MW00999739

BIG IDEAS

MATH®

ALGEBRA 1

A Bridge to Success

Ron Larson
Laurie Boswell

Erie, Pennsylvania
BigIdeasLearning.com

Big Ideas Learning, LLC
1762 Norcross Road
Erie, PA 16510-3838
USA

For product information and customer support, contact Big Ideas Learning
at **1-877-552-7766** or visit us at ***BigIdeasLearning.com***.

Cover Image
Nithon/Shutterstock.com

Printed in the U.S.A.

ISBN 13: 978-1-68033-114-1
ISBN 10: 1-68033-114-0

2 3 4 5 6 7 8 9 10 WEB 18 17 16 15

Authors

Ron Larson, Ph.D., is well known as the lead author of a comprehensive program for mathematics that spans middle school, high school, and college courses. He holds the distinction of Professor Emeritus from Penn State Erie, The Behrend College, where he taught for nearly 40 years. He received his Ph.D. in mathematics from the University of Colorado. Dr. Larson's numerous professional activities keep him actively involved in the mathematics education community and allow him to fully understand the needs of students, teachers, supervisors, and administrators.

Ron Larson

Laurie Boswell, Ed.D., is the Head of School and a mathematics teacher at the Riverside School in Lyndonville, Vermont. Dr. Boswell is a recipient of the Presidential Award for Excellence in Mathematics Teaching and has taught mathematics to students at all levels, from elementary through college. Dr. Boswell was a Tandy Technology Scholar and served on the NCTM Board of Directors from 2002 to 2005. She currently serves on the board of NCSM and is a popular national speaker.

Laurie Boswell

Dr. Ron Larson and **Dr. Laurie Boswell** began writing together in 1992. Since that time, they have authored over two dozen textbooks. In their collaboration, Ron is primarily responsible for the student edition while Laurie is primarily responsible for the teaching edition.

For the Student

Welcome to *Big Ideas Math Algebra 1*. From start to finish, this program was designed with you, the learner, in mind.

As you work through the chapters in your Algebra 1 course, you will be encouraged to think and to make conjectures while you persevere through challenging problems and exercises. You will make errors—and that is ok! Learning and understanding occur when you make errors and push through mental roadblocks to comprehend and solve new and challenging problems.

In this program, you will also be required to explain your thinking and your analysis of diverse problems and exercises. You will master content through engaging explorations that will provide deeper understanding, concise stepped-out examples, and rich thought-provoking exercises. Being actively involved in learning will help you develop mathematical reasoning and use it to solve math problems and work through other everyday challenges.

We wish you the best of luck as you explore Algebra 1. We are excited to be a part of your preparation for the challenges you will face in the remainder of your high school career and beyond.

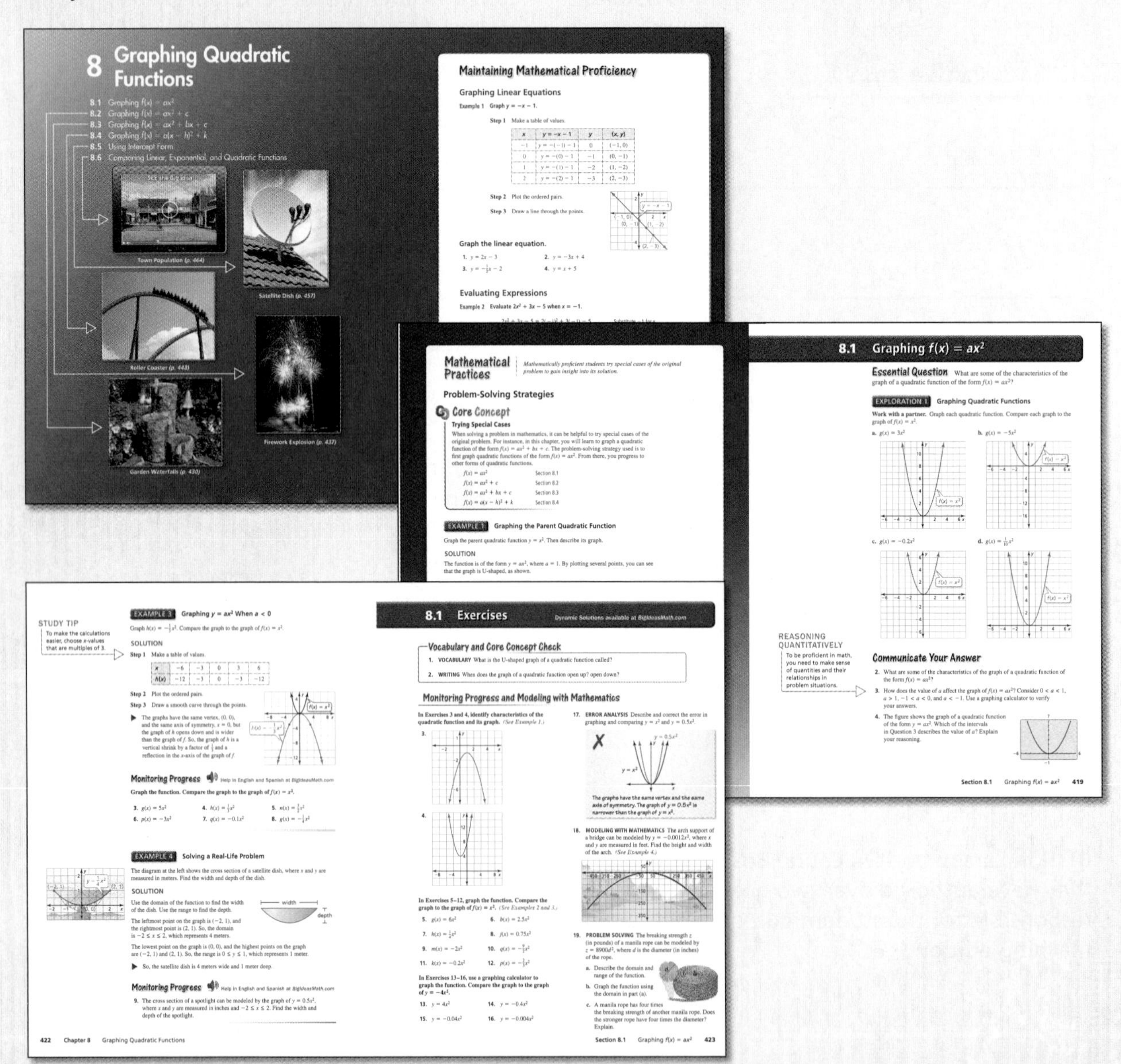

Big Ideas Math High School Research

Big Ideas Math Algebra 1, *Geometry*, and *Algebra 2* is a research-based program providing a rigorous, focused, and coherent curriculum for high school students. Ron Larson and Laurie Boswell utilized their expertise as well as the body of knowledge collected by additional expert mathematicians and researchers to develop each course.

The pedagogical approach to this program follows the best practices outlined in the most prominent and widely-accepted educational research and standards, including:

Achieve, ACT, and The College Board

Adding It Up: Helping Children Learn Mathematics
National Research Council ©2001

Common Core State Standards for Mathematics
National Governors Association Center for Best Practices and the Council of Chief State School Officers ©2010

Curriculum Focal Points and the *Principles and Standards for School Mathematics* ©2000
National Council of Teachers of Mathematics (NCTM)

Project Based Learning
The Buck Institute

Rigor/Relevance Framework™
International Center for Leadership in Education

Universal Design for Learning Guidelines
CAST ©2011

Big Ideas Math would like to express our gratitude to the mathematics education and instruction experts who served as consultants during the writing of *Big Ideas Math Algebra 1*, *Geometry*, and *Algebra 2*. Their input was an invaluable asset during the development of this program.

Kristen Karbon
Curriculum and Assessment Coordinator
Troy School District
Troy, Michigan

Jean Carwin
Math Specialist/TOSA
Snohomish School District
Snohomish, Washington

Carolyn Briles
Performance Tasks Consultant
Mathematics Teacher, Loudoun County Public Schools
Leesburg, Virginia

Bonnie Spence
Differentiated Instruction Consultant
Mathematics Lecturer, The University of Montana
Missoula, Montana

Connie Schrock, Ph.D.
Performance Tasks Consultant
Mathematics Professor, Emporia State University
Emporia, Kansas

We would also like to thank all of our reviewers who took the time to provide feedback during the final development phases. For a complete list of the *Big Ideas Math* program reviewers, please visit *www.BigIdeasLearning.com*.

Mathematical Practices

Make sense of problems and persevere in solving them.

- *Essential Questions* help students focus on core concepts as they analyze and work through each *Exploration*.
- Section opening *Explorations* allow students to struggle with new mathematical concepts and explain their reasoning in the *Communicate Your Answer* questions.

Reason abstractly and quantitatively.

- *Reasoning*, *Critical Thinking*, *Abstract Reasoning*, and *Problem Solving* exercises challenge students to apply their acquired knowledge and reasoning skills to solve each problem.
- *Thought Provoking* exercises test the reasoning skills of students as they analyze and interpret perplexing scenarios.

Construct viable arguments and critique the reasoning of others.

- Students must justify their responses to each *Essential Question* in the *Communicate Your Answer* questions at the end of each *Exploration* set.
- Students are asked to construct arguments and critique the reasoning of others in specialized exercises, including *Making an Argument*, *How Do You See It?*, *Drawing Conclusions*, *Reasoning*, *Error Analysis*, *Problem Solving*, and *Writing*.

Model with mathematics.

- Real-life scenarios are utilized in *Explorations*, *Examples*, *Exercises*, and *Assessments* so students have opportunities to apply the mathematical concepts they have learned to realistic situations.
- *Modeling with Mathematics* exercises allow students to interpret a problem in the context of a real-life situation, often utilizing tables, graphs, visual representations, and formulas.

Use appropriate tools strategically.

- Students are provided opportunities for selecting and utilizing the appropriate mathematical tool in *Using Tools* exercises. Students work with graphing calculators, dynamic geometry software, models, and more.
- A variety of tool papers and manipulatives are available for students to use in problems as strategically appropriate.

Attend to precision.

- *Vocabulary and Core Concept Check* exercises require students to use clear, precise mathematical language in their solutions and explanations.
- The many opportunities for cooperative learning in this program, including working with partners for each *Exploration*, support precise, explicit mathematical communication.

Look for and make use of structure.

- *Using Structure* exercises provide students with the opportunity to explore patterns and structure in mathematics.
- Students analyze structure in problems through *Justifying Steps* and *Analyzing Equations* exercises.

Look for and express regularity in repeated reasoning.

- Students are continually encouraged to evaluate the reasonableness of their solutions and their steps in the problem-solving process.
- Stepped-out *Examples* encourage students to maintain oversight of their problem-solving process and pay attention to the relevant details in each step.

Go to *BigIdeasLearning.com* for more information on the Mathematical Practices.

Mathematical Content

Chapter Coverage

1 2 3 4 5 6 7 8 9 10 11

Number and Quantity

- The Real Number System
- Quantities

1 2 3 4 5 6 7 8 9 10 11

Algebra

- Seeing Structure in Expressions
- Arithmetic with Polynomials and Rational Expressions
- Creating Equations
- Reasoning with Equations and Inequalities

1 2 3 4 5 6 7 8 9 10 11

Functions

- Interpreting Functions
- Building Functions
- Linear, Quadratic, and Exponential Models

1 2 3 4 5 6 7 8 9 10 11

Statistics and Probability

- Interpreting Categorical and Quantitative Data

1 Solving Linear Equations

See the Big Idea
Learn how boat navigators use dead reckoning to calculate their distance covered in a single direction.

Solving Linear Inequalities 2

See the Big Idea
Determine how designers decide on the number of electrical circuits needed in a house.

3 Graphing Linear Functions

See the Big Idea

Discover why unlike almost any other natural phenomenon, light travels at a constant speed.

Writing Linear Functions 4

See the Big Idea
Explore wind power and discover where the future of wind power will take us.

5 Solving Systems of Linear Equations

See the Big Idea

Learn how fisheries manage their complex ecosystems.

Exponential Functions and Sequences

See the Big Idea
Explore the variety of recursive sequences in language, art, music, nature, and games.

7 Polynomial Equations and Factoring

See the Big Idea

Explore whether seagulls and crows use the optimal height while dropping hard-shelled food to crack it open.

8 Graphing Quadratic Functions

See the Big Idea
Investigate the link between population growth and the classic exponential pay doubling application.

9 Solving Quadratic Equations

See the Big Idea
Explore the Parthenon and investigate how the use of the golden rectangle has evolved since its discovery.

10 Radical Functions and Equations

See the Big Idea

Explore how a tsunami's speed changes with depth and explore other factors in determining tsunami danger zones.

11 Data Analysis and Displays

See the Big Idea
Foray into the fashion differences between men and women.

How to Use Your Math Book

Get ready for each chapter by **Maintaining Mathematical Proficiency** and reviewing the **Mathematical Practices**. Begin each section by working through the **EXPLORATIONS** to **Communicate Your Answer** to the **Essential Question**. Each **Lesson** will explain **What You Will Learn** through **EXAMPLES**, **Core Concepts**, and **Core Vocabulary**.

Answer the **Monitoring Progress** questions as you work through each lesson. Look for **STUDY TIPS**, **COMMON ERRORS**, and suggestions for looking at a problem **ANOTHER WAY** throughout the lessons. We will also provide you with guidance for accurate mathematical **READING** and concept details you should **REMEMBER**.

Sharpen your newly acquired skills with **Exercises** at the end of every section. Halfway through each chapter you will be asked **What Did You Learn?** and you can use the Mid-Chapter **Quiz** to check your progress. You can also use the **Chapter Review** and **Chapter Test** to review and assess yourself after you have completed a chapter.

Apply what you learned in each chapter to a **Performance Task** and build your confidence for taking standardized tests with each chapter's **Cumulative Assessment**.

For extra practice in any chapter, use your *Online Resources*, *Skills Review Handbook*, or your *Student Journal*.

1 Solving Linear Equations

Density of Pyrite *(p. 41)*

Cheerleading Competition *(p. 29)*

Boat *(p. 22)*

Biking *(p. 14)*

Average Speed *(p. 6)*

Maintaining Mathematical Proficiency

Adding and Subtracting Integers

Example 1 **Evaluate $4 + (-12)$.**

$|-12| > |4|$. So, subtract $|4|$ from $|-12|$.

$$4 + (-12) = -8$$

Use the sign of -12.

Example 2 **Evaluate $-7 - (-16)$.**

$-7 - (-16) = -7 + 16$ Add the opposite of -16.

$= 9$ Add.

Add or subtract.

1. $-5 + (-2)$ **2.** $0 + (-13)$ **3.** $-6 + 14$

4. $19 - (-13)$ **5.** $-1 - 6$ **6.** $-5 - (-7)$

7. $17 + 5$ **8.** $8 + (-3)$ **9.** $11 - 15$

Multiplying and Dividing Integers

Example 3 **Evaluate $-3 \cdot (-5)$.**

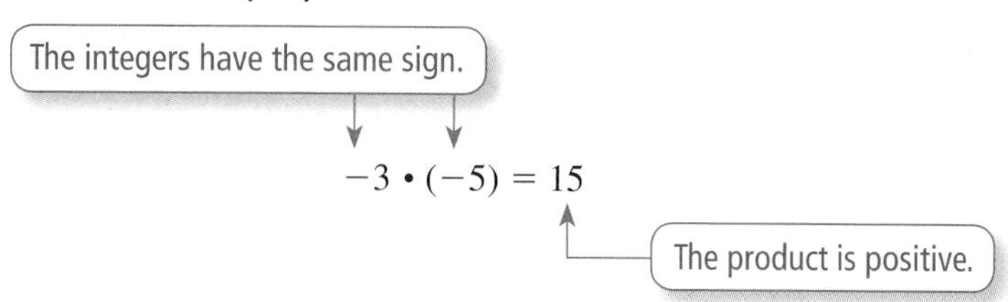

The integers have the same sign.
−3 • (−5) = 15
The product is positive.

Example 4 **Evaluate $15 \div (-3)$.**

The integers have different signs.

$$15 \div (-3) = -5$$

The quotient is negative.

Multiply or divide.

10. $-3(8)$ **11.** $-7 \cdot (-9)$ **12.** $4 \cdot (-7)$

13. $-24 \div (-6)$ **14.** $-16 \div 2$ **15.** $12 \div (-3)$

16. $6 \cdot 8$ **17.** $36 \div 6$ **18.** $-3(-4)$

19. **ABSTRACT REASONING** Summarize the rules for (a) adding integers, (b) subtracting integers, (c) multiplying integers, and (d) dividing integers. Give an example of each.

Mathematical Practices

Mathematically proficient students carefully specify units of measure.

Specifying Units of Measure

Core Concept

Operations and Unit Analysis

Addition and Subtraction

When you add or subtract quantities, they must have the same units of measure. The sum or difference will have the *same* unit of measure.

Example

3 ft

5 ft

Perimeter of rectangle

$= (3 \text{ ft}) + (5 \text{ ft}) + (3 \text{ ft}) + (5 \text{ ft})$

$= 16$ feet

When you add **feet**, you get **feet**.

Multiplication and Division

When you multiply or divide quantities, the product or quotient will have a *different* unit of measure.

Example Area of rectangle $= (3 \text{ ft}) \times (5 \text{ ft})$

$= 15$ square feet

When you multiply **feet**, you get feet squared, or **square feet**.

EXAMPLE 1 Specifying Units of Measure

You work 8 hours and earn \$72. What is your hourly wage?

SOLUTION

dollars per hour — dollars per hour

Hourly wage (\$ per h) $=$ \$72 $\div$ 8 h

$=$ \$9 per hour

The units on each side of the equation balance. Both are specified in dollars per hour.

▶ Your hourly wage is \$9 per hour.

Monitoring Progress

Solve the problem and specify the units of measure.

1. The population of the United States was about 280 million in 2000 and about 310 million in 2010. What was the annual rate of change in population from 2000 to 2010?
2. You drive 240 miles and use 8 gallons of gasoline. What was your car's gas mileage (in miles per gallon)?
3. A bathtub is in the shape of a rectangular prism. Its dimensions are 5 feet by 3 feet by 18 inches. The bathtub is three-fourths full of water and drains at a rate of 1 cubic foot per minute. About how long does it take for all the water to drain?

1.1 Solving Simple Equations

Essential Question How can you use simple equations to solve real-life problems?

EXPLORATION 1 Measuring Angles

Work with a partner. Use a protractor to measure the angles of each quadrilateral. Copy and complete the table to organize your results. (The notation $m\angle A$ denotes the measure of angle A.) How precise are your measurements?

a.

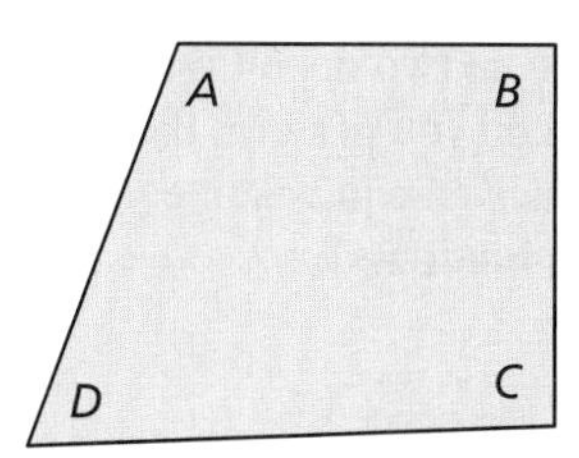

b.
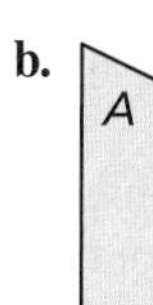

c.
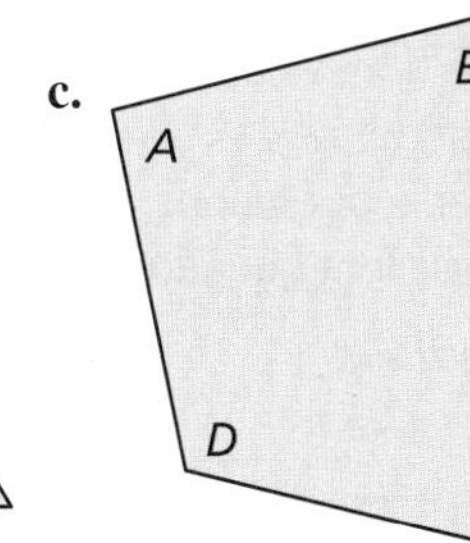

Quadrilateral	$m\angle A$ (degrees)	$m\angle B$ (degrees)	$m\angle C$ (degrees)	$m\angle D$ (degrees)	$m\angle A + m\angle B + m\angle C + m\angle D$
a.					
b.					
c.					

UNDERSTANDING MATHEMATICAL TERMS

A **conjecture** is an unproven statement about a general mathematical concept. After the statement is proven, it is called a **rule** or a **theorem**.

EXPLORATION 2 Making a Conjecture

Work with a partner. Use the completed table in Exploration 1 to write a conjecture about the sum of the angle measures of a quadrilateral. Draw three quadrilaterals that are different from those in Exploration 1 and use them to justify your conjecture.

EXPLORATION 3 Applying Your Conjecture

Work with a partner. Use the conjecture you wrote in Exploration 2 to write an equation for each quadrilateral. Then solve the equation to find the value of x. Use a protractor to check the reasonableness of your answer.

a.
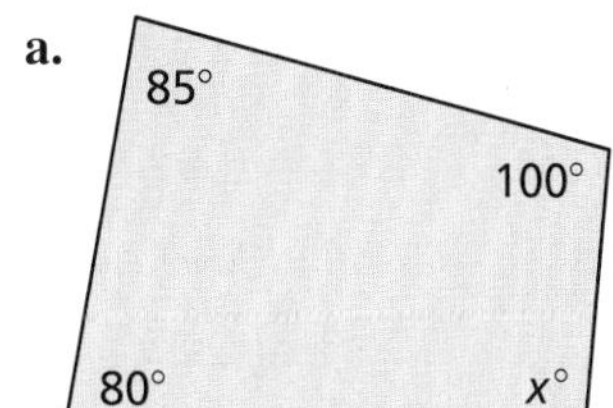

b.
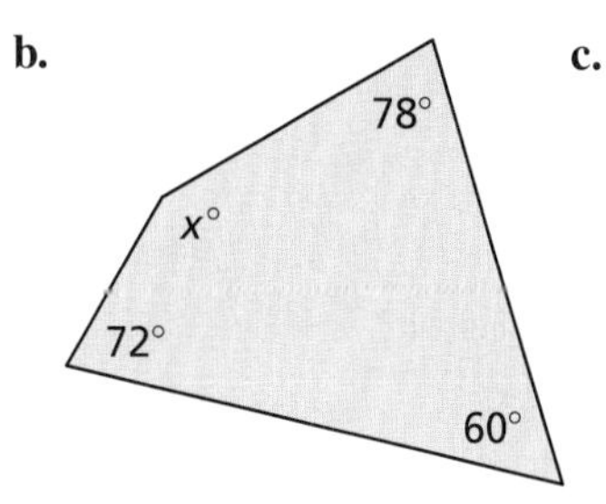

c.
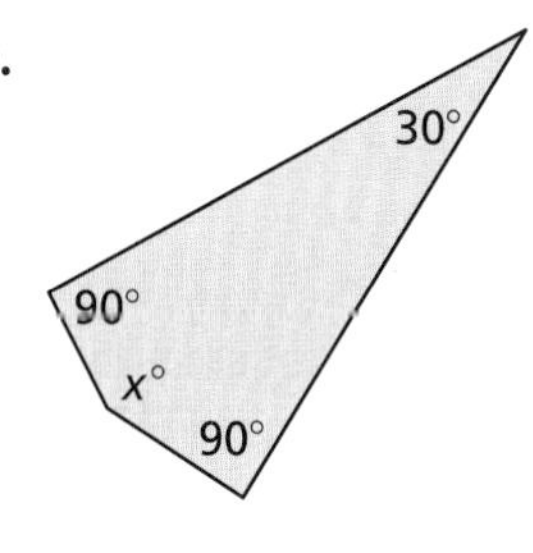

Communicate Your Answer

4. How can you use simple equations to solve real-life problems?
5. Draw your own quadrilateral and cut it out. Tear off the four corners of the quadrilateral and rearrange them to affirm the conjecture you wrote in Exploration 2. Explain how this affirms the conjecture.

1.1 Lesson

Core Vocabulary

conjecture, *p. 3*
rule, *p. 3*
theorem, *p. 3*
equation, *p. 4*
linear equation in one variable, *p. 4*
solution, *p. 4*
inverse operations, *p. 4*
equivalent equations, *p. 4*

Previous
expression

What You Will Learn

- Solve linear equations using addition and subtraction.
- Solve linear equations using multiplication and division.
- Use linear equations to solve real-life problems.

Solving Linear Equations by Adding or Subtracting

An **equation** is a statement that two expressions are equal. A **linear equation in one variable** is an equation that can be written in the form $ax + b = 0$, where a and b are constants and $a \neq 0$. A **solution** of an equation is a value that makes the equation true.

Inverse operations are two operations that undo each other, such as addition and subtraction. When you perform the same inverse operation on each side of an equation, you produce an equivalent equation. **Equivalent equations** are equations that have the same solution(s).

Core Concept

Addition Property of Equality

Words Adding the same number to each side of an equation produces an equivalent equation.

Algebra If $a = b$, then $a + c = b + c$.

Subtraction Property of Equality

Words Subtracting the same number from each side of an equation produces an equivalent equation.

Algebra If $a = b$, then $a - c = b - c$.

EXAMPLE 1 Solving Equations by Addition or Subtraction

Solve each equation. Justify each step. Check your answer.

a. $x - 3 = -5$ **b.** $0.9 = y + 2.8$

SOLUTION

a. $x - 3 = -5$ Write the equation.

Addition Property of Equality → $\underline{+3 \quad +3}$ Add 3 to each side.

$x = -2$ Simplify.

▶ The solution is $x = -2$.

Check

$x - 3 = -5$
$-2 - 3 \stackrel{?}{=} -5$
$-5 = -5$ ✓

b. $0.9 = y + 2.8$ Write the equation.

Subtraction Property of Equality → $\underline{-2.8 \quad -2.8}$ Subtract 2.8 from each side.

$-1.9 = y$ Simplify.

▶ The solution is $y = -1.9$.

Check

$0.9 = y + 2.8$
$0.9 \stackrel{?}{=} -1.9 + 2.8$
$0.9 = 0.9$ ✓

Monitoring Progress  Help in English and Spanish at *BigIdeasMath.com*

Solve the equation. Justify each step. Check your solution.

1. $n + 3 = -7$ **2.** $g - \frac{1}{3} = -\frac{2}{3}$ **3.** $-6.5 = p + 3.9$

Solving Linear Equations by Multiplying or Dividing

Core Concept

REMEMBER
Multiplication and division are inverse operations.

Multiplication Property of Equality

Words Multiplying each side of an equation by the same nonzero number produces an equivalent equation.

Algebra If $a = b$, then $a \cdot c = b \cdot c, c \neq 0$.

Division Property of Equality

Words Dividing each side of an equation by the same nonzero number produces an equivalent equation.

Algebra If $a = b$, then $a \div c = b \div c, c \neq 0$.

EXAMPLE 2 Solving Equations by Multiplication or Division

Solve each equation. Justify each step. Check your answer.

a. $-\frac{n}{5} = -3$ **b.** $\pi x = -2\pi$ **c.** $1.3z = 5.2$

SOLUTION

a. $-\frac{n}{5} = -3$ Write the equation.

Multiplication Property of Equality → $-5 \cdot \left(-\frac{n}{5}\right) = -5 \cdot (-3)$ Multiply each side by -5.

$n = 15$ Simplify.

▶ The solution is $n = 15$.

Check

$-\frac{n}{5} = -3$

$-\frac{15}{5} \stackrel{?}{=} -3$

$-3 = -3$ ✓

b. $\pi x = -2\pi$ Write the equation.

Division Property of Equality → $\frac{\pi x}{\pi} = \frac{-2\pi}{\pi}$ Divide each side by π.

$x = -2$ Simplify.

▶ The solution is $x = -2$.

Check

$\pi x = -2\pi$

$\pi(-2) \stackrel{?}{=} -2\pi$

$-2\pi = -2\pi$ ✓

c. $1.3z = 5.2$ Write the equation.

Division Property of Equality → $\frac{1.3z}{1.3} = \frac{5.2}{1.3}$ Divide each side by 1.3.

$z = 4$ Simplify.

▶ The solution is $z = 4$.

Check

$1.3z = 5.2$

$1.3(4) \stackrel{?}{=} 5.2$

$5.2 = 5.2$ ✓

Monitoring Progress

Help in English and Spanish at *BigIdeasMath.com*

Solve the equation. Justify each step. Check your solution.

4. $\frac{y}{3} = -6$ **5.** $9\pi = \pi x$ **6.** $0.05w = 1.4$

Solving Real-Life Problems

Core Concept

MODELING WITH MATHEMATICS

Mathematically proficient students routinely check that their solutions make sense in the context of a real-life problem.

Four-Step Approach to Problem Solving

1. **Understand the Problem** What is the unknown? What information is being given? What is being asked?
2. **Make a Plan** This plan might involve one or more of the problem-solving strategies shown on the next page.
3. **Solve the Problem** Carry out your plan. Check that each step is correct.
4. **Look Back** Examine your solution. Check that your solution makes sense in the original statement of the problem.

EXAMPLE 3 Modeling with Mathematics

In the 2012 Olympics, Usain Bolt won the 200-meter dash with a time of 19.32 seconds. Write and solve an equation to find his average speed to the nearest hundredth of a meter per second.

SOLUTION

REMEMBER

The formula that relates distance d, rate or speed r, and time t is

$$d = rt.$$

1. **Understand the Problem** You know the winning time and the distance of the race. You are asked to find the average speed to the nearest hundredth of a meter per second.
2. **Make a Plan** Use the Distance Formula to write an equation that represents the problem. Then solve the equation.
3. **Solve the Problem**

$d = r \cdot t$	Write the Distance Formula.
$200 = r \cdot 19.32$	Substitute 200 for d and 19.32 for t.
$\frac{200}{19.32} = \frac{19.32r}{19.32}$	Divide each side by 19.32.
$10.35 \approx r$	Simplify.

REMEMBER

The symbol $\approx$ means "approximately equal to."

▶ Bolt's average speed was about 10.35 meters per second.

4. **Look Back** Round Bolt's average speed to 10 meters per second. At this speed, it would take

$$\frac{200 \text{ m}}{10 \text{ m/sec}} = 20 \text{ seconds}$$

to run 200 meters. Because 20 is close to 19.32, your solution is reasonable.

Monitoring Progress  Help in English and Spanish at *BigIdeasMath.com*

7. Suppose Usain Bolt ran 400 meters at the same average speed that he ran the 200 meters. How long would it take him to run 400 meters? Round your answer to the nearest hundredth of a second.

Core Concept

Common Problem-Solving Strategies

Use a verbal model.	Guess, check, and revise.
Draw a diagram.	Sketch a graph or number line.
Write an equation.	Make a table.
Look for a pattern.	Make a list.
Work backward.	Break the problem into parts.

EXAMPLE 4 Modeling with Mathematics

On January 22, 1943, the temperature in Spearfish, South Dakota, fell from 54°F at 9:00 A.M. to -4°F at 9:27 A.M. How many degrees did the temperature fall?

SOLUTION

1. **Understand the Problem** You know the temperature before and after the temperature fell. You are asked to find how many degrees the temperature fell.
2. **Make a Plan** Use a verbal model to write an equation that represents the problem. Then solve the equation.
3. **Solve the Problem**

Words Temperature at 9:27 A.M. = Temperature at 9:00 A.M. − Number of degrees the temperature fell

Variable Let T be the number of degrees the temperature fell.

Equation $-4 = 54 - T$

$-4 = 54 - T$	Write the equation.
$-4 - 54 = 54 - 54 - T$	Subtract 54 from each side.
$-58 = -T$	Simplify.
$58 = T$	Divide each side by -1.

▶ The temperature fell 58°F.

REMEMBER
The distance between two points on a number line is always positive.

4. **Look Back** The temperature fell from 54 degrees *above* 0 to 4 degrees *below* 0. You can use a number line to check that your solution is reasonable.

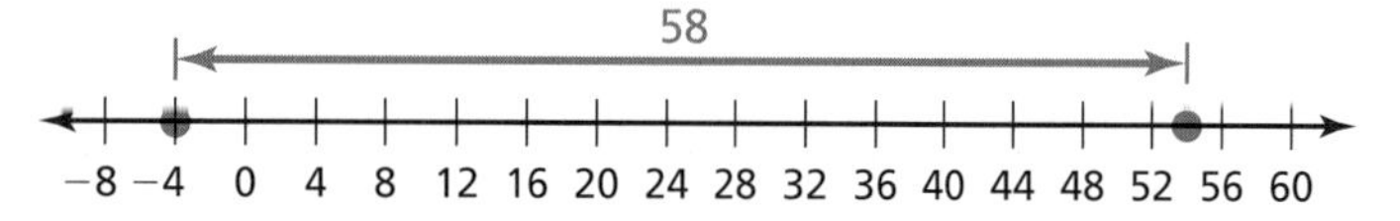

Monitoring Progress

Help in English and Spanish at *BigIdeasMath.com*

8. You thought the balance in your checking account was \$68. When your bank statement arrives, you realize that you forgot to record a check. The bank statement lists your balance as \$26. Write and solve an equation to find the amount of the check that you forgot to record.

1.1 Exercises

Dynamic Solutions available at *BigIdeasMath.com*

Vocabulary and Core Concept Check

1. **VOCABULARY** Which of the operations $+$, $-$, $\times$, and $\div$ are inverses of each other?

2. **VOCABULARY** Are the equations $-2x = 10$ and $-5x = 25$ equivalent? Explain.

3. **WRITING** Which property of equality would you use to solve the equation $14x = 56$? Explain.

4. **WHICH ONE DOESN'T BELONG?** Which expression does not belong with the other three? Explain your reasoning.

$8 = \frac{x}{2}$ $\quad$ $3 = x \div 4$ $\quad$ $x - 6 = 5$ $\quad$ $\frac{x}{3} = 9$

Monitoring Progress and Modeling with Mathematics

In Exercises 5–14, solve the equation. Justify each step. Check your solution. *(See Example 1.)*

5. $x + 5 = 8$

6. $m + 9 = 2$

7. $y - 4 = 3$

8. $s - 2 = 1$

9. $w + 3 = -4$

10. $n - 6 = -7$

11. $-14 = p - 11$

12. $0 = 4 + q$

13. $r + (-8) = 10$

14. $t - (-5) = 9$

15. **MODELING WITH MATHEMATICS** A discounted amusement park ticket costs $12.95 less than the original price p. Write and solve an equation to find the original price.

16. **MODELING WITH MATHEMATICS** You and a friend are playing a board game. Your final score x is 12 points less than your friend's final score. Write and solve an equation to find your final score.

	ROUND 9	ROUND 10	FINAL SCORE
Your Friend	22	12	195
You	9	25	?

USING TOOLS The sum of the angle measures of a quadrilateral is 360°. In Exercises 17–20, write and solve an equation to find the value of x. Use a protractor to check the reasonableness of your answer.

17.

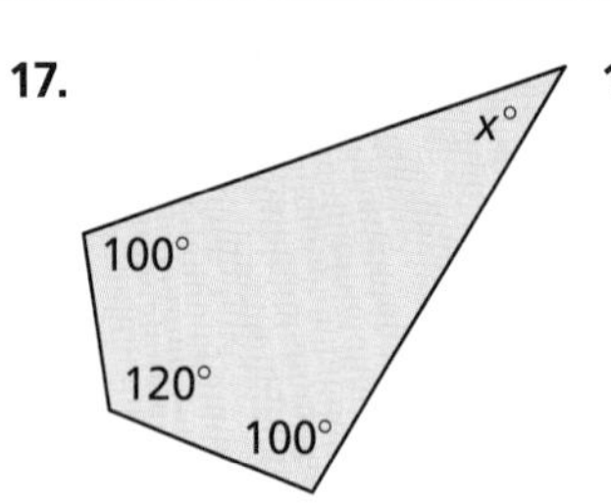

18.

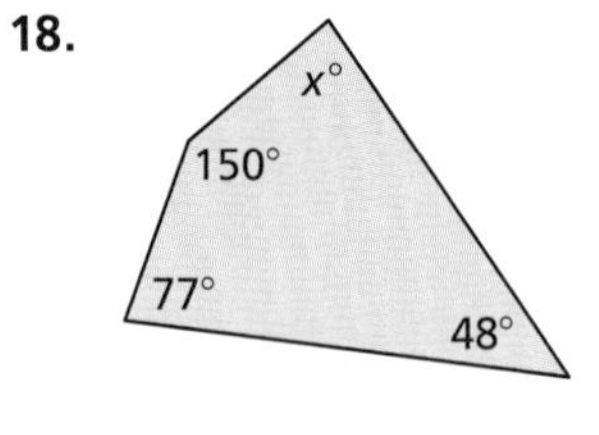

19.

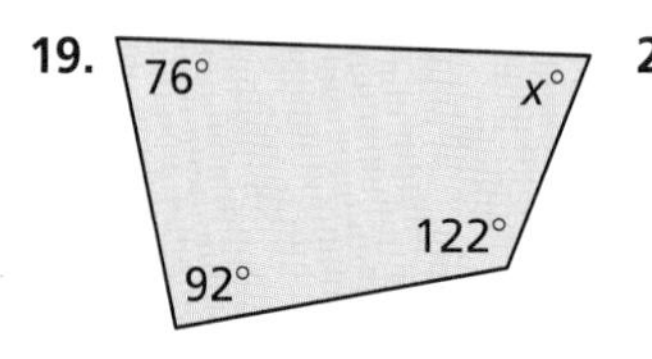

20.

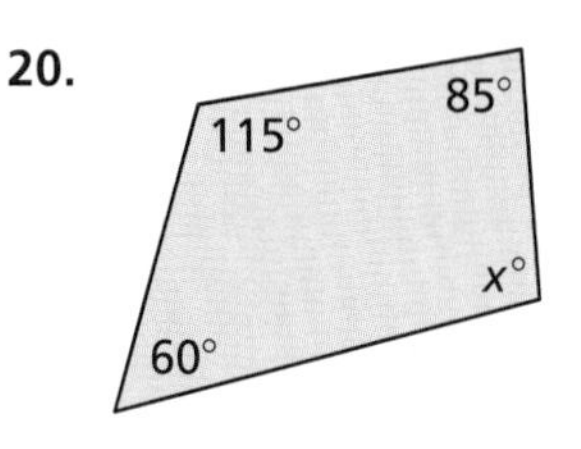

In Exercises 21–30, solve the equation. Justify each step. Check your solution. *(See Example 2.)*

21. $5g = 20$

22. $4q = 52$

23. $p \div 5 = 3$

24. $y \div 7 = 1$

25. $-8r = 64$

26. $x \div (-2) = 8$

27. $\frac{x}{6} = 8$

28. $\frac{w}{-3} = 6$

29. $-54 = 9s$

30. $-7 = \frac{t}{7}$

In Exercises 31–38, solve the equation. Check your solution.

31. $\frac{3}{2} + t = \frac{1}{2}$ **32.** $b - \frac{3}{16} = \frac{5}{16}$

33. $\frac{3}{7}m = 6$ **34.** $-\frac{2}{5}y = 4$

35. $5.2 = a - 0.4$ **36.** $f + 3\pi = 7\pi$

37. $-108\pi = 6\pi j$ **38.** $x \div (-2) = 1.4$

ERROR ANALYSIS In Exercises 39 and 40, describe and correct the error in solving the equation.

39. ✗

$$-0.8 + r = 12.6$$
$$r = 12.6 + (-0.8)$$
$$r = 11.8$$

40. ✗

$$-\frac{m}{3} = -4$$
$$3 \cdot \left(-\frac{m}{3}\right) = 3 \cdot (-4)$$
$$m = -12$$

41. ANALYZING RELATIONSHIPS A baker orders 162 eggs. Each carton contains 18 eggs. Which equation can you use to find the number x of cartons? Explain your reasoning and solve the equation.

Ⓐ $162x = 18$ Ⓑ $\frac{x}{18} = 162$

Ⓒ $18x = 162$ Ⓓ $x + 18 = 162$

MODELING WITH MATHEMATICS In Exercises 42–44, write and solve an equation to answer the question. *(See Examples 3 and 4.)*

42. The temperature at 5 P.M. is 20°F. The temperature at 10 P.M. is −5°F. How many degrees did the temperature fall?

43. The length of an American flag is 1.9 times its width. What is the width of the flag?

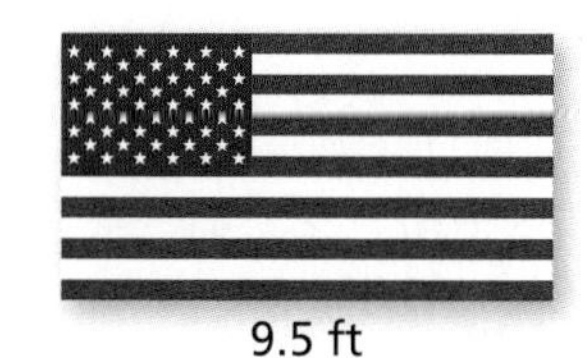

44. The balance of an investment account is $308 more than the balance 4 years ago. The current balance of the account is $4708. What was the balance 4 years ago?

45. REASONING Identify the property of equality that makes Equation 1 and Equation 2 equivalent.

Equation 1 $x - \frac{1}{2} = \frac{x}{4} + 3$

Equation 2 $4x - 2 = x + 12$

46. PROBLEM SOLVING Tatami mats are used as a floor covering in Japan. One possible layout uses four identical rectangular mats and one square mat, as shown. The area of the square mat is half the area of one of the rectangular mats.

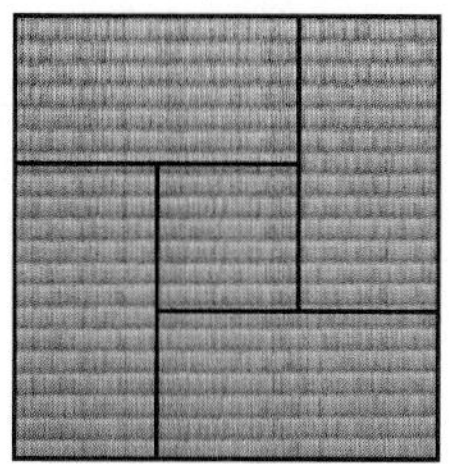

a. Write and solve an equation to find the area of one rectangular mat.

b. The length of a rectangular mat is twice the width. Use Guess, Check, and Revise to find the dimensions of one rectangular mat.

47. PROBLEM SOLVING You spend $30.40 on 4 CDs. Each CD costs the same amount and is on sale for 80% of the original price.

a. Write and solve an equation to find how much you spend on each CD.

b. The next day, the CDs are no longer on sale. You have $25. Will you be able to buy 3 more CDs? Explain your reasoning.

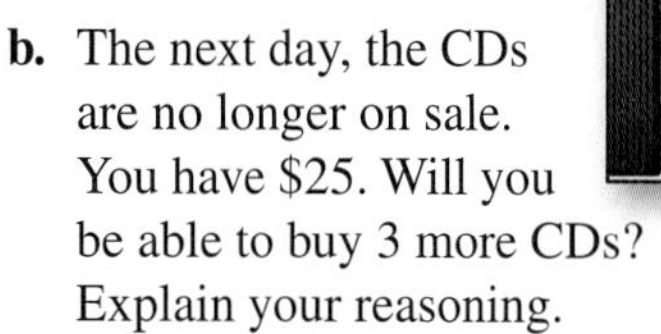

48. ANALYZING RELATIONSHIPS As c increases, does the value of x *increase*, *decrease*, or *stay the same* for each equation? Assume c is positive.

Equation	Value of x
$x - c = 0$	
$cx = 1$	
$cx = c$	
$\frac{x}{c} = 1$	

49. **USING STRUCTURE** Use the values −2, 5, 9, and 10 to complete each statement about the equation $ax = b - 5$.

a. When $a =$ ___ and $b =$ ___, x is a positive integer.

b. When $a =$ ___ and $b =$ ___, x is a negative integer.

50. **HOW DO YOU SEE IT?** The circle graph shows the percents of different animals sold at a local pet store in 1 year.

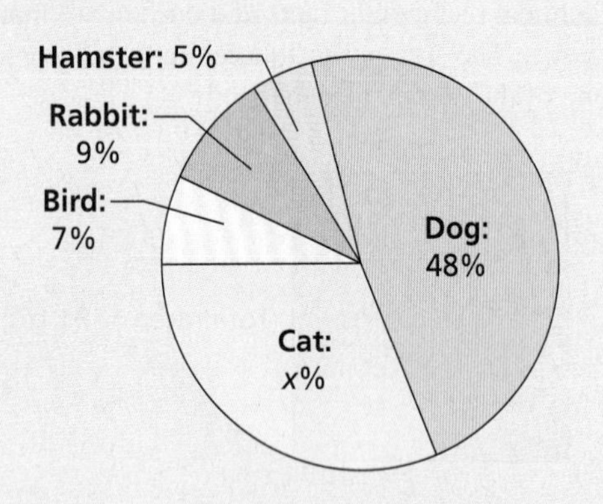

a. What percent is represented by the entire circle?

b. How does the equation $7 + 9 + 5 + 48 + x = 100$ relate to the circle graph? How can you use this equation to find the percent of cats sold?

51. **REASONING** One-sixth of the girls and two-sevenths of the boys in a school marching band are in the percussion section. The percussion section has 6 girls and 10 boys. How many students are in the marching band? Explain.

52. **THOUGHT PROVOKING** Write a real-life problem that can be modeled by an equation equivalent to the equation $5x = 30$. Then solve the equation and write the answer in the context of your real-life problem.

MATHEMATICAL CONNECTIONS In Exercises 53–56, find the height h or the area of the base B of the solid.

53.

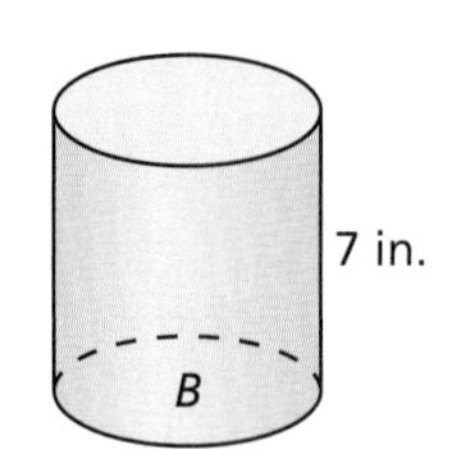

Volume $= 84\pi$ in.^{3}

54.

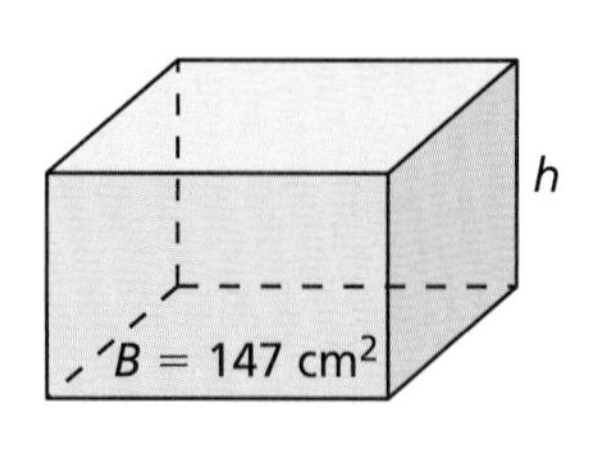

Volume $= 1323$ cm^3

55.

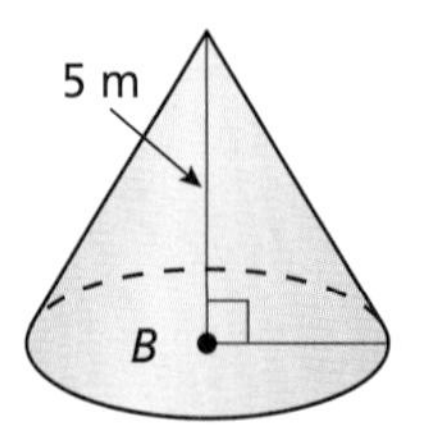

Volume $= 15\pi$ m^3

56.

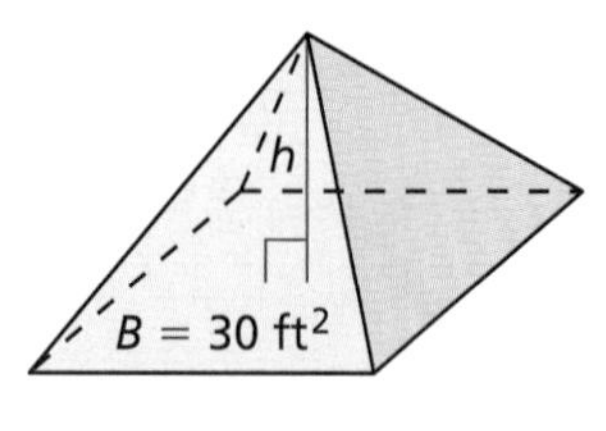

Volume $= 35$ ft^3

57. **MAKING AN ARGUMENT** In baseball, a player's batting average is calculated by dividing the number of hits by the number of at-bats. The table shows Player A's batting average and number of at-bats for three regular seasons.

Season	Batting average	At-bats
2010	.312	596
2011	.296	446
2012	.295	599

a. How many hits did Player A have in the 2011 regular season? Round your answer to the nearest whole number.

b. Player B had 33 fewer hits in the 2011 season than Player A but had a greater batting average. Your friend concludes that Player B had more at-bats in the 2011 season than Player A. Is your friend correct? Explain.

Maintaining Mathematical Proficiency

Reviewing what you learned in previous grades and lessons

Use the Distributive Property to simplify the expression. *(Skills Review Handbook)*

58. $8(y + 3)$

59. $\frac{5}{6}\left(x + \frac{1}{2} + 4\right)$

60. $5(m + 3 + n)$

61. $4(2p + 4q + 6)$

Copy and complete the statement. Round to the nearest hundredth, if necessary. *(Skills Review Handbook)*

62. $\frac{5 \text{ L}}{\text{min}} = \frac{\square \text{ L}}{\text{h}}$

63. $\frac{68 \text{ mi}}{\text{h}} \approx \frac{\square \text{ mi}}{\text{sec}}$

64. $\frac{7 \text{ gal}}{\text{min}} \approx \frac{\square \text{ qt}}{\text{sec}}$

65. $\frac{8 \text{ km}}{\text{min}} \approx \frac{\square \text{ mi}}{\text{h}}$

1.2 Solving Multi-Step Equations

Essential Question How can you use multi-step equations to solve real-life problems?

EXPLORATION 1 Solving for the Angle Measures of a Polygon

Work with a partner. The sum S of the angle measures of a polygon with n sides can be found using the formula $S = 180(n - 2)$. Write and solve an equation to find each value of x. Justify the steps in your solution. Then find the angle measures of each polygon. How can you check the reasonableness of your answers?

a.

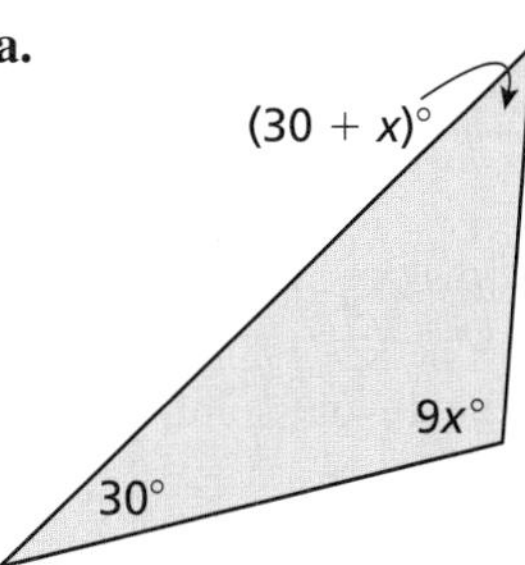

b.

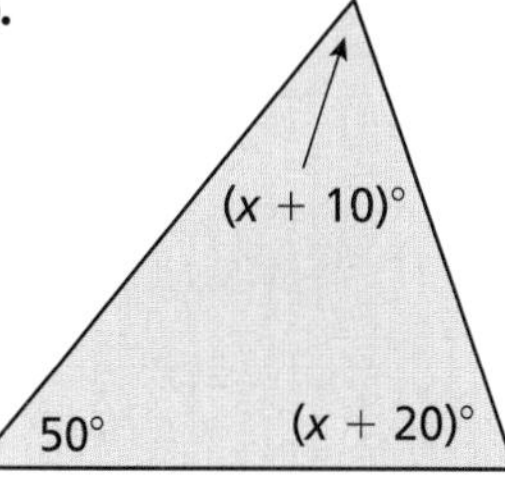

c.

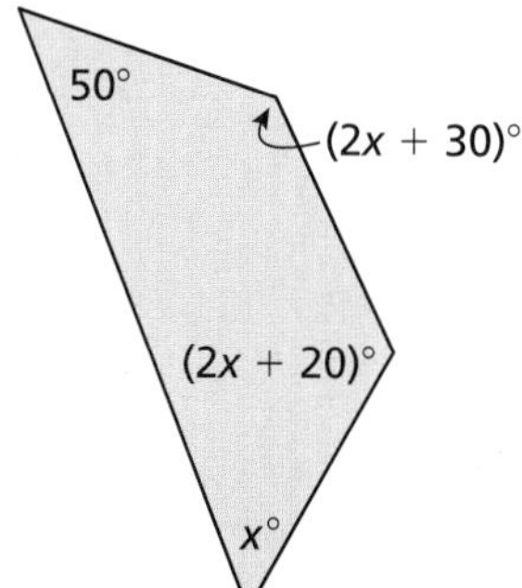

d.

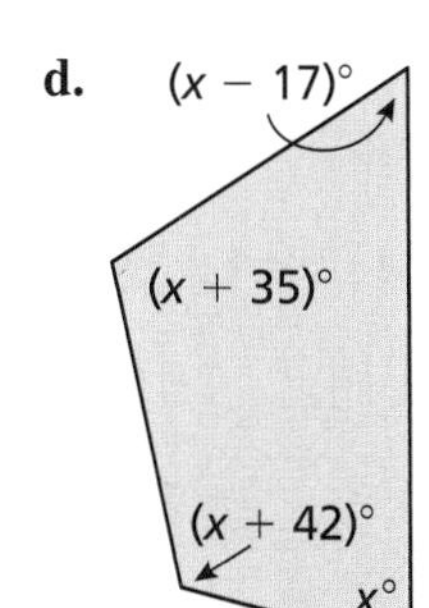

e.

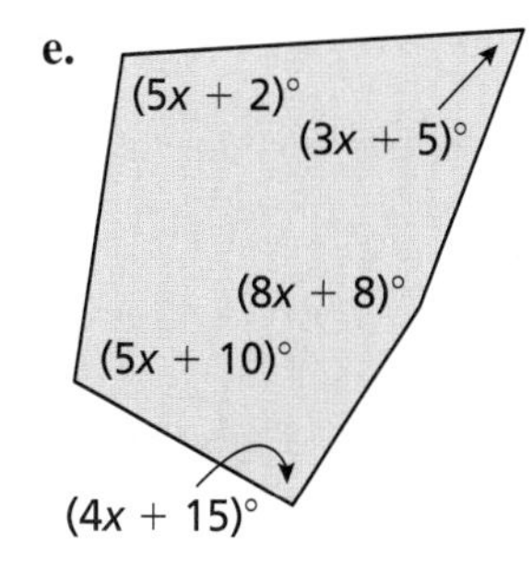

f.

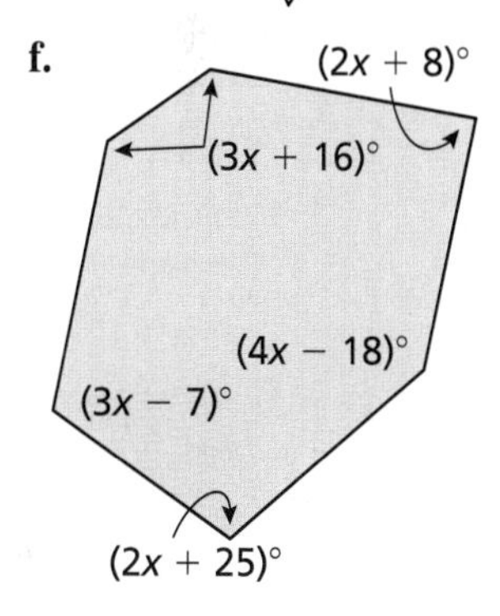

JUSTIFYING CONCLUSIONS

To be proficient in math, you need to be sure your answers make sense in the context of the problem. For instance, if you find the angle measures of a triangle, and they have a sum that is not equal to 180°, then you should check your work for mistakes.

EXPLORATION 2 Writing a Multi-Step Equation

Work with a partner.

a. Draw an irregular polygon.

b. Measure the angles of the polygon. Record the measurements on a separate sheet of paper.

c. Choose a value for x. Then, using this value, work backward to assign a variable expression to each angle measure, as in Exploration 1.

d. Trade polygons with your partner.

e. Solve an equation to find the angle measures of the polygon your partner drew. Do your answers seem reasonable? Explain.

Communicate Your Answer

3. How can you use multi-step equations to solve real-life problems?

4. In Exploration 1, you were given the formula for the sum S of the angle measures of a polygon with n sides. Explain why this formula works.

5. The sum of the angle measures of a polygon is 1080°. How many sides does the polygon have? Explain how you found your answer.

1.2 Lesson

Core Vocabulary

Previous
inverse operations
mean

What You Will Learn

- Solve multi-step linear equations using inverse operations.
- Use multi-step linear equations to solve real-life problems.
- Use unit analysis to model real-life problems.

Solving Multi-Step Linear Equations

Core Concept

Solving Multi-Step Equations

To solve a multi-step equation, simplify each side of the equation, if necessary. Then use inverse operations to isolate the variable.

EXAMPLE 1 Solving a Two-Step Equation

Solve $2.5x - 13 = 2$. Check your solution.

SOLUTION

	$2.5x - 13 = 2$	Write the equation.
Undo the subtraction.	$+13 \quad +13$	Add 13 to each side.
	$2.5x = 15$	Simplify.
Undo the multiplication.	$\frac{2.5x}{2.5} = \frac{15}{2.5}$	Divide each side by 2.5.
	$x = 6$	Simplify.

- The solution is $x = 6$.

Check

$2.5x - 13 = 2$

$2.5(6) - 13 \stackrel{?}{=} 2$

$2 = 2$ ✓

EXAMPLE 2 Combining Like Terms to Solve an Equation

Solve $-12 = 9x - 6x + 15$. Check your solution.

SOLUTION

	$-12 = 9x - 6x + 15$	Write the equation.
	$-12 = 3x + 15$	Combine like terms.
Undo the addition.	$-15 \quad -15$	Subtract 15 from each side.
	$-27 = 3x$	Simplify.
Undo the multiplication.	$\frac{-27}{3} = \frac{3x}{3}$	Divide each side by 3.
	$-9 = x$	Simplify.

- The solution is $x = -9$.

Check

$-12 = 9x - 6x + 15$

$-12 \stackrel{?}{=} 9(-9) - 6(-9) + 15$

$-12 = -12$ ✓

Monitoring Progress

Help in English and Spanish at *BigIdeasMath.com*

Solve the equation. Check your solution.

1. $-2n + 3 = 9$ **2.** $-21 = \frac{1}{2}c - 11$ **3.** $-2x - 10x + 12 = 18$

EXAMPLE 3 Using Structure to Solve a Multi-Step Equation

Solve $2(1 - x) + 3 = -8$. Check your solution.

SOLUTION

Method 1 One way to solve the equation is by using the Distributive Property.

$2(1 - x) + 3 = -8$	Write the equation.
$2(1) - 2(x) + 3 = -8$	Distributive Property
$2 - 2x + 3 = -8$	Multiply.
$-2x + 5 = -8$	Combine like terms.
$\underline{-5} \quad \underline{-5}$	Subtract 5 from each side.
$-2x = -13$	Simplify.
$\dfrac{-2x}{-2} = \dfrac{-13}{-2}$	Divide each side by -2.
$x = 6.5$	Simplify.

▶ The solution is $x = 6.5$.

Check

$$2(1 - x) + 3 = -8$$

$$2(1 - 6.5) + 3 \stackrel{?}{=} -8$$

$$-8 = -8 \checkmark$$

Method 2 Another way to solve the equation is by interpreting the expression $1 - x$ as a single quantity.

LOOKING FOR STRUCTURE

First solve for the expression $1 - x$, and then solve for x.

$2(1 - x) + 3 = -8$	Write the equation.
$\underline{-3} \quad \underline{-3}$	Subtract 3 from each side.
$2(1 - x) = -11$	Simplify.
$\dfrac{2(1 - x)}{2} = \dfrac{-11}{2}$	Divide each side by 2.
$1 - x = -5.5$	Simplify.
$\underline{-1} \quad \underline{-1}$	Subtract 1 from each side.
$-x = -6.5$	Simplify.
$\dfrac{-x}{-1} = \dfrac{-6.5}{-1}$	Divide each side by -1.
$x = 6.5$	Simplify.

▶ The solution is $x = 6.5$, which is the same solution obtained in Method 1.

Monitoring Progress

Help in English and Spanish at *BigIdeasMath.com*

Solve the equation. Check your solution.

4. $3(x + 1) + 6 = -9$

5. $15 = 5 + 4(2d - 3)$

6. $13 = -2(y - 4) + 3y$

7. $2x(5 - 3) - 3x = 5$

8. $-4(2m + 5) - 3m = 35$

9. $5(3 - x) + 2(3 - x) = 14$

Solving Real-Life Problems

EXAMPLE 4 Modeling with Mathematics

Use the table to find the number of miles x you need to bike on Friday so that the mean number of miles biked per day is 5.

Day	Miles
Monday	3.5
Tuesday	5.5
Wednesday	0
Thursday	5
Friday	x

SOLUTION

1. **Understand the Problem** You know how many miles you biked Monday through Thursday. You are asked to find the number of miles you need to bike on Friday so that the mean number of miles biked per day is 5.
2. **Make a Plan** Use the definition of mean to write an equation that represents the problem. Then solve the equation.
3. **Solve the Problem** The mean of a data set is the sum of the data divided by the number of data values.

$$\frac{3.5 + 5.5 + 0 + 5 + x}{5} = 5$$ Write the equation.

$$\frac{14 + x}{5} = 5$$ Combine like terms.

$$5 \cdot \frac{14 + x}{5} = 5 \cdot 5$$ Multiply each side by 5.

$$14 + x = 25$$ Simplify.

$$\underline{-14} \qquad \underline{-14}$$ Subtract 14 from each side.

$$x = 11$$ Simplify.

▶ You need to bike 11 miles on Friday.

4. **Look Back** Notice that on the days that you did bike, the values are close to the mean. Because you did not bike on Wednesday, you need to bike about twice the mean on Friday. Eleven miles is about twice the mean. So, your solution is reasonable.

Monitoring Progress  Help in English and Spanish at *BigIdeasMath.com*

10. The formula $d = \frac{1}{2}n + 26$ relates the nozzle pressure n (in pounds per square inch) of a fire hose and the maximum horizontal distance the water reaches d (in feet). How much pressure is needed to reach a fire 50 feet away?

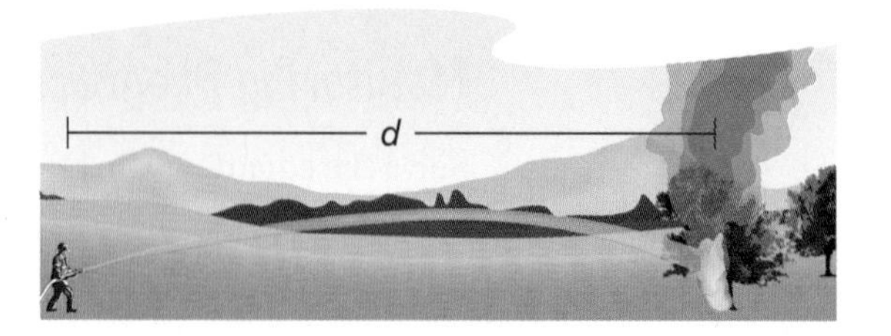

REMEMBER

When you add **miles** to **miles**, you get **miles**. But, when you divide **miles** by **days**, you get **miles per day**.

Using Unit Analysis to Model Real-Life Problems

When you write an equation to model a real-life problem, you should check that the units on each side of the equation balance. For instance, in Example 4, notice how the units balance.

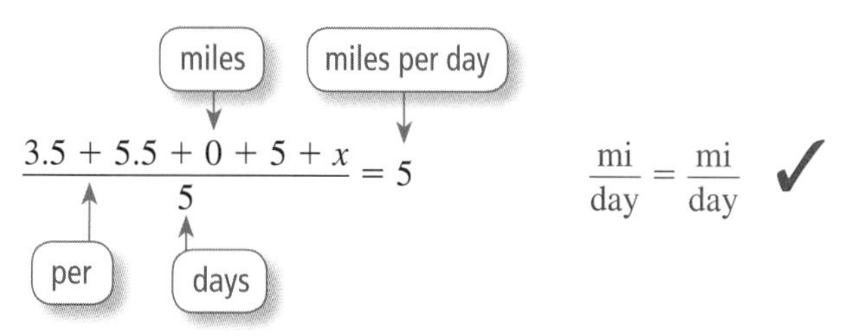

$$\frac{3.5 + 5.5 + 0 + 5 + x}{5} = 5 \qquad \frac{\text{mi}}{\text{day}} = \frac{\text{mi}}{\text{day}} \ ✓$$

EXAMPLE 5 Solving a Real-Life Problem

Your school's drama club charges \$4 per person for admission to a play. The club borrowed \$400 to pay for costumes and props. After paying back the loan, the club has a profit of \$100. How many people attended the play?

SOLUTION

1. **Understand the Problem** You know how much the club charges for admission. You also know how much the club borrowed and its profit. You are asked to find how many people attended the play.
2. **Make a Plan** Use a verbal model to write an equation that represents the problem. Then solve the equation.
3. **Solve the Problem**

REMEMBER

When you multiply **dollars per person** by **people**, you get **dollars**.

Words Ticket price • Number of people who attended − Amount of loan = Profit

Variable Let x be the number of people who attended.

Equation $\frac{\$4}{\text{person}} \cdot x \text{ people} - \$400 = \$100 \qquad \$ = \$$ ✓

$4x - 400 = 100$	Write the equation.
$4x - 400 + 400 = 100 + 400$	Add 400 to each side.
$4x = 500$	Simplify.
$\frac{4x}{4} = \frac{500}{4}$	Divide each side by 4.
$x = 125$	Simplify.

▶ So, 125 people attended the play.

4. **Look Back** To check that your solution is reasonable, multiply \$4 per person by 125 people. The result is \$500. After paying back the \$400 loan, the club has \$100, which is the profit.

Monitoring Progress  Help in English and Spanish at *BigIdeasMath.com*

11. You have 96 feet of fencing to enclose a rectangular pen for your dog. To provide sufficient running space for your dog to exercise, the pen should be three times as long as it is wide. Find the dimensions of the pen.

1.2 Exercises

Dynamic Solutions available at *BigIdeasMath.com*

Vocabulary and Core Concept Check

1. **COMPLETE THE SENTENCE** To solve the equation $2x + 3x = 20$, first combine $2x$ and $3x$ because they are __________.

2. **WRITING** Describe two ways to solve the equation $2(4x - 11) = 10$.

Monitoring Progress and Modeling with Mathematics

In Exercises 3–14, solve the equation. Check your solution. *(See Examples 1 and 2.)*

3. $3w + 7 = 19$
4. $2g - 13 = 3$
5. $11 = 12 - q$
6. $10 = 7 - m$
7. $5 = \frac{z}{-4} - 3$
8. $\frac{a}{3} + 4 = 6$
9. $\frac{h+6}{5} = 2$
10. $\frac{d-8}{-2} = 12$
11. $8y + 3y = 44$
12. $36 = 13n - 4n$
13. $12v + 10v + 14 = 80$
14. $6c - 8 - 2c = -16$

15. **MODELING WITH MATHEMATICS** The altitude a (in feet) of a plane t minutes after liftoff is given by $a = 3400t + 600$. How many minutes after liftoff is the plane at an altitude of 21,000 feet?

16. **MODELING WITH MATHEMATICS** A repair bill for your car is \$553. The parts cost \$265. The labor cost is \$48 per hour. Write and solve an equation to find the number of hours of labor spent repairing the car.

In Exercises 17–24, solve the equation. Check your solution. *(See Example 3.)*

17. $4(z + 5) = 32$
18. $-2(4g - 3) = 30$
19. $6 + 5(m + 1) = 26$
20. $5h + 2(11 - h) = -5$
21. $27 = 3c - 3(6 - 2c)$
22. $-3 = 12y - 5(2y - 7)$
23. $-3(3 + x) + 4(x - 6) = -4$
24. $5(r + 9) - 2(1 - r) = 1$

USING TOOLS In Exercises 25–28, find the value of the variable. Then find the angle measures of the polygon. Use a protractor to check the reasonableness of your answer.

25.

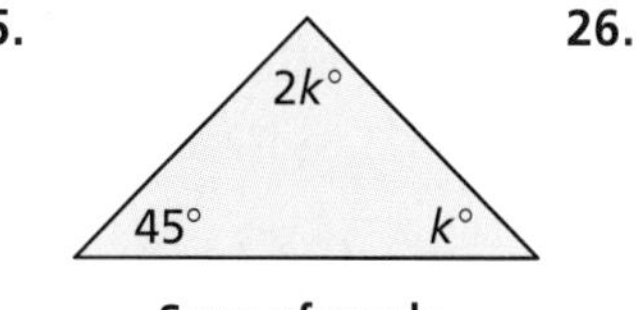

Sum of angle measures: 180°

26.

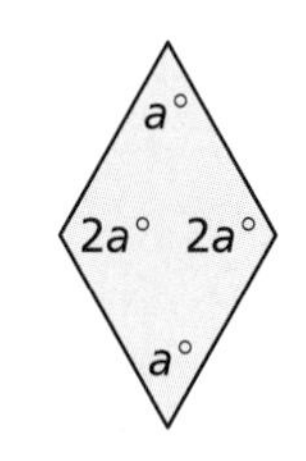

Sum of angle measures: 360°

27.

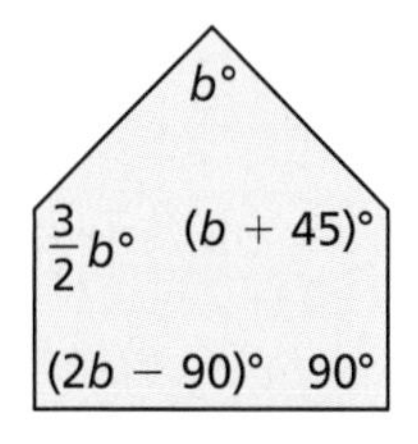

Sum of angle measures: 540°

28.

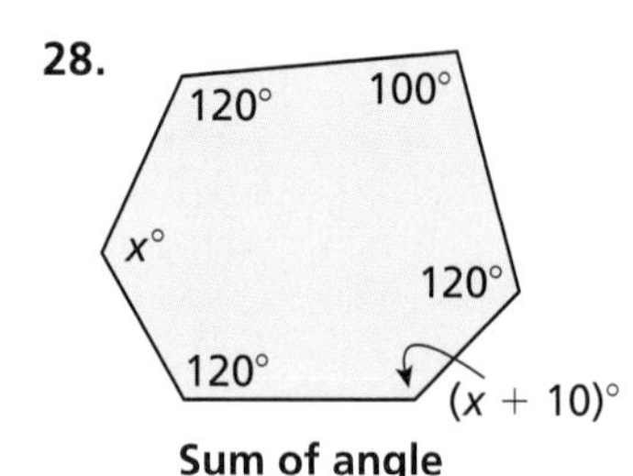

Sum of angle measures: 720°

In Exercises 29–34, write and solve an equation to find the number.

29. The sum of twice a number and 13 is 75.
30. The difference of three times a number and 4 is -19.
31. Eight plus the quotient of a number and 3 is -2.
32. The sum of twice a number and half the number is 10.
33. Six times the sum of a number and 15 is -42.
34. Four times the difference of a number and 7 is 12.

USING EQUATIONS In Exercises 35–37, write and solve an equation to answer the question. Check that the units on each side of the equation balance. *(See Examples 4 and 5.)*

35. During the summer, you work 30 hours per week at a gas station and earn \$8.75 per hour. You also work as a landscaper for \$11 per hour and can work as many hours as you want. You want to earn a total of \$400 per week. How many hours must you work as a landscaper?

36. The area of the surface of the swimming pool is 210 square feet. What is the length d of the deep end (in feet)?

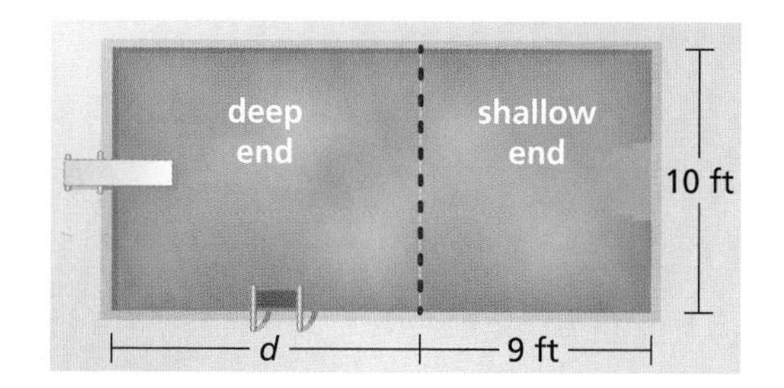

37. You order two tacos and a salad. The salad costs \$2.50. You pay 8% sales tax and leave a \$3 tip. You pay a total of \$13.80. How much does one taco cost?

JUSTIFYING STEPS In Exercises 38 and 39, justify each step of the solution.

38. $-\frac{1}{2}(5x - 8) - 1 = 6$ Write the equation.

$-\frac{1}{2}(5x - 8) = 7$ 

$5x - 8 = -14$ 

$5x = -6$ 

$x = -\frac{6}{5}$ 

39. $2(x + 3) + x = -9$ Write the equation.

$2(x) + 2(3) + x = -9$ 

$2x + 6 + x = -9$ 

$3x + 6 = -9$ 

$3x = -15$ 

$x = -5$

ERROR ANALYSIS In Exercises 40 and 41, describe and correct the error in solving the equation.

40.

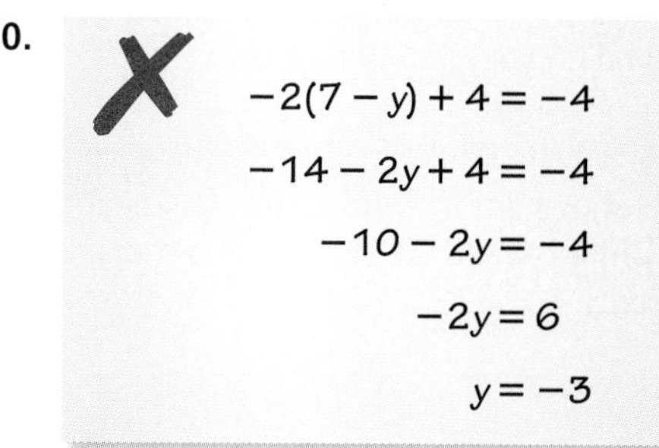

41.

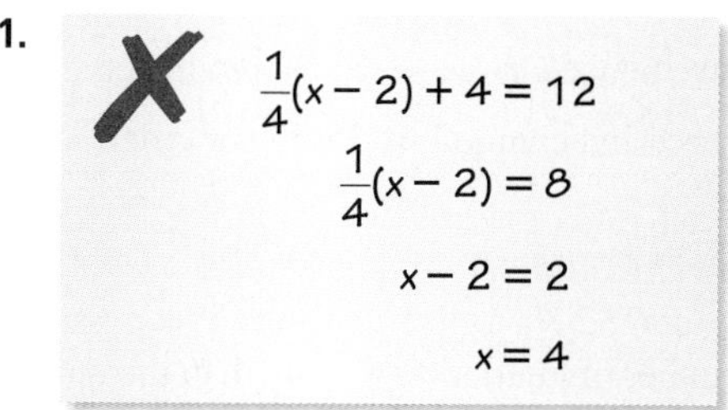

MATHEMATICAL CONNECTIONS In Exercises 42–44, write and solve an equation to answer the question.

42. The perimeter of the tennis court is 228 feet. What are the dimensions of the court?

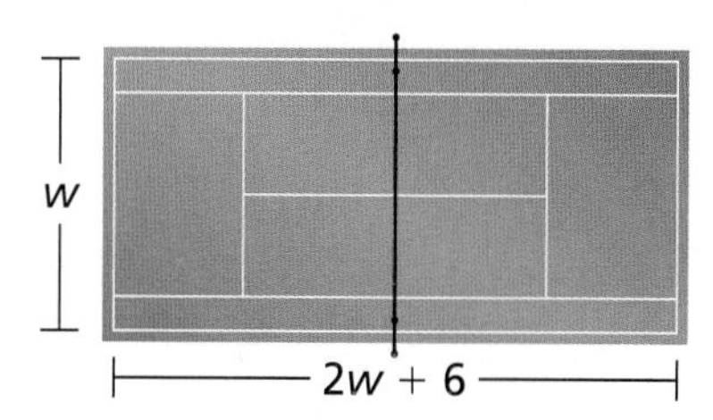

43. The perimeter of the Norwegian flag is 190 inches. What are the dimensions of the flag?

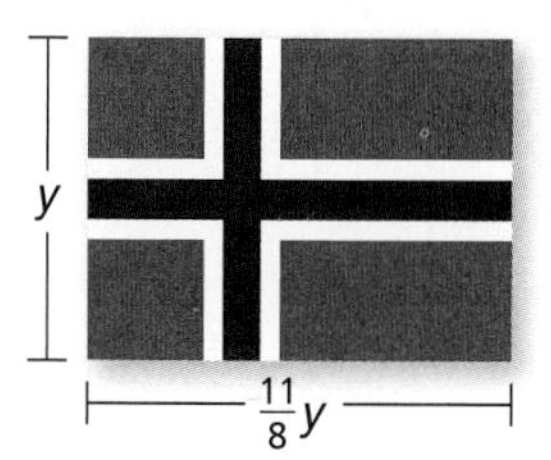

44. The perimeter of the school crossing sign is 102 inches. What is the length of each side?

45. **COMPARING METHODS** Solve the equation $2(4 - 8x) + 6 = -1$ using (a) Method 1 from Example 3 and (b) Method 2 from Example 3. Which method do you prefer? Explain.

46. **PROBLEM SOLVING** An online ticket agency charges the amounts shown for basketball tickets. The total cost for an order is $220.70. How many tickets are purchased?

Charge	Amount
Ticket price	$32.50 per ticket
Convenience charge	$3.30 per ticket
Processing charge	$5.90 per order

47. **MAKING AN ARGUMENT** You have quarters and dimes that total $2.80. Your friend says it is possible that the number of quarters is 8 more than the number of dimes. Is your friend correct? Explain.

48. **THOUGHT PROVOKING** You teach a math class and assign a weight to each component of the class. You determine final grades by totaling the products of the weights and the component scores. Choose values for the remaining weights and find the necessary score on the final exam for a student to earn an A (90%) in the class, if possible. Explain your reasoning.

Component	Student's score	Weight	Score × Weight
Class Participation	92%	0.20	92% × 0.20 = 18.4%
Homework	95%		
Midterm Exam	88%		
Final Exam			
Total		1	

49. **REASONING** An even integer can be represented by the expression $2n$, where n is any integer. Find three consecutive even integers that have a sum of 54. Explain your reasoning.

50. **HOW DO YOU SEE IT?** The scatter plot shows the attendance for each meeting of a gaming club.

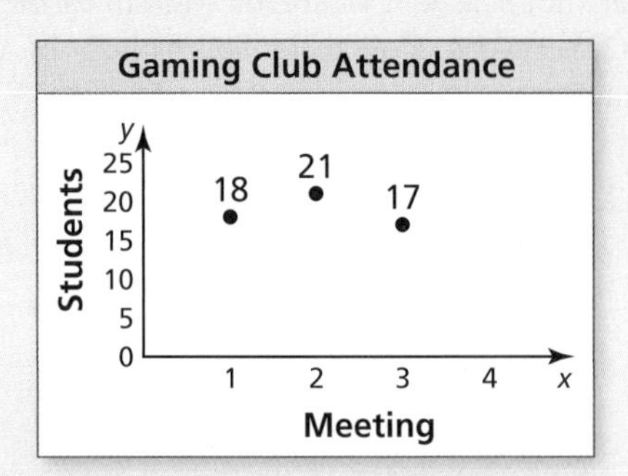

a. The mean attendance for the first four meetings is 20. Is the number of students who attended the fourth meeting greater than or less than 20? Explain.

b. Estimate the number of students who attended the fourth meeting.

c. Describe a way you can check your estimate in part (b).

REASONING In Exercises 51–56, the letters a, b, and c represent nonzero constants. Solve the equation for x.

51. $bx = -7$

52. $x + a = \frac{3}{4}$

53. $ax - b = 12.5$

54. $ax + b = c$

55. $2bx - bx = -8$

56. $cx - 4b = 5b$

Maintaining Mathematical Proficiency

Reviewing what you learned in previous grades and lessons

Simplify the expression. *(Skills Review Handbook)*

57. $4m + 5 - 3m$

58. $9 - 8b + 6b$

59. $6t + 3(1 - 2t) - 5$

Determine whether (a) $x = -1$ or (b) $x = 2$ is a solution of the equation. *(Skills Review Handbook)*

60. $x - 8 = -9$

61. $x + 1.5 = 3.5$

62. $2x - 1 = 3$

63. $3x + 4 = 1$

64. $x + 4 = 3x$

65. $-2(x - 1) = 1 - 3x$

1.3 Solving Equations with Variables on Both Sides

Essential Question How can you solve an equation that has variables on both sides?

EXPLORATION 1 **Perimeter**

Work with a partner. The two polygons have the same perimeter. Use this information to write and solve an equation involving x. Explain the process you used to find the solution. Then find the perimeter of each polygon.

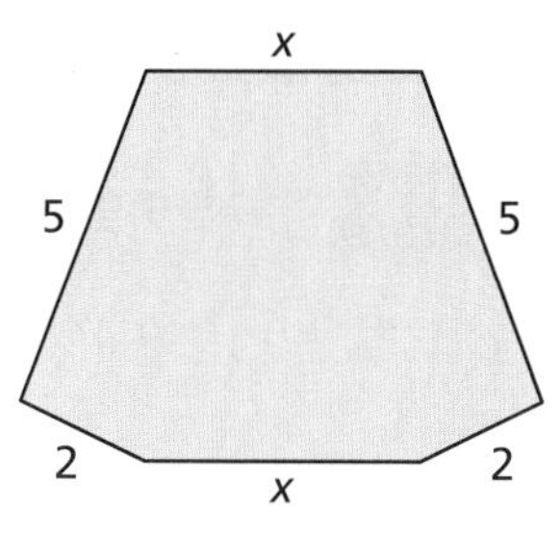

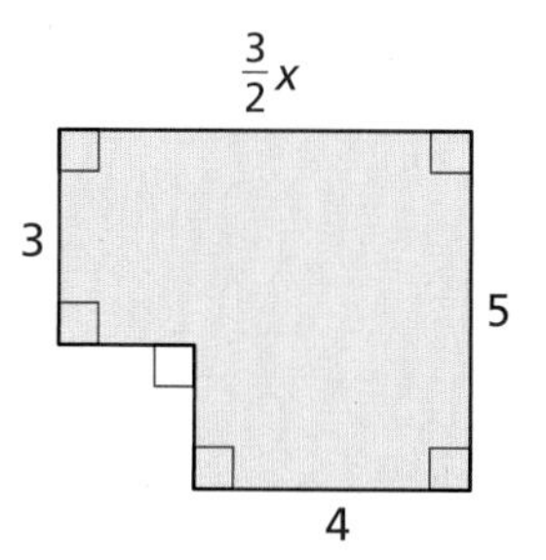

EXPLORATION 2 **Perimeter and Area**

Work with a partner.

- Each figure has the unusual property that the value of its perimeter (in feet) is equal to the value of its area (in square feet). Use this information to write an equation for each figure.
- Solve each equation for x. Explain the process you used to find the solution.
- Find the perimeter and area of each figure.

LOOKING FOR STRUCTURE

To be proficient in math, you need to visualize complex things, such as composite figures, as being made up of simpler, more manageable parts.

a.

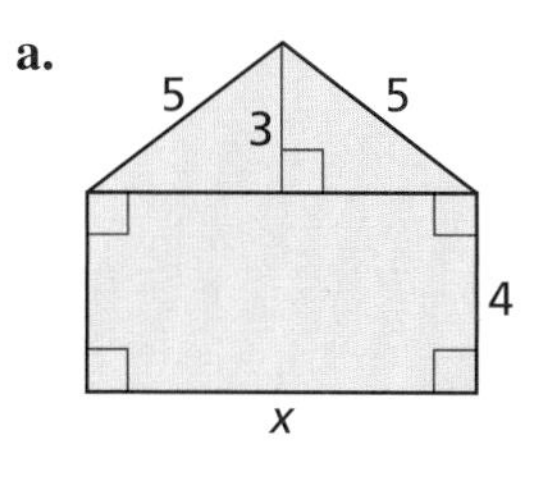

b.

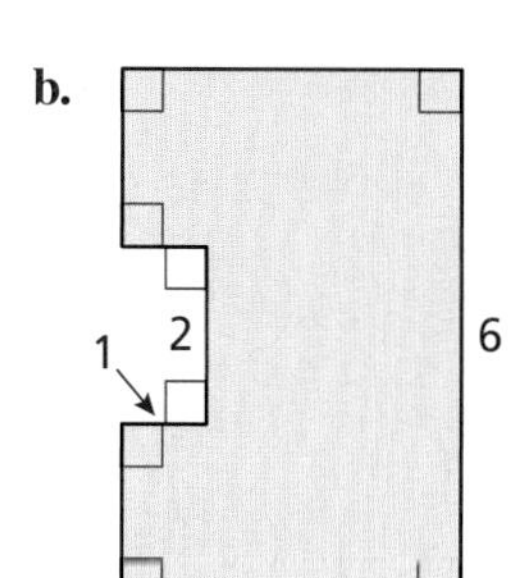

c.

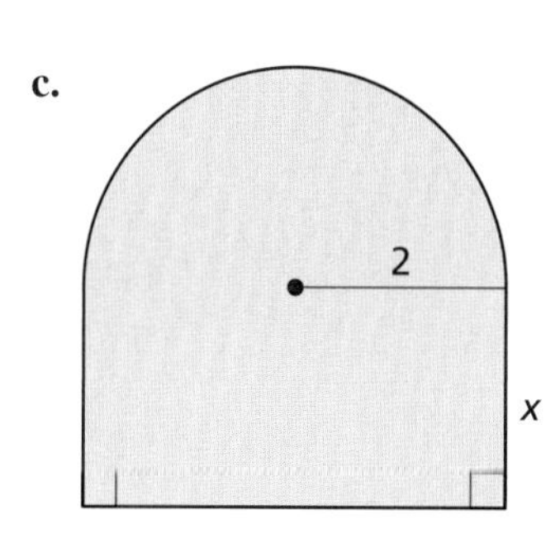

Communicate Your Answer

3. How can you solve an equation that has variables on both sides?
4. Write three equations that have the variable x on both sides. The equations should be different from those you wrote in Explorations 1 and 2. Have your partner solve the equations.

1.3 Lesson

Core Vocabulary

identity, *p. 21*

Previous

inverse operations

What You Will Learn

- Solve linear equations that have variables on both sides.
- Identify special solutions of linear equations.
- Use linear equations to solve real-life problems.

Solving Equations with Variables on Both Sides

Core Concept

Solving Equations with Variables on Both Sides

To solve an equation with variables on both sides, simplify one or both sides of the equation, if necessary. Then use inverse operations to collect the variable terms on one side, collect the constant terms on the other side, and isolate the variable.

EXAMPLE 1 Solving an Equation with Variables on Both Sides

Solve $10 - 4x = -9x$. Check your solution.

SOLUTION

$10 - 4x = -9x$	Write the equation.
$+4x \quad +4x$	Add $4x$ to each side.
$10 = -5x$	Simplify.
$\frac{10}{-5} = \frac{-5x}{-5}$	Divide each side by -5.
$-2 = x$	Simplify.

Check

$$10 - 4x = -9x$$
$$10 - 4(-2) \stackrel{?}{=} -9(-2)$$
$$18 = 18 \checkmark$$

▶ The solution is $x = -2$.

EXAMPLE 2 Solving an Equation with Grouping Symbols

Solve $3(3x - 4) = \frac{1}{4}(32x + 56)$.

SOLUTION

$3(3x - 4) = \frac{1}{4}(32x + 56)$	Write the equation.
$9x - 12 = 8x + 14$	Distributive Property
$+12 \quad +12$	Add 12 to each side.
$9x = 8x + 26$	Simplify.
$-8x \quad -8x$	Subtract $8x$ from each side.
$x = 26$	Simplify.

▶ The solution is $x = 26$.

Monitoring Progress  Help in English and Spanish at *BigIdeasMath.com*

Solve the equation. Check your solution.

1. $-2x = 3x + 10$ **2.** $\frac{1}{2}(6h - 4) = -5h + 1$ **3.** $-\frac{3}{4}(8n + 12) = 3(n - 3)$

Identifying Special Solutions of Linear Equations

Special Solutions of Linear Equations

Equations do not always have one solution. An equation that is true for all values of the variable is an **identity** and has *infinitely many solutions*. An equation that is not true for any value of the variable has *no solution*.

EXAMPLE 3 Identifying the Number of Solutions

REASONING

The equation $15x + 6 = 15x$ is not true because the number $15x$ cannot be equal to 6 more than itself.

Solve each equation.

a. $3(5x + 2) = 15x$

b. $-2(4y + 1) = -8y - 2$

SOLUTION

a.

$3(5x + 2) = 15x$	Write the equation.
$15x + 6 = 15x$	Distributive Property
$\underline{-15x} \qquad \underline{-15x}$	Subtract $15x$ from each side.
$6 = 0$ ✗	Simplify.

▶ The statement $6 = 0$ is never true. So, the equation has no solution.

READING

All real numbers are solutions of an identity.

b.

$-2(4y + 1) = -8y - 2$	Write the equation.
$-8y - 2 = -8y - 2$	Distributive Property
$\underline{+8y} \qquad \underline{+8y}$	Add $8y$ to each side.
$-2 = -2$	Simplify.

▶ The statement $-2 = -2$ is always true. So, the equation is an identity and has infinitely many solutions.

Monitoring Progress

Help in English and Spanish at *BigIdeasMath.com*

Solve the equation.

4. $4(1 - p) = -4p + 4$

5. $6m - m = \frac{5}{6}(6m - 10)$

6. $10k + 7 = -3 - 10k$

7. $3(2a - 2) = 2(3a - 3)$

Concept Summary

Steps for Solving Linear Equations

Here are several steps you can use to solve a linear equation. Depending on the equation, you may not need to use some steps.

Step 1 Use the Distributive Property to remove any grouping symbols.

Step 2 Simplify the expression on each side of the equation.

Step 3 Collect the variable terms on one side of the equation and the constant terms on the other side.

Step 4 Isolate the variable.

Step 5 Check your solution.

STUDY TIP

To check an identity, you can choose several different values of the variable.

Solving Real-Life Problems

EXAMPLE 4 Modeling with Mathematics

A boat leaves New Orleans and travels upstream on the Mississippi River for 4 hours. The return trip takes only 2.8 hours because the boat travels 3 miles per hour faster downstream due to the current. How far does the boat travel upstream?

SOLUTION

1. **Understand the Problem** You are given the amounts of time the boat travels and the difference in speeds for each direction. You are asked to find the distance the boat travels upstream.
2. **Make a Plan** Use the Distance Formula to write expressions that represent the problem. Because the distance the boat travels in both directions is the same, you can use the expressions to write an equation.
3. **Solve the Problem** Use the formula (distance) = (rate)(time).

Words Distance upstream = Distance downstream

Variable Let x be the speed (in miles per hour) of the boat traveling upstream.

Equation $\frac{x \text{ mi}}{1 \cancel{h}} \cdot 4 \cancel{h} = \frac{(x+3) \text{ mi}}{1 \cancel{h}} \cdot 2.8 \cancel{h}$ (mi = mi) ✓

$4x = 2.8(x + 3)$ Write the equation.

$4x = 2.8x + 8.4$ Distributive Property

$-2.8x \quad -2.8x$ Subtract $2.8x$ from each side.

$1.2x = 8.4$ Simplify.

$\frac{1.2x}{1.2} = \frac{8.4}{1.2}$ Divide each side by 1.2.

$x = 7$ Simplify.

▶ So, the boat travels 7 miles per hour upstream. To determine how far the boat travels upstream, multiply 7 miles per hour by 4 hours to obtain 28 miles.

4. **Look Back** To check that your solution is reasonable, use the formula for distance. Because the speed upstream is 7 miles per hour, the speed downstream is $7 + 3 = 10$ miles per hour. When you substitute each speed into the Distance Formula, you get the same distance for upstream and downstream.

Upstream

$$\text{Distance} = \frac{7 \text{ mi}}{1 \cancel{h}} \cdot 4 \cancel{h} = 28 \text{ mi}$$

Downstream

$$\text{Distance} = \frac{10 \text{ mi}}{1 \cancel{h}} \cdot 2.8 \cancel{h} = 28 \text{ mi}$$

Monitoring Progress

Help in English and Spanish at *BigIdeasMath.com*

8. A boat travels upstream on the Mississippi River for 3.5 hours. The return trip only takes 2.5 hours because the boat travels 2 miles per hour faster downstream due to the current. How far does the boat travel upstream?

1.3 Exercises

Dynamic Solutions available at *BigIdeasMath.com*

Vocabulary and Core Concept Check

1. **VOCABULARY** Is the equation $-2(4 - x) = 2x + 8$ an identity? Explain your reasoning.

2. **WRITING** Describe the steps in solving the linear equation $3(3x - 8) = 4x + 6$.

Monitoring Progress and Modeling with Mathematics

In Exercises 3–16, solve the equation. Check your solution. *(See Examples 1 and 2.)*

3. $15 - 2x = 3x$
4. $26 - 4s = 9s$
5. $5p - 9 = 2p + 12$
6. $8g + 10 = 35 + 3g$
7. $5t + 16 = 6 - 5t$
8. $-3r + 10 = 15r - 8$
9. $7 + 3x - 12x = 3x + 1$
10. $w - 2 + 2w = 6 + 5w$
11. $10(g + 5) = 2(g + 9)$
12. $-9(t - 2) = 4(t - 15)$
13. $\frac{2}{3}(3x + 9) = -2(2x + 6)$
14. $2(2t + 4) = \frac{3}{4}(24 - 8t)$
15. $10(2y + 2) - y = 2(8y - 8)$
16. $2(4x + 2) = 4x - 12(x - 1)$

17. **MODELING WITH MATHEMATICS** You and your friend drive toward each other. The equation $50h = 190 - 45h$ represents the number h of hours until you and your friend meet. When will you meet?

18. **MODELING WITH MATHEMATICS** The equation $1.5r + 15 = 2.25r$ represents the number r of movies you must rent to spend the same amount at each movie store. How many movies must you rent to spend the same amount at each movie store?

Membership Fee: $15

Membership Fee: Free

In Exercises 19–24, solve the equation. Determine whether the equation has *one solution*, *no solution*, or *infinitely many solutions*. *(See Example 3.)*

19. $3t + 4 = 12 + 3t$
20. $6d + 8 = 14 + 3d$
21. $2(h + 1) = 5h - 7$
22. $12y + 6 = 6(2y + 1)$
23. $3(4g + 6) = 2(6g + 9)$
24. $5(1 + 2m) = \frac{1}{2}(8 + 20m)$

ERROR ANALYSIS In Exercises 25 and 26, describe and correct the error in solving the equation.

25.

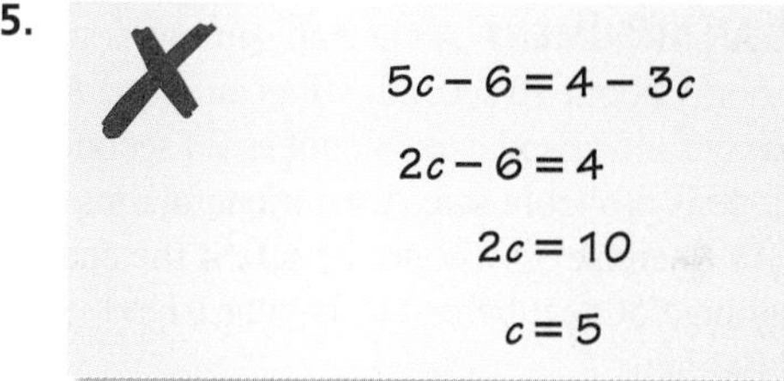

26.

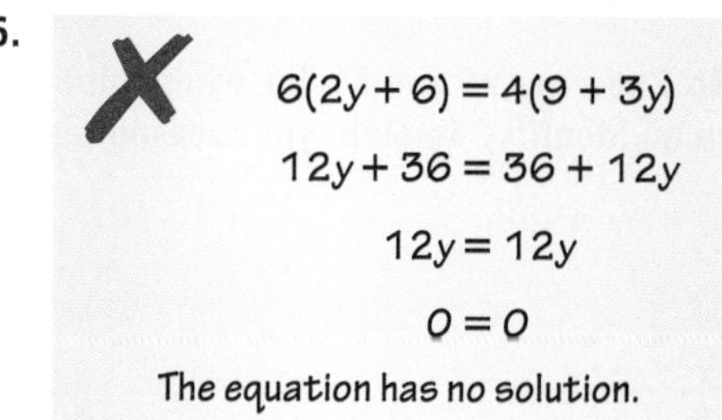

27. **MODELING WITH MATHEMATICS** Write and solve an equation to find the month when you would pay the same total amount for each Internet service.

	Installation fee	Price per month
Company A	$60.00	$42.95
Company B	$25.00	$49.95

28. PROBLEM SOLVING One serving of granola provides 4% of the protein you need daily. You must get the remaining 48 grams of protein from other sources. How many grams of protein do you need daily?

USING STRUCTURE In Exercises 29 and 30, find the value of r.

29. $8(x + 6) - 10 + r = 3(x + 12) + 5x$

30. $4(x - 3) - r + 2x = 5(3x - 7) - 9x$

MATHEMATICAL CONNECTIONS In Exercises 31 and 32, the value of the surface area of the cylinder is equal to the value of the volume of the cylinder. Find the value of x. Then find the surface area and volume of the cylinder.

31.

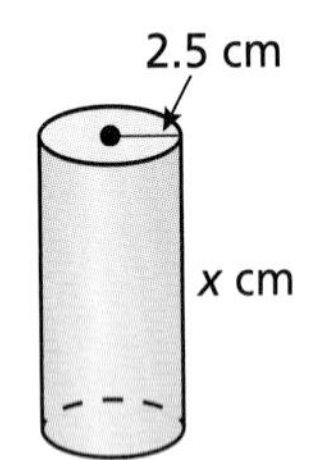

32.

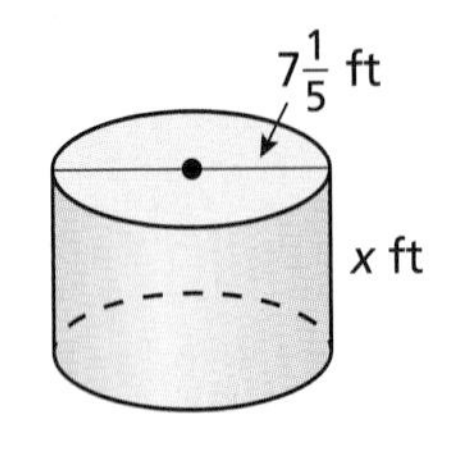

33. MODELING WITH MATHEMATICS A cheetah that is running 90 feet per second is 120 feet behind an antelope that is running 60 feet per second. How long will it take the cheetah to catch up to the antelope? *(See Example 4.)*

34. MAKING AN ARGUMENT A cheetah can run at top speed for only about 20 seconds. If an antelope is too far away for a cheetah to catch it in 20 seconds, the antelope is probably safe. Your friend claims the antelope in Exercise 33 will not be safe if the cheetah starts running 650 feet behind it. Is your friend correct? Explain.

REASONING In Exercises 35 and 36, for what value of a is the equation an identity? Explain your reasoning.

35. $a(2x + 3) = 9x + 15 + x$

36. $8x - 8 + 3ax = 5ax - 2a$

37. REASONING Two times the greater of two consecutive integers is 9 less than three times the lesser integer. What are the integers?

38. HOW DO YOU SEE IT? The table and the graph show information about students enrolled in Spanish and French classes at a high school.

	Students enrolled this year	Average rate of change
Spanish	355	9 fewer students each year
French	229	12 more students each year

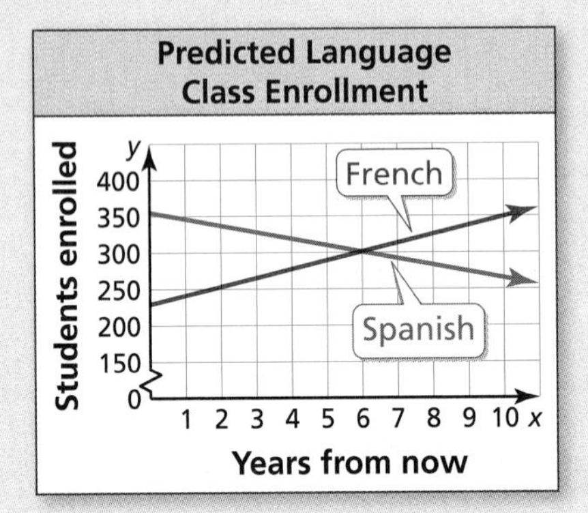

a. Use the graph to determine after how many years there will be equal enrollment in Spanish and French classes.

b. How does the equation $355 - 9x = 229 + 12x$ relate to the table and the graph? How can you use this equation to determine whether your answer in part (a) is reasonable?

39. WRITING EQUATIONS Give an example of a linear equation that has (a) no solution and (b) infinitely many solutions. Justify your answers.

40. THOUGHT PROVOKING Draw a different figure that has the same perimeter as the triangle shown. Explain why your figure has the same perimeter.

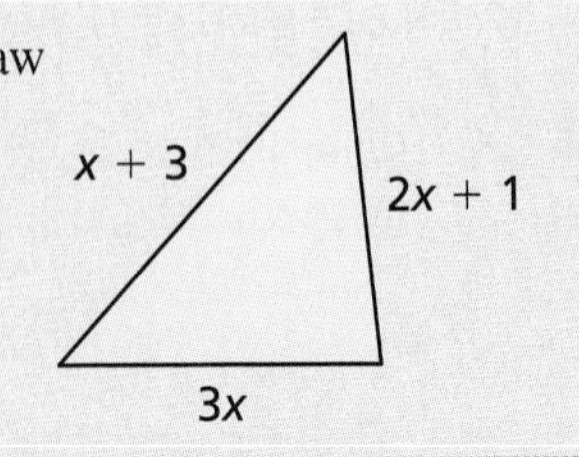

Maintaining Mathematical Proficiency

Reviewing what you learned in previous grades and lessons

Order the values from least to greatest. *(Skills Review Handbook)*

41. $9, |-4|, -4, 5, |2|$

42. $|-32|, 22, -16, -|21|, |-10|$

43. $-18, |-24|, -19, |-18|, |22|$

44. $-|-3|, |0|, -1, |2|, -2$

1.1–1.3 What Did You Learn?

Core Vocabulary

conjecture, *p. 3*
rule, *p. 3*
theorem, *p. 3*
equation, *p. 4*
linear equation in one variable, *p. 4*
solution, *p. 4*
inverse operations, *p. 4*
equivalent equations, *p. 4*
identity, *p. 21*

Core Concepts

Section 1.1

Addition Property of Equality, *p. 4*
Subtraction Property of Equality, *p. 4*
Multiplication Property of Equality, *p. 5*
Division Property of Equality, *p. 5*
Four-Step Approach to Problem Solving, *p. 6*
Common Problem-Solving Strategies, *p. 7*

Section 1.2

Solving Multi-Step Equations, *p. 12*
Unit Analysis, *p. 15*

Section 1.3

Solving Equations with Variables on Both Sides, *p. 20*
Special Solutions of Linear Equations, *p. 21*

Mathematical Practices

1. How did you make sense of the relationships between the quantities in Exercise 46 on page 9?
2. What is the limitation of the tool you used in Exercises 25–28 on page 16?
3. What definition did you use in your reasoning in Exercises 35 and 36 on page 24?

Study Skills

Completing Homework Efficiently

Before doing homework, review the Core Concepts and examples. Use the tutorials at *BigIdeasMath.com* for additional help.

Complete homework as though you are also preparing for a quiz. Memorize different types of problems, vocabulary, rules, and so on.

1.1–1.3 Quiz

Solve the equation. Justify each step. Check your solution. *(Section 1.1)*

1. $x + 9 = 7$

2. $8.6 = z - 3.8$

3. $60 = -12r$

4. $\frac{3}{4}p = 18$

Solve the equation. Check your solution. *(Section 1.2)*

5. $2m - 3 = 13$

6. $5 = 10 - v$

7. $5 = 7w + 8w + 2$

8. $-21a + 28a - 6 = -10.2$

9. $2k - 3(2k - 3) = 45$

10. $68 = \frac{1}{5}(20x + 50) + 2$

Solve the equation. *(Section 1.3)*

11. $3c + 1 = c + 1$

12. $-8 - 5n = 64 + 3n$

13. $2(8q - 5) = 4q$

14. $9(y - 4) - 7y = 5(3y - 2)$

15. $4(g + 8) = 7 + 4g$

16. $-4(-5h - 4) = 2(10h + 8)$

17. To estimate how many miles you are from a thunderstorm, count the seconds between when you see lightning and when you hear thunder. Then divide by 5. Write and solve an equation to determine how many seconds you would count for a thunderstorm that is 2 miles away. *(Section 1.1)*

18. You want to hang three equally-sized travel posters on a wall so that the posters on the ends are each 3 feet from the end of the wall. You want the spacing between posters to be equal. Write and solve an equation to determine how much space you should leave between the posters. *(Section 1.2)*

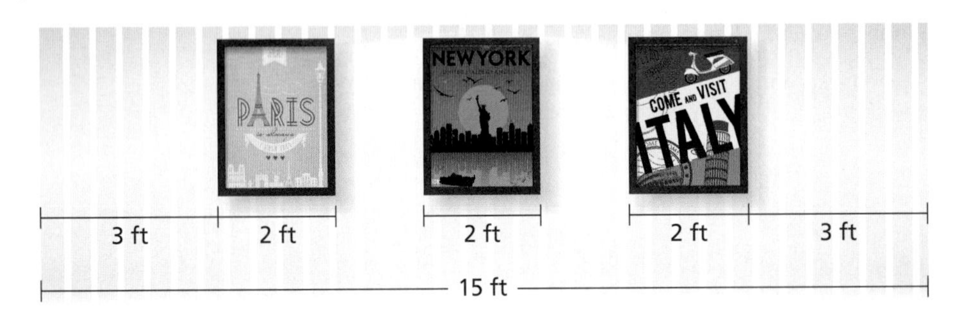

19. You want to paint a piece of pottery at an art studio. The total cost is the cost of the piece plus an hourly studio fee. There are two studios to choose from. *(Section 1.3)*

a. After how many hours of painting are the total costs the same at both studios? Justify your answer.

b. Studio B increases the hourly studio fee by $2. How does this affect your answer in part (a)? Explain.

1.4 Solving Absolute Value Equations

Essential Question How can you solve an absolute value equation?

EXPLORATION 1 Solving an Absolute Value Equation Algebraically

Work with a partner. Consider the absolute value equation

$|x + 2| = 3.$

a. Describe the values of $x + 2$ that make the equation true. Use your description to write two linear equations that represent the solutions of the absolute value equation.

b. Use the linear equations you wrote in part (a) to find the solutions of the absolute value equation.

c. How can you use linear equations to solve an absolute value equation?

MAKING SENSE OF PROBLEMS

To be proficient in math, you need to explain to yourself the meaning of a problem and look for entry points to its solution.

EXPLORATION 2 Solving an Absolute Value Equation Graphically

Work with a partner. Consider the absolute value equation

$|x + 2| = 3.$

a. On a real number line, locate the point for which $x + 2 = 0$.

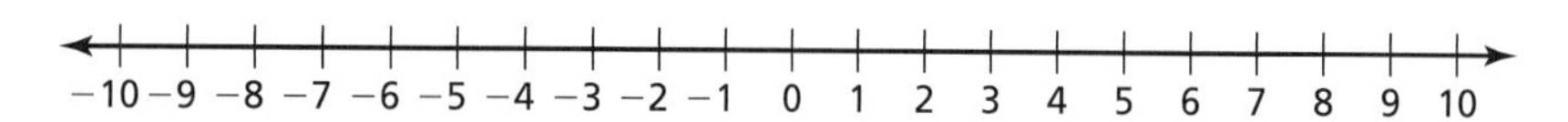

b. Locate the points that are 3 units from the point you found in part (a). What do you notice about these points?

c. How can you use a number line to solve an absolute value equation?

EXPLORATION 3 Solving an Absolute Value Equation Numerically

Work with a partner. Consider the absolute value equation

$|x + 2| = 3.$

a. Use a spreadsheet, as shown, to solve the absolute value equation.

b. Compare the solutions you found using the spreadsheet with those you found in Explorations 1 and 2. What do you notice?

c. How can you use a spreadsheet to solve an absolute value equation?

	A	B
1	x	|x + 2|
2	-6	4
3	-5	
4	-4	
5	-3	
6	-2	
7	-1	
8	0	
9	1	
10	2	
11		

abs(A2 + 2)

Communicate Your Answer

4. How can you solve an absolute value equation?

5. What do you like or dislike about the algebraic, graphical, and numerical methods for solving an absolute value equation? Give reasons for your answers.

1.4 Lesson

What You Will Learn

- Solve absolute value equations.
- Solve equations involving two absolute values.
- Identify special solutions of absolute value equations.

Core Vocabulary

absolute value equation, *p. 28*
extraneous solution, *p. 31*

Previous
absolute value
opposite

Solving Absolute Value Equations

An **absolute value equation** is an equation that contains an absolute value expression. You can solve these types of equations by solving two related linear equations.

Core Concept

Properties of Absolute Value

Let a and b be real numbers. Then the following properties are true.

1. $|a| \geq 0$ **2.** $|-a| = |a|$

3. $|ab| = |a||b|$ **4.** $\left|\frac{a}{b}\right| = \frac{|a|}{|b|}, b \neq 0$

Solving Absolute Value Equations

To solve $|ax + b| = c$ when $c \geq 0$, solve the related linear equations

$$ax + b = c \quad \text{or} \quad ax + b = -c.$$

When $c < 0$, the absolute value equation $|ax + b| = c$ has no solution because absolute value always indicates a number that is not negative.

EXAMPLE 1 Solving Absolute Value Equations

Solve each equation. Graph the solutions, if possible.

a. $|x - 4| = 6$ **b.** $|3x + 1| = -5$

SOLUTION

a. Write the two related linear equations for $|x - 4| = 6$. Then solve.

$x - 4 = 6$ *or* $x - 4 = -6$ Write related linear equations.

$x = 10$ $x = -2$ Add 4 to each side.

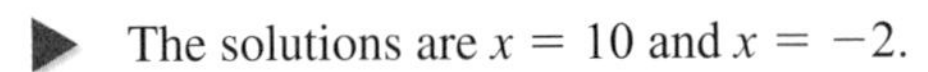

The solutions are $x = 10$ and $x = -2$.

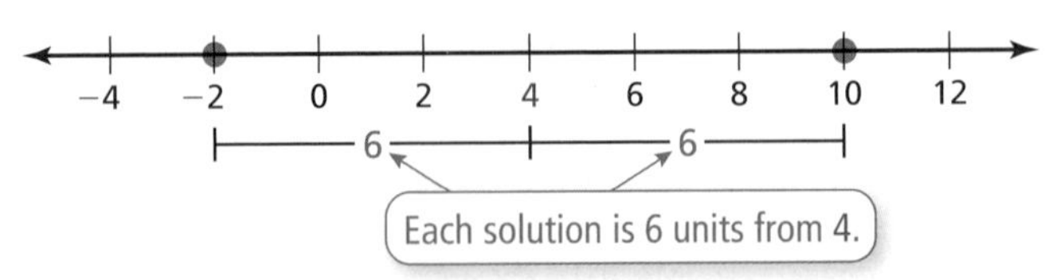

Property of Absolute Value

b. The absolute value of an expression must be greater than or equal to 0. The expression $|3x + 1|$ cannot equal -5.

So, the equation has no solution.

Monitoring Progress

Help in English and Spanish at *BigIdeasMath.com*

Solve the equation. Graph the solutions, if possible.

1. $|x| = 10$ **2.** $|x - 1| = 4$ **3.** $|3 + x| = -3$

EXAMPLE 2 Solving an Absolute Value Equation

Solve $|3x + 9| - 10 = -4$.

SOLUTION

> **ANOTHER WAY**
> Using the product property of absolute value, $|ab| = |a|\,|b|$, you could rewrite the equation as
> $$3|x + 3| - 10 = -4$$
> and then solve.

First isolate the absolute value expression on one side of the equation.

$\lvert 3x + 9\rvert - 10 = -4$	Write the equation.
$\lvert 3x + 9\rvert = 6$	Add 10 to each side.

Now write two related linear equations for $|3x + 9| = 6$. Then solve.

$3x + 9 = 6$	*or*	$3x + 9 = -6$	Write related linear equations.
$3x = -3$		$3x = -15$	Subtract 9 from each side.
$x = -1$		$x = -5$	Divide each side by 3.

▶ The solutions are $x = -1$ and $x = -5$.

EXAMPLE 3 Writing an Absolute Value Equation

In a cheerleading competition, the minimum length of a routine is 4 minutes. The maximum length of a routine is 5 minutes. Write an absolute value equation that represents the minimum and maximum lengths.

> **REASONING**
> Mathematically proficient students have the ability to decontextualize problem situations.

SOLUTION

1. **Understand the Problem** You know the minimum and maximum lengths. You are asked to write an absolute value equation that represents these lengths.
2. **Make a Plan** Consider the minimum and maximum lengths as solutions to an absolute value equation. Use a number line to find the halfway point between the solutions. Then use the halfway point and the distance to each solution to write an absolute value equation.
3. **Solve the Problem**

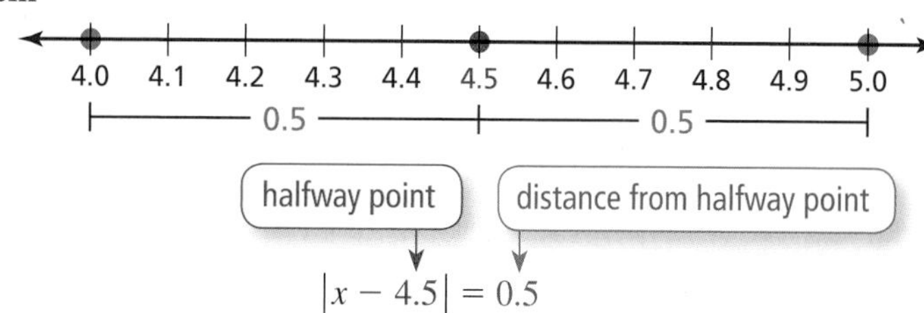

▶ The equation is $|x - 4.5| = 0.5$.

4. **Look Back** To check that your equation is reasonable, substitute the minimum and maximum lengths into the equation and simplify.

Minimum	**Maximum**
$\lvert 4 - 4.5\rvert = 0.5$ ✓	$\lvert 5 - 4.5\rvert = 0.5$ ✓

Monitoring Progress

Help in English and Spanish at *BigIdeasMath.com*

Solve the equation. Check your solutions.

4. $|x - 2| + 5 = 9$ **5.** $4|2x + 7| = 16$ **6.** $-2|5x - 1| - 3 = -11$

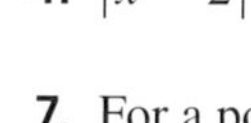

7. For a poetry contest, the minimum length of a poem is 16 lines. The maximum length is 32 lines. Write an absolute value equation that represents the minimum and maximum lengths.

Solving Equations with Two Absolute Values

If the absolute values of two algebraic expressions are equal, then they must either be equal to each other or be opposites of each other.

Solving Equations with Two Absolute Values

To solve $|ax + b| = |cx + d|$, solve the related linear equations

$$ax + b = cx + d \quad or \quad ax + b = -(cx + d).$$

EXAMPLE 4 **Solving Equations with Two Absolute Values**

Solve (a) $|3x - 4| = |x|$ and (b) $|4x - 10| = 2|3x + 1|$.

SOLUTION

a. Write the two related linear equations for $|3x - 4| = |x|$. Then solve.

$$3x - 4 = x \quad or \quad 3x - 4 = -x$$

$$\underline{-x} \quad \underline{-x} \qquad \underline{+x} \quad \underline{+x}$$

$$2x - 4 = 0 \qquad 4x - 4 = 0$$

$$\underline{+4} \quad \underline{+4} \qquad \underline{+4} \quad \underline{+4}$$

$$2x = 4 \qquad 4x = 4$$

$$\frac{2x}{2} = \frac{4}{2} \qquad \frac{4x}{4} = \frac{4}{4}$$

$$x = 2 \qquad x = 1$$

▶ The solutions are $x = 2$ and $x = 1$.

Check

$|3x - 4| = |x|$

$|3(2) - 4| \stackrel{?}{=} |2|$

$|2| \stackrel{?}{=} |2|$

$2 = 2$ ✓

$|3x - 4| = |x|$

$|3(1) - 4| \stackrel{?}{=} |1|$

$|-1| \stackrel{?}{=} |1|$

$1 = 1$ ✓

b. Write the two related linear equations for $|4x - 10| = 2|3x + 1|$. Then solve.

$$4x - 10 = 2(3x + 1) \quad or \quad 4x - 10 = 2[-(3x + 1)]$$

$$4x - 10 = 6x + 2 \qquad 4x - 10 = 2(-3x - 1)$$

$$\underline{-6x} \quad \underline{-6x} \qquad 4x - 10 = -6x - 2$$

$$-2x - 10 = 2 \qquad \underline{+6x} \quad \underline{+6x}$$

$$\underline{+10} \quad \underline{+10} \qquad 10x - 10 = -2$$

$$-2x = 12 \qquad \underline{+10} \quad \underline{+10}$$

$$\frac{-2x}{-2} = \frac{12}{-2} \qquad 10x = 8$$

$$x = -6 \qquad \frac{10x}{10} = \frac{8}{10}$$

$$x = 0.8$$

▶ The solutions are $x = -6$ and $x = 0.8$.

Monitoring Progress

Help in English and Spanish at *BigIdeasMath.com*

Solve the equation. Check your solutions.

8. $|x + 8| = |2x + 1|$

9. $3|x - 4| = |2x + 5|$

Identifying Special Solutions

When you solve an absolute value equation, it is possible for a solution to be *extraneous*. An **extraneous solution** is an apparent solution that must be rejected because it does not satisfy the original equation.

EXAMPLE 5 Identifying Extraneous Solutions

Solve $|2x + 12| = 4x$. Check your solutions.

SOLUTION

Write the two related linear equations for $|2x + 12| = 4x$. Then solve.

$2x + 12 = 4x$	*or*	$2x + 12 = -4x$	Write related linear equations.
$12 = 2x$		$12 = -6x$	Subtract $2x$ from each side.
$6 = x$		$-2 = x$	Solve for x.

Check the apparent solutions to see if either is extraneous.

Check

$$|2x + 12| = 4x$$

$$|2(6) + 12| \stackrel{?}{=} 4(6)$$

$$|24| \stackrel{?}{=} 24$$

$$24 = 24$$ 

$$|2x + 12| = 4x$$

$$|2(-2) + 12| \stackrel{?}{=} 4(-2)$$

$$|8| \stackrel{?}{=} -8$$

$$8 \neq -8 \ ✗$$

▶ The solution is $x = 6$. Reject $x = -2$ because it is extraneous.

When solving equations of the form $|ax + b| = |cx + d|$, it is possible that one of the related linear equations will not have a solution.

EXAMPLE 6 Solving an Equation with Two Absolute Values

Solve $|x + 5| = |x + 11|$.

SOLUTION

By equating the expression $x + 5$ and the opposite of $x + 11$, you obtain

$x + 5 = -(x + 11)$	Write related linear equation.
$x + 5 = -x - 11$	Distributive Property
$2x + 5 = -11$	Add x to each side.
$2x = -16$	Subtract 5 from each side.
$x = -8.$	Divide each side by 2.

However, by equating the expressions $x + 5$ and $x + 11$, you obtain

$x + 5 = x + 11$	Write related linear equation.
$x = x + 6$	Subtract 5 from each side.
$0 = 6$ ✗	Subtract x from each side.

which is a false statement. So, the original equation has only one solution.

REMEMBER

Always check your solutions in the original equation to make sure they are not extraneous.

▶ The solution is $x = -8$.

Monitoring Progress  Help in English and Spanish at *BigIdeasMath.com*

Solve the equation. Check your solutions.

10. $|x + 6| = 2x$

11. $|3x - 2| = x$

12. $|2 + x| = |x - 8|$

13. $|5x - 2| = |5x + 4|$

1.4 Exercises

Dynamic Solutions available at *BigIdeasMath.com*

Vocabulary and Core Concept Check

1. **VOCABULARY** What is an extraneous solution?

2. **WRITING** Without calculating, how do you know that the equation $|4x - 7| = -1$ has no solution?

Monitoring Progress and Modeling with Mathematics

In Exercises 3–10, simplify the expression.

3. $|-9|$

4. $-|15|$

5. $|14| - |-14|$

6. $|-3| + |3|$

7. $-|-5 \cdot (-7)|$

8. $|-0.8 \cdot 10|$

9. $\left|\frac{27}{-3}\right|$

10. $\left|-\frac{-12}{4}\right|$

In Exercises 11–24, solve the equation. Graph the solution(s), if possible. *(See Examples 1 and 2.)*

11. $|w| = 6$

12. $|r| = -2$

13. $|y| = -18$

14. $|x| = 13$

15. $|m + 3| = 7$

16. $|q - 8| = 14$

17. $|-3d| = 15$

18. $\left|\frac{t}{2}\right| = 6$

19. $|4b - 5| = 19$

20. $|x - 1| + 5 = 2$

21. $-4|8 - 5n| = 13$

22. $-3\left|1 - \frac{2}{3}v\right| = -9$

23. $3 = -2\left|\frac{1}{4}s - 5\right| + 3$

24. $9|4p + 2| + 8 = 35$

25. **WRITING EQUATIONS** The minimum distance from Earth to the Sun is 91.4 million miles. The maximum distance is 94.5 million miles. *(See Example 3.)*

    a. Represent these two distances on a number line.

    b. Write an absolute value equation that represents the minimum and maximum distances.

26. **WRITING EQUATIONS** The shoulder heights of the shortest and tallest miniature poodles are shown.

    a. Represent these two heights on a number line.

    b. Write an absolute value equation that represents these heights.

USING STRUCTURE In Exercises 27–30, match the absolute value equation with its graph without solving the equation.

27. $|x + 2| = 4$

28. $|x - 4| = 2$

29. $|x - 2| = 4$

30. $|x + 4| = 2$

A. 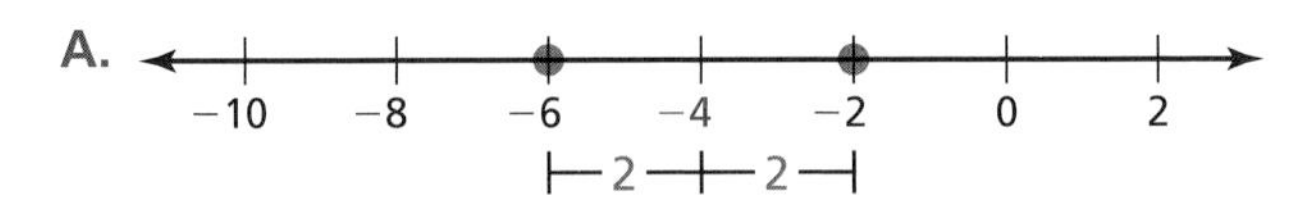

B. 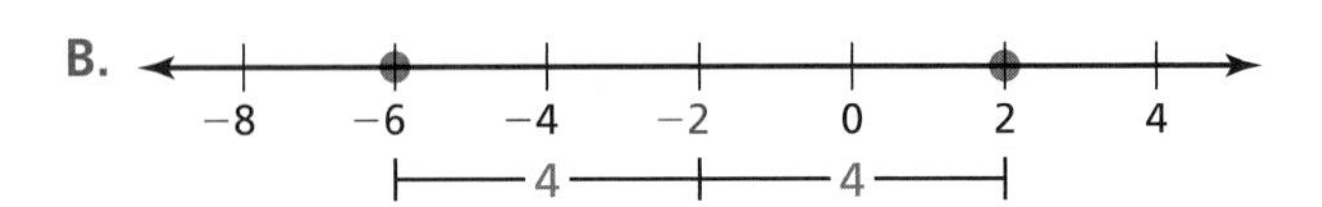

C. 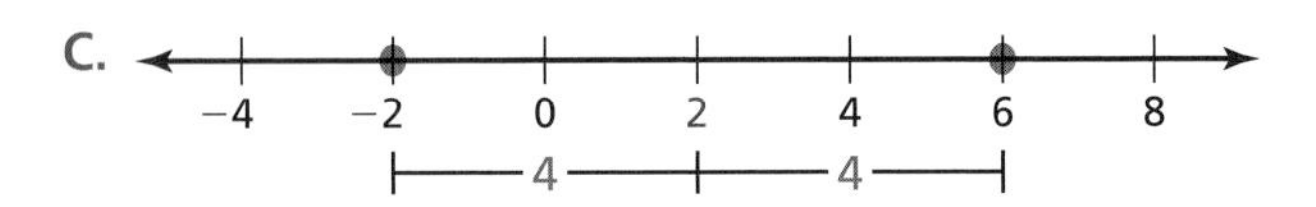

D. 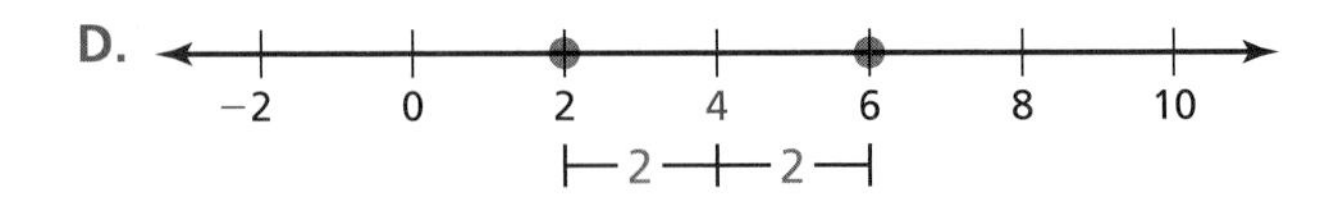

In Exercises 31–34, write an absolute value equation that has the given solutions.

31. $x = 8$ and $x = 18$

32. $x = -6$ and $x = 10$

33. $x = 2$ and $x = 9$

34. $x = -10$ and $x = -5$

In Exercises 35–44, solve the equation. Check your solutions. *(See Examples 4, 5, and 6.)*

35. $|4n - 15| = |n|$

36. $|2c + 8| = |10c|$

37. $|2b - 9| = |b - 6|$

38. $|3k - 2| = 2|k + 2|$

39. $4|p - 3| = |2p + 8|$

40. $2|4w - 1| = 3|4w + 2|$

41. $|3h + 1| = 7h$

42. $|6a - 5| = 4a$

43. $|f - 6| = |f + 8|$

44. $|3x - 4| = |3x - 5|$

45. MODELING WITH MATHEMATICS Starting from 300 feet away, a car drives toward you. It then passes by you at a speed of 48 feet per second. The distance d (in feet) of the car from you after t seconds is given by the equation $d = |300 - 48t|$. At what times is the car 60 feet from you?

46. MAKING AN ARGUMENT Your friend says that the absolute value equation $|3x + 8| - 9 = -5$ has no solution because the constant on the right side of the equation is negative. Is your friend correct? Explain.

47. MODELING WITH MATHEMATICS You randomly survey students about year-round school. The results are shown in the graph.

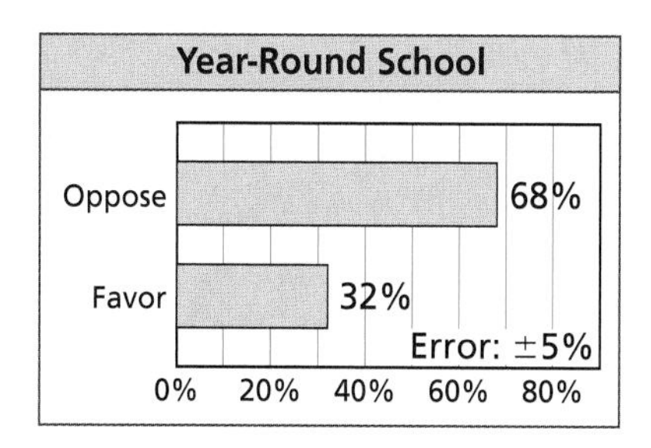

The error given in the graph means that the actual percent could be 5% more or 5% less than the percent reported by the survey.

a. Write and solve an absolute value equation to find the least and greatest percents of students who could be in favor of year-round school.

b. A classmate claims that $\frac{1}{3}$ of the student body is actually in favor of year-round school. Does this conflict with the survey data? Explain.

48. MODELING WITH MATHEMATICS The recommended weight of a soccer ball is 430 grams. The actual weight is allowed to vary by up to 20 grams.

a. Write and solve an absolute value equation to find the minimum and maximum acceptable soccer ball weights.

b. A soccer ball weighs 423 grams. Due to wear and tear, the weight of the ball decreases by 16 grams. Is the weight acceptable? Explain.

ERROR ANALYSIS In Exercises 49 and 50, describe and correct the error in solving the equation.

49.

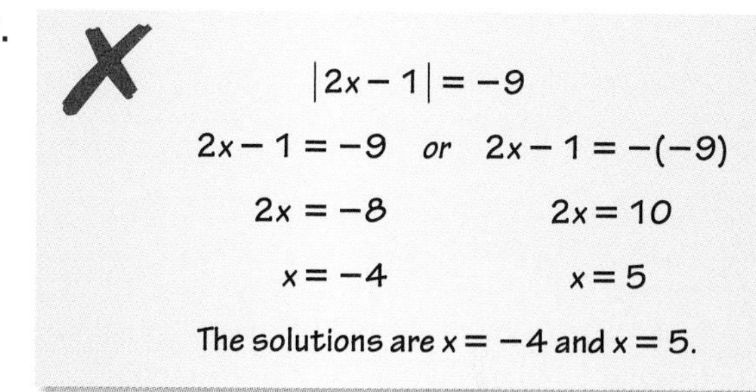

50.

✗

$|5x + 8| = x$

$5x + 8 = x$ or $5x + 8 = -x$

$4x + 8 = 0$ $\quad$ $6x + 8 = 0$

$4x = -8$ $\quad$ $6x = -8$

$x = -2$ $\quad$ $x = -\frac{4}{3}$

The solutions are $x = -2$ and $x = -\frac{4}{3}$.

51. ANALYZING EQUATIONS Without solving completely, place each equation into one of the three categories.

No solution	One solution	Two solutions

$|x - 2| + 6 = 0$ $\quad$ $|x + 3| - 1 = 0$

$|x + 8| + 2 = 7$ $\quad$ $|x - 1| + 4 = 4$

$|x - 6| - 5 = -9$ $\quad$ $|x + 5| - 8 = -8$

52. USING STRUCTURE Fill in the equation $|x - \square| = \square$ with a, b, c, or d so that the equation is graphed correctly.

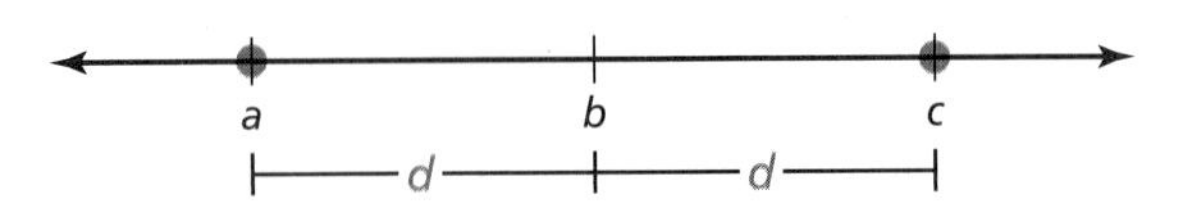

ABSTRACT REASONING In Exercises 53–56, complete the statement with *always*, *sometimes*, or *never*. Explain your reasoning.

53. If $x^2 = a^2$, then $|x|$ is ________ equal to $|a|$.

54. If a and b are real numbers, then $|a - b|$ is ________ equal to $|b - a|$.

55. For any real number p, the equation $|x - 4| = p$ will ________ have two solutions.

56. For any real number p, the equation $|x - p| = 4$ will ________ have two solutions.

57. WRITING Explain why absolute value equations can have no solution, one solution, or two solutions. Give an example of each case.

58. THOUGHT PROVOKING Describe a real-life situation that can be modeled by an absolute value equation with the solutions $x = 62$ and $x = 72$.

59. CRITICAL THINKING Solve the equation shown. Explain how you found your solution(s).

$$8|x + 2| - 6 = 5|x + 2| + 3$$

60. HOW DO YOU SEE IT? The circle graph shows the results of a survey of registered voters the day of an election.

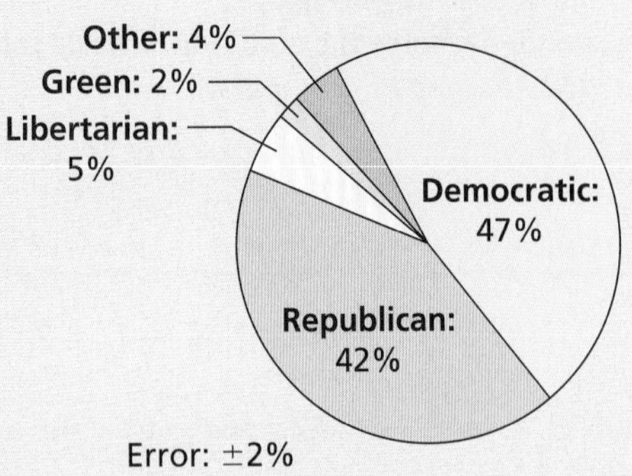

The error given in the graph means that the actual percent could be 2% more or 2% less than the percent reported by the survey.

a. What are the minimum and maximum percents of voters who could vote Republican? Green?

b. How can you use absolute value equations to represent your answers in part (a)?

c. One candidate receives 44% of the vote. Which party does the candidate belong to? Explain.

61. ABSTRACT REASONING How many solutions does the equation $a|x + b| + c = d$ have when $a > 0$ and $c = d$? when $a < 0$ and $c > d$? Explain your reasoning.

Maintaining Mathematical Proficiency

Reviewing what you learned in previous grades and lessons

Identify the property of equality that makes Equation 1 and Equation 2 equivalent. *(Section 1.1)*

62.

Equation 1	$3x + 8 = x - 1$
Equation 2	$3x + 9 = x$

63.

Equation 1	$4y = 28$
Equation 2	$y = 7$

Use a geometric formula to solve the problem. *(Skills Review Handbook)*

64. A square has an area of 81 square meters. Find the side length.

65. A circle has an area of 36π square inches. Find the radius.

66. A triangle has a height of 8 feet and an area of 48 square feet. Find the base.

67. A rectangle has a width of 4 centimeters and a perimeter of 26 centimeters. Find the length.

1.5 Rewriting Equations and Formulas

Essential Question How can you use a formula for one measurement to write a formula for a different measurement?

EXPLORATION 1 Using an Area Formula

Work with a partner.

a. Write a formula for the area A of a parallelogram.

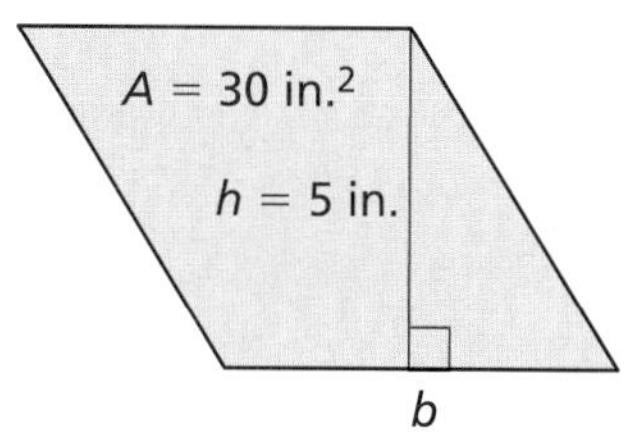

b. Substitute the given values into the formula. Then solve the equation for b. Justify each step.

c. Solve the formula in part (a) for b without first substituting values into the formula. Justify each step.

d. Compare how you solved the equations in parts (b) and (c). How are the processes similar? How are they different?

REASONING QUANTITATIVELY

To be proficient in math, you need to consider the given units. For instance, in Exploration 1, the area A is given in square inches and the height h is given in inches. A unit analysis shows that the units for the base b are also inches, which makes sense.

EXPLORATION 2 Using Area, Circumference, and Volume Formulas

Work with a partner. Write the indicated formula for each figure. Then write a new formula by solving for the variable whose value is not given. Use the new formula to find the value of the variable.

a. Area A of a trapezoid

$b_1 = 8$ cm

h $\quad A = 63$ cm^2

$b_2 = 10$ cm

b. Circumference C of a circle

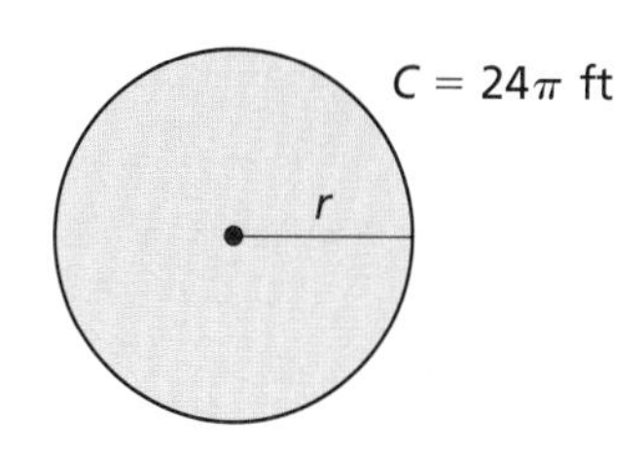

c. Volume V of a rectangular prism

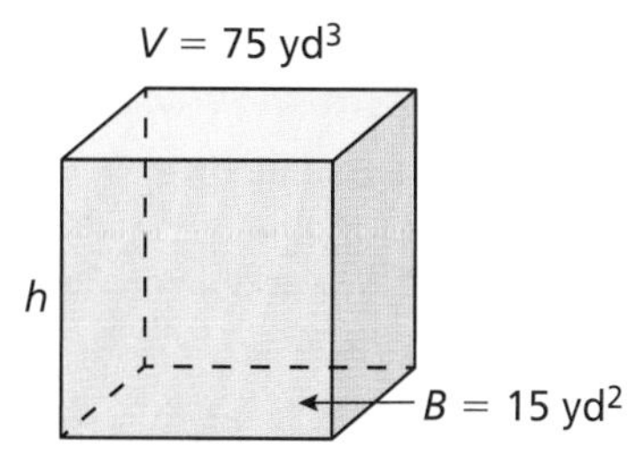

d. Volume V of a cone

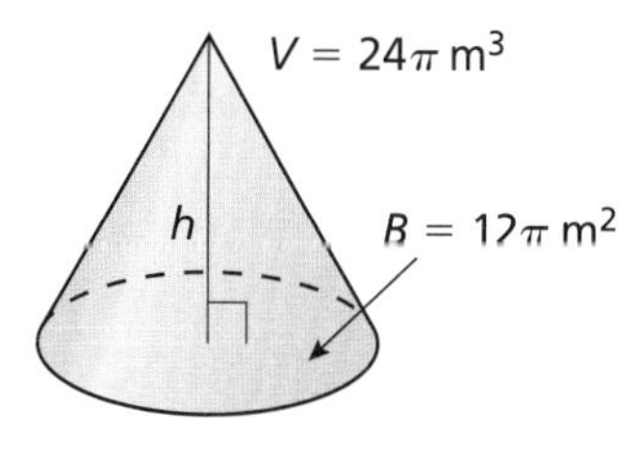

Communicate Your Answer

3. How can you use a formula for one measurement to write a formula for a different measurement? Give an example that is different from those given in Explorations 1 and 2.

1.5 Lesson

Core Vocabulary

literal equation, *p. 36*
formula, *p. 37*

Previous
surface area

What You Will Learn

- Rewrite literal equations.
- Rewrite and use formulas for area.
- Rewrite and use other common formulas.

Rewriting Literal Equations

An equation that has two or more variables is called a **literal equation**. To rewrite a literal equation, solve for one variable in terms of the other variable(s).

EXAMPLE 1 Rewriting a Literal Equation

Solve the literal equation $3y + 4x = 9$ for y.

SOLUTION

$3y + 4x = 9$	Write the equation.
$3y + 4x - 4x = 9 - 4x$	Subtract $4x$ from each side.
$3y = 9 - 4x$	Simplify.
$\frac{3y}{3} = \frac{9 - 4x}{3}$	Divide each side by 3.
$y = 3 - \frac{4}{3}x$	Simplify.

- The rewritten literal equation is $y = 3 - \frac{4}{3}x$.

EXAMPLE 2 Rewriting a Literal Equation

Solve the literal equation $y = 3x + 5xz$ for x.

SOLUTION

$y = 3x + 5xz$	Write the equation.
$y = x(3 + 5z)$	Distributive Property
$\frac{y}{3 + 5z} = \frac{x(3 + 5z)}{3 + 5z}$	Divide each side by $3 + 5z$.
$\frac{y}{3 + 5z} = x$	Simplify.

- The rewritten literal equation is $x = \frac{y}{3 + 5z}$.

REMEMBER
Division by 0 is undefined.

In Example 2, you must assume that $z \neq -\frac{3}{5}$ in order to divide by $3 + 5z$. In general, if you have to divide by a variable or variable expression when solving a literal equation, you should assume that the variable or variable expression does not equal 0.

Monitoring Progress  Help in English and Spanish at *BigIdeasMath.com*

Solve the literal equation for y.

1. $3y - x = 9$ **2.** $2x - 2y = 5$ **3.** $20 = 8x + 4y$

Solve the literal equation for x.

4. $y = 5x - 4x$ **5.** $2x + kx = m$ **6.** $3 + 5x - kx = y$

Rewriting and Using Formulas for Area

A **formula** shows how one variable is related to one or more other variables. A formula is a type of literal equation.

EXAMPLE 3 Rewriting a Formula for Surface Area

The formula for the surface area S of a rectangular prism is $S = 2\ell w + 2\ell h + 2wh$. Solve the formula for the length ℓ.

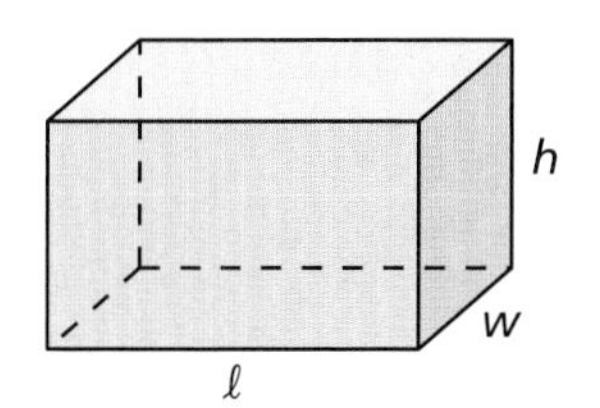

SOLUTION

$S = 2\ell w + 2\ell h + 2wh$	Write the equation.
$S - 2wh = 2\ell w + 2\ell h + 2wh - 2wh$	Subtract $2wh$ from each side.
$S - 2wh = 2\ell w + 2\ell h$	Simplify.
$S - 2wh = \ell(2w + 2h)$	Distributive Property
$\dfrac{S - 2wh}{2w + 2h} = \dfrac{\ell(2w + 2h)}{2w + 2h}$	Divide each side by $2w + 2h$.
$\dfrac{S - 2wh}{2w + 2h} = \ell$	Simplify.

▶ When you solve the formula for ℓ, you obtain $\ell = \dfrac{S - 2wh}{2w + 2h}$.

EXAMPLE 4 Using a Formula for Area

You own a rectangular lot that is 500 feet deep. It has an area of 100,000 square feet. To pay for a new water system, you are assessed \$5.50 per foot of lot frontage.

a. Find the frontage of your lot.

b. How much are you assessed for the new water system?

SOLUTION

a. In the formula for the area of a rectangle, let the width w represent the lot frontage.

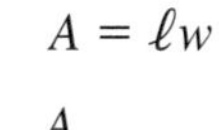

$A = \ell w$	Write the formula for area of a rectangle.
$\dfrac{A}{\ell} = w$	Divide each side by ℓ to solve for w.
$\dfrac{100{,}000}{500} = w$	Substitute 100,000 for A and 500 for ℓ.
$200 = w$	Simplify.

▶ The frontage of your lot is 200 feet.

b. Each foot of frontage costs \$5.50, and $\dfrac{\$5.50}{1\ \cancel{\text{ft}}} \cdot 200\ \cancel{\text{ft}} = \$1100$.

▶ So, your total assessment is \$1100.

Monitoring Progress  Help in English and Spanish at *BigIdeasMath.com*

Solve the formula for the indicated variable.

7. Area of a triangle: $A = \frac{1}{2}bh$; Solve for h.

8. Surface area of a cone: $S = \pi r^2 + \pi r\ell$; Solve for ℓ.

Rewriting and Using Other Common Formulas

Core Concept

Common Formulas

Temperature F = degrees Fahrenheit, C = degrees Celsius

$$C = \frac{5}{9}(F - 32)$$

Simple Interest I = interest, P = principal,
r = annual interest rate (decimal form),
t = time (years)

$$I = Prt$$

Distance d = distance traveled, r = rate, t = time

$$d = rt$$

EXAMPLE 5 Rewriting the Formula for Temperature

Solve the temperature formula for F.

SOLUTION

$C = \frac{5}{9}(F - 32)$	Write the temperature formula.
$\frac{9}{5}C = F - 32$	Multiply each side by $\frac{9}{5}$.
$\frac{9}{5}C + 32 = F - 32 + 32$	Add 32 to each side.
$\frac{9}{5}C + 32 = F$	Simplify.

▶ The rewritten formula is $F = \frac{9}{5}C + 32$.

EXAMPLE 6 Using the Formula for Temperature

Mercury
427°C

Venus
864°F

Which has the greater surface temperature: Mercury or Venus?

SOLUTION

Convert the Celsius temperature of Mercury to degrees Fahrenheit.

$F = \frac{9}{5}C + 32$	Write the rewritten formula from Example 5.
$= \frac{9}{5}(427) + 32$	Substitute 427 for C.
$= 800.6$	Simplify.

▶ Because 864°F is greater than 800.6°F, Venus has the greater surface temperature.

Monitoring Progress Help in English and Spanish at *BigIdeasMath.com*

9. A fever is generally considered to be a body temperature greater than 100°F. Your friend has a temperature of 37°C. Does your friend have a fever?

EXAMPLE 7 Using the Formula for Simple Interest

You deposit \$5000 in an account that earns simple interest. After 6 months, the account earns \$162.50 in interest. What is the annual interest rate?

COMMON ERROR

The unit of t is years. Be sure to convert months to years.

$\frac{1 \text{ yr}}{12 \text{ mo}} \cdot 6 \text{ mo} = 0.5 \text{ yr}$

SOLUTION

To find the annual interest rate, solve the simple interest formula for r.

$I = Prt$ Write the simple interest formula.

$\frac{I}{Pt} = r$ Divide each side by Pt to solve for r.

$\frac{162.50}{(5000)(0.5)} = r$ Substitute 162.50 for I, 5000 for P, and 0.5 for t.

$0.065 = r$ Simplify.

▶ The annual interest rate is 0.065, or 6.5%.

EXAMPLE 8 Solving a Real-Life Problem

A truck driver averages 60 miles per hour while delivering freight to a customer. On the return trip, the driver averages 50 miles per hour due to construction. The total driving time is 6.6 hours. How long does each trip take?

SOLUTION

Step 1 Rewrite the Distance Formula to write expressions that represent the two trip times. Solving the formula $d = rt$ for t, you obtain $t = \frac{d}{r}$. So, $\frac{d}{60}$ represents the delivery time, and $\frac{d}{50}$ represents the return trip time.

Step 2 Use these expressions and the total driving time to write and solve an equation to find the distance one way.

$\frac{d}{60} + \frac{d}{50} = 6.6$ The sum of the two trip times is 6.6 hours.

$\frac{11d}{300} = 6.6$ Add the left side using the LCD.

$11d = 1980$ Multiply each side by 300 and simplify.

$d = 180$ Divide each side by 11 and simplify.

The distance one way is 180 miles.

Step 3 Use the expressions from Step 1 to find the two trip times.

▶ So, the delivery takes $180 \text{ mi} \div \frac{60 \text{ mi}}{1 \text{ h}} = 3$ hours, and the return trip takes $180 \text{ mi} \div \frac{50 \text{ mi}}{1 \text{ h}} = 3.6$ hours.

Monitoring Progress

Help in English and Spanish at *BigIdeasMath.com*

10. How much money must you deposit in a simple interest account to earn \$500 in interest in 5 years at 4% annual interest?

11. A truck driver averages 60 miles per hour while delivering freight and 45 miles per hour on the return trip. The total driving time is 7 hours. How long does each trip take?

1.5 Exercises

Dynamic Solutions available at *BigIdeasMath.com*

Vocabulary and Core Concept Check

1. **VOCABULARY** Is $9r + 16 = \frac{\pi}{5}$ a literal equation? Explain.

2. **DIFFERENT WORDS, SAME QUESTION** Which is different? Find "both" answers.

Solve $3x + 6y = 24$ for x.	Solve $24 - 3x = 6y$ for x.
Solve $6y = 24 - 3x$ for y in terms of x.	Solve $24 - 6y = 3x$ for x in terms of y.

Monitoring Progress and Modeling with Mathematics

In Exercises 3–12, solve the literal equation for y. *(See Example 1.)*

3. $y - 3x = 13$
4. $2x + y = 7$
5. $2y - 18x = -26$
6. $20x + 5y = 15$
7. $9x - y = 45$
8. $6 - 3y = -6$
9. $4x - 5 = 7 + 4y$
10. $16x + 9 = 9y - 2x$
11. $2 + \frac{1}{6}y = 3x + 4$
12. $11 - \frac{1}{2}y = 3 + 6x$

In Exercises 13–22, solve the literal equation for x. *(See Example 2.)*

13. $y = 4x + 8x$
14. $m = 10x - x$
15. $a = 2x + 6xz$
16. $y = 3bx - 7x$
17. $y = 4x + rx + 6$
18. $z = 8 + 6x - px$
19. $sx + tx = r$
20. $a = bx + cx + d$
21. $12 - 5x - 4kx = y$
22. $x - 9 + 2wx = y$

23. **MODELING WITH MATHEMATICS** The total cost C (in dollars) to participate in a ski club is given by the literal equation $C = 85x + 60$, where x is the number of ski trips you take.

    a. Solve the equation for x.

    b. How many ski trips do you take if you spend a total of \$315? \$485?

24. **MODELING WITH MATHEMATICS** The penny size of a nail indicates the length of the nail. The penny size d is given by the literal equation $d = 4n - 2$, where n is the length (in inches) of the nail.

    a. Solve the equation for n.

    b. Use the equation from part (a) to find the lengths of nails with the following penny sizes: 3, 6, and 10.

n

ERROR ANALYSIS In Exercises 25 and 26, describe and correct the error in solving the equation for x.

25.

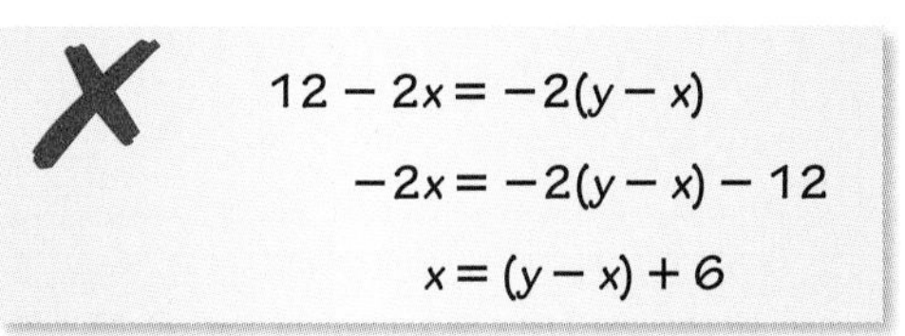

$12 - 2x = -2(y - x)$

$-2x = -2(y - x) - 12$

$x = (y - x) + 6$

26.

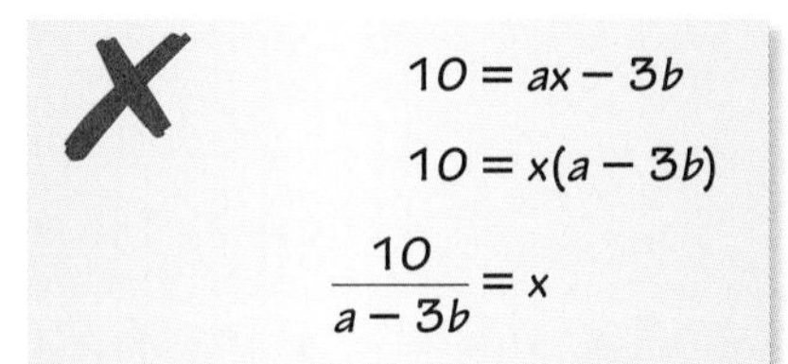

$10 = ax - 3b$

$10 = x(a - 3b)$

$\frac{10}{a - 3b} = x$

In Exercises 27–30, solve the formula for the indicated variable. *(See Examples 3 and 5.)*

27. Profit: $P = R - C$; Solve for C.

28. Surface area of a cylinder: $S = 2\pi r^2 + 2\pi rh$; Solve for h.

29. Area of a trapezoid: $A = \frac{1}{2}h(b_1 + b_2)$; Solve for b_2.

30. Average acceleration of an object: $a = \frac{v_1 - v_0}{t}$; Solve for v_1.

31. REWRITING A FORMULA A common statistic used in professional football is the quarterback rating. This rating is made up of four major factors. One factor is the completion rating given by the formula

$$R = 5\left(\frac{C}{A} - 0.3\right)$$

where C is the number of completed passes and A is the number of attempted passes. Solve the formula for C.

32. REWRITING A FORMULA Newton's law of gravitation is given by the formula

$$F = G\left(\frac{m_1 m_2}{d^2}\right)$$

where F is the force between two objects of masses m_1 and m_2, G is the gravitational constant, and d is the distance between the two objects. Solve the formula for m_1.

33. MODELING WITH MATHEMATICS The sale price S (in dollars) of an item is given by the formula $S = L - rL$, where L is the list price (in dollars) and r is the discount rate (in decimal form). *(See Examples 4 and 6.)*

a. Solve the formula for r.

b. The list price of the shirt is \$30. What is the discount rate?

34. MODELING WITH MATHEMATICS The density d of a substance is given by the formula $d = \frac{m}{V}$, where m is its mass and V is its volume.

Pyrite

Density: 5.01g/cm^3 Volume: 1.2 cm^3

a. Solve the formula for m.

b. Find the mass of the pyrite sample.

35. PROBLEM SOLVING You deposit \$2000 in an account that earns simple interest at an annual rate of 4%. How long must you leave the money in the account to earn \$500 in interest? *(See Example 7.)*

36. PROBLEM SOLVING A flight averages 460 miles per hour. The return flight averages 500 miles per hour due to a tailwind. The total flying time is 4.8 hours. How long is each flight? Explain. *(See Example 8.)*

37. USING STRUCTURE An athletic facility is building an indoor track. The track is composed of a rectangle and two semicircles, as shown.

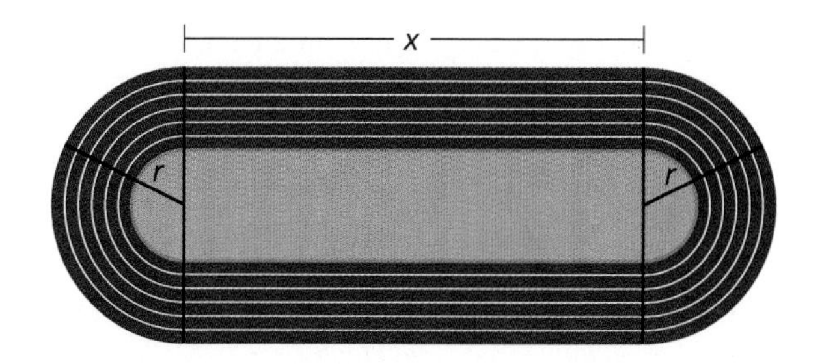

a. Write a formula for the perimeter of the indoor track.

b. Solve the formula for x.

c. The perimeter of the track is 660 feet, and r is 50 feet. Find x. Round your answer to the nearest foot.

38. MODELING WITH MATHEMATICS The distance d (in miles) you travel in a car is given by the two equations shown, where t is the time (in hours) and g is the number of gallons of gasoline the car uses.

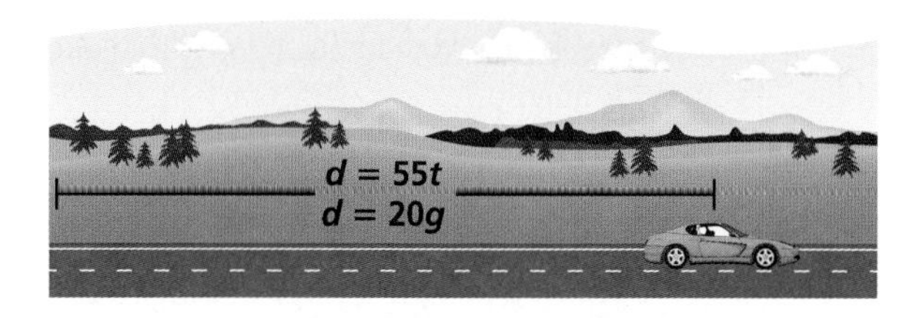

a. Write an equation that relates g and t.

b. Solve the equation for g.

c. You travel for 6 hours. How many gallons of gasoline does the car use? How far do you travel? Explain.

39. MODELING WITH MATHEMATICS One type of stone formation found in Carlsbad Caverns in New Mexico is called a column. This cylindrical stone formation connects to the ceiling and the floor of a cave.

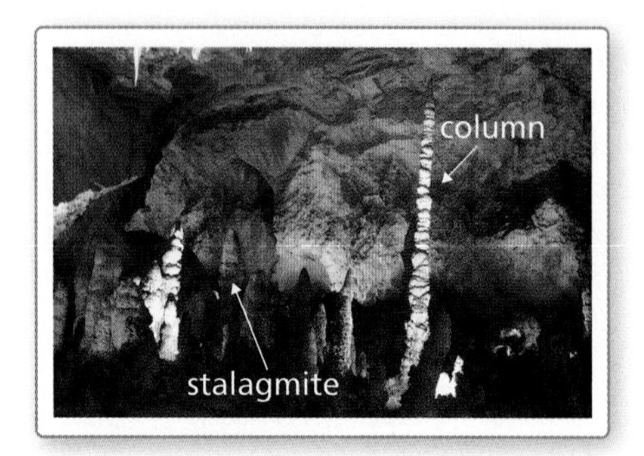

a. Rewrite the formula for the circumference of a circle, so that you can easily calculate the radius of a column given its circumference.

b. What is the radius (to the nearest tenth of a foot) of a column that has a circumference of 7 feet? 8 feet? 9 feet?

c. Explain how you can find the area of a cross section of a column when you know its circumference.

40. HOW DO YOU SEE IT? The rectangular prism shown has bases with equal side lengths.

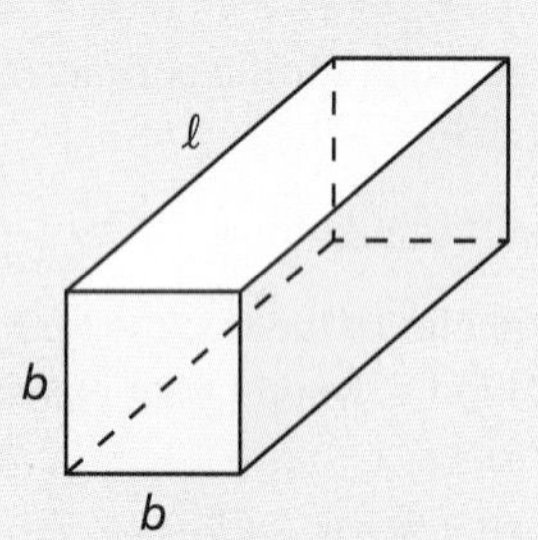

a. Use the figure to write a formula for the surface area S of the rectangular prism.

b. Your teacher asks you to rewrite the formula by solving for one of the side lengths, b or ℓ. Which side length would you choose? Explain your reasoning.

41. MAKING AN ARGUMENT Your friend claims that Thermometer A displays a greater temperature than Thermometer B. Is your friend correct? Explain your reasoning.

Thermometer A

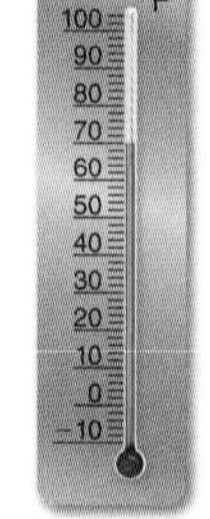

Thermometer B

42. THOUGHT PROVOKING Give a possible value for h. Justify your answer. Draw and label the figure using your chosen value of h.

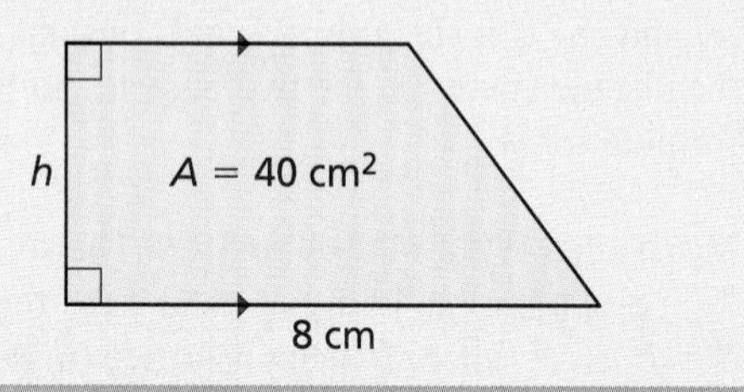

MATHEMATICAL CONNECTIONS In Exercises 43 and 44, write a formula for the area of the regular polygon. Solve the formula for the height h.

43.

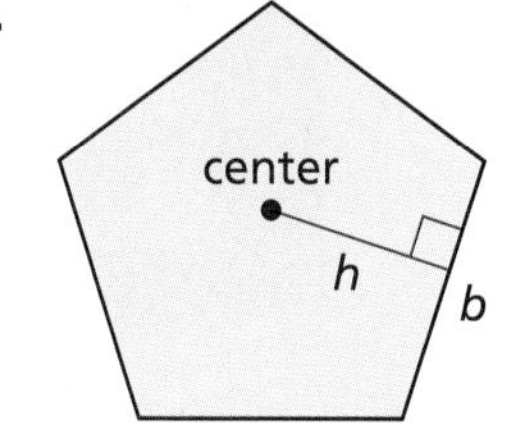

44.

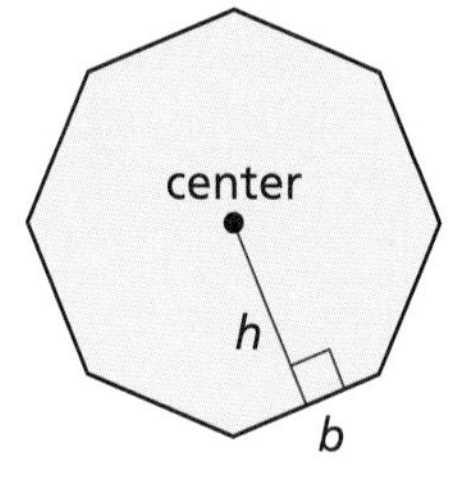

REASONING In Exercises 45 and 46, solve the literal equation for a.

45. $x = \dfrac{a + b + c}{ab}$

46. $y = x\left(\dfrac{ab}{a - b}\right)$

Maintaining Mathematical Proficiency

Reviewing what you learned in previous grades and lessons

Evaluate the expression. *(Skills Review Handbook)*

47. $15 - 5 + 5^2$ **48.** $18 \cdot 2 - 4^2 \div 8$ **49.** $3^3 + 12 \div 3 \cdot 5$ **50.** $2^5(5 - 6) + 9 \div 3$

Solve the equation. Graph the solutions, if possible. *(Section 1.4)*

51. $|x - 3| + 4 = 9$ **52.** $|3y - 12| - 7 = 2$ **53.** $2|2r + 4| = -16$ **54.** $-4|s + 9| = -24$

1.4–1.5 What Did You Learn?

Core Vocabulary

absolute value equation, *p. 28*
extraneous solution, *p. 31*
literal equation, *p. 36*
formula, *p. 37*

Core Concepts

Section 1.4

Properties of Absolute Value, *p. 28*
Solving Absolute Value Equations, *p. 28*
Solving Equations with Two Absolute Values, *p. 30*
Special Solutions of Absolute Value Equations, *p. 31*

Section 1.5

Rewriting Literal Equations, *p. 36*
Common Formulas, *p. 38*

Mathematical Practices

1. How did you decide whether your friend's argument in Exercise 46 on page 33 made sense?
2. How did you use the structure of the equation in Exercise 59 on page 34 to rewrite the equation?
3. What entry points did you use to answer Exercises 43 and 44 on page 42?

Performance Task

Magic of Mathematics

Have you ever watched a magician perform a number trick? You can use algebra to explain how these types of tricks work.

To explore the answers to these questions and more, go to ***BigIdeasMath.com***.

1 Chapter Review

Dynamic Solutions available at *BigIdeasMath.com*

1.1 Solving Simple Equations *(pp. 3–10)*

a. Solve $x - 5 = -9$. Justify each step.

	$x - 5 = -9$	Write the equation.
Addition Property of Equality →	$+5 \quad +5$	Add 5 to each side.
	$x = -4$	Simplify.

▶ The solution is $x = -4$.

b. Solve $4x = 12$. Justify each step.

	$4x = 12$	Write the equation.
Division Property of Equality →	$\frac{4x}{4} = \frac{12}{4}$	Divide each side by 4.
	$x = 3$	Simplify.

▶ The solution is $x = 3$.

Solve the equation. Justify each step. Check your solution.

1. $z + 3 = -6$

2. $2.6 = -0.2t$

3. $-\frac{n}{5} = -2$

1.2 Solving Multi-Step Equations *(pp. 11–18)*

Solve $-6x + 23 + 2x = 15$.

$-6x + 23 + 2x = 15$	Write the equation.
$-4x + 23 = 15$	Combine like terms.
$-4x = -8$	Subtract 23 from each side.
$x = 2$	Divide each side by -4.

▶ The solution is $x = 2$.

Solve the equation. Check your solution.

4. $3y + 11 = -16$

5. $6 = 1 - b$

6. $n + 5n + 7 = 43$

7. $-4(2z + 6) - 12 = 4$

8. $\frac{3}{2}(x - 2) - 5 = 19$

9. $6 = \frac{1}{5}w + \frac{7}{5}w - 4$

Find the value of x. Then find the angle measures of the polygon.

10.

110°
5x°
2x°

Sum of angle measures: 180°

11.

(x − 30)°
x°
x°
(x − 30)°
(x − 30)°

Sum of angle measures: 540°

1.3 Solving Equations with Variables on Both Sides (pp. 19–24)

Solve $2(y - 4) = -4(y + 8)$.

$2(y - 4) = -4(y + 8)$	Write the equation.
$2y - 8 = -4y - 32$	Distributive Property
$6y - 8 = -32$	Add $4y$ to each side.
$6y = -24$	Add 8 to each side.
$y = -4$	Divide each side by 6.

▶ The solution is $y = -4$.

Solve the equation.

12. $3n - 3 = 4n + 1$ **13.** $5(1 + x) = 5x + 5$ **14.** $3(n + 4) = \frac{1}{2}(6n + 4)$

1.4 Solving Absolute Value Equations (pp. 27–34)

a. Solve $|x - 5| = 3$.

$x - 5 = 3$	*or*	$x - 5 = -3$	Write related linear equations.
$+5 \quad +5$		$+5 \quad +5$	Add 5 to each side.
$x = 8$		$x = 2$	Simplify.

▶ The solutions are $x = 8$ and $x = 2$.

b. Solve $|2x + 6| = 4x$. Check your solutions.

$2x + 6 = 4x$	*or*	$2x + 6 = -4x$	Write related linear equations.
$-2x \quad -2x$		$-2x \quad -2x$	Subtract $2x$ from each side.
$6 = 2x$		$6 = -6x$	Simplify.
$\frac{6}{2} = \frac{2x}{2}$		$\frac{6}{-6} = \frac{-6x}{-6}$	Solve for x.
$3 = x$		$-1 = x$	Simplify.

Check the apparent solutions to see if either is extraneous.

▶ The solution is $x = 3$. Reject $x = -1$ because it is extraneous.

Check

$|2x + 6| = 4x$

$|2(3) + 6| \stackrel{?}{=} 4(3)$

$|12| \stackrel{?}{=} 12$

$12 = 12$ ✓

$|2x + 6| = 4x$

$|2(-1) + 6| \stackrel{?}{=} 4(-1)$

$|4| \stackrel{?}{=} -4$

$4 \neq -4$ ✗

Solve the equation. Check your solutions.

15. $|y + 3| = 17$ **16.** $-2|5w - 7| + 9 = -7$ **17.** $|x - 2| = |4 + x|$

18. The minimum sustained wind speed of a Category 1 hurricane is 74 miles per hour. The maximum sustained wind speed is 95 miles per hour. Write an absolute value equation that represents the minimum and maximum speeds.

1.5 Rewriting Equations and Formulas (pp. 35–42)

a. The slope-intercept form of a linear equation is $y = mx + b$. Solve the equation for m.

$y = mx + b$	Write the equation.
$y - b = mx + b - b$	Subtract b from each side.
$y - b = mx$	Simplify.
$\frac{y - b}{x} = \frac{mx}{x}$	Divide each side by x.
$\frac{y - b}{x} = m$	Simplify.

▶ When you solve the equation for m, you obtain $m = \frac{y - b}{x}$.

b. The formula for the surface area S of a cylinder is $S = 2\pi r^2 + 2\pi rh$. Solve the formula for the height h.

$S = 2\pi r^2 + 2\pi rh$	Write the equation.
$-2\pi r^2 \quad -2\pi r^2$	Subtract $2\pi r^2$ from each side.
$S - 2\pi r^2 = 2\pi rh$	Simplify.
$\frac{S - 2\pi r^2}{2\pi r} = \frac{2\pi rh}{2\pi r}$	Divide each side by $2\pi r$.
$\frac{S - 2\pi r^2}{2\pi r} = h$	Simplify.

▶ When you solve the formula for h, you obtain $h = \frac{S - 2\pi r^2}{2\pi r}$.

Solve the literal equation for y.

19. $2x - 4y = 20$

20. $8x - 3 = 5 + 4y$

21. $a = 9y + 3yx$

22. The volume V of a pyramid is given by the formula $V = \frac{1}{3}Bh$, where B is the area of the base and h is the height.

a. Solve the formula for h.

b. Find the height h of the pyramid.

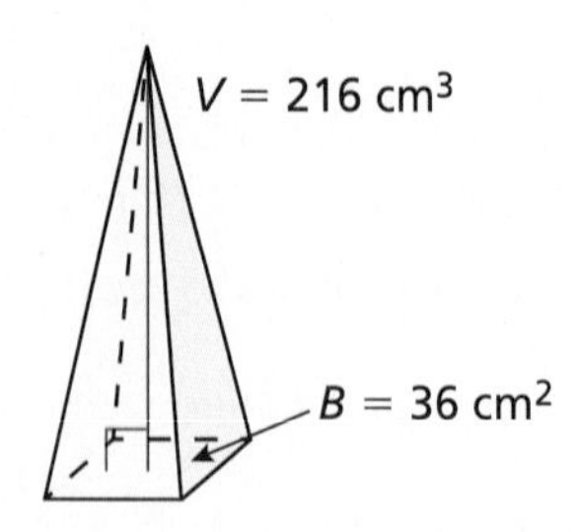

23. The formula $F = \frac{9}{5}(K - 273.15) + 32$ converts a temperature from kelvin K to degrees Fahrenheit F.

a. Solve the formula for K.

b. Convert 180°F to kelvin K. Round your answer to the nearest hundredth.

1 Chapter Test

Solve the equation. Justify each step. Check your solution.

1. $x - 7 = 15$

2. $\frac{2}{3}x + 5 = 3$

3. $11x + 1 = -1 + x$

Solve the equation.

4. $2|x - 3| - 5 = 7$

5. $|2x - 19| = 4x + 1$

6. $-2 + 5x - 7 = 3x - 9 + 2x$

7. $3(x + 4) - 1 = -7$

8. $|20 + 2x| = |4x + 4|$

9. $\frac{1}{3}(6x + 12) - 2(x - 7) = 19$

Describe the values of *c* for which the equation has no solution. Explain your reasoning.

10. $3x - 5 = 3x - c$

11. $|x - 7| = c$

12. A safety regulation states that the minimum height of a handrail is 30 inches. The maximum height is 38 inches. Write an absolute value equation that represents the minimum and maximum heights.

13. The perimeter P (in yards) of a soccer field is represented by the formula $P = 2\ell + 2w$, where ℓ is the length (in yards) and w is the width (in yards).

a. Solve the formula for w.

b. Find the width of the field.

c. About what percent of the field is inside the circle?

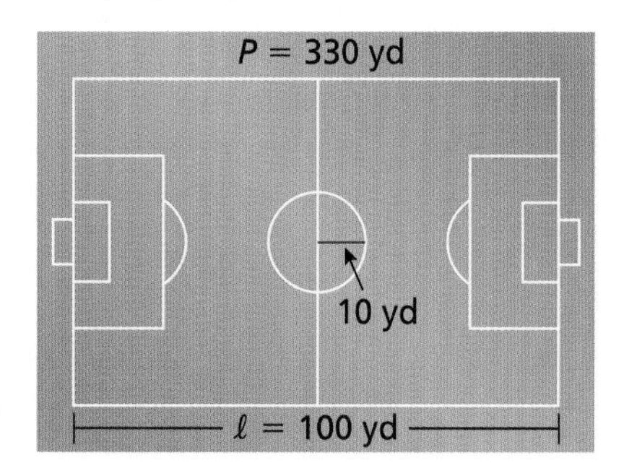

14. Your car needs new brakes. You call a dealership and a local mechanic for prices.

	Cost of parts	Labor cost per hour
Dealership	\$24	\$99
Local Mechanic	\$45	\$89

a. After how many hours are the total costs the same at both places? Justify your answer.

b. When do the repairs cost less at the dealership? at the local mechanic? Explain.

15. Consider the equation $|4x + 20| = 6x$. Without calculating, how do you know that $x = -2$ is an extraneous solution?

16. Your friend was solving the equation shown and was confused by the result "$-8 = -8$." Explain what this result means.

$$4(y - 2) - 2y = 6y - 8 - 4y$$
$$4y - 8 - 2y = 6y - 8 - 4y$$
$$2y - 8 = 2y - 8$$
$$-8 = -8$$

1 Cumulative Assessment

1. A mountain biking park has 48 trails, 37.5% of which are beginner trails. The rest are divided evenly between intermediate and expert trails. How many of each kind of trail are there?

Ⓐ 12 beginner, 18 intermediate, 18 expert

Ⓑ 18 beginner, 15 intermediate, 15 expert

Ⓒ 18 beginner, 12 intermediate, 18 expert

Ⓓ 30 beginner, 9 intermediate, 9 expert

2. Which of the equations are equivalent to $cx - a = b$?

$cx - a + b = 2b$ $\qquad 0 = cx - a + b$ $\qquad 2cx - 2a = \frac{b}{2}$

$x - a = \frac{b}{c}$ $\qquad x = \frac{a + b}{c}$ $\qquad b + a = cx$

3. Let N represent the number of solutions of the equation $3(x - a) = 3x - 6$. Complete each statement with the symbol <, >, or =.

a. When $a = 3$, N ____ 1.

b. When $a = -3$, N ____ 1.

c. When $a = 2$, N ____ 1.

d. When $a = -2$, N ____ 1.

e. When $a = x$, N ____ 1.

f. When $a = -x$, N ____ 1.

4. You are painting your dining room white and your living room blue. You spend \$132 on 5 cans of paint. The white paint costs \$24 per can, and the blue paint costs \$28 per can.

a. Use the numbers and symbols to write an equation that represents how many cans of each color you bought.

b. How much would you have saved by switching the colors of the dining room and living room? Explain.

5. Which of the equations are equivalent?

$6x + 6 = -14$ $\quad$ $8x + 6 = -2x - 14$

$5x + 3 = -7$ $\quad$ $7x + 3 = 2x - 13$

6. The perimeter of the triangle is 13 inches. What is the length of the shortest side?

Ⓐ 2 in.

Ⓑ 3 in.

Ⓒ 4 in.

Ⓓ 8 in.

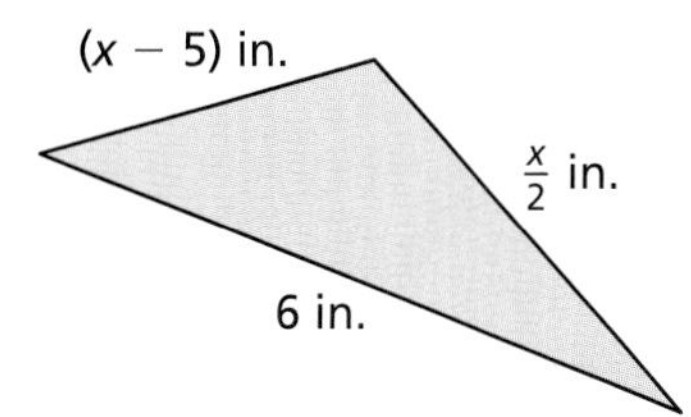

7. You pay \$45 per month for cable TV. Your friend buys a satellite TV receiver for \$99 and pays \$36 per month for satellite TV. Your friend claims that the expenses for a year of satellite TV are less than the expenses for a year of cable.

   a. Write and solve an equation to determine when you and your friend will have paid the same amount for TV services.

   b. Is your friend correct? Explain.

8. Place each equation into one of the four categories.

No solution	One solution	Two solutions	Infinitely many solutions

$|8x + 3| = 0$ $\quad$ $-6 = 5x - 9$ $\quad$ $3x - 12 = 3(x - 4) + 1$

$-2x + 4 = 2x + 4$ $\quad$ $0 = |x + 13| + 2$ $\quad$ $-4(x + 4) = -4x - 16$

$12x - 2x = 10x - 8$ $\quad$ $9 = 3|2x - 11|$ $\quad$ $7 - 2x = 3 - 2(x - 2)$

9. A car travels 1000 feet in 12.5 seconds. Which of the expressions do *not* represent the average speed of the car?

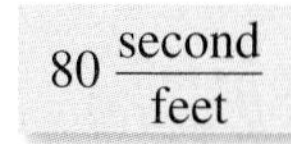

$80 \dfrac{\text{second}}{\text{feet}}$ $\quad$ $80 \dfrac{\text{feet}}{\text{second}}$

$\dfrac{80 \text{ feet}}{\text{second}}$ $\quad$ $\dfrac{\text{second}}{80 \text{ feet}}$

2 Solving Linear Inequalities

Camel Physiology *(p. 91)*

Mountain Plant Life *(p. 85)*

Digital Camera *(p. 70)*

Microwave Electricity *(p. 64)*

Natural Arch *(p. 59)*

Maintaining Mathematical Proficiency

Graphing Numbers on a Number Line

Example 1 **Graph each number.**

a. 3

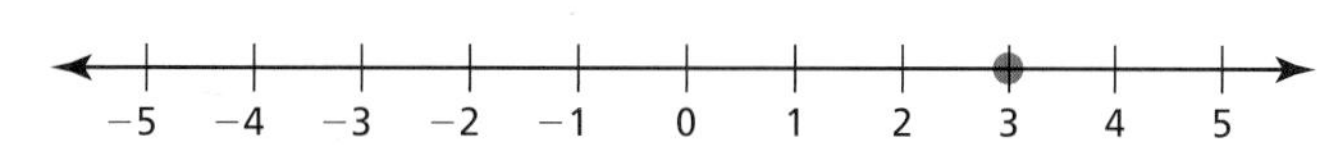

b. -5

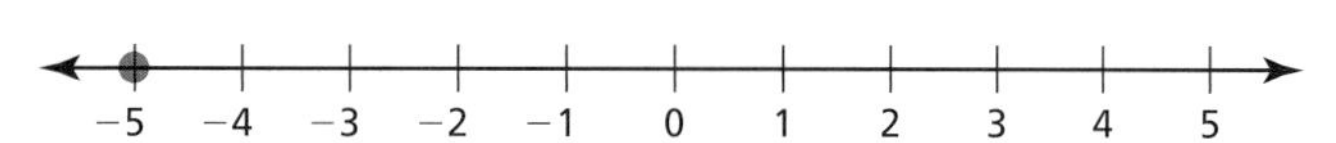

Example 2 **Graph each number.**

a. $|4|$

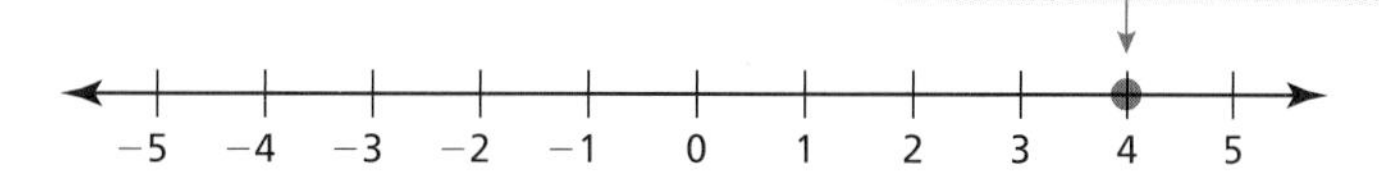
The absolute value of a positive number is positive.

b. $|-2|$

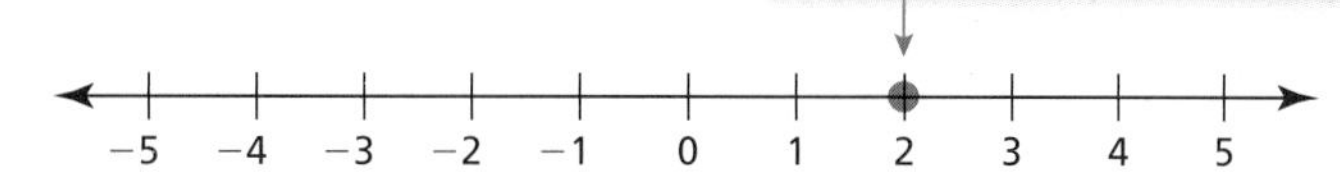
The absolute value of a negative number is positive.

Graph the number.

1. 6

2. $|2|$

3. $|-1|$

4. $2 + |-2|$

5. $1 - |-4|$

6. $-5 + |3|$

Comparing Real Numbers

Example 3 **Complete the statement -1 ▢ -5 with <, >, or =.**

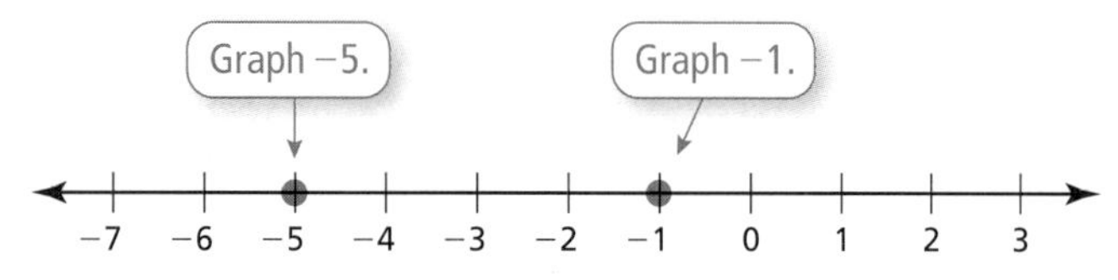
Graph −5.
Graph −1.

▶ -1 is to the right of -5. So, $-1 > -5$.

Complete the statement with <, >, or =.

7. 2 ▢ 9

8. -6 ▢ 5

9. -12 ▢ -4

10. -7 ▢ -13

11. $|-8|$ ▢ $|8|$

12. -10 ▢ $|-18|$

13. **ABSTRACT REASONING** A number a is to the left of a number b on the number line. How do the numbers $-a$ and $-b$ compare?

Mathematical Practices

Mathematically proficient students use technology tools to explore concepts.

Using a Graphing Calculator

Core Concept

Solving an Inequality in One Variable

You can use a graphing calculator to solve an inequality.

1. Enter the inequality into a graphing calculator.
2. Graph the inequality.
3. Use the graph to write the solution.

EXAMPLE 1 Using a Graphing Calculator

Use a graphing calculator to solve (a) $2x - 1 < x + 2$ and (b) $2x - 1 \leq x + 2$.

SOLUTION

a. Enter the inequality $2x - 1 < x + 2$ into a graphing calculator. Press *graph*.

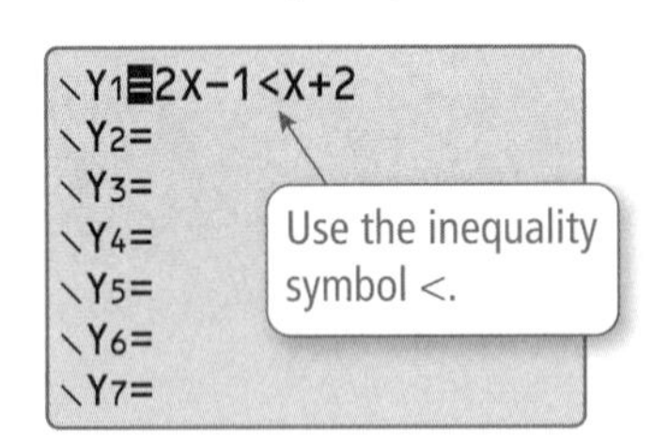

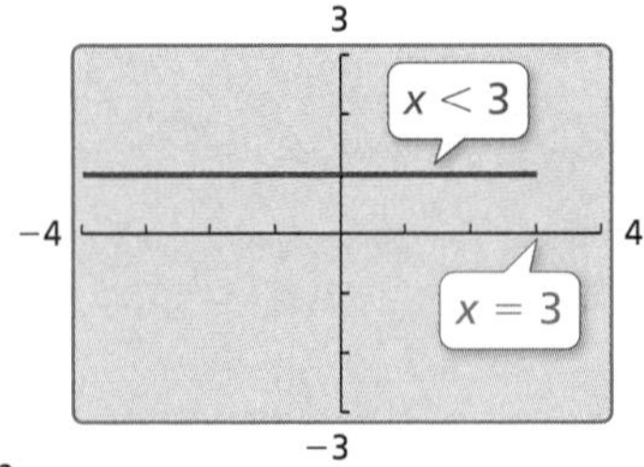

▶ The solution of the inequality is $x < 3$.

b. Enter the inequality $2x - 1 \leq x + 2$ into a graphing calculator. Press *graph*.

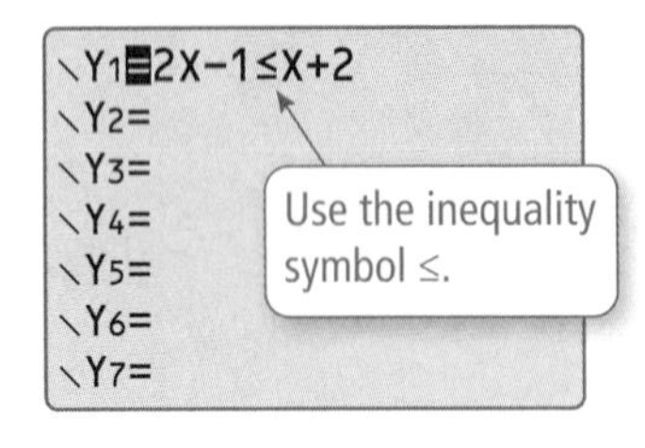

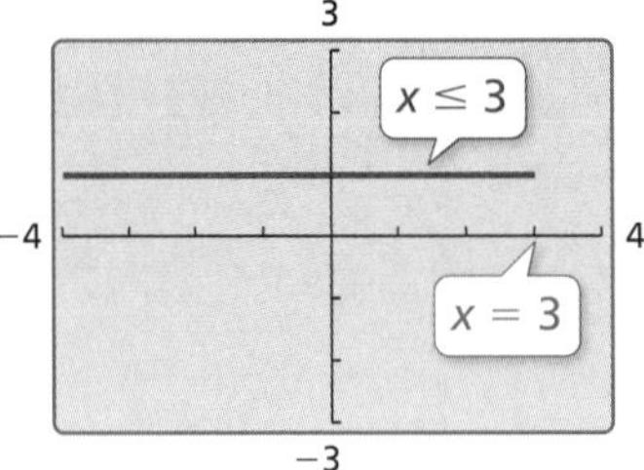

▶ The solution of the inequality is $x \leq 3$.

Notice that the graphing calculator does not distinguish between the solutions $x < 3$ and $x \leq 3$. You must distinguish between these yourself, based on the inequality symbol used in the original inequality.

Monitoring Progress

Use a graphing calculator to solve the inequality.

1. $2x + 3 < x - 1$ **2.** $-x - 1 > -2x + 2$ **3.** $\frac{1}{2}x + 1 \leq \frac{3}{2}x + 3$

2.1 Writing and Graphing Inequalities

Essential Question How can you use an inequality to describe a real-life statement?

EXPLORATION 1 **Writing and Graphing Inequalities**

Work with a partner. Write an inequality for each statement. Then sketch the graph of the numbers that make each inequality true.

a. Statement The temperature t in Sweden is at least $-10°C$.

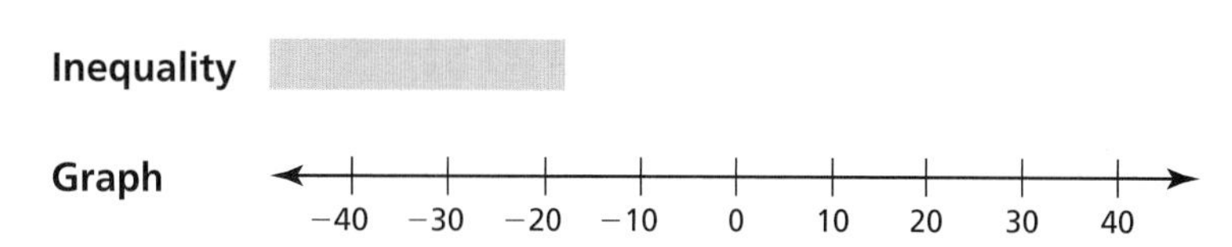

b. Statement The elevation e of Alabama is at most 2407 feet.

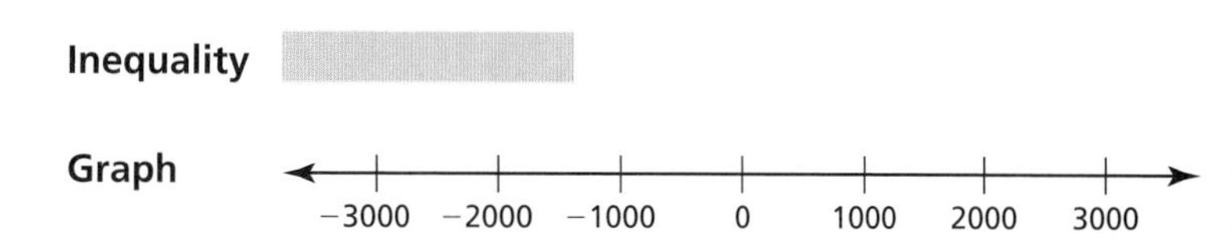

ATTENDING TO PRECISION

To be proficient in math, you need to communicate precisely. You also need to state the meanings of the symbols you use.

EXPLORATION 2 **Writing Inequalities**

Work with a partner. Write an inequality for each graph. Then, in words, describe all the values of x that make each inequality true.

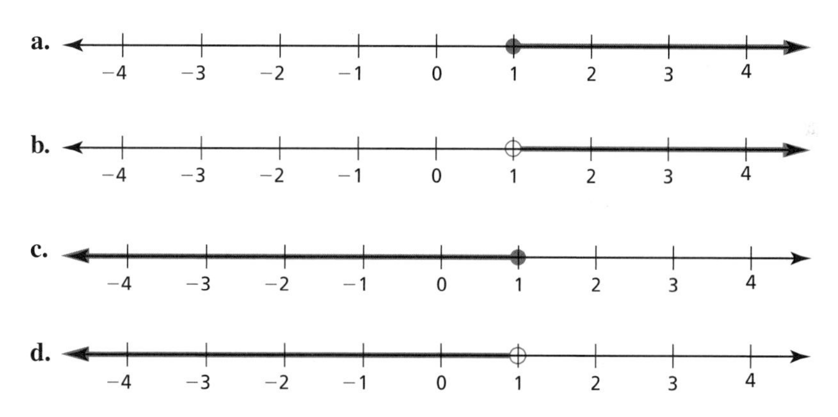

Communicate Your Answer

3. How can you use an inequality to describe a real-life statement?

4. Write a real-life statement that involves each inequality.

a. $x < 3.5$

b. $x \leq 6$

c. $x > -2$

d. $x \geq 10$

2.1 Lesson

Core Vocabulary

inequality, *p. 54*
solution of an inequality, *p. 55*
solution set, *p. 55*
graph of an inequality, *p. 56*

Previous
expression

What You Will Learn

- Write linear inequalities.
- Sketch the graphs of linear inequalities.
- Write linear inequalities from graphs.

Writing Linear Inequalities

An **inequality** is a mathematical sentence that compares expressions. An inequality contains the symbol $<$, $>$, $\leq$, or $\geq$. To write an inequality, look for the following phrases to determine what inequality symbol to use.

Inequality Symbols				
Symbol	$<$	$>$	$\leq$	$\geq$
Key Phrases	• is less than • is fewer than	• is greater than • is more than	• is less than or equal to • is at most • is no more than	• is greater than or equal to • is at least • is no less than

EXAMPLE 1 Writing Inequalities

Write each sentence as an inequality.

a. A number w minus 3.5 is less than or equal to -2.

b. Three is less than a number n plus 5.

c. Zero is greater than or equal to twice a number x plus 1.

SOLUTION

a. A number w minus 3.5 is less than or equal to -2.

$w - 3.5$ $\leq$ -2

▶ An inequality is $w - 3.5 \leq -2$.

READING
The inequality $3 < n + 5$ is the same as $n + 5 > 3$.

b. Three is less than a number n plus 5.

3 $<$ $n + 5$

▶ An inequality is $3 < n + 5$.

c. Zero is greater than or equal to twice a number x plus 1.

0 $\geq$ $2x + 1$

▶ An inequality is $0 \geq 2x + 1$.

Monitoring Progress

Help in English and Spanish at *BigIdeasMath.com*

Write the sentence as an inequality.

1. A number b is fewer than 30.4.

2. $-\frac{7}{10}$ is at least twice a number k minus 4.

Sketching the Graphs of Linear Inequalities

A **solution of an inequality** is a value that makes the inequality true. An inequality can have more than one solution. The set of all solutions of an inequality is called the **solution set**.

Value of x	$x + 5 \geq -2$	Is the inequality true?
-6	$-6 + 5 \overset{?}{\geq} -2$ $-1 \geq -2$ ✓	yes
-7	$-7 + 5 \overset{?}{\geq} -2$ $-2 \geq -2$ ✓	yes
-8	$-8 + 5 \overset{?}{\geq} -2$ $-3 \ngeq -2$ ✗	no

Recall that a diagonal line through an inequality symbol means the inequality is *not* true. For instance, the symbol $\ngeq$ means "is not greater than or equal to."

EXAMPLE 2 Checking Solutions

Tell whether -4 is a solution of each inequality.

a. $x + 8 < -3$

b. $-4.5x > -21$

SOLUTION

a. $x + 8 < -3$ Write the inequality.

$-4 + 8 \overset{?}{<} -3$ Substitute -4 for x.

$4 \nless -3$ ✗ Simplify.

4 is *not* less than -3.

▶ So, -4 is *not* a solution of the inequality.

b. $-4.5x > -21$ Write the inequality.

$-4.5(-4) \overset{?}{>} -21$ Substitute -4 for x.

$18 > -21$ ✓ Simplify.

18 is greater than -21.

▶ So, -4 is a solution of the inequality.

Monitoring Progress  Help in English and Spanish at *BigIdeasMath.com*

Tell whether -6 is a solution of the inequality.

3. $c + 4 < -1$

4. $10 \leq 3 - m$

5. $21 \div x \geq -3.5$

6. $4x - 25 > -2$

The **graph of an inequality** shows the solution set of the inequality on a number line. An open circle, ○, is used when a number is *not* a solution. A closed circle, ●, is used when a number is a solution. An arrow to the left or right shows that the graph continues in that direction.

EXAMPLE 3 Graphing Inequalities

Graph each inequality.

a. $y \leq -3$ **b.** $2 < x$ **c.** $x > 0$

SOLUTION

a. Test a number to the left of −3. $y = -4$ is a solution.

Test a number to the right of −3. $y = 0$ is *not* a solution.

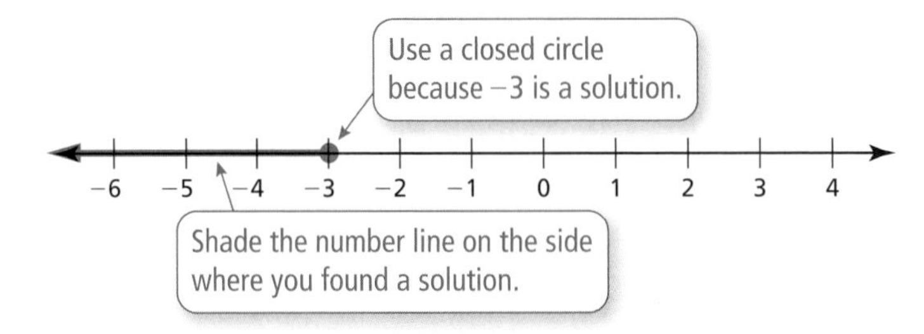

ANOTHER WAY

Another way to represent the solutions of an inequality is to use *set-builder notation*. In Example 3b, the solutions can be written as $\{x \mid x > 2\}$, which is read as "the set of all numbers x such that x is greater than 2."

b. Test a number to the left of 2. $x = 0$ is *not* a solution.

Test a number to the right of 2. $x = 4$ is a solution.

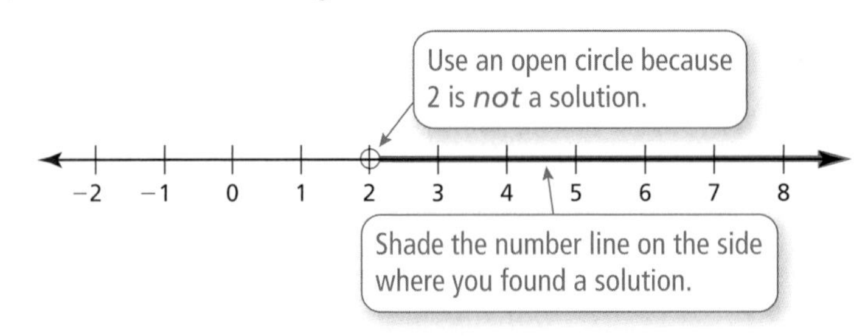

c. Just by looking at the inequality, you can see that it represents the set of all positive numbers.

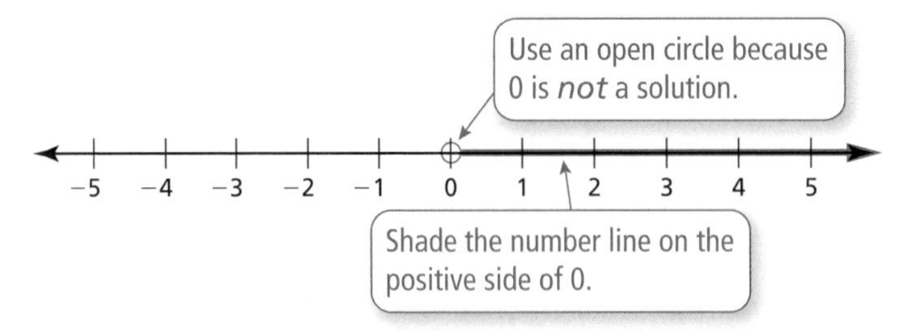

Monitoring Progress  Help in English and Spanish at *BigIdeasMath.com*

Graph the inequality.

7. $b > -8$

8. $1.4 \geq g$

9. $r < \frac{1}{2}$

10. $v \geq \sqrt{36}$

Writing Linear Inequalities from Graphs

EXAMPLE 4 **Writing Inequalities from Graphs**

The graphs show the height restrictions h (in inches) for two rides at an amusement park. Write an inequality that represents the height restriction of each ride.

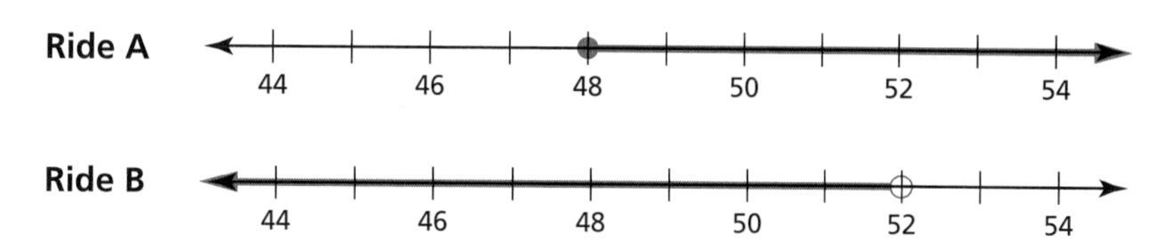

SOLUTION

Ride A

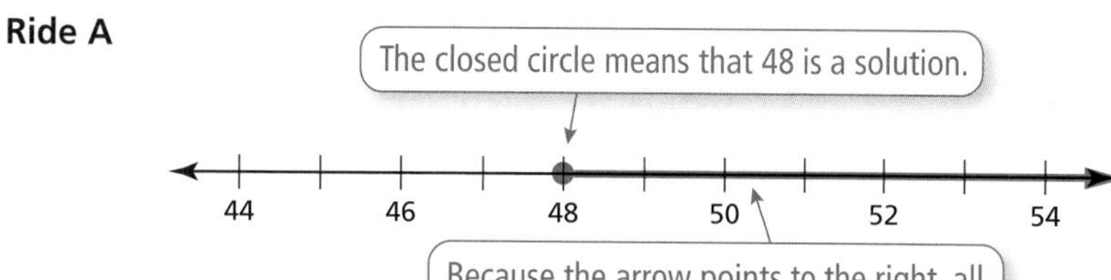

Ride B

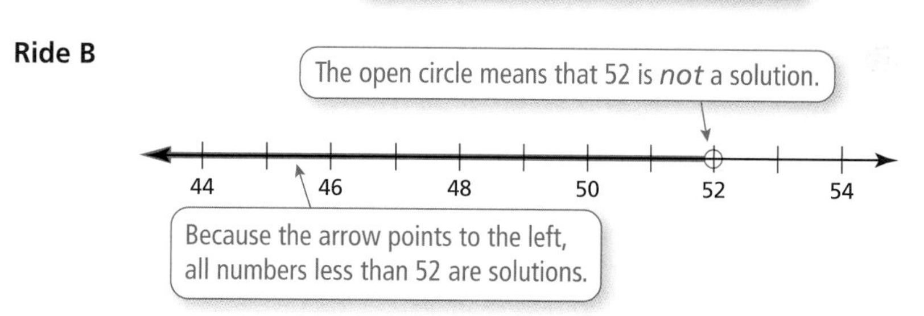

▶ So, $h \geq 48$ represents the height restriction for Ride A, and $h < 52$ represents the height restriction for Ride B.

Monitoring Progress  Help in English and Spanish at *BigIdeasMath.com*

11. Write an inequality that represents the graph.

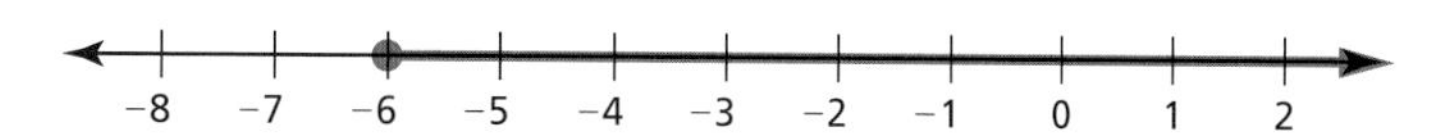

Concept Summary

Representing Linear Inequalities

Words	Algebra	Graph
x is less than 2	$x < 2$	
x is greater than 2	$x > 2$	
x is less than or equal to 2	$x \leq 2$	
x is greater than or equal to 2	$x \geq 2$	

2.1 Exercises

Dynamic Solutions available at *BigIdeasMath.com*

Vocabulary and Core Concept Check

1. **COMPLETE THE SENTENCE** A mathematical sentence using the symbols $<$, $>$, $\leq$, or $\geq$ is called a(n)________.

2. **VOCABULARY** Is 5 in the solution set of $x + 3 > 8$? Explain.

3. **ATTENDING TO PRECISION** Describe how to graph an inequality.

4. **DIFFERENT WORDS, SAME QUESTION** Which is different? Write "both" inequalities.

w is greater than or equal to -7.	w is no less than -7.
w is no more than -7.	w is at least -7.

Monitoring Progress and Modeling with Mathematics

In Exercises 5–12, write the sentence as an inequality. *(See Example 1.)*

5. A number x is greater than 3.

6. A number n plus 7 is less than or equal to 9.

7. Fifteen is no more than a number t divided by 5.

8. Three times a number w is less than 18.

9. One-half of a number y is more than 22.

10. Three is less than the sum of a number s and 4.

11. Thirteen is at least the difference of a number v and 1.

12. Four is no less than the quotient of a number x and 2.1.

13. **MODELING WITH MATHEMATICS** On a fishing trip, you catch two fish. The weight of the first fish is shown. The second fish weighs at least 0.5 pound more than the first fish. Write an inequality that represents the possible weights of the second fish.

14. **MODELING WITH MATHEMATICS** There are 430 people in a wave pool. Write an inequality that represents how many more people can enter the pool.

HOURS

Monday – Friday: 10 A.M. – 6 P.M.

Saturday – Sunday: 10 A.M. – 7 P.M.

Maximum Capacity: 600

In Exercises 15–24, tell whether the value is a solution of the inequality. *(See Example 2.)*

15. $r + 4 > 8;\ r = 2$

16. $5 - x < 8;\ x = -3$

17. $3s \leq 19;\ s = -6$

18. $17 \geq 2y;\ y = 7$

19. $-1 > -\frac{x}{2};\ x = 3$

20. $\frac{4}{z} \geq 3;\ z = 2$

21. $14 \geq -2n + 4;\ n = -5$

22. $-5 \div (2s) < -1;\ s = 10$

23. $20 \leq \frac{10}{2z} + 20;\ z = 5$

24. $\frac{3m}{6} - 2 > 3;\ m = 8$

25. **MODELING WITH MATHEMATICS** The tallest person who ever lived was approximately 8 feet 11 inches tall.

    a. Write an inequality that represents the heights of every other person who has ever lived.

    b. Is 9 feet a solution of the inequality? Explain.

26. DRAWING CONCLUSIONS The winner of a weight-lifting competition bench-pressed 400 pounds. The other competitors all bench-pressed at least 23 pounds less.

a. Write an inequality that represents the weights that the other competitors bench-pressed.

b. Was one of the other competitors able to bench-press 379 pounds? Explain.

ERROR ANALYSIS In Exercises 27 and 28, describe and correct the error in determining whether 8 is in the solution set of the inequality.

27.

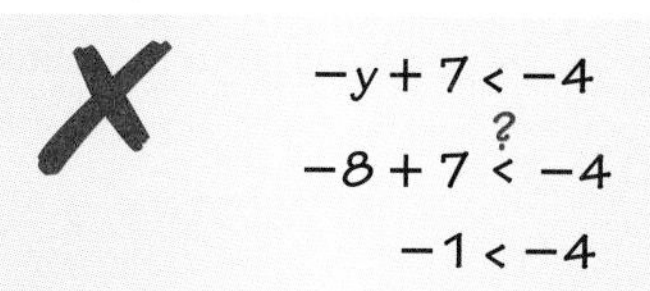

28.

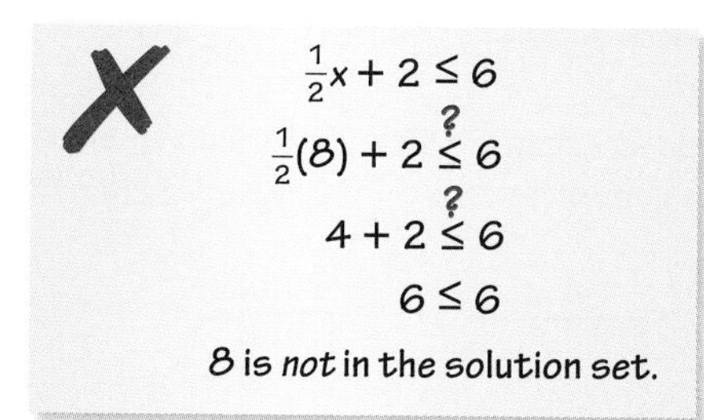

1/2 x + 2 ≤ 6
1/2(8) + 2 ≤? 6
4 + 2 ≤? 6
6 ≤ 6
8 is *not* in the solution set.

In Exercises 29–36, graph the inequality. *(See Example 3.)*

29. $x \geq 2$

30. $z \leq 5$

31. $-1 > t$

32. $-2 < w$

33. $v \leq -4$

34. $s < 1$

35. $\frac{1}{4} < p$

36. $r \geq -|5|$

In Exercises 37–40, write and graph an inequality for the given solution set.

37. $\{x \mid x < 7\}$

38. $\{n \mid n \geq -2\}$

39. $\{z \mid 1.3 \leq z\}$

40. $\{w \mid 5.2 > w\}$

In Exercises 41–44, write an inequality that represents the graph. *(See Example 4.)*

41.

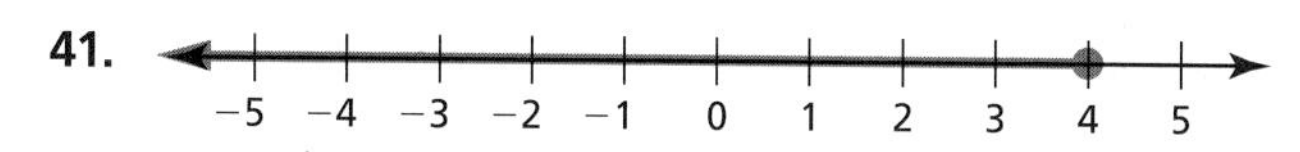

42.

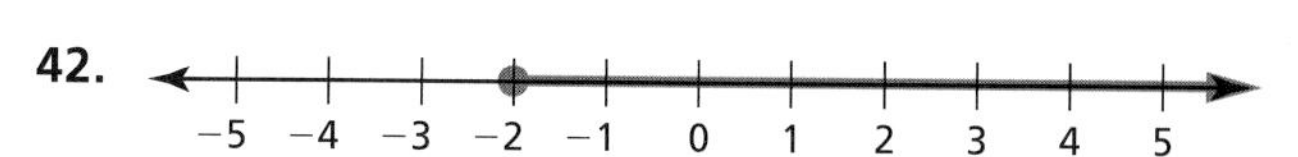

43.

−5 −4 −3 −2 −1 0 1 2 3 4 5

44.

−5 −4 −3 −2 −1 0 1 2 3 4 5

45. ANALYZING RELATIONSHIPS The water temperature of a swimming pool must be no less than 76°F. The temperature is currently 74°F. Which graph correctly shows how much the temperature needs to increase? Explain your reasoning.

Ⓐ

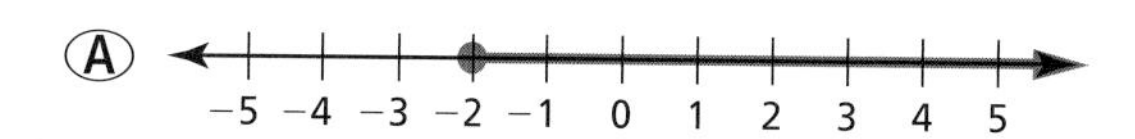

Ⓑ

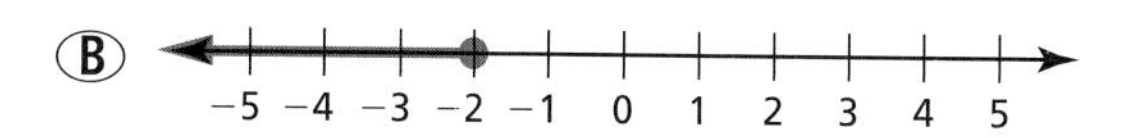

Ⓒ

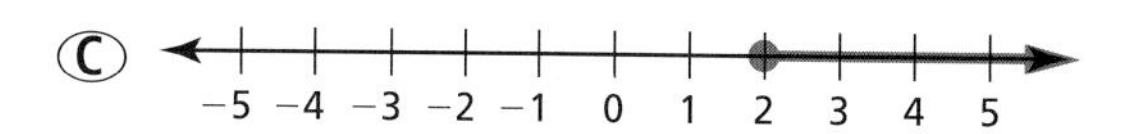

Ⓓ

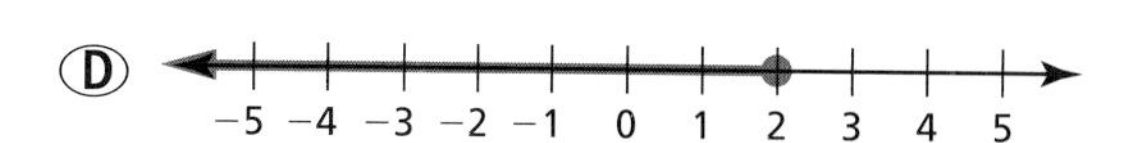

46. MODELING WITH MATHEMATICS According to a state law for vehicles traveling on state roads, the maximum total weight of a vehicle and its contents depends on the number of axles on the vehicle. For each type of vehicle, write and graph an inequality that represents the possible total weights w (in pounds) of the vehicle and its contents.

Maximum Total Weights

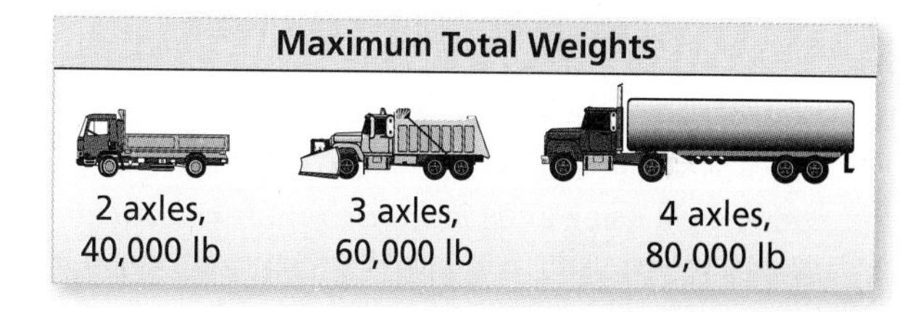

2 axles	3 axles	4 axles
40,000 lb	60,000 lb	80,000 lb

47. PROBLEM SOLVING The Xianren Bridge is located in Guangxi Province, China. This arch is the world's longest natural arch, with a length of 400 feet. Write and graph an inequality that represents the lengths ℓ (in *inches*) of all other natural arches.

48. THOUGHT PROVOKING A student works no more than 25 hours each week at a part-time job. Write an inequality that represents how many hours the student can work each day.

49. WRITING Describe a real-life situation modeled by the inequality $23 + x \leq 31$.

50. HOW DO YOU SEE IT? The graph represents the known melting points of all metallic elements (in degrees Celsius).

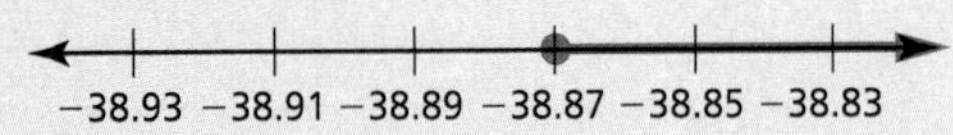

a. Write an inequality represented by the graph.

b. Is it possible for a metallic element to have a melting point of $-38.87°C$? Explain.

51. DRAWING CONCLUSIONS A one-way ride on a subway costs \$0.90. A monthly pass costs \$24. Write an inequality that represents how many one-way rides you can buy before it is cheaper to buy the monthly pass. Is it cheaper to pay the one-way fare for 25 rides? Explain.

Subway Prices	
One-way ride	\$0.90
Monthly pass	\$24.00

52. MAKING AN ARGUMENT The inequality $x \leq 1324$ represents the weights (in pounds) of all mako sharks ever caught using a rod and reel. Your friend says this means no one using a rod and reel has ever caught a mako shark that weighs 1324 pounds. Your cousin says this means someone using a rod and reel *has* caught a mako shark that weighs 1324 pounds. Who is correct? Explain your reasoning.

53. CRITICAL THINKING Describe a real-life situation that can be modeled by more than one inequality.

54. MODELING WITH MATHEMATICS In 1997, Superman's cape from the 1978 movie *Superman* was sold at an auction. The winning bid was \$17,000. Write and graph an inequality that represents the amounts all the losing bids.

MATHEMATICAL CONNECTIONS In Exercises 55–58, write an inequality that represents the missing dimension x.

55. The area is less than 42 square meters.

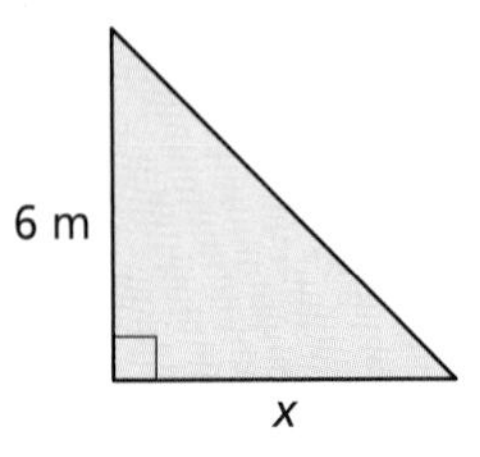

56. The area is greater than or equal to 8 square feet.

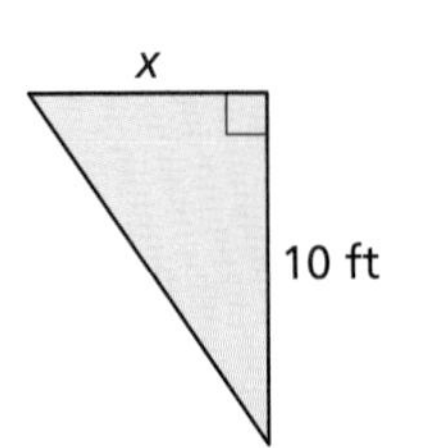

57. The area is less than 18 square centimeters.

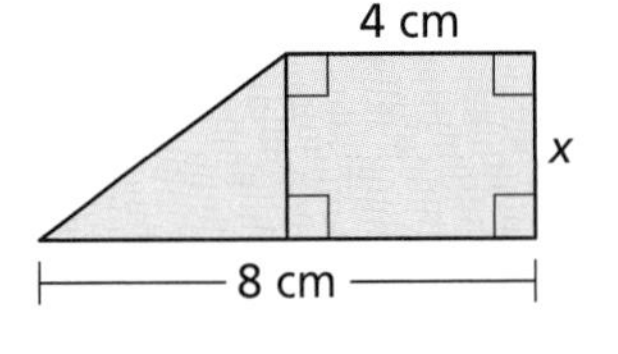

58. The area is greater than 12 square inches.

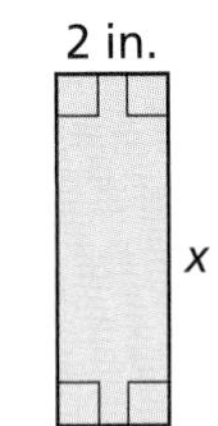

59. WRITING A runner finishes a 200-meter dash in 35 seconds. Let r represent any speed (in meters per second) faster than the runner's speed.

a. Write an inequality that represents r. Then graph the inequality.

b. Every point on the graph represents a speed faster than the runner's speed. Do you think every point could represent the speed of a runner? Explain.

Maintaining Mathematical Proficiency

Reviewing what you learned in previous grades and lessons

Solve the equation. Check your solution. *(Section 1.1)*

60. $x + 2 = 3$

61. $y - 9 = 5$

62. $6 = 4 + y$

63. $-12 = y - 11$

Solve the literal equation for x. *(Section 1.5)*

64. $v = x \cdot y \cdot z$

65. $s = 2r + 3x$

66. $w = 5 + 3(x - 1)$

67. $n = \dfrac{2x + 1}{2}$

2.2 Solving Inequalities Using Addition or Subtraction

Essential Question

How can you use addition or subtraction to solve an inequality?

EXPLORATION 1 Quarterback Passing Efficiency

Work with a partner. The National Collegiate Athletic Association (NCAA) uses the following formula to rank the passing efficiencies P of quarterbacks.

$$P = \frac{8.4Y + 100C + 330T - 200N}{A}$$

Y = total length of all completed passes (in Yards)

C = Completed passes

T = passes resulting in a Touchdown

N = iNtercepted passes

A = Attempted passes

M = incoMplete passes

To be proficient in math, you need to identify and analyze important relationships and then draw conclusions, using tools such as diagrams, flowcharts, and formulas.

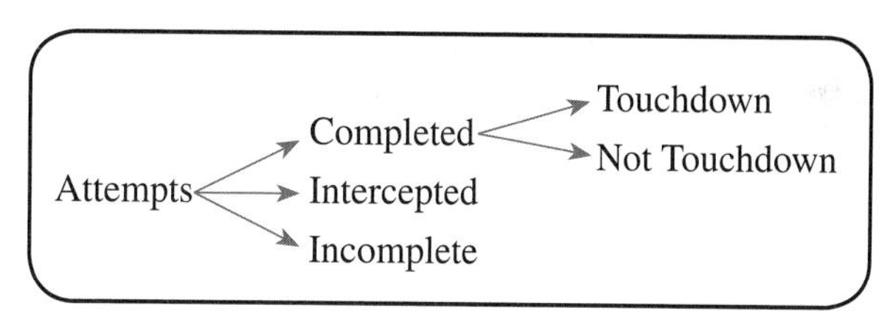

Determine whether each inequality must be true. Explain your reasoning.

a. $T < C$ **b.** $C + N \leq A$ **c.** $N < A$ **d.** $A - C \geq M$

EXPLORATION 2 Finding Solutions of Inequalities

Work with a partner. Use the passing efficiency formula to create a passing record that makes each inequality true. Record your results in the table. Then describe the values of P that make each inequality true.

Attempts	Completions	Yards	Touchdowns	Interceptions

a. $P < 0$

b. $P + 100 \geq 250$

c. $P - 250 > -80$

Communicate Your Answer

3. How can you use addition or subtraction to solve an inequality?

4. Solve each inequality.

a. $x + 3 < 4$ **b.** $x - 3 \geq 5$

c. $4 > x - 2$ **d.** $-2 \leq x + 1$

2.2 Lesson

Core Vocabulary

equivalent inequalities, *p. 62*

Previous
inequality

What You Will Learn

- Solve inequalities using addition.
- Solve inequalities using subtraction.
- Use inequalities to solve real-life problems.

Solving Inequalities Using Addition

Just as you used the properties of equality to produce equivalent equations, you can use the properties of inequality to produce equivalent inequalities. **Equivalent inequalities** are inequalities that have the same solutions.

Core Concept

Addition Property of Inequality

Words Adding the same number to each side of an inequality produces an equivalent inequality.

Numbers

$$\begin{array}{rcr} -3 & < & 2 \\ +4 & & +4 \\ \hline 1 & < & 6 \end{array} \qquad \begin{array}{rcr} -3 & \geq & -10 \\ +3 & & +3 \\ \hline 0 & \geq & -7 \end{array}$$

Algebra If $a > b$, then $a + c > b + c$. If $a \geq b$, then $a + c \geq b + c$.

If $a < b$, then $a + c < b + c$. If $a \leq b$, then $a + c \leq b + c$.

The diagram shows one way to visualize the Addition Property of Inequality when $c > 0$.

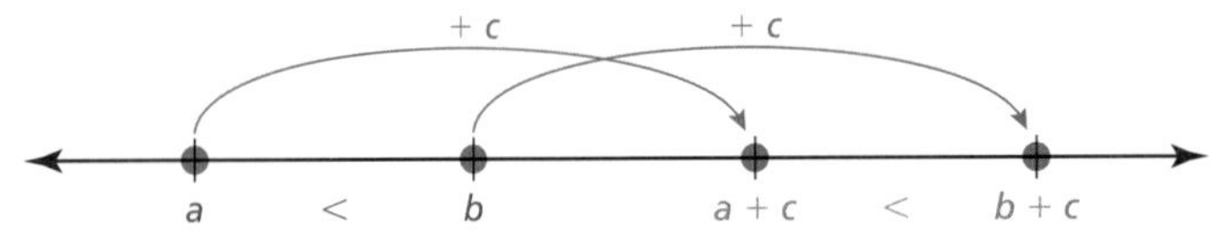

EXAMPLE 1 Solving an Inequality Using Addition

Solve $x - 6 \geq -10$. Graph the solution.

SOLUTION

$$\begin{array}{rcr} x - 6 & \geq & -10 \\ +6 & & +6 \\ \hline x & \geq & -4 \end{array}$$

Write the inequality.
Add 6 to each side.
Simplify.

▶ The solution is $x \geq -4$.

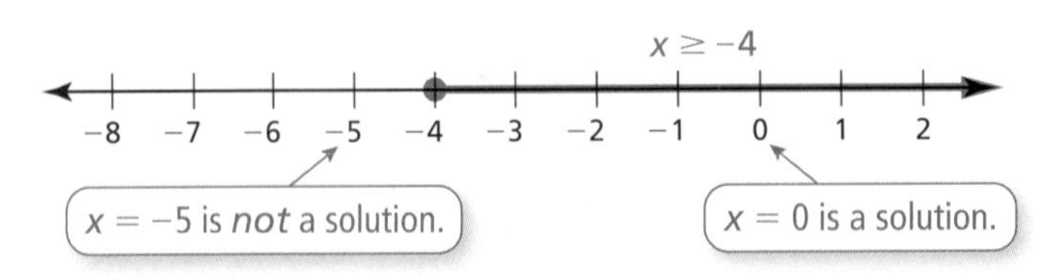

REMEMBER
To check this solution, substitute a few numbers to the left and right of −4 into the original inequality.

Monitoring Progress

Help in English and Spanish at *BigIdeasMath.com*

Solve the inequality. Graph the solution.

1. $b - 2 > -9$ **2.** $m - 3 \leq 5$ **3.** $\frac{1}{4} > y - \frac{1}{4}$

Solving Inequalities Using Subtraction

Core Concept

Subtraction Property of Inequality

Words Subtracting the same number from each side of an inequality produces an equivalent inequality.

Numbers

$$\begin{array}{rcr} -3 & \leq & 1 \\ \underline{-5} & & \underline{-5} \\ -8 & \leq & -4 \end{array} \qquad \begin{array}{rcr} 7 & > & -20 \\ \underline{-7} & & \underline{-7} \\ 0 & > & -27 \end{array}$$

Algebra If $a > b$, then $a - c > b - c$. If $a \geq b$, then $a - c \geq b - c$.

If $a < b$, then $a - c < b - c$. If $a \leq b$, then $a - c \leq b - c$.

The diagram shows one way to visualize the Subtraction Property of Inequality when $c > 0$.

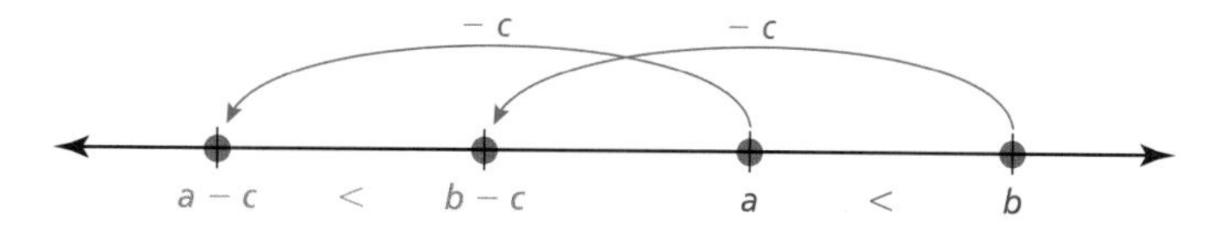

EXAMPLE 2 **Solving an Inequality Using Subtraction**

Solve each inequality. Graph the solution.

a. $y + 8 \leq 5$

b. $-8 < 1.4 + m$

SOLUTION

a. $y + 8 \leq 5$ Write the inequality.

Subtraction Property of Inequality → $\underline{-8} \quad \underline{-8}$ Subtract 8 from each side.

$y \leq -3$ Simplify.

▶ The solution is $y \leq -3$.

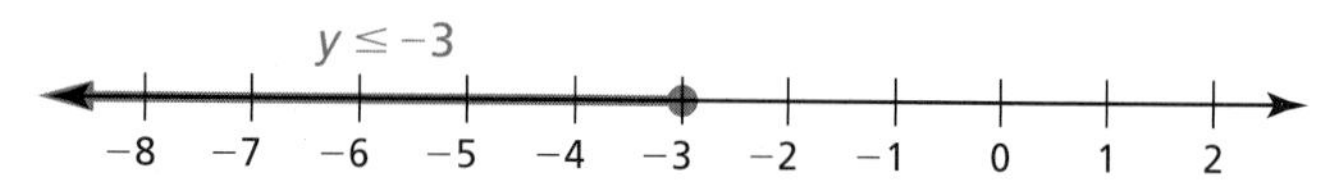

b. $-8 < 1.4 + m$ Write the inequality.

Subtraction Property of Inequality → $\underline{-1.4} \quad \underline{-1.4}$ Subtract 1.4 from each side.

$-9.4 < m$ Simplify.

▶ The solution is $m > -9.4$.

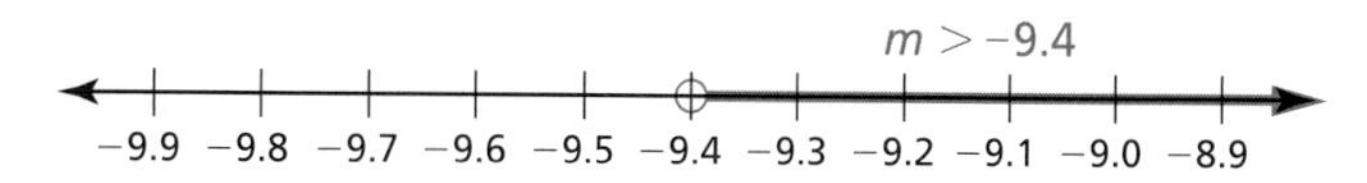

Monitoring Progress

Help in English and Spanish at *BigIdeasMath.com*

Solve the inequality. Graph the solution.

4. $k + 5 \leq -3$

5. $\frac{5}{6} \leq z + \frac{1}{6}$

6. $p + 0.7 > -2.3$

Solving Real-Life Problems

Modeling with Mathematics

A circuit overloads at 1800 watts of electricity. You plug a microwave oven that uses 1100 watts of electricity into the circuit.

a. Write and solve an inequality that represents how many watts you can add to the circuit without overloading the circuit.

b. In addition to the microwave oven, which of the following appliances can you plug into the circuit at the same time without overloading the circuit?

Appliance	Watts
Clock radio	50
Blender	300
Hot plate	1200
Toaster	800

SOLUTION

1. **Understand the Problem** You know that the microwave oven uses 1100 watts out of a possible 1800 watts. You are asked to write and solve an inequality that represents how many watts you can add without overloading the circuit. You also know the numbers of watts used by four other appliances. You are asked to identify the appliances you can plug in at the same time without overloading the circuit.
2. **Make a Plan** Use a verbal model to write an inequality. Then solve the inequality and identify other appliances that you can plug into the circuit at the same time without overloading the circuit.
3. **Solve the Problem**

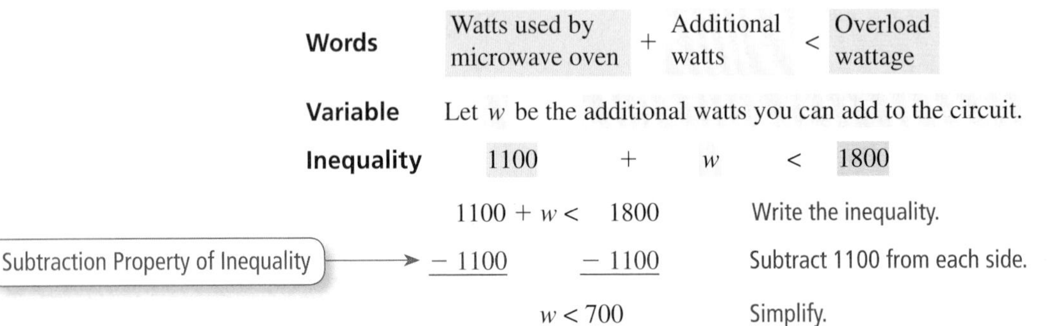

Words Watts used by microwave oven + Additional watts < Overload wattage

Variable Let w be the additional watts you can add to the circuit.

Inequality $1100 + w < 1800$

$1100 + w < 1800$ Write the inequality.

Subtraction Property of Inequality → $\underline{-1100} \quad \underline{-1100}$ Subtract 1100 from each side.

$w < 700$ Simplify.

▶ You can add up to 700 watts to the circuit, which means that you can also plug in the clock radio and the blender.

4. **Look Back** You can check that your answer is correct by adding the numbers of watts used by the microwave oven, clock radio, and blender.

$1100 + 50 + 300 = 1450$

The circuit will not overload because the total wattage is less than 1800 watts.

Monitoring Progress  Help in English and Spanish at *BigIdeasMath.com*

7. The microwave oven uses only 1000 watts of electricity. Does this allow you to have both the microwave oven and the toaster plugged into the circuit at the same time? Explain your reasoning.

2.2 Exercises

Dynamic Solutions available at *BigIdeasMath.com*

Vocabulary and Core Concept Check

1. **VOCABULARY** Why is the inequality $x \le 6$ equivalent to the inequality $x - 5 \le 6 - 5$?

2. **WRITING** Compare solving equations using addition with solving inequalities using addition.

Monitoring Progress and Modeling with Mathematics

In Exercises 3–6, tell which number you would add to or subtract from each side of the inequality to solve it.

3. $k + 11 < -3$
4. $v - 2 > 14$
5. $-1 \ge b - 9$
6. $-6 \le 17 + p$

In Exercises 7–20, solve the inequality. Graph the solution. *(See Examples 1 and 2.)*

7. $x - 4 < -5$
8. $1 \le s - 8$
9. $6 \ge m - 1$
10. $c - 12 > -4$
11. $r + 4 < 5$
12. $-8 \le 8 + y$
13. $9 + w > 7$
14. $15 \ge q + 3$
15. $h - (-2) \ge 10$
16. $-6 > t - (-13)$
17. $j + 9 - 3 < 8$
18. $1 - 12 + y \ge -5$
19. $10 \ge 3p - 2p - 7$
20. $18 - 5z + 6z > 3 + 6$

In Exercises 21–24, write the sentence as an inequality. Then solve the inequality.

21. A number plus 8 is greater than 11.

22. A number minus 3 is at least -5.

23. The difference of a number and 9 is fewer than 4.

24. Six is less than or equal to the sum of a number and 15.

25. **MODELING WITH MATHEMATICS** You are riding a train. Your carry-on bag can weigh no more than 50 pounds. Your bag weighs 38 pounds. *(See Example 3.)*

    a. Write and solve an inequality that represents how much weight you can add to your bag.

    b. Can you add both a 9-pound laptop and a 5-pound pair of boots to your bag without going over the weight limit? Explain.

26. **MODELING WITH MATHEMATICS** You order the hardcover book shown from a website that offers free shipping on orders of \$25 or more. Write and solve an inequality that represents how much more you must spend to get free shipping.

Price: \$19.76

ERROR ANALYSIS In Exercises 27 and 28, describe and correct the error in solving the inequality or graphing the solution.

27. ✗

$$-17 < x - 14$$
$$-17 + 14 < x - 14 + 14$$
$$-3 < x$$

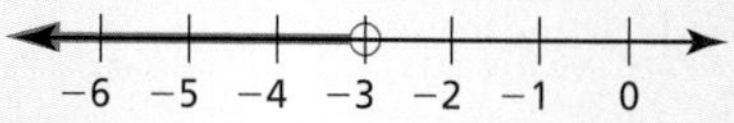

28. ✗

$$-10 + x \ge -9$$
$$-10 + 10 + x \ge -9$$
$$x \ge -9$$

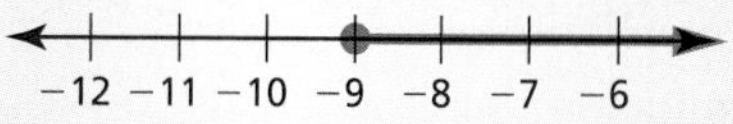

29. **PROBLEM SOLVING** An NHL hockey player has 59 goals so far in a season. What are the possible numbers of additional goals the player can score to match or break the NHL record of 92 goals in a season?

30. **MAKING AN ARGUMENT** In an aerial ski competition, you perform two acrobatic ski jumps. The scores on the two jumps are then added together.

Ski jump	Competitor's score	Your score
1	117.1	119.5
2	119.8	

a. Describe the score that you must earn on your second jump to beat your competitor.

b. Your coach says that you will beat your competitor if you score 118.4 points. A teammate says that you only need 117.5 points. Who is correct? Explain.

31. **REASONING** Which of the following inequalities are equivalent to the inequality $x - b < 3$, where b is a constant? Justify your answer.

Ⓐ $x - b - 3 < 0$ Ⓑ $0 > b - x + 3$

Ⓒ $x < 3 - b$ Ⓓ $-3 < b - x$

MATHEMATICAL CONNECTIONS In Exercises 32 and 33, write and solve an inequality to find the possible values of x.

32. Perimeter < 51.3 inches

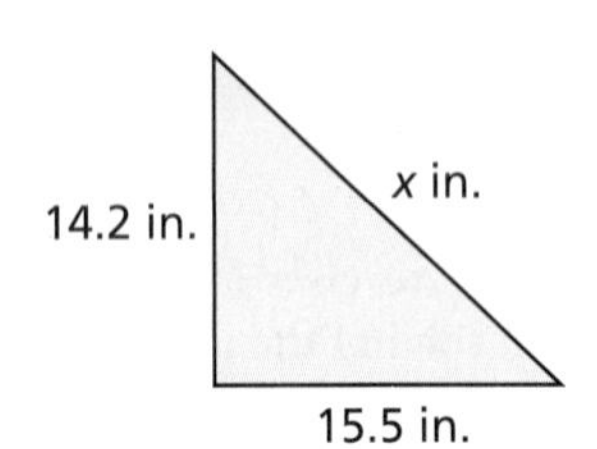

33. Perimeter ≤ 18.7 feet

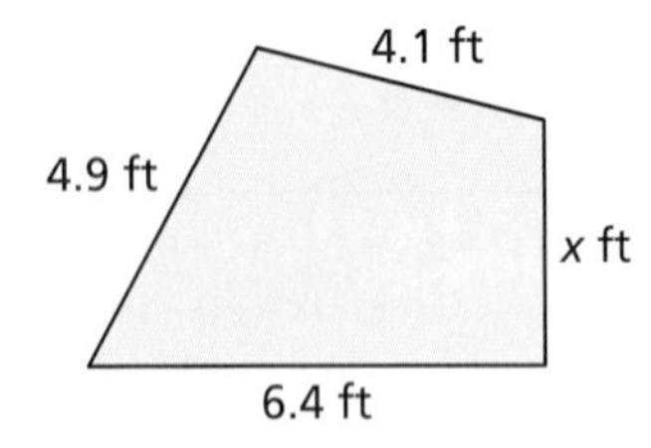

34. **THOUGHT PROVOKING** Write an inequality that has the solution shown in the graph. Describe a real-life situation that can be modeled by the inequality.

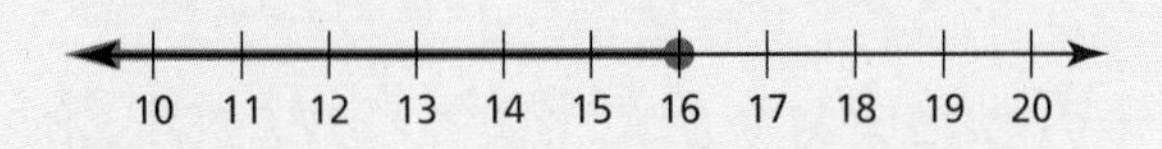

35. **WRITING** Is it possible to check all the numbers in the solution set of an inequality? When you solve the inequality $x - 11 \geq -3$, which numbers can you check to verify your solution? Explain your reasoning.

36. **HOW DO YOU SEE IT?** The diagram represents the numbers of students in a school with brown eyes, brown hair, or both.

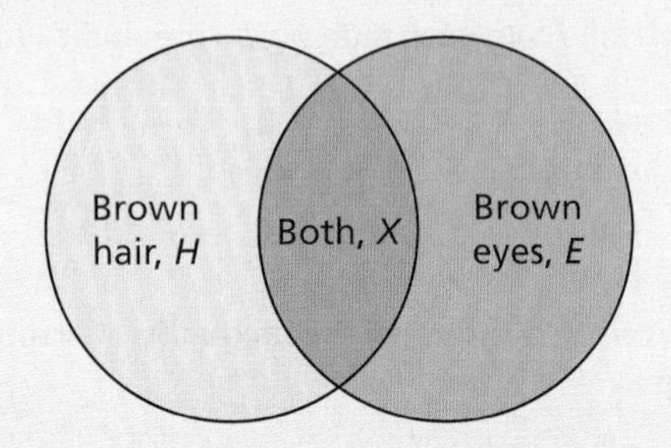

Determine whether each inequality must be true. Explain your reasoning.

a. $H \geq E$ b. $H + 10 \geq E$

c. $H \geq X$ d. $H + 10 \geq X$

e. $H > X$ f. $H + 10 > X$

37. **REASONING** Write and graph an inequality that represents the numbers that are *not* solutions of each inequality.

a. $x + 8 < 14$

b. $x - 12 \geq 5.7$

38. **PROBLEM SOLVING** Use the inequalities $c - 3 \geq d$, $b + 4 < a + 1$, and $a - 2 \leq d - 7$ to order a, b, c, and d from least to greatest.

Maintaining Mathematical Proficiency

Reviewing what you learned in previous grades and lessons

Find the product or quotient. *(Skills Review Handbook)*

39. $7 \cdot (-9)$ 40. $-11 \cdot (-12)$ 41. $-27 \div (-3)$ 42. $20 \div (-5)$

Solve the equation. Check your solution. *(Section 1.1)*

43. $6x = 24$ 44. $-3y = -18$ 45. $\frac{s}{-8} = 13$ 46. $\frac{n}{4} = -7.3$

2.3 Solving Inequalities Using Multiplication or Division

Essential Question How can you use division to solve an inequality?

EXPLORATION 1 Writing a Rule

Work with a partner.

a. Copy and complete the table. Decide which graph represents the solution of the inequality $6 < 3x$. Write the solution of the inequality.

x	-1	0	1	2	3	4	5
$3x$	-3						
$6 \overset{?}{<} 3x$	No						

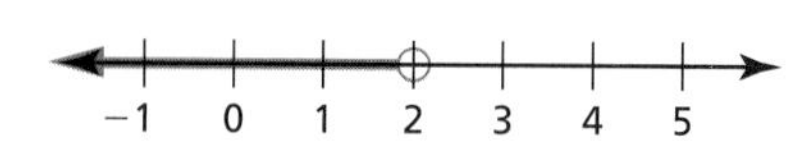

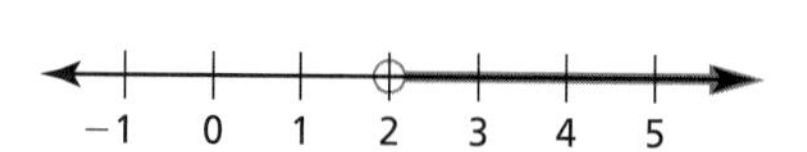

LOOKING FOR A PATTERN

To be proficient in math, you need to investigate relationships, observe patterns, and use your observations to write general rules.

b. Use a table to solve each inequality. Then write a rule that describes how to use division to solve the inequalities.

i. $2x < 4$ **ii.** $3 \geq 3x$ **iii.** $2x < 8$ **iv.** $6 \geq 3x$

EXPLORATION 2 Writing a Rule

Work with a partner.

a. Copy and complete the table. Decide which graph represents the solution of the inequality $6 < -3x$. Write the solution of the inequality.

x	-5	-4	-3	-2	-1	0	1
$-3x$							
$6 \overset{?}{<} -3x$							

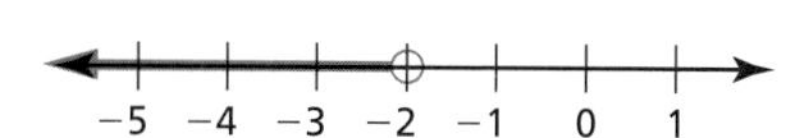

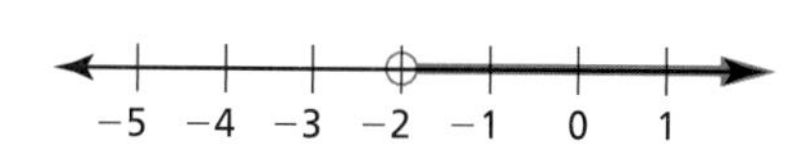

b. Use a table to solve each inequality. Then write a rule that describes how to use division to solve the inequalities.

i. $-2x < 4$ **ii.** $3 \geq -3x$ **iii.** $-2x < 8$ **iv.** $6 \geq -3x$

Communicate Your Answer

3. How can you use division to solve an inequality?

4. Use the rules you wrote in Explorations 1(b) and 2(b) to solve each inequality.

a. $7x < -21$ **b.** $12 \leq 4x$ **c.** $10 < -5x$ **d.** $-3x \leq 0$

2.3 Lesson

What You Will Learn

- Solve inequalities by multiplying or dividing by *positive* numbers.
- Solve inequalities by multiplying or dividing by *negative* numbers.
- Use inequalities to solve real-life problems.

Multiplying or Dividing by Positive Numbers

Core Concept

Multiplication and Division Properties of Inequality ($c > 0$)

Words Multiplying or dividing each side of an inequality by the same *positive* number produces an equivalent inequality.

Numbers

$$-6 < 8 \qquad\qquad 6 > -8$$

$$2 \cdot (-6) < 2 \cdot 8 \qquad\qquad \frac{6}{2} > \frac{-8}{2}$$

$$-12 < 16 \qquad\qquad 3 > -4$$

Algebra If $a > b$ and $c > 0$, then $ac > bc$. If $a > b$ and $c > 0$, then $\frac{a}{c} > \frac{b}{c}$.

If $a < b$ and $c > 0$, then $ac < bc$. If $a < b$ and $c > 0$, then $\frac{a}{c} < \frac{b}{c}$.

These properties are also true for $\leq$ and $\geq$.

EXAMPLE 1 Multiplying or Dividing by Positive Numbers

Solve (a) $\frac{x}{8} > -5$ and (b) $-24 \geq 3x$. Graph each solution.

SOLUTION

a. $\frac{x}{8} > -5$ Write the inequality.

Multiplication Property of Inequality → $8 \cdot \frac{x}{8} > 8 \cdot (-5)$ Multiply each side by 8.

$x > -40$ Simplify.

▶ The solution is $x > -40$.

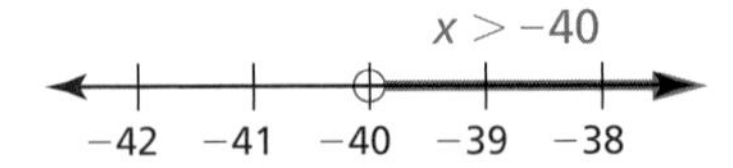

b. $-24 \geq 3x$ Write the inequality.

Division Property of Inequality → $\frac{-24}{3} \geq \frac{3x}{3}$ Divide each side by 3.

$-8 \geq x$ Simplify.

▶ The solution is $x \leq -8$.

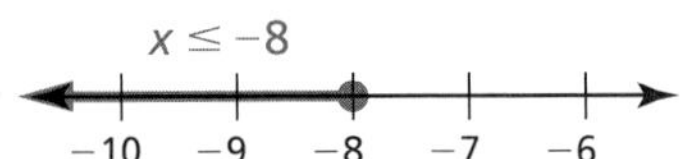

Monitoring Progress

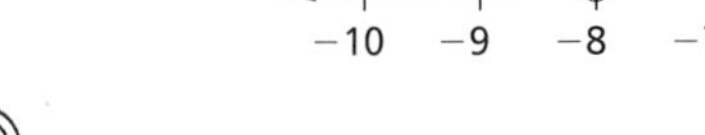

Solve the inequality. Graph the solution.

1. $\frac{n}{7} \geq -1$ **2.** $-6.4 \geq \frac{1}{5}w$ **3.** $4b \geq 36$ **4.** $-18 > 1.5q$

Multiplying or Dividing by Negative Numbers

Core Concept

Multiplication and Division Properties of Inequality ($c < 0$)

Words When multiplying or dividing each side of an inequality by the same *negative* number, the direction of the inequality symbol must be reversed to produce an equivalent inequality.

Numbers

$$-6 < 8 \qquad\qquad 6 > -8$$

$$-2 \cdot (-6) > -2 \cdot 8 \qquad\qquad \frac{6}{-2} < \frac{-8}{-2}$$

$$12 > -16 \qquad\qquad -3 < 4$$

Algebra If $a > b$ and $c < 0$, then $ac < bc$. If $a > b$ and $c < 0$, then $\frac{a}{c} < \frac{b}{c}$.

If $a < b$ and $c < 0$, then $ac > bc$. If $a < b$ and $c < 0$, then $\frac{a}{c} > \frac{b}{c}$.

These properties are also true for $\leq$ and $\geq$.

COMMON ERROR

A negative sign in an inequality does not necessarily mean you must reverse the inequality symbol, as shown in Example 1.

Only reverse the inequality symbol when you multiply or divide each side by a negative number.

EXAMPLE 2 Multiplying or Dividing by Negative Numbers

Solve each inequality. Graph each solution.

a. $2 < \frac{y}{-3}$

b. $-7y \leq -35$

SOLUTION

a. $2 < \frac{y}{-3}$ Write the inequality.

Multiplication Property of Inequality → $-3 \cdot 2 > -3 \cdot \frac{y}{-3}$ Multiply each side by -3. Reverse the inequality symbol.

$-6 > y$ Simplify.

▶ The solution is $y < -6$.

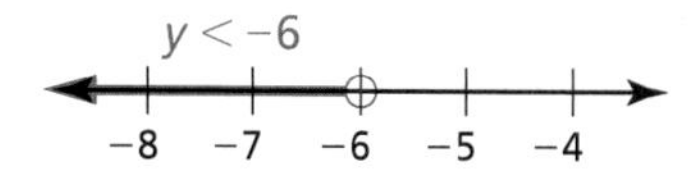

b. $-7y \leq -35$ Write the inequality.

Division Property of Inequality → $\frac{-7y}{-7} \geq \frac{-35}{-7}$ Divide each side by -7. Reverse the inequality symbol.

$y \geq 5$ Simplify.

▶ The solution is $y \geq 5$.

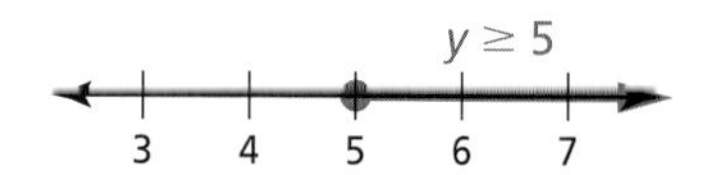

Monitoring Progress

Help in English and Spanish at *BigIdeasMath.com*

Solve the inequality. Graph the solution.

5. $\frac{p}{-4} < 7$

6. $\frac{x}{-5} \leq -5$

7. $1 \geq -\frac{1}{10}z$

8. $-9m > 63$

9. $-2r \geq -22$

10. $-0.4y \geq -12$

Solving Real-Life Problems

Modeling with Mathematics

You earn \$9.50 per hour at your summer job. Write and solve an inequality that represents the numbers of hours you need to work to buy a digital camera that costs \$247.

SOLUTION

1. **Understand the Problem** You know your hourly wage and the cost of the digital camera. You are asked to write and solve an inequality that represents the numbers of hours you need to work to buy the digital camera.
2. **Make a Plan** Use a verbal model to write an inequality. Then solve the inequality.
3. **Solve the Problem**

Words	Hourly wage	•	Hours worked	≥	Cost of camera
Variable	Let n be the number of hours worked.				
Inequality	9.5	•	n	≥	247

$$9.5n \geq 247 \quad \text{Write the inequality.}$$

Division Property of Inequality →
$$\frac{9.5n}{9.5} \geq \frac{247}{9.5} \quad \text{Divide each side by 9.5.}$$

$$n \geq 26 \quad \text{Simplify.}$$

▶ You need to work at least 26 hours for your gross pay to be at least \$247. If you have payroll deductions, such as Social Security taxes, you need to work more than 26 hours.

REMEMBER

Compatible numbers are numbers that are easy to compute mentally.

4. **Look Back** You can use estimation to check that your answer is reasonable.

\$247 ÷ \$9.50/h
↓ ↓
\$250 ÷ \$10/h = 25 h   Use compatible numbers.

Your hourly wage is about \$10 per hour. So, to earn about \$250, you need to work about 25 hours.

Unit Analysis Each time you set up an equation or inequality to represent a real-life problem, be sure to check that the units balance.

$$\frac{\$9.50}{\cancel{h}} \times 26\cancel{h} = \$247$$

Monitoring Progress

Help in English and Spanish at *BigIdeasMath.com*

11. You have at most \$3.65 to make copies. Each copy costs \$0.25. Write and solve an inequality that represents the numbers of copies you can make.

12. The maximum speed limit for a school bus is 55 miles per hour. Write and solve an inequality that represents the numbers of hours it takes to travel 165 miles in a school bus.

Vocabulary and Core Concept Check

1. **WRITING** Explain how solving $2x < -8$ is different from solving $-2x < 8$.

2. **OPEN-ENDED** Write an inequality that is solved using the Division Property of Inequality where the inequality symbol needs to be reversed.

Monitoring Progress and Modeling with Mathematics

In Exercises 3–10, solve the inequality. Graph the solution. *(See Example 1.)*

3. $4x < 8$

4. $3y \leq -9$

5. $-20 \leq 10n$

6. $35 < 7t$

7. $\frac{x}{2} > -2$

8. $\frac{a}{4} < 10.2$

9. $20 \geq \frac{4}{5}w$

10. $-16 \leq \frac{8}{3}t$

In Exercises 11–18, solve the inequality. Graph the solution. *(See Example 2.)*

11. $-6t < 12$

12. $-9y > 9$

13. $-10 \geq -2z$

14. $-15 \leq -3c$

15. $\frac{n}{-3} \geq 1$

16. $\frac{w}{-5} \leq 16$

17. $-8 < -\frac{1}{4}m$

18. $-6 > -\frac{2}{3}y$

19. **MODELING WITH MATHEMATICS** You have \$12 to buy five goldfish for your new fish tank. Write and solve an inequality that represents the prices you can pay per fish. *(See Example 3.)*

20. **MODELING WITH MATHEMATICS** A weather forecaster predicts that the temperature in Antarctica will decrease 8°F each hour for the next 6 hours. Write and solve an inequality to determine how many hours it will take for the temperature to drop at least 36°F.

USING TOOLS In Exercises 21–26, solve the inequality. Use a graphing calculator to verify your answer.

21. $36 < 3y$

22. $17v \geq 51$

23. $2 \leq -\frac{2}{9}x$

24. $4 > \frac{n}{-4}$

25. $2x > \frac{3}{4}$

26. $1.1y < 4.4$

ERROR ANALYSIS In Exercises 27 and 28, describe and correct the error in solving the inequality.

27.

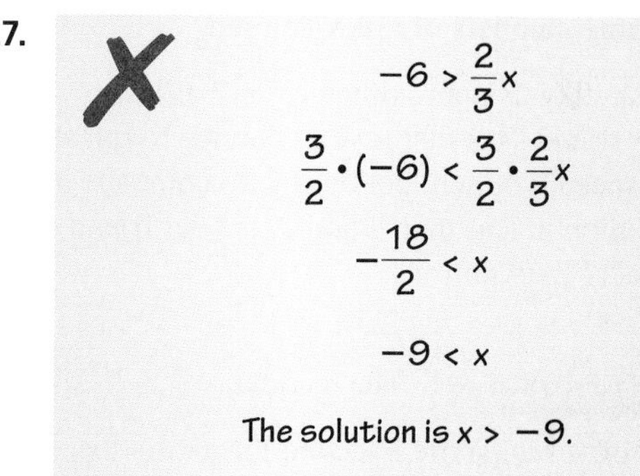

28.

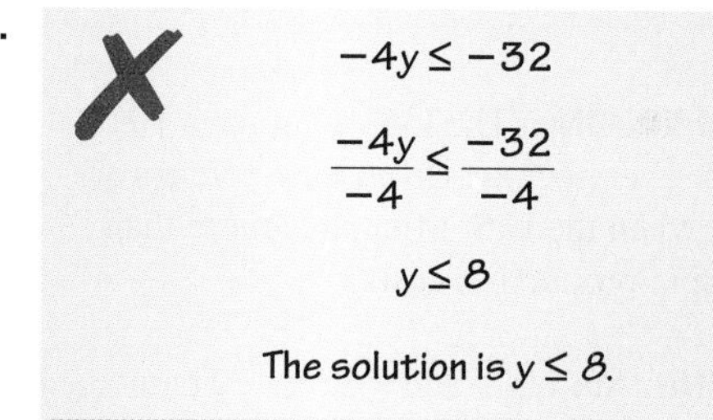

29. **ATTENDING TO PRECISION** You have \$700 to buy new carpet for your bedroom. Write and solve an inequality that represents the costs per square foot that you can pay for the new carpet. Specify the units of measure in each step.

14 ft

14 ft

30. HOW DO YOU SEE IT? Let $m > 0$. Match each inequality with its graph. Explain your reasoning.

a. $\frac{x}{m} < -1$ **b.** $\frac{x}{m} > 1$

c. $\frac{x}{m} < 1$ **d.** $-\frac{x}{m} < 1$

A. m

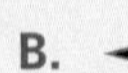

$-m$

D. $-m$

31. MAKING AN ARGUMENT You run for 2 hours at a speed no faster than 6.3 miles per hour.

a. Write and solve an inequality that represents the possible numbers of miles you run.

b. A marathon is approximately 26.2 miles. Your friend says that if you continue to run at this speed, you will not be able to complete a marathon in less than 4 hours. Is your friend correct? Explain.

32. THOUGHT PROVOKING The inequality $\frac{x}{4} \leq 5$ has a solution of $x = p$. Write a second inequality that also has a solution of $x = p$.

33. PROBLEM SOLVING The U.S. Mint pays \$0.02 to produce every penny. How many pennies are produced when the U.S. Mint pays more than \$6 million in production costs?

34. REASONING Are $x \leq \frac{2}{3}$ and $-3x \leq -2$ equivalent? Explain your reasoning.

35. ANALYZING RELATIONSHIPS Consider the number line shown.

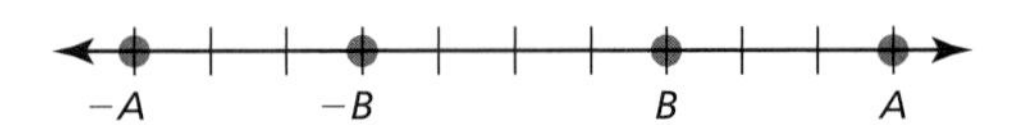

a. Write an inequality relating A and B.

b. Write an inequality relating $-A$ and $-B$.

c. Use the results from parts (a) and (b) to explain why the direction of the inequality symbol must be reversed when multiplying or dividing each side of an inequality by the same negative number.

36. REASONING Why might solving the inequality $\frac{4}{x} \geq 2$ by multiplying each side by x lead to an error? (*Hint:* Consider $x > 0$ and $x < 0$.)

37. MATHEMATICAL CONNECTIONS The radius of a circle is represented by the formula $r = \frac{C}{2\pi}$. Write and solve an inequality that represents the possible circumferences C of the circle.

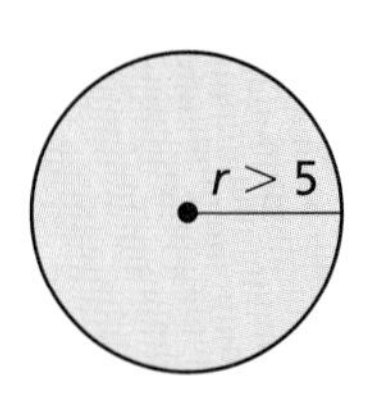

38. CRITICAL THINKING A water-skiing instructor recommends that a boat pulling a beginning skier has a speed less than 18 miles per hour. Write and solve an inequality that represents the possible distances d (in miles) that a beginner can travel in 45 minutes of practice time.

39. CRITICAL THINKING A local zoo employs 36 people to take care of the animals each day. At most, 24 of the employees work full time. Write and solve an inequality that represents the fraction of employees who work part time. Graph the solution.

Maintaining Mathematical Proficiency

Reviewing what you learned in previous grades and lessons

Solve the equation. Check your solution. *(Section 1.2 and Section 1.3)*

40. $5x + 3 = 13$

41. $\frac{1}{2}y - 8 = -10$

42. $-3n + 2 = 2n - 3$

43. $\frac{1}{2}z + 4 = \frac{5}{2}z - 8$

Tell which number is greater. *(Skills Review Handbook)*

44. 0.8, 85%

45. $\frac{16}{30}$, 50%

46. 120%, 0.12

47. 60%, $\frac{2}{3}$

2.4 Solving Multi-Step Inequalities

Essential Question How can you solve a multi-step inequality?

EXPLORATION 1 Solving a Multi-Step Inequality

Work with a partner.

- Use what you already know about solving equations and inequalities to solve each multi-step inequality. Justify each step.
- Match each inequality with its graph. Use a graphing calculator to check your answer.

JUSTIFYING STEPS

To be proficient in math, you need to justify each step in a solution and communicate your justification to others.

a. $2x + 3 \leq x + 5$ **b.** $-2x + 3 > x + 9$

c. $27 \geq 5x + 4x$ **d.** $-8x + 2x - 16 < -5x + 7x$

e. $3(x - 3) - 5x > -3x - 6$ **f.** $-5x - 6x \leq 8 - 8x - x$

A.

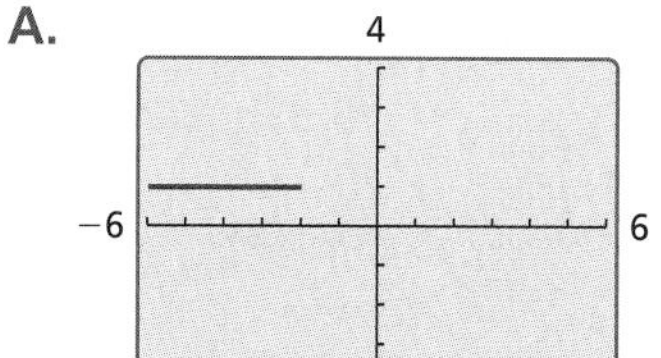

B.

4
−6
6
−4

C.

4
−6
6
−4

D.

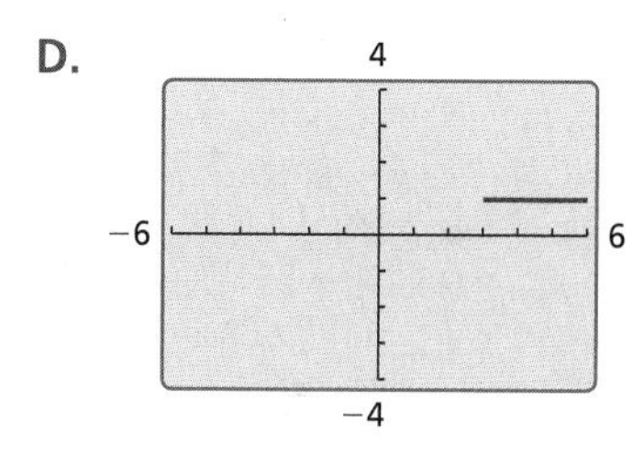

E.

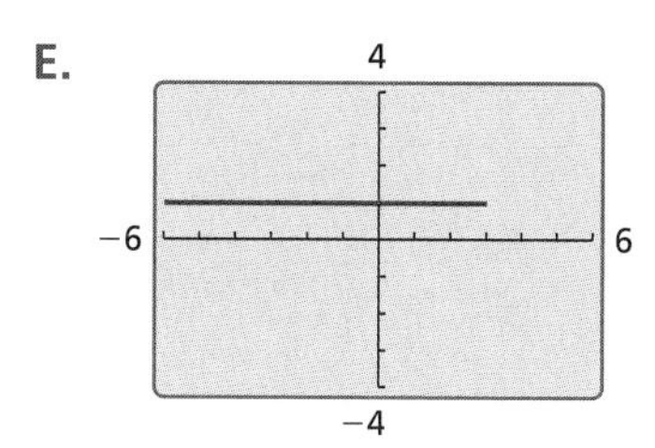

F.

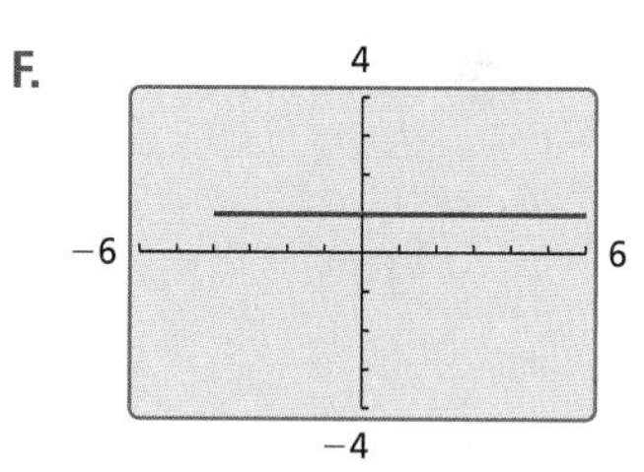

Communicate Your Answer

2. How can you solve a multi-step inequality?

3. Write two different multi-step inequalities whose solutions are represented by the graph.

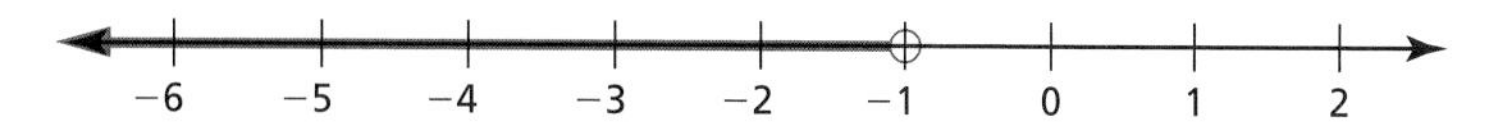

2.4 Lesson

What You Will Learn

- Solve multi-step inequalities.
- Use multi-step inequalities to solve real-life problems.

Solving Multi-Step Inequalities

To solve a multi-step inequality, simplify each side of the inequality, if necessary. Then use inverse operations to isolate the variable. Be sure to reverse the inequality symbol when multiplying or dividing by a negative number.

EXAMPLE 1 Solving Multi-Step Inequalities

Solve each inequality. Graph each solution.

a. $\frac{y}{-6} + 7 < 9$

b. $2v - 4 \geq 8$

SOLUTION

a.

$\frac{y}{-6} + 7 < 9$	Write the inequality.
$\underline{-7} \quad \underline{-7}$	Subtract 7 from each side.
$\frac{y}{-6} < 2$	Simplify.
$-6 \cdot \frac{y}{-6} > -6 \cdot 2$	Multiply each side by –6. Reverse the inequality symbol.
$y > -12$	Simplify.

▶ The solution is $y > -12$.

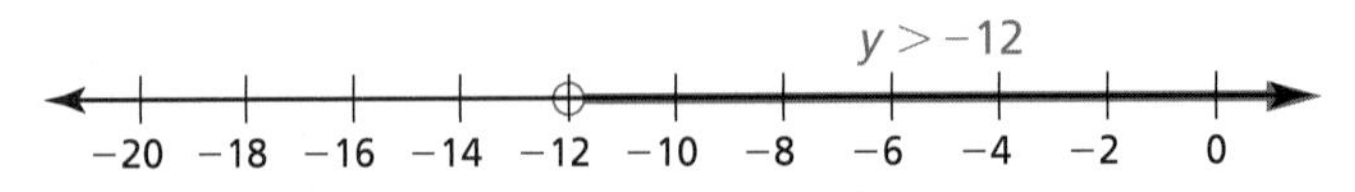

b.

$2v - 4 \geq 8$	Write the inequality.
$\underline{+4} \quad \underline{+4}$	Add 4 to each side.
$2v \geq 12$	Simplify.
$\frac{2v}{2} \geq \frac{12}{2}$	Divide each side by 2.
$v \geq 6$	Simplify.

▶ The solution is $v \geq 6$.

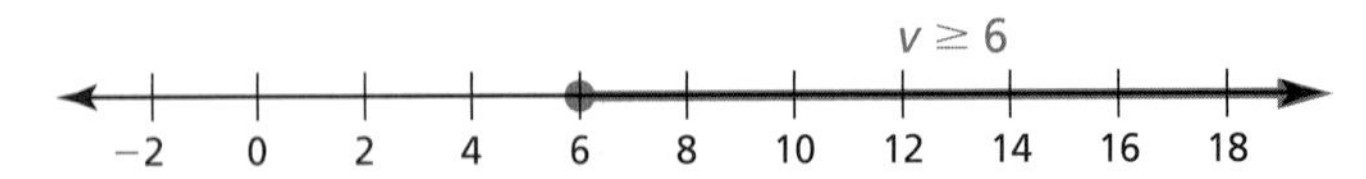

Monitoring Progress  Help in English and Spanish at *BigIdeasMath.com*

Solve the inequality. Graph the solution.

1. $4b - 1 < 7$

2. $8 - 9c \geq -28$

3. $\frac{n}{-2} + 11 > 12$

4. $6 \geq 5 - \frac{v}{3}$

EXAMPLE 2 Solving an Inequality with Variables on Both Sides

Solve $6x - 5 < 2x + 11$.

SOLUTION

$6x - 5 < 2x + 11$	Write the inequality.
$\underline{+5} \qquad \underline{+5}$	Add 5 to each side.
$6x < 2x + 16$	Simplify.
$\underline{-2x} \qquad \underline{-2x}$	Subtract 2*x* from each side.
$4x < 16$	Simplify.
$\frac{4x}{4} < \frac{16}{4}$	Divide each by 4.
$x < 4$	Simplify.

▶ The solution is $x < 4$.

When solving an inequality, if you obtain an equivalent inequality that is true, such as $-5 < 0$, the solutions of the inequality are *all real numbers*. If you obtain an equivalent inequality that is false, such as $3 \leq -2$, the inequality has *no solution.*

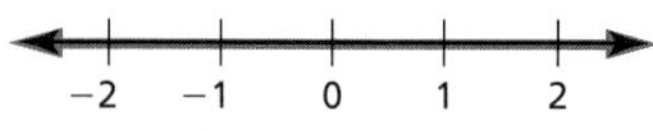

Graph of an inequality whose solutions are all real numbers

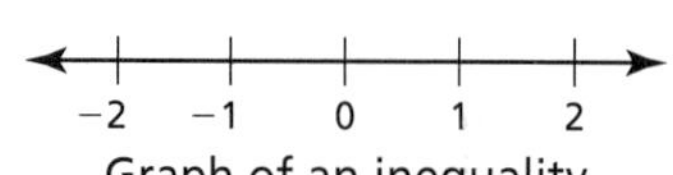

Graph of an inequality that has no solution

EXAMPLE 3 Inequalities with Special Solutions

Solve (a) $8b - 3 > 4(2b + 3)$ and (b) $2(5w - 1) \leq 7 + 10w$.

SOLUTION

LOOKING FOR STRUCTURE

When the variable terms on each side of an inequality are the same, the constant terms will determine whether the inequality is true or false.

a.

$8b - 3 > 4(2b + 3)$	Write the inequality.
$8b - 3 > 8b + 12$	Distributive Property
$\underline{-8b} \qquad \underline{-8b}$	Subtract 8*b* from each side.
$-3 > 12$ 	Simplify.

▶ The inequality $-3 > 12$ is false. So, there is no solution.

b.

$2(5w - 1) \leq 7 + 10w$	Write the inequality.
$10w - 2 \leq 7 + 10w$	Distributive Property
$\underline{-10w} \qquad \underline{-10w}$	Subtract 10*w* from each side.
$-2 \leq 7$	Simplify.

▶ The inequality $-2 \leq 7$ is true. So, all real numbers are solutions.

Monitoring Progress

Help in English and Spanish at *BigIdeasMath.com*

Solve the inequality.

5. $5x - 12 \leq 3x - 4$

6. $2(k - 5) < 2k + 5$

7. $-4(3n - 1) > -12n + 5.2$

8. $3(2a - 1) \geq 10a - 11$

Solving Real-Life Problems

EXAMPLE 4 Modeling with Mathematics

You need a mean score of at least 90 points to advance to the next round of the touch-screen trivia game. What scores in the fifth game will allow you to advance?

SOLUTION

REMEMBER

The mean in Example 4 is equal to the sum of the game scores divided by the number of games.

1. **Understand the Problem** You know the scores of your first four games. You are asked to find the scores in the fifth game that will allow you to advance.
2. **Make a Plan** Use the definition of the mean of a set of numbers to write an inequality. Then solve the inequality and answer the question.
3. **Solve the Problem** Let x be your score in the fifth game.

$$\frac{95 + 91 + 77 + 89 + x}{5} \geq 90 \qquad \text{Write an inequality.}$$

$$\frac{352 + x}{5} \geq 90 \qquad \text{Simplify.}$$

$$5 \cdot \frac{352 + x}{5} \geq 5 \cdot 90 \qquad \text{Multiply each side by 5.}$$

$$352 + x \geq 450 \qquad \text{Simplify.}$$

$$-352 \qquad -352 \qquad \text{Subtract 352 from each side.}$$

$$x \geq 98 \qquad \text{Simplify.}$$

▶ A score of at least 98 points will allow you to advance.

4. **Look Back** You can draw a diagram to check that your answer is reasonable. The horizontal bar graph shows the differences between the game scores and the desired mean of 90.

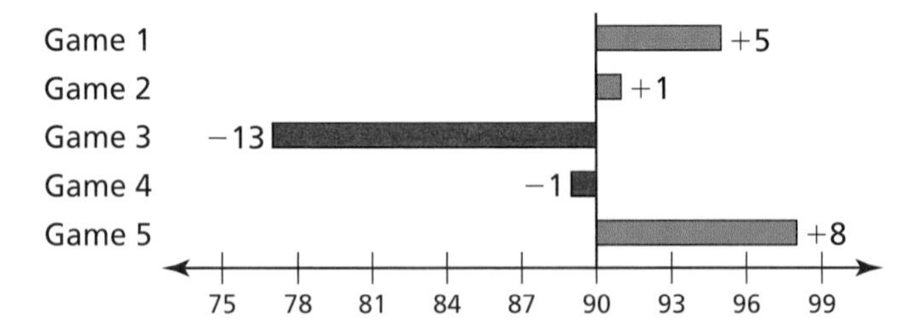

To have a mean of 90, the sum of the differences must be zero.

$5 + 1 - 13 - 1 + 8 = 0$ ✓

Monitoring Progress  Help in English and Spanish at *BigIdeasMath.com*

9. **WHAT IF?** You need a mean score of at least 85 points to advance to the next round. What scores in the fifth game will allow you to advance?

2.4 Exercises

Dynamic Solutions available at *BigIdeasMath.com*

Vocabulary and Core Concept Check

1. **WRITING** Compare solving multi-step inequalities and solving multi-step equations.
2. **WRITING** Without solving, how can you tell that the inequality $4x + 8 \le 4x - 3$ has no solution?

Monitoring Progress and Modeling with Mathematics

In Exercises 3–6, match the inequality with its graph.

3. $7b - 4 \le 10$
4. $4p + 4 \ge 12$
5. $-6g + 2 \ge 20$
6. $3(2 - f) \le 15$

A. −3 −2 −1 0 1 2 3 4 5 6 7

B. −3 −2 −1 0 1 2 3 4 5 6 7

C. −8 −7 −6 −5 −4 −3 −2 −1 0 1 2

D. −8 −7 −6 −5 −4 −3 −2 −1 0 1 2

In Exercises 7–16, solve the inequality. Graph the solution. *(See Example 1.)*

7. $2x - 3 > 7$
8. $5y + 9 \le 4$
9. $-9 \le 7 - 8v$
10. $2 > -3t - 10$
11. $\frac{w}{2} + 4 > 5$
12. $1 + \frac{m}{3} \le 6$
13. $\frac{p}{-8} + 9 > 13$
14. $3 + \frac{r}{-4} \le 6$
15. $6 \ge -6(a + 2)$
16. $18 \le 3(b - 4)$

In Exercises 17–28, solve the inequality. *(See Examples 2 and 3.)*

17. $4 - 2m > 7 - 3m$
18. $8n + 2 \le 8n - 9$
19. $-2d - 2 < 3d + 8$
20. $8 + 10f > 14 - 2f$
21. $8g - 5g - 4 \le -3 + 3g$
22. $3w - 5 > 2w + w - 7$
23. $6(\ell + 3) < 3(2\ell + 6)$
24. $2(5c - 7) \ge 10(c - 3)$
25. $4\left(\frac{1}{2}t - 2\right) > 2(t - 3)$
26. $15\left(\frac{1}{3}b + 3\right) \le 6(b + 9)$
27. $9j - 6 + 6j \ge 3(5j - 2)$
28. $6h - 6 + 2h < 2(4h - 3)$

ERROR ANALYSIS In Exercises 29 and 30, describe and correct the error in solving the inequality.

29.

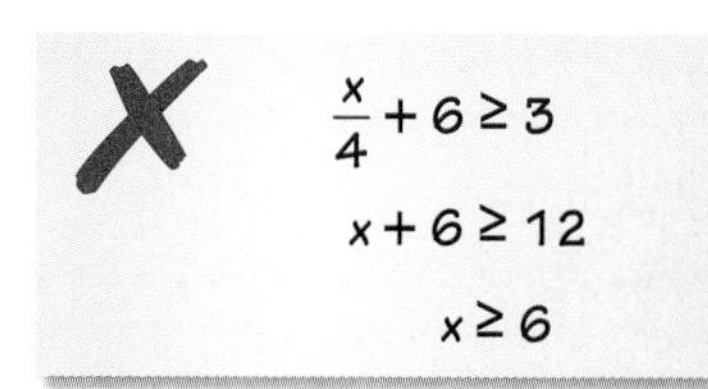

30.

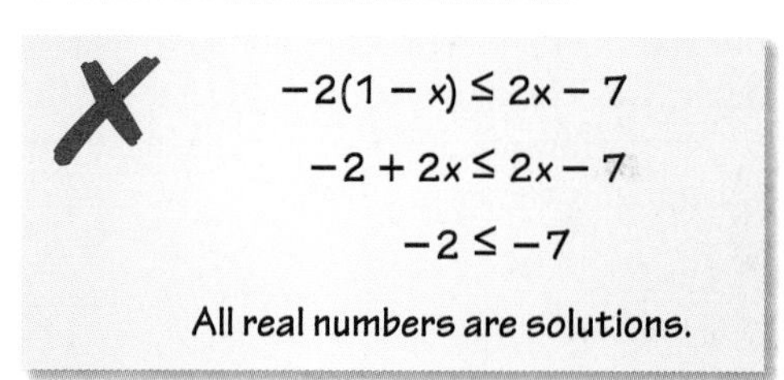

31. **MODELING WITH MATHEMATICS** Write and solve an inequality that represents how many \$20 bills you can withdraw from the account without going below the minimum balance. *(See Example 4.)*

32. MODELING WITH MATHEMATICS A woodworker wants to earn at least \$25 an hour making and selling cabinets. He pays \$125 for materials. Write and solve an inequality that represents how many hours the woodworker can spend building the cabinet.

33. MATHEMATICAL CONNECTIONS The area of the rectangle is greater than 60 square feet. Write and solve an inequality to find the possible values of x.

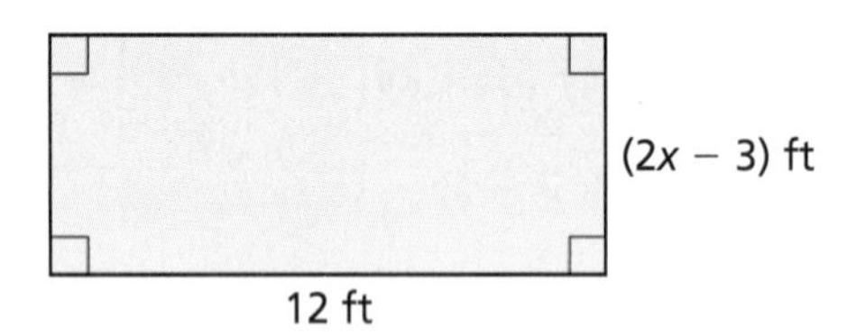

34. MAKING AN ARGUMENT Forest Park Campgrounds charges a \$100 membership fee plus \$35 per night. Woodland Campgrounds charges a \$20 membership fee plus \$55 per night. Your friend says that if you plan to camp for four or more nights, then you should choose Woodland Campgrounds. Is your friend correct? Explain.

35. PROBLEM SOLVING The height of one story of a building is about 10 feet. The bottom of the ladder on the fire truck must be at least 24 feet away from the building. How many stories can the ladder reach? Justify your answer.

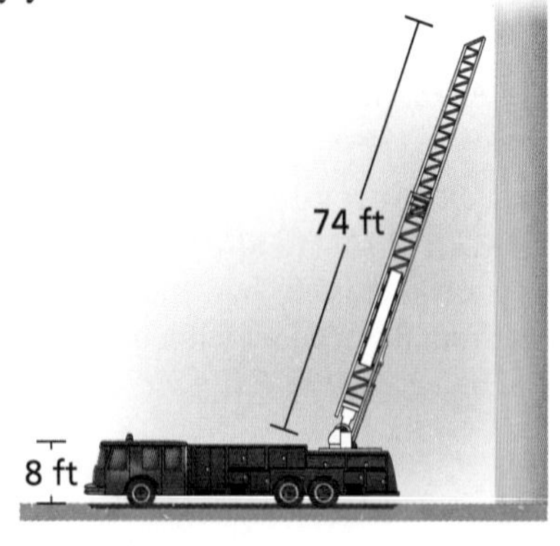

36. HOW DO YOU SEE IT? The graph shows your budget and the total cost of x gallons of gasoline and a car wash. You want to determine the possible amounts (in gallons) of gasoline you can buy within your budget.

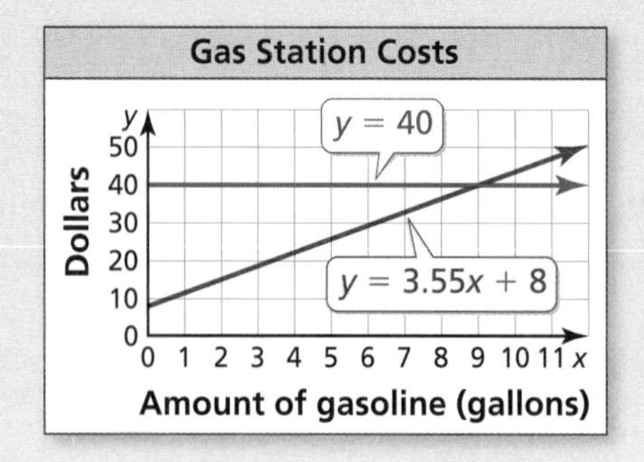

a. What is your budget?

b. How much does a gallon of gasoline cost? How much does a car wash cost?

c. Write an inequality that represents the possible amounts of gasoline you can buy.

d. Use the graph to estimate the solution of your inequality in part (c).

37. PROBLEM SOLVING For what values of r will the area of the shaded region be greater than or equal to $9(\pi - 2)$?

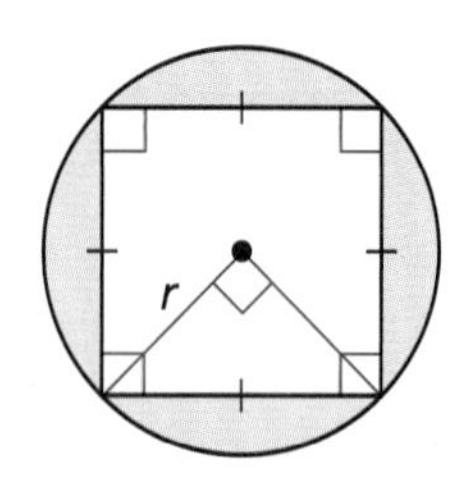

38. THOUGHT PROVOKING A runner's times (in minutes) in the four races he has completed are 25.5, 24.3, 24.8, and 23.5. The runner plans to run at least one more race and wants to have an average time less than 24 minutes. Write and solve an inequality to show how the runner can achieve his goal.

REASONING In Exercises 39 and 40, find the value of *a* for which the solution of the inequality is all real numbers.

39. $a(x + 3) < 5x + 15 - x$

40. $3x + 8 + 2ax \geq 3ax - 4a$

Maintaining Mathematical Proficiency

Reviewing what you learned in previous grades and lessons

Write the sentence as an inequality. *(Section 2.1)*

41. Six times a number y is less than or equal to 10.

42. A number p plus 7 is greater than 24.

43. The quotient of a number r and 7 is no more than 18.

2.1–2.4 What Did You Learn?

Core Vocabulary

inequality, *p. 54*
solution of an inequality, *p. 55*
solution set, *p. 55*
graph of an inequality, *p. 56*
equivalent inequalities, *p. 62*

Core Concepts

Section 2.1

Representing Linear Inequalities, *p. 57*

Section 2.2

Addition Property of Inequality, *p. 62*
Subtraction Property of Inequality, *p. 63*

Section 2.3

Multiplication and Division Properties of Inequality ($c > 0$), *p. 68*
Multiplication and Division Properties of Inequality ($c < 0$), *p. 69*

Section 2.4

Solving Multi-Step Inequalities, *p. 74*
Special Solutions of Linear Inequalities, *p. 75*

Mathematical Practices

1. Explain the meaning of the inequality symbol in your answer to Exercise 47 on page 59. How did you know which symbol to use?
2. In Exercise 30 on page 66, why is it important to check the reasonableness of your answer in part (a) before answering part (b)?
3. Explain how considering the units involved in Exercise 29 on page 71 helped you answer the question.

Study Skills

Analyzing Your Errors

Application Errors

What Happens: You can do numerical problems, but you struggle with problems that have context.

How to Avoid This Error: Do not just mimic the steps of solving an application problem. Explain out loud what the question is asking and why you are doing each step. After solving the problem, ask yourself, "Does my solution make sense?"

2.1–2.4 Quiz

Write the sentence as an inequality. *(Section 2.1)*

1. A number z minus 6 is greater than or equal to 11.
2. Twelve is no more than the sum of -1.5 times a number w and 4.

Write an inequality that represents the graph. *(Section 2.1)*

3. 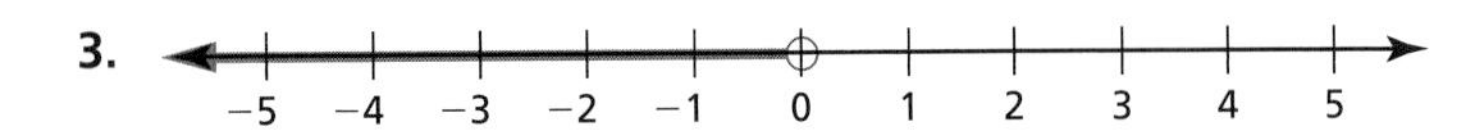

4. 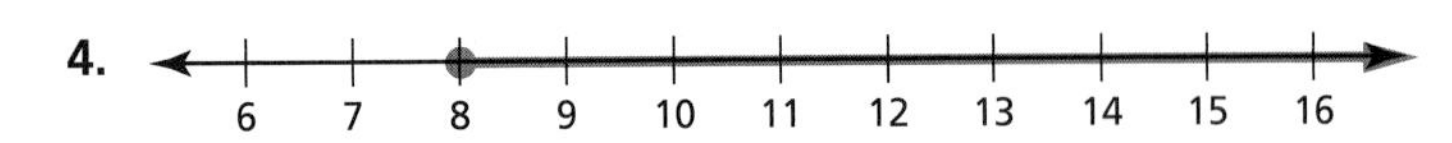

Solve the inequality. Graph the solution. *(Section 2.2 and Section 2.3)*

5. $9 + q \leq 15$
6. $z - (-7) < 5$
7. $-3 < y - 4$
8. $3p \geq 18$
9. $6 > \frac{w}{-2}$
10. $-20x > 5$

Solve the inequality. *(Section 2.4)*

11. $3y - 7 \geq 17$
12. $8(3g - 2) \leq 12(2g + 1)$
13. $6(2x - 1) \geq 3(4x + 1)$

14. Three requirements for a lifeguard training course are shown. *(Section 2.1)*
    a. Write and graph three inequalities that represent the requirements.
    b. You can swim 250 feet, tread water for 6 minutes, and swim 35 feet underwater without taking a breath. Do you satisfy the requirements of the course? Explain.

LIFEGUARDS NEEDED
Take Our Training Course NOW!!!
Lifeguard Training Requirements
- Swim at least 100 yards.
- Tread water for at least 5 minutes.
- Swim 10 yards or more underwater without taking a breath.

15. The maximum volume of an American white pelican's bill is about 700 cubic inches. A pelican scoops up 100 cubic inches of water. Write and solve an inequality that represents the additional volumes the pelican's bill can contain. *(Section 2.2)*

16. You save \$15 per week to purchase one of the bikes shown. *(Section 2.3 and Section 2.4)*
    a. Write and solve an inequality to find the numbers of weeks you need to save to purchase a bike.
    b. Your parents give you \$65 to help you buy the new bike. How does this affect you answer in part (a)? Use an inequality to justify your answer.

2.5 Solving Compound Inequalities

Essential Question How can you use inequalities to describe intervals on the real number line?

EXPLORATION 1 **Describing Intervals on the Real Number Line**

Work with a partner. In parts (a)–(d), use two inequalities to describe the interval.

REASONING ABSTRACTLY

To be proficient in math, you need to create a clear representation of the problem at hand.

a.

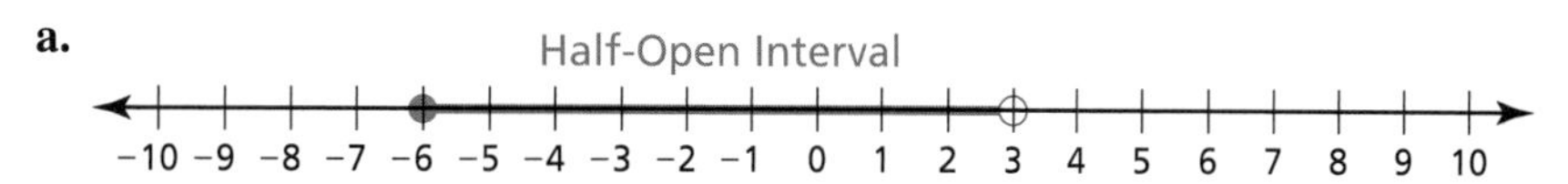

b.

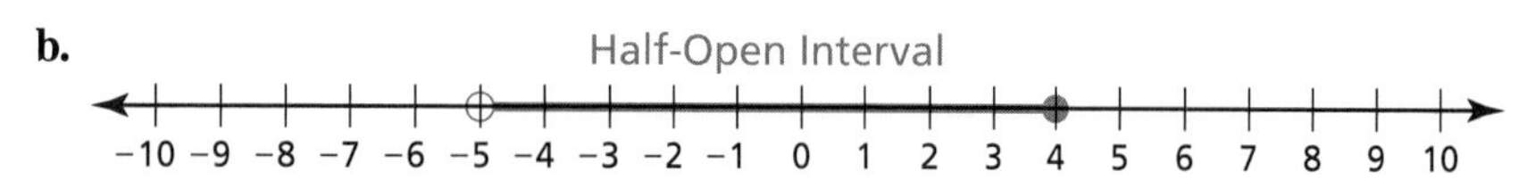

c.

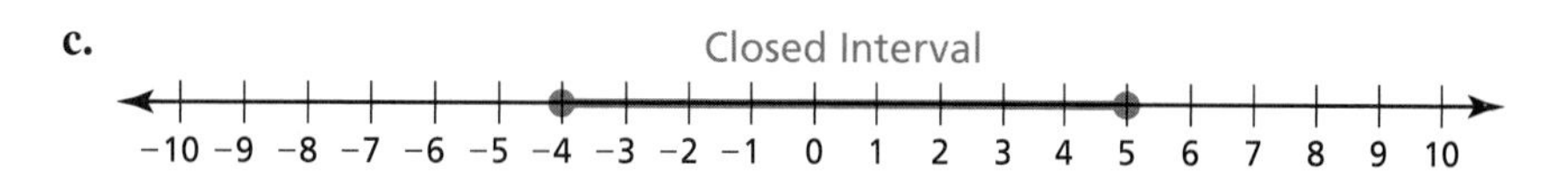

d.

e. Do you use "and" or "or" to connect the two inequalities in parts (a)–(d)? Explain.

EXPLORATION 2 **Describing Two Infinite Intervals**

Work with a partner. In parts (a)–(d), use two inequalities to describe the interval.

a.

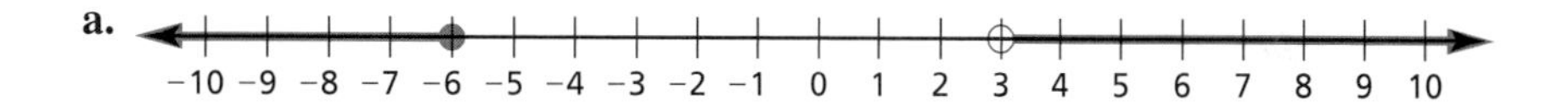

b. −10 −9 −8 −7 −6 −5 −4 −3 −2 −1 0 1 2 3 4 5 6 7 8 9 10

c.

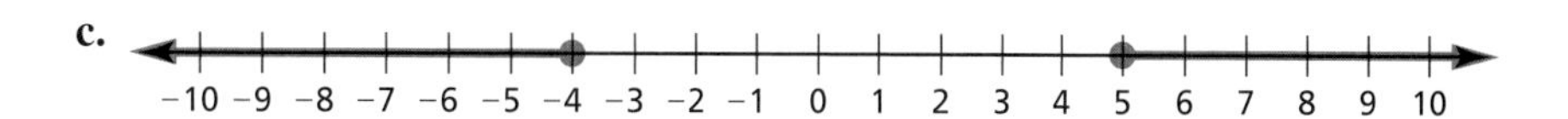

d. −10 −9 −8 −7 −6 −5 −4 −3 −2 −1 0 1 2 3 4 5 6 7 8 9 10

e. Do you use "and" or "or" to connect the two inequalities in parts (a)–(d)? Explain.

Communicate Your Answer

3. How can you use inequalities to describe intervals on the real number line?

2.5 Lesson

Core Vocabulary

compound inequality, *p. 82*

What You Will Learn

- Write and graph compound inequalities.
- Solve compound inequalities.
- Use compound inequalities to solve real-life problems.

Writing and Graphing Compound Inequalities

A **compound inequality** is an inequality formed by joining two inequalities with the word "and" or the word "or."

The graph of a compound inequality with "and" is the *intersection* of the graphs of the inequalities. The graph shows numbers that are solutions of *both* inequalities.

The graph of a compound inequality with "or" is the *union* of the graphs of the inequalities. The graph shows numbers that are solutions of *either* inequality.

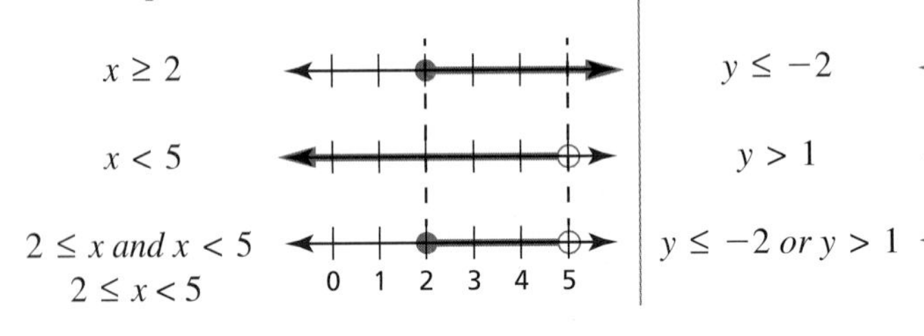

EXAMPLE 1 Writing and Graphing Compound Inequalities

Write each sentence as an inequality. Graph each inequality.

a. A number x is greater than -8 and less than or equal to 4.

b. A number y is at most 0 or at least 2.

REMEMBER

A compound inequality with "and" can be written as a single inequality. For example, you can write $x > -8$ and $x \leq 4$ as $-8 < x \leq 4$.

SOLUTION

a. A number x is greater than -8 and less than or equal to 4.

$x > -8$ *and* $x \leq 4$

▶ An inequality is $-8 < x \leq 4$.

−10 −8 −6 −4 −2 0 2 4 6

Graph the intersection of the graphs of $x > -8$ and $x \leq 4$.

b. A number y is at most 0 or at least 2.

$y \leq 0$ *or* $y \geq 2$

▶ An inequality is $y \leq 0$ *or* $y \geq 2$.

−2 −1 0 1 2 3 4 5 6

Graph the union of the graphs of $y \leq 0$ and $y \geq 2$.

Monitoring Progress

Help in English and Spanish at *BigIdeasMath.com*

Write the sentence as an inequality. Graph the inequality.

1. A number d is more than 0 and less than 10.
2. A number a is fewer than -6 or no less than -3.

Solving Compound Inequalities

You can solve a compound inequality by solving two inequalities separately. When a compound inequality with "and" is written as a single inequality, you can solve the inequality by performing the same operation on each expression.

LOOKING FOR STRUCTURE

To be proficient in math, you need to see complicated things as single objects or as being composed of several objects.

EXAMPLE 2 **Solving Compound Inequalities with "And"**

Solve each inequality. Graph each solution.

a. $-4 < x - 2 < 3$ **b.** $-3 < -2x + 1 \leq 9$

SOLUTION

a. Separate the compound inequality into two inequalities, then solve.

$-4 < x - 2$	*and*	$x - 2 < 3$	Write two inequalities.
$\underline{+2} \quad \underline{+2}$		$\underline{+2} \quad \underline{+2}$	Add 2 to each side.
$-2 < x$	*and*	$x < 5$	Simplify.

▶ The solution is $-2 < x < 5$.

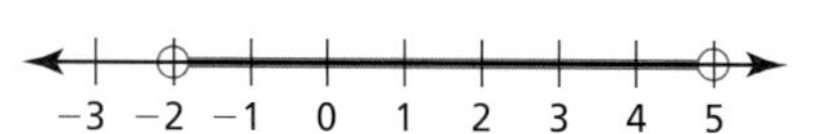

b.

$-3 < -2x + 1 \leq 9$	Write the inequality.
$\underline{-1} \quad \underline{-1} \quad \underline{-1}$	Subtract 1 from each expression.
$-4 < -2x \leq 8$	Simplify.
$\frac{-4}{-2} > \frac{-2x}{-2} \geq \frac{8}{-2}$	Divide each expression by -2. Reverse each inequality symbol.
$2 > x \geq -4$	Simplify.

▶ The solution is $-4 \leq x < 2$.

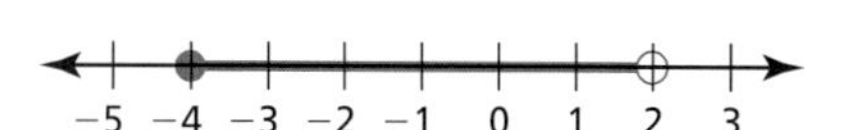

EXAMPLE 3 **Solving a Compound Inequality with "Or"**

Solve $3y - 5 < -8$ *or* $2y - 1 > 5$. Graph the solution.

SOLUTION

$3y - 5 < -8$	*or*	$2y - 1 > 5$	Write the inequality.
$\underline{+5} \quad \underline{+5}$		$\underline{+1} \quad \underline{+1}$	Addition Property of Inequality
$3y < -3$		$2y > 6$	Simplify.
$\frac{3y}{3} < \frac{-3}{3}$		$\frac{2y}{2} > \frac{6}{2}$	Division Property of Inequality
$y < -1$	*or*	$y > 3$	Simplify.

▶ The solution is $y < -1$ *or* $y > 3$.

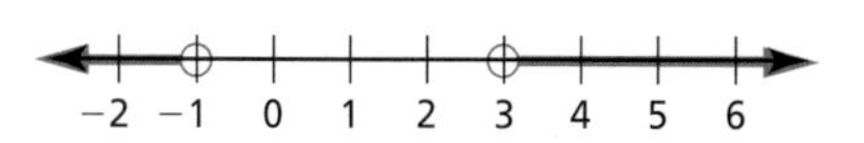

Monitoring Progress  Help in English and Spanish at *BigIdeasMath.com*

Solve the inequality. Graph the solution.

3. $5 \leq m + 4 < 10$

4. $-3 < 2k - 5 < 7$

5. $4c + 3 \leq -5$ *or* $c - 8 > -1$

6. $2p + 1 < -7$ *or* $3 - 2p \leq -1$

Solving Real-Life Problems

EXAMPLE 4 Modeling with Mathematics

Operating temperature: 0°C to 35°C

Electrical devices should operate effectively within a specified temperature range. Outside the operating temperature range, the device may fail.

a. Write and solve a compound inequality that represents the possible operating temperatures (in degrees Fahrenheit) of the smartphone.

b. Describe one situation in which the surrounding temperature could be below the operating range and one in which it could be above.

SOLUTION

1. **Understand the Problem** You know the operating temperature range in degrees Celsius. You are asked to write and solve a compound inequality that represents the possible operating temperatures (in degrees Fahrenheit) of the smartphone. Then you are asked to describe situations outside this range.

2. **Make a Plan** Write a compound inequality in degrees Celsius. Use the formula $C = \frac{5}{9}(F - 32)$ to rewrite the inequality in degrees Fahrenheit. Then solve the inequality and describe the situations.

3. **Solve the Problem** Let C be the temperature in degrees Celsius, and let F be the temperature in degrees Fahrenheit.

STUDY TIP

You can also solve the inequality by first multiplying each expression by $\frac{9}{5}$.

$0 \leq C \leq 35$	Write the inequality using C.
$0 \leq \frac{5}{9}(F - 32) \leq 35$	Substitute $\frac{5}{9}(F - 32)$ for C.
$9 \cdot 0 \leq 9 \cdot \frac{5}{9}(F - 32) \leq 9 \cdot 35$	Multiply each expression by 9.
$0 \leq 5(F - 32) \leq 315$	Simplify.
$0 \leq 5F - 160 \leq 315$	Distributive Property
$+160 \quad +160 \quad +160$	Add 160 to each expression.
$160 \leq 5F \leq 475$	Simplify.
$\frac{160}{5} \leq \frac{5F}{5} \leq \frac{475}{5}$	Divide each expression by 5.
$32 \leq F \leq 95$	Simplify.

▶ The solution is $32 \leq F \leq 95$. So, the operating temperature range of the smartphone is 32°F to 95°F. One situation when the surrounding temperature could be below this range is winter in Alaska. One situation when the surrounding temperature could be above this range is daytime in the Mojave Desert of the American Southwest.

4. **Look Back** You can use the formula $C = \frac{5}{9}(F - 32)$ to check that your answer is correct. Substitute 32 and 95 for F in the formula to verify that 0°C and 35°C are the minimum and maximum operating temperatures in degrees Celsius.

Monitoring Progress  Help in English and Spanish at *BigIdeasMath.com*

−40°C to 15°C

7. Write and solve a compound inequality that represents the temperature rating (in degrees Fahrenheit) of the winter boots.

2.5 Exercises

Dynamic Solutions available at *BigIdeasMath.com*

Vocabulary and Core Concept Check

1. **WRITING** Compare the graph of $-6 \le x \le -4$ with the graph of $x \le -6$ *or* $x \ge -4$.

2. **WHICH ONE DOESN'T BELONG?** Which compound inequality does *not* belong with the other three? Explain your reasoning.

$a > 4$ *or* $a < -3$	$a < -2$ *or* $a > 8$	$a > 7$ *or* $a < -5$	$a < 6$ *or* $a > -9$

Monitoring Progress and Modeling with Mathematics

In Exercises 3–6, write a compound inequality that is represented by the graph.

3. 

4. 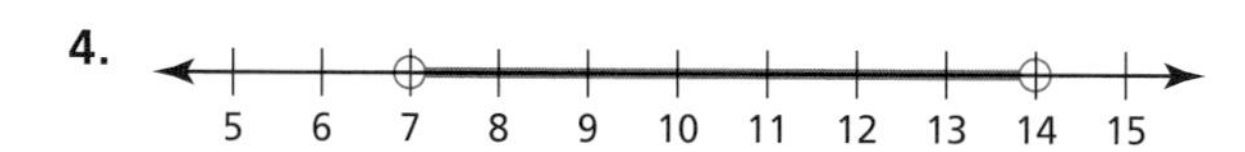

5.

6. 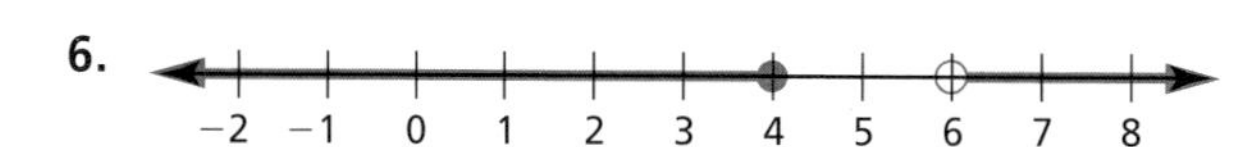

In Exercises 7–10, write the sentence as an inequality. Graph the inequality. *(See Example 1.)*

7. A number p is less than 6 and greater than 2.

8. A number n is less than or equal to -7 or greater than 12.

9. A number m is more than $-7\frac{2}{3}$ or at most -10.

10. A number r is no less than -1.5 and fewer than 9.5.

11. **MODELING WITH MATHEMATICS** Slitsnails are large mollusks that live in deep waters. They have been found in the range of elevations shown. Write and graph a compound inequality that represents this range.

12. **MODELING WITH MATHEMATICS** The life zones on Mount Rainier, a mountain in Washington, can be approximately classified by elevation, as follows.

Low-elevation forest: above 1700 feet to 2500 feet
Mid-elevation forest: above 2500 feet to 4000 feet
Subalpine: above 4000 feet to 6500 feet
Alpine: above 6500 feet to the summit

Elevation of Mount Rainier: 14,410 ft

Write a compound inequality that represents the elevation range for each type of plant life.

a. trees in the low-elevation forest zone

b. flowers in the subalpine and alpine zones

In Exercises 13–20, solve the inequality. Graph the solution. *(See Examples 2 and 3.)*

13. $6 < x + 5 \le 11$

14. $24 > -3r \ge -9$

15. $v + 8 < 3$ *or* $-8v < -40$

16. $-14 > w + 3$ *or* $3w \ge -27$

17. $2r + 3 < 7$ *or* $-r + 9 \le 2$

18. $-6 < 3n + 9 < 21$

19. $-12 < \frac{1}{2}(4x + 16) < 18$

20. $35 < 7(2 - b)$ *or* $\frac{1}{3}(15b - 12) \ge 21$

ERROR ANALYSIS In Exercises 21 and 22, describe and correct the error in solving the inequality or graphing the solution.

21.

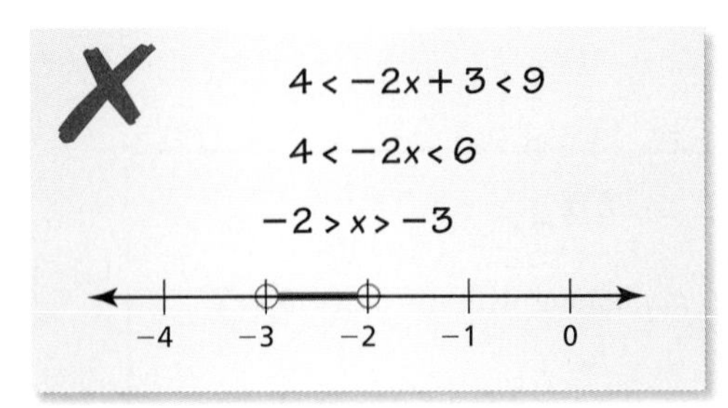

22.

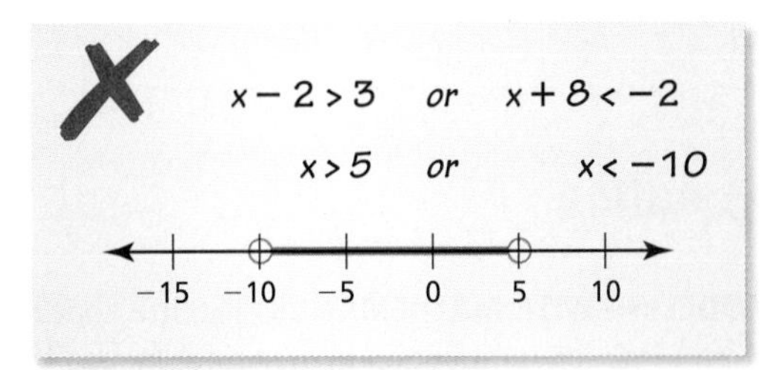

23. MODELING WITH MATHEMATICS Write and solve a compound inequality that represents the possible temperatures (in degrees Fahrenheit) of the interior of the iceberg. *(See Example 4.)*

24. PROBLEM SOLVING A ski shop sells skis with lengths ranging from 150 centimeters to 220 centimeters. The shop says the length of the skis should be about 1.16 times a skier's height (in centimeters). Write and solve a compound inequality that represents the heights of skiers the shop does *not* provide skis for.

In Exercises 25–30, solve the inequality. Graph the solution, if possible.

25. $22 < -3c + 4 < 14$

26. $2m - 1 \geq 5$ *or* $5m > -25$

27. $-y + 3 \leq 8$ *and* $y + 2 > 9$

28. $x - 8 \leq 4$ *or* $2x + 3 > 9$

29. $2n + 19 \leq 10 + n$ *or* $-3n + 3 < -2n + 33$

30. $3x - 18 < 4x - 23$ *and* $x - 16 < -22$

31. REASONING Fill in the compound inequality $4(x - 6)$ ▢ $2(x - 10)$ *and* $5(x + 2) \geq 2(x + 8)$ with <, ≤, >, or ≥ so that the solution is only one value.

32. THOUGHT PROVOKING Write a real-life story that can be modeled by the graph.

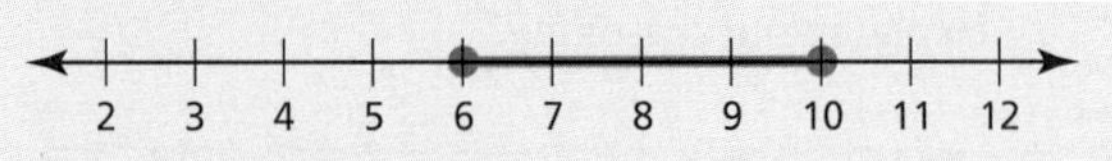

33. MAKING AN ARGUMENT The sum of the lengths of any two sides of a triangle is greater than the length of the third side. Use the triangle shown to write and solve three inequalities. Your friend claims the value of x can be 1. Is your friend correct? Explain.

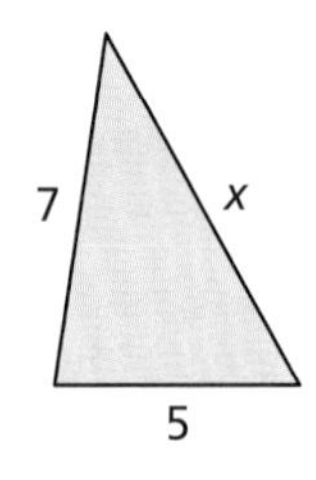

34. HOW DO YOU SEE IT? The graph shows the annual profits of a company from 2006 to 2013.

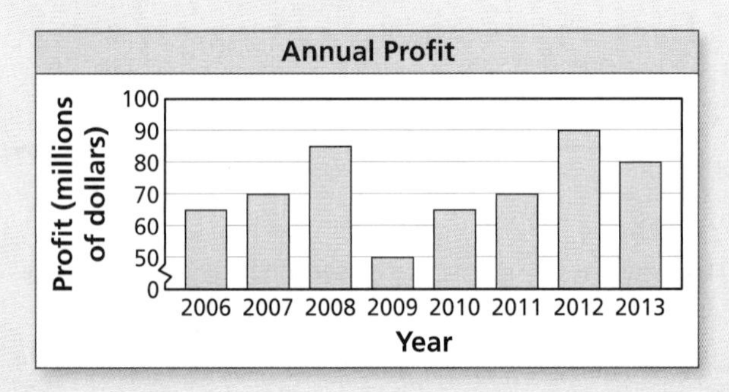

a. Write a compound inequality that represents the annual profits from 2006 to 2013.

b. You can use the formula $P = R - C$ to find the profit P, where R is the revenue and C is the cost. From 2006 to 2013, the company's annual cost was about \$125 million. Is it possible the company had an annual revenue of \$160 million from 2006 to 2013? Explain.

Maintaining Mathematical Proficiency

Reviewing what you learned in previous grades and lessons

Solve the equation. Graph the solutions, if possible. *(Section 1.4)*

35. $\left|\frac{d}{9}\right| = 6$

36. $7|5p - 7| = -21$

37. $|r + 2| = |3r - 4|$

38. $\left|\frac{1}{2}w - 6\right| = |w + 7|$

Find and interpret the mean absolute deviation of the data. *(Skills Review Handbook)*

39. 1, 1, 2, 5, 6, 8, 10, 12, 12, 13

40. 24, 26, 28, 28, 30, 30, 32, 32, 34, 36

2.6 Solving Absolute Value Inequalities

Essential Question

How can you solve an absolute value inequality?

EXPLORATION 1 Solving an Absolute Value Inequality Algebraically

Work with a partner. Consider the absolute value inequality

$|x + 2| \leq 3.$

a. Describe the values of $x + 2$ that make the inequality true. Use your description to write two linear inequalities that represent the solutions of the absolute value inequality.

b. Use the linear inequalities you wrote in part (a) to find the solutions of the absolute value inequality.

c. How can you use linear inequalities to solve an absolute value inequality?

MAKING SENSE OF PROBLEMS

To be proficient in math, you need to explain to yourself the meaning of a problem and look for entry points to its solution.

EXPLORATION 2 Solving an Absolute Value Inequality Graphically

Work with a partner. Consider the absolute value inequality

$|x + 2| \leq 3.$

a. On a real number line, locate the point for which $x + 2 = 0$.

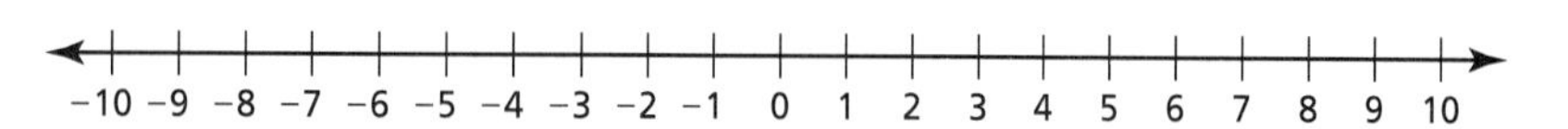

b. Locate the points that are within 3 units from the point you found in part (a). What do you notice about these points?

c. How can you use a number line to solve an absolute value inequality?

EXPLORATION 3 Solving an Absolute Value Inequality Numerically

Work with a partner. Consider the absolute value inequality

$|x + 2| \leq 3.$

a. Use a spreadsheet, as shown, to solve the absolute value inequality.

b. Compare the solutions you found using the spreadsheet with those you found in Explorations 1 and 2. What do you notice?

c. How can you use a spreadsheet to solve an absolute value inequality?

	A	B
1	x	|x + 2|
2	-6	4
3	-5	
4	-4	
5	-3	
6	-2	
7	-1	
8	0	
9	1	
10	2	
11		

abs(A2 + 2)

Communicate Your Answer

4. How can you solve an absolute value inequality?

5. What do you like or dislike about the algebraic, graphical, and numerical methods for solving an absolute value inequality? Give reasons for your answers.

2.6 Lesson

Core Vocabulary

absolute value inequality, *p. 88*
absolute deviation, *p. 90*

Previous
compound inequality
mean

What You Will Learn

- Solve absolute value inequalities.
- Use absolute value inequalities to solve real-life problems.

Solving Absolute Value Inequalities

An **absolute value inequality** is an inequality that contains an absolute value expression. For example, $|x| < 2$ and $|x| > 2$ are absolute value inequalities. Recall that $|x| = 2$ means the distance between x and 0 is 2.

The inequality $|x| < 2$ means the distance between x and 0 is *less than* 2.

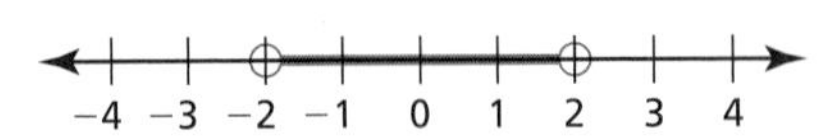

The graph of $|x| < 2$ is the graph of $x > -2$ *and* $x < 2$.

The inequality $|x| > 2$ means the distance between x and 0 is *greater than* 2.

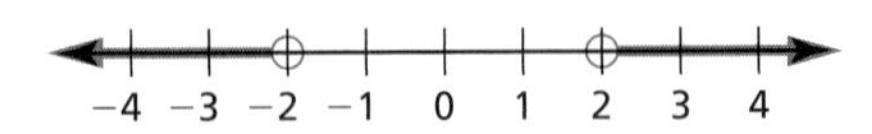

The graph of $|x| > 2$ is the graph of $x < -2$ *or* $x > 2$.

You can solve these types of inequalities by solving a compound inequality.

Core Concept

Solving Absolute Value Inequalities

To solve $|ax + b| < c$ for $c > 0$, solve the compound inequality

$$ax + b > -c \quad \textit{and} \quad ax + b < c.$$

To solve $|ax + b| > c$ for $c > 0$, solve the compound inequality

$$ax + b < -c \quad \textit{or} \quad ax + b > c.$$

In the inequalities above, you can replace $<$ with $\le$ and $>$ with $\ge$.

EXAMPLE 1 Solving Absolute Value Inequalities

Solve each inequality. Graph each solution, if possible.

a. $|x + 7| \le 2$ **b.** $|8x - 11| < 0$

SOLUTION

a. Use $|x + 7| \le 2$ to write a compound inequality. Then solve.

$x + 7 \ge -2$	*and*	$x + 7 \le 2$	Write a compound inequality.
$\underline{-7 \quad -7}$		$\underline{-7 \quad -7}$	Subtract 7 from each side.
$x \ge -9$	*and*	$x \le -5$	Simplify.

REMEMBER

A compound inequality with "and" can be written as a single inequality.

▶ The solution is $-9 \le x \le -5$.

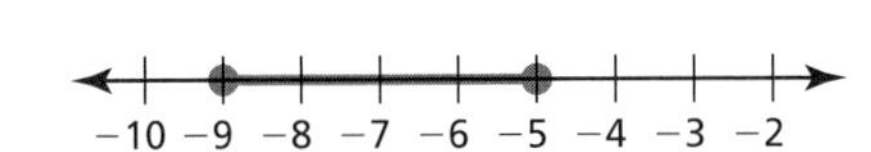

b. By definition, the absolute value of an expression must be greater than or equal to 0. The expression $|8x - 11|$ cannot be less than 0.

▶ So, the inequality has no solution.

Monitoring Progress  Help in English and Spanish at *BigIdeasMath.com*

Solve the inequality. Graph the solution, if possible.

1. $|x| \leq 3.5$ **2.** $|k - 3| < -1$ **3.** $|2w - 1| < 11$

EXAMPLE 2 Solving Absolute Value Inequalities

Solve each inequality. Graph each solution.

a. $|c - 1| \geq 5$ **b.** $|10 - m| \geq -2$ **c.** $4|2x - 5| + 1 > 21$

SOLUTION

a. Use $|c - 1| \geq 5$ to write a compound inequality. Then solve.

$c - 1 \leq -5$ *or* $c - 1 \geq 5$ Write a compound inequality.

$+1 \quad +1$ $\quad +1 \quad +1$ Add 1 to each side.

$c \leq -4$ *or* $c \geq 6$ Simplify.

▶ The solution is $c \leq -4$ *or* $c \geq 6$.

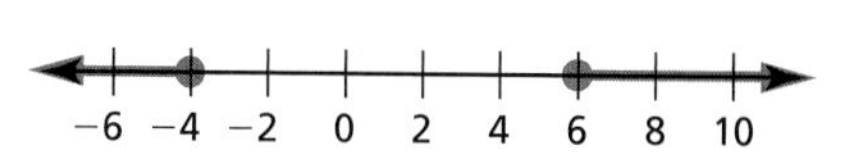

b. By definition, the absolute value of an expression must be greater than or equal to 0. The expression $|10 - m|$ will always be greater than -2.

▶ So, all real numbers are solutions.

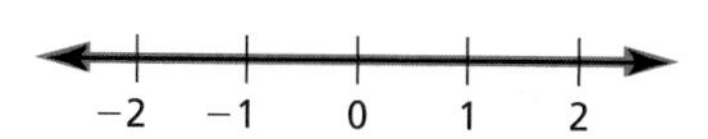

c. First isolate the absolute value expression on one side of the inequality.

$4|2x - 5| + 1 > 21$ Write the inequality.

$-1 \quad -1$ Subtract 1 from each side.

$4|2x - 5| > 20$ Simplify.

$\frac{4|2x - 5|}{4} > \frac{20}{4}$ Divide each side by 4.

$|2x - 5| > 5$ Simplify.

Then use $|2x - 5| > 5$ to write a compound inequality. Then solve.

$2x - 5 < -5$ *or* $2x - 5 > 5$ Write a compound inequality.

$+5 \quad +5$ $\quad +5 \quad +5$ Add 5 to each side.

$2x < 0$ $\quad 2x > 10$ Simplify.

$\frac{2x}{2} < \frac{0}{2}$ $\quad \frac{2x}{2} > \frac{10}{2}$ Divide each side by 2.

$x < 0$ *or* $x > 5$ Simplify.

▶ The solution is $x < 0$ *or* $x > 5$.

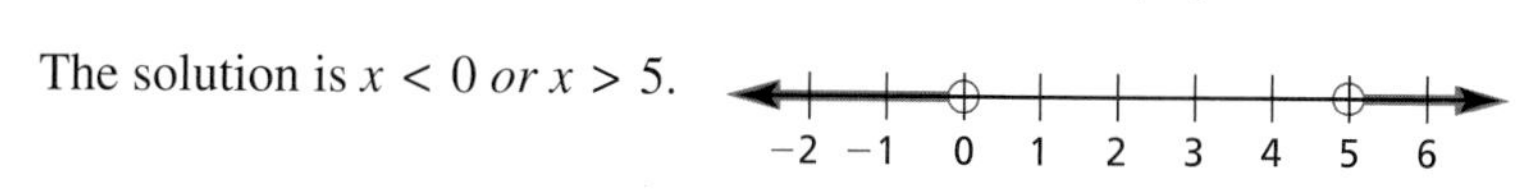

Monitoring Progress  Help in English and Spanish at *BigIdeasMath.com*

Solve the inequality. Graph the solution.

4. $|x + 3| > 8$ **5.** $|n + 2| - 3 \geq -6$ **6.** $3|d + 1| - 7 \geq -1$

Solving Real-Life Problems

The **absolute deviation** of a number x from a given value is the absolute value of the difference of x and the given value.

$$\text{absolute deviation} = |x - \text{given value}|$$

EXAMPLE 3 Modeling with Mathematics

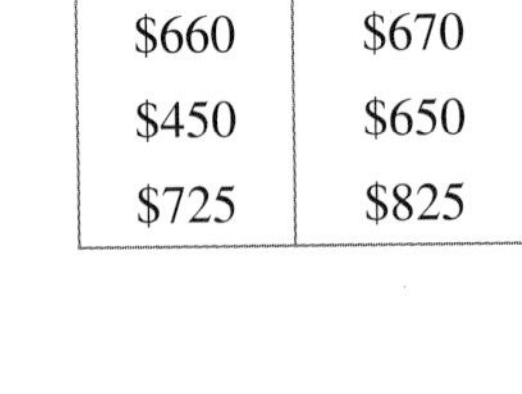

Computer prices	
\$890	\$750
\$650	\$370
\$660	\$670
\$450	\$650
\$725	\$825

You are buying a new computer. The table shows the prices of computers in a store advertisement. You are willing to pay the mean price with an absolute deviation of at most \$100. How many of the computer prices meet your condition?

SOLUTION

1. **Understand the Problem** You know the prices of 10 computers. You are asked to find how many computers are at most \$100 from the mean price.

2. **Make a Plan** Calculate the mean price by dividing the sum of the prices by the number of prices, 10. Use the absolute deviation and the mean price to write an absolute value inequality. Then solve the inequality and use it to answer the question.

3. **Solve the Problem**

   The mean price is $\frac{6640}{10} = \$664$. Let x represent a price you are willing to pay.

   $|x - 664| \le 100$ Write the absolute value inequality.

   $-100 \le x - 664 \le 100$ Write a compound inequality.

   $564 \le x \le 764$ Add 664 to each expression and simplify.

   ▶ The prices you will consider must be at least \$564 and at most \$764. Six prices meet your condition: \$750, \$650, \$660, \$670, \$650, and \$725.

STUDY TIP

The absolute deviation of at most \$100 from the mean, \$664, is given by the inequality $|x - 664| \le 100$.

4. **Look Back** You can check that your answer is correct by graphing the computer prices and the mean on a number line. Any point within 100 of 664 represents a price that you will consider.

Monitoring Progress

Help in English and Spanish at *BigIdeasMath.com*

7. **WHAT IF?** You are willing to pay the mean price with an absolute deviation of at most \$75. How many of the computer prices meet your condition?

Concept Summary

Solving Inequalities

One-Step and Multi-Step Inequalities

- Follow the steps for solving an equation. Reverse the inequality symbol when multiplying or dividing by a negative number.

Compound Inequalities

- If necessary, write the inequality as two separate inequalities. Then solve each inequality separately. Include *and* or *or* in the solution.

Absolute Value Inequalities

- If necessary, isolate the absolute value expression on one side of the inequality. Write the absolute value inequality as a compound inequality. Then solve the compound inequality.

2.6 Exercises

Dynamic Solutions available at *BigIdeasMath.com*

Vocabulary and Core Concept Check

1. **REASONING** Can you determine the solution of $|4x - 2| \geq -6$ without solving? Explain.

2. **WRITING** Describe how solving $|w - 9| \leq 2$ is different from solving $|w - 9| \geq 2$.

Monitoring Progress and Modeling with Mathematics

In Exercises 3–18, solve the inequality. Graph the solution, if possible. *(See Examples 1 and 2.)*

3. $|x| < 3$
4. $|y| \geq 4.5$
5. $|d + 9| > 3$
6. $|h - 5| \leq 10$
7. $|2s - 7| \geq -1$
8. $|4c + 5| > 7$
9. $|5p + 2| < -4$
10. $|9 - 4n| < 5$
11. $|6t - 7| - 8 \geq 3$
12. $|3j - 1| + 6 > 0$
13. $3|14 - m| > 18$
14. $-4|6b - 8| \leq 12$
15. $2|3w + 8| - 13 \leq -5$
16. $-3|2 - 4u| + 5 < -13$
17. $6|-f + 3| + 7 > 7$
18. $\frac{2}{3}|4v + 6| - 2 \leq 10$

19. **MODELING WITH MATHEMATICS** The rules for an essay contest say that entries can have 500 words with an absolute deviation of at most 30 words. Write and solve an absolute value inequality that represents the acceptable numbers of words. *(See Example 3.)*

20. **MODELING WITH MATHEMATICS** The normal body temperature of a camel is 37°C. This temperature varies by up to 3°C throughout the day. Write and solve an absolute value inequality that represents the range of normal body temperatures (in degrees Celsius) of a camel throughout the day.

ERROR ANALYSIS In Exercises 21 and 22, describe and correct the error in solving the absolute value inequality.

21.

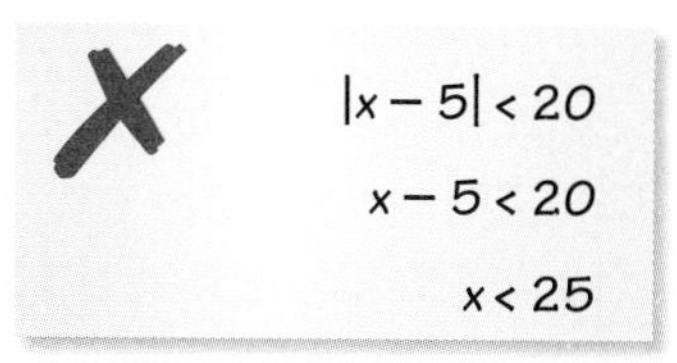

22.

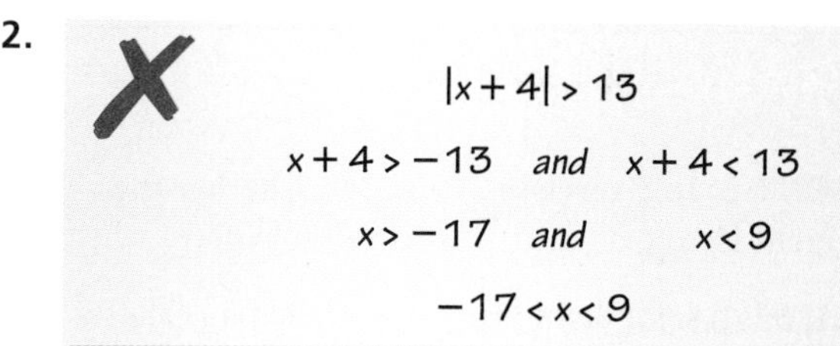

In Exercises 23–26, write the sentence as an absolute value inequality. Then solve the inequality.

23. A number is less than 6 units from 0.

24. A number is more than 9 units from 3.

25. Half of a number is at most 5 units from 14.

26. Twice a number is no less than 10 units from -1.

27. **PROBLEM SOLVING** An auto parts manufacturer throws out gaskets with weights that are not within 0.06 pound of the mean weight of the batch. The weights (in pounds) of the gaskets in a batch are 0.58, 0.63, 0.65, 0.53, and 0.61. Which gasket(s) should be thrown out?

28. **PROBLEM SOLVING** Six students measure the acceleration (in meters per second per second) of an object in free fall. The measured values are shown. The students want to state that the absolute deviation of each measured value x from the mean is at most d. Find the value of d.

    10.56, 9.52, 9.73, 9.80, 9.78, 10.91

MATHEMATICAL CONNECTIONS In Exercises 29 and 30, write an absolute value inequality that represents the situation. Then solve the inequality.

29. The difference between the areas of the figures is less than 2.

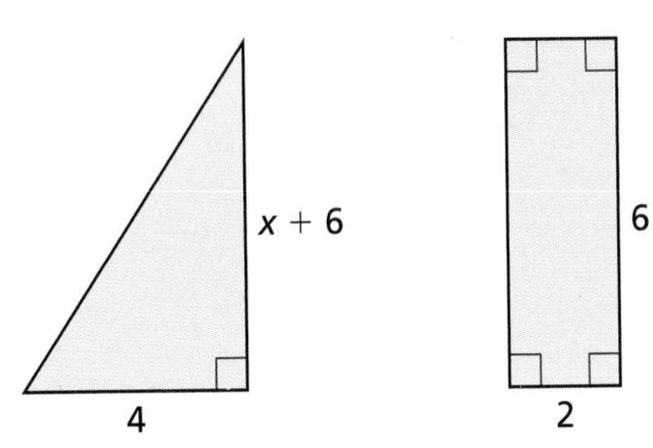

30. The difference between the perimeters of the figures is less than or equal to 3.

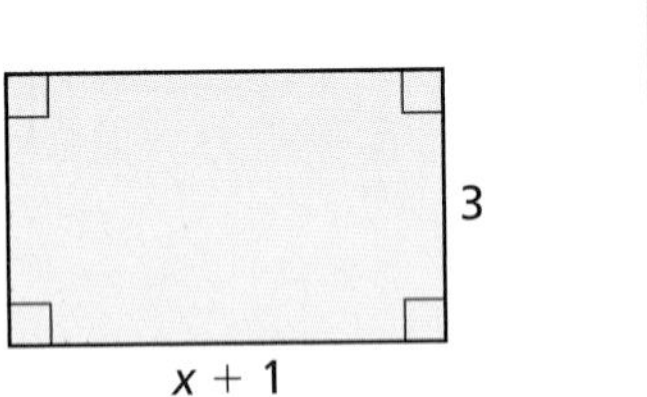

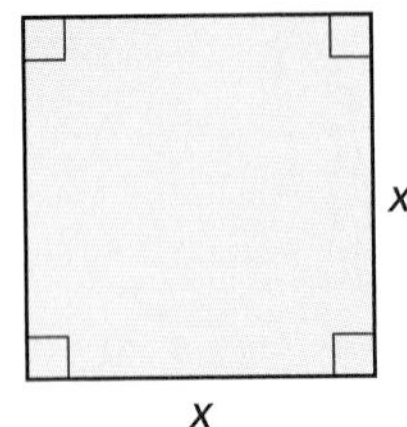

REASONING In Exercises 31–34, tell whether the statement is true or false. If it is false, explain why.

31. If a is a solution of $|x + 3| \leq 8$, then a is also a solution of $x + 3 \geq -8$.

32. If a is a solution of $|x + 3| > 8$, then a is also a solution of $x + 3 > 8$.

33. If a is a solution of $|x + 3| \geq 8$, then a is also a solution of $x + 3 \geq -8$.

34. If a is a solution of $x + 3 \leq -8$, then a is also a solution of $|x + 3| \geq 8$.

35. MAKING AN ARGUMENT One of your classmates claims that the solution of $|n| > 0$ is all real numbers. Is your classmate correct? Explain your reasoning.

36. THOUGHT PROVOKING Draw and label a geometric figure so that the perimeter P of the figure is a solution of the inequality $|P - 60| \leq 12$.

37. REASONING What is the solution of the inequality $|ax + b| < c$, where $c < 0$? What is the solution of the inequality $|ax + b| > c$, where $c < 0$? Explain.

38. HOW DO YOU SEE IT? Write an absolute value inequality for each graph.

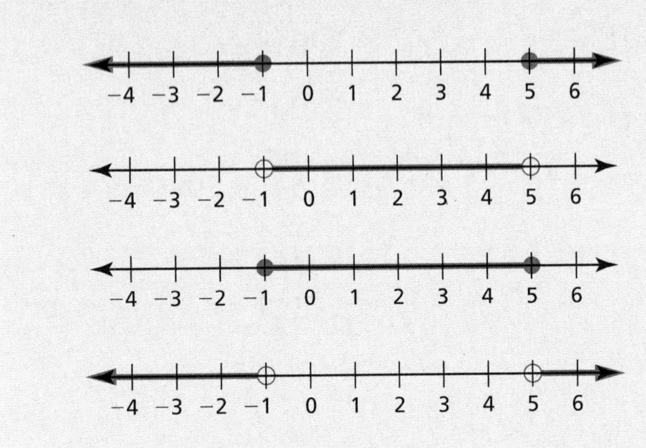

How did you decide which inequality symbol to use for each inequality?

39. WRITING Explain why the solution set of the inequality $|x| < 5$ is the *intersection* of two sets, while the solution set of the inequality $|x| > 5$ is the *union* of two sets.

40. PROBLEM SOLVING Solve the compound inequality below. Describe your steps.

$$|x - 3| < 4 \text{ and } |x + 2| > 8$$

Maintaining Mathematical Proficiency

Reviewing what you learned in previous grades and lessons

Plot the ordered pair in a coordinate plane. Describe the location of the point. *(Skills Review Handbook)*

41. $A(1, 3)$ **42.** $B(0, -3)$ **43.** $C(-4, -2)$ **44.** $D(-1, 2)$

Copy and complete the table. *(Skills Review Handbook)*

45.

x	0	1	2	3	4
5x + 1					

46.

x	−2	−1	0	1	2
−2x − 3					

2.5–2.6 What Did You Learn?

Core Vocabulary

compound inequality, *p. 82*
absolute value inequality, *p. 88*
absolute deviation, *p. 90*

Core Concepts

Section 2.5

Writing and Graphing Compound Inequalities, *p. 82*
Solving Compound Inequalities, *p. 83*

Section 2.6

Solving Absolute Value Inequalities, *p. 88*

Mathematical Practices

1. How can you use a diagram to help you solve Exercise 12 on page 85?
2. In Exercises 13 and 14 on page 85, how can you use structure to break down the compound inequality into two inequalities?
3. Describe the given information and the overall goal of Exercise 27 on page 91.
4. For false statements in Exercises 31–34 on page 92, use examples to show the statements are false.

Performance Task

Grading Calculations

You are not doing as well as you had hoped in one of your classes. So, you want to figure out the minimum grade you need on the final exam to receive the semester grade that you want. Is it still possible to get an A? How would you explain your calculations to a classmate?

To explore the answers to this question and more, go to ***BigIdeasMath.com***.

2 Chapter Review

Dynamic Solutions available at *BigIdeasMath.com*

2.1 Writing and Graphing Inequalities (pp. 53–60)

a. A number x plus 36 is no more than 40. Write this sentence as an inequality.

A number x plus 36 is no more than 40.

$x + 36 \quad \leq \quad 40$

▶ An inequality is $x + 36 \leq 40$.

b. Graph $w > -3$.

Test a number to the left of -3. $w = -4$ is *not* a solution.

Test a number to the right of -3. $w = 0$ is a solution.

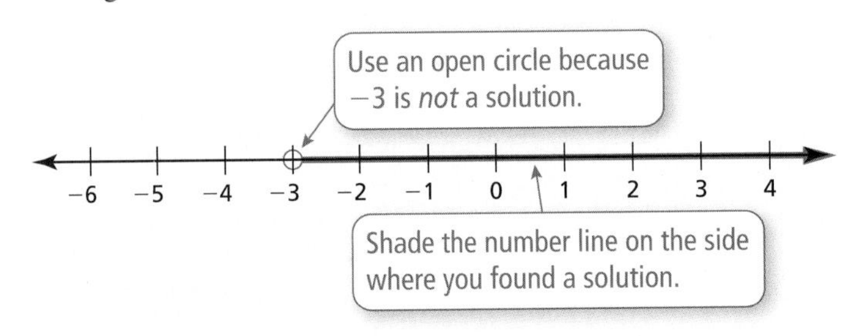

Use an open circle because −3 is *not* a solution.
−6 −5 −4 −3 −2 −1 0 1 2 3 4
Shade the number line on the side where you found a solution.

Write the sentence as an inequality.

1. A number d minus 2 is less than -1.
2. Ten is at least the product of a number h and 5.

Graph the inequality.

3. $x > 4$
4. $y \leq 2$
5. $-1 \geq z$

2.2 Solving Inequalities Using Addition or Subtraction (pp. 61–66)

Solve $x + 2.5 \leq -6$. Graph the solution.

$x + 2.5 \leq -6$ — Write the inequality.

Subtraction Property of Inequality → $-2.5 \quad -2.5$ — Subtract 2.5 from each side.

$x \leq -8.5$ — Simplify.

▶ The solution is $x \leq -8.5$.

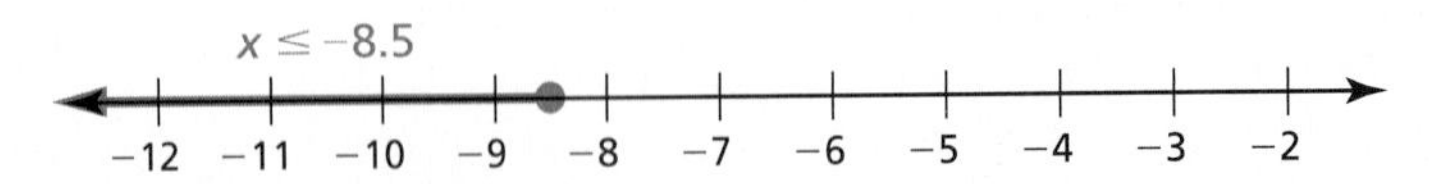

x ≤ −8.5
−12 −11 −10 −9 −8 −7 −6 −5 −4 −3 −2

Solve the inequality. Graph the solution.

6. $p + 4 < 10$
7. $r - 4 < -6$
8. $2.1 \geq m - 6.7$

2.3 Solving Inequalities Using Multiplication or Division (pp. 67–72)

Solve $\frac{n}{-10} > 5$. Graph the solution.

$\frac{n}{-10} > 5$	Write the inequality.
Multiplication Property of Inequality → $-10 \cdot \frac{n}{-10} < -10 \cdot 5$	Multiply each side by −10. Reverse the inequality symbol.
$n < -50$	Simplify.

▶ The solution is $n < -50$.

$n < -50$

−80 −70 −60 −50 −40 −30 −20 −10 0 10 20

Solve the inequality. Graph the solution.

9. $3x > -21$

10. $-4 \leq \frac{g}{5}$

11. $-\frac{3}{4}n \leq 3$

12. $\frac{s}{-8} \geq 11$

13. $36 < 2q$

14. $-1.2k > 6$

2.4 Solving Multi-Step Inequalities (pp. 73–78)

Solve $22 + 3y \geq 4$. Graph the solution.

$22 + 3y \geq 4$	Write the inequality.
$-22 \qquad -22$	Subtract 22 from each side.
$3y \geq -18$	Simplify.
$\frac{3y}{3} \geq \frac{-18}{3}$	Divide each side by 3.
$y \geq -6$	Simplify.

▶ The solution is $y \geq -6$.

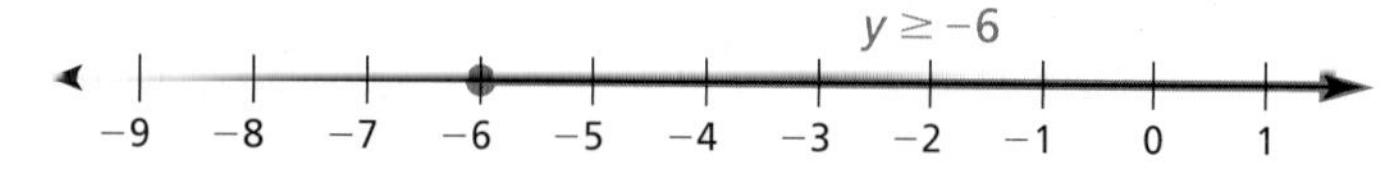

Solve the inequality. Graph the solution, if possible.

15. $3x - 4 > 11$

16. $-4 < \frac{b}{2} + 9$

17. $7 - 3n \leq n + 3$

18. $2(-4s + 2) \geq -5s - 10$

19. $6(2t + 9) \leq 12t - 1$

20. $3r - 8 > 3(r - 6)$

2.5 Solving Compound Inequalities (pp. 81–86)

Solve $-1 \le -2d + 7 \le 9$. Graph the solution.

$-1 \le -2d + 7 \le 9$	Write the inequality.
$\underline{-7} \quad \underline{-7} \quad \underline{-7}$	Subtract 7 from each expression.
$-8 \le -2d \le 2$	Simplify.
$\frac{-8}{-2} \ge \frac{-2d}{-2} \ge \frac{2}{-2}$	Divide each expression by -2. Reverse each inequality symbol.
$4 \ge d \ge -1$	Simplify.

▶ The solution is $-1 \le d \le 4$.

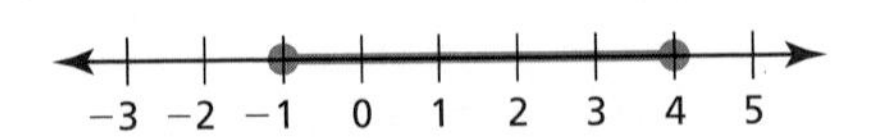

21. A number x is more than -6 and at most 8. Write this sentence as an inequality. Graph the inequality.

Solve the inequality. Graph the solution.

22. $19 \ge 3z + 1 \ge -5$

23. $\frac{r}{4} < -5$ *or* $-2r - 7 \le 3$

2.6 Solving Absolute Value Inequalities (pp. 87–92)

Solve $|2x + 11| + 3 > 8$. Graph the solution.

$\lvert 2x + 11\rvert + 3 > 8$	Write the inequality.
$\underline{-3} \quad \underline{-3}$	Subtract 3 from each side.
$\lvert 2x + 11\rvert > 5$	Simplify.
$2x + 11 < -5$ *or* $2x + 11 > 5$	Write a compound inequality.
$\underline{-11} \quad \underline{-11} \qquad \underline{-11} \quad \underline{-11}$	Subtract 11 from each side.
$2x < -16 \qquad 2x > -6$	Simplify.
$\frac{2x}{2} < \frac{-16}{2} \qquad \frac{2x}{2} > \frac{-6}{2}$	Divide each side by 2.
$x < -8$ *or* $x > -3$	Simplify.

▶ The solution is $x < -8$ *or* $x > -3$.

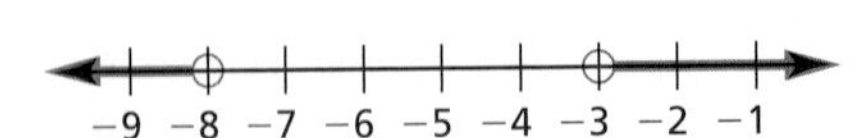

Solve the inequality. Graph the solution, if possible.

24. $|m| \ge 10$

25. $|k - 9| < -4$

26. $4|f - 6| \le 12$

27. $5|b + 8| - 7 > 13$

28. $|-3g - 2| + 1 < 6$

29. $|9 - 2j| + 10 \ge 2$

30. A safety regulation states that the height of a guardrail should be 106 centimeters with an absolute deviation of no more than 7 centimeters. Write and solve an absolute value inequality that represents the acceptable heights of a guardrail.

2 Chapter Test

Write the sentence as an inequality.

1. The sum of a number y and 9 is at least -1.
2. A number r is more than 0 or less than or equal to -8.
3. A number k is less than 3 units from 10.

Solve the inequality. Graph the solution, if possible.

4. $\frac{x}{2} - 5 \geq -9$
5. $-4s < 6s + 1$
6. $4p + 3 \geq 2(2p + 1)$
7. $-7 < 2c - 1 < 10$
8. $-2 \leq 4 - 3a \leq 13$
9. $-5 < 2 - h$ *or* $6h + 5 > 71$
10. $|2q + 8| > 4$
11. $-2|y - 3| - 5 \geq -4$
12. $4|-3b + 5| - 9 < 7$

13. You start a small baking business, and you want to earn a profit of at least \$250 in the first month. The expenses in the first month are \$155. What are the possible revenues that you need to earn to meet the profit goal?

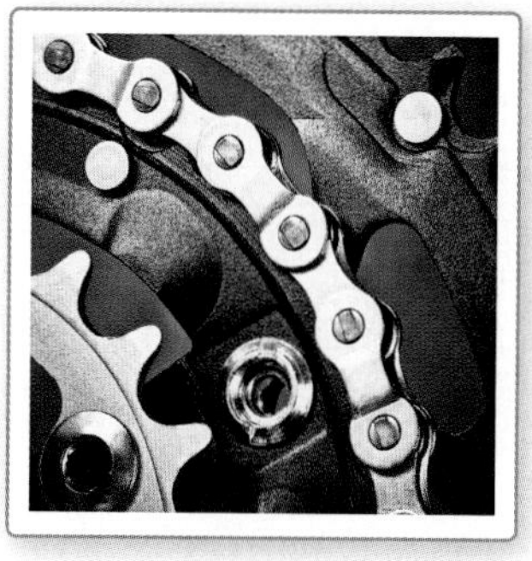

14. A manufacturer of bicycle parts requires that a bicycle chain have a width of 0.3 inch with an absolute deviation of at most 0.0003 inch. Write and solve an absolute value inequality that represents the acceptable widths.

15. Let a, b, c, and d be constants. Describe the possible solution sets of the inequality $ax + b < cx + d$.

Write and graph a compound inequality that represents the numbers that are *not* solutions of the inequality represented by the graph shown. Explain your reasoning.

16. 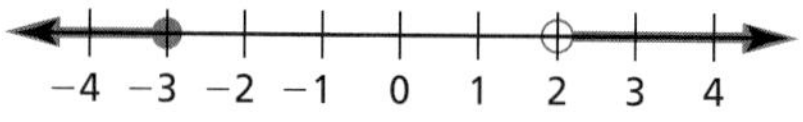

17. 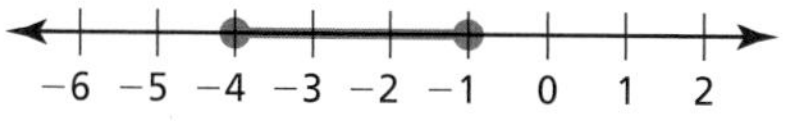

18. A state imposes a sales tax on items of clothing that cost more than \$175. The tax applies only to the difference of the price of the item and \$175.

    a. Use the receipt shown to find the tax rate (as a percent).

    b. A shopper has \$430 to spend on a winter coat. Write and solve an inequality to find the prices p of coats that the shopper can afford. Assume that $p \geq 175$.

    c. Another state imposes a 5% sales tax on the entire price of an item of clothing. For which prices would paying the 5% tax be cheaper than paying the tax described above? Write and solve an inequality to find your answer and list three prices that are solutions.

2 Cumulative Assessment

1. The expected attendance at a school event is 65 people. The actual attendance can vary by up to 30 people. Which equation can you use to find the minimum and maximum attendances?

   Ⓐ $|x - 65| = 30$ Ⓑ $|x + 65| = 30$

   Ⓒ $|x - 30| = 65$ Ⓓ $|x + 30| = 65$

2. Fill in values for a and b so that each statement is true for the inequality $ax + 4 \leq 3x + b$.

   a. When $a = 5$ and $b =$ _____, $x \leq -3$.

   b. When $a =$ _____ and $b =$ _____, the solution of the inequality is all real numbers.

   c. When $a =$ _____ and $b =$ _____, the inequality has no solution.

3. Place each inequality into one of the two categories.

At least one integer solution	No integer solutions

$5x - 6 + x \geq 2x - 8$

$x - 8 + 4x \leq 3(x - 3) + 2x$

$2(3x + 8) > 3(2x + 6)$

$9x - 3 < 12$ *or* $6x + 2 > -10$

$17 < 4x + 5 < 21$

$5(x - 1) \leq 5x - 3$

4. Admission to a play costs \$25. A season pass costs \$180.

   a. Write an inequality that represents the numbers x of plays you must attend for the season pass to be a better deal.

   b. Select the numbers of plays for which the season pass is *not* a better deal.

0	1	2	3	4
5	6	7	8	9
10	11	12	13	14

5. Select the values of a that make the solution of the equation $3(2x - 4) = 4(ax - 2)$ positive.

-2	-1	0	1	2	3	4	5

6. Fill in the compound inequality with $<$, $\leq$, $=$, $\geq$, or $>$ so the solution is shown in the graph.

$$4x - 18 \;\square\; -x - 3 \textit{ and } -3x - 9 \;\square\; -3$$

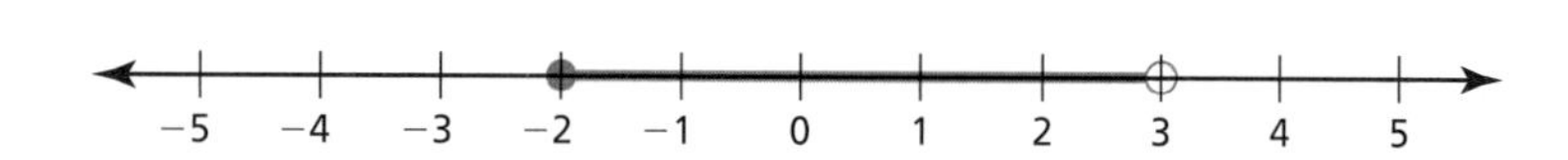

7. You have a \$250 gift card to use at a sporting goods store.

a. Write an inequality that represents the possible numbers x of pairs of socks you can buy when you buy 2 pairs of sneakers. Can you buy 8 pairs of socks? Explain.

b. Describe what the inequality $60 + 80x \leq 250$ represents in this context.

8. Consider the equation shown, where a, b, c, and d are integers.

$$ax + b = cx + d$$

Student A claims the equation will always have one solution. Student B claims the equation will always have no solution. Use the numbers shown to answer parts (a)–(c).

-1	0	1	2	3	4	5	6

a. Select values for a, b, c, and d to create an equation that supports Student A's claim.

b. Select values for a, b, c, and d to create an equation that supports Student B's claim.

c. Select values for a, b, c, and d to create an equation that shows both Student A and Student B are incorrect.

3 Graphing Linear Functions

Submersible *(p. 140)*

Basketball *(p. 134)*

Speed of Light *(p. 125)*

Coins *(p. 116)*

Taxi Ride *(p. 109)*

Maintaining Mathematical Proficiency

Plotting Points

Example 1 **Plot the point $A(-3, 4)$ in a coordinate plane. Describe the location of the point.**

Start at the origin. Move 3 units left and 4 units up. Then plot the point.
The point is in Quadrant II.

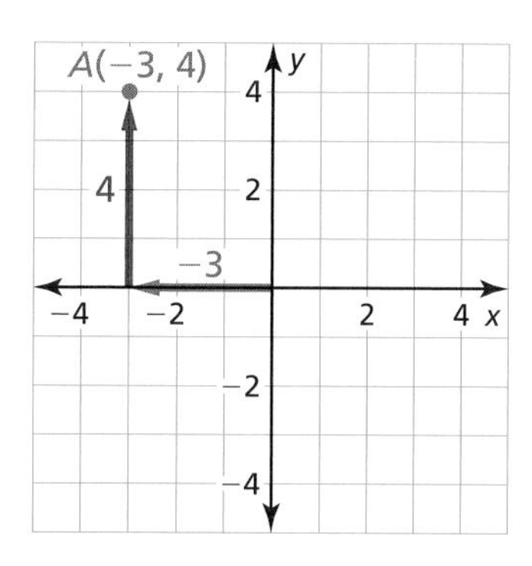

Plot the point in a coordinate plane. Describe the location of the point.

1. $A(3, 2)$
2. $B(-5, 1)$
3. $C(0, 3)$
4. $D(-1, -4)$
5. $E(-3, 0)$
6. $F(2, -1)$

Evaluating Expressions

Example 2 **Evaluate $4x - 5$ when $x = 3$.**

$4x - 5 = 4(3) - 5$ Substitute 3 for x.

$= 12 - 5$ Multiply.

$= 7$ Subtract.

Example 3 **Evaluate $-2x + 9$ when $x = -8$.**

$-2x + 9 = -2(-8) + 9$ Substitute -8 for x.

$= 16 + 9$ Multiply.

$= 25$ Add.

Evaluate the expression for the given value of x.

7. $3x - 4; x = 7$
8. $-5x + 8; x = 3$
9. $10x + 18; x = 5$
10. $-9x - 2; x = -4$
11. $24 - 8x; x = -2$
12. $15x + 9; x = -1$

13. ABSTRACT REASONING Let a and b be positive real numbers. Describe how to plot (a, b), $(-a, b)$, $(a, -b)$, and $(-a, -b)$.

Mathematical Practices

Mathematically proficient students use technological tools to explore concepts.

Using a Graphing Calculator

Core Concept

Standard and Square Viewing Windows

A typical graphing calculator screen has a height to width ratio of 2 to 3. This means that when you use the *standard viewing window* of -10 to 10 (on each axis), the graph will not be in its true perspective.

To see a graph in its true perspective, you need to use a *square viewing window*, in which the tick marks on the x-axis are spaced the same as the tick marks on the y-axis.

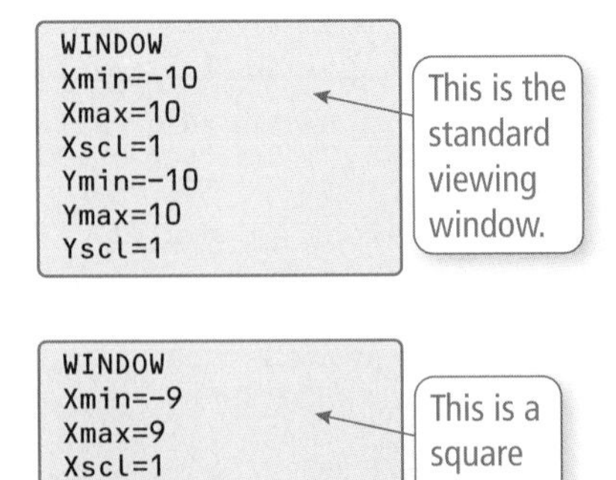

EXAMPLE 1 Using a Graphing Calculator

Use a graphing calculator to graph $y = 2x + 5$.

SOLUTION

Enter the equation $y = 2x + 5$ into your calculator. Then graph the equation. The standard viewing window does not show the graph in its true perspective. Notice that the tick marks on the y-axis are closer together than the tick marks on the x-axis. To see the graph in its true perspective, use a square viewing window.

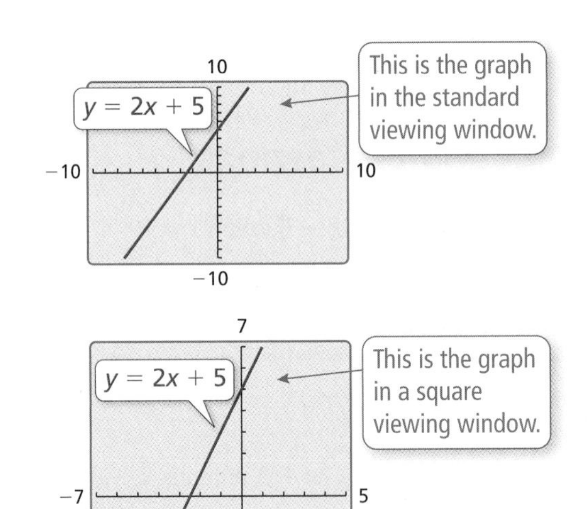

Monitoring Progress

Determine whether the viewing window is square. Explain.

1. $-8 \leq x \leq 7, -3 \leq y \leq 7$
2. $-6 \leq x \leq 6, -9 \leq y \leq 9$
3. $-18 \leq x \leq 18, -12 \leq y \leq 12$

Use a graphing calculator to graph the equation. Use a square viewing window.

4. $y = x + 3$
5. $y = -x - 2$
6. $y = 2x - 1$
7. $y = -2x + 1$
8. $y = -\frac{1}{3}x - 4$
9. $y = \frac{1}{2}x + 2$

10. How does the appearance of the slope of a line change between a standard viewing window and a square viewing window?

3.1 Functions

Essential Question

What is a function?

A **relation** pairs inputs with outputs. When a relation is given as ordered pairs, the x-coordinates are inputs and the y-coordinates are outputs. A relation that pairs each input with *exactly one* output is a **function**.

EXPLORATION 1 Describing a Function

Work with a partner. Functions can be described in many ways.

- by an equation
- by an input-output table
- using words
- by a graph
- as a set of ordered pairs

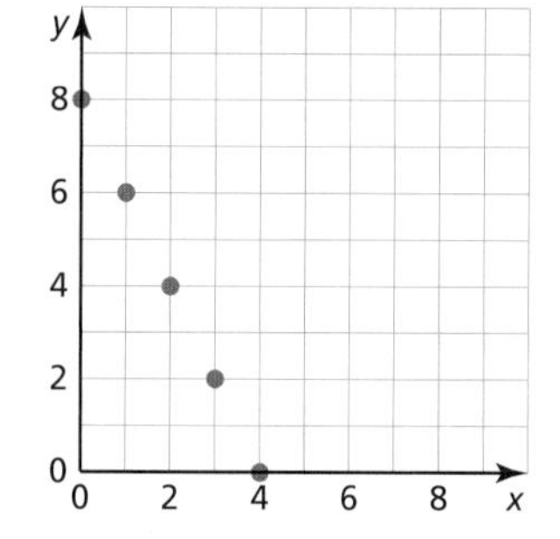

a. Explain why the graph shown represents a function.

b. Describe the function in two other ways.

ANALYZING RELATIONSHIPS

To be proficient in math, you need to analyze relationships mathematically to draw conclusions.

EXPLORATION 2 Identifying Functions

Work with a partner. Determine whether each relation represents a function. Explain your reasoning.

a.

Input, x	0	1	2	3	4
Output, y	8	8	8	8	8

b.

Input, x	8	8	8	8	8
Output, y	0	1	2	3	4

c. Input, x Output, y

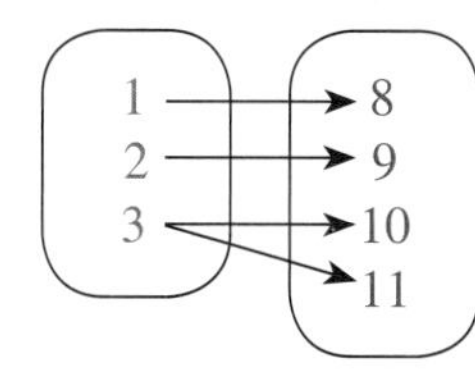

d.

e. $(-2, 5), (-1, 8), (0, 6), (1, 6), (2, 7)$

f. $(-2, 0), (-1, 0), (-1, 1), (0, 1), (1, 2), (2, 2)$

g. Each radio frequency x in a listening area has exactly one radio station y.

h. The same television station x can be found on more than one channel y.

i. $x = 2$

j. $y = 2x + 3$

Communicate Your Answer

3. What is a function? Give examples of relations, other than those in Explorations 1 and 2, that (a) are functions and (b) are not functions.

3.1 Lesson

Core Vocabulary

relation, *p. 104*
function, *p. 104*
domain, *p. 106*
range, *p. 106*
independent variable, *p. 107*
dependent variable, *p. 107*

Previous
ordered pair
mapping diagram

What You Will Learn

- Determine whether relations are functions.
- Find the domain and range of a function.
- Identify the independent and dependent variables of functions.

Determining Whether Relations Are Functions

A **relation** pairs inputs with outputs. When a relation is given as ordered pairs, the x-coordinates are inputs and the y-coordinates are outputs. A relation that pairs each input with *exactly one* output is a **function**.

EXAMPLE 1 Determining Whether Relations Are Functions

Determine whether each relation is a function. Explain.

a. $(-2, 2), (-1, 2), (0, 2), (1, 0), (2, 0)$

b. $(4, 0), (8, 7), (6, 4), (4, 3), (5, 2)$

c.

Input, x	-2	-1	0	0	1	2
Output, y	3	4	5	6	7	8

d. Input, x Output, y

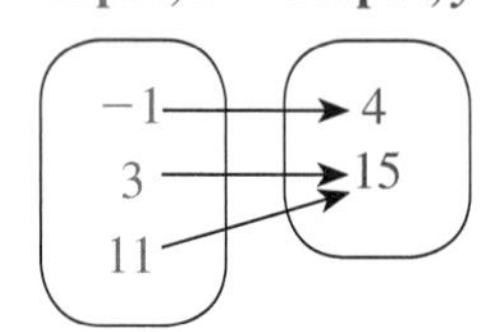

REMEMBER
A relation can be represented by a mapping diagram.

SOLUTION

a. Every input has exactly one output.

- So, the relation is a function.

b. The input 4 has two outputs, 0 and 3.

- So, the relation is *not* a function.

c. The input 0 has two outputs, 5 and 6.

- So, the relation is *not* a function.

d. Every input has exactly one output.

- So, the relation is a function.

Monitoring Progress  Help in English and Spanish at *BigIdeasMath.com*

Determine whether the relation is a function. Explain.

1. $(-5, 0), (0, 0), (5, 0), (5, 10)$

2. $(-4, 8), (-1, 2), (2, -4), (5, -10)$

3.

Input, x	Output, y
2	2.6
4	5.2
6	7.8

4. Input, x Output, y

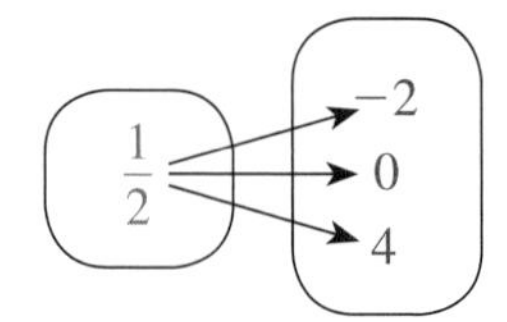

Vertical Line Test

Words A graph represents a function when no vertical line passes through more than one point on the graph.

Examples Function Not a function

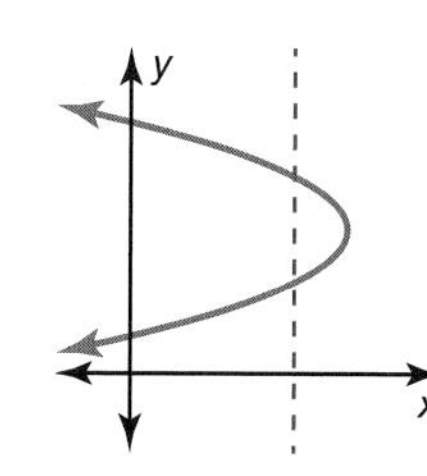

EXAMPLE 2 Using the Vertical Line Test

Determine whether each graph represents a function. Explain.

a.

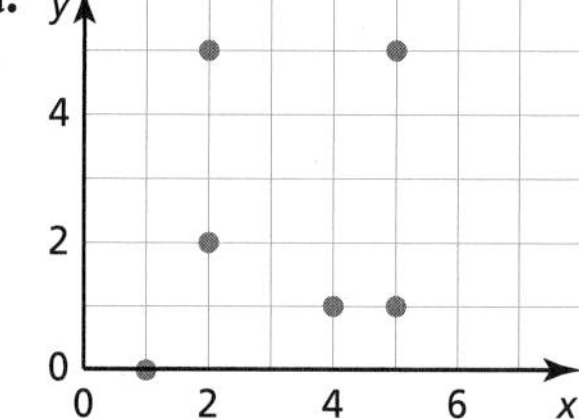

b.

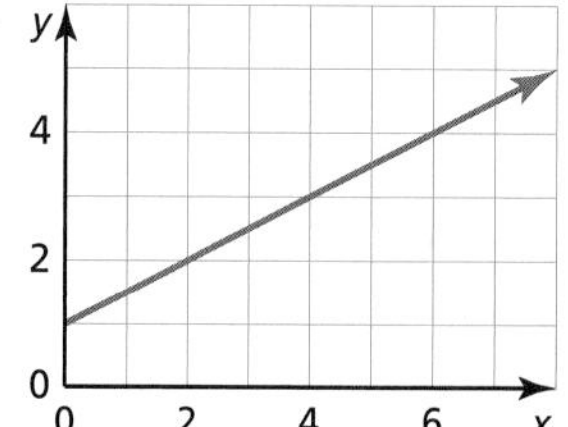

SOLUTION

a. You can draw a vertical line through (2, 2) and (2, 5).

▶ So, the graph does *not* represent a function.

b. No vertical line can be drawn through more than one point on the graph.

▶ So, the graph represents a function.

Monitoring Progress  Help in English and Spanish at *BigIdeasMath.com*

Determine whether the graph represents a function. Explain.

5.

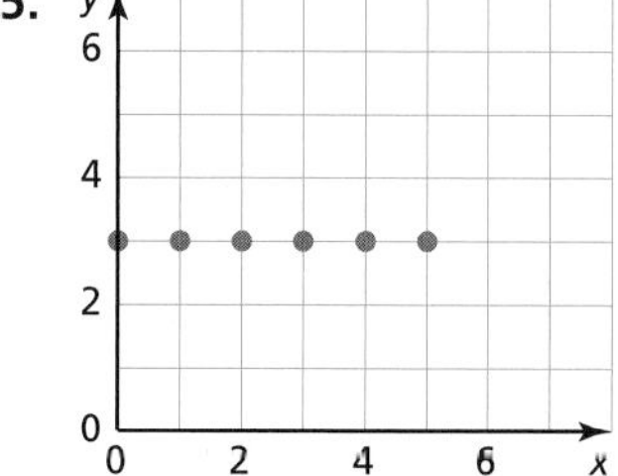

6.

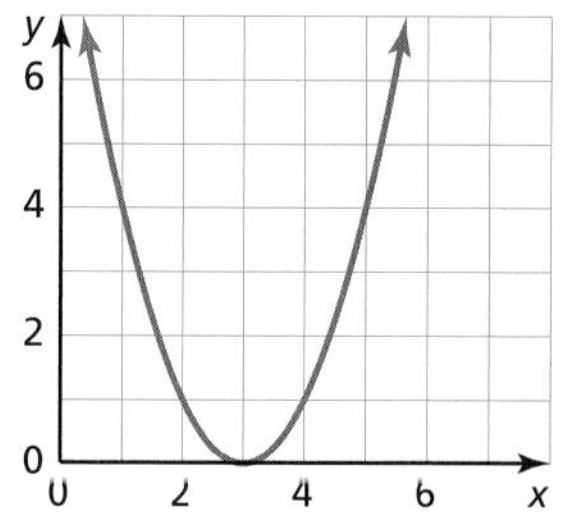

7.

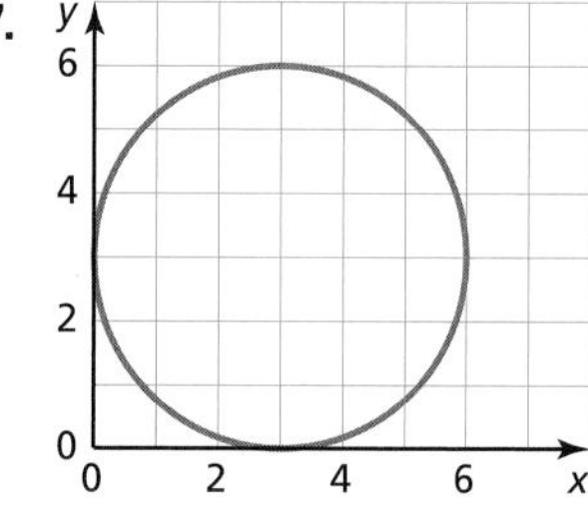

8.

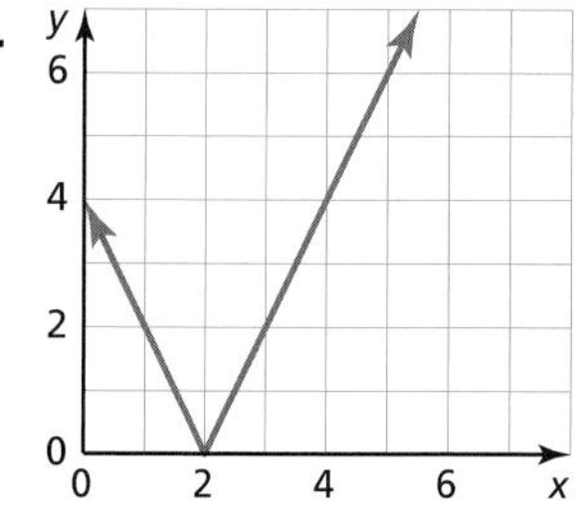

Finding the Domain and Range of a Function

Core Concept

The Domain and Range of a Function

The **domain** of a function is the set of all possible input values.

The **range** of a function is the set of all possible output values.

EXAMPLE 3 Finding the Domain and Range from a Graph

Find the domain and range of the function represented by the graph.

a.

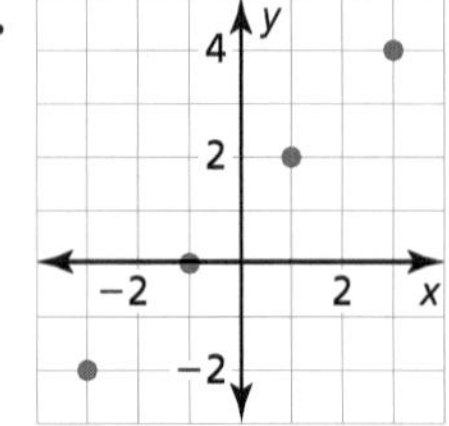

b.

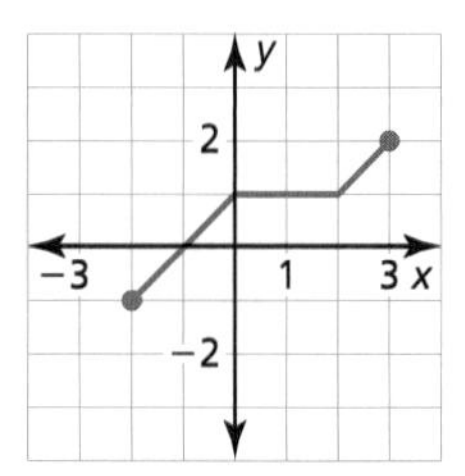

SOLUTION

a. Write the ordered pairs. Identify the inputs and outputs.

STUDY TIP

A relation also has a domain and a range.

▶ The domain is −3, −1, 1, and 3. The range is −2, 0, 2, and 4.

b. Identify the x- and y-values represented by the graph.

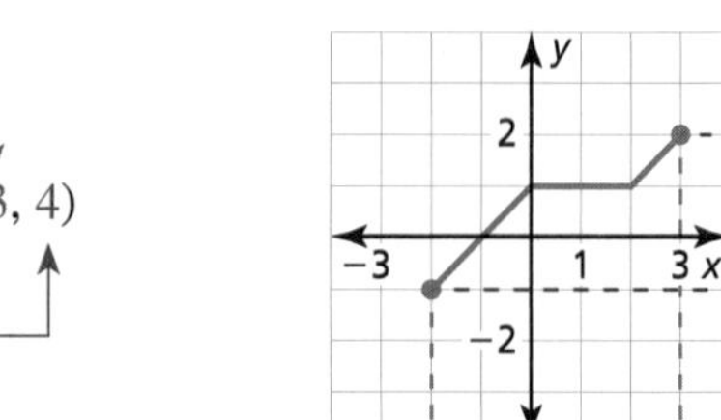

▶ The domain is $-2 \le x \le 3$. The range is $-1 \le y \le 2$.

Monitoring Progress

Help in English and Spanish at *BigIdeasMath.com*

Find the domain and range of the function represented by the graph.

9.

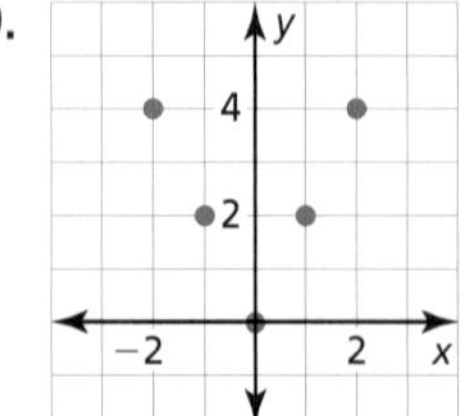

10.

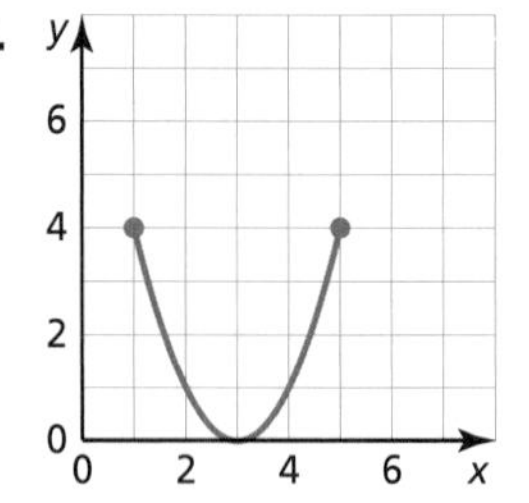

Identifying Independent and Dependent Variables

The variable that represents the input values of a function is the **independent variable** because it can be *any* value in the domain. The variable that represents the output values of a function is the **dependent variable** because it *depends* on the value of the independent variable. When an equation represents a function, the dependent variable is defined in terms of the independent variable. The statement "y is a function of x" means that y varies depending on the value of x.

$$y = -x + 10$$

dependent variable, y — independent variable, x

EXAMPLE 4 Identifying Independent and Dependent Variables

The function $y = -3x + 12$ represents the amount y (in fluid ounces) of juice remaining in a bottle after you take x gulps.

a. Identify the independent and dependent variables.

b. The domain is 0, 1, 2, 3, and 4. What is the range?

SOLUTION

a. The amount y of juice remaining depends on the number x of gulps.

▶ So, y is the dependent variable, and x is the independent variable.

b. Make an input-output table to find the range.

Input, x	$-3x + 12$	Output, y
0	$-3(0) + 12$	12
1	$-3(1) + 12$	9
2	$-3(2) + 12$	6
3	$-3(3) + 12$	3
4	$-3(4) + 12$	0

▶ The range is 12, 9, 6, 3, and 0.

Monitoring Progress  Help in English and Spanish at *BigIdeasMath.com*

11. The function $a = -4b + 14$ represents the number a of avocados you have left after making b batches of guacamole.

a. Identify the independent and dependent variables.

b. The domain is 0, 1, 2, and 3. What is the range?

12. The function $t = 19m + 65$ represents the temperature t (in degrees Fahrenheit) of an oven after preheating for m minutes.

a. Identify the independent and dependent variables.

b. A recipe calls for an oven temperature of 350°F. Describe the domain and range of the function.

 Dynamic Solutions available at *BigIdeasMath.com*

Vocabulary and Core Concept Check

1. **WRITING** How are independent variables and dependent variables different?

2. **DIFFERENT WORDS, SAME QUESTION** Which is different? Find "both" answers.

Find the range of the function represented by the table.

Find the inputs of the function represented by the table.

x	-1	0	1
y	7	5	-1

Find the x-values of the function represented by $(-1, 7)$, $(0, 5)$, and $(1, -1)$.

Find the domain of the function represented by $(-1, 7)$, $(0, 5)$, and $(1, -1)$.

Monitoring Progress and Modeling with Mathematics

In Exercises 3–8, determine whether the relation is a function. Explain. *(See Example 1.)*

3. $(1, -2)$, $(2, 1)$, $(3, 6)$, $(4, 13)$, $(5, 22)$

4. $(7, 4)$, $(5, -1)$, $(3, -8)$, $(1, -5)$, $(3, 6)$

5. Input, x Output, y

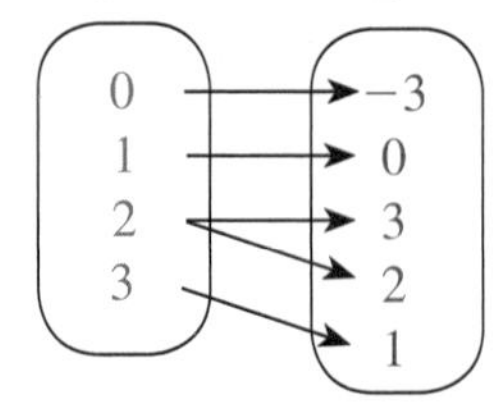

6. Input, x Output, y

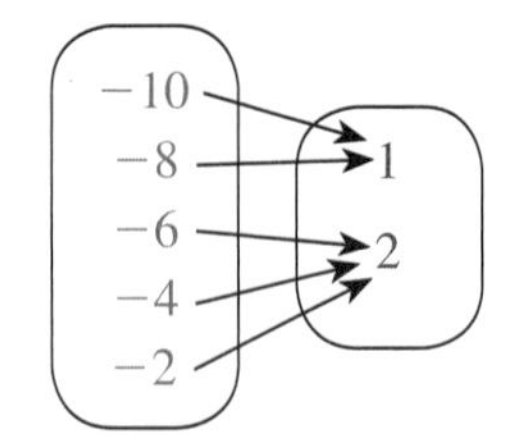

7.

Input, x	16	1	0	1	16
Output, y	-2	-1	0	1	2

8.

Input, x	-3	0	3	6	9
Output, y	11	5	-1	-7	-13

In Exercises 9–12, determine whether the graph represents a function. Explain. *(See Example 2.)*

9.

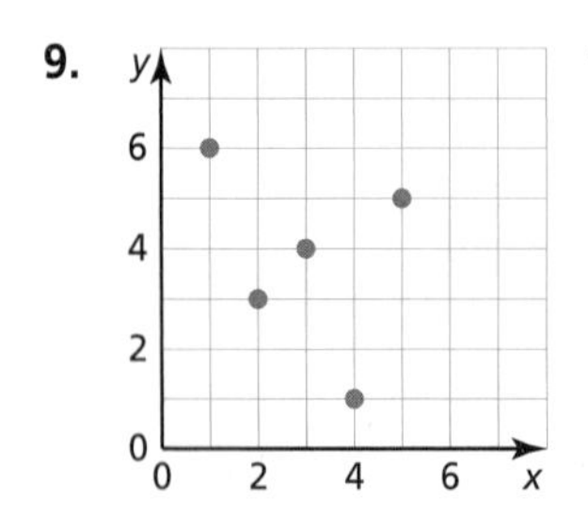

10.

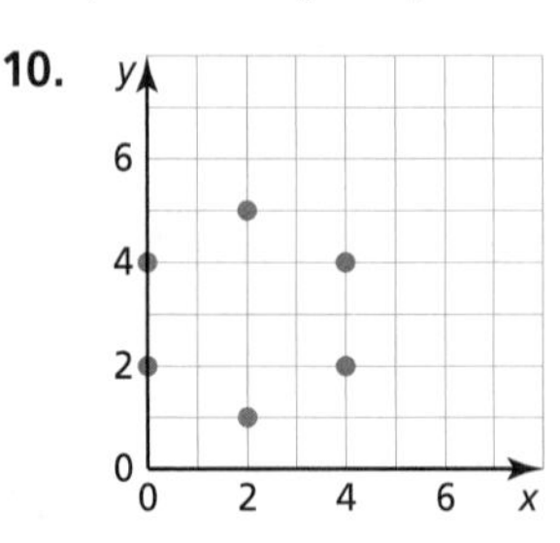

11. 12.

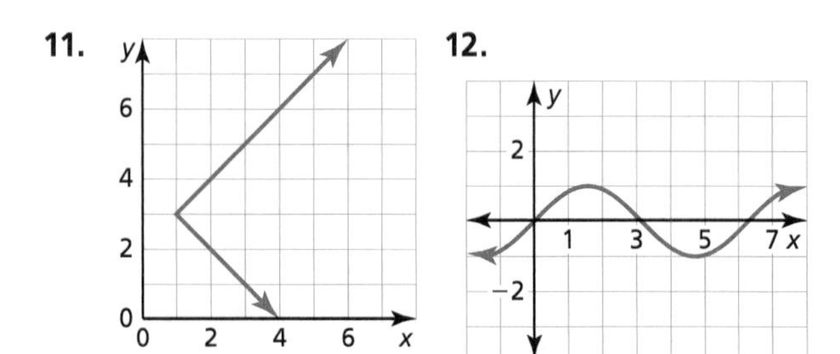

In Exercises 13–16, find the domain and range of the function represented by the graph. *(See Example 3.)*

13.

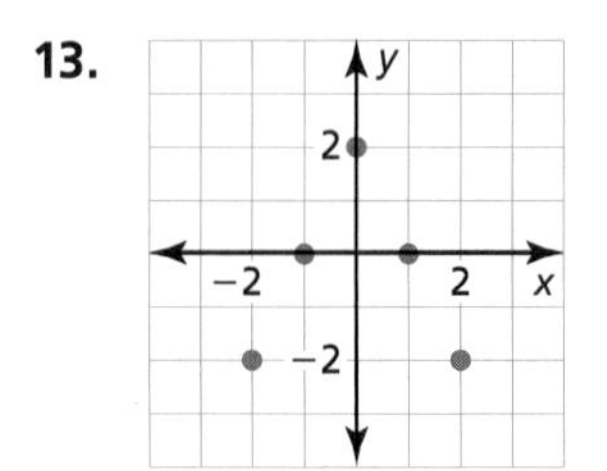

14.

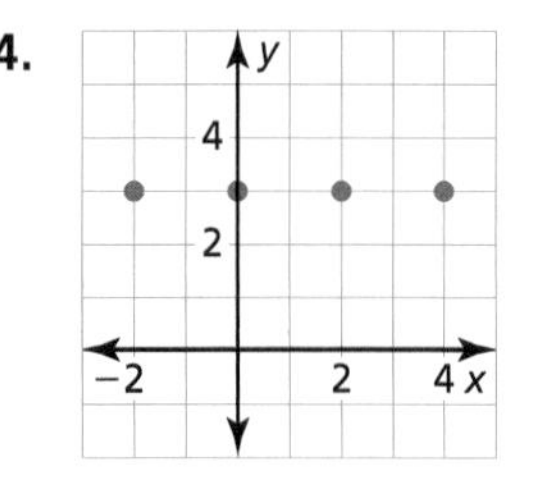

15.

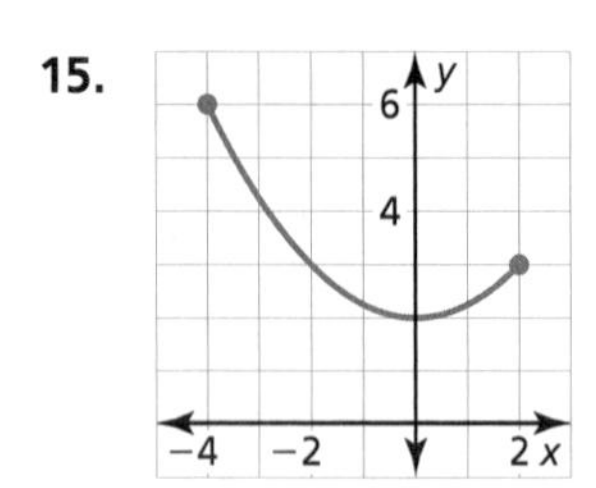

16.

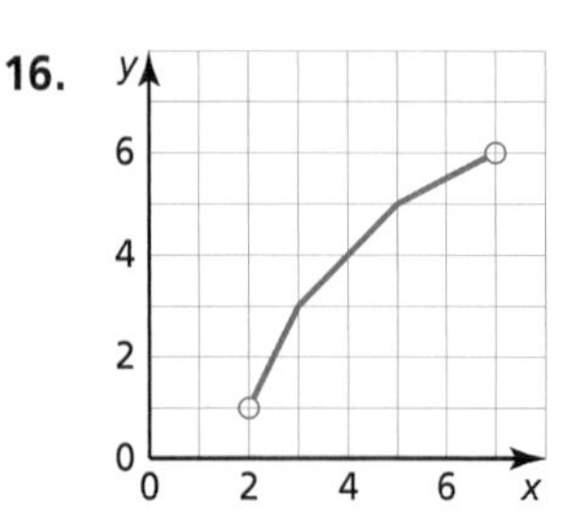

17. **MODELING WITH MATHEMATICS** The function $y = 25x + 500$ represents your monthly rent y (in dollars) when you pay x days late. *(See Example 4.)*

    a. Identify the independent and dependent variables.

    b. The domain is 0, 1, 2, 3, 4, and 5. What is the range?

18. MODELING WITH MATHEMATICS The function $y = 3.5x + 2.8$ represents the cost y (in dollars) of a taxi ride of x miles.

a. Identify the independent and dependent variables.

b. You have enough money to travel at most 20 miles in the taxi. Find the domain and range of the function.

ERROR ANALYSIS In Exercises 19 and 20, describe and correct the error in the statement about the relation shown in the table.

Input, x	1	2	3	4	5
Output, y	6	7	8	6	9

19.  The relation is *not* a function. One output is paired with two inputs.

20.  The relation is a function. The range is 1, 2, 3, 4, and 5.

ANALYZING RELATIONSHIPS In Exercises 21 and 22, identify the independent and dependent variables.

21. The number of quarters you put into a parking meter affects the amount of time you have on the meter.

22. The battery power remaining on your MP3 player is based on the amount of time you listen to it.

23. MULTIPLE REPRESENTATIONS The balance y (in dollars) of your savings account is a function of the month x.

Month, x	0	1	2	3	4
Balance (dollars), y	100	125	150	175	200

a. Describe this situation in words.

b. Write the function as a set of ordered pairs.

c. Plot the ordered pairs in a coordinate plane.

24. MULTIPLE REPRESENTATIONS The function $1.5x + 0.5y = 12$ represents the number of hardcover books x and softcover books y you can buy at a used book sale.

a. Solve the equation for y.

b. Make an input-output table to find ordered pairs for the function.

c. Plot the ordered pairs in a coordinate plane.

25. ATTENDING TO PRECISION The graph represents a function. Find the input value corresponding to an output of 2.

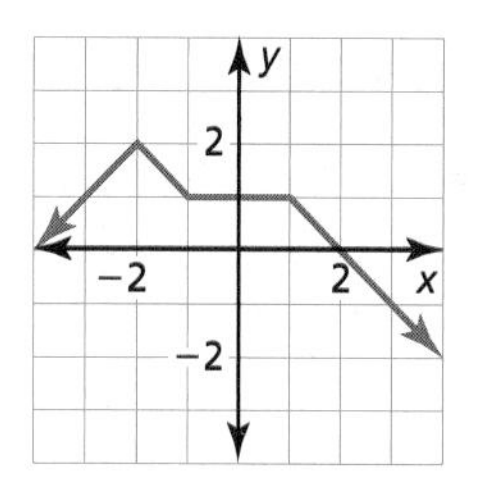

26. OPEN-ENDED Fill in the table so that when t is the independent variable, the relation is a function, and when t is the dependent variable, the relation is not a function.

t				
v				

27. ANALYZING RELATIONSHIPS You select items in a vending machine by pressing one letter and then one number.

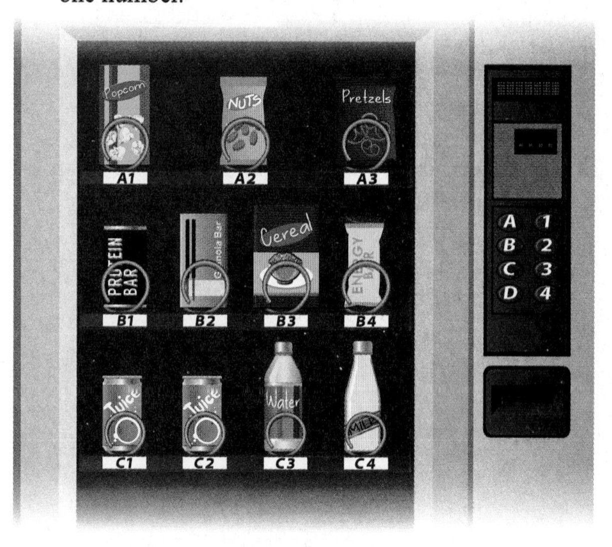

a. Explain why the relation that pairs letter-number combinations with food or drink items is a function.

b. Identify the independent and dependent variables.

c. Find the domain and range of the function.

28. HOW DO YOU SEE IT? The graph represents the height h of a projectile after t seconds.

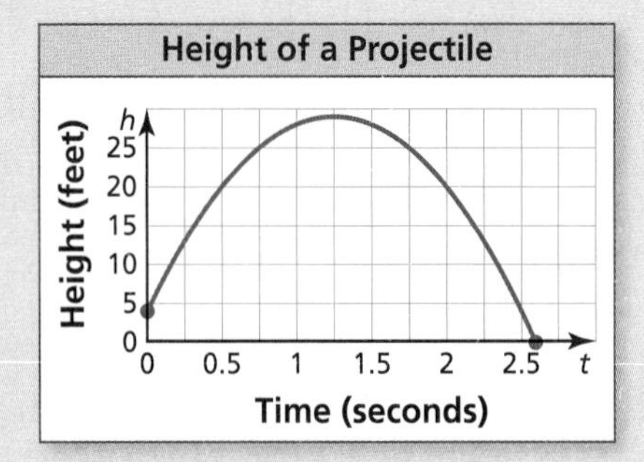

a. Explain why h is a function of t.

b. Approximate the height of the projectile after 0.5 second and after 1.25 seconds.

c. Approximate the domain of the function.

d. Is t a function of h? Explain.

29. MAKING AN ARGUMENT Your friend says that a line always represents a function. Is your friend correct? Explain.

30. THOUGHT PROVOKING Write a function in which the inputs and/or the outputs are not numbers. Identify the independent and dependent variables. Then find the domain and range of the function.

ATTENDING TO PRECISION In Exercises 31–34, determine whether the statement uses the word *function* in a way that is mathematically correct. Explain your reasoning.

31. The selling price of an item is a function of the cost of making the item.

32. The sales tax on a purchased item in a given state is a function of the selling price.

33. A function pairs each student in your school with a homeroom teacher.

34. A function pairs each chaperone on a school trip with 10 students.

REASONING In Exercises 35–38, tell whether the statement is true or false. If it is false, explain why.

35. Every function is a relation.

36. Every relation is a function.

37. When you switch the inputs and outputs of any function, the resulting relation is a function.

38. When the domain of a function has an infinite number of values, the range always has an infinite number of values.

39. MATHEMATICAL CONNECTIONS Consider the triangle shown.

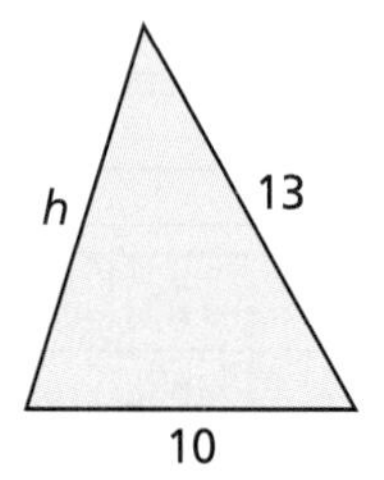

a. Write a function that represents the perimeter of the triangle.

b. Identify the independent and dependent variables.

c. Describe the domain and range of the function. (*Hint:* The sum of the lengths of any two sides of a triangle is greater than the length of the remaining side.)

REASONING In Exercises 40–43, find the domain and range of the function.

40. $y = |x|$

41. $y = -|x|$

42. $y = |x| - 6$

43. $y = 4 - |x|$

Maintaining Mathematical Proficiency

Reviewing what you learned in previous grades and lessons

Write the sentence as an inequality. *(Section 2.1)*

44. A number y is less than 16.

45. Three is no less than a number x.

46. Seven is at most the quotient of a number d and -5.

47. The sum of a number w and 4 is more than -12.

Evaluate the expression. *(Skills Review Handbook)*

48. 11^2

49. $(-3)^4$

50. -5^2

51. 2^5

3.2 Linear Functions

Essential Question

How can you determine whether a function is linear or nonlinear?

EXPLORATION 1 Finding Patterns for Similar Figures

Work with a partner. Copy and complete each table for the sequence of similar figures. (In parts (a) and (b), use the rectangle shown.) Graph the data in each table. Decide whether each pattern is linear or nonlinear. Justify your conclusion.

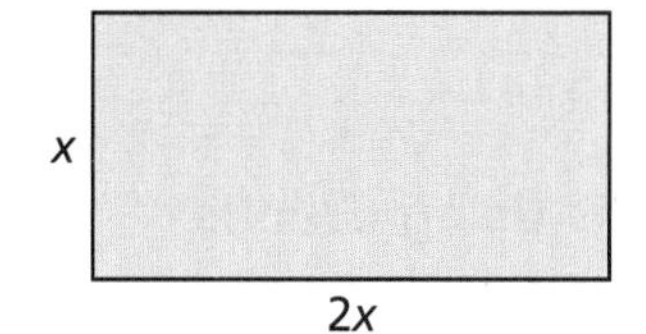

a. perimeters of similar rectangles

x	1	2	3	4	5
P					

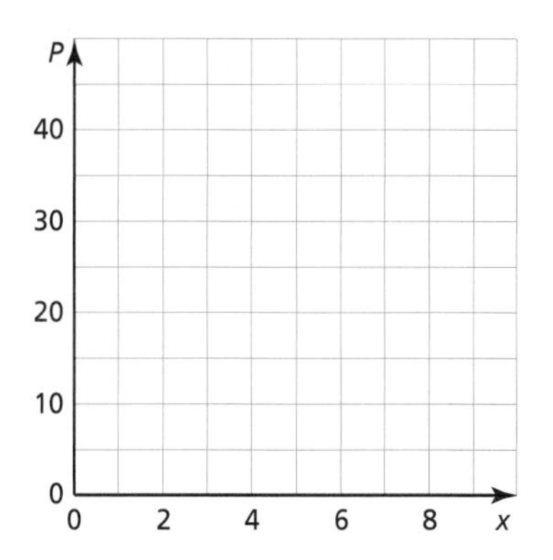

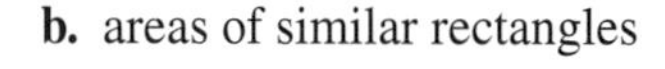

b. areas of similar rectangles

x	1	2	3	4	5
A					

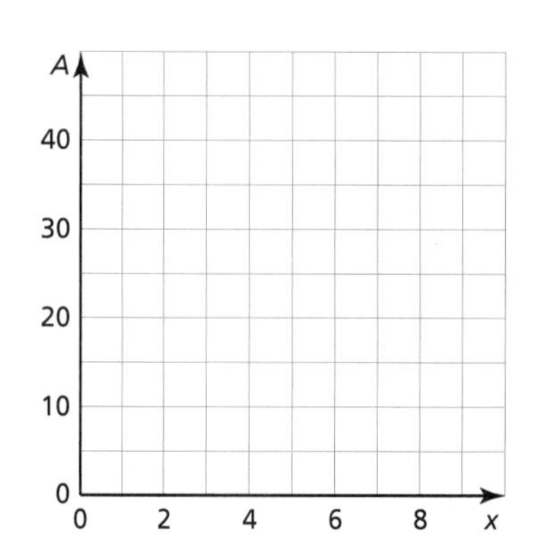

c. circumferences of circles of radius r

r	1	2	3	4	5
C					

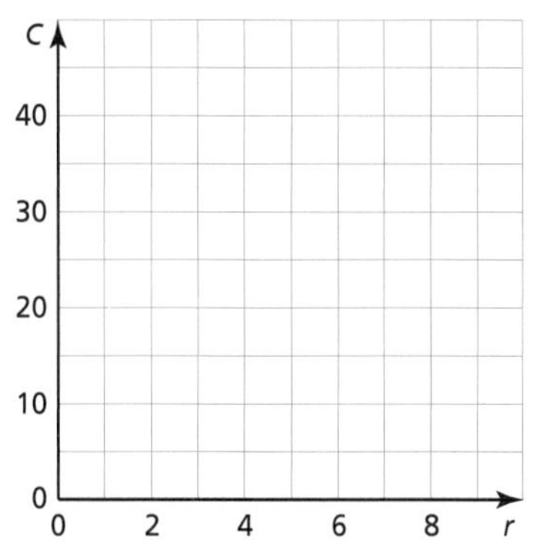

d. areas of circles of radius r

r	1	2	3	4	5
A					

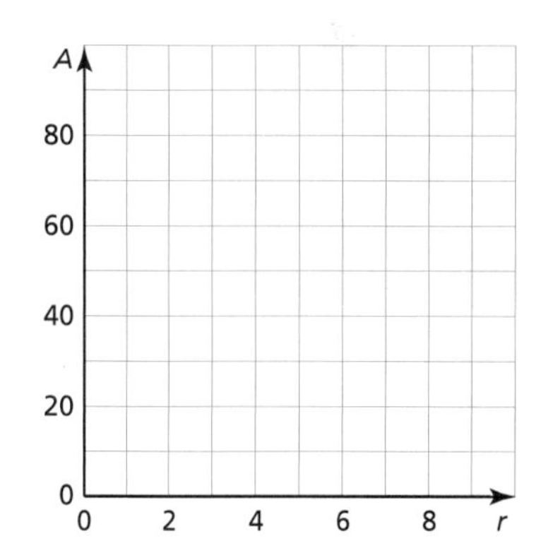

USING TOOLS STRATEGICALLY

To be proficient in math, you need to identify relationships using tools, such as tables and graphs.

Communicate Your Answer

2. How do you know that the patterns you found in Exploration 1 represent functions?

3. How can you determine whether a function is linear or nonlinear?

4. Describe two real-life patterns: one that is linear and one that is nonlinear. Use patterns that are different from those described in Exploration 1.

3.2 Lesson

Core Vocabulary

linear equation in two variables, *p. 112*
linear function, *p. 112*
nonlinear function, *p. 112*
solution of a linear equation in two variables, *p. 114*
discrete domain, *p. 114*
continuous domain, *p. 114*

Previous
whole number

What You Will Learn

- Identify linear functions using graphs, tables, and equations.
- Graph linear functions using discrete and continuous data.
- Write real-life problems to fit data.

Identifying Linear Functions

A **linear equation in two variables**, x and y, is an equation that can be written in the form $y = mx + b$, where m and b are constants. The graph of a linear equation is a line. Likewise, a **linear function** is a function whose graph is a nonvertical line. A linear function has a constant rate of change and can be represented by a linear equation in two variables. A **nonlinear function** does not have a constant rate of change. So, its graph is *not* a line.

EXAMPLE 1 Identifying Linear Functions Using Graphs

Does the graph represent a *linear* or *nonlinear* function? Explain.

a.

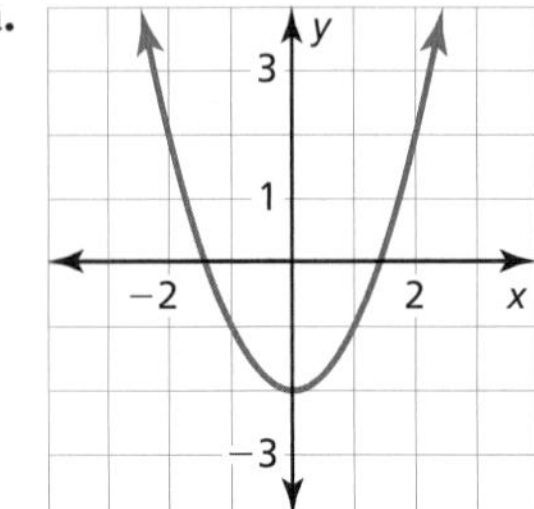

b.

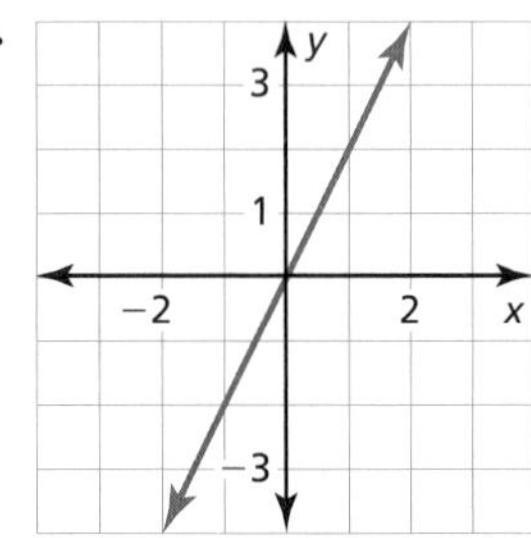

SOLUTION

a. The graph is *not* a line.

▶ So, the function is nonlinear.

b. The graph is a line.

▶ So, the function is linear.

EXAMPLE 2 Identifying Linear Functions Using Tables

Does the table represent a *linear* or *nonlinear* function? Explain.

a.

x	3	6	9	12
y	36	30	24	18

b.

x	1	3	5	7
y	2	9	20	35

SOLUTION

a.

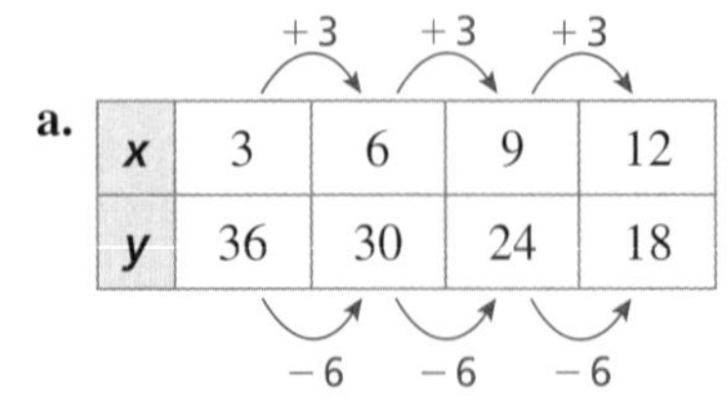

x	3	6	9	12
y	36	30	24	18

As x increases by 3, y decreases by 6. The rate of change is constant.

▶ So, the function is linear.

b.

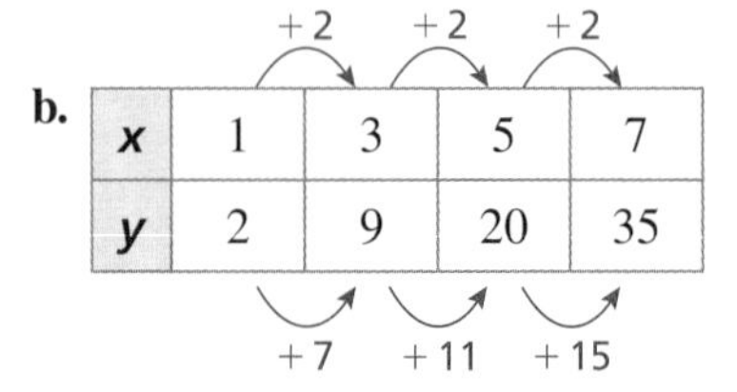

x	1	3	5	7
y	2	9	20	35

As x increases by 2, y increases by different amounts. The rate of change is *not* constant.

▶ So, the function is nonlinear.

REMEMBER

A constant rate of change describes a quantity that changes by equal amounts over equal intervals.

Monitoring Progress  Help in English and Spanish at *BigIdeasMath.com*

Does the graph or table represent a *linear* or *nonlinear* function? Explain.

1. 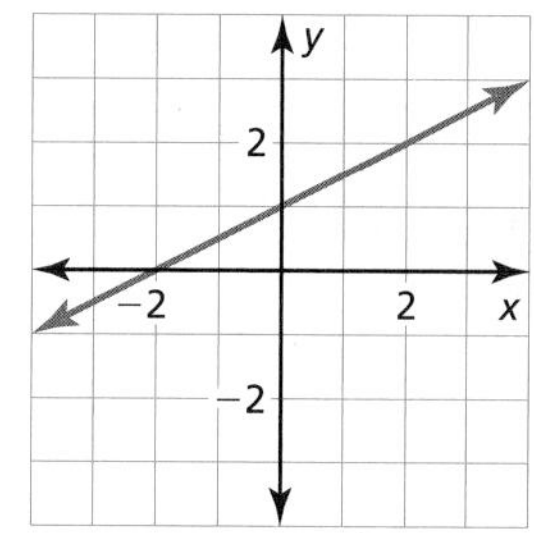

2. 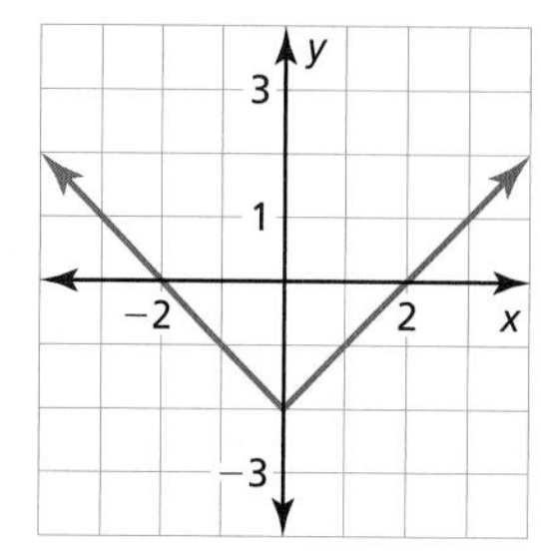

3.

x	0	1	2	3
y	3	5	7	9

4.

x	1	2	3	4
y	16	8	4	2

EXAMPLE 3 Identifying Linear Functions Using Equations

Which of the following equations represent linear functions? Explain.

$$y = 3.8,\ y = \sqrt{x},\ y = 3^x,\ y = \frac{2}{x},\ y = 6(x - 1), \text{ and } x^2 - y = 0$$

SOLUTION

You cannot rewrite the equations $y = \sqrt{x}$, $y = 3^x$, $y = \frac{2}{x}$, and $x^2 - y = 0$ in the form $y = mx + b$. So, these equations cannot represent linear functions.

▶ You can rewrite the equation $y = 3.8$ as $y = 0x + 3.8$ and the equation $y = 6(x - 1)$ as $y = 6x - 6$. So, they represent linear functions.

Monitoring Progress  Help in English and Spanish at *BigIdeasMath.com*

Does the equation represent a *linear* or *nonlinear* function? Explain.

5. $y = x + 9$

6. $y = \frac{3x}{5}$

7. $y = 5 - 2x^2$

Concept Summary

Representations of Functions

Words An output is 3 more than the input.

Equation $y = x + 3$

Input-Output Table

Input, x	Output, y
−1	2
0	3
1	4
2	5

Mapping Diagram

Input, x → Output, y: −1 → 2, 0 → 3, 1 → 4, 2 → 5

Graph

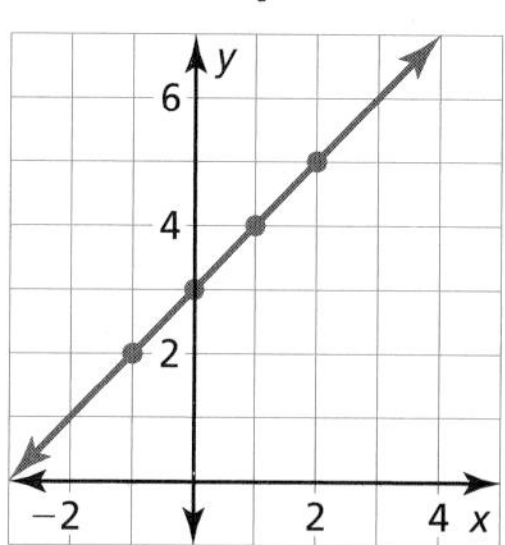

Graphing Linear Functions

A **solution of a linear equation in two variables** is an ordered pair (x, y) that makes the equation true. The graph of a linear equation in two variables is the set of points (x, y) in a coordinate plane that represents all solutions of the equation. Sometimes the points are distinct, and other times the points are connected.

Core Concept

Discrete and Continuous Domains

A **discrete domain** is a set of input values that consists of only certain numbers in an interval.

Example: Integers from 1 to 5

A **continuous domain** is a set of input values that consists of all numbers in an interval.

Example: All numbers from 1 to 5

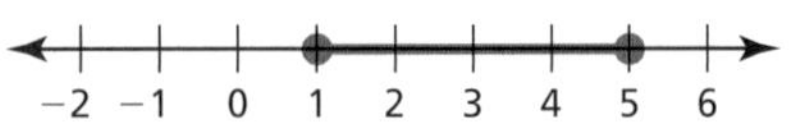

EXAMPLE 4 Graphing Discrete Data

The linear function $y = 15.95x$ represents the cost y (in dollars) of x tickets for a museum. Each customer can buy a maximum of four tickets.

a. Find the domain of the function. Is the domain discrete or continuous? Explain.

b. Graph the function using its domain.

STUDY TIP

The domain of a function depends on the real-life context of the function, not just the equation that represents the function.

SOLUTION

a. You cannot buy part of a ticket, only a certain number of tickets. Because x represents the number of tickets, it must be a whole number. The maximum number of tickets a customer can buy is four.

▶ So, the domain is 0, 1, 2, 3, and 4, and it is discrete.

b. Step 1 Make an input-output table to find the ordered pairs.

Input, x	$15.95x$	Output, y	(x, y)
0	15.95(0)	0	(0, 0)
1	15.95(1)	15.95	(1, 15.95)
2	15.95(2)	31.9	(2, 31.9)
3	15.95(3)	47.85	(3, 47.85)
4	15.95(4)	63.8	(4, 63.8)

Step 2 Plot the ordered pairs. The domain is discrete. So, the graph consists of individual points.

Museum Tickets

Cost (dollars) vs. Number of tickets: (0, 0), (1, 15.95), (2, 31.9), (3, 47.85), (4, 63.8)

Monitoring Progress  Help in English and Spanish at *BigIdeasMath.com*

8. The linear function $m = 50 - 9d$ represents the amount m (in dollars) of money you have after buying d DVDs. (a) Find the domain of the function. Is the domain discrete or continuous? Explain. (b) Graph the function using its domain.

EXAMPLE 5 Graphing Continuous Data

A cereal bar contains 130 calories. The number c of calories consumed is a function of the number b of bars eaten.

a. Does this situation represent a linear function? Explain.

b. Find the domain of the function. Is the domain discrete or continuous? Explain.

c. Graph the function using its domain.

STUDY TIP

When the domain of a linear function is not specified or cannot be obtained from a real-life context, it is understood to be all real numbers.

SOLUTION

a. As b increases by 1, c increases by 130. The rate of change is constant.

▶ So, this situation represents a linear function.

b. You can eat part of a cereal bar. The number b of bars eaten can be any value greater than or equal to 0.

▶ So, the domain is $b \geq 0$, and it is continuous.

c. Step 1 Make an input-output table to find ordered pairs.

Input, b	Output, c	(b, c)
0	0	(0, 0)
1	130	(1, 130)
2	260	(2, 260)
3	390	(3, 390)
4	520	(4, 520)

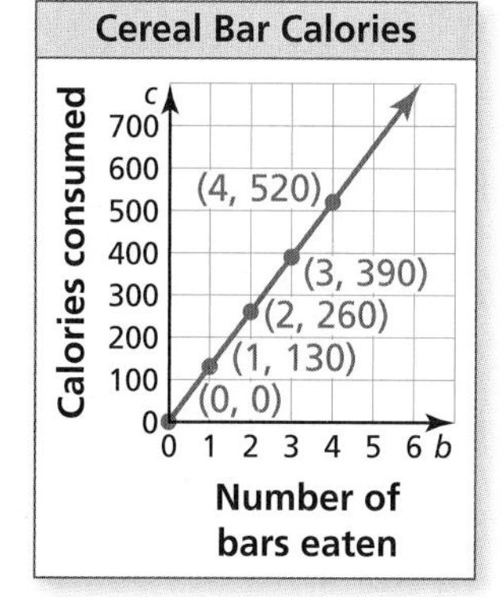

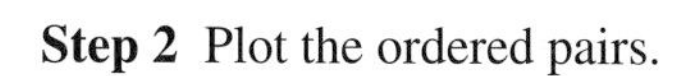

Step 2 Plot the ordered pairs.

Step 3 Draw a line through the points. The line should start at (0, 0) and continue to the right. Use an arrow to indicate that the line continues without end, as shown. The domain is continuous. So, the graph is a line with a domain of $b \geq 0$.

Monitoring Progress  Help in English and Spanish at *BigIdeasMath.com*

9. Is the domain discrete or continuous? Explain.

Input Number of stories, x	1	2	3
Output Height of building (feet), y	12	24	36

10. A 20-gallon bathtub is draining at a rate of 2.5 gallons per minute. The number g of gallons remaining is a function of the number m of minutes.

a. Does this situation represent a linear function? Explain.

b. Find the domain of the function. Is the domain discrete or continuous? Explain.

c. Graph the function using its domain.

Writing Real-Life Problems

EXAMPLE 6 Writing Real-Life Problems

Write a real-life problem to fit the data shown in each graph. Is the domain of each function *discrete* or *continuous*? Explain.

a. 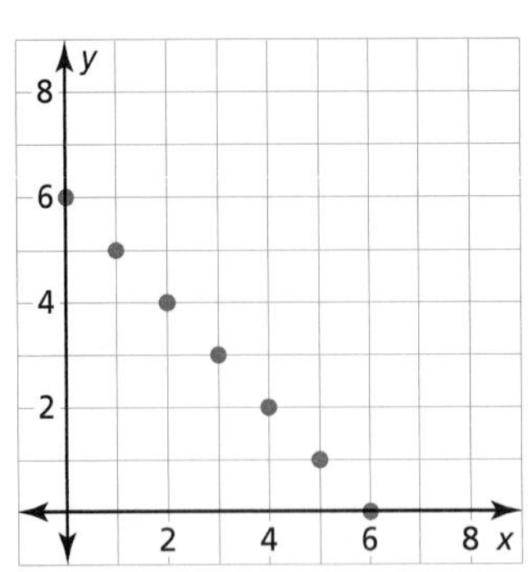

b. 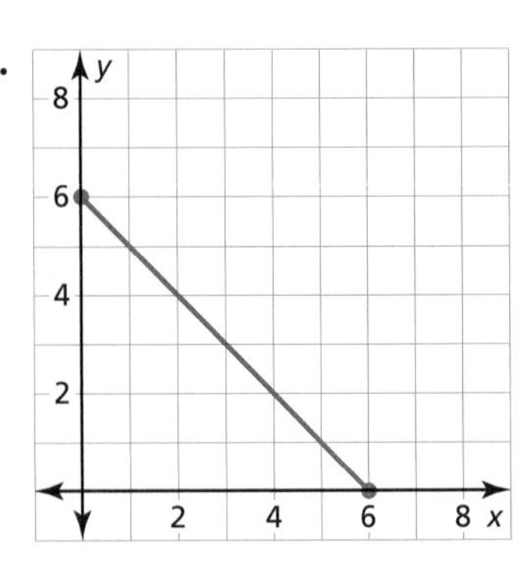

SOLUTION

a. You want to think of a real-life situation in which there are two variables, x and y. Using the graph, notice that the sum of the variables is always 6, and the value of each variable must be a whole number from 0 to 6.

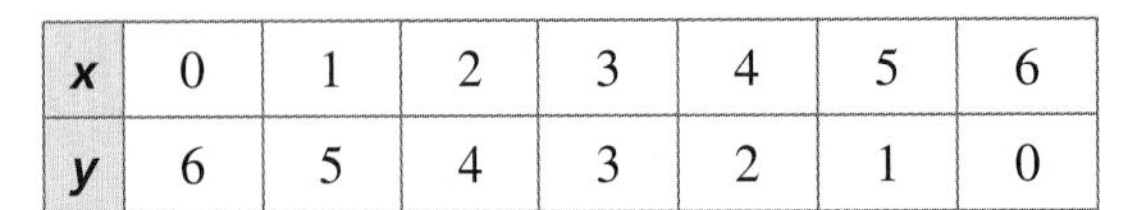

x	0	1	2	3	4	5	6
y	6	5	4	3	2	1	0

Discrete domain

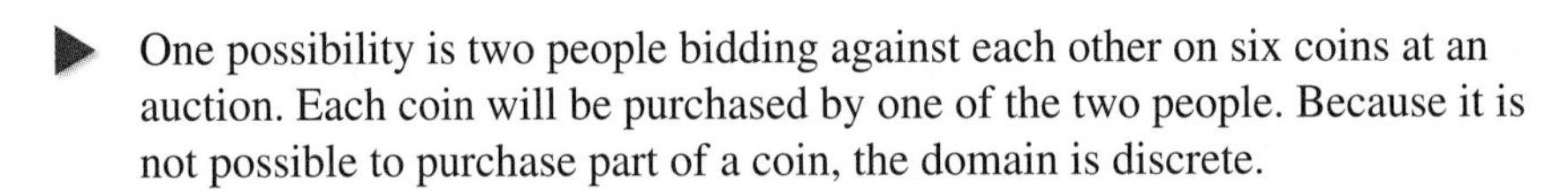

▶ One possibility is two people bidding against each other on six coins at an auction. Each coin will be purchased by one of the two people. Because it is not possible to purchase part of a coin, the domain is discrete.

b. You want to think of a real-life situation in which there are two variables, x and y. Using the graph, notice that the sum of the variables is always 6, and the value of each variable can be any real number from 0 to 6.

$$x + y = 6 \quad \text{or} \quad y = -x + 6$$

Continuous domain

▶ One possibility is two people bidding against each other on 6 ounces of gold dust at an auction. All the dust will be purchased by the two people. Because it is possible to purchase any portion of the dust, the domain is continuous.

Monitoring Progress  Help in English and Spanish at *BigIdeasMath.com*

Write a real-life problem to fit the data shown in the graph. Is the domain of the function *discrete* or *continuous*? Explain.

11. 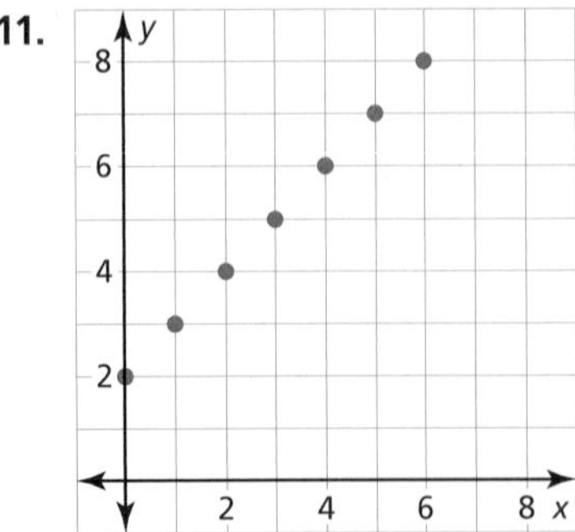

12. 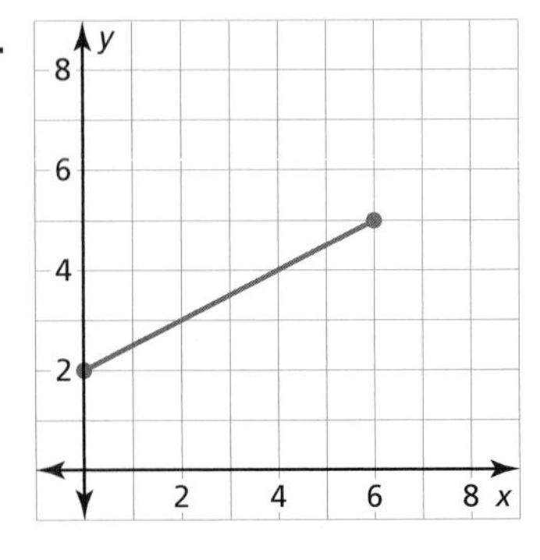

3.2 Exercises

Dynamic Solutions available at *BigIdeasMath.com*

Vocabulary and Core Concept Check

1. **COMPLETE THE SENTENCE** A linear equation in two variables is an equation that can be written in the form ________, where m and b are constants.

2. **VOCABULARY** Compare linear functions and nonlinear functions.

3. **VOCABULARY** Compare discrete domains and continuous domains.

4. **WRITING** How can you tell whether a graph shows a discrete domain or a continuous domain?

Monitoring Progress and Modeling with Mathematics

In Exercises 5–10, determine whether the graph represents a *linear* or *nonlinear* function. Explain. *(See Example 1.)*

5. 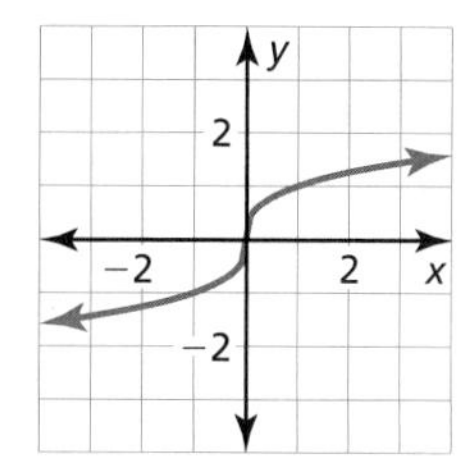

6. 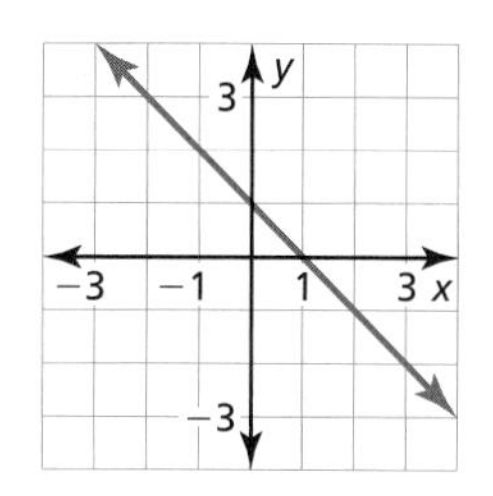

7. 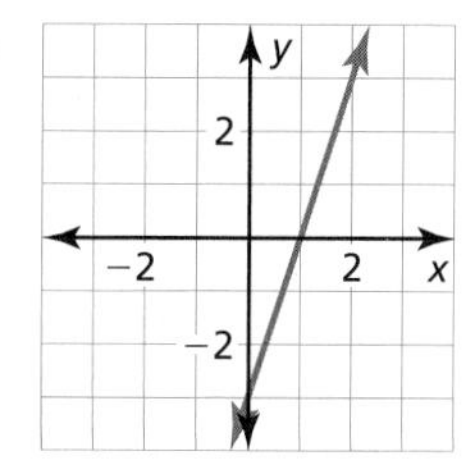

8. 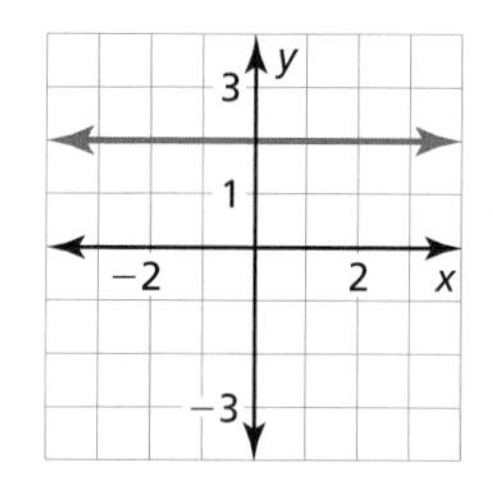

9. 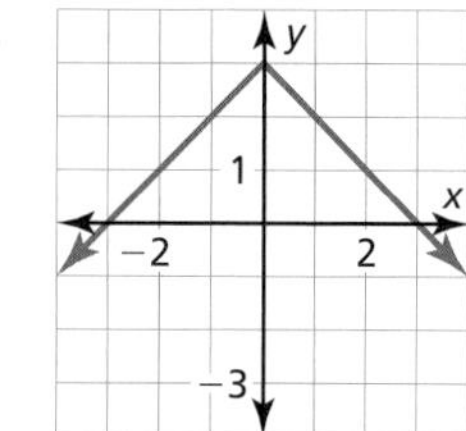

10. 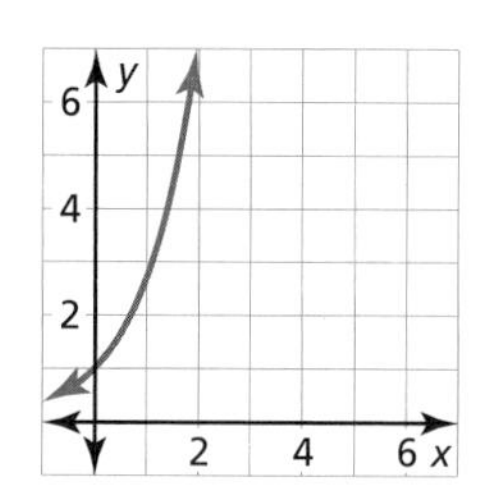

In Exercises 11–14, determine whether the table represents a *linear* or *nonlinear* function. Explain. *(See Example 2.)*

11.

x	1	2	3	4
y	5	10	15	20

12.

x	5	7	9	11
y	−9	−3	−1	3

13.

x	4	8	12	16
y	16	12	7	1

14.

x	−1	0	1	2
y	35	20	5	−10

ERROR ANALYSIS In Exercises 15 and 16, describe and correct the error in determining whether the table or graph represents a linear function.

15. 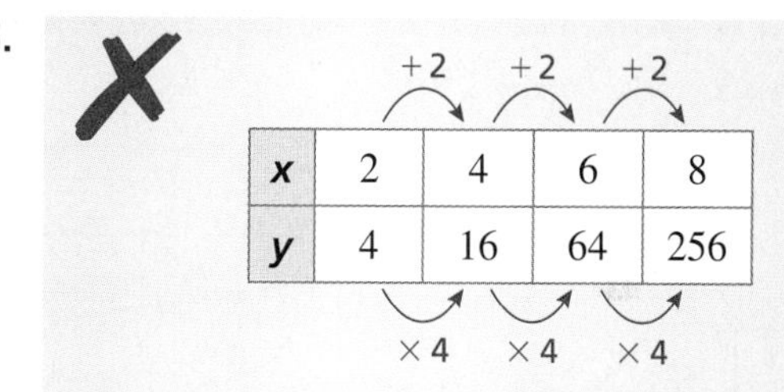

x	2	4	6	8
y	4	16	64	256

16. 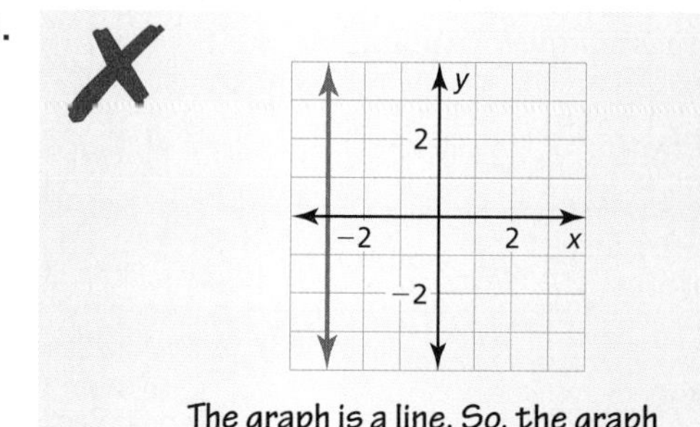

In Exercises 17–24, determine whether the equation represents a *linear* or *nonlinear* function. Explain. *(See Example 3.)*

17. $y = x^2 + 13$

18. $y = 7 - 3x$

19. $y = \sqrt[3]{8} - x$

20. $y = 4x(8 - x)$

21. $2 + \frac{1}{6}y = 3x + 4$

22. $y - x = 2x - \frac{2}{3}y$

23. $18x - 2y = 26$

24. $2x + 3y = 9xy$

25. CLASSIFYING FUNCTIONS Which of the following equations *do not* represent linear functions? Explain.

Ⓐ $12 = 2x^2 + 4y^2$

Ⓑ $y - x + 3 = x$

Ⓒ $x = 8$

Ⓓ $x = 9 - \frac{3}{4}y$

Ⓔ $y = \frac{5x}{11}$

Ⓕ $y = \sqrt{x} + 3$

26. USING STRUCTURE Fill in the table so it represents a linear function.

x	5	10	15	20	25
y	−1				11

In Exercises 27 and 28, find the domain of the function represented by the graph. Determine whether the domain is *discrete* or *continuous*. Explain.

27.

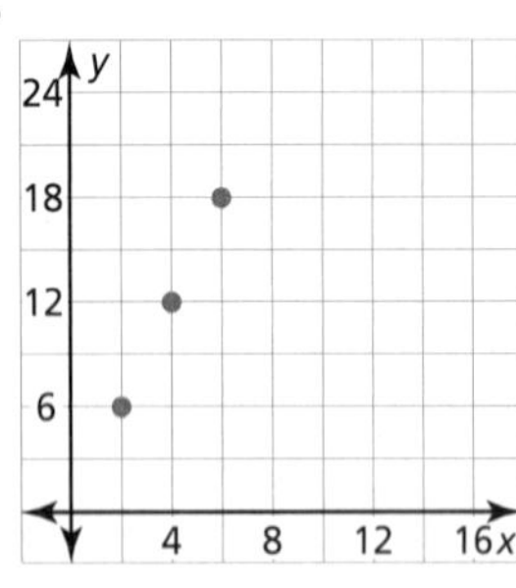

28.

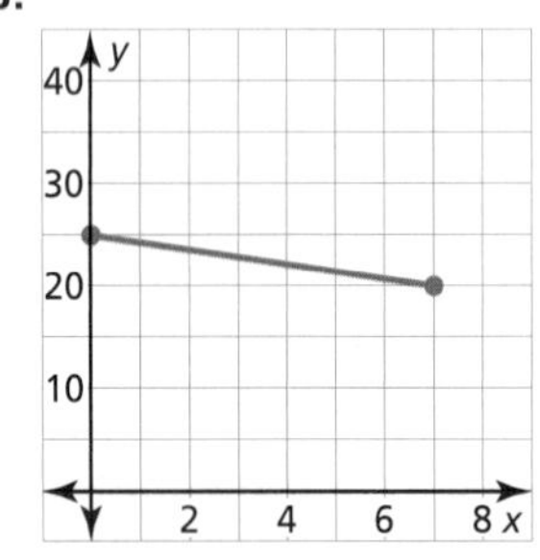

In Exercises 29–32, determine whether the domain is *discrete* or *continuous*. Explain.

29.

Input Bags, *x*	2	4	6
Output Marbles, *y*	20	40	60

30.

Input Years, *x*	1	2	3
Output Height of tree (feet), *y*	6	9	12

31.

Input Time (hours), *x*	3	6	9
Output Distance (miles), *y*	150	300	450

32.

Input Relay teams, *x*	0	1	2
Output Athletes, *y*	0	4	8

ERROR ANALYSIS In Exercises 33 and 34, describe and correct the error in the statement about the domain.

33.

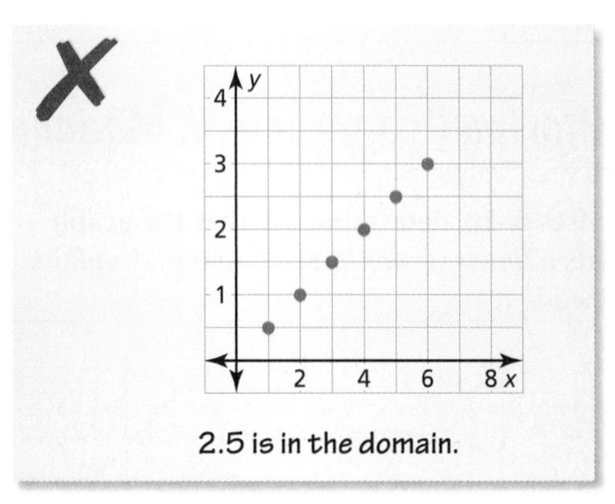
2.5 is in the domain.

34.

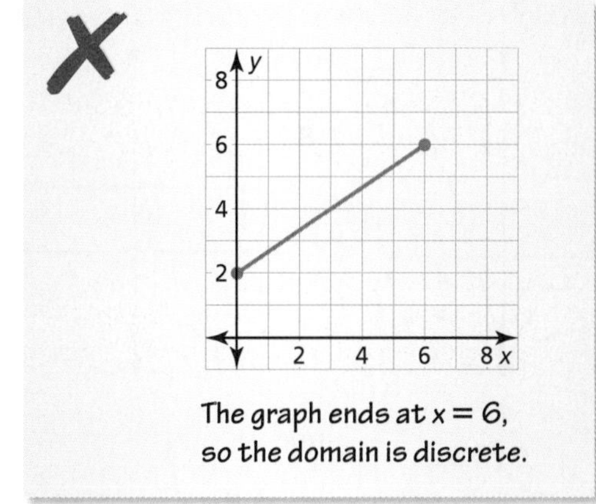
The graph ends at x = 6,
so the domain is discrete.

35. MODELING WITH MATHEMATICS The linear function $m = 55 - 8.5b$ represents the amount m (in dollars) of money that you have after buying b books. *(See Example 4.)*

a. Find the domain of the function. Is the domain discrete or continuous? Explain.

b. Graph the function using its domain.

36. **MODELING WITH MATHEMATICS** The number y of calories burned after x hours of rock climbing is represented by the linear function $y = 650x$.

a. Find the domain of the function. Is the domain discrete or continuous? Explain.

b. Graph the function using its domain.

37. **MODELING WITH MATHEMATICS** You are researching the speed of sound waves in dry air at 86°F. The table shows the distances d (in miles) sound waves travel in t seconds. *(See Example 5.)*

Time (seconds), t	Distance (miles), d
2	0.434
4	0.868
6	1.302
8	1.736
10	2.170

a. Does this situation represent a linear function? Explain.

b. Find the domain of the function. Is the domain discrete or continuous? Explain.

c. Graph the function using its domain.

38. **MODELING WITH MATHEMATICS** The function $y = 30 + 5x$ represents the cost y (in dollars) of having your dog groomed and buying x extra services.

a. Does this situation represent a linear function? Explain.

b. Find the domain of the function. Is the domain discrete or continuous? Explain.

c. Graph the function using its domain.

WRITING In Exercises 39–42, write a real-life problem to fit the data shown in the graph. Determine whether the domain of the function is *discrete* or *continuous*. Explain. *(See Example 6.)*

39.

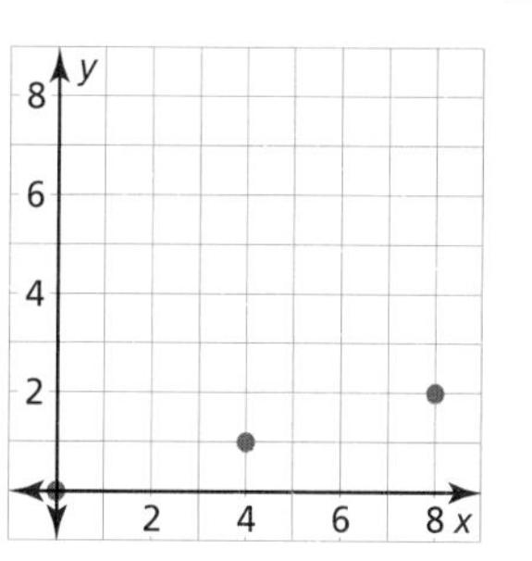

40.

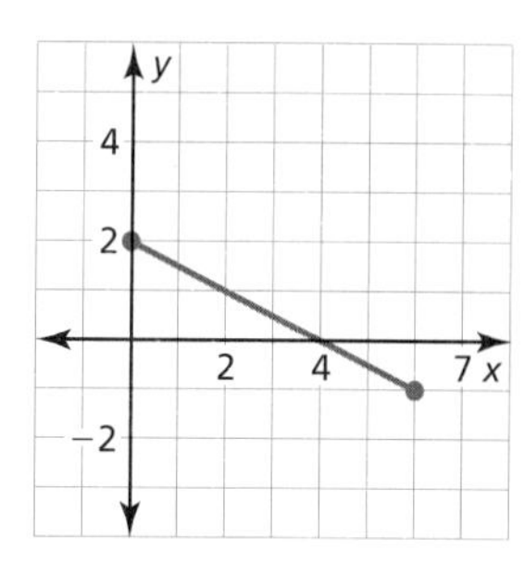

41.

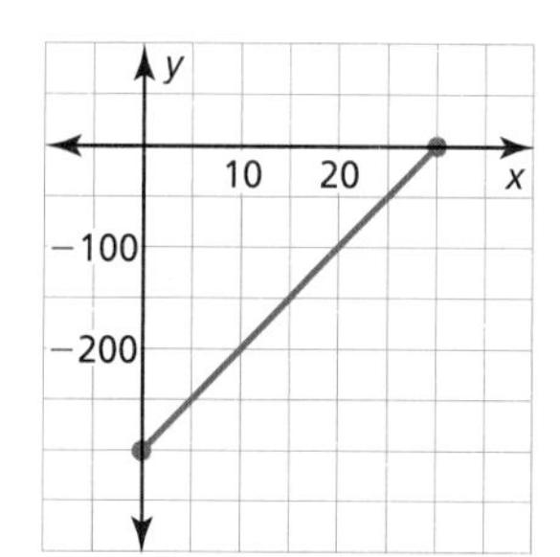

42.

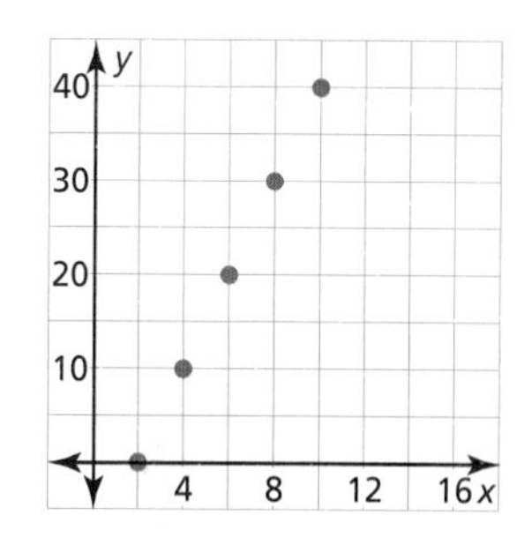

43. **USING STRUCTURE** The table shows your earnings y (in dollars) for working x hours.

a. What is the missing y-value that makes the table represent a linear function?

b. What is your hourly pay rate?

Time (hours), x	Earnings (dollars), y
4	40.80
5	
6	61.20
7	71.40

44. **MAKING AN ARGUMENT** The linear function $d = 50t$ represents the distance d (in miles) Car A is from a car rental store after t hours. The table shows the distances Car B is from the rental store.

Time (hours), t	Distance (miles), d
1	60
3	180
5	310

a. Does the table represent a linear or nonlinear function? Explain.

b. Your friend claims Car B is moving at a faster rate. Is your friend correct? Explain.

MATHEMATICAL CONNECTIONS **In Exercises 45–48, tell whether the volume of the solid is a linear or nonlinear function of the missing dimension(s). Explain.**

45.

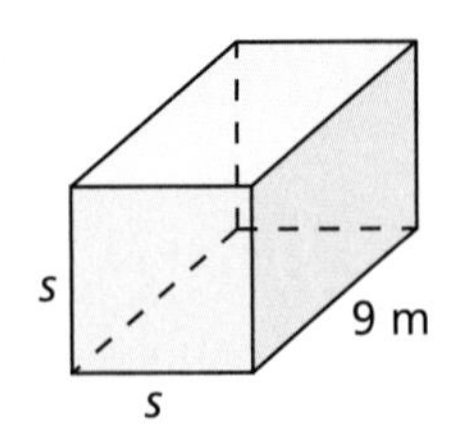

46.

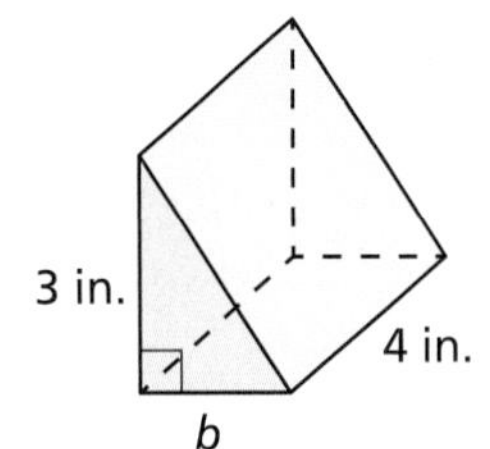

47.

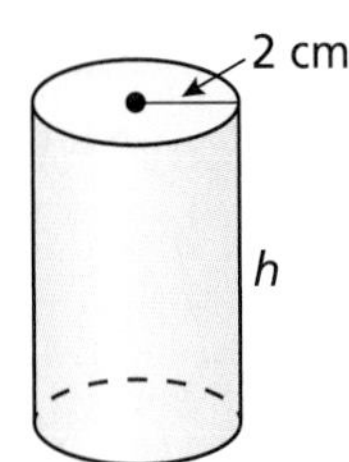

48.

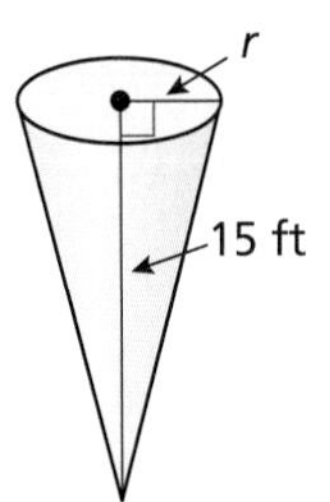

49. REASONING A water company fills two different-sized jugs. The first jug can hold x gallons of water. The second jug can hold y gallons of water. The company fills A jugs of the first size and B jugs of the second size. What does each expression represent? Does each expression represent a set of discrete or continuous values?

a. $x + y$

b. $A + B$

c. Ax

d. $Ax + By$

50. THOUGHT PROVOKING You go to a farmer's market to buy tomatoes. Graph a function that represents the cost of buying tomatoes. Explain your reasoning.

51. CLASSIFYING A FUNCTION Is the function represented by the ordered pairs linear or nonlinear? Explain your reasoning.

$$(0, 2), (3, 14), (5, 22), (9, 38), (11, 46)$$

52. HOW DO YOU SEE IT? You and your friend go running. The graph shows the distances you and your friend run.

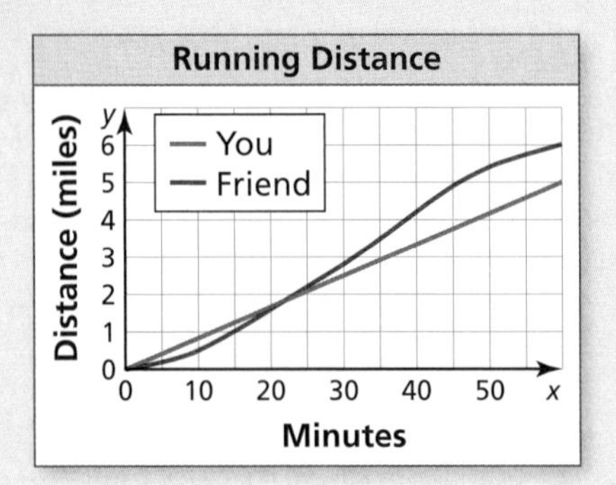

a. Describe your run and your friend's run. Who runs at a constant rate? How do you know? Why might a person not run at a constant rate?

b. Find the domain of each function. Describe the domains using the context of the problem.

WRITING **In Exercises 53 and 54, describe a real-life situation for the constraints.**

53. The function has at least one negative number in the domain. The domain is continuous.

54. The function gives at least one negative number as an output. The domain is discrete.

Maintaining Mathematical Proficiency

Reviewing what you learned in previous grades and lessons

Tell whether x and y show direct variation. Explain your reasoning. *(Skills Review Handbook)*

55.

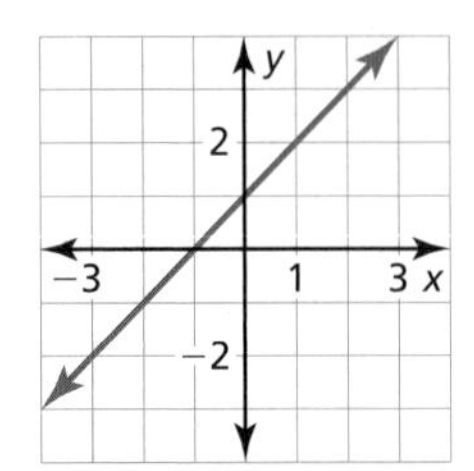

56.

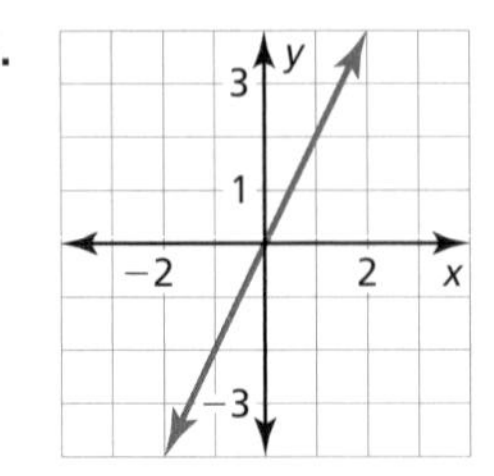

57.

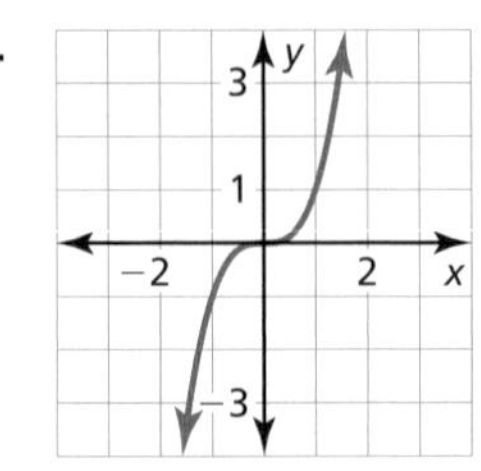

Evaluate the expression when $x = 2$. *(Skills Review Handbook)*

58. $6x + 8$

59. $10 - 2x + 8$

60. $4(x + 2 - 5x)$

61. $\frac{x}{2} + 5x - 7$

3.3 Function Notation

Essential Question

Essential Question How can you use function notation to represent a function?

The notation $f(x)$, called **function notation**, is another name for y. This notation is read as "the value of f at x" or "f of x." The parentheses do not imply multiplication. You can use letters other than f to name a function. The letters g, h, j, and k are often used to name functions.

EXPLORATION 1 Matching Functions with Their Graphs

Work with a partner. Match each function with its graph.

a. $f(x) = 2x - 3$ **b.** $g(x) = -x + 2$

c. $h(x) = x^2 - 1$ **d.** $j(x) = 2x^2 - 3$

ATTENDING TO PRECISION

To be proficient in math, you need to use clear definitions and state the meanings of the symbols you use.

A.

4
−6 6
−4

B.

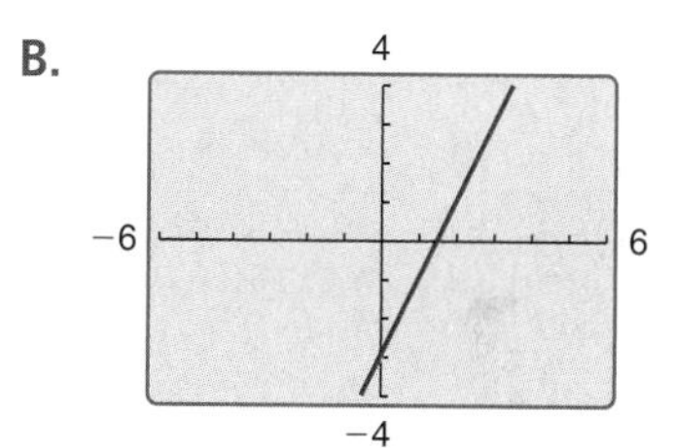

C.

4
−6 6
−4

D.

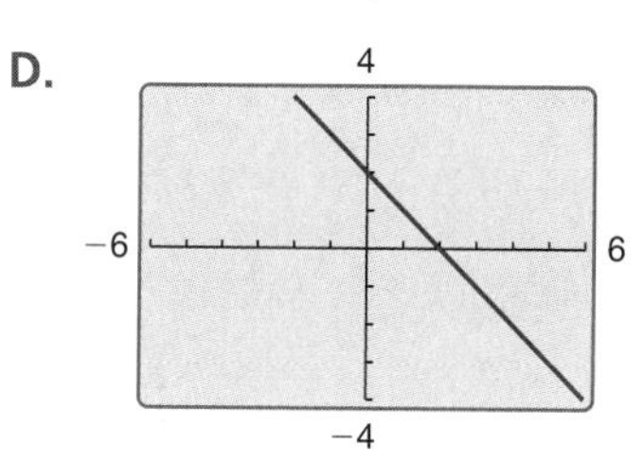

EXPLORATION 2 Evaluating a Function

Work with a partner. Consider the function

$f(x) = -x + 3.$

Locate the points $(x, f(x))$ on the graph. Explain how you found each point.

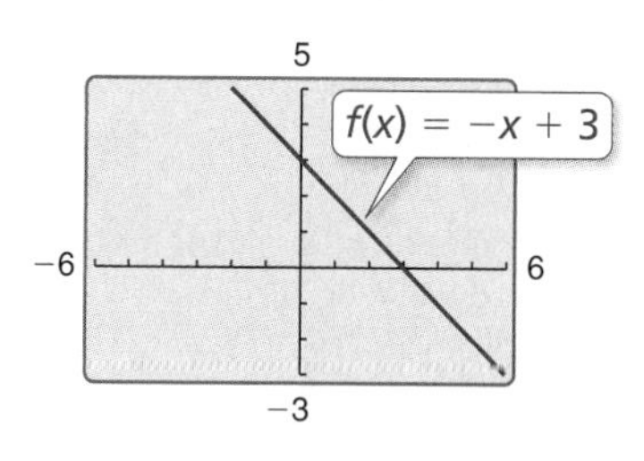

a. $(-1, f(-1))$

b. $(0, f(0))$

c. $(1, f(1))$

d. $(2, f(2))$

Communicate Your Answer

3. How can you use function notation to represent a function? How are standard notation and function notation similar? How are they different?

Standard Notation	*Function Notation*
$y = 2x + 5$	$f(x) = 2x + 5$

3.3 Lesson

Core Vocabulary

function notation, *p. 122*

Previous
linear function
quadrant

What You Will Learn

- Use function notation to evaluate and interpret functions.
- Use function notation to solve and graph functions.
- Solve real-life problems using function notation.

Using Function Notation to Evaluate and Interpret

You know that a linear function can be written in the form $y = mx + b$. By naming a linear function f, you can also write the function using **function notation**.

$f(x) = mx + b$ Function notation

The notation $f(x)$ is another name for y. If f is a function, and x is in its domain, then $f(x)$ represents the output of f corresponding to the input x. You can use letters other than f to name a function, such as g or h.

READING

The notation $f(x)$ is read as "the value of f at x" or "f of x." It does not mean "f times x."

EXAMPLE 1 Evaluating a Function

Evaluate $f(x) = -4x + 7$ when $x = 2$ and $x = -2$.

SOLUTION

$f(x) = -4x + 7$	Write the function.	$f(x) = -4x + 7$
$f(2) = -4(2) + 7$	Substitute for x.	$f(-2) = -4(-2) + 7$
$= -8 + 7$	Multiply.	$= 8 + 7$
$= -1$	Add.	$= 15$

▶ When $x = 2, f(x) = -1$, and when $x = -2, f(x) = 15$.

EXAMPLE 2 Interpreting Function Notation

Let $f(t)$ be the outside temperature (°F) t hours after 6 A.M. Explain the meaning of each statement.

a. $f(0) = 58$ **b.** $f(6) = n$ **c.** $f(3) < f(9)$

SOLUTION

a. The initial value of the function is 58. So, the temperature at 6 A.M. is 58°F.

b. The output of f when $t = 6$ is n. So, the temperature at noon (6 hours after 6 A.M.) is n°F.

c. The output of f when $t = 3$ is less than the output of f when $t = 9$. So, the temperature at 9 A.M. (3 hours after 6 A.M.) is less than the temperature at 3 P.M. (9 hours after 6 A.M.).

Monitoring Progress  Help in English and Spanish at *BigIdeasMath.com*

Evaluate the function when $x = -4$, 0, and 3.

1. $f(x) = 2x - 5$ **2.** $g(x) = -x - 1$

3. WHAT IF? In Example 2, let $f(t)$ be the outside temperature (°F) t hours after 9 A.M. Explain the meaning of each statement.

a. $f(4) = 75$ **b.** $f(m) = 70$ **c.** $f(2) = f(9)$ **d.** $f(6) > f(0)$

Using Function Notation to Solve and Graph

EXAMPLE 3 Solving for the Independent Variable

For $h(x) = \frac{2}{3}x - 5$, find the value of x for which $h(x) = -7$.

SOLUTION

$h(x) = \frac{2}{3}x - 5$	Write the function.
$-7 = \frac{2}{3}x - 5$	Substitute −7 for $h(x)$.
$+5 \qquad +5$	Add 5 to each side.
$-2 = \frac{2}{3}x$	Simplify.
$\frac{3}{2} \cdot (-2) = \frac{3}{2} \cdot \frac{2}{3}x$	Multiply each side by $\frac{3}{2}$.
$-3 = x$	Simplify.

▶ When $x = -3$, $h(x) = -7$.

EXAMPLE 4 Graphing a Linear Function in Function Notation

Graph $f(x) = 2x + 5$.

SOLUTION

Step 1 Make an input-output table to find ordered pairs.

x	−2	−1	0	1	2
$f(x)$	1	3	5	7	9

Step 2 Plot the ordered pairs.

Step 3 Draw a line through the points.

STUDY TIP

The graph of $y = f(x)$ consists of the points $(x, f(x))$.

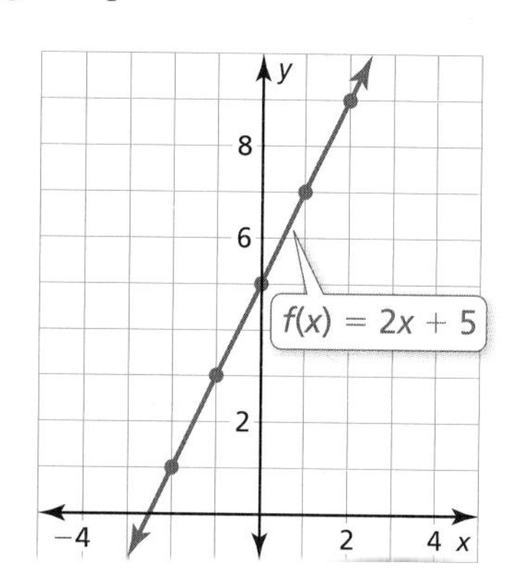

Monitoring Progress  Help in English and Spanish at *BigIdeasMath.com*

Find the value of x so that the function has the given value.

4. $f(x) = 6x + 9$; $f(x) = 21$

5. $g(x) = -\frac{1}{2}x + 3$; $g(x) = -1$

Graph the linear function.

6. $f(x) = 3x - 2$

7. $g(x) = -x + 4$

8. $h(x) = -\frac{3}{4}x - 1$

Solving Real-Life Problems

EXAMPLE 5 Modeling with Mathematics

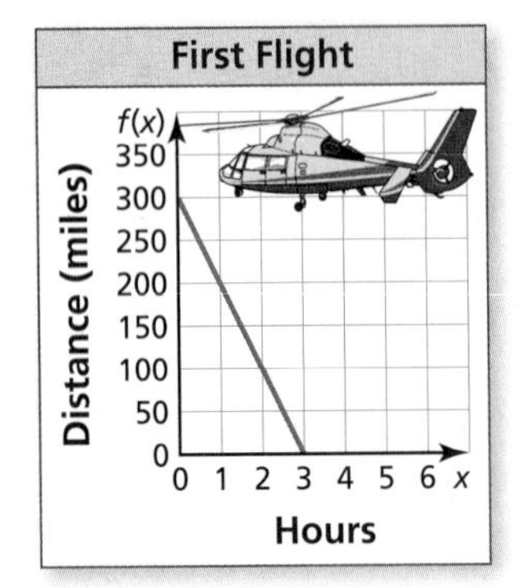

The graph shows the number of miles a helicopter is from its destination after x hours on its first flight. On its second flight, the helicopter travels 50 miles farther and increases its speed by 25 miles per hour. The function $f(x) = 350 - 125x$ represents the second flight, where $f(x)$ is the number of miles the helicopter is from its destination after x hours. Which flight takes less time? Explain.

SOLUTION

1. **Understand the Problem** You are given a graph of the first flight and an equation of the second flight. You are asked to compare the flight times to determine which flight takes less time.

2. **Make a Plan** Graph the function that represents the second flight. Compare the graph to the graph of the first flight. The x-value that corresponds to $f(x) = 0$ represents the flight time.

3. **Solve the Problem** Graph $f(x) = 350 - 125x$.

   Step 1 Make an input-output table to find the ordered pairs.

x	0	1	2	3
$f(x)$	350	225	100	−25

   Step 2 Plot the ordered pairs.

   Step 3 Draw a line through the points. Note that the function only makes sense when x and $f(x)$ are positive. So, only draw the line in the first quadrant.

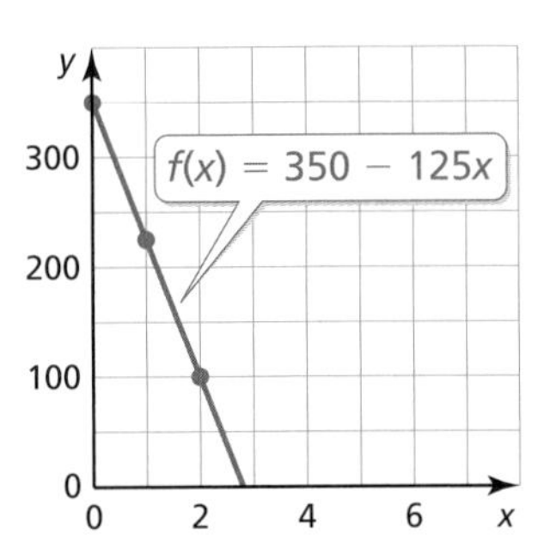

▶ From the graph of the first flight, you can see that when $f(x) = 0$, $x = 3$. From the graph of the second flight, you can see that when $f(x) = 0$, x is slightly less than 3. So, the second flight takes less time.

4. **Look Back** You can check that your answer is correct by finding the value of x for which $f(x) = 0$.

$f(x) = 350 - 125x$	Write the function.
$0 = 350 - 125x$	Substitute 0 for $f(x)$.
$-350 = -125x$	Subtract 350 from each side.
$2.8 = x$	Divide each side by −125.

So, the second flight takes 2.8 hours, which is less than 3.

Monitoring Progress  Help in English and Spanish at *BigIdeasMath.com*

9. **WHAT IF?** Let $f(x) = 250 - 75x$ represent the second flight, where $f(x)$ is the number of miles the helicopter is from its destination after x hours. Which flight takes less time? Explain.

3.3 Exercises

Dynamic Solutions available at *BigIdeasMath.com*

Vocabulary and Core Concept Check

1. **COMPLETE THE SENTENCE** When you write the function $y = 2x + 10$ as $f(x) = 2x + 10$, you are using ______________.

2. **REASONING** Your height can be represented by a function h, where the input is your age. What does $h(14)$ represent?

Monitoring Progress and Modeling with Mathematics

In Exercises 3–10, evaluate the function when $x = -2$, 0, and 5. *(See Example 1.)*

3. $f(x) = x + 6$
4. $g(x) = 3x$
5. $h(x) = -2x + 9$
6. $r(x) = -x - 7$
7. $p(x) = -3 + 4x$
8. $b(x) = 18 - 0.5x$
9. $v(x) = 12 - 2x - 5$
10. $n(x) = -1 - x + 4$

11. **INTERPRETING FUNCTION NOTATION** Let $c(t)$ be the number of customers in a restaurant t hours after 8 A.M. Explain the meaning of each statement. *(See Example 2.)*

 a. $c(0) = 0$ b. $c(3) = c(8)$

 c. $c(n) = 29$ d. $c(13) < c(12)$

12. **INTERPRETING FUNCTION NOTATION** Let $H(x)$ be the percent of U.S. households with Internet use x years after 1980. Explain the meaning of each statement.

 a. $H(23) = 55$ b. $H(4) = k$

 c. $H(27) \geq 61$

 d. $H(17) + H(21) \approx H(29)$

In Exercises 13–18, find the value of x so that the function has the given value. *(See Example 3.)*

13. $h(x) = -7x$; $h(x) = 63$

14. $t(x) = 3x$; $t(x) = 24$

15. $m(x) = 4x + 15$; $m(x) = 7$

16. $k(x) = 6x - 12$; $k(x) = 18$

17. $q(x) = \frac{1}{2}x - 3$; $q(x) = -4$

18. $j(x) = -\frac{4}{5}x + 7$; $j(x) = -5$

In Exercises 19 and 20, find the value of x so that $f(x) = 7$.

19.

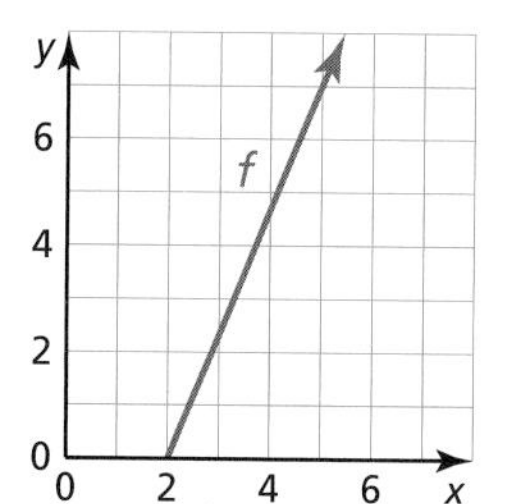

20.

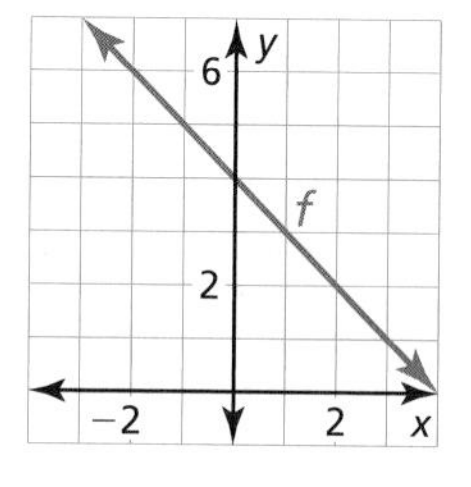

21. **MODELING WITH MATHEMATICS** The function $C(x) = 17.5x - 10$ represents the cost (in dollars) of buying x tickets to the orchestra with a \$10 coupon.

 a. How much does it cost to buy five tickets?

 b. How many tickets can you buy with \$130?

22. **MODELING WITH MATHEMATICS** The function $d(t) = 300{,}000t$ represents the distance (in kilometers) that light travels in t seconds.

 a. How far does light travel in 15 seconds?

 b. How long does it take light to travel 12 million kilometers?

In Exercises 23–28, graph the linear function. *(See Example 4.)*

23. $p(x) = 4x$
24. $h(x) = -5$
25. $d(x) = -\frac{1}{2}x - 3$
26. $w(x) = \frac{3}{5}x + 2$
27. $g(x) = -4 + 7x$
28. $f(x) = 3 - 6x$

29. **PROBLEM SOLVING** The graph shows the percent p (in decimal form) of battery power remaining in a laptop computer after t hours of use. A tablet computer initially has 75% of its battery power remaining and loses 12.5% per hour. Which computer's battery will last longer? Explain. *(See Example 5.)*

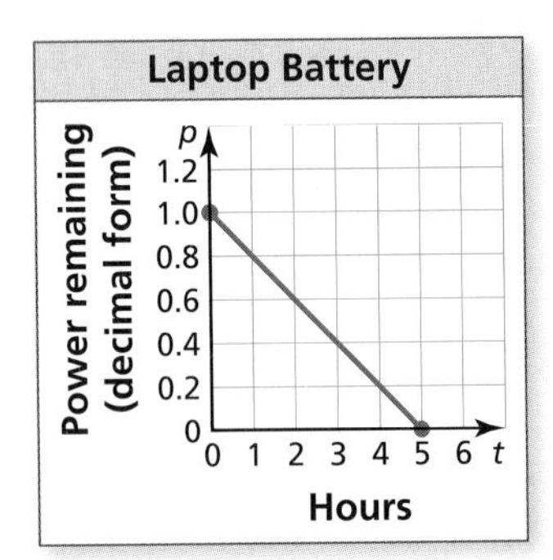

30. **PROBLEM SOLVING** The function $C(x) = 25x + 50$ represents the labor cost (in dollars) for Certified Remodeling to build a deck, where x is the number of hours of labor. The table shows sample labor costs from its main competitor, Master Remodeling. The deck is estimated to take 8 hours of labor. Which company would you hire? Explain.

Hours	Cost
2	\$130
4	\$160
6	\$190

31. **MAKING AN ARGUMENT** Let $P(x)$ be the number of people in the U.S. who own a cell phone x years after 1990. Your friend says that $P(x + 1) > P(x)$ for any x because $x + 1$ is always greater than x. Is your friend correct? Explain.

32. **THOUGHT PROVOKING** Let $B(t)$ be your bank account balance after t days. Describe a situation in which $B(0) < B(4) < B(2)$.

33. **MATHEMATICAL CONNECTIONS** Rewrite each geometry formula using function notation. Evaluate each function when $r = 5$ feet. Then explain the meaning of the result.

a. Diameter, $d = 2r$

b. Area, $A = \pi r^2$

c. Circumference, $C = 2\pi r$

34. **HOW DO YOU SEE IT?** The function $y = A(x)$ represents the attendance at a high school x weeks after a flu outbreak. The graph of the function is shown.

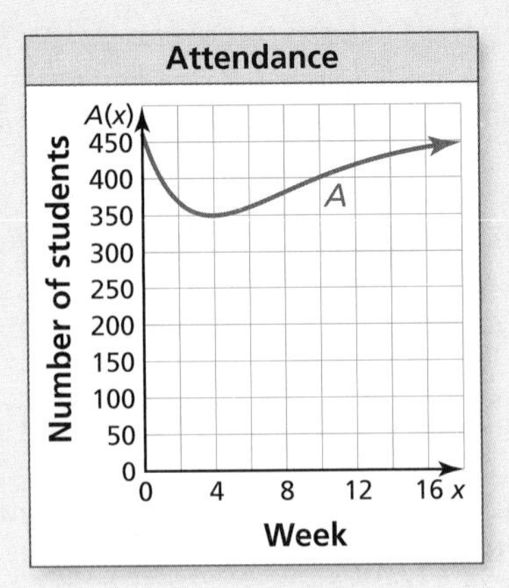

a. What happens to the school's attendance after the flu outbreak?

b. Estimate $A(13)$ and explain its meaning.

c. Use the graph to estimate the solution(s) of the equation $A(x) = 400$. Explain the meaning of the solution(s).

d. What was the least attendance? When did that occur?

e. How many students do you think are enrolled at this high school? Explain your reasoning.

35. **INTERPRETING FUNCTION NOTATION** Let f be a function. Use each statement to find the coordinates of a point on the graph of f.

a. $f(5)$ is equal to 9.

b. A solution of the equation $f(n) = -3$ is 5.

36. **REASONING** Given a function f, tell whether the statement

$$f(a + b) = f(a) + f(b)$$

is true or false for all inputs a and b. If it is false, explain why.

Maintaining Mathematical Proficiency

Reviewing what you learned in previous grades and lessons

Solve the inequality. Graph the solution. *(Section 2.5)*

37. $-2 \leq x - 11 \leq 6$

38. $5a < -35$ *or* $a - 14 > 1$

39. $-16 < 6k + 2 < 0$

40. $2d + 7 < -9$ *or* $4d - 1 > -3$

41. $5 \leq 3y + 8 < 17$

42. $4v + 9 \leq 5$ *or* $-3v \geq -6$

3.1–3.3 What Did You Learn?

Core Vocabulary

Core Concepts

Section 3.1

Section 3.2

Section 3.3

Mathematical Practices

1. How can you use technology to confirm your answers in Exercises 40–43 on page 110?
2. How can you use patterns to solve Exercise 43 on page 119?
3. How can you make sense of the quantities in the function in Exercise 21 on page 125?

Study Skills

Staying Focused during Class

As soon as class starts, quickly review your notes from the previous class and start thinking about math.

Repeat what you are writing in your head.

When a particular topic is difficult, ask for another example.

3.1–3.3 Quiz

Determine whether the relation is a function. Explain. *(Section 3.1)*

1.

Input, *x*	−1	0	1	2	3
Output, *y*	0	1	4	4	8

2. (−10, 2), (−8, 3), (−6, 5), (−8, 8), (−10, 6)

Find the domain and range of the function represented by the graph. *(Section 3.1)*

3.

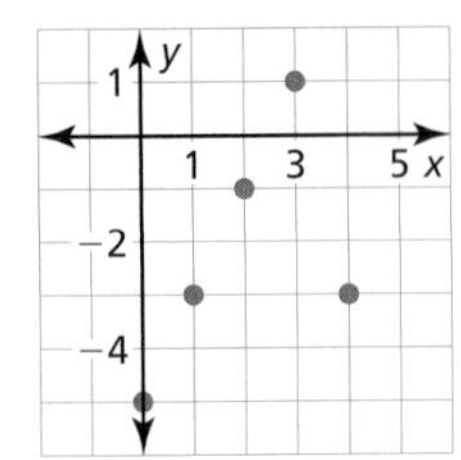

4.

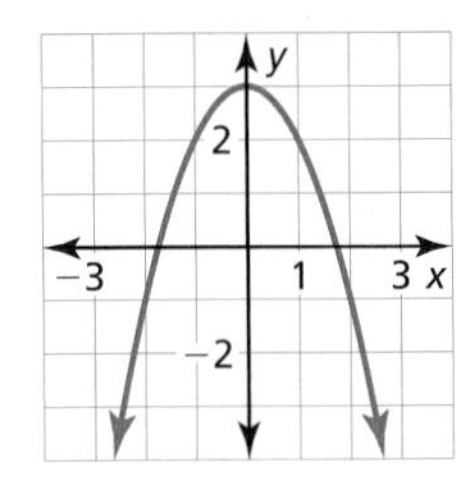

5.

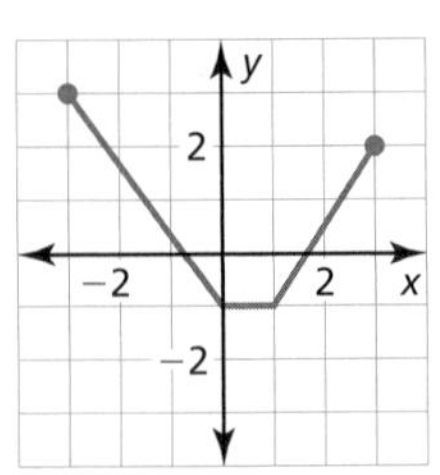

Determine whether the graph, table, or equation represents a *linear* or *nonlinear* function. Explain. *(Section 3.2)*

6.

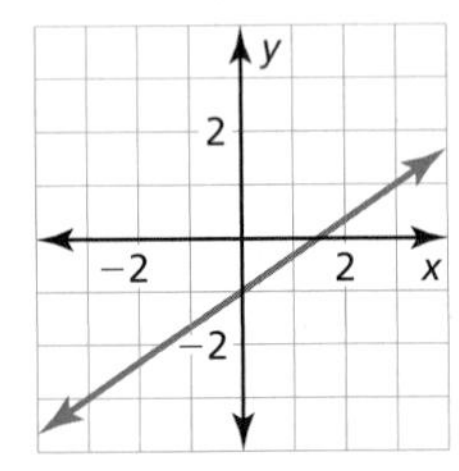

7.

x	*y*
−5	3
0	7
5	10

8. $y = x(2 - x)$

Determine whether the domain is *discrete* or *continuous*. Explain. *(Section 3.2)*

9.

Depth (feet), *x*	33	66	99
Pressure (ATM), *y*	2	3	4

10.

Hats, *x*	2	3	4
Cost (dollars), *y*	36	54	72

11. For $w(x) = -2x + 7$, find the value of x for which $w(x) = -3$. *(Section 3.3)*

Graph the linear function. *(Section 3.3)*

12. $g(x) = x + 3$

13. $p(x) = -3x - 1$

14. $m(x) = \frac{2}{3}x$

15. The function $m = 30 - 3r$ represents the amount m (in dollars) of money you have after renting r video games. *(Section 3.1 and Section 3.2)*

a. Identify the independent and dependent variables.

b. Find the domain and range of the function. Is the domain discrete or continuous? Explain.

c. Graph the function using its domain.

16. The function $d(x) = 1375 - 110x$ represents the distance (in miles) a high-speed train is from its destination after x hours. *(Section 3.3)*

a. How far is the train from its destination after 8 hours?

b. How long does the train travel before reaching its destination?

3.4 Graphing Linear Equations in Standard Form

Essential Question

How can you describe the graph of the equation $Ax + By = C$?

EXPLORATION 1 Using a Table to Plot Points

Work with a partner. You sold a total of $16 worth of tickets to a fundraiser. You lost track of how many of each type of ticket you sold. Adult tickets are $4 each. Child tickets are $2 each.

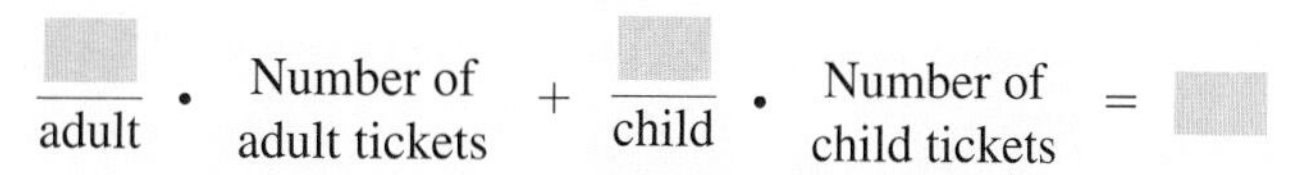

a. Let x represent the number of adult tickets. Let y represent the number of child tickets. Use the verbal model to write an equation that relates x and y.

b. Copy and complete the table to show the different combinations of tickets you might have sold.

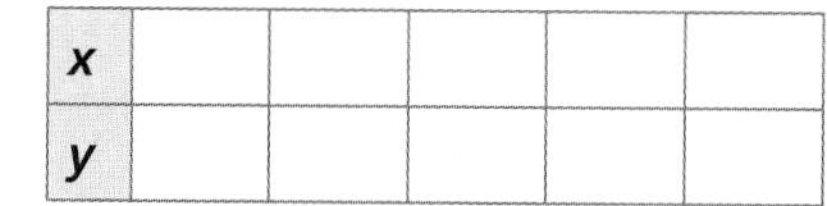

x					
y					

c. Plot the points from the table. Describe the pattern formed by the points.

d. If you remember how many adult tickets you sold, can you determine how many child tickets you sold? Explain your reasoning.

FINDING AN ENTRY POINT

To be proficient in math, you need to find an entry point into the solution of a problem. Determining what information you know, and what you can do with that information, can help you find an entry point.

EXPLORATION 2 Rewriting and Graphing an Equation

Work with a partner. You sold a total of $48 worth of cheese. You forgot how many pounds of each type of cheese you sold. Swiss cheese costs $8 per pound. Cheddar cheese costs $6 per pound.

$$\frac{\square}{\text{pound}} \cdot \text{Pounds of Swiss} + \frac{\square}{\text{pound}} \cdot \text{Pounds of cheddar} = \square$$

a. Let x represent the number of pounds of Swiss cheese. Let y represent the number of pounds of cheddar cheese. Use the verbal model to write an equation that relates x and y.

b. Solve the equation for y. Then use a graphing calculator to graph the equation. Given the real-life context of the problem, find the domain and range of the function.

c. The ***x*-intercept** of a graph is the x-coordinate of a point where the graph crosses the x-axis. The ***y*-intercept** of a graph is the y-coordinate of a point where the graph crosses the y-axis. Use the graph to determine the x- and y-intercepts.

d. How could you use the equation you found in part (a) to determine the x- and y-intercepts? Explain your reasoning.

e. Explain the meaning of the intercepts in the context of the problem.

Communicate Your Answer

3. How can you describe the graph of the equation $Ax + By = C$?

4. Write a real-life problem that is similar to those shown in Explorations 1 and 2.

3.4 Lesson

Core Vocabulary

standard form, *p. 130*
x-intercept, *p. 131*
y-intercept, *p. 131*

Previous
ordered pair
quadrant

What You Will Learn

- Graph equations of horizontal and vertical lines.
- Graph linear equations in standard form using intercepts.
- Use linear equations in standard form to solve real-life problems.

Horizontal and Vertical Lines

The **standard form** of a linear equation is $Ax + By = C$, where A, B, and C are real numbers and A and B are not both zero.

Consider what happens when $A = 0$ or when $B = 0$. When $A = 0$, the equation becomes $By = C$, or $y = \frac{C}{B}$. Because $\frac{C}{B}$ is a constant, you can write $y = b$. Similarly, when $B = 0$, the equation becomes $Ax = C$, or $x = \frac{C}{A}$, and you can write $x = a$.

Core Concept

Horizontal and Vertical Lines

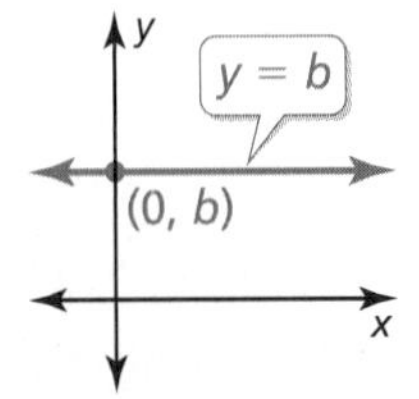

The graph of $y = b$ is a horizontal line. The line passes through the point $(0, b)$.

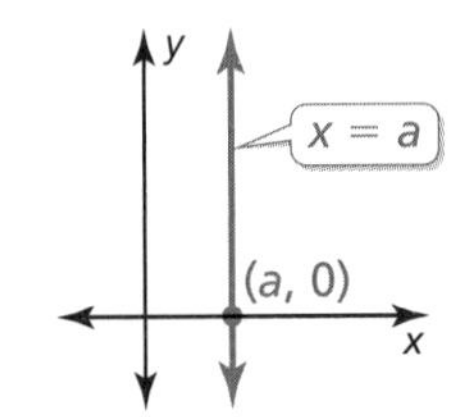

The graph of $x = a$ is a vertical line. The line passes through the point $(a, 0)$.

EXAMPLE 1 Horizontal and Vertical Lines

Graph (a) $y = 4$ and (b) $x = -2$.

SOLUTION

a. For every value of x, the value of y is 4. The graph of the equation $y = 4$ is a horizontal line 4 units above the x-axis.

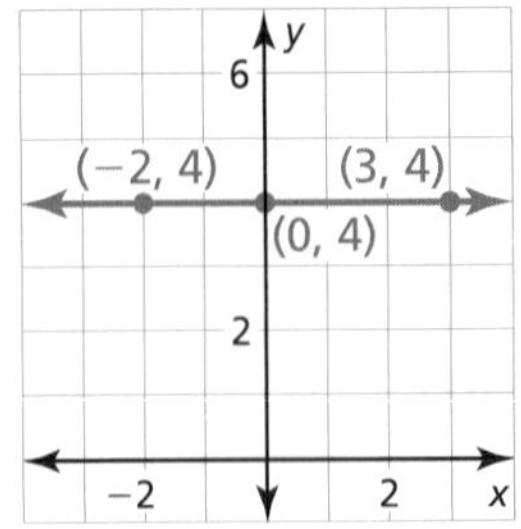

b. For every value of y, the value of x is -2. The graph of the equation $x = -2$ is a vertical line 2 units to the left of the y-axis.

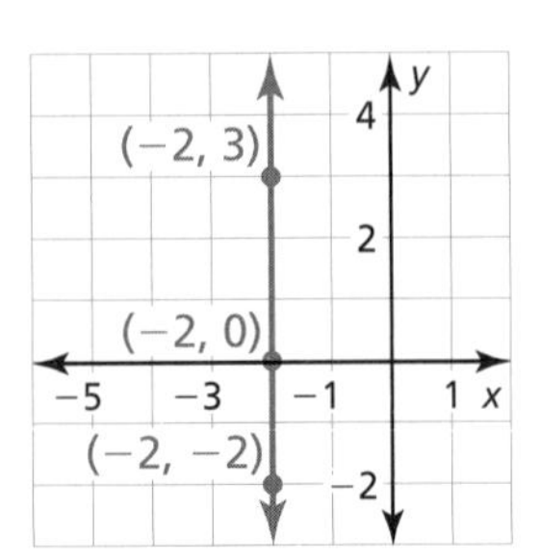

STUDY TIP

For every value of x, the ordered pair $(x, 4)$ is a solution of $y = 4$.

Monitoring Progress

Help in English and Spanish at *BigIdeasMath.com*

Graph the linear equation.

1. $y = -2.5$

2. $x = 5$

Using Intercepts to Graph Linear Equations

You can use the fact that two points determine a line to graph a linear equation. Two convenient points are the points where the graph crosses the axes.

Core Concept

Using Intercepts to Graph Equations

The **x-intercept** of a graph is the x-coordinate of a point where the graph crosses the x-axis. It occurs when $y = 0$.

The **y-intercept** of a graph is the y-coordinate of a point where the graph crosses the y-axis. It occurs when $x = 0$.

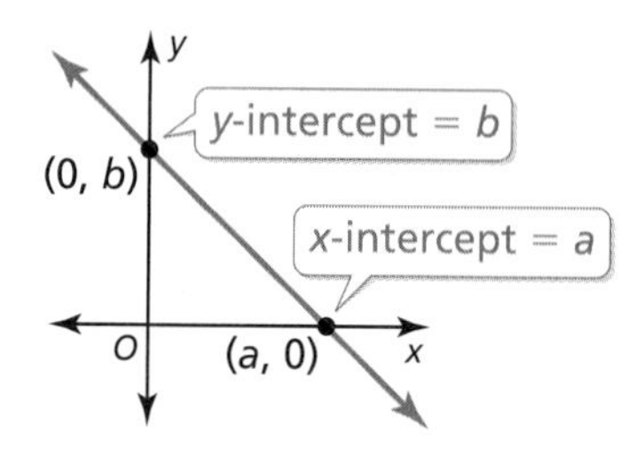

To graph the linear equation $Ax + By = C$, find the intercepts and draw the line that passes through the two intercepts.

- To find the x-intercept, let $y = 0$ and solve for x.
- To find the y-intercept, let $x = 0$ and solve for y.

EXAMPLE 2 Using Intercepts to Graph a Linear Equation

Use intercepts to graph the equation $3x + 4y = 12$.

SOLUTION

Step 1 Find the intercepts.

To find the x-intercept, substitute 0 for y and solve for x.

$3x + 4y = 12$ Write the original equation.

$3x + 4(0) = 12$ Substitute 0 for y.

$x = 4$ Solve for x.

To find the y-intercept, substitute 0 for x and solve for y.

$3x + 4y = 12$ Write the original equation.

$3(0) + 4y = 12$ Substitute 0 for x.

$y = 3$ Solve for y.

STUDY TIP

As a check, you can find a third solution of the equation and verify that the corresponding point is on the graph. To find a third solution, substitute any value for one of the variables and solve for the other variable.

Step 2 Plot the points and draw the line.

The x-intercept is 4, so plot the point (4, 0). The y-intercept is 3, so plot the point (0, 3). Draw a line through the points.

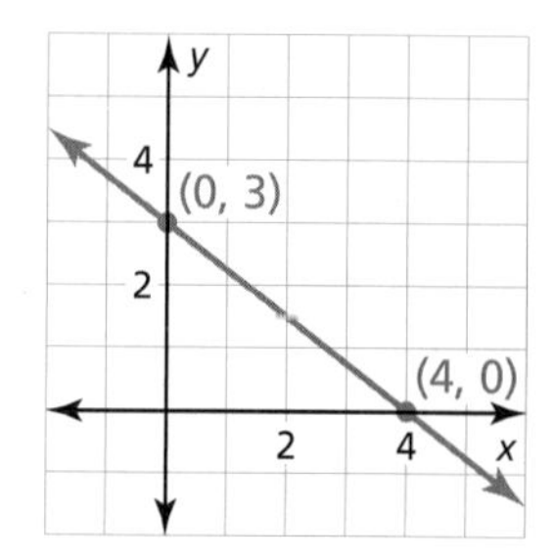

Monitoring Progress  Help in English and Spanish at *BigIdeasMath.com*

Use intercepts to graph the linear equation. Label the points corresponding to the intercepts.

3. $2x - y = 4$

4. $x + 3y = -9$

Solving Real-Life Problems

EXAMPLE 3 Modeling with Mathematics

You are planning an awards banquet for your school. You need to rent tables to seat 180 people. Tables come in two sizes. Small tables seat 6 people, and large tables seat 10 people. The equation $6x + 10y = 180$ models this situation, where x is the number of small tables and y is the number of large tables.

a. Graph the equation. Interpret the intercepts.

b. Find four possible solutions in the context of the problem.

SOLUTION

1. **Understand the Problem** You know the equation that models the situation. You are asked to graph the equation, interpret the intercepts, and find four solutions.

2. **Make a Plan** Use intercepts to graph the equation. Then use the graph to interpret the intercepts and find other solutions.

3. **Solve the Problem**

   a. Use intercepts to graph the equation. Neither x nor y can be negative, so only graph the equation in the first quadrant.

STUDY TIP

Although x and y represent whole numbers, it is convenient to draw a line segment that includes points whose coordinates are not whole numbers.

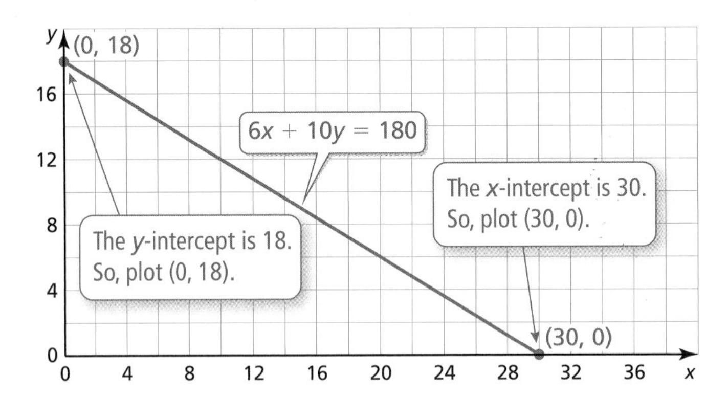

▶ The x-intercept shows that you can rent 30 small tables when you do not rent any large tables. The y-intercept shows that you can rent 18 large tables when you do not rent any small tables.

   b. Only whole-number values of x and y make sense in the context of the problem. Besides the intercepts, it appears that the line passes through the points (10, 12) and (20, 6). To verify that these points are solutions, check them in the equation, as shown.

Check

$$6x + 10y = 180$$
$$6(10) + 10(12) \stackrel{?}{=} 180$$
$$180 = 180 \checkmark$$

$$6x + 10y = 180$$
$$6(20) + 10(6) \stackrel{?}{=} 180$$
$$180 = 180 \checkmark$$

▶ So, four possible combinations of tables that will seat 180 people are 0 small and 18 large, 10 small and 12 large, 20 small and 6 large, and 30 small and 0 large.

4. **Look Back** The graph shows that as the number x of small tables increases, the number y of large tables decreases. This makes sense in the context of the problem. So, the graph is reasonable.

Monitoring Progress 

Help in English and Spanish at *BigIdeasMath.com*

5. **WHAT IF?** You decide to rent tables from a different company. The situation can be modeled by the equation $4x + 6y = 180$, where x is the number of small tables and y is the number of large tables. Graph the equation and interpret the intercepts.

3.4 Exercises

Dynamic Solutions available at *BigIdeasMath.com*

Vocabulary and Core Concept Check

1. **WRITING** How are x-intercepts and y-intercepts alike? How are they different?

2. **WHICH ONE DOESN'T BELONG?** Which point does not belong with the other three? Explain your reasoning.

(0, −3) (0, 0) (4, −3) (4, 0)

Monitoring Progress and Modeling with Mathematics

In Exercises 3–6, graph the linear equation. *(See Example 1.)*

3. $x = 4$
4. $y = 2$
5. $y = -3$
6. $x = -1$

In Exercises 7–12, find the x- and y-intercepts of the graph of the linear equation.

7. $2x + 3y = 12$
8. $3x + 6y = 24$
9. $-4x + 8y = -16$
10. $-6x + 9y = -18$
11. $3x - 6y = 2$
12. $-x + 8y = 4$

In Exercises 13–22, use intercepts to graph the linear equation. Label the points corresponding to the intercepts. *(See Example 2.)*

13. $5x + 3y = 30$
14. $4x + 6y = 12$
15. $-12x + 3y = 24$
16. $-2x + 6y = 18$
17. $-4x + 3y = -30$
18. $-2x + 7y = -21$
19. $-x + 2y = 7$
20. $3x - y = -5$
21. $-\frac{5}{2}x + y = 10$
22. $-\frac{1}{2}x + y = -4$

23. **MODELING WITH MATHEMATICS** A football team has an away game, and the bus breaks down. The coaches decide to drive the players to the game in cars and vans. Four players can ride in each car. Six players can ride in each van. There are 48 players on the team. The equation $4x + 6y = 48$ models this situation, where x is the number of cars and y is the number of vans. *(See Example 3.)*
    a. Graph the equation. Interpret the intercepts.
    b. Find four possible solutions in the context of the problem.

24. **MODELING WITH MATHEMATICS** You are ordering shirts for the math club at your school. Short-sleeved shirts cost \$10 each. Long-sleeved shirts cost \$12 each. You have a budget of \$300 for the shirts. The equation $10x + 12y = 300$ models the total cost, where x is the number of short-sleeved shirts and y is the number of long-sleeved shirts.
    a. Graph the equation. Interpret the intercepts.
    b. Twelve students decide they want short-sleeved shirts. How many long-sleeved shirts can you order?

ERROR ANALYSIS In Exercises 25 and 26, describe and correct the error in finding the intercepts of the graph of the equation.

25. ✗

$3x + 12y = 24$	$3x + 12y = 24$
$3x + 12(0) = 24$	$3(0) + 12y = 24$
$3x = 24$	$12y = 24$
$x = 8$	$y = 2$

The intercept is at (8, 2).

26. ✗

$4x + 10y = 20$	$4x + 10y = 20$
$4x + 10(0) = 20$	$4(0) + 10y = 20$
$4x = 20$	$10y = 20$
$x = 5$	$y = 2$

The x-intercept is at (0, 5), and the y-intercept is at (2, 0).

27. **MAKING AN ARGUMENT** You overhear your friend explaining how to find intercepts to a classmate. Your friend says, "When you want to find the x-intercept, just substitute 0 for x and continue to solve the equation." Is your friend's explanation correct? Explain.

28. **ANALYZING RELATIONSHIPS** You lose track of how many 2-point baskets and 3-point baskets a team makes in a basketball game. The team misses all the 1-point baskets and still scores 54 points. The equation $2x + 3y = 54$ models the total points scored, where x is the number of 2-point baskets made and y is the number of 3-point baskets made.

   a. Find and interpret the intercepts.

   b. Can the number of 3-point baskets made be odd? Explain your reasoning.

   c. Graph the equation. Find two more possible solutions in the context of the problem.

MULTIPLE REPRESENTATIONS **In Exercises 29–32, match the equation with its graph.**

29. $5x + 3y = 30$

30. $5x + 3y = -30$

31. $5x - 3y = 30$

32. $5x - 3y = -30$

A.
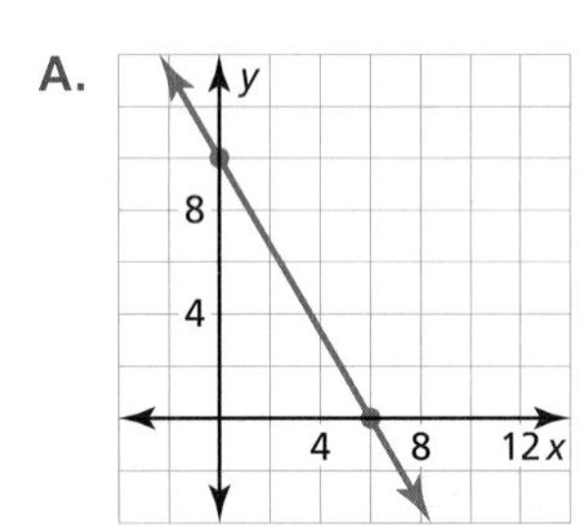

B.
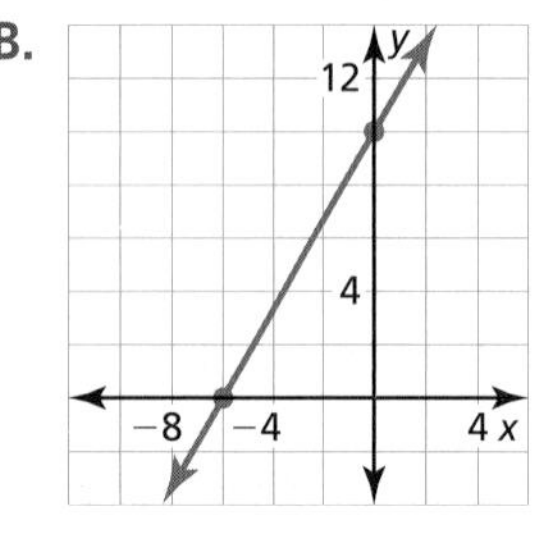

C.
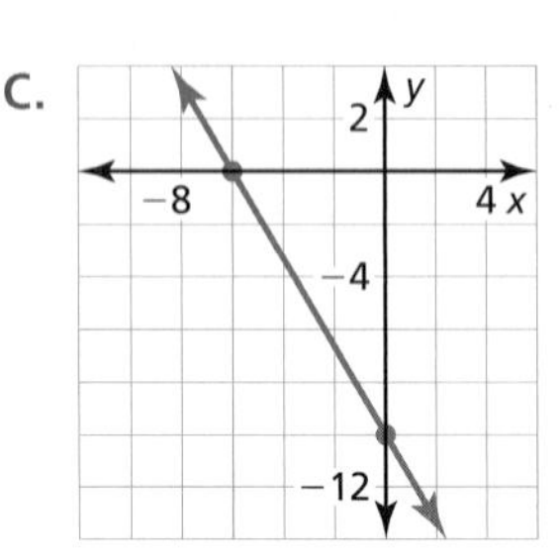

D.
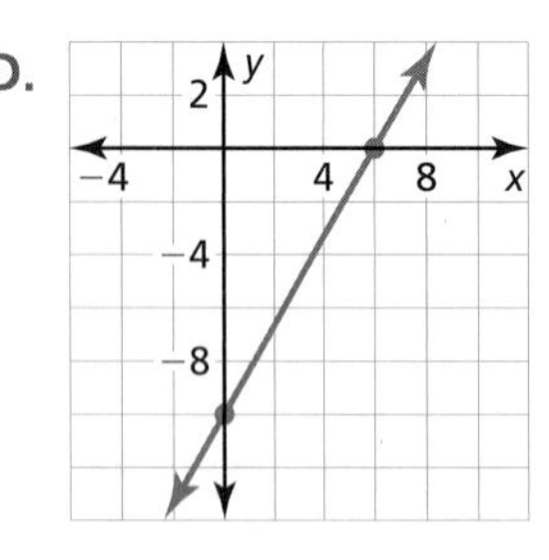

33. **MATHEMATICAL CONNECTIONS** Graph the equations $x = 5$, $x = 2$, $y = -2$, and $y = 1$. What enclosed shape do the lines form? Explain your reasoning.

34. **HOW DO YOU SEE IT?** You are organizing a class trip to an amusement park. The cost to enter the park is \$30. The cost to enter with a meal plan is \$45. You have a budget of \$2700 for the trip. The equation $30x + 45y = 2700$ models the total cost for the class to go on the trip, where x is the number of students who do not choose the meal plan and y is the number of students who do choose the meal plan.

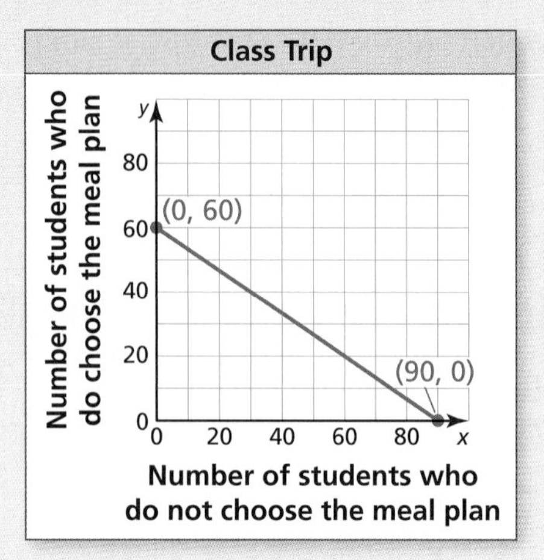

   a. Interpret the intercepts of the graph.

   b. Describe the domain and range in the context of the problem.

35. **REASONING** Use the values to fill in the equation $\square x + \square y = 30$ so that the x-intercept of the graph is -10 and the y-intercept of the graph is 5.

   -10 -3 1 5 6

36. **THOUGHT PROVOKING** Write an equation in standard form of a line whose intercepts are integers. Explain how you know the intercepts are integers.

37. **WRITING** Are the equations of horizontal and vertical lines written in standard form? Explain your reasoning.

38. **ABSTRACT REASONING** The x- and y-intercepts of the graph of the equation $3x + 5y = k$ are integers. Describe the values of k. Explain your reasoning.

Maintaining Mathematical Proficiency

Reviewing what you learned in previous grades and lessons

Simplify the expression. *(Skills Review Handbook)*

39. $\dfrac{2-(-2)}{4-(-4)}$

40. $\dfrac{14-18}{0-2}$

41. $\dfrac{-3-9}{8-(-7)}$

42. $\dfrac{12-17}{-5-(-2)}$

3.5 Graphing Linear Equations in Slope-Intercept Form

Essential Question

Essential Question How can you describe the graph of the equation $y = mx + b$?

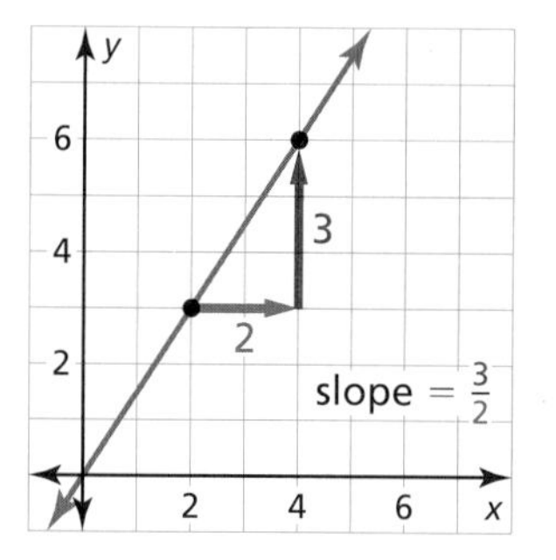

Slope is the rate of change between any two points on a line. It is the measure of the *steepness* of the line.

To find the slope of a line, find the ratio of the change in y (vertical change) to the change in x (horizontal change).

$$\text{slope} = \frac{\text{change in } y}{\text{change in } x}$$

EXPLORATION 1 Finding Slopes and y-Intercepts

Work with a partner. Find the slope and y-intercept of each line.

a.

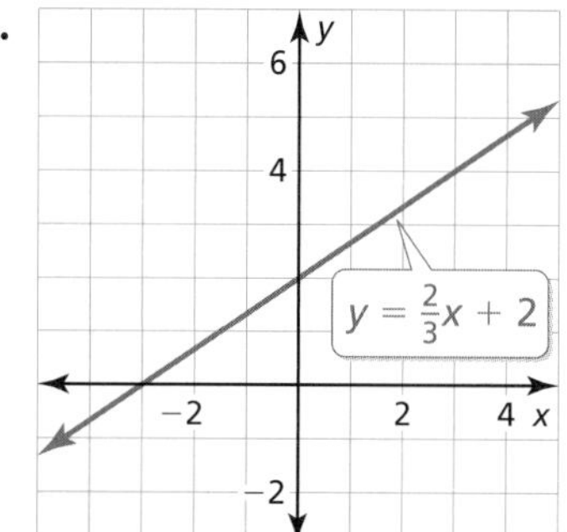

b.

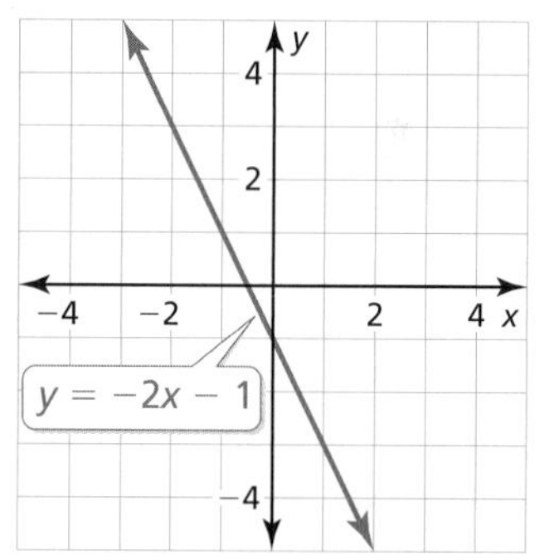

MAKING CONJECTURES

To be proficient in math, you first need to collect and organize data. Then make conjectures about the patterns you observe in the data.

EXPLORATION 2 Writing a Conjecture

Work with a partner. Graph each equation. Then copy and complete the table. Use the completed table to write a conjecture about the relationship between the graph of $y = mx + b$ and the values of m and b.

Equation	Description of graph	Slope of graph	y-Intercept
a. $y = -\frac{2}{3}x + 3$	Line	$-\frac{2}{3}$	3
b. $y = 2x - 2$			
c. $y - -x + 1$			
d. $y = x - 4$			

Communicate Your Answer

3. How can you describe the graph of the equation $y = mx + b$?

a. How does the value of m affect the graph of the equation?

b. How does the value of b affect the graph of the equation?

c. Check your answers to parts (a) and (b) by choosing one equation from Exploration 2 and (1) varying only m and (2) varying only b.

3.5 Lesson

Core Vocabulary

slope, *p. 136*
rise, *p. 136*
run, *p. 136*
slope-intercept form, *p. 138*
constant function, *p. 138*

Previous
dependent variable
independent variable

What You Will Learn

- Find the slope of a line.
- Use the slope-intercept form of a linear equation.
- Use slopes and y-intercepts to solve real-life problems.

The Slope of a Line

Core Concept

Slope

The **slope** m of a nonvertical line passing through two points (x_1, y_1) and (x_2, y_2) is the ratio of the **rise** (change in y) to the **run** (change in x).

$$\text{slope} = m = \frac{\text{rise}}{\text{run}} = \frac{\text{change in } y}{\text{change in } x} = \frac{y_2 - y_1}{x_2 - x_1}$$

(x_2, y_2) (x_1, y_1) rise $= y_2 - y_1$ run $= x_2 - x_1$ O x y

When the line rises from left to right, the slope is positive. When the line falls from left to right, the slope is negative.

EXAMPLE 1 **Finding the Slope of a Line**

Describe the slope of each line. Then find the slope.

a.

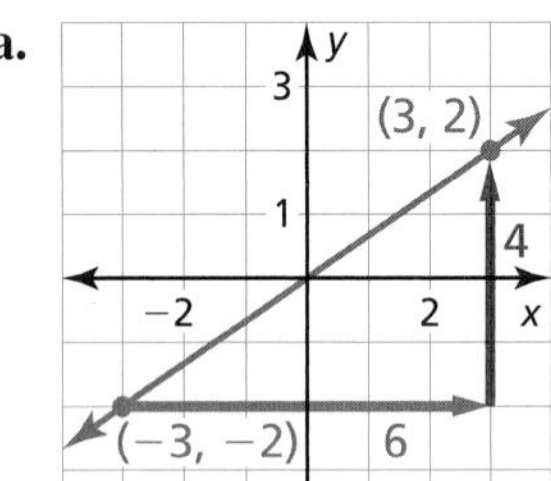

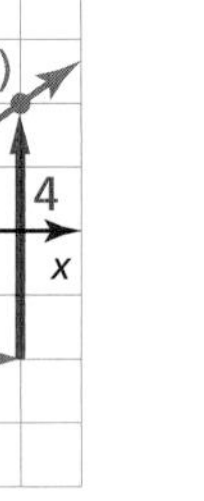

b.

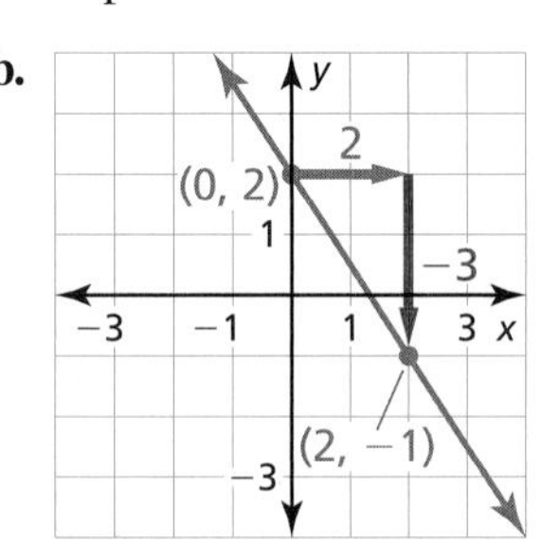

STUDY TIP

When finding slope, you can label either point as (x_1, y_1) and the other point as (x_2, y_2). The result is the same.

READING

In the slope formula, x_1 is read as "x sub one" and y_2 is read as "y sub two." The numbers 1 and 2 in x_1 and y_2 are called *subscripts*.

SOLUTION

a. The line rises from left to right. So, the slope is positive. Let $(x_1, y_1) = (-3, -2)$ and $(x_2, y_2) = (3, 2)$.

$$m = \frac{y_2 - y_1}{x_2 - x_1} = \frac{2 - (-2)}{3 - (-3)} = \frac{4}{6} = \frac{2}{3}$$

b. The line falls from left to right. So, the slope is negative. Let $(x_1, y_1) = (0, 2)$ and $(x_2, y_2) = (2, -1)$.

$$m = \frac{y_2 - y_1}{x_2 - x_1} = \frac{-1 - 2}{2 - 0} = \frac{-3}{2} = -\frac{3}{2}$$

Monitoring Progress

Help in English and Spanish at *BigIdeasMath.com*

Describe the slope of the line. Then find the slope.

1.

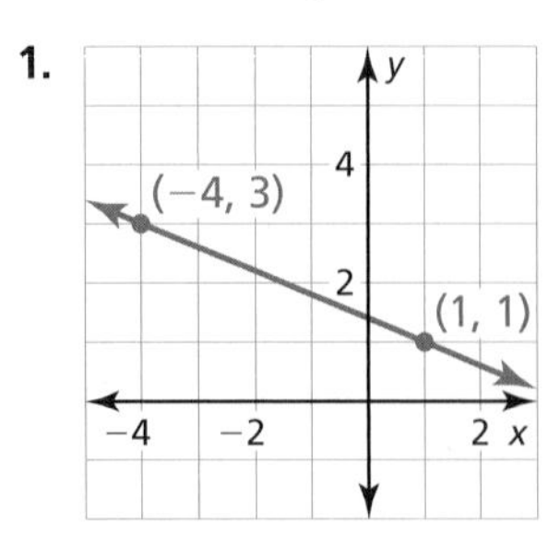

2.

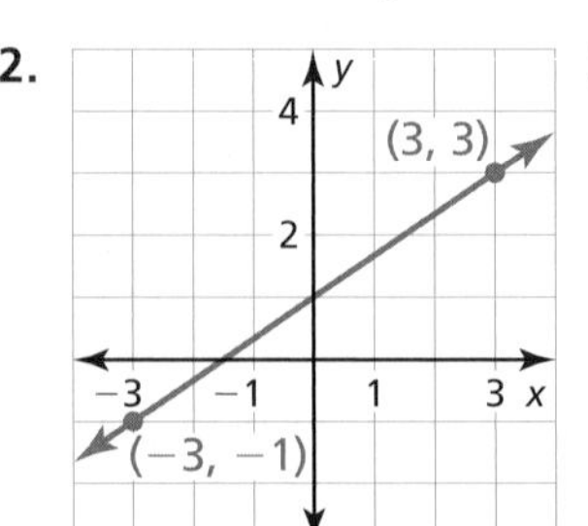

3.

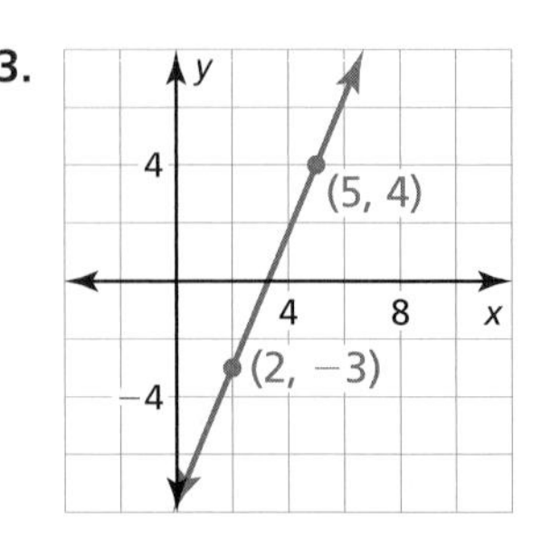

EXAMPLE 2 Finding Slope from a Table

The points represented by each table lie on a line. How can you find the slope of each line from the table? What is the slope of each line?

a.

x	y
4	20
7	14
10	8
13	2

b.

x	y
−1	2
1	2
3	2
5	2

c.

x	y
−3	−3
−3	0
−3	6
−3	9

STUDY TIP

As a check, you can plot the points represented by the table to verify that the line through them has a slope of −2.

SOLUTION

a. Choose any two points from the table and use the slope formula. Use the points $(x_1, y_1) = (4, 20)$ and $(x_2, y_2) = (7, 14)$.

$$m = \frac{y_2 - y_1}{x_2 - x_1} = \frac{14 - 20}{7 - 4} = \frac{-6}{3}, \text{ or } -2$$

▶ The slope is −2.

b. Note that there is no change in y. Choose any two points from the table and use the slope formula. Use the points $(x_1, y_1) = (-1, 2)$ and $(x_2, y_2) = (5, 2)$.

$$m = \frac{y_2 - y_1}{x_2 - x_1} = \frac{2 - 2}{5 - (-1)} = \frac{0}{6}, \text{ or } 0$$

The change in y is 0.

▶ The slope is 0.

c. Note that there is no change in x. Choose any two points from the table and use the slope formula. Use the points $(x_1, y_1) = (-3, 0)$ and $(x_2, y_2) = (-3, 6)$.

$$m = \frac{y_2 - y_1}{x_2 - x_1} = \frac{6 - 0}{-3 - (-3)} = \frac{6}{0} \text{ ✗}$$

The change in x is 0.

▶ Because division by zero is undefined, the slope of the line is undefined.

Monitoring Progress

Help in English and Spanish at *BigIdeasMath.com*

The points represented by the table lie on a line. How can you find the slope of the line from the table? What is the slope of the line?

4.

x	2	4	6	8
y	10	15	20	25

5.

x	5	5	5	5
y	−12	−9	−6	−3

Concept Summary

Slope

Positive slope — The line rises from left to right.

Negative slope — The line falls from left to right.

Slope of 0 — The line is horizontal.

Undefined slope — The line is vertical.

Using the Slope-Intercept Form of a Linear Equation

Core Concept

Slope-Intercept Form

Words A linear equation written in the form $y = mx + b$ is in **slope-intercept form**. The slope of the line is m, and the y-intercept of the line is b.

Algebra $y = mx + b$

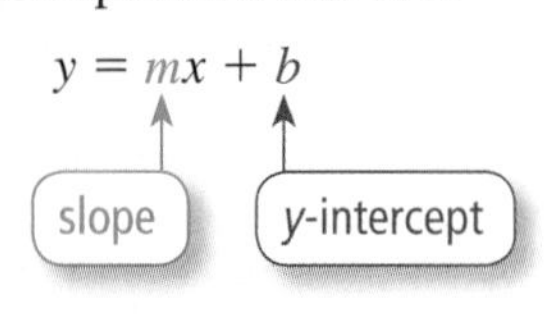

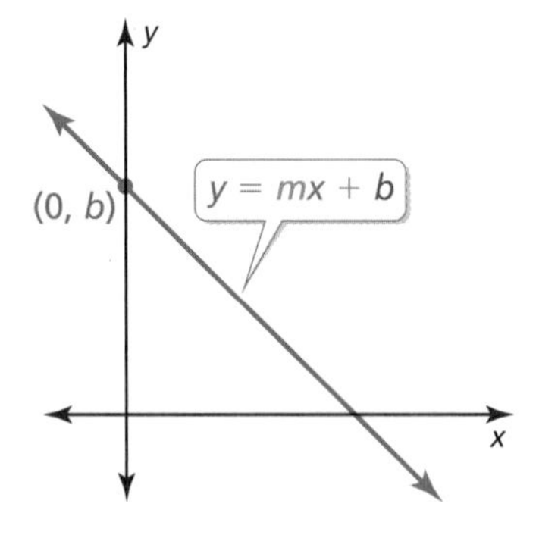

A linear equation written in the form $y = 0x + b$, or $y = b$, is a **constant function**. The graph of a constant function is a horizontal line.

EXAMPLE 3 Identifying Slopes and y-Intercepts

Find the slope and the y-intercept of the graph of each linear equation.

a. $y = 3x - 4$ **b.** $y = 6.5$ **c.** $-5x - y = -2$

SOLUTION

a. $y = mx + b$ Write the slope-intercept form.

slope y-intercept

$y = 3x + (-4)$ Rewrite the original equation in slope-intercept form.

▶ The slope is 3, and the y-intercept is -4.

STUDY TIP

For a constant function, every input has the same output. For instance, in Example 3b, every input has an output of 6.5.

b. The equation represents a constant function. The equation can also be written as $y = 0x + 6.5$.

▶ The slope is 0, and the y-intercept is 6.5.

c. Rewrite the equation in slope-intercept form by solving for y.

$$-5x - y = -2$$ Write the original equation.

$$+5x \qquad +5x$$ Add $5x$ to each side.

$$-y = 5x - 2$$ Simplify.

$$\frac{-y}{-1} = \frac{5x - 2}{-1}$$ Divide each side by -1.

$$y = -5x + 2$$ Simplify.

▶ The slope is -5, and the y-intercept is 2.

STUDY TIP

When you rewrite a linear equation in slope-intercept form, you are expressing y as a function of x.

Monitoring Progress  Help in English and Spanish at *BigIdeasMath.com*

Find the slope and the y-intercept of the graph of the linear equation.

6. $y = -6x + 1$ **7.** $y = 8$ **8.** $x + 4y = -10$

EXAMPLE 4 Using Slope-Intercept Form to Graph

STUDY TIP

You can use the slope to find points on a line in either direction. In Example 4, note that the slope can be written as $\frac{2}{-1}$. So, you could move 1 unit left and 2 units up from (0, 2) to find the point (–1, 4).

Graph $2x + y = 2$. Identify the x-intercept.

SOLUTION

Step 1 Rewrite the equation in slope-intercept form.

$$y = -2x + 2$$

Step 2 Find the slope and the y-intercept.

$$m = -2 \text{ and } b = 2$$

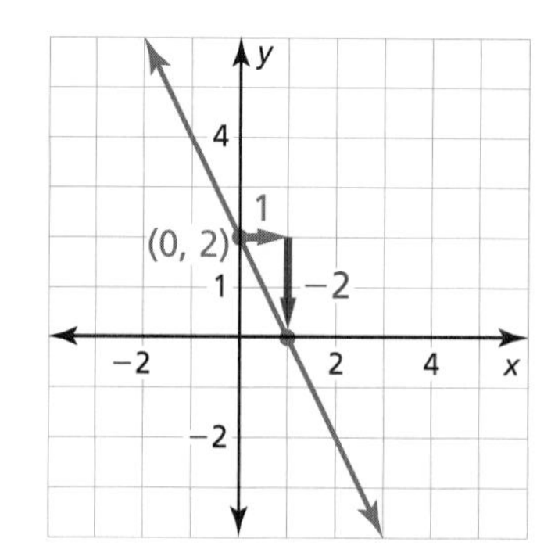

Step 3 The y-intercept is 2. So, plot (0, 2).

Step 4 Use the slope to find another point on the line.

$$\text{slope} = \frac{\text{rise}}{\text{run}} = \frac{-2}{1}$$

Plot the point that is 1 unit right and 2 units down from (0, 2). Draw a line through the two points.

▶ The line crosses the x-axis at (1, 0). So, the x-intercept is 1.

REMEMBER

You can also find the x-intercept by substituting 0 for y in the equation $2x + y = 2$ and solving for x.

EXAMPLE 5 Graphing from a Verbal Description

A linear function g models a relationship in which the dependent variable increases 3 units for every 1 unit the independent variable increases. Graph g when $g(0) = 3$. Identify the slope, y-intercept, and x-intercept of the graph.

SOLUTION

Because the function g is linear, it has a constant rate of change. Let x represent the independent variable and y represent the dependent variable.

Step 1 **Find the slope.** When the dependent variable increases by 3, the change in y is $+3$. When the independent variable increases by 1, the change in x is $+1$.

So, the slope is $\frac{3}{1}$, or 3.

Step 2 **Find the y-intercept.** The statement $g(0) = 3$ indicates that when $x = 0$, $y = 3$. So, the y-intercept is 3. Plot (0, 3).

Step 3 **Use the slope to find another point on the line.** A slope of 3 can be written as $\frac{-3}{-1}$. Plot the point that is 1 unit left and 3 units down from (0, 3). Draw a line through the two points. The line crosses the x-axis at $(-1, 0)$. So, the x-intercept is -1.

▶ The slope is 3, the y-intercept is 3, and the x-intercept is -1.

Monitoring Progress  Help in English and Spanish at *BigIdeasMath.com*

Graph the linear equation. Identify the x-intercept.

9. $y = 4x - 4$ **10.** $3x + y = -3$ **11.** $x + 2y = 6$

12. A linear function h models a relationship in which the dependent variable decreases 2 units for every 5 units the independent variable increases. Graph h when $h(0) = 4$. Identify the slope, y-intercept, and x-intercept of the graph.

Solving Real-Life Problems

In most real-life problems, slope is interpreted as a rate, such as miles per hour, dollars per hour, or people per year.

EXAMPLE 6 **Modeling with Mathematics**

A submersible that is exploring the ocean floor begins to ascend to the surface. The elevation h (in feet) of the submersible is modeled by the function $h(t) = 650t - 13{,}000$, where t is the time (in minutes) since the submersible began to ascend.

a. Graph the function and identify its domain and range.

b. Interpret the slope and the intercepts of the graph.

SOLUTION

1. **Understand the Problem** You know the function that models the elevation. You are asked to graph the function and identify its domain and range. Then you are asked to interpret the slope and intercepts of the graph.

2. **Make a Plan** Use the slope-intercept form of a linear equation to graph the function. Only graph values that make sense in the context of the problem. Examine the graph to interpret the slope and the intercepts.

3. **Solve the Problem**

a. The time t must be greater than or equal to 0. The elevation h is below sea level and must be less than or equal to 0. Use the slope of 650 and the h-intercept of $-13{,}000$ to graph the function in Quadrant IV.

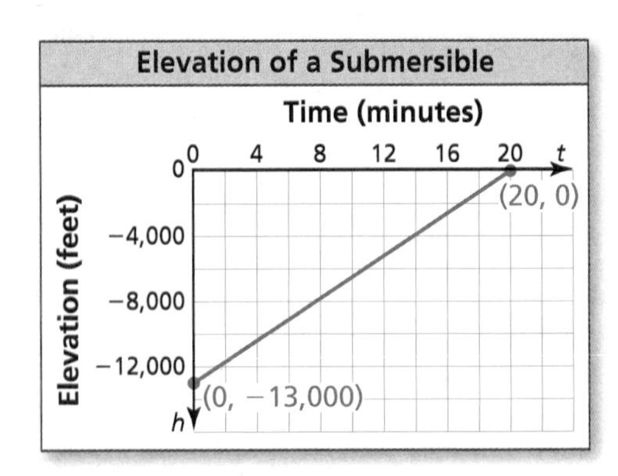

STUDY TIP

Because t is the independent variable, the horizontal axis is the t-axis and the graph will have a "t-intercept." Similarly, the vertical axis is the h-axis and the graph will have an "h-intercept."

▶ The domain is $0 \le t \le 20$, and the range is $-13{,}000 \le h \le 0$.

b. The slope is 650. So, the submersible ascends at a rate of 650 feet per minute. The h-intercept is $-13{,}000$. So, the elevation of the submersible after 0 minutes, or when the ascent begins, is $-13{,}000$ feet. The t-intercept is 20. So, the submersible takes 20 minutes to reach an elevation of 0 feet, or sea level.

4. **Look Back** You can check that your graph is correct by substituting the t-intercept for t in the function. If $h = 0$ when $t = 20$, the graph is correct.

$h = 650(20) - 13{,}000$ Substitute 20 for t in the original equation.

$h = 0$ ✓ Simplify.

Monitoring Progress  Help in English and Spanish at *BigIdeasMath.com*

13. **WHAT IF?** The elevation of the submersible is modeled by $h(t) = 500t - 10{,}000$. (a) Graph the function and identify its domain and range. (b) Interpret the slope and the intercepts of the graph.

Vocabulary and Core Concept Check

1. **COMPLETE THE SENTENCE** The ________ of a nonvertical line passing through two points is the ratio of the rise to the run.

2. **VOCABULARY** What is a constant function? What is the slope of a constant function?

3. **WRITING** What is the slope-intercept form of a linear equation? Explain why this form is called the slope-intercept form.

4. **WHICH ONE DOESN'T BELONG?** Which equation does *not* belong with the other three? Explain your reasoning.

$y = -5x - 1$ | $2x - y = 8$ | $y = x + 4$ | $y = -3x + 13$

Monitoring Progress and Modeling with Mathematics

In Exercises 5–8, describe the slope of the line. Then find the slope. *(See Example 1.)*

5. 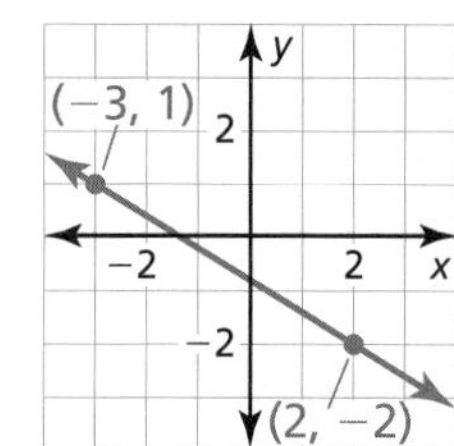

6. 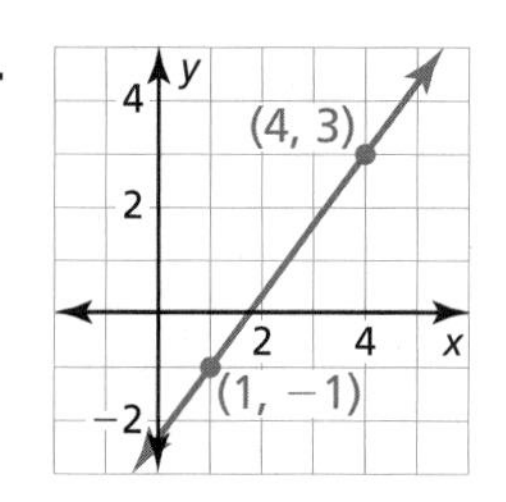

7. 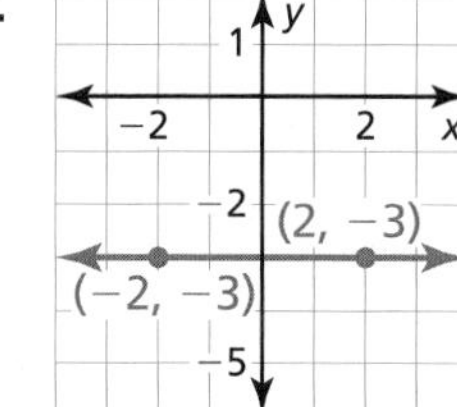

8. 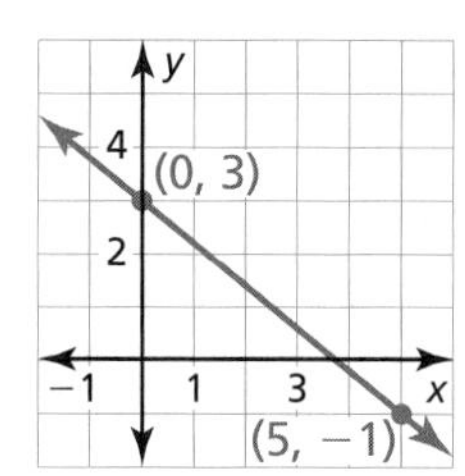

In Exercises 9–12, the points represented by the table lie on a line. Find the slope of the line. *(See Example 2.)*

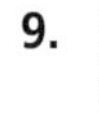

9.

x	−9	−5	−1	3
y	−2	0	2	4

10.

x	−1	2	5	8
y	−6	−6	−6	−6

11.

x	0	0	0	0
y	−4	0	4	8

12.

x	−4	−3	−2	−1
y	2	−5	−12	−19

13. **ANALYZING A GRAPH** The graph shows the distance y (in miles) that a bus travels in x hours. Find and interpret the slope of the line.

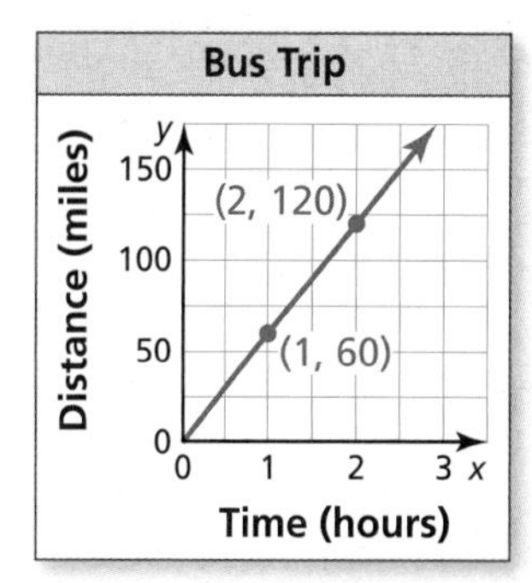

14. **ANALYZING A TABLE** The table shows the amount x (in hours) of time you spend at a theme park and the admission fee y (in dollars) to the park. The points represented by the table lie on a line. Find and interpret the slope of the line.

Time (hours), x	Admission (dollars), y
6	54.99
7	54.99
8	54.99

In Exercises 15–22, find the slope and the y-intercept of the graph of the linear equation. *(See Example 3.)*

15. $y = -3x + 2$

16. $y = 4x - 7$

17. $y = 6x$

18. $y = -1$

19. $-2x + y = 4$

20. $x + y = -6$

21. $-5x = 8 - y$

22. $0 = 1 - 2y + 14x$

ERROR ANALYSIS In Exercises 23 and 24, describe and correct the error in finding the slope and the y-intercept of the graph of the equation.

23. 

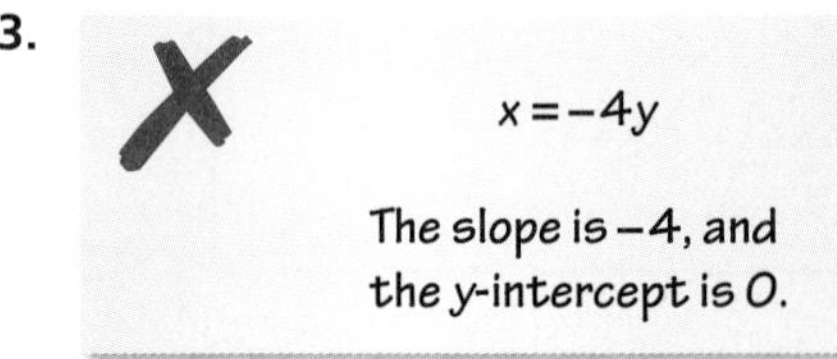

24.

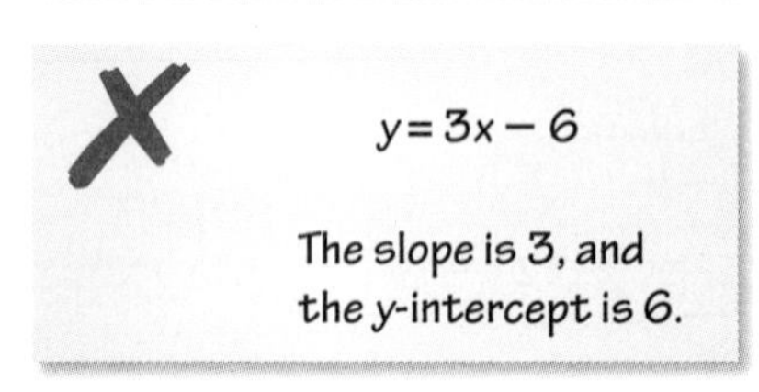

y = 3x − 6
The slope is 3, and
the y-intercept is 6.

In Exercises 25–32, graph the linear equation. Identify the x-intercept. *(See Example 4.)*

25. $y = -x + 7$

26. $y = \frac{1}{2}x + 3$

27. $y = 2x$

28. $y = -x$

29. $3x + y = -1$

30. $x + 4y = 8$

31. $-y + 5x = 0$

32. $2x - y + 6 = 0$

In Exercises 33 and 34, graph the function with the given description. Identify the slope, y-intercept, and x-intercept of the graph. *(See Example 5.)*

33. A linear function f models a relationship in which the dependent variable decreases 4 units for every 2 units the independent variable increases. The value of the function at 0 is -2.

34. A linear function h models a relationship in which the dependent variable increases 1 unit for every 5 units the independent variable decreases. The value of the function at 0 is 3.

35. GRAPHING FROM A VERBAL DESCRIPTION A linear function r models the growth of your right index fingernail. The length of the fingernail increases 0.7 millimeter every week. Graph r when $r(0) = 12$. Identify the slope and interpret the y-intercept of the graph.

36. GRAPHING FROM A VERBAL DESCRIPTION A linear function m models the amount of milk sold by a farm per month. The amount decreases 500 gallons for every \$1 increase in price. Graph m when $m(0) = 3000$. Identify the slope and interpret the x- and y-intercepts of the graph.

37. MODELING WITH MATHEMATICS The function shown models the depth d (in inches) of snow on the ground during the first 9 hours of a snowstorm, where t is the time (in hours) after the snowstorm begins. *(See Example 6.)*

a. Graph the function and identify its domain and range.

b. Interpret the slope and the d-intercept of the graph.

38. MODELING WITH MATHEMATICS The function $c(x) = 0.5x + 70$ represents the cost c (in dollars) of renting a truck from a moving company, where x is the number of miles you drive the truck.

a. Graph the function and identify its domain and range.

b. Interpret the slope and the c-intercept of the graph.

39. COMPARING FUNCTIONS A linear function models the cost of renting a truck from a moving company. The table shows the cost y (in dollars) when you drive the truck x miles. Graph the function and compare the slope and the y-intercept of the graph with the slope and the c-intercept of the graph in Exercise 38.

Miles, x	Cost (dollars), y
0	40
50	80
100	120

ERROR ANALYSIS In Exercises 40 and 41, describe and correct the error in graphing the function.

40. 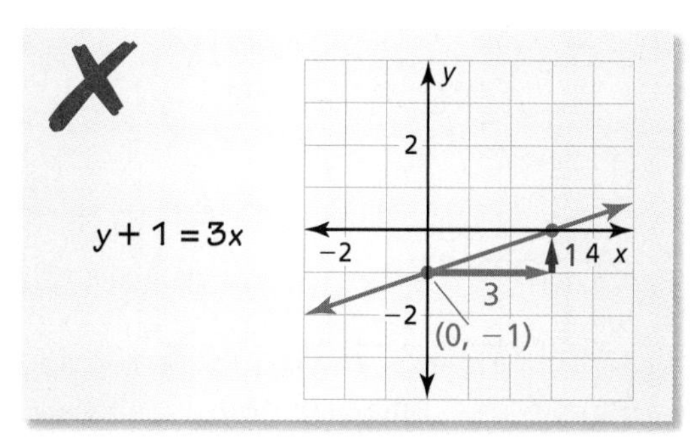

41. 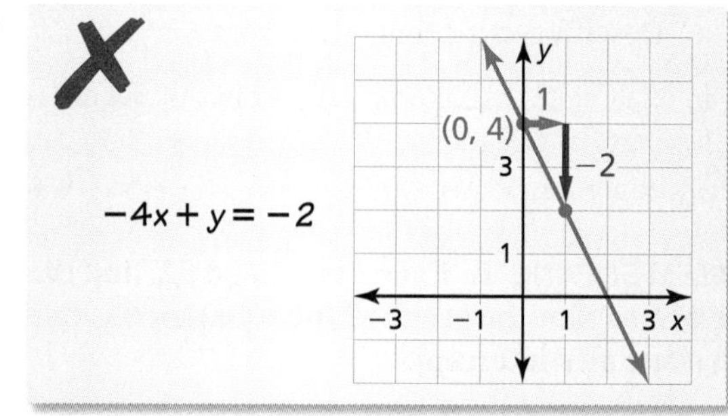

42. **MATHEMATICAL CONNECTIONS** Graph the four equations in the same coordinate plane.

$$3y = -x - 3$$

$$2y - 14 = 4x$$

$$4x - 3 - y = 0$$

$$x - 12 = -3y$$

a. What enclosed shape do you think the lines form? Explain.

b. Write a conjecture about the equations of parallel lines.

43. **MATHEMATICAL CONNECTIONS** The graph shows the relationship between the width y and the length x of a rectangle in inches. The perimeter of a second rectangle is 10 inches less than the perimeter of the first rectangle.

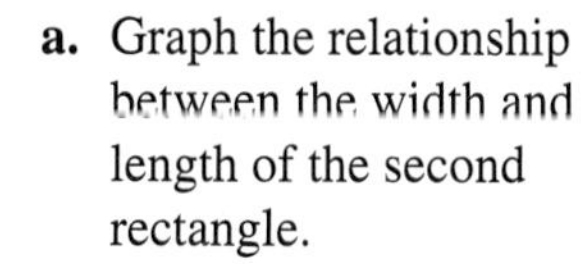

a. Graph the relationship between the width and length of the second rectangle.

b. How does the graph in part (a) compare to the the graph shown?

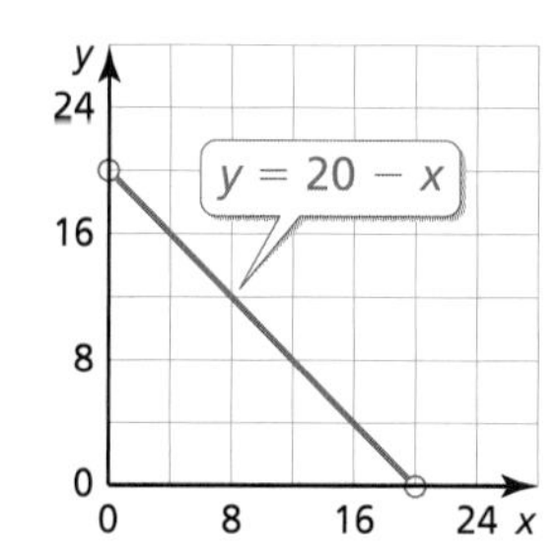

44. **MATHEMATICAL CONNECTIONS** The graph shows the relationship between the base length x and the side length (of the two equal sides) y of an isosceles triangle in meters. The perimeter of a second isosceles triangle is 8 meters more than the perimeter of the first triangle.

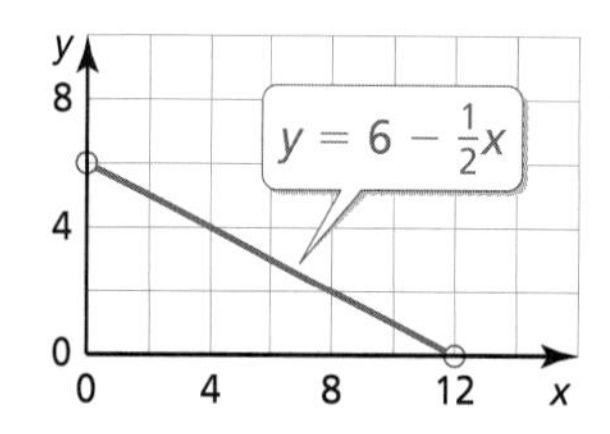

a. Graph the relationship between the base length and the side length of the second triangle.

b. How does the graph in part (a) compare to the graph shown?

45. **ANALYZING EQUATIONS** Determine which of the equations could be represented by each graph.

$y = -3x + 8$	$y = -x - \frac{4}{3}$
$y = -7x$	$y = 2x - 4$
$y = \frac{7}{4}x - \frac{1}{4}$	$y = \frac{1}{3}x + 5$
$y = -4x - 9$	$y = 6$

a. **b.**

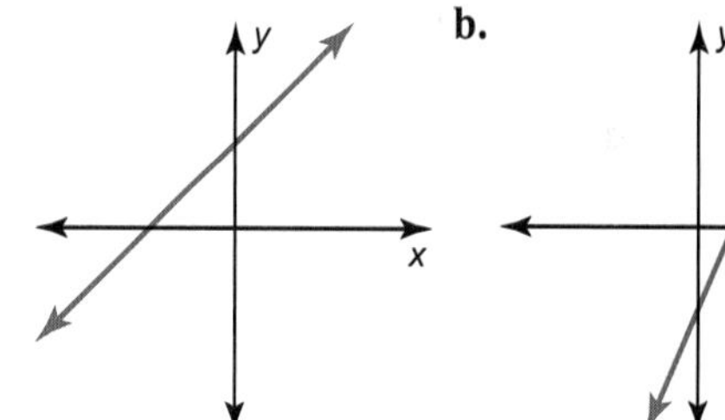

c. **d.**

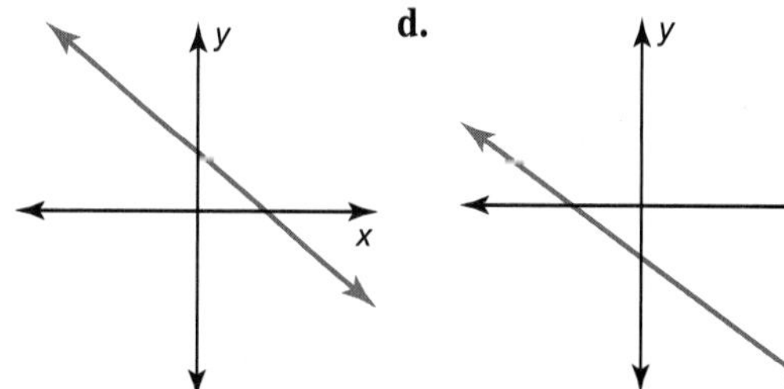

46. **MAKING AN ARGUMENT** Your friend says that you can write the equation of any line in slope-intercept form. Is your friend correct? Explain your reasoning.

47. WRITING Write the definition of the slope of a line in two different ways.

48. THOUGHT PROVOKING Your family goes on vacation to a beach 300 miles from your house. You reach your destination 6 hours after departing. Draw a graph that describes your trip. Explain what each part of your graph represents.

49. ANALYZING A GRAPH The graphs of the functions $g(x) = 6x + a$ and $h(x) = 2x + b$, where a and b are constants, are shown. They intersect at the point (p, q).

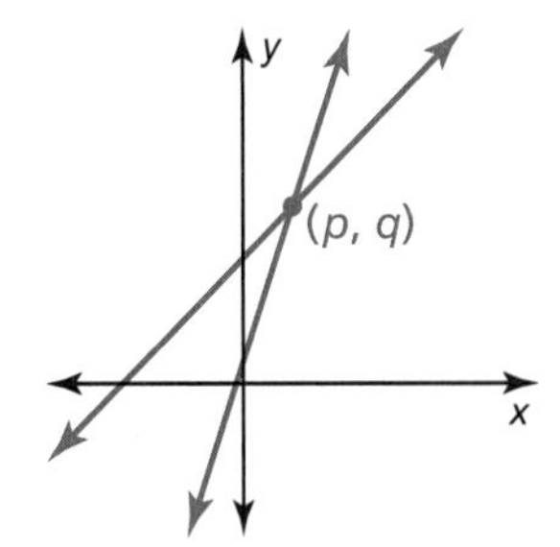

a. Label the graphs of g and h.

b. What do a and b represent?

c. Starting at the point (p, q), trace the graph of g until you get to the point with the x-coordinate $p + 2$. Mark this point C. Do the same with the graph of h. Mark this point D. How much greater is the y-coordinate of point C than the y-coordinate of point D?

50. HOW DO YOU SEE IT? You commute to school by walking and by riding a bus. The graph represents your commute.

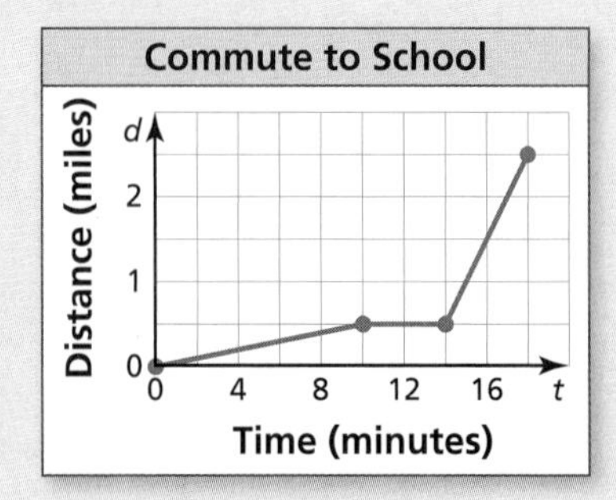

a. Describe your commute in words.

b. Calculate and interpret the slopes of the different parts of the graph.

PROBLEM SOLVING **In Exercises 51 and 52, find the value of k so that the graph of the equation has the given slope or y-intercept.**

51. $y = 4kx - 5;\ m = \frac{1}{2}$

52. $y = -\frac{1}{3}x + \frac{5}{6}k;\ b = -10$

53. ABSTRACT REASONING To show that the slope of a line is constant, let (x_1, y_1) and (x_2, y_2) be any two points on the line $y = mx + b$. Use the equation of the line to express y_1 in terms of x_1 and y_2 in terms of x_2. Then use the slope formula to show that the slope between the points is m.

Maintaining Mathematical Proficiency

Reviewing what you learned in previous grades and lessons

Find the coordinates of the figure after the transformation. *(Skills Review Handbook)*

54. Translate the rectangle 4 units left.

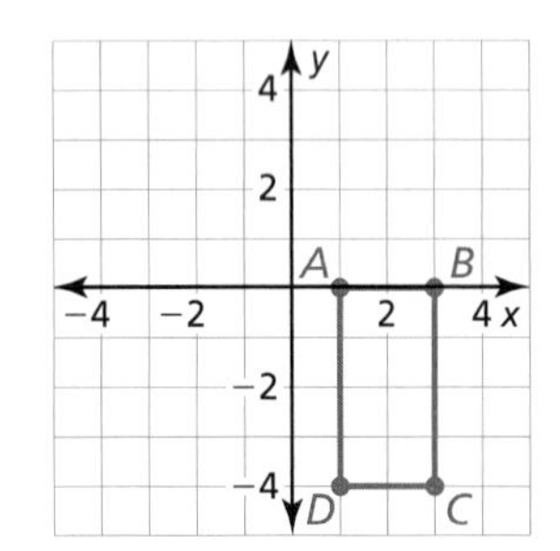

55. Dilate the triangle with respect to the origin using a scale factor of 2.

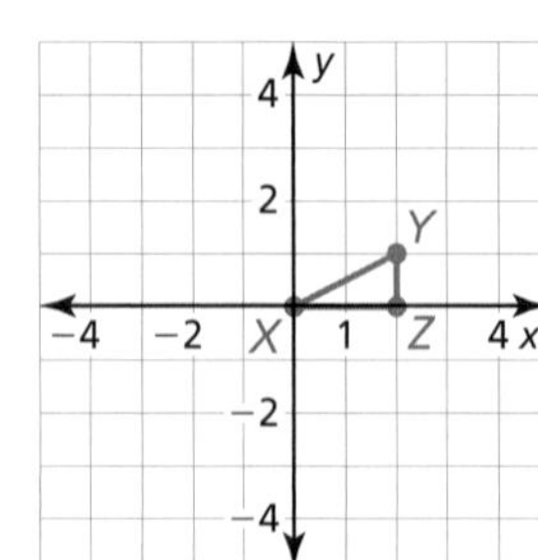

56. Reflect the trapezoid in the y-axis.

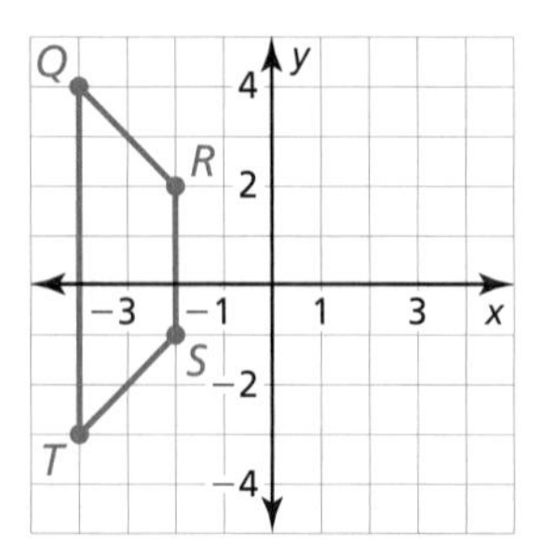

Determine whether the equation represents a *linear* or *nonlinear* function. Explain. *(Section 3.2)*

57. $y - 9 = \frac{2}{x}$

58. $x = 3 + 15y$

59. $\frac{x}{4} + \frac{y}{12} = 1$

60. $y = 3x^4 - 6$

3.6 Transformations of Graphs of Linear Functions

Essential Question How does the graph of the linear function $f(x) = x$ compare to the graphs of $g(x) = f(x) + c$ and $h(x) = f(cx)$?

EXPLORATION 1 Comparing Graphs of Functions

Work with a partner. The graph of $f(x) = x$ is shown. Sketch the graph of each function, along with f, on the same set of coordinate axes. Use a graphing calculator to check your results. What can you conclude?

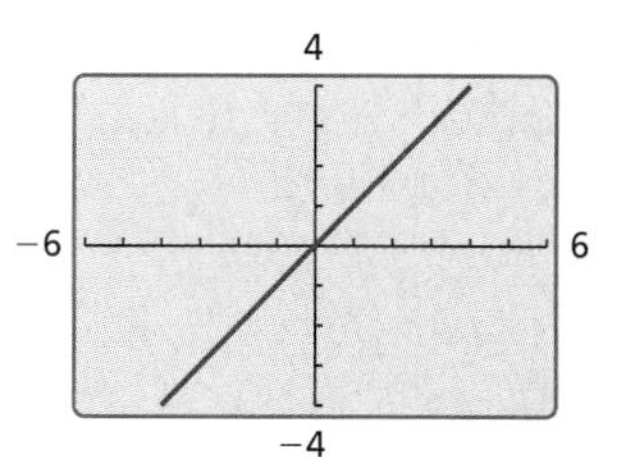

a. $g(x) = x + 4$ **b.** $g(x) = x + 2$

c. $g(x) = x - 2$ **d.** $g(x) = x - 4$

USING TOOLS STRATEGICALLY

To be proficient in math, you need to use the appropriate tools, including graphs, tables, and technology, to check your results.

EXPLORATION 2 Comparing Graphs of Functions

Work with a partner. Sketch the graph of each function, along with $f(x) = x$, on the same set of coordinate axes. Use a graphing calculator to check your results. What can you conclude?

a. $h(x) = \frac{1}{2}x$ **b.** $h(x) = 2x$ **c.** $h(x) = -\frac{1}{2}x$ **d.** $h(x) = -2x$

EXPLORATION 3 Matching Functions with Their Graphs

Work with a partner. Match each function with its graph. Use a graphing calculator to check your results. Then use the results of Explorations 1 and 2 to compare the graph of k to the graph of $f(x) = x$.

a. $k(x) = 2x - 4$ **b.** $k(x) = -2x + 2$

c. $k(x) = \frac{1}{2}x + 4$ **d.** $k(x) = -\frac{1}{2}x - 2$

A.

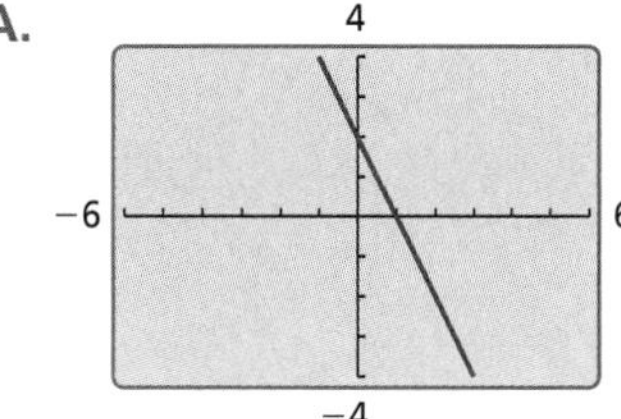

B.

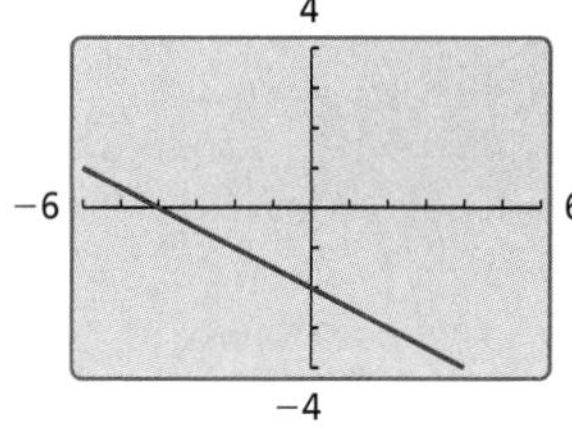

C.

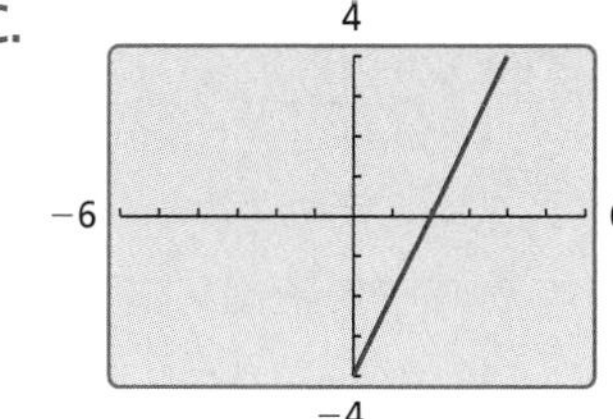

D.

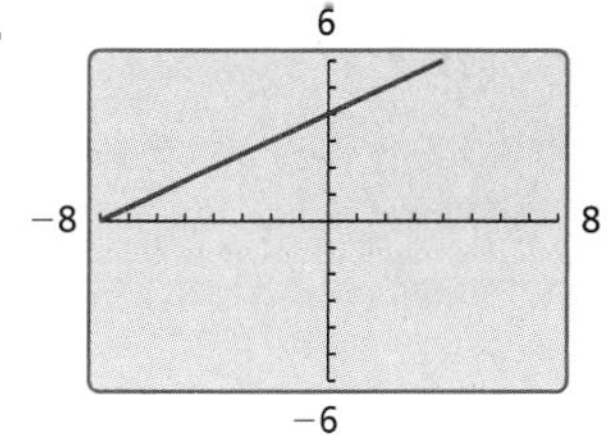

Communicate Your Answer

4. How does the graph of the linear function $f(x) = x$ compare to the graphs of $g(x) = f(x) + c$ and $h(x) = f(cx)$?

3.6 Lesson

Core Vocabulary

family of functions, *p. 146*
parent function, *p. 146*
transformation, *p. 146*
translation, *p. 146*
reflection, *p. 147*
horizontal shrink, *p. 148*
horizontal stretch, *p. 148*
vertical stretch, *p. 148*
vertical shrink, *p. 148*

Previous
linear function

What You Will Learn

- Translate and reflect graphs of linear functions.
- Stretch and shrink graphs of linear functions.
- Combine transformations of graphs of linear functions.

Translations and Reflections

A **family of functions** is a group of functions with similar characteristics. The most basic function in a family of functions is the **parent function**. For nonconstant linear functions, the parent function is $f(x) = x$. The graphs of all other nonconstant linear functions are *transformations* of the graph of the parent function. A **transformation** changes the size, shape, position, or orientation of a graph.

Core Concept

A **translation** is a transformation that shifts a graph horizontally or vertically but does not change the size, shape, or orientation of the graph.

Horizontal Translations

The graph of $y = f(x - h)$ is a horizontal translation of the graph of $y = f(x)$, where $h \neq 0$.

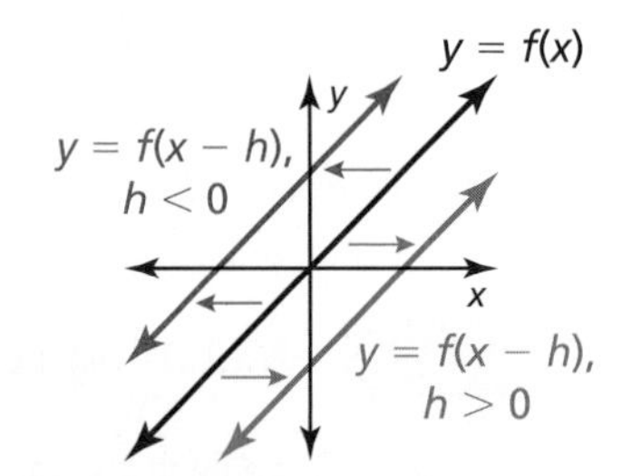

Subtracting h from the *inputs* before evaluating the function shifts the graph left when $h < 0$ and right when $h > 0$.

Vertical Translations

The graph of $y = f(x) + k$ is a vertical translation of the graph of $y = f(x)$, where $k \neq 0$.

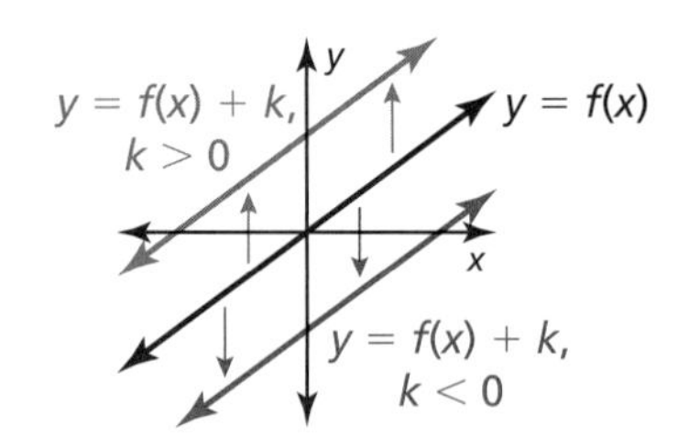

Adding k to the *outputs* shifts the graph down when $k < 0$ and up when $k > 0$.

EXAMPLE 1 Horizontal and Vertical Translations

Let $f(x) = 2x - 1$. Graph (a) $g(x) = f(x) + 3$ and (b) $t(x) = f(x + 3)$. Describe the transformations from the graph of f to the graphs of g and t.

SOLUTION

a. The function g is of the form $y = f(x) + k$, where $k = 3$. So, the graph of g is a vertical translation 3 units up of the graph of f.

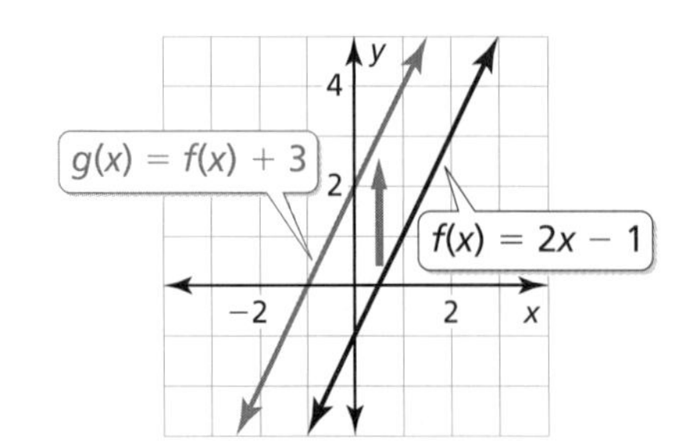

b. The function t is of the form $y = f(x - h)$, where $h = -3$. So, the graph of t is a horizontal translation 3 units left of the graph of f.

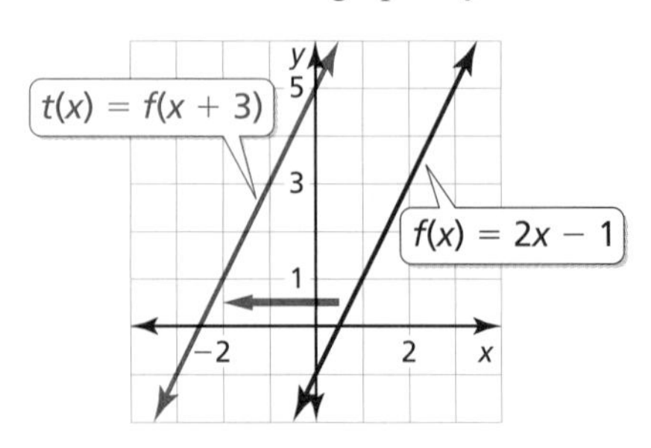

LOOKING FOR A PATTERN

In part (a), the output of g is equal to the output of f plus 3.

In part (b), the output of t is equal to the output of f when the input of f is 3 more than the input of t.

Core Concept

A **reflection** is a transformation that flips a graph over a line called the *line of reflection*.

Reflections in the x-axis

The graph of $y = -f(x)$ is a reflection in the x-axis of the graph of $y = f(x)$.

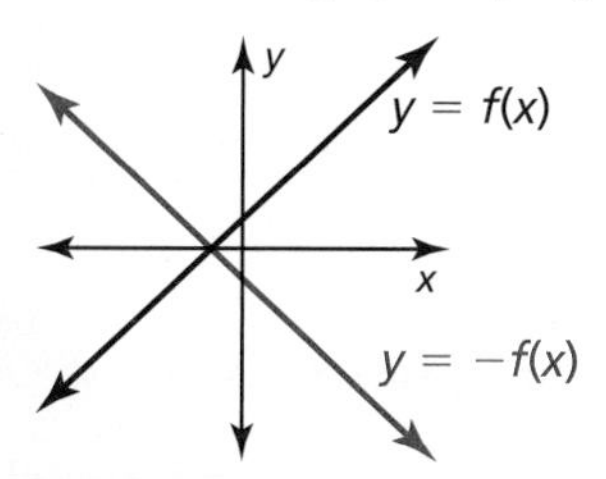

Multiplying the outputs by -1 changes their signs.

Reflections in the y-axis

The graph of $y = f(-x)$ is a reflection in the y-axis of the graph of $y = f(x)$.

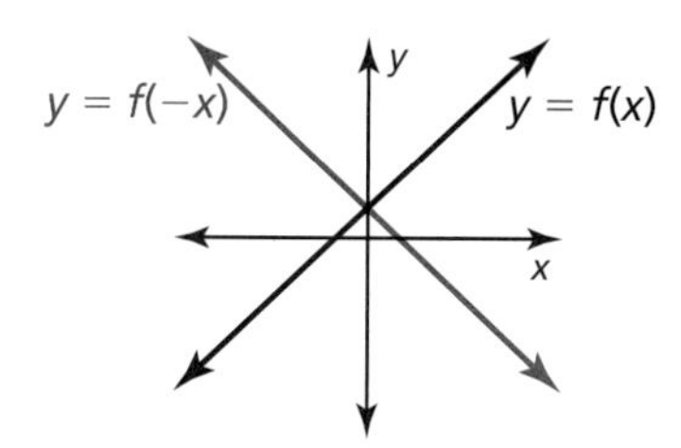

Multiplying the inputs by -1 changes their signs.

STUDY TIP

A reflected point is the same distance from the line of reflection as the original point but on the opposite side of the line.

EXAMPLE 2 Reflections in the x-axis and the y-axis

Let $f(x) = \frac{1}{2}x + 1$. Graph (a) $g(x) = -f(x)$ and (b) $t(x) = f(-x)$. Describe the transformations from the graph of f to the graphs of g and t.

SOLUTION

a. To find the outputs of g, multiply the outputs of f by -1. The graph of g consists of the points $(x, -f(x))$.

x	−4	−2	0
f(x)	−1	0	1
−f(x)	1	0	−1

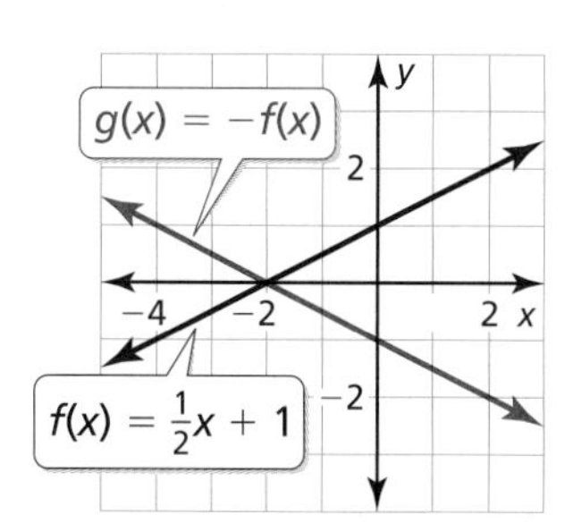

▶ The graph of g is a reflection in the x-axis of the graph of f.

b. To find the outputs of t, multiply the inputs by -1 and then evaluate f. The graph of t consists of the points $(x, f(-x))$.

x	−2	0	2
−x	2	0	−2
f(−x)	2	1	0

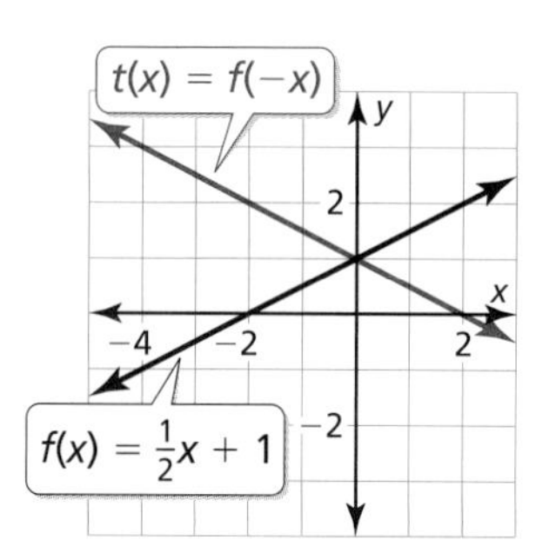

▶ The graph of t is a reflection in the y-axis of the graph of f.

Monitoring Progress

Help in English and Spanish at *BigIdeasMath.com*

Using f, graph (a) g and (b) h. Describe the transformations from the graph of f to the graphs of g and h.

1. $f(x) = 3x + 1$; $g(x) = f(x) - 2$; $h(x) = f(x - 2)$
2. $f(x) = -4x - 2$; $g(x) = -f(x)$; $h(x) = f(-x)$

Stretches and Shrinks

You can transform a function by multiplying all the x-coordinates (inputs) by the same factor a. When $a > 1$, the transformation is a **horizontal shrink** because the graph shrinks toward the y-axis. When $0 < a < 1$, the transformation is a **horizontal stretch** because the graph stretches away from the y-axis. In each case, the y-intercept stays the same.

You can also transform a function by multiplying all the y-coordinates (outputs) by the same factor a. When $a > 1$, the transformation is a **vertical stretch** because the graph stretches away from the x-axis. When $0 < a < 1$, the transformation is a **vertical shrink** because the graph shrinks toward the x-axis. In each case, the x-intercept stays the same.

STUDY TIP

The graphs of $y = f(-ax)$ and $y = -a \cdot f(x)$ represent a stretch or shrink *and* a reflection in the x- or y-axis of the graph of $y = f(x)$.

Core Concept

Horizontal Stretches and Shrinks

The graph of $y = f(ax)$ is a horizontal stretch or shrink by a factor of $\frac{1}{a}$ of the graph of $y = f(x)$, where $a > 0$ and $a \neq 1$.

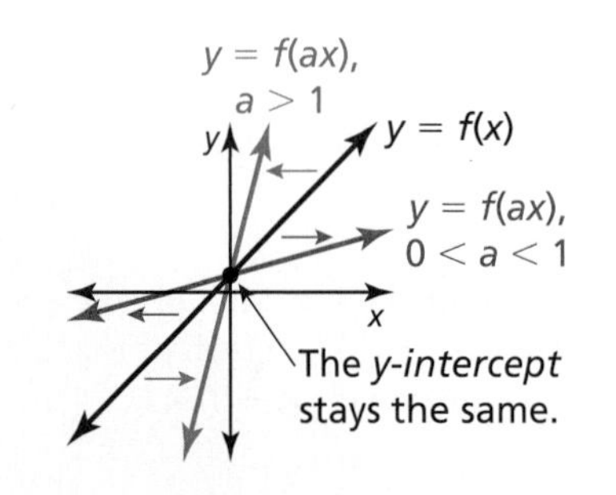

Vertical Stretches and Shrinks

The graph of $y = a \cdot f(x)$ is a vertical stretch or shrink by a factor of a of the graph of $y = f(x)$, where $a > 0$ and $a \neq 1$.

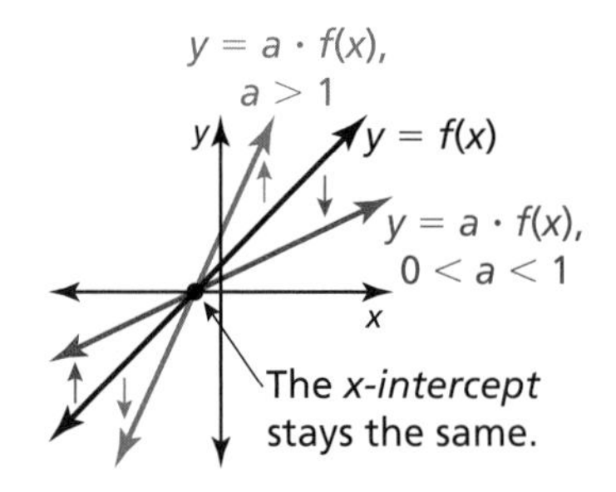

EXAMPLE 3 Horizontal and Vertical Stretches

Let $f(x) = x - 1$. Graph (a) $g(x) = f\left(\frac{1}{3}x\right)$ and (b) $h(x) = 3f(x)$. Describe the transformations from the graph of f to the graphs of g and h.

SOLUTION

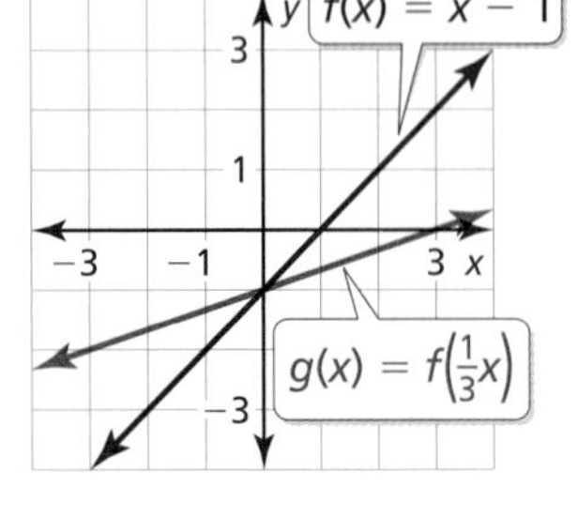

a. To find the outputs of g, multiply the inputs by $\frac{1}{3}$. Then evaluate f. The graph of g consists of the points $\left(x, f\left(\frac{1}{3}x\right)\right)$.

x	−3	0	3
$\frac{1}{3}(x)$	−1	0	1
$f\left(\frac{1}{3}x\right)$	−2	−1	0

▶ The graph of g is a horizontal stretch of the graph of f by a factor of $1 \div \frac{1}{3} = 3$.

b. To find the outputs of h, multiply the outputs of f by 3. The graph of h consists of the points $(x, 3f(x))$.

x	0	1	2
$f(x)$	−1	0	1
$3f(x)$	−3	0	3

▶ The graph of h is a vertical stretch of the graph of f by a factor of 3.

EXAMPLE 4 Horizontal and Vertical Shrinks

Let $f(x) = x + 2$. Graph (a) $g(x) = f(4x)$ and (b) $h(x) = \frac{1}{4}f(x)$. Describe the transformations from the graph of f to the graphs of g and h.

SOLUTION

a. To find the outputs of g, multiply the inputs by 4. Then evaluate f. The graph of g consists of the points $(x, f(4x))$.

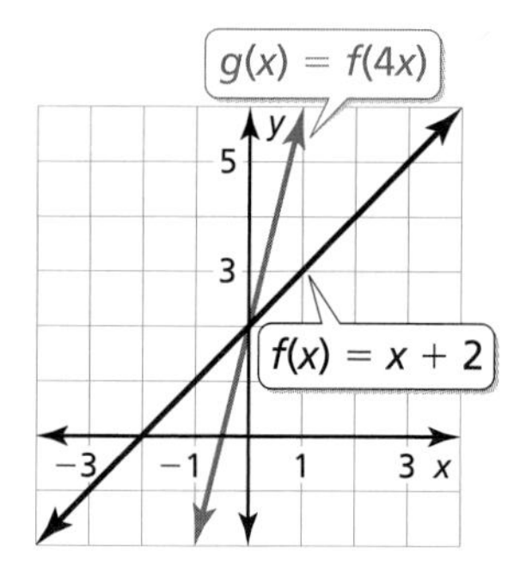

x	−1	0	1
4x	−4	0	4
f(4x)	−2	2	6

▶ The graph of g is a horizontal shrink of the graph of f by a factor of $\frac{1}{4}$.

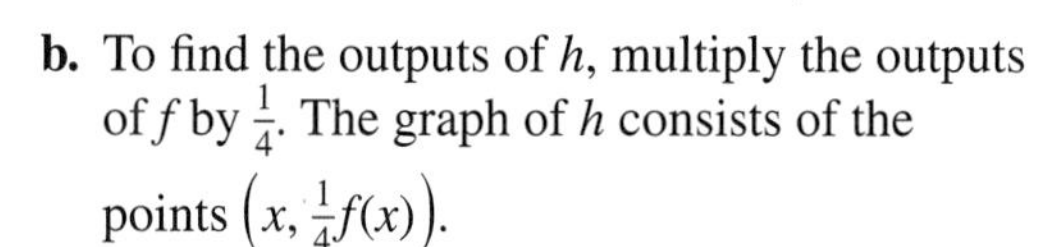

b. To find the outputs of h, multiply the outputs of f by $\frac{1}{4}$. The graph of h consists of the points $\left(x, \frac{1}{4}f(x)\right)$.

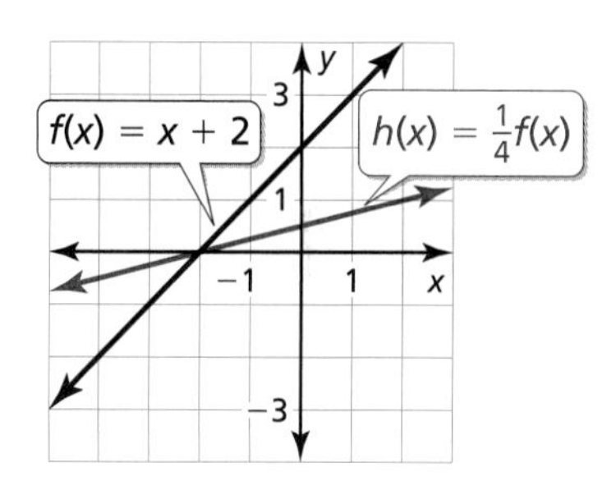

x	−2	0	2
f(x)	0	2	4
$\frac{1}{4}$f(x)	0	$\frac{1}{2}$	1

▶ The graph of h is a vertical shrink of the graph of f by a factor of $\frac{1}{4}$.

Monitoring Progress

Help in English and Spanish at *BigIdeasMath.com*

Using f, graph (a) g and (b) h. Describe the transformations from the graph of f to the graphs of g and h.

3. $f(x) = 4x - 2$; $g(x) = f\left(\frac{1}{2}x\right)$; $h(x) = 2f(x)$

4. $f(x) = -3x + 4$; $g(x) = f(2x)$; $h(x) = \frac{1}{2}f(x)$

Combining Transformations

STUDY TIP

You can perform transformations on the graph of *any* function f using these steps.

Core Concept

Transformations of Graphs

The graph of $y = a \cdot f(x - h) + k$ or the graph of $y = f(ax - h) + k$ can be obtained from the graph of $y = f(x)$ by performing these steps.

Step 1 Translate the graph of $y = f(x)$ horizontally h units.

Step 2 Use a to stretch or shrink the resulting graph from Step 1.

Step 3 Reflect the resulting graph from Step 2 when $a < 0$.

Step 4 Translate the resulting graph from Step 3 vertically k units.

EXAMPLE 5 Combining Transformations

Graph $f(x) = x$ and $g(x) = -2x + 3$. Describe the transformations from the graph of f to the graph of g.

SOLUTION

Note that you can rewrite g as $g(x) = -2f(x) + 3$.

ANOTHER WAY

You could also rewrite g as $g(x) = f(-2x) + 3$. In this case, the transformations from the graph of f to the graph of g will be different from those in Example 5.

Step 1 There is no horizontal translation from the graph of f to the graph of g.

Step 2 Stretch the graph of f vertically by a factor of 2 to get the graph of $h(x) = 2x$.

Step 3 Reflect the graph of h in the x-axis to get the graph of $r(x) = -2x$.

Step 4 Translate the graph of r vertically 3 units up to get the graph of $g(x) = -2x + 3$.

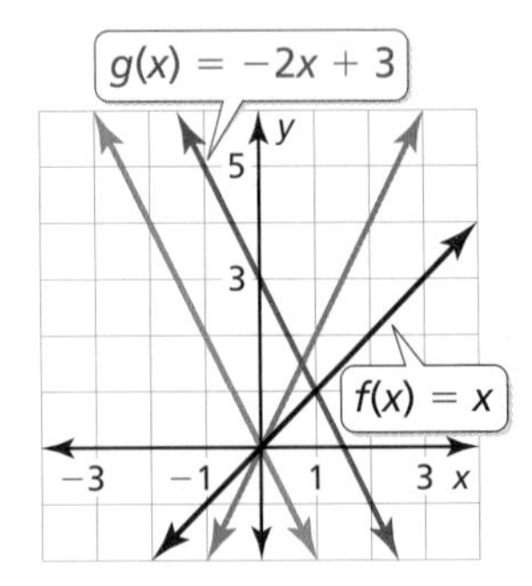

EXAMPLE 6 Solving a Real-Life Problem

A cable company charges customers \$60 per month for its service, with no installation fee. The cost to a customer is represented by $c(m) = 60m$, where m is the number of months of service. To attract new customers, the cable company reduces the monthly fee to \$30 but adds an installation fee of \$45. The cost to a new customer is represented by $r(m) = 30m + 45$, where m is the number of months of service. Describe the transformations from the graph of c to the graph of r.

SOLUTION

Note that you can rewrite r as $r(m) = \frac{1}{2}c(m) + 45$. In this form, you can use the order of operations to get the outputs of r from the outputs of c. First, multiply the outputs of c by $\frac{1}{2}$ to get $h(m) = 30m$. Then add 45 to the outputs of h to get $r(m) = 30m + 45$.

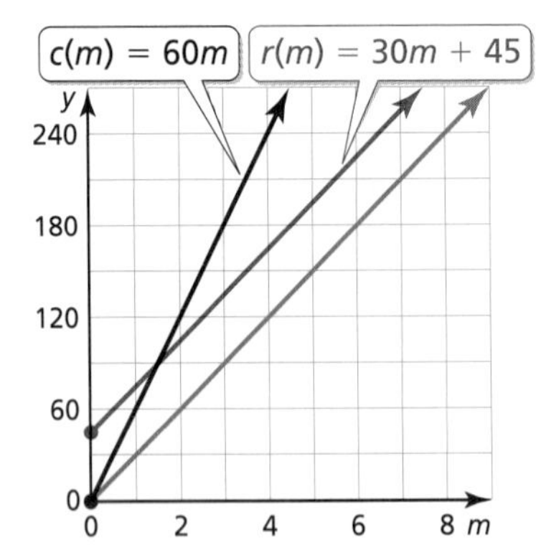

▶ The transformations are a vertical shrink by a factor of $\frac{1}{2}$ and then a vertical translation 45 units up.

Monitoring Progress  Help in English and Spanish at *BigIdeasMath.com*

5. Graph $f(x) = x$ and $h(x) = \frac{1}{4}x - 2$. Describe the transformations from the graph of f to the graph of h.

3.6 Exercises

Dynamic Solutions available at *BigIdeasMath.com*

Vocabulary and Core Concept Check

1. **WRITING** Describe the relationship between $f(x) = x$ and all other nonconstant linear functions.

2. **VOCABULARY** Name four types of transformations. Give an example of each and describe how it affects the graph of a function.

3. **WRITING** How does the value of a in the equation $y = f(ax)$ affect the graph of $y = f(x)$? How does the value of a in the equation $y = af(x)$ affect the graph of $y = f(x)$?

4. **REASONING** The functions f and g are linear functions. The graph of g is a vertical shrink of the graph of f. What can you say about the x-intercepts of the graphs of f and g? Is this always true? Explain.

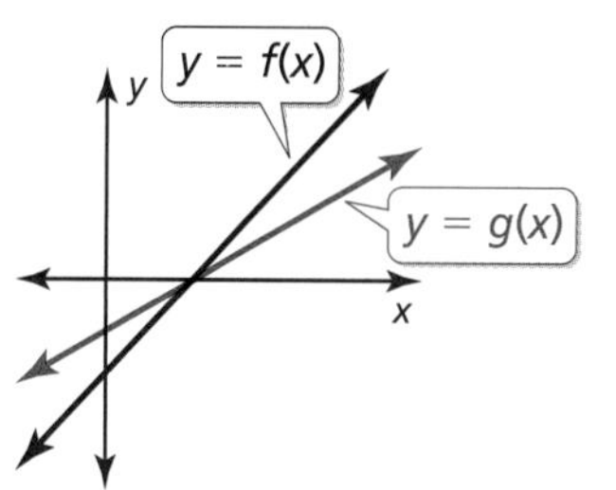

Monitoring Progress and Modeling with Mathematics

In Exercises 5–10, use the graphs of f and g to describe the transformation from the graph of f to the graph of g. *(See Example 1.)*

5.

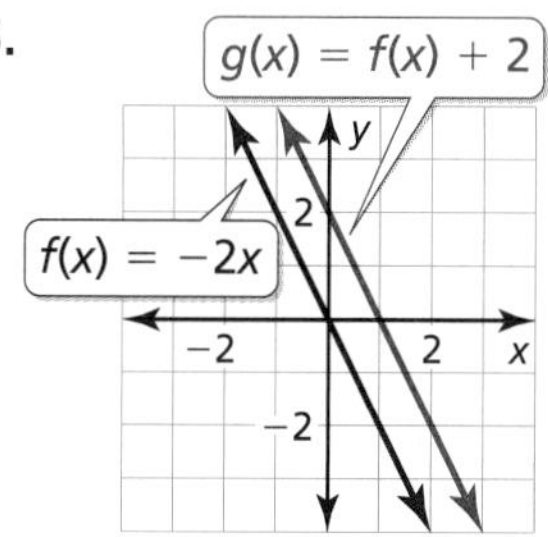

6.

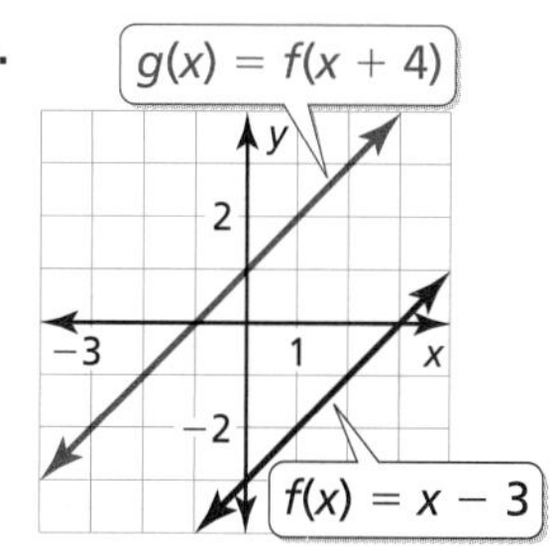

7. $f(x) = \frac{1}{3}x + 3$; $g(x) = f(x) - 3$

8. $f(x) = -3x + 4$; $g(x) = f(x) + 1$

9. $f(x) = -x - 2$; $g(x) = f(x + 5)$

10. $f(x) = \frac{1}{2}x - 5$; $g(x) = f(x - 3)$

11. **MODELING WITH MATHEMATICS** You and a friend start biking from the same location. Your distance d (in miles) after t minutes is given by the function $d(t) = \frac{1}{5}t$. Your friend starts biking 5 minutes after you. Your friend's distance f is given by the function $f(t) = d(t - 5)$. Describe the transformation from the graph of d to the graph of f.

12. **MODELING WITH MATHEMATICS** The total cost C (in dollars) to cater an event with p people is given by the function $C(p) = 18p + 50$. The set-up fee increases by \$25. The new total cost T is given by the function $T(p) = C(p) + 25$. Describe the transformation from the graph of C to the graph of T.

In Exercises 13–16, use the graphs of f and h to describe the transformation from the graph of f to the graph of h. *(See Example 2.)*

13.

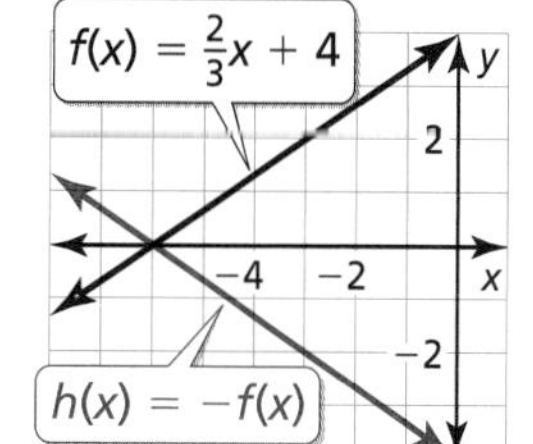

14.

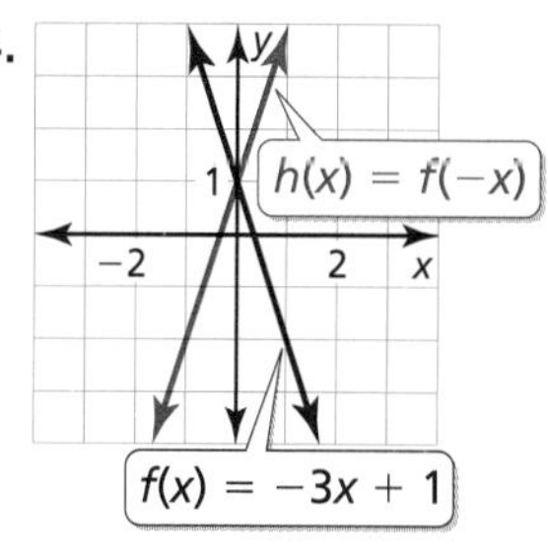

15. $f(x) = -5 - x$; $h(x) = f(-x)$

16. $f(x) = \frac{1}{4}x - 2$; $h(x) = -f(x)$

In Exercises 17–22, use the graphs of *f* and *r* to describe the transformation from the graph of *f* to the graph of *r*. *(See Example 3.)*

17.

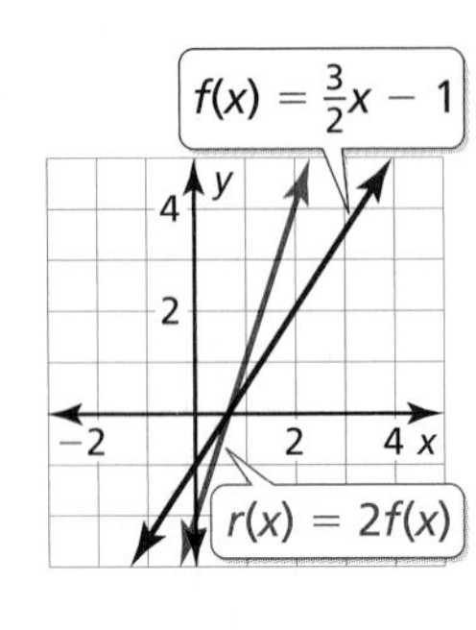

18.

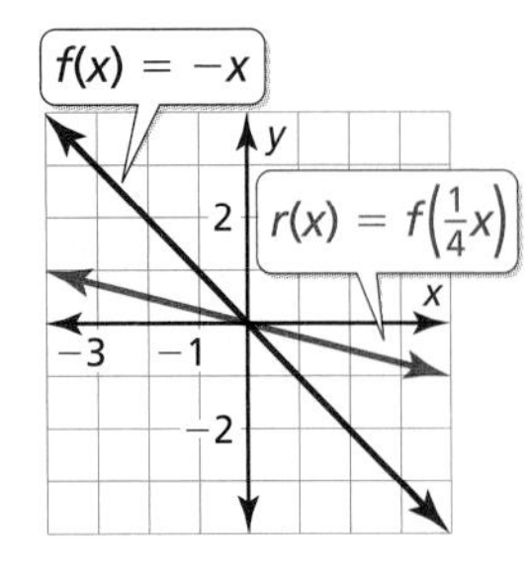

19. $f(x) = -2x - 4;\ r(x) = f\left(\frac{1}{2}x\right)$

20. $f(x) = 3x + 5;\ r(x) = f\left(\frac{1}{3}x\right)$

21. $f(x) = \frac{2}{3}x + 1;\ r(x) = 3f(x)$

22. $f(x) = -\frac{1}{4}x - 2;\ r(x) = 4f(x)$

In Exercises 23–28, use the graphs of *f* and *h* to describe the transformation from the graph of *f* to the graph of *h*. *(See Example 4.)*

23.

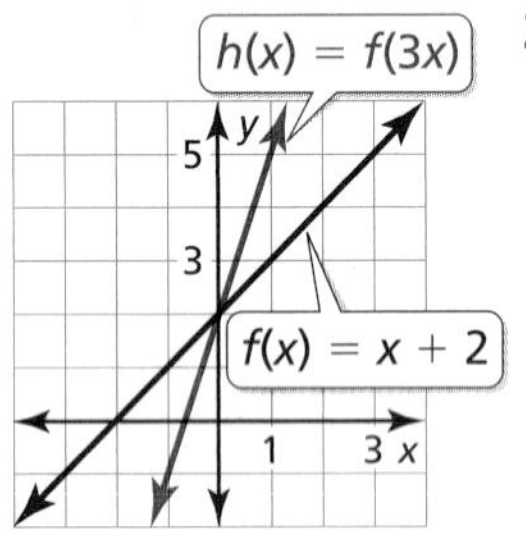

24.

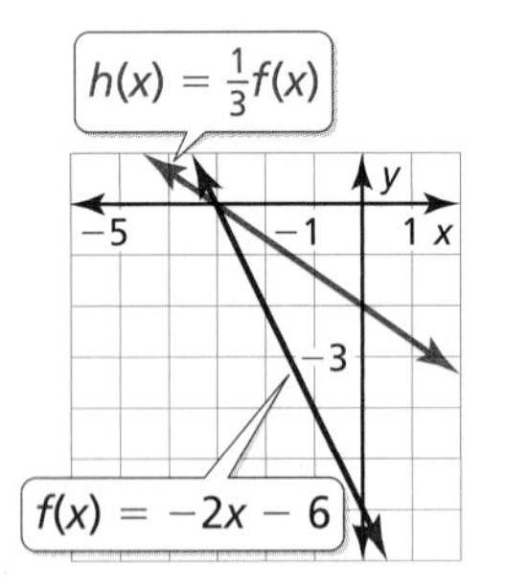

25. $f(x) = 3x - 12;\ h(x) = \frac{1}{6}f(x)$

26. $f(x) = -x + 1;\ h(x) = f(2x)$

27. $f(x) = -2x - 2;\ h(x) = f(5x)$

28. $f(x) = 4x + 8;\ h(x) = \frac{3}{4}f(x)$

In Exercises 29–34, use the graphs of *f* and *g* to describe the transformation from the graph of *f* to the graph of *g*.

29. $f(x) = x - 2;\ g(x) = \frac{1}{4}f(x)$

30. $f(x) = -4x + 8;\ g(x) = -f(x)$

31. $f(x) = -2x - 7;\ g(x) = f(x - 2)$

32. $f(x) = 3x + 8;\ g(x) = f\left(\frac{2}{3}x\right)$

33. $f(x) = x - 6;\ g(x) = 6f(x)$

34. $f(x) = -x;\ g(x) = f(x) - 3$

In Exercises 35–38, write a function *g* in terms of *f* so that the statement is true.

35. The graph of *g* is a horizontal translation 2 units right of the graph of *f*.

36. The graph of *g* is a reflection in the *y*-axis of the graph of *f*.

37. The graph of *g* is a vertical stretch by a factor of 4 of the graph of *f*.

38. The graph of *g* is a horizontal shrink by a factor of $\frac{1}{5}$ of the graph of *f*.

ERROR ANALYSIS In Exercises 39 and 40, describe and correct the error in graphing *g*.

39.

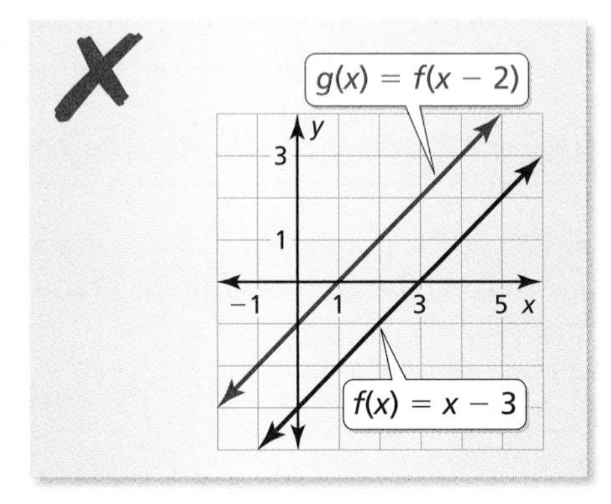

40.

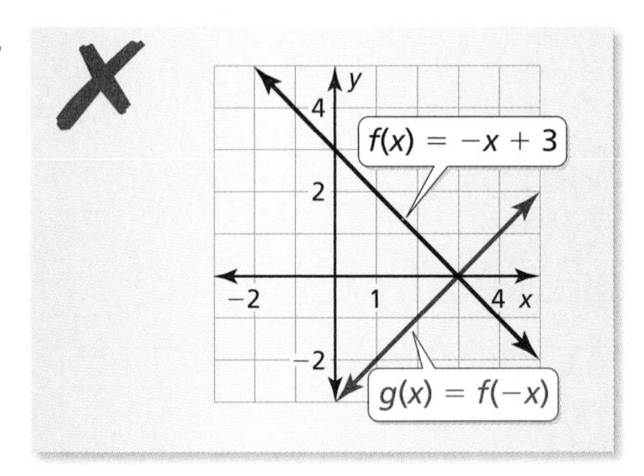

In Exercises 41–46, graph *f* and *h*. Describe the transformations from the graph of *f* to the graph of *h*. *(See Example 5.)*

41. $f(x) = x;\ h(x) = \frac{1}{3}x + 1$

42. $f(x) = x;\ h(x) = 4x - 2$

43. $f(x) = x;\ h(x) = -3x - 4$

44. $f(x) = x;\ h(x) = -\frac{1}{2}x + 3$

45. $f(x) = 2x;\ h(x) = 6x - 5$

46. $f(x) = 3x;\ h(x) = -3x - 7$

47. MODELING WITH MATHEMATICS The function $t(x) = -4x + 72$ represents the temperature from 5 P.M. to 11 P.M., where x is the number of hours after 5 P.M. The function $d(x) = 4x + 72$ represents the temperature from 10 A.M. to 4 P.M., where x is the number of hours after 10 A.M. Describe the transformation from the graph of t to the graph of d.

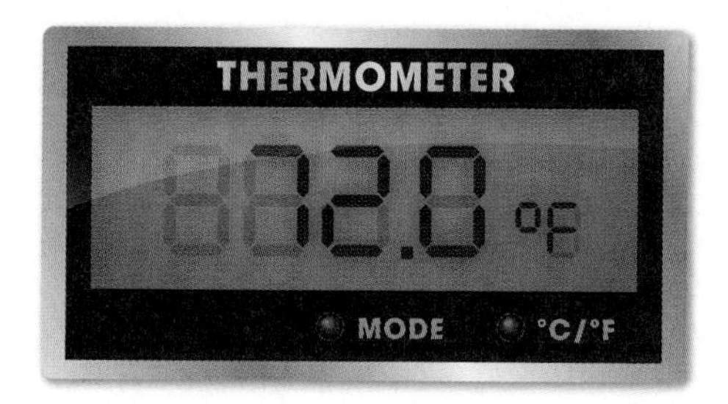

48. MODELING WITH MATHEMATICS A school sells T-shirts to promote school spirit. The school's profit is given by the function $P(x) = 8x - 150$, where x is the number of T-shirts sold. During the play-offs, the school increases the price of the T-shirts. The school's profit during the play-offs is given by the function $Q(x) = 16x - 200$, where x is the number of T-shirts sold. Describe the transformations from the graph of P to the graph of Q. *(See Example 6.)*

49. USING STRUCTURE The graph of $g(x) = a \cdot f(x - b) + c$ is a transformation of the graph of the linear function f. Select the word or value that makes each statement true.

reflection	translation	-1
stretch	shrink	0
left	right	1
y-axis	x-axis	

a. The graph of g is a vertical ______ of the graph of f when $a = 4$, $b = 0$, and $c = 0$.

b. The graph of g is a horizontal translation ______ of the graph of f when $a = 1$, $b = 2$, and $c = 0$.

c. The graph of g is a vertical translation 1 unit up of the graph of f when $a = 1$, $b = 0$, and $c =$ ____.

50. USING STRUCTURE The graph of $h(x) = a \cdot f(bx - c) + d$ is a transformation of the graph of the linear function f. Select the word or value that makes each statement true.

vertical	horizontal	0
stretch	shrink	$\frac{1}{5}$
y-axis	x-axis	5

a. The graph of h is a ______ shrink of the graph of f when $a = \frac{1}{3}$, $b = 1$, $c = 0$, and $d = 0$.

b. The graph of h is a reflection in the ______ of the graph of f when $a = 1$, $b = -1$, $c = 0$, and $d = 0$.

c. The graph of h is a horizontal stretch of the graph of f by a factor of 5 when $a = 1$, $b =$ _____, $c = 0$, and $d = 0$.

51. ANALYZING GRAPHS Which of the graphs are related by only a translation? Explain.

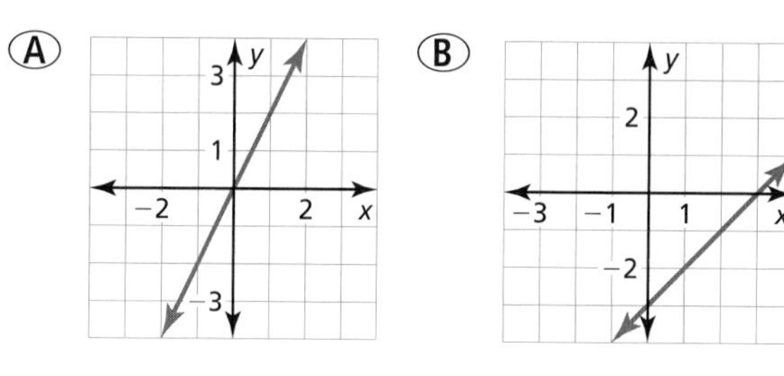

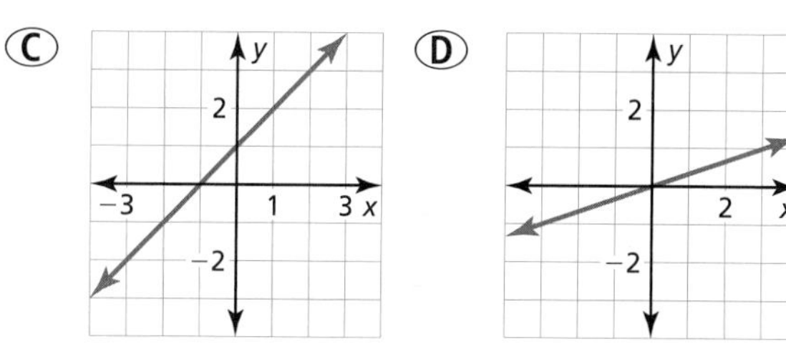

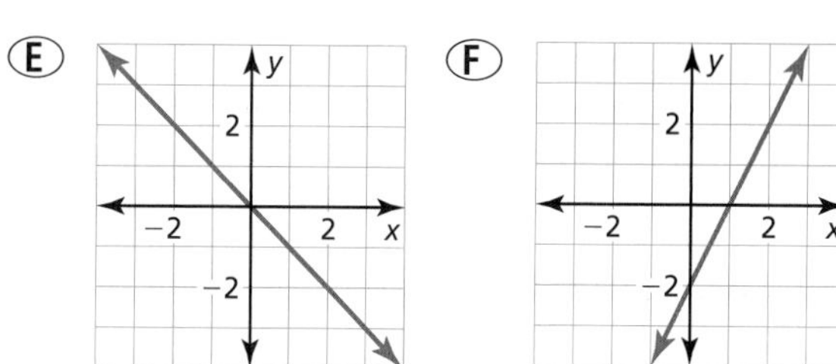

52. ANALYZING RELATIONSHIPS A swimming pool is filled with water by a hose at a rate of 1020 gallons per hour. The amount v (in gallons) of water in the pool after t hours is given by the function $v(t) = 1020t$. How does the graph of v change in each situation?

a. A larger hose is found. Then the pool is filled at a rate of 1360 gallons per hour.

b. Before filling up the pool with a hose, a water truck adds 2000 gallons of water to the pool.

53. ANALYZING RELATIONSHIPS You have \$50 to spend on fabric for a blanket. The amount m (in dollars) of money you have after buying y yards of fabric is given by the function $m(y) = -9.98y + 50$. How does the graph of m change in each situation?

a. You receive an additional \$10 to spend on the fabric.

b. The fabric goes on sale, and each yard now costs \$4.99.

54. THOUGHT PROVOKING Write a function g whose graph passes through the point (4, 2) and is a transformation of the graph of $f(x) = x$.

In Exercises 55–60, graph f and g. Write g in terms of f. Describe the transformation from the graph of f to the graph of g.

55. $f(x) = 2x - 5$; $g(x) = 2x - 8$

56. $f(x) = 4x + 1$; $g(x) = -4x - 1$

57. $f(x) = 3x + 9$; $g(x) = 3x + 15$

58. $f(x) = -x - 4$; $g(x) = x - 4$

59. $f(x) = x + 2$; $g(x) = \frac{2}{3}x + 2$

60. $f(x) = x - 1$; $g(x) = 3x - 3$

61. REASONING The graph of $f(x) = x + 5$ is a vertical translation 5 units up of the graph of $f(x) = x$. How can you obtain the graph of $f(x) = x + 5$ from the graph of $f(x) = x$ using a horizontal translation?

62. HOW DO YOU SEE IT? Match each function with its graph. Explain your reasoning.

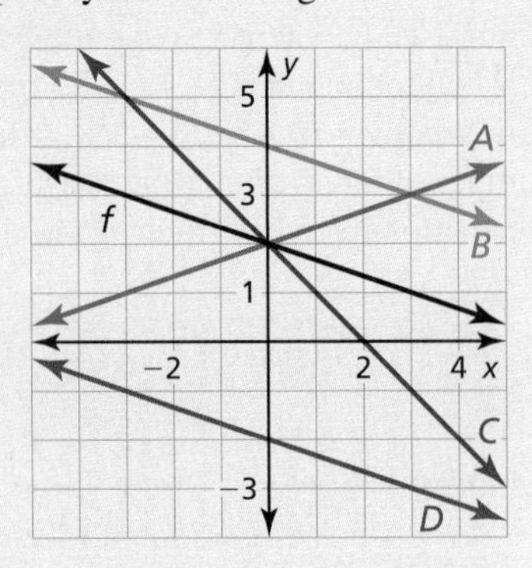

a. $a(x) = f(-x)$ b. $g(x) = f(x) - 4$

c. $h(x) = f(x) + 2$ d. $k(x) = f(3x)$

REASONING In Exercises 63–66, find the value of r.

63.

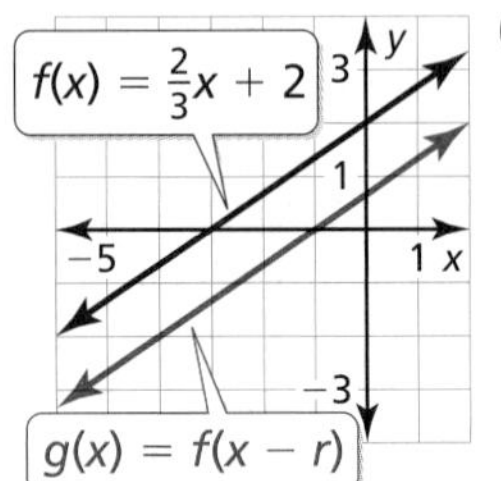

64.

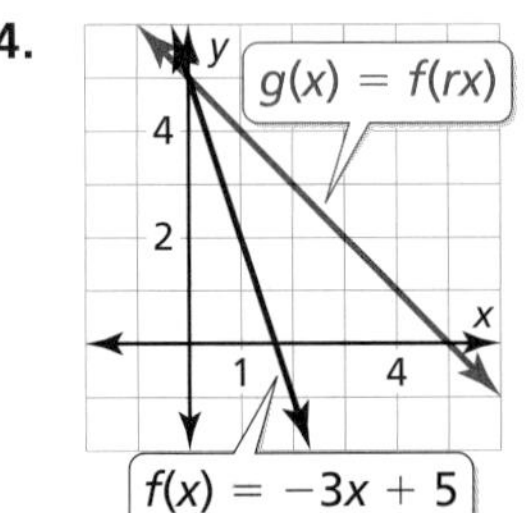

65.

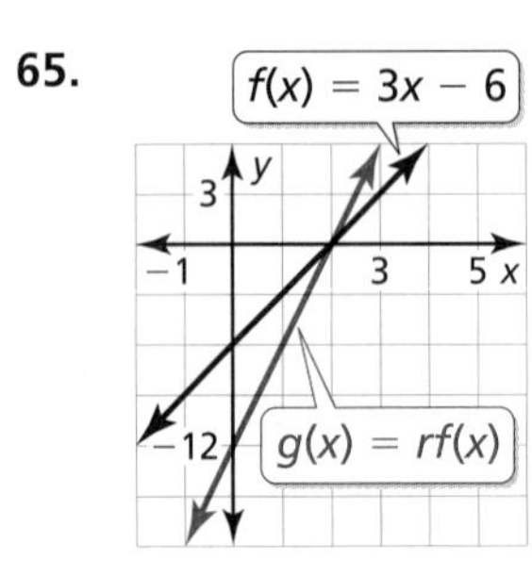

66.

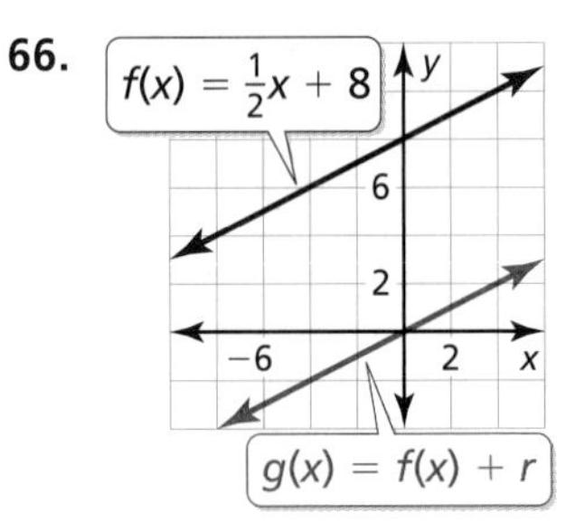

67. CRITICAL THINKING When is the graph of $y = f(x) + w$ the same as the graph of $y = f(x + w)$ for linear functions? Explain your reasoning.

Maintaining Mathematical Proficiency

Reviewing what you learned in previous grades and lessons

Solve the formula for the indicated variable. *(Section 1.5)*

68. Solve for h.

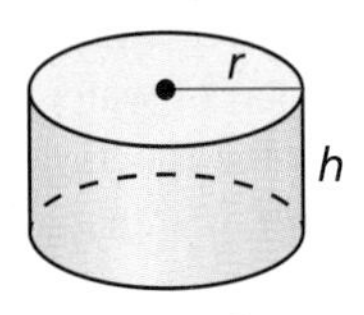

$V = \pi r^2 h$

69. Solve for w.

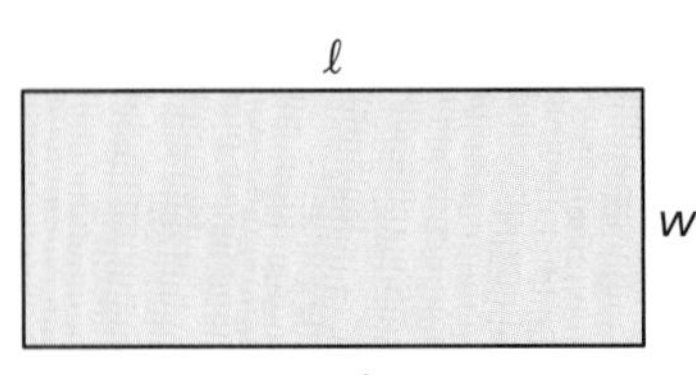

$P = 2\ell + 2w$

Solve the inequality. Graph the solution, if possible. *(Section 2.6)*

70. $|x - 3| \leq 14$ **71.** $|2x + 4| > 16$ **72.** $5|x + 7| < 25$ **73.** $-2|x + 1| \geq 18$

3.7 Graphing Absolute Value Functions

Essential Question How do the values of a, h, and k affect the graph of the absolute value function $g(x) = a|x - h| + k$?

The parent absolute value function is

$f(x) = |x|$. Parent absolute value function

The graph of f is V-shaped.

EXPLORATION 1 Identifying Graphs of Absolute Value Functions

Work with a partner. Match each absolute value function with its graph. Then use a graphing calculator to verify your answers.

a. $g(x) = -|x - 2|$ **b.** $g(x) = |x - 2| + 2$ **c.** $g(x) = -|x + 2| - 2$

d. $g(x) = |x - 2| - 2$ **e.** $g(x) = 2|x - 2|$ **f.** $g(x) = -|x + 2| + 2$

A.
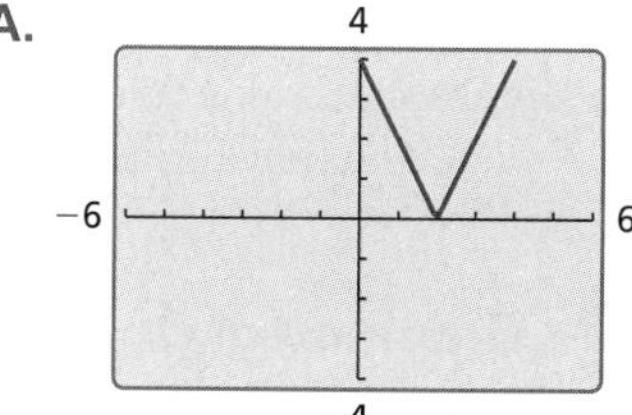

B.
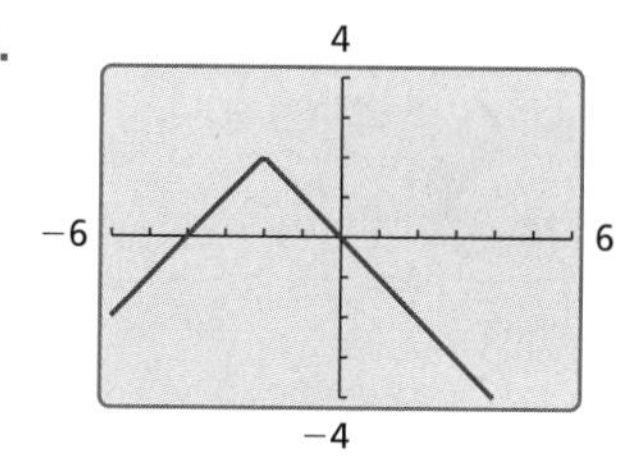

C.
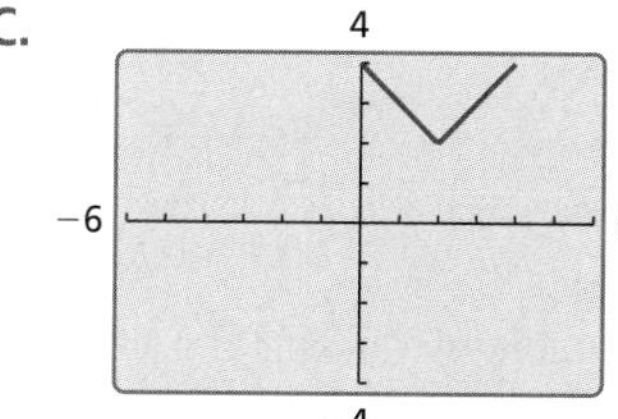

D.
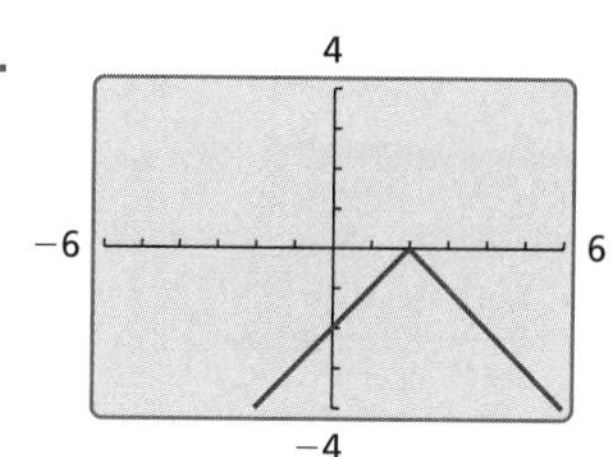

E.
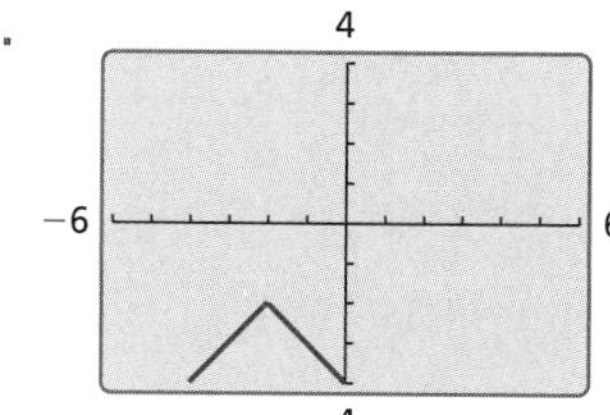

F.
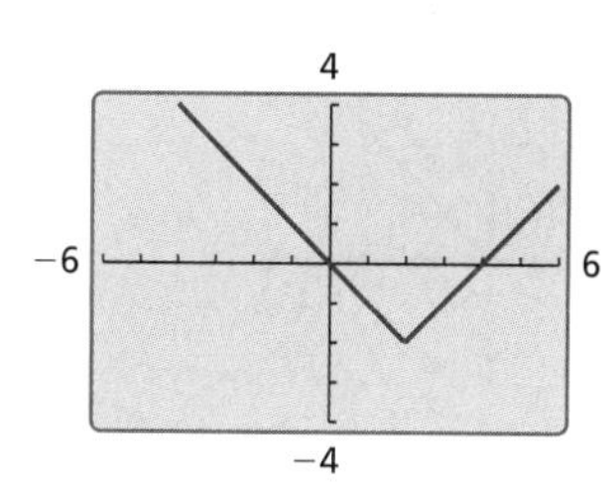

LOOKING FOR STRUCTURE

To be proficient in math, you need to look closely to discern a pattern or structure.

Communicate Your Answer

2. How do the values of a, h, and k affect the graph of the absolute value function

$g(x) = a|x - h| + k$?

3. Write the equation of the absolute value function whose graph is shown. Use a graphing calculator to verify your equation.

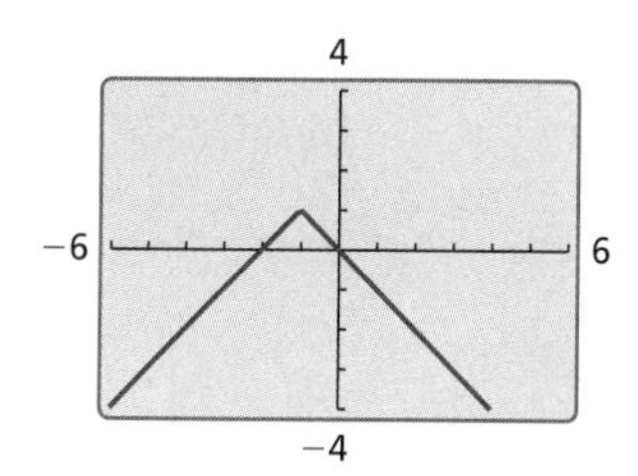

3.7 Lesson

What You Will Learn

- Translate graphs of absolute value functions.
- Stretch, shrink, and reflect graphs of absolute value functions.
- Combine transformations of graphs of absolute value functions.

Core Vocabulary

absolute value function, *p. 156*
vertex, *p. 156*
vertex form, *p. 158*

Previous
domain
range

Translating Graphs of Absolute Value Functions

Core Concept

Absolute Value Function

An **absolute value function** is a function that contains an absolute value expression. The parent absolute value function is $f(x) = |x|$. The graph of $f(x) = |x|$ is V-shaped and symmetric about the y-axis. The **vertex** is the point where the graph changes direction. The vertex of the graph of $f(x) = |x|$ is (0, 0).

The domain of $f(x) = |x|$ is all real numbers.
The range is $y \geq 0$.

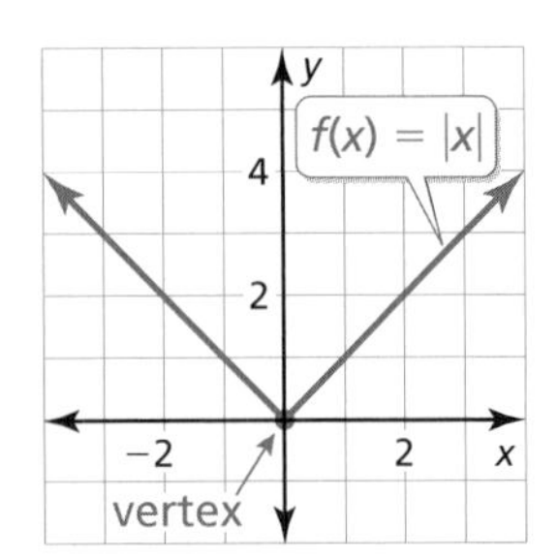

The graphs of all other absolute value functions are transformations of the graph of the parent function $f(x) = |x|$. The transformations presented in Section 3.6 also apply to absolute value functions.

EXAMPLE 1 **Graphing $g(x) = |x| + k$ and $g(x) = |x - h|$**

Graph each function. Compare each graph to the graph of $f(x) = |x|$. Describe the domain and range.

a. $g(x) = |x| + 3$

b. $m(x) = |x - 2|$

SOLUTION

a. Step 1 Make a table of values.

x	−2	−1	0	1	2
$g(x)$	5	4	3	4	5

Step 2 Plot the ordered pairs.

Step 3 Draw the V-shaped graph.

▶ The function g is of the form $y = f(x) + k$, where $k = 3$. So, the graph of g is a vertical translation 3 units up of the graph of f. The domain is all real numbers. The range is $y \geq 3$.

b. Step 1 Make a table of values.

x	0	1	2	3	4
$m(x)$	2	1	0	1	2

Step 2 Plot the ordered pairs.

Step 3 Draw the V-shaped graph.

▶ The function m is of the form $y = f(x - h)$, where $h = 2$. So, the graph of m is a horizontal translation 2 units right of the graph of f. The domain is all real numbers. The range is $y \geq 0$.

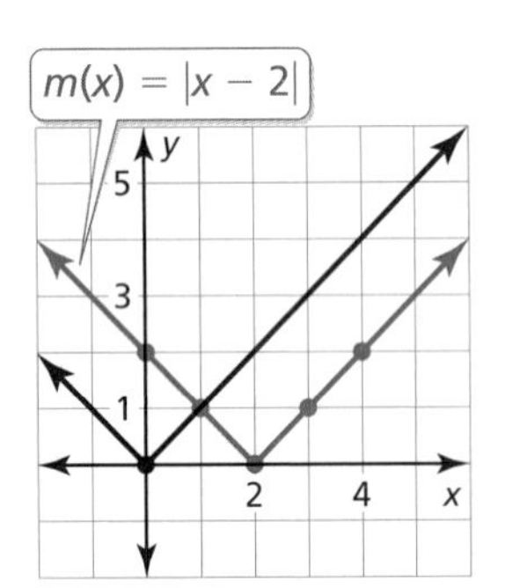

Monitoring Progress  Help in English and Spanish at *BigIdeasMath.com*

Graph the function. Compare the graph to the graph of $f(x) = |x|$. Describe the domain and range.

1. $h(x) = |x| - 1$

2. $n(x) = |x + 4|$

Stretching, Shrinking, and Reflecting

EXAMPLE 2 **Graphing $g(x) = a|x|$**

Graph each function. Compare each graph to the graph of $f(x) = |x|$. Describe the domain and range.

a. $q(x) = 2|x|$

b. $p(x) = -\frac{1}{2}|x|$

SOLUTION

a. Step 1 Make a table of values.

x	-2	-1	0	1	2
$q(x)$	4	2	0	2	4

Step 2 Plot the ordered pairs.

Step 3 Draw the V-shaped graph.

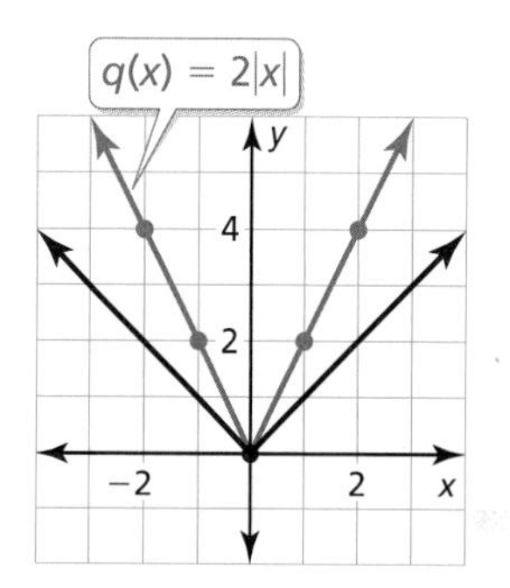

STUDY TIP

A vertical stretch of the graph of $f(x) = |x|$ is *narrower* than the graph of $f(x) = |x|$.

▶ The function q is of the form $y = a \cdot f(x)$, where $a = 2$. So, the graph of q is a vertical stretch of the graph of f by a factor of 2. The domain is all real numbers. The range is $y \geq 0$.

b. Step 1 Make a table of values.

x	-2	-1	0	1	2
$p(x)$	-1	$-\frac{1}{2}$	0	$-\frac{1}{2}$	-1

Step 2 Plot the ordered pairs.

Step 3 Draw the V-shaped graph.

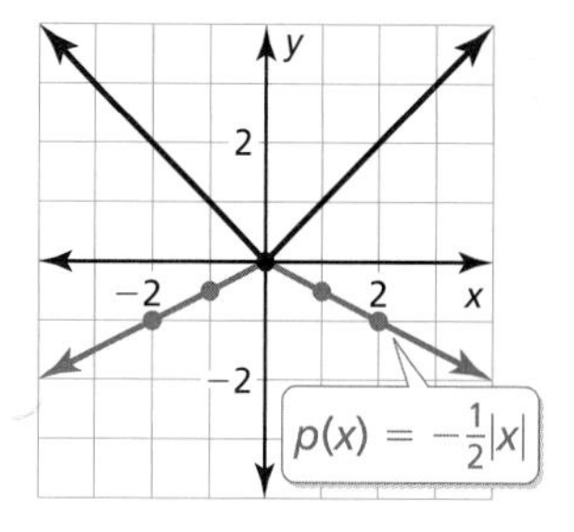

STUDY TIP

A vertical shrink of the graph of $f(x) = |x|$ is *wider* than the graph of $f(x) = |x|$.

▶ The function p is of the form $y = -a \cdot f(x)$, where $a = \frac{1}{2}$. So, the graph of p is a vertical shrink of the graph of f by a factor of $\frac{1}{2}$ and a reflection in the x-axis. The domain is all real numbers. The range is $y \leq 0$.

Monitoring Progress

Help in English and Spanish at *BigIdeasMath.com*

Graph the function. Compare the graph to the graph of $f(x) = |x|$. Describe the domain and range.

3. $t(x) = -3|x|$

4. $v(x) = \frac{1}{4}|x|$

Core Concept

Vertex Form of an Absolute Value Function

An absolute value function written in the form $g(x) = a|x - h| + k$, where $a \neq 0$, is in **vertex form**. The vertex of the graph of g is (h, k).

Any absolute value function can be written in vertex form, and its graph is symmetric about the line $x = h$.

STUDY TIP

The function g is *not* in vertex form because the x variable does not have a coefficient of 1.

EXAMPLE 3 Graphing $f(x) = |x - h| + k$ and $g(x) = f(ax)$

Graph $f(x) = |x + 2| - 3$ and $g(x) = |2x + 2| - 3$. Compare the graph of g to the graph of f.

SOLUTION

Step 1 Make a table of values for each function.

x	−4	−3	−2	−1	0	1	2
$f(x)$	−1	−2	−3	−2	−1	0	1

x	−2	−1.5	−1	−0.5	0	0.5	1
$g(x)$	−1	−2	−3	−2	−1	0	1

Step 2 Plot the ordered pairs.

Step 3 Draw the V-shaped graph of each function. Notice that the vertex of the graph of f is $(-2, -3)$ and the graph is symmetric about $x = -2$.

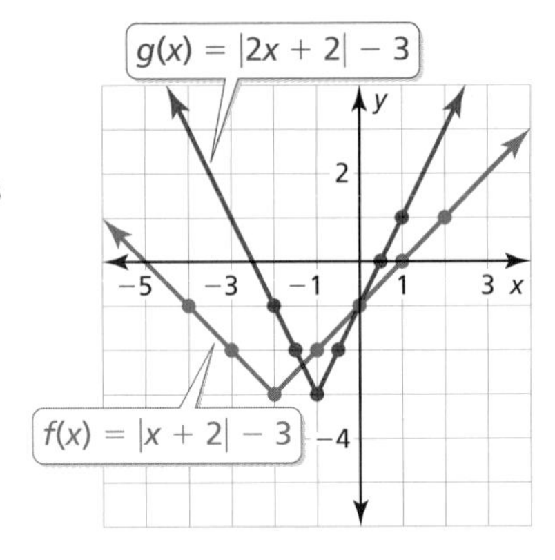

▶ Note that you can rewrite g as $g(x) = f(2x)$, which is of the form $y = f(ax)$, where $a = 2$. So, the graph of g is a horizontal shrink of the graph of f by a factor of $\frac{1}{2}$. The y-intercept is the same for both graphs. The points on the graph of f move halfway closer to the y-axis, resulting in the graph of g. When the input values of f are 2 times the input values of g, the output values of f and g are the same.

Monitoring Progress  Help in English and Spanish at *BigIdeasMath.com*

5. Graph $f(x) = |x - 1|$ and $g(x) = \left|\frac{1}{2}x - 1\right|$. Compare the graph of g to the graph of f.

6. Graph $f(x) = |x + 2| + 2$ and $g(x) = |-4x + 2| + 2$. Compare the graph of g to the graph of f.

Combining Transformations

EXAMPLE 4 **Graphing $g(x) = a|x - h| + k$**

> **REMEMBER**
> You can obtain the graph of $y = a \cdot f(x - h) + k$ from the graph of $y = f(x)$ using the steps you learned in Section 3.6.

Let $g(x) = -2|x - 1| + 3$. (a) Describe the transformations from the graph of $f(x) = |x|$ to the graph of g. (b) Graph g.

SOLUTION

a. Step 1 Translate the graph of f horizontally 1 unit right to get the graph of $t(x) = |x - 1|$.

Step 2 Stretch the graph of t vertically by a factor of 2 to get the graph of $h(x) = 2|x - 1|$.

Step 3 Reflect the graph of h in the x-axis to get the graph of $r(x) = -2|x - 1|$.

Step 4 Translate the graph of r vertically 3 units up to get the graph of $g(x) = -2|x - 1| + 3$.

b. Method 1

Step 1 Make a table of values.

x	−1	0	1	2	3
$g(x)$	−1	1	3	1	−1

Step 2 Plot the ordered pairs.

Step 3 Draw the V-shaped graph.

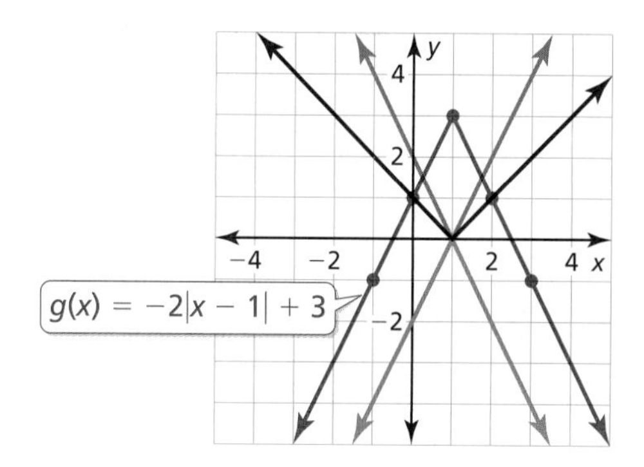

Method 2

Step 1 Identify and plot the vertex. $(h, k) = (1, 3)$

Step 2 Plot another point on the graph, such as $(2, 1)$. Because the graph is symmetric about the line $x = 1$, you can use symmetry to plot a third point, $(0, 1)$.

Step 3 Draw the V-shaped graph.

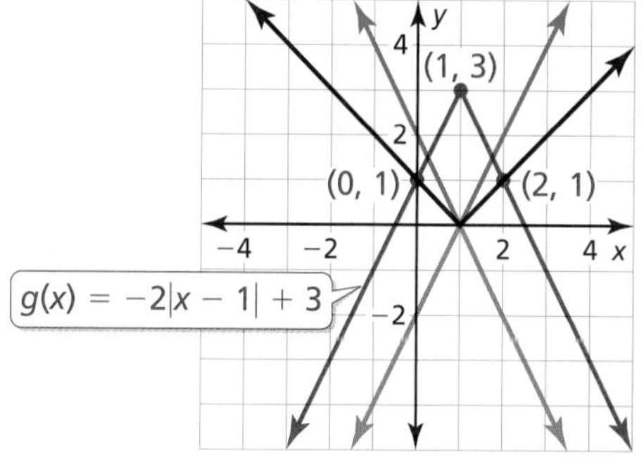

Monitoring Progress  Help in English and Spanish at *BigIdeasMath.com*

7. Let $g(x) = \left|-\frac{1}{2}x + 2\right| + 1$. (a) Describe the transformations from the graph of $f(x) = |x|$ to the graph of g. (b) Graph g.

3.7 Exercises

Dynamic Solutions available at *BigIdeasMath.com*

Vocabulary and Core Concept Check

1. **COMPLETE THE SENTENCE** The point $(1, -4)$ is the _______ of the graph of $f(x) = -3|x - 1| - 4$.

2. **USING STRUCTURE** How do you know whether the graph of $f(x) = a|x - h| + k$ is a vertical stretch or a vertical shrink of the graph of $f(x) = |x|$?

3. **WRITING** Describe three different types of transformations of the graph of an absolute value function.

4. **REASONING** The graph of which function has the same y-intercept as the graph of $f(x) = |x - 2| + 5$? Explain.

   $g(x) = |3x - 2| + 5$ $\quad$ $h(x) = 3|x - 2| + 5$

Monitoring Progress and Modeling with Mathematics

In Exercises 5–12, graph the function. Compare the graph to the graph of $f(x) = |x|$. Describe the domain and range. *(See Examples 1 and 2.)*

5. $d(x) = |x| - 4$
6. $r(x) = |x| + 5$
7. $m(x) = |x + 1|$
8. $v(x) = |x - 3|$
9. $p(x) = \frac{1}{3}|x|$
10. $j(x) = 3|x|$
11. $a(x) = -5|x|$
12. $q(x) = -\frac{3}{2}|x|$

In Exercises 13–16, graph the function. Compare the graph to the graph of $f(x) = |x - 6|$.

13. $h(x) = |x - 6| + 2$
14. $n(x) = \frac{1}{2}|x - 6|$
15. $k(x) = -3|x - 6|$
16. $g(x) = |x - 1|$

In Exercises 17 and 18, graph the function. Compare the graph to the graph of $f(x) = |x + 3| - 2$.

17. $y(x) = |x + 4| - 2$
18. $b(x) = |x + 3| + 3$

In Exercises 19–22, compare the graphs. Find the value of h, k, or a.

19.

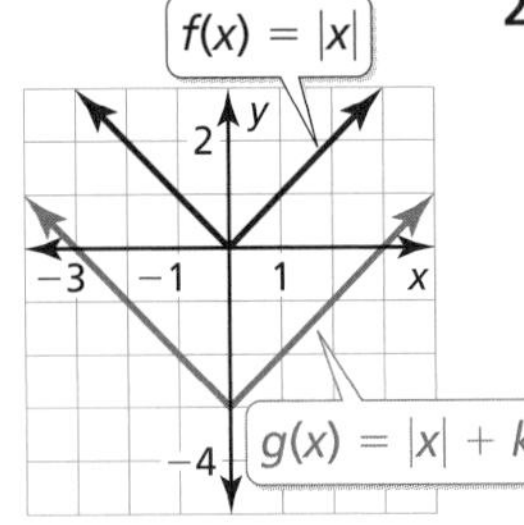

20.

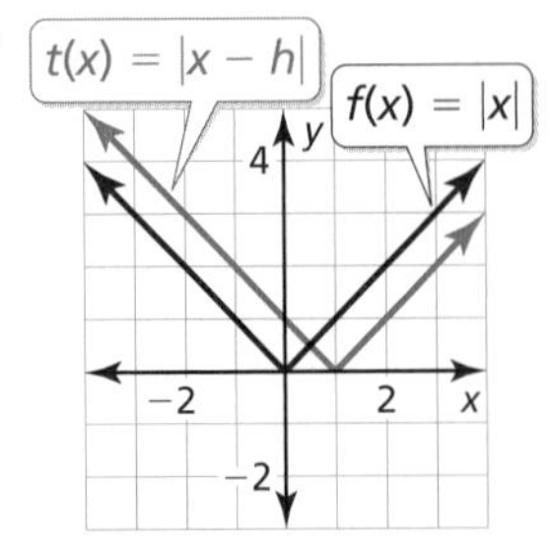

21.

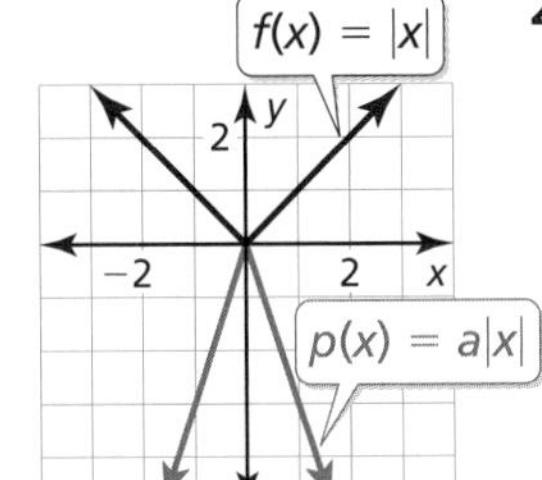

22.

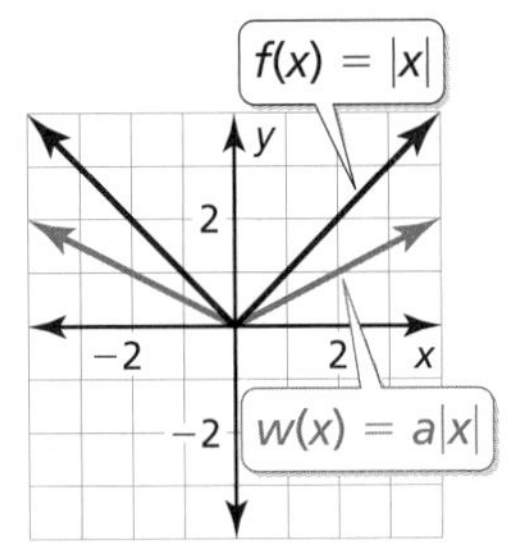

In Exercises 23–26, write an equation that represents the given transformation(s) of the graph of $g(x) = |x|$.

23. vertical translation 7 units down
24. horizontal translation 10 units left
25. vertical shrink by a factor of $\frac{1}{4}$
26. vertical stretch by a factor of 3 and a reflection in the x-axis

In Exercises 27–32, graph and compare the two functions. *(See Example 3.)*

27. $f(x) = |x - 4|$; $g(x) = |3x - 4|$
28. $h(x) = |x + 5|$; $t(x) = |2x + 5|$
29. $p(x) = |x + 1| - 2$; $q(x) = \left|\frac{1}{4}x + 1\right| - 2$
30. $w(x) = |x - 3| + 4$; $y(x) = |5x - 3| + 4$
31. $a(x) = |x + 2| + 3$; $b(x) = |-4x + 2| + 3$
32. $u(x) = |x - 1| + 2$; $v(x) = \left|-\frac{1}{2}x - 1\right| + 2$

In Exercises 33–40, describe the transformations from the graph of $f(x) = |x|$ to the graph of the given function. Then graph the given function. *(See Example 4.)*

33. $r(x) = |x + 2| - 6$

34. $c(x) = |x + 4| + 4$

35. $d(x) = -|x - 3| + 5$

36. $v(x) = -3|x + 1| + 4$

37. $m(x) = \frac{1}{2}|x + 4| - 1$

38. $s(x) = |2x - 2| - 3$

39. $j(x) = |-x + 1| - 5$

40. $n(x) = \left|-\frac{1}{3}x + 1\right| + 2$

41. MODELING WITH MATHEMATICS The number of pairs of shoes sold s (in thousands) increases and then decreases as described by the function $s(t) = -2|t - 15| + 50$, where t is the time (in weeks).

a. Graph the function.

b. What is the greatest number of pairs of shoes sold in 1 week?

42. MODELING WITH MATHEMATICS On the pool table shown, you bank the five ball off the side represented by the x-axis. The path of the ball is described by the function $p(x) = \frac{4}{3}\left|x - \frac{5}{4}\right|$.

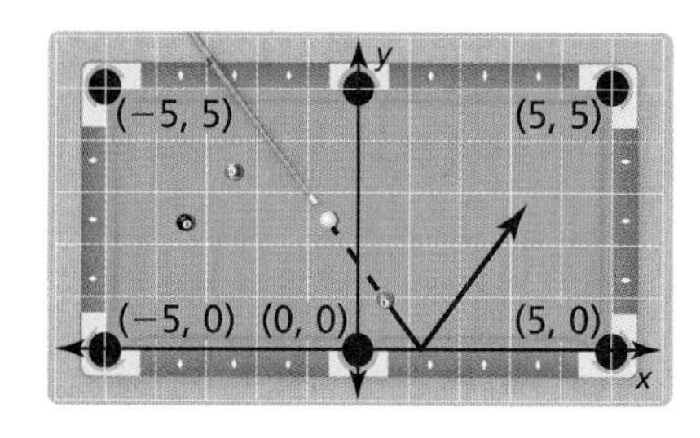

a. At what point does the five ball bank off the side?

b. Do you make the shot? Explain your reasoning.

43. USING TRANSFORMATIONS The points $A\left(-\frac{1}{2}, 3\right)$, $B(1, 0)$, and $C(-4, -2)$ lie on the graph of the absolute value function f. Find the coordinates of the points corresponding to A, B, and C on the graph of each function.

a. $g(x) = f(x) - 5$

b. $h(x) = f(x - 3)$

c. $j(x) = -f(x)$

d. $k(x) = 4f(x)$

44. USING STRUCTURE Explain how the graph of each function compares to the graph of $y = |x|$ for positive and negative values of k, h, and a.

a. $y = |x| + k$

b. $y = |x - h|$

c. $y = a|x|$

d. $y = |ax|$

ERROR ANALYSIS In Exercises 45 and 46, describe and correct the error in graphing the function.

45.

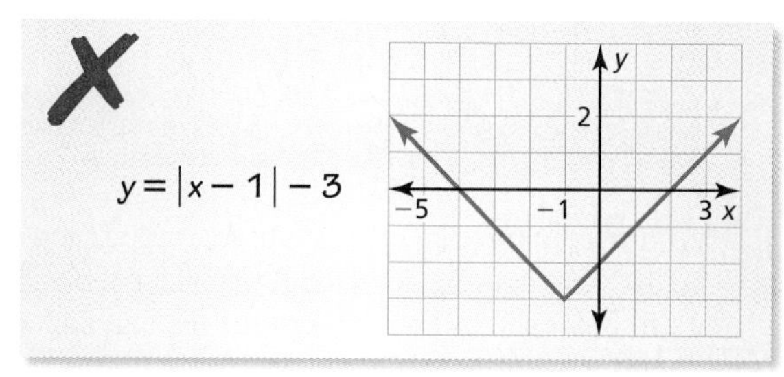

46.

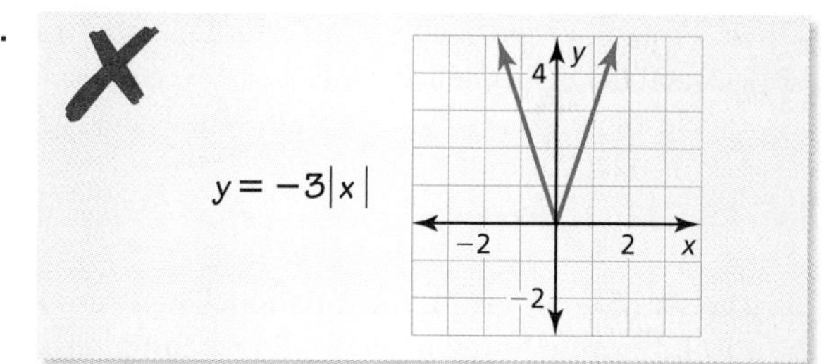

MATHEMATICAL CONNECTIONS In Exercises 47 and 48, write an absolute value function whose graph forms a square with the given graph.

47.

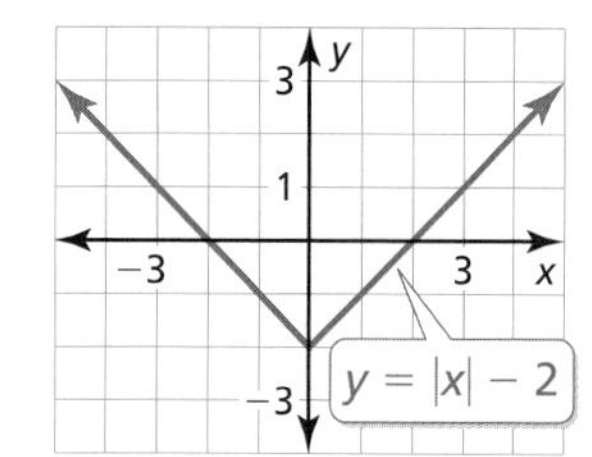

48.

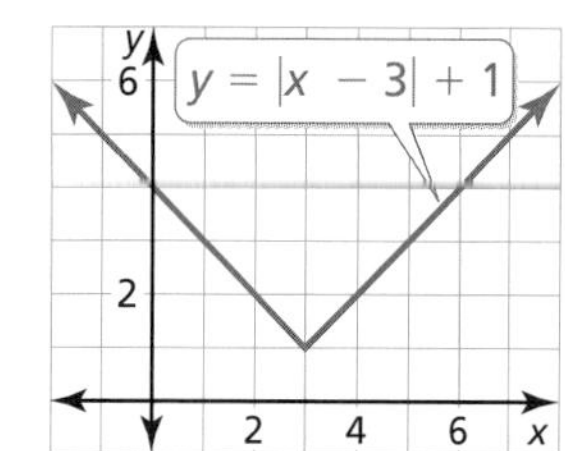

49. WRITING Compare the graphs of $p(x) = |x - 6|$ and $q(x) = |x| - 6$.

50. **HOW DO YOU SEE IT?** The object of a computer game is to break bricks by deflecting a ball toward them using a paddle. The graph shows the current path of the ball and the location of the last brick.

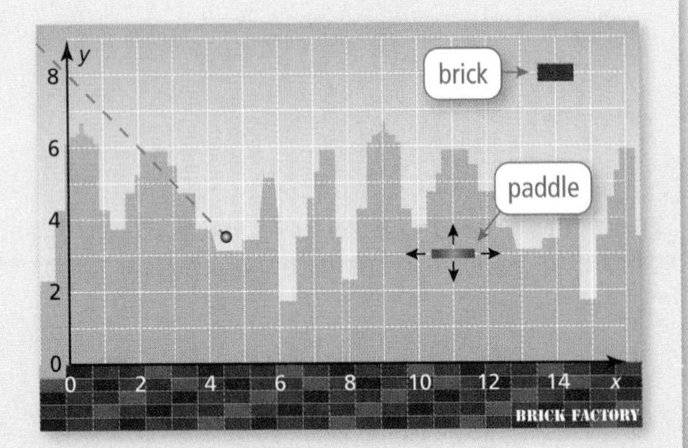

a. You can move the paddle up, down, left, and right. At what coordinates should you place the paddle to break the last brick? Assume the ball deflects at a right angle.

b. You move the paddle to the coordinates in part (a), and the ball is deflected. How can you write an absolute value function that describes the path of the ball?

In Exercises 51–54, graph the function. Then rewrite the absolute value function as two linear functions, one that has the domain $x < 0$ and one that has the domain $x \geq 0$.

51. $y = |x|$

52. $y = |x| - 3$

53. $y = -|x| + 9$

54. $y = -4|x|$

In Exercises 55–58, graph and compare the two functions.

55. $f(x) = |x - 1| + 2$; $g(x) = 4|x - 1| + 8$

56. $s(x) = |2x - 5| - 6$; $t(x) = \frac{1}{2}|2x - 5| - 3$

57. $v(x) = -2|3x + 1| + 4$; $w(x) = 3|3x + 1| - 6$

58. $c(x) = 4|x + 3| - 1$; $d(x) = -\frac{4}{3}|x + 3| + \frac{1}{3}$

59. **REASONING** Describe the transformations from the graph of $g(x) = -2|x + 1| + 4$ to the graph of $h(x) = |x|$. Explain your reasoning.

60. **THOUGHT PROVOKING** Graph an absolute value function f that represents the route a wide receiver runs in a football game. Let the x-axis represent distance (in yards) across the field horizontally. Let the y-axis represent distance (in yards) down the field. Be sure to limit the domain so the route is realistic.

61. **SOLVING BY GRAPHING** Graph $y = 2|x + 2| - 6$ and $y = -2$ in the same coordinate plane. Use the graph to solve the equation $2|x + 2| - 6 = -2$. Check your solutions.

62. **MAKING AN ARGUMENT** Let p be a positive constant. Your friend says that because the graph of $y = |x| + p$ is a *positive* vertical translation of the graph of $y = |x|$, the graph of $y = |x + p|$ is a *positive* horizontal translation of the graph of $y = |x|$. Is your friend correct? Explain.

63. **ABSTRACT REASONING** Write the vertex of the absolute value function $f(x) = |ax - h| + k$ in terms of a, h, and k.

Maintaining Mathematical Proficiency

Reviewing what you learned in previous grades and lessons

Solve the inequality. *(Section 2.4)*

64. $8a - 7 \leq 2(3a - 1)$

65. $-3(2p + 4) > -6p - 5$

66. $4(3h + 1.5) \geq 6(2h - 2)$

67. $-4(x + 6) < 2(2x - 9)$

Find the slope of the line. *(Section 3.5)*

68.

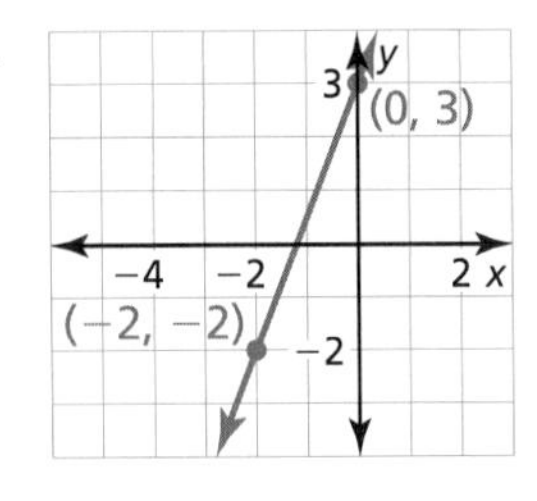

69.

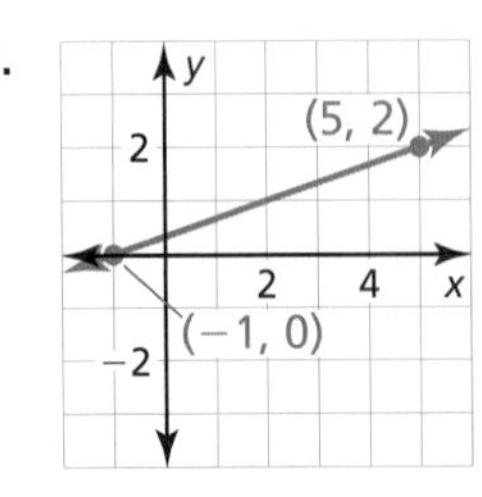

70.

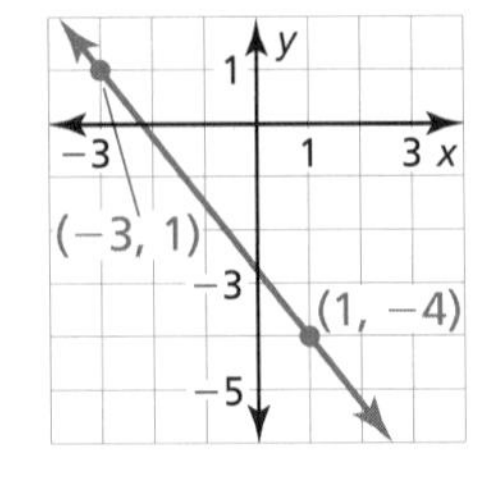

3.4–3.7 What Did You Learn?

Core Vocabulary

standard form, *p. 130*
x-intercept, *p. 131*
y-intercept, *p. 131*
slope, *p. 136*
rise, *p. 136*
run, *p. 136*
slope-intercept form, *p. 138*
constant function, *p. 138*
family of functions, *p. 146*
parent function, *p. 146*
transformation, *p. 146*
translation, *p. 146*
reflection, *p. 147*
horizontal shrink, *p. 148*
horizontal stretch, *p. 148*
vertical stretch, *p. 148*
vertical shrink, *p. 148*
absolute value function, *p. 156*
vertex, *p. 156*
vertex form, *p. 158*

Core Concepts

Section 3.4

Horizontal and Vertical Lines, *p. 130*
Using Intercepts to Graph Equations, *p. 131*

Section 3.5

Slope, *p. 136*
Slope-Intercept Form, *p. 138*

Section 3.6

Horizontal Translations, *p. 146*
Vertical Translations, *p. 146*
Reflections in the x-axis, *p. 147*
Reflections in the y-axis, *p. 147*
Horizontal Stretches and Shrinks, *p. 148*
Vertical Stretches and Shrinks, *p. 148*
Transformations of Graphs, *p. 149*

Section 3.7

Absolute Value Function, *p. 156*
Vertex Form of an Absolute Value Function, *p. 158*

Mathematical Practices

1. Explain how you determined what units of measure to use for the horizontal and vertical axes in Exercise 37 on page 142.
2. Explain your plan for solving Exercise 48 on page 153.

Performance Task

The Cost of a T-Shirt

You receive bids for making T-shirts for your class fundraiser from four companies. To present the pricing information, one company uses a table, one company uses a written description, one company uses an equation, and one company uses a graph. How will you compare the different representations and make the final choice?

To explore the answers to this question and more, go to ***BigIdeasMath.com***.

3 Chapter Review

Dynamic Solutions available at *BigIdeasMath.com*

3.1 Functions (pp. 103–110)

Determine whether the relation is a function. Explain.

Input, *x*	2	5	7	9	14
Output, *y*	5	11	19	12	3

Every input has exactly one output.

▶ So, the relation is a function.

Determine whether the relation is a function. Explain.

1. (0, 1), (5, 6), (7, 9)

2.

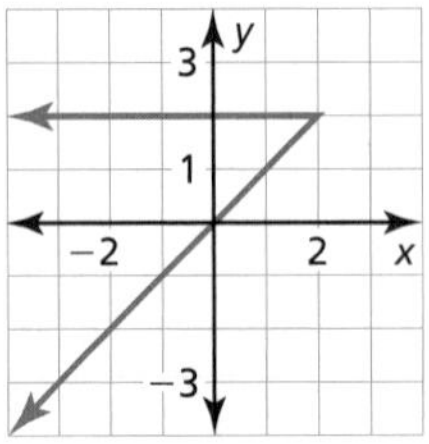

3. **Input, *x*** **Output, *y***

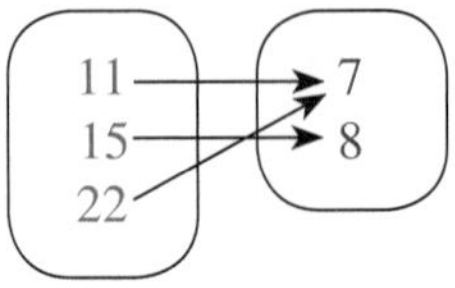

4. The function $y = 10x + 100$ represents the amount y (in dollars) of money in your bank account after you babysit for x hours.

a. Identify the independent and dependent variables.

b. You babysit for 4 hours. Find the domain and range of the function.

3.2 Linear Functions (pp. 111–120)

Does the table or equation represent a *linear* or *nonlinear* function? Explain.

a.

+4 +4 +4

x	6	10	14	18
y	5	9	14	20

+4 +5 +6

As x increases by 4, y increases by different amounts. The rate of change is *not* constant.

▶ So, the function is nonlinear.

b. $y = 3x - 4$

The equation is in the form $y = mx + b$.

▶ So, the equation represents a linear function.

Does the table or graph represent a *linear* or *nonlinear* function? Explain.

5.

x	2	7	12	17
y	2	−1	−4	−7

6.

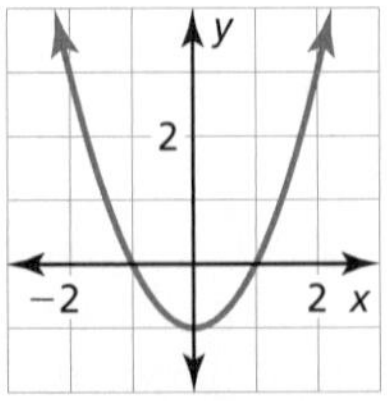

7. The function $y = 60 - 8x$ represents the amount y (in dollars) of money you have after buying x movie tickets. (a) Find the domain of the function. Is the domain discrete or continuous? Explain. (b) Graph the function using its domain.

3.3 Function Notation (pp. 121–126)

a. Evaluate $f(x) = 3x - 9$ when $x = 2$.

$f(x) = 3x - 9$	Write the function.
$f(2) = 3(2) - 9$	Substitute 2 for *x*.
$= 6 - 9$	Multiply.
$= -3$	Subtract.

▶ When $x = 2$, $f(x) = -3$.

b. For $f(x) = 4x$, find the value of x for which $f(x) = 12$.

$f(x) = 4x$	Write the function.
$12 = 4x$	Substitute 12 for $f(x)$.
$3 = x$	Divide each side by 4.

▶ When $x = 3$, $f(x) = 12$.

Evaluate the function when $x = -3$, 0, and 5.

8. $f(x) = x + 8$

9. $g(x) = 4 - 3x$

Find the value of x so that the function has the given value.

10. $k(x) = 7x$; $k(x) = 49$

11. $r(x) = -5x - 1$; $r(x) = 19$

Graph the linear function.

12. $g(x) = -2x - 3$

13. $h(x) = \frac{2}{3}x + 4$

3.4 Graphing Linear Equations in Standard Form (pp. 129–134)

Use intercepts to graph the equation $2x + 3y = 6$.

Step 1 Find the intercepts.

To find the *x*-intercept, substitute 0 for *y* and solve for *x*.

$$2x + 3y = 6$$
$$2x + 3(0) = 6$$
$$x = 3$$

To find the *y*-intercept, substitute 0 for *x* and solve for *y*.

$$2x + 3y = 6$$
$$2(0) + 3y = 6$$
$$y = 2$$

Step 2 Plot the points and draw the line.
The *x*-intercept is 3, so plot the point (3, 0).
The *y*-intercept is 2, so plot the point (0, 2).
Draw a line through the points.

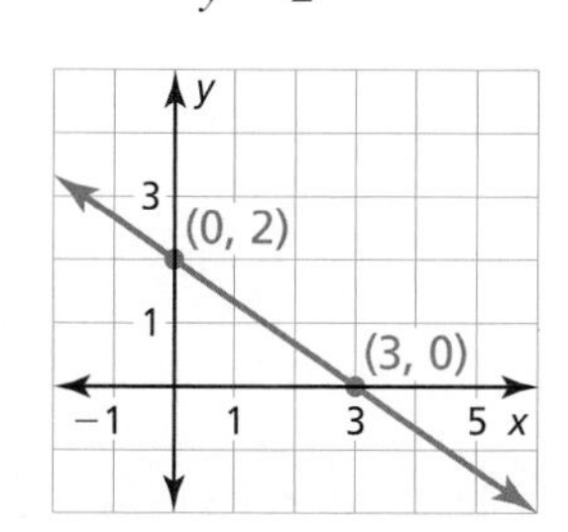

Graph the linear equation.

14. $8x - 4y = 16$

15. $-12x - 3y = 36$

16. $y = -5$

17. $x = 6$

3.5 Graphing Linear Equations in Slope-Intercept Form (pp. 135–144)

a. The points represented by the table lie on a line. How can you find the slope of the line from the table? What is the slope of the line?

x	y
1	−7
4	2
7	11
10	20

Choose any two points from the table and use the slope formula. Use the points $(x_1, y_1) = (1, -7)$ and $(x_2, y_2) = (4, 2)$.

$$\text{slope} = \frac{y_2 - y_1}{x_2 - x_1} = \frac{2 - (-7)}{4 - 1} = \frac{9}{3}, \text{ or } 3$$

▶ The slope is 3.

b. Graph $-\frac{1}{2}x + y = 1$. Identify the x-intercept.

Step 1 Rewrite the equation in slope-intercept form.

$$y = \frac{1}{2}x + 1$$

Step 2 Find the slope and the y-intercept.

$$m = \frac{1}{2} \text{ and } b = 1$$

Step 3 The y-intercept is 1. So, plot (0, 1).

Step 4 Use the slope to find another point on the line.

$$\text{slope} = \frac{\text{rise}}{\text{run}} = \frac{1}{2}$$

Plot the point that is 2 units right and 1 unit up from (0, 1). Draw a line through the two points.

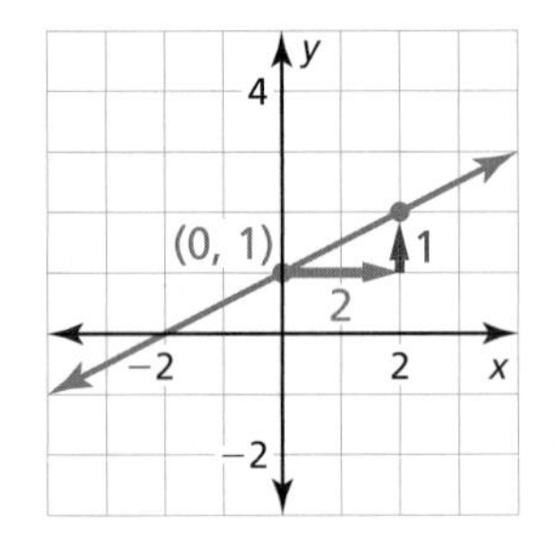

▶ The line crosses the x-axis at (−2, 0). So, the x-intercept is −2.

The points represented by the table lie on a line. Find the slope of the line.

18.

x	y
6	9
11	15
16	21
21	27

19.

x	y
3	−5
3	−2
3	5
3	8

20.

x	y
−4	−1
−3	−1
1	−1
9	−1

Graph the linear equation. Identify the x-intercept.

21. $y = 2x + 4$

22. $-5x + y = -10$

23. $x + 3y = 9$

24. A linear function h models a relationship in which the dependent variable decreases 2 units for every 3 units the independent variable increases. Graph h when $h(0) = 2$. Identify the slope, y-intercept, and x-intercept of the graph.

3.6 Transformations of Graphs of Linear Functions (pp. 145–154)

a. Let $f(x) = -\frac{1}{2}x + 2$. Graph $t(x) = f(x - 3)$. Describe the transformation from the graph of f to the graph of t.

The function t is of the form $y = f(x - h)$, where $h = 3$. So, the graph of t is a horizontal translation 3 units right of the graph of f.

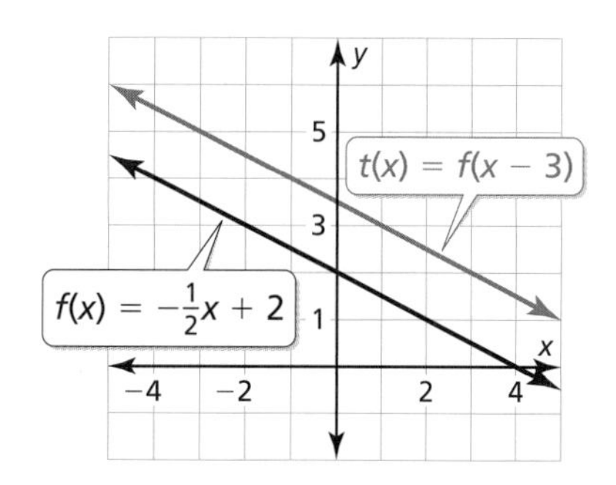

b. Graph $f(x) = x$ and $g(x) = -3x - 2$. Describe the transformations from the graph of f to the graph of g.

Note that you can rewrite g as $g(x) = -3f(x) - 2$.

Step 1 There is no horizontal translation from the graph of f to the graph of g.

Step 2 Stretch the graph of f vertically by a factor of 3 to get the graph of $h(x) = 3x$.

Step 3 Reflect the graph of h in the x-axis to get the graph of $r(x) = -3x$.

Step 4 Translate the graph of r vertically 2 units down to get the graph of $g(x) = -3x - 2$.

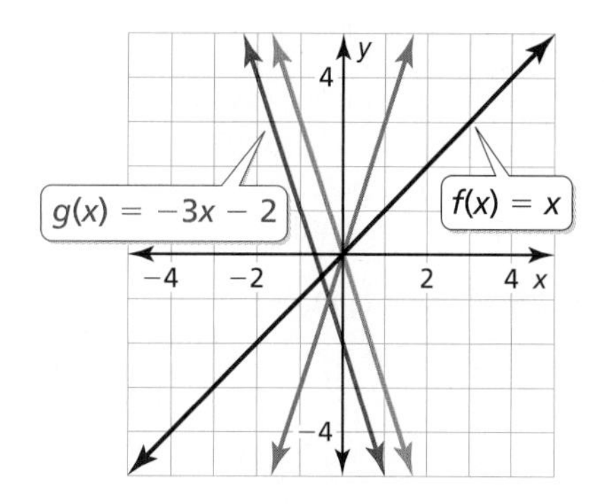

Let $f(x) = 3x + 4$. Graph f and h. Describe the transformation from the graph of f to the graph of h.

25. $h(x) = f(x + 3)$

26. $h(x) = f(x) + 1$

27. $h(x) = f(-x)$

28. $h(x) = -f(x)$

29. $h(x) = 3f(x)$

30. $h(x) = f(6x)$

31. Graph $f(x) = x$ and $g(x) = 5x + 1$. Describe the transformations from the graph of f to the graph of g.

3.7 Graphing Absolute Value Functions (pp. 155–162)

Let $g(x) = -3|x + 1| + 2$. (a) Describe the transformations from the graph of $f(x) = |x|$ to the graph of g. (b) Graph g.

a. Step 1 Translate the graph of f horizontally 1 unit left to get the graph of $t(x) = |x + 1|$.

Step 2 Stretch the graph of t vertically by a factor of 3 to get the graph of $h(x) = 3|x + 1|$.

Step 3 Reflect the graph of h in the x-axis to get the graph of $r(x) = -3|x + 1|$.

Step 4 Translate the graph of r vertically 2 units up to get the graph of $g(x) = -3|x + 1| + 2$.

b. Method 1

Step 1 Make a table of values.

x	−3	−2	−1	0	1
$g(x)$	−4	−1	2	−1	−4

Step 2 Plot the ordered pairs.

Step 3 Draw the V-shaped graph.

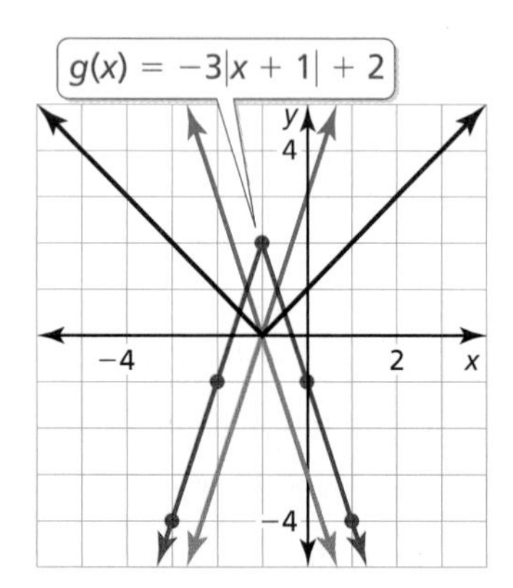

Method 2

Step 1 Identify and plot the vertex.

$$(h, k) = (-1, 2)$$

Step 2 Plot another point on the graph such as $(0, -1)$. Because the graph is symmetric about the line $x = -1$, you can use symmetry to plot a third point, $(-2, -1)$.

Step 3 Draw the V-shaped graph.

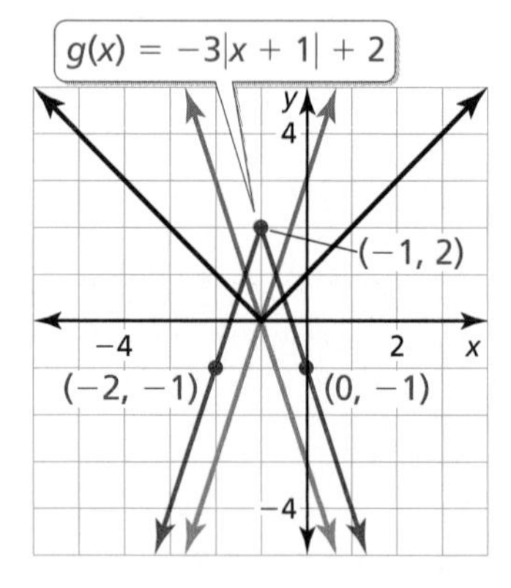

Graph the function. Compare the graph to the graph of $f(x) = |x|$. Describe the domain and range.

32. $m(x) = |x| + 6$ **33.** $p(x) = |x - 4|$ **34.** $q(x) = 4|x|$ **35.** $r(x) = -\frac{1}{4}|x|$

36. Graph $f(x) = |x - 2| + 4$ and $g(x) = |3x - 2| + 4$. Compare the graph of g to the graph of f.

37. Let $g(x) = \frac{1}{3}|x - 1| - 2$. (a) Describe the transformations from the graph of $f(x) = |x|$ to the graph of g. (b) Graph g.

3 Chapter Test

Determine whether the relation is a function. If the relation is a function, determine whether the function is *linear* or *nonlinear*. Explain.

1.

x	−1	0	1	2
y	6	5	9	14

2. $y = -2x + 3$

3. $x = -2$

Graph the equation and identify the intercept(s). If the equation is linear, find the slope of the line.

4. $2x - 3y = 6$

5. $y = 4.5$

6. $y = |x - 1| - 2$

Find the domain and range of the function represented by the graph. Determine whether the domain is *discrete* or *continuous*. Explain.

7.

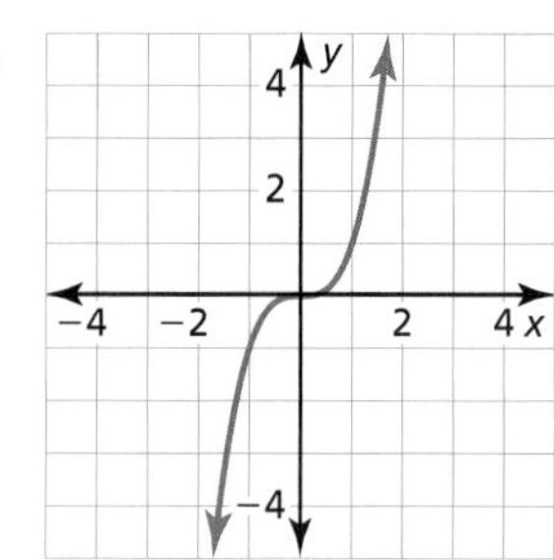

8.

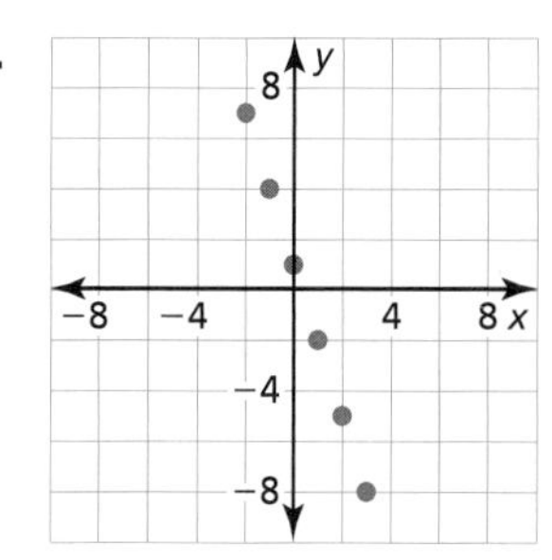

Graph *f* and *g*. Describe the transformations from the graph of *f* to the graph of *g*.

9. $f(x) = x$; $g(x) = -x + 3$

10. $f(x) = |x|$; $g(x) = |2x + 4|$

11. Function A represents the amount of money in a jar based on the number of quarters in the jar. Function B represents your distance from home over time. Compare the domains.

12. A mountain climber is scaling a 500-foot cliff. The graph shows the elevation of the climber over time.

a. Find and interpret the slope and the y-intercept of the graph.

b. Explain two ways to find $f(3)$. Then find $f(3)$ and interpret its meaning.

c. How long does it take the climber to reach the top of the cliff? Justify your answer.

Mountain Climbing

Elevation (feet)

Time (hours)

$f(x) = 125x + 50$

13. Without graphing, compare the slopes and the intercepts of the graphs of the functions $f(x) = x + 1$ and $g(x) = f(2x)$.

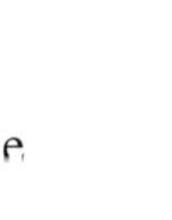

14. A rock band releases a new single. Weekly sales s (in thousands of dollars) increase and then decrease as described by the function $s(t) = -2|t - 20| + 40$, where t is the time (in weeks).

a. Identify the independent and dependent variables.

b. Graph s. Describe the transformations from the graph of $f(x) = |x|$ to the graph of s.

3 Cumulative Assessment

1. You claim you can create a table of values that represents a linear function. Your friend claims he can create a table of values that represents a nonlinear function. Using the given numbers, what values can you use for x (the input) and y (the output) to support your claim? What values can your friend use?

	Your claim			
x				
y				

	Friend's claim			
x				
y				

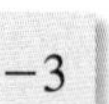

−4 −3 −2 −1 0

1 2 3 4 5

2. A car rental company charges an initial fee of \$42 and a daily fee of \$12.

a. Use the numbers and symbols to write a function that represents this situation.

x $f(x)$ 42 12 138

$+$ $-$ $\times$ $\div$ $=$

b. The bill is \$138. How many days did you rent the car?

3. Fill in values for a and b so that each statement is true for the inequality $ax - b > 0$.

a. When $a =$ _____ and $b =$ _____, $x > \frac{b}{a}$.

b. When $a =$ _____ and $b =$ _____, $x < \frac{b}{a}$.

4. Fill in the inequality with $<$, $\leq$, $>$, or $\geq$ so that the solution of the inequality is represented by the graph.

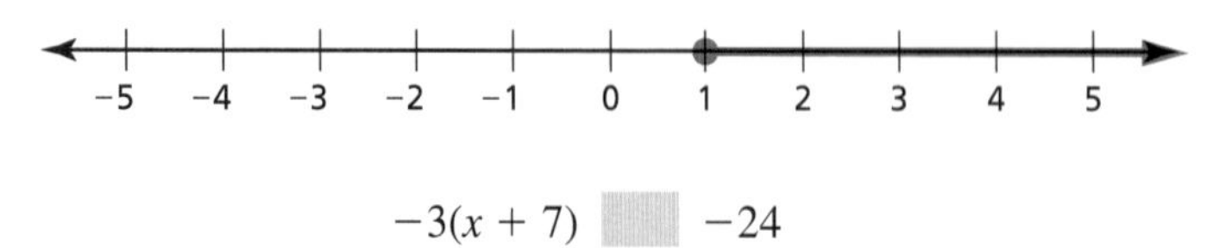

$-3(x + 7) \;\square\; -24$

5. Use the numbers to fill in the coefficients of $ax + by = 40$ so that when you graph the function, the x-intercept is -10 and the y-intercept is 8.

-10 -8 -5 -4 4 5 8 10

6. Solve each equation. Then classify each equation based on the solution. Explain your reasoning.

a. $2x - 9 = 5x - 33$

b. $5x - 6 = 10x + 10$

c. $2(8x - 3) = 4(4x + 7)$

d. $-7x + 5 = 2(x - 10.1)$

e. $6(2x + 4) = 4(4x + 10)$

f. $8(3x + 4) = 2(12x + 16)$

7. The table shows the cost of bologna at a deli. Plot the points represented by the table in a coordinate plane. Decide whether you should connect the points with a line. Explain your reasoning.

Pounds, x	0.5	1	1.5	2
Cost, y	\$3	\$6	\$9	\$12

8. The graph of g is a horizontal translation right, then a vertical stretch, then a vertical translation down of the graph of $f(x) = x$. Use the numbers and symbols to create g.

-3 -1 $-\frac{1}{2}$ 0 $\frac{1}{2}$ 1 3

x $g(x)$ $+$ $-$ $\times$ $\div$ $=$

9. What is the sum of the integer solutions of the compound inequality $2|x - 5| < 16$?

Ⓐ 72 Ⓑ 75 Ⓒ 85 Ⓓ 88

10. Your bank offers a text alert service that notifies you when your checking account balance drops below a specific amount. You set it up so you are notified when your balance drops below \$700. The balance is currently \$3000. You only use your account for paying your rent (no other deposits or deductions occur). Your rent each month is \$625.

a. Write an inequality that represents the number of months m you can pay your rent without receiving a text alert.

b. What is the maximum number of months you can pay your rent without receiving a text alert?

c. Suppose you start paying rent in June. Select all the months you can pay your rent without making a deposit.

June July August September October

4 Writing Linear Functions

Karaoke Machine *(p. 220)*

Old Faithful Geyser *(p. 204)*

Helicopter Rescue *(p. 190)*

School Spirit *(p. 184)*

Renewable Energy *(p. 178)*

Maintaining Mathematical Proficiency

Using a Coordinate Plane

Example 1 **What ordered pair corresponds to point *A*?**

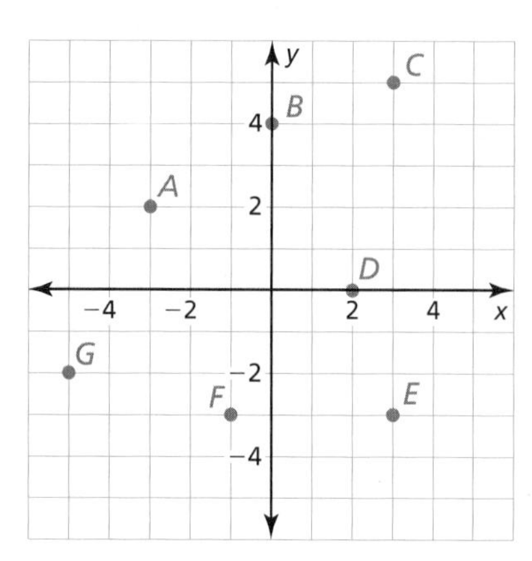

Point A is 3 units to the left of the origin and 2 units up. So, the x-coordinate is -3 and the y-coordinate is 2.

▶ The ordered pair $(-3, 2)$ corresponds to point A.

Use the graph to answer the question.

1. What ordered pair corresponds to point G?

2. What ordered pair corresponds to point D?

3. Which point is located in Quadrant I?

4. Which point is located in Quadrant IV?

Rewriting Equations

Example 2 **Solve the equation $3x - 2y = 8$ for y.**

$$3x - 2y = 8$$ Write the equation.

$$3x - 2y - 3x = 8 - 3x$$ Subtract $3x$ from each side.

$$-2y = 8 - 3x$$ Simplify.

$$\frac{-2y}{-2} = \frac{8 - 3x}{-2}$$ Divide each side by -2.

$$y = -4 + \frac{3}{2}x$$ Simplify.

Solve the equation for *y*.

5. $x - y = 5$

6. $6x + 3y = -1$

7. $0 = 2y - 8x + 10$

8. $-x + 4y - 28 = 0$

9. $2y + 1 - x = 7x$

10. $y - 4 = 3x + 5y$

11. **ABSTRACT REASONING** Both coordinates of the point (x, y) are multiplied by a negative number. How does this change the location of the point? Be sure to consider points originally located in all four quadrants.

Mathematical Practices

Mathematically proficient students try simpler forms of the original problem.

Problem-Solving Strategies

Core Concept

Solve a Simpler Problem

When solving a real-life problem, if the numbers in the problem seem complicated, then try solving a simpler form of the problem. After you have solved the simpler problem, look for a general strategy. Then apply that strategy to the original problem.

EXAMPLE 1 Using a Problem-Solving Strategy

In the deli section of a grocery store, a half pound of sliced roast beef costs \$3.19. You buy 1.81 pounds. How much do you pay?

SOLUTION

Step 1 Solve a simpler problem.

Suppose the roast beef costs \$3 per half pound, and you buy 2 pounds.

$$\text{Total cost} = \frac{\$3}{1/2 \text{ lb}} \cdot 2 \text{ lb}$$ Use unit analysis to write a verbal model.

$$= \frac{\$6}{1 \cancel{\text{lb}}} \cdot 2 \cancel{\text{lb}}$$ Rewrite \$3 per $\frac{1}{2}$ pound as \$6 per pound.

$$= \$12$$ Simplify.

▶ In the simpler problem, you pay \$12.

Step 2 Apply the strategy to the original problem.

$$\text{Total cost} = \frac{\$3.19}{1/2 \text{ lb}} \cdot 1.81 \text{ lb}$$ Use unit analysis to write a verbal model.

$$= \frac{\$6.38}{1 \cancel{\text{lb}}} \cdot 1.81 \cancel{\text{lb}}$$ Rewrite \$3.19 per $\frac{1}{2}$ pound as \$6.38 per pound.

$$= \$11.55$$ Simplify.

▶ In the original problem, you pay \$11.55. ← Your answer is reasonable because you bought about 2 pounds.

Monitoring Progress

1. You work $37\frac{1}{2}$ hours and earn \$352.50. What is your hourly wage?
2. You drive 1244.5 miles and use 47.5 gallons of gasoline. What is your car's gas mileage (in miles per gallon)?
3. You drive 236 miles in 4.6 hours. At the same rate, how long will it take you to drive 450 miles?

4.1 Writing Equations in Slope-Intercept Form

Essential Question Given the graph of a linear function, how can you write an equation of the line?

EXPLORATION 1 Writing Equations in Slope-Intercept Form

Work with a partner.

- Find the slope and y-intercept of each line.
- Write an equation of each line in slope-intercept form.
- Use a graphing calculator to verify your equation.

a.

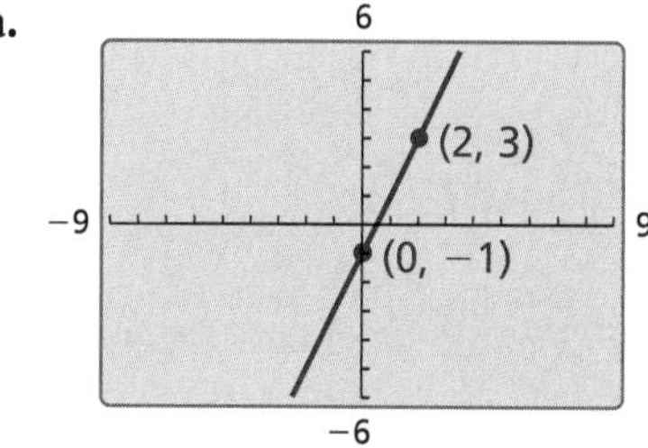

b.

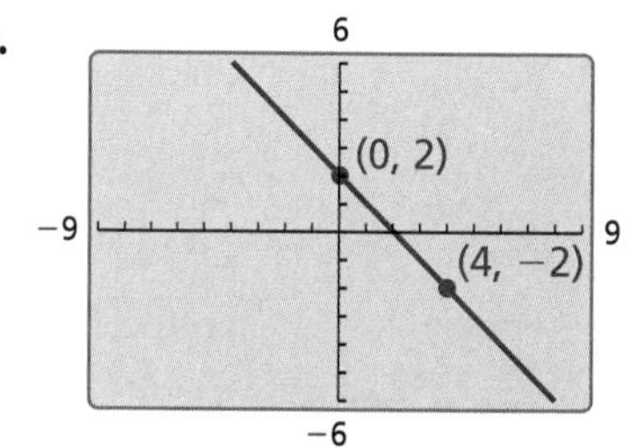

c.

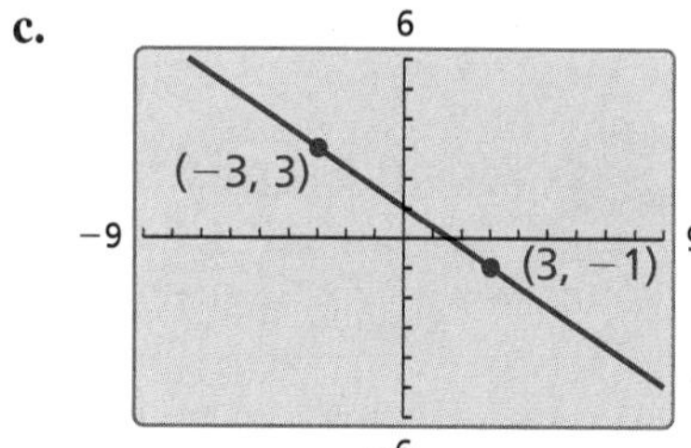

d.

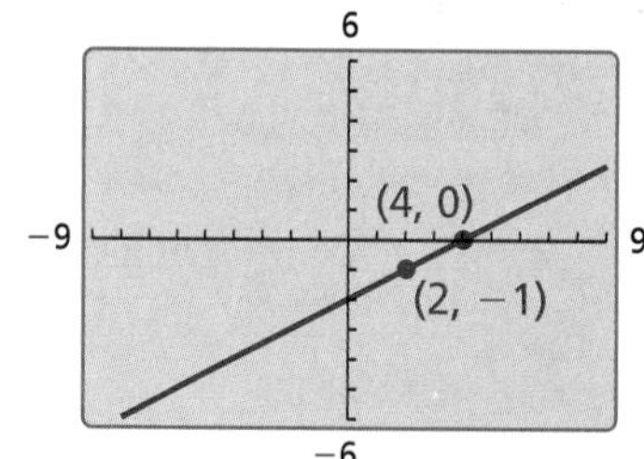

INTERPRETING MATHEMATICAL RESULTS

To be proficient in math, you need to routinely interpret your results in the context of the situation. The reason for studying mathematics is to enable you to model and solve real-life problems.

EXPLORATION 2 Mathematical Modeling

Work with a partner. The graph shows the cost of a smartphone plan.

a. What is the y-intercept of the line? Interpret the y-intercept in the context of the problem.

b. Approximate the slope of the line. Interpret the slope in the context of the problem.

c. Write an equation that represents the cost as a function of data usage.

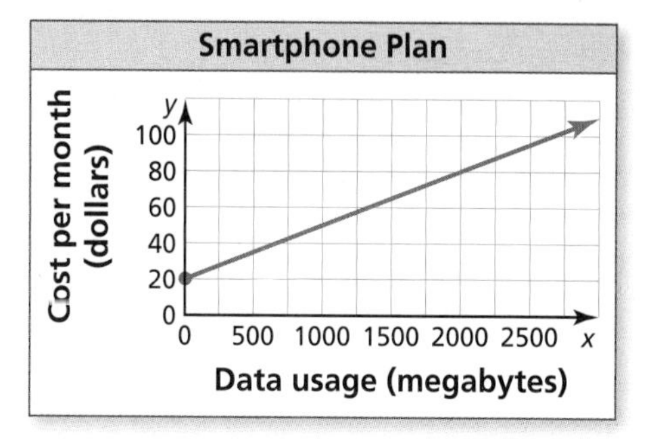

Communicate Your Answer

3. Given the graph of a linear function, how can you write an equation of the line?

4. Give an example of a graph of a linear function that is different from those above. Then use the graph to write an equation of the line.

4.1 Lesson

What You Will Learn

- Write equations in slope-intercept form.
- Use linear equations to solve real-life problems.

Core Vocabulary

linear model, *p. 178*

Previous
slope-intercept form
function
rate

Writing Equations in Slope-Intercept Form

EXAMPLE 1 Using Slopes and y-Intercepts to Write Equations

Write an equation of each line with the given slope and y-intercept.

a. slope $= -3$; y-intercept $= \frac{1}{2}$ **b.** slope $= 0$; y-intercept $= -2$

SOLUTION

a. $y = mx + b$ Write the slope-intercept form.

$y = -3x + \frac{1}{2}$ Substitute -3 for m and $\frac{1}{2}$ for b.

▶ An equation is $y = -3x + \frac{1}{2}$.

b. $y = mx + b$ Write the slope-intercept form.

$y = 0x + (-2)$ Substitute 0 for m and -2 for b.

$y = -2$ Simplify.

▶ An equation is $y = -2$.

EXAMPLE 2 Using Graphs to Write Equations

Write an equation of each line in slope-intercept form.

a.

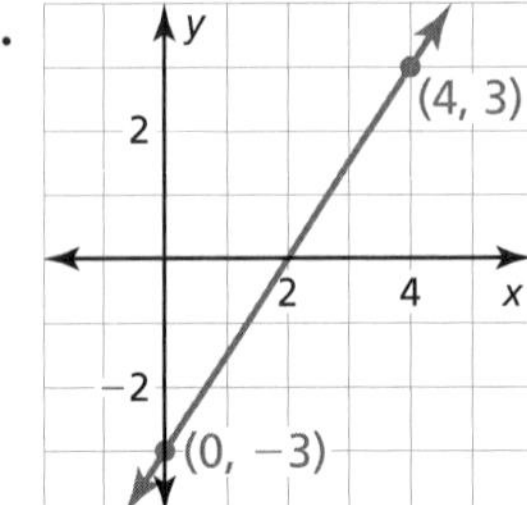

b.

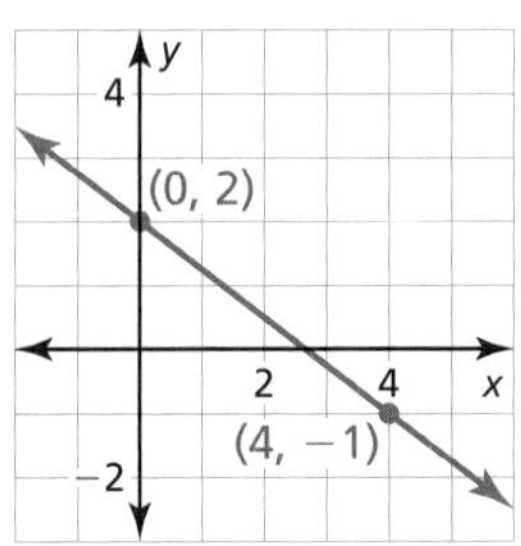

SOLUTION

STUDY TIP
You can use any two points on a line to find the slope.

a. Find the slope and y-intercept.

Let $(x_1, y_1) = (0, -3)$ and $(x_2, y_2) = (4, 3)$.

$$m = \frac{y_2 - y_1}{x_2 - x_1} = \frac{3 - (-3)}{4 - 0} = \frac{6}{4}, \text{ or } \frac{3}{2}$$

Because the line crosses the y-axis at $(0, -3)$, the y-intercept is -3.

▶ So, the equation is $y = \frac{3}{2}x - 3$.

b. Find the slope and y-intercept.

Let $(x_1, y_1) = (0, 2)$ and $(x_2, y_2) = (4, -1)$.

$$m = \frac{y_2 - y_1}{x_2 - x_1} = \frac{-1 - 2}{4 - 0} = \frac{-3}{4}, \text{ or } -\frac{3}{4}$$

Because the line crosses the y-axis at $(0, 2)$, the y-intercept is 2.

STUDY TIP
After writing an equation, check that the given points are solutions of the equation.

▶ So, the equation is $y = -\frac{3}{4}x + 2$.

EXAMPLE 3 Using Points to Write Equations

Write an equation of each line that passes through the given points.

a. $(-3, 5), (0, -1)$

b. $(0, -5), (8, -5)$

SOLUTION

a. Find the slope and y-intercept.

$$m = \frac{-1 - 5}{0 - (-3)} = -2$$

Because the line crosses the y-axis at $(0, -1)$, the y-intercept is -1.

▶ So, an equation is $y = -2x - 1$.

b. Find the slope and y-intercept.

$$m = \frac{-5 - (-5)}{8 - 0} = 0$$

Because the line crosses the y-axis at $(0, -5)$, the y-intercept is -5.

▶ So, an equation is $y = -5$.

REMEMBER

If f is a function and x is in its domain, then $f(x)$ represents the output of f corresponding to the input x.

EXAMPLE 4 Writing a Linear Function

Write a linear function f with the values $f(0) = 10$ and $f(6) = 34$.

SOLUTION

Step 1 Write $f(0) = 10$ as $(0, 10)$ and $f(6) = 34$ as $(6, 34)$.

Step 2 Find the slope of the line that passes through $(0, 10)$ and $(6, 34)$.

$$m = \frac{34 - 10}{6 - 0} = \frac{24}{6}, \text{ or } 4$$

Step 3 Write an equation of the line. Because the line crosses the y-axis at $(0, 10)$, the y-intercept is 10.

$y = mx + b$ Write the slope-intercept form.

$y = 4x + 10$ Substitute 4 for m and 10 for b.

▶ A function is $f(x) = 4x + 10$.

Monitoring Progress  Help in English and Spanish at *BigIdeasMath.com*

Write an equation of the line with the given slope and y-intercept.

1. slope = 7; y-intercept = 2
2. slope = $\frac{1}{3}$; y-intercept = -1

Write an equation of the line in slope-intercept form.

3. 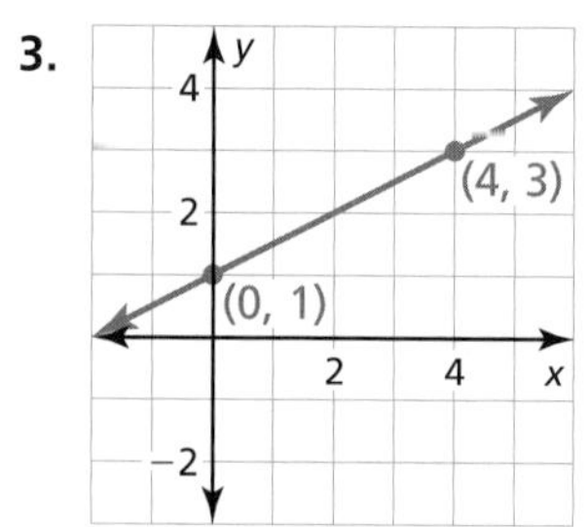

4. 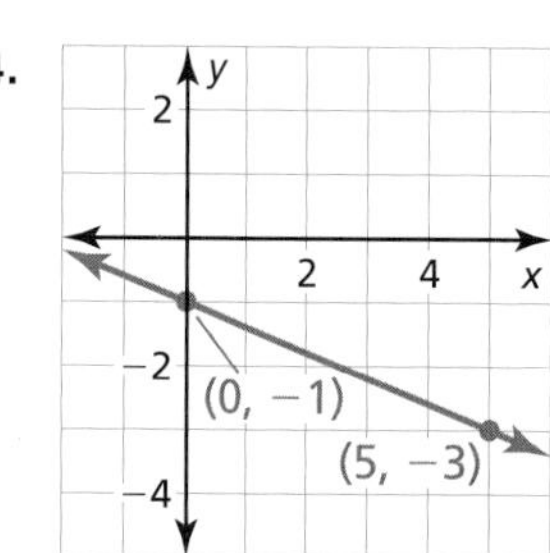

5. Write an equation of the line that passes through $(0, -2)$ and $(4, 10)$.
6. Write a linear function g with the values $g(0) = 9$ and $g(8) = 7$.

Solving Real-Life Problems

A **linear model** is a linear function that models a real-life situation. When a quantity y changes at a constant rate with respect to a quantity x, you can use the equation $y = mx + b$ to model the relationship. The value of m is the constant rate of change, and the value of b is the initial, or starting, value of y.

EXAMPLE 5 Modeling with Mathematics

Excluding hydropower, U.S. power plants used renewable energy sources to generate 105 million megawatt hours of electricity in 2007. By 2012, the amount of electricity generated had increased to 219 million megawatt hours. Write a linear model that represents the number of megawatt hours generated by non-hydropower renewable energy sources as a function of the number of years since 2007. Use the model to predict the number of megawatt hours that will be generated in 2017.

SOLUTION

1. **Understand the Problem** You know the amounts of electricity generated in two distinct years. You are asked to write a linear model that represents the amount of electricity generated each year since 2007 and then predict a future amount.

2. **Make a Plan** Break the problem into parts and solve each part. Then combine the results to help you solve the original problem.

   Part 1 Define the variables. Find the initial value and the rate of change.

   Part 2 Write a linear model and predict the amount in 2017.

3. **Solve the Problem**

   Part 1 Let x represent the time (in years) since 2007 and let y represent the number of megawatt hours (in millions). Because time x is defined in years since 2007, 2007 corresponds to $x = 0$ and 2012 corresponds to $x = 5$. Let $(x_1, y_1) = (0, 105)$ and $(x_2, y_2) = (5, 219)$. The initial value is the y-intercept b, which is 105. The rate of change is the slope m.

$$m = \frac{y_2 - y_1}{x_2 - x_1} = \frac{219 - 105}{5 - 0} = \frac{114}{5} = 22.8$$

   Part 2

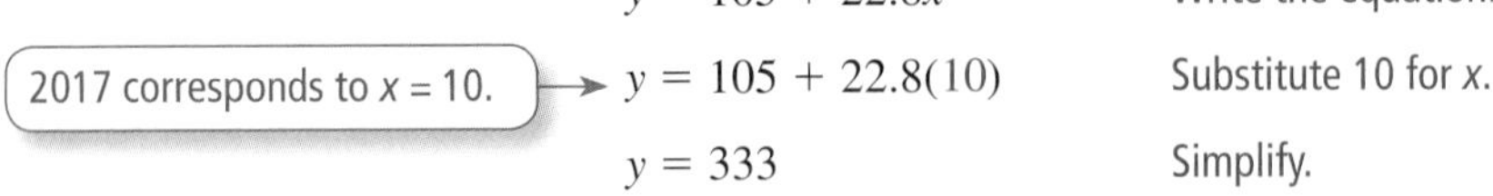

$$y \quad = \quad 105 \quad + \quad 22.8 \quad \cdot \quad x$$

$y = 105 + 22.8x$ Write the equation.

2017 corresponds to $x = 10$. → $y = 105 + 22.8(10)$ Substitute 10 for x.

$y = 333$ Simplify.

▶ The linear model is $y = 22.8x + 105$. The model predicts non-hydropower renewable energy sources will generate 333 million megawatt hours in 2017.

4. **Look Back** To check that your model is correct, verify that (0, 105) and (5, 219) are solutions of the equation.

Monitoring Progress  Help in English and Spanish at *BigIdeasMath.com*

7. The corresponding data for electricity generated by hydropower are 248 million megawatt hours in 2007 and 277 million megawatt hours in 2012. Write a linear model that represents the number of megawatt hours generated by hydropower as a function of the number of years since 2007.

4.1 Exercises

Dynamic Solutions available at *BigIdeasMath.com*

Vocabulary and Core Concept Check

1. **COMPLETE THE SENTENCE** A linear function that models a real-life situation is called a __________.

2. **WRITING** Explain how you can use slope-intercept form to write an equation of a line given its slope and y-intercept.

Monitoring Progress and Modeling with Mathematics

In Exercises 3–8, write an equation of the line with the given slope and y-intercept. *(See Example 1.)*

3. slope: 2
   y-intercept: 9

4. slope: 0
   y-intercept: 5

5. slope: -3
   y-intercept: 0

6. slope: -7
   y-intercept: 1

7. slope: $\frac{2}{3}$
   y-intercept: -8

8. slope: $-\frac{3}{4}$
   y-intercept: -6

In Exercises 9–12, write an equation of the line in slope-intercept form. *(See Example 2.)*

9.

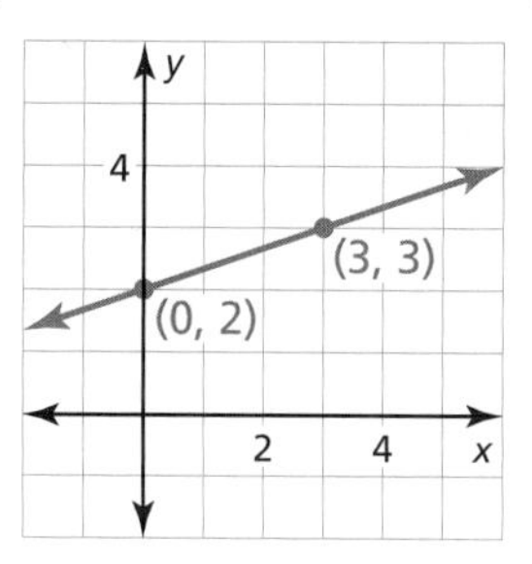

10.

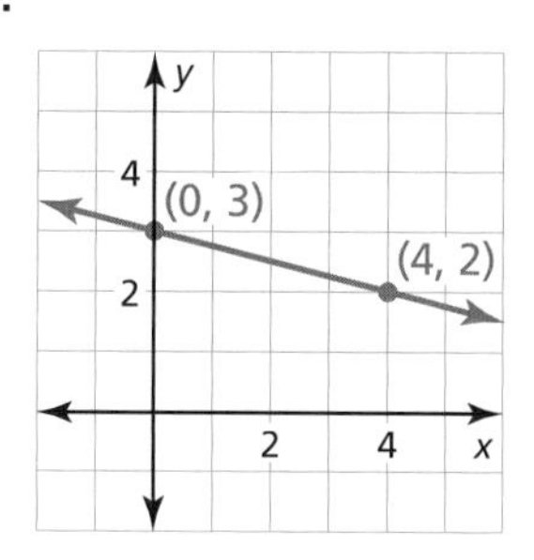

11.

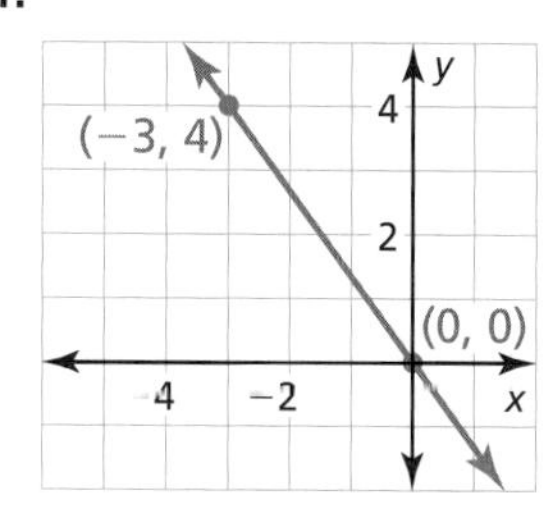

12.

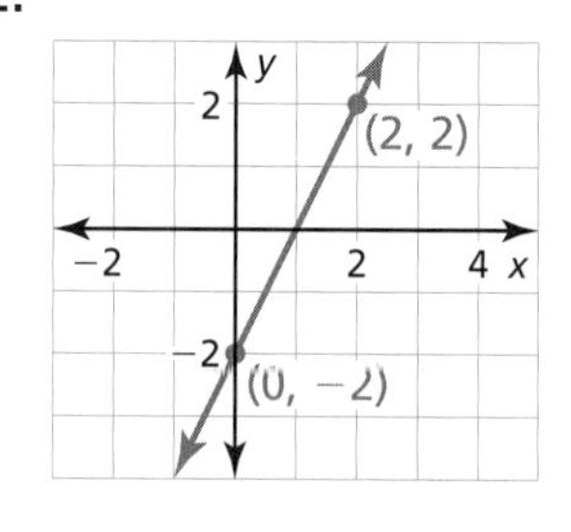

In Exercises 13–18, write an equation of the line that passes through the given points. *(See Example 3.)*

13. (3, 1), (0, 10)

14. (2, 7), (0, −5)

15. (2, −4), (0, −4)

16. (−6, 0), (0, −24)

17. (0, 5), (−1.5, 1)

18. (0, 3), (−5, 2.5)

In Exercises 19–24, write a linear function f with the given values. *(See Example 4.)*

19. $f(0) = 2, f(2) = 4$

20. $f(0) = 7, f(3) = 1$

21. $f(4) = -3, f(0) = -2$

22. $f(5) = -1, f(0) = -5$

23. $f(-2) = 6, f(0) = -4$

24. $f(0) = 3, f(-6) = 3$

In Exercises 25 and 26, write a linear function f with the given values.

25.

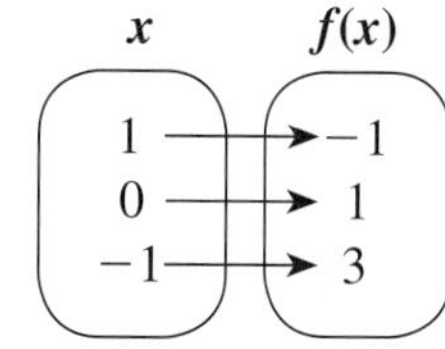

26.

x	$f(x)$
−4	−2
−2	−1
0	0

27. **ERROR ANALYSIS** Describe and correct the error in writing an equation of the line with a slope of 2 and a y-intercept of 7.

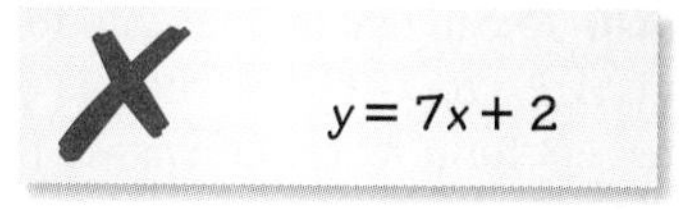

28. **ERROR ANALYSIS** Describe and correct the error in writing an equation of the line shown.

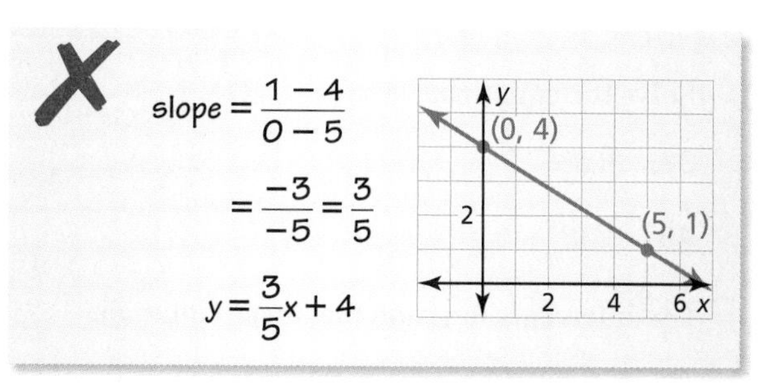

29. **MODELING WITH MATHEMATICS** In 1960, the world record for the men's mile was 3.91 minutes. In 1980, the record time was 3.81 minutes. *(See Example 5.)*

   a. Write a linear model that represents the world record (in minutes) for the men's mile as a function of the number of years since 1960.

   b. Use the model to estimate the record time in 2000 and predict the record time in 2020.

30. **MODELING WITH MATHEMATICS** A recording studio charges musicians an initial fee of \$50 to record an album. Studio time costs an additional \$75 per hour.

   a. Write a linear model that represents the total cost of recording an album as a function of studio time (in hours).

   b. Is it less expensive to purchase 12 hours of recording time at the studio or a \$750 music software program that you can use to record on your own computer? Explain.

31. **WRITING** A line passes through the points $(0, -2)$ and $(0, 5)$. Is it possible to write an equation of the line in slope-intercept form? Justify your answer.

32. **THOUGHT PROVOKING** Describe a real-life situation involving a linear function whose graph passes through the points.

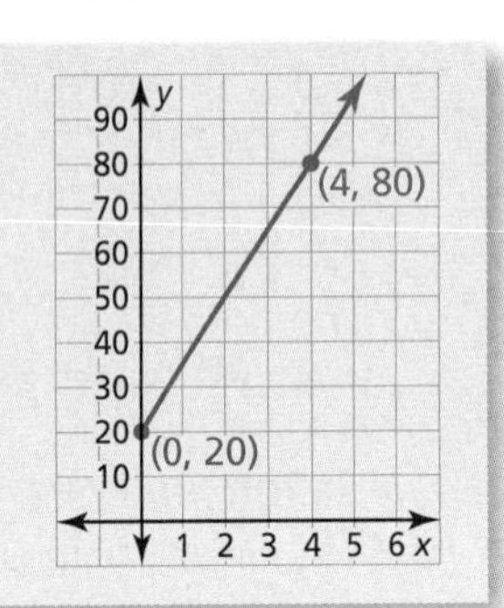

33. **REASONING** Recall that the standard form of a linear equation is $Ax + By = C$. Rewrite this equation in slope-intercept form. Use your answer to find the slope and y-intercept of the graph of the equation $-6x + 5y = 9$.

34. **MAKING AN ARGUMENT** Your friend claims that given $f(0)$ and any other value of a linear function f, you can write an equation in slope-intercept form that represents the function. Your cousin disagrees, claiming that the two points could lie on a vertical line. Who is correct? Explain.

35. **ANALYZING A GRAPH** Line ℓ is a reflection in the x-axis of line k. Write an equation that represents line k.

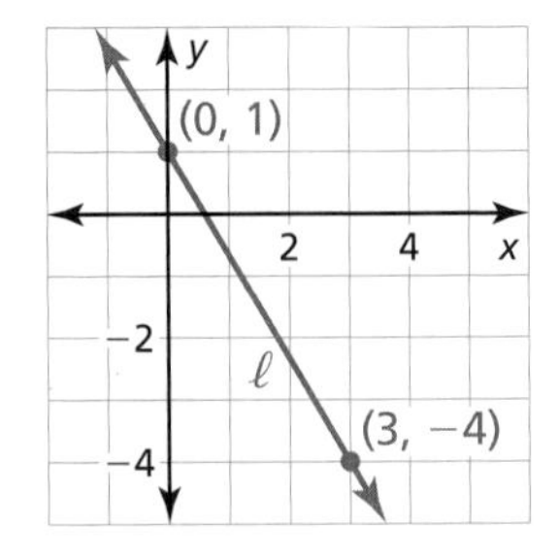

36. **HOW DO YOU SEE IT?** The graph shows the approximate U.S. box office revenues (in billions of dollars) from 2000 to 2012, where $x = 0$ represents the year 2000.

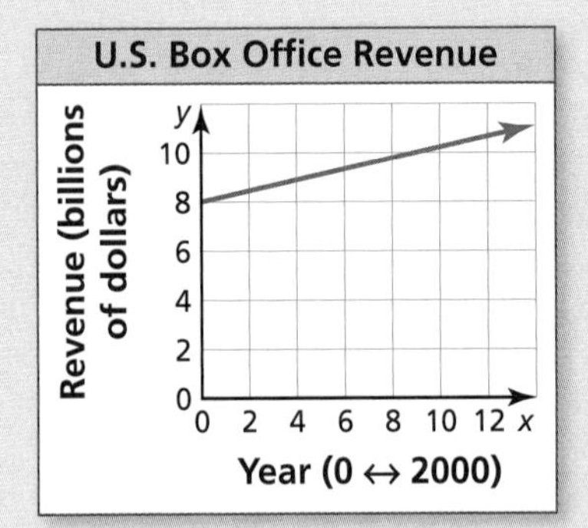

   a. Estimate the slope and y-intercept of the graph.

   b. Interpret your answers in part (a) in the context of the problem.

   c. How can you use your answers in part (a) to predict the U.S. box office revenue in 2018?

37. **ABSTRACT REASONING** Show that the equation of the line that passes through the points $(0, b)$ and $(1, b + m)$ is $y = mx + b$. Explain how you can be sure that the point $(-1, b - m)$ also lies on the line.

Maintaining Mathematical Proficiency

Reviewing what you learned in previous grades and lessons

Solve the equation. *(Section 1.3)*

38. $3(x - 15) = x + 11$

39. $-4y - 10 = 4(y - 3)$

40. $2(3d + 3) = 7 + 6d$

41. $-5(4 - 3n) = 10(n - 2)$

Use intercepts to graph the linear equation. *(Section 3.4)*

42. $-4x + 2y = 16$

43. $3x + 5y = -15$

44. $x - 6y = 24$

45. $-7x - 2y = -21$

4.2 Writing Equations in Point-Slope Form

Essential Question

How can you write an equation of a line when you are given the slope and a point on the line?

EXPLORATION 1 Writing Equations of Lines

Work with a partner.

- Sketch the line that has the given slope and passes through the given point.
- Find the y-intercept of the line.
- Write an equation of the line.

a. $m = \frac{1}{2}$

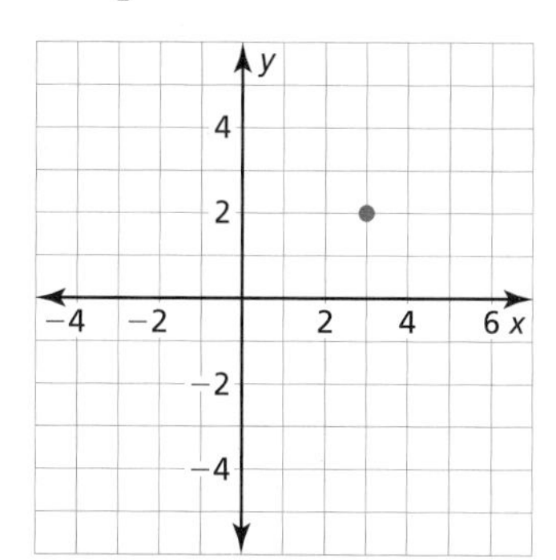

b. $m = -2$

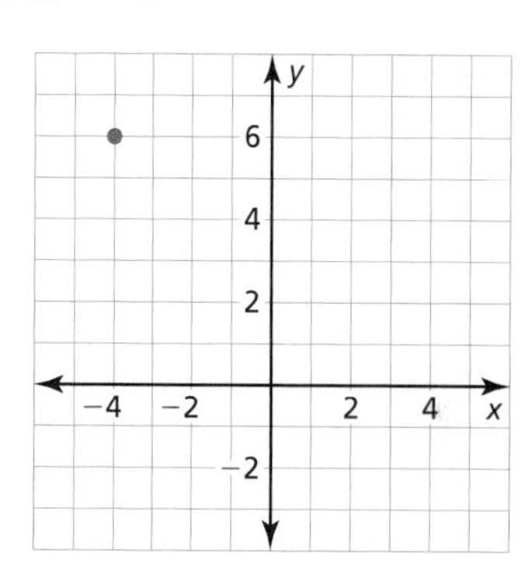

EXPLORATION 2 Writing a Formula

Work with a partner.

The point (x_1, y_1) is a given point on a nonvertical line. The point (x, y) is any other point on the line. Write an equation that represents the slope m of the line. Then rewrite this equation by multiplying each side by the difference of the x-coordinates to obtain the **point-slope form** of a linear equation.

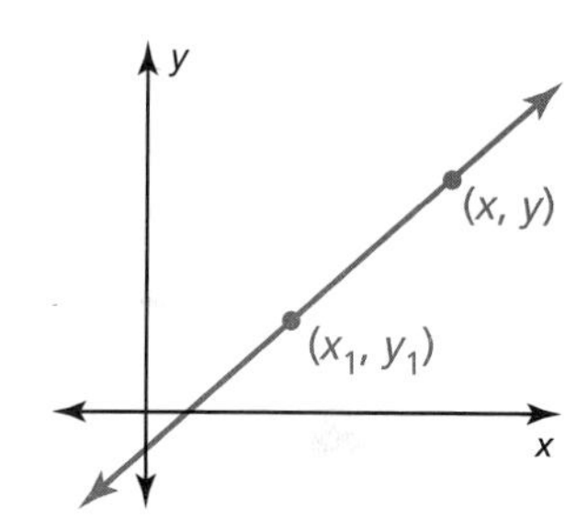

USING A GRAPHING CALCULATOR

To be proficient in math, you need to understand the feasibility, appropriateness, and limitations of the technological tools at your disposal. For instance, in real-life situations such as the one given in Exploration 3, it may not be feasible to use a square viewing window on a graphing calculator.

EXPLORATION 3 Writing an Equation

Work with a partner.

For four months, you have saved \$25 per month. You now have \$175 in your savings account.

a. Use your result from Exploration 2 to write an equation that represents the balance A after t months.

b. Use a graphing calculator to verify your equation.

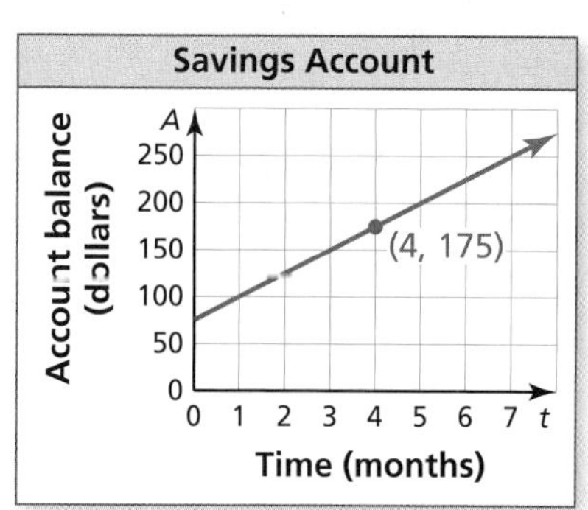

Communicate Your Answer

4. How can you write an equation of a line when you are given the slope and a point on the line?

5. Give an example of how to write an equation of a line when you are given the slope and a point on the line. Your example should be different from those above.

4.2 Lesson

Core Vocabulary

point-slope form, *p. 182*

Previous
slope-intercept form
function
linear model
rate

What You Will Learn

- Write an equation of a line given its slope and a point on the line.
- Write an equation of a line given two points on the line.
- Use linear equations to solve real-life problems.

Writing Equations of Lines in Point-Slope Form

Given a point on a line and the slope of the line, you can write an equation of the line. Consider the line that passes through (2, 3) and has a slope of $\frac{1}{2}$. Let (x, y) be another point on the line where $x \neq 2$. You can write an equation relating x and y using the slope formula with $(x_1, y_1) = (2, 3)$ and $(x_2, y_2) = (x, y)$.

$m = \dfrac{y_2 - y_1}{x_2 - x_1}$ Write the slope formula.

$\dfrac{1}{2} = \dfrac{y - 3}{x - 2}$ Substitute values.

$\dfrac{1}{2}(x - 2) = y - 3$ Multiply each side by $(x - 2)$.

The equation in *point-slope form* is $y - 3 = \frac{1}{2}(x - 2)$.

Core Concept

Point-Slope Form

Words A linear equation written in the form $y - y_1 = m(x - x_1)$ is in **point-slope form**. The line passes through the point (x_1, y_1), and the slope of the line is m.

Algebra $y - y_1 = m(x - x_1)$

slope: m; passes through (x_1, y_1)

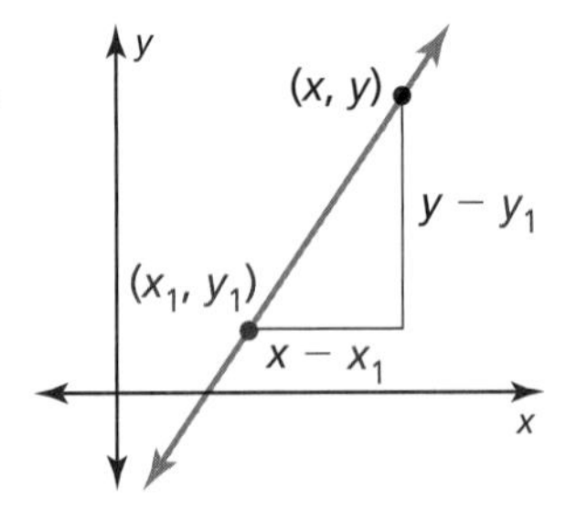

EXAMPLE 1 Using a Slope and a Point to Write an Equation

Write an equation in point-slope form of the line that passes through the point (−8, 3) and has a slope of $\frac{1}{4}$.

SOLUTION

$y - y_1 = m(x - x_1)$ Write the point-slope form.

$y - 3 = \frac{1}{4}[x - (-8)]$ Substitute $\frac{1}{4}$ for m, −8 for x_1, and 3 for y_1.

$y - 3 = \frac{1}{4}(x + 8)$ Simplify.

▶ The equation is $y - 3 = \frac{1}{4}(x + 8)$.

Check

$y - 3 = \frac{1}{4}(x + 8)$

$3 - 3 \stackrel{?}{=} \frac{1}{4}(-8 + 8)$

$0 = 0$ ✓

Monitoring Progress  Help in English and Spanish at *BigIdeasMath.com*

Write an equation in point-slope form of the line that passes through the given point and has the given slope.

1. $(3, -1)$; $m = -2$

2. $(4, 0)$; $m = -\frac{2}{3}$

Writing Equations of Lines Given Two Points

When you are given two points on a line, you can write an equation of the line using the following steps.

Step 1 Find the slope of the line.

Step 2 Use the slope and one of the points to write an equation of the line in point-slope form.

> **ANOTHER WAY**
>
> You can use either of the given points to write an equation of the line.
>
> Use $m = -2$ and $(3, -2)$.
>
> $y - (-2) = -2(x - 3)$
>
> $y + 2 = -2x + 6$
>
> $y = -2x + 4$

EXAMPLE 2 Using Two Points to Write an Equation

Write an equation in slope-intercept form of the line shown.

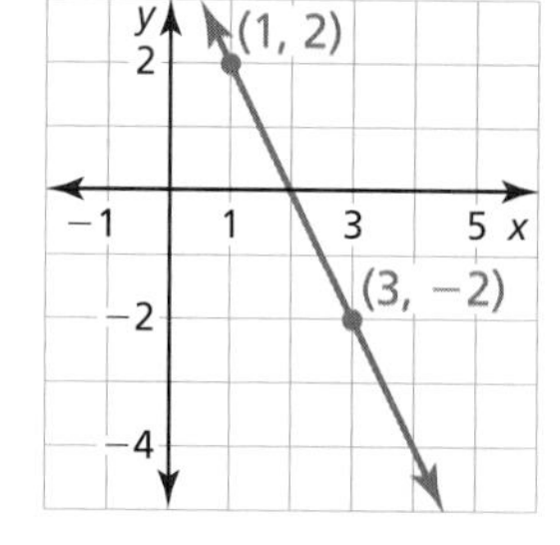

SOLUTION

Step 1 Find the slope of the line.

$$m = \frac{-2 - 2}{3 - 1} = \frac{-4}{2}, \text{ or } -2$$

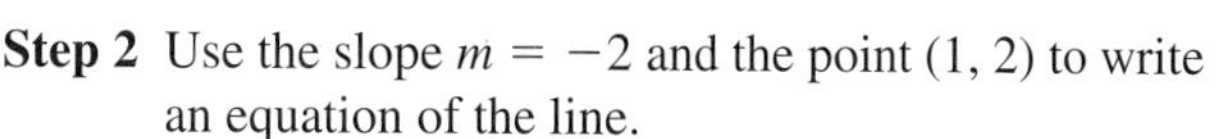

Step 2 Use the slope $m = -2$ and the point $(1, 2)$ to write an equation of the line.

$y - y_1 = m(x - x_1)$	Write the point-slope form.
$y - 2 = -2(x - 1)$	Substitute -2 for m, 1 for x_1, and 2 for y_1.
$y - 2 = -2x + 2$	Distributive Property
$y = -2x + 4$	Write in slope-intercept form.

▶ The equation is $y = -2x + 4$.

EXAMPLE 3 Writing a Linear Function

Write a linear function f with the values $f(4) = -2$ and $f(8) = 4$.

SOLUTION

Note that you can rewrite $f(4) = -2$ as $(4, -2)$ and $f(8) = 4$ as $(8, 4)$.

Step 1 Find the slope of the line that passes through $(4, -2)$ and $(8, 4)$.

$$m = \frac{4 - (-2)}{8 - 4} = \frac{6}{4}, \text{ or } 1.5$$

Step 2 Use the slope $m = 1.5$ and the point $(8, 4)$ to write an equation of the line.

$y - y_1 = m(x - x_1)$	Write the point-slope form.
$y - 4 = 1.5(x - 8)$	Substitute 1.5 for m, 8 for x_1, and 4 for y_1.
$y - 4 = 1.5x - 12$	Distributive Property
$y = 1.5x - 8$	Write in slope-intercept form.

▶ A function is $f(x) = 1.5x - 8$.

Monitoring Progress  Help in English and Spanish at *BigIdeasMath.com*

Write an equation in slope-intercept form of the line that passes through the given points.

3. $(1, 4), (3, 10)$

4. $(-4, -1), (8, -4)$

5. Write a linear function g with the values $g(2) = 3$ and $g(6) = 5$.

Solving Real-Life Problems

Modeling with Mathematics

The student council is ordering customized foam hands to promote school spirit. The table shows the cost of ordering different numbers of foam hands. Can the situation be modeled by a linear equation? Explain. If possible, write a linear model that represents the cost as a function of the number of foam hands.

Number of foam hands	4	6	8	10	12
Cost (dollars)	34	46	58	70	82

SOLUTION

1. **Understand the Problem** You know five data pairs from the table. You are asked whether the data are linear. If so, write a linear model that represents the cost.

2. **Make a Plan** Find the rate of change for consecutive data pairs in the table. If the rate of change is constant, use the point-slope form to write an equation. Rewrite the equation in slope-intercept form so that the cost is a function of the number of foam hands.

3. **Solve the Problem**

   Step 1 Find the rate of change for consecutive data pairs in the table.

   $$\frac{46-34}{6-4}=6,\ \frac{58-46}{8-6}=6,\ \frac{70-58}{10-8}=6,\ \frac{82-70}{12-10}=6$$

   Because the rate of change is constant, the data are linear. So, use the point-slope form to write an equation that represents the data.

   Step 2 Use the constant rate of change (slope) $m = 6$ and the data pair (4, 34) to write an equation. Let C be the cost (in dollars) and n be the number of foam hands.

   $C - C_1 = m(n - n_1)$ — Write the point-slope form.

   $C - 34 = 6(n - 4)$ — Substitute 6 for m, 4 for n_1, and 34 for C_1.

   $C - 34 = 6n - 24$ — Distributive Property

   $C = 6n + 10$ — Write in slope-intercept form.

   ▶ Because the cost increases at a constant rate, the situation can be modeled by a linear equation. The linear model is $C = 6n + 10$.

4. **Look Back** To check that your model is correct, verify that the other data pairs are solutions of the equation.

   $46 = 6(6) + 10$ ✓ $58 = 6(8) + 10$ ✓

   $70 = 6(10) + 10$ ✓ $82 = 6(12) + 10$ ✓

Monitoring Progress

Help in English and Spanish at *BigIdeasMath.com*

6. You pay an installation fee and a monthly fee for Internet service. The table shows the total cost for different numbers of months. Can the situation be modeled by a linear equation? Explain. If possible, write a linear model that represents the total cost as a function of the number of months.

Number of months	Total cost (dollars)
3	176
6	302
9	428
12	554

4.2 Exercises

Dynamic Solutions available at *BigIdeasMath.com*

Vocabulary and Core Concept Check

1. **USING STRUCTURE** Without simplifying, identify the slope of the line given by the equation $y - 5 = -2(x + 5)$. Then identify one point on the line.

2. **WRITING** Explain how you can use the slope formula to write an equation of the line that passes through $(3, -2)$ and has a slope of 4.

Monitoring Progress and Modeling with Mathematics

In Exercises 3–10, write an equation in point-slope form of the line that passes through the given point and has the given slope. *(See Example 1.)*

3. $(2, 1)$; $m = 2$
4. $(3, 5)$; $m = -1$
5. $(7, -4)$; $m = -6$
6. $(-8, -2)$; $m = 5$
7. $(9, 0)$; $m = -3$
8. $(0, 2)$; $m = 4$
9. $(-6, 6)$; $m = \frac{3}{2}$
10. $(5, -12)$; $m = -\frac{2}{5}$

In Exercises 11–14, write an equation in slope-intercept form of the line shown. *(See Example 2.)*

11.

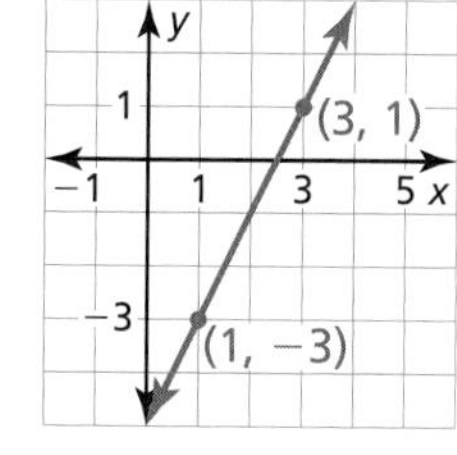

12.

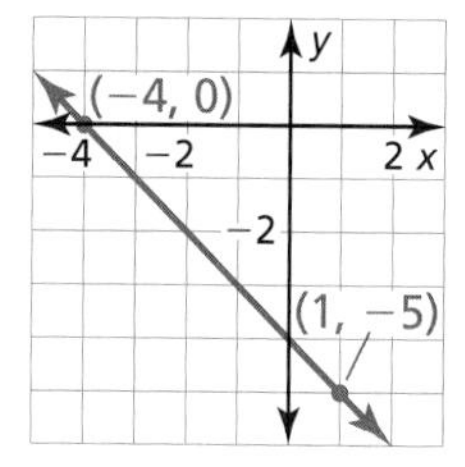

13.

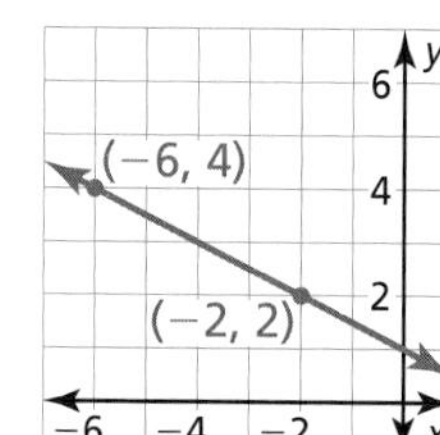

14.

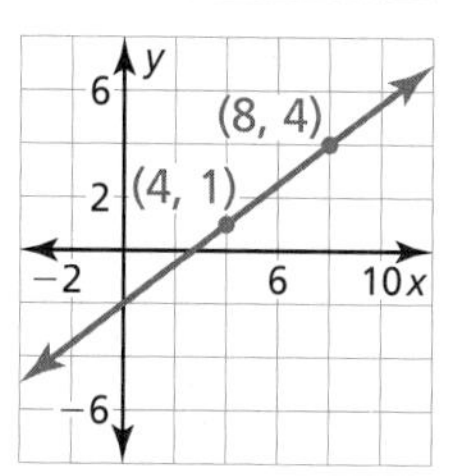

In Exercises 15–20, write an equation in slope-intercept form of the line that passes through the given points.

15. $(7, 2), (2, 12)$
16. $(6, -2), (12, 1)$
17. $(6, -1), (3, -7)$
18. $(-2, 5), (-4, -5)$
19. $(1, -9), (-3, -9)$
20. $(-5, 19), (5, 13)$

In Exercises 21–26, write a linear function f with the given values. *(See Example 3.)*

21. $f(2) = -2, f(1) = 1$
22. $f(5) = 7, f(-2) = 0$
23. $f(-4) = 2, f(6) = -3$
24. $f(-10) = 4, f(-2) = 4$
25. $f(-3) = 1, f(13) = 5$
26. $f(-9) = 10, f(-1) = -2$

In Exercises 27–30, tell whether the data in the table can be modeled by a linear equation. Explain. If possible, write a linear equation that represents y as a function of x. *(See Example 4.)*

27.

x	2	4	6	8	10
y	−1	5	15	29	47

28.

x	−3	−1	1	3	5
y	16	10	4	−2	−8

29.

x	y
0	1.2
1	1.4
2	1.6
4	2

30.

x	y
1	18
2	15
4	12
8	9

31. **ERROR ANALYSIS** Describe and correct the error in writing a linear function g with the values $g(5) = 4$ and $g(3) = 10$.

✗

$$m = \frac{10 - 4}{3 - 5} = \frac{6}{-2} = -3$$

$$y - y_1 = mx - x_1$$
$$y - 4 = -3x - 5$$
$$y = -3x - 1$$

A function is $g(x) = -3x - 1$.

32. ERROR ANALYSIS Describe and correct the error in writing an equation of the line that passes through the points (1, 2) and (4, 3).

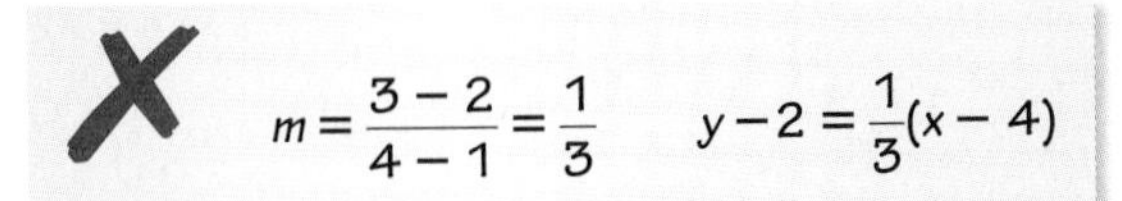

33. MODELING WITH MATHEMATICS You are designing a sticker to advertise your band. A company charges $225 for the first 1000 stickers and $80 for each additional 1000 stickers.

a. Write an equation that represents the total cost (in dollars) of the stickers as a function of the number (in thousands) of stickers ordered.

b. Find the total cost of 9000 stickers.

34. MODELING WITH MATHEMATICS You pay a processing fee and a daily fee to rent a beach house. The table shows the total cost of renting the beach house for different numbers of days.

Days	2	4	6	8
Total cost (dollars)	246	450	654	858

a. Can the situation be modeled by a linear equation? Explain.

b. What is the processing fee? the daily fee?

c. You can spend no more than $1200 on the beach house rental. What is the maximum number of days you can rent the beach house?

35. WRITING Describe two ways to graph the equation $y - 1 = \frac{3}{2}(x - 4)$.

36. THOUGHT PROVOKING The graph of a linear function passes through the point $(12, -5)$ and has a slope of $\frac{2}{5}$. Represent this function in two other ways.

37. REASONING You are writing an equation of the line that passes through two points that are not on the y-axis. Would you use slope-intercept form or point-slope form to write the equation? Explain.

38. HOW DO YOU SEE IT? The graph shows two points that lie on the graph of a linear function.

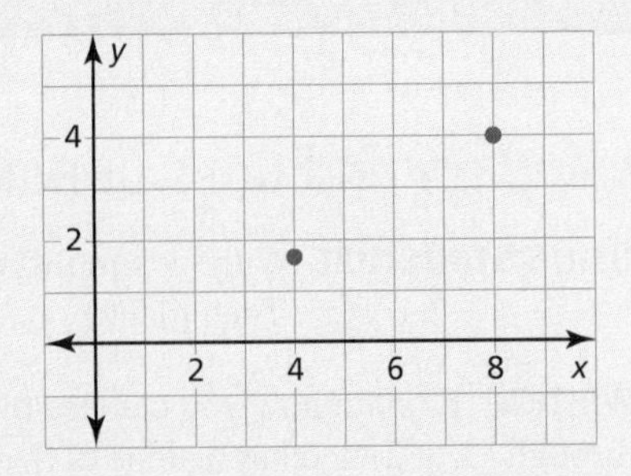

a. Does the y-intercept of the graph of the linear function appear to be positive or negative? Explain.

b. Estimate the coordinates of the two points. How can you use your estimates to confirm your answer in part (a)?

39. CONNECTION TO TRANSFORMATIONS Compare the graph of $y = 2x$ to the graph of $y - 1 = 2(x + 3)$. Make a conjecture about the graphs of $y = mx$ and $y - k = m(x - h)$.

40. COMPARING FUNCTIONS Three siblings each receive money for a holiday and then spend it at a constant weekly rate. The graph describes Sibling A's spending, the table describes Sibling B's spending, and the equation $y = -22.5x + 90$ describes Sibling C's spending. The variable y represents the amount of money left after x weeks.

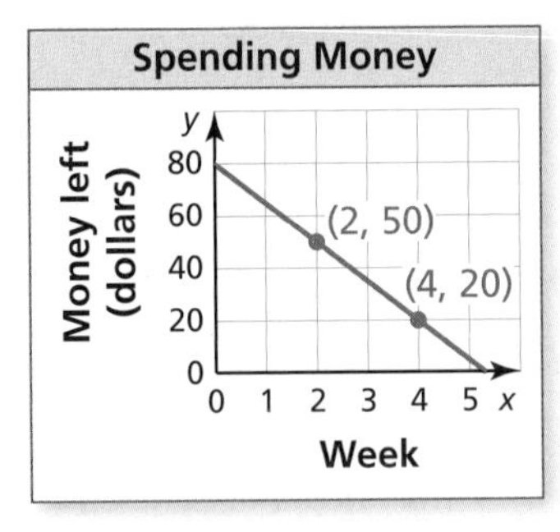

Week, x	Money left, y
1	$100
2	$75
3	$50
4	$25

a. Which sibling received the most money? the least money?

b. Which sibling spends money at the fastest rate? the slowest rate?

c. Which sibling runs out of money first? last?

Maintaining Mathematical Proficiency

Reviewing what you learned in previous grades and lessons

Write the reciprocal of the number. *(Skills Review Handbook)*

41. 5 **42.** -8 **43.** $-\frac{2}{7}$ **44.** $\frac{3}{2}$

4.3 Writing Equations of Parallel and Perpendicular Lines

Essential Question How can you recognize lines that are parallel or perpendicular?

EXPLORATION 1 **Recognizing Parallel Lines**

Work with a partner. Write each linear equation in slope-intercept form. Then use a graphing calculator to graph the three equations in the same square viewing window. (The graph of the first equation is shown.) Which two lines appear parallel? How can you tell?

a. $3x + 4y = 6$
$3x + 4y = 12$
$4x + 3y = 12$

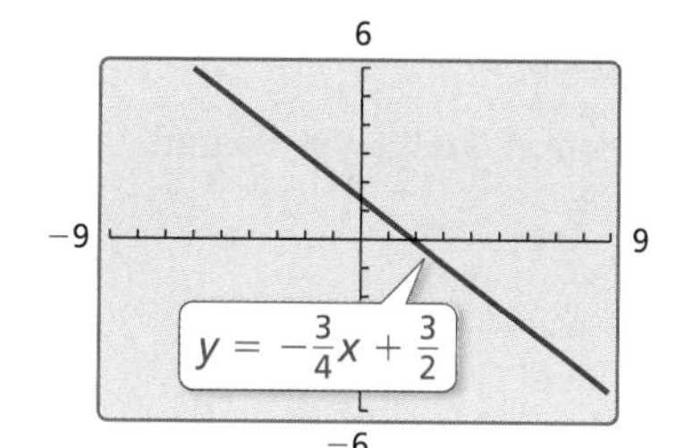

b. $5x + 2y = 6$
$2x + y = 3$
$2.5x + y = 5$

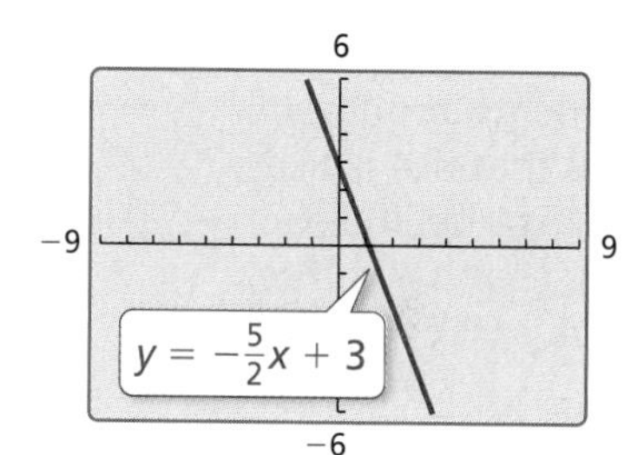

USING TOOLS STRATEGICALLY

To be proficient in math, you need to use a graphing calculator and other available technological tools, as appropriate, to help you explore relationships and deepen your understanding of concepts.

EXPLORATION 2 **Recognizing Perpendicular Lines**

Work with a partner. Write each linear equation in slope-intercept form. Then use a graphing calculator to graph the three equations in the same square viewing window. (The graph of the first equation is shown.) Which two lines appear perpendicular? How can you tell?

a. $3x + 4y = 6$
$3x - 4y = 12$
$4x - 3y = 12$

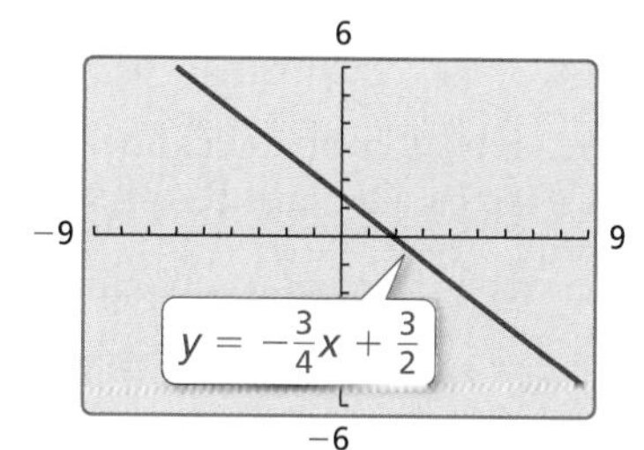

b. $2x + 5y = 10$
$-2x + y = 3$
$2.5x - y = 5$

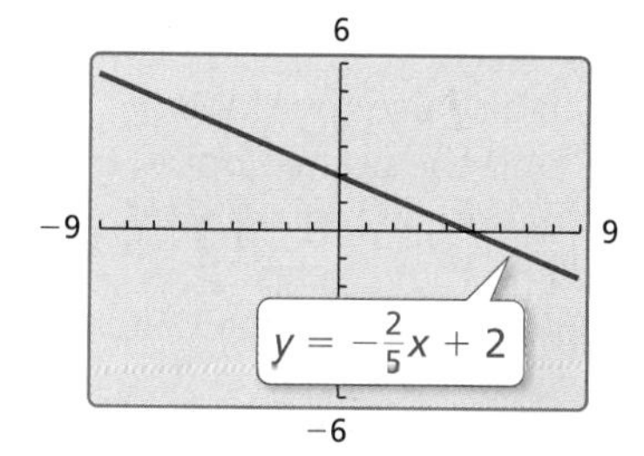

Communicate Your Answer

3. How can you recognize lines that are parallel or perpendicular?

4. Compare the slopes of the lines in Exploration 1. How can you use slope to determine whether two lines are parallel? Explain your reasoning.

5. Compare the slopes of the lines in Exploration 2. How can you use slope to determine whether two lines are perpendicular? Explain your reasoning.

4.3 Lesson

Core Vocabulary

parallel lines, *p. 188*
perpendicular lines, *p. 189*

Previous
reciprocal

What You Will Learn

- Identify and write equations of parallel lines.
- Identify and write equations of perpendicular lines.
- Use parallel and perpendicular lines in real-life problems.

Identifying and Writing Equations of Parallel Lines

Core Concept

Parallel Lines and Slopes

Two lines in the same plane that never intersect are **parallel lines**. Two distinct nonvertical lines are parallel if and only if they have the same slope.

All vertical lines are parallel.

READING

The phrase "*A if and only if B*" is a way of writing two conditional statements at once. It means that if *A* is true, then *B* is true. It also means that if *B* is true, then *A* is true.

EXAMPLE 1 Identifying Parallel Lines

Determine which of the lines are parallel.

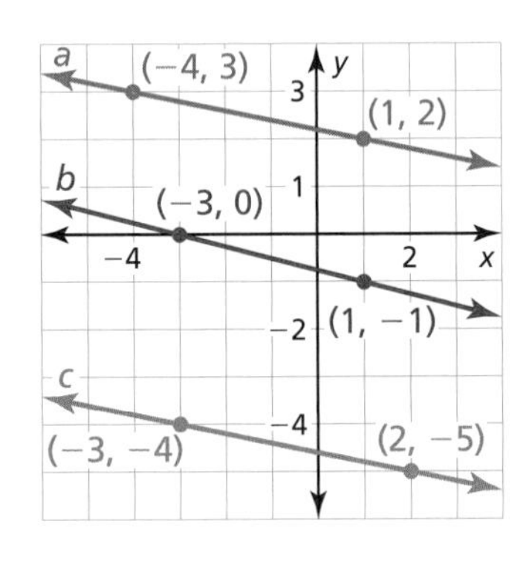

SOLUTION

Find the slope of each line.

Line a: $m = \dfrac{2-3}{1-(-4)} = -\dfrac{1}{5}$

Line b: $m = \dfrac{-1-0}{1-(-3)} = -\dfrac{1}{4}$

Line c: $m = \dfrac{-5-(-4)}{2-(-3)} = -\dfrac{1}{5}$

▶ Lines a and c have the same slope, so they are parallel.

EXAMPLE 2 Writing an Equation of a Parallel Line

Write an equation of the line that passes through $(5, -4)$ and is parallel to the line $y = 2x + 3$.

SOLUTION

Step 1 Find the slope of the parallel line. The graph of the given equation has a slope of 2. So, the parallel line that passes through $(5, -4)$ also has a slope of 2.

Step 2 Use the slope-intercept form to find the y-intercept of the parallel line.

$y = mx + b$	Write the slope-intercept form.
$-4 = 2(5) + b$	Substitute 2 for m, 5 for x, and -4 for y.
$-14 = b$	Solve for b.

▶ Using $m = 2$ and $b = -14$, an equation of the parallel line is $y = 2x - 14$.

ANOTHER WAY

You can also use the slope $m = 2$ and the point-slope form to write an equation of the line that passes through $(5, -4)$.

$$y - y_1 = m(x - x_1)$$
$$y - (-4) = 2(x - 5)$$
$$y = 2x - 14$$

Monitoring Progress

Help in English and Spanish at *BigIdeasMath.com*

1. Line a passes through $(-5, 3)$ and $(-6, -1)$. Line b passes through $(3, -2)$ and $(2, -7)$. Are the lines parallel? Explain.
2. Write an equation of the line that passes through $(-4, 2)$ and is parallel to the line $y = \frac{1}{4}x + 1$.

Identifying and Writing Equations of Perpendicular Lines

REMEMBER

The product of a nonzero number m and its negative reciprocal is -1:

$$m\left(-\frac{1}{m}\right) = -1.$$

Core Concept

Perpendicular Lines and Slopes

Two lines in the same plane that intersect to form right angles are **perpendicular lines.** Nonvertical lines are perpendicular if and only if their slopes are negative reciprocals.

Vertical lines are perpendicular to horizontal lines.

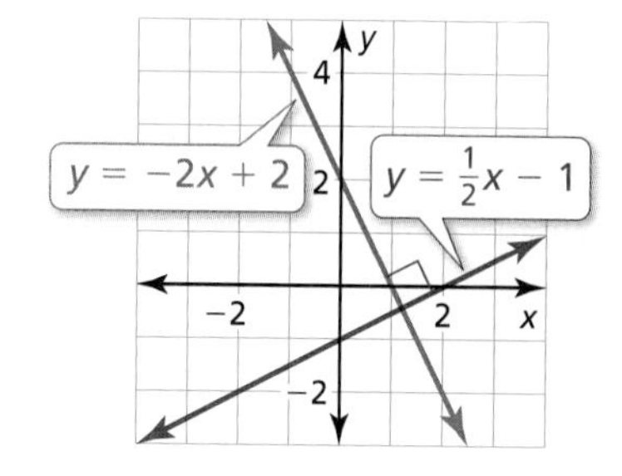

EXAMPLE 3 **Identifying Parallel and Perpendicular Lines**

Determine which of the lines, if any, are parallel or perpendicular.

Line a: $y = 4x + 2$ Line b: $x + 4y = 3$ Line c: $-8y - 2x = 16$

SOLUTION

Write the equations in slope-intercept form. Then compare the slopes.

Line a: $y = 4x + 2$ Line b: $y = -\frac{1}{4}x + \frac{3}{4}$ Line c: $y = -\frac{1}{4}x - 2$

▶ Lines b and c have slopes of $-\frac{1}{4}$, so they are parallel. Line a has a slope of 4, the negative reciprocal of $-\frac{1}{4}$, so it is perpendicular to lines b and c.

EXAMPLE 4 **Writing an Equation of a Perpendicular Line**

Write an equation of the line that passes through $(-3, 1)$ and is perpendicular to the line $y = \frac{1}{2}x + 3$.

ANOTHER WAY

You can also use the slope $m = -2$ and the slope-intercept form to write an equation of the line that passes through $(-3, 1)$.

$y = mx + b$

$1 = -2(-3) + b$

$-5 = b$

So, $y = -2x - 5$.

SOLUTION

Step 1 Find the slope of the perpendicular line. The graph of the given equation has a slope of $\frac{1}{2}$. Because the slopes of perpendicular lines are negative reciprocals, the slope of the perpendicular line that passes through $(-3, 1)$ is -2.

Step 2 Use the slope $m = -2$ and the point-slope form to write an equation of the perpendicular line that passes through $(-3, 1)$.

$y - y_1 = m(x - x_1)$	Write the point-slope form.
$y - 1 = -2[x - (-3)]$	Substitute -2 for m, -3 for x_1, and 1 for y_1.
$y - 1 = -2x - 6$	Simplify.
$y = -2x - 5$	Write in slope-intercept form.

▶ An equation of the perpendicular line is $y = -2x - 5$.

Monitoring Progress

Help in English and Spanish at *BigIdeasMath.com*

3. Determine which of the lines, if any, are parallel or perpendicular. Explain.

Line a: $2x + 6y = -3$ Line b: $y = 3x - 8$ Line c: $-6y + 18x = 9$

4. Write an equation of the line that passes through $(-3, 5)$ and is perpendicular to the line $y = -3x - 1$.

Writing Equations for Real-Life Problems

EXAMPLE 5 Writing an Equation of a Perpendicular Line

The position of a helicopter search and rescue crew is shown in the graph. The shortest flight path to the shoreline is one that is perpendicular to the shoreline. Write an equation that represents this path.

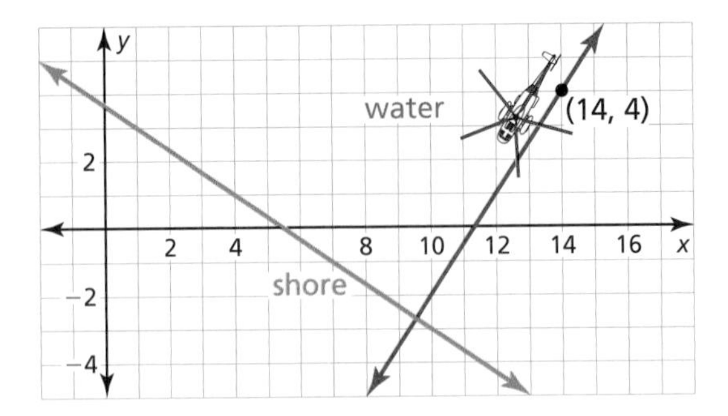

SOLUTION

1. **Understand the Problem** You can see the line that represents the shoreline. You know the coordinates of the helicopter. You are asked to write an equation that represents the shortest flight path to the shoreline.

2. **Make a Plan** Find the slope of the line that represents the shoreline. Use the negative reciprocal of this slope, the coordinates of the helicopter, and the point-slope form to write an equation.

3. **Solve the Problem**

Step 1 Find the slope of the line that represents the shoreline. The line passes through points (1, 3) and (4, 1). So, the slope is

$$m = \frac{1 - 3}{4 - 1} = -\frac{2}{3}.$$

Because the shoreline and shortest flight path are perpendicular, the slopes of their respective graphs are negative reciprocals. So, the slope of the graph of the shortest flight path is $\frac{3}{2}$.

Step 2 Use the slope $m = \frac{3}{2}$ and the point-slope form to write an equation of the shortest flight path that passes through (14, 4).

$y - y_1 = m(x - x_1)$	Write the point-slope form.
$y - 4 = \frac{3}{2}(x - 14)$	Substitute $\frac{3}{2}$ for m, 14 for x_1, and 4 for y_1.
$y - 4 = \frac{3}{2}x - 21$	Distributive Property
$y = \frac{3}{2}x - 17$	Write in slope-intercept form.

▶ An equation that represents the shortest flight path is $y = \frac{3}{2}x - 17$.

4. **Look Back** To check that your equation is correct, verify that (14, 4) is a solution of the equation.

$$4 = \frac{3}{2}(14) - 17 \checkmark$$

Monitoring Progress  Help in English and Spanish at *BigIdeasMath.com*

5. In Example 5, a boat is traveling parallel to the shoreline and passes through (9, 3). Write an equation that represents the path of the boat.

Vocabulary and Core Concept Check

1. **COMPLETE THE SENTENCE** Two distinct nonvertical lines that have the same slope are ________.

2. **VOCABULARY** Two lines are perpendicular. The slope of one line is $-\frac{5}{7}$. What is the slope of the other line? Justify your answer.

Monitoring Progress and Modeling with Mathematics

In Exercises 3–8, determine which of the lines, if any, are parallel. Explain. *(See Example 1.)*

3.

4.

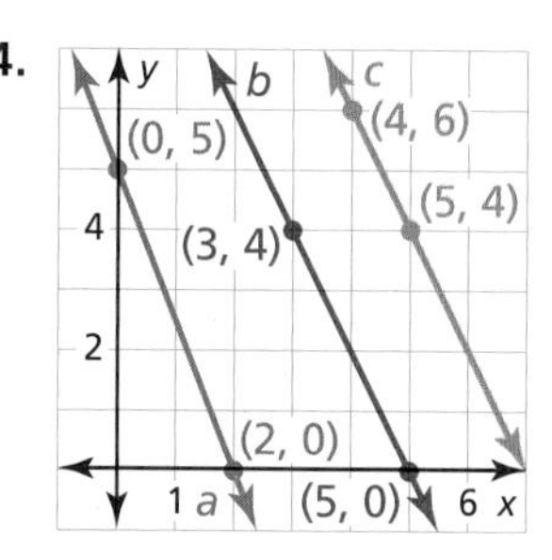

5. Line a passes through $(-1, -2)$ and $(1, 0)$.
Line b passes through $(4, 2)$ and $(2, -2)$.
Line c passes through $(0, 2)$ and $(-1, 1)$.

6. Line a passes through $(-1, 3)$ and $(1, 9)$.
Line b passes through $(-2, 12)$ and $(-1, 14)$.
Line c passes through $(3, 8)$ and $(6, 10)$.

7. Line a: $4y + x = 8$
Line b: $2y + x = 4$
Line c: $2y = -3x + 6$

8. Line a: $3y - x = 6$
Line b: $3y = x + 18$
Line c: $3y - 2x = 9$

In Exercises 9–12, write an equation of the line that passes through the given point and is parallel to the given line. *(See Example 2.)*

9. $(-1, 3)$; $y = 2x + 2$

10. $(1, 2)$; $y = -5x + 4$

11. $(18, 2)$; $3y - x = -12$

12. $(2, -5)$; $2y = 3x + 10$

In Exercises 13–18, determine which of the lines, if any, are parallel or perpendicular. Explain. *(See Example 3.)*

13.

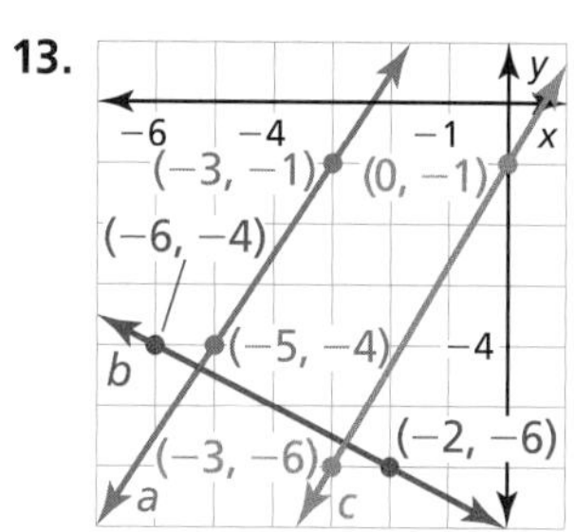

14.

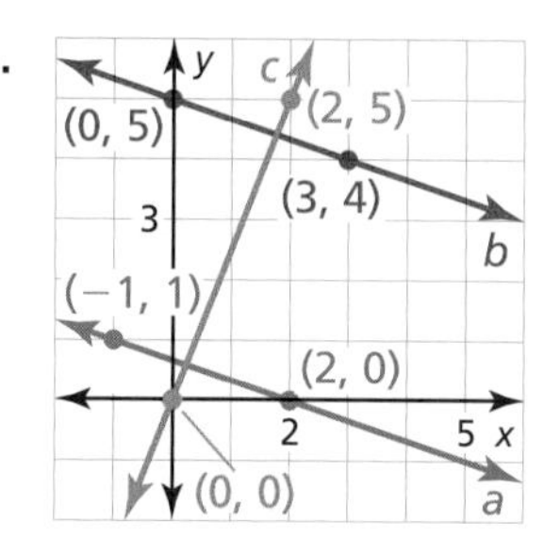

15. Line a passes through $(-2, 1)$ and $(0, 3)$.
Line b passes through $(4, 1)$ and $(6, 4)$.
Line c passes through $(1, 3)$ and $(4, 1)$.

16. Line a passes through $(2, 10)$ and $(4, 13)$.
Line b passes through $(4, 9)$ and $(6, 12)$.
Line c passes through $(2, 10)$ and $(4, 9)$.

17. Line a: $4x - 3y = 2$
Line b: $y = \frac{4}{3}x + 2$
Line c: $4y + 3x = 4$

18. Line a: $y = 6x - 2$
Line b: $6y = -x$
Line c: $y + 6x = 1$

In Exercises 19–22, write an equation of the line that passes through the given point and is perpendicular to the given line. *(See Example 4.)*

19. $(7, 10)$; $y = \frac{1}{2}x - 9$

20. $(-4, -1)$; $y = \frac{4}{3}x + 6$

21. $(-3, 3)$; $2y = 8x - 6$

22. $(8, 1)$; $2y + 4x = 12$

In Exercises 23 and 24, write an equation of the line that passes through the given point and is (a) parallel and (b) perpendicular to the given line.

23.

24.

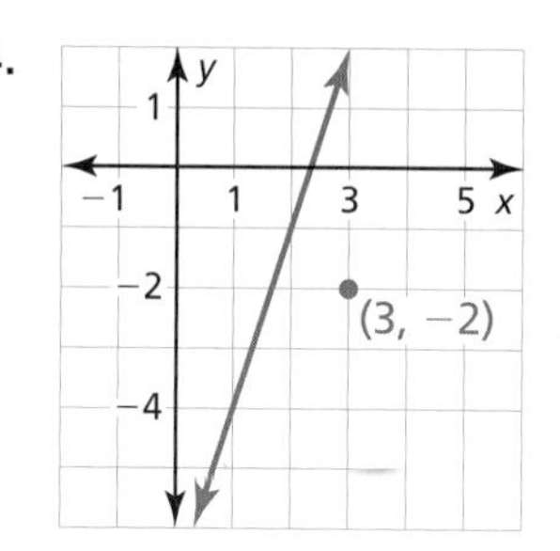

25. **ERROR ANALYSIS** Describe and correct the error in writing an equation of the line that passes through $(1, 3)$ and is parallel to the line $y = \frac{1}{4}x + 2$.

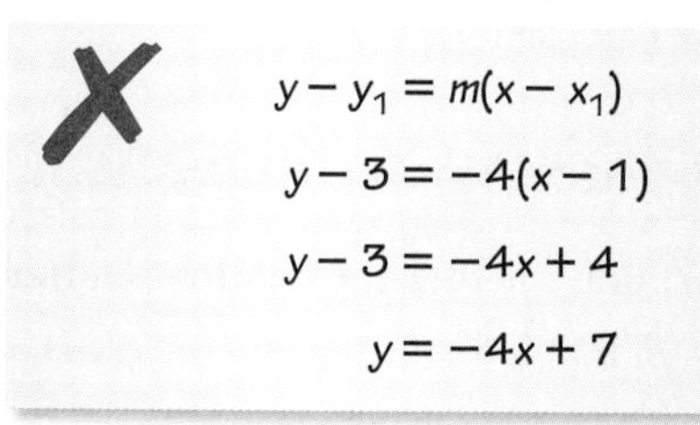

26. **ERROR ANALYSIS** Describe and correct the error in writing an equation of the line that passes through $(4, -5)$ and is perpendicular to the line $y = \frac{1}{3}x + 5$.

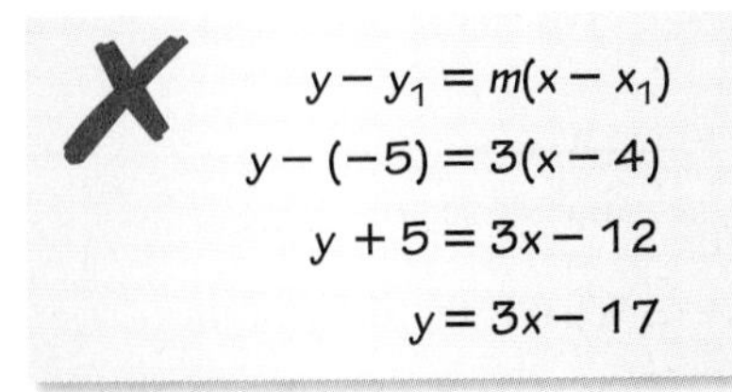

27. **MODELING WITH MATHEMATICS** A city water department is proposing the construction of a new water pipe, as shown. The new pipe will be perpendicular to the old pipe. Write an equation that represents the new pipe. *(See Example 5.)*

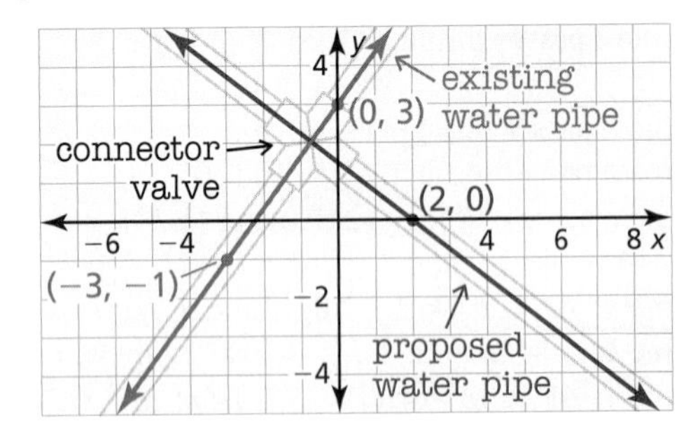

28. **MODELING WITH MATHEMATICS** A parks and recreation department is constructing a new bike path. The path will be parallel to the railroad tracks shown and pass through the parking area at the point (4, 5). Write an equation that represents the path.

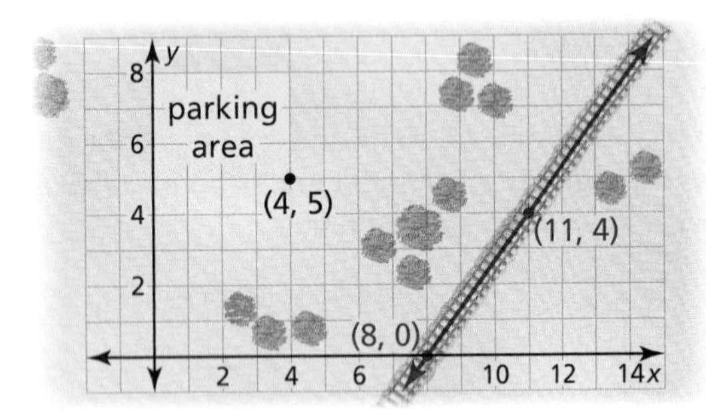

29. **MATHEMATICAL CONNECTIONS** The vertices of a quadrilateral are $A(2, 2)$, $B(6, 4)$, $C(8, 10)$, and $D(4, 8)$.

   a. Is quadrilateral $ABCD$ a parallelogram? Explain.

   b. Is quadrilateral $ABCD$ a rectangle? Explain.

30. **USING STRUCTURE** For what value of a are the graphs of $6y = -2x + 4$ and $2y = ax - 5$ parallel? perpendicular?

31. **MAKING AN ARGUMENT** A hockey puck leaves the blade of a hockey stick, bounces off a wall, and travels in a new direction, as shown. Your friend claims the path of the puck forms a right angle. Is your friend correct? Explain.

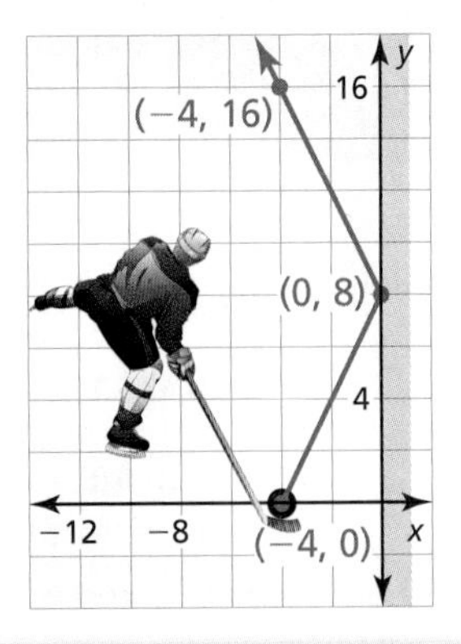

32. **HOW DO YOU SEE IT?** A softball academy charges students an initial registration fee plus a monthly fee. The graph shows the total amounts paid by two students over a 4-month period. The lines are parallel.

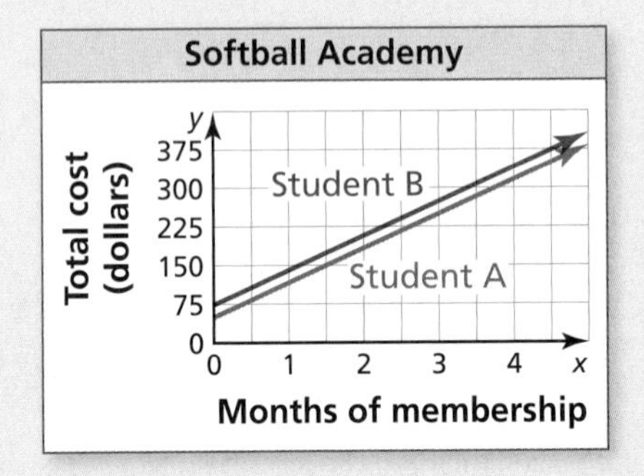

   a. Did one of the students pay a greater registration fee? Explain.

   b. Did one of the students pay a greater monthly fee? Explain.

REASONING In Exercises 33–35, determine whether the statement is *always*, *sometimes*, or *never* true. Explain your reasoning.

33. Two lines with positive slopes are perpendicular.

34. A vertical line is parallel to the y-axis.

35. Two lines with the same y-intercept are perpendicular.

36. **THOUGHT PROVOKING** You are designing a new logo for your math club. Your teacher asks you to include at least one pair of parallel lines and at least one pair of perpendicular lines. Sketch your logo in a coordinate plane. Write the equations of the parallel and perpendicular lines.

Maintaining Mathematical Proficiency

Reviewing what you learned in previous grades and lessons

Determine whether the relation is a function. Explain. *(Section 3.1)*

37. (3, 6), (4, 8), (5, 10), (6, 9), (7, 14)

38. (−1, 6), (1, 4), (−1, 2), (1, 6), (−1, 5)

4.1–4.3 What Did You Learn?

Core Vocabulary

linear model, *p. 178*
point-slope form, *p. 182*
parallel lines, *p. 188*
perpendicular lines, *p. 189*

Core Concepts

Section 4.1

Using Slope-Intercept Form, *p. 176*

Section 4.2

Using Point-Slope Form, *p. 182*

Section 4.3

Parallel Lines and Slopes, *p. 188*
Perpendicular Lines and Slopes, *p. 189*

Mathematical Practices

1. How can you explain to yourself the meaning of the graph in Exercise 36 on page 180?
2. How did you use the structure of the equations in Exercise 39 on page 186 to make a conjecture?
3. How did you use the diagram in Exercise 31 on page 192 to determine whether your friend was correct?

Study Skills

Getting Actively Involved in Class

If you do not understand something at all and do not even know how to phrase a question, just ask for clarification. You might say something like, "Could you please explain the steps in this problem one more time?"

If your teacher asks for someone to go up to the board, volunteer. The student at the board often receives additional attention and instruction to complete the problem.

4.1–4.3 Quiz

Write an equation of the line in slope-intercept form. *(Section 4.1)*

1.

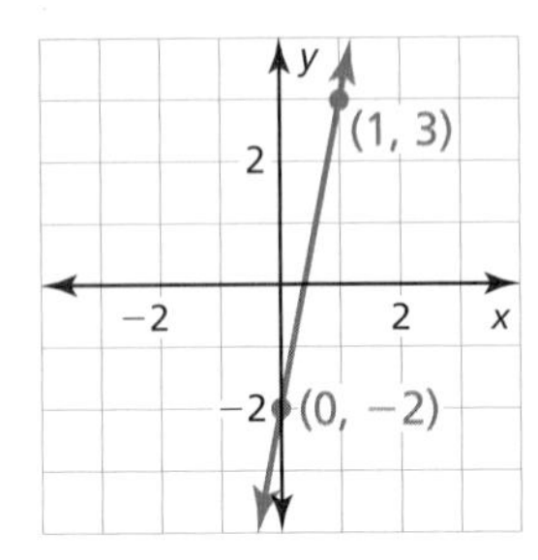

2.

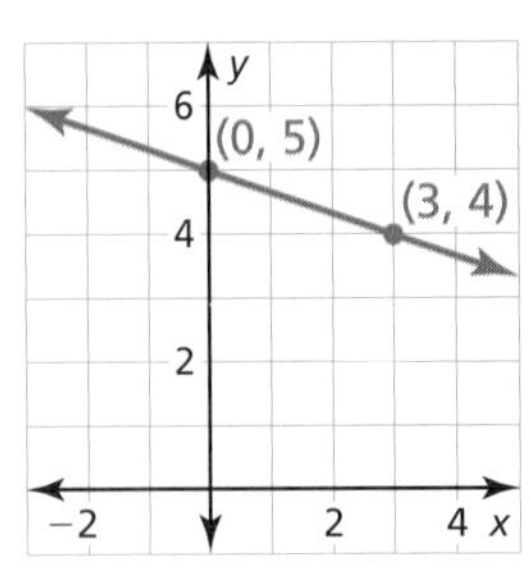

3.

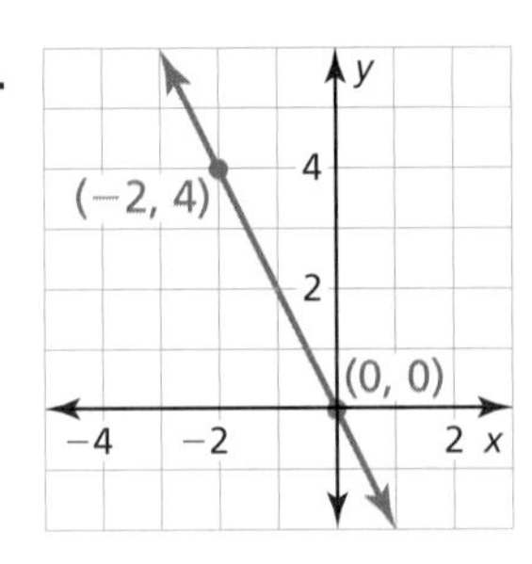

Write an equation in point-slope form of the line that passes through the given points. *(Section 4.2)*

4. $(-2, 5), (1, -1)$

5. $(-3, -2), (2, -1)$

6. $(1, 0), (4, 4)$

Write a linear function f with the given values. *(Section 4.1 and Section 4.2)*

7. $f(0) = 2, f(5) = -3$

8. $f(-1) = -6, f(4) = -6$

9. $f(-3) = -2, f(-2) = 3$

Determine which of the lines, if any, are parallel or perpendicular. Explain. *(Section 4.3)*

10. Line a passes through $(-2, 2)$ and $(2, 1)$.
Line b passes through $(1, -8)$ and $(3, 0)$.
Line c passes through $(-4, -3)$ and $(0, -2)$.

11. Line a: $2x + 6y = -12$
Line b: $y = \frac{3}{2}x - 5$
Line c: $3x - 2y = -4$

Write an equation of the line that passes through the given point and is (a) parallel and (b) perpendicular to the given line. *(Section 4.3)*

12.

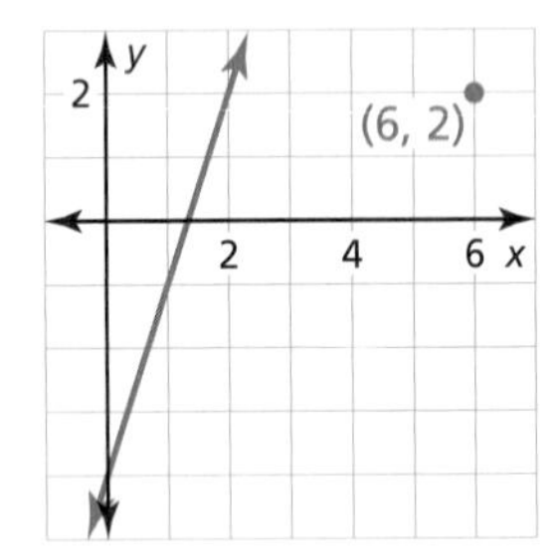

13.

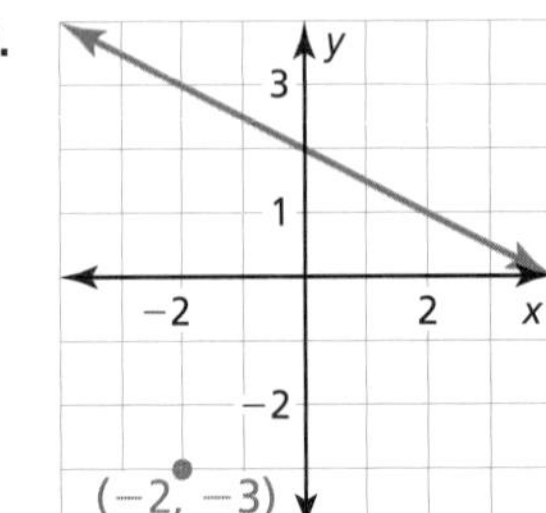

14.

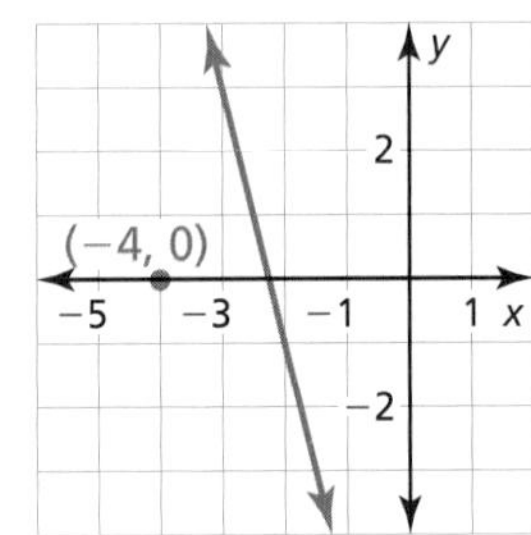

15. A website hosting company charges an initial fee of \$48 to set up a website. The company charges \$44 per month to maintain the website. *(Section 4.1)*

a. Write a linear model that represents the total cost of setting up and maintaining a website as a function of the number of months it is maintained.

b. Find the total cost of setting up a website and maintaining it for 6 months.

c. A different website hosting company charges \$62 per month to maintain a website, but there is no initial set-up fee. You have \$620. At which company can you set up and maintain a website for the greatest amount of time? Explain.

16. The table shows the amount of water remaining in a water tank as it drains. Can the situation be modeled by a linear equation? Explain. If possible, write a linear model that represents the amount of water remaining in the tank as a function of time. *(Section 4.2)*

Time (minutes)	8	10	12	14	16
Water (gallons)	155	150	145	140	135

4.4 Scatter Plots and Lines of Fit

Essential Question How can you use a scatter plot and a line of fit to make conclusions about data?

A **scatter plot** is a graph that shows the relationship between two data sets. The two data sets are graphed as ordered pairs in a coordinate plane.

EXPLORATION 1 Finding a Line of Fit

Work with a partner. A survey was taken of 179 married couples. Each person was asked his or her age. The scatter plot shows the results.

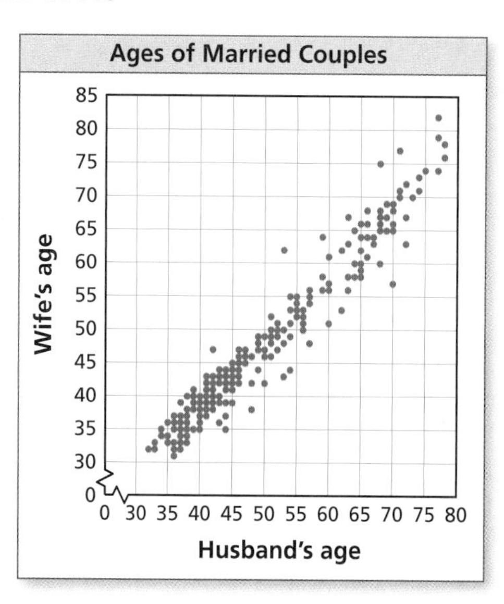

a. Draw a line that approximates the data. Write an equation of the line. Explain the method you used.

b. What conclusions can you make from the equation you wrote? Explain your reasoning.

REASONING QUANTITATIVELY

To be proficient in math, you need to make sense of quantities and their relationships in problem situations.

EXPLORATION 2 Finding a Line of Fit

Work with a partner. The scatter plot shows the median ages of American women at their first marriage for selected years from 1960 through 2010.

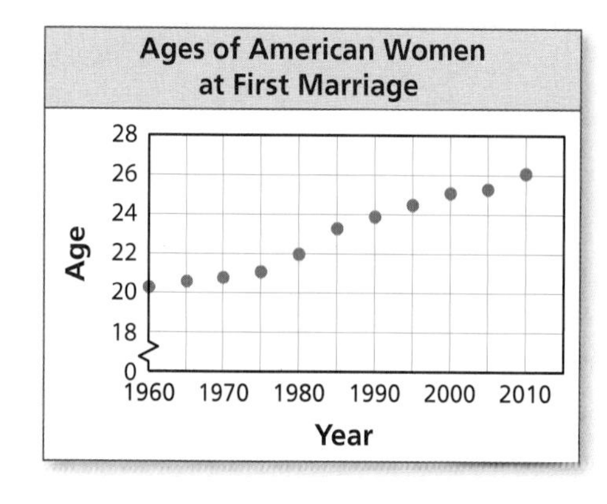

a. Draw a line that approximates the data. Write an equation of the line. Let x represent the number of years since 1960. Explain the method you used.

b. What conclusions can you make from the equation you wrote?

c. Use your equation to predict the median age of American women at their first marriage in the year 2020.

Communicate Your Answer

3. How can you use a scatter plot and a line of fit to make conclusions about data?

4. Use the Internet or some other reference to find a scatter plot of real-life data that is different from those given above. Then draw a line that approximates the data and write an equation of the line. Explain the method you used.

4.4 Lesson

What You Will Learn

- Interpret scatter plots.
- Identify correlations between data sets.
- Use lines of fit to model data.

Core Vocabulary

scatter plot, *p. 196*
correlation, *p. 197*
line of fit, *p. 198*

Interpreting Scatter Plots

Core Concept

Scatter Plot

A **scatter plot** is a graph that shows the relationship between two data sets. The two data sets are graphed as ordered pairs in a coordinate plane. Scatter plots can show trends in the data.

EXAMPLE 1 Interpreting a Scatter Plot

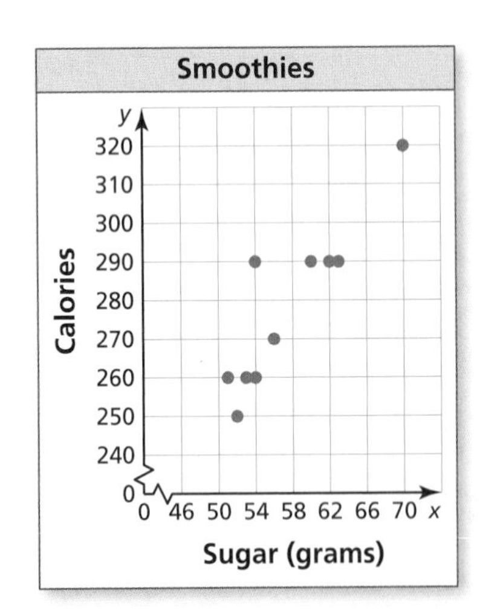

The scatter plot shows the amounts x (in grams) of sugar and the numbers y of calories in 10 smoothies.

a. How many calories are in the smoothie that contains 56 grams of sugar?

b. How many grams of sugar are in the smoothie that contains 320 calories?

c. What tends to happen to the number of calories as the number of grams of sugar increases?

SOLUTION

a. Draw a horizontal line from the point that has an x-value of 56. It crosses the y-axis at 270.

▶ So, the smoothie has 270 calories.

b. Draw a vertical line from the point that has a y-value of 320. It crosses the x-axis at 70.

▶ So, the smoothie has 70 grams of sugar.

c. Looking at the graph, the plotted points go up from left to right.

▶ So, as the number of grams of sugar increases, the number of calories increases.

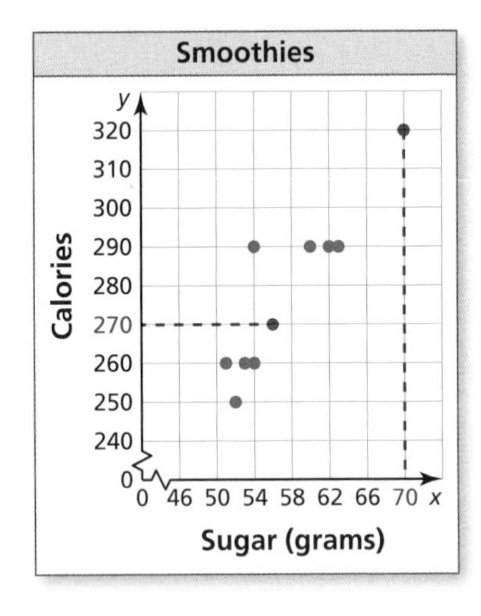

Monitoring Progress  Help in English and Spanish at *BigIdeasMath.com*

1. How many calories are in the smoothie that contains 51 grams of sugar?
2. How many grams of sugar are in the smoothie that contains 250 calories?

STUDY TIP

You can think of a positive correlation as having a positive slope and a negative correlation as having a negative slope.

Identifying Correlations between Data Sets

A **correlation** is a relationship between data sets. You can use a scatter plot to describe the correlation between data.

Positive Correlation

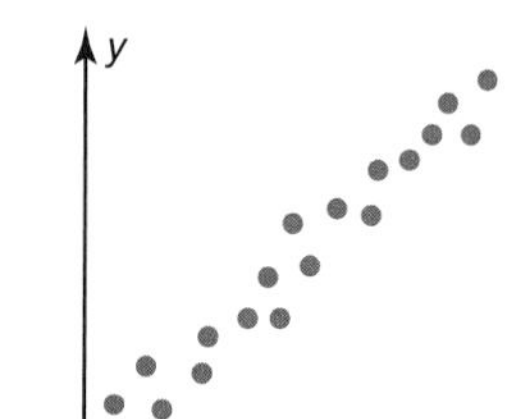

As x increases, y increases.

Negative Correlation

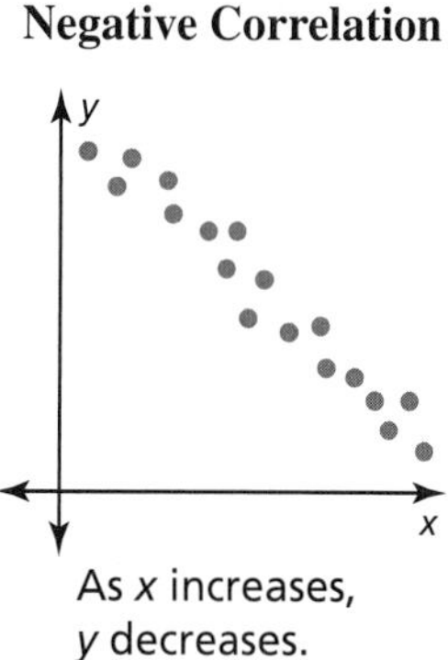

As x increases, y decreases.

No Correlation

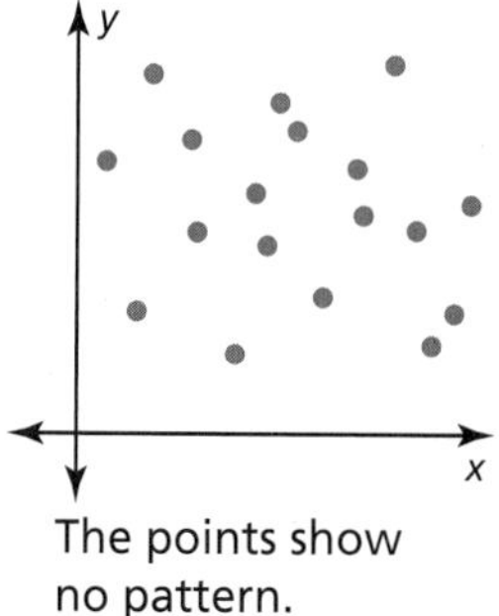

The points show no pattern.

EXAMPLE 2 Identifying Correlations

Tell whether the data show a *positive*, a *negative*, or *no* correlation.

a. age and vehicles owned

b. temperature and coat sales at a store

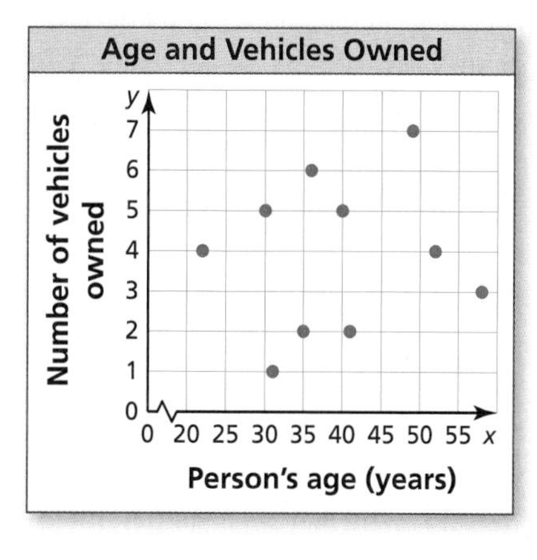

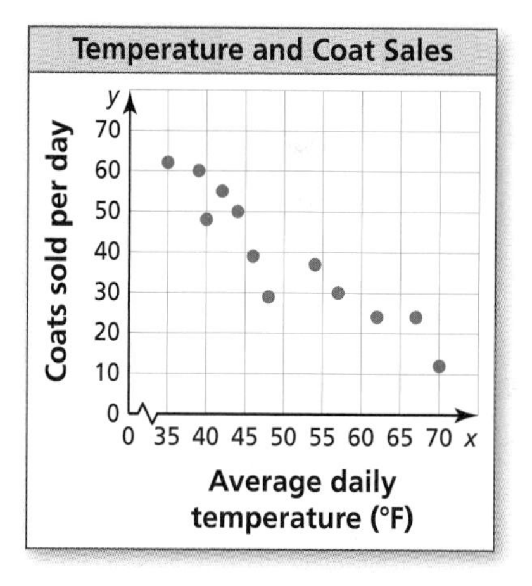

SOLUTION

a. The points show no pattern. The number of vehicles owned does not depend on a person's age.

▶ So, the scatter plot shows no correlation.

b. As the average temperature increases, the number of coats sold decreases.

▶ So, the scatter plot shows a negative correlation.

Monitoring Progress

Help in English and Spanish at *BigIdeasMath.com*

Make a scatter plot of the data. Tell whether the data show a *positive*, a *negative*, or *no* correlation.

3.

Temperature (°F), x	82	78	68	87	75	71	92	84
Attendees (thousands), y	4.5	4.0	1.7	5.5	3.8	2.9	4.7	5.3

4.

Age of a car (years), x	1	2	3	4	5	6	7	8
Value (thousands), y	\$24	\$21	\$19	\$18	\$15	\$12	\$8	\$7

Using Lines of Fit to Model Data

When data show a positive or negative correlation, you can model the *trend* in the data using a line of fit. A **line of fit** is a line drawn on a scatter plot that is close to most of the data points.

STUDY TIP

A line of fit is also called a *trend line*.

Core Concept

Using a Line of Fit to Model Data

Step 1 Make a scatter plot of the data.

Step 2 Decide whether the data can be modeled by a line.

Step 3 Draw a line that appears to fit the data closely. There should be approximately as many points above the line as below it.

Step 4 Write an equation using two points on the line. The points do not have to represent actual data pairs, but they must lie on the line of fit.

EXAMPLE 3 Finding a Line of Fit

The table shows the weekly sales of a DVD and the number of weeks since its release. Write an equation that models the DVD sales as a function of the number of weeks since its release. Interpret the slope and y-intercept of the line of fit.

Week, x	1	2	3	4	5	6	7	8
Sales (millions), y	\$19	\$15	\$13	\$11	\$10	\$8	\$7	\$5

SOLUTION

Step 1 Make a scatter plot of the data.

Step 2 Decide whether the data can be modeled by a line. Because the scatter plot shows a negative correlation, you can fit a line to the data.

Step 3 Draw a line that appears to fit the data closely.

Step 4 Write an equation using two points on the line. Use (5, 10) and (6, 8).

The slope of the line is $m = \dfrac{8 - 10}{6 - 5} = -2$.

Use the slope $m = -2$ and the point (6, 8) to write an equation of the line.

$y - y_1 = m(x - x_1)$ Write the point-slope form.

$y - 8 = -2(x - 6)$ Substitute -2 for m, 6 for x_1, and 8 for y_1.

$y = -2x + 20$ Solve for y.

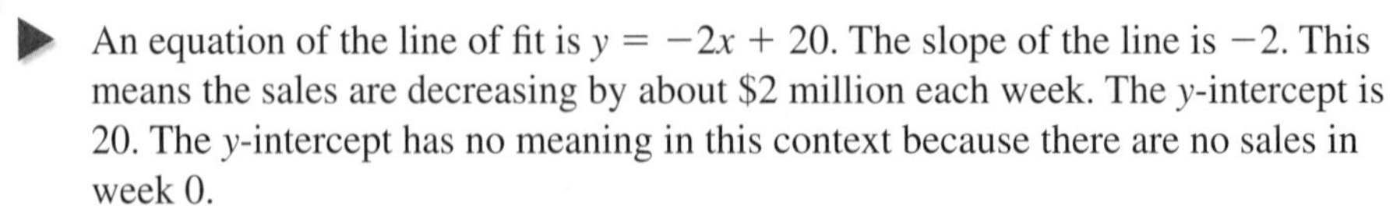

An equation of the line of fit is $y = -2x + 20$. The slope of the line is -2. This means the sales are decreasing by about \$2 million each week. The y-intercept is 20. The y-intercept has no meaning in this context because there are no sales in week 0.

Monitoring Progress  Help in English and Spanish at *BigIdeasMath.com*

5. The following data pairs show the monthly income x (in dollars) and the monthly car payment y (in dollars) of six people: (2100, 410), (1650, 315), (1950, 405), (1500, 295), (2250, 440), and (1800, 375). Write an equation that models the monthly car payment as a function of the monthly income. Interpret the slope and y-intercept of the line of fit.

4.4 Exercises

Dynamic Solutions available at *BigIdeasMath.com*

Vocabulary and Core Concept Check

1. **COMPLETE THE SENTENCE** When data show a positive correlation, the dependent variable tends to ____________ as the independent variable increases.

2. **VOCABULARY** What is a line of fit?

Monitoring Progress and Modeling with Mathematics

In Exercises 3–6, use the scatter plot to fill in the missing coordinate of the ordered pair.

3. (16, ▒▒)

4. (3, ▒▒)

5. (▒▒, 12)

6. (▒▒, 17)

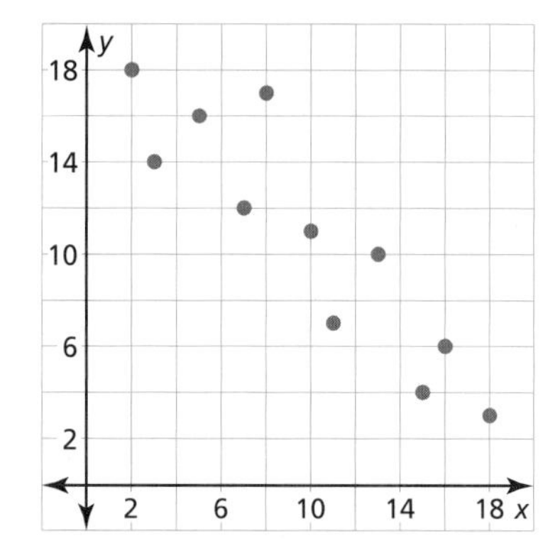

7. **INTERPRETING A SCATTER PLOT** The scatter plot shows the hard drive capacities (in gigabytes) and the prices (in dollars) of 10 laptops. *(See Example 1.)*

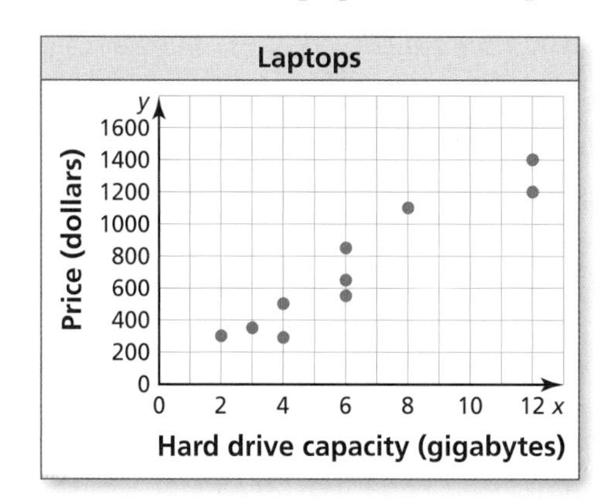

a. What is the price of the laptop with a hard drive capacity of 8 gigabytes?

b. What is the hard drive capacity of the $1200 laptop?

c. What tends to happen to the price as the hard drive capacity increases?

8. **INTERPRETING A SCATTER PLOT** The scatter plot shows the earned run averages and the winning percentages of eight pitchers on a baseball team.

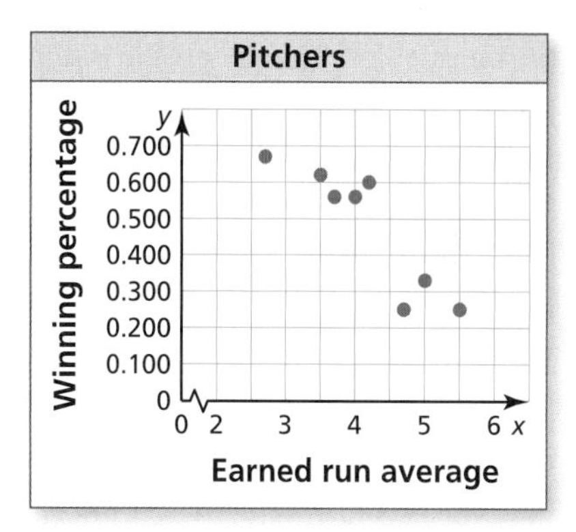

a. What is the winning percentage of the pitcher with an earned run average of 4.2?

b. What is the earned run average of the pitcher with a winning percentage of 0.33?

c. What tends to happen to the winning percentage as the earned run average increases?

In Exercises 9–12, tell whether *x* and *y* show a *positive*, a *negative*, or *no* correlation. *(See Example 2.)*

9.

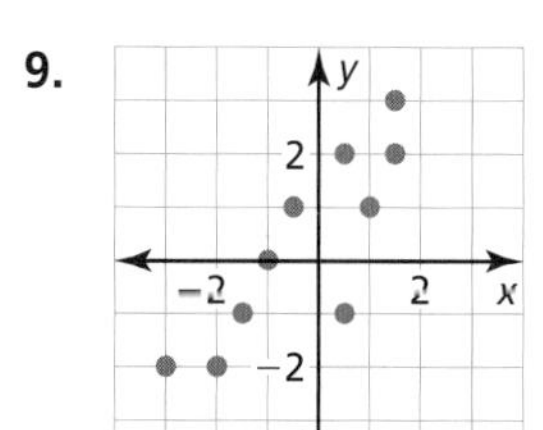

10.

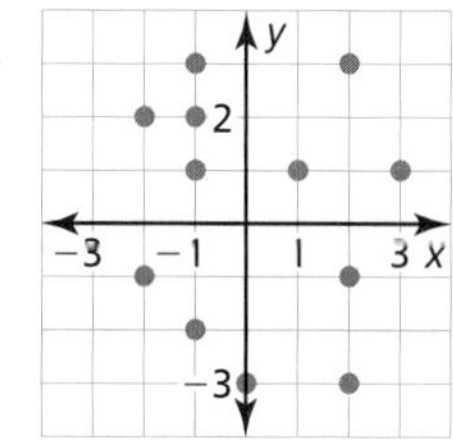

11.

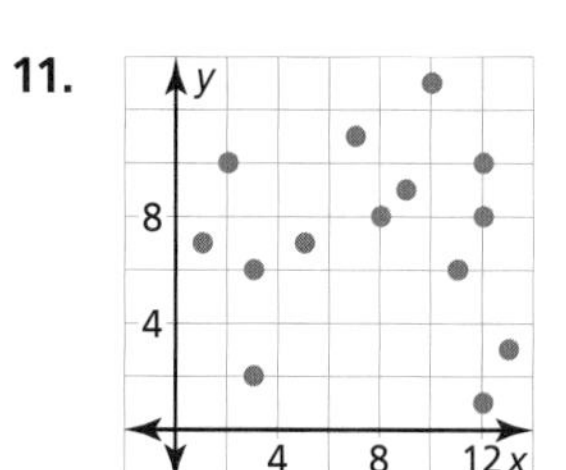

12.

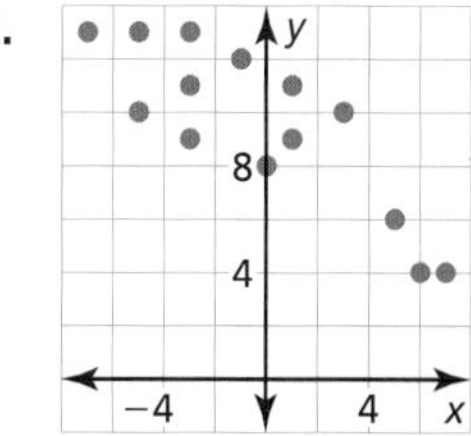

In Exercises 13 and 14, make a scatter plot of the data. Tell whether x and y show a *positive*, a *negative*, or *no* correlation.

13.

x	3.1	2.2	2.5	3.7	3.9	1.5	2.7	2.0
y	1	0	1	2	0	2	3	2

14.

x	3	4	5	6	7	8	9	10
y	67	67	50	33	25	21	19	4

15. MODELING WITH MATHEMATICS The table shows the world birth rates y (number of births per 1000 people) x years since 1960. *(See Example 3.)*

x	0	10	20	30	40	50
y	35.4	33.6	28.3	27.0	22.4	20.0

a. Write an equation that models the birthrate as a function of the number of years since 1960.

b. Interpret the slope and y-intercept of the line of fit.

16. MODELING WITH MATHEMATICS The table shows the total earnings y (in dollars) of a food server who works x hours.

x	0	1	2	3	4	5	6
y	0	18	40	62	77	85	113

a. Write an equation that models the server's earnings as a function of the number of hours the server works.

b. Interpret the slope and y-intercept of the line of fit.

17. OPEN-ENDED Give an example of a real-life data set that shows a negative correlation.

18. MAKING AN ARGUMENT Your friend says that the data in the table show a negative correlation because the dependent variable y is decreasing. Is your friend correct? Explain.

x	14	12	10	8	6	4	2
y	4	1	0	−1	−2	−4	−5

19. USING TOOLS Use a ruler or a yardstick to find the heights and arm spans of five people.

a. Make a scatter plot using the data you collected. Then draw a line of fit for the data.

b. Interpret the slope and y-intercept of the line of fit.

20. THOUGHT PROVOKING A line of fit for a scatter plot is given by the equation $y = 5x + 20$. Describe a real-life data set that could be represented by the scatter plot.

21. WRITING When is data best displayed in a scatter plot, rather than another type of display, such as a bar graph or circle graph?

22. HOW DO YOU SEE IT? The scatter plot shows part of a data set and a line of fit for the data set. Four data points are missing. Choose possible coordinates for these data points.

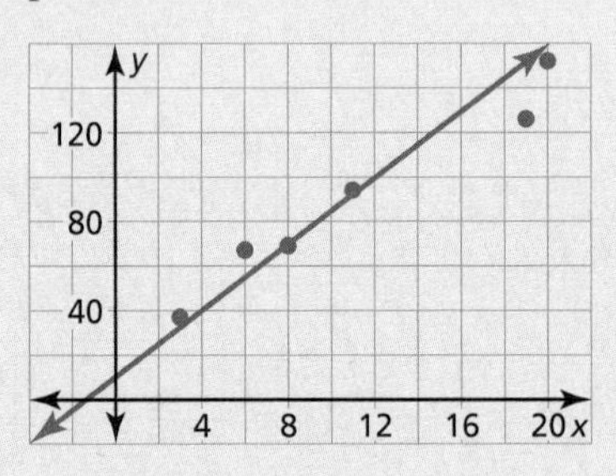

23. REASONING A data set has no correlation. Is it possible to find a line of fit for the data? Explain.

24. ANALYZING RELATIONSHIPS Make a scatter plot of the data in the tables. Describe the relationship between the variables. Is it possible to fit a line to the data? If so, write an equation of the line. If not, explain why.

x	−12	−9	−7	−4	−3	−1
y	150	76	50	15	10	1

x	2	5	6	7	9	15
y	5	22	37	52	90	226

Maintaining Mathematical Proficiency

Reviewing what you learned in previous grades and lessons

Evaluate the function when $x = -3, 0,$ and 4. *(Section 3.3)*

25. $g(x) = 6x$

26. $h(x) = -10x$

27. $f(x) = 5x - 8$

28. $v(x) = 14 - 3x$

4.5 Analyzing Lines of Fit

Essential Question How can you *analytically* find a line of best fit for a scatter plot?

EXPLORATION 1 Finding a Line of Best Fit

Work with a partner.
The scatter plot shows the median ages of American women at their first marriage for selected years from 1960 through 2010. In Exploration 2 in Section 4.4, you approximated a line of fit graphically. To find the line of *best* fit, you can use a computer, spreadsheet, or graphing calculator that has a *linear regression* feature.

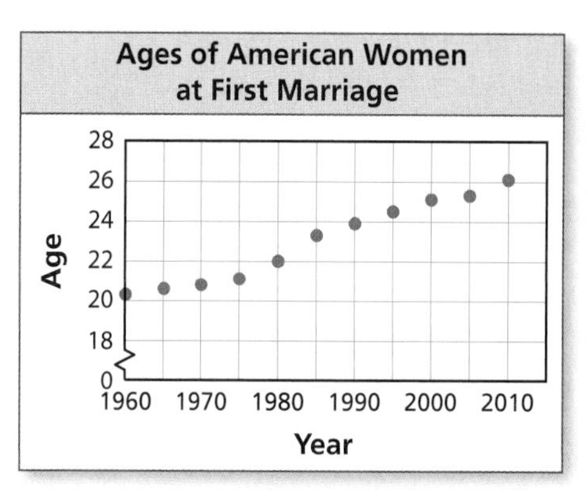

a. The data from the scatter plot is shown in the table. Note that 0, 5, 10, and so on represent the numbers of years since 1960. What does the ordered pair (25, 23.3) represent?

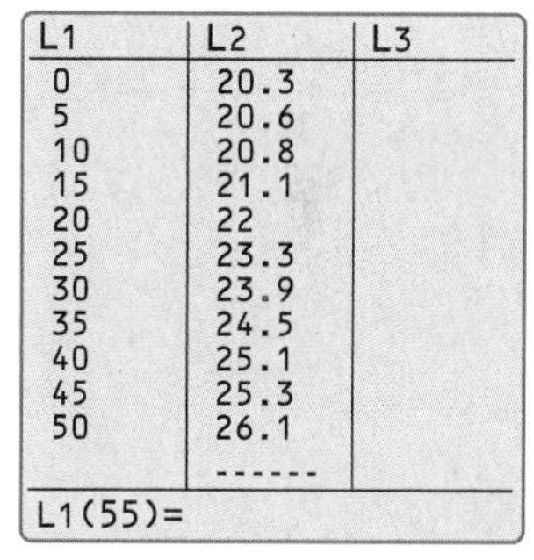

L1	L2	L3
0	20.3	
5	20.6	
10	20.8	
15	21.1	
20	22	
25	23.3	
30	23.9	
35	24.5	
40	25.1	
45	25.3	
50	26.1	

L1(55)=

b. Use the *linear regression* feature to find an equation of the line of best fit. You should obtain results such as those shown below.

```
LinReg
 y=ax+b
 a=.1261818182
 b=19.84545455
 r²=.9738676804
 r=.986847344
```

c. Write an equation of the line of best fit. Compare your result with the equation you obtained in Exploration 2 in Section 4.4.

CONSTRUCTING VIABLE ARGUMENTS

To be proficient in math, you need to reason inductively about data.

Communicate Your Answer

2. How can you *analytically* find a line of best fit for a scatter plot?

3. The data set relates the number of chirps per second for striped ground crickets and the outside temperature in degrees Fahrenheit. Make a scatter plot of the data. Then find an equation of the line of best fit. Use your result to estimate the outside temperature when there are 19 chirps per second.

Chirps per second	20.0	16.0	19.8	18.4	17.1
Temperature (°F)	88.6	71.6	93.3	84.3	80.6

Chirps per second	14.7	15.4	16.2	15.0	14.4
Temperature (°F)	69.7	69.4	83.3	79.6	76.3

4.5 Lesson

What You Will Learn

- Use residuals to determine how well lines of fit model data.
- Use technology to find lines of best fit.
- Distinguish between correlation and causation.

Core Vocabulary

residual, *p. 202*
linear regression, *p. 203*
line of best fit, *p. 203*
correlation coefficient, *p. 203*
interpolation, *p. 205*
extrapolation, *p. 205*
causation, *p. 205*

Analyzing Residuals

One way to determine how well a line of fit models a data set is to analyze *residuals*.

Core Concept

Residuals

A **residual** is the difference of the y-value of a data point and the corresponding y-value found using the line of fit. A residual can be positive, negative, or zero.

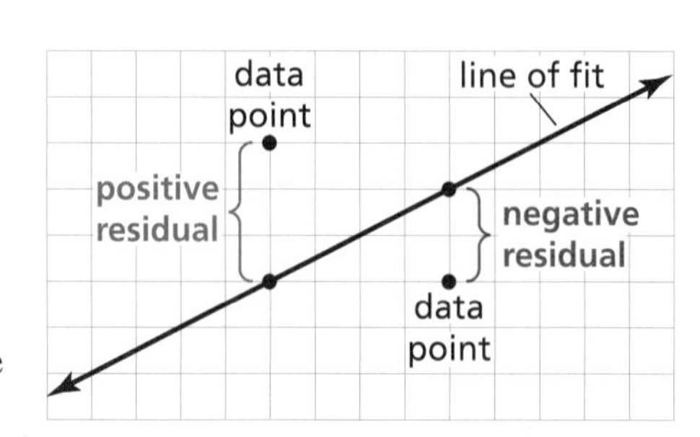

A scatter plot of the residuals shows how well a model fits a data set. If the model is a good fit, then the absolute values of the residuals are relatively small, and the residual points will be more or less evenly dispersed about the horizontal axis. If the model is not a good fit, then the residual points will form some type of pattern that suggests the data are not linear. Wildly scattered residual points suggest that the data might have no correlation.

EXAMPLE 1 Using Residuals

In Example 3 in Section 4.4, the equation $y = -2x + 20$ models the data in the table shown. Is the model a good fit?

Week, x	Sales (millions), y
1	\$19
2	\$15
3	\$13
4	\$11
5	\$10
6	\$8
7	\$7
8	\$5

SOLUTION

Step 1 Calculate the residuals. Organize your results in a table.

Step 2 Use the points (x, residual) to make a scatter plot.

x	y	y-Value from model	Residual
1	19	18	$19 - 18 = 1$
2	15	16	$15 - 16 = -1$
3	13	14	$13 - 14 = -1$
4	11	12	$11 - 12 = -1$
5	10	10	$10 - 10 = 0$
6	8	8	$8 - 8 = 0$
7	7	6	$7 - 6 = 1$
8	5	4	$5 - 4 = 1$

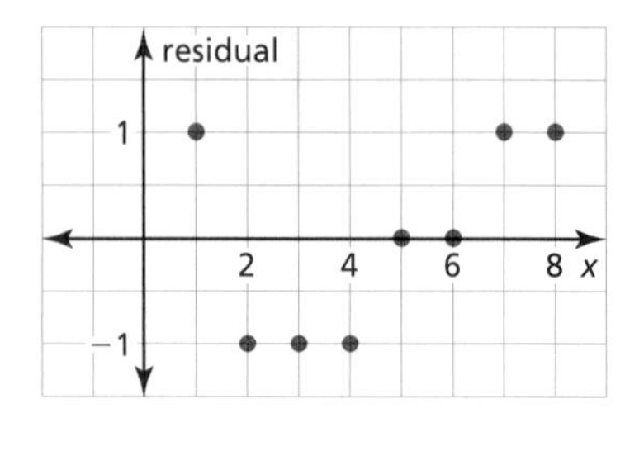

▶ The points are evenly dispersed about the horizontal axis. So, the equation $y = -2x + 20$ is a good fit.

EXAMPLE 2 Using Residuals

The table shows the ages x and salaries y (in thousands of dollars) of eight employees at a company. The equation $y = 0.2x + 38$ models the data. Is the model a good fit?

Age, x	35	37	41	43	45	47	53	55
Salary, y	42	44	47	50	52	51	49	45

SOLUTION

Step 1 Calculate the residuals. Organize your results in a table.

Step 2 Use the points (x, residual) to make a scatter plot.

x	y	y-Value from model	Residual
35	42	45.0	$42 - 45.0 = -3.0$
37	44	45.4	$44 - 45.4 = -1.4$
41	47	46.2	$47 - 46.2 = 0.8$
43	50	46.6	$50 - 46.6 = 3.4$
45	52	47.0	$52 - 47.0 = 5.0$
47	51	47.4	$51 - 47.4 = 3.6$
53	49	48.6	$49 - 48.6 = 0.4$
55	45	49.0	$45 - 49.0 = -4.0$

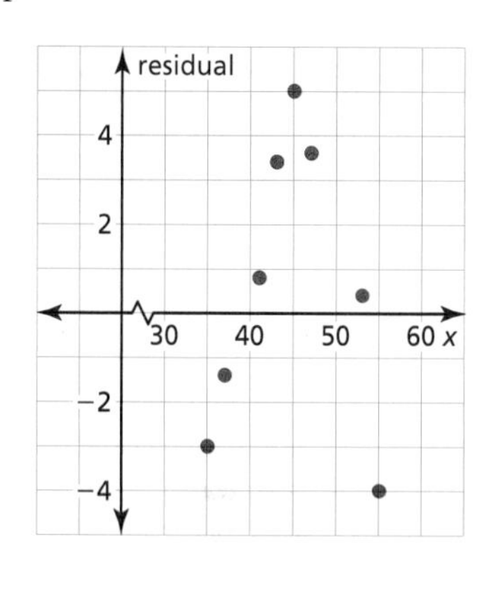

The residual points form a ∩-shaped pattern, which suggests the data are not linear. So, the equation $y = 0.2x + 38$ does not model the data well.

Monitoring Progress

Help in English and Spanish at *BigIdeasMath.com*

1. The table shows the attendances y (in thousands) at an amusement park from 2005 to 2014, where $x = 0$ represents the year 2005. The equation $y = -9.8x + 850$ models the data. Is the model a good fit?

Year, x	0	1	2	3	4	5	6	7	8	9
Attendance, y	850	845	828	798	800	792	785	781	775	760

STUDY TIP

You know how to use two points to find an equation of a line of fit. When finding an equation of the line of best fit, every point in the data set is used.

Finding Lines of Best Fit

Graphing calculators use a method called **linear regression** to find a precise line of fit called a **line of best fit**. This line best models a set of data. A calculator often gives a value r, called the **correlation coefficient**. This value tells whether the correlation is positive or negative and how closely the equation models the data. Values of r range from -1 to 1. When r is close to 1 or -1, there is a strong correlation between the variables. As r, gets closer to 0, the correlation becomes weaker.

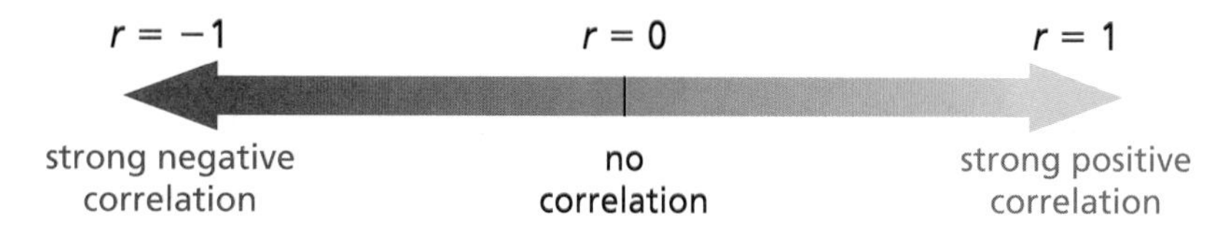

EXAMPLE 3 Finding a Line of Best Fit Using Technology

The table shows the durations x (in minutes) of several eruptions of the geyser Old Faithful and the times y (in minutes) until the next eruption. (a) Use a graphing calculator to find an equation of the line of best fit. Then plot the data and graph the equation in the same viewing window. (b) Identify and interpret the correlation coefficient. (c) Interpret the slope and y-intercept of the line of best fit.

Duration, *x*	2.0	3.7	4.2	1.9	3.1	2.5	4.4	3.9
Time, *y*	60	83	84	58	72	62	85	85

SOLUTION

a. Step 1 Enter the data from the table into two lists.

L1	L2	L3 1
2	60	
3.7	83	
4.2	84	
1.9	58	
3.1	72	
2.5	62	
4.4	85	

L1(1)=2

Step 2 Use the *linear regression* feature. The values in the equation can be rounded to obtain $y = 12.0x + 35$.

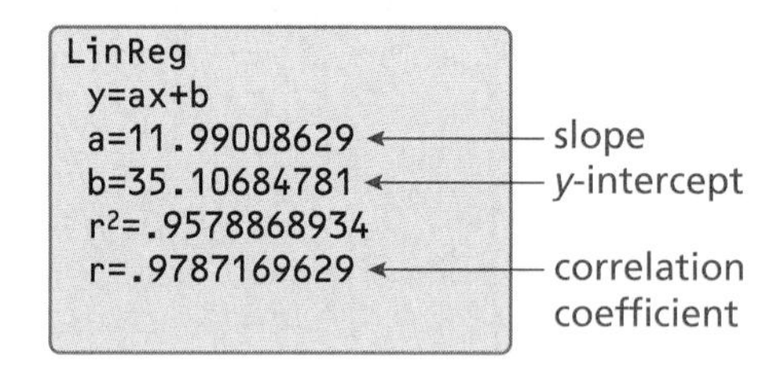

Step 3 Enter the equation $y = 12.0x + 35$ into the calculator. Then plot the data and graph the equation in the same viewing window.

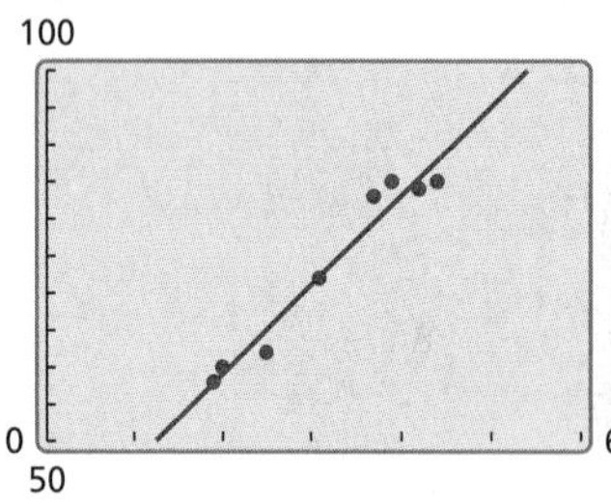

PRECISION

Be sure to analyze the data values to select an appropriate viewing window for your graph.

b. The correlation coefficient is about 0.979. This means that the relationship between the durations and the times until the next eruption has a strong positive correlation and the equation closely models the data, as shown in the graph.

c. The slope of the line is 12. This means the time until the next eruption increases by about 12 minutes for each minute the duration increases. The y-intercept is 35, but it has no meaning in this context because the duration cannot be 0 minutes.

Monitoring Progress

Help in English and Spanish at *BigIdeasMath.com*

2. Use the data in Monitoring Progress Question 1. (a) Use a graphing calculator to find an equation of the line of best fit. Then plot the data and graph the equation in the same viewing window. (b) Identify and interpret the correlation coefficient. (c) Interpret the slope and y-intercept of the line of best fit.

Using a graph or its equation to *approximate* a value between two known values is called **interpolation**. Using a graph or its equation to *predict* a value outside the range of known values is called **extrapolation**. In general, the farther removed a value is from the known values, the less confidence you can have in the accuracy of the prediction.

STUDY TIP

To approximate or predict an unknown value, you can evaluate the model algebraically or graph the model with a graphing calculator and use the *trace* feature.

EXAMPLE 4 Interpolating and Extrapolating Data

Refer to Example 3. Use the equation of the line of best fit.

a. Approximate the duration before a time of 77 minutes.

b. Predict the time after an eruption lasting 5.0 minutes.

SOLUTION

a. $y = 12.0x + 35$ Write the equation.

$77 = 12.0x + 35$ Substitute 77 for y.

$3.5 = x$ Solve for x.

▶ An eruption lasts about 3.5 minutes before a time of 77 minutes.

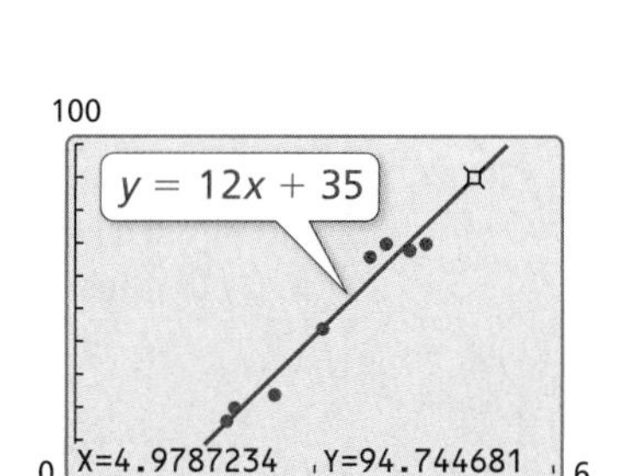

b. Use a graphing calculator to graph the equation. Use the *trace* feature to find the value of y when $x \approx 5.0$, as shown.

▶ A time of about 95 minutes will follow an eruption of 5.0 minutes.

Monitoring Progress  Help in English and Spanish at *BigIdeasMath.com*

3. Refer to Monitoring Progress Question 2. Use the equation of the line of best fit to predict the attendance at the amusement park in 2017.

Correlation and Causation

When a change in one variable causes a change in another variable, it is called **causation**. Causation produces a strong correlation between the two variables. The converse is *not* true. In other words, correlation does not imply causation.

READING

A causal relationship exists when one variable causes a change in another variable.

EXAMPLE 5 Identifying Correlation and Causation

Tell whether a correlation is likely in the situation. If so, tell whether there is a causal relationship. Explain your reasoning.

a. time spent exercising and the number of calories burned

b. the number of banks and the population of a city

SOLUTION

a. There is a positive correlation and a causal relationship because the more time you spend exercising, the more calories you burn.

b. There may be a positive correlation but no causal relationship. Building more banks will not cause the population to increase.

Monitoring Progress  Help in English and Spanish at *BigIdeasMath.com*

4. Is there a correlation between time spent playing video games and grade point average? If so, is there a causal relationship? Explain your reasoning.

4.5 Exercises

Dynamic Solutions available at *BigIdeasMath.com*

Vocabulary and Core Concept Check

1. **VOCABULARY** When is a residual positive? When is it negative?

2. **WRITING** Explain how you can use residuals to determine how well a line of fit models a data set.

3. **VOCABULARY** Compare interpolation and extrapolation.

4. **WHICH ONE DOESN'T BELONG?** Which correlation coefficient does *not* belong with the other three? Explain your reasoning.

$r = -0.98$ $r = 0.96$ $r = -0.09$ $r = 0.97$

Monitoring Progress and Modeling with Mathematics

In Exercises 5–8, use residuals to determine whether the model is a good fit for the data in the table. Explain. *(See Examples 1 and 2.)*

5. $y = 4x - 5$

x	−4	−3	−2	−1	0	1	2	3	4
y	−18	−13	−10	−7	−2	0	6	10	15

6. $y = 6x + 4$

x	1	2	3	4	5	6	7	8	9
y	13	14	23	26	31	42	45	52	62

7. $y = -1.3x + 1$

x	−8	−6	−4	−2	0	2	4	6	8
y	9	10	5	8	−1	1	−4	−12	−7

8. $y = -0.5x - 2$

x	4	6	8	10	12	14	16	18	20
y	−1	−3	−6	−8	−10	−10	−10	−9	−9

9. **ANALYZING RESIDUALS** The table shows the growth y (in inches) of an elk's antlers during week x. The equation $y = -0.7x + 6.8$ models the data. Is the model a good fit? Explain.

Week, *x*	1	2	3	4	5
Growth, *y*	6.0	5.5	4.7	3.9	3.3

10. **ANALYZING RESIDUALS** The table shows the approximate numbers y (in thousands) of movie tickets sold from January to June for a theater. In the table, $x = 1$ represents January. The equation $y = 1.3x + 27$ models the data. Is the model a good fit? Explain.

Month, *x*	Ticket sales, *y*
1	27
2	28
3	36
4	28
5	32
6	35

In Exercises 11–14, use a graphing calculator to find an equation of the line of best fit for the data. Identify and interpret the correlation coefficient.

11.

x	0	1	2	3	4	5	6	7
y	−8	−5	−2	−1	−1	2	5	8

12.

x	−4	−2	0	2	4	6	8	10
y	17	7	8	1	5	−2	2	−8

13.

x	−15	−10	−5	0	5	10	15	20
y	−4	2	7	16	22	30	37	43

14.

x	5	6	7	8	9	10	11	12
y	12	−2	8	3	−1	−4	6	0

ERROR ANALYSIS In Exercises 15 and 16, describe and correct the error in interpreting the graphing calculator display.

```
LinReg
 y=ax+b
 a=-4.47
 b=23.16
 r²=.9989451055
 r=-.9994724136
```

15. 

An equation of the line of best fit is $y = 23.16x - 4.47$.

16. 

The data have a strong positive correlation.

17. **MODELING WITH MATHEMATICS** The table shows the total numbers y of people who reported an earthquake x minutes after it ended. *(See Example 3.)*

Minutes, x	People, y
1	10
2	100
3	400
4	900
5	1400
6	1800
7	2100

a. Use a graphing calculator to find an equation of the line of best fit. Then plot the data and graph the equation in the same viewing window.

b. Identify and interpret the correlation coefficient.

c. Interpret the slope and y-intercept of the line of best fit.

18. **MODELING WITH MATHEMATICS** The table shows the numbers y of people who volunteer at an animal shelter on each day x.

Day, x	1	2	3	4	5	6	7	8
People, y	9	5	13	11	10	11	19	12

a. Use a graphing calculator to find an equation of the line of best fit. Then plot the data and graph the equation in the same viewing window.

b. Identify and interpret the correlation coefficient.

c. Interpret the slope and y-intercept of the line of best fit.

19. **MODELING WITH MATHEMATICS** The table shows the mileages x (in thousands of miles) and the selling prices y (in thousands of dollars) of several used automobiles of the same year and model. *(See Example 4.)*

Mileage, x	22	14	18	30	8	24
Price, y	16	17	17	14	18	15

a. Use a graphing calculator to find an equation of the line of best fit.

b. Identify and interpret the correlation coefficient.

c. Interpret the slope and y-intercept of the line of best fit.

d. Approximate the mileage of an automobile that costs $15,500.

e. Predict the price of an automobile with 6000 miles.

20. **MODELING WITH MATHEMATICS** The table shows the lengths x and costs y of several sailboats.

Length (feet), x	Cost (thousands of dollars), y
27	94
18	56
25	58
32	123
18	60
26	87
36	145

a. Use a graphing calculator to find an equation of the line of best fit.

b. Identify and interpret the correlation coefficient.

c. Interpret the slope and y-intercept of the line of best fit.

d. Approximate the cost of a sailboat that is 20 feet long.

e. Predict the length of a sailboat that costs $147,000.

In Exercises 21–24, tell whether a correlation is likely in the situation. If so, tell whether there is a causal relationship. Explain your reasoning. *(See Example 5.)*

21. the amount of time spent talking on a cell phone and the remaining battery life

22. the height of a toddler and the size of the toddler's vocabulary

23. the number of hats you own and the size of your head

24. the weight of a dog and the length of its tail

25. **OPEN-ENDED** Describe a data set that has a strong correlation but does not have a causal relationship.

26. **HOW DO YOU SEE IT?** Match each graph with its correlation coefficient. Explain your reasoning.

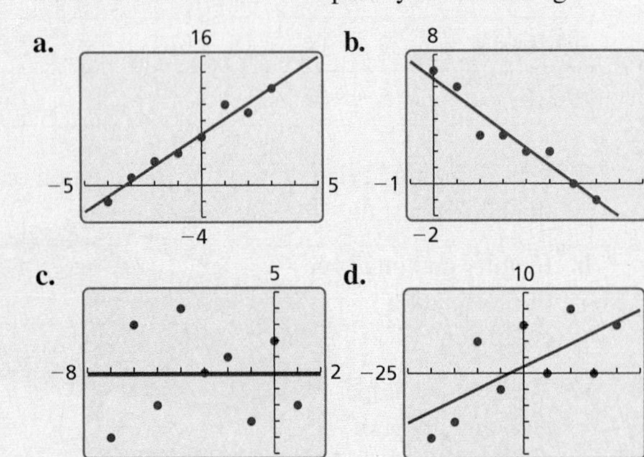

A. $r = 0$ B. $r = 0.98$

C. $r = -0.97$ D. $r = 0.69$

27. **ANALYZING RELATIONSHIPS** The table shows the grade point averages y of several students and the numbers x of hours they spend watching television each week.

Hours, x	Grade point average, y
10	3.0
5	3.4
3	3.5
12	2.7
20	2.1
15	2.8
8	3.0
4	3.7
16	2.5

a. Use a graphing calculator to find an equation of the line of best fit. Identify and interpret the correlation coefficient.

b. Interpret the slope and y-intercept of the line of best fit.

c. Another student watches about 14 hours of television each week. Approximate the student's grade point average.

d. Do you think there is a causal relationship between time spent watching television and grade point average? Explain.

28. **MAKING AN ARGUMENT** A student spends 2 hours watching television each week and has a grade point average of 2.4. Your friend says including this information in the data set in Exercise 27 will weaken the correlation. Is your friend correct? Explain.

29. **USING MODELS** Refer to Exercise 17.

a. Predict the total numbers of people who reported an earthquake 9 minutes and 15 minutes after it ended.

b. The table shows the actual data. Describe the accuracy of your extrapolations in part (a).

Minutes, x	9	15
People, y	2750	3200

30. **THOUGHT PROVOKING** A data set consists of the numbers x of people at Beach 1 and the numbers y of people at Beach 2 recorded daily for 1 week. Sketch a possible graph of the data set. Describe the situation shown in the graph and give a possible correlation coefficient. Determine whether there is a causal relationship. Explain.

31. **COMPARING METHODS** The table shows the numbers y (in billions) of text messages sent each year in a five-year period, where $x = 1$ represents the first year in the five-year period.

Year, x	1	2	3	4	5
Text messages (billions), y	241	601	1360	1806	2206

a. Use a graphing calculator to find an equation of the line of best fit. Identify and interpret the correlation coefficient.

b. Is there a causal relationship? Explain your reasoning.

c. Calculate the residuals. Then make a scatter plot of the residuals and interpret the results.

d. Compare the methods you used in parts (a) and (c) to determine whether the model is a good fit. Which method do you prefer? Explain.

Maintaining Mathematical Proficiency

Reviewing what you learned in previous grades and lessons

Determine whether the table represents a *linear* or *nonlinear* function. Explain. *(Section 3.2)*

32.

x	5	6	7	8
y	−4	4	−4	4

33.

x	2	4	6	8
y	13	8	3	−2

4.6 Arithmetic Sequences

Essential Question How can you use an arithmetic sequence to describe a pattern?

An **arithmetic sequence** is an ordered list of numbers in which the difference between each pair of consecutive **terms**, or numbers in the list, is the same.

EXPLORATION 1 **Describing a Pattern**

Work with a partner. Use the figures to complete the table. Plot the points given by your completed table. Describe the pattern of the y-values.

LOOKING FOR A PATTERN

To be proficient in math, you need to look closely to discern patterns and structure.

a. $n = 1$ $n = 2$ $n = 3$ $n = 4$ $n = 5$

Number of stars, n	1	2	3	4	5
Number of sides, y					

y: 60, 50, 40, 30, 20, 10, 0; n: 0 1 2 3 4 5

b. $n = 1$ $n = 2$ $n = 3$ $n = 4$ $n = 5$

n	1	2	3	4	5
Number of circles, y					

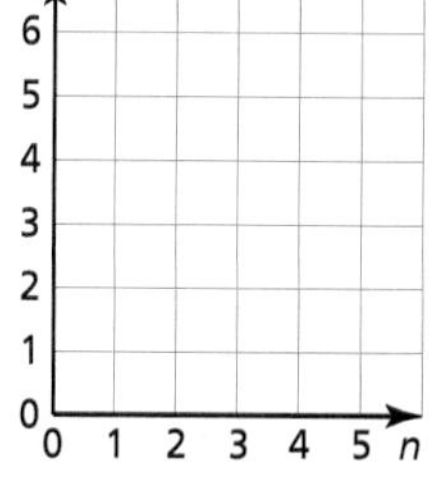

c. $n = 1$ $n = 2$ $n = 3$ $n = 4$ $n = 5$

Number of rows, n	1	2	3	4	5
Number of dots, y					

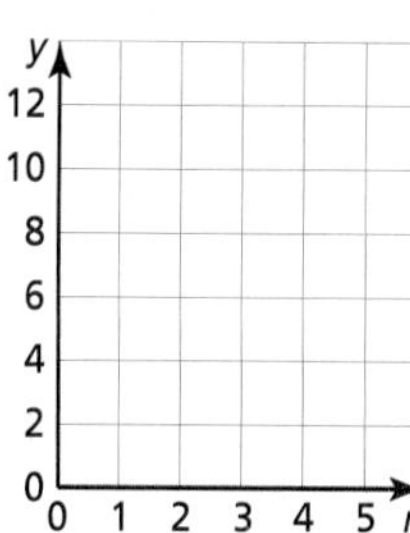

Communicate Your Answer

2. How can you use an arithmetic sequence to describe a pattern? Give an example from real life.

3. In chemistry, water is called H_2O because each molecule of water has two hydrogen atoms and one oxygen atom. Describe the pattern shown below. Use the pattern to determine the number of atoms in 23 molecules.

$n = 1$ $n = 2$ $n = 3$ $n = 4$ $n = 5$

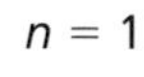

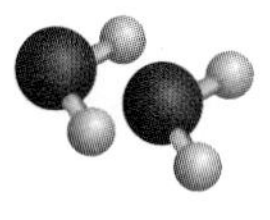
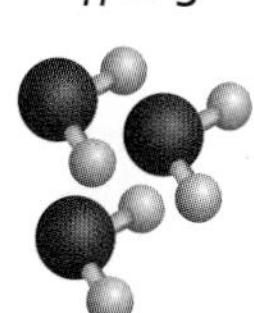
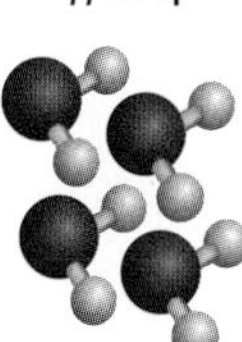
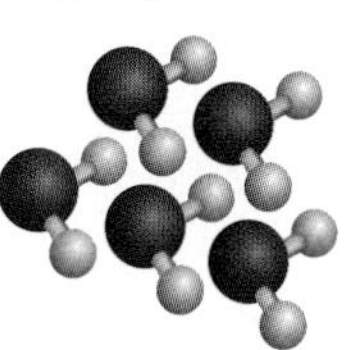

4.6 Lesson

Core Vocabulary

sequence, *p. 210*
term, *p. 210*
arithmetic sequence, *p. 210*
common difference, *p. 210*

Previous
point-slope form
function notation

What You Will Learn

- Write the terms of arithmetic sequences.
- Graph arithmetic sequences.
- Write arithmetic sequences as functions.

Writing the Terms of Arithmetic Sequences

A **sequence** is an ordered list of numbers. Each number in a sequence is called a **term**. Each term a_n has a specific position n in the sequence.

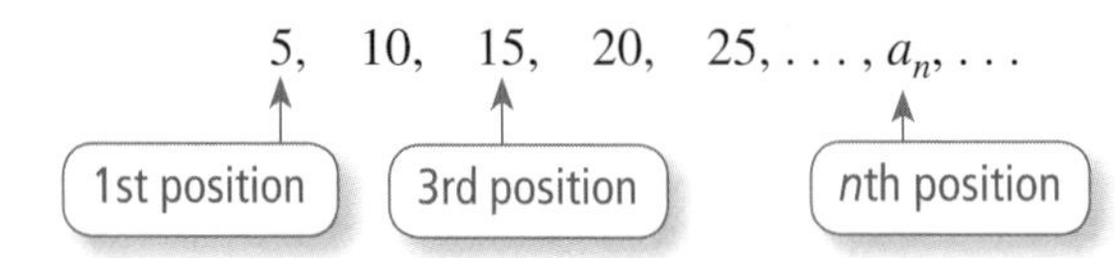

Core Concept

Arithmetic Sequence

In an **arithmetic sequence**, the difference between each pair of consecutive terms is the same. This difference is called the **common difference**. Each term is found by adding the common difference to the previous term.

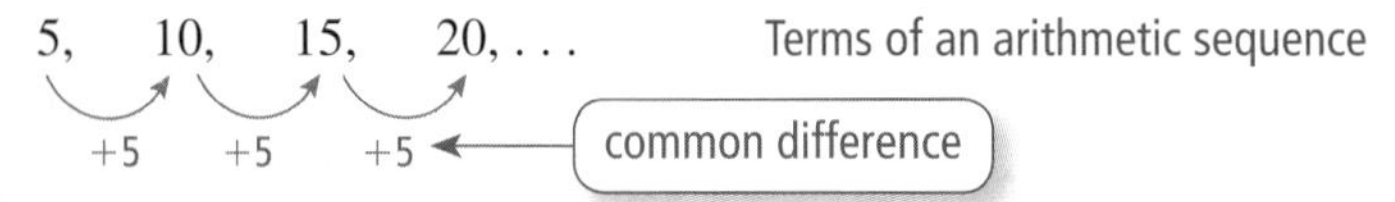

READING

An ellipsis (. . .) is a series of dots that indicates an intentional omission of information. In mathematics, the . . . notation means "and so forth." The ellipsis indicates that there are more terms in the sequence that are not shown.

EXAMPLE 1 Extending an Arithmetic Sequence

Write the next three terms of the arithmetic sequence.

$-7, -14, -21, -28, \ldots$

SOLUTION

Use a table to organize the terms and find the pattern.

Position	1	2	3	4
Term	−7	−14	−21	−28

+(−7) +(−7) +(−7)

Each term is 7 less than the previous term. So, the common difference is −7.

Add −7 to a term to find the next term.

Position	1	2	3	4	5	6	7
Term	−7	−14	−21	−28	−35	−42	−49

+(−7) +(−7) +(−7)

▶ The next three terms are −35, −42, and −49.

Monitoring Progress  Help in English and Spanish at *BigIdeasMath.com*

Write the next three terms of the arithmetic sequence.

1. $-12, 0, 12, 24, \ldots$ **2.** $0.2, 0.6, 1, 1.4, \ldots$ **3.** $4, 3\frac{3}{4}, 3\frac{1}{2}, 3\frac{1}{4}, \ldots$

Graphing Arithmetic Sequences

To graph a sequence, let a term's position number n in the sequence be the x-value. The term a_n is the corresponding y-value. Plot the ordered pairs (n, a_n).

EXAMPLE 2 Graphing an Arithmetic Sequence

Graph the arithmetic sequence 4, 8, 12, 16, ... What do you notice?

SOLUTION

Make a table. Then plot the ordered pairs (n, a_n).

Position, n	Term, a_n
1	4
2	8
3	12
4	16

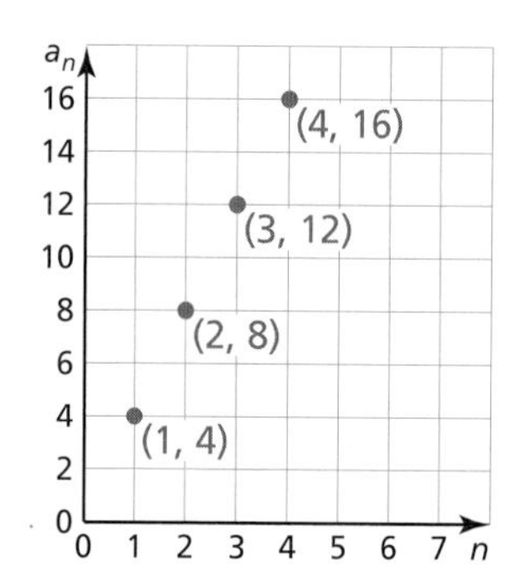

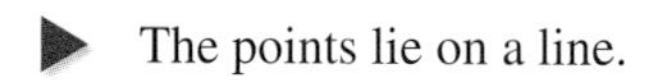
The points lie on a line.

EXAMPLE 3 Identifying an Arithmetic Sequence from a Graph

Does the graph represent an arithmetic sequence? Explain.

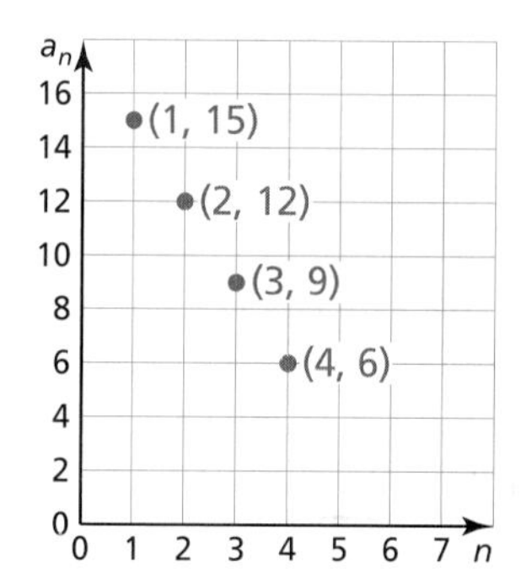

SOLUTION

Make a table to organize the ordered pairs. Then determine whether there is a common difference.

Position, n	1	2	3	4
Term, a_n	15	12	9	6

+(−3) +(−3) +(−3)

Each term is 3 less than the previous term. So, the common difference is −3.

▶ Consecutive terms have a common difference of −3. So, the graph represents the arithmetic sequence 15, 12, 9, 6, ...

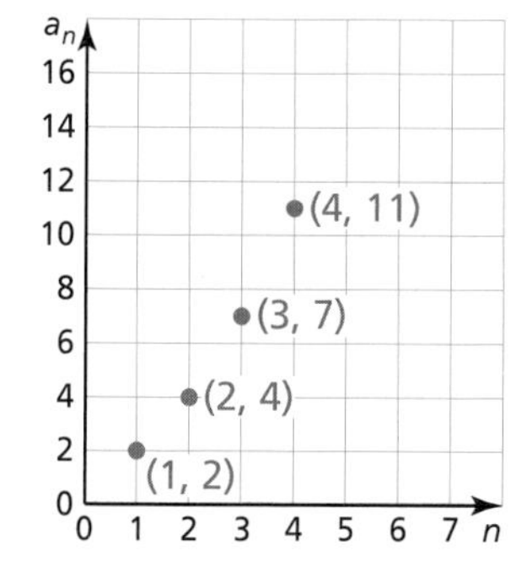

Monitoring Progress

Help in English and Spanish at *BigIdeasMath.com*

Graph the arithmetic sequence. What do you notice?

4. 3, 6, 9, 12, . . . **5.** 4, 2, 0, −2, . . . **6.** 1, 0.8, 0.6, 0.4, . . .

7. Does the graph shown represent an arithmetic sequence? Explain.

Writing Arithmetic Sequences as Functions

Because consecutive terms of an arithmetic sequence have a common difference, the sequence has a constant rate of change. So, the points represented by any arithmetic sequence lie on a line. You can use the first term and the common difference to write a linear function that describes an arithmetic sequence. Let $a_1 = 4$ and $d = 3$.

Position, n	Term, a_n	Written using a_1 and d	Numbers
1	first term, a_1	a_1	4
2	second term, a_2	$a_1 + d$	$4 + 3 = 7$
3	third term, a_3	$a_1 + 2d$	$4 + 2(3) = 10$
4	fourth term, a_4	$a_1 + 3d$	$4 + 3(3) = 13$
⋮	⋮	⋮	⋮
n	nth term, a_n	$a_1 + (n - 1)d$	$4 + (n - 1)(3)$

ANOTHER WAY

An *arithmetic sequence* is a linear function whose domain is the set of positive integers. You can think of d as the slope and $(1, a_1)$ as a point on the graph of the function. An equation in point-slope form for the function is

$$a_n - a_1 = d(n - 1).$$

This equation can be rewritten as

$$a_n = a_1 + (n - 1)d.$$

Core Concept

Equation for an Arithmetic Sequence

Let a_n be the nth term of an arithmetic sequence with first term a_1 and common difference d. The nth term is given by

$$a_n = a_1 + (n - 1)d.$$

EXAMPLE 4 Finding the nth Term of an Arithmetic Sequence

Write an equation for the nth term of the arithmetic sequence 14, 11, 8, 5, ... Then find a_{50}.

STUDY TIP

Notice that the equation in Example 4 is of the form $y = mx + b$, where y is replaced by a_n and x is replaced by n.

SOLUTION

The first term is 14, and the common difference is -3.

$a_n = a_1 + (n - 1)d$ — Equation for an arithmetic sequence

$a_n = 14 + (n - 1)(-3)$ — Substitute 14 for a_1 and -3 for d.

$a_n = -3n + 17$ — Simplify.

Use the equation to find the 50th term.

$a_n = -3n + 17$ — Write the equation.

$a_{50} = -3(50) + 17$ — Substitute 50 for n.

$= -133$ — Simplify.

▶ The 50th term of the arithmetic sequence is -133.

Monitoring Progress  Help in English and Spanish at *BigIdeasMath.com*

Write an equation for the nth term of the arithmetic sequence. Then find a_{25}.

8. 4, 5, 6, 7, . . .

9. 8, 16, 24, 32, . . .

10. 1, 0, -1, -2, . . .

You can rewrite the equation for an arithmetic sequence with first term a_1 and common difference d in function notation by replacing a_n with $f(n)$.

$$f(n) = a_1 + (n - 1)d$$

The domain of the function is the set of positive integers.

EXAMPLE 5 Writing Real-Life Functions

Online bidding for a purse increases by $5 for each bid after the $60 initial bid.

Bid number	1	2	3	4
Bid amount	$60	$65	$70	$75

a. Write a function that represents the arithmetic sequence.

b. Graph the function.

c. The winning bid is $105. How many bids were there?

SOLUTION

a. The first term is 60, and the common difference is 5.

$f(n) = a_1 + (n - 1)d$	Function for an arithmetic sequence
$f(n) = 60 + (n - 1)5$	Substitute 60 for a_1 and 5 for d.
$f(n) = 5n + 55$	Simplify.

▶ The function $f(n) = 5n + 55$ represents the arithmetic sequence.

b. Make a table. Then plot the ordered pairs (n, a_n).

REMEMBER

The domain is the set of positive integers.

Bid number, n	Bid amount, a_n
1	60
2	65
3	70
4	75

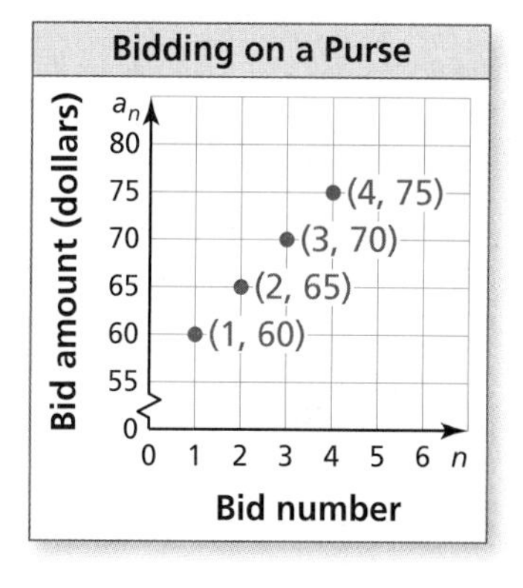

c. Use the function to find the value of n for which $f(n) = 105$.

$f(n) = 5n + 55$	Write the function.
$105 = 5n + 55$	Substitute 105 for $f(n)$.
$10 = n$	Solve for n.

▶ There were 10 bids.

Monitoring Progress  Help in English and Spanish at *BigIdeasMath.com*

11. A carnival charges $2 for each game after you pay a $5 entry fee.

Games	Total cost
1	$7
2	$9
3	$11
4	$13

a. Write a function that represents the arithmetic sequence.

b. Graph the function.

c. How many games can you play when you take $29 to the carnival?

4.6 Exercises

Dynamic Solutions available at *BigIdeasMath.com*

Vocabulary and Core Concept Check

1. **WRITING** Describe the graph of an arithmetic sequence.

2. **DIFFERENT WORDS, SAME QUESTION** Consider the arithmetic sequence represented by the graph. Which is different? Find "both" answers.

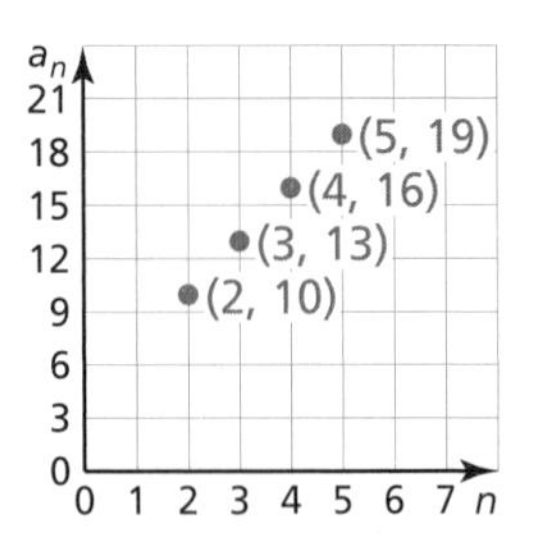

Find the slope of the linear function.	Find the difference between consecutive terms of the arithmetic sequence.
Find the difference between the terms a_2 and a_4.	Find the common difference of the arithmetic sequence.

Monitoring Progress and Modeling with Mathematics

In Exercises 3 and 4, write the next three terms of the arithmetic sequence.

3. First term: 2
   Common difference: 13

4. First term: 18
   Common difference: −6

In Exercises 5–10, find the common difference of the arithmetic sequence.

5. 13, 18, 23, 28, . . .
6. 175, 150, 125, 100, . . .
7. −16, −12, −8, −4, . . .
8. $4, 3\frac{2}{3}, 3\frac{1}{3}, 3, \ldots$
9. 6.5, 5, 3.5, 2, . . .
10. −16, −7, 2, 11, . . .

In Exercises 11–16, write the next three terms of the arithmetic sequence. *(See Example 1.)*

11. 19, 22, 25, 28, . . .
12. 1, 12, 23, 34, . . .
13. 16, 21, 26, 31, . . .
14. 60, 30, 0, −30, . . .
15. 1.3, 1, 0.7, 0.4, . . .
16. $\frac{5}{6}, \frac{2}{3}, \frac{1}{2}, \frac{1}{3}, \ldots$

In Exercises 17–22, graph the arithmetic sequence. *(See Example 2.)*

17. 4, 12, 20, 28, . . .
18. −15, 0, 15, 30, . . .
19. −1, −3, −5, −7, . . .
20. 2, 19, 36, 53, . . .
21. $0, 4\frac{1}{2}, 9, 13\frac{1}{2}, \ldots$
22. 6, 5.25, 4.5, 3.75, . . .

In Exercises 23–26, determine whether the graph represents an arithmetic sequence. Explain. *(See Example 3.)*

23. 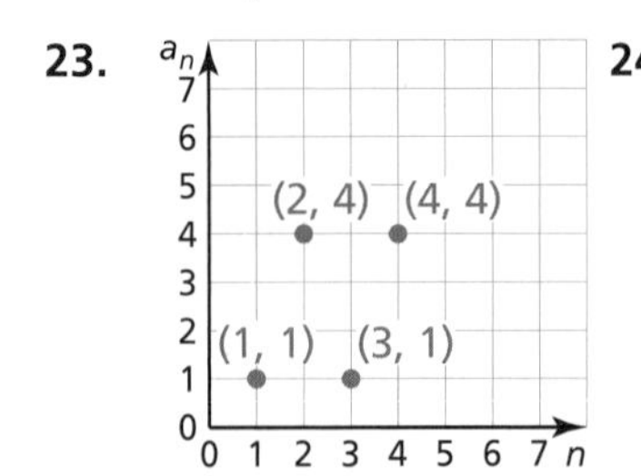

24. 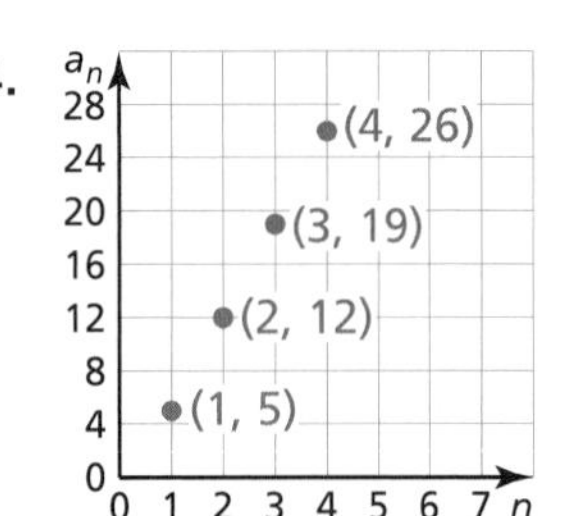

25. 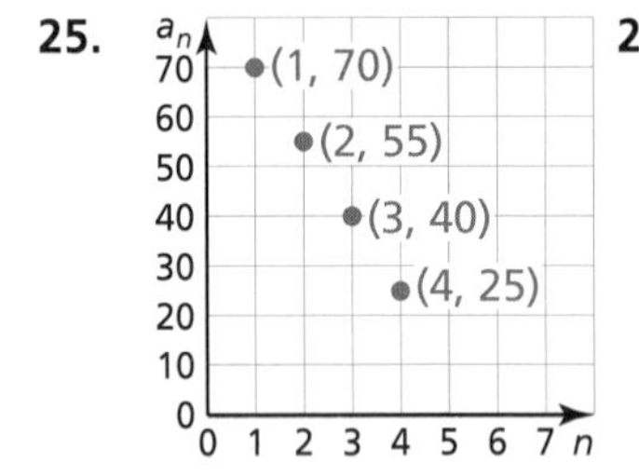

26. 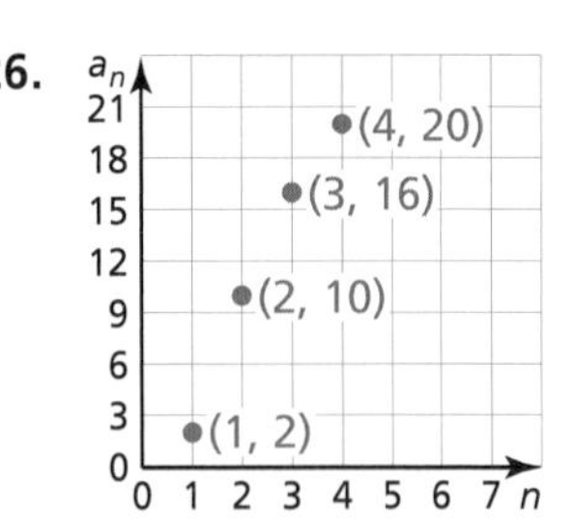

In Exercises 27–30, determine whether the sequence is arithmetic. If so, find the common difference.

27. 13, 26, 39, 52, . . .
28. 5, 9, 14, 20, . . .
29. 48, 24, 12, 6, . . .
30. 87, 81, 75, 69, . . .

31. **FINDING A PATTERN** Write a sequence that represents the number of smiley faces in each group. Is the sequence arithmetic? Explain.

32. FINDING A PATTERN Write a sequence that represents the sum of the numbers in each roll. Is the sequence arithmetic? Explain.

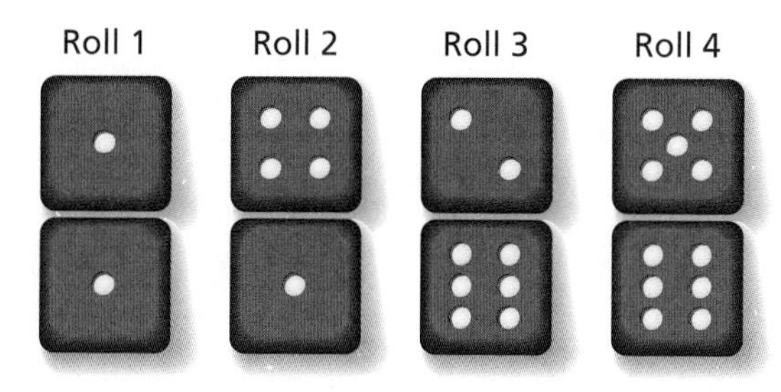

In Exercises 33–38, write an equation for the nth term of the arithmetic sequence. Then find a_{10}. *(See Example 4.)*

33. $-5, -4, -3, -2, \ldots$

34. $-6, -9, -12, -15, \ldots$

35. $\frac{1}{2}, 1, 1\frac{1}{2}, 2, \ldots$

36. $100, 110, 120, 130, \ldots$

37. $10, 0, -10, -20, \ldots$

38. $\frac{3}{7}, \frac{4}{7}, \frac{5}{7}, \frac{6}{7}, \ldots$

39. ERROR ANALYSIS Describe and correct the error in finding the common difference of the arithmetic sequence.

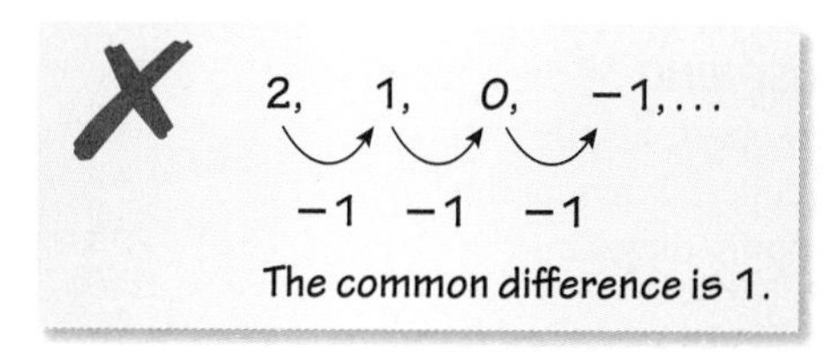

40. ERROR ANALYSIS Describe and correct the error in writing an equation for the nth term of the arithmetic sequence.

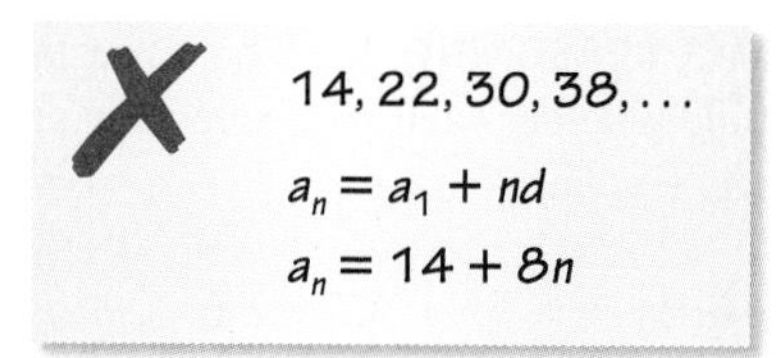

41. NUMBER SENSE The first term of an arithmetic sequence is 3. The common difference of the sequence is 1.5 times the first term. Write the next three terms of the sequence. Then graph the sequence.

42. NUMBER SENSE The first row of a dominoes display has 10 dominoes. Each row after the first has two more dominoes than the row before it. Write the first five terms of the sequence that represents the number of dominoes in each row. Then graph the sequence.

REPEATED REASONING In Exercises 43 and 44, (a) draw the next three figures in the sequence and (b) describe the 20th figure in the sequence.

43.

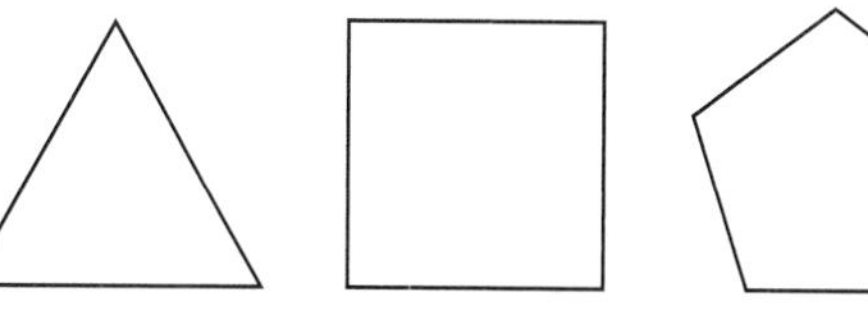

44.

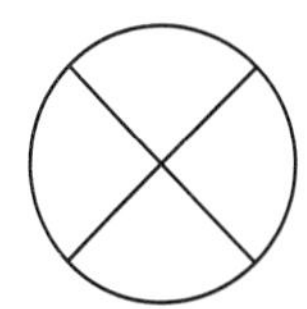

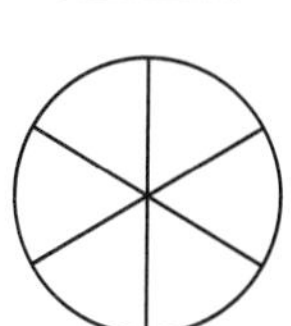

45. MODELING WITH MATHEMATICS The total number of babies born in a country each minute after midnight January 1st can be estimated by the sequence shown in the table. *(See Example 5.)*

Minutes after midnight January 1st	1	2	3	4
Total babies born	5	10	15	20

a. Write a function that represents the arithmetic sequence.

b. Graph the function.

c. Estimate how many minutes after midnight January 1st it takes for 100 babies to be born.

46. MODELING WITH MATHEMATICS The amount of money a movie earns each week after its release can be approximated by the sequence shown in the graph.

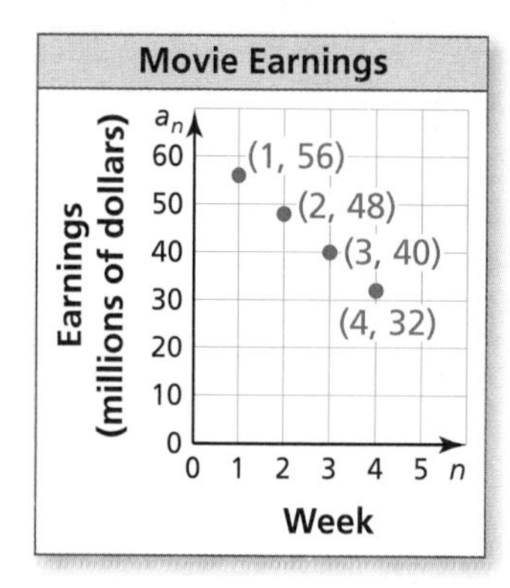

a. Write a function that represents the arithmetic sequence.

b. In what week does the movie earn $16 million?

c. How much money does the movie earn overall?

MATHEMATICAL CONNECTIONS In Exercises 47 and 48, each small square represents 1 square inch. Determine whether the areas of the figures form an arithmetic sequence. If so, write a function *f* that represents the arithmetic sequence and find *f*(30).

47.

48.

49. REASONING Is the domain of an arithmetic sequence discrete or continuous? Is the range of an arithmetic sequence discrete or continuous?

50. MAKING AN ARGUMENT Your friend says that the range of a function that represents an arithmetic sequence always contains only positive numbers or only negative numbers. Your friend claims this is true because the domain is the set of positive integers and the output values either constantly increase or constantly decrease. Is your friend correct? Explain.

51. OPEN-ENDED Write the first four terms of two different arithmetic sequences with a common difference of -3. Write an equation for the nth term of each sequence.

52. THOUGHT PROVOKING Describe an arithmetic sequence that models the numbers of people in a real-life situation.

53. REPEATED REASONING Firewood is stacked in a pile. The bottom row has 20 logs, and the top row has 14 logs. Each row has one more log than the row above it. How many logs are in the pile?

54. HOW DO YOU SEE IT? The bar graph shows the costs of advertising in a magazine.

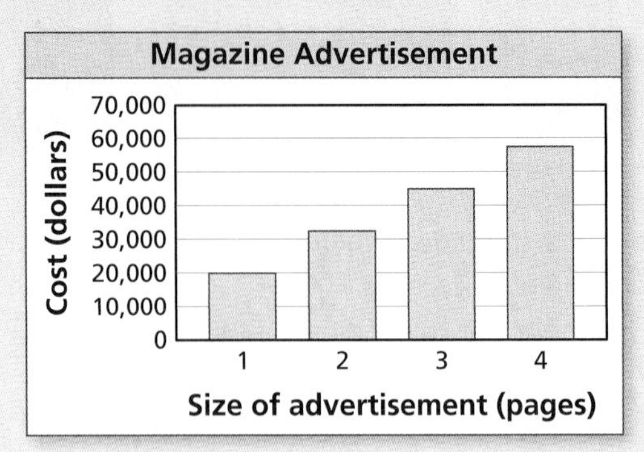

a. Does the graph represent an arithmetic sequence? Explain.

b. Explain how you would estimate the cost of a six-page advertisement in the magazine.

55. REASONING Write a function *f* that represents the arithmetic sequence shown in the mapping diagram.

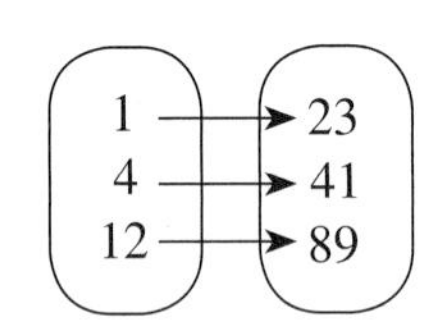

56. PROBLEM SOLVING A train stops at a station every 12 minutes starting at 6:00 A.M. You arrive at the station at 7:29 A.M. How long must you wait for the train?

57. ABSTRACT REASONING Let x be a constant. Determine whether each sequence is an arithmetic sequence. Explain.

a. $x + 6, 3x + 6, 5x + 6, 7x + 6, \ldots$

b. $x + 1, 3x + 1, 9x + 1, 27x + 1, \ldots$

Maintaining Mathematical Proficiency

Reviewing what you learned in previous grades and lessons

Solve the inequality. Graph the solution. *(Section 2.2)*

58. $x + 8 \geq -9$ **59.** $15 < b - 4$ **60.** $t - 21 < -12$ **61.** $7 + y \leq 3$

Graph the function. Compare the graph to the graph of $f(x) = |x|$. Describe the domain and range. *(Section 3.7)*

62. $h(x) = 3|x|$ **63.** $v(x) = |x - 5|$

64. $g(x) = |x| + 1$ **65.** $r(x) = -2|x|$

4.7 Piecewise Functions

Essential Question How can you describe a function that is represented by more than one equation?

EXPLORATION 1 Writing Equations for a Function

Work with a partner.

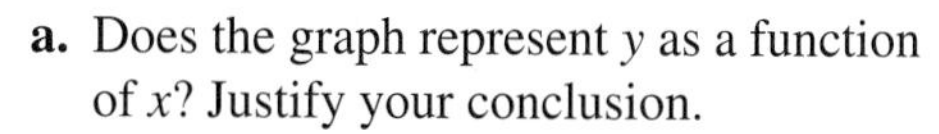

CONSTRUCTING VIABLE ARGUMENTS

To be proficient in math, you need to justify your conclusions and communicate them to others.

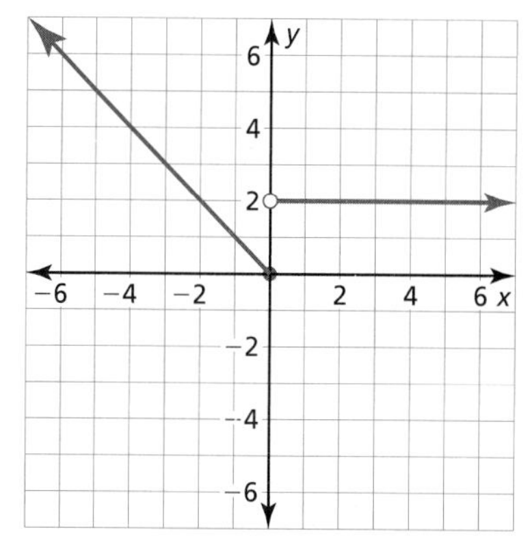

a. Does the graph represent y as a function of x? Justify your conclusion.

b. What is the value of the function when $x = 0$? How can you tell?

c. Write an equation that represents the values of the function when $x \leq 0$.

$f(x) =$ ______, if $x \leq 0$

d. Write an equation that represents the values of the function when $x > 0$.

$f(x) =$ ______, if $x > 0$

e. Combine the results of parts (c) and (d) to write a single description of the function.

$$f(x) = \begin{cases} \_\_\_\_\_\_, & \text{if } x \leq 0 \\ \_\_\_\_\_\_, & \text{if } x > 0 \end{cases}$$

EXPLORATION 2 Writing Equations for a Function

Work with a partner.

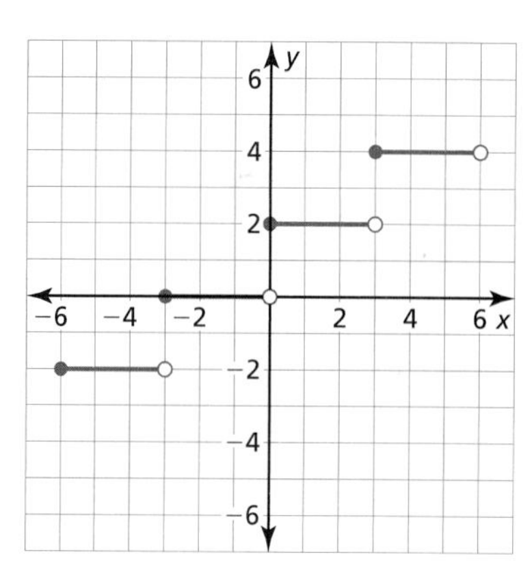

a. Does the graph represent y as a function of x? Justify your conclusion.

b. Describe the values of the function for the following intervals.

$$f(x) = \begin{cases} \_\_\_\_\_\_, & \text{if } -6 \leq x < -3 \\ \_\_\_\_\_\_, & \text{if } -3 \leq x < 0 \\ \_\_\_\_\_\_, & \text{if } 0 \leq x < 3 \\ \_\_\_\_\_\_, & \text{if } 3 \leq x < 6 \end{cases}$$

Communicate Your Answer

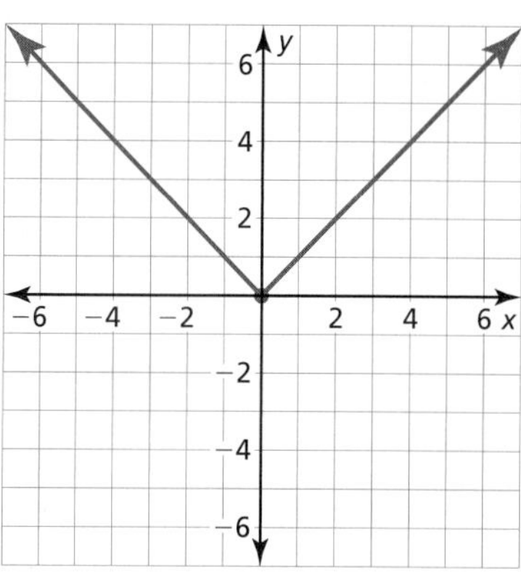

3. How can you describe a function that is represented by more than one equation?

4. Use two equations to describe the function represented by the graph.

4.7 Lesson

Core Vocabulary

piecewise function, *p. 218*
step function, *p. 220*

Previous
absolute value function
vertex form
vertex

What You Will Learn

- Evaluate piecewise functions.
- Graph and write piecewise functions.
- Graph and write step functions.
- Write absolute value functions.

Evaluating Piecewise Functions

Core Concept

Piecewise Function

A **piecewise function** is a function defined by two or more equations. Each "piece" of the function applies to a different part of its domain. An example is shown below.

$$f(x) = \begin{cases} x - 2, & \text{if } x \leq 0 \\ 2x + 1, & \text{if } x > 0 \end{cases}$$

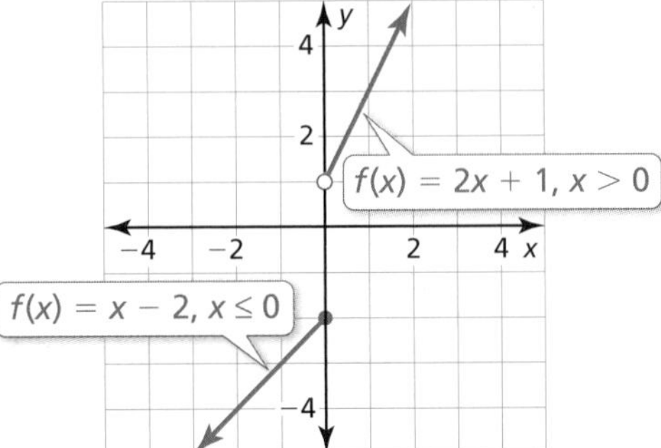

- The expression $x - 2$ represents the value of f when x is less than or equal to 0.
- The expression $2x + 1$ represents the value of f when x is greater than 0.

EXAMPLE 1 Evaluating a Piecewise Function

Evaluate the function f above when (a) $x = 0$ and (b) $x = 4$.

SOLUTION

a. $f(x) = x - 2$ — Because $0 \leq 0$, use the first equation.

$f(0) = 0 - 2$ — Substitute 0 for x.

$f(0) = -2$ — Simplify.

▶ The value of f is -2 when $x = 0$.

b. $f(x) = 2x + 1$ — Because $4 > 0$, use the second equation.

$f(4) = 2(4) + 1$ — Substitute 4 for x.

$f(4) = 9$ — Simplify.

▶ The value of f is 9 when $x = 4$.

Monitoring Progress  Help in English and Spanish at *BigIdeasMath.com*

Evaluate the function.

$$f(x) = \begin{cases} 3, & \text{if } x < -2 \\ x + 2, & \text{if } -2 \leq x \leq 5 \\ 4x, & \text{if } x > 5 \end{cases}$$

1. $f(-8)$
2. $f(-2)$
3. $f(0)$
4. $f(3)$
5. $f(5)$
6. $f(10)$

Graphing and Writing Piecewise Functions

EXAMPLE 2 Graphing a Piecewise Function

Graph $y = \begin{cases} -x - 4, & \text{if } x < 0 \\ x, & \text{if } x \geq 0 \end{cases}$. Describe the domain and range.

SOLUTION

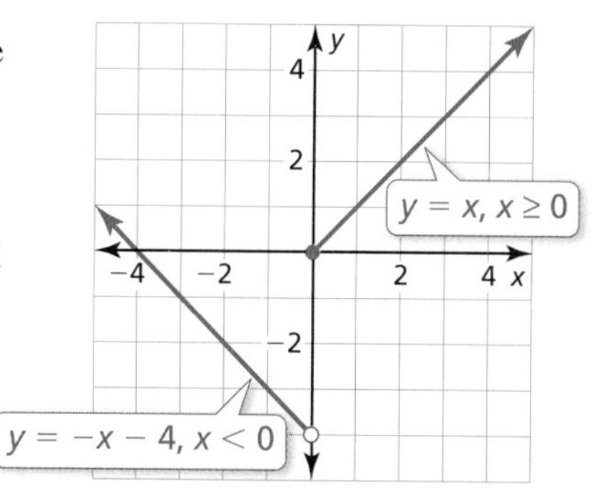

Step 1 Graph $y = -x - 4$ for $x < 0$. Because x is not equal to 0, use an open circle at $(0, -4)$.

Step 2 Graph $y = x$ for $x \geq 0$. Because x is greater than or equal to 0, use a closed circle at $(0, 0)$.

▶ The domain is all real numbers. The range is $y > -4$.

Monitoring Progress

Help in English and Spanish at *BigIdeasMath.com*

Graph the function. Describe the domain and range.

7. $y = \begin{cases} x + 1, & \text{if } x \leq 0 \\ -x, & \text{if } x > 0 \end{cases}$

8. $y = \begin{cases} x - 2, & \text{if } x < 0 \\ 4x, & \text{if } x \geq 0 \end{cases}$

EXAMPLE 3 Writing a Piecewise Function

Write a piecewise function for the graph.

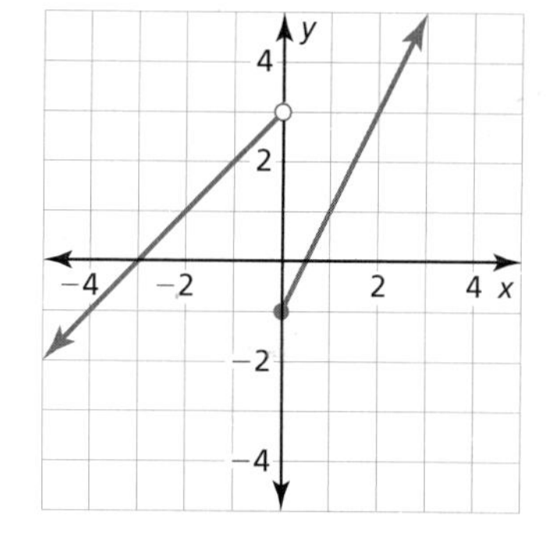

SOLUTION

Each "piece" of the function is linear.

Left Piece When $x < 0$, the graph is the line given by $y = x + 3$.

Right Piece When $x \geq 0$, the graph is the line given by $y = 2x - 1$.

▶ So, a piecewise function for the graph is

$$f(x) = \begin{cases} x + 3, & \text{if } x < 0 \\ 2x - 1, & \text{if } x \geq 0 \end{cases}.$$

Monitoring Progress

Help in English and Spanish at *BigIdeasMath.com*

Write a piecewise function for the graph.

9.

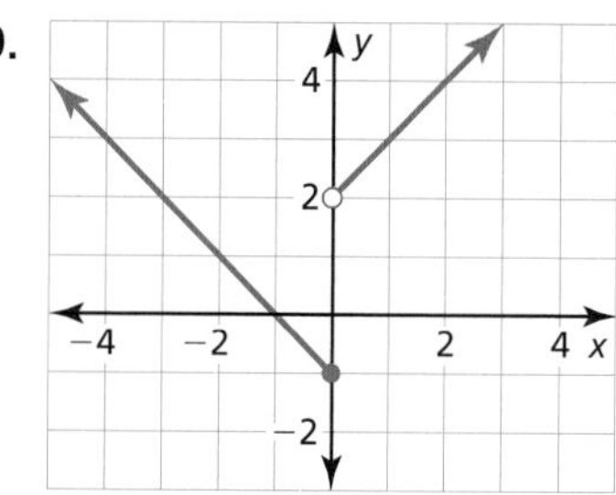

10.

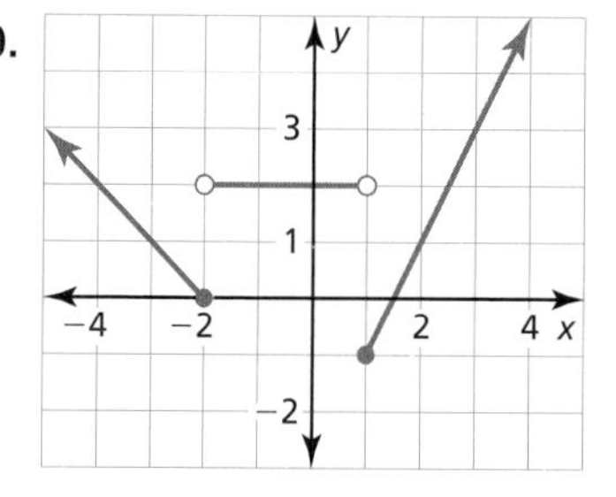

Graphing and Writing Step Functions

A **step function** is a piecewise function defined by a constant value over each part of its domain. The graph of a step function consists of a series of line segments.

STUDY TIP

The graph of a step function looks like a staircase.

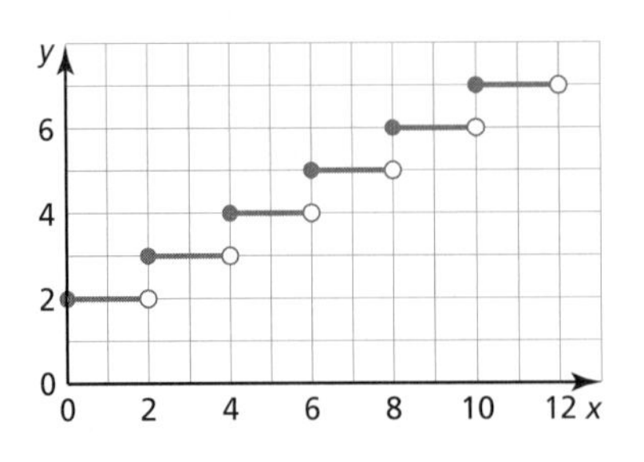

$$f(x) = \begin{cases} 2, & \text{if } 0 \le x < 2 \\ 3, & \text{if } 2 \le x < 4 \\ 4, & \text{if } 4 \le x < 6 \\ 5, & \text{if } 6 \le x < 8 \\ 6, & \text{if } 8 \le x < 10 \\ 7, & \text{if } 10 \le x < 12 \end{cases}$$

EXAMPLE 4 Graphing and Writing a Step Function

You rent a karaoke machine for 5 days. The rental company charges $50 for the first day and $25 for each additional day. Write and graph a step function that represents the relationship between the number x of days and the total cost y (in dollars) of renting the karaoke machine.

SOLUTION

Step 1 Use a table to organize the information.

Number of days	Total cost (dollars)
$0 < x \le 1$	50
$1 < x \le 2$	75
$2 < x \le 3$	100
$3 < x \le 4$	125
$4 < x \le 5$	150

Step 2 Write the step function.

$$f(x) = \begin{cases} 50, & \text{if } 0 < x \le 1 \\ 75, & \text{if } 1 < x \le 2 \\ 100, & \text{if } 2 < x \le 3 \\ 125, & \text{if } 3 < x \le 4 \\ 150, & \text{if } 4 < x \le 5 \end{cases}$$

Step 3 Graph the step function.

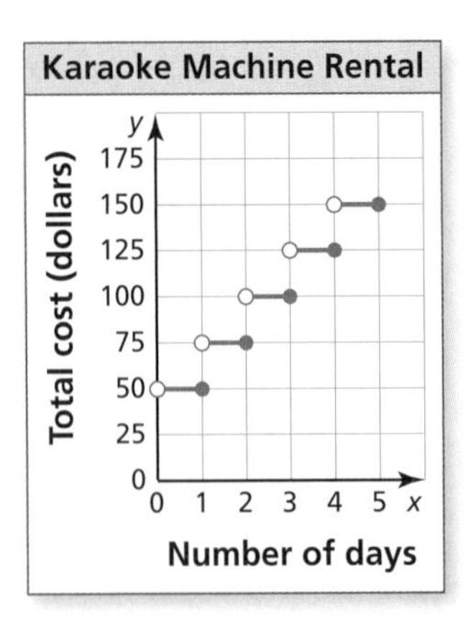

Monitoring Progress  Help in English and Spanish at *BigIdeasMath.com*

11. A landscaper rents a wood chipper for 4 days. The rental company charges $100 for the first day and $50 for each additional day. Write and graph a step function that represents the relationship between the number x of days and the total cost y (in dollars) of renting the chipper.

Writing Absolute Value Functions

The absolute value function $f(x) = |x|$ can be written as a piecewise function.

$$f(x) = \begin{cases} -x, & \text{if } x < 0 \\ x, & \text{if } x \geq 0 \end{cases}$$

Similarly, the vertex form of an absolute value function $g(x) = a|x - h| + k$ can be written as a piecewise function.

$$g(x) = \begin{cases} a[-(x - h)] + k, & \text{if } x - h < 0 \\ a(x - h) + k, & \text{if } x - h \geq 0 \end{cases}$$

REMEMBER

The vertex form of an absolute value function is $g(x) = a|x - h| + k$, where $a \neq 0$. The vertex of the graph of g is (h, k).

EXAMPLE 5 Writing an Absolute Value Function

In holography, light from a laser beam is split into two beams, a reference beam and an object beam. Light from the object beam reflects off an object and is recombined with the reference beam to form images on film that can be used to create three-dimensional images.

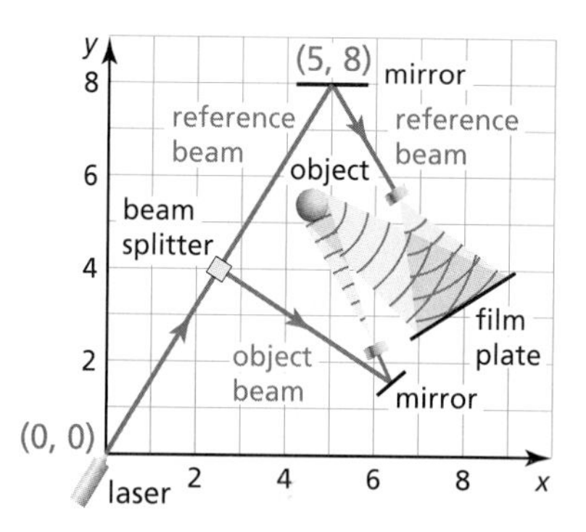

a. Write an absolute value function that represents the path of the reference beam.

b. Write the function in part (a) as a piecewise function.

SOLUTION

a. The vertex of the path of the reference beam is (5, 8). So, the function has the form $g(x) = a|x - 5| + 8$. Substitute the coordinates of the point (0, 0) into the equation and solve for a.

$g(x) = a|x - 5| + 8$ — Vertex form of the function

$0 = a|0 - 5| + 8$ — Substitute 0 for x and 0 for $g(x)$.

$-1.6 = a$ — Solve for a.

▶ So, the function $g(x) = -1.6|x - 5| + 8$ represents the path of the reference beam.

b. Write $g(x) = -1.6|x - 5| + 8$ as a piecewise function.

$$g(x) = \begin{cases} -1.6[-(x - 5)] + 8, & \text{if } x - 5 < 0 \\ -1.6(x - 5) + 8, & \text{if } x - 5 \geq 0 \end{cases}$$

Simplify each expression and solve the inequalities.

▶ So, a piecewise function for $g(x) = -1.6|x - 5| + 8$ is

$$g(x) = \begin{cases} 1.6x, & \text{if } x < 5 \\ -1.6x + 16, & \text{if } x \geq 5 \end{cases}.$$

STUDY TIP

Recall that the graph of an absolute value function is symmetric about the line $x = h$. So, it makes sense that the piecewise definition "splits" the function at $x = 5$.

Monitoring Progress

Help in English and Spanish at *BigIdeasMath.com*

12. WHAT IF? The reference beam originates at (3, 0) and reflects off a mirror at (5, 4).

a. Write an absolute value function that represents the path of the reference beam.

b. Write the function in part (a) as a piecewise function.

4.7 Exercises

Dynamic Solutions available at *BigIdeasMath.com*

Vocabulary and Core Concept Check

1. **VOCABULARY** Compare piecewise functions and step functions.

2. **WRITING** Use a graph to explain why you can write the absolute value function $y = |x|$ as a piecewise function.

Monitoring Progress and Modeling with Mathematics

In Exercises 3–12, evaluate the function. *(See Example 1.)*

$$f(x) = \begin{cases} 5x - 1, & \text{if } x < -2 \\ x + 3, & \text{if } x \geq -2 \end{cases}$$

$$g(x) = \begin{cases} -x + 4, & \text{if } x \leq -1 \\ 3, & \text{if } -1 < x < 2 \\ 2x - 5, & \text{if } x \geq 2 \end{cases}$$

3. $f(-3)$
4. $f(-2)$
5. $f(0)$
6. $f(5)$
7. $g(-4)$
8. $g(-1)$
9. $g(0)$
10. $g(1)$
11. $g(2)$
12. $g(5)$

13. **MODELING WITH MATHEMATICS** On a trip, the total distance (in miles) you travel in x hours is represented by the piecewise function

$$d(x) = \begin{cases} 55x, & \text{if } 0 \leq x \leq 2 \\ 65x - 20, & \text{if } 2 < x \leq 5 \end{cases}.$$

How far do you travel in 4 hours?

14. **MODELING WITH MATHEMATICS** The total cost (in dollars) of ordering x custom shirts is represented by the piecewise function

$$c(x) = \begin{cases} 17x + 20, & \text{if } 0 \leq x < 25 \\ 15.80x + 20, & \text{if } 25 \leq x < 50 \\ 14x + 20, & \text{if } x \geq 50 \end{cases}.$$

Determine the total cost of ordering 26 shirts.

CUSTOM SHIRTS

Quantity	Price/Shirt
0-24	\$17.00
25-49	\$15.80
50+	\$14.00

plus a \$20 processing fee on all orders

In Exercises 15–20, graph the function. Describe the domain and range. *(See Example 2.)*

15. $y = \begin{cases} -x, & \text{if } x < 2 \\ x - 6, & \text{if } x \geq 2 \end{cases}$

16. $y = \begin{cases} 2x, & \text{if } x \leq -3 \\ -2x, & \text{if } x > -3 \end{cases}$

17. $y = \begin{cases} -3x - 2, & \text{if } x \leq -1 \\ x + 2, & \text{if } x > -1 \end{cases}$

18. $y = \begin{cases} x + 8, & \text{if } x < 4 \\ 4x - 4, & \text{if } x \geq 4 \end{cases}$

19. $y = \begin{cases} 1, & \text{if } x < -3 \\ x - 1, & \text{if } -3 \leq x \leq 3 \\ -2x + 4, & \text{if } x > 3 \end{cases}$

20. $y = \begin{cases} 2x + 1, & \text{if } x \leq -1 \\ -x + 2, & \text{if } -1 < x < 2 \\ -3, & \text{if } x \geq 2 \end{cases}$

21. **ERROR ANALYSIS** Describe and correct the error in finding $f(5)$ when $f(x) = \begin{cases} 2x - 3, & \text{if } x < 5 \\ x + 8, & \text{if } x \geq 5 \end{cases}$.

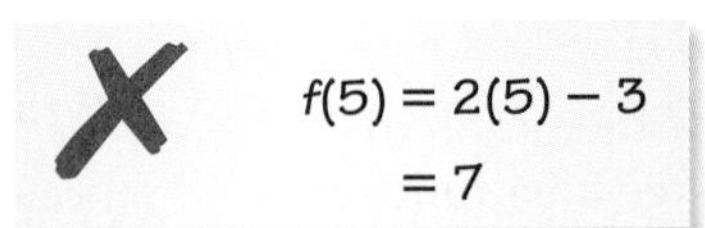

22. **ERROR ANALYSIS** Describe and correct the error in graphing $y = \begin{cases} x + 6, & \text{if } x \leq -2 \\ 1, & \text{if } x > -2 \end{cases}$.

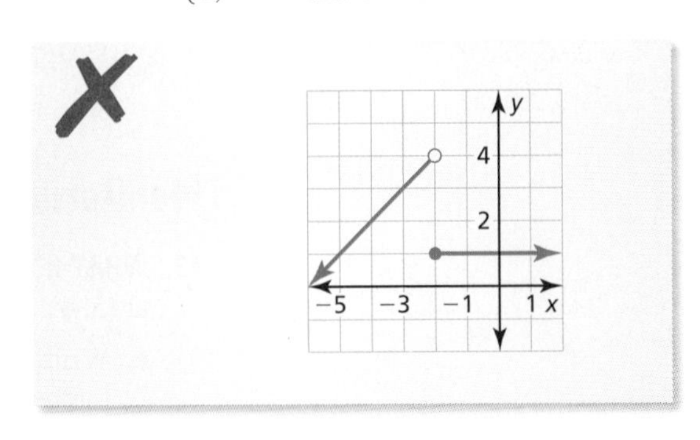

In Exercises 23–30, write a piecewise function for the graph. *(See Example 3.)*

23.

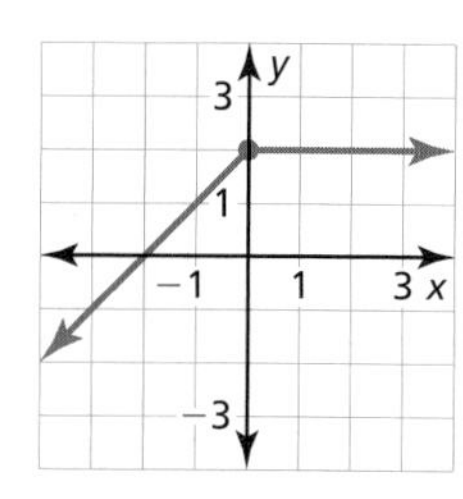

24.

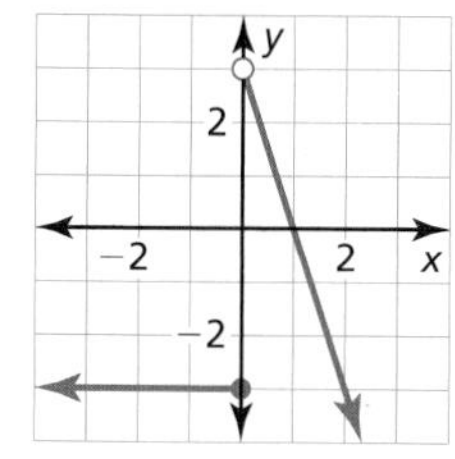

25.

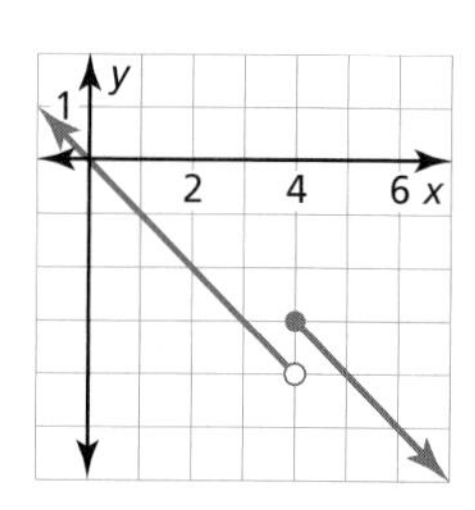

26.

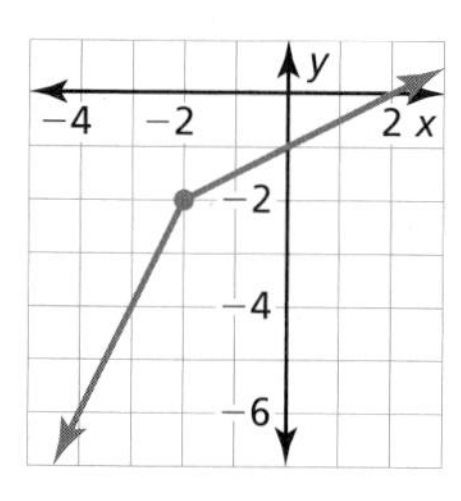

27.

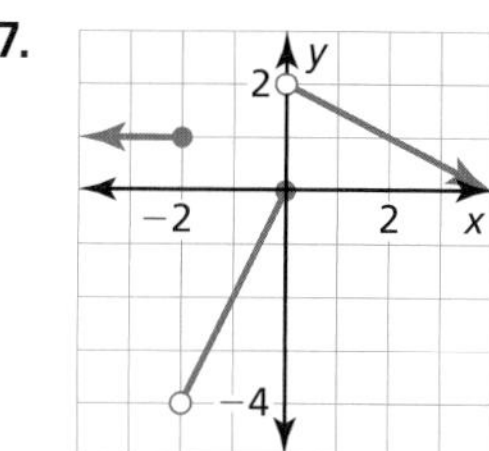

28.

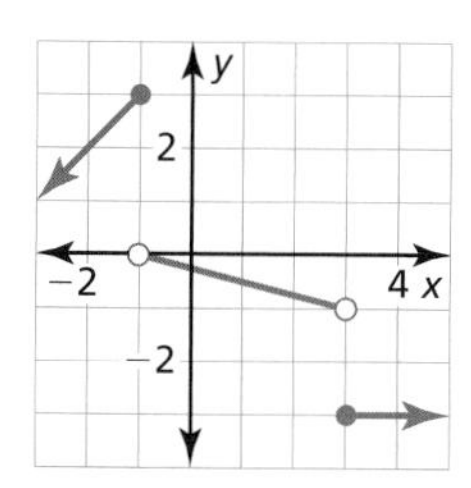

29.

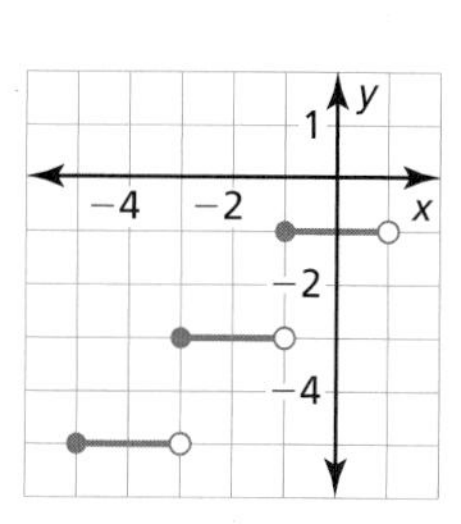

30.

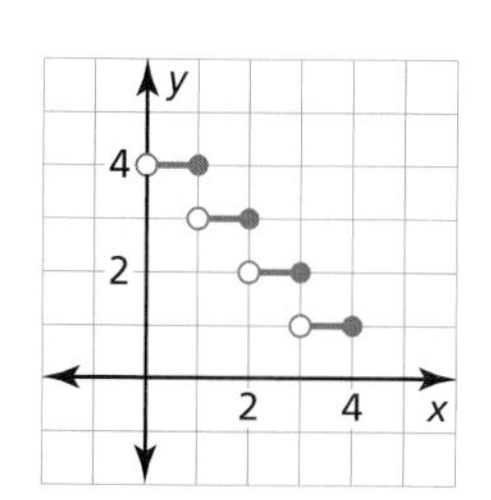

In Exercises 31–34, graph the step function. Describe the domain and range.

31. $f(x) = \begin{cases} 3, & \text{if } 0 \le x < 2 \\ 4, & \text{if } 2 \le x < 4 \\ 5, & \text{if } 4 \le x < 6 \\ 6, & \text{if } 6 \le x < 8 \end{cases}$

32. $f(x) = \begin{cases} -4, & \text{if } 1 < x \le 2 \\ -6, & \text{if } 2 < x \le 3 \\ -8, & \text{if } 3 < x \le 4 \\ -10, & \text{if } 4 < x \le 5 \end{cases}$

33. $f(x) = \begin{cases} 9, & \text{if } 1 < x \le 2 \\ 6, & \text{if } 2 < x \le 4 \\ 5, & \text{if } 4 < x \le 9 \\ 1, & \text{if } 9 < x \le 12 \end{cases}$

34. $f(x) = \begin{cases} -2, & \text{if } -6 \le x < -5 \\ -1, & \text{if } -5 \le x < -3 \\ 0, & \text{if } -3 \le x < -2 \\ 1, & \text{if } -2 \le x < 0 \end{cases}$

35. MODELING WITH MATHEMATICS The cost to join an intramural sports league is \$180 per team and includes the first five team members. For each additional team member, there is a \$30 fee. You plan to have nine people on your team. Write and graph a step function that represents the relationship between the number p of people on your team and the total cost of joining the league. *(See Example 4.)*

36. MODELING WITH MATHEMATICS The rates for a parking garage are shown. Write and graph a step function that represents the relationship between the number x of hours a car is parked in the garage and the total cost of parking in the garage for 1 day.

Daily Parking Garage
Rates
\$4 per hour
\$15 daily maximum

In Exercises 37–46, write the absolute value function as a piecewise function.

37. $y = |x| + 1$

38. $y = |x| - 3$

39. $y = |x - 2|$

40. $y = |x + 5|$

41. $y = 2|x + 3|$

42. $y = 4|x - 1|$

43. $y = -5|x - 8|$

44. $y = -3|x + 6|$

45. $y = -|x - 3| + 2$

46. $y = 7|x + 1| - 5$

47. MODELING WITH MATHEMATICS You are sitting on a boat on a lake. You can get a sunburn from the sunlight that hits you directly and also from the sunlight that reflects off the water. *(See Example 5.)*

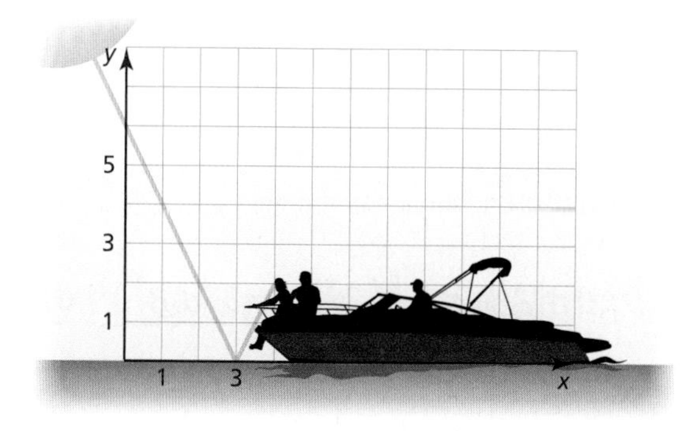

a. Write an absolute value function that represents the path of the sunlight that reflects off the water.

b. Write the function in part (a) as a piecewise function.

48. MODELING WITH MATHEMATICS You are trying to make a hole in one on the miniature golf green.

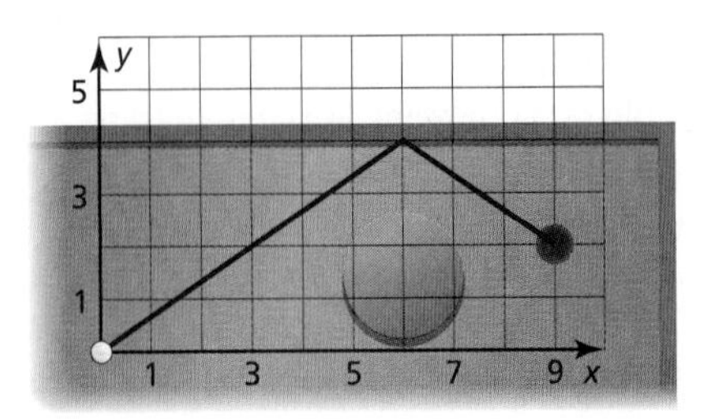

a. Write an absolute value function that represents the path of the golf ball.

b. Write the function in part (a) as a piecewise function.

49. REASONING The piecewise function f consists of two linear "pieces." The graph of f is shown.

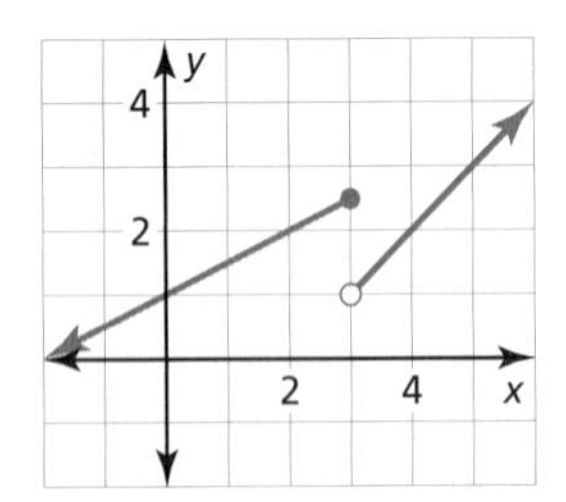

a. What is the value of $f(-10)$?

b. What is the value of $f(8)$?

50. CRITICAL THINKING Describe how the graph of each piecewise function changes when $<$ is replaced with $\leq$ and $\geq$ is replaced with $>$. Do the domain and range change? Explain.

a. $f(x) = \begin{cases} x + 2, & \text{if } x < 2 \\ -x - 1, & \text{if } x \geq 2 \end{cases}$

b. $f(x) = \begin{cases} \frac{1}{2}x + \frac{3}{2}, & \text{if } x < 1 \\ -x + 3, & \text{if } x \geq 1 \end{cases}$

51. USING STRUCTURE Graph

$$y = \begin{cases} -x + 2, & \text{if } x \leq -2 \\ |x|, & \text{if } x > -2 \end{cases}.$$

Describe the domain and range.

52. HOW DO YOU SEE IT? The graph shows the total cost C of making x photocopies at a copy shop.

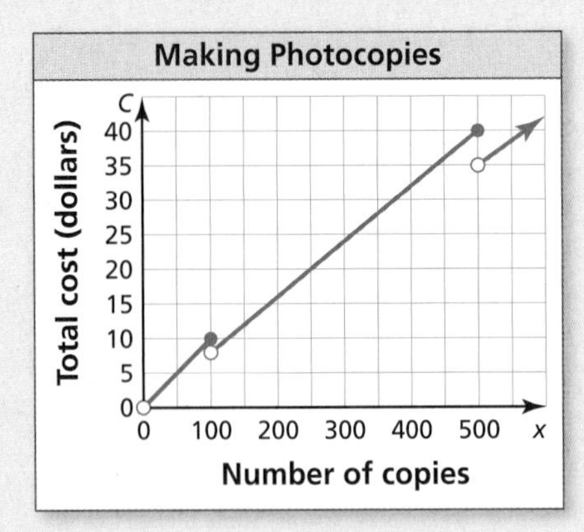

a. Does it cost more money to make 100 photocopies or 101 photocopies? Explain.

b. You have \$40 to make photocopies. Can you buy more than 500 photocopies? Explain.

53. USING STRUCTURE The output y of the *greatest integer function* is the greatest integer less than or equal to the input value x. This function is written as $f(x) = [\![x]\!]$. Graph the function for $-4 \leq x < 4$. Is it a piecewise function? a step function? Explain.

54. THOUGHT PROVOKING Explain why

$$y = \begin{cases} 2x - 2, & \text{if } x \leq 3 \\ -3, & \text{if } x \geq 3 \end{cases}$$

does not represent a function. How can you redefine y so that it does represent a function?

55. MAKING AN ARGUMENT During a 9-hour snowstorm, it snows at a rate of 1 inch per hour for the first 2 hours, 2 inches per hour for the next 6 hours, and 1 inch per hour for the final hour.

a. Write and graph a piecewise function that represents the depth of the snow during the snowstorm.

b. Your friend says 12 inches of snow accumulated during the storm. Is your friend correct? Explain.

Maintaining Mathematical Proficiency

Reviewing what you learned in previous grades and lessons

Write the sentence as an inequality. Graph the inequality. *(Section 2.5)*

56. A number r is greater than -12 and no more than 13.

57. A number t is less than or equal to 4 or no less than 18.

Graph f and h. Describe the transformations from the graph of f to the graph of h. *(Section 3.6)*

58. $f(x) = x$; $h(x) = 4x + 3$

59. $f(x) = x$; $h(x) = -x - 8$

60. $f(x) = x$; $h(x) = -\frac{1}{2}x + 5$

4.4–4.7 What Did You Learn?

Core Vocabulary

scatter plot, *p. 196*
correlation, *p. 197*
line of fit, *p. 198*
residual, *p. 202*
linear regression, *p. 203*
line of best fit, *p. 203*
correlation coefficient, *p. 203*
interpolation, *p. 205*
extrapolation, *p. 205*
causation, *p. 205*
sequence, *p. 210*
term, *p. 210*
arithmetic sequence, *p. 210*
common difference, *p. 210*
piecewise function, *p. 218*
step function, *p. 220*

Core Concepts

Section 4.4

Scatter Plot, *p. 196*
Identifying Correlations, *p. 197*
Using a Line of Fit to Model Data, *p. 198*

Section 4.5

Residuals, *p. 202*
Lines of Best Fit, *p. 203*
Correlation and Causation, *p. 205*

Section 4.6

Arithmetic Sequence, *p. 210*
Equation for an Arithmetic Sequence, *p. 212*

Section 4.7

Piecewise Function, *p. 218*
Step Function, *p. 220*
Writing Absolute Value Functions, *p. 221*

Mathematical Practices

1. What resources can you use to help you answer Exercise 17 on page 200?
2. What calculations are repeated in Exercises 11–16 on page 214? When finding a term such as a_{50}, is there a general method or shortcut you can use instead of repeating calculations?
3. Describe the definitions you used when you explained your answer in Exercise 53 on page 224.

Performance Task

Any Beginning

With so many ways to represent a linear relationship, where do you start? Use what you know to move between equations, graphs, tables, and contexts.

To explore the answers to this question and more, go to ***BigIdeasMath.com***.

4 Chapter Review

Dynamic Solutions available at *BigIdeasMath.com*

4.1 Writing Equations in Slope-Intercept Form (pp. 175–180)

Write an equation of the line in slope-intercept form.

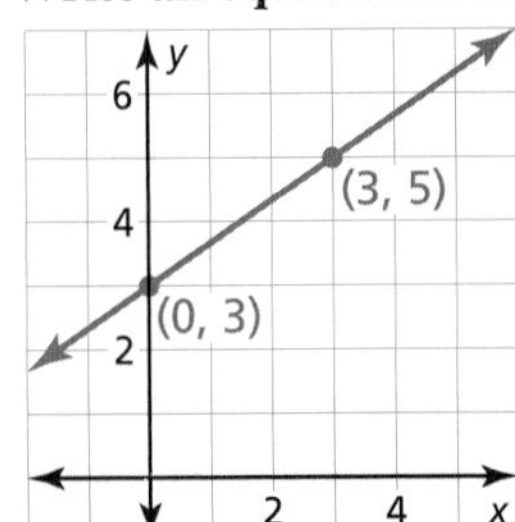

Find the slope and y-intercept.

Let $(x_1, y_1) = (0, 3)$ and $(x_2, y_2) = (3, 5)$.

$$m = \frac{y_2 - y_1}{x_2 - x_1} = \frac{5 - 3}{3 - 0} = \frac{2}{3}$$

Because the line crosses the y-axis at $(0, 3)$, the y-intercept is 3.

▶ So, the equation is $y = \frac{2}{3}x + 3$.

1. Write an equation of the line in slope-intercept form.

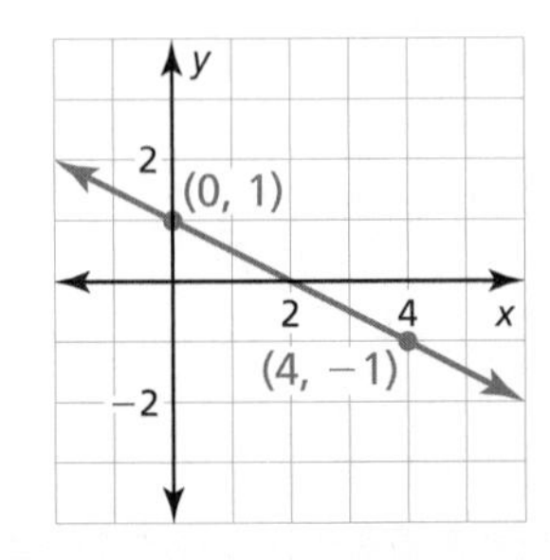

4.2 Writing Equations in Point-Slope Form (pp. 181–186)

Write an equation in point-slope form of the line that passes through the point $(-1, -8)$ and has a slope of 3.

$y - y_1 = m(x - x_1)$	Write the point-slope form.
$y - (-8) = 3[x - (-1)]$	Substitute 3 for m, -1 for x_1, and -8 for y_1.
$y + 8 = 3(x + 1)$	Simplify.

▶ The equation is $y + 8 = 3(x + 1)$.

2. Write an equation in point-slope form of the line that passes through the point $(4, 7)$ and has a slope of -1.

Write a linear function f with the given values.

3. $f(10) = 5, f(2) = -3$ **4.** $f(3) = -4, f(5) = -4$ **5.** $f(6) = 8, f(9) = 3$

4.3 Writing Equations of Parallel and Perpendicular Lines (pp. 187–192)

Determine which of the lines, if any, are parallel or perpendicular.

Line a: $y = 2x + 3$ **Line b: $2y + x = 5$** **Line c: $4y - 8x = -4$**

Write the equations in slope-intercept form. Then compare the slopes.

Line a: $y = 2x + 3$ Line b: $y = -\frac{1}{2}x + \frac{5}{2}$ Line c: $y = 2x - 1$

▶ Lines a and c have slopes of 2, so they are parallel. Line b has a slope of $-\frac{1}{2}$, the negative reciprocal of 2, so it is perpendicular to lines a and c.

Determine which of the lines, if any, are parallel or perpendicular. Explain.

6. Line a passes through (0, 4) and (4, 3).
Line b passes through (0, 1) and (4, 0).
Line c passes through (2, 0) and (4, 4).

7. Line a: $2x - 7y = 14$
Line b: $y = \frac{7}{2}x - 8$
Line c: $2x + 7y = -21$

8. Write an equation of the line that passes through (1, 5) and is parallel to the line $y = -4x + 2$.

9. Write an equation of the line that passes through $(2, -3)$ and is perpendicular to the line $y = -2x - 3$.

4.4 Scatter Plots and Lines of Fit *(pp. 195–200)*

The scatter plot shows the roasting times (in hours) and weights (in pounds) of seven turkeys. Tell whether the data show a *positive*, a *negative*, or *no* correlation.

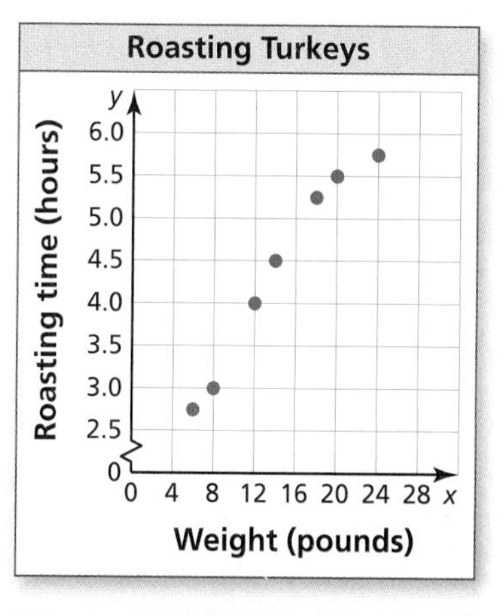

As the weight of a turkey increases, the roasting time increases.

▶ So, the scatter plot shows a positive correlation.

Use the scatter plot in the example.

10. What is the roasting time for a 12-pound turkey?

11. Write an equation that models the roasting time as a function of the weight of a turkey. Interpret the slope and y-intercept of the line of fit.

4.5 Analyzing Lines of Fit *(pp. 201–208)*

The table shows the heights x (in inches) and shoe sizes y of several students. Use a graphing calculator to find an equation of the line of best fit. Identify and interpret the correlation coefficient.

Height, x	Shoe size, y
64	9
62	7
70	12
63	8
72	13
68	9.5
66	9
74	13.5
68	10
59	6.5

Step 1 Enter the data from the table into two lists.

Step 2 Use the *linear regression* feature.

```
LinReg
 y=ax+b
 a=.4989919355
 b=-23.4828629
 r²=.9477256904
 r=.9735120392
```

▶ An equation of the line of best fit is $y = 0.50x - 23.5$. The correlation coefficient is about 0.974. This means that the relationship between the heights and the shoe sizes has a strong positive correlation and the equation closely models the data.

12. Make a scatter plot of the residuals to verify that the model in the example is a good fit.

13. Use the data in the example. (a) Approximate the height of a student whose shoe size is 9. (b) Predict the shoe size of a student whose height is 60 inches.

14. Is there a causal relationship in the data in the example? Explain.

4.6 Arithmetic Sequences (pp. 209–216)

Write an equation for the nth term of the arithmetic sequence $-3, -5, -7, -9, \ldots$. Then find a_{20}.

The first term is −3, and the common difference is −2.

$a_n = a_1 + (n - 1)d$ — Equation for an arithmetic sequence

$a_n = -3 + (n - 1)(-2)$ — Substitute −3 for a_1 and −2 for d.

$a_n = -2n - 1$ — Simplify.

Use the equation to find the 20th term.

$a_{20} = -2(20) - 1$ — Substitute 20 for n.

$= -41$ — Simplify.

▶ The 20th term of the arithmetic sequence is −41.

Write an equation for the nth term of the arithmetic sequence. Then find a_{30}.

15. 11, 10, 9, 8, . . . **16.** 6, 12, 18, 24, . . . **17.** −9, −6, −3, 0, . . .

4.7 Piecewise Functions (pp. 217–224)

Graph $y = \begin{cases} \frac{3}{2}x + 3, & \text{if } x \le 0 \\ -2x, & \text{if } x > 0 \end{cases}$. Describe the domain and range.

Step 1 Graph $y = \frac{3}{2}x + 3$ for $x \le 0$. Because x is less than or equal to 0, use a closed circle at (0, 3).

Step 2 Graph $y = -2x$ for $x > 0$. Because x is not equal to 0, use an open circle at (0, 0).

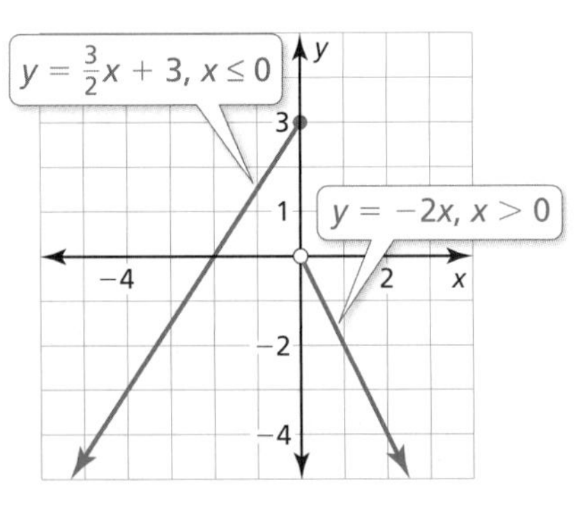

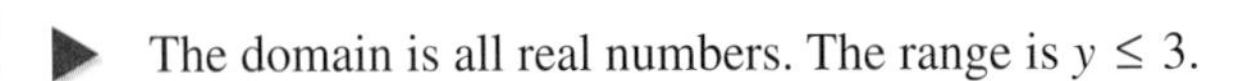

▶ The domain is all real numbers. The range is $y \le 3$.

18. Evaluate the function in the example when (a) $x = 0$ and (b) $x = 5$.

Graph the function. Describe the domain and range.

19. $y = \begin{cases} x + 6, & \text{if } x \le 0 \\ -3x, & \text{if } x > 0 \end{cases}$ **20.** $y = \begin{cases} 4x + 2, & \text{if } x < -4 \\ 2x - 6, & \text{if } x \ge -4 \end{cases}$

Write the absolute value function as a piecewise function.

21. $y = |x| + 15$ **22.** $y = 4|x + 5|$ **23.** $y = 2|x + 2| - 3$

24. You are organizing a school fair and rent a popcorn machine for 3 days. The rental company charges \$65 for the first day and \$35 for each additional day. Write and graph a step function that represents the relationship between the number x of days and the total cost y (in dollars) of renting the popcorn machine.

4 Chapter Test

Graph the function. Describe the domain and range.

1. $y = \begin{cases} 2x + 4, & \text{if } x \le -1 \\ \frac{1}{3}x - 1, & \text{if } x > -1 \end{cases}$

2. $y = \begin{cases} 1, & \text{if } 0 \le x < 3 \\ 0, & \text{if } 3 \le x < 6 \\ -1, & \text{if } 6 \le x < 9 \\ -2, & \text{if } 9 \le x < 12 \end{cases}$

Write an equation in slope-intercept form of the line with the given characteristics.

3. slope $= \frac{2}{5}$; y-intercept $= -7$

4. passes through $(0, 6)$ and $(3, -3)$

5. parallel to the line $y = 3x - 1$; passes through $(-2, -8)$

6. perpendicular to the line $y = \frac{1}{4}x - 9$; passes through $(1, 1)$

Write an equation in point-slope form of the line with the given characteristics.

7. slope $= 10$; passes through $(6, 2)$

8. passes through $(-3, 2)$ and $(6, -1)$

9. The first row of an auditorium has 42 seats. Each row after the first has three more seats than the row before it.

a. Find the number of seats in Row 25.

b. Which row has 90 seats?

10. The table shows the amount x (in dollars) spent on advertising for a neighborhood festival and the attendance y of the festival for several years.

Advertising (dollars), x	Yearly attendance, y
500	400
1000	550
1500	550
2000	800
2500	650
3000	800
3500	1050
4000	1100

a. Make a scatter plot of the data. Describe the correlation.

b. Write an equation that models the attendance as a function of the amount spent on advertising.

c. Interpret the slope and y-intercept of the line of fit.

11. Consider the data in the table in Exercise 10.

a. Use a graphing calculator to find an equation of the line of best fit.

b. Identify and interpret the correlation coefficient.

c. What would you expect the scatter plot of the residuals to look like?

d. Is there a causal relationship in the data? Explain your reasoning.

e. Predict the amount that must be spent on advertising to get 2000 people to attend the festival.

12. Let a, b, c, and d be constants. Determine which of the lines, if any, are parallel or perpendicular. Explain.

Line 1: $y - c = ax$ Line 2: $ay = -x - b$ Line 3: $ax + y = d$

13. Write a piecewise function defined by three equations that has a domain of all real numbers and a range of $-3 < y \le 1$.

4 Cumulative Assessment

1. Which function represents the arithmetic sequence shown in the graph?

Ⓐ $f(n) = 15 + 3n$

Ⓑ $f(n) = 4 - 3n$

Ⓒ $f(n) = 27 - 3n$

Ⓓ $f(n) = 24 - 3n$

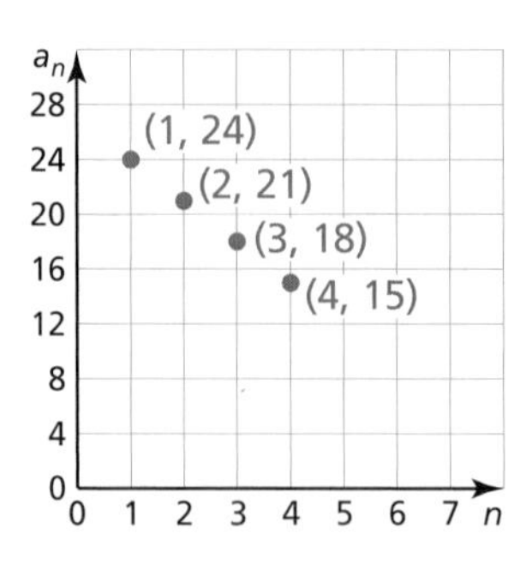

2. Which of the inequalities are equivalent?

$3x + 6 \leq 8 + 2x$ $\quad$ $5x - 5 \geq 7x - 9$ $\quad$ $12 - 3x \leq 18$ $\quad$ $-2 - \frac{3}{2}x \geq -3 - x$

3. Complete the table for the four situations below. Explain your reasoning.

a. the price of a pair of pants and the number sold

b. the number of cell phones and the number of taxis in a city

c. a person's IQ and the time it takes the person to run 50 meters

d. the amount of time spent studying and the score earned

Situation	Correlation		Causation	
	Yes	No	Yes	No
a.				
b.				
c.				
d.				

4. Consider the function $f(x) = x - 1$. Select the functions that are shown in the graph. Explain your reasoning.

$g(x) = f(x + 2)$ $\quad$ $h(x) = f(3x)$

$k(x) = f(x) + 4$ $\quad$ $p(x) = f(-x)$

$r(x) = 3f(x)$ $\quad$ $q(x) = -f(x)$

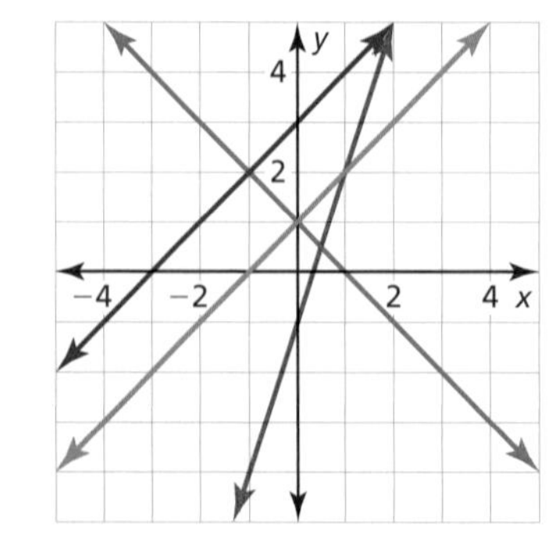

5. Use the numbers to fill in values for m and b in the equation $y = mx + b$ so that its graph passes through the points $(6, 1)$ and $(-2, -3)$.

-5	-2	-1	$-\frac{1}{2}$	0	$\frac{1}{2}$	1	2	5

6. Fill in the piecewise function with $-$, $+$, $<$, $\leq$, $>$, or $\geq$ so that the function is represented by the graph.

$$y = \begin{cases} 2x \ \square \ 3, & \text{if } x \ \square \ 0 \\ 2x \ \square \ 3, & \text{if } x \ \square \ 0 \end{cases}$$

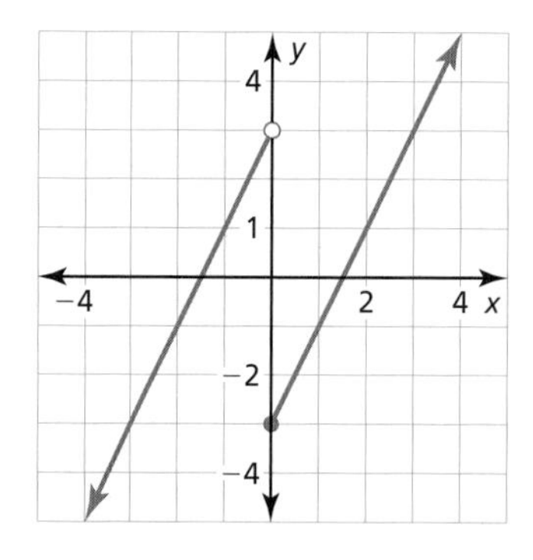

7. You claim that you can create a relation that is a function, and your friend claims that she can create a relation that is not a function. Using the given numbers, create a relation of five ordered pairs that supports your claim. What relation of five ordered pairs can your friend use to support her claim?

-4	-3	-2	-1	0	1	2	3	4

8. You have two coupons you can use at a restaurant. Write and solve an equation to determine how much your total bill must be for both coupons to save you the same amount of money.

9. The table shows the daily high temperatures x (in degrees Fahrenheit) and the numbers y of frozen fruit bars sold on eight randomly selected days. The equation $y = 3x - 50$ models the data.

Temperature (°F), x	54	60	68	72	78	84	92	98
Frozen fruit bars, y	40	120	180	260	280	260	220	180

a. Select the points that appear on a scatter plot of the residuals.

$(92, -6)$	$(78, 96)$	$(60, -10)$	$(84, 58)$	$(98, -64)$
$(72, 94)$	$(54, -72)$	$(96, 78)$	$(60, 10)$	$(68, 26)$

b. Determine whether the model is a good fit for the data. Explain your reasoning.

5 Solving Systems of Linear Equations

Fishing *(p. 279)*

Pets *(p. 266)*

Delivery Vans *(p. 250)*

Drama Club *(p. 244)*

Roofing Contractor *(p. 238)*

Maintaining Mathematical Proficiency

Graphing Linear Functions

Example 1 **Graph $3 + y = \frac{1}{2}x$.**

Step 1 Rewrite the equation in slope-intercept form.

$$y = \frac{1}{2}x - 3$$

Step 2 Find the slope and the y-intercept.

$$m = \frac{1}{2} \text{ and } b = -3$$

Step 3 The y-intercept is -3. So, plot $(0, -3)$.

Step 4 Use the slope to find another point on the line.

$$\text{slope} = \frac{\text{rise}}{\text{run}} = \frac{1}{2}$$

Plot the point that is 2 units right and 1 unit up from $(0, -3)$. Draw a line through the two points.

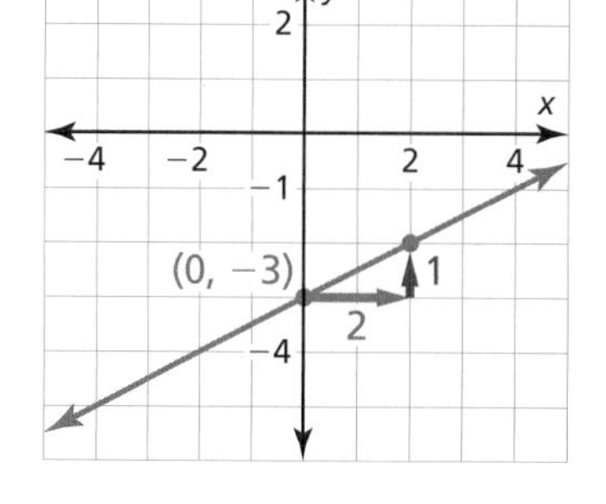

Graph the equation.

1. $y + 4 = x$ **2.** $6x - y = -1$ **3.** $4x + 5y = 20$ **4.** $-2y + 12 = -3x$

Solving and Graphing Linear Inequalities

Example 2 **Solve $2x - 17 \leq 8x - 5$. Graph the solution.**

$2x - 17 \leq 8x - 5$	Write the inequality.
$\underline{+5} \qquad \underline{+5}$	Add 5 to each side.
$2x - 12 \leq 8x$	Simplify.
$\underline{-2x} \qquad \underline{-2x}$	Subtract $2x$ from each side.
$-12 \leq 6x$	Simplify.
$\frac{-12}{6} \leq \frac{6x}{6}$	Divide each side by 6.
$-2 \leq x$	Simplify.

▶ The solution is $x \geq -2$.

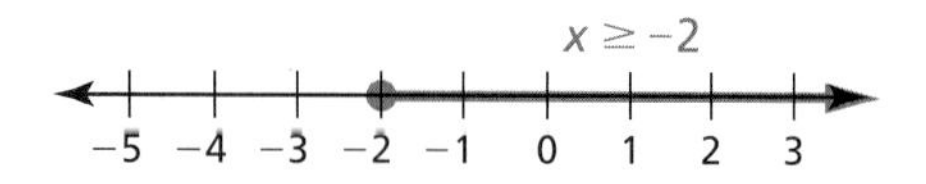

Solve the inequality. Graph the solution.

5. $m + 4 > 9$ **6.** $24 \leq -6t$ **7.** $2a - 5 \leq 13$

8. $-5z + 1 < -14$ **9.** $4k - 16 < k + 2$ **10.** $7w + 12 \geq 2w - 3$

11. **ABSTRACT REASONING** The graphs of the linear functions g and h have different slopes. The value of both functions at $x = a$ is b. When g and h are graphed in the same coordinate plane, what happens at the point (a, b)?

Mathematical Practices

Mathematically proficient students use technological tools to explore concepts.

Using a Graphing Calculator

Core Concept

Finding the Point of Intersection

You can use a graphing calculator to find the point of intersection, if it exists, of the graphs of two linear equations.

1. Enter the equations into a graphing calculator.
2. Graph the equations in an appropriate viewing window, so that the point of intersection is visible.
3. Use the *intersect* feature of the graphing calculator to find the point of intersection.

EXAMPLE 1 Using a Graphing Calculator

Use a graphing calculator to find the point of intersection, if it exists, of the graphs of the two linear equations.

$y = -\frac{1}{2}x + 2$ Equation 1

$y = 3x - 5$ Equation 2

SOLUTION

The slopes of the lines are not the same, so you know that the lines intersect. Enter the equations into a graphing calculator. Then graph the equations in an appropriate viewing window.

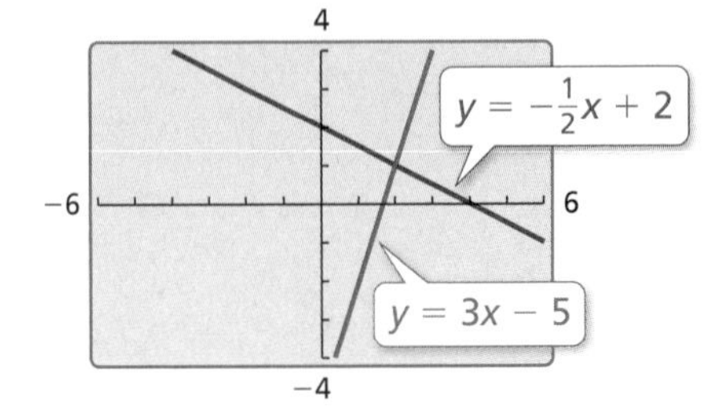

Use the *intersect* feature to find the point of intersection of the lines.

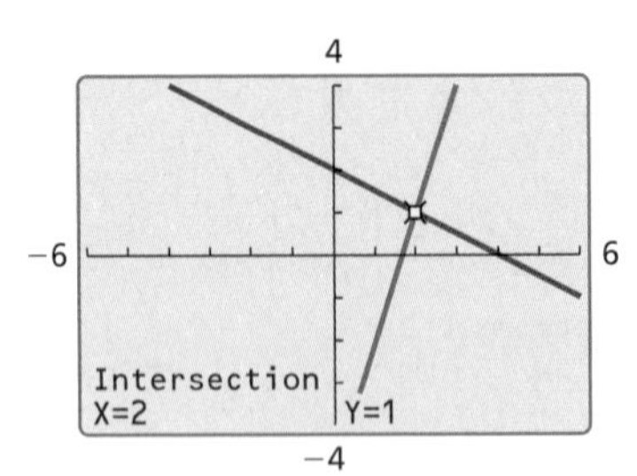

▶ The point of intersection is (2, 1).

Monitoring Progress

Use a graphing calculator to find the point of intersection of the graphs of the two linear equations.

1. $y = -2x - 3$
   $y = \frac{1}{2}x - 3$
2. $y = -x + 1$
   $y = x - 2$
3. $3x - 2y = 2$
   $2x - y = 2$

5.1 Solving Systems of Linear Equations by Graphing

Essential Question How can you solve a system of linear equations?

EXPLORATION 1 **Writing a System of Linear Equations**

Work with a partner. Your family opens a bed-and-breakfast. They spend \$600 preparing a bedroom to rent. The cost to your family for food and utilities is \$15 per night. They charge \$75 per night to rent the bedroom.

a. Write an equation that represents the costs.

Cost, C (in dollars) = \$15 per night • Number of nights, x + \$600

b. Write an equation that represents the revenue (income).

Revenue, R (in dollars) = \$75 per night • Number of nights, x

c. A set of two (or more) linear equations is called a **system of linear equations**. Write the system of linear equations for this problem.

MODELING WITH MATHEMATICS

To be proficient in math, you need to identify important quantities in real-life problems and map their relationships using tools such as diagrams, tables, and graphs.

EXPLORATION 2 **Using a Table or Graph to Solve a System**

Work with a partner. Use the cost and revenue equations from Exploration 1 to determine how many nights your family needs to rent the bedroom before recovering the cost of preparing the bedroom. This is the *break-even point*.

a. Copy and complete the table.

x (nights)	0	1	2	3	4	5	6	7	8	9	10	11
C (dollars)												
R (dollars)												

b. How many nights does your family need to rent the bedroom before breaking even?

c. In the same coordinate plane, graph the cost equation and the revenue equation from Exploration 1.

d. Find the point of intersection of the two graphs. What does this point represent? How does this compare to the break-even point in part (b)? Explain.

Communicate Your Answer

3. How can you solve a system of linear equations? How can you check your solution?

4. Solve each system by using a table or sketching a graph. Explain why you chose each method. Use a graphing calculator to check each solution.

a. $y = -4.3x - 1.3$
$y = 1.7x + 4.7$

b. $y = x$
$y = -3x + 8$

c. $y = -x - 1$
$y = 3x + 5$

5.1 Lesson

Core Vocabulary

system of linear equations, *p. 236*
solution of a system of linear equations, *p. 236*

Previous
linear equation
ordered pair

What You Will Learn

- Check solutions of systems of linear equations.
- Solve systems of linear equations by graphing.
- Use systems of linear equations to solve real-life problems.

Systems of Linear Equations

A **system of linear equations** is a set of two or more linear equations in the same variables. An example is shown below.

$x + y = 7$ Equation 1

$2x - 3y = -11$ Equation 2

A **solution of a system of linear equations** in two variables is an ordered pair that is a solution of each equation in the system.

EXAMPLE 1 Checking Solutions

Tell whether the ordered pair is a solution of the system of linear equations.

a. $(2, 5)$; $x + y = 7$ Equation 1; $2x - 3y = -11$ Equation 2

b. $(-2, 0)$; $y = -2x - 4$ Equation 1; $y = x + 4$ Equation 2

SOLUTION

a. Substitute 2 for x and 5 for y in each equation.

Equation 1	Equation 2
$x + y = 7$	$2x - 3y = -11$
$2 + 5 \stackrel{?}{=} 7$	$2(2) - 3(5) \stackrel{?}{=} -11$
$7 = 7$ ✓	$-11 = -11$ ✓

▶ Because the ordered pair (2, 5) is a solution of each equation, it is a solution of the linear system.

READING

A system of linear equations is also called a *linear system.*

b. Substitute -2 for x and 0 for y in each equation.

Equation 1	Equation 2
$y = -2x - 4$	$y = x + 4$
$0 \stackrel{?}{=} -2(-2) - 4$	$0 \stackrel{?}{=} -2 + 4$
$0 = 0$ ✓	$0 \neq 2$ ✗

▶ The ordered pair $(-2, 0)$ is a solution of the first equation, but it is not a solution of the second equation. So, $(-2, 0)$ is *not* a solution of the linear system.

Monitoring Progress

Help in English and Spanish at *BigIdeasMath.com*

Tell whether the ordered pair is a solution of the system of linear equations.

1. $(1, -2)$; $2x + y = 0$, $-x + 2y = 5$

2. $(1, 4)$; $y = 3x + 1$, $y = -x + 5$

Solving Systems of Linear Equations by Graphing

The solution of a system of linear equations is the point of intersection of the graphs of the equations.

Core Concept

Solving a System of Linear Equations by Graphing

Step 1 Graph each equation in the same coordinate plane.

Step 2 Estimate the point of intersection.

Step 3 Check the point from Step 2 by substituting for x and y in each equation of the original system.

REMEMBER

Note that the linear equations are in slope-intercept form. You can use the method presented in Section 3.5 to graph the equations.

EXAMPLE 2 Solving a System of Linear Equations by Graphing

Solve the system of linear equations by graphing.

$y = -2x + 5$ Equation 1

$y = 4x - 1$ Equation 2

SOLUTION

Step 1 Graph each equation.

Step 2 Estimate the point of intersection. The graphs appear to intersect at (1, 3).

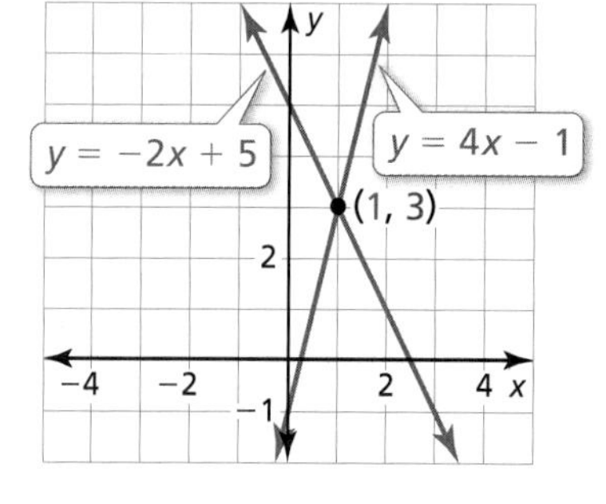

Step 3 Check your point from Step 2.

Equation 1	Equation 2
$y = -2x + 5$	$y = 4x - 1$
$3 \stackrel{?}{=} -2(1) + 5$	$3 \stackrel{?}{=} 4(1) - 1$
$3 = 3$ ✓	$3 = 3$ ✓

▶ The solution is (1, 3).

Check

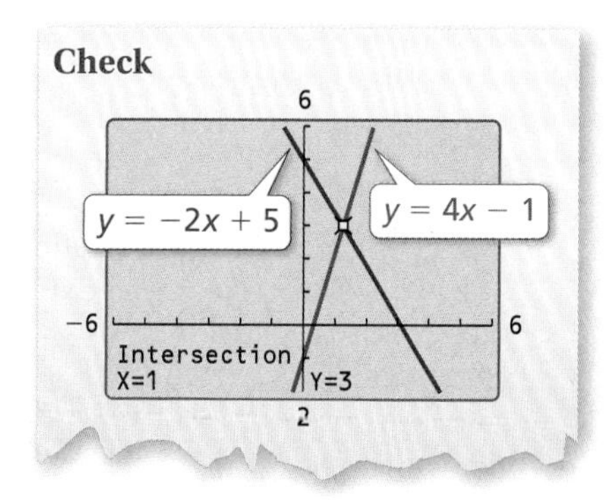

Monitoring Progress

Help in English and Spanish at *BigIdeasMath.com*

Solve the system of linear equations by graphing.

3. $y = x - 2$
$y = -x + 4$

4. $y = \frac{1}{2}x + 3$
$y = -\frac{3}{2}x - 5$

5. $2x + y = 5$
$3x - 2y = 4$

Solving Real-Life Problems

EXAMPLE 3 Modeling with Mathematics

A roofing contractor buys 30 bundles of shingles and 4 rolls of roofing paper for \$1040. In a second purchase (at the same prices), the contractor buys 8 bundles of shingles for \$256. Find the price per bundle of shingles and the price per roll of roofing paper.

SOLUTION

1. **Understand the Problem** You know the total price of each purchase and how many of each item were purchased. You are asked to find the price of each item.
2. **Make a Plan** Use a verbal model to write a system of linear equations that represents the problem. Then solve the system of linear equations.
3. **Solve the Problem**

Words $30 \cdot$ Price per bundle $+ 4 \cdot$ Price per roll $= 1040$

$8 \cdot$ Price per bundle $+ 0 \cdot$ Price per roll $= 256$

Variables Let x be the price (in dollars) per bundle and let y be the price (in dollars) per roll.

System $30x + 4y = 1040$ Equation 1

$8x = 256$ Equation 2

Step 1 Graph each equation. Note that only the first quadrant is shown because x and y must be positive.

Step 2 Estimate the point of intersection. The graphs appear to intersect at (32, 20).

Step 3 Check your point from Step 2.

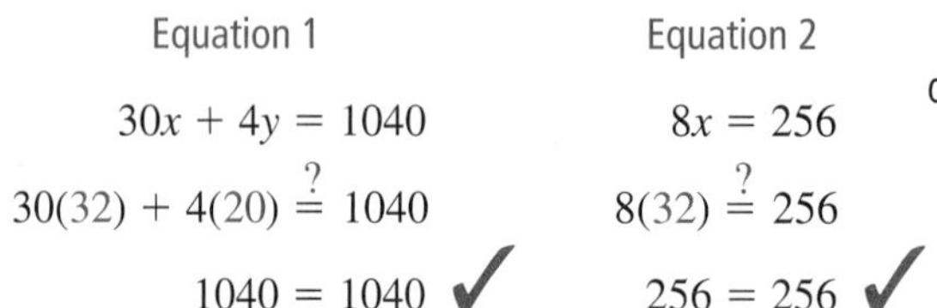

Equation 1	Equation 2
$30x + 4y = 1040$	$8x = 256$
$30(32) + 4(20) \stackrel{?}{=} 1040$	$8(32) \stackrel{?}{=} 256$
$1040 = 1040$ ✓	$256 = 256$ ✓

y = −7.5x + 260
x = 32
(32, 20)
y: 0, 80, 160, 240, 320
x: 0, 8, 16, 24, 32

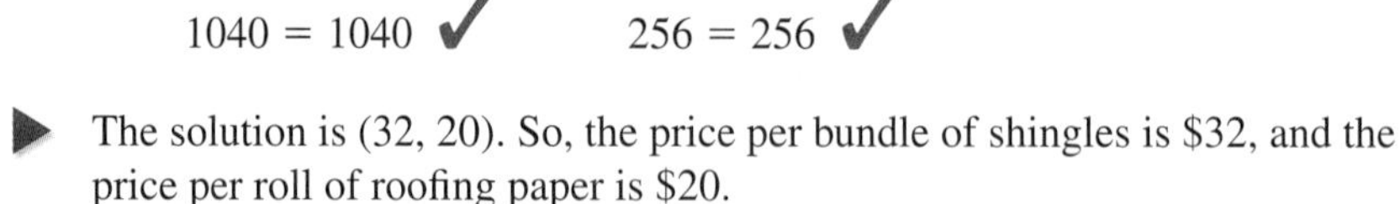

▶ The solution is (32, 20). So, the price per bundle of shingles is \$32, and the price per roll of roofing paper is \$20.

4. **Look Back** You can use estimation to check that your solution is reasonable. A bundle of shingles costs about \$30. So, 30 bundles of shingles and 4 rolls of roofing paper (at \$20 per roll) cost about $30(30) + 4(20) = \$980$, and 8 bundles of shingles costs about $8(30) = \$240$. These prices are close to the given values, so the solution seems reasonable.

Monitoring Progress

Help in English and Spanish at *BigIdeasMath.com*

6. You have a total of 18 math and science exercises for homework. You have six more math exercises than science exercises. How many exercises do you have in each subject?

5.1 Exercises

Dynamic Solutions available at *BigIdeasMath.com*

Vocabulary and Core Concept Check

1. **VOCABULARY** Do the equations $5y - 2x = 18$ and $6x = -4y - 10$ form a system of linear equations? Explain.

2. **DIFFERENT WORDS, SAME QUESTION** Consider the system of linear equations $-4x + 2y = 4$ and $4x - y = -6$. Which is different? Find "both" answers.

Solve the system of linear equations.	Solve each equation for y.
Find the point of intersection of the graphs of the equations.	Find an ordered pair that is a solution of each equation in the system.

Monitoring Progress and Modeling with Mathematics

In Exercises 3–8, tell whether the ordered pair is a solution of the system of linear equations. *(See Example 1.)*

3. $(2, 6);\ \begin{array}{l} x + y = 8 \\ 3x - y = 0 \end{array}$

4. $(8, 2);\ \begin{array}{l} x - y = 6 \\ 2x - 10y = 4 \end{array}$

5. $(-1, 3);\ \begin{array}{l} y = -7x - 4 \\ y = 8x + 5 \end{array}$

6. $(-4, -2);\ \begin{array}{l} y = 2x + 6 \\ y = -3x - 14 \end{array}$

7. $(-2, 1);\ \begin{array}{l} 6x + 5y = -7 \\ 2x - 4y = -8 \end{array}$

8. $(5, -6);\ \begin{array}{l} 6x + 3y = 12 \\ 4x + y = 14 \end{array}$

In Exercises 9–12, use the graph to solve the system of linear equations. Check your solution.

9. $x - y = 4$
   $4x + y = 1$

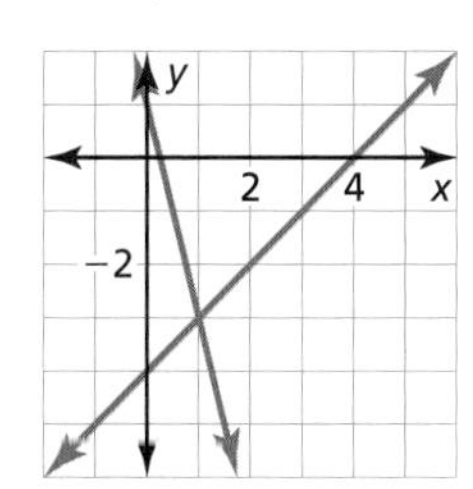

10. $x + y = 5$
    $y - 2x = -4$

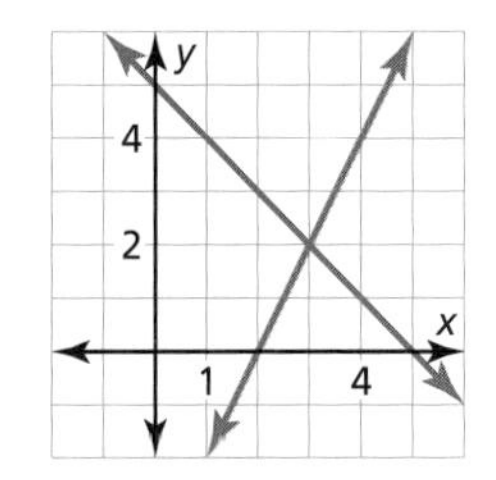

11. $6y + 3x = 18$
    $-x + 4y = 24$

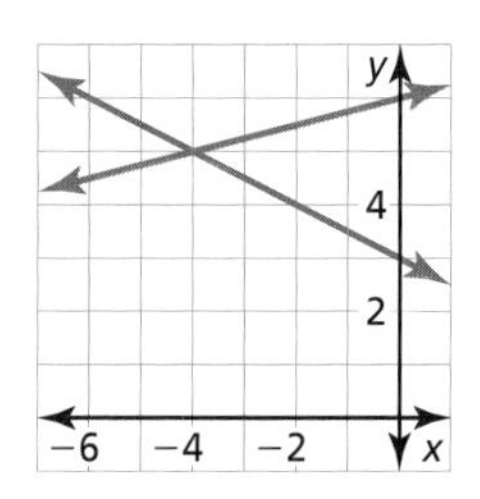

12. $2x - y = -2$
    $2x + 4y = 8$

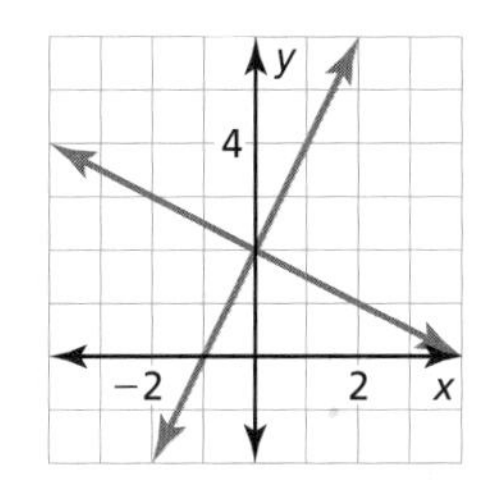

In Exercises 13–20, solve the system of linear equations by graphing. *(See Example 2.)*

13. $y = -x + 7$
    $y = x + 1$

14. $y = -x + 4$
    $y = 2x - 8$

15. $y = \frac{1}{3}x + 2$
    $y = \frac{2}{3}x + 5$

16. $y = \frac{3}{4}x - 4$
    $y = -\frac{1}{2}x + 11$

17. $9x + 3y = -3$
    $2x - y = -4$

18. $4x - 4y = 20$
    $y = -5$

19. $x - 4y = -4$
    $-3x - 4y = 12$

20. $3y + 4x = 3$
    $x + 3y = -6$

ERROR ANALYSIS In Exercises 21 and 22, describe and correct the error in solving the system of linear equations.

21.

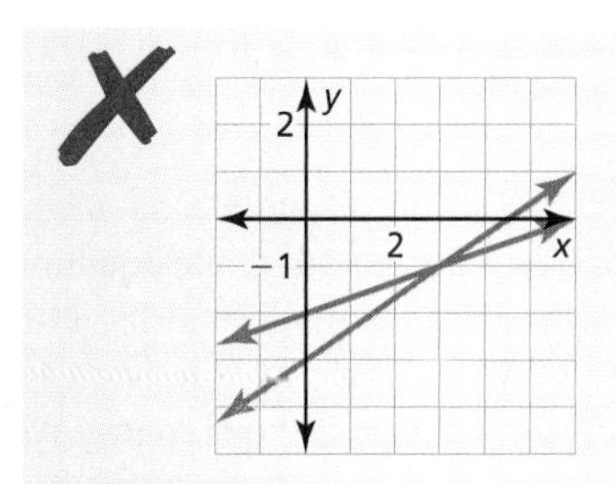

The solution of the linear system $x - 3y = 6$ and $2x - 3y = 3$ is $(3, -1)$.

22.

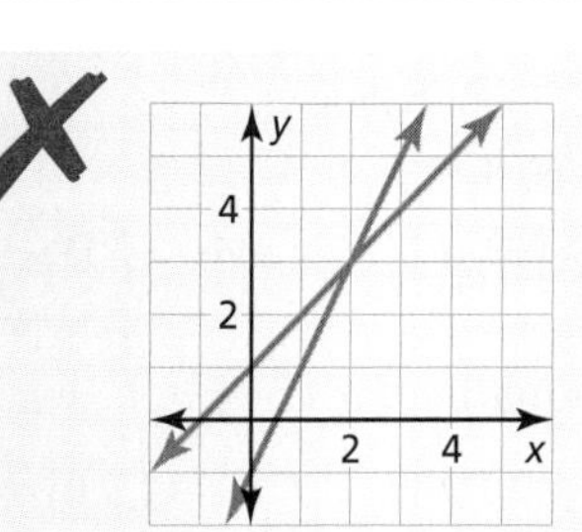

The solution of the linear system $y = 2x - 1$ and $y = x + 1$ is $x = 2$.

USING TOOLS **In Exercises 23–26, use a graphing calculator to solve the system of linear equations.**

23. $0.2x + 0.4y = 4$
$-0.6x + 0.6y = -3$

24. $-1.6x - 3.2y = -24$
$2.6x + 2.6y = 26$

25. $-7x + 6y = 0$
$0.5x + y = 2$

26. $4x - y = 1.5$
$2x + y = 1.5$

27. MODELING WITH MATHEMATICS You have 40 minutes to exercise at the gym, and you want to burn 300 calories total using both machines. How much time should you spend on each machine? *(See Example 3.)*

Elliptical Trainer

8 calories per minute

Stationary Bike

6 calories per minute

28. MODELING WITH MATHEMATICS You sell small and large candles at a craft fair. You collect \$144 selling a total of 28 candles. How many of each type of candle did you sell?

29. MATHEMATICAL CONNECTIONS Write a linear equation that represents the area and a linear equation that represents the perimeter of the rectangle. Solve the system of linear equations by graphing. Interpret your solution.

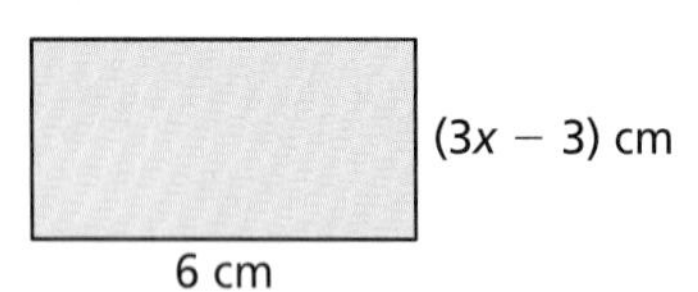

30. THOUGHT PROVOKING Your friend's bank account balance (in dollars) is represented by the equation $y = 25x + 250$, where x is the number of months. Graph this equation. After 6 months, you want to have the same account balance as your friend. Write a linear equation that represents your account balance. Interpret the slope and y-intercept of the line that represents your account balance.

31. COMPARING METHODS Consider the equation $x + 2 = 3x - 4$.

a. Solve the equation using algebra.

b. Solve the system of linear equations $y = x + 2$ and $y = 3x - 4$ by graphing.

c. How is the linear system and the solution in part (b) related to the original equation and the solution in part (a)?

32. HOW DO YOU SEE IT? A teacher is purchasing binders for students. The graph shows the total costs of ordering x binders from three different companies.

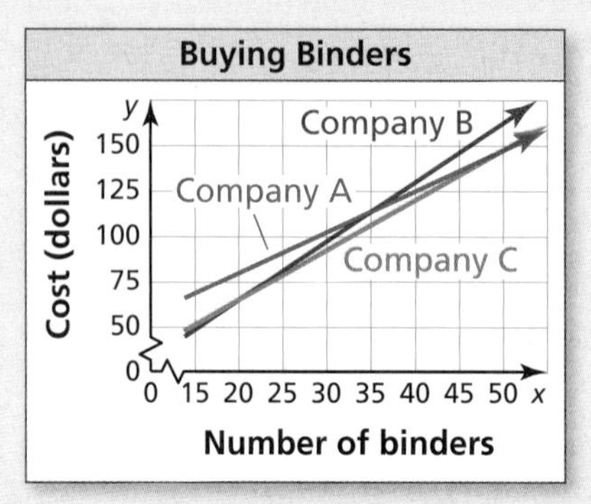

a. For what numbers of binders are the costs the same at two different companies? Explain.

b. How do your answers in part (a) relate to systems of linear equations?

33. MAKING AN ARGUMENT You and a friend are going hiking but start at different locations. You start at the trailhead and walk 5 miles per hour. Your friend starts 3 miles from the trailhead and walks 3 miles per hour.

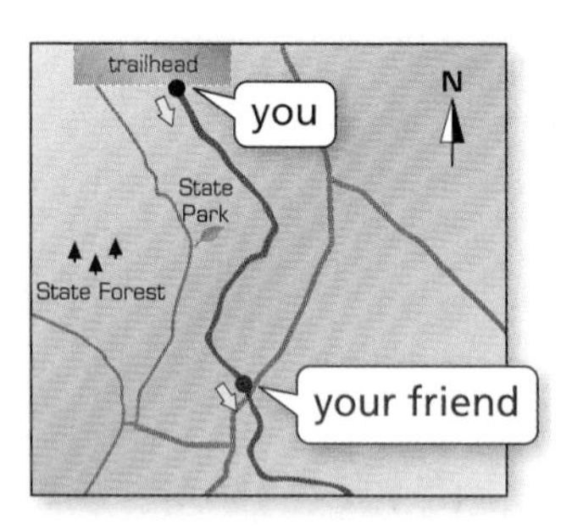

a. Write and graph a system of linear equations that represents this situation.

b. Your friend says that after an hour of hiking you will both be at the same location on the trail. Is your friend correct? Use the graph from part (a) to explain your answer.

Maintaining Mathematical Proficiency

Reviewing what you learned in previous grades and lessons

Solve the literal equation for *y*. *(Section 1.5)*

34. $10x + 5y = 5x + 20$

35. $9x + 18 = 6y - 3x$

36. $\frac{3}{4}x + \frac{1}{4}y = 5$

5.2 Solving Systems of Linear Equations by Substitution

Essential Question How can you use substitution to solve a system of linear equations?

EXPLORATION 1 Using Substitution to Solve Systems

Work with a partner. Solve each system of linear equations using two methods.

Method 1 Solve for x first.

Solve for x in one of the equations. Substitute the expression for x into the other equation to find y. Then substitute the value of y into one of the original equations to find x.

Method 2 Solve for y first.

Solve for y in one of the equations. Substitute the expression for y into the other equation to find x. Then substitute the value of x into one of the original equations to find y.

Is the solution the same using both methods? Explain which method you would prefer to use for each system.

a. $x + y = -7$
$-5x + y = 5$

b. $x - 6y = -11$
$3x + 2y = 7$

c. $4x + y = -1$
$3x - 5y = -18$

EXPLORATION 2 Writing and Solving a System of Equations

Work with a partner.

a. Write a random ordered pair with integer coordinates. One way to do this is to use a graphing calculator. The ordered pair generated at the right is $(-2, -3)$.

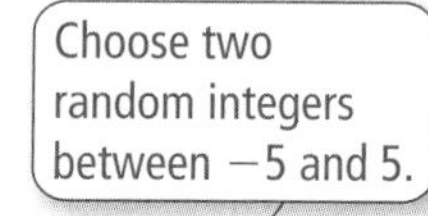
Choose two
random integers
between −5 and 5.

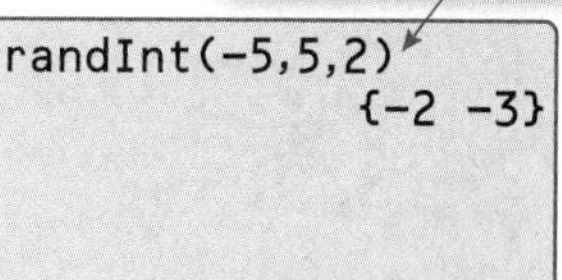
randInt(-5,5,2)
{-2 -3}

b. Write a system of linear equations that has your ordered pair as its solution.

c. Exchange systems with your partner and use one of the methods from Exploration 1 to solve the system. Explain your choice of method.

ATTENDING TO PRECISION

To be proficient in math, you need to communicate precisely with others.

Communicate Your Answer

3. How can you use substitution to solve a system of linear equations?

4. Use one of the methods from Exploration 1 to solve each system of linear equations. Explain your choice of method. Check your solutions.

a. $x + 2y = -7$
$2x - y = -9$

b. $x - 2y = -6$
$2x + y = -2$

c. $-3x + 2y = -10$
$-2x + y = -6$

d. $3x + 2y = 13$
$x - 3y = -3$

e. $3x - 2y = 9$
$-x - 3y = 8$

f. $3x - y = -6$
$4x + 5y = 11$

5.2 Lesson

Core Vocabulary

Previous
system of linear equations
solution of a system of linear equations

What You Will Learn

- Solve systems of linear equations by substitution.
- Use systems of linear equations to solve real-life problems.

Solving Linear Systems by Substitution

Another way to solve a system of linear equations is to use substitution.

Core Concept

Solving a System of Linear Equations by Substitution

Step 1 Solve one of the equations for one of the variables.

Step 2 Substitute the expression from Step 1 into the other equation and solve for the other variable.

Step 3 Substitute the value from Step 2 into one of the original equations and solve.

EXAMPLE 1 Solving a System of Linear Equations by Substitution

Solve the system of linear equations by substitution.

$y = -2x - 9$ Equation 1

$6x - 5y = -19$ Equation 2

SOLUTION

Step 1 Equation 1 is already solved for y.

Step 2 Substitute $-2x - 9$ for y in Equation 2 and solve for x.

$6x - 5y = -19$	Equation 2
$6x - 5(-2x - 9) = -19$	Substitute $-2x - 9$ for y.
$6x + 10x + 45 = -19$	Distributive Property
$16x + 45 = -19$	Combine like terms.
$16x = -64$	Subtract 45 from each side.
$x = -4$	Divide each side by 16.

Step 3 Substitute -4 for x in Equation 1 and solve for y.

$y = -2x - 9$	Equation 1
$= -2(-4) - 9$	Substitute -4 for x.
$= 8 - 9$	Multiply.
$= -1$	Subtract.

The solution is $(-4, -1)$.

Check

Equation 1

$y = -2x - 9$

$-1 \stackrel{?}{=} -2(-4) - 9$

$-1 = -1$ ✓

Equation 2

$6x - 5y = -19$

$6(-4) - 5(-1) \stackrel{?}{=} -19$

$-19 = -19$ ✓

Monitoring Progress  Help in English and Spanish at *BigIdeasMath.com*

Solve the system of linear equations by substitution. Check your solution.

1. $y = 3x + 14$
$y = -4x$

2. $3x + 2y = 0$
$y = \frac{1}{2}x - 1$

3. $x = 6y - 7$
$4x + y = -3$

EXAMPLE 2 Solving a System of Linear Equations by Substitution

ANOTHER WAY

You could also begin by solving for x in Equation 1, solving for y in Equation 2, or solving for x in Equation 2.

Solve the system of linear equations by substitution.

$-x + y = 3$ Equation 1

$3x + y = -1$ Equation 2

SOLUTION

Step 1 Solve for y in Equation 1.

$y = x + 3$ Revised Equation 1

Step 2 Substitute $x + 3$ for y in Equation 2 and solve for x.

$3x + y = -1$ Equation 2

$3x + (x + 3) = -1$ Substitute $x + 3$ for y.

$4x + 3 = -1$ Combine like terms.

$4x = -4$ Subtract 3 from each side.

$x = -1$ Divide each side by 4.

Step 3 Substitute -1 for x in Equation 1 and solve for y.

$-x + y = 3$ Equation 1

$-(-1) + y = 3$ Substitute -1 for x.

$y = 2$ Subtract 1 from each side.

▶ The solution is $(-1, 2)$.

Algebraic Check

Equation 1

$-x + y = 3$

$-(-1) + 2 \stackrel{?}{=} 3$

$3 = 3$ ✓

Equation 2

$3x + y = -1$

$3(-1) + 2 \stackrel{?}{=} -1$

$-1 = -1$ ✓

Graphical Check

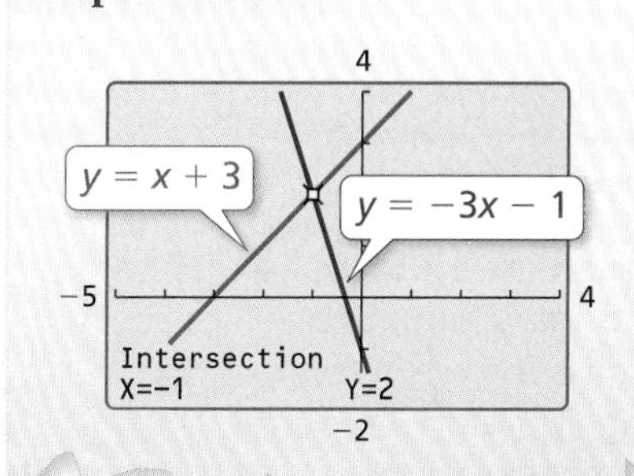

Monitoring Progress  Help in English and Spanish at BigIdeasMath.com

Solve the system of linear equations by substitution. Check your solution.

4. $x + y = -2$
$-3x + y = 6$

5. $-x + y = -4$
$4x - y = 10$

6. $2x - y = -5$
$3x - y = 1$

7. $x - 2y = 7$
$3x - 2y = 3$

Solving Real-Life Problems

EXAMPLE 3 Modeling with Mathematics

A drama club earns \$1040 from a production. A total of 64 adult tickets and 132 student tickets are sold. An adult ticket costs twice as much as a student ticket. Write a system of linear equations that represents this situation. What is the price of each type of ticket?

SOLUTION

1. **Understand the Problem** You know the amount earned, the total numbers of adult and student tickets sold, and the relationship between the price of an adult ticket and the price of a student ticket. You are asked to write a system of linear equations that represents the situation and find the price of each type of ticket.

2. **Make a Plan** Use a verbal model to write a system of linear equations that represents the problem. Then solve the system of linear equations.

3. **Solve the Problem**

Words 64 • Adult ticket price + 132 • Student ticket price = 1040

Adult ticket price = 2 • Student ticket price

Variables Let x be the price (in dollars) of an adult ticket and let y be the price (in dollars) of a student ticket.

System

$64x + 132y = 1040$ Equation 1

$x = 2y$ Equation 2

Step 1 Equation 2 is already solved for x.

Step 2 Substitute $2y$ for x in Equation 1 and solve for y.

$64x + 132y = 1040$ Equation 1

$64(2y) + 132y = 1040$ Substitute $2y$ for x.

$260y = 1040$ Simplify.

$y = 4$ Simplify.

STUDY TIP

You can use either of the original equations to solve for x. However, using Equation 2 requires fewer calculations.

Step 3 Substitute 4 for y in Equation 2 and solve for x.

$x = 2y$ Equation 2

$x = 2(4)$ Substitute 4 for y.

$x = 8$ Simplify.

▶ The solution is (8, 4). So, an adult ticket costs \$8 and a student ticket costs \$4.

4. **Look Back** To check that your solution is correct, substitute the values of x and y into both of the original equations and simplify.

$64(8) + 132(4) = 1040$ $8 = 2(4)$

$1040 = 1040$ ✓ $8 = 8$ ✓

Monitoring Progress

Help in English and Spanish at *BigIdeasMath.com*

8. There are a total of 64 students in a drama club and a yearbook club. The drama club has 10 more students than the yearbook club. Write a system of linear equations that represents this situation. How many students are in each club?

5.2 Exercises

Dynamic Solutions available at *BigIdeasMath.com*

Vocabulary and Core Concept Check

1. **WRITING** Describe how to solve a system of linear equations by substitution.

2. **NUMBER SENSE** When solving a system of linear equations by substitution, how do you decide which variable to solve for in Step 1?

Monitoring Progress and Modeling with Mathematics

In Exercises 3–8, tell which equation you would choose to solve for one of the variables. Explain.

3. $x + 4y = 30$
   $x - 2y = 0$

4. $3x - y = 0$
   $2x + y = -10$

5. $5x + 3y = 11$
   $5x - y = 5$

6. $3x - 2y = 19$
   $x + y = 8$

7. $x - y = -3$
   $4x + 3y = -5$

8. $3x + 5y = 25$
   $x - 2y = -6$

In Exercises 9–16, solve the sytem of linear equations by substitution. Check your solution. *(See Examples 1 and 2.)*

9. $x = 17 - 4y$
   $y = x - 2$

10. $6x - 9 = y$
    $y = -3x$

11. $x = 16 - 4y$
    $3x + 4y = 8$

12. $-5x + 3y = 51$
    $y = 10x - 8$

13. $2x = 12$
    $x - 5y = -29$

14. $2x - y = 23$
    $x - 9 = -1$

15. $5x + 2y = 9$
    $x + y = -3$

16. $11x - 7y = -14$
    $x - 2y = -4$

17. **ERROR ANALYSIS** Describe and correct the error in solving for one of the variables in the linear system $8x + 2y = -12$ and $5x - y = 4$.

✗

Step 1 $5x - y = 4$
$-y = -5x + 4$
$y = 5x - 4$

Step 2 $5x - (5x - 4) = 4$
$5x - 5x + 4 = 4$
$4 = 4$

18. **ERROR ANALYSIS** Describe and correct the error in solving for one of the variables in the linear system $4x + 2y = 6$ and $3x + y = 9$.

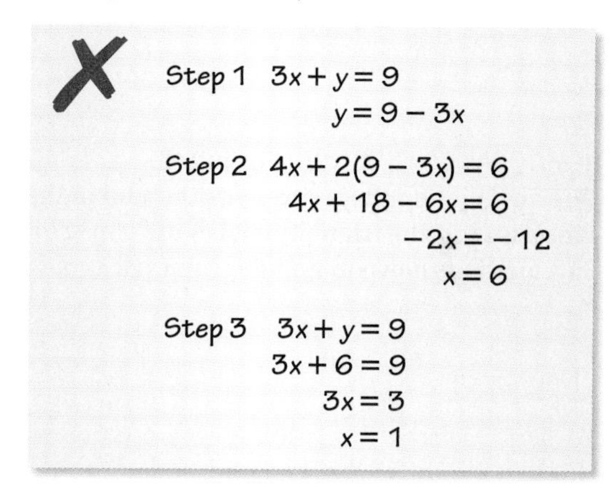

19. **MODELING WITH MATHEMATICS** A farmer plants corn and wheat on a 180-acre farm. The farmer wants to plant three times as many acres of corn as wheat. Write a system of linear equations that represents this situation. How many acres of each crop should the farmer plant? *(See Example 3.)*

20. **MODELING WITH MATHEMATICS** A company that offers tubing trips down a river rents tubes for a person to use and "cooler" tubes to carry food and water. A group spends \$270 to rent a total of 15 tubes. Write a system of linear equations that represents this situation. How many of each type of tube does the group rent?

In Exercises 21–24, write a system of linear equations that has the ordered pair as its solution.

21. $(3, 5)$

22. $(-2, 8)$

23. $(-4, -12)$

24. $(15, -25)$

25. **PROBLEM SOLVING** A math test is worth 100 points and has 38 problems. Each problem is worth either 5 points or 2 points. How many problems of each point value are on the test?

26. **PROBLEM SOLVING** An investor owns shares of Stock A and Stock B. The investor owns a total of 200 shares with a total value of $4000. How many shares of each stock does the investor own?

Stock	Price
A	$9.50
B	$27.00

MATHEMATICAL CONNECTIONS **In Exercises 27 and 28, (a) write an equation that represents the sum of the angle measures of the triangle and (b) use your equation and the equation shown to find the values of *x* and *y*.**

27.

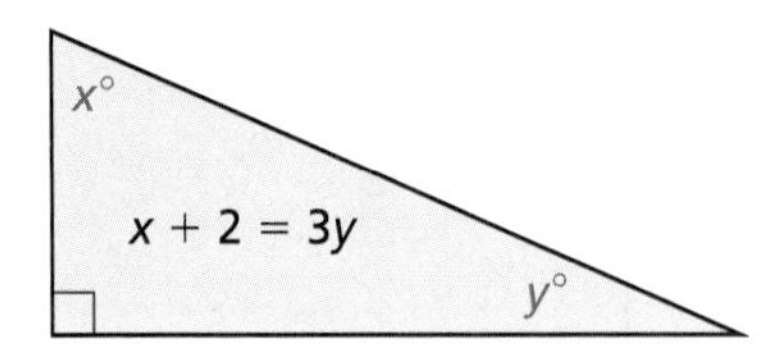

28.

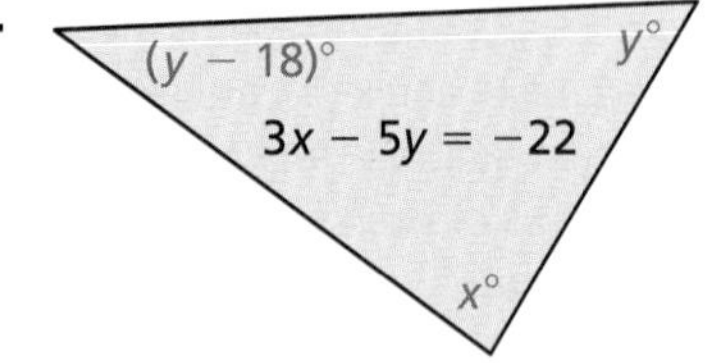

29. **REASONING** Find the values of a and b so that the solution of the linear system is $(-9, 1)$.

$ax + by = -31$ Equation 1
$ax - by = -41$ Equation 2

30. **MAKING AN ARGUMENT** Your friend says that given a linear system with an equation of a horizontal line and an equation of a vertical line, you cannot solve the system by substitution. Is your friend correct? Explain.

31. **OPEN-ENDED** Write a system of linear equations in which $(3, -5)$ is a solution of Equation 1 but not a solution of Equation 2, and $(-1, 7)$ is a solution of the system.

32. **HOW DO YOU SEE IT?** The graphs of two linear equations are shown.

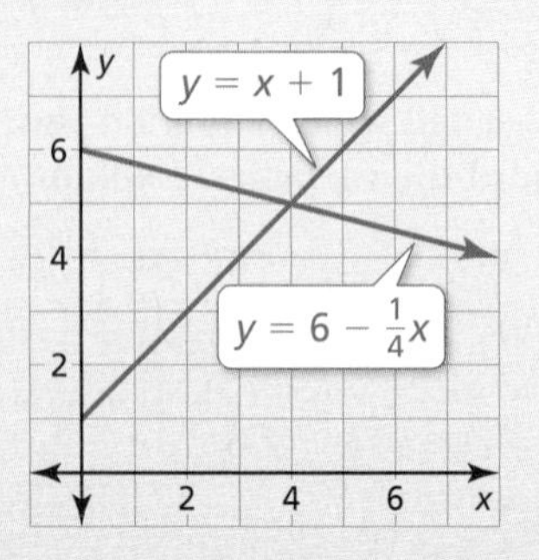

a. At what point do the lines appear to intersect?

b. Could you solve a system of linear equations by substitution to check your answer in part (a)? Explain.

33. **REPEATED REASONING** A radio station plays a total of 272 pop, rock, and hip-hop songs during a day. The number of pop songs is 3 times the number of rock songs. The number of hip-hop songs is 32 more than the number of rock songs. How many of each type of song does the radio station play?

34. **THOUGHT PROVOKING** You have $2.65 in coins. Write a system of equations that represents this situation. Use variables to represent the number of each type of coin.

35. **NUMBER SENSE** The sum of the digits of a two-digit number is 11. When the digits are reversed, the number increases by 27. Find the original number.

Maintaining Mathematical Proficiency

Reviewing what you learned in previous grades and lessons

Find the sum or difference. (*Skills Review Handbook*)

36. $(x - 4) + (2x - 7)$

37. $(5y - 12) + (-5y - 1)$

38. $(t - 8) - (t + 15)$

39. $(6d + 2) - (3d - 3)$

40. $4(m + 2) + 3(6m - 4)$

41. $2(5v + 6) - 6(-9v + 2)$

5.3 Solving Systems of Linear Equations by Elimination

Essential Question

How can you use elimination to solve a system of linear equations?

EXPLORATION 1 Writing and Solving a System of Equations

Work with a partner. You purchase a drink and a sandwich for $4.50. Your friend purchases a drink and five sandwiches for $16.50. You want to determine the price of a drink and the price of a sandwich.

a. Let x represent the price (in dollars) of one drink. Let y represent the price (in dollars) of one sandwich. Write a system of equations for the situation. Use the following verbal model.

Number of drinks	$\cdot$	Price per drink	$+$	Number of sandwiches	$\cdot$	Price per sandwich	$=$	Total price

Label one of the equations Equation 1 and the other equation Equation 2.

b. Subtract Equation 1 from Equation 2. Explain how you can use the result to solve the system of equations. Then find and interpret the solution.

CHANGING COURSE

To be proficient in math, you need to monitor and evaluate your progress and change course using a different solution method, if necessary.

EXPLORATION 2 Using Elimination to Solve Systems

Work with a partner. Solve each system of linear equations using two methods.

Method 1 Subtract. Subtract Equation 2 from Equation 1. Then use the result to solve the system.

Method 2 Add. Add the two equations. Then use the result to solve the system.

Is the solution the same using both methods? Which method do you prefer?

a. $3x - y = 6$
$3x + y = 0$

b. $2x + y = 6$
$2x - y = 2$

c. $x - 2y = -7$
$x + 2y = 5$

EXPLORATION 3 Using Elimination to Solve a System

Work with a partner.

$2x + y = 7$ Equation 1

$x + 5y = 17$ Equation 2

a. Can you eliminate a variable by adding or subtracting the equations as they are? If not, what do you need to do to one or both equations so that you can?

b. Solve the system individually. Then exchange solutions with your partner and compare and check the solutions.

Communicate Your Answer

4. How can you use elimination to solve a system of linear equations?

5. When can you add or subtract the equations in a system to solve the system? When do you have to multiply first? Justify your answers with examples.

6. In Exploration 3, why can you multiply an equation in the system by a constant and not change the solution of the system? Explain your reasoning.

5.3 Lesson

Core Vocabulary

Previous
coefficient

What You Will Learn

- Solve systems of linear equations by elimination.
- Use systems of linear equations to solve real-life problems.

Solving Linear Systems by Elimination

Core Concept

Solving a System of Linear Equations by Elimination

Step 1 Multiply, if necessary, one or both equations by a constant so at least one pair of like terms has the same or opposite coefficients.

Step 2 Add or subtract the equations to eliminate one of the variables.

Step 3 Solve the resulting equation.

Step 4 Substitute the value from Step 3 into one of the original equations and solve for the other variable.

You can use elimination to solve a system of equations because replacing one equation in the system with the sum of that equation and a multiple of the other produces a system that has the same solution. Here is why.

System 1

$a = b$	Equation 1
$c = d$	Equation 2

System 2

$a + kc = b + kd$	Equation 3
$c = d$	Equation 2

Consider System 1. In this system, a and c are algebraic expressions, and b and d are constants. Begin by multiplying each side of Equation 2 by a constant k. By the Multiplication Property of Equality, $kc = kd$. You can rewrite Equation 1 as Equation 3 by adding kc on the left and kd on the right. You can rewrite Equation 3 as Equation 1 by subtracting kc on the left and kd on the right. Because you can rewrite either system as the other, System 1 and System 2 have the same solution.

EXAMPLE 1 Solving a System of Linear Equations by Elimination

Solve the system of linear equations by elimination.

$3x + 2y = 4$	Equation 1
$3x - 2y = -4$	Equation 2

SOLUTION

Step 1 Because the coefficients of the y-terms are opposites, you do not need to multiply either equation by a constant.

Step 2 Add the equations.

$3x + 2y = 4$	Equation 1
$\underline{3x - 2y = -4}$	Equation 2
$6x \qquad = 0$	Add the equations.

Step 3 Solve for x.

$6x = 0$	Resulting equation from Step 2
$x = 0$	Divide each side by 6.

Step 4 Substitute 0 for x in one of the original equations and solve for y.

$3x + 2y = 4$	Equation 1
$3(0) + 2y = 4$	Substitute 0 for x.
$y = 2$	Solve for y.

▶ The solution is (0, 2).

Check

Equation 1

$$3x + 2y = 4$$
$$3(0) + 2(2) \stackrel{?}{=} 4$$
$$4 = 4 \checkmark$$

Equation 2

$$3x - 2y = -4$$
$$3(0) - 2(2) \stackrel{?}{=} -4$$
$$-4 = -4 \checkmark$$

EXAMPLE 2 Solving a System of Linear Equations by Elimination

Solve the system of linear equations by elimination.

$-10x + 3y = 1$ Equation 1

$-5x - 6y = 23$ Equation 2

ANOTHER WAY

To use subtraction to eliminate one of the variables, multiply Equation 2 by 2 and then subtract the equations.

$$\begin{array}{r} -10x + 3y = 1 \\ \underline{-(-10x - 12y = 46)} \\ 15y = -45 \end{array}$$

SOLUTION

Step 1 Multiply Equation 2 by -2 so that the coefficients of the x-terms are opposites.

$-10x + 3y = 1$ → $-10x + 3y = 1$ Equation 1

$-5x - 6y = 23$ **Multiply by −2.** → $10x + 12y = -46$ Revised Equation 2

Step 2 Add the equations.

$$\begin{array}{rl} -10x + 3y = 1 & \text{Equation 1} \\ \underline{10x + 12y = -46} & \text{Revised Equation 2} \\ 15y = -45 & \text{Add the equations.} \end{array}$$

Step 3 Solve for y.

$15y = -45$ Resulting equation from Step 2

$y = -3$ Divide each side by 15.

Step 4 Substitute -3 for y in one of the original equations and solve for x.

$-5x - 6y = 23$ Equation 2

$-5x - 6(-3) = 23$ Substitute -3 for y.

$-5x + 18 = 23$ Multiply.

$-5x = 5$ Subtract 18 from each side.

$x = -1$ Divide each side by -5.

▶ The solution is $(-1, -3)$.

Check

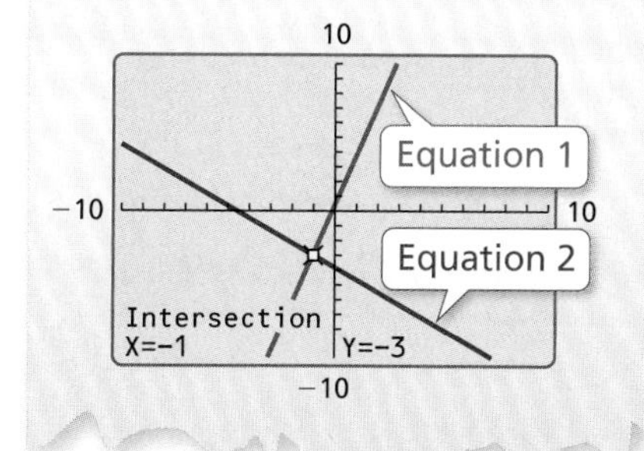

Monitoring Progress

Help in English and Spanish at *BigIdeasMath.com*

Solve the system of linear equations by elimination. Check your solution.

1. $3x + 2y = 7$
$-3x + 4y = 5$

2. $x - 3y = 24$
$3x + y = 12$

3. $x + 4y = 22$
$4x + y = 13$

Concept Summary

Methods for Solving Systems of Linear Equations

Method	When to Use
Graphing *(Lesson 5.1)*	To estimate solutions
Substitution *(Lesson 5.2)*	When one of the variables in one of the equations has a coefficient of 1 or -1
Elimination *(Lesson 5.3)*	When at least one pair of like terms has the same or opposite coefficients
Elimination (Multiply First) *(Lesson 5.3)*	When one of the variables cannot be eliminated by adding or subtracting the equations

Solving Real-Life Problems

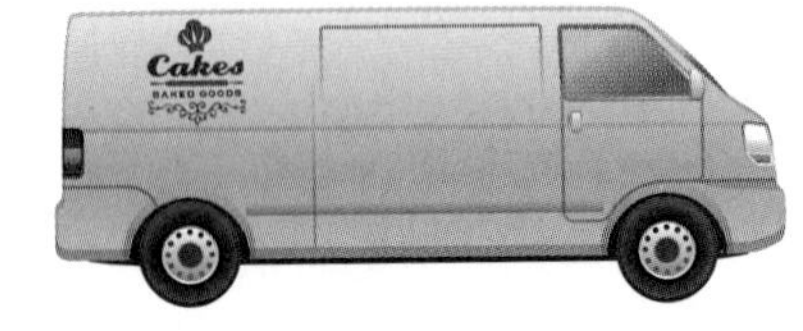

EXAMPLE 3 Modeling with Mathematics

A business with two locations buys seven large delivery vans and five small delivery vans. Location A receives five large vans and two small vans for a total cost of \$235,000. Location B receives two large vans and three small vans for a total cost of \$160,000. What is the cost of each type of van?

SOLUTION

1. **Understand the Problem** You know how many of each type of van each location receives. You also know the total cost of the vans for each location. You are asked to find the cost of each type of van.
2. **Make a Plan** Use a verbal model to write a system of linear equations that represents the problem. Then solve the system of linear equations.
3. **Solve the Problem**

Words 5 • Cost of large van + 2 • Cost of small van = 235,000

2 • Cost of large van + 3 • Cost of small van = 160,000

Variables Let x be the cost (in dollars) of a large van and let y be the cost (in dollars) of a small van.

System $5x + 2y = 235{,}000$ Equation 1

$2x + 3y = 160{,}000$ Equation 2

Step 1 Multiply Equation 1 by -3. Multiply Equation 2 by 2.

$5x + 2y = 235{,}000$ **Multiply by −3.** $-15x - 6y = -705{,}000$ Revised Equation 1

$2x + 3y = 160{,}000$ **Multiply by 2.** $4x + 6y = 320{,}000$ Revised Equation 2

Step 2 Add the equations.

$-15x - 6y = -705{,}000$ Revised Equation 1

$4x + 6y = 320{,}000$ Revised Equation 2

$-11x = -385{,}000$ Add the equations.

Step 3 Solving the equation $-11x = -385{,}000$ gives $x = 35{,}000$.

Step 4 Substitute 35,000 for x in one of the original equations and solve for y.

$5x + 2y = 235{,}000$ Equation 1

$5(35{,}000) + 2y = 235{,}000$ Substitute 35,000 for x.

$y = 30{,}000$ Solve for y.

▶ The solution is (35,000, 30,000). So, a large van costs \$35,000 and a small van costs \$30,000.

4. **Look Back** Check to make sure your solution makes sense with the given information. For Location A, the total cost is $5(35{,}000) + 2(30{,}000) = \$235{,}000$. For Location B, the total cost is $2(35{,}000) + 3(30{,}000) = \$160{,}000$. So, the solution makes sense.

STUDY TIP

In Example 3, both equations are multiplied by a constant so that the coefficients of the y-terms are opposites.

Monitoring Progress  Help in English and Spanish at *BigIdeasMath.com*

4. Solve the system in Example 3 by eliminating x.

5.3 Exercises

Dynamic Solutions available at *BigIdeasMath.com*

Vocabulary and Core Concept Check

1. **OPEN-ENDED** Give an example of a system of linear equations that can be solved by first adding the equations to eliminate one variable.

2. **WRITING** Explain how to solve the system of linear equations by elimination.

   $2x - 3y = -4$ Equation 1
   $-5x + 9y = 7$ Equation 2

Monitoring Progress and Modeling with Mathematics

In Exercises 3–10, solve the system of linear equations by elimination. Check your solution. *(See Example 1.)*

3. $x + 2y = 13$
   $-x + y = 5$

4. $9x + y = 2$
   $-4x - y = -17$

5. $5x + 6y = 50$
   $x - 6y = -26$

6. $-x + y = 4$
   $x + 3y = 4$

7. $-3x - 5y = -7$
   $-4x + 5y = 14$

8. $4x - 9y = -21$
   $-4x - 3y = 9$

9. $-y - 10 = 6x$
   $5x + y = -10$

10. $3x - 30 = y$
    $7y - 6 = 3x$

In Exercises 11–18, solve the system of linear equations by elimination. Check your solution. *(See Examples 2 and 3.)*

11. $x + y = 2$
    $2x + 7y = 9$

12. $8x - 5y = 11$
    $4x - 3y = 5$

13. $11x - 20y = 28$
    $3x + 4y = 36$

14. $10x - 9y = 46$
    $-2x + 3y = 10$

15. $4x - 3y = 8$
    $5x - 2y = -11$

16. $-2x - 5y = 9$
    $3x + 11y = 4$

17. $9x + 2y = 39$
    $6x + 13y = -9$

18. $12x - 7y = -2$
    $8x + 11y = 30$

19. **ERROR ANALYSIS** Describe and correct the error in solving for one of the variables in the linear system $5x - 7y = 16$ and $x + 7y = 8$.

    ✗
    $5x - 7y = 16$
    $x + 7y = 8$
    $4x = 24$
    $x = 6$

20. **ERROR ANALYSIS** Describe and correct the error in solving for one of the variables in the linear system $4x + 3y = 8$ and $x - 2y = -13$.

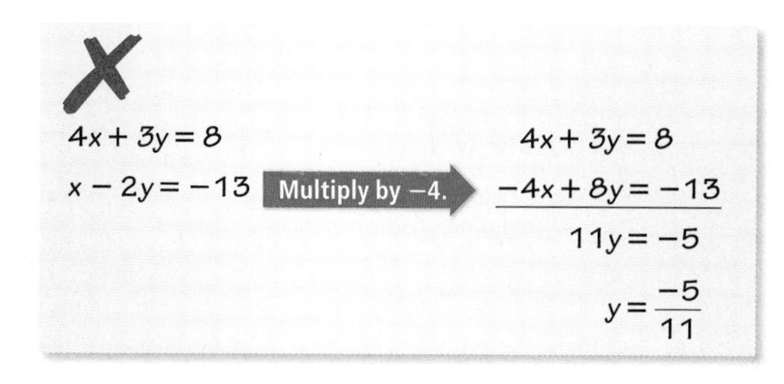

21. **MODELING WITH MATHEMATICS** A service center charges a fee of x dollars for an oil change plus y dollars per quart of oil used. A sample of its sales record is shown. Write a system of linear equations that represents this situation. Find the fee and cost per quart of oil.

	A	B	C
1	Customer	Oil Tank Size (quarts)	Total Cost
2	A	5	$22.45
3	B	7	$25.45

22. **MODELING WITH MATHEMATICS** A music website charges x dollars for individual songs and y dollars for entire albums. Person A pays $25.92 to download 6 individual songs and 2 albums. Person B pays $33.93 to download 4 individual songs and 3 albums. Write a system of linear equations that represents this situation. How much does the website charge to download a song? an entire album?

In Exercises 23–26, solve the system of linear equations using any method. Explain why you chose the method.

23. $3x + 2y = 4$
$2y = 8 - 5x$

24. $-6y + 2 = -4x$
$y - 2 = x$

25. $y - x = 2$
$y = -\frac{1}{4}x + 7$

26. $3x + y = \frac{1}{3}$
$2x - 3y = \frac{8}{3}$

27. WRITING For what values of a can you solve the linear system $ax + 3y = 2$ and $4x + 5y = 6$ by elimination without multiplying first? Explain.

28. HOW DO YOU SEE IT? The circle graph shows the results of a survey in which 50 students were asked about their favorite meal.

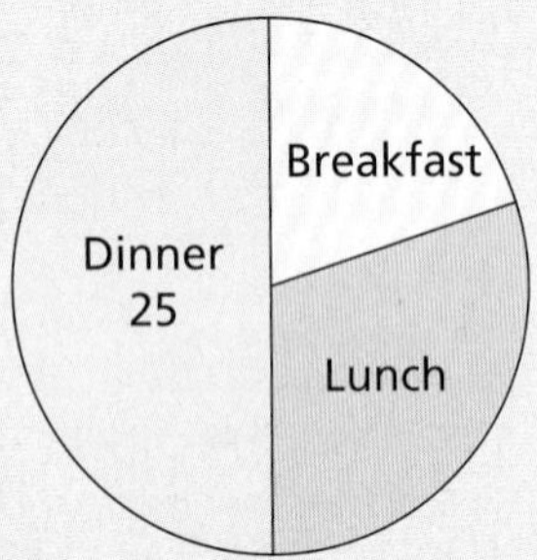

a. Estimate the numbers of students who chose breakfast and lunch.

b. The number of students who chose lunch was 5 more than the number of students who chose breakfast. Write a system of linear equations that represents the numbers of students who chose breakfast and lunch.

c. Explain how you can solve the linear system in part (b) to check your answers in part (a).

29. MAKING AN ARGUMENT Your friend says that any system of equations that can be solved by elimination can be solved by substitution in an equal or fewer number of steps. Is your friend correct? Explain.

30. THOUGHT PROVOKING Write a system of linear equations that can be added to eliminate a variable *or* subtracted to eliminate a variable.

31. MATHEMATICAL CONNECTIONS A rectangle has a perimeter of 18 inches. A new rectangle is formed by doubling the width w and tripling the length ℓ, as shown. The new rectangle has a perimeter P of 46 inches.

$P = 46$ in. | $2w$ | 3ℓ

a. Write and solve a system of linear equations to find the length and width of the original rectangle.

b. Find the length and width of the new rectangle.

32. CRITICAL THINKING Refer to the discussion of System 1 and System 2 on page 248. Without solving, explain why the two systems shown have the same solution.

System 1		**System 2**	
$3x - 2y = 8$	Equation 1	$5x = 20$	Equation 3
$x + y = 6$	Equation 2	$x + y = 6$	Equation 2

33. PROBLEM SOLVING You are making 6 quarts of fruit punch for a party. You have bottles of 100% fruit juice and 20% fruit juice. How many quarts of each type of juice should you mix to make 6 quarts of 80% fruit juice?

34. PROBLEM SOLVING A motorboat takes 40 minutes to travel 20 miles downstream. The return trip takes 60 minutes. What is the speed of the current?

35. CRITICAL THINKING Solve for x, y, and z in the system of equations. Explain your steps.

$x + 7y + 3z = 29$ Equation 1
$3z + x - 2y = -7$ Equation 2
$5y = 10 - 2x$ Equation 3

Maintaining Mathematical Proficiency

Reviewing what you learned in previous grades and lessons

Solve the equation. Determine whether the equation has *one solution*, *no solution*, or *infinitely many solutions*. *(Section 1.3)*

36. $5d - 8 = 1 + 5d$

37. $9 + 4t = 12 - 4t$

38. $3n + 2 = 2(n - 3)$

39. $-3(4 - 2v) = 6v - 12$

Write an equation of the line that passes through the given point and is parallel to the given line. *(Section 4.3)*

40. $(4, -1);\ y = -2x + 7$

41. $(0, 6);\ y = 5x - 3$

42. $(-5, -2);\ y = \frac{2}{3}x + 1$

5.4 Solving Special Systems of Linear Equations

Essential Question

Can a system of linear equations have no solution or infinitely many solutions?

EXPLORATION 1 Using a Table to Solve a System

Work with a partner. You invest \$450 for equipment to make skateboards. The materials for each skateboard cost \$20. You sell each skateboard for \$20.

a. Write the cost and revenue equations. Then copy and complete the table for your cost C and your revenue R.

x (skateboards)	0	1	2	3	4	5	6	7	8	9	10
C (dollars)											
R (dollars)											

b. When will your company break even? What is wrong?

MODELING WITH MATHEMATICS

To be proficient in math, you need to interpret mathematical results in real-life contexts.

EXPLORATION 2 Writing and Analyzing a System

Work with a partner. A necklace and matching bracelet have two types of beads. The necklace has 40 small beads and 6 large beads and weighs 10 grams. The bracelet has 20 small beads and 3 large beads and weighs 5 grams. The threads holding the beads have no significant weight.

a. Write a system of linear equations that represents the situation. Let x be the weight (in grams) of a small bead and let y be the weight (in grams) of a large bead.

b. Graph the system in the coordinate plane shown. What do you notice about the two lines?

c. Can you find the weight of each type of bead? Explain your reasoning.

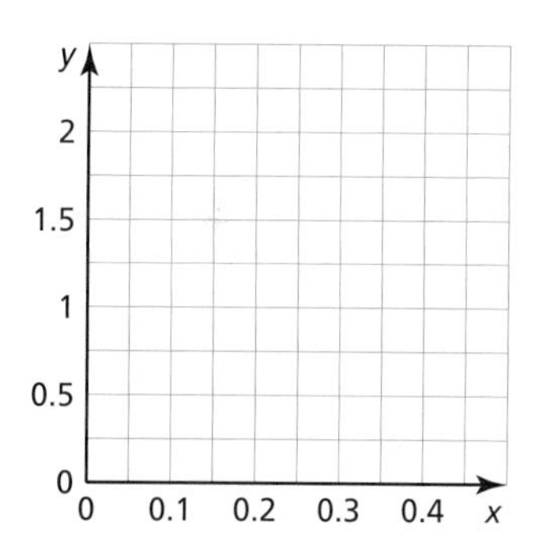

Communicate Your Answer

3. Can a system of linear equations have no solution or infinitely many solutions? Give examples to support your answers.

4. Does the system of linear equations represented by each graph have *no solution*, *one solution*, or *infinitely many solutions*? Explain.

a.

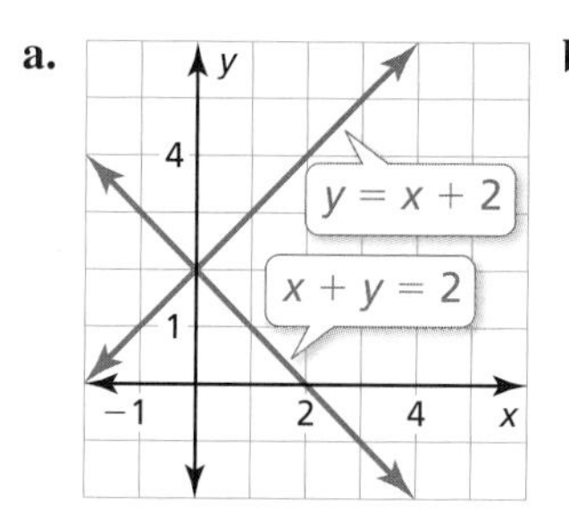

b.

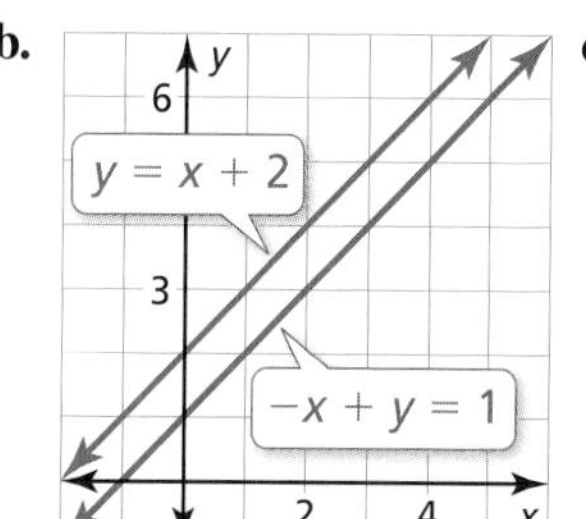

c.

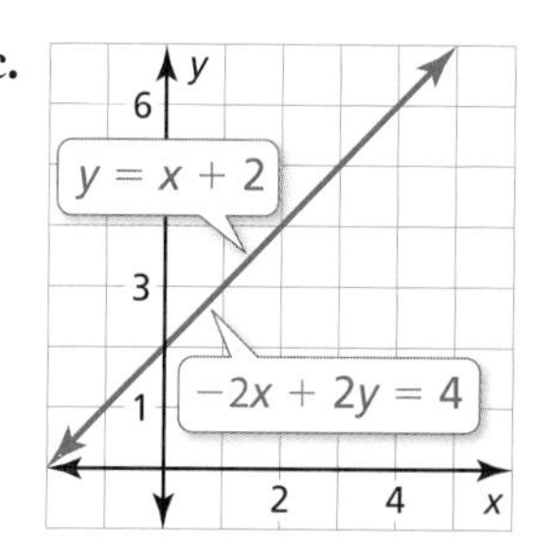

5.4 Lesson

Core Vocabulary

Previous
parallel

What You Will Learn

- Determine the numbers of solutions of linear systems.
- Use linear systems to solve real-life problems.

The Numbers of Solutions of Linear Systems

Core Concept

Solutions of Systems of Linear Equations

A system of linear equations can have *one solution*, *no solution*, or *infinitely many solutions*.

One solution | **No solution** | **Infinitely many solutions**

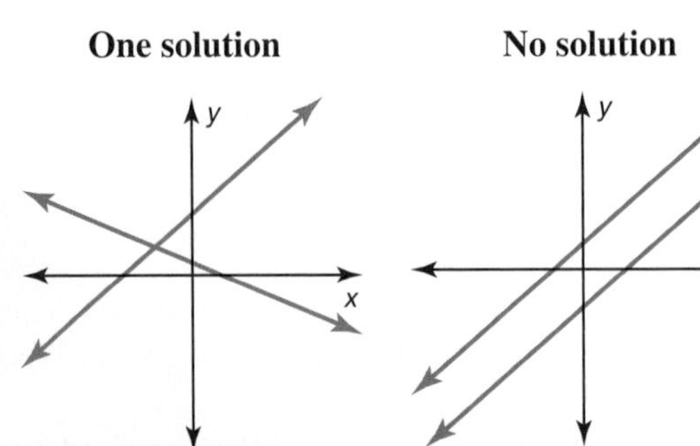

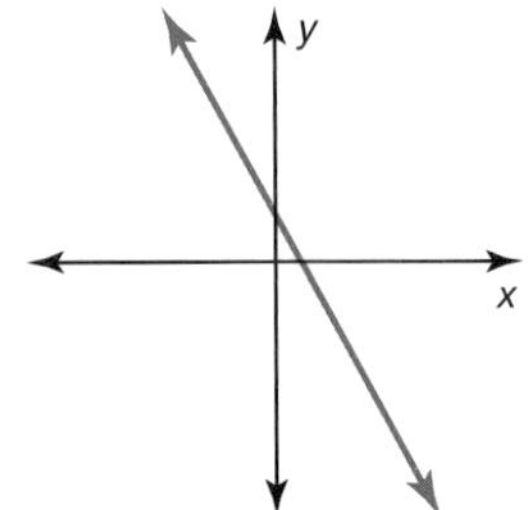

The lines intersect. | The lines are parallel. | The lines are the same.

ANOTHER WAY

You can solve some linear systems by inspection. In Example 1, notice you can rewrite the system as

$-2x + y = 1$
$-2x + y = -5.$

This system has no solution because $-2x + y$ cannot be equal to both 1 and -5.

EXAMPLE 1 Solving a System: No Solution

Solve the system of linear equations.

$y = 2x + 1$ Equation 1
$y = 2x - 5$ Equation 2

SOLUTION

Method 1 Solve by graphing.

Graph each equation.

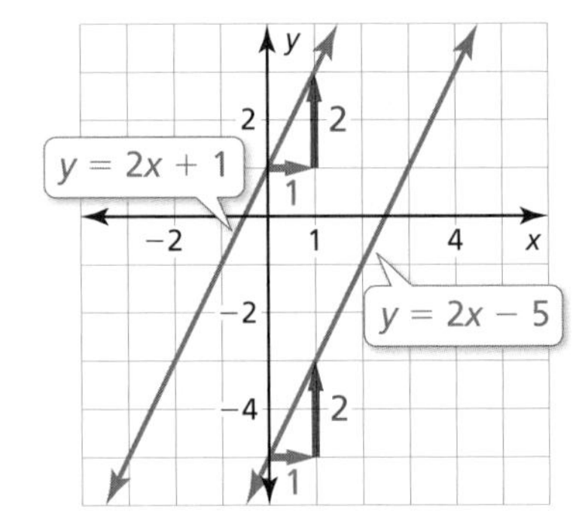

The lines have the same slope and different y-intercepts. So, the lines are parallel.

Because parallel lines do not intersect, there is no point that is a solution of both equations.

▶ So, the system of linear equations has no solution.

Method 2 Solve by substitution.

Substitute $2x - 5$ for y in Equation 1.

$y = 2x + 1$ Equation 1
$2x - 5 = 2x + 1$ Substitute $2x - 5$ for y.
$-5 = 1$  Subtract $2x$ from each side.

▶ The equation $-5 = 1$ is never true. So, the system of linear equations has no solution.

STUDY TIP

A linear system with no solution is called an *inconsistent system.*

EXAMPLE 2 Solving a System: Infinitely Many Solutions

Solve the system of linear equations.

$-2x + y = 3$ Equation 1

$-4x + 2y = 6$ Equation 2

SOLUTION

Method 1 Solve by graphing.

Graph each equation.

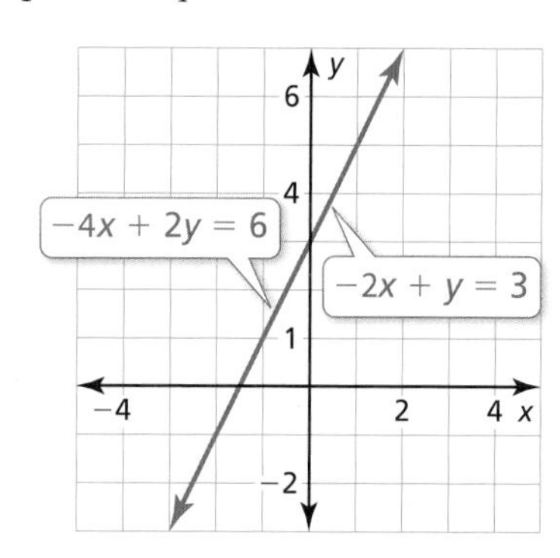

The lines have the same slope and the same y-intercept. So, the lines are the same. Because the lines are the same, all points on the line are solutions of both equations.

▶ So, the system of linear equations has infinitely many solutions.

Method 2 Solve by elimination.

Step 1 Multiply Equation 1 by −2.

$-2x + y = 3$ **Multiply by −2.** → $4x - 2y = -6$ Revised Equation 1

$-4x + 2y = 6$ → $-4x + 2y = 6$ Equation 2

Step 2 Add the equations.

$4x - 2y = -6$ Revised Equation 1

$\underline{-4x + 2y = 6}$ Equation 2

$0 = 0$ Add the equations.

STUDY TIP

A linear system with infinitely many solutions is called a *consistent dependent system.*

▶ The equation $0 = 0$ is always true. So, the solutions are all the points on the line $-2x + y = 3$. The system of linear equations has infinitely many solutions.

Monitoring Progress  Help in English and Spanish at *BigIdeasMath.com*

Solve the system of linear equations.

1. $x + y = 3$
$2x + 2y = 6$

2. $y = -x + 3$
$2x + 2y = 4$

3. $x + y = 3$
$x + 2y = 4$

4. $y = -10x + 2$
$10x + y = 10$

Solving Real-Life Problems

Modeling with Mathematics

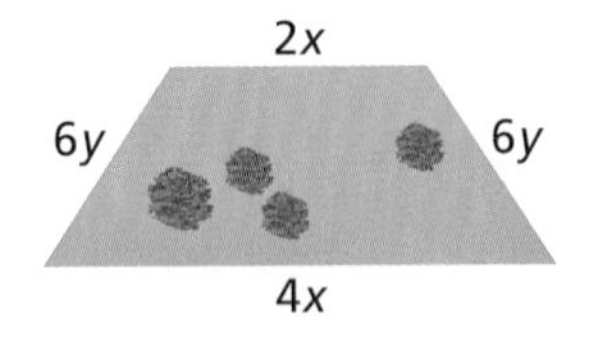

18y

9x 9x

18y

The perimeter of the trapezoidal piece of land is 48 kilometers. The perimeter of the rectangular piece of land is 144 kilometers. Write and solve a system of linear equations to find the values of x and y.

SOLUTION

1. **Understand the Problem** You know the perimeter of each piece of land and the side lengths in terms of x or y. You are asked to write and solve a system of linear equations to find the values of x and y.
2. **Make a Plan** Use the figures and the definition of perimeter to write a system of linear equations that represents the problem. Then solve the system of linear equations.
3. **Solve the Problem**

Perimeter of trapezoid

$2x + 4x + 6y + 6y = 48$

$6x + 12y = 48$ Equation 1

Perimeter of rectangle

$9x + 9x + 18y + 18y = 144$

$18x + 36y = 144$ Equation 2

System $6x + 12y = 48$ Equation 1

$18x + 36y = 144$ Equation 2

Method 1 Solve by graphing.

Graph each equation.

The lines have the same slope and the same y-intercept. So, the lines are the same.

In this context, x and y must be positive. Because the lines are the same, all the points on the line in Quadrant I are solutions of both equations.

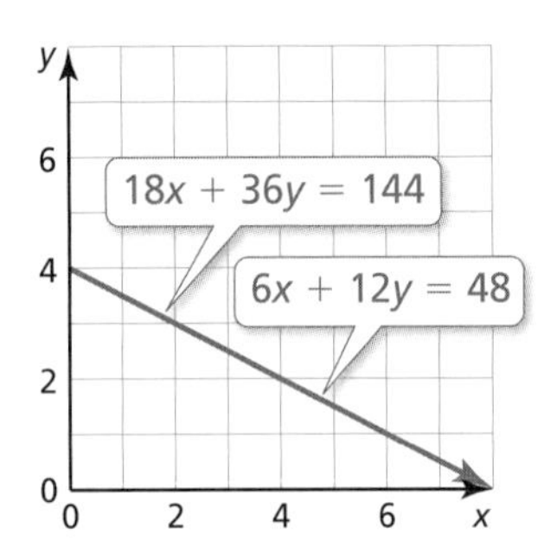

▶ So, the system of linear equations has infinitely many solutions.

Method 2 Solve by elimination.

Multiply Equation 1 by -3 and add the equations.

$6x + 12y = 48$	Multiply by −3.	$-18x - 36y = -144$	Revised Equation 1
$18x + 36y = 144$		$18x + 36y = 144$	Equation 2
		$0 = 0$	Add the equations.

▶ The equation $0 = 0$ is always true. In this context, x and y must be positive. So, the solutions are all the points on the line $6x + 12y = 48$ in Quadrant I. The system of linear equations has infinitely many solutions.

4. **Look Back** Choose a few of the ordered pairs (x, y) that are solutions of Equation 1. You should find that no matter which ordered pairs you choose, they will also be solutions of Equation 2. So, *infinitely many solutions* seems reasonable.

Monitoring Progress  Help in English and Spanish at *BigIdeasMath.com*

5. **WHAT IF?** What happens to the solution in Example 3 when the perimeter of the trapezoidal piece of land is 96 kilometers? Explain.

5.4 Exercises

Dynamic Solutions available at *BigIdeasMath.com*

Vocabulary and Core Concept Check

1. **REASONING** Is it possible for a system of linear equations to have exactly two solutions? Explain.

2. **WRITING** Compare the graph of a system of linear equations that has infinitely many solutions and the graph of a system of linear equations that has no solution.

Monitoring Progress and Modeling with Mathematics

In Exercises 3–8, match the system of linear equations with its graph. Then determine whether the system has *one solution*, *no solution*, or *infinitely many solutions*.

3. $-x + y = 1$
   $x - y = 1$

4. $2x - 2y = 4$
   $-x + y = -2$

5. $2x + y = 4$
   $-4x - 2y = -8$

6. $x - y = 0$
   $5x - 2y = 6$

7. $-2x + 4y = 1$
   $3x - 6y = 9$

8. $5x + 3y = 17$
   $x - 3y = -2$

A.
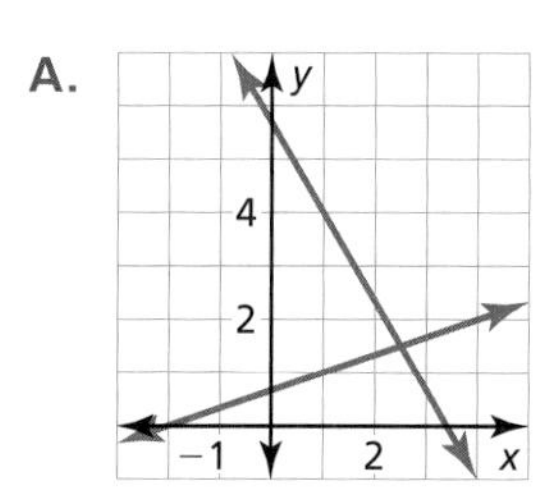

B.
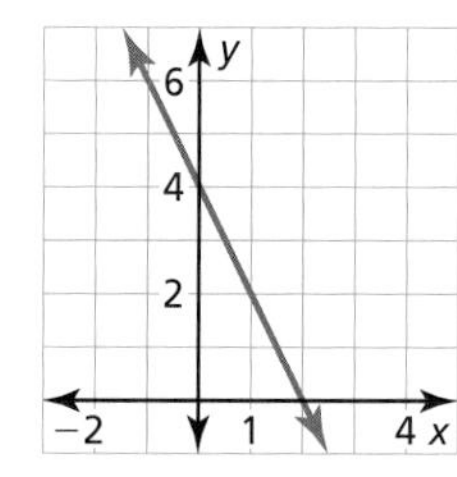

C.
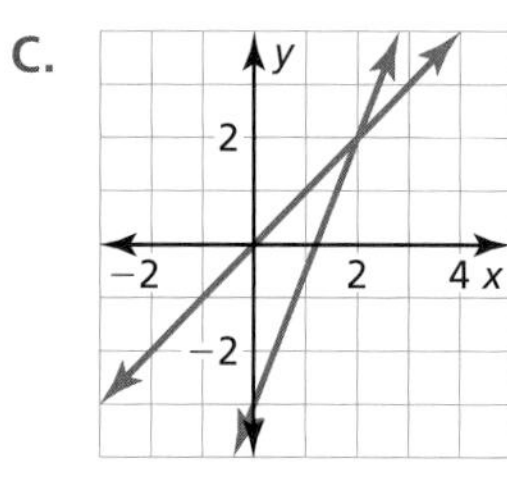

D.
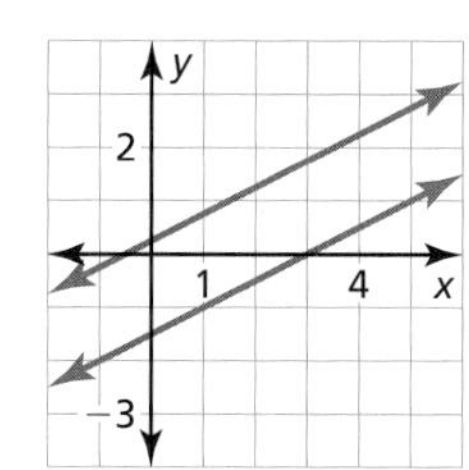

E.
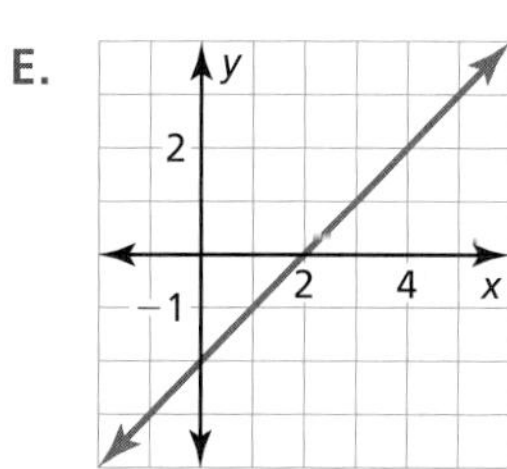

F.
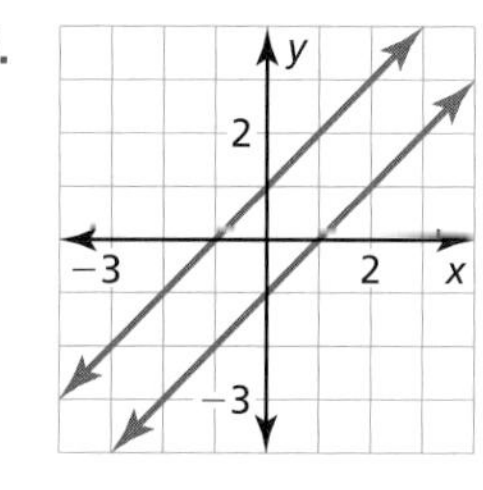

In Exercises 9–16, solve the system of linear equations. *(See Examples 1 and 2.)*

9. $y = -2x - 4$
   $y = 2x - 4$

10. $y = -6x - 8$
    $y = -6x + 8$

11. $3x - y = 6$
    $-3x + y = -6$

12. $-x + 2y = 7$
    $x - 2y = 7$

13. $4x + 4y = -8$
    $-2x - 2y = 4$

14. $15x - 5y = -20$
    $-3x + y = 4$

15. $9x - 15y = 24$
    $6x - 10y = -16$

16. $3x - 2y = -5$
    $4x + 5y = 47$

In Exercises 17–22, use only the slopes and y-intercepts of the graphs of the equations to determine whether the system of linear equations has *one solution*, *no solution*, or *infinitely many solutions*. Explain.

17. $y = 7x + 13$
    $-21x + 3y = 39$

18. $y = -6x - 2$
    $12x + 2y = -6$

19. $4x + 3y = 27$
    $4x - 3y = -27$

20. $-7x + 7y = 1$
    $2x - 2y = -18$

21. $-18x + 6y = 24$
    $3x - y = -2$

22. $2x - 2y = 16$
    $3x - 6y = 30$

ERROR ANALYSIS In Exercises 23 and 24, describe and correct the error in solving the system of linear equations.

23.
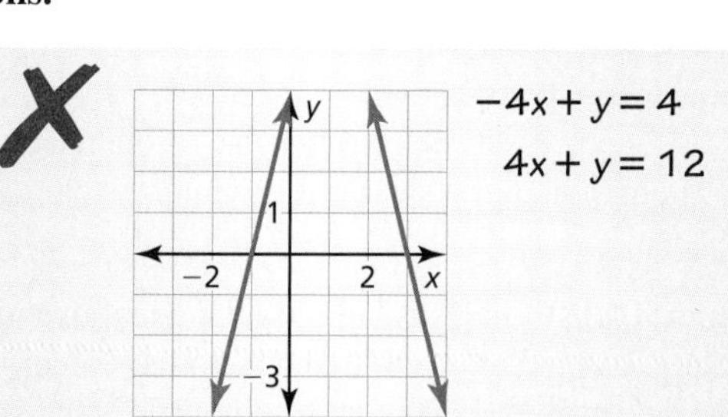

The lines do not intersect. So, the system has no solution.

24.
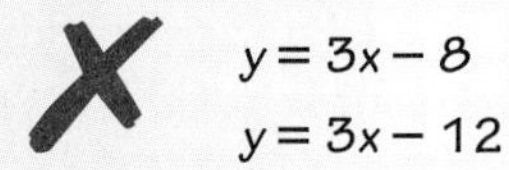

$y = 3x - 8$
$y = 3x - 12$

The lines have the same slope. So, the system has infinitely many solutions.

25. **MODELING WITH MATHEMATICS** A small bag of trail mix contains 3 cups of dried fruit and 4 cups of almonds. A large bag contains $4\frac{1}{2}$ cups of dried fruit and 6 cups of almonds. Write and solve a system of linear equations to find the price of 1 cup of dried fruit and 1 cup of almonds. *(See Example 3.)*

$9

$6

26. **MODELING WITH MATHEMATICS** In a canoe race, Team A is traveling 6 miles per hour and is 2 miles ahead of Team B. Team B is also traveling 6 miles per hour. The teams continue traveling at their current rates for the remainder of the race. Write a system of linear equations that represents this situation. Will Team B catch up to Team A? Explain.

27. **PROBLEM SOLVING** A train travels from New York City to Washington, D.C., and then back to New York City. The table shows the number of tickets purchased for each leg of the trip. The cost per ticket is the same for each leg of the trip. Is there enough information to determine the cost of one coach ticket? Explain.

Destination	Coach tickets	Business class tickets	Money collected (dollars)
Washington, D.C.	150	80	22,860
New York City	170	100	27,280

28. **THOUGHT PROVOKING** Write a system of three linear equations in two variables so that any two of the equations have exactly one solution, but the entire system of equations has no solution.

29. **REASONING** In a system of linear equations, one equation has a slope of 2 and the other equation has a slope of $-\frac{1}{3}$. How many solutions does the system have? Explain.

30. **HOW DO YOU SEE IT?** The graph shows information about the last leg of a 4×200-meter relay for three relay teams. Team A's runner ran about 7.8 meters per second, Team B's runner ran about 7.8 meters per second, and Team C's runner ran about 8.8 meters per second.

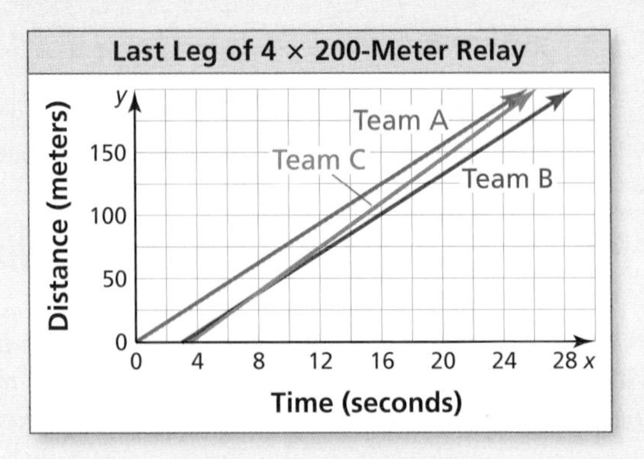

a. Estimate the distance at which Team C's runner passed Team B's runner.

b. If the race was longer, could Team C's runner have passed Team A's runner? Explain.

c. If the race was longer, could Team B's runner have passed Team A's runner? Explain.

31. **ABSTRACT REASONING** Consider the system of linear equations $y = ax + 4$ and $y = bx - 2$, where a and b are real numbers. Determine whether each statement is *always*, *sometimes*, or *never* true. Explain your reasoning.

a. The system has infinitely many solutions.

b. The system has no solution.

c. When $a < b$, the system has one solution.

32. **MAKING AN ARGUMENT** One admission to an ice skating rink costs x dollars, and renting a pair of ice skates costs y dollars. Your friend says she can determine the exact cost of one admission and one skate rental. Is your friend correct? Explain.

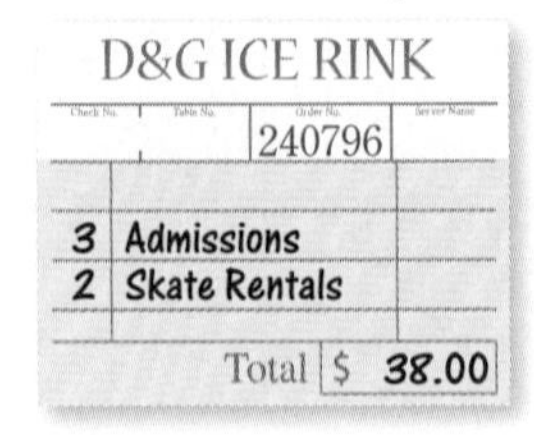

Maintaining Mathematical Proficiency

Reviewing what you learned in previous grades and lessons

Solve the equation. Check your solutions. *(Section 1.4)*

33. $|2x + 6| = |x|$

34. $|3x - 45| = |12x|$

35. $|x - 7| = |2x - 8|$

36. $|2x + 1| = |3x - 11|$

5.1–5.4 What Did You Learn?

Core Vocabulary

system of linear equations, *p. 236*

solution of a system of linear equations, *p. 236*

Core Concepts

Section 5.1

Solving a System of Linear Equations by Graphing, *p. 237*

Section 5.2

Solving a System of Linear Equations by Substitution, *p. 242*

Section 5.3

Solving a System of Linear Equations by Elimination, *p. 248*

Section 5.4

Solutions of Systems of Linear Equations, *p. 254*

Mathematical Practices

1. Describe the given information in Exercise 33 on page 246 and your plan for finding the solution.
2. Describe another real-life situation similar to Exercise 22 on page 251 and the mathematics that you can apply to solve the problem.
3. What question(s) can you ask your friend to help her understand the error in the statement she made in Exercise 32 on page 258?

Study Skills

Analyzing Your Errors

Study Errors

What Happens: You do not study the right material or you do not learn it well enough to remember it on a test without resources such as notes.

How to Avoid This Error: Take a practice test. Work with a study group. Discuss the topics on the test with your teacher. Do not try to learn a whole chapter's worth of material in one night.

5.1–5.4 Quiz

Use the graph to solve the system of linear equations. Check your solution. *(Section 5.1)*

1. $y = -\frac{1}{3}x + 2$
$y = x - 2$

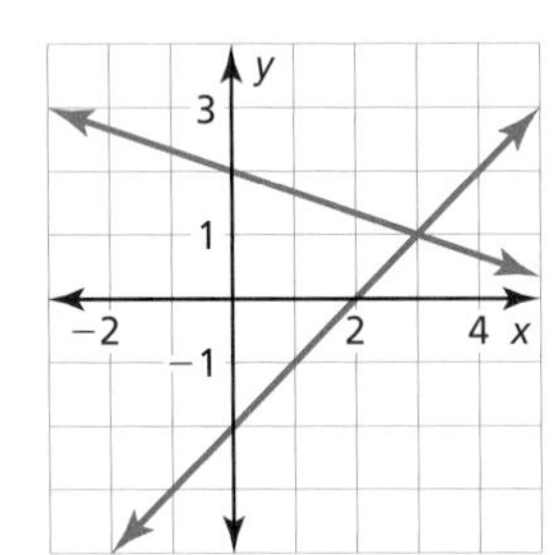

2. $y = \frac{1}{2}x - 1$
$y = 4x + 6$

3. $y = 1$
$y = 2x + 1$

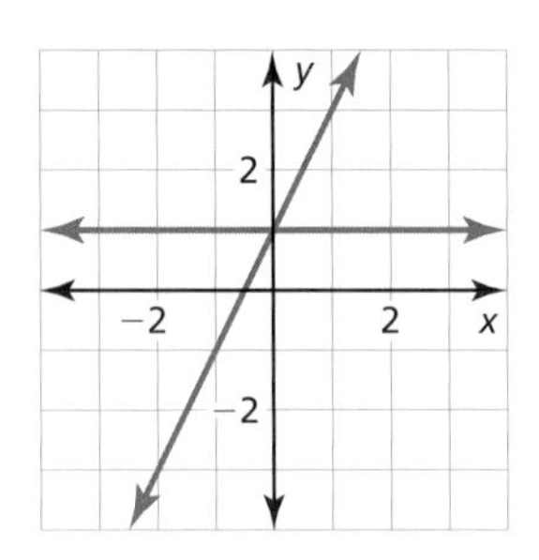

Solve the system of linear equations by substitution. Check your solution. *(Section 5.2)*

4. $y = x - 4$
$-2x + y = 18$

5. $2y + x = -4$
$y - x = -5$

6. $3x - 5y = 13$
$x + 4y = 10$

Solve the system of linear equations by elimination. Check your solution. *(Section 5.3)*

7. $x + y = 4$
$-3x - y = -8$

8. $x + 3y = 1$
$5x + 6y = 14$

9. $2x - 3y = -5$
$5x + 2y = 16$

Solve the system of linear equations. *(Section 5.4)*

10. $x - y = 1$
$x - y = 6$

11. $6x + 2y = 16$
$2x - y = 2$

12. $3x - 3y = -2$
$-6x + 6y = 4$

13. You plant a spruce tree that grows 4 inches per year and a hemlock tree that grows 6 inches per year. The initial heights are shown. *(Section 5.1)*

a. Write a system of linear equations that represents this situation.

b. Solve the system by graphing. Interpret your solution.

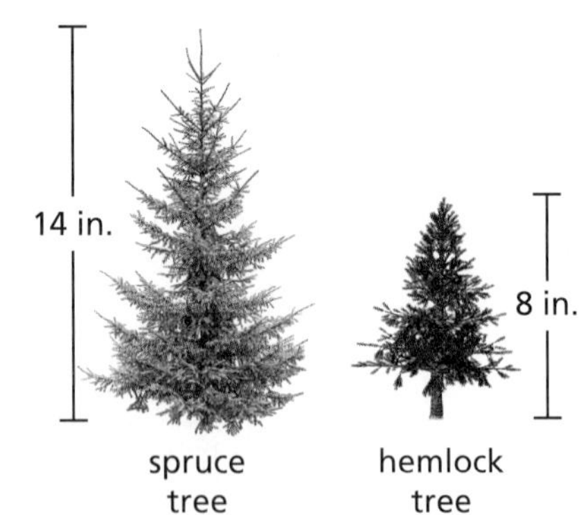

14. It takes you 3 hours to drive to a concert 135 miles away. You drive 55 miles per hour on highways and 40 miles per hour on the rest of the roads. *(Section 5.1, Section 5.2, and Section 5.3)*

a. How much time do you spend driving at each speed?

b. How many miles do you drive on highways? the rest of the roads?

15. In a football game, all of the home team's points are from 7-point touchdowns and 3-point field goals. The team scores six times. Write and solve a system of linear equations to find the numbers of touchdowns and field goals that the home team scores. *(Section 5.1, Section 5.2, and Section 5.3)*

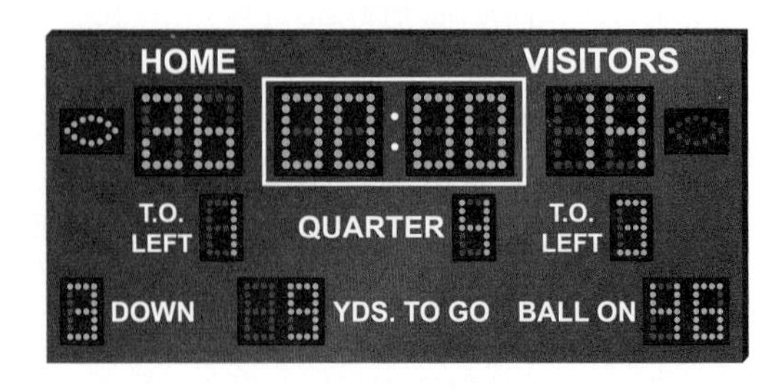

5.5 Solving Equations by Graphing

Essential Question How can you use a system of linear equations to solve an equation with variables on both sides?

Previously, you learned how to use algebra to solve equations with variables on both sides. Another way is to use a system of linear equations.

EXPLORATION 1 **Solving an Equation by Graphing**

Work with a partner. Solve $2x - 1 = -\frac{1}{2}x + 4$ by graphing.

a. Use the left side to write a linear equation. Then use the right side to write another linear equation.

b. Graph the two linear equations from part (a). Find the x-value of the point of intersection. Check that the x-value is the solution of

$$2x - 1 = -\frac{1}{2}x + 4.$$

c. Explain why this "graphical method" works.

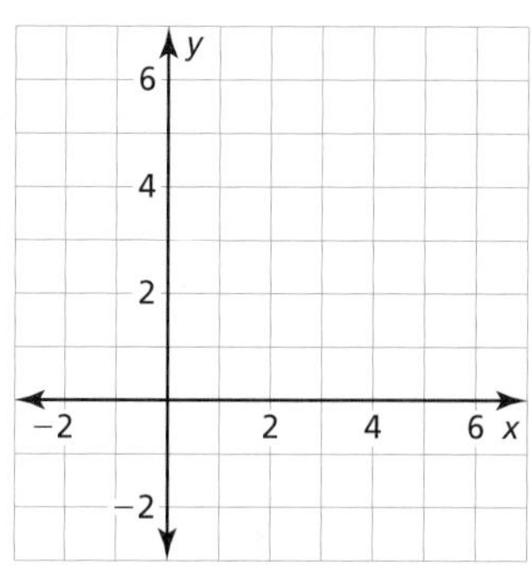

USING TOOLS STRATEGICALLY

To be proficient in math, you need to consider the available tools, which may include pencil and paper or a graphing calculator, when solving a mathematical problem.

EXPLORATION 2 **Solving Equations Algebraically and Graphically**

Work with a partner. Solve each equation using two methods.

Method 1 Use an algebraic method.

Method 2 Use a graphical method.

Is the solution the same using both methods?

a. $\frac{1}{2}x + 4 = -\frac{1}{4}x + 1$

b. $\frac{2}{3}x + 4 = \frac{1}{3}x + 3$

c. $-\frac{2}{3}x - 1 = \frac{1}{3}x - 4$

d. $\frac{4}{5}x + \frac{7}{5} = 3x - 3$

e. $-x + 2.5 = 2x - 0.5$

f. $-3x + 1.5 = x + 1.5$

Communicate Your Answer

3. How can you use a system of linear equations to solve an equation with variables on both sides?

4. Compare the algebraic method and the graphical method for solving a linear equation with variables on both sides. Describe the advantages and disadvantages of each method.

5.5 Lesson

Core Vocabulary

Previous
absolute value equation

What You Will Learn

- Solve linear equations by graphing.
- Solve absolute value equations by graphing.
- Use linear equations to solve real-life problems.

Solving Linear Equations by Graphing

You can use a system of linear equations to solve an equation with variables on both sides.

Core Concept

Solving Linear Equations by Graphing

Step 1 To solve the equation $ax + b = cx + d$, write two linear equations.

$$ax + b = cx + d$$

$y = ax + b$ and $y = cx + d$

Step 2 Graph the system of linear equations. The x-value of the solution of the system of linear equations is the solution of the equation $ax + b = cx + d$.

EXAMPLE 1 Solving an Equation by Graphing

Solve $-x + 1 = 2x - 5$ by graphing. Check your solution.

SOLUTION

Step 1 Write a system of linear equations using each side of the original equation.

$$-x + 1 = 2x - 5$$

$y = -x + 1$ and $y = 2x - 5$

Step 2 Graph the system.

$y = -x + 1$ Equation 1

$y = 2x - 5$ Equation 2

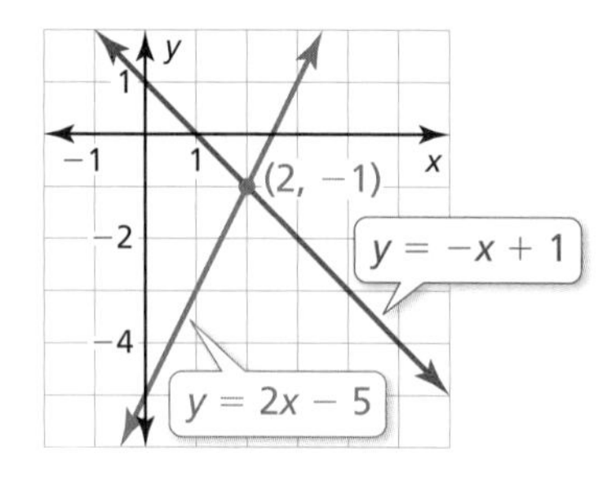

The graphs intersect at $(2, -1)$.

Check

$$-x + 1 = 2x - 5$$

$$-(2) + 1 \stackrel{?}{=} 2(2) - 5$$

$$-1 = -1 \checkmark$$

▶ So, the solution of the equation is $x = 2$.

Monitoring Progress

Help in English and Spanish at *BigIdeasMath.com*

Solve the equation by graphing. Check your solution.

1. $\frac{1}{2}x - 3 = 2x$

2. $-4 + 9x = -3x + 2$

Solving Absolute Value Equations by Graphing

EXAMPLE 2 Solving an Absolute Value Equation by Graphing

Solve $|x + 1| = |2x - 4|$ by graphing. Check your solutions.

SOLUTION

Recall that an absolute value equation of the form $|ax + b| = |cx + d|$ has two related equations.

$ax + b = cx + d$ Equation 1

$ax + b = -(cx + d)$ Equation 2

So, the related equations of $|x + 1| = |2x - 4|$ are as follows.

$x + 1 = 2x - 4$ Equation 1

$x + 1 = -(2x - 4)$ Equation 2

Apply the steps for solving an equation by graphing to each of the related equations.

Step 1 Write a system of linear equations for each related equation.

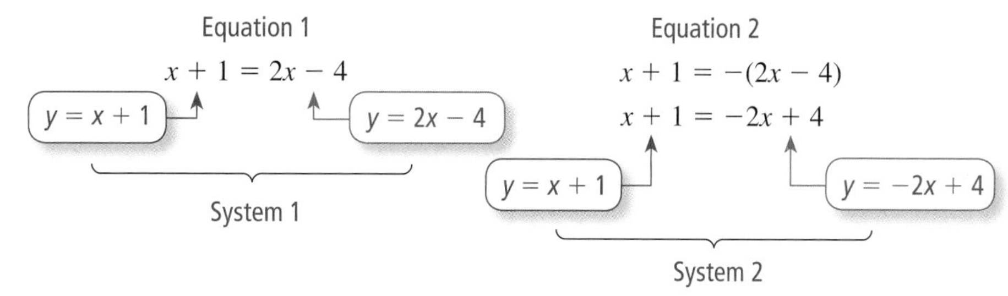

Step 2 Graph each system.

System 1

$y = x + 1$

$y = 2x - 4$

The graphs intersect at (5, 6).

System 2

$y = x + 1$

$y = -2x + 4$

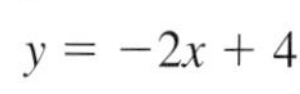

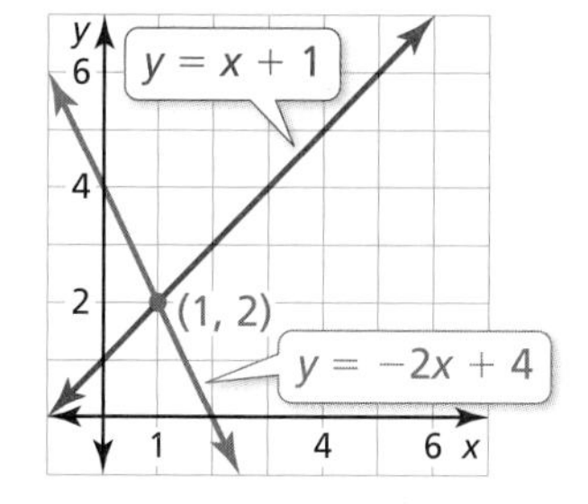

The graphs intersect at (1, 2).

▶ So, the solutions of the equation are $x = 5$ and $x = 1$.

Check

$|x + 1| = |2x - 4|$

$|5 + 1| \stackrel{?}{=} |2(5) - 4|$

$|6| \stackrel{?}{=} |6|$

$6 = 6$ ✓

$|x + 1| = |2x - 4|$

$|1 + 1| \stackrel{?}{=} |2(1) - 4|$

$|2| \stackrel{?}{=} |-2|$

$2 = 2$ ✓

Monitoring Progress

Help in English and Spanish at *BigIdeasMath.com*

Solve the equation by graphing. Check your solutions.

3. $|2x + 2| = |x - 2|$

4. $|x - 6| = |-x + 4|$

Solving Real-Life Problems

EXAMPLE 3 **Modeling with Mathematics**

Your family needs to rent a car for a week while on vacation. Company A charges \$3.25 per mile plus a flat fee of \$125 per week. Company B charges \$3 per mile plus a flat fee of \$150 per week. After how many miles of travel are the total costs the same at both companies?

SOLUTION

1. **Understand the Problem** You know the costs of renting a car from two companies. You are asked to determine how many miles of travel will result in the same total costs at both companies.
2. **Make a Plan** Use a verbal model to write an equation that represents the problem. Then solve the equation by graphing.
3. **Solve the Problem**

Words

Company A: Cost per mile • Miles + Flat fee = Company B: Cost per mile • Miles + Flat fee

Variable Let x be the number of miles traveled.

Equation $3.25x + 125 = 3x + 150$

Solve the equation by graphing.

Step 1 Write a system of linear equations using each side of the original equation.

$$3.25x + 125 = 3x + 150$$

$y = 3.25x + 125$ $\quad$ $y = 3x + 150$

Step 2 Use a graphing calculator to graph the system.

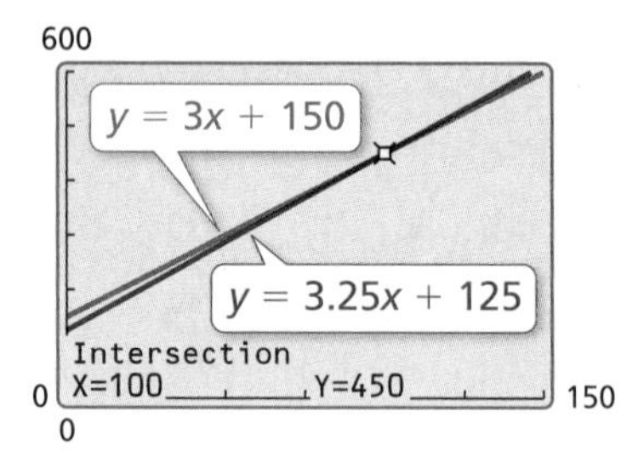

Because the graphs intersect at (100, 450), the solution of the equation is $x = 100$.

▶ So, the total costs are the same after 100 miles.

4. **Look Back** One way to check your solution is to solve the equation algebraically, as shown.

Check

$3.25x + 125 = 3x + 150$

$0.25x + 125 = 150$

$0.25x = 25$

$x = 100$

Monitoring Progress

Help in English and Spanish at *BigIdeasMath.com*

5. **WHAT IF?** Company C charges \$3.30 per mile plus a flat fee of \$115 per week. After how many miles are the total costs the same at Company A and Company C?

5.5 Exercises

Dynamic Solutions available at *BigIdeasMath.com*

Vocabulary and Core Concept Check

1. **REASONING** The graphs of the equations $y = 3x - 20$ and $y = -2x + 10$ intersect at the point $(6, -2)$. Without solving, find the solution of the equation $3x - 20 = -2x + 10$.

2. **WRITING** Explain how to rewrite the absolute value equation $|2x - 4| = |-5x + 1|$ as two systems of linear equations.

Monitoring Progress and Modeling with Mathematics

In Exercises 3–6, use the graph to solve the equation. Check your solution.

3. $-2x + 3 = x$ 

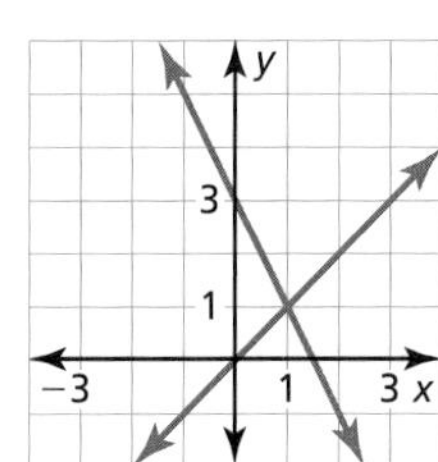

4. $-3 = 4x + 1$

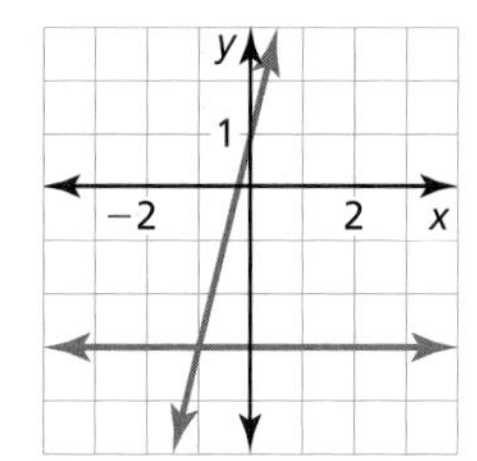

5. $-x - 1 = \frac{1}{3}x + 3$

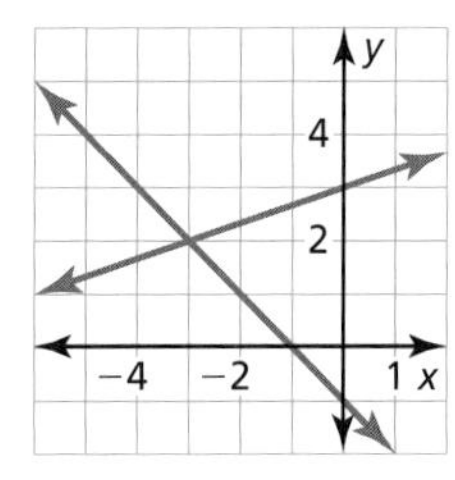

6. $-\frac{3}{2}x - 2 = -4x + 3$

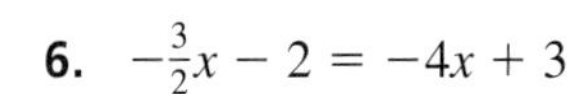

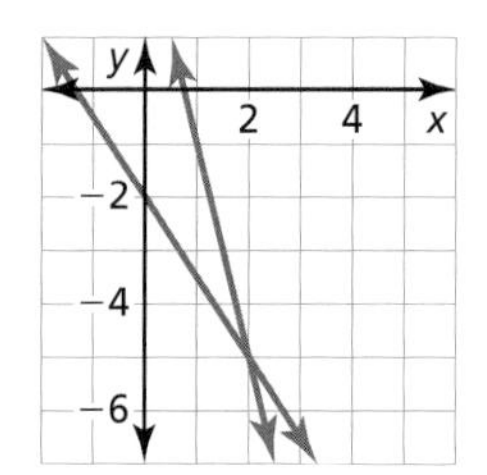

In Exercises 7–14, solve the equation by graphing. Check your solution. *(See Example 1.)*

7. $x + 4 = -x$
8. $4x = x + 3$
9. $x + 5 = -2x - 4$
10. $-2x + 6 = 5x - 1$
11. $\frac{1}{2}x - 2 = 9 - 5x$
12. $-5 + \frac{1}{4}x = 3x + 6$
13. $5x - 7 = 2(x + 1)$
14. $-6(x + 4) = -3x - 6$

In Exercises 15–20, solve the equation by graphing. Determine whether the equation has ***one solution, no solution,*** **or** ***infinitely many solutions.***

15. $3x - 1 = -x + 7$
16. $5x - 4 = 5x + 1$
17. $-4(2 - x) = 4x - 8$
18. $-2x - 3 = 2(x - 2)$
19. $-x - 5 = -\frac{1}{3}(3x + 5)$
20. $\frac{1}{2}(8x + 3) = 4x + \frac{3}{2}$

In Exercises 21 and 22, use the graphs to solve the equation. Check your solutions.

21. $|x - 4| = |3x|$

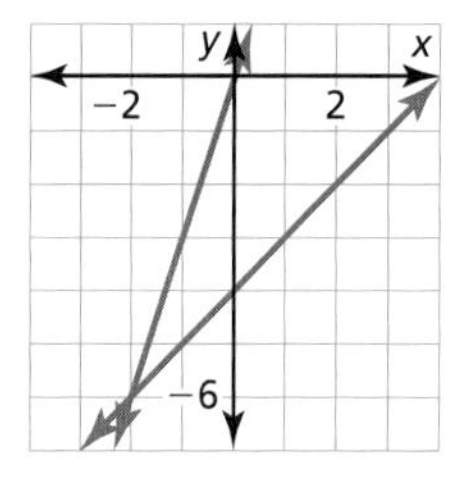

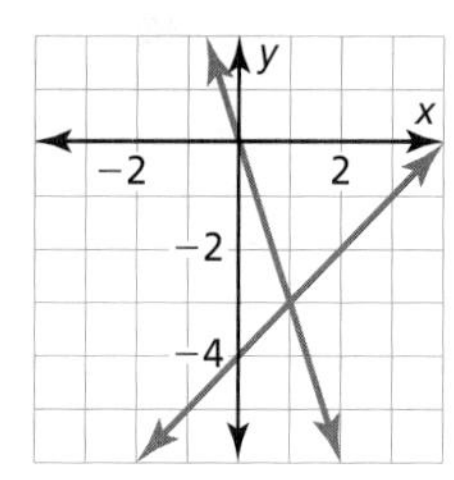

22. $|2x + 4| = |x - 1|$

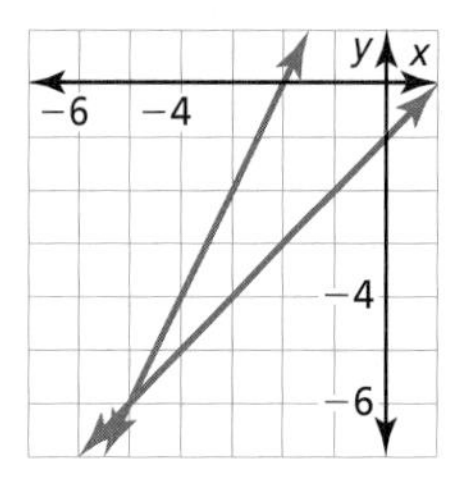

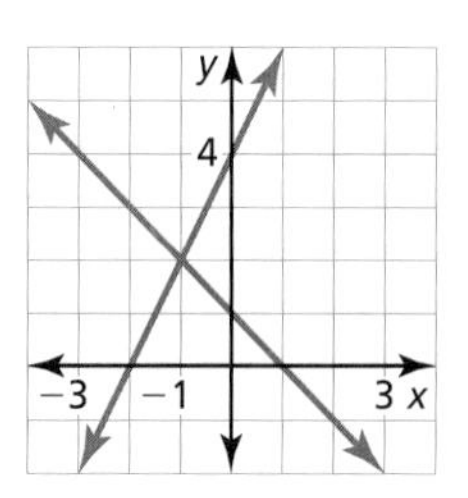

In Exercises 23–30, solve the equation by graphing. Check your solutions. *(See Example 2.)*

23. $|2x| = |x + 3|$
24. $|2x - 6| = |x|$
25. $|-x + 4| = |2x - 2|$
26. $|x + 2| = |-3x + 6|$
27. $|x + 1| = |x - 5|$
28. $|2x + 5| = |-2x + 1|$
29. $|x - 3| = 2|x|$
30. $4|x + 2| = |2x + 7|$

USING TOOLS In Exercises 31 and 32, use a graphing calculator to solve the equation.

31. $0.7x + 0.5 = -0.2x - 1.3$

32. $2.1x + 0.6 = -1.4x + 6.9$

33. MODELING WITH MATHEMATICS You need to hire a catering company to serve meals to guests at a wedding reception. Company A charges \$500 plus \$20 per guest. Company B charges \$800 plus \$16 per guest. For how many guests are the total costs the same at both companies? *(See Example 3.)*

34. MODELING WITH MATHEMATICS Your dog is 16 years old in dog years. Your cat is 28 years old in cat years. For every human year, your dog ages by 7 dog years and your cat ages by 4 cat years. In how many human years will both pets be the same age in their respective types of years?

35. MODELING WITH MATHEMATICS You and a friend race across a field to a fence and back. Your friend has a 50-meter head start. The equations shown represent you and your friend's distances d (in meters) from the fence t seconds after the race begins. Find the time at which you catch up to your friend.

You: $d = |-5t + 100|$

Your friend: $d = \left|-3\frac{1}{3}t + 50\right|$

36. MAKING AN ARGUMENT The graphs of $y = -x + 4$ and $y = 2x - 8$ intersect at the point (4, 0). So, your friend says the solution of the equation $-x + 4 = 2x - 8$ is (4, 0). Is your friend correct? Explain.

37. OPEN-ENDED Find values for m and b so that the solution of the equation $mx + b = -2x - 1$ is $x = -3$.

38. HOW DO YOU SEE IT? The graph shows the total revenue and expenses of a company x years after it opens for business.

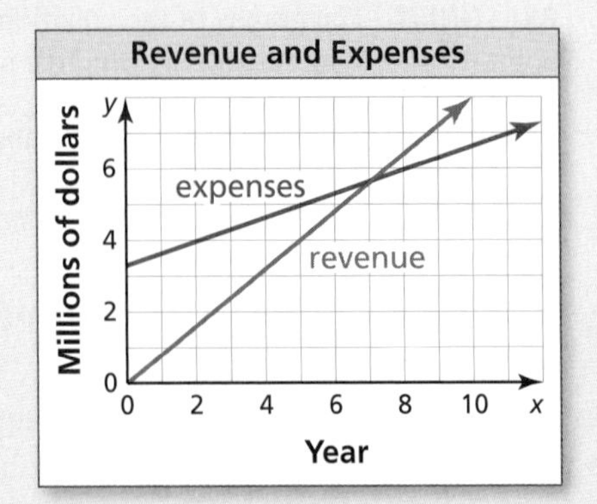

a. Estimate the point of intersection of the graphs.

b. Interpret your answer in part (a).

39. MATHEMATICAL CONNECTIONS The value of the perimeter of the triangle (in feet) is equal to the value of the area of the triangle (in square feet). Use a graph to find x.

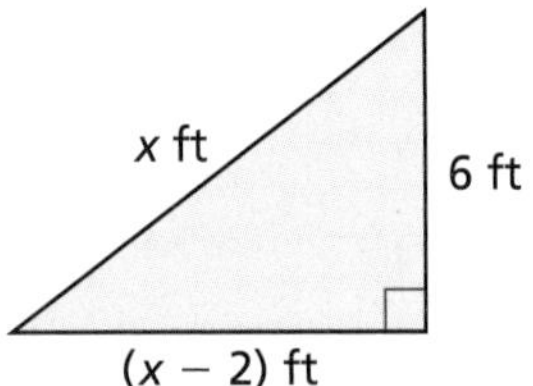

40. THOUGHT PROVOKING A car has an initial value of \$20,000 and decreases in value at a rate of \$1500 per year. Describe a different car that will be worth the same amount as this car in exactly 5 years. Specify the initial value and the rate at which the value decreases.

41. ABSTRACT REASONING Use a graph to determine the sign of the solution of the equation $ax + b = cx + d$ in each situation.

a. $0 < b < d$ and $a < c$ b. $d < b < 0$ and $a < c$

Maintaining Mathematical Proficiency

Reviewing what you learned in previous grades and lessons

Graph the inequality. *(Section 2.1)*

42. $y > 5$ **43.** $x \leq -2$ **44.** $n \geq 9$ **45.** $c < -6$

Use the graphs of f and g to describe the transformation from the graph of f to the graph of g. *(Section 3.6)*

46. $f(x) = x - 5$; $g(x) = f(x + 2)$

47. $f(x) = 6x$; $g(x) = -f(x)$

48. $f(x) = -2x + 1$; $g(x) = f(4x)$

49. $f(x) = \frac{1}{2}x - 2$; $g(x) = f(x - 1)$

5.6 Graphing Linear Inequalities in Two Variables

Essential Question

How can you graph a linear inequality in two variables?

A **solution of a linear inequality in two variables** is an ordered pair (x, y) that makes the inequality true. The **graph of a linear inequality** in two variables shows all the solutions of the inequality in a coordinate plane.

EXPLORATION 1 Writing a Linear Inequality in Two Variables

Work with a partner.

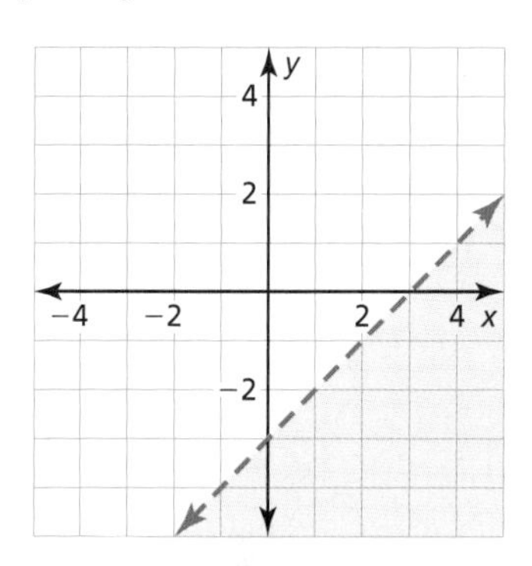

a. Write an equation represented by the dashed line.

b. The solutions of an inequality are represented by the shaded region. In words, describe the solutions of the inequality.

c. Write an inequality represented by the graph. Which inequality symbol did you use? Explain your reasoning.

EXPLORATION 2 Using a Graphing Calculator

USING TOOLS STRATEGICALLY

To be proficient in math, you need to use technological tools to explore and deepen your understanding of concepts.

Work with a partner. Use a graphing calculator to graph $y \geq \frac{1}{4}x - 3$.

a. Enter the equation $y = \frac{1}{4}x - 3$ into your calculator.

b. The inequality has the symbol $\geq$. So, the region to be shaded is above the graph of $y = \frac{1}{4}x - 3$, as shown. Verify this by testing a point in this region, such as $(0, 0)$, to make sure it is a solution of the inequality.

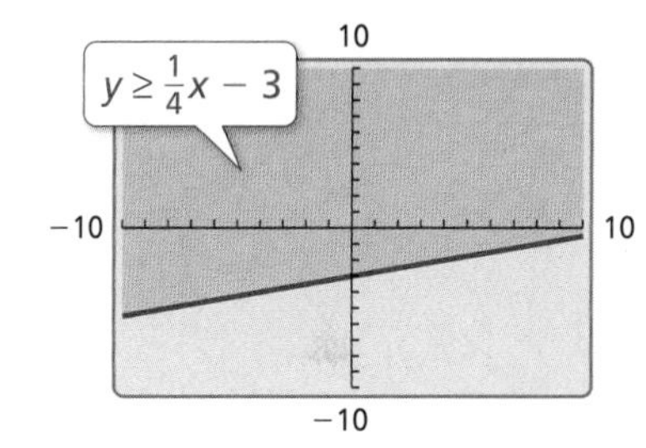

Because the inequality symbol is *greater than or equal to*, the line is solid and not dashed. Some graphing calculators always use a solid line when graphing inequalities. In this case, you have to determine whether the line should be solid or dashed, based on the inequality symbol used in the original inequality.

EXPLORATION 3 Graphing Linear Inequalities in Two Variables

Work with a partner. Graph each linear inequality in two variables. Explain your steps. Use a graphing calculator to check your graphs.

a. $y > x + 5$ **b.** $y \leq -\frac{1}{2}x + 1$ **c.** $y \geq -x - 5$

Communicate Your Answer

4. How can you graph a linear inequality in two variables?

5. Give an example of a real-life situation that can be modeled using a linear inequality in two variables.

5.6 Lesson

Core Vocabulary

linear inequality in two variables, *p. 268*
solution of a linear inequality in two variables, *p. 268*
graph of a linear inequality, *p. 268*
half-planes, *p. 268*

Previous
ordered pair

What You Will Learn

- Check solutions of linear inequalities.
- Graph linear inequalities in two variables.
- Use linear inequalities to solve real-life problems.

Linear Inequalities

A **linear inequality in two variables**, x and y, can be written as

$$ax + by < c \qquad ax + by \leq c \qquad ax + by > c \qquad ax + by \geq c$$

where a, b, and c are real numbers. A **solution of a linear inequality in two variables** is an ordered pair (x, y) that makes the inequality true.

EXAMPLE 1 Checking Solutions

Tell whether the ordered pair is a solution of the inequality.

a. $2x + y < -3; (-1, 9)$ **b.** $x - 3y \geq 8; (2, -2)$

SOLUTION

a. $2x + y < -3$ Write the inequality.

$2(-1) + 9 \overset{?}{<} -3$ Substitute -1 for x and 9 for y.

$7 \nless -3$ ✗ Simplify. 7 is *not* less than -3.

▶ So, $(-1, 9)$ is *not* a solution of the inequality.

b. $x - 3y \geq 8$ Write the inequality.

$2 - 3(-2) \overset{?}{\geq} 8$ Substitute 2 for x and -2 for y.

$8 \geq 8$ ✓ Simplify. 8 is equal to 8.

▶ So, $(2, -2)$ is a solution of the inequality.

Monitoring Progress

Help in English and Spanish at *BigIdeasMath.com*

Tell whether the ordered pair is a solution of the inequality.

1. $x + y > 0; (-2, 2)$ **2.** $4x - y \geq 5; (0, 0)$

3. $5x - 2y \leq -1; (-4, -1)$ **4.** $-2x - 3y < 15; (5, -7)$

Graphing Linear Inequalities in Two Variables

The **graph of a linear inequality** in two variables shows all the solutions of the inequality in a coordinate plane.

READING

A dashed boundary line means that points on the line are *not* solutions. A solid boundary line means that points on the line are solutions.

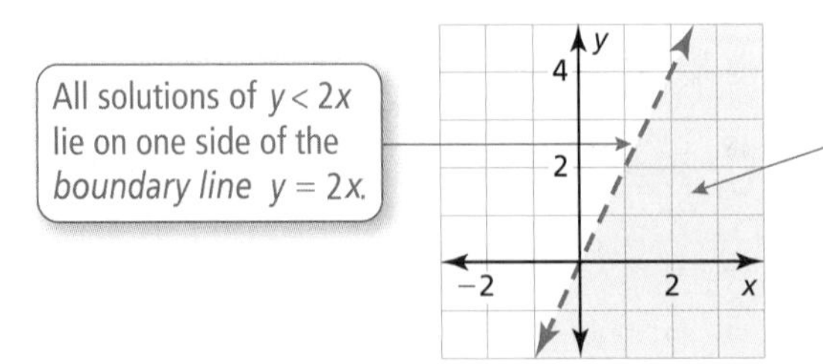
All solutions of y < 2x lie on one side of the boundary line y = 2x.
y
4
2
−2
2
x
The boundary line divides the coordinate plane into two half-planes. The shaded half-plane is the graph of y < 2x.

Core Concept

Graphing a Linear Inequality in Two Variables

Step 1 Graph the boundary line for the inequality. Use a dashed line for $<$ or $>$. Use a solid line for $\leq$ or $\geq$.

Step 2 Test a point that is not on the boundary line to determine whether it is a solution of the inequality.

Step 3 When the test point is a solution, shade the half-plane that contains the point. When the test point is *not* a solution, shade the half-plane that does *not* contain the point.

STUDY TIP

It is often convenient to use the origin as a test point. However, you must choose a different test point when the origin is on the boundary line.

EXAMPLE 2 Graphing a Linear Inequality in One Variable

Graph $y \leq 2$ in a coordinate plane.

SOLUTION

Step 1 Graph $y = 2$. Use a solid line because the inequality symbol is $\leq$.

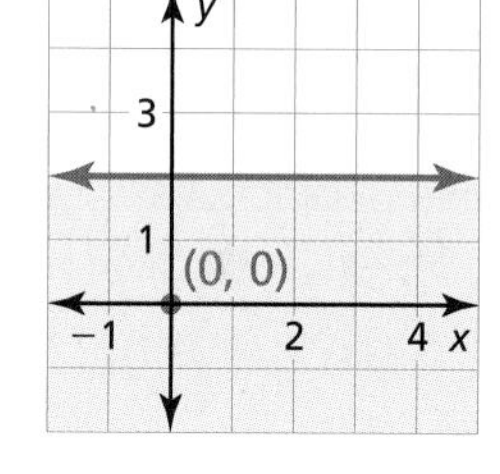

Step 2 Test (0, 0).

$y \leq 2$ Write the inequality.

$0 \leq 2$ ✓ Substitute.

Step 3 Because (0, 0) is a solution, shade the half-plane that contains (0, 0).

EXAMPLE 3 Graphing a Linear Inequality in Two Variables

Graph $-x + 2y > 2$ in a coordinate plane.

SOLUTION

Step 1 Graph $-x + 2y = 2$, or $y = \frac{1}{2}x + 1$. Use a dashed line because the inequality symbol is $>$.

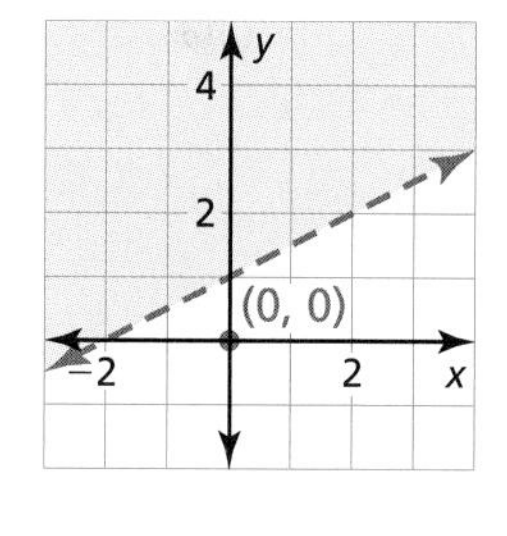

Step 2 Test (0, 0).

$-x + 2y > 2$ Write the inequality.

$-(0) + 2(0) \overset{?}{>} 2$ Substitute.

$0 \not> 2$ ✗ Simplify.

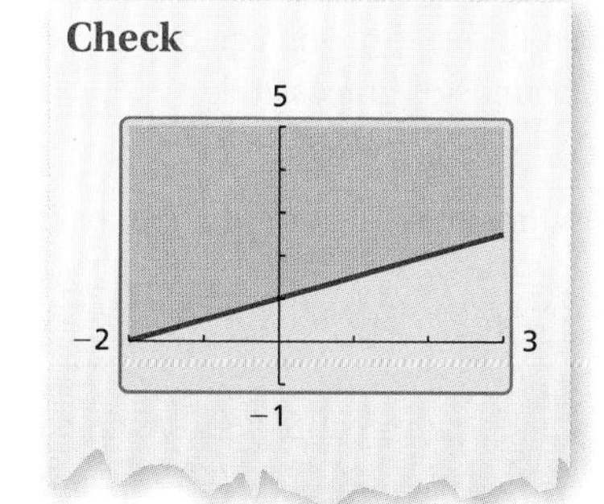

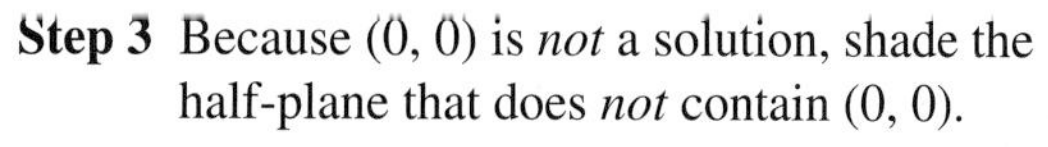

Step 3 Because (0, 0) is *not* a solution, shade the half-plane that does *not* contain (0, 0).

Monitoring Progress

Help in English and Spanish at *BigIdeasMath.com*

Graph the inequality in a coordinate plane.

5. $y > -1$

6. $x \leq -4$

7. $x + y \leq -4$

8. $x - 2y < 0$

Solving Real-Life Problems

EXAMPLE 4 Modeling with Mathematics

You can spend at most \$10 on grapes and apples for a fruit salad. Grapes cost \$2.50 per pound, and apples cost \$1 per pound. Write and graph an inequality that represents the amounts of grapes and apples you can buy. Identify and interpret two solutions of the inequality.

SOLUTION

1. **Understand the Problem** You know the most that you can spend and the prices per pound for grapes and apples. You are asked to write and graph an inequality and then identify and interpret two solutions.
2. **Make a Plan** Use a verbal model to write an inequality that represents the problem. Then graph the inequality. Use the graph to identify two solutions. Then interpret the solutions.
3. **Solve the Problem**

Words	Cost per pound of grapes	•	Pounds of grapes	+	Cost per pound of apples	•	Pounds of apples	≤	Amount you can spend

Variables Let x be pounds of grapes and y be pounds of apples.

Inequality $2.50 \cdot x + 1 \cdot y \leq 10$

Step 1 Graph $2.5x + y = 10$, or $y = -2.5x + 10$. Use a solid line because the inequality symbol is ≤. Restrict the graph to positive values of x and y because negative values do not make sense in this real-life context.

Step 2 Test (0, 0).

$2.5x + y \leq 10$	Write the inequality.
$2.5(0) + 0 \overset{?}{\leq} 10$	Substitute.
$0 \leq 10$ ✓	Simplify.

Step 3 Because (0, 0) is a solution, shade the half-plane that contains (0, 0).

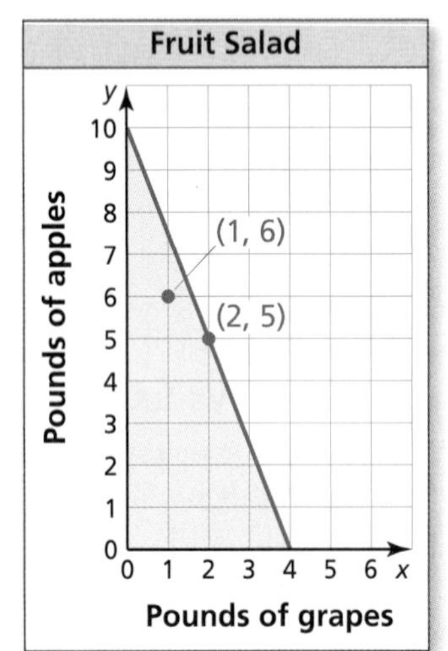

▶ One possible solution is (1, 6) because it lies in the shaded half-plane. Another possible solution is (2, 5) because it lies on the solid line. So, you can buy 1 pound of grapes and 6 pounds of apples, or 2 pounds of grapes and 5 pounds of apples.

4. **Look Back** Check your solutions by substituting them into the original inequality, as shown.

Check

$2.5x + y \leq 10$

$2.5(1) + 6 \overset{?}{\leq} 10$

$8.5 \leq 10$ ✓

$2.5x + y \leq 10$

$2.5(2) + 5 \overset{?}{\leq} 10$

$10 \leq 10$ ✓

Monitoring Progress

Help in English and Spanish at *BigIdeasMath.com*

9. You can spend at most \$12 on red peppers and tomatoes for salsa. Red peppers cost \$4 per pound, and tomatoes cost \$3 per pound. Write and graph an inequality that represents the amounts of red peppers and tomatoes you can buy. Identify and interpret two solutions of the inequality.

 Dynamic Solutions available at *BigIdeasMath.com*

Vocabulary and Core Concept Check

1. **VOCABULARY** How can you tell whether an ordered pair is a solution of a linear inequality?

2. **WRITING** Compare the graph of a linear inequality in two variables with the graph of a linear equation in two variables.

Monitoring Progress and Modeling with Mathematics

In Exercises 3–10, tell whether the ordered pair is a solution of the inequality. *(See Example 1.)*

3. $x + y < 7$; $(2, 3)$
4. $x - y \leq 0$; $(5, 2)$
5. $x + 3y \geq -2$; $(-9, 2)$
6. $8x + y > -6$; $(-1, 2)$
7. $-6x + 4y \leq 6$; $(-3, -3)$
8. $3x - 5y \geq 2$; $(-1, -1)$
9. $-x - 6y > 12$; $(-8, 2)$
10. $-4x - 8y < 15$; $(-6, 3)$

In Exercises 11–16, tell whether the ordered pair is a solution of the inequality whose graph is shown.

11. $(0, -1)$
12. $(-1, 3)$
13. $(1, 4)$
14. $(0, 0)$
15. $(3, 3)$
16. $(2, 1)$

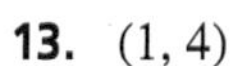

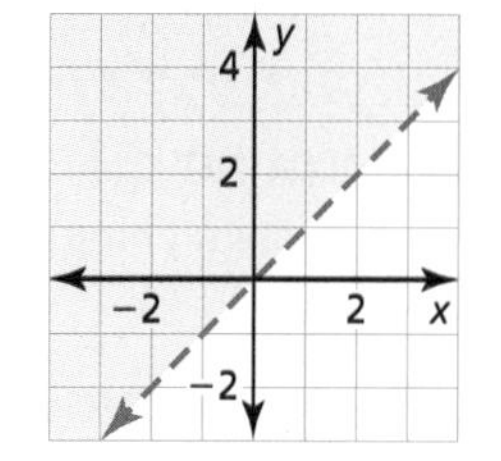

17. **MODELING WITH MATHEMATICS** A carpenter has at most \$250 to spend on lumber. The inequality $8x + 12y \leq 250$ represents the numbers x of 2-by-8 boards and the numbers y of 4-by-4 boards the carpenter can buy. Can the carpenter buy twelve 2-by-8 boards and fourteen 4-by-4 boards? Explain.

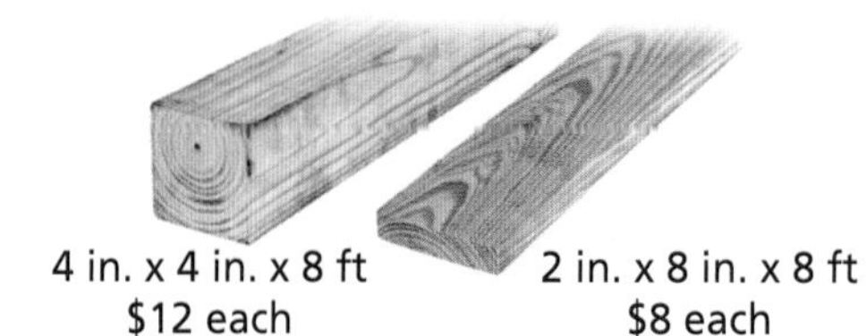

4 in. x 4 in. x 8 ft
\$12 each

2 in. x 8 in. x 8 ft
\$8 each

18. **MODELING WITH MATHEMATICS** The inequality $3x + 2y \geq 93$ represents the numbers x of multiple-choice questions and the numbers y of matching questions you can answer correctly to receive an A on a test. You answer 20 multiple-choice questions and 18 matching questions correctly. Do you receive an A on the test? Explain.

In Exercises 19–24, graph the inequality in a coordinate plane. *(See Example 2.)*

19. $y \leq 5$
20. $y > 6$
21. $x < 2$
22. $x \geq -3$
23. $y > -7$
24. $x < 9$

In Exercises 25–30, graph the inequality in a coordinate plane. *(See Example 3.)*

25. $y > -2x - 4$
26. $y \leq 3x - 1$
27. $-4x + y < -7$
28. $3x - y \geq 5$
29. $5x - 2y \leq 6$
30. $-x + 4y > -12$

ERROR ANALYSIS In Exercises 31 and 32, describe and correct the error in graphing the inequality.

31. $y < -x + 1$

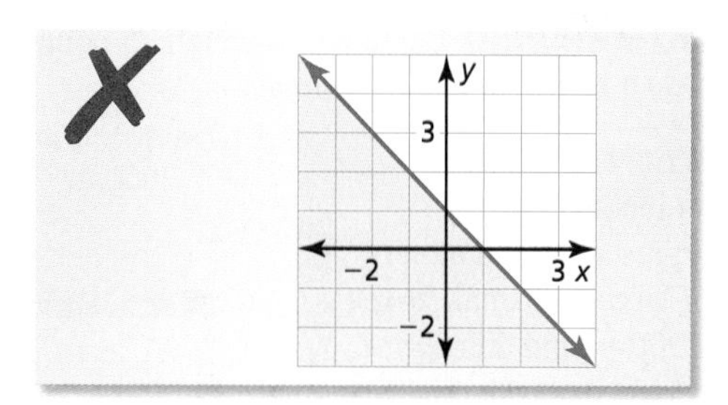

32. $y \leq 3x - 2$

33. MODELING WITH MATHEMATICS You have at most \$20 to spend at an arcade. Arcade games cost \$0.75 each, and snacks cost \$2.25 each. Write and graph an inequality that represents the numbers of games you can play and snacks you can buy. Identify and interpret two solutions of the inequality. *(See Example 4.)*

34. MODELING WITH MATHEMATICS A drama club must sell at least \$1500 worth of tickets to cover the expenses of producing a play. Write and graph an inequality that represents how many adult and student tickets the club must sell. Identify and interpret two solutions of the inequality.

School Play
Adults: \$10
Students: \$6

In Exercises 35–38, write an inequality that represents the graph.

35.

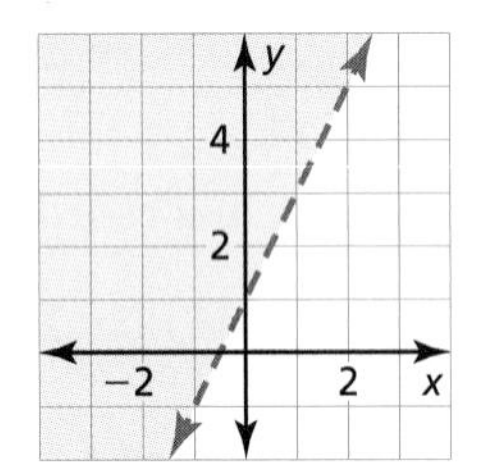

36.

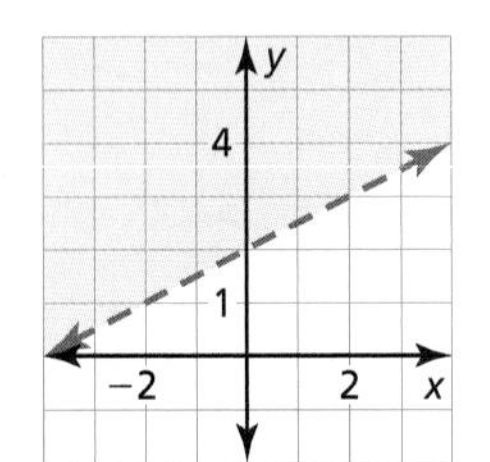

37.

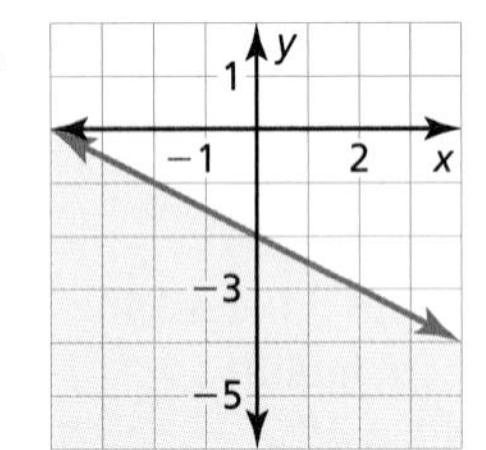

38.

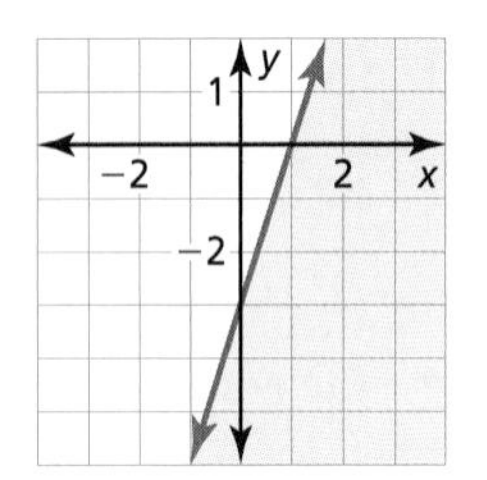

39. PROBLEM SOLVING Large boxes weigh 75 pounds, and small boxes weigh 40 pounds.

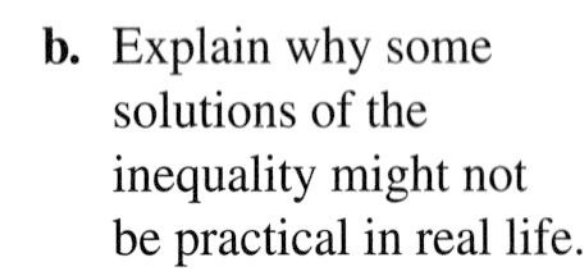

a. Write and graph an inequality that represents the numbers of large and small boxes a 200-pound delivery person can take on the elevator.

b. Explain why some solutions of the inequality might not be practical in real life.

40. HOW DO YOU SEE IT? Match each inequality with its graph.

a. $3x - 2y \leq 6$ **b.** $3x - 2y < 6$

c. $3x - 2y > 6$ **d.** $3x - 2y \geq 6$

A.

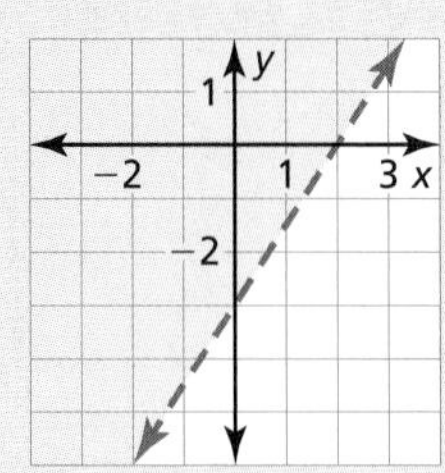

B.

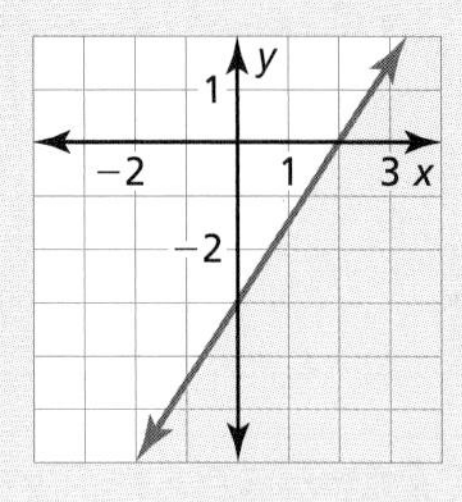

C.

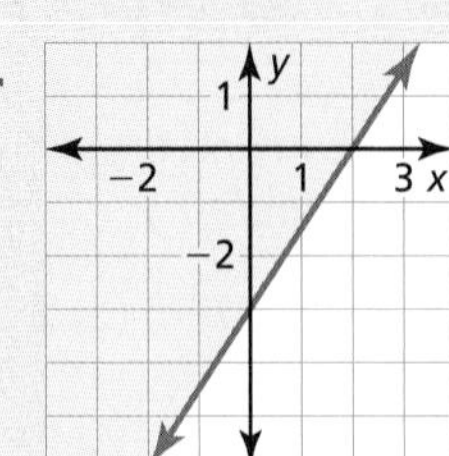

D.

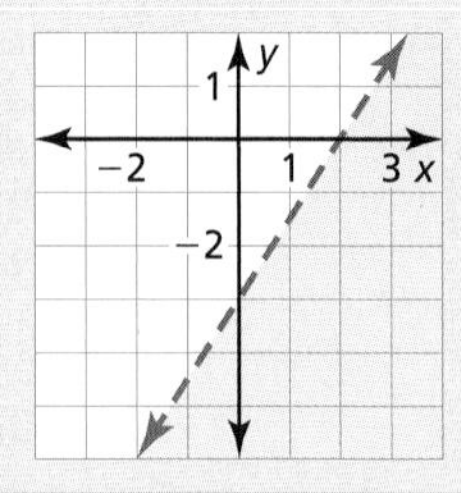

41. REASONING When graphing a linear inequality in two variables, why must you choose a test point that is *not* on the boundary line?

42. THOUGHT PROVOKING Write a linear inequality in two variables that has the following two properties.

- $(0, 0)$, $(0, -1)$, and $(0, 1)$ are not solutions.
- $(1, 1)$, $(3, -1)$, and $(-1, 3)$ are solutions.

43. WRITING Can you always use $(0, 0)$ as a test point when graphing an inequality? Explain.

CRITICAL THINKING In Exercises 44 and 45, write and graph an inequality whose graph is described by the given information.

44. The points $(2, 5)$ and $(-3, -5)$ lie on the boundary line. The points $(6, 5)$ and $(-2, -3)$ are solutions of the inequality.

45. The points $(-7, -16)$ and $(1, 8)$ lie on the boundary line. The points $(-7, 0)$ and $(3, 14)$ are *not* solutions of the inequality.

Maintaining Mathematical Proficiency

Reviewing what you learned in previous grades and lessons

Write the next three terms of the arithmetic sequence. *(Section 4.6)*

46. $0, 8, 16, 24, 32, \ldots$

47. $-5, -8, -11, -14, -17, \ldots$

48. $-\frac{3}{2}, -\frac{1}{2}, \frac{1}{2}, \frac{3}{2}, \frac{5}{2}, \ldots$

5.7 Systems of Linear Inequalities

Essential Question

How can you graph a system of linear inequalities?

EXPLORATION 1 Graphing Linear Inequalities

Work with a partner. Match each linear inequality with its graph. Explain your reasoning.

$2x + y \leq 4$ Inequality 1

$2x - y \leq 0$ Inequality 2

A.

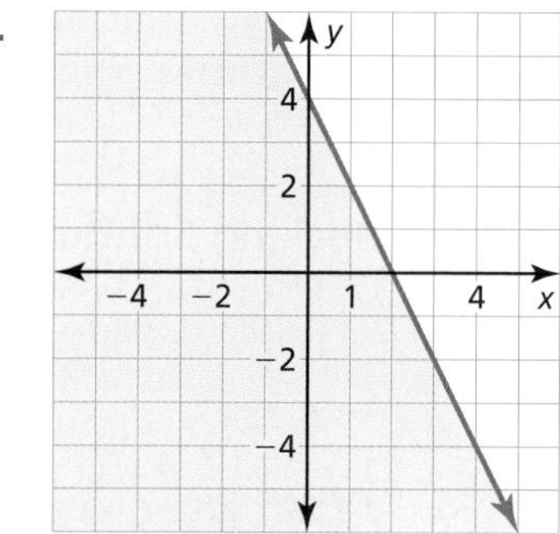

B.

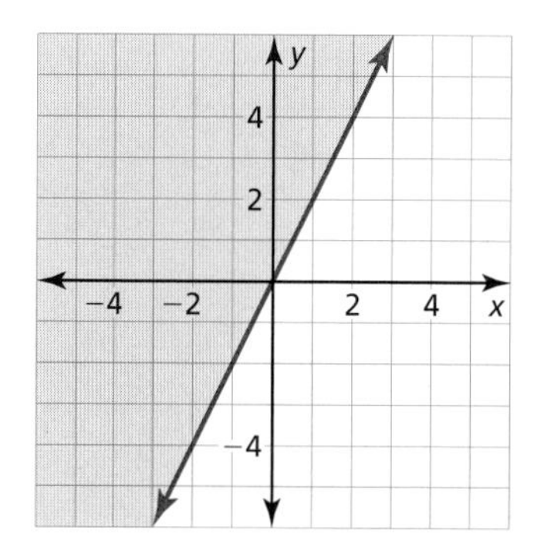

EXPLORATION 2 Graphing a System of Linear Inequalities

Work with a partner. Consider the linear inequalities given in Exploration 1.

$2x + y \leq 4$ Inequality 1

$2x - y \leq 0$ Inequality 2

a. Use two different colors to graph the inequalities in the same coordinate plane. What is the result?

b. Describe each of the shaded regions of the graph. What does the unshaded region represent?

MAKING SENSE OF PROBLEMS

To be proficient in math, you need to explain to yourself the meaning of a problem.

Communicate Your Answer

3. How can you graph a system of linear inequalities?

4. When graphing a system of linear inequalities, which region represents the solution of the system?

5. Do you think all systems of linear inequalities have a solution? Explain your reasoning.

6. Write a system of linear inequalities represented by the graph.

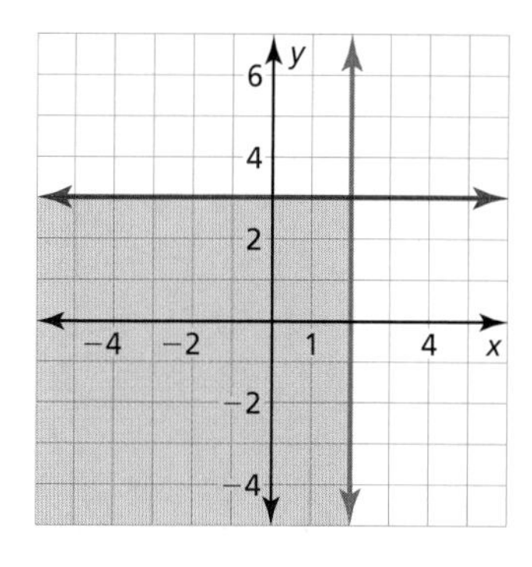

5.7 Lesson

Core Vocabulary

system of linear inequalities, *p. 274*
solution of a system of linear inequalities, *p. 274*
graph of a system of linear inequalities, *p. 275*

Previous
linear inequality in two variables

What You Will Learn

- Check solutions of systems of linear inequalities.
- Graph systems of linear inequalities.
- Write systems of linear inequalities.
- Use systems of linear inequalities to solve real-life problems.

Systems of Linear Inequalities

A **system of linear inequalities** is a set of two or more linear inequalities in the same variables. An example is shown below.

$y < x + 2$ Inequality 1

$y \geq 2x - 1$ Inequality 2

A **solution of a system of linear inequalities** in two variables is an ordered pair that is a solution of each inequality in the system.

EXAMPLE 1 Checking Solutions

Tell whether each ordered pair is a solution of the system of linear inequalities.

$y < 2x$ Inequality 1

$y \geq x + 1$ Inequality 2

a. $(3, 5)$ **b.** $(-2, 0)$

SOLUTION

a. Substitute 3 for x and 5 for y in each inequality.

Inequality 1	Inequality 2
$y < 2x$	$y \geq x + 1$
$5 \overset{?}{<} 2(3)$	$5 \overset{?}{\geq} 3 + 1$
$5 < 6$ ✓	$5 \geq 4$ ✓

▶ Because the ordered pair (3, 5) is a solution of each inequality, it is a solution of the system.

b. Substitute -2 for x and 0 for y in each inequality.

Inequality 1	Inequality 2
$y < 2x$	$y \geq x + 1$
$0 \overset{?}{<} 2(-2)$	$0 \overset{?}{\geq} -2 + 1$
$0 \not< -4$ ✗	$0 \geq -1$ ✓

▶ Because $(-2, 0)$ is not a solution of each inequality, it is *not* a solution of the system.

Monitoring Progress

Help in English and Spanish at *BigIdeasMath.com*

Tell whether the ordered pair is a solution of the system of linear inequalities.

1. $(-1, 5)$; $y < 5$, $y > x - 4$

2. $(1, 4)$; $y \geq 3x + 1$, $y > x - 1$

Graphing Systems of Linear Inequalities

The **graph of a system of linear inequalities** is the graph of all the solutions of the system.

Core Concept

Graphing a System of Linear Inequalities

Step 1 Graph each inequality in the same coordinate plane.

Step 2 Find the intersection of the half-planes that are solutions of the inequalities. This intersection is the graph of the system.

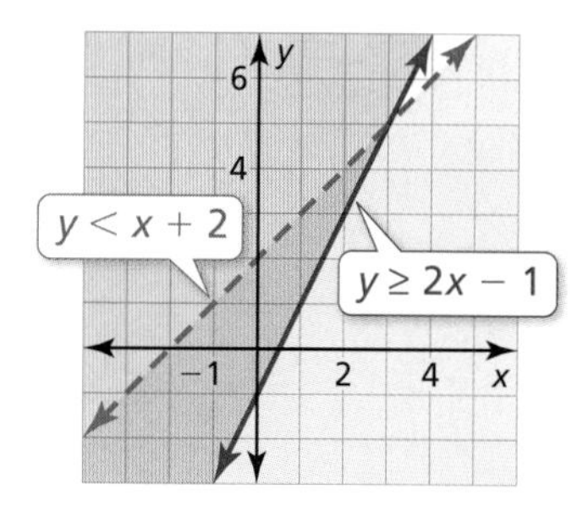

Check

Verify that $(-3, 1)$ is a solution of each inequality.

Inequality 1

$y \leq 3$

$1 \leq 3$ ✓

Inequality 2

$y > x + 2$

$1 \stackrel{?}{>} -3 + 2$

$1 > -1$ ✓

EXAMPLE 2 Graphing a System of Linear Inequalities

Graph the system of linear inequalities.

$y \leq 3$ Inequality 1

$y > x + 2$ Inequality 2

SOLUTION

Step 1 Graph each inequality.

Step 2 Find the intersection of the half-planes. One solution is $(-3, 1)$.

The solution is the purple-shaded region.

EXAMPLE 3 Graphing a System of Linear Inequalities: No Solution

Graph the system of linear inequalities.

$2x + y < -1$ Inequality 1

$2x + y > 3$ Inequality 2

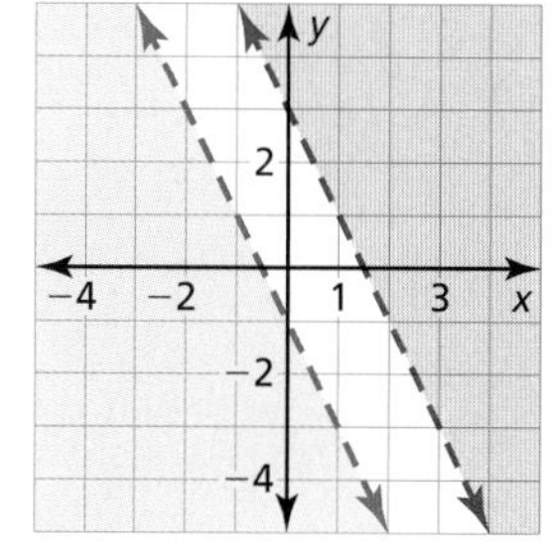

SOLUTION

Step 1 Graph each inequality.

Step 2 Find the intersection of the half-planes. Notice that the lines are parallel, and the half-planes do not intersect.

▶ So, the system has no solution.

Monitoring Progress  Help in English and Spanish at *BigIdeasMath.com*

Graph the system of linear inequalities.

3. $y \geq -x + 4$
$x + y \leq 0$

4. $y > 2x - 3$
$y \geq \frac{1}{2}x + 1$

5. $-2x + y < 4$
$2x + y > 4$

Writing Systems of Linear Inequalities

EXAMPLE 4 Writing a System of Linear Inequalities

Write a system of linear inequalities represented by the graph.

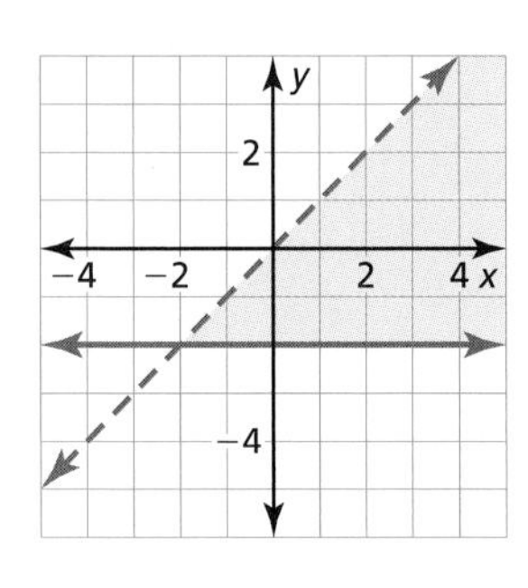

SOLUTION

Inequality 1 The horizontal boundary line passes through (0, −2). So, an equation of the line is $y = -2$. The shaded region is *above* the *solid* boundary line, so the inequality is $y \geq -2$.

Inequality 2 The slope of the other boundary line is 1, and the y-intercept is 0. So, an equation of the line is $y = x$. The shaded region is *below* the *dashed* boundary line, so the inequality is $y < x$.

▶ The system of linear inequalities represented by the graph is

$y \geq -2$ Inequality 1

$y < x.$ Inequality 2

EXAMPLE 5 Writing a System of Linear Inequalities

Write a system of linear inequalities represented by the graph.

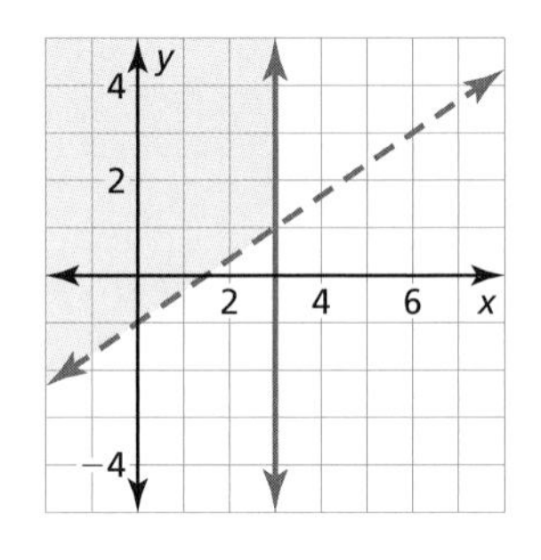

SOLUTION

Inequality 1 The vertical boundary line passes through (3, 0). So, an equation of the line is $x = 3$. The shaded region is to the *left* of the *solid* boundary line, so the inequality is $x \leq 3$.

Inequality 2 The slope of the other boundary line is $\frac{2}{3}$, and the y-intercept is −1. So, an equation of the line is $y = \frac{2}{3}x - 1$. The shaded region is *above* the *dashed* boundary line, so the inequality is $y > \frac{2}{3}x - 1$.

▶ The system of linear inequalities represented by the graph is

$x \leq 3$ Inequality 1

$y > \frac{2}{3}x - 1.$ Inequality 2

Monitoring Progress  Help in English and Spanish at *BigIdeasMath.com*

Write a system of linear inequalities represented by the graph.

6. 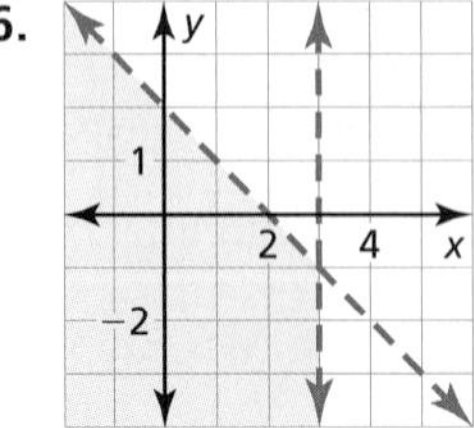

7. 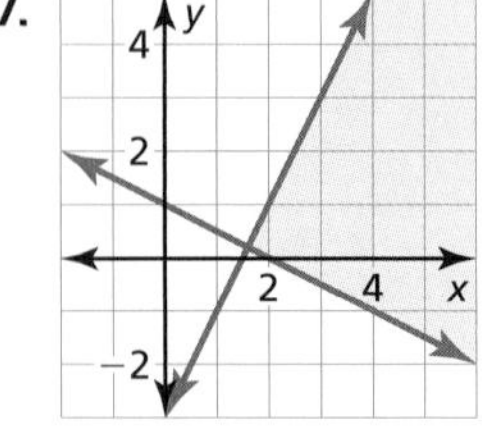

Solving Real-Life Problems

EXAMPLE 6 Modeling with Mathematics

You have at most 8 hours to spend at the mall and at the beach. You want to spend at least 2 hours at the mall and more than 4 hours at the beach. Write and graph a system that represents the situation. How much time can you spend at each location?

SOLUTION

1. **Understand the Problem** You know the total amount of time you can spend at the mall and at the beach. You also know how much time you want to spend at each location. You are asked to write and graph a system that represents the situation and determine how much time you can spend at each location.

2. **Make a Plan** Use the given information to write a system of linear inequalities. Then graph the system and identify an ordered pair in the solution region.

3. **Solve the Problem** Let x be the number of hours at the mall and let y be the number of hours at the beach.

$x + y \leq 8$	at most 8 hours at the mall and at the beach
$x \geq 2$	at least 2 hours at the mall
$y > 4$	more than 4 hours at the beach

Graph the system.

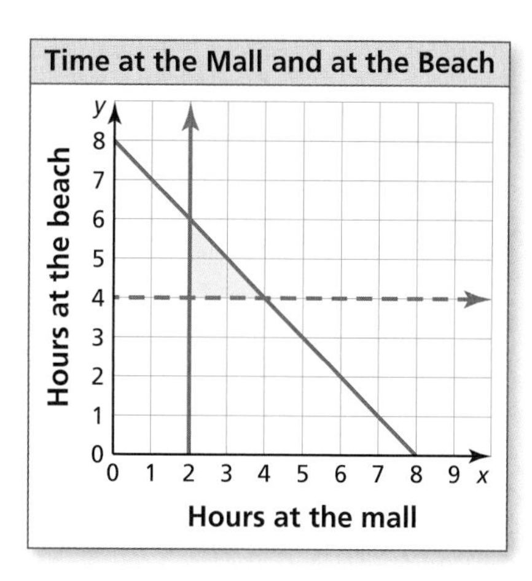

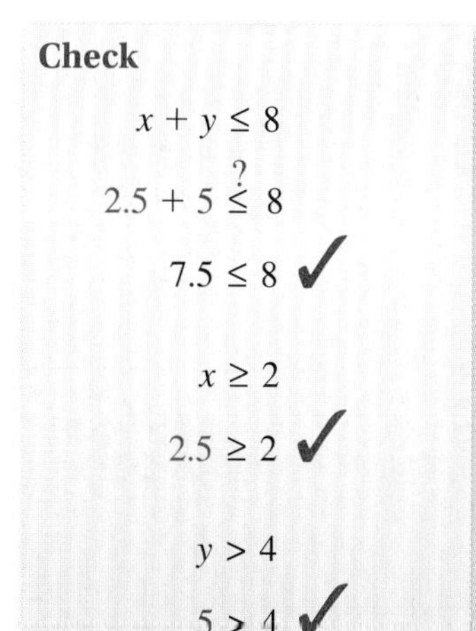

One ordered pair in the solution region is (2.5, 5).

▶ So, you can spend 2.5 hours at the mall and 5 hours at the beach.

4. **Look Back** Check your solution by substituting it into the inequalities in the system, as shown.

Monitoring Progress

Help in English and Spanish at *BigIdeasMath.com*

8. Name another solution of Example 6.

9. **WHAT IF?** You want to spend at least 3 hours at the mall. How does this change the system? Is (2.5, 5) still a solution? Explain.

5.7 Exercises

Dynamic Solutions available at *BigIdeasMath.com*

Vocabulary and Core Concept Check

1. **VOCABULARY** How can you verify that an ordered pair is a solution of a system of linear inequalities?

2. **WHICH ONE DOESN'T BELONG?** Use the graph shown. Which of the ordered pairs does *not* belong with the other three? Explain your reasoning.

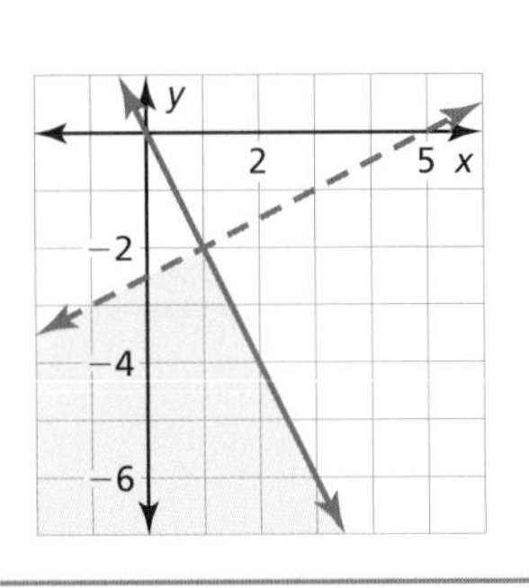

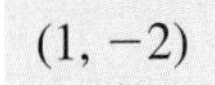
(1, −2) (0, −4) (−1, −6) (2, −4)

Monitoring Progress and Modeling with Mathematics

In Exercises 3–6, tell whether the ordered pair is a solution of the system of linear inequalities.

3. (−4, 3)

4. (−3, −1)

5. (−2, 5)

6. (1, 1)

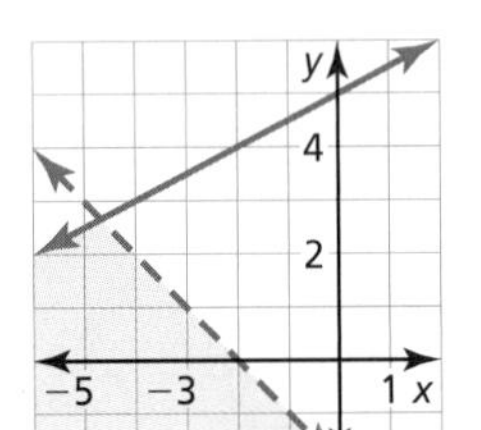

In Exercises 7–10, tell whether the ordered pair is a solution of the system of linear inequalities. *(See Example 1.)*

7. $(-5, 2);\ y < 4$, $y > x + 3$

8. $(1, -1);\ y > -2$, $y > x - 5$

9. $(0, 0);\ y \le x + 7$, $y \ge 2x + 3$

10. $(4, -3);\ y \le -x + 1$, $y \le 5x - 2$

In Exercises 11–20, graph the system of linear inequalities. *(See Examples 2 and 3.)*

11. $y > -3$
    $y \ge 5x$

12. $y < -1$
    $x > 4$

13. $y < -2$
    $y > 2$

14. $y < x - 1$
    $y \ge x + 1$

15. $y \ge -5$
    $y - 1 < 3x$

16. $x + y > 4$
    $y \ge \frac{3}{2}x - 9$

17. $x + y > 1$
    $-x - y < -3$

18. $2x + y \le 5$
    $y + 2 \ge -2x$

19. $x < 4$
    $y > 1$
    $y \ge -x + 1$

20. $x + y \le 10$
    $x - y \ge 2$
    $y > 2$

In Exercises 21–26, write a system of linear inequalities represented by the graph. *(See Examples 4 and 5.)*

21.
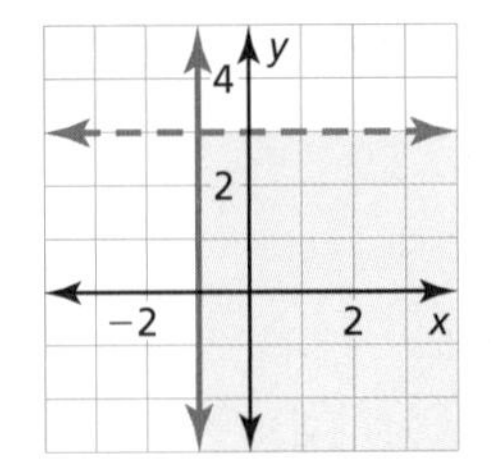

22.
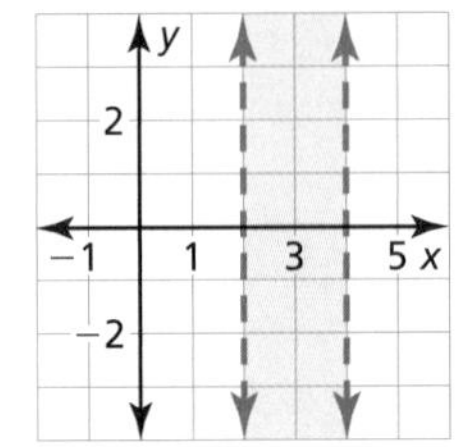

23.
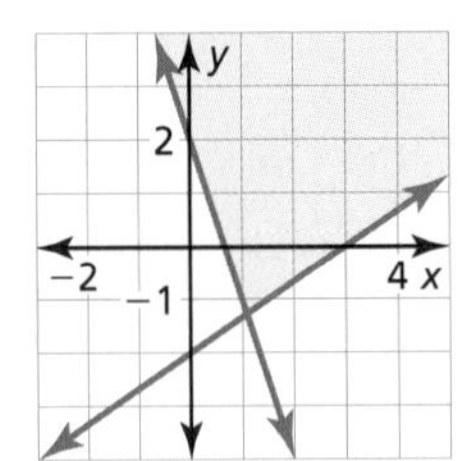

24.
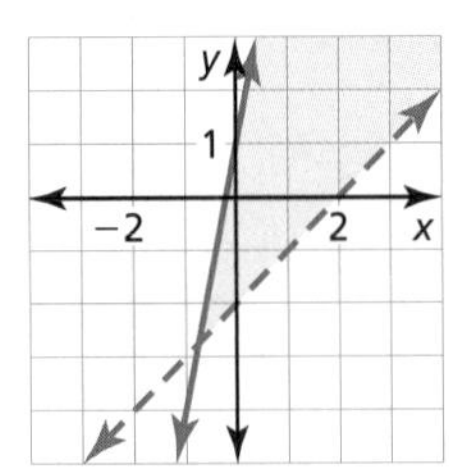

25.
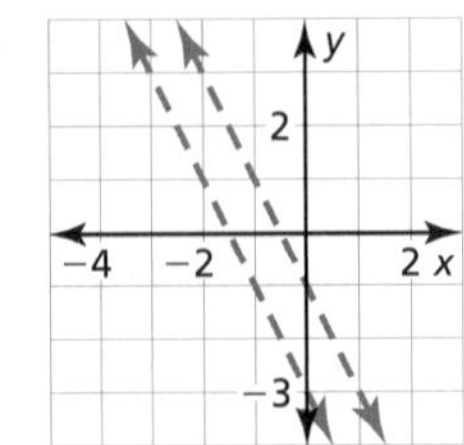

26.
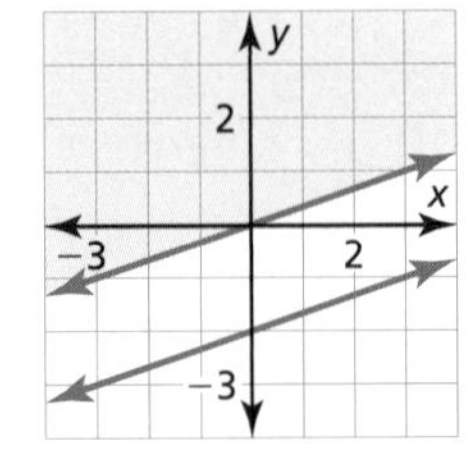

ERROR ANALYSIS In Exercises 27 and 28, describe and correct the error in graphing the system of linear inequalities.

27. 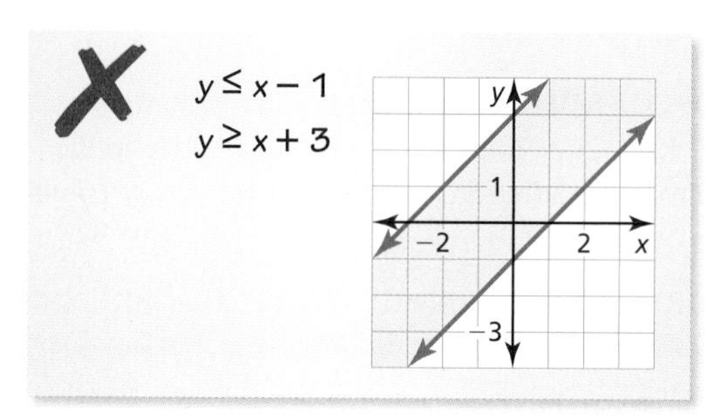

28. 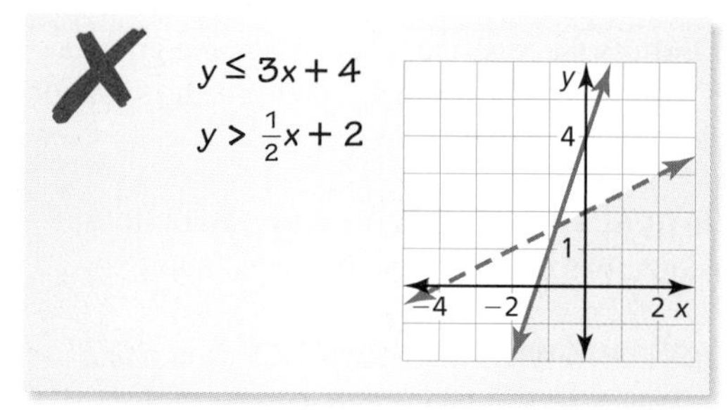

29. MODELING WITH MATHEMATICS You can spend at most \$21 on fruit. Blueberries cost \$4 per pound, and strawberries cost \$3 per pound. You need at least 3 pounds of fruit to make muffins. *(See Example 6.)*

a. Write and graph a system of linear inequalities that represents the situation.

b. Identify and interpret a solution of the system.

c. Use the graph to determine whether you can buy 4 pounds of blueberries and 1 pound of strawberries.

30. MODELING WITH MATHEMATICS You earn \$10 per hour working as a manager at a grocery store. You are required to work at the grocery store at least 8 hours per week. You also teach music lessons for \$15 per hour. You need to earn at least \$120 per week, but you do not want to work more than 20 hours per week.

a. Write and graph a system of linear inequalities that represents the situation.

b. Identify and interpret a solution of the system.

c. Use the graph to determine whether you can work 8 hours at the grocery store and teach 1 hour of music lessons.

31. MODELING WITH MATHEMATICS You are fishing for surfperch and rockfish, which are species of bottomfish. Gaming laws allow you to catch no more than 15 surfperch per day, no more than 10 rockfish per day, and no more than 20 total bottomfish per day.

a. Write and graph a system of linear inequalities that represents the situation.

b. Use the graph to determine whether you can catch 11 surfperch and 9 rockfish in 1 day.

surfperch

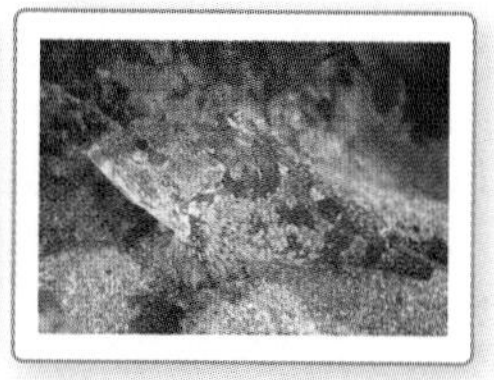
rockfish

32. REASONING Describe the intersection of the half-planes of the system shown.

$$x - y \leq 4$$
$$x - y \geq 4$$

33. MATHEMATICAL CONNECTIONS The following points are the vertices of a shaded rectangle.

$(-1, 1), (6, 1), (6, -3), (-1, -3)$

a. Write a system of linear inequalities represented by the shaded rectangle.

b. Find the area of the rectangle.

34. MATHEMATICAL CONNECTIONS The following points are the vertices of a shaded triangle.

$(2, 5), (6, -3), (-2, -3)$

a. Write a system of linear inequalities represented by the shaded triangle.

b. Find the area of the triangle.

35. PROBLEM SOLVING You plan to spend less than half of your monthly \$2000 paycheck on housing and savings. You want to spend at least 10% of your paycheck on savings and at most 30% of it on housing. How much money can you spend on savings and housing?

36. PROBLEM SOLVING On a road trip with a friend, you drive about 70 miles per hour, and your friend drives about 60 miles per hour. The plan is to drive less than 15 hours and at least 600 miles each day. Your friend will drive more hours than you. How many hours can you and your friend each drive in 1 day?

37. WRITING How are solving systems of linear inequalities and solving systems of linear equations similar? How are they different?

38. HOW DO YOU SEE IT? The graphs of two linear equations are shown.

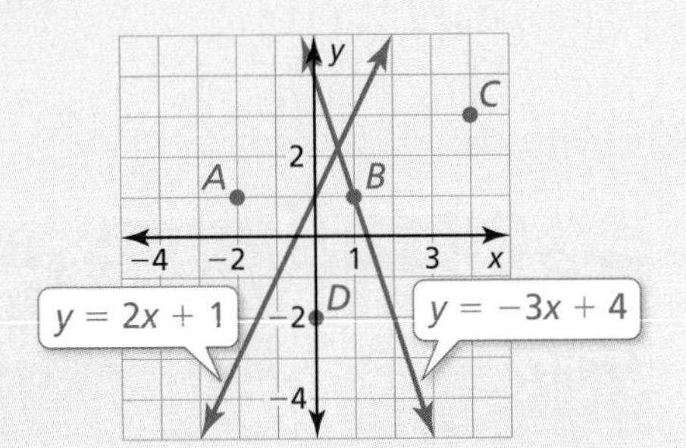

Replace the equal signs with inequality symbols to create a system of linear inequalities that has point C as a solution, but not points A, B, and D. Explain your reasoning.

$y \;\square\; -3x + 4$

$y \;\square\; 2x + 1$

39. USING STRUCTURE Write a system of linear inequalities that is equivalent to $|y| < x$, where $x > 0$. Graph the system.

40. MAKING AN ARGUMENT Your friend says that a system of linear inequalities in which the boundary lines are parallel must have no solution. Is your friend correct? Explain.

41. CRITICAL THINKING Is it possible for the solution set of a system of linear inequalities to be all real numbers? Explain your reasoning.

OPEN-ENDED In Exercises 42–44, write a system of linear inequalities with the given characteristic.

42. All solutions are in Quadrant I.

43. All solutions have one positive coordinate and one negative coordinate.

44. There are no solutions.

45. OPEN-ENDED One inequality in a system is $-4x + 2y > 6$. Write another inequality so the system has (a) no solution and (b) infinitely many solutions.

46. THOUGHT PROVOKING You receive a gift certificate for a clothing store and plan to use it to buy T-shirts and sweatshirts. Describe a situation in which you can buy 9 T-shirts and 1 sweatshirt, but you cannot buy 3 T-shirts and 8 sweatshirts. Write and graph a system of linear inequalities that represents the situation.

47. CRITICAL THINKING Write a system of linear inequalities that has exactly one solution.

48. MODELING WITH MATHEMATICS You make necklaces and key chains to sell at a craft fair. The table shows the amounts of time and money it takes to make a necklace and a key chain, and the amounts of time and money you have available for making them.

	Necklace	Key chain	Available
Time to make (hours)	0.5	0.25	20
Cost to make (dollars)	2	3	120

a. Write and graph a system of four linear inequalities that represents the number x of necklaces and the number y of key chains that you can make.

b. Find the vertices (corner points) of the graph of the system.

c. You sell each necklace for \$10 and each key chain for \$8. The revenue R is given by the equation $R = 10x + 8y$. Find the revenue corresponding to each ordered pair in part (b). Which vertex results in the maximum revenue?

Maintaining Mathematical Proficiency

Reviewing what you learned in previous grades and lessons

Write the product using exponents. *(Skills Review Handbook)*

49. $4 \cdot 4 \cdot 4 \cdot 4 \cdot 4$

50. $(-13) \cdot (-13) \cdot (-13)$

51. $x \cdot x \cdot x \cdot x \cdot x \cdot x$

Write an equation of the line with the given slope and y-intercept. *(Section 4.1)*

52. slope: 1
y-intercept: -6

53. slope: -3
y-intercept: 5

54. slope: $-\frac{1}{4}$
y-intercept: -1

55. slope: $\frac{4}{3}$
y-intercept: 0

5.5–5.7 What Did You Learn?

Core Vocabulary

linear inequality in two variables, *p. 268*
solution of a linear inequality in two variables, *p. 268*
graph of a linear inequality, *p. 268*
half-planes, *p. 268*
system of linear inequalities, *p. 274*
solution of a system of linear inequalities, *p. 274*
graph of a system of linear inequalities, *p. 275*

Core Concepts

Section 5.5

Solving Linear Equations by Graphing, *p. 262*
Solving Absolute Value Equations by Graphing, *p. 263*

Section 5.6

Graphing a Linear Inequality in Two Variables, *p. 269*

Section 5.7

Graphing a System of Linear Inequalities, *p. 275*
Writing a System of Linear Inequalities, *p. 276*

Mathematical Practices

1. Why do the equations in Exercise 35 on page 266 contain absolute value expressions?
2. Why is it important to be precise when answering part (a) of Exercise 39 on page 272?
3. Describe the overall step-by-step process you used to solve Exercise 35 on page 279.

Performance Task

Prize Patrol

You have been selected to drive a prize patrol cart and place prizes on the competing teams' predetermined paths. You know the teams' routes and you can only make one pass. Where will you place the prizes so that each team will have a chance to find a prize on their route?

To explore the answers to these questions and more, go to ***BigIdeasMath.com***.

5 Chapter Review

Dynamic Solutions available at *BigIdeasMath.com*

5.1 Solving Systems of Linear Equations by Graphing *(pp. 235–240)*

Solve the system by graphing. $y = x - 2$ Equation 1

$y = -3x + 2$ Equation 2

Step 1 Graph each equation.

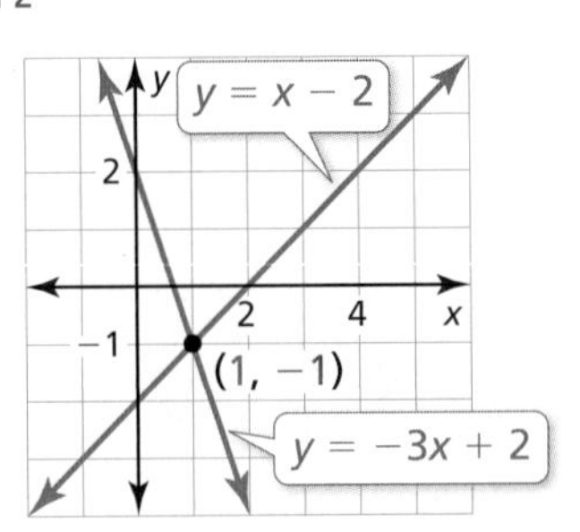

Step 2 Estimate the point of intersection.
The graphs appear to intersect at $(1, -1)$.

Step 3 Check your point from Step 2.

Equation 1	Equation 2
$y = x - 2$	$y = -3x + 2$
$-1 \stackrel{?}{=} 1 - 2$	$-1 \stackrel{?}{=} -3(1) + 2$
$-1 = -1$ ✓	$-1 = -1$ ✓

▶ The solution is $(1, -1)$.

Solve the system of linear equations by graphing.

1. $y = -3x + 1$
$y = x - 7$

2. $y = -4x + 3$
$4x - 2y = 6$

3. $5x + 5y = 15$
$2x - 2y = 10$

5.2 Solving Systems of Linear Equations by Substitution *(pp. 241–246)*

Solve the system by substitution. $-2x + y = -8$ Equation 1

$7x + y = 10$ Equation 2

Step 1 Solve for y in Equation 1.

$y = 2x - 8$ Revised Equation 1

Step 2 Substitute $2x - 8$ for y in Equation 2 and solve for x.

$7x + y = 10$	Equation 2
$7x + (2x - 8) = 10$	Substitute $2x - 8$ for y.
$9x - 8 = 10$	Combine like terms.
$9x = 18$	Add 8 to each side.
$x = 2$	Divide each side by 9.

Step 3 Substituting 2 for x in Equation 1 and solving for y gives $y = -4$.

▶ The solution is $(2, -4)$.

Solve the system of linear equations by substitution. Check your solution.

4. $3x + y = -9$
$y = 5x + 7$

5. $x + 4y = 6$
$x - y = 1$

6. $2x + 3y = 4$
$y + 3x = 6$

7. You spend \$20 total on tubes of paint and disposable brushes for an art project. Tubes of paint cost \$4.00 each and paintbrushes cost \$0.50 each. You purchase twice as many brushes as tubes of paint. How many brushes and tubes of paint do you purchase?

5.3 Solving Systems of Linear Equations by Elimination (pp. 247–252)

Solve the system by elimination. $4x + 6y = -8$ Equation 1
$x - 2y = -2$ Equation 2

Step 1 Multiply Equation 2 by 3 so that the coefficients of the y-terms are opposites.

$4x + 6y = -8$ → $4x + 6y = -8$ Equation 1

$x - 2y = -2$ **Multiply by 3.** → $3x - 6y = -6$ Revised Equation 2

Step 2 Add the equations.

$4x + 6y = -8$ Equation 1

$3x - 6y = -6$ Revised Equation 2

$7x = -14$ Add the equations.

Step 3 Solve for x.

$7x = -14$ Resulting equation from Step 2

$x = -2$ Divide each side by 7.

Step 4 Substitute -2 for x in one of the original equations and solve for y.

$4x + 6y = -8$ Equation 1

$4(-2) + 6y = -8$ Substitute -2 for x.

$-8 + 6y = -8$ Multiply.

$y = 0$ Solve for y.

▶ The solution is $(-2, 0)$.

Check

Equation 1

$4x + 6y = -8$

$4(-2) + 6(0) \stackrel{?}{=} -8$

$-8 = -8$ ✓

Equation 2

$x - 2y = -2$

$(-2) - 2(0) \stackrel{?}{=} -2$

$-2 = -2$ ✓

Solve the system of linear equations by elimination. Check your solution.

8. $9x - 2y = 34$
$5x + 2y = -6$

9. $x + 6y = 28$
$2x - 3y = -19$

10. $8x - 7y = -3$
$6x - 5y = -1$

5.4 Solving Special Systems of Linear Equations (pp. 253–258)

Solve the system. $4x + 2y = -14$ Equation 1
$y = -2x - 6$ Equation 2

Solve by substitution. Substitute $-2x - 6$ for y in Equation 1.

$4x + 2y = -14$ Equation 1

$4x + 2(-2x - 6) = -14$ Substitute $-2x - 6$ for y.

$4x - 4x - 12 = -14$ Distributive Property

$-12 = -14$ ✗ Combine like terms.

▶ The equation $-12 = -14$ is never true. So, the system has no solution.

Solve the system of linear equations.

11. $x = y + 2$
$-3x + 3y = 6$

12. $3x - 6y = -9$
$-5x + 10y = 10$

13. $-4x + 4y = 32$
$3x + 24 = 3y$

5.5 Solving Equations by Graphing (pp. 261–266)

Solve $3x - 1 = -2x + 4$ by graphing. Check your solution.

Step 1 Write a system of linear equations using each side of the original equation.

$y = 3x - 1$ → $3x - 1 = -2x + 4$ ← $y = -2x + 4$

Step 2 Graph the system.

$y = 3x - 1$ Equation 1

$y = -2x + 4$ Equation 2

The graphs intersect at (1, 2).

▶ So, the solution of the equation is $x = 1$.

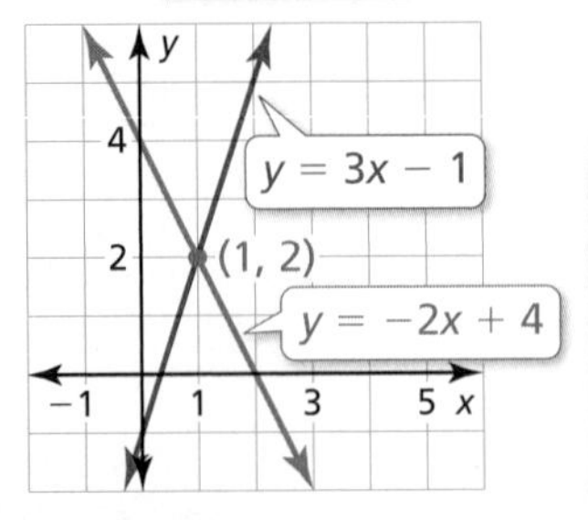

Check

$3x - 1 = -2x + 4$

$3(1) - 1 \stackrel{?}{=} -2(1) + 4$

$2 = 2$ ✓

Solve the equation by graphing. Check your solution(s).

14. $\frac{1}{3}x + 5 = -2x - 2$ **15.** $|x + 1| = |-x - 9|$ **16.** $|2x - 8| = |x + 5|$

5.6 Graphing Linear Inequalities in Two Variables (pp. 267–272)

Graph $4x + 2y \geq -6$ in a coordinate plane.

Step 1 Graph $4x + 2y = -6$, or $y = -2x - 3$. Use a solid line because the inequality symbol is $\geq$.

Step 2 Test (0, 0).

$4x + 2y \geq -6$ Write the inequality.

$4(0) + 2(0) \stackrel{?}{\geq} -6$ Substitute.

$0 \geq -6$ ✓ Simplify.

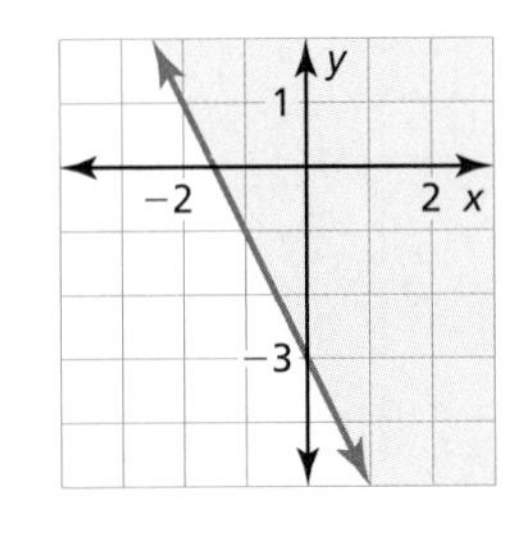

Step 3 Because (0, 0) is a solution, shade the half-plane that contains (0, 0).

Graph the inequality in a coordinate plane.

17. $y > -4$ **18.** $-9x + 3y \geq 3$ **19.** $5x + 10y < 40$

5.7 Systems of Linear Inequalities (pp. 273–280)

Graph the system. $y < x - 2$ Inequality 1

$y \geq 2x - 4$ Inequality 2

Step 1 Graph each inequality.

Step 2 Find the intersection of the half-planes. One solution is (0, −3).

The solution is the purple-shaded region.

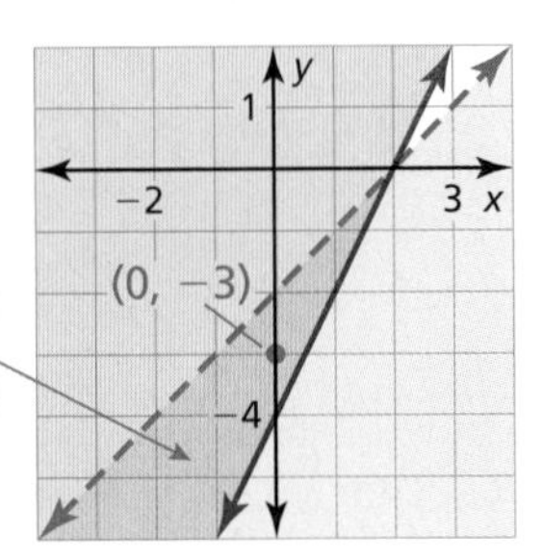

Graph the system of linear inequalities.

20. $y \leq x - 3$
$y \geq x + 1$

21. $y > -2x + 3$
$y \geq \frac{1}{4}x - 1$

22. $x + 3y > 6$
$2x + y < 7$

5 Chapter Test

Solve the system of linear equations using any method. Explain why you chose the method.

1. $8x + 3y = -9$
$-8x + y = 29$

2. $\frac{1}{2}x + y = -6$
$y = \frac{3}{5}x + 5$

3. $y = 4x + 4$
$-8x + 2y = 8$

4. $x = y - 11$
$x - 3y = 1$

5. $6x - 4y = 9$
$9x - 6y = 15$

6. $y = 5x - 7$
$-4x + y = -1$

7. Write a system of linear inequalities so the points $(1, 2)$ and $(4, -3)$ are solutions of the system, but the point $(-2, 8)$ is not a solution of the system.

8. How is solving the equation $|2x + 1| = |x - 7|$ by graphing similar to solving the equation $4x + 3 = -2x + 9$ by graphing? How is it different?

Graph the system of linear inequalities.

9. $y > \frac{1}{2}x + 4$
$2y \le x + 4$

10. $x + y < 1$
$5x + y > 4$

11. $y \ge -\frac{2}{3}x + 1$
$-3x + y > -2$

12. You pay \$45.50 for 10 gallons of gasoline and 2 quarts of oil at a gas station. Your friend pays \$22.75 for 5 gallons of the same gasoline and 1 quart of the same oil.

a. Is there enough information to determine the cost of 1 gallon of gasoline and 1 quart of oil? Explain.

b. The receipt shown is for buying the same gasoline and same oil. Is there now enough information to determine the cost of 1 gallon of gasoline and 1 quart of oil? Explain.

c. Determine the cost of 1 gallon of gasoline and 1 quart of oil.

WELCOME

DATE 11/12/13 16:25
PUMP # 03
PRODUCT: REGUNL
GALLONS: 8.00
2 QUARTS OIL
TOTAL: \$38.40

THANK YOU
HAVE A NICE DAY

13. Describe the advantages and disadvantages of solving a system of linear equations by graphing.

14. You have at most \$60 to spend on trophies and medals to give as prizes for a contest.

a. Write and graph an inequality that represents the numbers of trophies and medals you can buy. Identify and interpret a solution of the inequality.

b. You want to purchase at least 6 items. Write and graph a system that represents the situation. How many of each item can you buy?

15. Compare the slopes and y-intercepts of the graphs of the equations in the linear system $8x + 4y = 12$ and $3y = -6x - 15$ to determine whether the system has one solution, no solution, or infinitely many solutions. Explain.

5 Cumulative Assessment

1. The graph of which equation is shown?

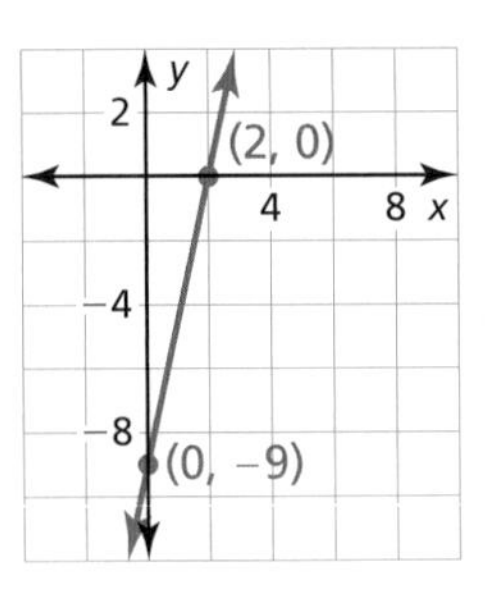

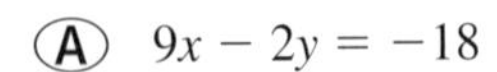

Ⓐ $9x - 2y = -18$

Ⓑ $-9x - 2y = 18$

Ⓒ $9x + 2y = 18$

Ⓓ $-9x + 2y = -18$

2. A van rental company rents out 6-, 8-, 12-, and 16-passenger vans. The function $C(x) = 100 + 5x$ represents the cost C (in dollars) of renting an x-passenger van for a day. Choose the numbers that are in the range of the function.

130	140	150	160	170	180	190	200

3. Fill in the system of linear inequalities with <, ≤, >, or ≥ so that the graph represents the system.

y ▭ $3x - 2$

y ▭ $-x + 5$

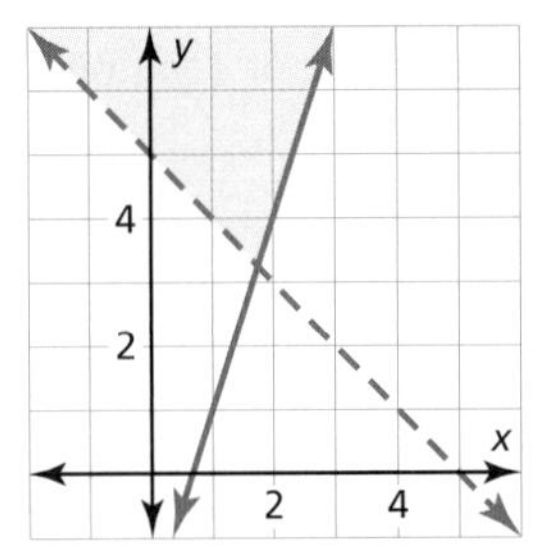

4. Your friend claims to be able to fill in each box with a constant so that when you set each side of the equation equal to y and graph the resulting equations, the lines will intersect exactly once. Do you support your friend's claim? Explain.

$$4x + \square = 4x + \square$$

5. Select the phrases you should use when describing the transformations from the graph of f to the graph of g.

reflection in the x-axis	reflection in the y-axis
horizontal translation	vertical translation
horizontal stretch	vertical stretch
horizontal shrink	vertical shrink

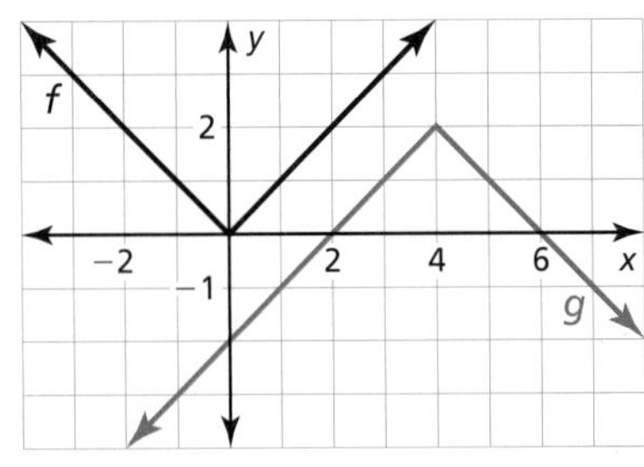

6. Which two equations form a system of linear equations that has no solution?

$y = 3x + 2$	$y = \frac{1}{3}x + 2$	$y = 2x + 3$	$y = 3x + \frac{1}{2}$

7. Fill in a value for a so that each statement is true for the equation $ax - 8 = 4 - x$.

a. When $a =$ ▢, the solution is $x = -2$.

b. When $a =$ ▢, the solution is $x = 12$.

c. When $a =$ ▢, the solution is $x = 3$.

8. Which ordered pair is a solution of the linear inequality whose graph is shown?

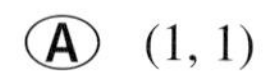

Ⓐ (1, 1)

Ⓑ (−1, 1)

Ⓒ (−1, −1)

Ⓓ (1, −1)

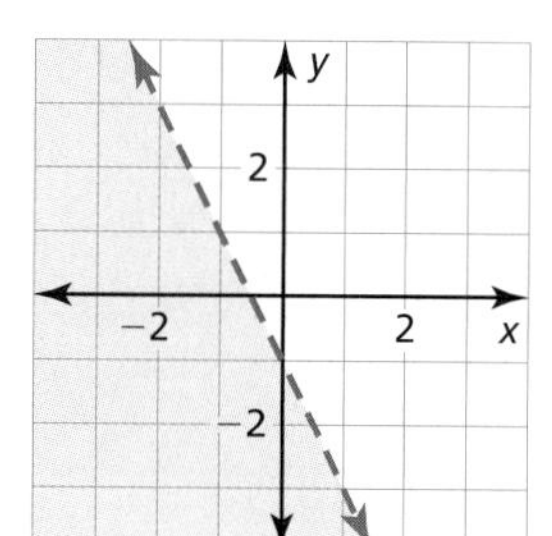

9. Which of the systems of linear equations are equivalent?

$4x - 5y = 3$ $2x + 15y = -1$	$4x - 5y = 3$ $-4x - 30y = 2$	$4x - 5y = 3$ $4x + 30y = -1$	$12x - 15y = 9$ $2x + 15y = -1$

10. The value of x is more than 9. Which of the inequalities correctly describe the triangle? The perimeter (in feet) is represented by P, and the area (in square feet) is represented by A.

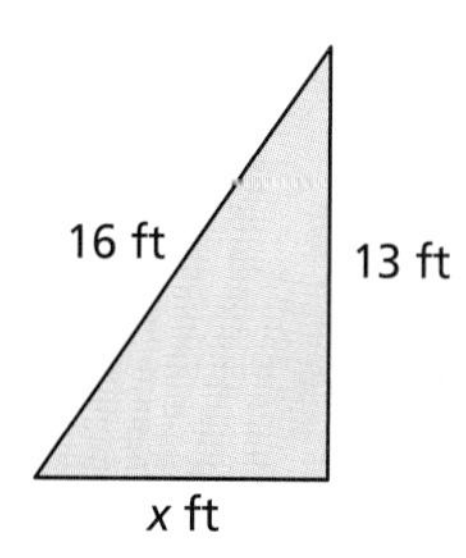

$P < 29$	$A > 117$	$P > 38$	$A > 58.5$	$A > 104$

6 Exponential Functions and Sequences

Fibonacci and Flowers *(p. 343)*

Soup Kitchen *(p. 338)*

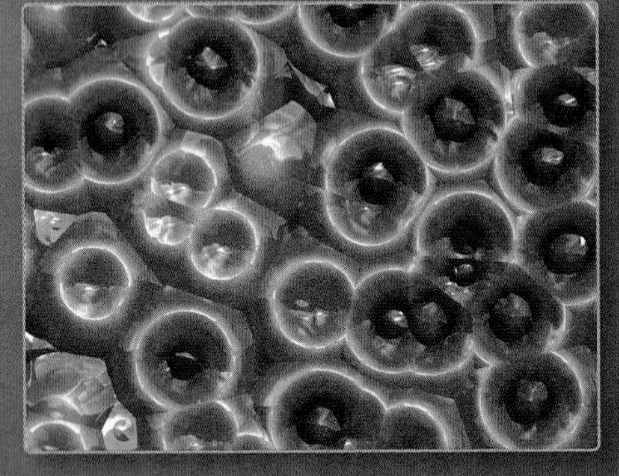

Bacterial Culture *(p. 330)*

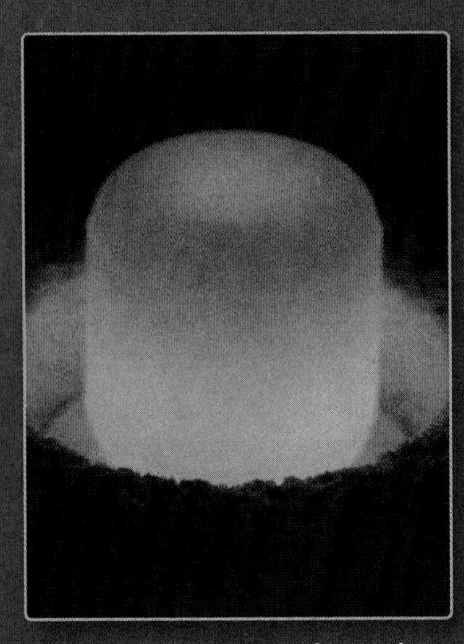

Plutonium Decay *(p. 321)*

Coyote Population *(p. 311)*

Maintaining Mathematical Proficiency

Using Order of Operations

Example 1 **Evaluate $10^2 \div (30 \div 3) - 4(3 - 9) + 5^1$.**

First:	Parentheses	$10^2 \div (30 \div 3) - 4(3 - 9) + 5^1 = 10^2 \div 10 - 4(-6) + 5^1$
Second:	Exponents	$= 100 \div 10 - 4(-6) + 5$
Third:	Multiplication and Division (from left to right)	$= 10 + 24 + 5$
Fourth:	Addition and Subtraction (from left to right)	$= 39$

Evaluate the expression.

1. $12\left(\frac{14}{2}\right) - 3^3 + 15 - 9^2$

2. $5^2 \cdot 8 \div 2^2 + 20 \cdot 3 - 4$

3. $-7 + 16 \div 2^4 + (10 - 4^2)$

Finding Square Roots

Example 2 **Find $-\sqrt{81}$.**

▶ $-\sqrt{81}$ represents the negative square root. Because $9^2 = 81$, $-\sqrt{81} = -\sqrt{9^2} = -9$.

Find the square root(s).

4. $\sqrt{64}$

5. $-\sqrt{4}$

6. $-\sqrt{25}$

7. $\pm\sqrt{121}$

Writing Equations for Arithmetic Sequences

Example 3 **Write an equation for the *n*th term of the arithmetic sequence 5, 15, 25, 35, ...**

The first term is 5, and the common difference is 10.

$a_n = a_1 + (n - 1)d$	Equation for an arithmetic sequence
$a_n = 5 + (n - 1)(10)$	Substitute 5 for a_1 and 10 for d.
$a_n = 10n - 5$	Simplify.

Write an equation for the *n*th term of the arithmetic sequence.

8. 12, 14, 16, 18, . . .

9. 6, 3, 0, −3, . . .

10. 22, 15, 8, 1, . . .

11. **ABSTRACT REASONING** Recall that a perfect square is a number with integers as its square roots. Is the product of two perfect squares always a perfect square? Is the quotient of two perfect squares always a perfect square? Explain your reasoning.

Mathematical Practices

Mathematically proficient students look closely to find a pattern.

Problem-Solving Strategies

Core Concept

Finding a Pattern

When solving a real-life problem, look for a pattern in the data. The pattern could include repeating items, numbers, or events. After you find the pattern, describe it and use it to solve the problem.

EXAMPLE 1 Using a Problem-Solving Strategy

The volumes of seven chambers of a chambered nautilus are given. Find the volume of Chamber 10.

Chamber 7: 1.207 cm^3
Chamber 6: 1.135 cm^3
Chamber 5: 1.068 cm^3
Chamber 4: 1.005 cm^3
Chamber 3: 0.945 cm^3
Chamber 2: 0.889 cm^3
Chamber 1: 0.836 cm^3

SOLUTION

To find a pattern, try dividing each volume by the volume of the previous chamber.

$$\frac{0.889}{0.836} \approx 1.063 \qquad \frac{0.945}{0.889} \approx 1.063 \qquad \frac{1.005}{0.945} \approx 1.063$$

$$\frac{1.068}{1.005} \approx 1.063 \qquad \frac{1.135}{1.068} \approx 1.063 \qquad \frac{1.207}{1.135} \approx 1.063$$

From this, you can see that the volume of each chamber is about 6.3% greater than the volume of the previous chamber. To find the volume of Chamber 10, multiply the volume of Chamber 7 by 1.063 three times.

$1.207(1.063) \approx \mathbf{1.283}$ (volume of Chamber 8)

$1.283(1.063) \approx \mathbf{1.364}$ (volume of Chamber 9)

$1.364(1.063) \approx \mathbf{1.450}$ (volume of Chamber 10)

▶ The volume of Chamber 10 is about 1.450 cubic centimeters.

Monitoring Progress

1. A rabbit population over 8 consecutive years is given by 50, 80, 128, 205, 328, 524, 839, 1342. Find the population in the tenth year.
2. The sums of the numbers in the first eight rows of Pascal's Triangle are 1, 2, 4, 8, 16, 32, 64, 128. Find the sum of the numbers in the tenth row.

6.1 Properties of Exponents

Essential Question How can you write general rules involving properties of exponents?

EXPLORATION 1 **Writing Rules for Properties of Exponents**

Work with a partner.

a. What happens when you multiply two powers with the same base? Write the product of the two powers as a single power. Then write a *general rule* for finding the product of two powers with the same base.

i. $(2^2)(2^3) =$ ______ **ii.** $(4^1)(4^5) =$ ______

iii. $(5^3)(5^5) =$ ______ **iv.** $(x^2)(x^6) =$ ______

b. What happens when you divide two powers with the same base? Write the quotient of the two powers as a single power. Then write a *general rule* for finding the quotient of two powers with the same base.

i. $\frac{4^3}{4^2} =$ ______ **ii.** $\frac{2^5}{2^2} =$ ______

iii. $\frac{x^6}{x^3} =$ ______ **iv.** $\frac{3^4}{3^4} =$ ______

c. What happens when you find a power of a power? Write the expression as a single power. Then write a *general rule* for finding a power of a power.

i. $(2^2)^4 =$ ______ **ii.** $(7^3)^2 =$ ______

iii. $(y^3)^3 =$ ______ **iv.** $(x^4)^2 =$ ______

d. What happens when you find a power of a product? Write the expression as the product of two powers. Then write a *general rule* for finding a power of a product.

i. $(2 \cdot 5)^2 =$ ______ **ii.** $(5 \cdot 4)^3 =$ ______

iii. $(6a)^2 =$ ______ **iv.** $(3x)^2 =$ ______

e. What happens when you find a power of a quotient? Write the expression as the quotient of two powers. Then write a *general rule* for finding a power of a quotient.

i. $\left(\frac{2}{3}\right)^2 =$ ______ **ii.** $\left(\frac{4}{3}\right)^3 =$ ______

iii. $\left(\frac{x}{2}\right)^3 =$ ______ **iv.** $\left(\frac{a}{b}\right)^4 =$ ______

WRITING GENERAL RULES

To be proficient in math, you need to understand and use stated assumptions, definitions, and previously established results in writing general rules.

Communicate Your Answer

2. How can you write general rules involving properties of exponents?

3. There are 3^3 small cubes in the cube below. Write an expression for the number of small cubes in the large cube at the right.

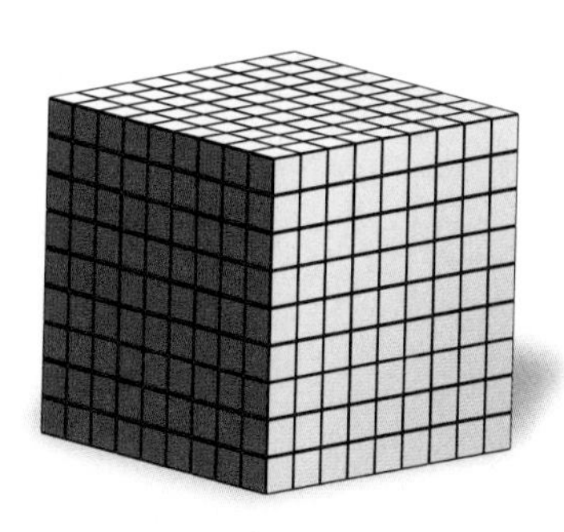

6.1 Lesson

Core Vocabulary

Previous
power
exponent
base
scientific notation

What You Will Learn

- Use zero and negative exponents.
- Use the properties of exponents.
- Solve real-life problems involving exponents.

Using Zero and Negative Exponents

Core Concept

Zero Exponent

Words For any nonzero number a, $a^0 = 1$. The power 0^0 is undefined.

Numbers $4^0 = 1$ **Algebra** $a^0 = 1$, where $a \neq 0$

Negative Exponents

Words For any integer n and any nonzero number a, a^{-n} is the reciprocal of a^n.

Numbers $4^{-2} = \frac{1}{4^2}$ **Algebra** $a^{-n} = \frac{1}{a^n}$, where $a \neq 0$

EXAMPLE 1 Using Zero and Negative Exponents

Evaluate each expression.

a. 6.7^0 **b.** $(-2)^{-4}$

SOLUTION

a. $6.7^0 = 1$ Definition of zero exponent

b. $(-2)^{-4} = \frac{1}{(-2)^4}$ Definition of negative exponent

$= \frac{1}{16}$ Simplify.

EXAMPLE 2 Simplifying an Expression

Simplify the expression $\frac{4x^0}{y^{-3}}$. Write your answer using only positive exponents.

SOLUTION

$\frac{4x^0}{y^{-3}} = 4x^0y^3$ Definition of negative exponent

$= 4y^3$ Definition of zero exponent

Monitoring Progress  Help in English and Spanish at *BigIdeasMath.com*

Evaluate the expression.

1. $(-9)^0$ **2.** 3^{-3} **3.** $\frac{-5^0}{2^{-2}}$

4. Simplify the expression $\frac{3^{-2}x^{-5}}{y^0}$. Write your answer using only positive exponents.

Using the Properties of Exponents

REMEMBER

The expression x^3 is called a *power*. The *base*, x, is used as a factor 3 times because the *exponent* is 3.

Core Concept

Product of Powers Property

Let a be a real number, and let m and n be integers.

Words To multiply powers with the same base, add their exponents.

Numbers $4^6 \cdot 4^3 = 4^{6+3} = 4^9$ **Algebra** $a^m \cdot a^n = a^{m+n}$

Quotient of Powers Property

Let a be a nonzero real number, and let m and n be integers.

Words To divide powers with the same base, subtract their exponents.

Numbers $\frac{4^6}{4^3} = 4^{6-3} = 4^3$ **Algebra** $\frac{a^m}{a^n} = a^{m-n}$, where $a \neq 0$

Power of a Power Property

Let a be a real number, and let m and n be integers.

Words To find a power of a power, multiply the exponents.

Numbers $(4^6)^3 = 4^{6 \cdot 3} = 4^{18}$ **Algebra** $(a^m)^n = a^{mn}$

EXAMPLE 3 Using Properties of Exponents

Simplify each expression. Write your answer using only positive exponents.

a. $3^2 \cdot 3^6$ **b.** $\frac{(-4)^2}{(-4)^7}$ **c.** $(z^4)^{-3}$

SOLUTION

a. $3^2 \cdot 3^6 = 3^{2+6}$ Product of Powers Property

$= 3^8 = 6561$ Simplify.

b. $\frac{(-4)^2}{(-4)^7} = (-4)^{2-7}$ Quotient of Powers Property

$= (-4)^{-5}$ Simplify.

$= \frac{1}{(-4)^5} = -\frac{1}{1024}$ Definition of negative exponent

c. $(z^4)^{-3} = z^{4 \cdot (-3)}$ Power of a Power Property

$= z^{-12}$ Simplify.

$= \frac{1}{z^{12}}$ Definition of negative exponent

Monitoring Progress  Help in English and Spanish at *BigIdeasMath.com*

Simplify the expression. Write your answer using only positive exponents.

5. $10^4 \cdot 10^{-6}$ **6.** $x^9 \cdot x^{-9}$ **7.** $\frac{-5^8}{-5^4}$

8. $\frac{y^6}{y^7}$ **9.** $(6^{-2})^{-1}$ **10.** $(w^{12})^5$

Power of a Product Property

Let a and b be real numbers, and let m be an integer.

Words To find a power of a product, find the power of each factor and multiply.

Numbers $(3 \cdot 2)^5 = 3^5 \cdot 2^5$ **Algebra** $(ab)^m = a^m b^m$

Power of a Quotient Property

Let a and b be real numbers with $b \neq 0$, and let m be an integer.

Words To find the power of a quotient, find the power of the numerator and the power of the denominator and divide.

Numbers $\left(\frac{3}{2}\right)^5 = \frac{3^5}{2^5}$ **Algebra** $\left(\frac{a}{b}\right)^m = \frac{a^m}{b^m}$, where $b \neq 0$

EXAMPLE 4 Using Properties of Exponents

Simplify each expression. Write your answer using only positive exponents.

a. $(-1.5y)^2$ **b.** $\left(\frac{a}{-10}\right)^3$ **c.** $\left(\frac{3d}{2}\right)^4$ **d.** $\left(\frac{2x}{3}\right)^{-5}$

SOLUTION

a. $(-1.5y)^2 = (-1.5)^2 \cdot y^2$ Power of a Product Property

$= 2.25y^2$ Simplify.

b. $\left(\frac{a}{-10}\right)^3 = \frac{a^3}{(-10)^3}$ Power of a Quotient Property

$= -\frac{a^3}{1000}$ Simplify.

c. $\left(\frac{3d}{2}\right)^4 = \frac{(3d)^4}{2^4}$ Power of a Quotient Property

$= \frac{3^4 d^4}{2^4}$ Power of a Product Property

$= \frac{81d^4}{16}$ Simplify.

d. $\left(\frac{2x}{3}\right)^{-5} = \frac{(2x)^{-5}}{3^{-5}}$ Power of a Quotient Property

$= \frac{3^5}{(2x)^5}$ Definition of negative exponent

$= \frac{3^5}{2^5 x^5}$ Power of a Product Property

$= \frac{243}{32x^5}$ Simplify.

ANOTHER WAY

Because the exponent is negative, you could find the reciprocal of the base first. Then simplify.

$$\left(\frac{2x}{3}\right)^{-5} = \left(\frac{3}{2x}\right)^5 = \frac{243}{32x^5}$$

Monitoring Progress  Help in English and Spanish at *BigIdeasMath.com*

Simplify the expression. Write your answer using only positive exponents.

11. $(10y)^{-3}$ **12.** $\left(-\frac{4}{n}\right)^5$ **13.** $\left(\frac{1}{2k^2}\right)^5$ **14.** $\left(\frac{6c}{7}\right)^{-2}$

Solving Real-Life Problems

EXAMPLE 5 Simplifying a Real-Life Expression

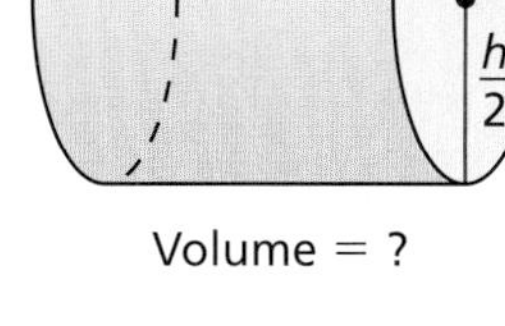

Volume = ?

$2\pi r^3$	$\pi h^3 2^{-2}$	$\pi h 4^{-1}$
$\dfrac{\pi h^2}{4}$	$\dfrac{\pi h^3}{4}$	$\dfrac{\pi h^3}{2}$

Which of the expressions shown represent the volume of the cylinder, where r is the radius and h is the height?

SOLUTION

$V = \pi r^2 h$ — Formula for the volume of a cylinder

$= \pi\left(\dfrac{h}{2}\right)^2(h)$ — Substitute $\dfrac{h}{2}$ for r.

$= \pi\left(\dfrac{h^2}{2^2}\right)(h)$ — Power of a Quotient Property

$= \dfrac{\pi h^3}{4}$ — Simplify.

Any expression equivalent to $\dfrac{\pi h^3}{4}$ represents the volume of the cylinder.

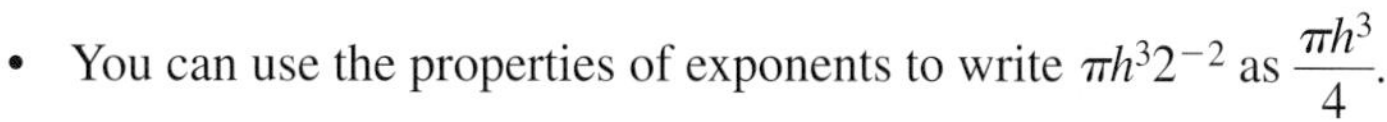

- You can use the properties of exponents to write $\pi h^3 2^{-2}$ as $\dfrac{\pi h^3}{4}$.
- Note $h = 2r$. When you substitute $2r$ for h in $\dfrac{\pi h^3}{4}$, you can write $\dfrac{\pi(2r)^3}{4}$ as $2\pi r^3$.
- None of the other expressions are equivalent to $\dfrac{\pi h^3}{4}$.

▶ The expressions $2\pi r^3$, $\pi h^3 2^{-2}$, and $\dfrac{\pi h^3}{4}$ represent the volume of the cylinder.

REMEMBER

A number is written in scientific notation when it is of the form $a \times 10^b$, where $1 \le a < 10$ and b is an integer.

EXAMPLE 6 Solving a Real-Life Problem

A jellyfish emits about 1.25×10^8 particles of light, or photons, in 6.25×10^{-4} second. How many photons does the jellyfish emit each second? Write your answer in scientific notation and in standard form.

SOLUTION

Divide to find the unit rate.

$\dfrac{1.25 \times 10^8}{6.25 \times 10^{-4}}$ (photons / seconds) — Write the rate.

$= \dfrac{1.25}{6.25} \times \dfrac{10^8}{10^{-4}}$ — Rewrite.

$= 0.2 \times 10^{12}$ — Simplify.

$= 2 \times 10^{11}$ — Write in scientific notation.

▶ The jellyfish emits 2×10^{11}, or 200,000,000,000 photons per second.

Monitoring Progress

Help in English and Spanish at *BigIdeasMath.com*

15. Write two expressions that represent the area of a base of the cylinder in Example 5.

16. It takes the Sun about 2.3×10^8 years to orbit the center of the Milky Way. It takes Pluto about 2.5×10^2 years to orbit the Sun. How many times does Pluto orbit the Sun while the Sun completes one orbit around the center of the Milky Way? Write your answer in scientific notation.

6.1 Exercises

Dynamic Solutions available at *BigIdeasMath.com*

Vocabulary and Core Concept Check

1. **VOCABULARY** Which definitions or properties would you use to simplify the expression $(4^8 \cdot 4^{-4})^{-2}$? Explain.

2. **WRITING** Explain when and how to use the Power of a Product Property.

3. **WRITING** Explain when and how to use the Quotient of Powers Property.

4. **DIFFERENT WORDS, SAME QUESTION** Which is different? Find "both" answers.

Simplify $3^3 \cdot 3^6$.	Simplify 3^{3+6}.	Simplify $3^{6 \cdot 3}$.	Simplify $3^6 \cdot 3^3$.

Monitoring Progress and Modeling with Mathematics

In Exercises 5–12, evaluate the expression. *(See Example 1.)*

5. $(-7)^0$
6. 4^0
7. 5^{-4}
8. $(-2)^{-5}$
9. $\frac{2^{-4}}{4^0}$
10. $\frac{5^{-1}}{-9^0}$
11. $\frac{-3^{-3}}{6^{-2}}$
12. $\frac{(-8)^{-2}}{3^{-4}}$

In Exercises 13–22, simplify the expression. Write your answer using only positive exponents. *(See Example 2.)*

13. x^{-7}
14. y^0
15. $9x^0y^{-3}$
16. $15c^{-8}d^0$
17. $\frac{2^{-2}m^{-3}}{n^0}$
18. $\frac{10^0r^{-11}s}{3^2}$
19. $\frac{4^{-3}a^0}{b^{-7}}$
20. $\frac{p^{-8}}{7^{-2}q^{-9}}$
21. $\frac{2^2y^{-6}}{8^{-1}z^0x^{-7}}$
22. $\frac{13x^{-5}y^0}{5^{-3}z^{-10}}$

In Exercises 23–32, simplify the expression. Write your answer using only positive exponents. *(See Example 3.)*

23. $\frac{5^6}{5^2}$
24. $\frac{(-6)^8}{(-6)^5}$
25. $(-9)^2 \cdot (-9)^2$
26. $4^{-5} \cdot 4^5$
27. $(p^6)^4$
28. $(s^{-5})^3$
29. $6^{-8} \cdot 6^5$
30. $-7 \cdot (-7)^{-4}$
31. $\frac{x^5}{x^4} \cdot x$
32. $\frac{z^8 \cdot z^2}{z^5}$

33. **USING PROPERTIES** A microscope magnifies an object 10^5 times. The length of an object is 10^{-7} meter. What is its magnified length?

34. **USING PROPERTIES** The area of the rectangular computer chip is $112a^3b^2$ square microns. What is the length?

width = $8ab$ microns

ERROR ANALYSIS In Exercises 35 and 36, describe and correct the error in simplifying the expression.

35.

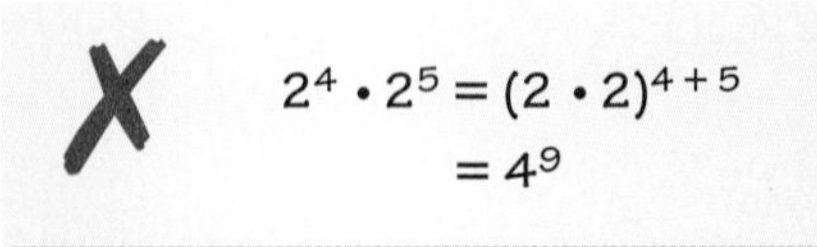

$$2^4 \cdot 2^5 = (2 \cdot 2)^{4+5} = 4^9$$

36.

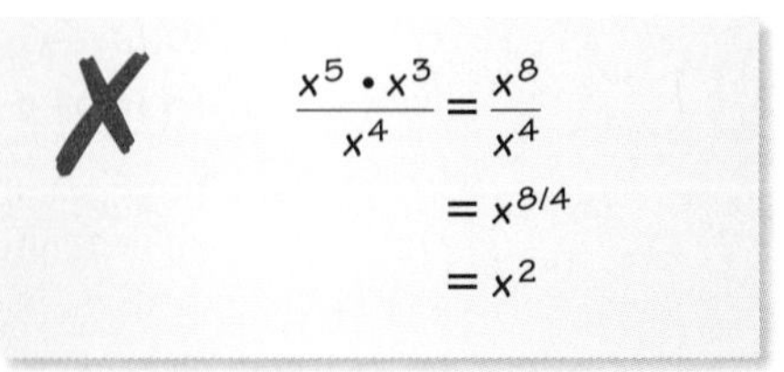

$$\frac{x^5 \cdot x^3}{x^4} = \frac{x^8}{x^4} = x^{8/4} = x^2$$

In Exercises 37–44, simplify the expression. Write your answer using only positive exponents. *(See Example 4.)*

37. $(-5z)^3$

38. $(4x)^{-4}$

39. $\left(\frac{6}{n}\right)^{-2}$

40. $\left(\frac{-t}{3}\right)^2$

41. $(3s^8)^{-5}$

42. $(-5p^3)^3$

43. $\left(-\frac{w^3}{6}\right)^{-2}$

44. $\left(\frac{1}{2r^6}\right)^{-6}$

45. USING PROPERTIES Which of the expressions represent the volume of the sphere? Explain. *(See Example 5.)*

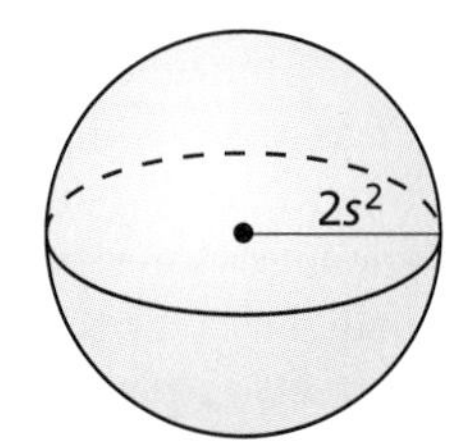

Ⓐ $\left(\frac{3s^2}{2^4\pi s^8}\right)^{-1}$

Ⓑ $(2^5\pi s^6)(3^{-1})$

Ⓒ $\frac{32\pi s^6}{3}$

Ⓓ $(2s)^5 \cdot \frac{\pi s}{3}$

Ⓔ $\left(\frac{3\pi s^6}{32}\right)^{-1}$

Ⓕ $\frac{32}{3}\pi s^5$

46. MODELING WITH MATHEMATICS Diffusion is the movement of molecules from one location to another. The time t (in seconds) it takes molecules to diffuse a distance of x centimeters is given by $t = \frac{x^2}{2D}$, where D is the diffusion coefficient. The diffusion coefficient for a drop of ink in water is about 10^{-5} square centimeters per second. How long will it take the ink to diffuse 1 micrometer (10^{-4} centimeter)?

In Exercises 47–50, simplify the expression. Write your answer using only positive exponents.

47. $\left(\frac{2x^{-2}y^3}{3xy^{-4}}\right)^4$

48. $\left(\frac{4s^5t^{-7}}{-2s^{-2}t^4}\right)^3$

49. $\left(\frac{3m^{-5}n^2}{4m^{-2}n^0}\right)^2 \cdot \left(\frac{mn^4}{9n}\right)^2$

50. $\left(\frac{3x^3y^0}{x^{-2}}\right)^4 \cdot \left(\frac{y^2x^{-4}}{5xy^{-8}}\right)^3$

In Exercises 51–54, evaluate the expression. Write your answer in scientific notation and standard form.

51. $(3 \times 10^2)(1.5 \times 10^{-5})$

52. $(6.1 \times 10^{-3})(8 \times 10^9)$

53. $\frac{(6.4 \times 10^7)}{(1.6 \times 10^5)}$

54. $\frac{(3.9 \times 10^{-5})}{(7.8 \times 10^{-8})}$

55. PROBLEM SOLVING In 2012, on average, about 9.46×10^{-1} pound of potatoes was produced for every 2.3×10^{-5} acre harvested. How many pounds of potatoes on average were produced for each acre harvested? Write your answer in scientific notation and in standard form. *(See Example 6.)*

56. PROBLEM SOLVING The speed of light is approximately 3×10^5 kilometers per second. How long does it take sunlight to reach Jupiter? Write your answer in scientific notation and in standard form.

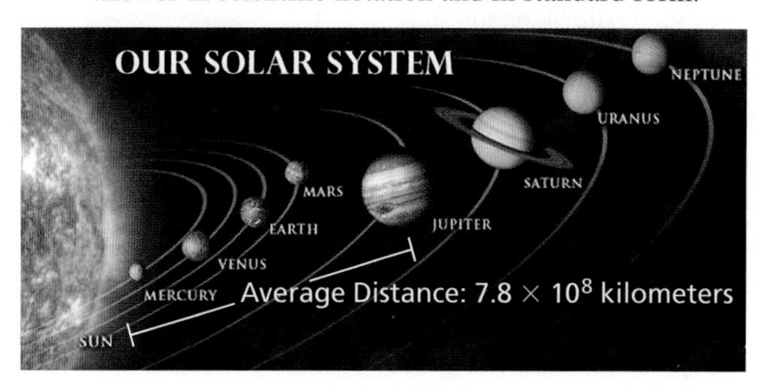

57. MATHEMATICAL CONNECTIONS Consider Cube A and Cube B.

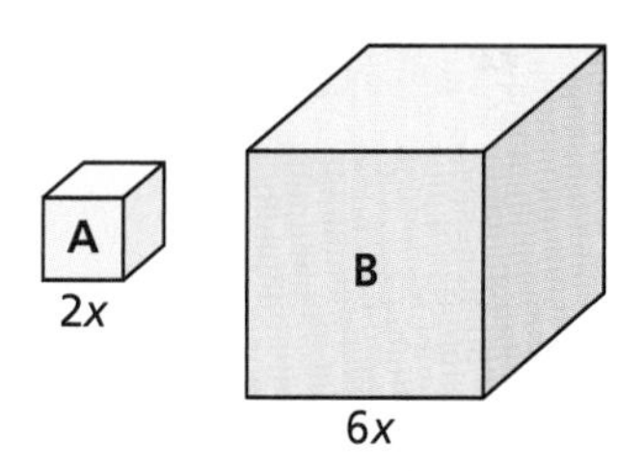

a. Which property of exponents should you use to simplify an expression for the volume of each cube?

b. How can you use the Power of a Quotient Property to find how many times greater the volume of Cube B is than the volume of Cube A?

58. PROBLEM SOLVING A byte is a unit used to measure a computer's memory. The table shows the numbers of bytes in several units of measure.

Unit	kilobyte	megabyte	gigabyte	terabyte
Number of bytes	2^{10}	2^{20}	2^{30}	2^{40}

a. How many kilobytes are in 1 terabyte?

b. How many megabytes are in 16 gigabytes?

c. Another unit used to measure a computer's memory is a bit. There are 8 bits in a byte. How can you convert the number of bytes in each unit of measure given in the table to bits? Can you still use a base of 2? Explain.

REWRITING EXPRESSIONS In Exercises 59–62, rewrite the expression as a power of a product.

59. $8a^3b^3$

60. $16r^2s^2$

61. $64w^{18}z^{12}$

62. $81x^4y^8$

63. USING STRUCTURE The probability of rolling a 6 on a number cube is $\frac{1}{6}$. The probability of rolling a 6 twice in a row is $\left(\frac{1}{6}\right)^2 = \frac{1}{36}$.

a. Write an expression that represents the probability of rolling a 6 n times in a row.

b. What is the probability of rolling a 6 four times in a row?

c. What is the probability of flipping heads on a coin five times in a row? Explain.

64. HOW DO YOU SEE IT? The shaded part of Figure n represents the portion of a piece of paper visible after folding the paper in half n times.

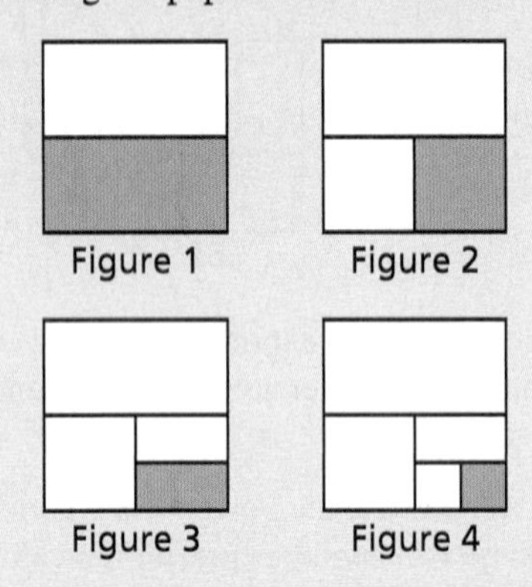

a. What fraction of the original piece of paper is each shaded part?

b. Rewrite each fraction from part (a) in the form 2^x.

65. REASONING Find x and y when $\frac{b^x}{b^y} = b^9$ and $\frac{b^x \cdot b^2}{b^{3y}} = b^{13}$. Explain how you found your answer.

66. THOUGHT PROVOKING Write expressions for r and h so that the volume of the cone can be represented by the expression $27\pi x^8$. Find r and h.

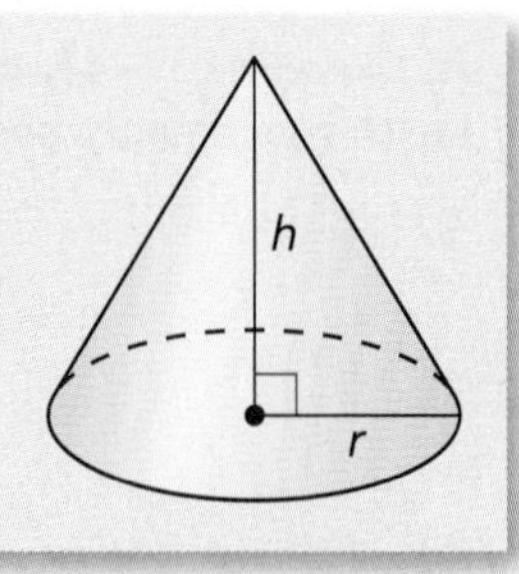

67. MAKING AN ARGUMENT One of the smallest plant seeds comes from an orchid, and one of the largest plant seeds comes from a double coconut palm. A seed from an orchid has a mass of 10^{-6} gram. The mass of a seed from a double coconut palm is 10^{10} times the mass of the seed from the orchid. Your friend says that the seed from the double coconut palm has a mass of about 1 kilogram. Is your friend correct? Explain.

68. CRITICAL THINKING Your school is conducting a survey. Students can answer the questions in either part with "agree" or "disagree."

Part 1: 13 questions		
Part 2: 10 questions		
Part 1: Classroom	Agree	Disagree
1. I come prepared for class.	O	O
2. I enjoy my assignments.	O	O

a. What power of 2 represents the number of different ways that a student can answer all the questions in Part 1?

b. What power of 2 represents the number of different ways that a student can answer all the questions on the entire survey?

c. The survey changes, and students can now answer "agree," "disagree," or "no opinion." How does this affect your answers in parts (a) and (b)?

69. ABSTRACT REASONING Compare the values of a^n and a^{-n} when $n < 0$, when $n = 0$, and when $n > 0$ for (a) $a > 1$ and (b) $0 < a < 1$. Explain your reasoning.

Maintaining Mathematical Proficiency

Reviewing what you learned in previous grades and lessons

Find the square root(s). *(Skills Review Handbook)*

70. $\sqrt{25}$

71. $-\sqrt{100}$

72. $\pm\sqrt{\frac{1}{64}}$

Classify the real number in as many ways as possible. *(Skills Review Handbook)*

73. 12

74. $\frac{65}{9}$

75. $\frac{\pi}{4}$

6.2 Radicals and Rational Exponents

Essential Question How can you write and evaluate an nth root of a number?

Recall that you cube a number as follows.

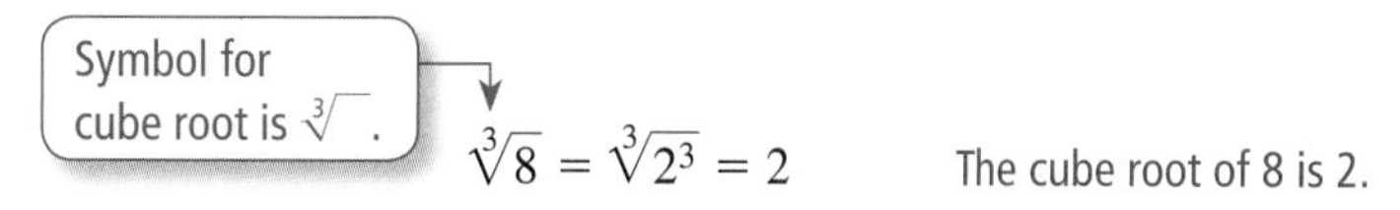

$$2^3 = 2 \cdot 2 \cdot 2 = 8$$

To "undo" cubing a number, take the cube root of the number.

Symbol for cube root is $\sqrt[3]{}$.

$$\sqrt[3]{8} = \sqrt[3]{2^3} = 2$$

The cube root of 8 is 2.

EXPLORATION 1 Finding Cube Roots

Work with a partner. Use a cube root symbol to write the side length of each cube. Then find the cube root. Check your answers by multiplying. Which cube is the largest? Which two cubes are the same size? Explain your reasoning.

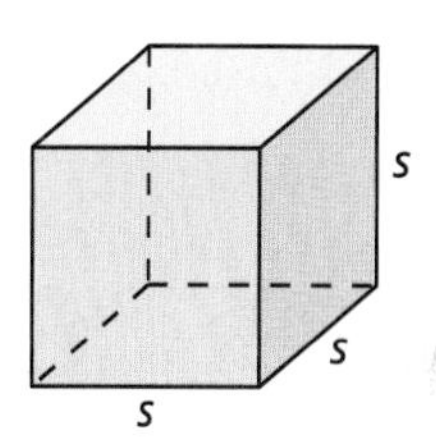

a. Volume $= 27\text{ ft}^3$ **b.** Volume $= 125\text{ cm}^3$ **c.** Volume $= 3375\text{ in.}^3$

d. Volume $= 3.375\text{ m}^3$ **e.** Volume $= 1\text{ yd}^3$ **f.** Volume $= \frac{125}{8}\text{ mm}^3$

JUSTIFYING CONCLUSIONS

To be proficient in math, you need to justify your conclusions and communicate them to others.

EXPLORATION 2 Estimating nth Roots

Work with a partner. Estimate each positive nth root. Then match each nth root with the point on the number line. Justify your answers.

a. $\sqrt[4]{25}$ **b.** $\sqrt{0.5}$ **c.** $\sqrt[5]{2.5}$

d. $\sqrt[3]{65}$ **e.** $\sqrt[3]{55}$ **f.** $\sqrt[6]{20{,}000}$

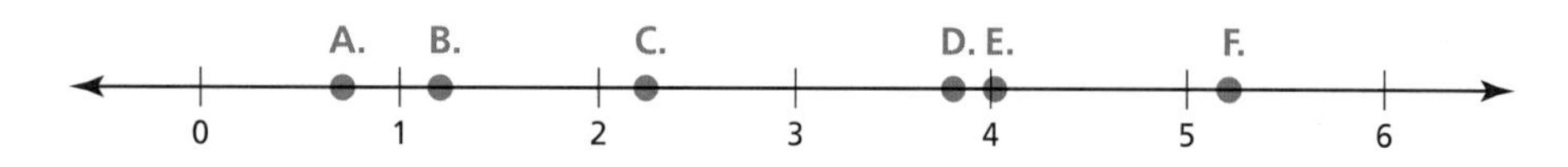

Communicate Your Answer

3. How can you write and evaluate an nth root of a number?

4. The body mass m (in kilograms) of a dinosaur that walked on two feet can be modeled by

$$m = (0.00016)C^{2.73}$$

where C is the circumference (in millimeters) of the dinosaur's femur. The mass of a *Tyrannosaurus rex* was 4000 kilograms. Use a calculator to approximate the circumference of its femur.

6.2 Lesson

Core Vocabulary

nth root of a, *p. 300*
radical, *p. 300*
index of a radical, *p. 300*

Previous
square root

What You Will Learn

- Find nth roots.
- Evaluate expressions with rational exponents.
- Solve real-life problems involving rational exponents.

Finding nth Roots

You can extend the concept of a square root to other types of roots. For example, 2 is a cube root of 8 because $2^3 = 8$, and 3 is a fourth root of 81 because $3^4 = 81$. In general, for an integer n greater than 1, if $b^n = a$, then b is an **nth root of a**. An nth root of a is written as $\sqrt[n]{a}$, where the expression $\sqrt[n]{a}$ is called a **radical** and n is the **index** of the radical.

You can also write an nth root of a as a power of a. If you assume the Power of a Power Property applies to rational exponents, then the following is true.

$$(a^{1/2})^2 = a^{(1/2) \cdot 2} = a^1 = a$$
$$(a^{1/3})^3 = a^{(1/3) \cdot 3} = a^1 = a$$
$$(a^{1/4})^4 = a^{(1/4) \cdot 4} = a^1 = a$$

Because $a^{1/2}$ is a number whose square is a, you can write $\sqrt{a} = a^{1/2}$. Similarly, $\sqrt[3]{a} = a^{1/3}$ and $\sqrt[4]{a} = a^{1/4}$. In general, $\sqrt[n]{a} = a^{1/n}$ for any integer n greater than 1.

READING

$\pm\sqrt[n]{a}$ represents both the positive and negative nth roots of a.

Core Concept

Real nth Roots of a

Let n be an integer greater than 1, and let a be a real number.

- If n is odd, then a has one real nth root: $\sqrt[n]{a} = a^{1/n}$
- If n is even and $a > 0$, then a has two real nth roots: $\pm\sqrt[n]{a} = \pm a^{1/n}$
- If n is even and $a = 0$, then a has one real nth root: $\sqrt[n]{0} = 0$
- If n is even and $a < 0$, then a has no real nth roots.

The nth roots of a number may be real numbers or *imaginary numbers*. You will study imaginary numbers in a future course.

EXAMPLE 1 Finding nth Roots

Find the indicated real nth root(s) of a.

a. $n = 3, a = -27$

b. $n = 4, a = 16$

SOLUTION

a. The index $n = 3$ is odd, so -27 has one real cube root. Because $(-3)^3 = -27$, the cube root of -27 is $\sqrt[3]{-27} = -3$, or $(-27)^{1/3} = -3$.

b. The index $n = 4$ is even, and $a > 0$. So, 16 has two real fourth roots. Because $2^4 = 16$ and $(-2)^4 = 16$, the fourth roots of 16 are $\pm\sqrt[4]{16} = \pm 2$, or $\pm 16^{1/4} = \pm 2$.

Monitoring Progress

Help in English and Spanish at *BigIdeasMath.com*

Find the indicated real nth root(s) of a.

1. $n = 3, a = -125$

2. $n = 6, a = 64$

Evaluating Expressions with Rational Exponents

Recall that the radical $\sqrt{a}$ indicates the positive square root of a. Similarly, an nth root of a, $\sqrt[n]{a}$, with an *even* index indicates the positive nth root of a.

REMEMBER

The expression under the radical sign is the radicand.

EXAMPLE 2 **Evaluating *n*th Root Expressions**

Evaluate each expression.

a. $\sqrt[3]{-8}$ **b.** $-\sqrt[3]{8}$ **c.** $16^{1/4}$ **d.** $(-16)^{1/4}$

SOLUTION

a. $\sqrt[3]{-8} = \sqrt[3]{(-2) \cdot (-2) \cdot (-2)}$ — Rewrite the expression showing factors.

$= -2$ — Evaluate the cube root.

b. $-\sqrt[3]{8} = -\left(\sqrt[3]{2 \cdot 2 \cdot 2}\right)$ — Rewrite the expression showing factors.

$= -(2)$ — Evaluate the cube root.

$= -2$ — Simplify.

c. $16^{1/4} = \sqrt[4]{16}$ — Rewrite the expression in radical form.

$= \sqrt[4]{2 \cdot 2 \cdot 2 \cdot 2}$ — Rewrite the expression showing factors.

$= 2$ — Evaluate the fourth root.

d. $(-16)^{1/4}$ is not a real number because there is no real number that can be multiplied by itself four times to produce -16.

A rational exponent does not have to be of the form $1/n$. Other rational numbers such as $3/2$ can also be used as exponents. You can use the properties of exponents to evaluate or simplify expressions involving rational exponents.

STUDY TIP

You can rewrite $27^{2/3}$ as $27^{(1/3) \cdot 2}$ and then use the Power of a Power Property to show that

$27^{(1/3) \cdot 2} = (27^{1/3})^2$.

Core Concept

Rational Exponents

Let $a^{1/n}$ be an nth root of a, and let m be a positive integer.

Algebra $a^{m/n} = (a^{1/n})^m = (\sqrt[n]{a})^m$

Numbers $27^{2/3} = (27^{1/3})^2 = \left(\sqrt[3]{27}\right)^2$

EXAMPLE 3 **Evaluating Expressions with Rational Exponents**

Evaluate (a) $16^{3/4}$ and (b) $27^{4/3}$.

SOLUTION

a. $16^{3/4} = (16^{1/4})^3$ — Rational exponents

$= 2^3$ — Evaluate the nth root.

$= 8$ — Evaluate the power.

b. $27^{4/3} = (27^{1/3})^4$

$= 3^4$

$= 81$

Monitoring Progress

Help in English and Spanish at *BigIdeasMath.com*

Evaluate the expression.

3. $\sqrt[3]{-125}$ **4.** $(-64)^{2/3}$ **5.** $9^{5/2}$ **6.** $256^{3/4}$

Solving Real-Life Problems

EXAMPLE 4 Solving a Real-Life Problem

Volume = 113 cubic feet

The radius r of a sphere is given by the equation $r = \left(\frac{3V}{4\pi}\right)^{1/3}$, where V is the volume of the sphere. Find the radius of the beach ball to the nearest foot. Use 3.14 for π.

SOLUTION

1. **Understand the Problem** You know the equation that represents the radius of a sphere in terms of its volume. You are asked to find the radius for a given volume.
2. **Make a Plan** Substitute the given volume into the equation. Then evaluate to find the radius.
3. **Solve the Problem**

$$r = \left(\frac{3V}{4\pi}\right)^{1/3}$$ Write the equation.

$$= \left(\frac{3(113)}{4(3.14)}\right)^{1/3}$$ Substitute 113 for V and 3.14 for π.

$$= \left(\frac{339}{12.56}\right)^{1/3}$$ Multiply.

$$\approx 3$$ Use a calculator.

▶ The radius of the beach ball is about 3 feet.

4. **Look Back** To check that your answer is reasonable, compare the size of the ball to the size of the woman pushing the ball. The ball appears to be slightly taller than the woman. The average height of a woman is between 5 and 6 feet. So, a radius of 3 feet, or height of 6 feet, seems reasonable for the beach ball.

EXAMPLE 5 Solving a Real-Life Problem

To calculate the annual inflation rate r (in decimal form) of an item that increases in value from P to F over a period of n years, you can use the equation $r = \left(\frac{F}{P}\right)^{1/n} - 1$. Find the annual inflation rate to the nearest tenth of a percent of a house that increases in value from \$200,000 to \$235,000 over a period of 5 years.

SOLUTION

$$r = \left(\frac{F}{P}\right)^{1/n} - 1$$ Write the equation.

$$= \left(\frac{235{,}000}{200{,}000}\right)^{1/5} - 1$$ Substitute 235,000 for F, 200,000 for P, and 5 for n.

$$= 1.175^{1/5} - 1$$ Divide.

$$\approx 0.03278$$ Use a calculator.

▶ The annual inflation rate is about 3.3%.

REMEMBER
To write a decimal as a percent, move the decimal point two places to the right. Then add a percent symbol.

Monitoring Progress  Help in English and Spanish at *BigIdeasMath.com*

7. **WHAT IF?** In Example 4, the volume of the beach ball is 17,000 cubic inches. Find the radius to the nearest inch. Use 3.14 for π.
8. The average cost of college tuition increases from \$8500 to \$13,500 over a period of 8 years. Find the annual inflation rate to the nearest tenth of a percent.

6.2 Exercises

Dynamic Solutions available at *BigIdeasMath.com*

Vocabulary and Core Concept Check

1. **WRITING** Explain how to evaluate $81^{1/4}$.

2. **WHICH ONE DOESN'T BELONG?** Which expression does *not* belong with the other three? Explain your reasoning.

$\left(\sqrt[3]{27}\right)^2$ | $27^{2/3}$ | 3^2 | $\left(\sqrt[2]{27}\right)^3$

Monitoring Progress and Modeling with Mathematics

In Exercises 3 and 4, rewrite the expression in rational exponent form.

3. $\sqrt{10}$
4. $\sqrt[5]{34}$

In Exercises 5 and 6, rewrite the expression in radical form.

5. $15^{1/3}$
6. $140^{1/8}$

In Exercises 7–10, find the indicated real *n*th root(s) of *a*. *(See Example 1.)*

7. $n = 2, a = 36$
8. $n = 4, a = 81$
9. $n = 3, a = 1000$
10. $n = 9, a = -512$

MATHEMATICAL CONNECTIONS In Exercises 11 and 12, find the dimensions of the cube. Check your answer.

11. Volume = 64 in.^{3}

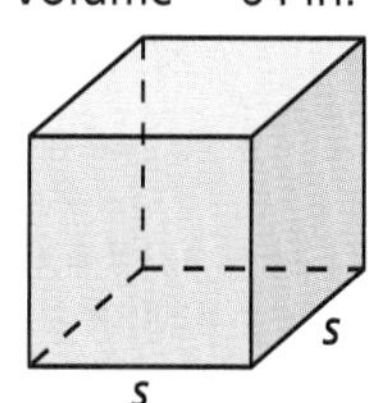

12. Volume = 216 cm^3

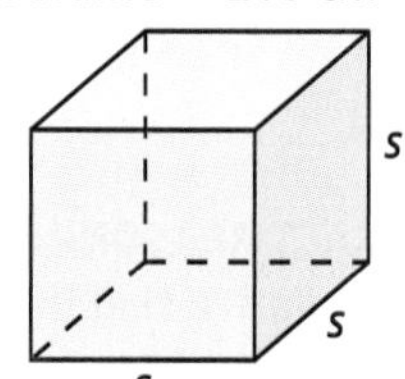

In Exercises 13–18, evaluate the expression. *(See Example 2.)*

13. $\sqrt[4]{256}$
14. $\sqrt[3]{-216}$
15. $\sqrt[3]{-343}$
16. $-\sqrt[5]{1024}$
17. $128^{1/7}$
18. $(-64)^{1/2}$

In Exercises 19 and 20, rewrite the expression in rational exponent form.

19. $\left(\sqrt[5]{8}\right)^4$
20. $\left(\sqrt[5]{-21}\right)^6$

In Exercises 21 and 22, rewrite the expression in radical form.

21. $(-4)^{2/7}$
22. $9^{5/2}$

In Exercises 23–28, evaluate the expression. *(See Example 3.)*

23. $32^{3/5}$
24. $125^{2/3}$
25. $(-36)^{3/2}$
26. $(-243)^{2/5}$
27. $(-128)^{5/7}$
28. $343^{4/3}$

29. **ERROR ANALYSIS** Describe and correct the error in rewriting the expression in rational exponent form.

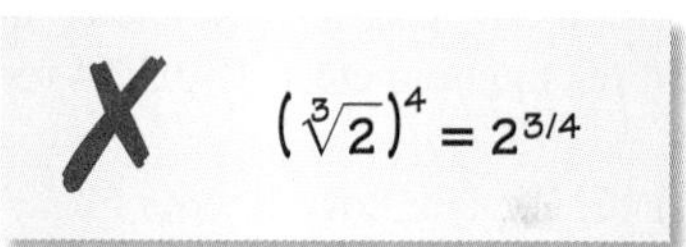

30. **ERROR ANALYSIS** Describe and correct the error in evaluating the expression.

$$
\begin{aligned}
(-81)^{3/4} &= [(-81)^{1/4}]^3 \\
&= (-3)^3 \\
&= -27
\end{aligned}
$$

In Exercises 31–34, evaluate the expression.

31. $\left(\frac{1}{1000}\right)^{1/3}$
32. $\left(\frac{1}{64}\right)^{1/6}$
33. $(27)^{-2/3}$
34. $(9)^{-5/2}$

35. PROBLEM SOLVING A math club is having a bake sale. Find the area of the bake sale sign.

36. PROBLEM SOLVING The volume of a cube-shaped box is 27^5 cubic millimeters. Find the length of one side of the box.

37. MODELING WITH MATHEMATICS The radius r of the base of a cone is given by the equation

$$r = \left(\frac{3V}{\pi h}\right)^{1/2}$$

where V is the volume of the cone and h is the height of the cone. Find the radius of the paper cup to the nearest inch. Use 3.14 for π. *(See Example 4.)*

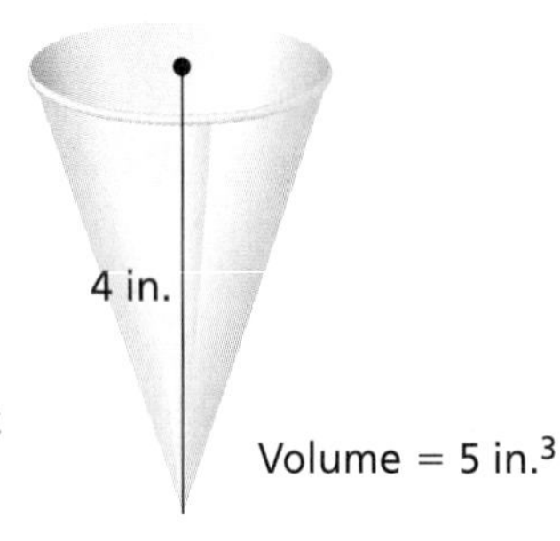

38. MODELING WITH MATHEMATICS The volume of a sphere is given by the equation $V = \frac{1}{6\sqrt{\pi}}S^{3/2}$, where S is the surface area of the sphere. Find the volume of a sphere, to the nearest cubic meter, that has a surface area of 60 square meters. Use 3.14 for π.

39. WRITING Explain how to write $(\sqrt[n]{a})^m$ in rational exponent form.

40. HOW DO YOU SEE IT? Write an expression in rational exponent form that represents the side length of the square.

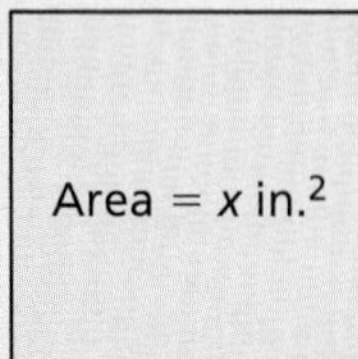

In Exercises 41 and 42, use the formula $r = \left(\frac{F}{P}\right)^{1/n} - 1$ to find the annual inflation rate to the nearest tenth of a percent. *(See Example 5.)*

41. A farm increases in value from \$800,000 to \$1,100,000 over a period of 6 years.

42. The cost of a gallon of gas increases from \$1.46 to \$3.53 over a period of 10 years.

43. REASONING For what values of x is $x = x^{1/5}$?

44. MAKING AN ARGUMENT Your friend says that for a real number a and a positive integer n, the value of $\sqrt[n]{a}$ is always positive and the value of $-\sqrt[n]{a}$ is always negative. Is your friend correct? Explain.

In Exercises 45–48, simplify the expression.

45. $(y^{1/6})^3 \cdot \sqrt{x}$

46. $(y \cdot y^{1/3})^{3/2}$

47. $x \cdot \sqrt[3]{y^6} + y^2 \cdot \sqrt[3]{x^3}$

48. $(x^{1/3} \cdot y^{1/2})^9 \cdot \sqrt{y}$

49. PROBLEM SOLVING The formula for the volume of a regular dodecahedron is $V \approx 7.66\ell^3$, where ℓ is the length of an edge. The volume of the dodecahedron is 20 cubic feet. Estimate the edge length.

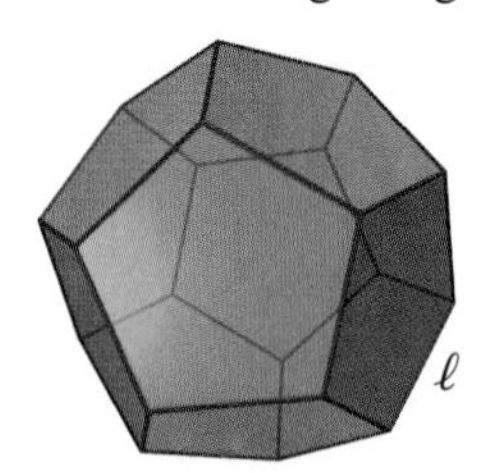

50. THOUGHT PROVOKING Find a formula (for instance, from geometry or physics) that contains a radical. Rewrite the formula using rational exponents.

ABSTRACT REASONING In Exercises 51–56, let x be a nonnegative real number. Determine whether the statement is *always*, *sometimes*, or *never* true. Justify your answer.

51. $(x^{1/3})^3 = x$

52. $x^{1/3} = x^{-3}$

53. $x^{1/3} = \sqrt[3]{x}$

54. $x^{1/3} = x^3$

55. $\frac{x^{2/3}}{x^{1/3}} = \sqrt[3]{x}$

56. $x = x^{1/3} \cdot x^3$

Maintaining Mathematical Proficiency

Reviewing what you learned in previous grades and lessons

Evaluate the function when $x = -3$, 0, and 8. *(Section 3.3)*

57. $f(x) = 2x - 10$

58. $w(x) = -5x - 1$

59. $h(x) = 13 - x$

60. $g(x) = 8x + 16$

6.3 Exponential Functions

Essential Question What are some of the characteristics of the graph of an exponential function?

EXPLORATION 1 **Exploring an Exponential Function**

Work with a partner. Copy and complete each table for the *exponential function* $y = 16(2)^x$. In each table, what do you notice about the values of x? What do you notice about the values of y?

x	$y = 16(2)^x$
0	
1	
2	
3	
4	
5	

x	$y = 16(2)^x$
0	
2	
4	
6	
8	
10	

JUSTIFYING CONCLUSIONS

To be proficient in math, you need to justify your conclusions and communicate them to others.

EXPLORATION 2 **Exploring an Exponential Function**

Work with a partner. Repeat Exploration 1 for the exponential function $y = 16\left(\frac{1}{2}\right)^x$. Do you think the statement below is true for *any* exponential function? Justify your answer.

"As the independent variable x changes by a constant amount, the dependent variable y is multiplied by a constant factor."

EXPLORATION 3 **Graphing Exponential Functions**

Work with a partner. Sketch the graphs of the functions given in Explorations 1 and 2. How are the graphs similar? How are they different?

Communicate Your Answer

4. What are some of the characteristics of the graph of an exponential function?

5. Sketch the graph of each exponential function. Does each graph have the characteristics you described in Question 4? Explain your reasoning.

a. $y = 2^x$ **b.** $y = 2(3)^x$ **c.** $y = 3(1.5)^x$

d. $y = \left(\frac{1}{2}\right)^x$ **e.** $y = 3\left(\frac{1}{2}\right)^x$ **f.** $y = 2\left(\frac{3}{4}\right)^x$

6.3 Lesson

Core Vocabulary

exponential function, *p. 306*

Previous
independent variable
dependent variable
parent function

What You Will Learn

- Identify and evaluate exponential functions.
- Graph exponential functions.
- Solve real-life problems involving exponential functions.

Identifying and Evaluating Exponential Functions

An **exponential function** is a nonlinear function of the form $y = ab^x$, where $a \neq 0$, $b \neq 1$, and $b > 0$. As the independent variable x changes by a constant amount, the dependent variable y is multiplied by a constant factor, which means consecutive y-values form a constant ratio.

EXAMPLE 1 Identifying Functions

Does each table represent a *linear* or an *exponential* function? Explain.

a.

x	0	1	2	3
y	2	4	6	8

b.

x	0	1	2	3
y	4	8	16	32

SOLUTION

a. $+1$, $+1$, $+1$

x	0	1	2	3
y	2	4	6	8

$+2$, $+2$, $+2$

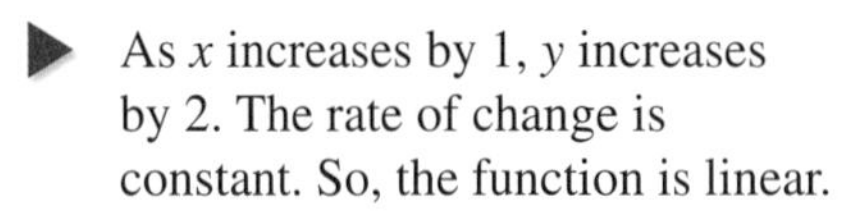

- As x increases by 1, y increases by 2. The rate of change is constant. So, the function is linear.

b.

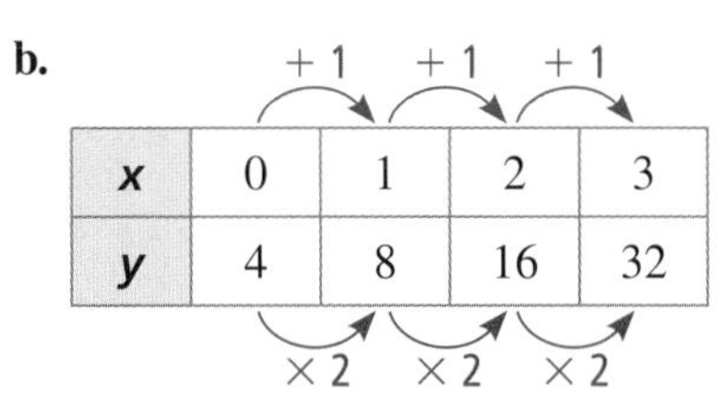

$+1$, $+1$, $+1$

x	0	1	2	3
y	4	8	16	32

$\times 2$, $\times 2$, $\times 2$

- As x increases by 1, y is multiplied by 2. So, the function is exponential.

STUDY TIP

In Example 1b, consecutive y-values form a constant ratio.

$\frac{8}{4} = 2, \frac{16}{8} = 2, \frac{32}{16} = 2$

EXAMPLE 2 Evaluating Exponential Functions

Evaluate each function for the given value of x.

a. $y = -2(5)^x$; $x = 3$

b. $y = 3(0.5)^x$; $x = -2$

SOLUTION

a.
$y = -2(5)^x$ — Write the function.
$= -2(5)^3$ — Substitute for x.
$= -2(125)$ — Evaluate the power.
$= -250$ — Multiply.

b.
$y = 3(0.5)^x$
$= 3(0.5)^{-2}$
$= 3(4)$
$= 12$

Monitoring Progress

Help in English and Spanish at *BigIdeasMath.com*

Does the table represent a *linear* or an *exponential* function? Explain.

1.

x	0	1	2	3
y	8	4	2	1

2.

x	−4	0	4	8
y	1	0	−1	−2

Evaluate the function when $x = -2$, 0, and $\frac{1}{2}$.

3. $y = 2(9)^x$

4. $y = 1.5(2)^x$

Graphing Exponential Functions

The graph of a function $y = ab^x$ is a vertical stretch or shrink by a factor of $|a|$ of the graph of the parent function $y = b^x$. When $a < 0$, the graph is also reflected in the x-axis. The y-intercept of the graph of $y = ab^x$ is a.

Core Concept

STUDY TIP

The graph of $y = ab^x$ approaches the x-axis but never intersects it.

Graphing $y = ab^x$ When $b > 1$

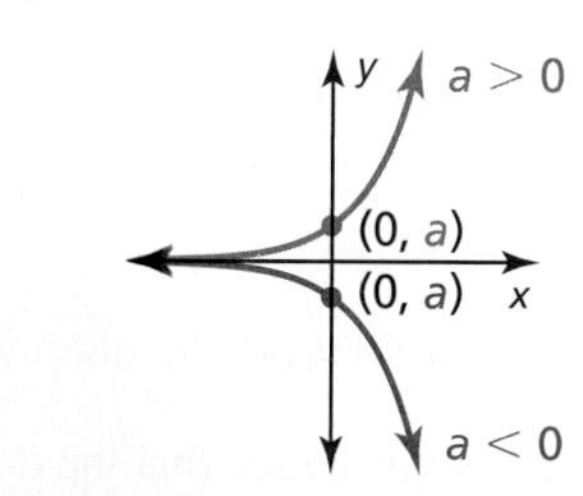

Graphing $y = ab^x$ When $0 < b < 1$

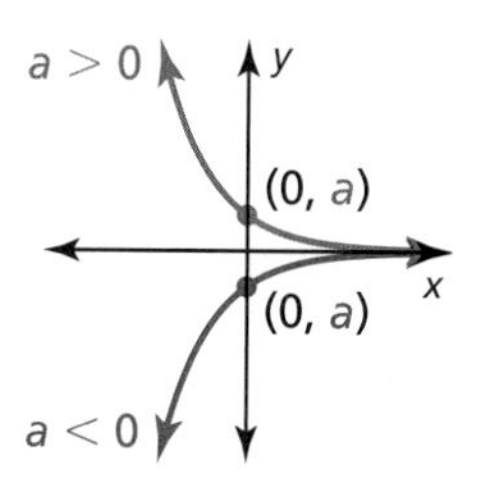

EXAMPLE 3 Graphing $y = ab^x$ When $b > 1$

Graph $f(x) = 4(2)^x$. Compare the graph to the graph of the parent function. Describe the domain and range of f.

SOLUTION

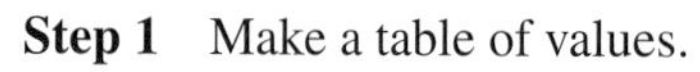

Step 1 Make a table of values.

Step 2 Plot the ordered pairs.

Step 3 Draw a smooth curve through the points.

x	-2	-1	0	1	2
$f(x)$	1	2	4	8	16

▶ The parent function is $g(x) = 2^x$. The graph of f is a vertical stretch by a factor of 4 of the graph of g. The y-intercept of the graph of f, 4, is above the y-intercept of the graph of g, 1. From the graph of f, you can see that the domain is all real numbers and the range is $y > 0$.

EXAMPLE 4 Graphing $y = ab^x$ When $0 < b < 1$

Graph $f(x) = -\left(\frac{1}{2}\right)^x$. Compare the graph to the graph of the parent function. Describe the domain and range of f.

SOLUTION

Step 1 Make a table of values.

Step 2 Plot the ordered pairs.

Step 3 Draw a smooth curve through the points.

x	-2	-1	0	1	2
$f(x)$	-4	-2	-1	$-\frac{1}{2}$	$-\frac{1}{4}$

▶ The parent function is $g(x) = \left(\frac{1}{2}\right)^x$. The graph of f is a reflection in the x-axis of the graph of g. The y-intercept of the graph of f, -1, is below the y-intercept of the graph of g, 1. From the graph of f, you can see that the domain is all real numbers and the range is $y < 0$.

Monitoring Progress

Help in English and Spanish at *BigIdeasMath.com*

Graph the function. Compare the graph to the graph of the parent function. Describe the domain and range of f.

5. $f(x) = -2(4)^x$

6. $f(x) = 2\left(\frac{1}{4}\right)^x$

To graph a function of the form $y = ab^{x-h} + k$, begin by graphing $y = ab^x$. Then translate the graph horizontally h units and vertically k units.

EXAMPLE 5 Graphing $y = ab^{x-h} + k$

Graph $y = 4(2)^{x-3} + 2$. Describe the domain and range.

SOLUTION

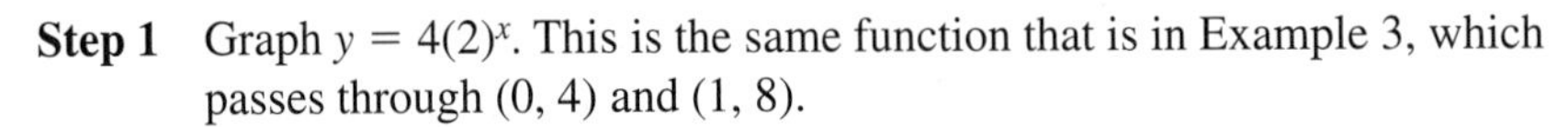

Step 1 Graph $y = 4(2)^x$. This is the same function that is in Example 3, which passes through (0, 4) and (1, 8).

Step 2 Translate the graph 3 units right and 2 units up. The graph passes through (3, 6) and (4, 10).

Notice that the graph approaches the line $y = 2$ but does not intersect it.

▶ From the graph, you can see that the domain is all real numbers and the range is $y > 2$.

EXAMPLE 6 Comparing Exponential Functions

An exponential function g models a relationship in which the dependent variable is multiplied by 1.5 for every 1 unit the independent variable x increases. Graph g when $g(0) = 4$. Compare g and the function f from Example 3 over the interval $x = 0$ to $x = 2$.

SOLUTION

You know (0, 4) is on the graph of g. To find points to the right of (0, 4), multiply $g(x)$ by 1.5 for every 1 unit increase in x. To find points to the left of (0, 4), divide $g(x)$ by 1.5 for every 1 unit decrease in x.

Step 1 Make a table of values.

x	−1	0	1	2	3
$g(x)$	2.7	4	6	9	13.5

Step 2 Plot the ordered pairs.

Step 3 Draw a smooth curve through the points.

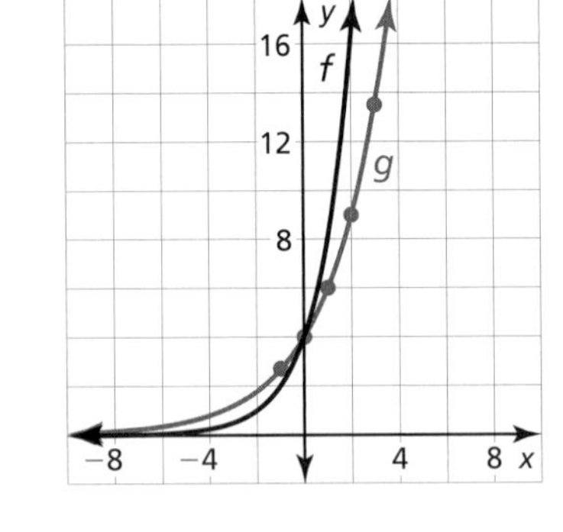

STUDY TIP

Note that f is increasing faster than g to the right of $x = 0$.

▶ Both functions have the same value when $x = 0$, but the value of f is greater than the value of g over the rest of the interval.

Monitoring Progress

Help in English and Spanish at *BigIdeasMath.com*

Graph the function. Describe the domain and range.

7. $y = -2(3)^{x+2} - 1$

8. $f(x) = (0.25)^x + 3$

9. WHAT IF? In Example 6, the dependent variable of g is multiplied by 3 for every 1 unit the independent variable x increases. Graph g when $g(0) = 4$. Compare g and the function f from Example 3 over the interval $x = 0$ to $x = 2$.

Solving Real-Life Problems

For an exponential function of the form $y = ab^x$, the y-values change by a factor of b as x increases by 1. You can use this fact to write an exponential function when you know the y-intercept, a. The table represents the exponential function $y = 2(5)^x$.

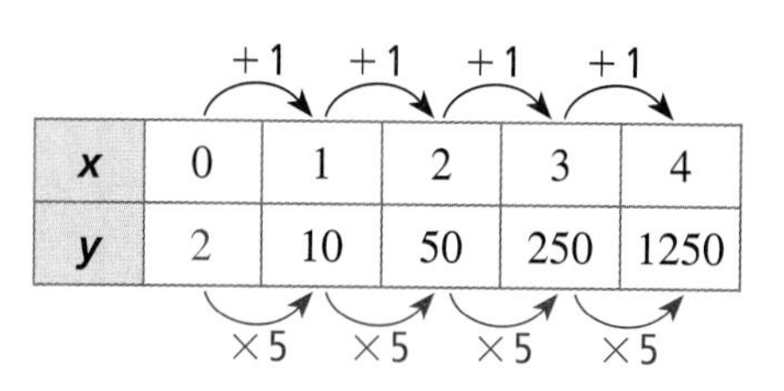

x	0	1	2	3	4
y	2	10	50	250	1250

(+1 between consecutive x-values; ×5 between consecutive y-values)

EXAMPLE 7 Modeling with Mathematics

Bacterial Population

Population vs. Day: points (0, 3), (1, 12), (2, 48), (3, 192), (4, 768)

The graph represents a bacterial population y after x days.

a. Write an exponential function that represents the population.

b. Find the population after 12 hours and after 5 days.

SOLUTION

1. **Understand the Problem** You have a graph of the population that shows some data points. You are asked to write an exponential function that represents the population and find the population after different amounts of time.

2. **Make a Plan** Use the graph to make a table of values. Use the table and the y-intercept to write an exponential function. Then evaluate the function to find the populations.

3. **Solve the Problem**

   a. Use the graph to make a table of values.

x	0	1	2	3	4
y	3	12	48	192	768

(+1 between consecutive x-values; ×4 between consecutive y-values)

The y-intercept is 3. The y-values increase by a factor of 4 as x increases by 1.

▶ So, the population can be modeled by $y = 3(4)^x$.

b. ***Population after 12 hours***

12 hours $= \frac{1}{2}$ day

Population after 12 hours		Population after 5 days
$y = 3(4)^x$	Write the function.	$y = 3(4)^x$
$= 3(4)^{1/2}$	Substitute for x.	$= 3(4)^5$
$= 3(2)$	Evaluate the power.	$= 3(1024)$
$= 6$	Multiply.	$= 3072$

▶ There are 6 bacteria after 12 hours and 3072 bacteria after 5 days.

4. **Look Back** The graph resembles an exponential function of the form $y = ab^x$, where $b > 1$ and $a > 0$. So, the exponential function $y = 3(4)^x$ is reasonable.

Monitoring Progress

Help in English and Spanish at *BigIdeasMath.com*

10. A bacterial population y after x days can be represented by an exponential function whose graph passes through (0, 100) and (1, 200). (a) Write a function that represents the population. (b) Find the population after 6 days. (c) Does this bacterial population grow faster than the bacterial population in Example 7? Explain.

6.3 Exercises

Dynamic Solutions available at *BigIdeasMath.com*

Vocabulary and Core Concept Check

1. **OPEN-ENDED** Sketch an increasing exponential function whose graph has a y-intercept of 2.
2. **REASONING** Why is a the y-intercept of the graph of the function $y = ab^x$?
3. **WRITING** Compare the graph of $y = 2(5)^x$ with the graph of $y = 5^x$.
4. **WHICH ONE DOESN'T BELONG?** Which equation does *not* belong with the other three? Explain your reasoning.

$y = 3^x$	$f(x) = 2(4)^x$	$f(x) = (-3)^x$	$y = 5(3)^x$

Monitoring Progress and Modeling with Mathematics

In Exercises 5–10, determine whether the equation represents an exponential function. Explain.

5. $y = 4(7)^x$
6. $y = -6x$
7. $y = 2x^3$
8. $y = -3^x$
9. $y = 9(-5)^x$
10. $y = \frac{1}{2}(1)^x$

In Exercises 11–14, determine whether the table represents a *linear* or an *exponential* function. Explain. *(See Example 1.)*

11.

x	y
1	−2
2	0
3	2
4	4

12.

x	y
1	6
2	12
3	24
4	48

13.

x	−1	0	1	2	3
y	0.25	1	4	16	64

14.

x	−3	0	3	6	9
y	10	1	−8	−17	−26

In Exercises 15–20, evaluate the function for the given value of x. *(See Example 2.)*

15. $y = 3^x;\ x = 2$
16. $f(x) = 3(2)^x;\ x = -1$
17. $y = -4(5)^x;\ x = 2$
18. $f(x) = 0.5^x;\ x = -3$
19. $f(x) = \frac{1}{3}(6)^x;\ x = 3$
20. $y = \frac{1}{4}(4)^x;\ x = \frac{3}{2}$

USING STRUCTURE In Exercises 21–24, match the function with its graph.

21. $f(x) = 2(0.5)^x$
22. $y = -2(0.5)^x$
23. $y = 2(2)^x$
24. $f(x) = -2(2)^x$

A.

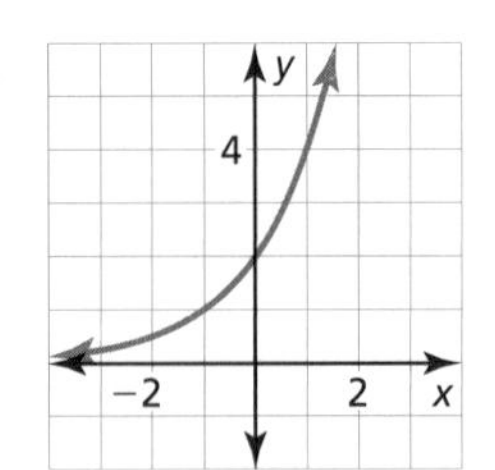

B.

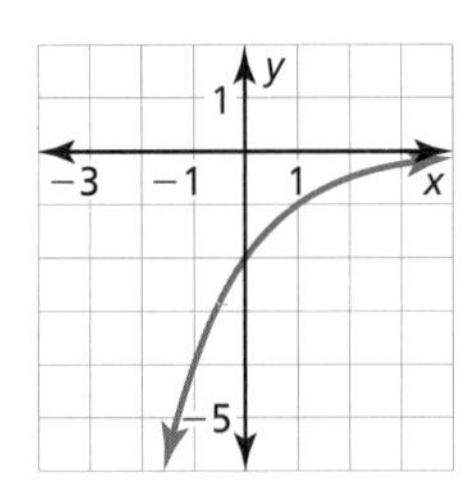

C.

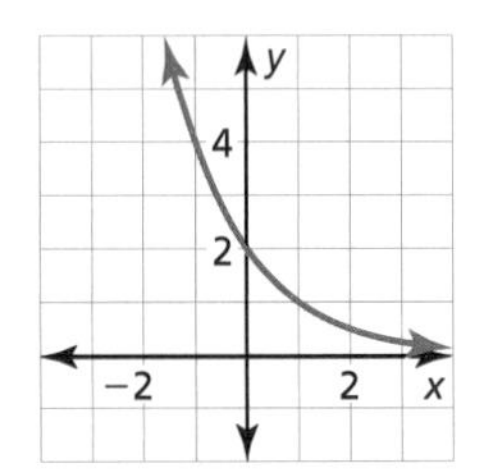

D.

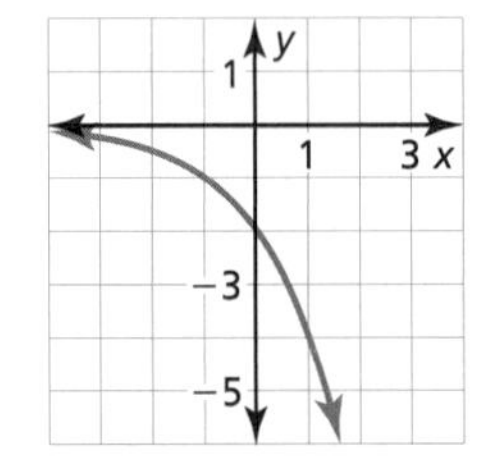

In Exercises 25–30, graph the function. Compare the graph to the graph of the parent function. Describe the domain and range of f. *(See Examples 3 and 4.)*

25. $f(x) = 3(0.5)^x$
26. $f(x) = -4^x$
27. $f(x) = -2(7)^x$
28. $f(x) = 6\left(\frac{1}{3}\right)^x$
29. $f(x) = \frac{1}{2}(8)^x$
30. $f(x) = \frac{3}{2}(0.25)^x$

In Exercises 31–36, graph the function. Describe the domain and range. *(See Example 5.)*

31. $f(x) = 3^x - 1$
32. $f(x) = 4^{x+3}$

33. $y = 5^{x-2} + 7$

34. $y = -\left(\frac{1}{2}\right)^{x+1} - 3$

35. $y = -8(0.75)^{x+2} - 2$

36. $f(x) = 3(6)^{x-1} - 5$

In Exercises 37–40, compare the graphs. Find the value of h, k, or a.

37.

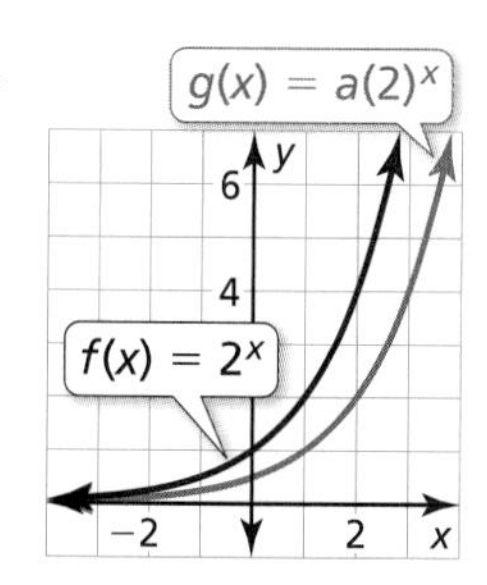

38.

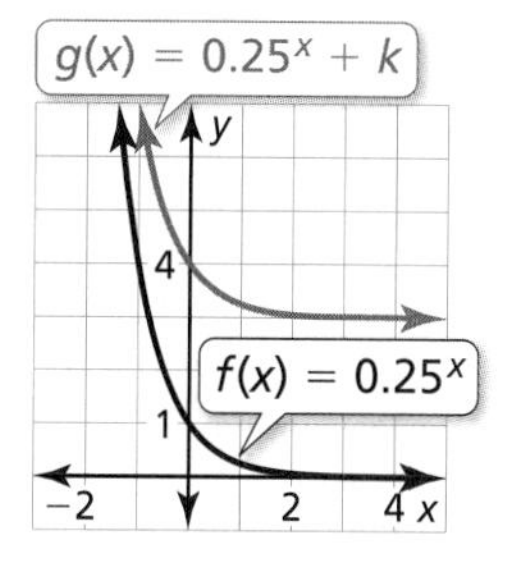

39.

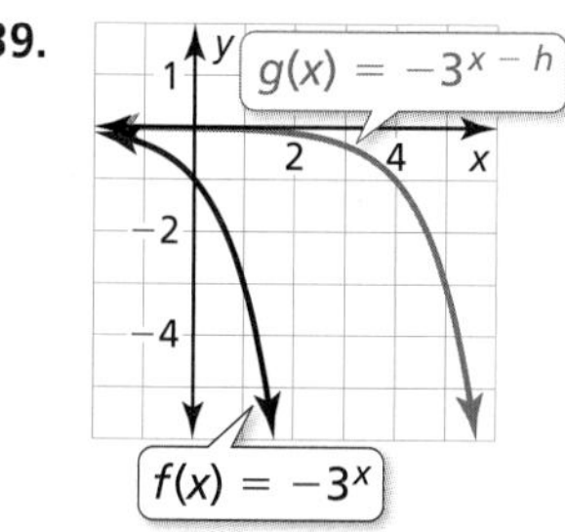

40.

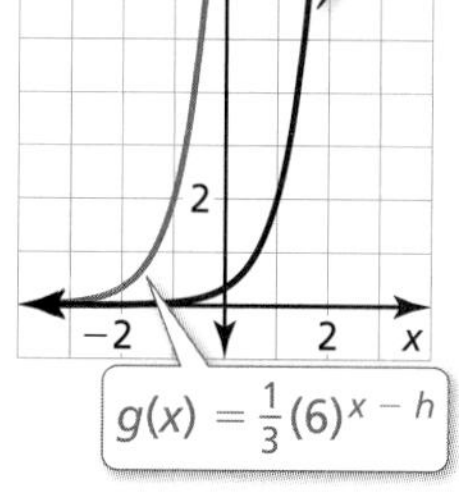

41. ERROR ANALYSIS Describe and correct the error in evaluating the function.

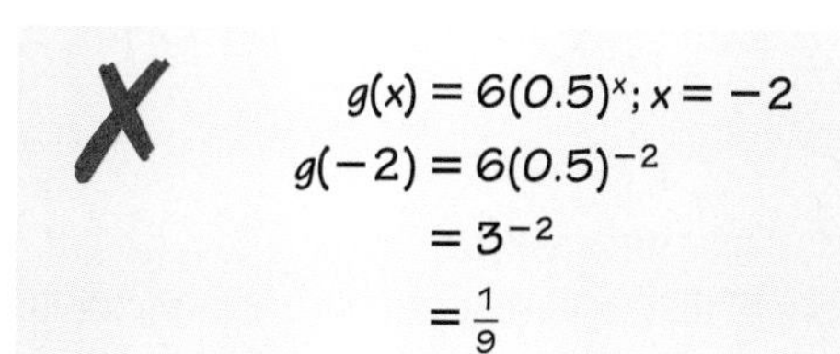

42. ERROR ANALYSIS Describe and correct the error in finding the domain and range of the function.

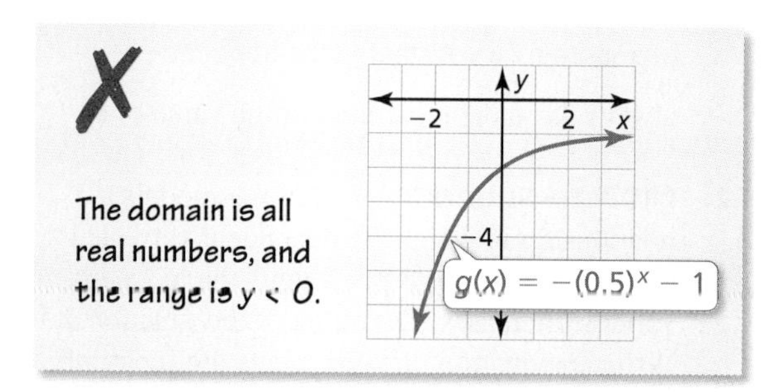

In Exercises 43 and 44, graph the function with the given description. Compare the function to $f(x) = 0.5(4)^x$ over the interval $x = 0$ to $x = 2$. *(See Example 6.)*

43. An exponential function g models a relationship in which the dependent variable is multiplied by 2.5 for every 1 unit the independent variable x increases. The value of the function at 0 is 8.

44. An exponential function h models a relationship in which the dependent variable is multiplied by $\frac{1}{2}$ for every 1 unit the independent variable x increases. The value of the function at 0 is 32.

45. MODELING WITH MATHEMATICS You graph an exponential function on a calculator. You zoom in repeatedly to 25% of the screen size. The function $y = 0.25^x$ represents the percent (in decimal form) of the original screen display that you see, where x is the number of times you zoom in.

a. Graph the function. Describe the domain and range.

b. Find and interpret the y-intercept.

c. You zoom in twice. What percent of the original screen do you see?

46. MODELING WITH MATHEMATICS A population y of coyotes in a national park triples every 20 years. The function $y = 15(3)^x$ represents the population, where x is the number of 20-year periods.

a. Graph the function. Describe the domain and range.

b. Find and interpret the y-intercept.

c. How many coyotes are in the national park in 40 years?

In Exercises 47–50, write an exponential function represented by the table or graph. *(See Example 7.)*

47.

x	0	1	2	3
y	2	14	98	686

48.

x	0	1	2	3
y	−50	−10	−2	−0.4

49. **50.**

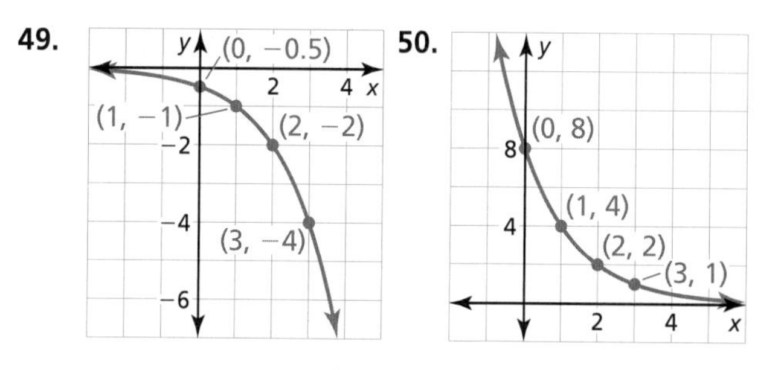

51. **MODELING WITH MATHEMATICS** The graph represents the number y of visitors to a new art gallery after x months.

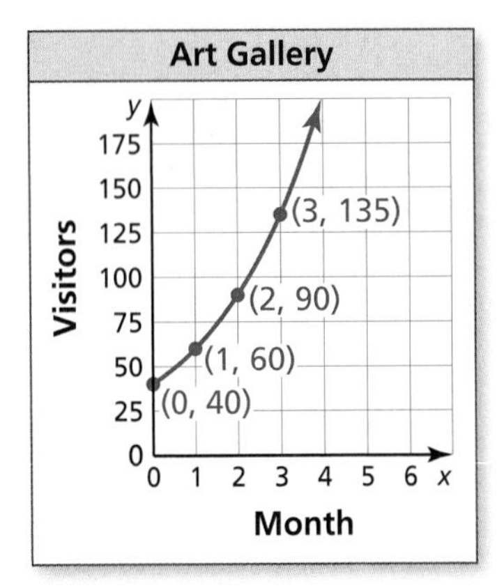

a. Write an exponential function that represents this situation.

b. Approximate the number of visitors after 5 months.

52. **PROBLEM SOLVING** A sales report shows that 3300 gas grills were purchased from a chain of hardware stores last year. The store expects grill sales to increase 6% each year. About how many grills does the store expect to sell in Year 6? Use an equation to justify your answer.

53. **WRITING** Graph the function $f(x) = -2^x$. Then graph $g(x) = -2^x - 3$. How are the y-intercept, domain, and range affected by the translation?

54. **MAKING AN ARGUMENT** Your friend says that the table represents an exponential function because y is multiplied by a constant factor. Is your friend correct? Explain.

x	0	1	3	6
y	2	10	50	250

55. **WRITING** Describe the effect of a on the graph of $y = a \cdot 2^x$ when a is positive and when a is negative.

56. **OPEN-ENDED** Write a function whose graph is a horizontal translation of the graph of $h(x) = 4^x$.

57. **USING STRUCTURE** The graph of g is a translation 4 units up and 3 units right of the graph of $f(x) = 5^x$. Write an equation for g.

58. **HOW DO YOU SEE IT?** The exponential function $y = V(x)$ represents the projected value of a stock x weeks after a corporation loses an important legal battle. The graph of the function is shown.

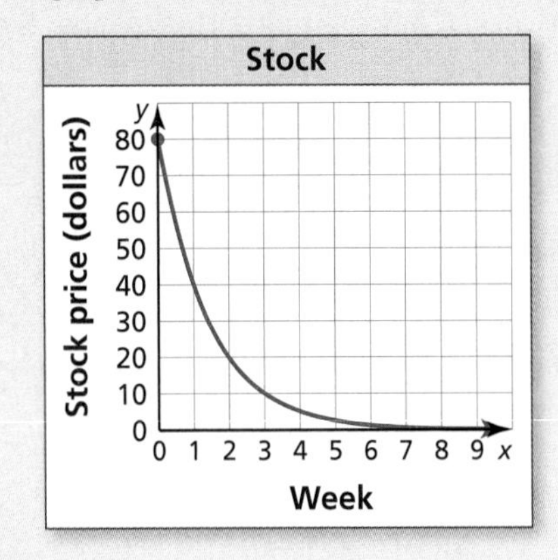

a. After how many weeks will the stock be worth \$20?

b. Describe the change in the stock price from Week 1 to Week 3.

59. **USING GRAPHS** The graph represents the exponential function f. Find $f(7)$.

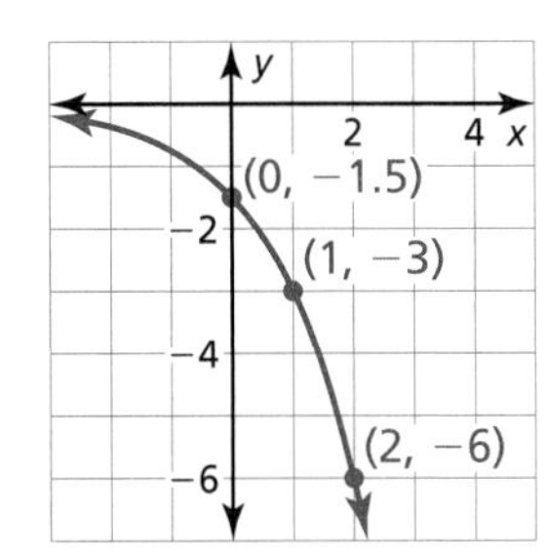

60. **THOUGHT PROVOKING** Write a function of the form $y = ab^x$ that represents a real-life population. Explain the meaning of each of the constants a and b in the real-life context.

61. **REASONING** Let $f(x) = ab^x$. Show that when x is increased by a constant k, the quotient $\frac{f(x+k)}{f(x)}$ is always the same regardless of the value of x.

62. **PROBLEM SOLVING** A function g models a relationship in which the dependent variable is multiplied by 4 for every 2 units the independent variable increases. The value of the function at 0 is 5. Write an equation that represents the function.

63. **PROBLEM SOLVING** Write an exponential function f so that the slope from the point $(0, f(0))$ to the point $(2, f(2))$ is equal to 12.

Maintaining Mathematical Proficiency

Reviewing what you learned in previous grades and lessons

Write the percent as a decimal. *(Skills Review Handbook)*

64. 4%   65. 35%   66. 128%   67. 250%

6.4 Exponential Growth and Decay

Essential Question What are some of the characteristics of exponential growth and exponential decay functions?

EXPLORATION 1 Predicting a Future Event

Work with a partner. It is estimated, that in 1782, there were about 100,000 nesting pairs of bald eagles in the United States. By the 1960s, this number had dropped to about 500 nesting pairs. In 1967, the bald eagle was declared an endangered species in the United States. With protection, the nesting pair population began to increase. Finally, in 2007, the bald eagle was removed from the list of endangered and threatened species.

MODELING WITH MATHEMATICS

To be proficient in math, you need to apply the mathematics you know to solve problems arising in everyday life.

Describe the pattern shown in the graph. Is it exponential growth? Assume the pattern continues. When will the population return to that of the late 1700s? Explain your reasoning.

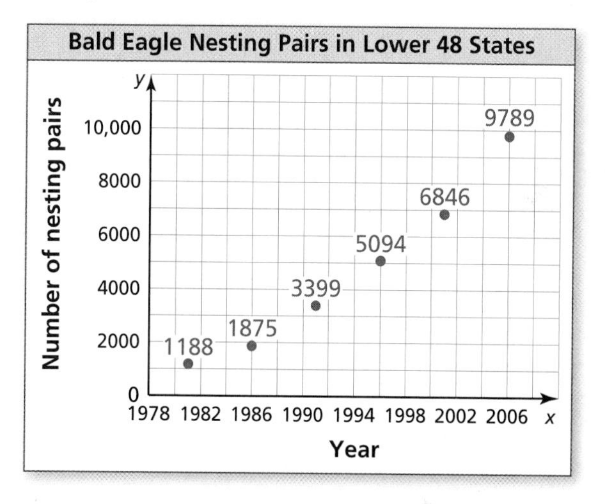

EXPLORATION 2 Describing a Decay Pattern

Work with a partner. A forensic pathologist was called to estimate the time of death of a person. At midnight, the body temperature was 80.5°F and the room temperature was a constant 60°F. One hour later, the body temperature was 78.5°F.

a. By what percent did the difference between the body temperature and the room temperature drop during the hour?

b. Assume that the original body temperature was 98.6°F. Use the percent decrease found in part (a) to make a table showing the decreases in body temperature. Use the table to estimate the time of death.

Communicate Your Answer

3. What are some of the characteristics of exponential growth and exponential decay functions?

4. Use the Internet or some other reference to find an example of each type of function. Your examples should be different than those given in Explorations 1 and 2.

a. exponential growth

b. exponential decay

6.4 Lesson

Core Vocabulary

exponential growth, *p. 314*
exponential growth function, *p. 314*
exponential decay, *p. 315*
exponential decay function, *p. 315*
compound interest, *p. 317*

What You Will Learn

- Use and identify exponential growth and decay functions.
- Interpret and rewrite exponential growth and decay functions.
- Solve real-life problems involving exponential growth and decay.

Exponential Growth and Decay Functions

Exponential growth occurs when a quantity increases by the same factor over equal intervals of time.

Core Concept

Exponential Growth Functions

A function of the form $y = a(1 + r)^t$, where $a > 0$ and $r > 0$, is an **exponential growth function**.

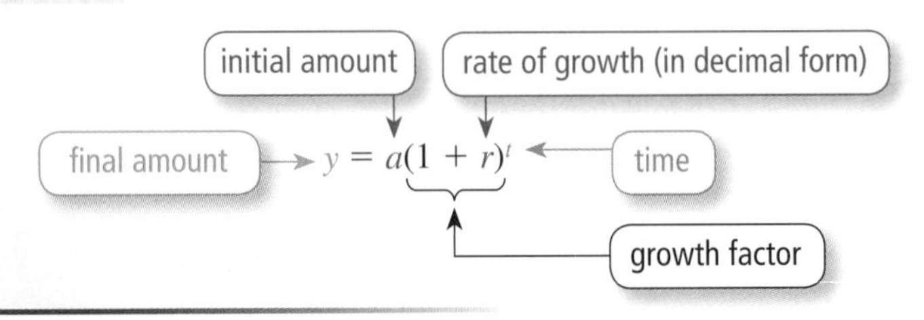

STUDY TIP

Notice that an exponential growth function is of the form $y = ab^x$, where b is replaced by $1 + r$ and x is replaced by t.

EXAMPLE 1 Using an Exponential Growth Function

The inaugural attendance of an annual music festival is 150,000. The attendance y increases by 8% each year.

a. Write an exponential growth function that represents the attendance after t years.

b. How many people will attend the festival in the fifth year? Round your answer to the nearest thousand.

SOLUTION

a. The initial amount is 150,000, and the rate of growth is 8%, or 0.08.

$y = a(1 + r)^t$ — Write the exponential growth function.

$= 150{,}000(1 + 0.08)^t$ — Substitute 150,000 for a and 0.08 for r.

$= 150{,}000(1.08)^t$ — Add.

▶ The festival attendance can be represented by $y = 150{,}000(1.08)^t$.

b. The value $t = 4$ represents the fifth year because $t = 0$ represents the first year.

$y = 150{,}000(1.08)^t$ — Write the exponential growth function.

$= 150{,}000(1.08)^4$ — Substitute 4 for t.

$\approx 204{,}073$ — Use a calculator.

▶ About 204,000 people will attend the festival in the fifth year.

Monitoring Progress  Help in English and Spanish at *BigIdeasMath.com*

1. A website has 500,000 members in 2010. The number y of members increases by 15% each year. (a) Write an exponential growth function that represents the website membership t years after 2010. (b) How many members will there be in 2016? Round your answer to the nearest ten thousand.

Exponential decay occurs when a quantity decreases by the same factor over equal intervals of time.

Core Concept

Exponential Decay Functions

A function of the form $y = a(1 - r)^t$, where $a > 0$ and $0 < r < 1$, is an **exponential decay function**.

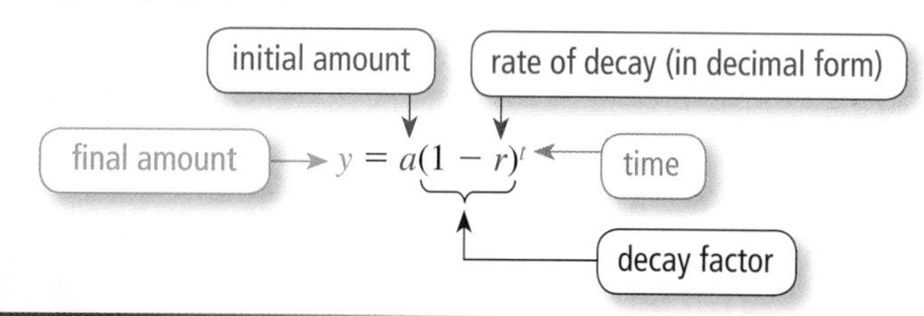

STUDY TIP

Notice that an exponential decay function is of the form $y = ab^x$, where b is replaced by $1 - r$ and x is replaced by t.

For exponential growth, the value inside the parentheses is greater than 1 because r is added to 1. For exponential decay, the value inside the parentheses is less than 1 because r is subtracted from 1.

EXAMPLE 2 Identifying Exponential Growth and Decay

Determine whether each table represents an *exponential growth function*, an *exponential decay function*, or *neither*.

a.

x	y
0	270
1	90
2	30
3	10

b.

x	0	1	2	3
y	5	10	20	40

SOLUTION

a.

x	y
0	270
1	90
2	30
3	10

x: $+1$, $+1$, $+1$; y: $\times\frac{1}{3}$, $\times\frac{1}{3}$, $\times\frac{1}{3}$

▶ As x increases by 1, y is multiplied by $\frac{1}{3}$. So, the table represents an exponential decay function.

b.

x	0	1	2	3
y	5	10	20	40

x: $+1$, $+1$, $+1$; y: $\times 2$, $\times 2$, $\times 2$

▶ As x increases by 1, y is multiplied by 2. So, the table represents an exponential growth function.

Monitoring Progress  Help in English and Spanish at *BigIdeasMath.com*

Determine whether the table represents an *exponential growth function*, an *exponential decay function*, or *neither*. Explain.

2.

x	0	1	2	3
y	64	16	4	1

3.

x	1	3	5	7
y	4	11	18	25

Interpreting and Rewriting Exponential Functions

EXAMPLE 3 Interpreting Exponential Functions

Determine whether each function represents *exponential growth* or *exponential decay*. Identify the percent rate of change.

a. $y = 5(1.07)^t$ **b.** $f(t) = 0.2(0.98)^t$

SOLUTION

a. The function is of the form $y = a(1 + r)^t$, where $1 + r > 1$, so it represents exponential growth. Use the growth factor $1 + r$ to find the rate of growth.

$1 + r = 1.07$ Write an equation.

$r = 0.07$ Solve for r.

▶ So, the function represents exponential growth and the rate of growth is 7%.

b. The function is of the form $y = a(1 - r)^t$, where $1 - r < 1$, so it represents exponential decay. Use the decay factor $1 - r$ to find the rate of decay.

$1 - r = 0.98$ Write an equation.

$r = 0.02$ Solve for r.

▶ So, the function represents exponential decay and the rate of decay is 2%.

STUDY TIP

You can rewrite exponential expressions and functions using the properties of exponents. Changing the form of an exponential function can reveal important attributes of the function.

EXAMPLE 4 Rewriting Exponential Functions

Rewrite each function to determine whether it represents *exponential growth* or *exponential decay*.

a. $y = 100(0.96)^{t/4}$ **b.** $f(t) = (1.1)^{t-3}$

SOLUTION

a. $y = 100(0.96)^{t/4}$ Write the function.

$= 100(0.96^{1/4})^t$ Power of a Power Property

$\approx 100(0.99)^t$ Evaluate the power.

▶ So, the function represents exponential decay.

b. $f(t) = (1.1)^{t-3}$ Write the function.

$= \dfrac{(1.1)^t}{(1.1)^3}$ Quotient of Powers Property

$\approx 0.75(1.1)^t$ Evaluate the power and simplify.

▶ So, the function represents exponential growth.

Monitoring Progress  Help in English and Spanish at *BigIdeasMath.com*

Determine whether the function represents *exponential growth* or *exponential decay*. Identify the percent rate of change.

4. $y = 2(0.92)^t$ **5.** $f(t) = (1.2)^t$

Rewrite the function to determine whether it represents *exponential growth* or *exponential decay*.

6. $f(t) = 3(1.02)^{10t}$ **7.** $y = (0.95)^{t+2}$

Solving Real-Life Problems

Exponential growth functions are used in real-life situations involving *compound interest*. Although interest earned is expressed as an *annual* rate, the interest is usually compounded more frequently than once per year. So, the formula $y = a(1 + r)^t$ must be modified for compound interest problems.

Core Concept

STUDY TIP

For interest compounded yearly, you can substitute 1 for n in the formula to get $y = P(1 + r)^t$.

Compound Interest

Compound interest is the interest earned on the principal *and* on previously earned interest. The balance y of an account earning compound interest is

$$y = P\left(1 + \frac{r}{n}\right)^{nt}.$$

P = principal (initial amount)
r = annual interest rate (in decimal form)
t = time (in years)
n = number of times interest is compounded per year

EXAMPLE 5 Writing a Function

You deposit \$100 in a savings account that earns 6% annual interest compounded monthly. Write a function that represents the balance after t years.

SOLUTION

$y = P\left(1 + \frac{r}{n}\right)^{nt}$ — Write the compound interest formula.

$= 100\left(1 + \frac{0.06}{12}\right)^{12t}$ — Substitute 100 for P, 0.06 for r, and 12 for n.

$= 100(1.005)^{12t}$ — Simplify.

EXAMPLE 6 Solving a Real-Life Problem

The table shows the balance of a money market account over time.

a. Write a function that represents the balance after t years.

b. Graph the functions from part (a) and from Example 5 in the same coordinate plane. Compare the account balances.

Year, t	Balance
0	\$100
1	\$110
2	\$121
3	\$133.10
4	\$146.41
5	\$161.05

SOLUTION

a. From the table, you know the initial balance is \$100, and it increases 10% each year. So, $P = 100$ and $r = 0.1$.

$y = P(1 + r)^t$ — Write the compound interest formula when $n = 1$.

$= 100(1 + 0.1)^t$ — Substitute 100 for P and 0.1 for r.

$= 100(1.1)^t$ — Add.

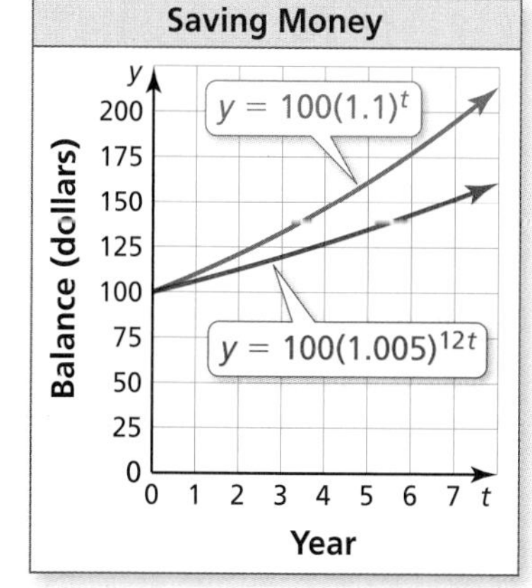

b. The money market account earns 10% interest each year, and the savings account earns 6% interest each year. So, the balance of the money market account increases faster.

Monitoring Progress  Help in English and Spanish at *BigIdeasMath.com*

8. You deposit \$500 in a savings account that earns 9% annual interest compounded monthly. Write and graph a function that represents the balance y (in dollars) after t years.

EXAMPLE 7 Solving a Real-Life Problem

The value of a car is \$21,500. It loses 12% of its value every year. (a) Write a function that represents the value y (in dollars) of the car after t years. (b) Find the approximate monthly percent decrease in value. (c) Graph the function from part (a). Use the graph to estimate the value of the car after 6 years.

SOLUTION

1. **Understand the Problem** You know the value of the car and its annual percent decrease in value. You are asked to write a function that represents the value of the car over time and approximate the monthly percent decrease in value. Then graph the function and use the graph to estimate the value of the car in the future.

2. **Make a Plan** Use the initial amount and the annual percent decrease in value to write an exponential decay function. Note that the annual percent decrease represents the rate of decay. Rewrite the function using the properties of exponents to approximate the monthly percent decrease (rate of decay). Then graph the original function and use the graph to estimate the y-value when the t-value is 6.

STUDY TIP

In real life, the percent decrease in value of an asset is called the *depreciation rate.*

3. **Solve the Problem**

   a. The initial value is \$21,500, and the rate of decay is 12%, or 0.12.

$y = a(1 - r)^t$	Write the exponential decay function.
$= 21{,}500(1 - 0.12)^t$	Substitute 21,500 for a and 0.12 for r.
$= 21{,}500(0.88)^t$	Subtract.

   ▶ The value of the car can be represented by $y = 21{,}500(0.88)^t$.

   b. Use the fact that $t = \frac{1}{12}(12t)$ and the properties of exponents to rewrite the function in a form that reveals the monthly rate of decay.

$y = 21{,}500(0.88)^t$	Write the original function.
$= 21{,}500(0.88)^{(1/12)(12t)}$	Rewrite the exponent.
$= 21{,}500(0.88^{1/12})^{12t}$	Power of a Power Property
$\approx 21{,}500(0.989)^{12t}$	Evaluate the power.

   Use the decay factor $1 - r \approx 0.989$ to find the rate of decay $r \approx 0.011$.

   ▶ So, the monthly percent decrease is about 1.1%.

   c. From the graph, you can see that the y-value is about 10,000 when $t = 6$.

   ▶ So, the value of the car is about \$10,000 after 6 years.

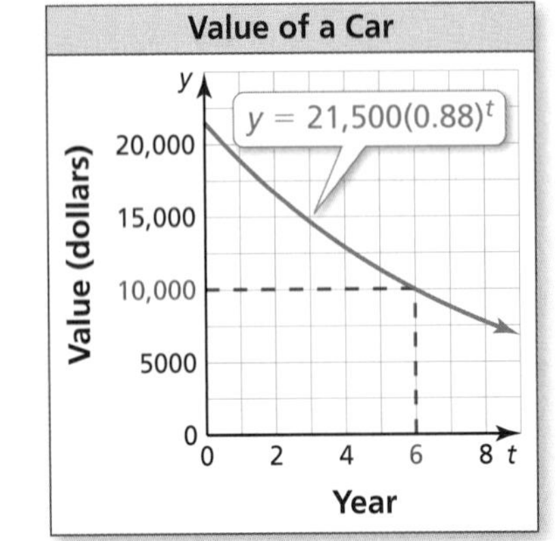

4. **Look Back** To check that the monthly percent decrease is reasonable, multiply it by 12 to see if it is close in value to the annual percent decrease of 12%.

   $1.1\% \times 12 = 13.2\%$ — 13.2% is close to 12%, so 1.1% is reasonable.

   When you evaluate $y = 21{,}500(0.88)^t$ for $t = 6$, you get about \$9985. So, \$10,000 is a reasonable estimation.

Monitoring Progress  Help in English and Spanish at *BigIdeasMath.com*

9. **WHAT IF?** The car loses 9% of its value every year. (a) Write a function that represents the value y (in dollars) of the car after t years. (b) Find the approximate monthly percent decrease in value. (c) Graph the function from part (a). Use the graph to estimate the value of the car after 12 years. Round your answer to the nearest thousand.

6.4 Exercises

Dynamic Solutions available at *BigIdeasMath.com*

Vocabulary and Core Concept Check

1. **COMPLETE THE SENTENCE** In the exponential growth function $y = a(1 + r)^t$, the quantity r is called the ________.

2. **VOCABULARY** What is the decay factor in the exponential decay function $y = a(1 - r)^t$?

3. **VOCABULARY** Compare exponential growth and exponential decay.

4. **WRITING** When does the function $y = ab^x$ represent exponential growth? exponential decay?

Monitoring Progress and Modeling with Mathematics

In Exercises 5–12, identify the initial amount a and the rate of growth r (as a percent) of the exponential function. Evaluate the function when $t = 5$. Round your answer to the nearest tenth.

5. $y = 350(1 + 0.75)^t$
6. $y = 10(1 + 0.4)^t$
7. $y = 25(1.2)^t$
8. $y = 12(1.05)^t$
9. $f(t) = 1500(1.074)^t$
10. $h(t) = 175(1.028)^t$
11. $g(t) = 6.72(2)^t$
12. $p(t) = 1.8^t$

In Exercises 13–16, write a function that represents the situation.

13. Sales of $10,000 increase by 65% each year.

14. Your starting annual salary of $35,000 increases by 4% each year.

15. A population of 210,000 increases by 12.5% each year.

16. An item costs $4.50, and its price increases by 3.5% each year.

17. **MODELING WITH MATHEMATICS** The population of a city has been increasing by 2% annually. The sign shown is from the year 2000. *(See Example 1.)*

    a. Write an exponential growth function that represents the population t years after 2000.

    b. What will the population be in 2020? Round your answer to the nearest thousand.

18. **MODELING WITH MATHEMATICS** A young channel catfish weighs about 0.1 pound. During the next 8 weeks, its weight increases by about 23% each week.

    a. Write an exponential growth function that represents the weight of the catfish after t weeks during the 8-week period.

    b. About how much will the catfish weigh after 4 weeks? Round your answer to the nearest thousandth.

In Exercises 19–26, identify the initial amount a and the rate of decay r (as a percent) of the exponential function. Evaluate the function when $t = 3$. Round your answer to the nearest tenth.

19. $y = 575(1 - 0.6)^t$
20. $y = 8(1 - 0.15)^t$
21. $g(t) = 240(0.75)^t$
22. $f(t) = 475(0.5)^t$
23. $w(t) = 700(0.995)^t$
24. $h(t) = 1250(0.865)^t$
25. $y = \left(\frac{7}{8}\right)^t$
26. $y = 0.5\left(\frac{3}{4}\right)^t$

In Exercises 27–30, write a function that represents the situation.

27. A population of 100,000 decreases by 2% each year.

28. A $900 sound system decreases in value by 9% each year.

29. A stock valued at $100 decreases in value by 9.5% each year.

30. A company profit of \$20,000 decreases by 13.4% each year.

31. **ERROR ANALYSIS** The growth rate of a bacterial culture is 150% each hour. Initially, there are 10 bacteria. Describe and correct the error in finding the number of bacteria in the culture after 8 hours.

$b(t) = 10(1.5)^t$

$b(8) = 10(1.5)^8 \approx 256.3$

After 8 hours, there are about 256 bacteria in the culture.

32. **ERROR ANALYSIS** You purchase a car in 2010 for \$25,000. The value of the car decreases by 14% annually. Describe and correct the error in finding the value of the car in 2015.

$v(t) = 25{,}000(1.14)^t$

$v(5) = 25{,}000(1.14)^5 \approx 48{,}135$

The value of the car in 2015 is about \$48,000.

In Exercises 33–38, determine whether the table represents an *exponential growth function*, an *exponential decay function*, or *neither*. Explain. *(See Example 2.)*

33.

x	y
−1	50
0	10
1	2
2	0.4

34.

x	y
0	32
1	28
2	24
3	20

35.

x	y
0	35
1	29
2	23
3	17

36.

x	y
1	17
2	51
3	153
4	459

37.

x	y
5	2
10	8
15	32
20	128

38.

x	y
3	432
5	72
7	12
9	2

39. **ANALYZING RELATIONSHIPS** The table shows the value of a camper t years after it is purchased.

t	Value
1	\$37,000
2	\$29,600
3	\$23,680
4	\$18,944

a. Determine whether the table represents an exponential growth function, an exponential decay function, or neither.

b. What is the value of the camper after 5 years?

40. **ANALYZING RELATIONSHIPS** The table shows the total numbers of visitors to a website t days after it is online.

t	42	43	44	45
Visitors	11,000	12,100	13,310	14,641

a. Determine whether the table represents an exponential growth function, an exponential decay function, or neither.

b. How many people will have visited the website after it is online 47 days?

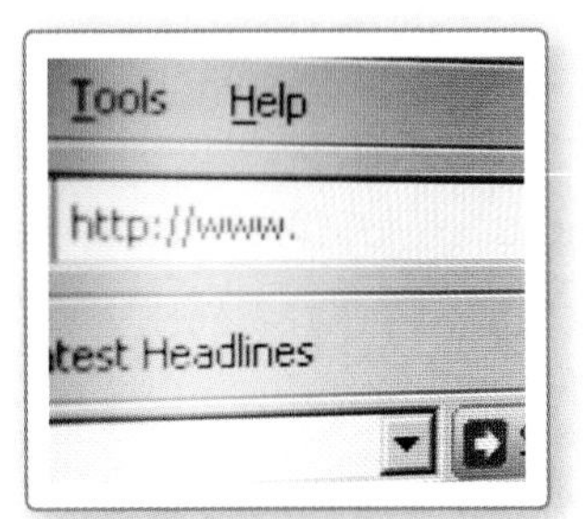

In Exercises 41–48, determine whether each function represents *exponential growth* or *exponential decay*. Identify the percent rate of change. *(See Example 3.)*

41. $y = 4(0.8)^t$

42. $y = 15(1.1)^t$

43. $y = 30(0.95)^t$

44. $y = 5(1.08)^t$

45. $r(t) = 0.4(1.06)^t$

46. $s(t) = 0.65(0.48)^t$

47. $g(t) = 2\left(\frac{5}{4}\right)^t$

48. $m(t) = \left(\frac{4}{5}\right)^t$

In Exercises 49–56, rewrite the function to determine whether it represents *exponential growth* or *exponential decay*. *(See Example 4.)*

49. $y = (0.9)^{t-4}$

50. $y = (1.4)^{t+8}$

51. $y = 2(1.06)^{9t}$

52. $y = 5(0.82)^{t/5}$

53. $x(t) = (1.45)^{t/2}$

54. $f(t) = 0.4(1.16)^{t-1}$

55. $b(t) = 4(0.55)^{t+3}$

56. $r(t) = (0.88)^{4t}$

In Exercises 57–60, write a function that represents the balance after t years. *(See Example 5.)*

57. \$2000 deposit that earns 5% annual interest compounded quarterly

58. \$1400 deposit that earns 10% annual interest compounded semiannually

59. \$6200 deposit that earns 8.4% annual interest compounded monthly

60. \$3500 deposit that earns 9.2% annual interest compounded quarterly

61. PROBLEM SOLVING The cross-sectional area of a tree 4.5 feet from the ground is called its *basal area*. The table shows the basal areas (in square inches) of Tree A over time. *(See Example 6.)*

Year, t	0	1	2	3	4
Basal area, A	120	132	145.2	159.7	175.7

a. Write functions that represent the basal areas of the trees after t years.

b. Graph the functions from part (a) in the same coordinate plane. Compare the basal areas.

62. PROBLEM SOLVING You deposit \$300 into an investment account that earns 12% annual interest compounded quarterly. The graph shows the balance of a savings account over time.

a. Write functions that represent the balances of the accounts after t years.

b. Graph the functions from part (a) in the same coordinate plane. Compare the account balances.

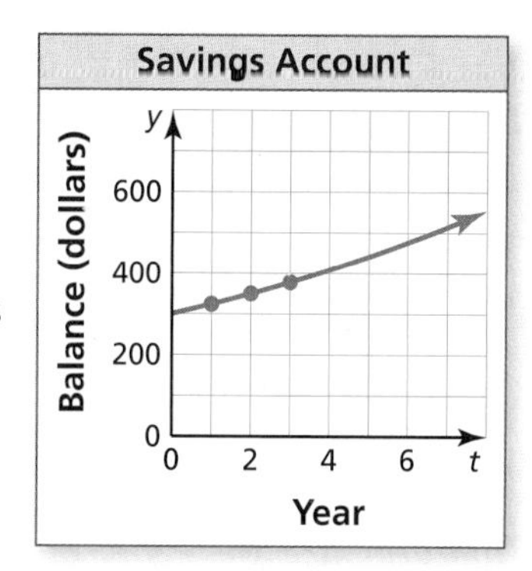

63. PROBLEM SOLVING A city has a population of 25,000. The population is expected to increase by 5.5% annually for the next decade. *(See Example 7.)*

a. Write a function that represents the population y after t years.

b. Find the approximate monthly percent increase in population.

c. Graph the function from part (a). Use the graph to estimate the population after 4 years.

64. PROBLEM SOLVING Plutonium-238 is a material that generates steady heat due to decay and is used in power systems for some spacecraft. The function $y = a(0.5)^{t/x}$ represents the amount y of a substance remaining after t years, where a is the initial amount and x is the length of the half-life (in years).

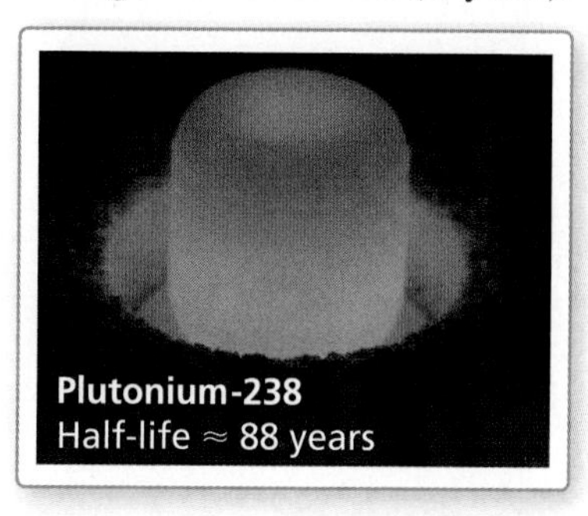

a. A scientist is studying a 3-gram sample. Write a function that represents the amount y of plutonium-238 after t years.

b. What is the yearly percent decrease of plutonium-238?

c. Graph the function from part (a). Use the graph to estimate the amount remaining after 12 years.

65. COMPARING FUNCTIONS The three given functions describe the amount y of ibuprofen (in milligrams) in a person's bloodstream t hours after taking the dosage.

$$y \approx 800(0.71)^t$$
$$y \approx 800(0.9943)^{60t}$$
$$y \approx 800(0.843)^{2t}$$

a. Show that these expressions are approximately equivalent.

b. Describe the information given by each of the functions.

66. **COMBINING FUNCTIONS** You deposit \$9000 in a savings account that earns 3.6% annual interest compounded monthly. You also save \$40 per month in a safe at home. Write a function $C(t) = b(t) + h(t)$, where $b(t)$ represents the balance of your savings account and $h(t)$ represents the amount in your safe after t years. What does $C(t)$ represent?

67. **NUMBER SENSE** During a flu epidemic, the number of sick people triples every week. What is the growth rate as a percent? Explain your reasoning.

68. **HOW DO YOU SEE IT?** Match each situation with its graph. Explain your reasoning.

    a. A bacterial population doubles each hour.

    b. The value of a computer decreases by 18% each year.

    c. A deposit earns 11% annual interest compounded yearly.

    d. A radioactive element decays 5.5% each year.

A. 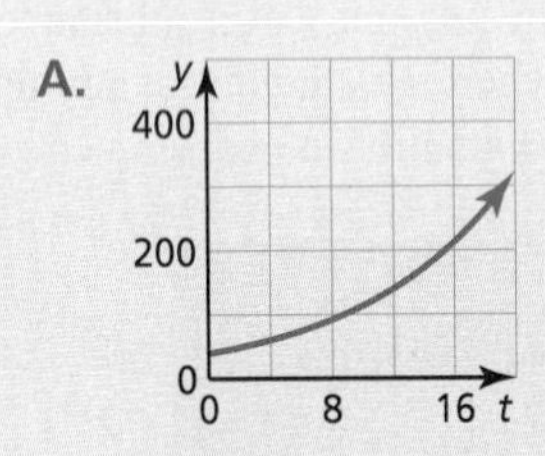

B. 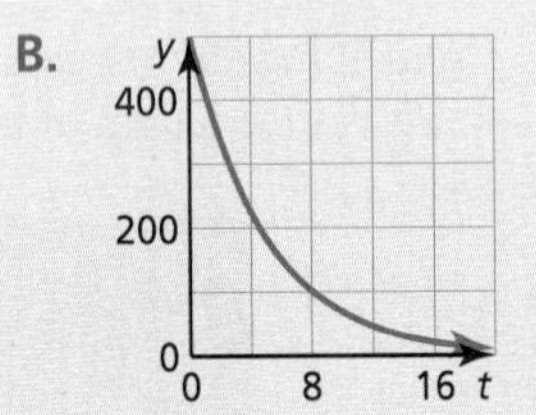

C. 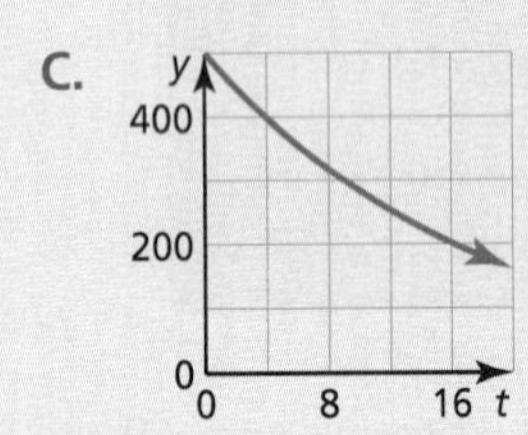

D. 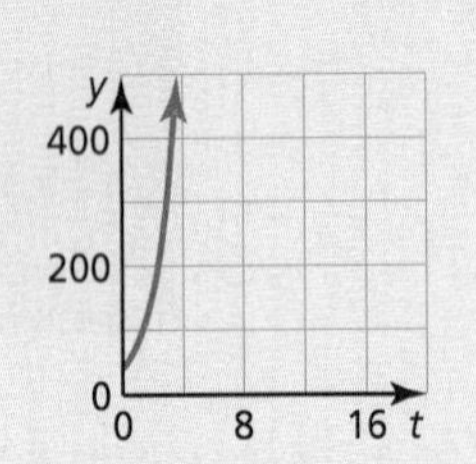

69. **WRITING** Give an example of an equation in the form $y = ab^x$ that does not represent an exponential growth function or an exponential decay function. Explain your reasoning.

70. **THOUGHT PROVOKING** Describe two account options into which you can deposit \$1000 and earn compound interest. Write a function that represents the balance of each account after t years. Which account would you rather use? Explain your reasoning.

71. **MAKING AN ARGUMENT** A store is having a sale on sweaters. On the first day, the prices of the sweaters are reduced by 20%. The prices will be reduced another 20% each day until the sweaters are sold. Your friend says the sweaters will be free on the fifth day. Is your friend correct? Explain.

72. **COMPARING FUNCTIONS** The graphs of f and g are shown.

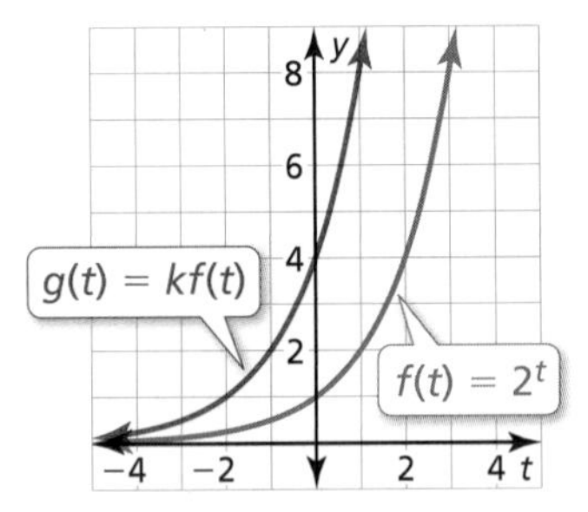

    a. Explain why f is an exponential growth function. Identify the rate of growth.

    b. Describe the transformation from the graph of f to the graph of g. Determine the value of k.

    c. The graph of g is the same as the graph of $h(t) = f(t + r)$. Use properties of exponents to find the value of r.

Maintaining Mathematical Proficiency

Reviewing what you learned in previous grades and lessons

Solve the equation. Check your solution. *(Section 1.3)*

73. $8x + 12 = 4x$

74. $5 - t = 7t + 21$

75. $6(r - 2) = 2r + 8$

Find the slope and the y-intercept of the graph of the linear equation. *(Section 3.5)*

76. $y = -6x + 7$

77. $y = \frac{1}{4}x + 7$

78. $3y = 6x - 12$

79. $2y + x = 8$

6.1–6.4 What Did You Learn?

Core Vocabulary

*n*th root of *a*, *p. 300*
radical, *p. 300*
index of a radical, *p. 300*
exponential function, *p. 306*
exponential growth, *p. 314*
exponential growth function, *p. 314*
exponential decay, *p. 315*
exponential decay function, *p. 315*
compound interest, *p. 317*

Core Concepts

Section 6.1

Zero Exponent, *p. 292*
Negative Exponents, *p. 292*
Product of Powers Property, *p. 293*
Quotient of Powers Property, *p. 293*
Power of a Power Property, *p. 293*
Power of a Product Property, *p. 294*
Power of a Quotient Property, *p. 294*

Section 6.2

Real *n*th Roots of *a*, *p. 300*
Rational Exponents, *p. 301*

Section 6.3

Graphing $y = ab^x$ When $b > 1$, *p. 307*
Graphing $y = ab^x$ When $0 < b < 1$, *p. 307*

Section 6.4

Exponential Growth Functions, *p. 314*
Exponential Decay Functions, *p. 315*
Compound Interest, *p. 317*

Mathematical Practices

1. How did you apply what you know to simplify the complicated situation in Exercise 56 on page 297?
2. How can you use previously established results to construct an argument in Exercise 44 on page 304?
3. How is the form of the function you wrote in Exercise 66 on page 322 related to the forms of other types of functions you have learned about in this course?

Study Skills

Analyzing Your Errors

Misreading Directions

- **What Happens:** You incorrectly read or do not understand directions.
- **How to Avoid This Error:** Read the instructions for exercises at least twice and make sure you understand what they mean. Make this a habit and use it when taking tests.

6.1–6.4 Quiz

Simplify the expression. Write your answer using only positive exponents. *(Section 6.1)*

1. $3^2 \cdot 3^4$

2. $(k^4)^{-3}$

3. $\left(\dfrac{4r^2}{3s^5}\right)^3$

4. $\left(\dfrac{2x^0}{4x^{-2}y^4}\right)^2$

Evaluate the expression. *(Section 6.2)*

5. $\sqrt[3]{27}$

6. $\left(\dfrac{1}{16}\right)^{1/4}$

7. $512^{2/3}$

8. $(\sqrt{4})^5$

Graph the function. Describe the domain and range. *(Section 6.3)*

9. $y = 5^x$

10. $y = -2\left(\dfrac{1}{6}\right)^x$

11. $y = 6(2)^{x-4} - 1$

Determine whether the table represents an *exponential growth function*, an *exponential decay function*, or *neither*. Explain. *(Section 6.4)*

12.

x	0	1	2	3
y	7	21	63	189

13.

x	1	2	3	4
y	14,641	1331	121	11

Determine whether the function represents *exponential growth* or *exponential decay*. Identify the percent rate of change. *(Section 6.4)*

14. $y = 3(1.88)^t$

15. $f(t) = \dfrac{1}{3}(1.26)^t$

16. $f(t) = 80\left(\dfrac{3}{5}\right)^t$

17. The table shows several units of mass. *(Section 6.1)*

Unit of mass	kilogram	hectogram	dekagram	decigram	centigram	milligram	microgram	nanogram
Mass (in grams)	10^3	10^2	10^1	10^{-1}	10^{-2}	10^{-3}	10^{-6}	10^{-9}

a. How many times larger is a kilogram than a nanogram? Write your answer using only positive exponents.

b. How many times smaller is a milligram than a hectogram? Write your answer using only positive exponents.

c. Which is greater, 10,000 milligrams or 1000 decigrams? Explain your reasoning.

18. You store blankets in a cedar chest. What is the volume of the cedar chest? *(Section 6.2)*

19. The function $f(t) = 5(4)^t$ represents the number of frogs in a pond after t years. *(Section 6.3 and Section 6.4)*

a. Does the function represent *exponential growth* or *exponential decay*? Explain.

b. Graph the function. Describe the domain and range.

c. What is the yearly percent change? the approximate monthly percent change?

d. How many frogs are in the pond after 4 years?

6.5 Solving Exponential Equations

Essential Question How can you solve an exponential equation graphically?

EXPLORATION 1 **Solving an Exponential Equation Graphically**

Work with a partner. Use a graphing calculator to solve the exponential equation $2.5^{x-3} = 6.25$ graphically. Describe your process and explain how you determined the solution.

EXPLORATION 2 **The Number of Solutions of an Exponential Equation**

Work with a partner.

a. Use a graphing calculator to graph the equation $y = 2^x$.

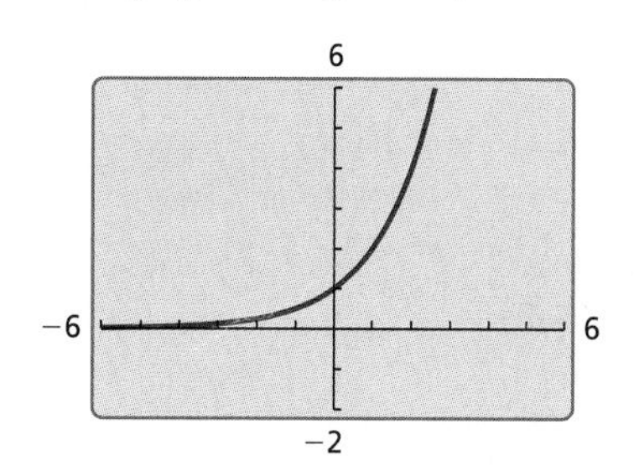

b. In the same viewing window, graph a linear equation (if possible) that does not intersect the graph of $y = 2^x$.

c. In the same viewing window, graph a linear equation (if possible) that intersects the graph of $y = 2^x$ in more than one point.

d. Is it possible for an exponential equation to have no solution? more than one solution? Explain your reasoning.

USING APPROPRIATE TOOLS

To be proficient in math, you need to use technological tools to explore and deepen your understanding of concepts.

EXPLORATION 3 **Solving Exponential Equations Graphically**

Work with a partner. Use a graphing calculator to solve each equation.

a. $2^x = \frac{1}{2}$ **b.** $2^{x+1} = 0$ **c.** $2^x = \sqrt{2}$

d. $3^x = 9$ **e.** $3^{x-1} = 0$ **f.** $4^{2x} = 2$

g. $2^{x/2} = \frac{1}{4}$ **h.** $3^{x+2} = \frac{1}{9}$ **i.** $2^{x-2} = \frac{3}{2}x - 2$

Communicate Your Answer

4. How can you solve an exponential equation graphically?

5. A population of 30 mice is expected to double each year. The number p of mice in the population each year is given by $p = 30(2^n)$. In how many years will there be 960 mice in the population?

6.5 Lesson

Core Vocabulary

exponential equation, *p. 326*

What You Will Learn

- Solve exponential equations with the same base.
- Solve exponential equations with unlike bases.
- Solve exponential equations by graphing.

Solving Exponential Equations with the Same Base

Exponential equations are equations in which variable expressions occur as exponents.

Core Concept

Property of Equality for Exponential Equations

Words Two powers with the *same positive base b*, where $b \neq 1$, are equal if and only if their exponents are equal.

Numbers If $2^x = 2^5$, then $x = 5$. If $x = 5$, then $2^x = 2^5$.

Algebra If $b > 0$ and $b \neq 1$, then $b^x = b^y$ if and only if $x = y$.

EXAMPLE 1 Solving Exponential Equations with the Same Base

Solve each equation.

a. $3^{x+1} = 3^5$ **b.** $6 = 6^{2x-3}$ **c.** $10^{3x} = 10^{2x+3}$

SOLUTION

a.

$3^{x+1} = 3^5$	Write the equation.
$x + 1 = 5$	Equate the exponents.
$-1 \quad -1$	Subtract 1 from each side.
$x = 4$	Simplify.

b.

$6 = 6^{2x-3}$	Write the equation.
$1 = 2x - 3$	Equate the exponents.
$+3 \quad +3$	Add 3 to each side.
$4 = 2x$	Simplify.
$\frac{4}{2} = \frac{2x}{2}$	Divide each side by 2.
$2 = x$	Simplify.

Check

$6 = 6^{2x-3}$

$6 \stackrel{?}{=} 6^{2(2)-3}$

$6 = 6$ ✓

c.

$10^{3x} = 10^{2x+3}$	Write the equation.
$3x = 2x + 3$	Equate the exponents.
$-2x \quad -2x$	Subtract $2x$ from each side.
$x = 3$	Simplify.

Monitoring Progress  Help in English and Spanish at *BigIdeasMath.com*

Solve the equation. Check your solution.

1. $2^{2x} = 2^6$ **2.** $5^{2x} = 5^{x+1}$ **3.** $7^{3x+5} = 7^{x+1}$

Solving Exponential Equations with Unlike Bases

To solve some exponential equations, you must first rewrite each side of the equation using the same base.

EXAMPLE 2 Solving Exponential Equations with Unlike Bases

Solve (a) $5^x = 125$, (b) $4^x = 2^{x-3}$, and (c) $9^{x+2} = 27^x$.

SOLUTION

a. $5^x = 125$ Write the equation.

$5^x = 5^3$ Rewrite 125 as 5^3.

$x = 3$ Equate the exponents.

b. $4^x = 2^{x-3}$ Write the equation.

$(2^2)^x = 2^{x-3}$ Rewrite 4 as 2^2.

$2^{2x} = 2^{x-3}$ Power of a Power Property

$2x = x - 3$ Equate the exponents.

$x = -3$ Solve for x.

Check

$$4^x = 2^{x-3}$$
$$4^{-3} \stackrel{?}{=} 2^{-3-3}$$
$$\frac{1}{64} = \frac{1}{64} \checkmark$$

c. $9^{x+2} = 27^x$ Write the equation.

$(3^2)^{x+2} = (3^3)^x$ Rewrite 9 as 3^2 and 27 as 3^3.

$3^{2x+4} = 3^{3x}$ Power of a Power Property

$2x + 4 = 3x$ Equate the exponents.

$4 = x$ Solve for x.

Check

$$9^{x+2} = 27^x$$
$$9^{4+2} \stackrel{?}{=} 27^4$$
$$531{,}441 = 531{,}441 \checkmark$$

EXAMPLE 3 Solving Exponential Equations When $0 < b < 1$

Solve (a) $\left(\frac{1}{2}\right)^x = 4$ and (b) $4^{x+1} = \frac{1}{64}$.

SOLUTION

a. $\left(\frac{1}{2}\right)^x = 4$ Write the equation.

$(2^{-1})^x = 2^2$ Rewrite $\frac{1}{2}$ as 2^{-1} and 4 as 2^2.

$2^{-x} = 2^2$ Power of a Power Property

$-x = 2$ Equate the exponents.

$x = -2$ Solve for x.

b. $4^{x+1} = \frac{1}{64}$ Write the equation.

$4^{x+1} = \frac{1}{4^3}$ Rewrite 64 as 4^3.

$4^{x+1} = 4^{-3}$ Definition of negative exponent

$x + 1 = -3$ Equate the exponents.

$x = -4$ Solve for x.

Check

$$4^{x+1} = \frac{1}{64}$$
$$4^{-4+1} \stackrel{?}{=} \frac{1}{64}$$
$$\frac{1}{64} = \frac{1}{64} \checkmark$$

Monitoring Progress

Help in English and Spanish at *BigIdeasMath.com*

Solve the equation. Check your solution.

4. $4^x = 256$ **5.** $9^{2x} = 3^{x-6}$ **6.** $4^{3x} = 8^{x+1}$ **7.** $\left(\frac{1}{3}\right)^{x-1} = 27$

Solving Exponential Equations by Graphing

Sometimes, it is impossible to rewrite each side of an exponential equation using the same base. You can solve these types of equations by graphing each side and finding the point(s) of intersection. Exponential equations can have no solution, one solution, or more than one solution depending on the number of points of intersection.

EXAMPLE 4 **Solving Exponential Equations by Graphing**

Use a graphing calculator to solve (a) $\left(\frac{1}{2}\right)^{x-1} = 7$ and (b) $3^{x+2} = x + 1$.

SOLUTION

a. Step 1 Write a system of equations using each side of the equation.

$y = \left(\frac{1}{2}\right)^{x-1}$ Equation 1

$y = 7$ Equation 2

Step 2 Enter the equations into a calculator. Then graph the equations in a viewing window that shows where the graphs could intersect.

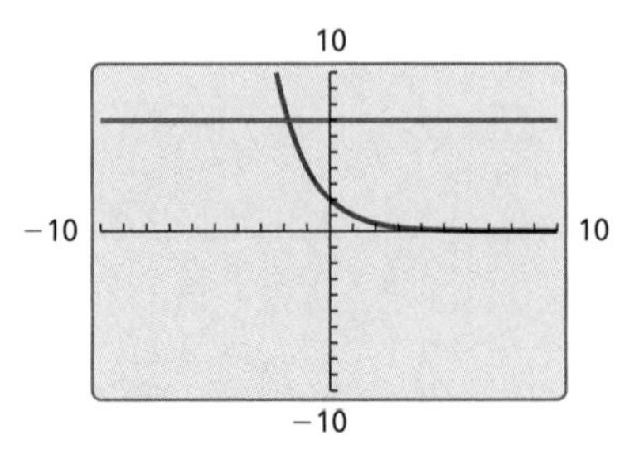

Step 3 Use the *intersect* feature to find the point of intersection. The graphs intersect at about $(-1.81, 7)$.

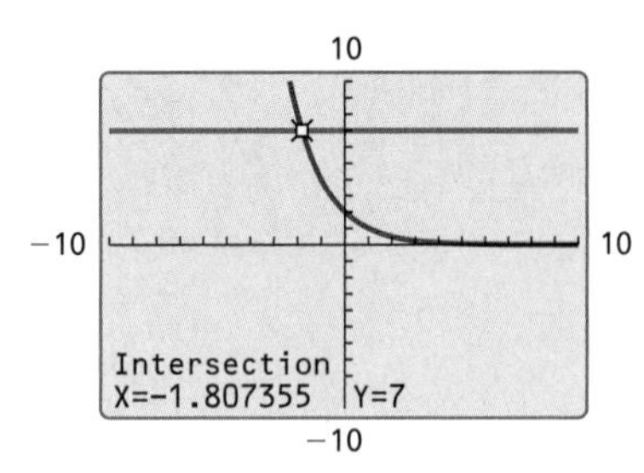

Check

$\left(\frac{1}{2}\right)^{x-1} = 7$

$\left(\frac{1}{2}\right)^{-1.81-1} \stackrel{?}{=} 7$

$7.01 \approx 7$ ✓

▶ So, the solution is $x \approx -1.81$.

b. Step 1 Write a system of equations using each side of the equation.

$y = 3^{x+2}$ Equation 1

$y = x + 1$ Equation 2

Step 2 Enter the equations into a calculator. Then graph the equations in a viewing window that shows where the graphs could intersect.

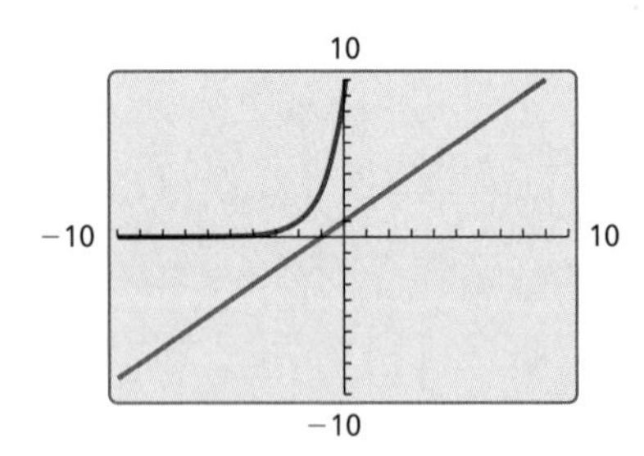

▶ The graphs do not intersect. So, the equation has no solution.

Monitoring Progress  Help in English and Spanish at *BigIdeasMath.com*

Use a graphing calculator to solve the equation.

8. $2^x = 1.8$ **9.** $4^{x-3} = x + 2$ **10.** $\left(\frac{1}{4}\right)^x = -2x - 3$

6.5 Exercises

Dynamic Solutions available at *BigIdeasMath.com*

Vocabulary and Core Concept Check

1. **WRITING** Describe how to solve an exponential equation with unlike bases.

2. **WHICH ONE DOESN'T BELONG?** Which equation does *not* belong with the other three? Explain your reasoning.

$2^x = 4^{x+6}$	$5^{3x+8} = 5^{2x}$	$3^4 = x + 4^2$	$2^{x-7} = 2^7$

Monitoring Progress and Modeling with Mathematics

In Exercises 3–12, solve the equation. Check your solution. *(See Examples 1 and 2.)*

3. $4^{5x} = 4^{10}$
4. $7^{x-4} = 7^8$
5. $3^{9x} = 3^{7x+8}$
6. $2^{4x} = 2^{x+9}$
7. $2^x = 64$
8. $3^x = 243$
9. $7^{x-5} = 49^x$
10. $216^x = 6^{x+10}$
11. $64^{2x+4} = 16^{5x}$
12. $27^x = 9^{x-2}$

In Exercises 13–18, solve the equation. Check your solution. *(See Example 3.)*

13. $\left(\frac{1}{5}\right)^x = 125$
14. $\left(\frac{1}{4}\right)^x = 256$
15. $\frac{1}{128} = 2^{5x+3}$
16. $3^{4x-9} = \frac{1}{243}$
17. $36^{-3x+3} = \left(\frac{1}{216}\right)^{x+1}$
18. $\left(\frac{1}{27}\right)^{4-x} = 9^{2x-1}$

ERROR ANALYSIS In Exercises 19 and 20, describe and correct the error in solving the exponential equation.

19. 

$$5^{3x+2} = 25^{x-8}$$
$$3x + 2 = x - 8$$
$$x = -5$$

20. 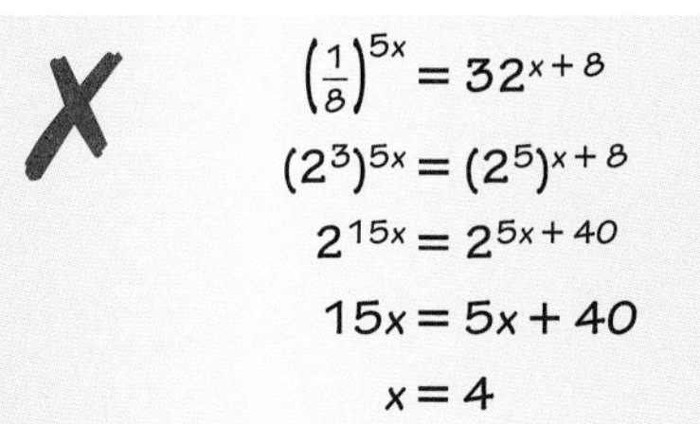

In Exercises 21–24, match the equation with the graph that can be used to solve it. Then solve the equation.

21. $2^x = 6$
22. $4^{2x-5} = 6$
23. $5^{x+2} = 6$
24. $3^{-x-1} = 6$

A. 
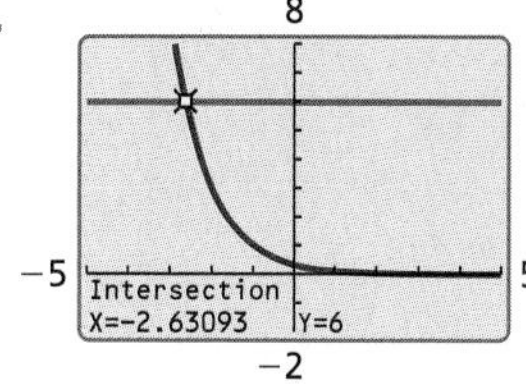

B. 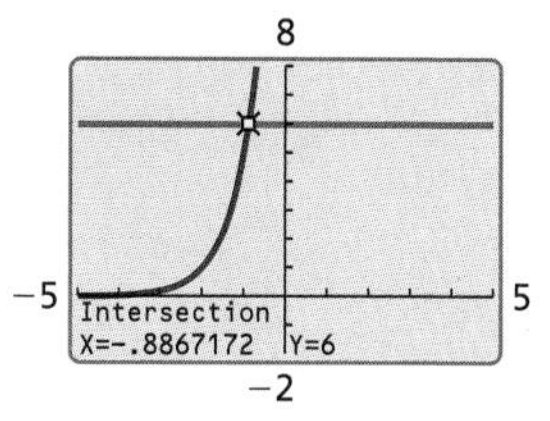

C. 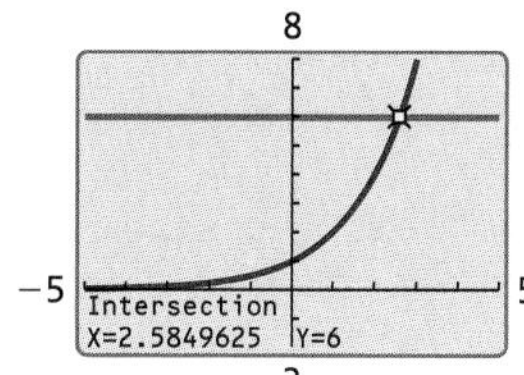

D. 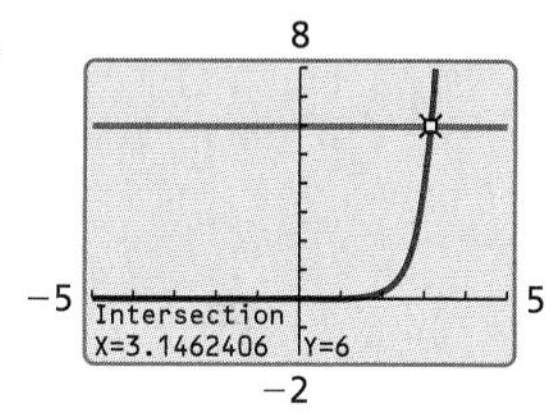

In Exercises 25–36, use a graphing calculator to solve the equation. *(See Example 4.)*

25. $6^{x+2} = 12$
26. $5^{x-4} = 8$
27. $\left(\frac{1}{2}\right)^{7x+1} = -9$
28. $\left(\frac{1}{3}\right)^{x+3} = 10$
29. $2^{x+6} = 2x + 15$
30. $3x - 2 = 5^{x-1}$
31. $\frac{1}{2}x - 1 = \left(\frac{1}{3}\right)^{2x-1}$
32. $2^{-x+1} = -\frac{3}{4}x + 3$
33. $5^x = -4^{-x+4}$
34. $7^{x-2} = 2^{-x}$
35. $2^{-x-3} = 3^{x+1}$
36. $5^{-2x+3} = -6^{x+5}$

In Exercises 37–40, solve the equation by using the Property of Equality for Exponential Equations.

37. $30 \cdot 5^{x+3} = 150$

38. $12 \cdot 2^{x-7} = 24$

39. $4(3^{-2x-4}) = 36$

40. $2(4^{2x+1}) = 128$

41. MODELING WITH MATHEMATICS You scan a photo into a computer at four times its original size. You continue to increase its size repeatedly by 100% using the computer. The new size of the photo y in comparison to its original size after x enlargements on the computer is represented by $y = 2^{x+2}$. How many times must the photo be enlarged on the computer so the new photo is 32 times the original size?

42. MODELING WITH MATHEMATICS A bacterial culture quadruples in size every hour. You begin observing the number of bacteria 3 hours after the culture is prepared. The amount y of bacteria x hours after the culture is prepared is represented by $y = 192(4^{x-3})$. When will there be 200,000 bacteria?

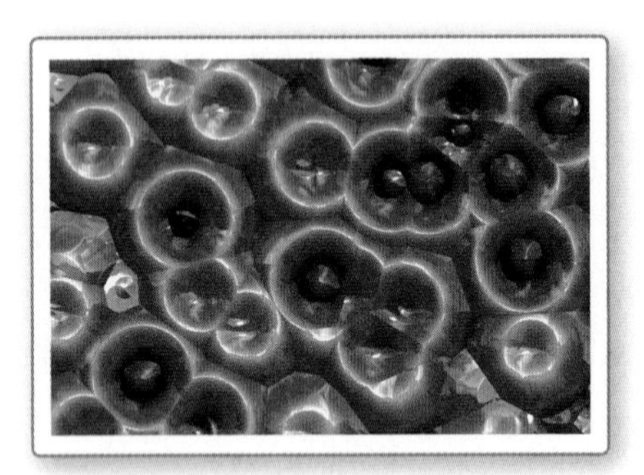

In Exercises 43–46, solve the equation.

43. $3^{3x+6} = 27^{x+2}$

44. $3^{4x+3} = 81^{x}$

45. $4^{x+3} = 2^{2(x+1)}$

46. $5^{8(x-1)} = 625^{2x-2}$

47. NUMBER SENSE Explain how you can use mental math to solve the equation $8^{x-4} = 1$.

48. PROBLEM SOLVING There are a total of 128 teams at the start of a citywide 3-on-3 basketball tournament. Half the teams are eliminated after each round. Write and solve an exponential equation to determine after which round there are 16 teams left.

49. PROBLEM SOLVING You deposit \$500 in a savings account that earns 6% annual interest compounded yearly. Write and solve an exponential equation to determine when the balance of the account will be \$800.

50. HOW DO YOU SEE IT? The graph shows the annual attendance at two different events. Each event began in 2004.

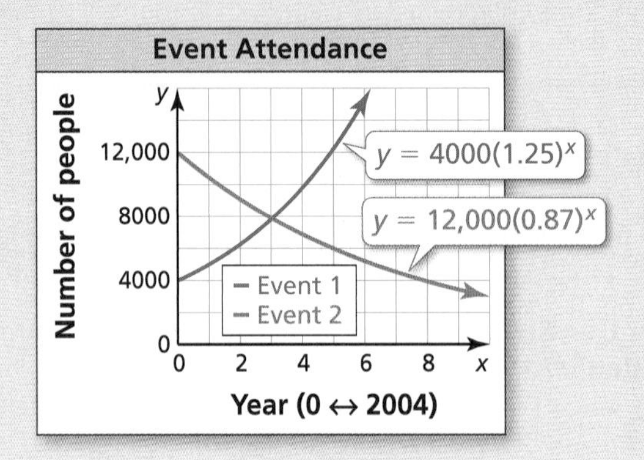

a. Estimate when the events will have about the same attendance.

b. Explain how you can verify your answer in part (a).

51. REASONING Explain why the Property of Equality for Exponential Equations does not work when $b = 1$. Give an example to justify your answer.

52. THOUGHT PROVOKING Is it possible for an exponential equation to have two different solutions? If not, explain your reasoning. If so, give an example.

USING STRUCTURE In Exercises 53–58, solve the equation.

53. $8^{x-2} = \sqrt{8}$

54. $\sqrt{5} = 5^{x+4}$

55. $\left(\sqrt[5]{7}\right)^{x} = 7^{2x+3}$

56. $12^{2x-1} = \left(\sqrt[4]{12}\right)^{x}$

57. $\left(\sqrt[3]{6}\right)^{2x} = \left(\sqrt{6}\right)^{x+6}$

58. $\left(\sqrt[5]{3}\right)^{5x-10} = \left(\sqrt[8]{3}\right)^{4x}$

59. MAKING AN ARGUMENT Consider the equation $\left(\frac{1}{a}\right)^{x} = b$, where $a > 1$ and $b > 1$. Your friend says the value of x will always be negative. Is your friend correct? Explain.

Maintaining Mathematical Proficiency

Reviewing what you learned in previous grades and lessons

Determine whether the sequence is arithmetic. If so, find the common difference. *(Section 4.6)*

60. $-20, -26, -32, -38, \ldots$

61. $9, 18, 36, 72, \ldots$

62. $-5, -8, -12, -17, \ldots$

63. $10, 20, 30, 40, \ldots$

6.6 Geometric Sequences

Essential Question How can you use a geometric sequence to describe a pattern?

In a **geometric sequence**, the ratio between each pair of consecutive terms is the same. This ratio is called the **common ratio**.

EXPLORATION 1 Describing Calculator Patterns

Work with a partner. Enter the keystrokes on a calculator and record the results in the table. Describe the pattern.

a. Step 1 2 =

Step 2 × 2 =

Step 3 × 2 =

Step 4 × 2 =

Step 5 × 2 =

Step	1	2	3	4	5
Calculator display					

b. Step 1 6 4 =

Step 2 × . 5 =

Step 3 × . 5 =

Step 4 × . 5 =

Step 5 × . 5 =

Step	1	2	3	4	5
Calculator display					

c. Use a calculator to make your own sequence. Start with any number and multiply by 3 each time. Record your results in the table.

Step	1	2	3	4	5
Calculator display					

d. Part (a) involves a geometric sequence with a common ratio of 2. What is the common ratio in part (b)? part (c)?

LOOKING FOR REGULARITY IN REPEATED REASONING

To be proficient in math, you need to notice when calculations are repeated and look both for general methods and for shortcuts.

EXPLORATION 2 Folding a Sheet of Paper

Work with a partner. A sheet of paper is about 0.1 millimeter thick.

a. How thick will it be when you fold it in half once? twice? three times?

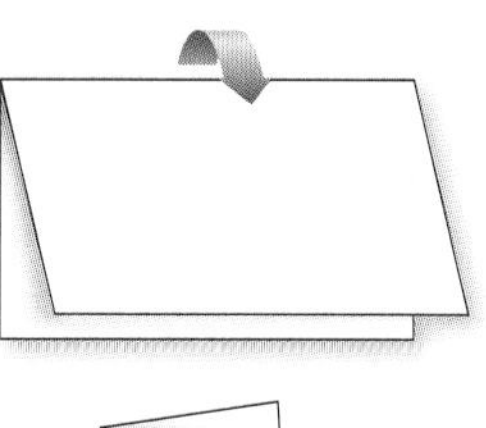

b. What is the greatest number of times you can fold a piece of paper in half? How thick is the result?

c. Do you agree with the statement below? Explain your reasoning.

"If it were possible to fold the paper in half 15 times, it would be taller than you."

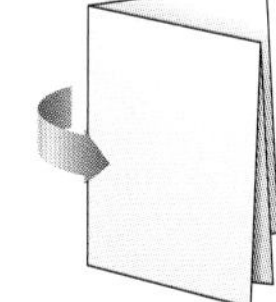

Communicate Your Answer

3. How can you use a geometric sequence to describe a pattern?

4. Give an example of a geometric sequence from real life other than paper folding.

6.6 Lesson

Core Vocabulary

geometric sequence, *p. 332*
common ratio, *p. 332*

Previous
arithmetic sequence
common difference

What You Will Learn

- Identify geometric sequences.
- Extend and graph geometric sequences.
- Write geometric sequences as functions.

Identifying Geometric Sequences

Core Concept

Geometric Sequence

In a **geometric sequence**, the ratio between each pair of consecutive terms is the same. This ratio is called the **common ratio**. Each term is found by multiplying the previous term by the common ratio.

$$1, \quad 5, \quad 25, \quad 125, \ldots$$ Terms of a geometric sequence

$\times 5 \quad \times 5 \quad \times 5$ ← common ratio

EXAMPLE 1 Identifying Geometric Sequences

Decide whether each sequence is *arithmetic*, *geometric*, or *neither*. Explain your reasoning.

a. 120, 60, 30, 15, . . . **b.** 2, 6, 11, 17, . . .

SOLUTION

a. Find the ratio between each pair of consecutive terms.

$$120 \quad 60 \quad 30 \quad 15$$

$$\frac{60}{120} = \frac{1}{2} \quad \frac{30}{60} = \frac{1}{2} \quad \frac{15}{30} = \frac{1}{2}$$

The ratios are the same. The common ratio is $\frac{1}{2}$.

▶ So, the sequence is geometric.

b. Find the ratio between each pair of consecutive terms.

$$2 \quad 6 \quad 11 \quad 17$$

$$\frac{6}{2} = 3 \quad \frac{11}{6} = 1\frac{5}{6} \quad \frac{17}{11} = 1\frac{6}{11}$$

There is no common ratio, so the sequence is *not* geometric.

Find the difference between each pair of consecutive terms.

$$2 \quad 6 \quad 11 \quad 17$$

$$6 - 2 = 4 \quad 11 - 6 = 5 \quad 17 - 11 = 6$$

There is no common difference, so the sequence is *not* arithmetic.

▶ So, the sequence is *neither* geometric nor arithmetic.

Monitoring Progress

Help in English and Spanish at *BigIdeasMath.com*

Decide whether the sequence is *arithmetic*, *geometric*, or *neither*. Explain your reasoning.

1. 5, 1, −3, −7, . . . **2.** 1024, 128, 16, 2, . . . **3.** 2, 6, 10, 16, . . .

Extending and Graphing Geometric Sequences

EXAMPLE 2 Extending Geometric Sequences

Write the next three terms of each geometric sequence.

a. 3, 6, 12, 24, . . .

b. 64, −16, 4, −1, . . .

SOLUTION

Use tables to organize the terms and extend each sequence.

a.

Position	1	2	3	4	5	6	7
Term	3	6	12	24	48	96	192

$\times 2$ $\times 2$ $\times 2$ $\times 2$ $\times 2$ $\times 2$

Each term is twice the previous term. So, the common ratio is 2.

Multiply a term by 2 to find the next term.

▶ The next three terms are 48, 96, and 192.

b.

Position	1	2	3	4	5	6	7
Term	64	−16	4	−1	$\frac{1}{4}$	$-\frac{1}{16}$	$\frac{1}{64}$

$\times\left(-\frac{1}{4}\right)$ $\times\left(-\frac{1}{4}\right)$ $\times\left(-\frac{1}{4}\right)$ $\times\left(-\frac{1}{4}\right)$ $\times\left(-\frac{1}{4}\right)$ $\times\left(-\frac{1}{4}\right)$

Multiply a term by $-\frac{1}{4}$ to find the next term.

▶ The next three terms are $\frac{1}{4}$, $-\frac{1}{16}$, and $\frac{1}{64}$.

LOOKING FOR STRUCTURE

When the terms of a geometric sequence alternate between positive and negative terms, or vice versa, the common ratio is negative.

EXAMPLE 3 Graphing a Geometric Sequence

Graph the geometric sequence 32, 16, 8, 4, 2, ... What do you notice?

SOLUTION

Make a table. Then plot the ordered pairs (n, a_n).

Position, n	1	2	3	4	5
Term, a_n	32	16	8	4	2

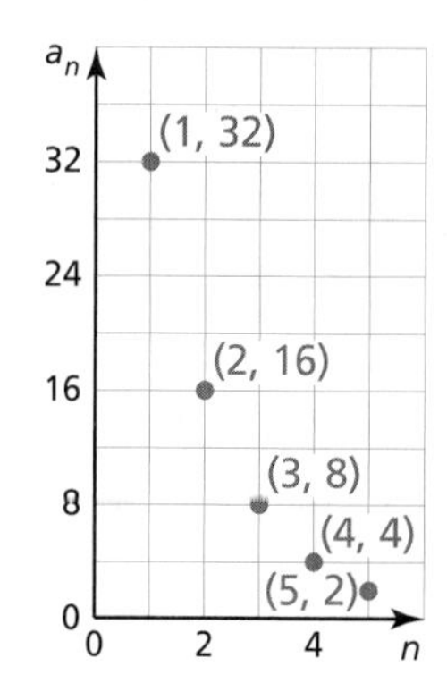

▶ The points appear to lie on an exponential curve.

STUDY TIP

The points of any geometric sequence with a *positive* common ratio lie on an exponential curve.

Monitoring Progress  Help in English and Spanish at *BigIdeasMath.com*

Write the next three terms of the geometric sequence. Then graph the sequence.

4. 1, 3, 9, 27, . . .

5. 2500, 500, 100, 20, . . .

6. 80, −40, 20, −10, . . .

7. −2, 4, −8, 16, . . .

Writing Geometric Sequences as Functions

Because consecutive terms of a geometric sequence have a common ratio, you can use the first term a_1 and the common ratio r to write an exponential function that describes a geometric sequence. Let $a_1 = 1$ and $r = 5$.

Position, n	Term, a_n	Written using a_1 and r	Numbers
1	first term, a_1	a_1	1
2	second term, a_2	a_1r	$1 \cdot 5 = 5$
3	third term, a_3	a_1r^2	$1 \cdot 5^2 = 25$
4	fourth term, a_4	a_1r^3	$1 \cdot 5^3 = 125$
⋮	⋮	⋮	⋮
n	nth term, a_n	a_1r^{n-1}	$1 \cdot 5^{n-1}$

STUDY TIP

Notice that the equation $a_n = a_1r^{n-1}$ is of the form $y = ab^x$.

Core Concept

Equation for a Geometric Sequence

Let a_n be the nth term of a geometric sequence with first term a_1 and common ratio r. The nth term is given by

$$a_n = a_1r^{n-1}.$$

EXAMPLE 4 Finding the nth Term of a Geometric Sequence

Write an equation for the nth term of the geometric sequence 2, 12, 72, 432, ... Then find a_{10}.

SOLUTION

The first term is 2, and the common ratio is 6.

$a_n = a_1r^{n-1}$ Equation for a geometric sequence

$a_n = 2(6)^{n-1}$ Substitute 2 for a_1 and 6 for r.

Use the equation to find the 10th term.

$a_n = 2(6)^{n-1}$ Write the equation.

$a_{10} = 2(6)^{10-1}$ Substitute 10 for n.

$= 20{,}155{,}392$ Simplify.

▶ The 10th term of the geometric sequence is 20,155,392.

Monitoring Progress  Help in English and Spanish at *BigIdeasMath.com*

Write an equation for the nth term of the geometric sequence. Then find a_7.

8. 1, −5, 25, −125, . . .

9. 13, 26, 52, 104, . . .

10. 432, 72, 12, 2, . . .

11. 4, 10, 25, 62.5, . . .

You can rewrite the equation for a geometric sequence with first term a_1 and common ratio r in function notation by replacing a_n with $f(n)$.

$$f(n) = a_1 r^{n-1}$$

The domain of the function is the set of positive integers.

EXAMPLE 5 Modeling with Mathematics

Clicking the *zoom-out* button on a mapping website doubles the side length of the square map. After how many clicks on the *zoom-out* button is the side length of the map 640 miles?

Zoom-out clicks	1	2	3
Map side length (miles)	5	10	20

SOLUTION

1. **Understand the Problem** You know that the side length of the square map doubles after each click on the *zoom-out* button. So, the side lengths of the map represent the terms of a geometric sequence. You need to find the number of clicks it takes for the side length of the map to be 640 miles.

2. **Make a Plan** Begin by writing a function f for the nth term of the geometric sequence. Then find the value of n for which $f(n) = 640$.

3. **Solve the Problem** The first term is 5, and the common ratio is 2.

$f(n) = a_1 r^{n-1}$ Function for a geometric sequence

$f(n) = 5(2)^{n-1}$ Substitute 5 for a_1 and 2 for r.

The function $f(n) = 5(2)^{n-1}$ represents the geometric sequence. Use this function to find the value of n for which $f(n) = 640$. So, use the equation $640 = 5(2)^{n-1}$ to write a system of equations.

$y = 5(2)^{n-1}$ Equation 1

$y = 640$ Equation 2

Then use a graphing calculator to graph the equations and find the point of intersection. The point of intersection is (8, 640).

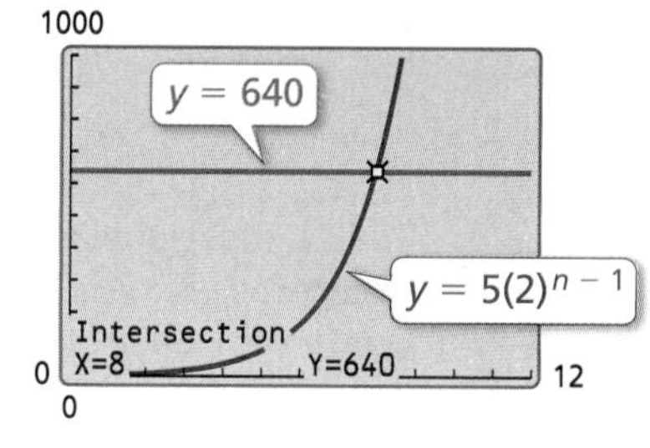

▶ So, after eight clicks, the side length of the map is 640 miles.

4. **Look Back** Find the value of n for which $f(n) = 640$ algebraically.

$640 = 5(2)^{n-1}$ Write the equation.

$128 = (2)^{n-1}$ Divide each side by 5.

$2^7 = (2)^{n-1}$ Rewrite 128 as 2^7.

$7 = n - 1$ Equate the exponents.

$8 = n$ ✓   Add 1 to each side.

USING APPROPRIATE TOOLS STRATEGICALLY

You can also use the *table* feature of a graphing calculator to find the value of n for which $f(n) = 640$.

X	Y1	Y2
3	20	640
4	40	640
5	80	640
6	160	640
7	320	640
8	640	640
9	1280	640

X=8

Monitoring Progress  Help in English and Spanish at *BigIdeasMath.com*

12. **WHAT IF?** After how many clicks on the *zoom-out* button is the side length of the map 2560 miles?

6.6 Exercises

Dynamic Solutions available at *BigIdeasMath.com*

Vocabulary and Core Concept Check

1. **WRITING** Compare the two sequences.

   2, 4, 6, 8, 10, . . .   2, 4, 8, 16, 32, . . .

2. **CRITICAL THINKING** Why do the points of a geometric sequence lie on an exponential curve only when the common ratio is positive?

Monitoring Progress and Modeling with Mathematics

In Exercises 3–8, find the common ratio of the geometric sequence.

3. 4, 12, 36, 108, . . .
4. 36, 6, 1, $\frac{1}{6}$, . . .
5. $\frac{3}{8}$, -3, 24, -192, . . .
6. 0.1, 1, 10, 100, . . .
7. 128, 96, 72, 54, . . .
8. -162, 54, -18, 6, . . .

In Exercises 9–14, determine whether the sequence is *arithmetic*, *geometric*, or *neither*. Explain your reasoning. *(See Example 1.)*

9. -8, 0, 8, 16, . . .
10. -1, 4, -7, 10, . . .
11. 9, 14, 20, 27, . . .
12. $\frac{3}{49}$, $\frac{3}{7}$, 3, 21, . . .
13. 192, 24, 3, $\frac{3}{8}$, . . .
14. -25, -18, -11, -4, . . .

In Exercises 15–18, determine whether the graph represents an *arithmetic sequence*, a *geometric sequence*, or *neither*. Explain your reasoning.

15.
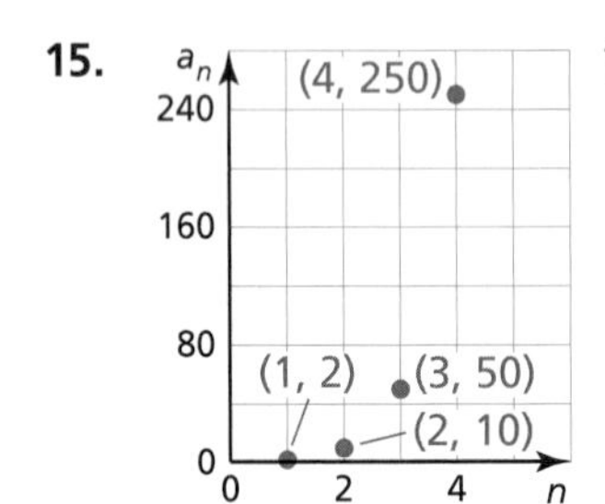

16.
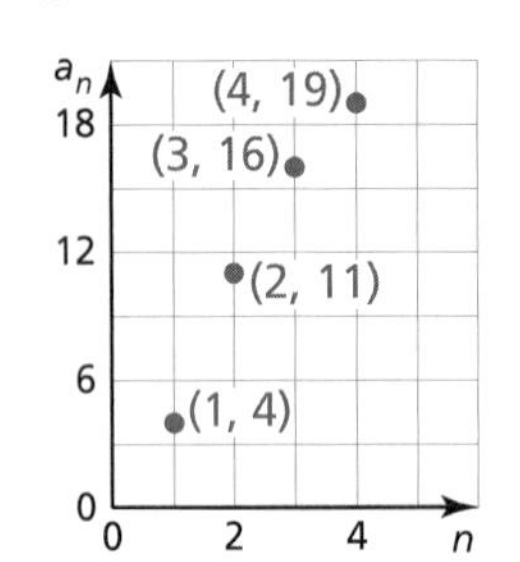

17.
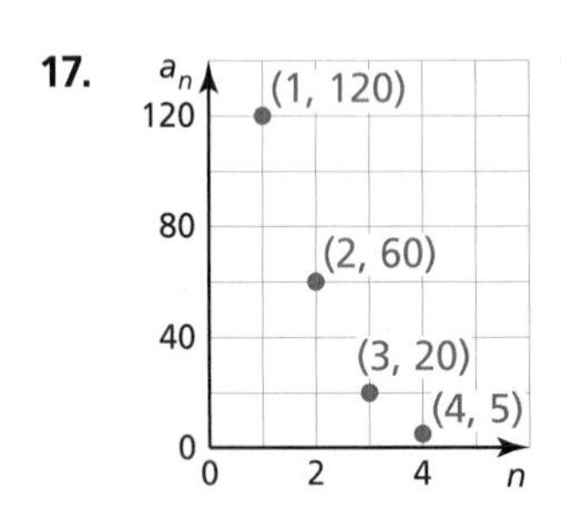

18.
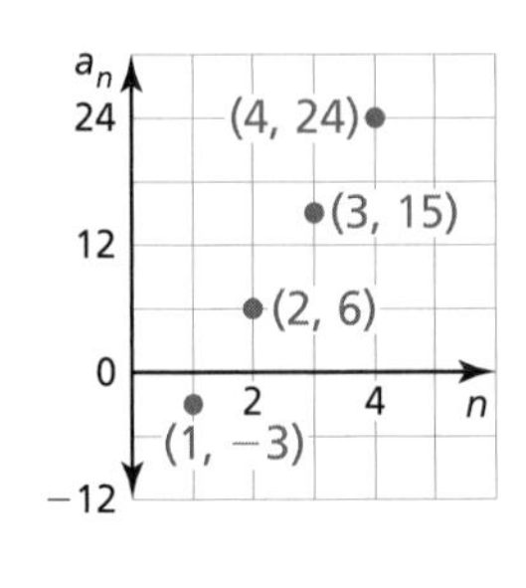

In Exercises 19–24, write the next three terms of the geometric sequence. Then graph the sequence. *(See Examples 2 and 3.)*

19. 5, 20, 80, 320, . . .
20. -3, 12, -48, 192, . . .
21. 81, -27, 9, -3, . . .
22. -375, -75, -15, -3, . . .
23. 32, 8, 2, $\frac{1}{2}$, . . .
24. $\frac{16}{9}$, $\frac{8}{3}$, 4, 6, . . .

In Exercises 25–32, write an equation for the *n*th term of the geometric sequence. Then find a_6. *(See Example 4.)*

25. 2, 8, 32, 128, . . .
26. 0.6, -3, 15, -75, . . .
27. $-\frac{1}{8}$, $-\frac{1}{4}$, $-\frac{1}{2}$, -1, . . .
28. 0.1, 0.9, 8.1, 72.9, . . .

29.

n	1	2	3	4
a_n	7640	764	76.4	7.64

30.

n	1	2	3	4
a_n	-192	48	-12	3

31.
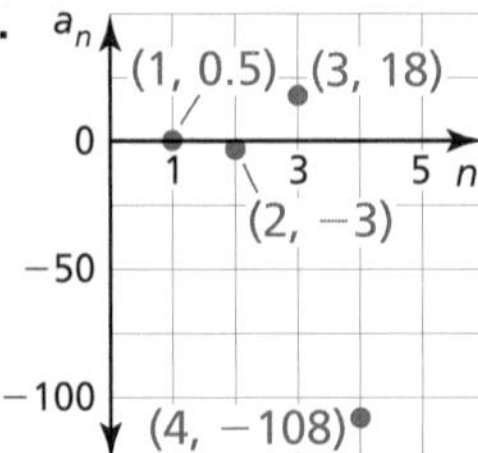

32.
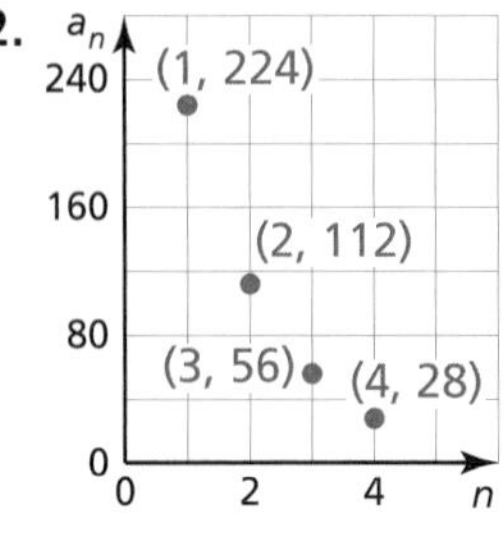

33. **PROBLEM SOLVING** A badminton tournament begins with 128 teams. After the first round, 64 teams remain. After the second round, 32 teams remain. How many teams remain after the third, fourth, and fifth rounds?

34. PROBLEM SOLVING The graphing calculator screen displays an area of 96 square units. After you zoom out once, the area is 384 square units. After you zoom out a second time, the area is 1536 square units. What is the screen area after you zoom out four times?

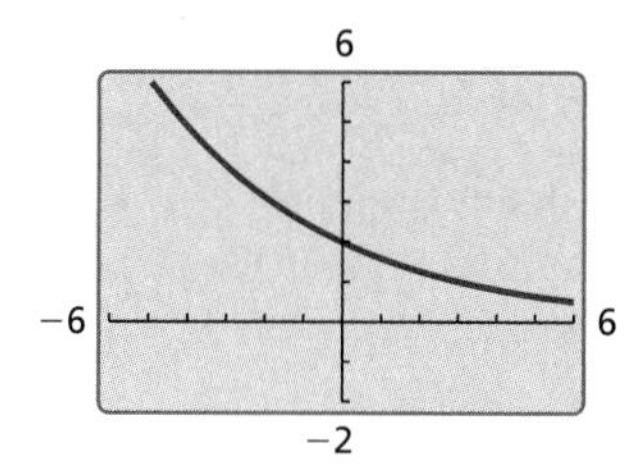

35. ERROR ANALYSIS Describe and correct the error in writing the next three terms of the geometric sequence.

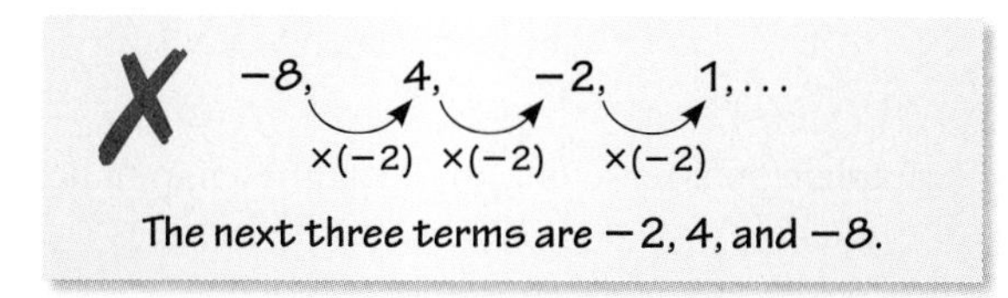

36. ERROR ANALYSIS Describe and correct the error in writing an equation for the *n*th term of the geometric sequence.

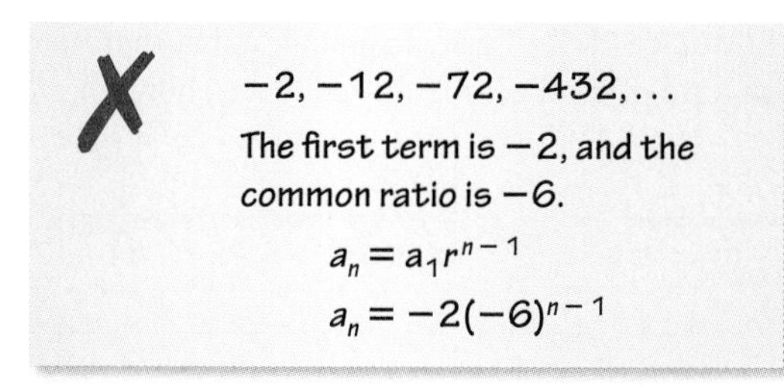

37. MODELING WITH MATHEMATICS The distance (in millimeters) traveled by a swinging pendulum decreases after each swing, as shown in the table. *(See Example 5.)*

Swing	1	2	3
Distance (in millimeters)	625	500	400

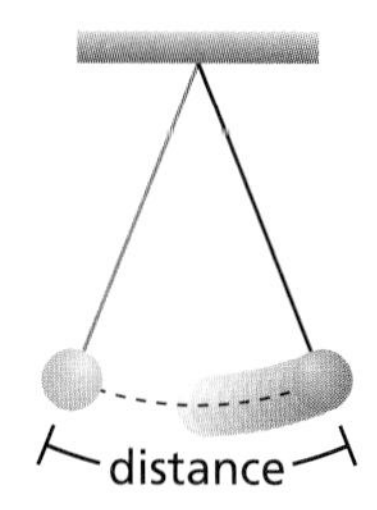

a. Write a function that represents the distance the pendulum swings on its *n*th swing.

b. After how many swings is the distance 256 millimeters?

38. MODELING WITH MATHEMATICS You start a chain email and send it to six friends. The next day, each of your friends forwards the email to six people. The process continues for a few days.

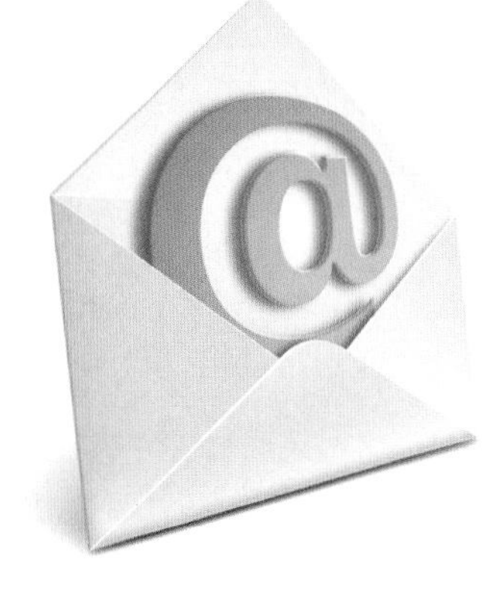

a. Write a function that represents the number of people who have received the email after *n* days.

b. After how many days will 1296 people have received the email?

MATHEMATICAL CONNECTIONS In Exercises 39 and 40, (a) write a function that represents the sequence of figures and (b) describe the 10th figure in the sequence.

39.

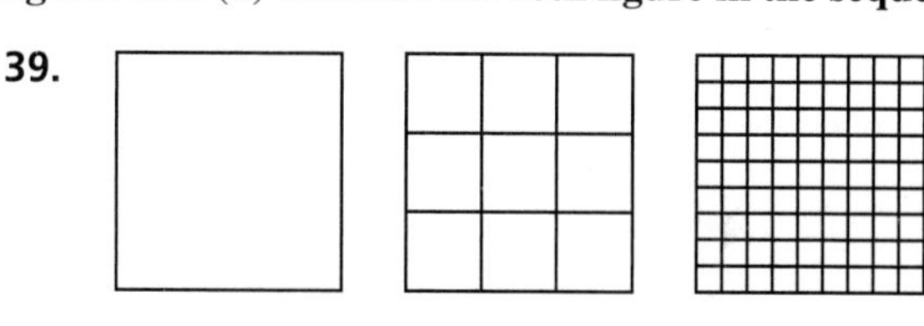

40.

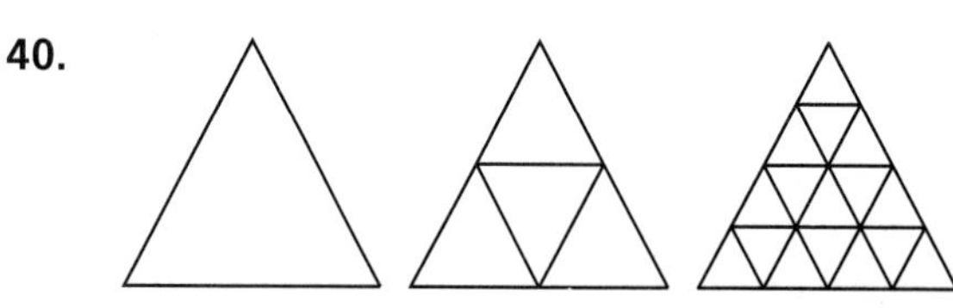

41. REASONING Write a sequence that represents the number of teams that have been eliminated after *n* rounds of the badminton tournament in Exercise 33. Determine whether the sequence is *arithmetic*, *geometric*, or *neither*. Explain your reasoning.

42. REASONING Write a sequence that represents the perimeter of the graphing calculator screen in Exercise 34 after you zoom out *n* times. Determine whether the sequence is *arithmetic*, *geometric*, or *neither*. Explain your reasoning.

43. WRITING Compare the graphs of arithmetic sequences to the graphs of geometric sequences.

44. MAKING AN ARGUMENT You are given two consecutive terms of a sequence.

$$\ldots, -8, 0, \ldots$$

Your friend says that the sequence is not geometric. A classmate says that is impossible to know given only two terms. Who is correct? Explain.

45. **CRITICAL THINKING** Is the sequence shown an *arithmetic* sequence? a *geometric* sequence? Explain your reasoning.

$$3, 3, 3, 3, \ldots$$

46. **HOW DO YOU SEE IT?** Without performing any calculations, match each equation with its graph. Explain your reasoning.

a. $a_n = 20\left(\frac{4}{3}\right)^{n-1}$

b. $a_n = 20\left(\frac{3}{4}\right)^{n-1}$

A.

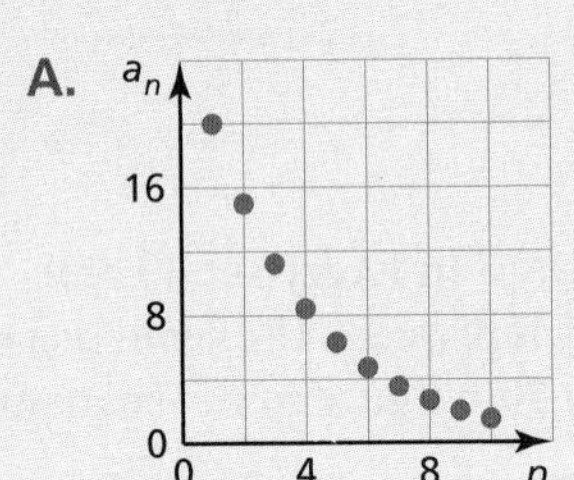

B.

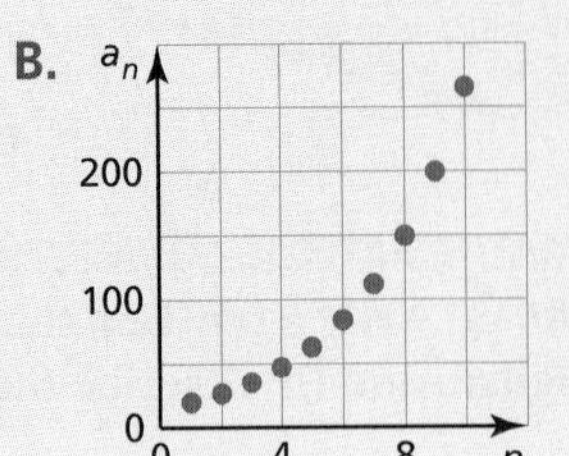

47. **REASONING** What is the 9th term of the geometric sequence where $a_3 = 81$ and $r = 3$?

48. **OPEN-ENDED** Write a sequence that has a pattern but is not arithmetic or geometric. Describe the pattern.

49. **ATTENDING TO PRECISION** Are the terms of a geometric sequence independent or dependent? Explain your reasoning.

50. **DRAWING CONCLUSIONS** A college student makes a deal with her parents to live at home instead of living on campus. She will pay her parents \$0.01 for the first day of the month, \$0.02 for the second day, \$0.04 for the third day, and so on.

a. Write an equation that represents the nth term of the geometric sequence.

b. What will she pay on the 25th day?

c. Did the student make a good choice or should she have chosen to live on campus? Explain.

51. **REPEATED REASONING** A soup kitchen makes 16 gallons of soup. Each day, a quarter of the soup is served and the rest is saved for the next day.

a. Write the first five terms of the sequence of the number of fluid ounces of soup left each day.

b. Write an equation that represents the nth term of the sequence.

c. When is all the soup gone? Explain.

52. **THOUGHT PROVOKING** Find the sum of the terms of the geometric sequence.

$$1, \frac{1}{2}, \frac{1}{4}, \frac{1}{8}, \ldots, \frac{1}{2^{n-1}}, \ldots$$

Explain your reasoning. Write a different infinite geometric sequence that has the same sum.

53. **OPEN-ENDED** Write a geometric sequence in which $a_2 < a_1 < a_3$.

54. **NUMBER SENSE** Write an equation that represents the nth term of each geometric sequence shown.

n	1	2	3	4
a_n	2	6	18	54

n	1	2	3	4
b_n	1	5	25	125

a. Do the terms $a_1 - b_1, a_2 - b_2, a_3 - b_3, \ldots$ form a geometric sequence? If so, how does the common ratio relate to the common ratios of the sequences above?

b. Do the terms $\frac{a_1}{b_1}, \frac{a_2}{b_2}, \frac{a_3}{b_3}, \ldots$ form a geometric sequence? If so, how does the common ratio relate to the common ratios of the sequences above?

Maintaining Mathematical Proficiency

Reviewing what you learned in previous grades and lessons

Use residuals to determine whether the model is a good fit for the data in the table. Explain. *(Section 4.5)*

55. $y = 3x - 8$

x	0	1	2	3	4	5	6
y	−10	−2	−1	2	1	7	10

56. $y = -5x + 1$

x	−3	−2	−1	0	1	2	3
y	6	4	6	1	2	−4	−3

6.7 Recursively Defined Sequences

Essential Question How can you define a sequence recursively?

A **recursive rule** gives the beginning term(s) of a sequence and a *recursive equation* that tells how a_n is related to one or more preceding terms.

EXPLORATION 1 Describing a Pattern

Work with a partner. Consider a hypothetical population of rabbits. Start with one breeding pair. After each month, each breeding pair produces another breeding pair. The total number of rabbits each month follows the exponential pattern 2, 4, 8, 16, 32, ... Now suppose that in the first month after each pair is born, the pair is too young to reproduce. Each pair produces another pair after it is 2 months old. Find the total number of pairs in months 6, 7, and 8.

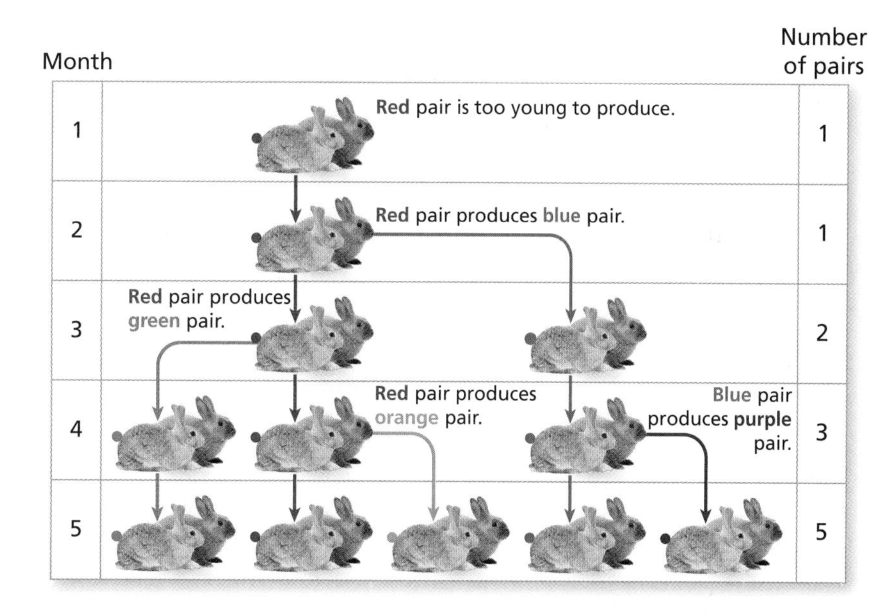

RECOGNIZING PATTERNS

To be proficient in math, you need to look closely to discern a pattern or structure.

EXPLORATION 2 Using a Recursive Equation

Work with a partner. Consider the following recursive equation.

$$a_n = a_{n-1} + a_{n-2}$$

Each term in the sequence is the sum of the two preceding terms.

Copy and complete the table. Compare the results with the sequence of the number of pairs in Exploration 1.

a_1	a_2	a_3	a_4	a_5	a_6	a_7	a_8
1	1						

Communicate Your Answer

3. How can you define a sequence recursively?
4. Use the Internet or some other reference to determine the mathematician who first described the sequences in Explorations 1 and 2.

6.7 Lesson

Core Vocabulary

explicit rule, *p. 340*
recursive rule, *p. 340*

Previous
arithmetic sequence
geometric sequence

What You Will Learn

- Write terms of recursively defined sequences.
- Write recursive rules for sequences.
- Translate between recursive rules and explicit rules.
- Write recursive rules for special sequences.

Writing Terms of Recursively Defined Sequences

So far in this book, you have defined arithmetic and geometric sequences *explicitly*. An **explicit rule** gives a_n as a function of the term's position number n in the sequence. For example, an explicit rule for the arithmetic sequence 3, 5, 7, 9, . . . is $a_n = 3 + 2(n - 1)$, or $a_n = 2n + 1$.

Now, you will define arithmetic and geometric sequences *recursively*. A **recursive rule** gives the beginning term(s) of a sequence and a *recursive equation* that tells how a_n is related to one or more preceding terms.

Core Concept

Recursive Equation for an Arithmetic Sequence

$a_n = a_{n-1} + d$, where d is the common difference

Recursive Equation for a Geometric Sequence

$a_n = r \cdot a_{n-1}$, where r is the common ratio

EXAMPLE 1 Writing Terms of Recursively Defined Sequences

Write the first six terms of each sequence. Then graph each sequence.

a. $a_1 = 2, a_n = a_{n-1} + 3$

b. $a_1 = 1, a_n = 3a_{n-1}$

SOLUTION

You are given the first term. Use the recursive equation to find the next five terms.

a. $a_1 = 2$

$a_2 = a_1 + 3 = 2 + 3 = 5$

$a_3 = a_2 + 3 = 5 + 3 = 8$

$a_4 = a_3 + 3 = 8 + 3 = 11$

$a_5 = a_4 + 3 = 11 + 3 = 14$

$a_6 = a_5 + 3 = 14 + 3 = 17$

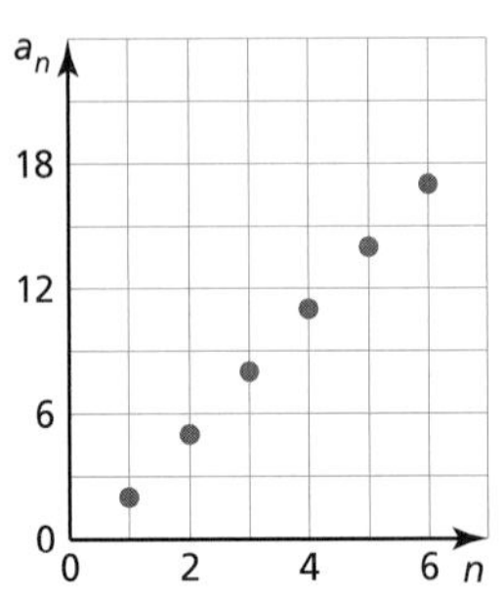

b. $a_1 = 1$

$a_2 = 3a_1 = 3(1) = 3$

$a_3 = 3a_2 = 3(3) = 9$

$a_4 = 3a_3 = 3(9) = 27$

$a_5 = 3a_4 = 3(27) = 81$

$a_6 = 3a_5 = 3(81) = 243$

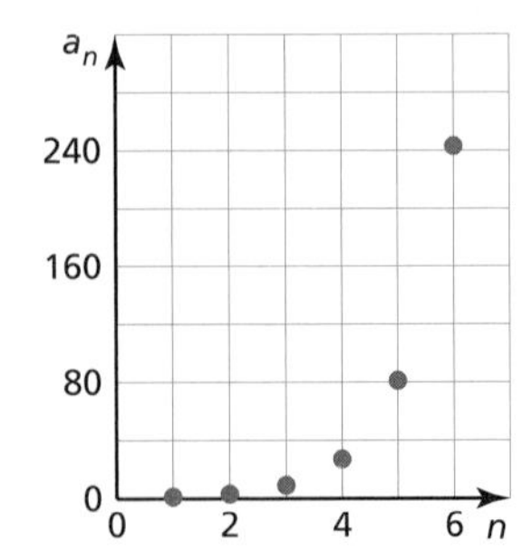

STUDY TIP

A sequence is a discrete function. So, the points on the graph are not connected.

Monitoring Progress  Help in English and Spanish at *BigIdeasMath.com*

Write the first six terms of the sequence. Then graph the sequence.

1. $a_1 = 0,\ a_n = a_{n-1} - 8$

2. $a_1 = -7.5,\ a_n = a_{n-1} + 2.5$

3. $a_1 = -36,\ a_n = \frac{1}{2}a_{n-1}$

4. $a_1 = 0.7,\ a_n = 10a_{n-1}$

Writing Recursive Rules

EXAMPLE 2 Writing Recursive Rules

Write a recursive rule for each sequence.

a. $-30, -18, -6, 6, 18, \ldots$

b. $500, 100, 20, 4, 0.8, \ldots$

SOLUTION

Use a table to organize the terms and find the pattern.

a.

Position, n	1	2	3	4	5
Term, a_n	-30	-18	-6	6	18

$+12 \quad +12 \quad +12 \quad +12$

The sequence is arithmetic, with first term $a_1 = -30$ and common difference $d = 12$.

$a_n = a_{n-1} + d$ Recursive equation for an arithmetic sequence

$a_n = a_{n-1} + 12$ Substitute 12 for d.

▶ So, a recursive rule for the sequence is $a_1 = -30,\ a_n = a_{n-1} + 12$.

COMMON ERROR

When writing a recursive rule for a sequence, you need to write both the beginning term(s) and the recursive equation.

b.

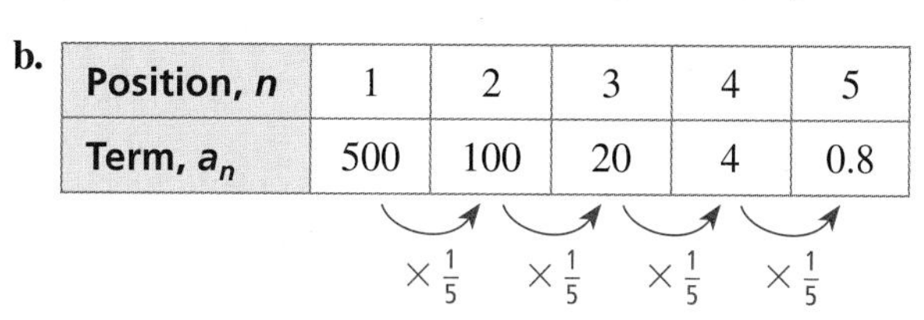

Position, n	1	2	3	4	5
Term, a_n	500	100	20	4	0.8

$\times\frac{1}{5} \quad \times\frac{1}{5} \quad \times\frac{1}{5} \quad \times\frac{1}{5}$

The sequence is geometric, with first term $a_1 = 500$ and common ratio $r = \frac{1}{5}$.

$a_n = r \cdot a_{n-1}$ Recursive equation for a geometric sequence

$a_n = \frac{1}{5}a_{n-1}$ Substitute $\frac{1}{5}$ for r.

▶ So, a recursive rule for the sequence is $a_1 = 500,\ a_n = \frac{1}{5}a_{n-1}$.

Monitoring Progress  Help in English and Spanish at *BigIdeasMath.com*

Write a recursive rule for the sequence.

5. $8, 3, -2, -7, -12, \ldots$

6. $1.3, 2.6, 3.9, 5.2, 6.5, \ldots$

7. $4, 20, 100, 500, 2500, \ldots$

8. $128, -32, 8, -2, 0.5, \ldots$

9. Write a recursive rule for the height of the sunflower over time.

Translating between Recursive and Explicit Rules

EXAMPLE 3 Translating from Recursive Rules to Explicit Rules

Write an explicit rule for each recursive rule.

a. $a_1 = 25, a_n = a_{n-1} - 10$

b. $a_1 = 19.6, a_n = -0.5a_{n-1}$

SOLUTION

a. The recursive rule represents an arithmetic sequence, with first term $a_1 = 25$ and common difference $d = -10$.

$a_n = a_1 + (n - 1)d$ — Explicit rule for an arithmetic sequence

$a_n = 25 + (n - 1)(-10)$ — Substitute 25 for a_1 and -10 for d.

$a_n = -10n + 35$ — Simplify.

▶ An explicit rule for the sequence is $a_n = -10n + 35$.

b. The recursive rule represents a geometric sequence, with first term $a_1 = 19.6$ and common ratio $r = -0.5$.

$a_n = a_1 r^{n-1}$ — Explicit rule for a geometric sequence

$a_n = 19.6(-0.5)^{n-1}$ — Substitute 19.6 for a_1 and -0.5 for r.

▶ An explicit rule for the sequence is $a_n = 19.6(-0.5)^{n-1}$.

EXAMPLE 4 Translating from Explicit Rules to Recursive Rules

Write a recursive rule for each explicit rule.

a. $a_n = -2n + 3$

b. $a_n = -3(2)^{n-1}$

SOLUTION

a. The explicit rule represents an arithmetic sequence, with first term $a_1 = -2(1) + 3 = 1$ and common difference $d = -2$.

$a_n = a_{n-1} + d$ — Recursive equation for an arithmetic sequence

$a_n = a_{n-1} + (-2)$ — Substitute -2 for d.

▶ So, a recursive rule for the sequence is $a_1 = 1, a_n = a_{n-1} - 2$.

b. The explicit rule represents a geometric sequence, with first term $a_1 = -3$ and common ratio $r = 2$.

$a_n = r \cdot a_{n-1}$ — Recursive equation for a geometric sequence

$a_n = 2a_{n-1}$ — Substitute 2 for r.

▶ So, a recursive rule for the sequence is $a_1 = -3, a_n = 2a_{n-1}$.

Monitoring Progress  Help in English and Spanish at *BigIdeasMath.com*

Write an explicit rule for the recursive rule.

10. $a_1 = -45, a_n = a_{n-1} + 20$

11. $a_1 = 13, a_n = -3a_{n-1}$

Write a recursive rule for the explicit rule.

12. $a_n = -n + 1$

13. $a_n = -2.5(4)^{n-1}$

Writing Recursive Rules for Special Sequences

You can write recursive rules for sequences that are neither arithmetic nor geometric. One way is to look for patterns in the sums of consecutive terms.

EXAMPLE 5 Writing Recursive Rules for Other Sequences

Use the sequence shown.

$$1, 1, 2, 3, 5, 8, \ldots$$

a. Write a recursive rule for the sequence.

b. Write the next three terms of the sequence.

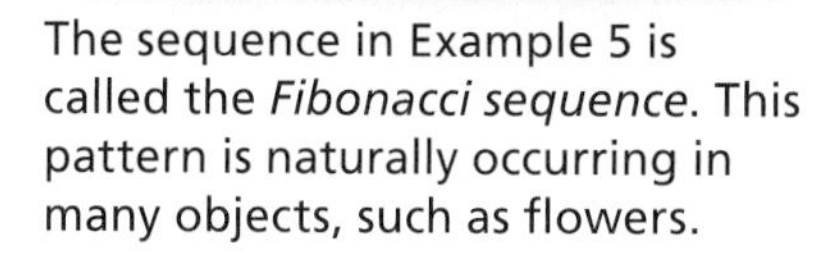

The sequence in Example 5 is called the *Fibonacci sequence*. This pattern is naturally occurring in many objects, such as flowers.

SOLUTION

a. Find the difference and ratio between each pair of consecutive terms.

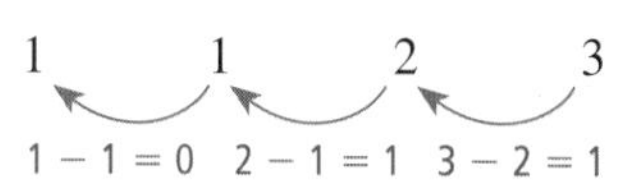

There is no common difference, so the sequence is *not* arithmetic.

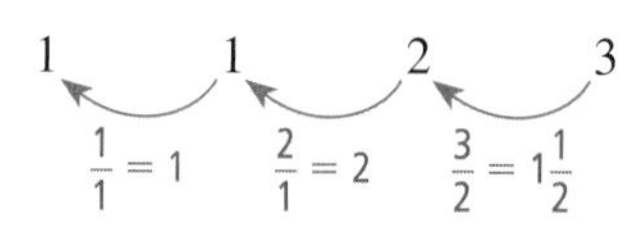

There is no common ratio, so the sequence is *not* geometric.

Find the sum of each pair of consecutive terms.

$a_1 + a_2 = 1 + 1 = 2$ 2 is the third term.

$a_2 + a_3 = 1 + 2 = 3$ 3 is the fourth term.

$a_3 + a_4 = 2 + 3 = 5$ 5 is the fifth term.

$a_4 + a_5 = 3 + 5 = 8$ 8 is the sixth term.

Beginning with the third term, each term is the sum of the two previous terms. A recursive equation for the sequence is $a_n = a_{n-2} + a_{n-1}$.

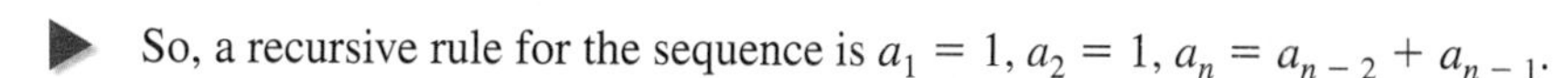

▶ So, a recursive rule for the sequence is $a_1 = 1$, $a_2 = 1$, $a_n = a_{n-2} + a_{n-1}$.

b. Use the recursive equation $a_n = a_{n-2} + a_{n-1}$ to find the next three terms.

$$a_7 = a_5 + a_6 = 5 + 8 = 13 \qquad a_8 = a_6 + a_7 = 8 + 13 = 21 \qquad a_9 = a_7 + a_8 = 13 + 21 = 34$$

▶ The next three terms are 13, 21, and 34.

Monitoring Progress

Help in English and Spanish at *BigIdeasMath.com*

Write a recursive rule for the sequence. Then write the next three terms of the sequence.

14. 5, 6, 11, 17, 28, . . .

15. −3, −4, −7, −11, −18, . . .

16. 1, 1, 0, −1, −1, 0, 1, 1, . . .

17. 4, 3, 1, 2, −1, 3, −4, . . .

6.7 Exercises

Dynamic Solutions available at *BigIdeasMath.com*

Vocabulary and Core Concept Check

1. **COMPLETE THE SENTENCE** A recursive rule gives the beginning term(s) of a sequence and a(n) ____________ that tells how a_n is related to one or more preceding terms.

2. **WHICH ONE DOESN'T BELONG?** Which rule does *not* belong with the other three? Explain your reasoning.

$a_1 = -1, a_n = 5a_{n-1}$	$a_n = 6n - 2$	$a_1 = -3, a_n = a_{n-1} + 1$	$a_1 = 9, a_n = 4a_{n-1}$

Monitoring Progress and Modeling with Mathematics

In Exercises 3–6, determine whether the recursive rule represents an *arithmetic sequence* or a *geometric sequence*.

3. $a_1 = 2, a_n = 7a_{n-1}$

4. $a_1 = 18, a_n = a_{n-1} + 1$

5. $a_1 = 5, a_n = a_{n-1} - 4$

6. $a_1 = 3, a_n = -6a_{n-1}$

In Exercises 7–12, write the first six terms of the sequence. Then graph the sequence. *(See Example 1.)*

7. $a_1 = 0, a_n = a_{n-1} + 2$

8. $a_1 = 10, a_n = a_{n-1} - 5$

9. $a_1 = 2, a_n = 3a_{n-1}$

10. $a_1 = 8, a_n = 1.5a_{n-1}$

11. $a_1 = 80, a_n = -\frac{1}{2}a_{n-1}$

12. $a_1 = -7, a_n = -4a_{n-1}$

In Exercises 13–20, write a recursive rule for the sequence. *(See Example 2.)*

13.

n	1	2	3	4
a_n	7	16	25	34

14.

n	1	2	3	4
a_n	8	24	72	216

15. 243, 81, 27, 9, 3, . . .

16. 3, 11, 19, 27, 35, . . .

17. 0, −3, −6, −9, −12, . . .

18. 5, −20, 80, −320, 1280, . . .

19.

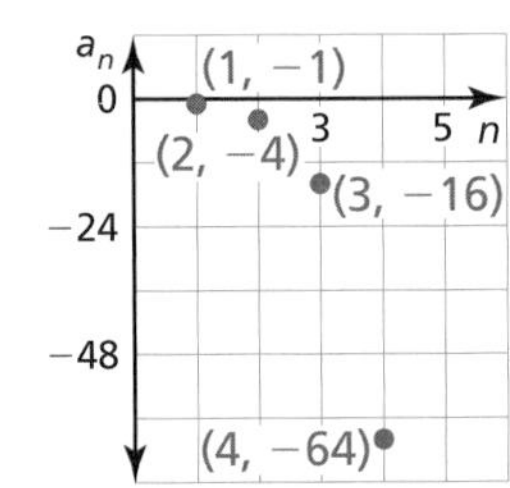

20.

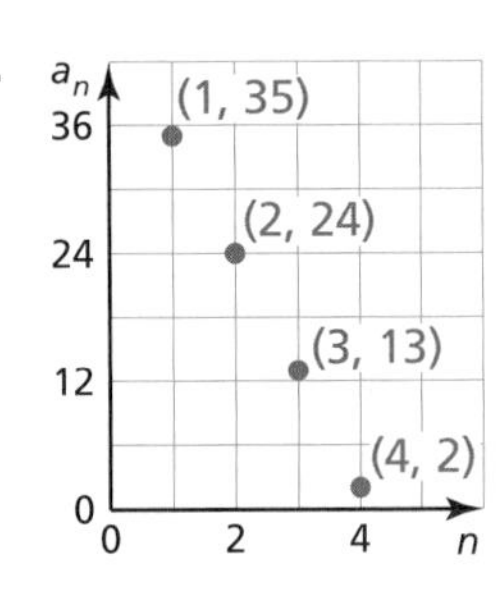

21. **MODELING WITH MATHEMATICS** Write a recursive rule for the number of bacterial cells over time.

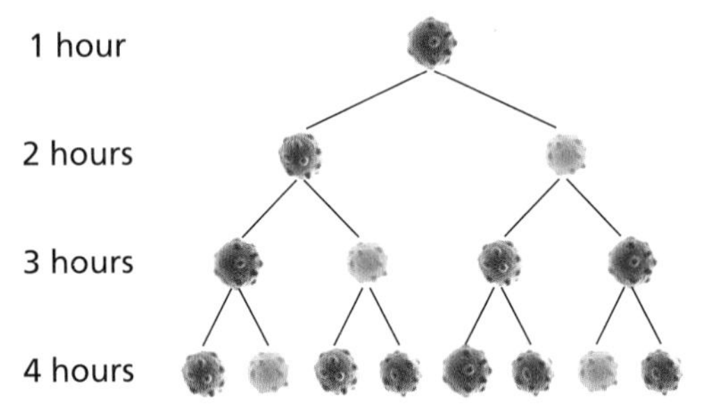

22. **MODELING WITH MATHEMATICS** Write a recursive rule for the length of the deer antler over time.

1 day: $4\frac{1}{2}$ in.

2 days: $4\frac{3}{4}$ in.

3 days: 5 in.

4 days: $5\frac{1}{4}$ in.

In Exercises 23–28, write an explicit rule for the recursive rule. *(See Example 3.)*

23. $a_1 = -3, a_n = a_{n-1} + 3$

24. $a_1 = 8, a_n = a_{n-1} - 12$

25. $a_1 = 16, a_n = 0.5a_{n-1}$

26. $a_1 = -2, a_n = 9a_{n-1}$

27. $a_1 = 4, a_n = a_{n-1} + 17$

28. $a_1 = 5, a_n = -5a_{n-1}$

In Exercises 29–34, write a recursive rule for the explicit rule. *(See Example 4.)*

29. $a_n = 7(3)^{n-1}$

30. $a_n = -4n + 2$

31. $a_n = 1.5n + 3$

32. $a_n = 6n - 20$

33. $a_n = (-5)^{n-1}$

34. $a_n = -81\left(\frac{2}{3}\right)^{n-1}$

In Exercises 35–38, graph the first four terms of the sequence with the given description. Write a recursive rule and an explicit rule for the sequence.

35. The first term of a sequence is 5. Each term of the sequence is 15 more than the preceding term.

36. The first term of a sequence is 16. Each term of the sequence is half the preceding term.

37. The first term of a sequence is −1. Each term of the sequence is −3 times the preceding term.

38. The first term of a sequence is 19. Each term of the sequence is 13 less than the preceding term.

In Exercises 39–44, write a recursive rule for the sequence. Then write the next two terms of the sequence. *(See Example 5.)*

39. 1, 3, 4, 7, 11, . . .

40. 10, 9, 1, 8, −7, 15, . . .

41. 2, 4, 2, −2, −4, −2, . . .

42. 6, 1, 7, 8, 15, 23, . . .

43.

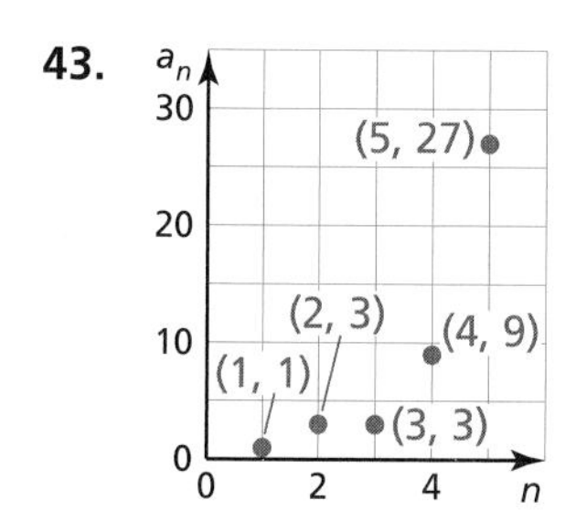

44.

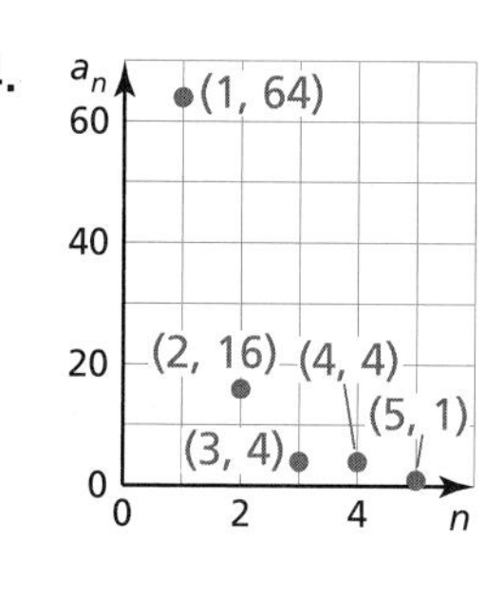

45. ERROR ANALYSIS Describe and correct the error in writing an explicit rule for the recursive rule $a_1 = 6$, $a_n = a_{n-1} - 12$.

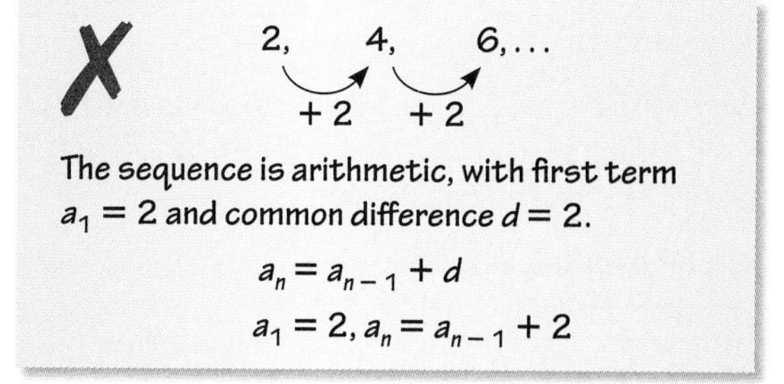

46. ERROR ANALYSIS Describe and correct the error in writing a recursive rule for the sequence 2, 4, 6, 10, 16, ...

2, 4, 6, . . .
+2 +2

The sequence is arithmetic, with first term $a_1 = 2$ and common difference $d = 2$.

$a_n = a_{n-1} + d$

$a_1 = 2, a_n = a_{n-1} + 2$

In Exercises 47–51, the function *f* represents a sequence. Find the 2nd, 5th, and 10th terms of the sequence.

47. $f(1) = 3, f(n) = f(n - 1) + 7$

48. $f(1) = -1, f(n) = 6f(n - 1)$

49. $f(1) = 8, f(n) = -f(n - 1)$

50. $f(1) = 4, f(2) = 5, f(n) = f(n - 2) + f(n - 1)$

51. $f(1) = 10, f(2) = 15, f(n) = f(n - 1) - f(n - 2)$

52. MODELING WITH MATHEMATICS The X-ray shows the lengths (in centimeters) of bones in a human hand.

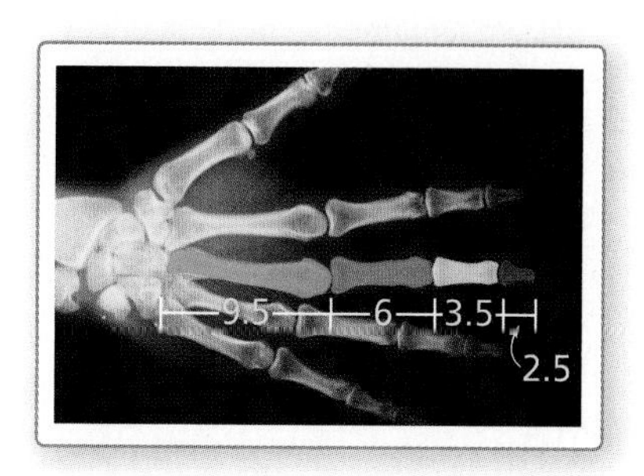

a. Write a recursive rule for the lengths of the bones.

b. Measure the lengths of different sections of your hand. Can the lengths be represented by a recursively defined sequence? Explain.

53. USING TOOLS You can use a spreadsheet to generate the terms of a sequence.

A2 ▼ = =A1+2

	A	B	C
1	3		
2	5		
3			
4			

a. To generate the terms of the sequence $a_1 = 3$, $a_n = a_{n-1} + 2$, enter the value of a_1, 3, into cell A1. Then enter "=A1+2" into cell A2, as shown. Use the *fill down* feature to generate the first 10 terms of the sequence.

b. Use a spreadsheet to generate the first 10 terms of the sequence $a_1 = 3$, $a_n = 4a_{n-1}$. (*Hint:* Enter "=4*A1" into cell A2.)

c. Use a spreadsheet to generate the first 10 terms of the sequence $a_1 = 4$, $a_2 = 7$, $a_n = a_{n-1} - a_{n-2}$. (*Hint:* Enter "=A2-A1" into cell A3.)

54. HOW DO YOU SEE IT? Consider Squares 1–6 in the diagram.

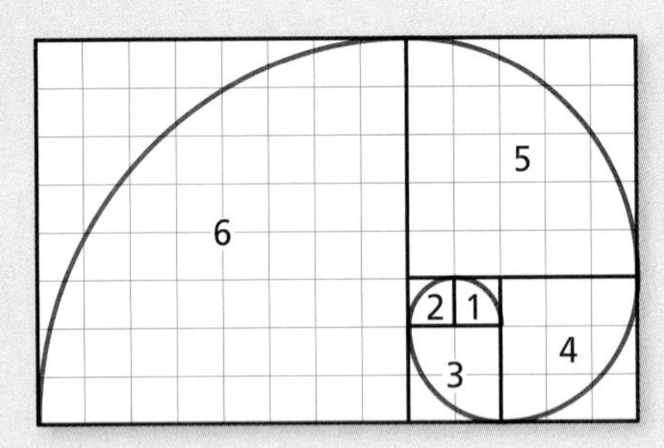

a. Write a sequence in which each term a_n is the side length of square n.

b. What is the name of this sequence? What is the next term of this sequence?

c. Use the term in part (b) to add another square to the diagram and extend the spiral.

55. REASONING Write the first 5 terms of the sequence $a_1 = 5$, $a_n = 3a_{n-1} + 4$. Determine whether the sequence is *arithmetic*, *geometric*, or *neither*. Explain your reasoning.

56. THOUGHT PROVOKING Describe the pattern for the numbers in Pascal's Triangle, shown below. Write a recursive rule that gives the mth number in the nth row.

1
1 1
1 2 1
1 3 3 1
1 4 6 4 1
1 5 10 10 5 1

57. REASONING The explicit rule $a_n = a_1 + (n - 1)d$ defines an arithmetic sequence.

a. Explain why $a_{n-1} = a_1 + [(n - 1) - 1]d$.

b. Justify each step in showing that a recursive equation for the sequence is $a_n = a_{n-1} + d$.

$$a_n = a_1 + (n - 1)d$$
$$= a_1 + [(n - 1) + 0]d$$
$$= a_1 + [(n - 1) - 1 + 1]d$$
$$= a_1 + [((n - 1) - 1) + 1]d$$
$$= a_1 + [(n - 1) - 1]d + d$$
$$= a_{n-1} + d$$

58. MAKING AN ARGUMENT Your friend claims that the sequence

$-5, 5, -5, 5, -5, \ldots$

cannot be represented by a recursive rule. Is your friend correct? Explain.

59. PROBLEM SOLVING Write a recursive rule for the sequence.

$3, 7, 15, 31, 63, \ldots$

Maintaining Mathematical Proficiency

Reviewing what you learned in previous grades and lessons

Simplify the expression. (*Skills Review Handbook*)

60. $5x + 12x$ **61.** $9 - 6y - 14$ **62.** $2d - 7 - 8d$ **63.** $3 - 3m + 11m$

Write a linear function f with the given values. (*Section 4.2*)

64. $f(2) = 6, f(-1) = -3$ **65.** $f(-2) = 0, f(6) = -4$

66. $f(-3) = 5, f(-1) = 5$ **67.** $f(3) = -1, f(-4) = -15$

6.5–6.7 What Did You Learn?

Core Vocabulary

exponential equation, *p. 326*
geometric sequence, *p. 332*
common ratio, *p. 332*
explicit rule, *p. 340*
recursive rule, *p. 340*

Core Concepts

Section 6.5

Property of Equality for Exponential Equations, *p. 326*
Solving Exponential Equations by Graphing, *p. 328*

Section 6.6

Geometric Sequence, *p. 332*
Equation for a Geometric Sequence, *p. 334*

Section 6.7

Recursive Equation for an Arithmetic Sequence, *p. 340*
Recursive Equation for a Geometric Sequence, *p. 340*

Mathematical Practices

1. How did you decide on an appropriate level of precision for your answer in Exercise 49 on page 330?
2. Explain how writing a function in Exercise 39 part (a) on page 337 created a shortcut for answering part (b).
3. How did you choose an appropriate tool in Exercise 52 part (b) on page 345?

Performance Task

The New Car

There is so much more to buying a new car than the purchase price. Interest rates, depreciation, and inflation are all factors. So, what is the real cost of your new car?

To explore the answers to this question and more, go to ***BigIdeasMath.com***.

6 Chapter Review

Dynamic Solutions available at *BigIdeasMath.com*

6.1 Properties of Exponents *(pp. 291–298)*

Simplify $\left(\frac{x}{4}\right)^{-4}$. Write your answer using only positive exponents.

$$\left(\frac{x}{4}\right)^{-4} = \frac{x^{-4}}{4^{-4}} \qquad \text{Power of a Quotient Property}$$
$$= \frac{4^4}{x^4} \qquad \text{Definition of negative exponent}$$
$$= \frac{256}{x^4} \qquad \text{Simplify.}$$

Simplify the expression. Write your answer using only positive exponents.

1. $y^3 \cdot y^{-5}$ **2.** $\frac{x^4}{x^7}$ **3.** $(x^0y^2)^3$ **4.** $\left(\frac{2x^2}{5y^4}\right)^{-2}$

6.2 Radicals and Rational Exponents *(pp. 299–304)*

Evaluate $512^{1/3}$.

$$512^{1/3} = \sqrt[3]{512} \qquad \text{Rewrite the expression in radical form.}$$
$$= \sqrt[3]{8 \cdot 8 \cdot 8} \qquad \text{Rewrite the expression showing factors.}$$
$$= 8 \qquad \text{Evaluate the cube root.}$$

Evaluate the expression.

5. $\sqrt[3]{8}$ **6.** $\sqrt[5]{-243}$ **7.** $625^{3/4}$ **8.** $(-25)^{1/2}$

6.3 Exponential Functions *(pp. 305–312)*

Graph $f(x) = 9(3)^x$.

Step 1 Make a table of values.

x	−2	−1	0	1	2
$f(x)$	1	3	9	27	81

Step 2 Plot the ordered pairs.

Step 3 Draw a smooth curve through the points.

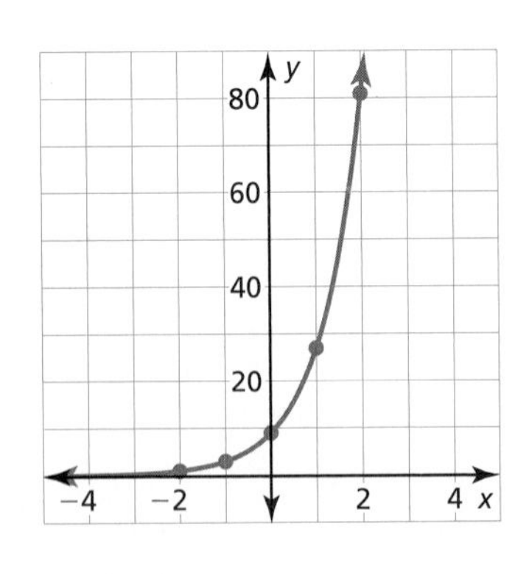

Graph the function. Describe the domain and range.

9. $f(x) = -4\left(\frac{1}{4}\right)^x$ **10.** $f(x) = 3^{x+2}$ **11.** $f(x) = 2^{x-4} - 3$

12. Write and graph an exponential function f represented by the table. Then compare the graph to the graph of $g(x) = \left(\frac{1}{2}\right)^x$.

x	0	1	2	3
y	2	1	0.5	0.25

6.4 Exponential Growth and Decay (pp. 313–322)

Rewrite the function $y = 10(0.65)^{t/8}$ to determine whether it represents *exponential growth* or *exponential decay*. Identify the percent rate of change.

$y = 10(0.65)^{t/8}$ — Write the function.

$= 10(0.65^{1/8})^t$ — Power of a Power Property

$\approx 10(0.95)^t$ — Evaluate the power.

The function is of the form $y = a(1 - r)^t$, where $1 - r < 1$, so it represents exponential decay. Use the decay factor $1 - r$ to find the rate of decay.

$1 - r = 0.95$ — Write an equation.

$r = 0.05$ — Solve for r.

▶ So, the function represents exponential decay, and the rate of decay is 5%.

Determine whether the table represents an *exponential growth function,* an *exponential decay function,* or *neither.* Explain.

13.

x	0	1	2	3
y	3	6	12	24

14.

x	1	2	3	4
y	162	108	72	48

Rewrite the function to determine whether it represents *exponential growth* or *exponential decay.* Identify the percent rate of change.

15. $f(t) = 4(1.25)^{t+3}$

16. $y = (1.06)^{8t}$

17. $f(t) = 6(0.84)^{t-4}$

18. You deposit $750 in a savings account that earns 5% annual interest compounded quarterly. (a) Write a function that represents the balance after t years. (b) What is the balance of the account after 4 years?

19. The value of a TV is $1500. Its value decreases by 14% each year. (a) Write a function that represents the value y (in dollars) of the TV after t years. (b) Find the approximate monthly percent decrease in value. (c) Graph the function from part (a). Use the graph to estimate the value of the TV after 3 years.

6.5 Solving Exponential Equations (pp. 325–330)

Solve $\frac{1}{9} = 3^{x+6}$.

$\frac{1}{9} = 3^{x+6}$ — Write the equation.

$3^{-2} = 3^{x+6}$ — Rewrite $\frac{1}{9}$ as 3^{-2}.

$-2 = x + 6$ — Equate the exponents.

$x = -8$ — Solve for x.

Solve the equation.

20. $5^x = 5^{3x-2}$

21. $3^{x-2} = 1$

22. $-4 = 6^{4x-3}$

23. $\left(\frac{1}{3}\right)^{2x+3} = 5$

24. $\left(\frac{1}{16}\right)^{3x} = 64^{2(x+8)}$

25. $27^{2x+2} = 81^{x+4}$

6.6 Geometric Sequences (pp. 331–338)

Write the next three terms of the geometric sequence 2, 6, 18, 54, ...

Use a table to organize the terms and extend the sequence.

Position	1	2	3	4	5	6	7
Term	2	6	18	54	162	486	1458

×3 ×3 ×3 ×3 ×3 ×3

Each term is 3 times the previous term. So, the common ratio is 3.

Multiply a term by 3 to find the next term.

▶ The next three terms are 162, 486, and 1458.

Decide whether the sequence is *arithmetic*, *geometric*, or *neither*. Explain your reasoning. If the sequence is geometric, write the next three terms and graph the sequence.

26. 3, 12, 48, 192, . . .

27. 9, −18, 27, −36, . . .

28. 375, −75, 15, −3, . . .

Write an equation for the *n*th term of the geometric sequence. Then find a_9.

29. 1, 4, 16, 64, . . .

30. 5, −10, 20, −40, . . .

31. 486, 162, 54, 18, . . .

6.7 Recursively Defined Sequences (pp. 339–346)

Write a recursive rule for the sequence 5, 12, 19, 26, 33, ...

Use a table to organize the terms and find the pattern.

Position, n	1	2	3	4	5
Term, a_n	5	12	19	26	33

+7 +7 +7 +7

The sequence is arithmetic, with first term $a_1 = 5$ and common difference $d = 7$.

$a_n = a_{n-1} + d$ Recursive equation for an arithmetic sequence

$a_n = a_{n-1} + 7$ Substitute 7 for d.

▶ So, a recursive rule for the sequence is $a_1 = 5$, $a_n = a_{n-1} + 7$.

Write the first six terms of the sequence. Then graph the sequence.

32. $a_1 = 4, a_n = a_{n-1} + 5$

33. $a_1 = -4, a_n = -3a_{n-1}$

34. $a_1 = 32, a_n = \frac{1}{4}a_{n-1}$

Write a recursive rule for the sequence.

35. 3, 8, 13, 18, 23, . . .

36. 3, 6, 12, 24, 48, . . .

37. 7, 6, 13, 19, 32, . . .

38. The first term of a sequence is 8. Each term of the sequence is 5 times the preceding term. Graph the first four terms of the sequence. Write a recursive rule and an explicit rule for the sequence.

6 Chapter Test

Evaluate the expression.

1. $-\sqrt[4]{16}$

2. $729^{1/6}$

3. $(-32)^{7/5}$

Simplify the expression. Write your answer using only positive exponents.

4. $z^{-2} \cdot z^4$

5. $\dfrac{b^{-5}}{a^0 b^{-8}}$

6. $\left(\dfrac{2c^4}{5}\right)^{-3}$

Write and graph a function that represents the situation.

7. Your starting annual salary of $42,500 increases by 3% each year.

8. You deposit $500 in an account that earns 6.5% annual interest compounded yearly.

Write an explicit rule and a recursive rule for the sequence.

9.

n	1	2	3	4
a_n	−6	8	22	36

10.

n	1	2	3	4
a_n	400	100	25	6.25

Solve the equation. Check your solution.

11. $2^x = \dfrac{1}{128}$

12. $256^{x+2} = 16^{3x-1}$

13. Graph $f(x) = 2(6)^x$. Compare the graph to the graph of $g(x) = 6^x$. Describe the domain and range of f.

Use the equation to complete the statement "a ▢ b" with the symbol <, >, or =. Do not attempt to solve the equation.

14. $\dfrac{5^a}{5^b} = 5^{-3}$

15. $9^a \cdot 9^{-b} = 1$

16. The first two terms of a sequence are $a_1 = 3$ and $a_2 = -12$. Let a_3 be the third term when the sequence is arithmetic and let b_3 be the third term when the sequence is geometric. Find $a_3 - b_3$.

17. At sea level, Earth's atmosphere exerts a pressure of 1 atmosphere. Atmospheric pressure P (in atmospheres) decreases with altitude. It can be modeled by $P = (0.99988)^a$, where a is the altitude (in meters).

a. Identify the initial amount, decay factor, and decay rate.

b. Use a graphing calculator to graph the function. Use the graph to estimate the atmospheric pressure at an altitude of 5000 feet.

18. You follow the training schedule from your coach.

a. Write an explicit rule and a recursive rule for the geometric sequence.

b. On what day do you run approximately 3 kilometers?

Training On Your Own

Day 1: Run 1 km.

Each day after Day 1: Run 20% farther than the previous day.

6 Cumulative Assessment

1. Fill in the exponent of x with a number to simplify the expression.

$$\frac{x^{5/3} \cdot x^{-1} \cdot \sqrt[3]{x}}{x^{-2} \cdot x^{0}} = x^{\square}$$

2. The graph of the exponential function f is shown. Find $f(-7)$.

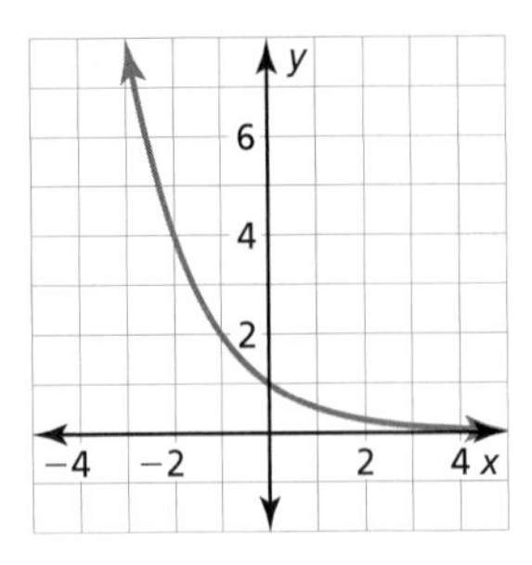

3. Student A claims he can form a linear system from the equations shown that has infinitely many solutions. Student B claims she can form a linear system from the equations shown that has one solution. Student C claims he can form a linear system from the equations shown that has no solution.

$3x + y = 12$	$3x + 2y = 12$	$6x + 2y = 6$
$3y + 9x = 36$	$2y - 6x = 12$	$9x - 3y = -18$

a. Select two equations to support Student A's claim.

b. Select two equations to support Student B's claim.

c. Select two equations to support Student C's claim.

4. Fill in the inequality with $<$, $\leq$, $>$, or $\geq$ so that the system of linear inequalities has no solution.

Inequality 1 $y - 2x \leq 4$

Inequality 2 $6x - 3y \;\square\; -12$

5. The second term of a sequence is 7. Each term of the sequence is 10 more than the preceding term. Fill in values to write a recursive rule and an explicit rule for the sequence.

$a_1 = \square$, $a_n = a_{n-1} + \square$

$a_n = \square n - \square$

6. A data set consists of the heights y (in feet) of a hot-air balloon t minutes after it begins its descent. An equation of the line of best fit is $y = 870 - 14.8t$. Which of the following is a correct interpretation of the line of best fit?

Ⓐ The initial height of the hot-air balloon is 870 feet. The slope has no meaning in this context.

Ⓑ The initial height of the hot-air balloon is 870 feet, and it descends 14.8 feet per minute.

Ⓒ The initial height of the hot-air balloon is 870 feet, and it ascends 14.8 feet per minute.

Ⓓ The hot-air balloon descends 14.8 feet per minute. The y-intercept has no meaning in this context.

7. Select all the functions whose x-value is an integer when $f(x) = 10$.

$f(x) = 3x - 2$	$f(x) = -2x + 4$	$f(x) = \frac{3}{2}x + 4$
$f(x) = -3x + 5$	$f(x) = \frac{1}{2}x - 6$	$f(x) = 4x + 14$

8. Place each function into one of the three categories. For exponential functions, state whether the function represents *exponential growth*, *exponential decay*, or *neither.*

Exponential	Linear	Neither
$f(x) = -2(8)^x$	$f(x) = 15 - x$	$f(x) = \frac{1}{2}(3)^x$
$f(x) = 6x^2 + 9$	$f(x) = 4(1.6)^{x/10}$	$f(x) = x(18 - x)$
$f(x) = 3\left(\frac{1}{6}\right)^x$	$f(x) = -3(4x + 1 - x)$	$f(x) = \sqrt[4]{16} + 2x$

9. How does the graph shown compare to the graph of $f(x) = 2^x$?

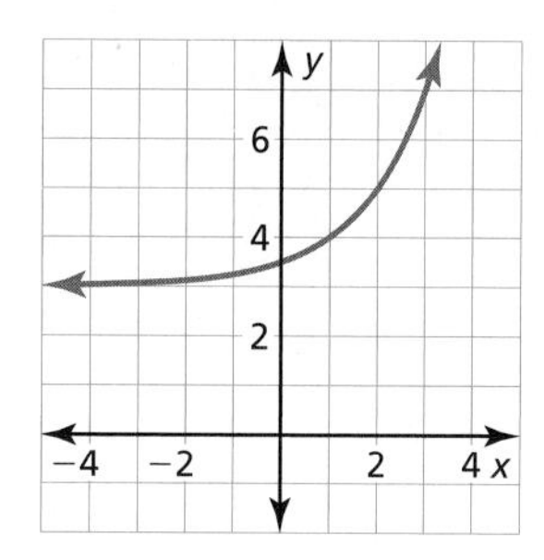

7 Polynomial Equations and Factoring

Height of a Falling Object *(p. 400)*

Game Reserve *(p. 394)*

Photo Cropping *(p. 390)*

Gateway Arch *(p. 382)*

Framing a Photo *(p. 370)*

Maintaining Mathematical Proficiency

Simplifying Algebraic Expressions

Example 1 **Simplify $6x + 5 - 3x - 4$.**

$6x + 5 - 3x - 4 = 6x - 3x + 5 - 4$	Commutative Property of Addition
$= (6 - 3)x + 5 - 4$	Distributive Property
$= 3x + 1$	Simplify.

Example 2 **Simplify $-8(y - 3) + 2y$.**

$-8(y - 3) + 2y = -8(y) - (-8)(3) + 2y$	Distributive Property
$= -8y + 24 + 2y$	Multiply.
$= -8y + 2y + 24$	Commutative Property of Addition
$= (-8 + 2)y + 24$	Distributive Property
$= -6y + 24$	Simplify.

Simplify the expression.

1. $3x - 7 + 2x$

2. $4r + 6 - 9r - 1$

3. $-5t + 3 - t - 4 + 8t$

4. $3(s - 1) + 5$

5. $2m - 7(3 - m)$

6. $4(h + 6) - (h - 2)$

Finding the Greatest Common Factor

Example 3 **Find the greatest common factor (GCF) of 42 and 70.**

To find the GCF of two numbers, first write the prime factorization of each number. Then find the product of the common prime factors.

$42 = 2 \cdot 3 \cdot 7$

$70 = 2 \cdot 5 \cdot 7$

▶ The GCF of 42 and 70 is $2 \cdot 7 = 14$.

Find the greatest common factor.

7. 20, 36

8. 42, 63

9. 54, 81

10. 72, 84

11. 28, 64

12. 30, 77

13. **ABSTRACT REASONING** Is it possible for two integers to have no common factors? Explain your reasoning.

Mathematical Practices

Mathematically proficient students consider concrete models when solving a mathematics problem.

Using Models

Using Algebra Tiles

When solving a problem, it can be helpful to use a model. For instance, you can use algebra tiles to model algebraic expressions and operations with algebraic expressions.

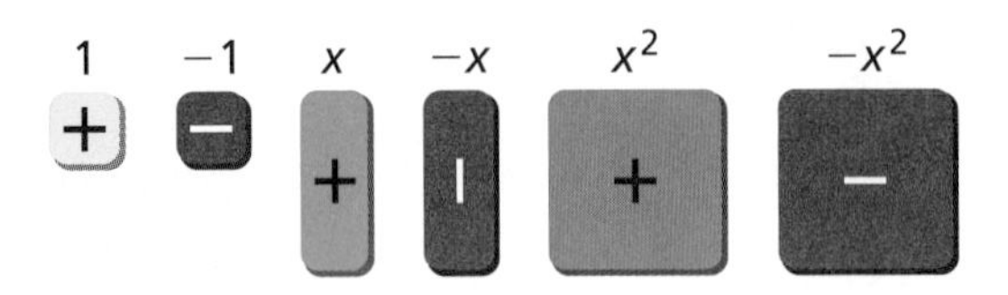

EXAMPLE 1 **Writing Expressions Modeled by Algebra Tiles**

Write the algebraic expression modeled by the algebra tiles.

a.

b. 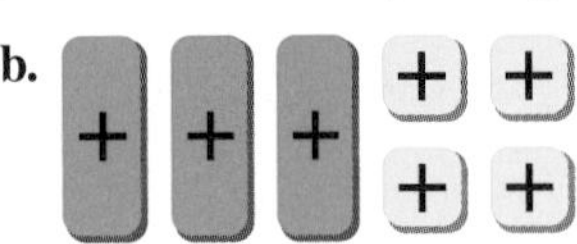

c. 

SOLUTION

a. The algebraic expression is x^2.

b. The algebraic expression is $3x + 4$.

c. The algebraic expression is $x^2 - x + 2$.

Monitoring Progress

Write the algebraic expression modeled by the algebra tiles.

1.

2. 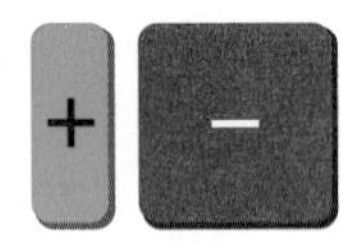

3. 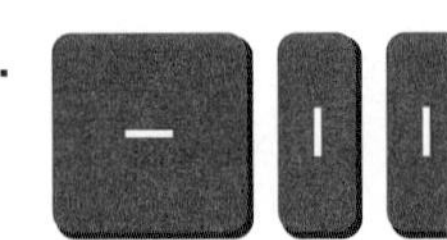

4. 

5. 

6. 

7. 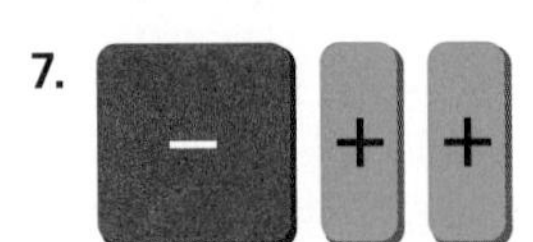

8.

9. 

7.1 Adding and Subtracting Polynomials

Essential Question How can you add and subtract polynomials?

EXPLORATION 1 Adding Polynomials

Work with a partner. Write the expression modeled by the algebra tiles in each step.

Step 1

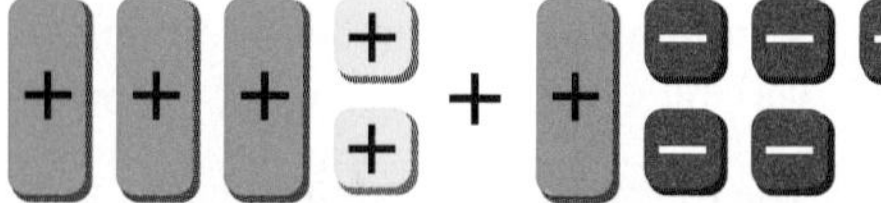

$(3x + 2) + (x - 5)$

Step 2

Step 3

Step 4

EXPLORATION 2 Subtracting Polynomials

Work with a partner. Write the expression modeled by the algebra tiles in each step.

Step 1

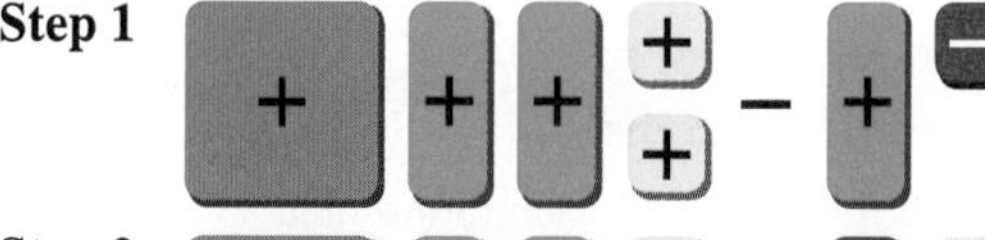

$(x^2 + 2x + 2) - (x - 1)$

Step 2

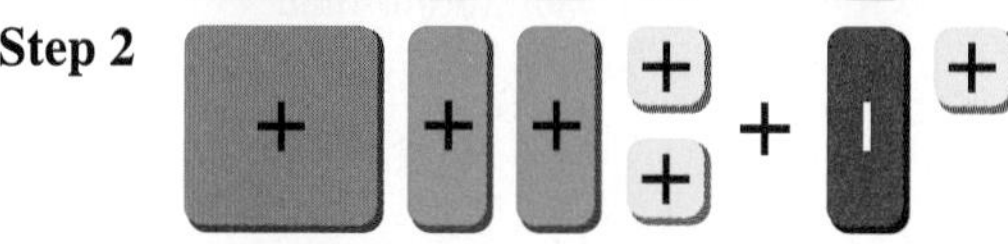

Step 3

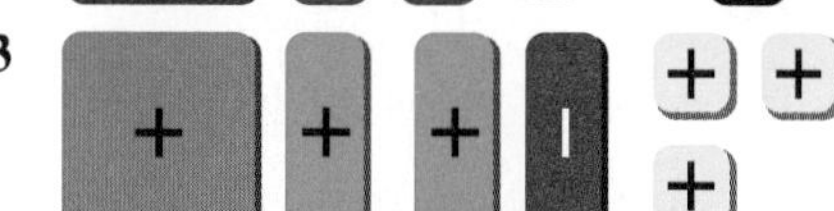

Step 4

Step 5

REASONING ABSTRACTLY

To be proficient in math, you need to represent a given situation using symbols.

Communicate Your Answer

3. How can you add and subtract polynomials?

4. Use your methods in Question 3 to find each sum or difference.

a. $(x^2 + 2x - 1) + (2x^2 - 2x + 1)$ **b.** $(4x + 3) + (x - 2)$

c. $(x^2 + 2) - (3x^2 + 2x + 5)$ **d.** $(2x - 3x) - (x^2 - 2x + 4)$

7.1 Lesson

Core Vocabulary

monomial, *p. 358*
degree of a monomial, *p. 358*
polynomial, *p. 359*
binomial, *p. 359*
trinomial, *p. 359*
degree of a polynomial, *p. 359*
standard form, *p. 359*
leading coefficient, *p. 359*
closed, *p. 360*

What You Will Learn

- Find the degrees of monomials.
- Classify polynomials.
- Add and subtract polynomials.
- Solve real-life problems.

Finding the Degrees of Monomials

A **monomial** is a number, a variable, or the product of a number and one or more variables with whole number exponents.

The **degree of a monomial** is the sum of the exponents of the variables in the monomial. The degree of a nonzero constant term is 0. The constant 0 does not have a degree.

Monomial	Degree
10	0
$3x$	1
$\frac{1}{2}ab^2$	$1 + 2 = 3$
$-1.8m^5$	5

Not a monomial	Reason
$5 + x$	A sum is not a monomial.
$\frac{2}{n}$	A monomial cannot have a variable in the denominator.
4^a	A monomial cannot have a variable exponent.
x^{-1}	The variable must have a whole number exponent.

EXAMPLE 1 Finding the Degrees of Monomials

Find the degree of each monomial.

a. $5x^2$ **b.** $-\frac{1}{2}xy^3$ **c.** $8x^3y^3$ **d.** -3

SOLUTION

a. The exponent of x is 2.

▶ So, the degree of the monomial is 2.

b. The exponent of x is 1, and the exponent of y is 3.

▶ So, the degree of the monomial is $1 + 3$, or 4.

c. The exponent of x is 3, and the exponent of y is 3.

▶ So, the degree of the monomial is $3 + 3$, or 6.

d. You can rewrite -3 as $-3x^0$.

▶ So, the degree of the monomial is 0.

Monitoring Progress  Help in English and Spanish at *BigIdeasMath.com*

Find the degree of the monomial.

1. $-3x^4$ **2.** $7c^3d^2$ **3.** $\frac{5}{3}y$ **4.** -20.5

Classifying Polynomials

Polynomials

A **polynomial** is a monomial or a sum of monomials. Each monomial is called a *term* of the polynomial. A polynomial with two terms is a **binomial**. A polynomial with three terms is a **trinomial**.

Binomial	Trinomial
$5x + 2$	$x^2 + 5x + 2$

The **degree of a polynomial** is the greatest degree of its terms. A polynomial in one variable is in **standard form** when the exponents of the terms decrease from left to right. When you write a polynomial in standard form, the coefficient of the first term is the **leading coefficient**.

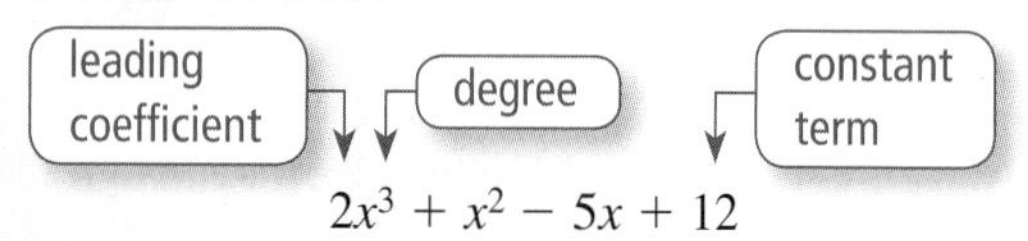

$$2x^3 + x^2 - 5x + 12$$

EXAMPLE 2 Writing a Polynomial in Standard Form

Write $15x - x^3 + 3$ in standard form. Identify the degree and leading coefficient of the polynomial.

SOLUTION

Consider the degree of each term of the polynomial.

Degree is 3.

Degree is 1. → $15x - x^3 + 3$ ← Degree is 0.

▶ You can write the polynomial in standard form as $-x^3 + 15x + 3$. The greatest degree is 3, so the degree of the polynomial is 3, and the leading coefficient is -1.

EXAMPLE 3 Classifying Polynomials

Write each polynomial in standard form. Identify the degree and classify each polynomial by the number of terms.

a. $-3z^4$ **b.** $4 + 5x^2 - x$ **c.** $8q + q^5$

SOLUTION

Polynomial	Standard Form	Degree	Type of Polynomial
a. $-3z^4$	$-3z^4$	4	monomial
b. $4 + 5x^2 - x$	$5x^2 - x + 4$	2	trinomial
c. $8q + q^5$	$q^5 + 8q$	5	binomial

Monitoring Progress

Help in English and Spanish at *BigIdeasMath.com*

Write the polynomial in standard form. Identify the degree and leading coefficient of the polynomial. Then classify the polynomial by the number of terms.

5. $4 - 9z$ **6.** $t^2 - t^3 - 10t$ **7.** $2.8x + x^3$

Adding and Subtracting Polynomials

A set of numbers is **closed** under an operation when the operation performed on any two numbers in the set results in a number that is also in the set. For example, the set of integers is closed under addition, subtraction, and multiplication. This means that if a and b are two integers, then $a + b$, $a - b$, and ab are also integers.

The set of polynomials is closed under addition and subtraction. So, the sum or difference of any two polynomials is also a polynomial.

To add polynomials, add like terms. You can use a vertical or a horizontal format.

EXAMPLE 4 Adding Polynomials

Find the sum.

a. $(2x^3 - 5x^2 + x) + (2x^2 + x^3 - 1)$ **b.** $(3x^2 + x - 6) + (x^2 + 4x + 10)$

SOLUTION

a. Vertical format: Align like terms vertically and add.

$$\begin{array}{r} 2x^3 - 5x^2 + x \\ + \quad x^3 + 2x^2 - 1 \\ \hline 3x^3 - 3x^2 + x - 1 \end{array}$$

▶ The sum is $3x^3 - 3x^2 + x - 1$.

STUDY TIP

When a power of the variable appears in one polynomial but not the other, leave a space in that column, or write the term with a coefficient of 0.

b. Horizontal format: Group like terms and simplify.

$$\begin{aligned}(3x^2 + x - 6) + (x^2 + 4x + 10) &= (3x^2 + x^2) + (x + 4x) + (-6 + 10) \\ &= 4x^2 + 5x + 4\end{aligned}$$

▶ The sum is $4x^2 + 5x + 4$.

To subtract a polynomial, add its opposite. To find the opposite of a polynomial, multiply each of its terms by -1.

EXAMPLE 5 Subtracting Polynomials

Find the difference.

a. $(4n^2 + 5) - (-2n^2 + 2n - 4)$ **b.** $(4x^2 - 3x + 5) - (3x^2 - x - 8)$

SOLUTION

a. Vertical format: Align like terms vertically and subtract.

$$\begin{array}{r} 4n^2 + 5 \\ - \; (-2n^2 + 2n - 4) \\ \hline \end{array} \quad \Rightarrow \quad \begin{array}{r} 4n^2 + 5 \\ + \quad 2n^2 - 2n + 4 \\ \hline 6n^2 - 2n + 9 \end{array}$$

▶ The difference is $6n^2 - 2n + 9$.

COMMON ERROR

Remember to multiply *each* term of the polynomial by -1 when you write the subtraction as addition.

b. Horizontal format: Group like terms and simplify.

$$\begin{aligned}(4x^2 - 3x + 5) - (3x^2 - x - 8) &= 4x^2 - 3x + 5 - 3x^2 + x + 8 \\ &= (4x^2 - 3x^2) + (-3x + x) + (5 + 8) \\ &= x^2 - 2x + 13\end{aligned}$$

▶ The difference is $x^2 - 2x + 13$.

Monitoring Progress Help in English and Spanish at *BigIdeasMath.com*

Find the sum or difference.

8. $(b - 10) + (4b - 3)$

9. $(x^2 - x - 2) + (7x^2 - x)$

10. $(p^2 + p + 3) - (-4p^2 - p + 3)$

11. $(-k + 5) - (3k^2 - 6)$

Solving Real-Life Problems

EXAMPLE 6 Solving a Real-Life Problem

A penny is thrown straight down from a height of 200 feet. At the same time, a paintbrush is dropped from a height of 100 feet. The polynomials represent the heights (in feet) of the objects after t seconds.

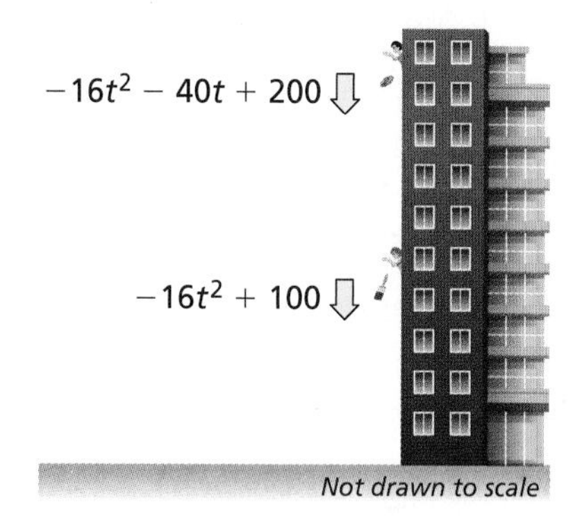

a. Write a polynomial that represents the distance between the penny and the paintbrush after t seconds.

b. Interpret the coefficients of the polynomial in part (a).

SOLUTION

a. To find the distance between the objects after t seconds, subtract the polynomials.

Penny	$-16t^2 - 40t + 200$		$-16t^2 - 40t + 200$
Paintbrush	$-\ (-16t^2 \quad + 100)$		$+\ \ 16t^2 \quad - 100$
			$-40t + 100$

▶ The polynomial $-40t + 100$ represents the distance between the objects after t seconds.

b. When $t = 0$, the distance between the objects is $-40(0) + 100 = 100$ feet. So, the constant term 100 represents the distance between the penny and the paintbrush when both objects begin to fall.

As the value of t increases by 1, the value of $-40t + 100$ decreases by 40. This means that the objects become 40 feet closer to each other each second. So, -40 represents the amount that the distance between the objects changes each second.

Monitoring Progress Help in English and Spanish at *BigIdeasMath.com*

12. WHAT IF? The polynomial $-16t^2 - 25t + 200$ represents the height of the penny after t seconds.

a. Write a polynomial that represents the distance between the penny and the paintbrush after t seconds.

b. Interpret the coefficients of the polynomial in part (a).

7.1 Exercises

Dynamic Solutions available at *BigIdeasMath.com*

Vocabulary and Core Concept Check

1. **VOCABULARY** When is a polynomial in one variable in standard form?

2. **OPEN-ENDED** Write a trinomial in one variable of degree 5 in standard form.

3. **VOCABULARY** How can you determine whether a set of numbers is closed under an operation?

4. **WHICH ONE DOESN'T BELONG?** Which expression does *not* belong with the other three? Explain your reasoning.

$a^3 + 4a$	$x^2 - 8^x$	$b - 2^{-1}$	$-\frac{\pi}{3} + 6y^8z$

Monitoring Progress and Modeling with Mathematics

In Exercises 5–12, find the degree of the monomial. *(See Example 1.)*

5. $4g$
6. $23x^4$
7. $-1.75k^2$
8. $-\frac{4}{9}$
9. s^8t
10. $8m^2n^4$
11. $9xy^3z^7$
12. $-3q^4rs^6$

In Exercises 13–20, write the polynomial in standard form. Identify the degree and leading coefficient of the polynomial. Then classify the polynomial by the number of terms. *(See Examples 2 and 3.)*

13. $6c^2 + 2c^4 - c$
14. $4w^{11} - w^{12}$
15. $7 + 3p^2$
16. $8d - 2 - 4d^3$
17. $3t^8$
18. $5z + 2z^3 + 3z^4$
19. $\pi r^2 - \frac{5}{7}r^8 + 2r^5$
20. $\sqrt{7}n^4$

21. **MODELING WITH MATHEMATICS** The expression $\frac{4}{3}\pi r^3$ represents the volume of a sphere with radius r. Why is this expression a monomial? What is its degree?

22. **MODELING WITH MATHEMATICS** The amount of money you have after investing \$400 for 8 years and \$600 for 6 years at the same interest rate is represented by $400x^8 + 600x^6$, where x is the growth factor. Classify the polynomial by the number of terms. What is its degree?

In Exercises 23–30, find the sum. *(See Example 4.)*

23. $(5y + 4) + (-2y + 6)$
24. $(-8x - 12) + (9x + 4)$
25. $(2n^2 - 5n - 6) + (-n^2 - 3n + 11)$
26. $(-3p^3 + 5p^2 - 2p) + (-p^3 - 8p^2 - 15p)$
27. $(3g^2 - g) + (3g^2 - 8g + 4)$
28. $(9r^2 + 4r - 7) + (3r^2 - 3r)$
29. $(4a - a^3 - 3) + (2a^3 - 5a^2 + 8)$
30. $(s^3 - 2s - 9) + (2s^2 - 6s^3 + s)$

In Exercises 31–38, find the difference. *(See Example 5.)*

31. $(d - 9) - (3d - 1)$
32. $(6x + 9) - (7x + 1)$
33. $(y^2 - 4y + 9) - (3y^2 - 6y - 9)$
34. $(4m^2 - m + 2) - (-3m^2 + 10m + 4)$
35. $(k^3 - 7k + 2) - (k^2 - 12)$
36. $(-r - 10) - (-4r^3 + r^2 + 7r)$

37. $(t^4 - t^2 + t) - (12 - 9t^2 - 7t)$

38. $(4d - 6d^3 + 3d^2) - (10d^3 + 7d - 2)$

ERROR ANALYSIS In Exercises 39 and 40, describe and correct the error in finding the sum or difference.

39.

✗
$$\begin{aligned}(x^2 + x) - (2x^2 - 3x) &= x^2 + x - 2x^2 - 3x \\ &= (x^2 - 2x^2) + (x - 3x) \\ &= -x^2 - 2x\end{aligned}$$

40.

✗
$$\begin{array}{r} x^3 - 4x^2 + 3 \\ + \ -3x^3 + 8x \ - 2 \\ \hline -2x^3 + 4x^2 + 1 \end{array}$$

41. **MODELING WITH MATHEMATICS** The cost (in dollars) of making b bracelets is represented by $4 + 5b$. The cost (in dollars) of making b necklaces is represented by $8b + 6$. Write a polynomial that represents how much more it costs to make b necklaces than b bracelets.

42. **MODELING WITH MATHEMATICS** The number of individual memberships at a fitness center in m months is represented by $142 + 12m$. The number of family memberships at the fitness center in m months is represented by $52 + 6m$. Write a polynomial that represents the total number of memberships at the fitness center.

In Exercises 43–46, find the sum or difference.

43. $(2s^2 - 5st - t^2) - (s^2 + 7st - t^2)$

44. $(a^2 - 3ab + 2b^2) + (-4a^2 + 5ab - b^2)$

45. $(c^2 - 6d^2) + (c^2 - 2cd + 2d^2)$

46. $(-x^2 + 9xy) - (x^2 + 6xy - 8y^2)$

REASONING In Exercises 47–50, complete the statement with *always*, *sometimes*, or *never*. Explain your reasoning.

47. The terms of a polynomial are ________ monomials.

48. The difference of two trinomials is ________ a trinomial.

49. A binomial is ________ a polynomial of degree 2.

50. The sum of two polynomials is ________ a polynomial.

MODELING WITH MATHEMATICS The polynomial $-16t^2 + v_0t + s_0$ represents the height (in feet) of an object, where v_0 is the initial vertical velocity (in feet per second), s_0 is the initial height of the object (in feet), and t is the time (in seconds). In Exercises 51 and 52, write a polynomial that represents the height of the object. Then find the height of the object after 1 second.

51. You throw a water balloon from a building.

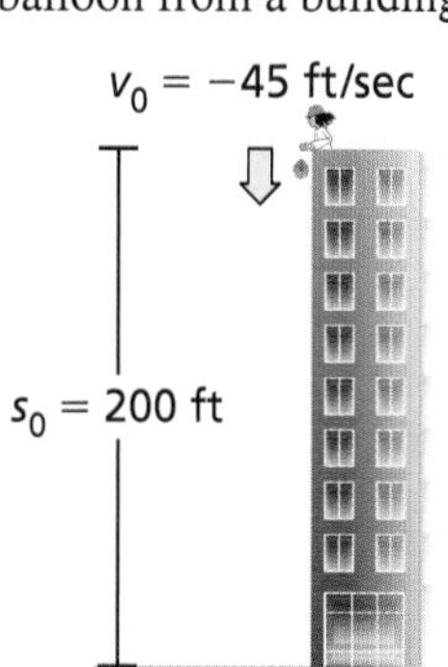

Not drawn to scale

52. You bounce a tennis ball on a racket.

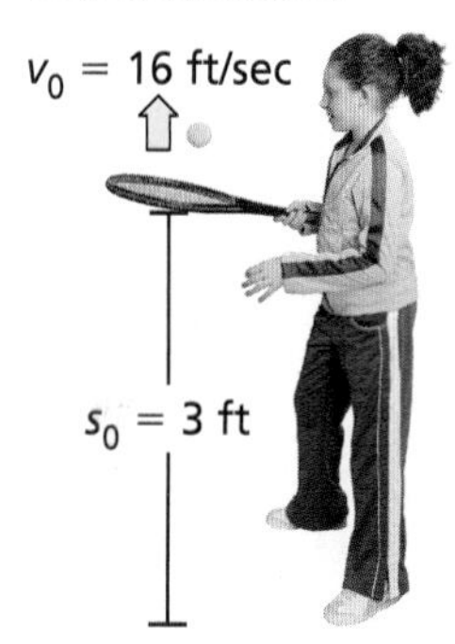

53. **MODELING WITH MATHEMATICS** You drop a ball from a height of 98 feet. At the same time, your friend throws a ball upward. The polynomials represent the heights (in feet) of the balls after t seconds. *(See Example 6.)*

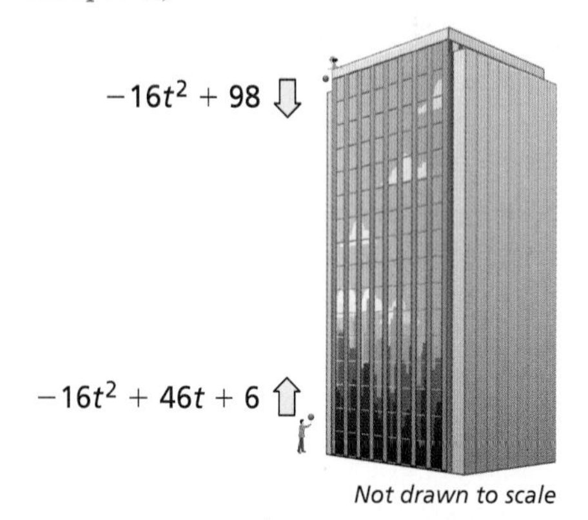

Not drawn to scale

a. Write a polynomial that represents the distance between your ball and your friend's ball after t seconds.

b. Interpret the coefficients of the polynomial in part (a).

54. **MODELING WITH MATHEMATICS** During a 7-year period, the amounts (in millions of dollars) spent each year on buying new vehicles N and used vehicles U by United States residents are modeled by the equations

$$N = -0.028t^3 + 0.06t^2 + 0.1t + 17$$
$$U = -0.38t^2 + 1.5t + 42$$

where $t = 1$ represents the first year in the 7-year period.

a. Write a polynomial that represents the total amount spent each year on buying new and used vehicles in the 7-year period.

b. How much is spent on buying new and used vehicles in the fifth year?

55. **MATHEMATICAL CONNECTIONS** Write the polynomial in standard form that represents the perimeter of the quadrilateral.

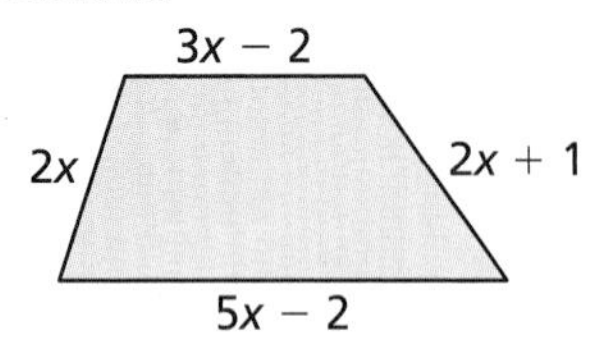

56. **HOW DO YOU SEE IT?** The right side of the equation of each line is a polynomial.

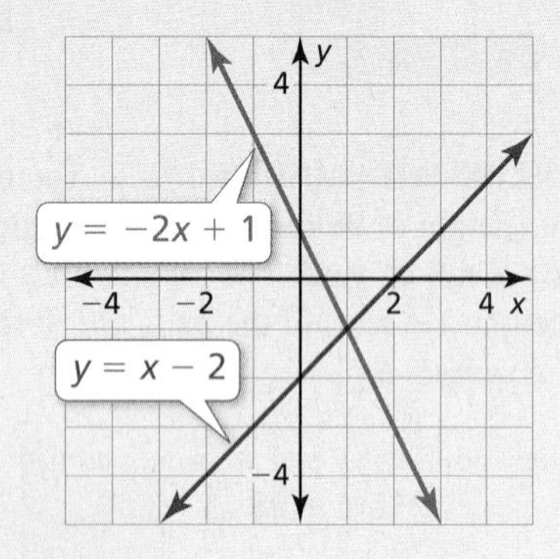

a. The absolute value of the difference of the two polynomials represents the vertical distance between points on the lines with the same x-value. Write this expression.

b. When does the expression in part (a) equal 0? How does this value relate to the graph?

57. **MAKING AN ARGUMENT** Your friend says that when adding polynomials, the order in which you add does not matter. Is your friend correct? Explain.

58. **THOUGHT PROVOKING** Write two polynomials whose sum is x^2 and whose difference is 1.

59. **REASONING** Determine whether the set is closed under the given operation. Explain.

a. the set of negative integers; multiplication

b. the set of whole numbers; addition

60. **PROBLEM SOLVING** You are building a multi-level deck.

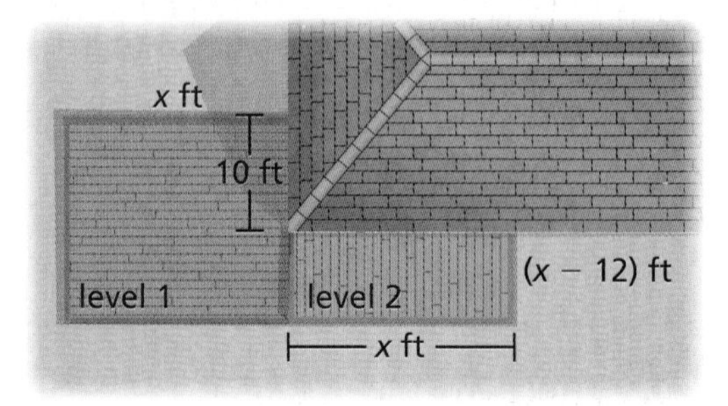

a. For each level, write a polynomial in standard form that represents the area of that level. Then write the polynomial in standard form that represents the total area of the deck.

b. What is the total area of the deck when $x = 20$?

c. A gallon of deck sealant covers 400 square feet. How many gallons of sealant do you need to cover the deck in part (b) once? Explain.

61. **PROBLEM SOLVING** A hotel installs a new swimming pool and a new hot tub.

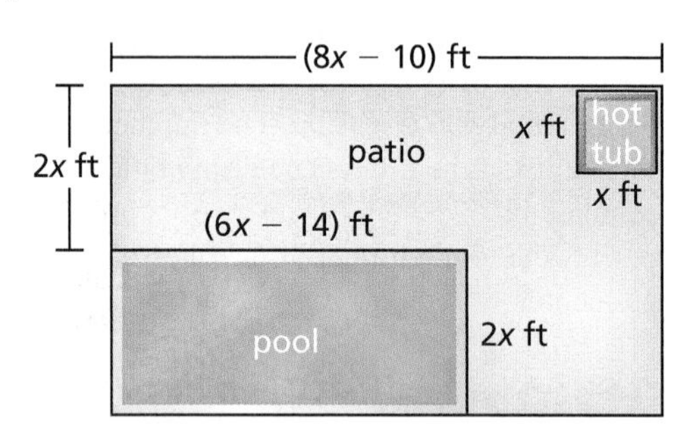

a. Write the polynomial in standard form that represents the area of the patio.

b. The patio will cost $10 per square foot. Determine the cost of the patio when $x = 9$.

Maintaining Mathematical Proficiency

Reviewing what you learned in previous grades and lessons

Simplify the expression. *(Skills Review Handbook)*

62. $2(x - 1) + 3(x + 2)$

63. $8(4y - 3) + 2(y - 5)$

64. $5(2r + 1) - 3(-4r + 2)$

7.2 Multiplying Polynomials

Essential Question How can you multiply two polynomials?

EXPLORATION 1 Multiplying Monomials Using Algebra Tiles

Work with a partner. Write each product. Explain your reasoning.

a. + • + =

b. + • − =

c. − • − =

d. + • + =

e. + • − =

f. − • + =

g. − • − =

h. + • + =

i. + • − =

j. − • − =

REASONING ABSTRACTLY

To be proficient in math, you need to reason abstractly and quantitatively. You need to pause as needed to recall the meanings of the symbols, operations, and quantities involved.

EXPLORATION 2 Multiplying Binomials Using Algebra Tiles

Work with a partner. Write the product of two binomials modeled by each rectangular array of algebra tiles. In parts (c) and (d), first draw the rectangular array of algebra tiles that models each product.

a. $(x + 3)(x - 2) =$

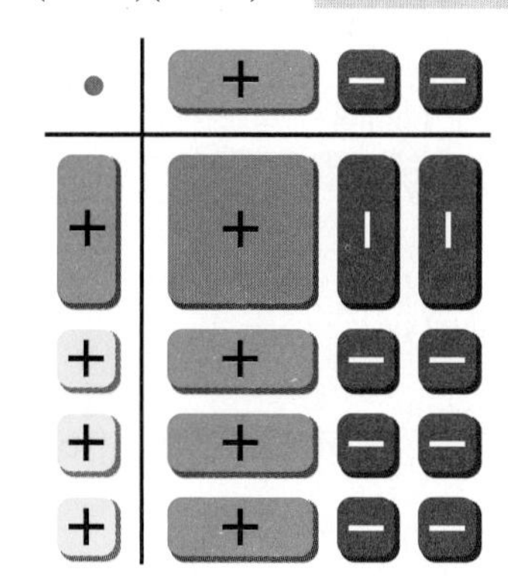

b. $(2x - 1)(2x + 1) =$

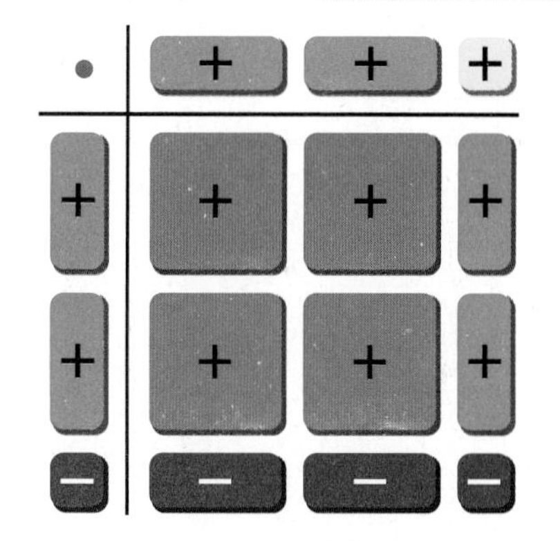

c. $(x + 2)(2x - 1) =$

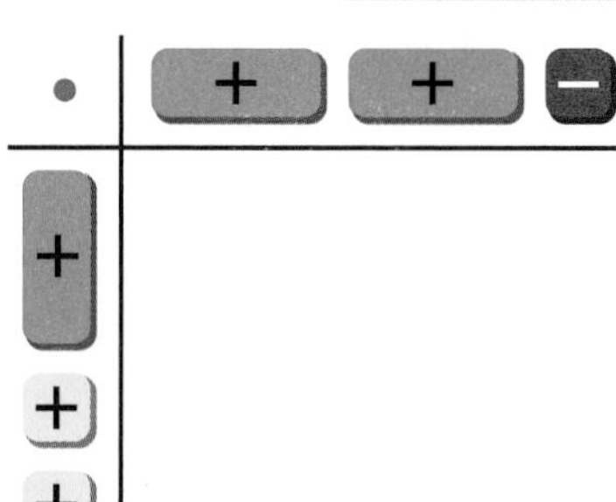

d. $(-x - 2)(x - 3) =$

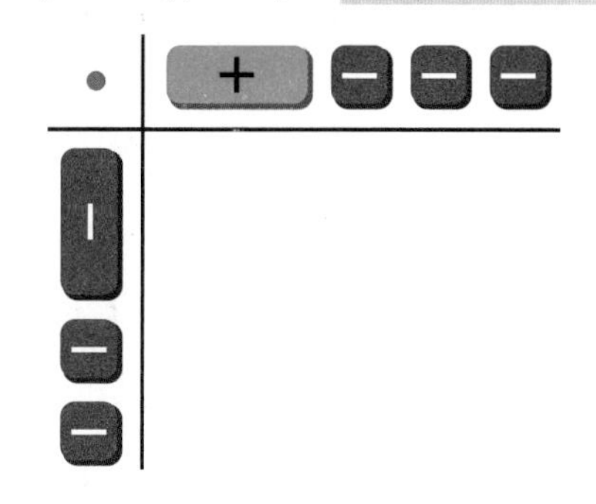

Communicate Your Answer

3. How can you multiply two polynomials?

4. Give another example of multiplying two binomials using algebra tiles that is similar to those in Exploration 2.

7.2 Lesson

Core Vocabulary

FOIL Method, *p. 367*

Previous

polynomial
closed
binomial
trinomial

What You Will Learn

- Multiply binomials.
- Use the FOIL Method.
- Multiply binomials and trinomials.

Multiplying Binomials

The product of two polynomials is always a polynomial. So, like the set of integers, the set of polynomials is closed under multiplication. You can use the Distributive Property to multiply two binomials.

EXAMPLE 1 Multiplying Binomials Using the Distributive Property

Find (a) $(x + 2)(x + 5)$ and (b) $(x + 3)(x - 4)$.

SOLUTION

a. Use the horizontal method.

$(x + 2)(x + 5) = x(x + 5) + 2(x + 5)$ — Distribute $(x + 5)$ to each term of $(x + 2)$.

$= x(x) + x(5) + 2(x) + 2(5)$ — Distributive Property

$= x^2 + 5x + 2x + 10$ — Multiply.

$= x^2 + 7x + 10$ — Combine like terms.

▶ The product is $x^2 + 7x + 10$.

b. Use the vertical method.

$$\begin{array}{rr} & x + 3 \\ \times & x - 4 \\ \hline & -4x - 12 \\ & x^2 + 3x \quad \\ \hline & x^2 - x - 12 \end{array}$$

Align like terms vertically.
Multiply $-4(x + 3)$. → $-4x - 12$ — Distributive Property
Multiply $x(x + 3)$. → $x^2 + 3x$ — Distributive Property
$x^2 - x - 12$ — Combine like terms.

▶ The product is $x^2 - x - 12$.

EXAMPLE 2 Multiplying Binomials Using a Table

Find $(2x - 3)(x + 5)$.

SOLUTION

Step 1 Write each binomial as a sum of terms.

$$(2x - 3)(x + 5) = [2x + (-3)](x + 5)$$

Step 2 Make a table of products.

	$2x$	-3
x	$2x^2$	$-3x$
5	$10x$	-15

▶ The product is $2x^2 - 3x + 10x - 15$, or $2x^2 + 7x - 15$.

Monitoring Progress

Help in English and Spanish at *BigIdeasMath.com*

Use the Distributive Property to find the product.

1. $(y + 4)(y + 1)$

2. $(z - 2)(z + 6)$

Use a table to find the product.

3. $(p + 3)(p - 8)$

4. $(r - 5)(2r - 1)$

Using the FOIL Method

The **FOIL Method** is a shortcut for multiplying two binomials.

Core Concept

FOIL Method

To multiply two binomials using the FOIL Method, find the sum of the products of the

First terms, $(x + 1)(x + 2)$ → $x(x) = x^2$

Outer terms, $(x + 1)(x + 2)$ → $x(2) = 2x$

Inner terms, and $(x + 1)(x + 2)$ → $1(x) = x$

Last terms. $(x + 1)(x + 2)$ → $1(2) = 2$

$$(x + 1)(x + 2) = x^2 + 2x + x + 2 = \mathbf{x^2 + 3x + 2}$$

EXAMPLE 3 Multiplying Binomials Using the FOIL Method

Find each product.

a. $(x - 3)(x - 6)$

b. $(2x + 1)(3x - 5)$

SOLUTION

a. Use the FOIL Method.

First Outer Inner Last

$(x - 3)(x - 6) = x(x) + x(-6) + (-3)(x) + (-3)(-6)$ FOIL Method

$= x^2 + (-6x) + (-3x) + 18$ Multiply.

$= x^2 - 9x + 18$ Combine like terms.

▶ The product is $x^2 - 9x + 18$.

b. Use the FOIL Method.

First Outer Inner Last

$(2x + 1)(3x - 5) = 2x(3x) + 2x(-5) + 1(3x) + 1(-5)$ FOIL Method

$= 6x^2 + (-10x) + 3x + (-5)$ Multiply.

$= 6x^2 - 7x - 5$ Combine like terms.

▶ The product is $6x^2 - 7x - 5$.

Monitoring Progress

Help in English and Spanish at *BigIdeasMath.com*

Use the FOIL Method to find the product.

5. $(m - 3)(m - 7)$

6. $(x - 4)(x + 2)$

7. $\left(2u + \frac{1}{2}\right)\left(u - \frac{3}{2}\right)$

8. $(n + 2)(n^2 + 3)$

Multiplying Binomials and Trinomials

EXAMPLE 4 Multiplying a Binomial and a Trinomial

Find $(x + 5)(x^2 - 3x - 2)$.

SOLUTION

$$
\begin{array}{rrrrl}
 & x^2 & -\,3x & -\,2 & \\
\times & & x & +\,5 & \text{Align like terms vertically.} \\
\hline
 & 5x^2 & -\,15x & -\,10 & \text{Distributive Property} \\
x^3 & -\,3x^2 & -\,2x & & \text{Distributive Property} \\
\hline
x^3 & +\,2x^2 & -\,17x & -\,10 & \text{Combine like terms.}
\end{array}
$$

Multiply $5(x^2 - 3x - 2)$. → $5x^2 - 15x - 10$

Multiply $x(x^2 - 3x - 2)$. → $x^3 - 3x^2 - 2x$

▶ The product is $x^3 + 2x^2 - 17x - 10$.

EXAMPLE 5 Solving a Real-Life Problem

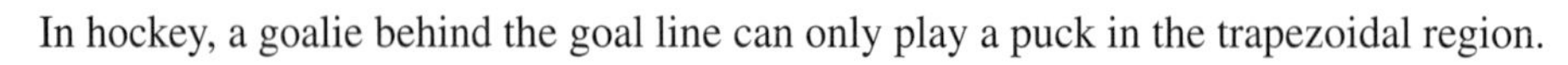

In hockey, a goalie behind the goal line can only play a puck in the trapezoidal region.

a. Write a polynomial that represents the area of the trapezoidal region.

b. Find the area of the trapezoidal region when the shorter base is 18 feet.

SOLUTION

a. $\frac{1}{2}h(b_1 + b_2) = \frac{1}{2}(x - 7)[x + (x + 10)]$ Substitute.

$= \frac{1}{2}(x - 7)(2x + 10)$ Combine like terms.

F O I L

$= \frac{1}{2}[2x^2 + 10x + (-14x) + (-70)]$ FOIL Method

$= \frac{1}{2}(2x^2 - 4x - 70)$ Combine like terms.

$= x^2 - 2x - 35$ Distributive Property

▶ A polynomial that represents the area of the trapezoidal region is $x^2 - 2x - 35$.

b. Find the value of $x^2 - 2x - 35$ when $x = 18$.

$x^2 - 2x - 35 = 18^2 - 2(18) - 35$ Substitute 18 for x.

$= 324 - 36 - 35$ Simplify.

$= 253$ Subtract.

▶ The area of the trapezoidal region is 253 square feet.

Monitoring Progress  Help in English and Spanish at *BigIdeasMath.com*

Find the product.

9. $(x + 1)(x^2 + 5x + 8)$

10. $(n - 3)(n^2 - 2n + 4)$

11. WHAT IF? In Example 5(a), how does the polynomial change when the longer base is extended by 1 foot? Explain.

7.2 Exercises

Dynamic Solutions available at *BigIdeasMath.com*

Vocabulary and Core Concept Check

1. **VOCABULARY** Describe two ways to find the product of two binomials.

2. **WRITING** Explain how the letters of the word FOIL can help you to remember how to multiply two binomials.

Monitoring Progress and Modeling with Mathematics

In Exercises 3–10, use the Distributive Property to find the product. *(See Example 1.)*

3. $(x + 1)(x + 3)$
4. $(y + 6)(y + 4)$
5. $(z - 5)(z + 3)$
6. $(a + 8)(a - 3)$
7. $(g - 7)(g - 2)$
8. $(n - 6)(n - 4)$
9. $(3m + 1)(m + 9)$
10. $(5s + 6)(s - 2)$

In Exercises 11–18, use a table to find the product. *(See Example 2.)*

11. $(x + 3)(x + 2)$
12. $(y + 10)(y - 5)$
13. $(h - 8)(h - 9)$
14. $(c - 6)(c - 5)$
15. $(3k - 1)(4k + 9)$
16. $(5g + 3)(g + 8)$
17. $(-3 + 2j)(4j - 7)$
18. $(5d - 12)(-7 + 3d)$

ERROR ANALYSIS In Exercises 19 and 20, describe and correct the error in finding the product of the binomials.

19. ✗

$$(t - 2)(t + 5) = t - 2(t + 5)$$
$$= t - 2t - 10$$
$$= -t - 10$$

20. ✗

$(x - 5)(3x + 1)$

	$3x$	1
x	$3x^2$	x
5	$15x$	5

$(x - 5)(3x + 1) = 3x^2 + 16x + 5$

In Exercises 21–30, use the FOIL Method to find the product. *(See Example 3.)*

21. $(b + 3)(b + 7)$
22. $(w + 9)(w + 6)$
23. $(k + 5)(k - 1)$
24. $(x - 4)(x + 8)$
25. $\left(q - \frac{3}{4}\right)\left(q + \frac{1}{4}\right)$
26. $\left(z - \frac{5}{3}\right)\left(z - \frac{2}{3}\right)$
27. $(9 - r)(2 - 3r)$
28. $(8 - 4x)(2x + 6)$
29. $(w + 5)(w^2 + 3w)$
30. $(v - 3)(v^2 + 8v)$

MATHEMATICAL CONNECTIONS In Exercises 31–34, write a polynomial that represents the area of the shaded region.

31.

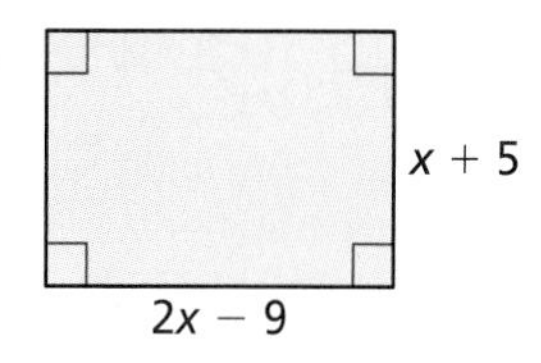

32.

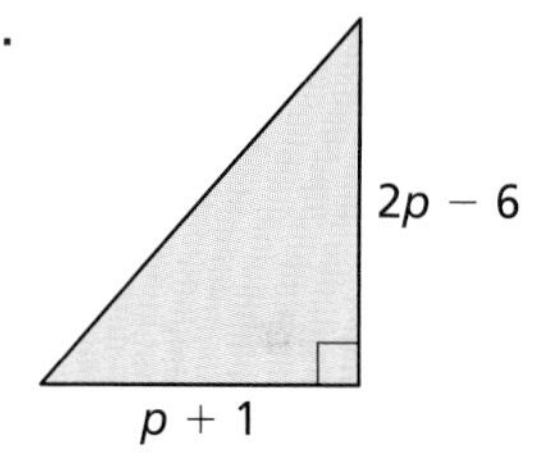

33.

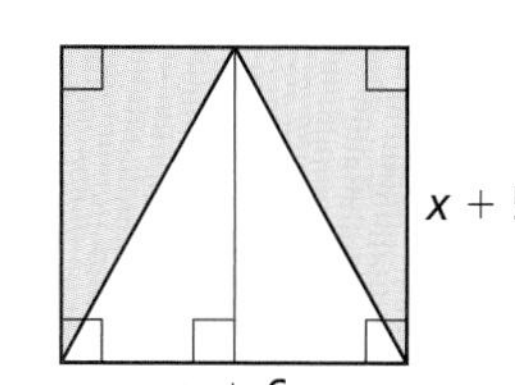

34.

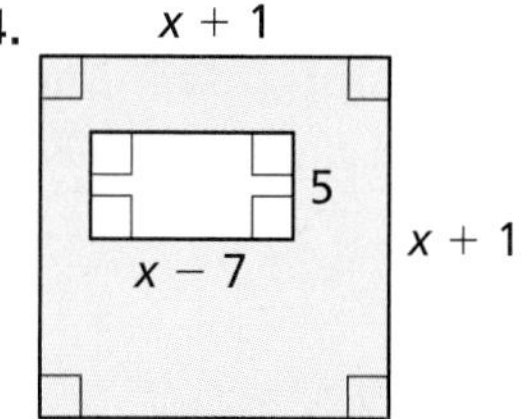

In Exercises 35–42, find the product. *(See Example 4.)*

35. $(x + 4)(x^2 + 3x + 2)$
36. $(f + 1)(f^2 + 4f + 8)$
37. $(y + 3)(y^2 + 8y - 2)$
38. $(t - 2)(t^2 - 5t + 1)$
39. $(4 - b)(5b^2 + 5b - 4)$
40. $(d + 6)(2d^2 - d + 7)$
41. $(3e^2 - 5e + 7)(6e + 1)$
42. $(6v^2 + 2v - 9)(4 - 5v)$

43. **MODELING WITH MATHEMATICS** The football field is rectangular. *(See Example 5.)*

(10x + 10) ft

(4x + 20) ft

a. Write a polynomial that represents the area of the football field.

b. Find the area of the football field when the width is 160 feet.

44. **MODELING WITH MATHEMATICS** You design a frame to surround a rectangular photo. The width of the frame is the same on every side, as shown.

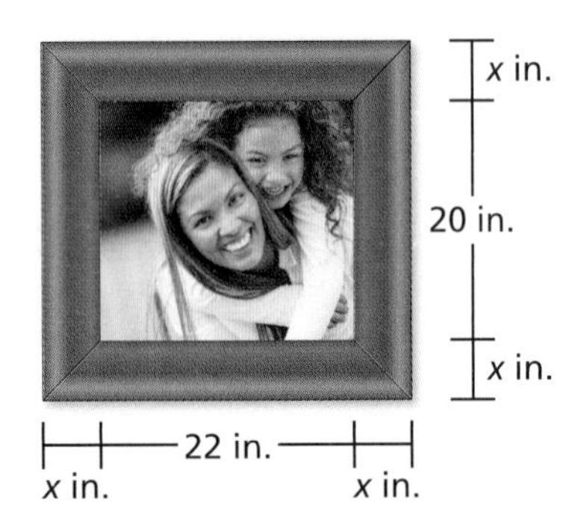

a. Write a polynomial that represents the combined area of the photo and the frame.

b. Find the combined area of the photo and the frame when the width of the frame is 4 inches.

45. **WRITING** When multiplying two binomials, explain how the degree of the product is related to the degree of each binomial.

46. **THOUGHT PROVOKING** Write two polynomials that are not monomials whose product is a trinomial of degree 3.

47. **MAKING AN ARGUMENT** Your friend says the FOIL Method can be used to multiply two trinomials. Is your friend correct? Explain your reasoning.

48. **HOW DO YOU SEE IT?** The table shows one method of finding the product of two binomials.

	$-4x$	3
$-8x$	a	b
-9	c	d

a. Write the two binomials being multiplied.

b. Determine whether a, b, c, and d will be positive or negative when $x > 0$.

49. **COMPARING METHODS** You use the Distributive Property to multiply $(x + 3)(x - 5)$. Your friend uses the FOIL Method to multiply $(x - 5)(x + 3)$. Should your answers be equivalent? Justify your answer.

50. **USING STRUCTURE** The shipping container is a rectangular prism. Write a polynomial that represents the volume of the container.

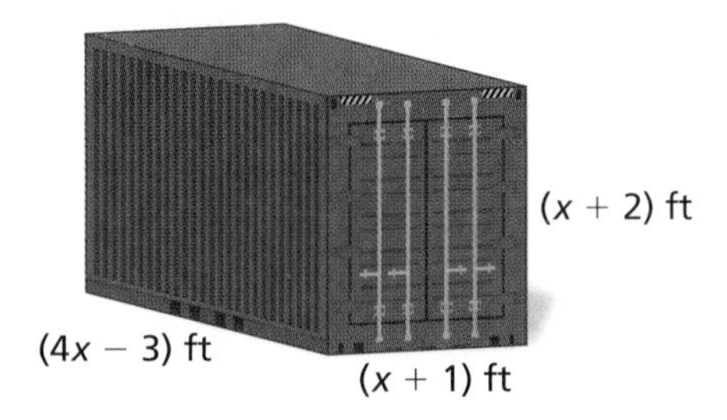

51. **ABSTRACT REASONING** The product of $(x + m)(x + n)$ is $x^2 + bx + c$.

a. What do you know about m and n when $c > 0$?

b. What do you know about m and n when $c < 0$?

Maintaining Mathematical Proficiency

Reviewing what you learned in previous grades and lessons

Write the absolute value function as a piecewise function. *(Section 4.7)*

52. $y = |x| + 4$

53. $y = 6|x - 3|$

54. $y = -4|x + 2|$

Simplify the expression. Write your answer using only positive exponents. *(Section 6.1)*

55. $10^2 \cdot 10^9$

56. $\frac{x^5 \cdot x}{x^8}$

57. $(3z^6)^{-3}$

58. $\left(\frac{2y^4}{y^3}\right)^{-2}$

7.3 Special Products of Polynomials

Essential Question What are the patterns in the special products $(a + b)(a - b)$, $(a + b)^2$, and $(a - b)^2$?

EXPLORATION 1 Finding a Sum and Difference Pattern

Work with a partner. Write the product of two binomials modeled by each rectangular array of algebra tiles.

a. $(x + 2)(x - 2) =$ ______

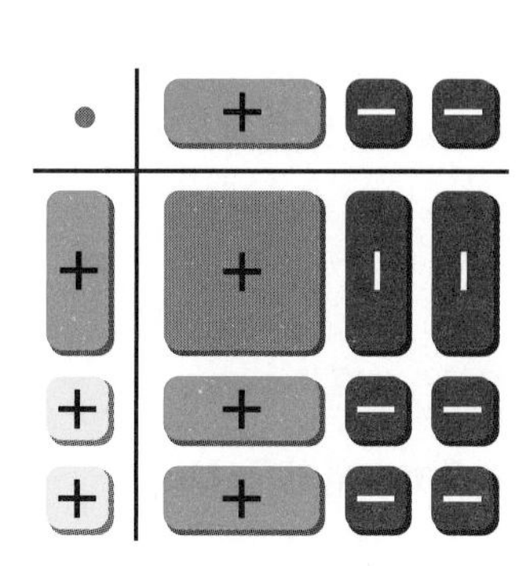

b. $(2x - 1)(2x + 1) =$ ______

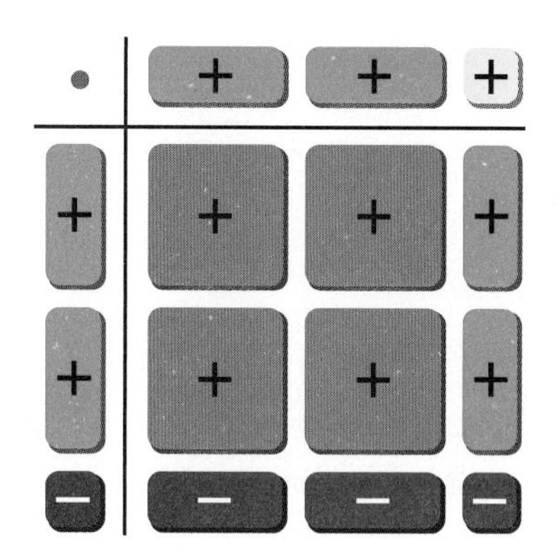

EXPLORATION 2 Finding the Square of a Binomial Pattern

Work with a partner. Draw the rectangular array of algebra tiles that models each product of two binomials. Write the product.

a. $(x + 2)^2 =$ ______

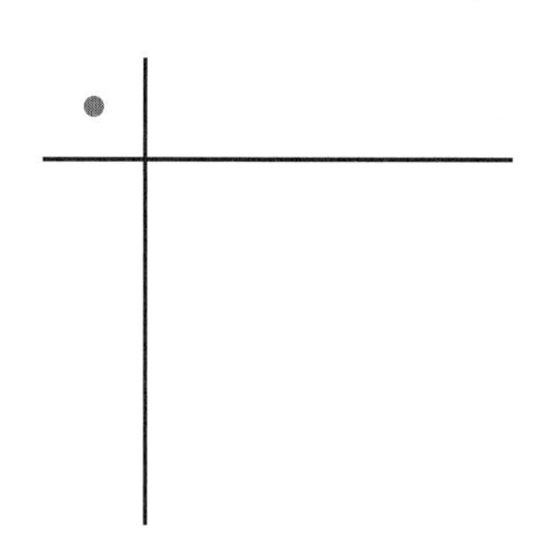

b. $(2x - 1)^2 =$ ______

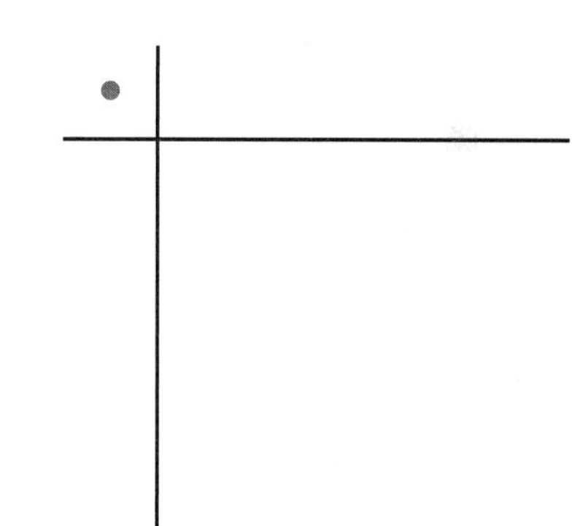

LOOKING FOR STRUCTURE

To be proficient in math, you need to look closely to discern a pattern or structure.

Communicate Your Answer

3. What are the patterns in the special products $(a + b)(a - b)$, $(a + b)^2$, and $(a - b)^2$?

4. Use the appropriate special product pattern to find each product. Check your answers using algebra tiles.

a. $(x + 3)(x - 3)$ **b.** $(x - 4)(x + 4)$ **c.** $(3x + 1)(3x - 1)$

d. $(x + 3)^2$ **e.** $(x - 2)^2$ **f.** $(3x + 1)^2$

7.3 Lesson

What You Will Learn

- Use the square of a binomial pattern.
- Use the sum and difference pattern.
- Use special product patterns to solve real-life problems.

Core Vocabulary

Previous
binomial

Using the Square of a Binomial Pattern

The diagram shows a square with a side length of $(a + b)$ units. You can see that the area of the square is

$$(a + b)^2 = a^2 + 2ab + b^2.$$

	a	b
a	a^2	ab
b	ab	b^2

This is one version of a pattern called the square of a binomial. To find another version of this pattern, use algebra: replace b with $-b$.

$(a + (-b))^2 = a^2 + 2a(-b) + (-b)^2$ — Replace b with $-b$ in the pattern above.

$(a - b)^2 = a^2 - 2ab + b^2$ — Simplify.

Core Concept

Square of a Binomial Pattern

Algebra	Example
$(a + b)^2 = a^2 + 2ab + b^2$	$(x + 5)^2 = (x)^2 + 2(x)(5) + (5)^2$ $= x^2 + 10x + 25$
$(a - b)^2 = a^2 - 2ab + b^2$	$(2x - 3)^2 = (2x)^2 - 2(2x)(3) + (3)^2$ $= 4x^2 - 12x + 9$

LOOKING FOR STRUCTURE

When you use special product patterns, remember that a and b can be numbers, variables, or variable expressions.

EXAMPLE 1 Using the Square of a Binomial Pattern

Find each product.

a. $(3x + 4)^2$

b. $(5x - 2y)^2$

SOLUTION

a. $(3x + 4)^2 = (3x)^2 + 2(3x)(4) + 4^2$ — Square of a binomial pattern

$= 9x^2 + 24x + 16$ — Simplify.

▶ The product is $9x^2 + 24x + 16$.

b. $(5x - 2y)^2 = (5x)^2 - 2(5x)(2y) + (2y)^2$ — Square of a binomial pattern

$= 25x^2 - 20xy + 4y^2$ — Simplify.

▶ The product is $25x^2 - 20xy + 4y^2$.

Monitoring Progress

Help in English and Spanish at *BigIdeasMath.com*

Find the product.

1. $(x + 7)^2$ **2.** $(7x - 3)^2$ **3.** $(4x - y)^2$ **4.** $(3m + n)^2$

Using the Sum and Difference Pattern

To find the product $(x + 2)(x - 2)$, you can multiply the two binomials using the FOIL Method.

$(x + 2)(x - 2) = x^2 - 2x + 2x - 4$	FOIL Method
$= x^2 - 4$	Combine like terms.

This suggests a pattern for the product of the sum and difference of two terms.

Core Concept

Sum and Difference Pattern

Algebra	**Example**
$(a + b)(a - b) = a^2 - b^2$	$(x + 3)(x - 3) = x^2 - 9$

EXAMPLE 2 Using the Sum and Difference Pattern

Find each product.

a. $(t + 5)(t - 5)$

b. $(3x + y)(3x - y)$

SOLUTION

a. $(t + 5)(t - 5) = t^2 - 5^2$	Sum and difference pattern
$= t^2 - 25$	Simplify.

▶ The product is $t^2 - 25$.

b. $(3x + y)(3x - y) = (3x)^2 - y^2$	Sum and difference pattern
$= 9x^2 - y^2$	Simplify.

▶ The product is $9x^2 - y^2$.

The special product patterns can help you use mental math to find certain products of numbers.

EXAMPLE 3 Using Special Product Patterns and Mental Math

Use special product patterns to find the product $26 \cdot 34$.

SOLUTION

Notice that 26 is 4 less than 30, while 34 is 4 more than 30.

$26 \cdot 34 = (30 - 4)(30 + 4)$	Write as product of difference and sum.
$= 30^2 - 4^2$	Sum and difference pattern
$= 900 - 16$	Evaluate powers.
$= 884$	Simplify.

▶ The product is 884.

Monitoring Progress

Help in English and Spanish at *BigIdeasMath.com*

Find the product.

5. $(x + 10)(x - 10)$

6. $(2x + 1)(2x - 1)$

7. $(x + 3y)(x - 3y)$

8. Describe how to use special product patterns to find 21^2.

Solving Real-Life Problems

EXAMPLE 4 **Modeling with Mathematics**

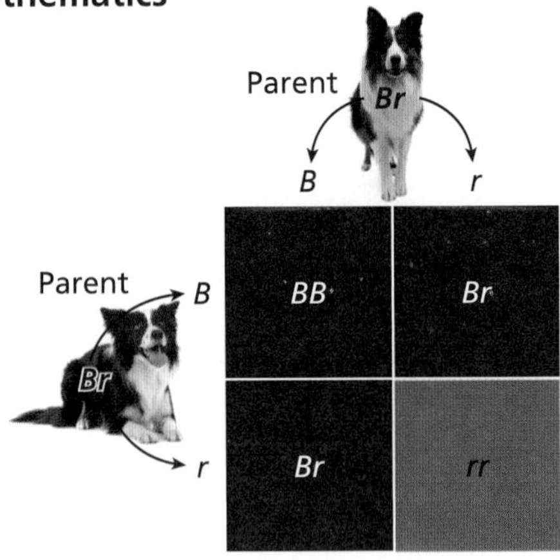

A combination of two genes determines the color of the dark patches of a border collie's coat. An offspring inherits one patch color gene from each parent. Each parent has two color genes, and the offspring has an equal chance of inheriting either one.

The gene B is for black patches, and the gene r is for red patches. Any gene combination with a B results in black patches. Suppose each parent has the same gene combination Br. The Punnett square shows the possible gene combinations of the offspring and the resulting patch colors.

a. What percent of the possible gene combinations result in black patches?

b. Show how you could use a polynomial to model the possible gene combinations.

SOLUTION

a. Notice that the Punnett square shows four possible gene combinations of the offspring. Of these combinations, three result in black patches.

▶ So, 75% of the possible gene combinations result in black patches.

b. Model the gene from each parent with $0.5B + 0.5r$. There is an equal chance that the offspring inherits a black or a red gene from each parent.

You can model the possible gene combinations of the offspring with $(0.5B + 0.5r)^2$. Notice that this product also represents the area of the Punnett square.

Expand the product to find the possible patch colors of the offspring.

$$(0.5B + 0.5r)^2 = (0.5B)^2 + 2(0.5B)(0.5r) + (0.5r)^2$$
$$= 0.25B^2 + 0.5Br + 0.25r^2$$

Consider the coefficients in the polynomial.

$$0.25B^2 + 0.5Br + 0.25r^2$$

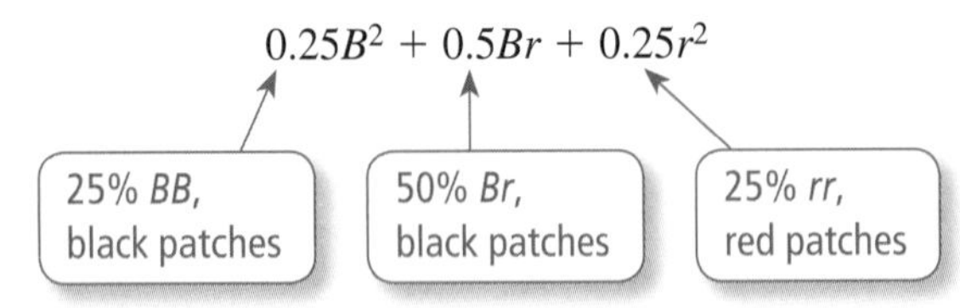

The coefficients show that 25% + 50% = 75% of the possible gene combinations result in black patches.

Monitoring Progress  Help in English and Spanish at *BigIdeasMath.com*

9. Each of two dogs has one black gene (B) and one white gene (W). The Punnett square shows the possible gene combinations of an offspring and the resulting colors.

	B	*W*
B	*BB* black	*BW* gray
W	*BW* gray	*WW* white

a. What percent of the possible gene combinations result in black?

b. Show how you could use a polynomial to model the possible gene combinations of the offspring.

7.3 Exercises

Dynamic Solutions available at *BigIdeasMath.com*

Vocabulary and Core Concept Check

1. **WRITING** Explain how to use the square of a binomial pattern.

2. **WHICH ONE DOESN'T BELONG?** Which expression does *not* belong with the other three? Explain your reasoning.

$(x + 1)(x - 1)$ $(3x + 2)(3x - 2)$ $(x + 2)(x - 3)$ $(2x + 5)(2x - 5)$

Monitoring Progress and Modeling with Mathematics

In Exercises 3–10, find the product. *(See Example 1.)*

3. $(x + 8)^2$
4. $(a - 6)^2$
5. $(2f - 1)^2$
6. $(5p + 2)^2$
7. $(-7t + 4)^2$
8. $(-12 - n)^2$
9. $(2a + b)^2$
10. $(6x - 3y)^2$

MATHEMATICAL CONNECTIONS In Exercises 11–14, write a polynomial that represents the area of the square.

11. 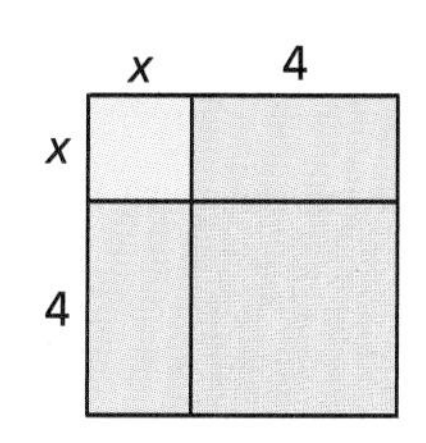

12. 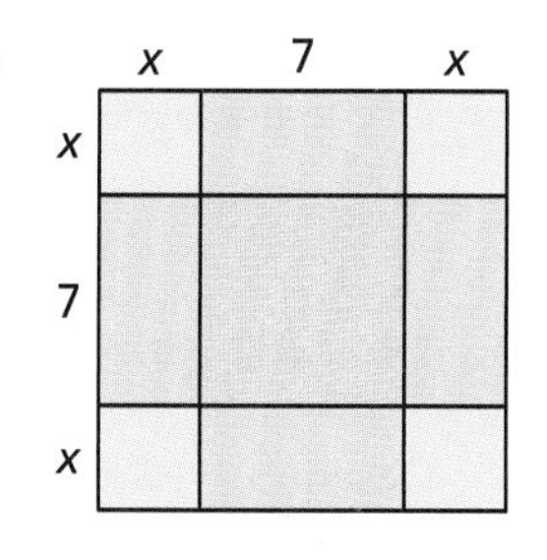

13. 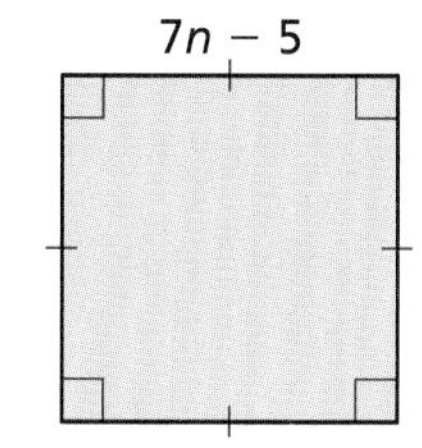

14. 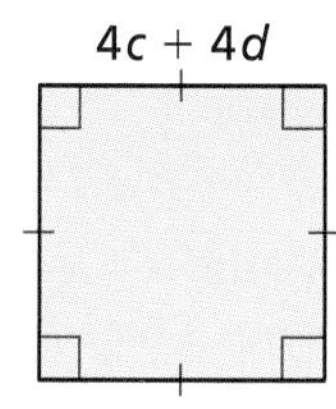

In Exercises 15–24, find the product. *(See Example 2.)*

15. $(t - 7)(t + 7)$
16. $(m + 6)(m - 6)$
17. $(4x + 1)(4x - 1)$
18. $(2k - 4)(2k + 4)$
19. $(8 + 3a)(8 - 3a)$
20. $\left(\frac{1}{2} - c\right)\left(\frac{1}{2} + c\right)$
21. $(p - 10q)(p + 10q)$
22. $(7m + 8n)(7m - 8n)$
23. $(-y + 4)(-y - 4)$
24. $(-5g - 2h)(-5g + 2h)$

In Exercises 25–30, use special product patterns to find the product. *(See Example 3.)*

25. $16 \cdot 24$
26. $33 \cdot 27$
27. 42^2
28. 29^2
29. 30.5^2
30. $10\frac{1}{3} \cdot 9\frac{2}{3}$

ERROR ANALYSIS In Exercises 31 and 32, describe and correct the error in finding the product.

31. 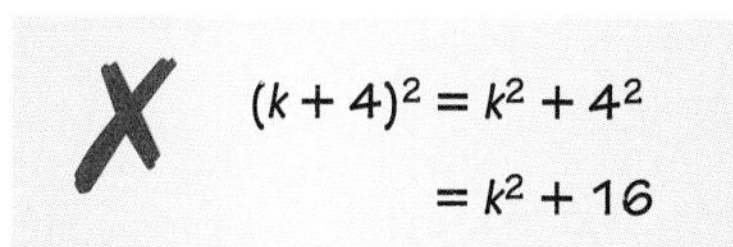

32. 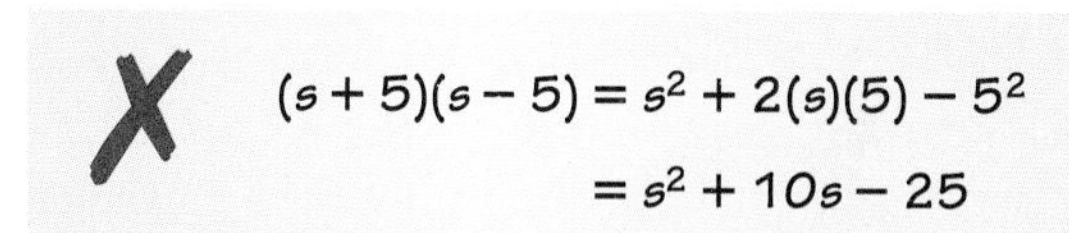

33. **MODELING WITH MATHEMATICS** A contractor extends a house on two sides.

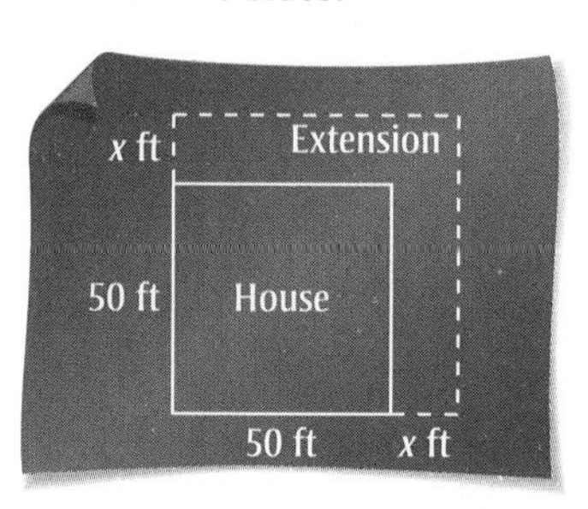

a. The area of the house after the renovation is represented by $(x + 50)^2$. Find this product.

b. Use the polynomial in part (a) to find the area when $x = 15$. What is the area of the extension?

34. MODELING WITH MATHEMATICS A square-shaped parking lot with 100-foot sides is reduced by x feet on one side and extended by x feet on an adjacent side.

a. The area of the new parking lot is represented by $(100 - x)(100 + x)$. Find this product.

b. Does the area of the parking lot increase, decrease, or stay the same? Explain.

c. Use the polynomial in part (a) to find the area of the new parking lot when $x = 21$.

35. MODELING WITH MATHEMATICS In deer, the gene N is for normal coloring and the gene a is for no coloring, or albino. Any gene combination with an N results in normal coloring. The Punnett square shows the possible gene combinations of an offspring and the resulting colors from parents that both have the gene combination Na. *(See Example 4.)*

a. What percent of the possible gene combinations result in albino coloring?

b. Show how you could use a polynomial to model the possible gene combinations of the offspring.

Parent A: *Na* → *N*, *a*; Parent B: *Na* → *N*, *a*

	N	*a*
N	*NN* normal	*Na* normal
a	*Na* normal	*aa* albino

36. MODELING WITH MATHEMATICS Your iris controls the amount of light that enters your eye by changing the size of your pupil.

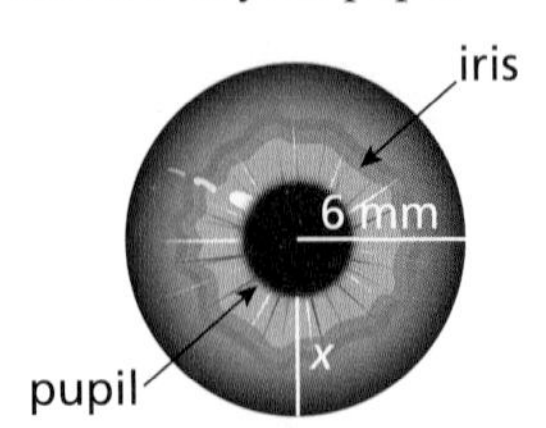

a. Write a polynomial that represents the area of your pupil. Write your answer in terms of π.

b. The width x of your iris decreases from 4 millimeters to 2 millimeters when you enter a dark room. How many times greater is the area of your pupil after entering the room than before entering the room? Explain.

37. CRITICAL THINKING Write two binomials that have the product $x^2 - 121$. Explain.

38. HOW DO YOU SEE IT? In pea plants, any gene combination with a green gene (G) results in a green pod. The Punnett square shows the possible gene combinations of the offspring of two Gy pea plants and the resulting pod colors.

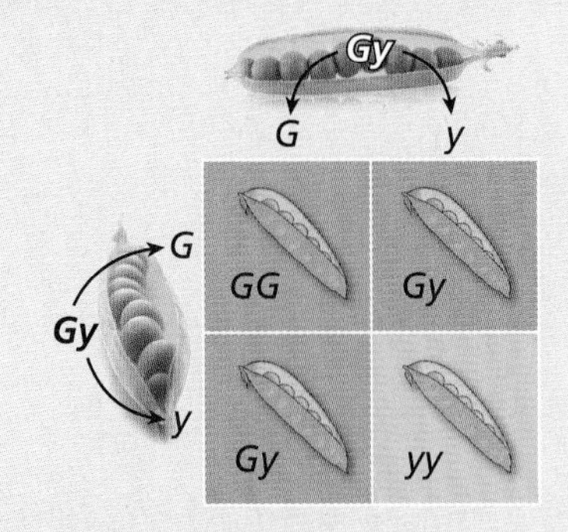

G = Green gene
y = Yellow gene

A polynomial that models the possible gene combinations of the offspring is

$$(0.5G + 0.5y)^2 = 0.25G^2 + 0.5Gy + 0.25y^2.$$

Describe two ways to determine the percent of possible gene combinations that result in green pods.

In Exercises 39–42, find the product.

39. $(x^2 + 1)(x^2 - 1)$

40. $(y^3 + 4)^2$

41. $(2m^2 - 5n^2)^2$

42. $(r^3 - 6t^4)(r^3 + 6t^4)$

43. MAKING AN ARGUMENT Your friend claims to be able to use a special product pattern to determine that $\left(4\frac{1}{3}\right)^2$ is equal to $16\frac{1}{9}$. Is your friend correct? Explain.

44. THOUGHT PROVOKING The area (in square meters) of the surface of an artificial lake is represented by x^2. Describe three ways to modify the dimensions of the lake so that the new area can be represented by the three types of special product patterns discussed in this section.

45. REASONING Find k so that $9x^2 - 48x + k$ is the square of a binomial.

46. REPEATED REASONING Find $(x + 1)^3$ and $(x + 2)^3$. Find a pattern in the terms and use it to write a pattern for the cube of a binomial $(a + b)^3$.

47. PROBLEM SOLVING Find two numbers a and b such that $(a + b)(a - b) < (a - b)^2 < (a + b)^2$.

Maintaining Mathematical Proficiency

Reviewing what you learned in previous grades and lessons

Factor the expression using the GCF. *(Skills Review Handbook)*

48. $12y - 18$

49. $9r + 27$

50. $49s + 35t$

51. $15x - 10y$

7.4 Solving Polynomial Equations in Factored Form

Essential Question How can you solve a polynomial equation?

EXPLORATION 1 Matching Equivalent Forms of an Equation

Work with a partner. An equation is considered to be in *factored form* when the product of the factors is equal to 0. Match each factored form of the equation with its equivalent standard form and nonstandard form.

	Factored Form		Standard Form		Nonstandard Form
a.	$(x-1)(x-3)=0$	**A.**	$x^2-x-2=0$	**1.**	$x^2-5x=-6$
b.	$(x-2)(x-3)=0$	**B.**	$x^2+x-2=0$	**2.**	$(x-1)^2=4$
c.	$(x+1)(x-2)=0$	**C.**	$x^2-4x+3=0$	**3.**	$x^2-x=2$
d.	$(x-1)(x+2)=0$	**D.**	$x^2-5x+6=0$	**4.**	$x(x+1)=2$
e.	$(x+1)(x-3)=0$	**E.**	$x^2-2x-3=0$	**5.**	$x^2-4x=-3$

USING TOOLS STRATEGICALLY

To be proficient in math, you need to consider using tools such as a table or a spreadsheet to organize your results.

EXPLORATION 2 Writing a Conjecture

Work with a partner. Substitute 1, 2, 3, 4, 5, and 6 for x in each equation and determine whether the equation is true. Organize your results in a table. Write a conjecture describing what you discovered.

a. $(x-1)(x-2)=0$

b. $(x-2)(x-3)=0$

c. $(x-3)(x-4)=0$

d. $(x-4)(x-5)=0$

e. $(x-5)(x-6)=0$

f. $(x-6)(x-1)=0$

EXPLORATION 3 Special Properties of 0 and 1

Work with a partner. The numbers 0 and 1 have special properties that are shared by no other numbers. For each of the following, decide whether the property is true for 0, 1, both, or neither. Explain your reasoning.

a. When you add ___ to a number n, you get n.

b. If the product of two numbers is ___, then at least one of the numbers is 0.

c. The square of ___ is equal to itself.

d. When you multiply a number n by ___, you get n.

e. When you multiply a number n by ___, you get 0.

f. The opposite of ___ is equal to itself.

Communicate Your Answer

4. How can you solve a polynomial equation?

5. One of the properties in Exploration 3 is called the Zero-Product Property. It is one of the most important properties in all of algebra. Which property is it? Why do you think it is called the Zero-Product Property? Explain how it is used in algebra and why it is so important.

7.4 Lesson

Core Vocabulary

factored form, *p. 378*
Zero-Product Property, *p. 378*
roots, *p. 378*
repeated roots, *p. 379*

Previous
polynomial
standard form
greatest common factor (GCF)
monomial

What You Will Learn

- Use the Zero-Product Property.
- Factor polynomials using the GCF.
- Use the Zero-Product Property to solve real-life problems.

Using the Zero-Product Property

A polynomial is in **factored form** when it is written as a product of factors.

Standard form	Factored form
$x^2 + 2x$	$x(x + 2)$
$x^2 + 5x - 24$	$(x - 3)(x + 8)$

When one side of an equation is a polynomial in factored form and the other side is 0, use the **Zero-Product Property** to solve the polynomial equation. The solutions of a polynomial equation are also called **roots**.

Core Concept

Zero-Product Property

Words If the product of two real numbers is 0, then at least one of the numbers is 0.

Algebra If a and b are real numbers and $ab = 0$, then $a = 0$ or $b = 0$.

EXAMPLE 1 Solving Polynomial Equations

Solve each equation.

a. $2x(x - 4) = 0$ **b.** $(x - 3)(x - 9) = 0$

SOLUTION

a. $2x(x - 4) = 0$ Write equation.

$2x = 0$ *or* $x - 4 = 0$ Zero-Product Property

$x = 0$ *or* $x = 4$ Solve for *x*.

▶ The roots are $x = 0$ and $x = 4$.

b. $(x - 3)(x - 9) = 0$ Write equation.

$x - 3 = 0$ *or* $x - 9 = 0$ Zero-Product Property

$x = 3$ *or* $x = 9$ Solve for *x*.

▶ The roots are $x = 3$ and $x = 9$.

Check

To check the solutions of Example 1(a), substitute each solution in the original equation.

$$2(0)(0 - 4) \stackrel{?}{=} 0$$
$$0(-4) \stackrel{?}{=} 0$$
$$0 = 0 \checkmark$$
$$2(4)(4 - 4) \stackrel{?}{=} 0$$
$$8(0) \stackrel{?}{=} 0$$
$$0 = 0 \checkmark$$

Monitoring Progress

Help in English and Spanish at *BigIdeasMath.com*

Solve the equation. Check your solutions.

1. $x(x - 1) = 0$
2. $3t(t + 2) = 0$
3. $(z - 4)(z - 6) = 0$

When two or more roots of an equation are the same number, the equation has **repeated roots**.

EXAMPLE 2 Solving Polynomial Equations

Solve each equation.

a. $(2x + 7)(2x - 7) = 0$ **b.** $(x - 1)^2 = 0$ **c.** $(x + 1)(x - 3)(x - 2) = 0$

SOLUTION

a. $(2x + 7)(2x - 7) = 0$ Write equation.

$2x + 7 = 0$ *or* $2x - 7 = 0$ Zero-Product Property

$x = -\frac{7}{2}$ *or* $x = \frac{7}{2}$ Solve for x.

▶ The roots are $x = -\frac{7}{2}$ and $x = \frac{7}{2}$.

b. $(x - 1)^2 = 0$ Write equation.

$(x - 1)(x - 1) = 0$ Expand equation.

$x - 1 = 0$ *or* $x - 1 = 0$ Zero-Product Property

$x = 1$ *or* $x = 1$ Solve for x.

▶ The equation has repeated roots of $x = 1$.

STUDY TIP

You can extend the Zero-Product Property to products of more than two real numbers.

c. $(x + 1)(x - 3)(x - 2) = 0$ Write equation.

$x + 1 = 0$ *or* $x - 3 = 0$ *or* $x - 2 = 0$ Zero-Product Property

$x = -1$ *or* $x = 3$ *or* $x = 2$ Solve for x.

▶ The roots are $x = -1$, $x = 3$, and $x = 2$.

Monitoring Progress  Help in English and Spanish at *BigIdeasMath.com*

Solve the equation. Check your solutions.

4. $(3s + 5)(5s + 8) = 0$ **5.** $(b + 7)^2 = 0$ **6.** $(d - 2)(d + 6)(d + 8) = 0$

Factoring Polynomials Using the GCF

To solve a polynomial equation using the Zero-Product Property, you may need to *factor* the polynomial, or write it as a product of other polynomials. Look for the *greatest common factor* (GCF) of the terms of the polynomial. This is a monomial that divides evenly into each term.

EXAMPLE 3 Finding the Greatest Common Monomial Factor

Factor out the greatest common monomial factor from $4x^4 + 24x^3$.

SOLUTION

The GCF of 4 and 24 is 4. The GCF of x^4 and x^3 is x^3. So, the greatest common monomial factor of the terms is $4x^3$.

▶ So, $4x^4 + 24x^3 = 4x^3(x + 6)$.

Monitoring Progress  Help in English and Spanish at *BigIdeasMath.com*

7. Factor out the greatest common monomial factor from $8y^2 - 24y$.

EXAMPLE 4 Solving Equations by Factoring

Solve (a) $2x^2 + 8x = 0$ and (b) $6n^2 = 15n$.

SOLUTION

a.

$2x^2 + 8x = 0$	Write equation.
$2x(x + 4) = 0$	Factor left side.
$2x = 0$ *or* $x + 4 = 0$	Zero-Product Property
$x = 0$ *or* $x = -4$	Solve for x.

▶ The roots are $x = 0$ and $x = -4$.

b.

$6n^2 = 15n$	Write equation.
$6n^2 - 15n = 0$	Subtract $15n$ from each side.
$3n(2n - 5) = 0$	Factor left side.
$3n = 0$ *or* $2n - 5 = 0$	Zero-Product Property
$n = 0$ *or* $n = \frac{5}{2}$	Solve for n.

▶ The roots are $n = 0$ and $n = \frac{5}{2}$.

Monitoring Progress

Help in English and Spanish at *BigIdeasMath.com*

Solve the equation. Check your solutions.

8. $a^2 + 5a = 0$ **9.** $3s^2 - 9s = 0$ **10.** $4x^2 = 2x$

Solving Real-Life Problems

EXAMPLE 5 Modeling with Mathematics

You can model the arch of a fireplace using the equation $y = -\frac{1}{9}(x + 18)(x - 18)$, where x and y are measured in inches. The x-axis represents the floor. Find the width of the arch at floor level.

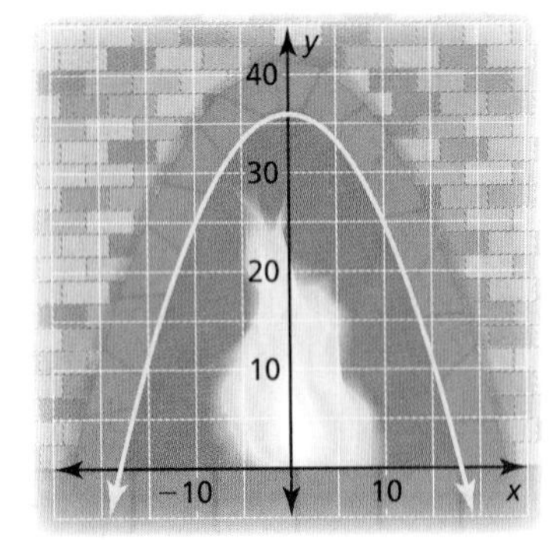

SOLUTION

Use the x-coordinates of the points where the arch meets the floor to find the width. At floor level, $y = 0$. So, substitute 0 for y and solve for x.

$y = -\frac{1}{9}(x + 18)(x - 18)$	Write equation.
$0 = -\frac{1}{9}(x + 18)(x - 18)$	Substitute 0 for y.
$0 = (x + 18)(x - 18)$	Multiply each side by -9.
$x + 18 = 0$ *or* $x - 18 = 0$	Zero-Product Property
$x = -18$ *or* $x = 18$	Solve for x.

The width is the distance between the x-coordinates, -18 and 18.

▶ So, the width of the arch at floor level is $|-18 - 18| = 36$ inches.

Monitoring Progress

Help in English and Spanish at *BigIdeasMath.com*

11. You can model the entrance to a mine shaft using the equation $y = -\frac{1}{2}(x + 4)(x - 4)$, where x and y are measured in feet. The x-axis represents the ground. Find the width of the entrance at ground level.

Dynamic Solutions available at *BigIdeasMath.com*

Vocabulary and Core Concept Check

1. **WRITING** Explain how to use the Zero-Product Property to find the solutions of the equation $3x(x - 6) = 0$.

2. **DIFFERENT WORDS, SAME QUESTION** Which is different? Find *both* answers.

Solve the equation $(2k + 4)(k - 3) = 0$.	Find the values of k for which $2k + 4 = 0$ *or* $k - 3 = 0$.
Find the value of k for which $(2k + 4) + (k - 3) = 0$.	Find the roots of the equation $(2k + 4)(k - 3) = 0$.

Monitoring Progress and Modeling with Mathematics

In Exercises 3–8, solve the equation. *(See Example 1.)*

3. $x(x + 7) = 0$
4. $r(r - 10) = 0$
5. $12t(t - 5) = 0$
6. $-2v(v + 1) = 0$
7. $(s - 9)(s - 1) = 0$
8. $(y + 2)(y - 6) = 0$

In Exercises 9–20, solve the equation. *(See Example 2.)*

9. $(2a - 6)(3a + 15) = 0$
10. $(4q + 3)(q + 2) = 0$
11. $(5m + 4)^2 = 0$
12. $(h - 8)^2 = 0$
13. $(3 - 2g)(7 - g) = 0$
14. $(2 - 4d)(2 + 4d) = 0$
15. $z(z + 2)(z - 1) = 0$
16. $5p(2p - 3)(p + 7) = 0$
17. $(r - 4)^2(r + 8) = 0$
18. $w(w - 6)^2 = 0$
19. $(15 - 5c)(5c + 5)(-c + 6) = 0$
20. $(2 - n)\left(6 + \frac{2}{3}n\right)(n - 2) = 0$

In Exercises 21–24, find the x-coordinates of the points where the graph crosses the x-axis.

21.

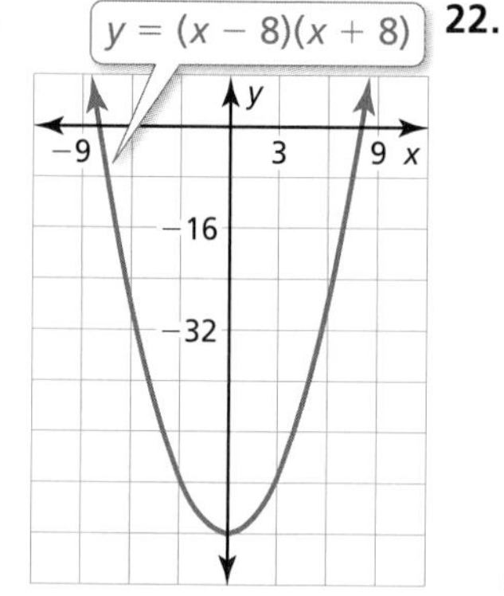

22.

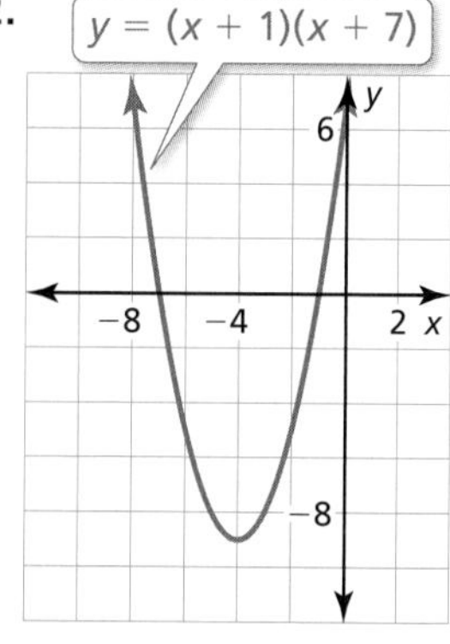

23.

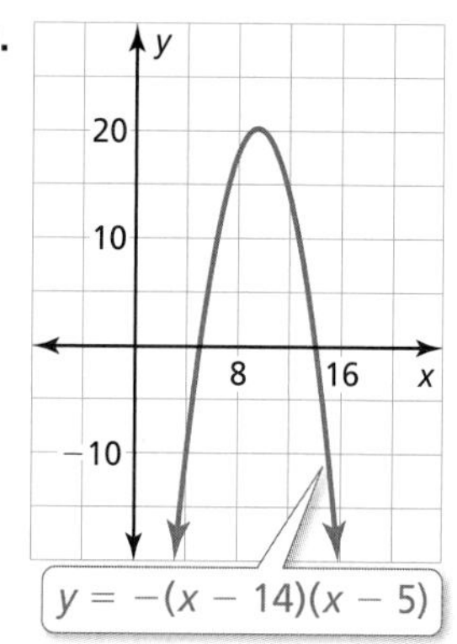

24.

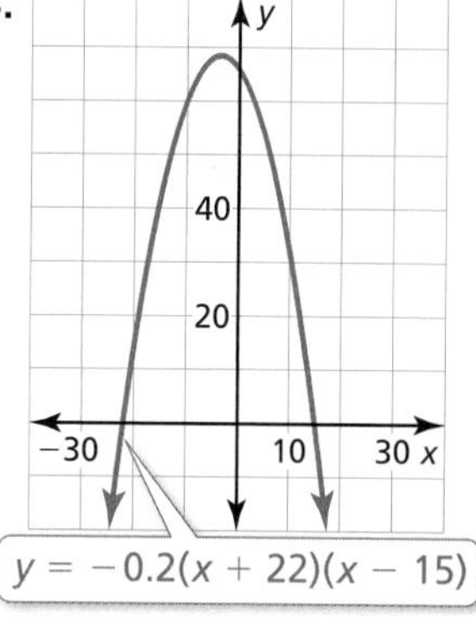

In Exercises 25–30, factor the polynomial. *(See Example 3.)*

25. $5z^2 + 45z$
26. $6d^2 - 21d$
27. $3y^3 - 9y^2$
28. $20x^3 + 30x^2$
29. $5n^6 + 2n^5$
30. $12a^4 + 8a$

In Exercises 31–36, solve the equation. *(See Example 4.)*

31. $4p^2 - p = 0$
32. $6m^2 + 12m = 0$
33. $25c + 10c^2 = 0$
34. $18q - 2q^2 = 0$
35. $3n^2 = 9n$
36. $-28r = 4r^2$

37. **ERROR ANALYSIS** Describe and correct the error in solving the equation.

$6x(x + 5) = 0$

$x + 5 = 0$

$x = -5$

The root is $x = -5$.

38. ERROR ANALYSIS Describe and correct the error in solving the equation.

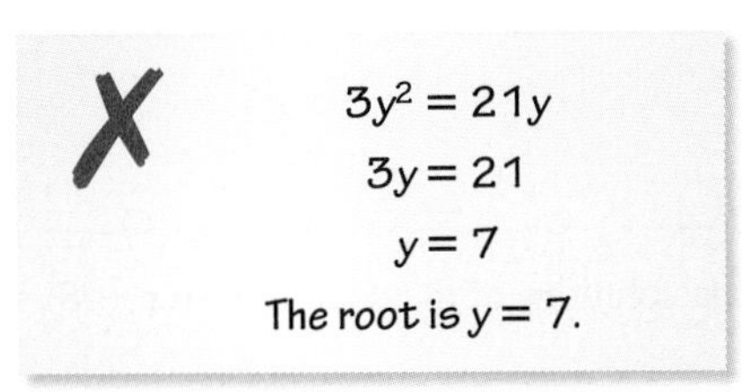

39. MODELING WITH MATHEMATICS The entrance of a tunnel can be modeled by $y = -\frac{11}{50}(x - 4)(x - 24)$, where x and y are measured in feet. The x-axis represents the ground. Find the width of the tunnel at ground level. *(See Example 5.)*

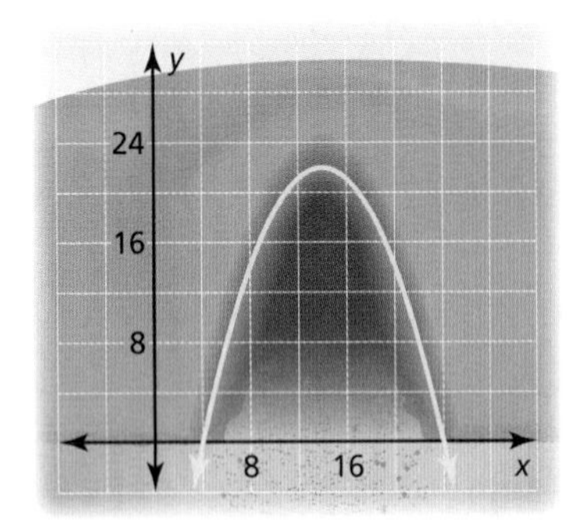

40. MODELING WITH MATHEMATICS The Gateway Arch in St. Louis can be modeled by $y = -\frac{2}{315}(x + 315)(x - 315)$, where x and y are measured in feet. The x-axis represents the ground.

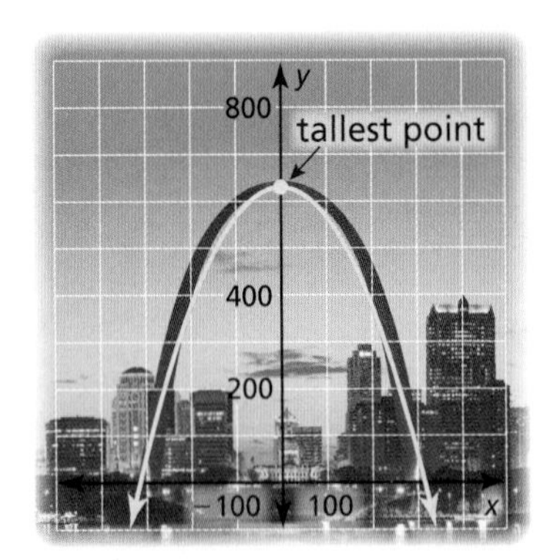

a. Find the width of the arch at ground level.

b. How tall is the arch?

41. MODELING WITH MATHEMATICS A penguin leaps out of the water while swimming. This action is called porpoising. The height y (in feet) of a porpoising penguin can be modeled by $y = -16x^2 + 4.8x$, where x is the time (in seconds) since the penguin leaped out of the water. Find the roots of the equation when $y = 0$. Explain what the roots mean in this situation.

42. HOW DO YOU SEE IT? Use the graph to fill in each blank in the equation with the symbol + or −. Explain your reasoning.

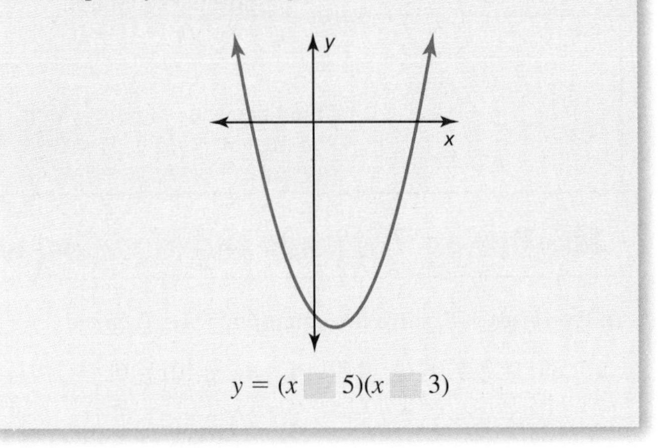

$y = (x \;\_\; 5)(x \;\_\; 3)$

43. CRITICAL THINKING How many x-intercepts does the graph of $y = (2x + 5)(x - 9)^2$ have? Explain.

44. MAKING AN ARGUMENT Your friend says that the graph of the equation $y = (x - a)(x - b)$ always has two x-intercepts for any values of a and b. Is your friend correct? Explain.

45. CRITICAL THINKING Does the equation $(x^2 + 3)(x^4 + 1) = 0$ have any real roots? Explain.

46. THOUGHT PROVOKING Write a polynomial equation of degree 4 whose only roots are $x = 1$, $x = 2$, and $x = 3$.

47. REASONING Find the values of x in terms of y that are solutions of each equation.

a. $(x + y)(2x - y) = 0$

b. $(x^2 - y^2)(4x + 16y) = 0$

48. PROBLEM SOLVING Solve the equation $(4^{x-5} - 16)(3^x - 81) = 0$.

Maintaining Mathematical Proficiency

Reviewing what you learned in previous grades and lessons

List the factor pairs of the number. *(Skills Review Handbook)*

49. 10

50. 18

51. 30

52. 48

7.1–7.4 What Did You Learn?

Core Vocabulary

monomial, *p. 358*
degree of a monomial, *p. 358*
polynomial, *p. 359*
binomial, *p. 359*
trinomial, *p. 359*
degree of a polynomial, *p. 359*
standard form, *p. 359*
leading coefficient, *p. 359*
closed, *p. 360*
FOIL Method, *p. 367*
factored form, *p. 378*
Zero-Product Property, *p. 378*
roots, *p. 378*
repeated roots, *p. 379*

Core Concepts

Section 7.1

Polynomials, *p. 359*
Adding Polynomials, *p. 360*
Subtracting Polynomials, *p. 360*

Section 7.2

Multiplying Binomials, *p. 366*
FOIL Method, *p. 367*
Multiplying Binomials and Trinomials, *p. 368*

Section 7.3

Square of a Binomial Pattern, *p. 372*
Sum and Difference Pattern, *p. 373*

Section 7.4

Zero-Product Property, *p. 378*
Factoring Polynomials Using the GCF, *p. 379*

Mathematical Practices

1. Explain how you wrote the polynomial in Exercise 11 on page 375. Is there another method you can use to write the same polynomial?
2. Find a shortcut for exercises like Exercise 7 on page 381 when the variable has a coefficient of 1. Does your shortcut work when the coefficient is *not* 1?

Study Skills

Preparing for a Test

- Review examples of each type of problem that could appear on the test.
- Review the homework problems your teacher assigned.
- Take a practice test.

7.1–7.4 Quiz

Write the polynomial in standard form. Identify the degree and leading coefficient of the polynomial. Then classify the polynomial by the number of terms. *(Section 7.1)*

1. $-8q^3$

2. $9 + d^2 - 3d$

3. $\frac{2}{3}m^4 - \frac{5}{6}m^6$

4. $-1.3z + 3z^4 + 7.4z^2$

Find the sum or difference. *(Section 7.1)*

5. $(2x^2 + 5) + (-x^2 + 4)$

6. $(-3n^2 + n) - (2n^2 - 7)$

7. $(-p^2 + 4p) - (p^2 - 3p + 15)$

8. $(a^2 - 3ab + b^2) + (-a^2 + ab + b^2)$

Find the product. *(Section 7.2 and Section 7.3)*

9. $(w + 6)(w + 7)$

10. $(3 - 4d)(2d - 5)$

11. $(y + 9)(y^2 + 2y - 3)$

12. $(3z - 5)(3z + 5)$

13. $(t + 5)^2$

14. $(2q - 6)^2$

Solve the equation. *(Section 7.4)*

15. $5x^2 - 15x = 0$

16. $(8 - g)(8 - g) = 0$

17. $(3p + 7)(3p - 7)(p + 8) = 0$

18. $-3y(y - 8)(2y + 1) = 0$

19. You are making a blanket with a fringe border of equal width on each side. *(Section 7.1 and Section 7.2)*

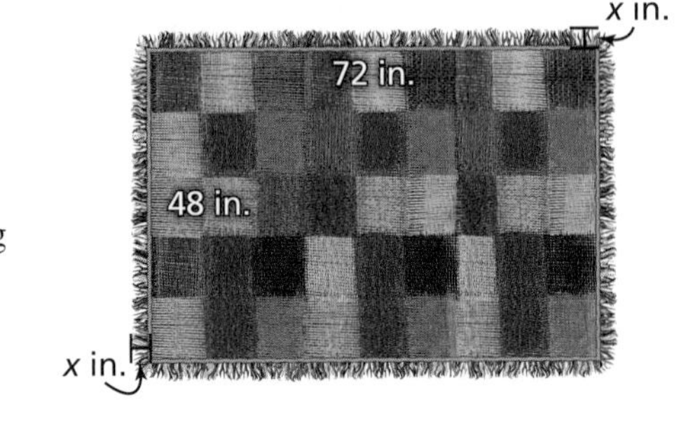

a. Write a polynomial that represents the perimeter of the blanket including the fringe.

b. Write a polynomial that represents the area of the blanket including the fringe.

c. Find the perimeter and the area of the blanket including the fringe when the width of the fringe is 4 inches.

20. You are saving money to buy an electric guitar. You deposit $1000 in an account that earns interest compounded annually. The expression $1000(1 + r)^2$ represents the balance after 2 years, where r is the annual interest rate in decimal form. *(Section 7.3)*

a. Write the polynomial in standard form that represents the balance of your account after 2 years.

b. The interest rate is 3%. What is the balance of your account after 2 years?

c. The guitar costs $1100. Do you have enough money in your account *after 3 years*? Explain.

21. The front of a storage bunker can be modeled by $y = -\frac{5}{216}(x - 72)(x + 72)$, where x and y are measured in inches. The x-axis represents the ground. Find the width of the bunker at ground level. *(Section 7.4)*

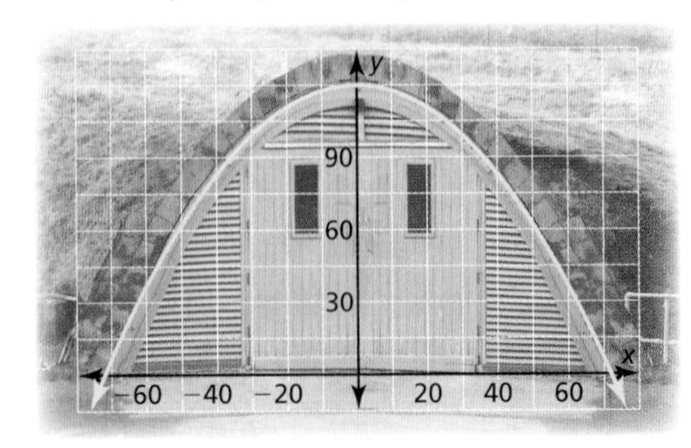

7.5 Factoring $x^2 + bx + c$

Essential Question How can you use algebra tiles to factor the trinomial $x^2 + bx + c$ into the product of two binomials?

EXPLORATION 1 Finding Binomial Factors

Work with a partner. Use algebra tiles to write each polynomial as the product of two binomials. Check your answer by multiplying.

Sample $x^2 + 5x + 6$

Step 1 Arrange algebra tiles that model $x^2 + 5x + 6$ into a rectangular array.

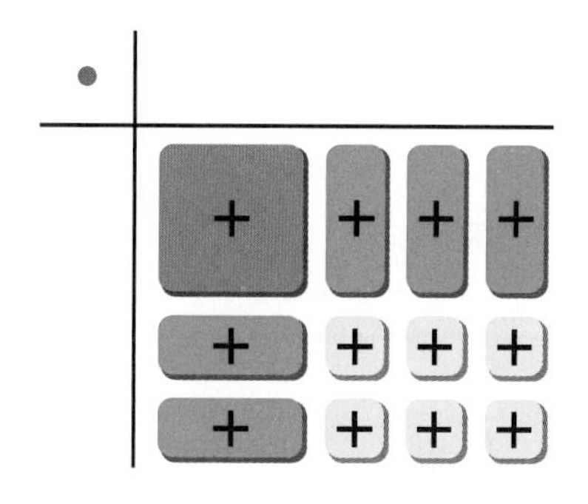

Step 2 Use additional algebra tiles to model the dimensions of the rectangle.

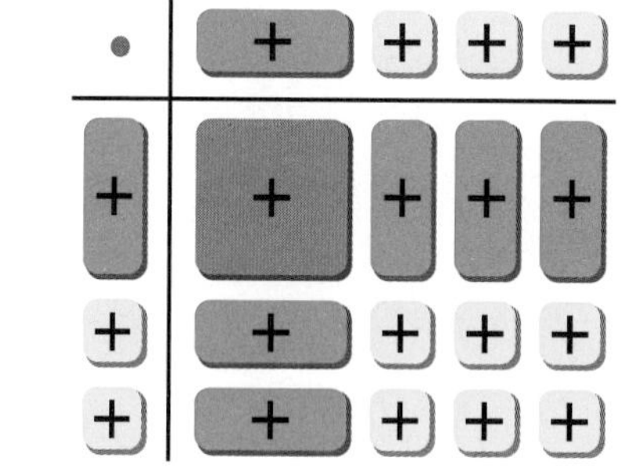

Step 3 Write the polynomial in factored form using the dimensions of the rectangle.

width — length

$$\text{Area} = x^2 + 5x + 6 = (x + 2)(x + 3)$$

a. $x^2 - 3x + 2 =$

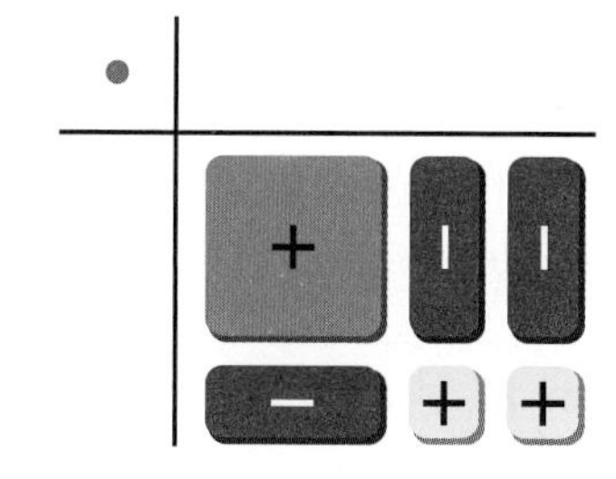

b. $x^2 + 5x + 4 =$

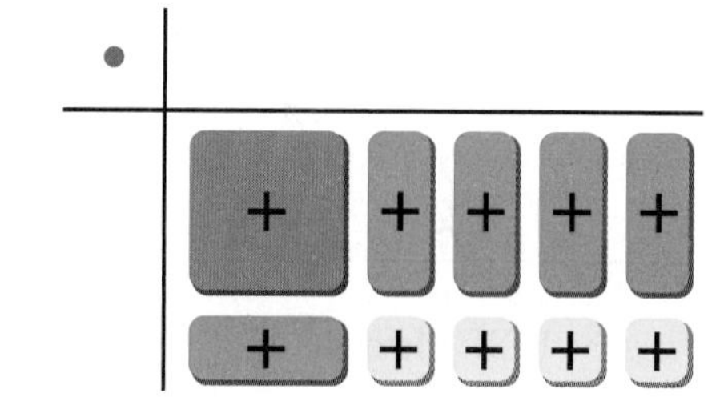

c. $x^2 - 7x + 12 =$

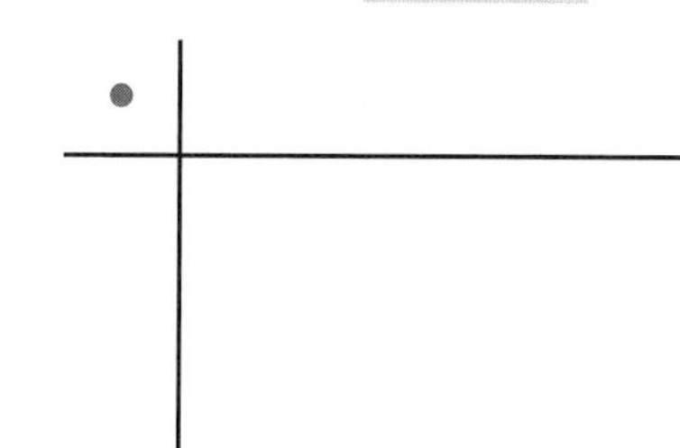

d. $x^2 + 7x + 12 =$

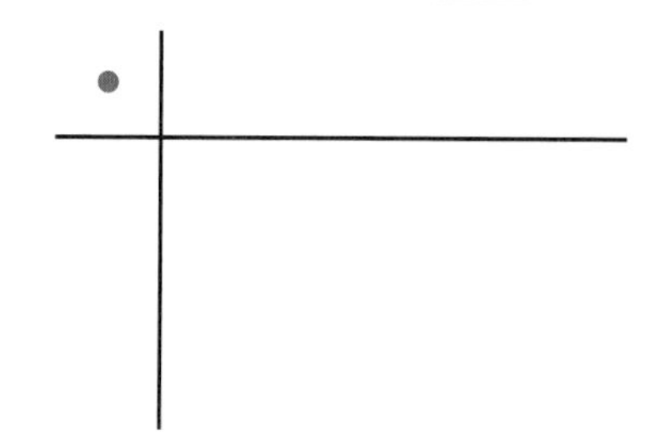

REASONING ABSTRACTLY

To be proficient in math, you need to understand a situation abstractly and represent it symbolically.

Communicate Your Answer

2. How can you use algebra tiles to factor the trinomial $x^2 + bx + c$ into the product of two binomials?

3. Describe a strategy for factoring the trinomial $x^2 + bx + c$ that does not use algebra tiles.

7.5 Lesson

Core Vocabulary

Previous
polynomial
FOIL Method
Zero-Product Property

What You Will Learn

- Factor $x^2 + bx + c$.
- Use factoring to solve real-life problems.

Factoring $x^2 + bx + c$

Writing a polynomial as a product of factors is called *factoring*. To factor $x^2 + bx + c$ as $(x + p)(x + q)$, you need to find p and q such that $p + q = b$ and $pq = c$.

$$(x + p)(x + q) = x^2 + px + qx + pq$$
$$= x^2 + (p + q)x + pq$$

Core Concept

Factoring $x^2 + bx + c$ When c Is Positive

Algebra $x^2 + bx + c = (x + p)(x + q)$ when $p + q = b$ and $pq = c$.

When c is positive, p and q have the same sign as b.

Examples $x^2 + 6x + 5 = (x + 1)(x + 5)$

$x^2 - 6x + 5 = (x - 1)(x - 5)$

EXAMPLE 1 Factoring $x^2 + bx + c$ When b and c Are Positive

Factor $x^2 + 10x + 16$.

SOLUTION

Notice that $b = 10$ and $c = 16$.

- Because c is positive, the factors p and q must have the same sign so that pq is positive.
- Because b is also positive, p and q must each be positive so that $p + q$ is positive.

Find two positive integer factors of 16 whose sum is 10.

Factors of 16	Sum of factors
1, 16	17
2, 8	10
4, 4	8

The values of p and q are 2 and 8.

Check

Use the FOIL Method.

$(x + 2)(x + 8)$

$= x^2 + 8x + 2x + 16$

$= x^2 + 10x + 16$ ✓

▶ So, $x^2 + 10x + 16 = (x + 2)(x + 8)$.

Monitoring Progress  Help in English and Spanish at *BigIdeasMath.com*

Factor the polynomial.

1. $x^2 + 7x + 6$
2. $x^2 + 9x + 8$

EXAMPLE 2 Factoring $x^2 + bx + c$ When b Is Negative and c Is Positive

Factor $x^2 - 8x + 12$.

SOLUTION

Notice that $b = -8$ and $c = 12$.

- Because c is positive, the factors p and q must have the same sign so that pq is positive.
- Because b is negative, p and q must each be negative so that $p + q$ is negative.

Find two negative integer factors of 12 whose sum is -8.

Factors of 12	$-1, -12$	$-2, -6$	$-3, -4$
Sum of factors	-13	-8	-7

The values of p and q are -2 and -6.

▶ So, $x^2 - 8x + 12 = (x - 2)(x - 6)$.

Check

Use the FOIL Method.

$(x - 2)(x - 6)$

$= x^2 - 6x - 2x + 12$

$= x^2 - 8x + 12$ ✓

Core Concept

Factoring $x^2 + bx + c$ When c Is Negative

Algebra $x^2 + bx + c = (x + p)(x + q)$ when $p + q = b$ and $pq = c$.

When c is negative, p and q have different signs.

Example $x^2 - 4x - 5 = (x + 1)(x - 5)$

EXAMPLE 3 Factoring $x^2 + bx + c$ When c Is Negative

Factor $x^2 + 4x - 21$.

SOLUTION

Notice that $b = 4$ and $c = -21$. Because c is negative, the factors p and q must have different signs so that pq is negative.

Find two integer factors of -21 whose sum is 4.

Factors of -21	$-21, 1$	$-1, 21$	$-7, 3$	$-3, 7$
Sum of factors	-20	20	-4	4

The values of p and q are -3 and 7.

▶ So, $x^2 + 4x - 21 = (x - 3)(x + 7)$.

Check

Use the FOIL Method.

$(x - 3)(x + 7)$

$= x^2 + 7x - 3x - 21$

$= x^2 + 4x - 21$ ✓

Monitoring Progress

Help in English and Spanish at *BigIdeasMath.com*

Factor the polynomial.

3. $w^2 - 4w + 3$

4. $n^2 - 12n + 35$

5. $x^2 - 14x + 24$

6. $x^2 + 2x - 15$

7. $y^2 + 13y - 30$

8. $v^2 - v - 42$

Solving Real-Life Problems

EXAMPLE 4 Solving a Real-Life Problem

40 m

s m

30 m

s m

A farmer plants a rectangular pumpkin patch in the northeast corner of a square plot of land. The area of the pumpkin patch is 600 square meters. What is the area of the square plot of land?

SOLUTION

1. **Understand the Problem** You are given the area of the pumpkin patch, the difference of the side length of the square plot and the length of the pumpkin patch, and the difference of the side length of the square plot and the width of the pumpkin patch.

2. **Make a Plan** The length of the pumpkin patch is $(s - 30)$ meters and the width is $(s - 40)$ meters. Write and solve an equation to find the side length s. Then use the solution to find the area of the square plot of land.

3. **Solve the Problem** Use the equation for the area of a rectangle to write and solve an equation to find the side length s of the square plot of land.

$600 = (s - 30)(s - 40)$	Write an equation.
$600 = s^2 - 70s + 1200$	Multiply.
$0 = s^2 - 70s + 600$	Subtract 600 from each side.
$0 = (s - 10)(s - 60)$	Factor the polynomial.
$s - 10 = 0$ *or* $s - 60 = 0$	Zero-Product Property
$s = 10$ *or* $s = 60$	Solve for s.

STUDY TIP

The diagram shows that the side length is more than 40 meters, so a side length of 10 meters does not make sense in this situation. The side length is 60 meters.

▶ So, the area of the square plot of land is $60(60) = 3600$ square meters.

4. **Look Back** Use the diagram to check that you found the correct side length. Using $s = 60$, the length of the pumpkin patch is $60 - 30 = 30$ meters and the width is $60 - 40 = 20$ meters. So, the area of the pumpkin patch is 600 square meters. This matches the given information and confirms the side length is 60 meters, which gives an area of 3600 square meters.

Monitoring Progress  Help in English and Spanish at *BigIdeasMath.com*

9. **WHAT IF?** The area of the pumpkin patch is 200 square meters. What is the area of the square plot of land?

Concept Summary

Factoring $x^2 + bx + c$ as $(x + p)(x + q)$

The diagram shows the relationships between the signs of b and c and the signs of p and q.

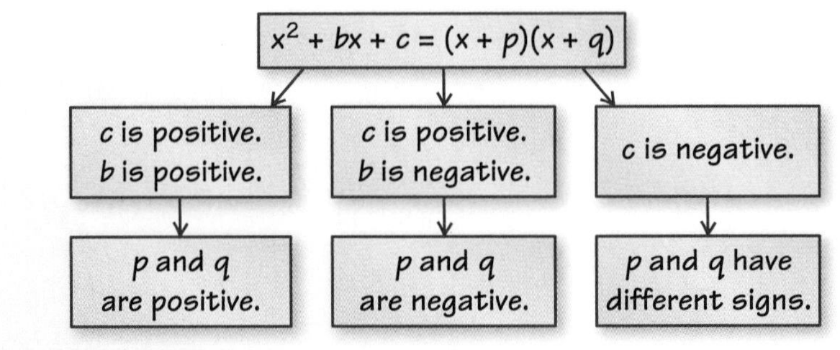

7.5 Exercises

Dynamic Solutions available at *BigIdeasMath.com*

Vocabulary and Core Concept Check

1. **WRITING** You are factoring $x^2 + 11x - 26$. What do the signs of the terms tell you about the factors? Explain.

2. **OPEN-ENDED** Write a trinomial that can be factored as $(x + p)(x + q)$, where p and q are positive.

Monitoring Progress and Modeling with Mathematics

In Exercises 3–8, factor the polynomial. *(See Example 1.)*

3. $x^2 + 8x + 7$
4. $z^2 + 10z + 21$
5. $n^2 + 9n + 20$
6. $s^2 + 11s + 30$
7. $h^2 + 11h + 18$
8. $y^2 + 13y + 40$

In Exercises 9–14, factor the polynomial. *(See Example 2.)*

9. $v^2 - 5v + 4$
10. $x^2 - 13x + 22$
11. $d^2 - 5d + 6$
12. $k^2 - 10k + 24$
13. $w^2 - 17w + 72$
14. $j^2 - 13j + 42$

In Exercises 15–24, factor the polynomial. *(See Example 3.)*

15. $x^2 + 3x - 4$
16. $z^2 + 7z - 18$
17. $n^2 + 4n - 12$
18. $s^2 + 3s - 40$
19. $y^2 + 2y - 48$
20. $h^2 + 6h - 27$
21. $x^2 - x - 20$
22. $m^2 - 6m - 7$
23. $-6t - 16 + t^2$
24. $-7y + y^2 - 30$

25. **MODELING WITH MATHEMATICS** A projector displays an image on a wall. The area (in square feet) of the projection is represented by $x^2 - 8x + 15$.

    a. Write a binomial that represents the height of the projection.

    b. Find the perimeter of the projection when the height of the wall is 8 feet.

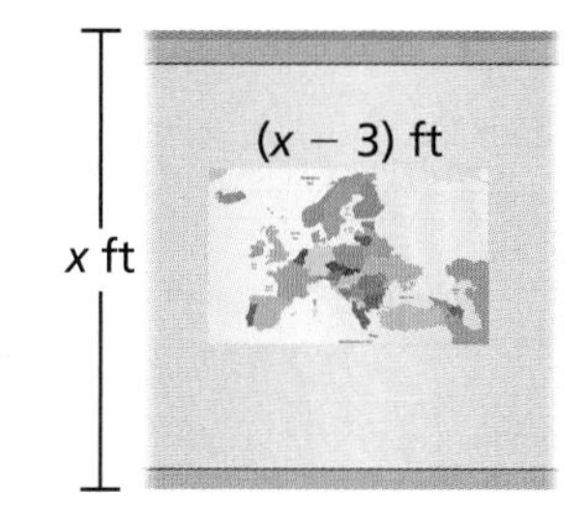

26. **MODELING WITH MATHEMATICS** A dentist's office and parking lot are on a rectangular piece of land. The area (in square meters) of the land is represented by $x^2 + x - 30$.

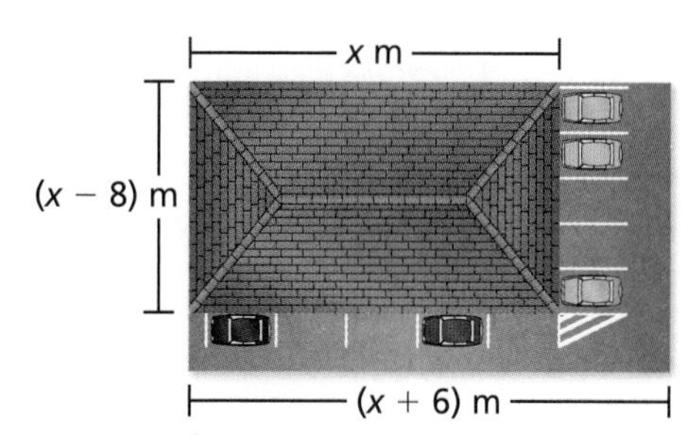

    a. Write a binomial that represents the width of the land.

    b. Find the area of the land when the length of the dentist's office is 20 meters.

ERROR ANALYSIS In Exercises 27 and 28, describe and correct the error in factoring the polynomial.

27. 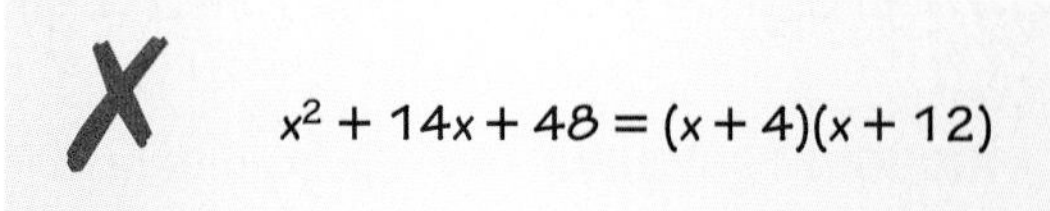

28. 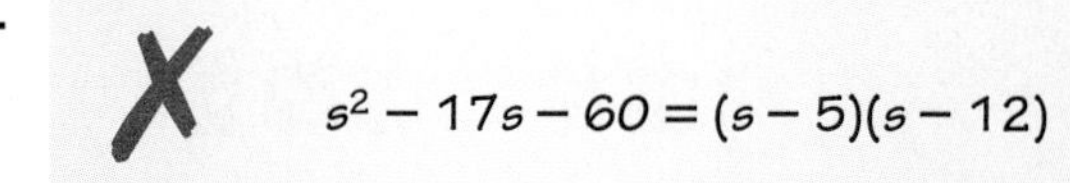

In Exercises 29–38, solve the equation.

29. $m^2 + 3m + 2 = 0$
30. $n^2 - 9n + 18 = 0$
31. $x^2 + 5x - 14 = 0$
32. $v^2 + 11v - 26 = 0$
33. $t^2 + 15t = -36$
34. $n^2 - 5n = 24$
35. $a^2 + 5a - 20 = 30$
36. $y^2 - 2y - 8 = 7$
37. $m^2 + 10 = 15m - 34$
38. $b^2 + 5 = 8b - 10$

39. **MODELING WITH MATHEMATICS** You trimmed a large square picture so that you could fit it into a frame. The area of the cut picture is 20 square inches. What is the area of the original picture? *(See Example 4.)*

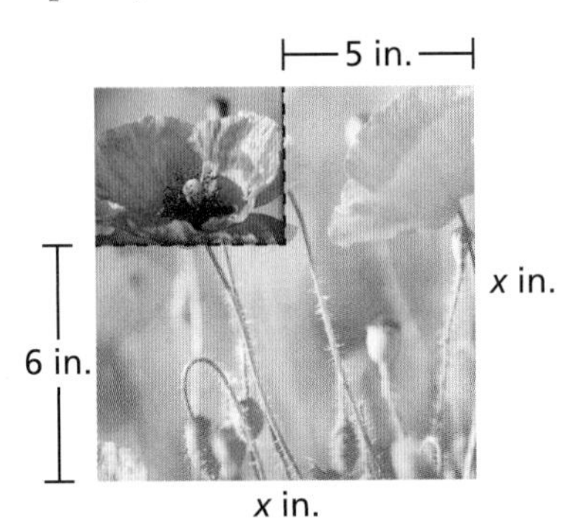

40. **MODELING WITH MATHEMATICS** A web browser is open on your computer screen.

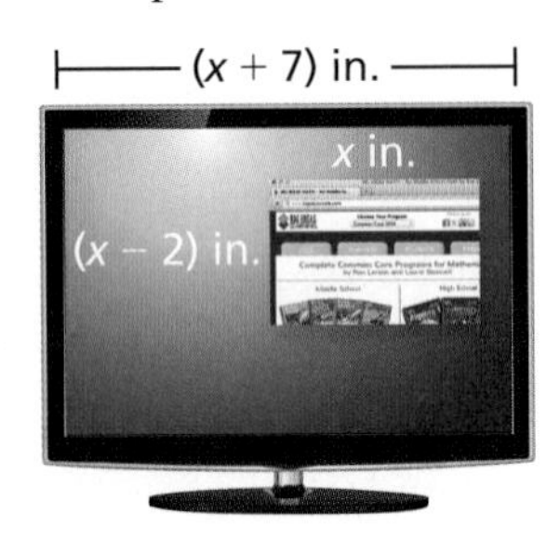

a. The area of the browser window is 24 square inches. Find the length of the browser window x.

b. The browser covers $\frac{3}{13}$ of the screen. What are the dimensions of the screen?

41. **MAKING AN ARGUMENT** Your friend says there are six integer values of b for which the trinomial $x^2 + bx - 12$ has two binomial factors of the form $(x + p)$ and $(x + q)$. Is your friend correct? Explain.

42. **THOUGHT PROVOKING** Use algebra tiles to factor each polynomial modeled by the tiles. Show your work.

a.

b.

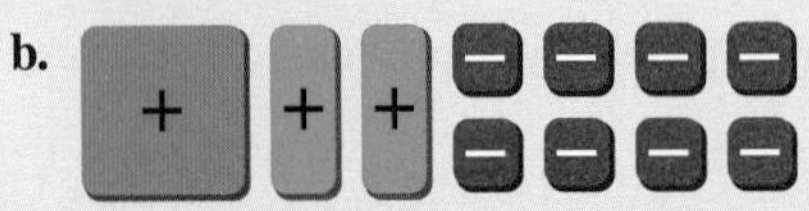

MATHEMATICAL CONNECTIONS In Exercises 43 and 44, find the dimensions of the polygon with the given area.

43. Area = 44 ft²

44. Area = 35 m²

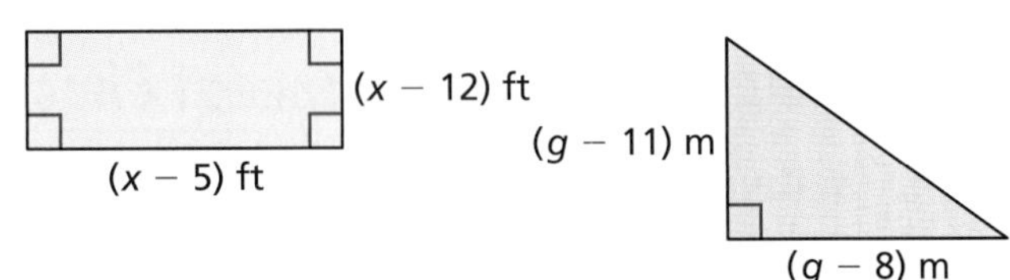

45. **REASONING** Write an equation of the form $x^2 + bx + c = 0$ that has the solutions $x = -4$ and $x = 6$. Explain how you found your answer.

46. **HOW DO YOU SEE IT?** The graph of $y = x^2 + x - 6$ is shown.

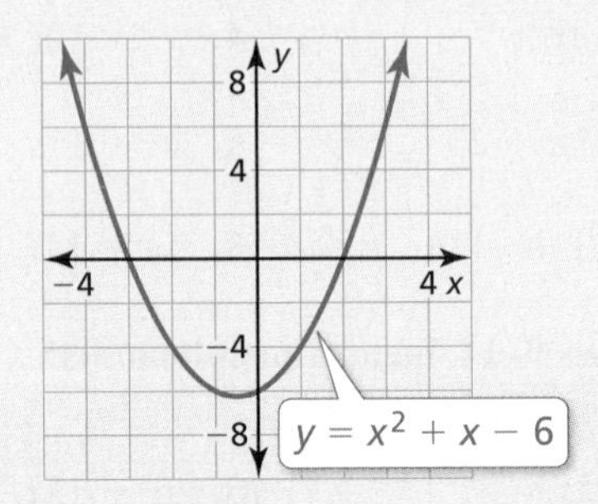

a. Explain how you can use the graph to factor the polynomial $x^2 + x - 6$.

b. Factor the polynomial.

47. **PROBLEM SOLVING** Road construction workers are paving the area shown.

a. Write an expression that represents the area being paved.

b. The area being paved is 280 square meters. Write and solve an equation to find the width of the road x.

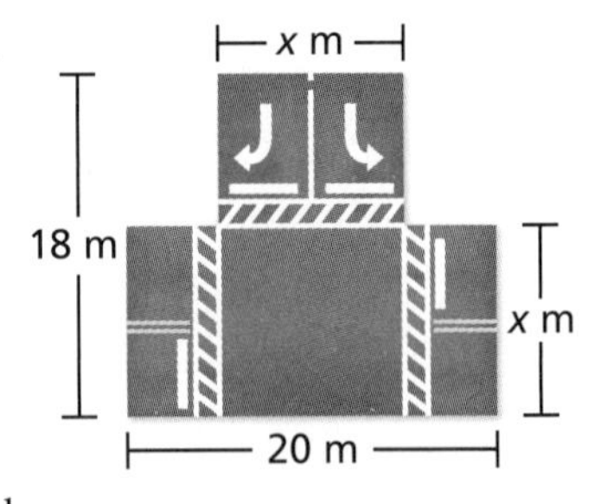

USING STRUCTURE In Exercises 48–51, factor the polynomial.

48. $x^2 + 6xy + 8y^2$

49. $r^2 + 7rs + 12s^2$

50. $a^2 + 11ab - 26b^2$

51. $x^2 - 2xy - 35y^2$

Maintaining Mathematical Proficiency

Reviewing what you learned in previous grades and lessons

Solve the equation. Check your solution. *(Section 1.1)*

52. $p - 9 = 0$

53. $z + 12 = -5$

54. $6 = \frac{c}{-7}$

55. $4k = 0$

7.6 Factoring $ax^2 + bx + c$

Essential Question How can you use algebra tiles to factor the trinomial $ax^2 + bx + c$ into the product of two binomials?

EXPLORATION 1 **Finding Binomial Factors**

Work with a partner. Use algebra tiles to write each polynomial as the product of two binomials. Check your answer by multiplying.

Sample $2x^2 + 5x + 2$

Step 1 Arrange algebra tiles that model $2x^2 + 5x + 2$ into a rectangular array.

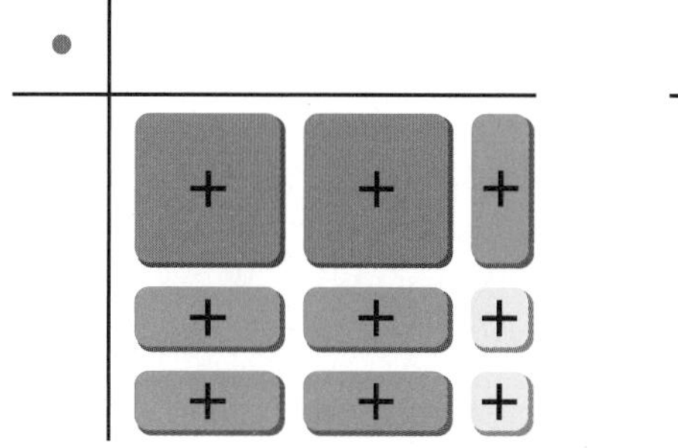

Step 2 Use additional algebra tiles to model the dimensions of the rectangle.

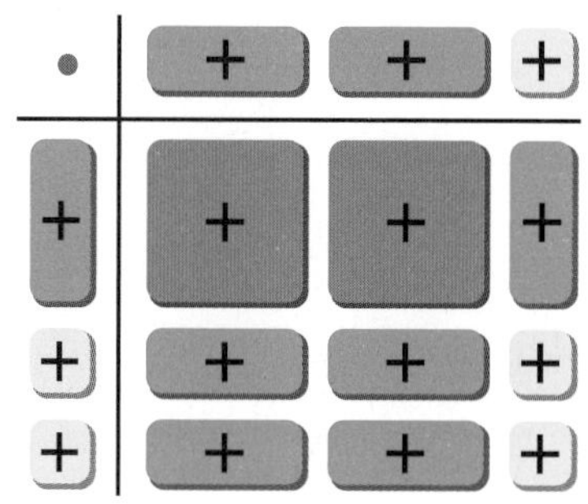

Step 3 Write the polynomial in factored form using the dimensions of the rectangle.

width length

$$\text{Area} = 2x^2 + 5x + 2 = (x + 2)(2x + 1)$$

a. $3x^2 + 5x + 2 =$ ______

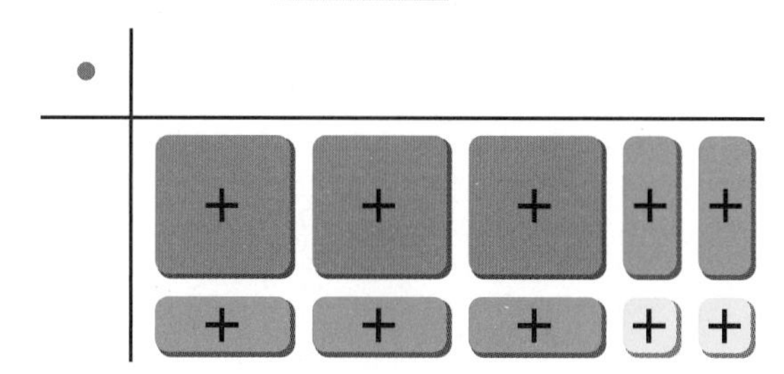

b. $4x^2 + 4x - 3 =$ ______

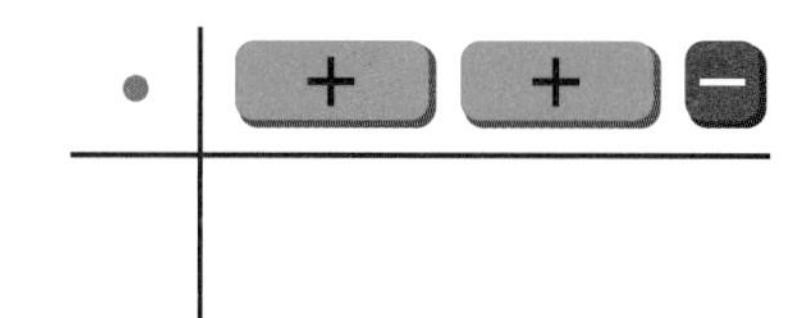

c. $2x^2 - 11x + 5 =$ ______

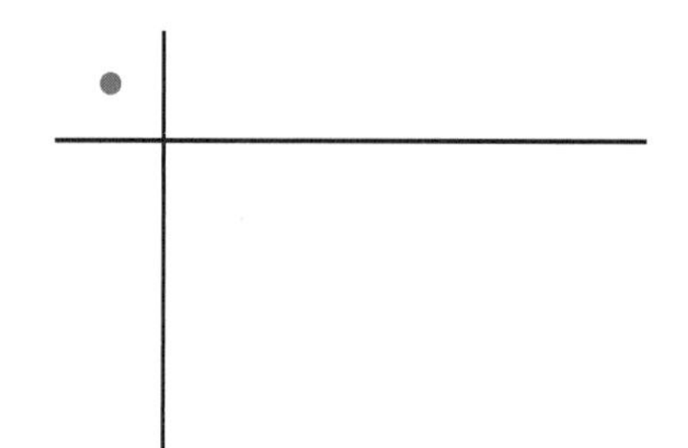

USING TOOLS STRATEGICALLY

To be proficient in math, you need to consider the available tools, including concrete models, when solving a mathematical problem.

Communicate Your Answer

2. How can you use algebra tiles to factor the trinomial $ax^2 + bx + c$ into the product of two binomials?

3. Is it possible to factor the trinomial $2x^2 + 2x + 1$? Explain your reasoning.

7.6 Lesson

Core Vocabulary

Previous
polynomial
greatest common factor (GCF)
Zero-Product Property

What You Will Learn

- Factor $ax^2 + bx + c$.
- Use factoring to solve real-life problems.

Factoring $ax^2 + bx + c$

In Section 7.5, you factored polynomials of the form $ax^2 + bx + c$, where $a = 1$. To factor polynomials of the form $ax^2 + bx + c$, where $a \neq 1$, first look for the GCF of the terms of the polynomial and then factor further if possible.

EXAMPLE 1 Factoring Out the GCF

Factor $5x^2 + 15x + 10$.

SOLUTION

Notice that the GCF of the terms $5x^2$, $15x$, and 10 is 5.

$5x^2 + 15x + 10 = 5(x^2 + 3x + 2)$ Factor out GCF.

$= 5(x + 1)(x + 2)$ Factor $x^2 + 3x + 2$.

▶ So, $5x^2 + 15x + 10 = 5(x + 1)(x + 2)$.

When there is no GCF, consider the possible factors of a and c.

EXAMPLE 2 Factoring $ax^2 + bx + c$ When ac Is Positive

Factor each polynomial.

a. $4x^2 + 13x + 3$ **b.** $3x^2 - 7x + 2$

SOLUTION

a. There is no GCF, so you need to consider the possible factors of a and c. Because b and c are both positive, the factors of c must be positive. Use a table to organize information about the factors of a and c.

STUDY TIP
You must consider the order of the factors of 3, because the middle terms formed by the possible factorizations are different.

Factors of 4	Factors of 3	Possible factorization	Middle term	
1, 4	1, 3	$(x + 1)(4x + 3)$	$3x + 4x = 7x$	✗
1, 4	3, 1	$(x + 3)(4x + 1)$	$x + 12x = 13x$	✓
2, 2	1, 3	$(2x + 1)(2x + 3)$	$6x + 2x = 8x$	✗

▶ So, $4x^2 + 13x + 3 = (x + 3)(4x + 1)$.

b. There is no GCF, so you need to consider the possible factors of a and c. Because b is negative and c is positive, both factors of c must be negative. Use a table to organize information about the factors of a and c.

Factors of 3	Factors of 2	Possible factorization	Middle term	
1, 3	$-1, -2$	$(x - 1)(3x - 2)$	$-2x - 3x = -5x$	✗
1, 3	$-2, -1$	$(x - 2)(3x - 1)$	$-x - 6x = -7x$	✓

▶ So, $3x^2 - 7x + 2 = (x - 2)(3x - 1)$.

EXAMPLE 3 Factoring $ax^2 + bx + c$ When ac Is Negative

Factor $2x^2 - 5x - 7$.

SOLUTION

There is no GCF, so you need to consider the possible factors of a and c. Because c is negative, the factors of c must have different signs. Use a table to organize information about the factors of a and c.

Factors of 2	Factors of −7	Possible factorization	Middle term	
1, 2	1, −7	$(x + 1)(2x - 7)$	$-7x + 2x = -5x$	✓
1, 2	7, −1	$(x + 7)(2x - 1)$	$-x + 14x = 13x$	✗
1, 2	−1, 7	$(x - 1)(2x + 7)$	$7x - 2x = 5x$	✗
1, 2	−7, 1	$(x - 7)(2x + 1)$	$x - 14x = -13x$	✗

▶ So, $2x^2 - 5x - 7 = (x + 1)(2x - 7)$.

STUDY TIP

When a is negative, factor -1 from each term of $ax^2 + bx + c$. Then factor the resulting trinomial as in the previous examples.

EXAMPLE 4 Factoring $ax^2 + bx + c$ When a Is Negative

Factor $-4x^2 - 8x + 5$.

SOLUTION

Step 1 Factor -1 from each term of the trinomial.

$$-4x^2 - 8x + 5 = -(4x^2 + 8x - 5)$$

Step 2 Factor the trinomial $4x^2 + 8x - 5$. Because c is negative, the factors of c must have different signs. Use a table to organize information about the factors of a and c.

Factors of 4	Factors of −5	Possible factorization	Middle term	
1, 4	1, −5	$(x + 1)(4x - 5)$	$-5x + 4x = -x$	✗
1, 4	5, −1	$(x + 5)(4x - 1)$	$-x + 20x = 19x$	✗
1, 4	−1, 5	$(x - 1)(4x + 5)$	$5x - 4x = x$	✗
1, 4	−5, 1	$(x - 5)(4x + 1)$	$x - 20x = -19x$	✗
2, 2	1, −5	$(2x + 1)(2x - 5)$	$-10x + 2x = -8x$	✗
2, 2	−1, 5	$(2x - 1)(2x + 5)$	$10x - 2x = 8x$	✓

▶ So, $-4x^2 - 8x + 5 = -(2x - 1)(2x + 5)$.

Monitoring Progress

Help in English and Spanish at *BigIdeasMath.com*

Factor the polynomial.

1. $8x^2 - 56x + 48$
2. $14x^2 + 31x + 15$
3. $2x^2 - 7x + 5$
4. $3x^2 - 14x + 8$
5. $4x^2 - 19x - 5$
6. $6x^2 + x - 12$
7. $-2y^2 - 5y - 3$
8. $-5m^2 + 6m - 1$
9. $-3x^2 - x + 2$

Solving Real-Life Problems

EXAMPLE 5 Solving a Real-Life Problem

The length of a rectangular game reserve is 1 mile longer than twice the width. The area of the reserve is 55 square miles. What is the width of the reserve?

SOLUTION

Use the formula for the area of a rectangle to write an equation for the area of the reserve. Let w represent the width. Then $2w + 1$ represents the length. Solve for w.

$w(2w + 1) = 55$ Area of the reserve

$2w^2 + w = 55$ Distributive Property

$2w^2 + w - 55 = 0$ Subtract 55 from each side.

Factor the left side of the equation. There is no GCF, so you need to consider the possible factors of a and c. Because c is negative, the factors of c must have different signs. Use a table to organize information about the factors of a and c.

Factors of 2	Factors of −55	Possible factorization	Middle term	
1, 2	1, −55	$(w + 1)(2w - 55)$	$-55w + 2w = -53w$	✗
1, 2	55, −1	$(w + 55)(2w - 1)$	$-w + 110w = 109w$	✗
1, 2	−1, 55	$(w - 1)(2w + 55)$	$55w - 2w = 53w$	✗
1, 2	−55, 1	$(w - 55)(2w + 1)$	$w - 110w = -109w$	✗
1, 2	5, −11	$(w + 5)(2w - 11)$	$-11w + 10w = -w$	✗
1, 2	11, −5	$(w + 11)(2w - 5)$	$-5w + 22w = 17w$	✗
1, 2	−5, 11	$(w - 5)(2w + 11)$	$11w - 10w = w$	✓
1, 2	−11, 5	$(w - 11)(2w + 5)$	$5w - 22w = -17w$	✗

So, you can rewrite $2w^2 + w - 55$ as $(w - 5)(2w + 11)$. Write the equation with the left side factored and continue solving for w.

$(w - 5)(2w + 11) = 0$ Rewrite equation with left side factored.

$w - 5 = 0$ *or* $2w + 11 = 0$ Zero-Product Property

$w = 5$ *or* $w = -\frac{11}{2}$ Solve for w.

Check

Use mental math.

The width is 5 miles, so the length is $5(2) + 1 = 11$ miles and the area is $5(11) = 55$ square miles. ✓

A negative width does not make sense, so you should use the positive solution.

▶ So, the width of the reserve is 5 miles.

Monitoring Progress  Help in English and Spanish at *BigIdeasMath.com*

10. WHAT IF? The area of the reserve is 136 square miles. How wide is the reserve?

7.6 Exercises

Dynamic Solutions available at *BigIdeasMath.com*

Vocabulary and Core Concept Check

1. **REASONING** What is the greatest common factor of the terms of $3y^2 - 21y + 36$?

2. **WRITING** Compare factoring $6x^2 - x - 2$ with factoring $x^2 - x - 2$.

Monitoring Progress and Modeling with Mathematics

In Exercises 3–8, factor the polynomial. *(See Example 1.)*

3. $3x^2 + 3x - 6$
4. $8v^2 + 8v - 48$
5. $4k^2 + 28k + 48$
6. $6y^2 - 24y + 18$
7. $7b^2 - 63b + 140$
8. $9r^2 - 36r - 45$

In Exercises 9–16, factor the polynomial. *(See Examples 2 and 3.)*

9. $3h^2 + 11h + 6$
10. $8m^2 + 30m + 7$
11. $6x^2 - 5x + 1$
12. $10w^2 - 31w + 15$
13. $3n^2 + 5n - 2$
14. $4z^2 + 4z - 3$
15. $8g^2 - 10g - 12$
16. $18v^2 - 15v - 18$

In Exercises 17–22, factor the polynomial. *(See Example 4.)*

17. $-3t^2 + 11t - 6$
18. $-7v^2 - 25v - 12$
19. $-4c^2 + 19c + 5$
20. $-8h^2 - 13h + 6$
21. $-15w^2 - w + 28$
22. $-22d^2 + 29d - 9$

ERROR ANALYSIS In Exercises 23 and 24, describe and correct the error in factoring the polynomial.

23. ✗
$$2x^2 - 2x - 24 = 2(x^2 - 2x - 24)$$
$$= 2(x - 6)(x + 4)$$

24. ✗
$$6x^2 - 7x - 3 = (3x - 3)(2x + 1)$$

In Exercises 25–28, solve the equation.

25. $5x^2 - 5x - 30 = 0$
26. $2k^2 - 5k - 18 = 0$
27. $-12n^2 - 11n = -15$
28. $14b^2 - 2 = -3b$

In Exercises 29–32, find the x-coordinates of the points where the graph crosses the x-axis.

29.

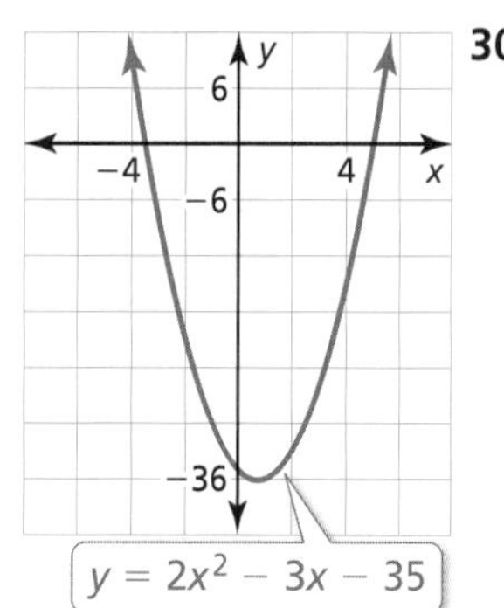

30.

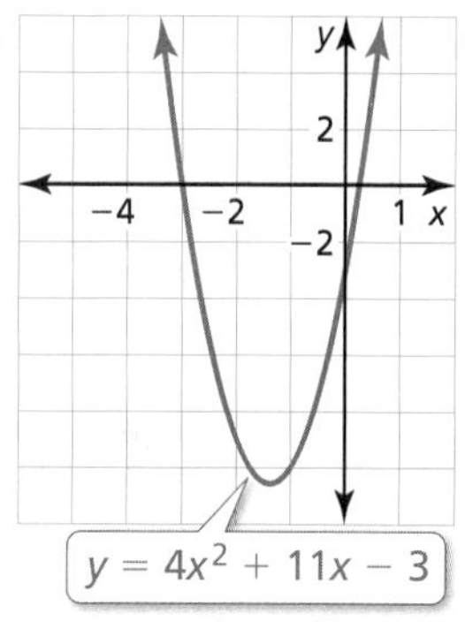

31.

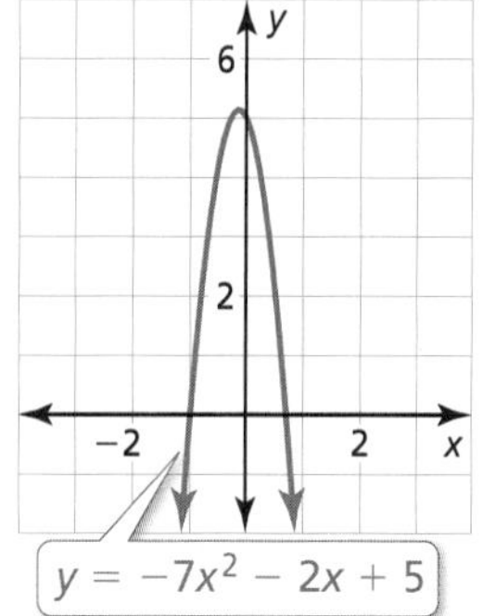

32.

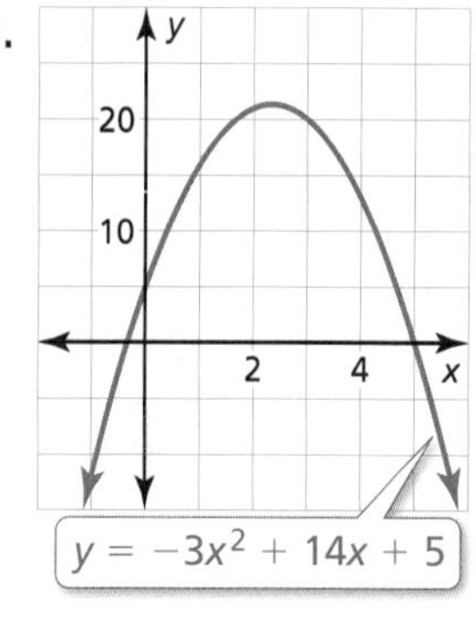

33. **MODELING WITH MATHEMATICS** The area (in square feet) of the school sign can be represented by $15x^2 - x - 2$.

   a. Write an expression that represents the length of the sign.

   b. Describe two ways to find the area of the sign when $x = 3$.

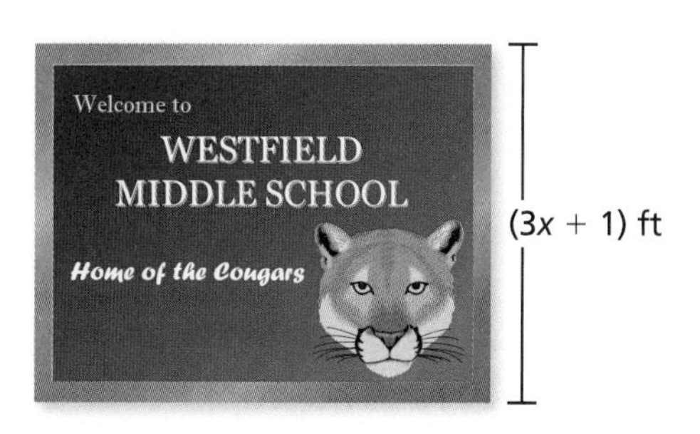

34. MODELING WITH MATHEMATICS The height h (in feet) above the water of a cliff diver is modeled by $h = -16t^2 + 8t + 80$, where t is the time (in seconds). How long is the diver in the air?

35. MODELING WITH MATHEMATICS The Parthenon in Athens, Greece, is an ancient structure that has a rectangular base. The length of the base of the Parthenon is 8 meters more than twice its width. The area of the base is about 2170 square meters. Find the length and width of the base. *(See Example 5.)*

36. MODELING WITH MATHEMATICS The length of a rectangular birthday party invitation is 1 inch less than twice its width. The area of the invitation is 15 square inches. Will the invitation fit in the envelope shown without being folded? Explain.

$3\frac{5}{8}$ in.

$5\frac{1}{8}$ in.

37. OPEN-ENDED Write a binomial whose terms have a GCF of $3x$.

38. HOW DO YOU SEE IT? Without factoring, determine which of the graphs represents the function $g(x) = 21x^2 + 37x + 12$ and which represents the function $h(x) = 21x^2 - 37x + 12$. Explain your reasoning.

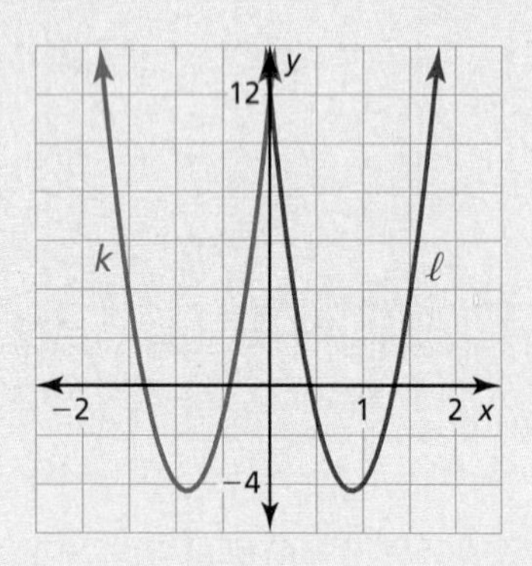

39. REASONING When is it not possible to factor $ax^2 + bx + c$, where $a \neq 1$? Give an example.

40. MAKING AN ARGUMENT Your friend says that to solve the equation $5x^2 + x - 4 = 2$, you should start by factoring the left side as $(5x - 4)(x + 1)$. Is your friend correct? Explain.

41. REASONING For what values of t can $2x^2 + tx + 10$ be written as the product of two binomials?

42. THOUGHT PROVOKING Use algebra tiles to factor each polynomial modeled by the tiles. Show your work.

a.

b.

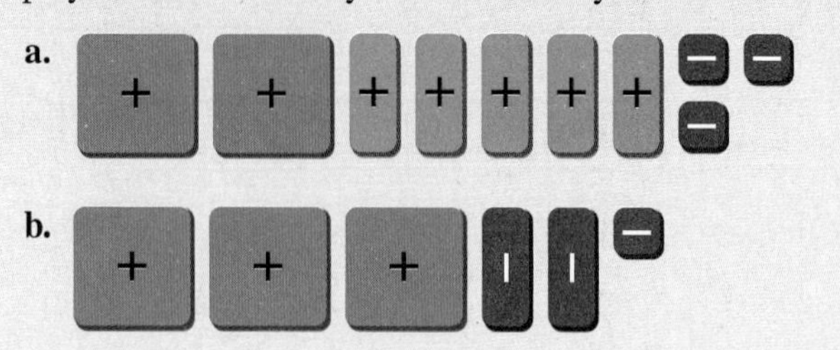

43. MATHEMATICAL CONNECTIONS The length of a rectangle is 1 inch more than twice its width. The value of the area of the rectangle (in square inches) is 5 more than the value of the perimeter of the rectangle (in inches). Find the width.

44. PROBLEM SOLVING A rectangular swimming pool is bordered by a concrete patio. The width of the patio is the same on every side. The area of the surface of the pool is equal to the area of the patio. What is the width of the patio?

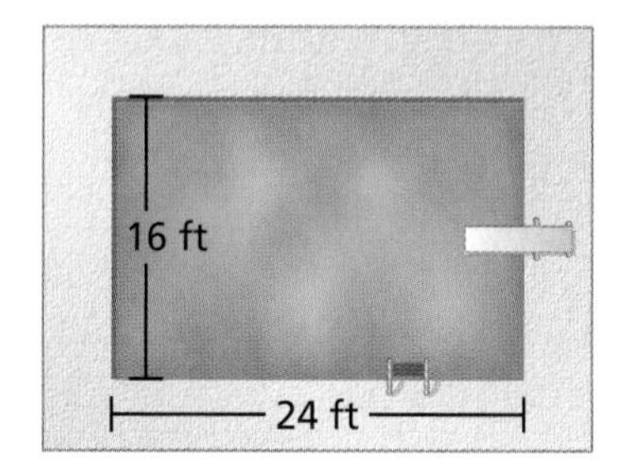

In Exercises 45–48, factor the polynomial.

45. $4k^2 + 7jk - 2j^2$

46. $6x^2 + 5xy - 4y^2$

47. $-6a^2 + 19ab - 14b^2$

48. $18m^3 + 39m^2n - 15mn^2$

Maintaining Mathematical Proficiency

Reviewing what you learned in previous grades and lessons

Find the square root(s). *(Skills Review Handbook)*

49. $\pm\sqrt{64}$

50. $\sqrt{4}$

51. $-\sqrt{225}$

52. $\pm\sqrt{81}$

Solve the system of linear equations by substitution. Check your solution. *(Section 5.2)*

53. $y = 3 + 7x$
$y - x = -3$

54. $2x = y + 2$
$-x + 3y = 14$

55. $5x - 2y = 14$
$-7 = -2x + y$

56. $-x - 8 = -y$
$9y - 12 + 3x = 0$

7.7 Factoring Special Products

Essential Question How can you recognize and factor special products?

EXPLORATION 1 **Factoring Special Products**

Work with a partner. Use algebra tiles to write each polynomial as the product of two binomials. Check your answer by multiplying. State whether the product is a "special product" that you studied in Section 7.3.

LOOKING FOR STRUCTURE

To be proficient in math, you need to see complicated things as single objects or as being composed of several objects.

a. $4x^2 - 1 =$ 

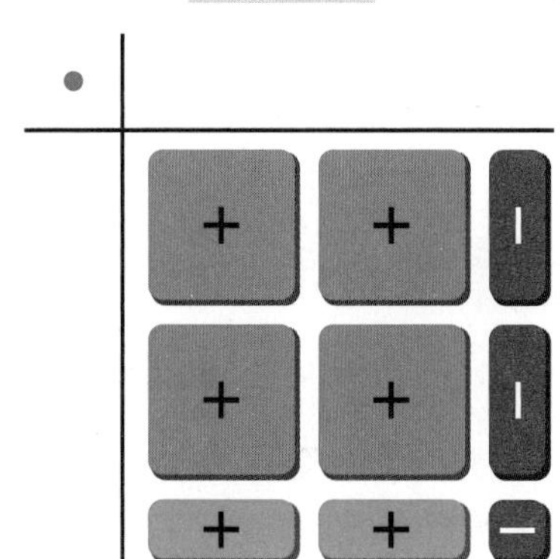

b. $4x^2 - 4x + 1 =$

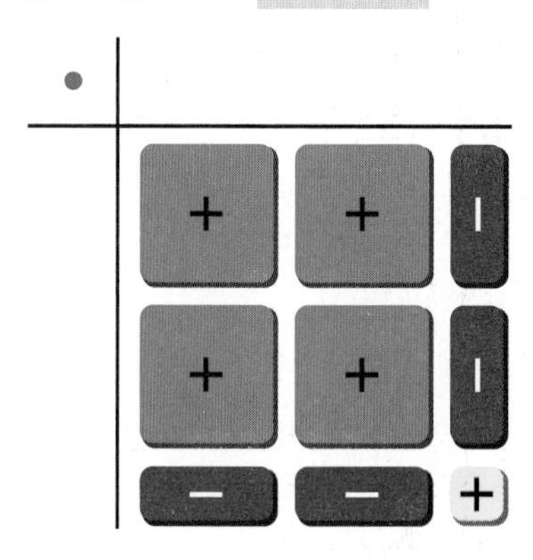

c. $4x^2 + 4x + 1 =$

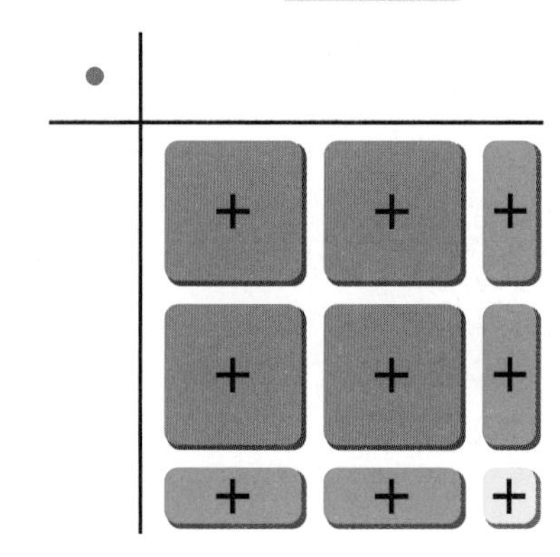

d. $4x^2 - 6x + 2 =$

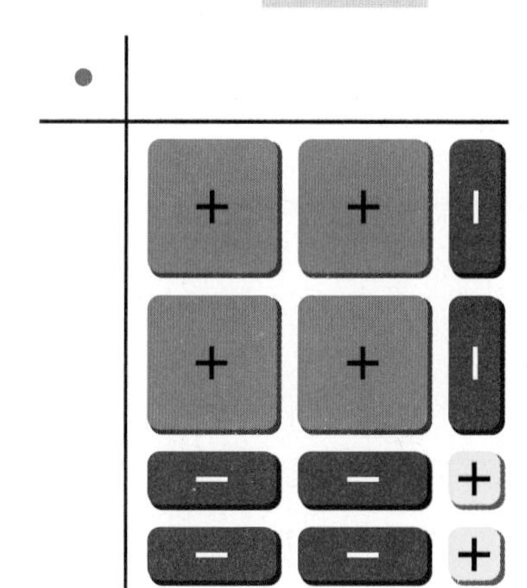

EXPLORATION 2 **Factoring Special Products**

Work with a partner. Use algebra tiles to complete the rectangular array at the left in three different ways, so that each way represents a different special product. Write each special product in standard form and in factored form.

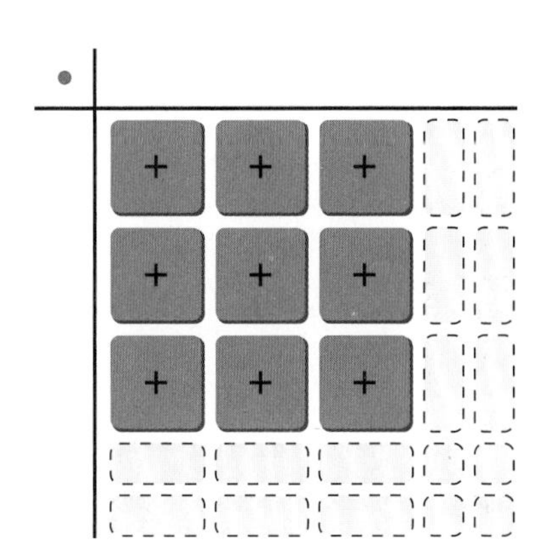

Communicate Your Answer

3. How can you recognize and factor special products? Describe a strategy for recognizing which polynomials can be factored as special products.

4. Use the strategy you described in Question 3 to factor each polynomial.

a. $25x^2 + 10x + 1$ **b.** $25x^2 - 10x + 1$ **c.** $25x^2 - 1$

7.7 Lesson

Core Vocabulary

Previous
polynomial
trinomial

What You Will Learn

- Factor the difference of two squares.
- Factor perfect square trinomials.
- Use factoring to solve real-life problems.

Factoring the Difference of Two Squares

You can use special product patterns to factor polynomials.

Core Concept

Difference of Two Squares Pattern

Algebra	**Example**
$a^2 - b^2 = (a + b)(a - b)$	$x^2 - 9 = x^2 - 3^2 = (x + 3)(x - 3)$

EXAMPLE 1 Factoring the Difference of Two Squares

Factor (a) $x^2 - 25$ and (b) $4z^2 - 1$.

SOLUTION

a. $x^2 - 25 = x^2 - 5^2$ — Write as $a^2 - b^2$.

$= (x + 5)(x - 5)$ — Difference of two squares pattern

▶ So, $x^2 - 25 = (x + 5)(x - 5)$.

b. $4z^2 - 1 = (2z)^2 - 1^2$ — Write as $a^2 - b^2$.

$= (2z + 1)(2z - 1)$ — Difference of two squares pattern

▶ So, $4z^2 - 1 = (2z + 1)(2z - 1)$.

EXAMPLE 2 Evaluating a Numerical Expression

Use a special product pattern to evaluate the expression $54^2 - 48^2$.

SOLUTION

Notice that $54^2 - 48^2$ is a difference of two squares. So, you can rewrite the expression in a form that it is easier to evaluate using the difference of two squares pattern.

$54^2 - 48^2 = (54 + 48)(54 - 48)$ — Difference of two squares pattern

$= 102(6)$ — Simplify.

$= 612$ — Multiply.

▶ So, $54^2 - 48^2 = 612$.

Monitoring Progress  Help in English and Spanish at *BigIdeasMath.com*

Factor the polynomial.

1. $x^2 - 36$ **2.** $100 - m^2$ **3.** $9n^2 - 16$ **4.** $16h^2 - 49$

Use a special product pattern to evaluate the expression.

5. $36^2 - 34^2$ **6.** $47^2 - 44^2$ **7.** $55^2 - 50^2$ **8.** $28^2 - 24^2$

Factoring Perfect Square Trinomials

Perfect Square Trinomial Pattern

Algebra	Example
$a^2 + 2ab + b^2 = (a + b)^2$	$x^2 + 6x + 9 = x^2 + 2(x)(3) + 3^2$ $= (x + 3)^2$
$a^2 - 2ab + b^2 = (a - b)^2$	$x^2 - 6x + 9 = x^2 - 2(x)(3) + 3^2$ $= (x - 3)^2$

EXAMPLE 3 Factoring Perfect Square Trinomials

Factor each polynomial.

a. $n^2 + 8n + 16$

b. $4x^2 - 12x + 9$

SOLUTION

a. $n^2 + 8n + 16 = n^2 + 2(n)(4) + 4^2$ — Write as $a^2 + 2ab + b^2$.

$= (n + 4)^2$ — Perfect square trinomial pattern

▶ So, $n^2 + 8n + 16 = (n + 4)^2$.

b. $4x^2 - 12x + 9 = (2x)^2 - 2(2x)(3) + 3^2$ — Write as $a^2 - 2ab + b^2$.

$= (2x - 3)^2$ — Perfect square trinomial pattern

▶ So, $4x^2 - 12x + 9 = (2x - 3)^2$.

EXAMPLE 4 Solving a Polynomial Equation

Solve $x^2 + \frac{2}{3}x + \frac{1}{9} = 0$.

SOLUTION

$x^2 + \frac{2}{3}x + \frac{1}{9} = 0$	Write equation.
$9x^2 + 6x + 1 = 0$	Multiply each side by 9.
$(3x)^2 + 2(3x)(1) + 1^2 = 0$	Write left side as $a^2 + 2ab + b^2$.
$(3x + 1)^2 = 0$	Perfect square trinomial pattern
$3x + 1 = 0$	Zero-Product Property
$x = -\frac{1}{3}$	Solve for x.

▶ The solution is $x = -\frac{1}{3}$.

LOOKING FOR STRUCTURE

Equations of the form $(x + a)^2 = 0$ always have repeated roots of $x = -a$.

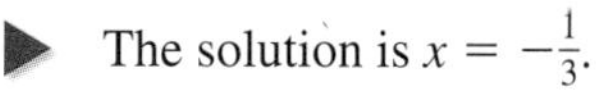

Help in English and Spanish at *BigIdeasMath.com*

Factor the polynomial.

9. $m^2 - 2m + 1$ **10.** $d^2 - 10d + 25$ **11.** $9z^2 + 36z + 36$

Solve the equation.

12. $a^2 + 6a + 9 = 0$ **13.** $w^2 - \frac{7}{3}w + \frac{49}{36} = 0$ **14.** $n^2 - 81 = 0$

Solving Real-Life Problems

EXAMPLE 5 **Modeling with Mathematics**

A bird picks up a golf ball and drops it while flying. The function represents the height y (in feet) of the golf ball t seconds after it is dropped. The ball hits the top of a 32-foot-tall pine tree. After how many seconds does the ball hit the tree?

$y = 81 - 16t^2$

SOLUTION

1. **Understand the Problem** You are given the height of the golf ball as a function of the amount of time after it is dropped and the height of the tree that the golf ball hits. You are asked to determine how many seconds it takes for the ball to hit the tree.

2. **Make a Plan** Use the function for the height of the golf ball. Substitute the height of the tree for y and solve for the time t.

3. **Solve the Problem** Substitute 32 for y and solve for t.

$y = 81 - 16t^2$	Write equation.
$32 = 81 - 16t^2$	Substitute 32 for y.
$0 = 49 - 16t^2$	Subtract 32 from each side.
$0 = 7^2 - (4t)^2$	Write as $a^2 - b^2$.
$0 = (7 + 4t)(7 - 4t)$	Difference of two squares pattern
$7 + 4t = 0$ *or* $7 - 4t = 0$	Zero-Product Property
$t = -\frac{7}{4}$ *or* $t = \frac{7}{4}$	Solve for t.

A negative time does not make sense in this situation.

▶ So, the golf ball hits the tree after $\frac{7}{4}$, or 1.75 seconds.

4. **Look Back** Check your solution, as shown, by substituting $t = \frac{7}{4}$ into the equation $32 = 81 - 16t^2$. Then verify that a time of $\frac{7}{4}$ seconds gives a height of 32 feet.

Check

$$32 = 81 - 16t^2$$
$$32 \stackrel{?}{=} 81 - 16\left(\frac{7}{4}\right)^2$$
$$32 \stackrel{?}{=} 81 - 16\left(\frac{49}{16}\right)$$
$$32 \stackrel{?}{=} 81 - 49$$
$$32 = 32 \checkmark$$

Monitoring Progress  Help in English and Spanish at *BigIdeasMath.com*

15. **WHAT IF?** The golf ball does not hit the pine tree. After how many seconds does the ball hit the ground?

7.7 Exercises

Dynamic Solutions available at *BigIdeasMath.com*

Vocabulary and Core Concept Check

1. **REASONING** Can you use the perfect square trinomial pattern to factor $y^2 + 16y + 64$? Explain.

2. **WHICH ONE DOESN'T BELONG?** Which polynomial does *not* belong with the other three? Explain your reasoning.

$n^2 - 4$	$g^2 - 6g + 9$	$r^2 + 12r + 36$	$k^2 + 25$

Monitoring Progress and Modeling with Mathematics

In Exercises 3–8, factor the polynomial. *(See Example 1.)*

3. $m^2 - 49$
4. $z^2 - 81$
5. $64 - 81d^2$
6. $25 - 4x^2$
7. $225a^2 - 36b^2$
8. $16x^2 - 169y^2$

In Exercises 9–14, use a special product pattern to evaluate the expression. *(See Example 2.)*

9. $12^2 - 9^2$
10. $19^2 - 11^2$
11. $78^2 - 72^2$
12. $54^2 - 52^2$
13. $53^2 - 47^2$
14. $39^2 - 36^2$

In Exercises 15–22, factor the polynomial. *(See Example 3.)*

15. $h^2 + 12h + 36$
16. $p^2 + 30p + 225$
17. $y^2 - 22y + 121$
18. $x^2 - 4x + 4$
19. $a^2 - 28a + 196$
20. $m^2 + 24m + 144$
21. $25n^2 + 20n + 4$
22. $49a^2 - 14a + 1$

ERROR ANALYSIS In Exercises 23 and 24, describe and correct the error in factoring the polynomial.

23.

$$n^2 - 64 = n^2 - 8^2$$
$$= (n - 8)^2$$

24.

$$y^2 - 6y + 9 = y^2 - 2(y)(3) + 3^2$$
$$= (y - 3)(y + 3)$$

25. **MODELING WITH MATHEMATICS** The area (in square centimeters) of a square coaster can be represented by $d^2 + 8d + 16$.
    a. Write an expression that represents the side length of the coaster.
    b. Write an expression for the perimeter of the coaster.

26. **MODELING WITH MATHEMATICS** The polynomial represents the area (in square feet) of the square playground.

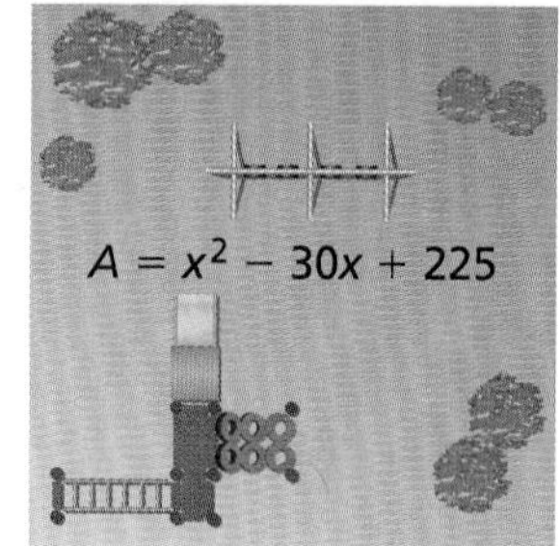

    a. Write a polynomial that represents the side length of the playground.
    b. Write an expression for the perimeter of the playground.

In Exercises 27–34, solve the equation. *(See Example 4.)*

27. $z^2 - 4 = 0$
28. $4x^2 = 49$
29. $k^2 - 16k + 64 = 0$
30. $s^2 + 20s + 100 = 0$
31. $n^2 + 9 = 6n$
32. $y^2 = 12y - 36$
33. $y^2 + \frac{1}{2}y = -\frac{1}{16}$
34. $-\frac{4}{3}x + \frac{4}{9} = -x^2$

In Exercises 35–40, factor the polynomial.

35. $3z^2 - 27$
36. $2m^2 - 50$
37. $4y^2 - 16y + 16$
38. $8k^2 + 80k + 200$
39. $50y^2 + 120y + 72$
40. $27m^2 - 36m + 12$

41. MODELING WITH MATHEMATICS While standing on a ladder, you drop a paintbrush. The function represents the height y (in feet) of the paintbrush t seconds after it is dropped. After how many seconds does the paintbrush land on the ground? *(See Example 5.)*

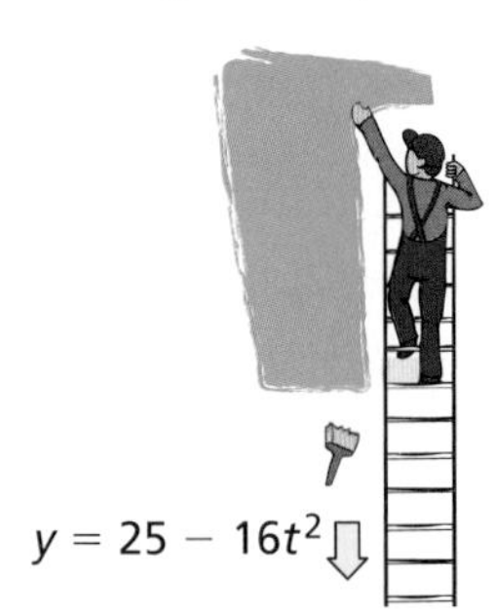

42. MODELING WITH MATHEMATICS The function represents the height y (in feet) of a grasshopper jumping straight up from the ground t seconds after the start of the jump. After how many seconds is the grasshopper 1 foot off the ground?

43. REASONING Tell whether the polynomial can be factored. If not, change the constant term so that the polynomial is a perfect square trinomial.

a. $w^2 + 18w + 84$ **b.** $y^2 - 10y + 23$

44. THOUGHT PROVOKING Use algebra tiles to factor each polynomial modeled by the tiles. Show your work.

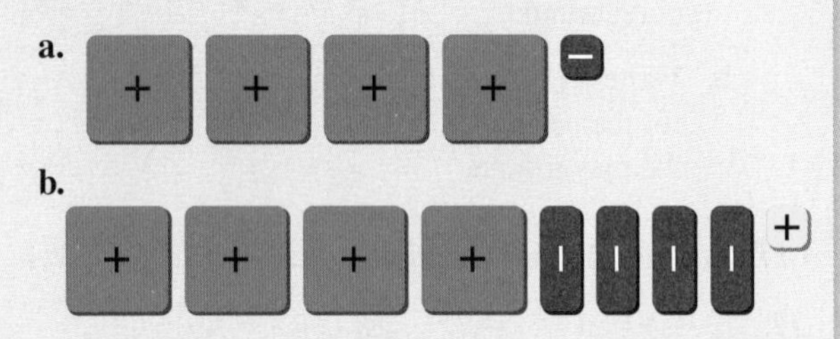

45. COMPARING METHODS Describe two methods you can use to simplify $(2x - 5)^2 - (x - 4)^2$. Which one would you use? Explain.

46. HOW DO YOU SEE IT? The figure shows a large square with an area of a^2 that contains a smaller square with an area of b^2.

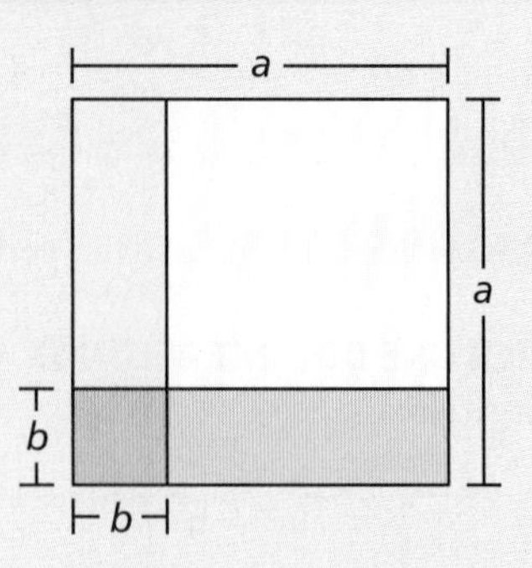

a. Describe the regions that represent $a^2 - b^2$. How can you rearrange these regions to show that $a^2 - b^2 = (a + b)(a - b)$?

b. How can you use the figure to show that $(a - b)^2 = a^2 - 2ab + b^2$?

47. PROBLEM SOLVING You hang nine identical square picture frames on a wall.

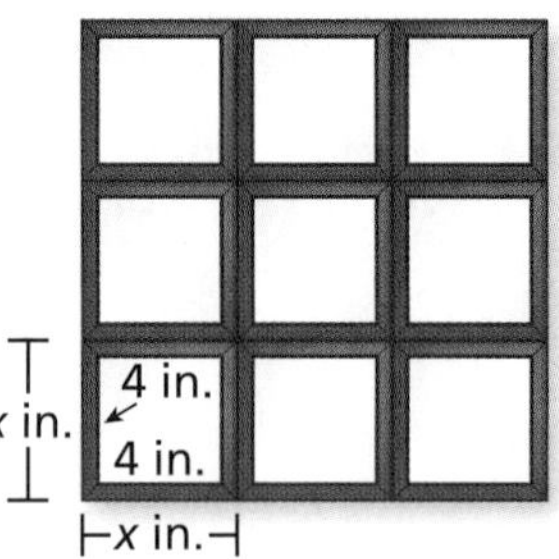

a. Write a polynomial that represents the area of the picture frames, not including the pictures.

b. The area in part (a) is 81 square inches. What is the side length of one of the picture frames? Explain your reasoning.

48. MATHEMATICAL CONNECTIONS The composite solid is made up of a cube and a rectangular prism.

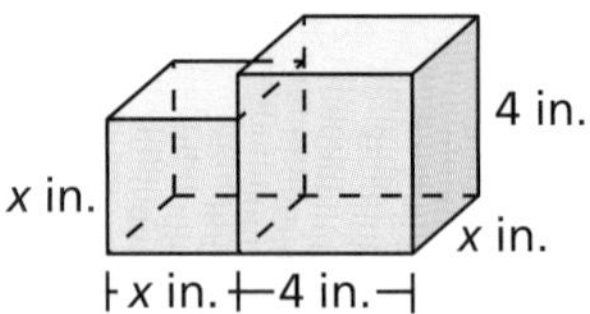

a. Write a polynomial that represents the volume of the composite solid.

b. The volume of the composite solid is equal to $25x$. What is the value of x? Explain your reasoning.

Maintaining Mathematical Proficiency

Reviewing what you learned in previous grades and lessons

Write the prime factorization of the number. *(Skills Review Handbook)*

49. 50 **50.** 44 **51.** 85 **52.** 96

Graph the inequality in a coordinate plane. *(Section 5.6)*

53. $y \leq 4x - 1$ **54.** $y > -\frac{1}{2}x + 3$ **55.** $4y - 12 \geq 8x$ **56.** $3y + 3 < x$

7.8 Factoring Polynomials Completely

Essential Question How can you factor a polynomial completely?

EXPLORATION 1 **Writing a Product of Linear Factors**

Work with a partner. Write the product represented by the algebra tiles. Then multiply to write the polynomial in standard form.

a. (+ +)(+ +)(− −)

b. (+ + +)(+ +)(−)

c. (+ + + +)(+)(+ +)

d. (+ +)(+ −)(+)

e. (− +)(+ +)(−)

f. (− −)(+ +)(− −)

REASONING ABSTRACTLY

To be proficient in math, you need to know and flexibly use different properties of operations and objects.

EXPLORATION 2 **Matching Standard and Factored Forms**

Work with a partner. Match the standard form of the polynomial with the equivalent factored form. Explain your strategy.

a. $x^3 + x^2$ | **A.** $x(x + 1)(x - 1)$

b. $x^3 - x$ | **B.** $x(x - 1)^2$

c. $x^3 + x^2 - 2x$ | **C.** $x(x + 1)^2$

d. $x^3 - 4x^2 + 4x$ | **D.** $x(x + 2)(x - 1)$

e. $x^3 - 2x^2 - 3x$ | **E.** $x(x - 1)(x - 2)$

f. $x^3 - 2x^2 + x$ | **F.** $x(x + 2)(x - 2)$

g. $x^3 - 4x$ | **G.** $x(x - 2)^2$

h. $x^3 + 2x^2$ | **H.** $x(x + 2)^2$

i. $x^3 - x^2$ | **I.** $x^2(x - 1)$

j. $x^3 - 3x^2 + 2x$ | **J.** $x^2(x + 1)$

k. $x^3 + 2x^2 - 3x$ | **K.** $x^2(x - 2)$

l. $x^3 - 4x^2 + 3x$ | **L.** $x^2(x + 2)$

m. $x^3 - 2x^2$ | **M.** $x(x + 3)(x - 1)$

n. $x^3 + 4x^2 + 4x$ | **N.** $x(x + 1)(x - 3)$

o. $x^3 + 2x^2 + x$ | **O.** $x(x - 1)(x - 3)$

Communicate Your Answer

3. How can you factor a polynomial completely?

4. Use your answer to Question 3 to factor each polynomial completely.

a. $x^3 + 4x^2 + 3x$ **b.** $x^3 - 6x^2 + 9x$ **c.** $x^3 + 6x^2 + 9x$

7.8 Lesson

Core Vocabulary

factoring by grouping, *p. 404*
factored completely, *p. 404*

Previous
polynomial
binomial

What You Will Learn

- Factor polynomials by grouping.
- Factor polynomials completely.
- Use factoring to solve real-life problems.

Factoring Polynomials by Grouping

You have used the Distributive Property to factor out a greatest common monomial from a polynomial. Sometimes, you can factor out a common binomial. You may be able to use the Distributive Property to factor polynomials with four terms, as described below.

Core Concept

Factoring by Grouping

To factor a polynomial with four terms, group the terms into pairs. Factor the GCF out of each pair of terms. Look for and factor out the common binomial factor. This process is called **factoring by grouping**.

EXAMPLE 1 Factoring by Grouping

Factor each polynomial by grouping.

a. $x^3 + 3x^2 + 2x + 6$ **b.** $x^2 + y + x + xy$

SOLUTION

a. $x^3 + 3x^2 + 2x + 6 = (x^3 + 3x^2) + (2x + 6)$ Group terms with common factors.

Common binomial factor is $x + 3$. → $= x^2(x + 3) + 2(x + 3)$ Factor out GCF of each pair of terms.

$= (x + 3)(x^2 + 2)$ Factor out $(x + 3)$.

▶ So, $x^3 + 3x^2 + 2x + 6 = (x + 3)(x^2 + 2)$.

b. $x^2 + y + x + xy = x^2 + x + xy + y$ Rewrite polynomial.

$= (x^2 + x) + (xy + y)$ Group terms with common factors.

Common binomial factor is $x + 1$. → $= x(x + 1) + y(x + 1)$ Factor out GCF of each pair of terms.

$= (x + 1)(x + y)$ Factor out $(x + 1)$.

▶ So, $x^2 + y + x + xy = (x + 1)(x + y)$.

Monitoring Progress

Help in English and Spanish at *BigIdeasMath.com*

Factor the polynomial by grouping.

1. $a^3 + 3a^2 + a + 3$ **2.** $y^2 + 2x + yx + 2y$

Factoring Polynomials Completely

You have seen that the polynomial $x^2 - 1$ can be factored as $(x + 1)(x - 1)$. This polynomial is factorable. Notice that the polynomial $x^2 + 1$ cannot be written as the product of polynomials with integer coefficients. This polynomial is unfactorable. A factorable polynomial with integer coefficients is **factored completely** when it is written as a product of unfactorable polynomials with integer coefficients.

Concept Summary

Guidelines for Factoring Polynomials Completely

To factor a polynomial completely, you should try each of these steps.

1. Factor out the greatest common monomial factor. $3x^2 + 6x = 3x(x + 2)$
2. Look for a difference of two squares or a perfect square trinomial. $x^2 + 4x + 4 = (x + 2)^2$
3. Factor a trinomial of the form $ax^2 + bx + c$ into a product of binomial factors. $3x^2 - 5x - 2 = (3x + 1)(x - 2)$
4. Factor a polynomial with four terms by grouping. $x^3 + x - 4x^2 - 4 = (x^2 + 1)(x - 4)$

EXAMPLE 2 Factoring Completely

Factor (a) $3x^3 + 6x^2 - 18x$ and (b) $7x^4 - 28x^2$.

SOLUTION

a. $3x^3 + 6x^2 - 18x = 3x(x^2 + 2x - 6)$ Factor out $3x$.

$x^2 + 2x - 6$ is unfactorable, so the polynomial is factored completely.

▶ So, $3x^3 + 6x^2 - 18x = 3x(x^2 + 2x - 6)$.

b. $7x^4 - 28x^2 = 7x^2(x^2 - 4)$ Factor out $7x^2$.

$= 7x^2(x^2 - 2^2)$ Write as $a^2 - b^2$.

$= 7x^2(x + 2)(x - 2)$ Difference of two squares pattern

▶ So, $7x^4 - 28x^2 = 7x^2(x + 2)(x - 2)$.

EXAMPLE 3 Solving an Equation by Factoring Completely

Solve $2x^3 + 8x^2 = 10x$.

SOLUTION

$2x^3 + 8x^2 = 10x$ Original equation

$2x^3 + 8x^2 - 10x = 0$ Subtract $10x$ from each side.

$2x(x^2 + 4x - 5) = 0$ Factor out $2x$.

$2x(x + 5)(x - 1) = 0$ Factor $x^2 + 4x - 5$.

$2x = 0$ *or* $x + 5 = 0$ *or* $x - 1 = 0$ Zero-Product Property

$x = 0$ *or* $x = -5$ *or* $x = 1$ Solve for x.

▶ The roots are $x = -5$, $x = 0$, and $x = 1$.

Monitoring Progress  Help in English and Spanish at *BigIdeasMath.com*

Factor the polynomial completely.

3. $3x^3 - 12x$ **4.** $2y^3 - 12y^2 + 18y$ **5.** $m^3 - 2m^2 - 8m$

Solve the equation.

6. $w^3 - 8w^2 + 16w = 0$ **7.** $x^3 - 25x = 0$ **8.** $c^3 - 7c^2 + 12c = 0$

Solving Real-Life Problems

EXAMPLE 4 Modeling with Mathematics

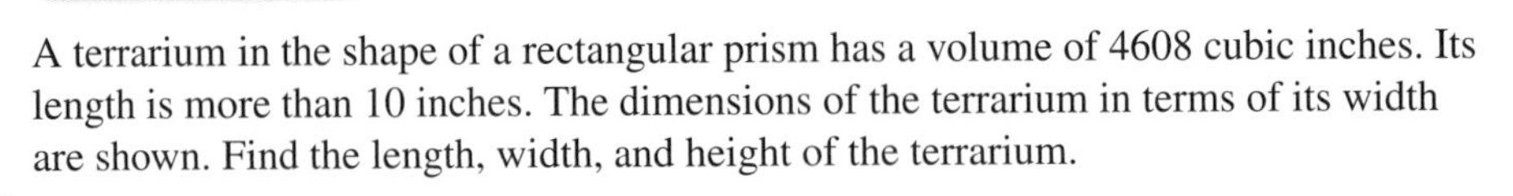

A terrarium in the shape of a rectangular prism has a volume of 4608 cubic inches. Its length is more than 10 inches. The dimensions of the terrarium in terms of its width are shown. Find the length, width, and height of the terrarium.

SOLUTION

1. **Understand the Problem** You are given the volume of a terrarium in the shape of a rectangular prism and a description of the length. The dimensions are written in terms of its width. You are asked to find the length, width, and height of the terrarium.

2. **Make a Plan** Use the formula for the volume of a rectangular prism to write and solve an equation for the width of the terrarium. Then substitute that value in the expressions for the length and height of the terrarium.

3. **Solve the Problem**

$\text{Volume} = \text{length} \cdot \text{width} \cdot \text{height}$	Volume of a rectangular prism
$4608 = (36 - w)(w)(w + 4)$	Write equation.
$4608 = 32w^2 + 144w - w^3$	Multiply.
$0 = 32w^2 + 144w - w^3 - 4608$	Subtract 4608 from each side.
$0 = (-w^3 + 32w^2) + (144w - 4608)$	Group terms with common factors.
$0 = -w^2(w - 32) + 144(w - 32)$	Factor out GCF of each pair of terms.
$0 = (w - 32)(-w^2 + 144)$	Factor out $(w - 32)$.
$0 = -1(w - 32)(w^2 - 144)$	Factor -1 from $-w^2 + 144$.
$0 = -1(w - 32)(w - 12)(w + 12)$	Difference of two squares pattern
$w - 32 = 0$ *or* $w - 12 = 0$ *or* $w + 12 = 0$	Zero-Product Property
$w = 32$ *or* $w = 12$ *or* $w = -12$	Solve for w.

Disregard $w = -12$ because a negative width does not make sense. You know that the length is more than 10 inches. Test the solutions of the equation, 12 and 32, in the expression for the length.

$\text{length} = 36 - w = 36 - 12 = 24$ ✓ *or* $\text{length} = 36 - w = 36 - 32 = 4$ ✗

The solution 12 gives a length of 24 inches, so 12 is the correct value of w.

Use $w = 12$ to find the height, as shown.

$\text{height} = w + 4 = 12 + 4 = 16$

▶ The width is 12 inches, the length is 24 inches, and the height is 16 inches.

Check

$V = \ell wh$

$4608 \stackrel{?}{=} 24(12)(16)$

$4608 = 4608$ ✓

4. **Look Back** Check your solution. Substitute the values for the length, width, and height when the width is 12 inches into the formula for volume. The volume of the terrarium should be 4608 cubic inches.

Monitoring Progress  Help in English and Spanish at *BigIdeasMath.com*

9. A box in the shape of a rectangular prism has a volume of 72 cubic feet. The box has a length of x feet, a width of $(x - 1)$ feet, and a height of $(x + 9)$ feet. Find the dimensions of the box.

7.8 Exercises

Dynamic Solutions available at *BigIdeasMath.com*

Vocabulary and Core Concept Check

1. **VOCABULARY** What does it mean for a polynomial to be factored completely?

2. **WRITING** Explain how to choose which terms to group together when factoring by grouping.

Monitoring Progress and Modeling with Mathematics

In Exercises 3–10, factor the polynomial by grouping. *(See Example 1.)*

3. $x^3 + x^2 + 2x + 2$
4. $y^3 - 9y^2 + y - 9$
5. $3z^3 + 2z - 12z^2 - 8$
6. $2s^3 - 27 - 18s + 3s^2$
7. $x^2 + xy + 8x + 8y$
8. $q^2 + q + 5pq + 5p$
9. $m^2 - 3m + mn - 3n$
10. $2a^2 + 8ab - 3a - 12b$

In Exercises 11–22, factor the polynomial completely. *(See Example 2.)*

11. $2x^3 - 2x$
12. $36a^4 - 4a^2$
13. $2c^2 - 7c + 19$
14. $m^2 - 5m - 35$
15. $6g^3 - 24g^2 + 24g$
16. $-15d^3 + 21d^2 - 6d$
17. $3r^5 + 3r^4 - 90r^3$
18. $5w^4 - 40w^3 + 80w^2$
19. $-4c^4 + 8c^3 - 28c^2$
20. $8t^2 + 8t - 72$
21. $b^3 - 5b^2 - 4b + 20$
22. $h^3 + 4h^2 - 25h - 100$

In Exercises 23–28, solve the equation. *(See Example 3.)*

23. $5n^3 - 30n^2 + 40n = 0$
24. $k^4 - 100k^2 = 0$
25. $x^3 + x^2 = 4x + 4$
26. $2t^5 + 2t^4 - 144t^3 = 0$
27. $12s - 3s^3 = 0$
28. $4y^3 - 7y^2 + 28 = 16y$

In Exercises 29–32, find the x-coordinates of the points where the graph crosses the x-axis.

29.
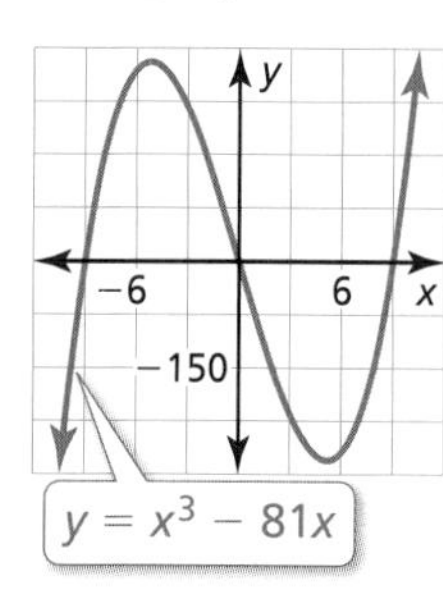

30.
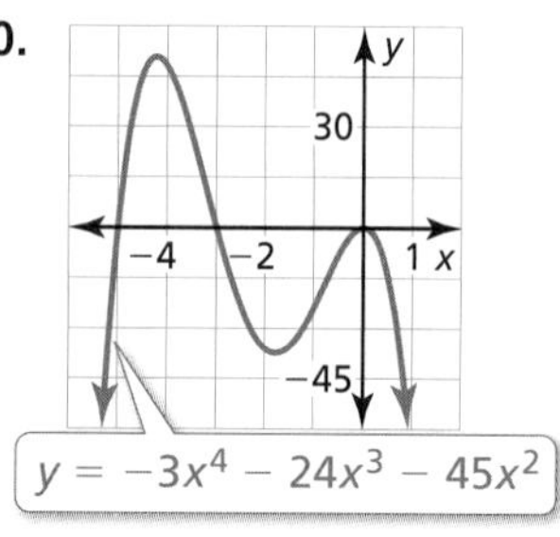

31.
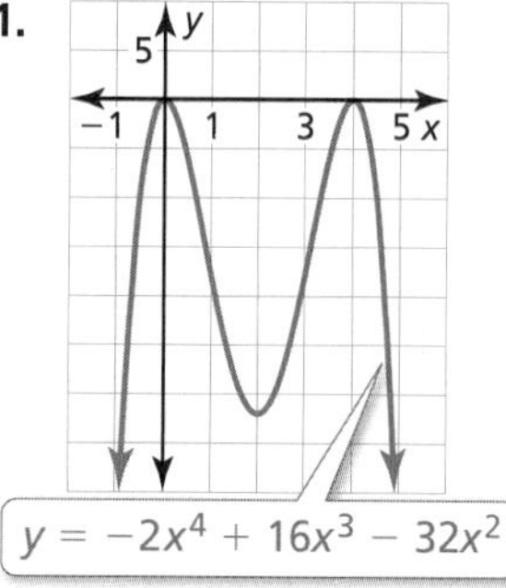

32.
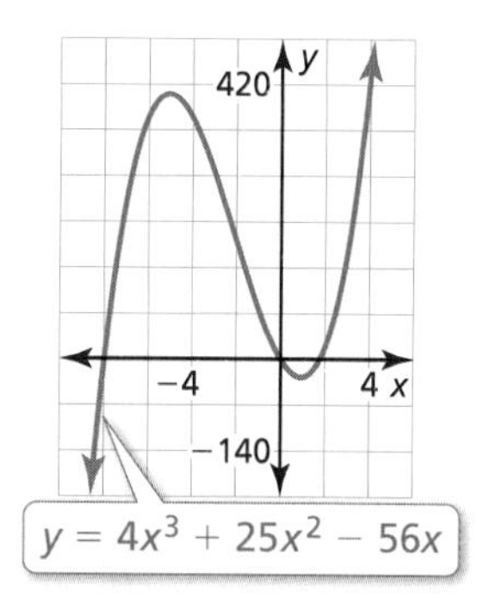

ERROR ANALYSIS In Exercises 33 and 34, describe and correct the error in factoring the polynomial completely.

33. ✗ $a^3 + 8a^2 - 6a - 48 = a^2(a + 8) + 6(a + 8)$
$= (a + 8)(a^2 + 6)$

34. ✗ $x^3 - 6x^2 - 9x + 54 = x^2(x - 6) - 9(x - 6)$
$= (x - 6)(x^2 - 9)$

35. **MODELING WITH MATHEMATICS** You are building a birdhouse in the shape of a rectangular prism that has a volume of 128 cubic inches. The dimensions of the birdhouse in terms of its width are shown. *(See Example 4.)*

a. Write a polynomial that represents the volume of the birdhouse.

b. What are the dimensions of the birdhouse?

36. MODELING WITH MATHEMATICS A gift bag shaped like a rectangular prism has a volume of 1152 cubic inches. The dimensions of the gift bag in terms of its width are shown. The height is greater than the width. What are the dimensions of the gift bag?

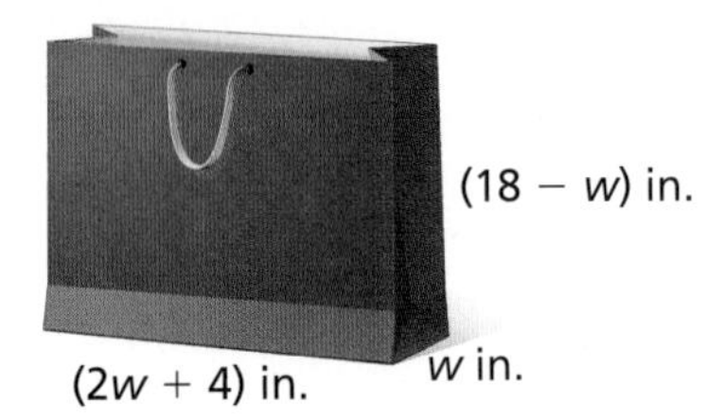

In Exercises 37–40, factor the polynomial completely.

37. $x^3 + 2x^2y - x - 2y$ **38.** $8b^3 - 4b^2a - 18b + 9a$

39. $4s^2 - s + 12st - 3t$

40. $6m^3 - 12mn + m^2n - 2n^2$

41. WRITING Is it possible to find three real solutions of the equation $x^3 + 2x^2 + 3x + 6 = 0$? Explain your reasoning.

42. HOW DO YOU SEE IT? How can you use the factored form of the polynomial $x^4 - 2x^3 - 9x^2 + 18x = x(x - 3)(x + 3)(x - 2)$ to find the x-intercepts of the graph of the function?

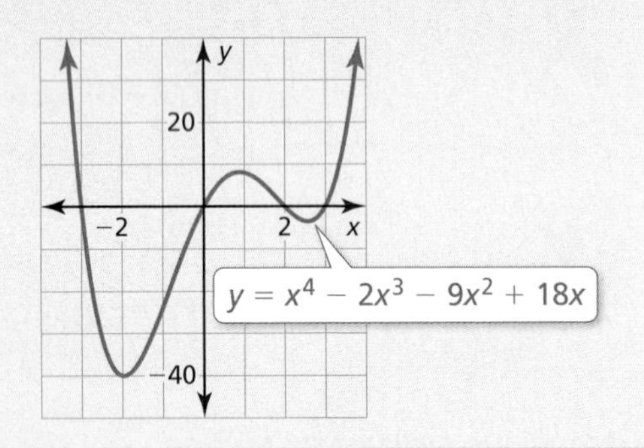

43. OPEN-ENDED Write a polynomial of degree 3 that satisfies each of the given conditions.

a. is not factorable **b.** can be factored by grouping

44. MAKING AN ARGUMENT Your friend says that if a trinomial cannot be factored as the product of two binomials, then the trinomial is factored completely. Is your friend correct? Explain.

45. PROBLEM SOLVING The volume (in cubic feet) of a room in the shape of a rectangular prism is represented by $12z^3 - 27z$. Find expressions that could represent the dimensions of the room.

46. MATHEMATICAL CONNECTIONS The width of a box in the shape of a rectangular prism is 4 inches more than the height h. The length is the difference of 9 inches and the height.

a. Write a polynomial that represents the volume of the box in terms of its height (in inches).

b. The volume of the box is 180 cubic inches. What are the possible dimensions of the box?

c. Which dimensions result in a box with the least possible surface area? Explain your reasoning.

47. MATHEMATICAL CONNECTIONS The volume of a cylinder is given by $V = \pi r^2h$, where r is the radius of the base of the cylinder and h is the height of the cylinder. Find the dimensions of the cylinder.

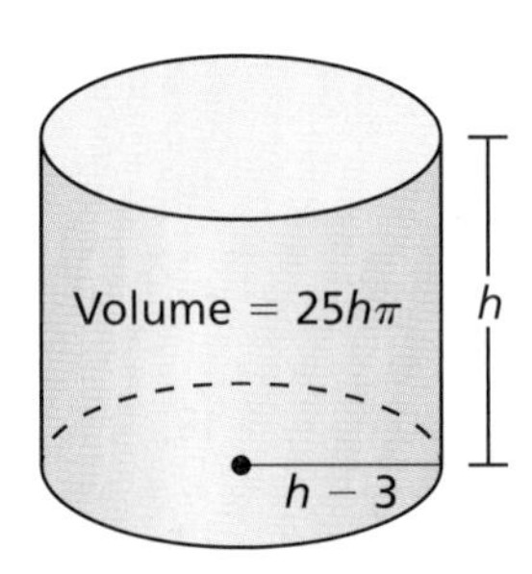

48. THOUGHT PROVOKING Factor the polynomial $x^5 - x^4 - 5x^3 + 5x^2 + 4x - 4$ completely.

49. REASONING Find a value for w so that the equation has (a) two solutions and (b) three solutions. Explain your reasoning.

$$5x^3 + wx^2 + 80x = 0$$

Maintaining Mathematical Proficiency

Reviewing what you learned in previous grades and lessons

Solve the system of linear equations by graphing. *(Section 5.1)*

50. $y = x - 4$
$y = -2x + 2$

51. $y = \frac{1}{2}x + 2$
$y = 3x - 3$

52. $5x - y = 12$
$\frac{1}{4}x + y = 9$

53. $x = 3y$
$y - 10 = 2x$

Graph the function. Describe the domain and range. *(Section 6.3)*

54. $f(x) = 5^x$ **55.** $y = 9\left(\frac{1}{3}\right)^x$ **56.** $y = -3(0.5)^x$ **57.** $f(x) = -3(4)^x$

7.5–7.8 What Did You Learn?

Core Vocabulary

factoring by grouping, *p. 404*
factored completely, *p. 404*

Core Concepts

Section 7.5

Factoring $x^2 + bx + c$ When c Is Positive, *p. 386*
Factoring $x^2 + bx + c$ When c Is Negative, *p. 387*

Section 7.6

Factoring $ax^2 + bx + c$ When ac Is Positive, *p. 392*
Factoring $ax^2 + bx + c$ When ac Is Negative, *p. 393*

Section 7.7

Difference of Two Squares Pattern, *p. 398*
Perfect Square Trinomial Pattern, *p. 399*

Section 7.8

Factoring by Grouping, *p. 404*
Factoring Polynomials Completely, *p. 404*

Mathematical Practices

1. How are the solutions of Exercise 29 on page 389 related to the graph of $y = m^2 + 3m + 2$?
2. The equation in part (b) of Exercise 47 on page 390 has two solutions. Are both solutions of the equation reasonable in the context of the problem? Explain your reasoning.

Performance Task

The View Matters

The way an equation or expression is written can help you interpret and solve problems. Which representation would you rather have when trying to solve for specific information? Why?

To explore the answers to these questions and more, go to ***BigIdeasMath.com***.

7 Chapter Review

Dynamic Solutions available at *BigIdeasMath.com*

7.1 Adding and Subtracting Polynomials *(pp. 357–364)*

Find $(2x^3 + 6x^2 - x) - (-3x^3 - 2x^2 - 9x)$.

$$\begin{aligned}(2x^3 + 6x^2 - x) - (-3x^3 - 2x^2 - 9x) &= (2x^3 + 6x^2 - x) + (3x^3 + 2x^2 + 9x)\\ &= (2x^3 + 3x^3) + (6x^2 + 2x^2) + (-x + 9x)\\ &= 5x^3 + 8x^2 + 8x\end{aligned}$$

Write the polynomial in standard form. Identify the degree and leading coefficient of the polynomial. Then classify the polynomial by the number of terms.

1. $6 + 2x^2$ **2.** $-3p^3 + 5p^6 - 4$ **3.** $9x^7 - 6x^2 + 13x^5$ **4.** $-12y + 8y^3$

Find the sum or difference.

5. $(3a + 7) + (a - 1)$

6. $(x^2 + 6x - 5) + (2x^2 + 15)$

7. $(-y^2 + y + 2) - (y^2 - 5y - 2)$

8. $(p + 7) - (6p^2 + 13p)$

7.2 Multiplying Polynomials *(pp. 365–370)*

Find $(x + 7)(x - 9)$.

$(x + 7)(x - 9) = x(x - 9) + 7(x - 9)$	Distribute $(x - 9)$ to each term of $(x + 7)$.
$= x(x) + x(-9) + 7(x) + 7(-9)$	Distributive Property
$= x^2 + (-9x) + 7x + (-63)$	Multiply.
$= x^2 - 2x - 63$	Combine like terms.

Find the product.

9. $(x + 6)(x - 4)$ **10.** $(y - 5)(3y + 8)$ **11.** $(x + 4)(x^2 + 7x)$ **12.** $(-3y + 1)(4y^2 - y - 7)$

7.3 Special Products of Polynomials *(pp. 371–376)*

Find each product.

a. $(6x + 4y)^2$

$(6x + 4y)^2 = (6x)^2 + 2(6x)(4y) + (4y)^2$	Square of a binomial pattern
$= 36x^2 + 48xy + 16y^2$	Simplify.

b. $(2x + 3y)(2x - 3y)$

$(2x + 3y)(2x - 3y) = (2x)^2 - (3y)^2$	Sum and difference pattern
$= 4x^2 - 9y^2$	Simplify.

Find the product.

13. $(x + 9)(x - 9)$ **14.** $(2y + 4)(2y - 4)$ **15.** $(p + 4)^2$ **16.** $(-1 + 2d)^2$

7.4 Solving Polynomial Equations in Factored Form (pp. 377–382)

Solve $(x + 6)(x - 8) = 0$.

$(x + 6)(x - 8) = 0$	Write equation.
$x + 6 = 0$ *or* $x - 8 = 0$	Zero-Product Property
$x = -6$ *or* $x = 8$	Solve for x.

Solve the equation.

17. $x^2 + 5x = 0$ **18.** $(z + 3)(z - 7) = 0$ **19.** $(b + 13)^2 = 0$ **20.** $2y(y - 9)(y + 4) = 0$

7.5 Factoring $x^2 + bx + c$ (pp. 385–390)

Factor $x^2 + 6x - 27$.

Notice that $b = 6$ and $c = -27$. Because c is negative, the factors p and q must have different signs so that pq is negative.

Find two integer factors of -27 whose sum is 6.

Factors of −27	−27, 1	−1, 27	−9, 3	−3, 9
Sum of factors	−26	26	−6	6

The values of p and q are -3 and 9.

▶ So, $x^2 + 6x - 27 = (x - 3)(x + 9)$.

Factor the polynomial.

21. $p^2 + 2p - 35$ **22.** $b^2 + 18b + 80$ **23.** $z^2 - 4z - 21$ **24.** $x^2 - 11x + 28$

7.6 Factoring $ax^2 + bx + c$ (pp. 391–396)

Factor $5x^2 + 36x + 7$.

There is no GCF, so you need to consider the possible factors of a and c. Because b and c are both positive, the factors of c must be positive. Use a table to organize information about the factors of a and c.

Factors of 5	Factors of 7	Possible factorization	Middle term	
1, 5	1, 7	$(x + 1)(5x + 7)$	$7x + 5x = 12x$	✗
1, 5	7, 1	$(x + 7)(5x + 1)$	$x + 35x = 36x$	✓

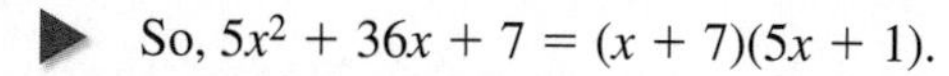

▶ So, $5x^2 + 36x + 7 = (x + 7)(5x + 1)$.

Factor the polynomial.

25. $3t^2 + 16t - 12$ **26.** $-5y^2 - 22y - 8$ **27.** $6x^2 + 17x + 7$

28. $-2y^2 + 7y - 6$ **29.** $3z^2 + 26z - 9$ **30.** $10a^2 - 13a - 3$

7.7 Factoring Special Products (pp. 397–402)

Factor each polynomial.

a. $x^2 - 16$

$x^2 - 16 = x^2 - 4^2$ — Write as $a^2 - b^2$.

$= (x + 4)(x - 4)$ — Difference of two squares pattern

b. $25x^2 - 30x + 9$

$25x^2 - 30x + 9 = (5x)^2 - 2(5x)(3) + 3^2$ — Write as $a^2 - 2ab + b^2$.

$= (5x - 3)^2$ — Perfect square trinomial pattern

Factor the polynomial.

31. $x^2 - 9$ **32.** $y^2 - 100$ **33.** $z^2 - 6z + 9$ **34.** $m^2 + 16m + 64$

7.8 Factoring Polynomials Completely (pp. 403–408)

Factor each polynomial completely.

a. $x^3 + 4x^2 - 3x - 12$

$x^3 + 4x^2 - 3x - 12 = (x^3 + 4x^2) + (-3x - 12)$ — Group terms with common factors.

$= x^2(x + 4) + (-3)(x + 4)$ — Factor out GCF of each pair of terms.

$= (x + 4)(x^2 - 3)$ — Factor out $(x + 4)$.

b. $2x^4 - 8x^2$

$2x^4 - 8x^2 = 2x^2(x^2 - 4)$ — Factor out $2x^2$.

$= 2x^2(x^2 - 2^2)$ — Write as $a^2 - b^2$.

$= 2x^2(x + 2)(x - 2)$ — Difference of two squares pattern

c. $2x^3 + 18x^2 - 72x$

$2x^3 + 18x^2 - 72x = 2x(x^2 + 9x - 36)$ — Factor out $2x$.

$= 2x(x + 12)(x - 3)$ — Factor $x^2 + 9x - 36$.

Factor the polynomial completely.

35. $n^3 - 9n$ **36.** $x^2 - 3x + 4ax - 12a$ **37.** $2x^4 + 2x^3 - 20x^2$

Solve the equation.

38. $3x^3 - 9x^2 - 54x = 0$ **39.** $16x^2 - 36 = 0$ **40.** $z^3 + 3z^2 - 25z - 75 = 0$

41. A box in the shape of a rectangular prism has a volume of 96 cubic feet. The box has a length of $(x + 8)$ feet, a width of x feet, and a height of $(x - 2)$ feet. Find the dimensions of the box.

7 Chapter Test

Find the sum or difference. Then identify the degree of the sum or difference and classify it by the number of terms.

1. $(-2p + 4) - (p^2 - 6p + 8)$

2. $(9c^6 - 5b^4) - (4c^6 - 5b^4)$

3. $(4s^4 + 2st + t) + (2s^4 - 2st - 4t)$

Find the product.

4. $(h - 5)(h - 8)$

5. $(2w - 3)(3w + 5)$

6. $(z + 11)(z - 11)$

7. Explain how you can determine whether a polynomial is a perfect square trinomial.

8. Is 18 a polynomial? Explain your reasoning.

Factor the polynomial completely.

9. $s^2 - 15s + 50$

10. $h^3 + 2h^2 - 9h - 18$

11. $-5k^2 - 22k + 15$

Solve the equation.

12. $(n - 1)(n + 6)(n + 5) = 0$

13. $d^2 + 14d + 49 = 0$

14. $6x^4 + 8x^2 = 26x^3$

15. The expression $\pi(r - 3)^2$ represents the area covered by the hour hand on a clock in one rotation, where r is the radius of the entire clock. Write a polynomial in standard form that represents the area covered by the hour hand in one rotation.

16. A magician's stage has a trapdoor.

a. The total area (in square feet) of the stage can be represented by $x^2 + 27x + 176$. Write an expression for the width of the stage.

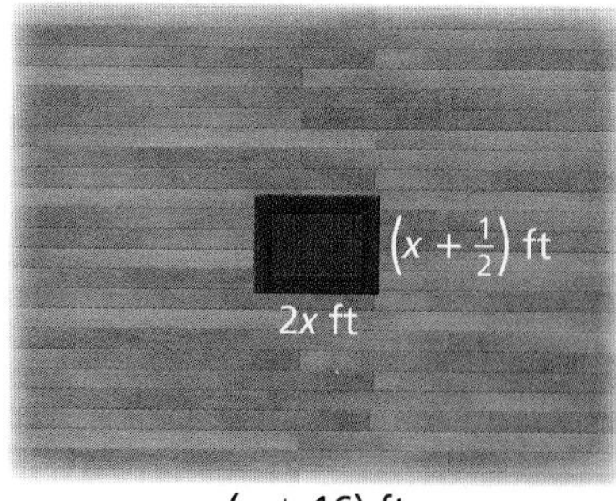

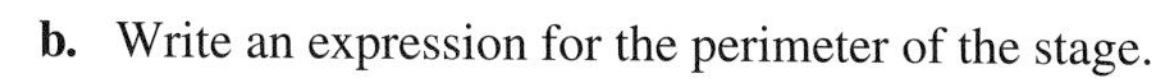

b. Write an expression for the perimeter of the stage.

c. The area of the trapdoor is 10 square feet. Find the value of x.

d. The magician wishes to have the area of the stage be at least 20 times the area of the trapdoor. Does this stage satisfy his requirement? Explain.

17. Write a polynomial equation in factored form that has three positive roots.

18. You are jumping on a trampoline. For one jump, your height y (in feet) above the trampoline after t seconds can be represented by $y = -16t^2 + 24t$. How many seconds are you in the air?

19. A cardboard box in the shape of a rectangular prism has the dimensions shown.

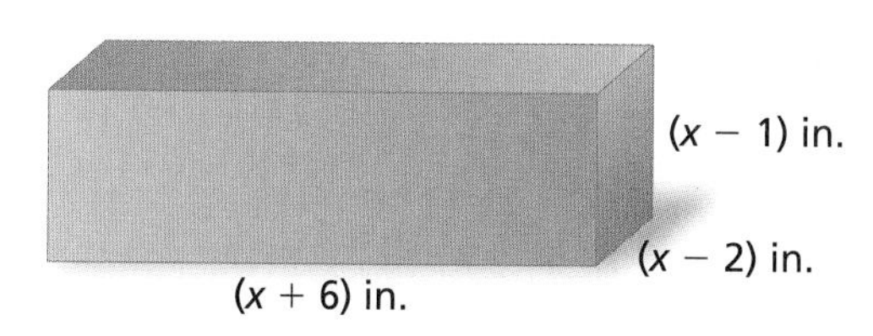

a. Write a polynomial that represents the volume of the box.

b. The volume of the box is 60 cubic inches. What are the length, width, and height of the box?

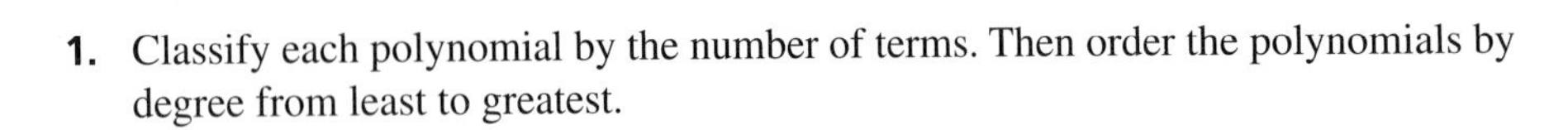

7 Cumulative Assessment

1. Classify each polynomial by the number of terms. Then order the polynomials by degree from least to greatest.

a. $-4x^3$

b. $6y - 3y^5$

c. $c^2 + 2 + c$

d. $-10d^4 + 7d^2$

e. $-5z^{11} + 8z^{12}$

f. $3b^6 - 12b^8 + 4b^4$

2. Which exponential function is increasing the fastest over the interval $x = 0$ to $x = 2$?

Ⓐ $f(x) = 4(2.5)^x$

Ⓑ

x	0	1	2	3	4
g(x)	8	12	18	27	40.5

Ⓒ 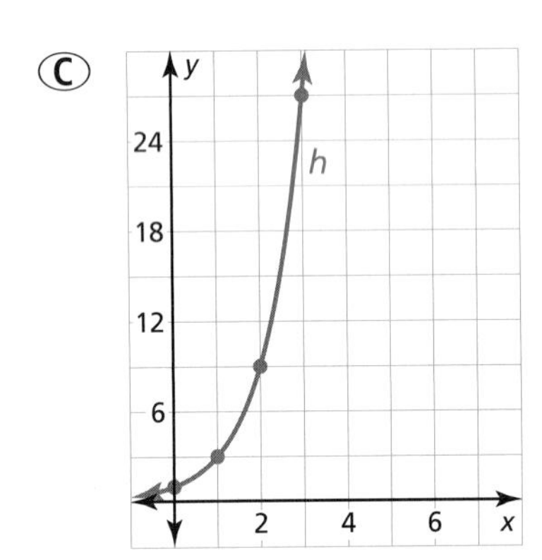

Ⓓ An exponential function j models a relationship in which the dependent variable is multiplied by 6 for every 1 unit the independent variable increases. The value of the function at 0 is 2.

3. Find all solutions of the equation $x^3 + 6x^2 - 4x = 24$.

−6   −4   −2   −1   0   1   2   4   6   24

4. The table shows the distances you travel over a 6-hour period. Create an equation that models the distance traveled as a function of the number of hours.

Hours, x	Distance (miles), y
1	62
2	123
3	184
4	245
5	306
6	367

5. Consider the equation $y = -\frac{1}{3}x + 2$.

a. Graph the equation in a coordinate plane.

b. Does the equation represent a linear or nonlinear function?

c. Is the domain discrete or continuous?

6. Which expressions are equivalent to $-2x + 15x^2 - 8$?

$15x^2 - 2x - 8$	$(5x + 4)(3x + 2)$
$(5x - 4)(3x + 2)$	$15x^2 + 2x - 8$
$(3x - 2)(5x - 4)$	$(3x + 2)(5x - 4)$

7. The graph shows the function $f(x) = 2(3)^x$.

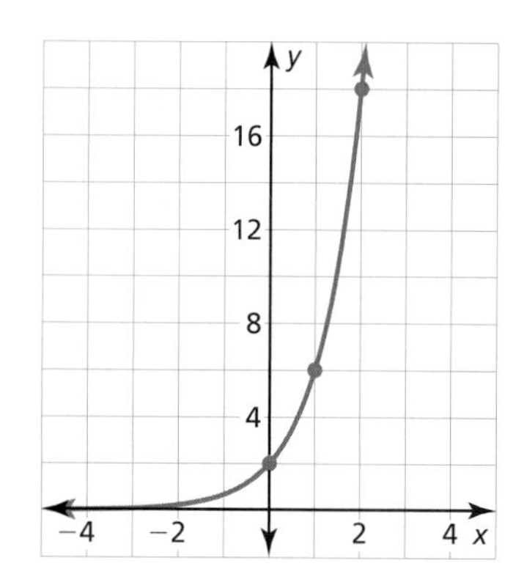

a. Is the function increasing or decreasing for increasing values of x?

b. Identify any x- and y-intercepts.

8. Which polynomial represents the product of $2x - 4$ and $x^2 + 6x - 2$?

Ⓐ $2x^3 + 8x^2 - 4x + 8$

Ⓑ $2x^3 + 8x^2 - 28x + 8$

Ⓒ $2x^3 + 8$

Ⓓ $2x^3 - 24x - 2$

9. You are playing miniature golf on the hole shown.

a. Write a polynomial that represents the area of the golf hole.

b. Write a polynomial that represents the perimeter of the golf hole.

c. Find the perimeter of the golf hole when the area is 216 square feet.

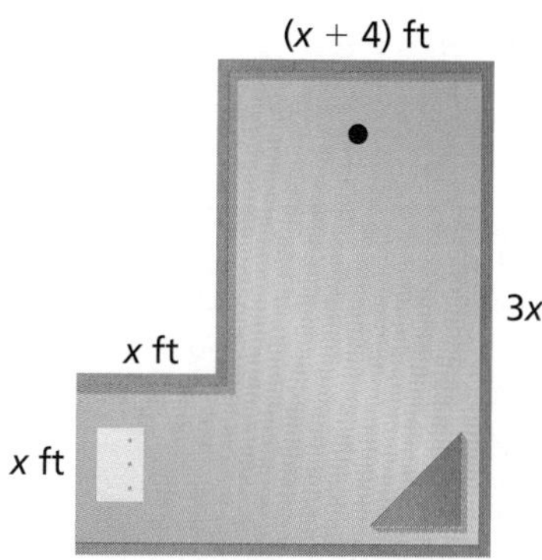

8 Graphing Quadratic Functions

Town Population *(p. 464)*

Satellite Dish *(p. 457)*

Roller Coaster *(p. 448)*

Firework Explosion *(p. 437)*

Garden Waterfalls *(p. 430)*

Maintaining Mathematical Proficiency

Graphing Linear Equations

Example 1 **Graph $y = -x - 1$.**

Step 1 Make a table of values.

x	$y = -x - 1$	y	(x, y)
-1	$y = -(-1) - 1$	0	$(-1, 0)$
0	$y = -(0) - 1$	-1	$(0, -1)$
1	$y = -(1) - 1$	-2	$(1, -2)$
2	$y = -(2) - 1$	-3	$(2, -3)$

Step 2 Plot the ordered pairs.

Step 3 Draw a line through the points.

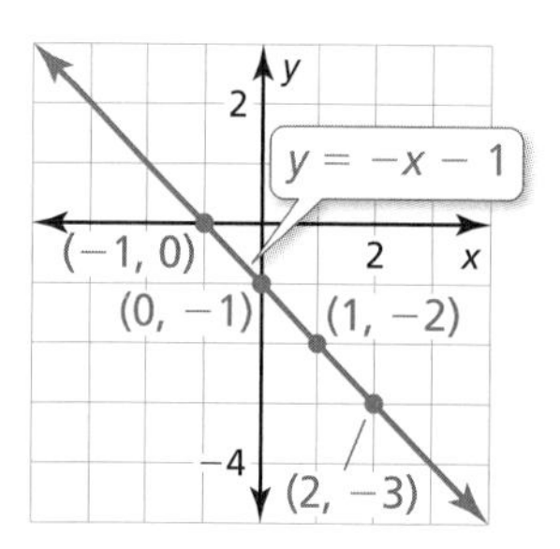

Graph the linear equation.

1. $y = 2x - 3$

2. $y = -3x + 4$

3. $y = -\frac{1}{2}x - 2$

4. $y = x + 5$

Evaluating Expressions

Example 2 **Evaluate $2x^2 + 3x - 5$ when $x = -1$.**

$$
\begin{aligned}
2x^2 + 3x - 5 &= 2(-1)^2 + 3(-1) - 5 && \text{Substitute } -1 \text{ for } x.\\
&= 2(1) + 3(-1) - 5 && \text{Evaluate the power.}\\
&= 2 - 3 - 5 && \text{Multiply.}\\
&= -6 && \text{Subtract.}
\end{aligned}
$$

Evaluate the expression when $x = -2$.

5. $5x^2 - 9$

6. $3x^2 + x - 2$

7. $-x^2 + 4x + 1$

8. $x^2 + 8x + 5$

9. $-2x^2 - 4x + 3$

10. $-4x^2 + 2x - 6$

11. ABSTRACT REASONING Complete the table. Find a pattern in the differences of consecutive y-values. Use the pattern to write an expression for y when $x = 6$.

x	1	2	3	4	5
$y = ax^2$					

Mathematical Practices

Mathematically proficient students try special cases of the original problem to gain insight into its solution.

Problem-Solving Strategies

Core Concept

Trying Special Cases

When solving a problem in mathematics, it can be helpful to try special cases of the original problem. For instance, in this chapter, you will learn to graph a quadratic function of the form $f(x) = ax^2 + bx + c$. The problem-solving strategy used is to first graph quadratic functions of the form $f(x) = ax^2$. From there, you progress to other forms of quadratic functions.

$f(x) = ax^2$	Section 8.1
$f(x) = ax^2 + c$	Section 8.2
$f(x) = ax^2 + bx + c$	Section 8.3
$f(x) = a(x - h)^2 + k$	Section 8.4

EXAMPLE 1 Graphing the Parent Quadratic Function

Graph the parent quadratic function $y = x^2$. Then describe its graph.

SOLUTION

The function is of the form $y = ax^2$, where $a = 1$. By plotting several points, you can see that the graph is U-shaped, as shown.

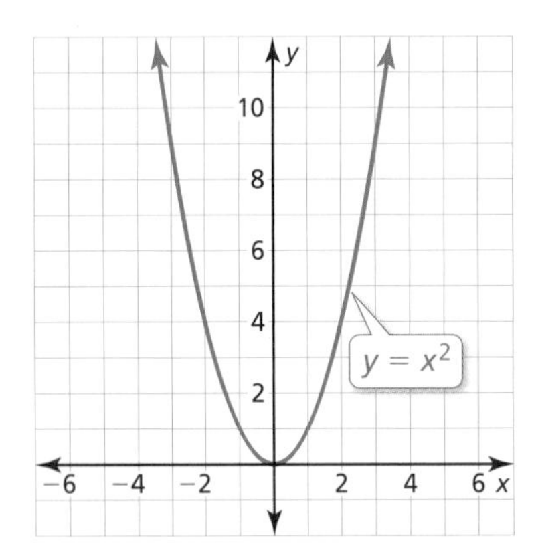

▶ The graph opens up, and the lowest point is at the origin.

Monitoring Progress

Graph the quadratic function. Then describe its graph.

1. $y = -x^2$
2. $y = 2x^2$
3. $f(x) = 2x^2 + 1$
4. $f(x) = 2x^2 - 1$
5. $f(x) = \frac{1}{2}x^2 + 4x + 3$
6. $f(x) = \frac{1}{2}x^2 - 4x + 3$
7. $y = -2(x + 1)^2 + 1$
8. $y = -2(x - 1)^2 + 1$
9. How are the graphs in Monitoring Progress Questions 1−8 similar? How are they different?

8.1 Graphing $f(x) = ax^2$

Essential Question What are some of the characteristics of the graph of a quadratic function of the form $f(x) = ax^2$?

EXPLORATION 1 Graphing Quadratic Functions

Work with a partner. Graph each quadratic function. Compare each graph to the graph of $f(x) = x^2$.

a. $g(x) = 3x^2$

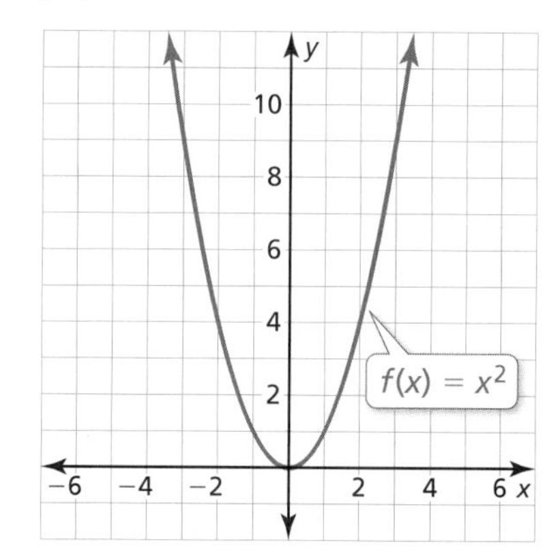

b. $g(x) = -5x^2$

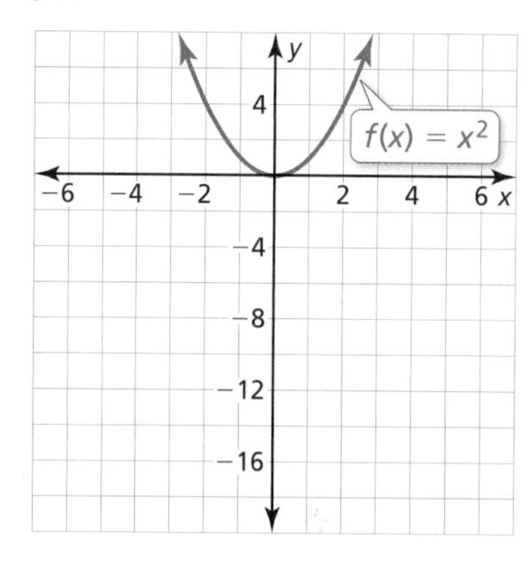

c. $g(x) = -0.2x^2$

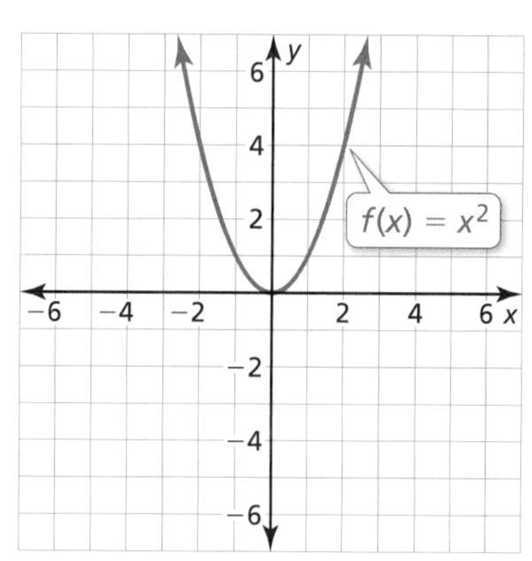

d. $g(x) = \frac{1}{10}x^2$

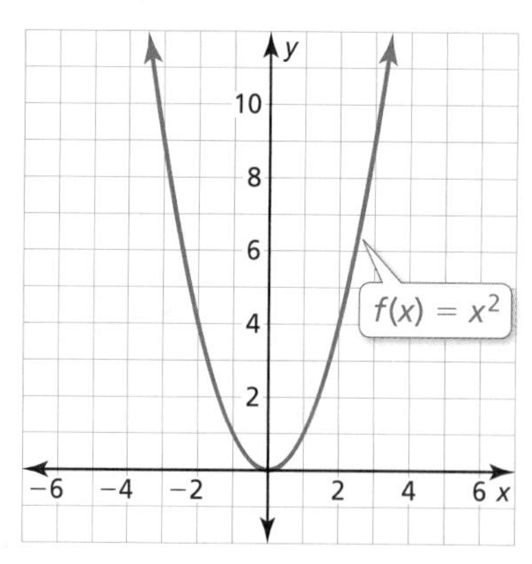

REASONING QUANTITATIVELY

To be proficient in math, you need to make sense of quantities and their relationships in problem situations.

Communicate Your Answer

2. What are some of the characteristics of the graph of a quadratic function of the form $f(x) = ax^2$?

3. How does the value of a affect the graph of $f(x) = ax^2$? Consider $0 < a < 1$, $a > 1$, $-1 < a < 0$, and $a < -1$. Use a graphing calculator to verify your answers.

4. The figure shows the graph of a quadratic function of the form $y = ax^2$. Which of the intervals in Question 3 describes the value of a? Explain your reasoning.

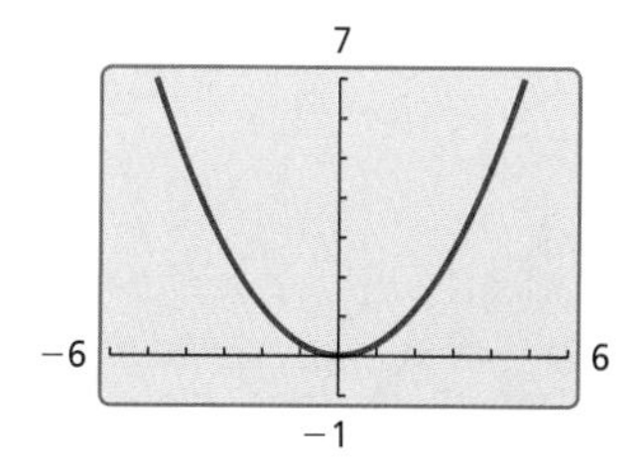

8.1 Lesson

Core Vocabulary

quadratic function, *p. 420*
parabola, *p. 420*
vertex, *p. 420*
axis of symmetry, *p. 420*

Previous
domain
range
vertical shrink
vertical stretch
reflection

What You Will Learn

- Identify characteristics of quadratic functions.
- Graph and use quadratic functions of the form $f(x) = ax^2$.

Identifying Characteristics of Quadratic Functions

A **quadratic function** is a nonlinear function that can be written in the standard form $y = ax^2 + bx + c$, where $a \neq 0$. The U-shaped graph of a quadratic function is called a **parabola**. In this lesson, you will graph quadratic functions, where b and c equal 0.

REMEMBER

The notation $f(x)$ is another name for y.

Core Concept

Characteristics of Quadratic Functions

The *parent quadratic function* is $f(x) = x^2$. The graphs of all other quadratic functions are *transformations* of the graph of the parent quadratic function.

The lowest point on a parabola that opens up or the highest point on a parabola that opens down is the **vertex**. The vertex of the graph of $f(x) = x^2$ is (0, 0).

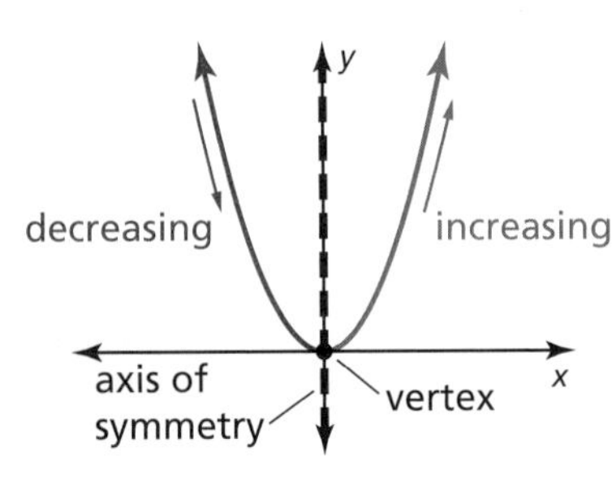

The vertical line that divides the parabola into two symmetric parts is the **axis of symmetry**. The axis of symmetry passes through the vertex. For the graph of $f(x) = x^2$, the axis of symmetry is the y-axis, or $x = 0$.

EXAMPLE 1 Identifying Characteristics of a Quadratic Function

Consider the graph of the quadratic function.

Using the graph, you can identify characteristics such as the vertex, axis of symmetry, and the behavior of the graph, as shown.

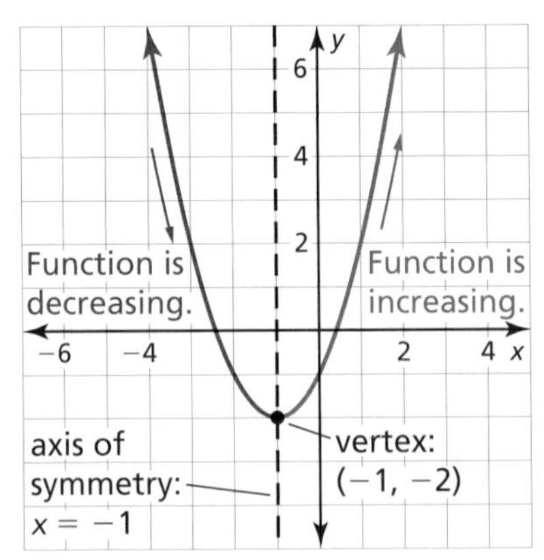

You can also determine the following:

- The domain is all real numbers.
- The range is all real numbers greater than or equal to -2.
- When $x < -1$, y increases as x decreases.
- When $x > -1$, y increases as x increases.

Monitoring Progress

Help in English and Spanish at *BigIdeasMath.com*

Identify characteristics of the quadratic function and its graph.

1.

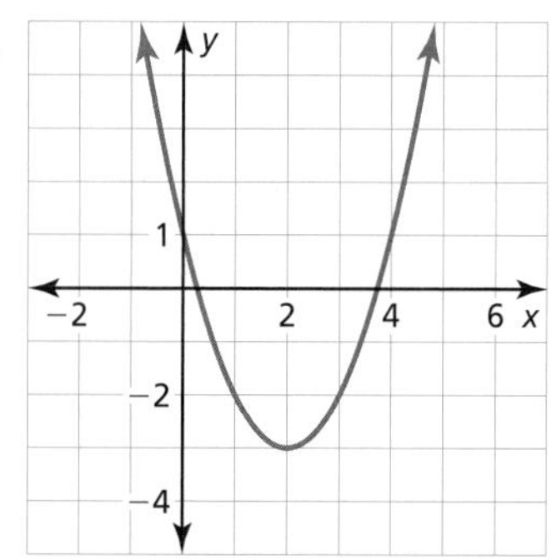

2.

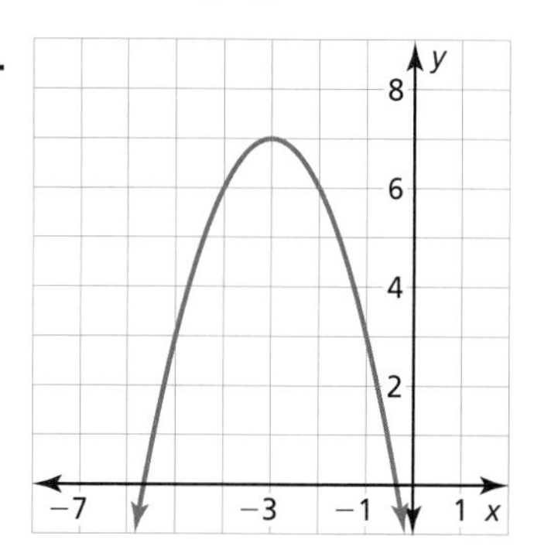

REMEMBER

The graph of $y = a \cdot f(x)$ is a vertical stretch or shrink by a factor of a of the graph of $y = f(x)$.

The graph of $y = -f(x)$ is a reflection in the x-axis of the graph of $y = f(x)$.

Graphing and Using $f(x) = ax^2$

Core Concept

Graphing $f(x) = ax^2$ When $a > 0$

- When $0 < a < 1$, the graph of $f(x) = ax^2$ is a vertical shrink of the graph of $f(x) = x^2$.
- When $a > 1$, the graph of $f(x) = ax^2$ is a vertical stretch of the graph of $f(x) = x^2$.

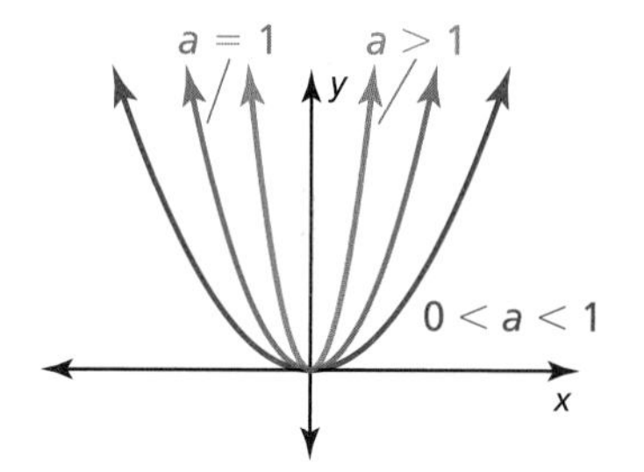

Graphing $f(x) = ax^2$ When $a < 0$

- When $-1 < a < 0$, the graph of $f(x) = ax^2$ is a vertical shrink with a reflection in the x-axis of the graph of $f(x) = x^2$.
- When $a < -1$, the graph of $f(x) = ax^2$ is a vertical stretch with a reflection in the x-axis of the graph of $f(x) = x^2$.

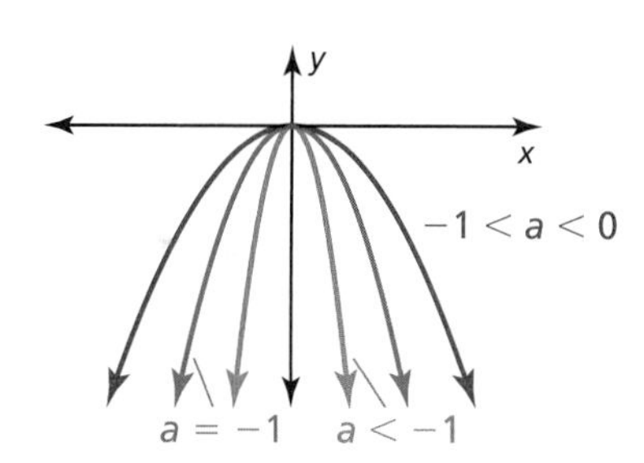

EXAMPLE 2 Graphing $y = ax^2$ When $a > 0$

Graph $g(x) = 2x^2$. Compare the graph to the graph of $f(x) = x^2$.

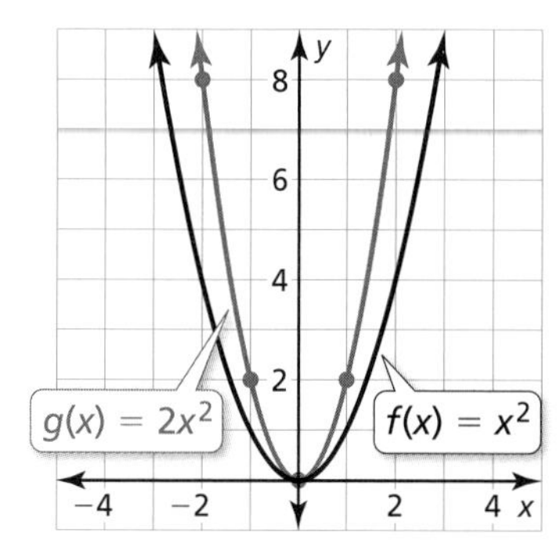

SOLUTION

Step 1 Make a table of values.

x	−2	−1	0	1	2
$g(x)$	8	2	0	2	8

Step 2 Plot the ordered pairs.

Step 3 Draw a smooth curve through the points.

▶ Both graphs open up and have the same vertex, (0, 0), and the same axis of symmetry, $x = 0$. The graph of g is narrower than the graph of f because the graph of g is a vertical stretch by a factor of 2 of the graph of f.

EXAMPLE 3 Graphing $y = ax^2$ When $a < 0$

STUDY TIP

To make the calculations easier, choose x-values that are multiples of 3.

Graph $h(x) = -\frac{1}{3}x^2$. Compare the graph to the graph of $f(x) = x^2$.

SOLUTION

Step 1 Make a table of values.

x	-6	-3	0	3	6
$h(x)$	-12	-3	0	-3	-12

Step 2 Plot the ordered pairs.

Step 3 Draw a smooth curve through the points.

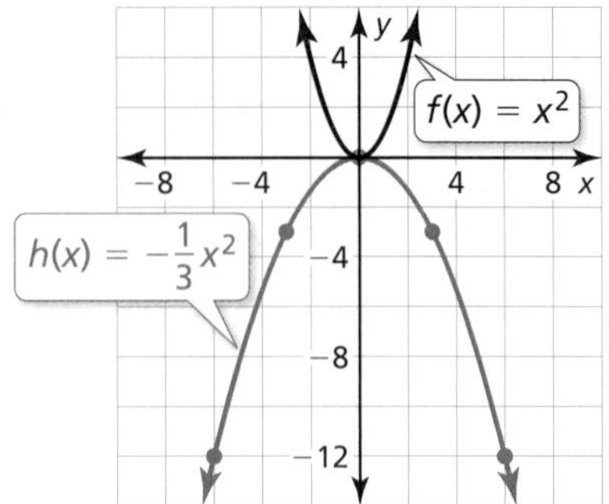

The graphs have the same vertex, $(0, 0)$, and the same axis of symmetry, $x = 0$, but the graph of h opens down and is wider than the graph of f. So, the graph of h is a vertical shrink by a factor of $\frac{1}{3}$ and a reflection in the x-axis of the graph of f.

Monitoring Progress

Help in English and Spanish at *BigIdeasMath.com*

Graph the function. Compare the graph to the graph of $f(x) = x^2$.

3. $g(x) = 5x^2$

4. $h(x) = \frac{1}{3}x^2$

5. $n(x) = \frac{3}{2}x^2$

6. $p(x) = -3x^2$

7. $q(x) = -0.1x^2$

8. $g(x) = -\frac{1}{4}x^2$

EXAMPLE 4 Solving a Real-Life Problem

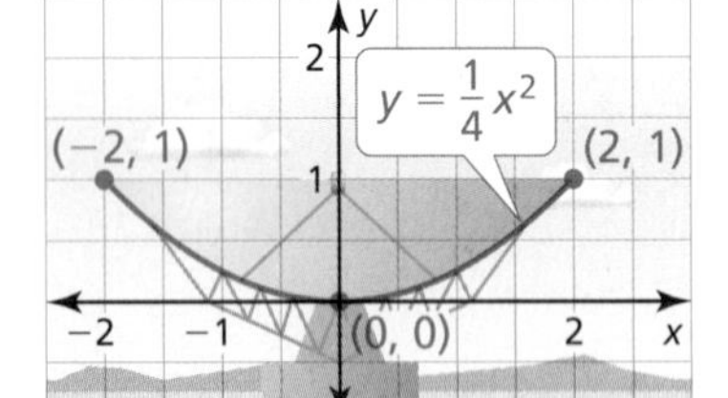

The diagram at the left shows the cross section of a satellite dish, where x and y are measured in meters. Find the width and depth of the dish.

SOLUTION

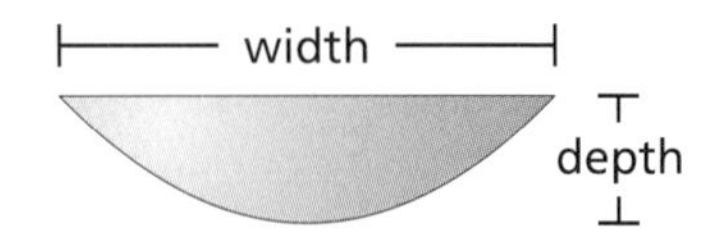

Use the domain of the function to find the width of the dish. Use the range to find the depth.

The leftmost point on the graph is $(-2, 1)$, and the rightmost point is $(2, 1)$. So, the domain is $-2 \le x \le 2$, which represents 4 meters.

The lowest point on the graph is $(0, 0)$, and the highest points on the graph are $(-2, 1)$ and $(2, 1)$. So, the range is $0 \le y \le 1$, which represents 1 meter.

So, the satellite dish is 4 meters wide and 1 meter deep.

Monitoring Progress

Help in English and Spanish at *BigIdeasMath.com*

9. The cross section of a spotlight can be modeled by the graph of $y = 0.5x^2$, where x and y are measured in inches and $-2 \le x \le 2$. Find the width and depth of the spotlight.

8.1 Exercises

Dynamic Solutions available at *BigIdeasMath.com*

Vocabulary and Core Concept Check

1. **VOCABULARY** What is the U-shaped graph of a quadratic function called?

2. **WRITING** When does the graph of a quadratic function open up? open down?

Monitoring Progress and Modeling with Mathematics

In Exercises 3 and 4, identify characteristics of the quadratic function and its graph. *(See Example 1.)*

3.

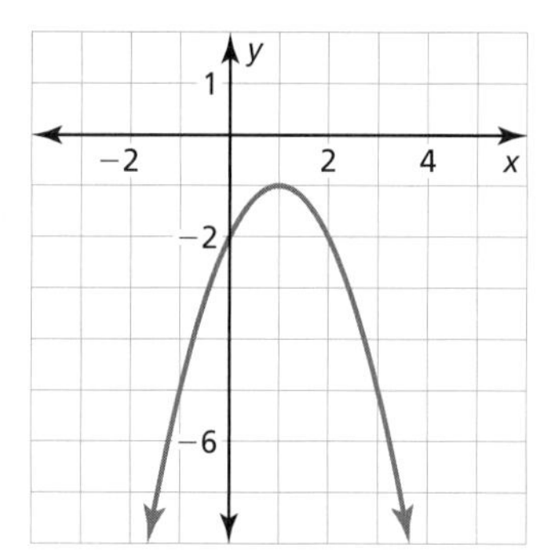

4.

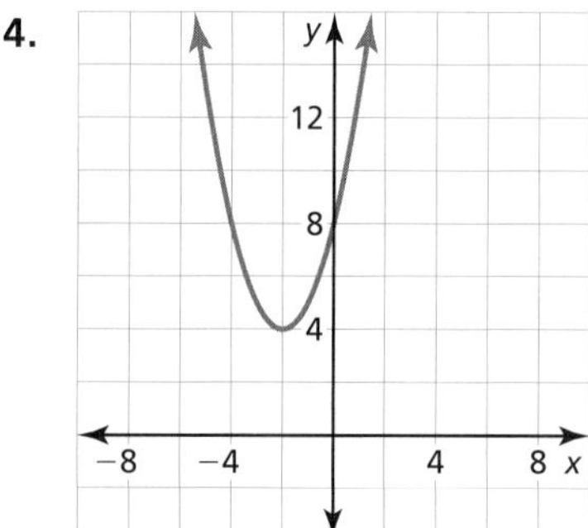

In Exercises 5–12, graph the function. Compare the graph to the graph of $f(x) = x^2$. *(See Examples 2 and 3.)*

5. $g(x) = 6x^2$

6. $b(x) = 2.5x^2$

7. $h(x) = \frac{1}{4}x^2$

8. $j(x) = 0.75x^2$

9. $m(x) = -2x^2$

10. $q(x) = -\frac{9}{2}x^2$

11. $k(x) = -0.2x^2$

12. $p(x) = -\frac{2}{3}x^2$

In Exercises 13–16, use a graphing calculator to graph the function. Compare the graph to the graph of $y = -4x^2$.

13. $y = 4x^2$

14. $y = -0.4x^2$

15. $y = -0.04x^2$

16. $y = -0.004x^2$

17. **ERROR ANALYSIS** Describe and correct the error in graphing and comparing $y = x^2$ and $y = 0.5x^2$.

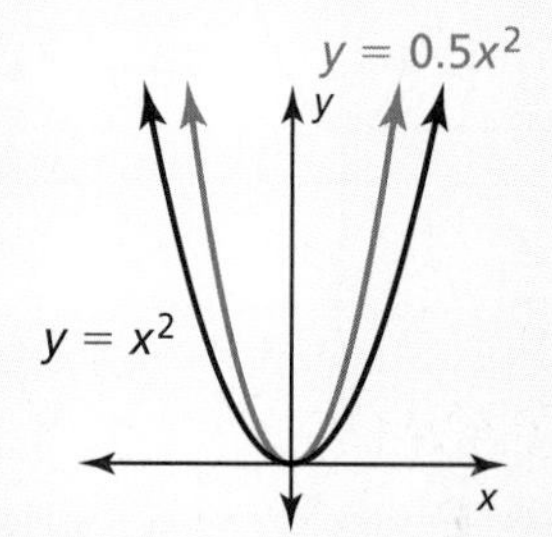

The graphs have the same vertex and the same axis of symmetry. The graph of $y = 0.5x^2$ is narrower than the graph of $y = x^2$.

18. **MODELING WITH MATHEMATICS** The arch support of a bridge can be modeled by $y = -0.0012x^2$, where x and y are measured in feet. Find the height and width of the arch. *(See Example 4.)*

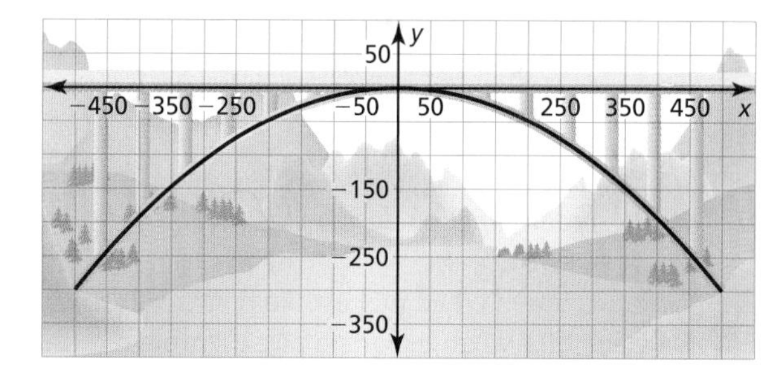

19. **PROBLEM SOLVING** The breaking strength z (in pounds) of a manila rope can be modeled by $z = 8900d^2$, where d is the diameter (in inches) of the rope.

    a. Describe the domain and range of the function.

    b. Graph the function using the domain in part (a).

    c. A manila rope has four times the breaking strength of another manila rope. Does the stronger rope have four times the diameter? Explain.

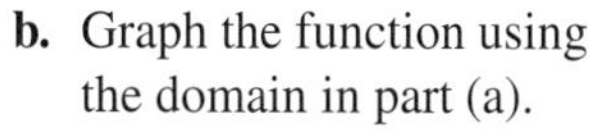

20. HOW DO YOU SEE IT? Describe the possible values of a.

a.

$g(x) = ax^2$ $f(x) = x^2$

b.

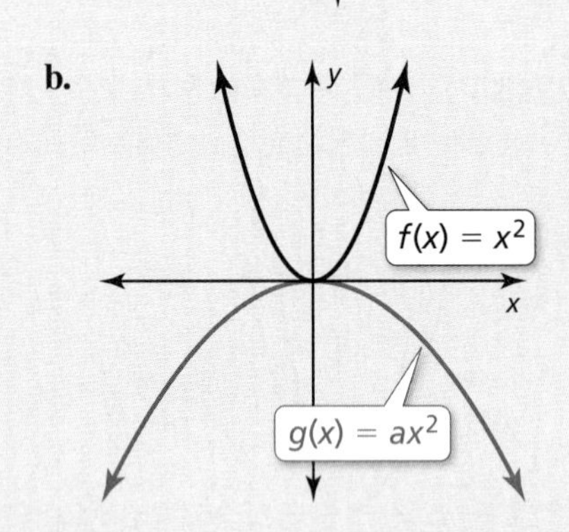

ANALYZING GRAPHS **In Exercises 21–23, use the graph.**

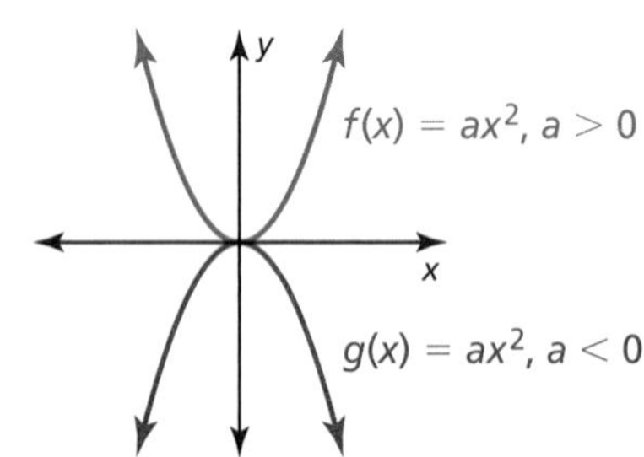

21. When is each function increasing?

22. When is each function decreasing?

23. Which function could include the point $(-2, 3)$? Find the value of a when the graph passes through $(-2, 3)$.

24. REASONING Is the x-intercept of the graph of $y = ax^2$ always 0? Justify your answer.

25. REASONING A parabola opens up and passes through $(-4, 2)$ and $(6, -3)$. How do you know that $(-4, 2)$ is not the vertex?

ABSTRACT REASONING **In Exercises 26–29, determine whether the statement is *always*, *sometimes*, or *never* true. Explain your reasoning.**

26. The graph of $f(x) = ax^2$ is narrower than the graph of $g(x) = x^2$ when $a > 0$.

27. The graph of $f(x) = ax^2$ is narrower than the graph of $g(x) = x^2$ when $|a| > 1$.

28. The graph of $f(x) = ax^2$ is wider than the graph of $g(x) = x^2$ when $0 < |a| < 1$.

29. The graph of $f(x) = ax^2$ is wider than the graph of $g(x) = dx^2$ when $|a| > |d|$.

30. THOUGHT PROVOKING Draw the isosceles triangle shown. Divide each leg into eight congruent segments. Connect the highest point of one leg with the lowest point of the other leg. Then connect the second highest point of one leg to the second lowest point of the other leg. Continue this process. Write a quadratic equation whose graph models the shape that appears.

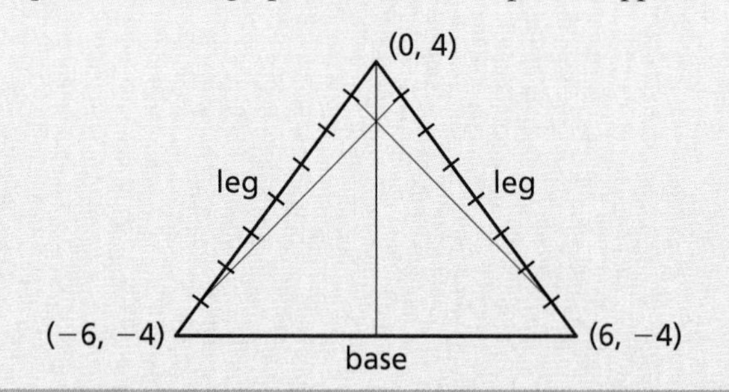

31. MAKING AN ARGUMENT The diagram shows the parabolic cross section of a swirling glass of water, where x and y are measured in centimeters.

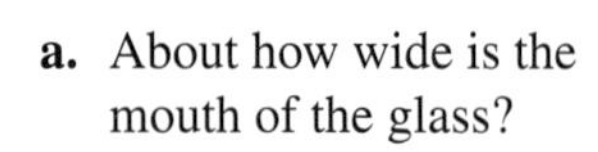

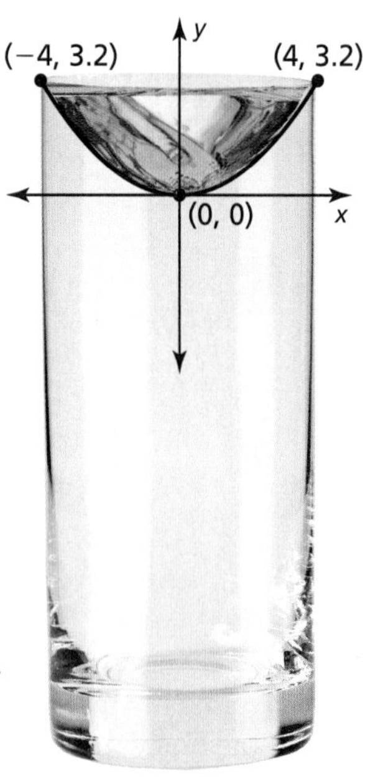

a. About how wide is the mouth of the glass?

b. Your friend claims that the rotational speed of the water would have to increase for the cross section to be modeled by $y = 0.1x^2$. Is your friend correct? Explain your reasoning.

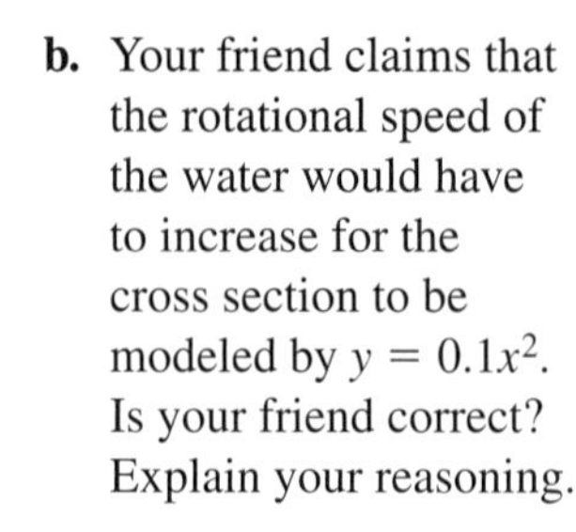

Maintaining Mathematical Proficiency

Reviewing what you learned in previous grades and lessons

Evaluate the expression when $n = 3$ and $x = -2$. *(Skills Review Handbook)*

32. $n^2 + 5$ **33.** $3x^2 - 9$ **34.** $-4n^2 + 11$ **35.** $n + 2x^2$

8.2 Graphing $f(x) = ax^2 + c$

Essential Question How does the value of c affect the graph of $f(x) = ax^2 + c$?

EXPLORATION 1 Graphing $y = ax^2 + c$

Work with a partner. Sketch the graphs of the functions in the same coordinate plane. What do you notice?

a. $f(x) = x^2$ and $g(x) = x^2 + 2$

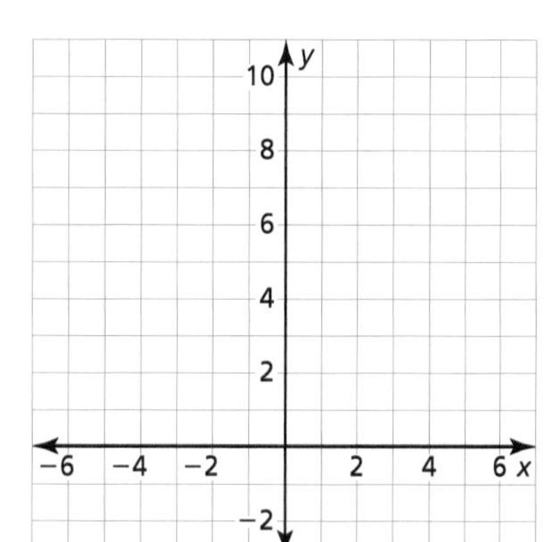

b. $f(x) = 2x^2$ and $g(x) = 2x^2 - 2$

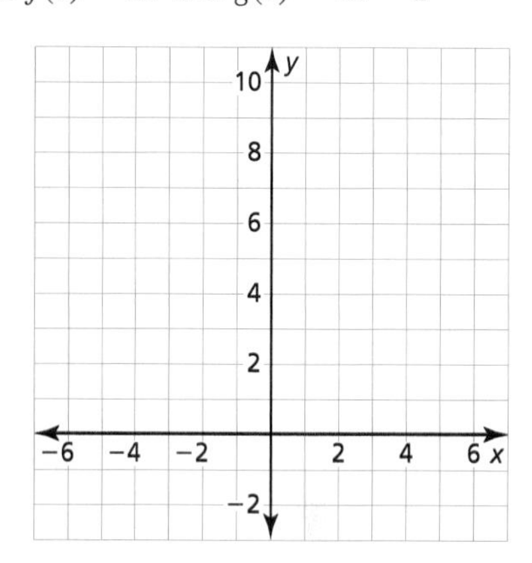

EXPLORATION 2 Finding x-Intercepts of Graphs

Work with a partner. Graph each function. Find the x-intercepts of the graph. Explain how you found the x-intercepts.

a. $y = x^2 - 7$

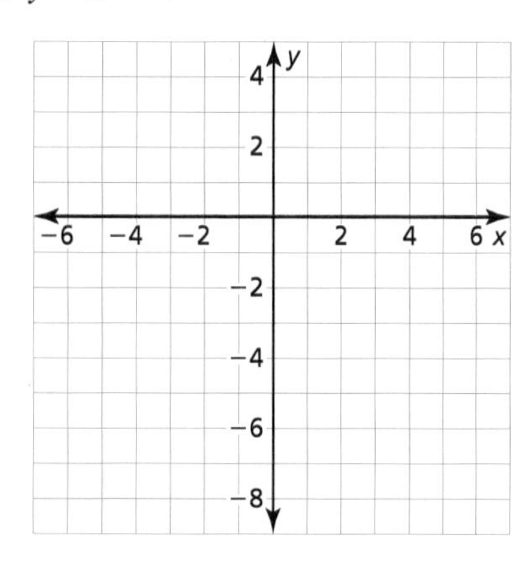

b. $y = -x^2 + 1$

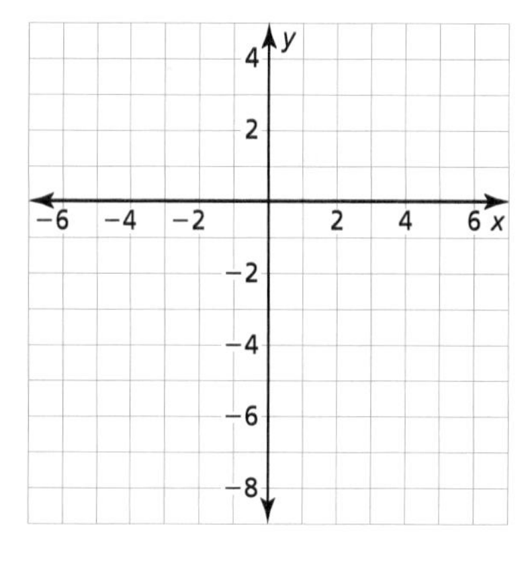

USING TOOLS STRATEGICALLY

To be proficient in math, you need to consider the available tools, such as a graphing calculator, when solving a mathematical problem.

Communicate Your Answer

3. How does the value of c affect the graph of $f(x) = ax^2 + c$?

4. Use a graphing calculator to verify your answers to Question 3.

5. The figure shows the graph of a quadratic function of the form $y = ax^2 + c$. Describe possible values of a and c. Explain your reasoning.

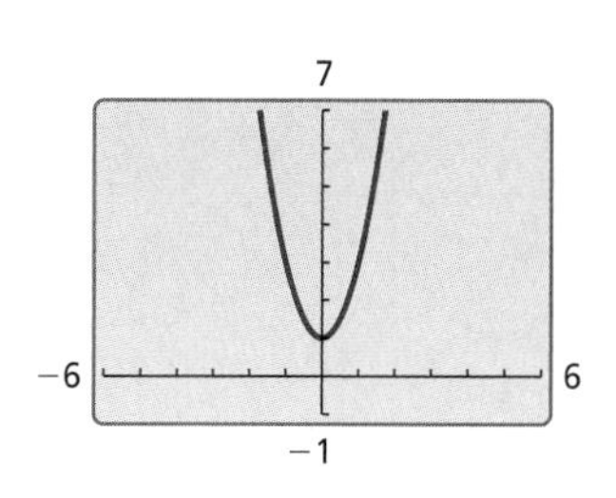

8.2 Lesson

Core Vocabulary

zero of a function, *p. 428*

Previous
translation
vertex of a parabola
axis of symmetry
vertical stretch
vertical shrink

What You Will Learn

- Graph quadratic functions of the form $f(x) = ax^2 + c$.
- Solve real-life problems involving functions of the form $f(x) = ax^2 + c$.

Graphing $f(x) = ax^2 + c$

Core Concept

Graphing $f(x) = ax^2 + c$

- When $c > 0$, the graph of $f(x) = ax^2 + c$ is a vertical translation c units up of the graph of $f(x) = ax^2$.
- When $c < 0$, the graph of $f(x) = ax^2 + c$ is a vertical translation $|c|$ units down of the graph of $f(x) = ax^2$.

The vertex of the graph of $f(x) = ax^2 + c$ is $(0, c)$, and the axis of symmetry is $x = 0$.

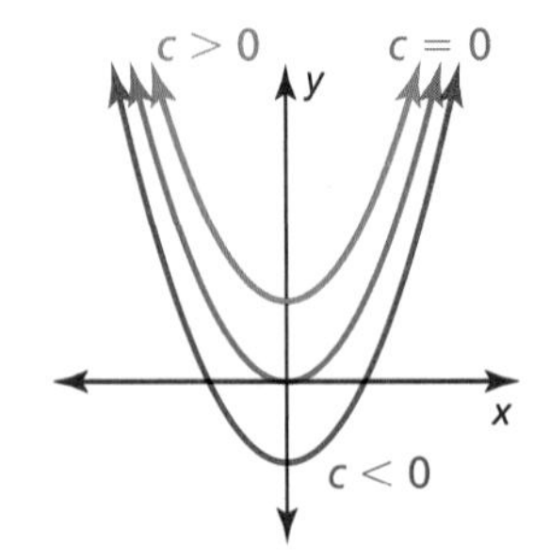

EXAMPLE 1 **Graphing $y = x^2 + c$**

Graph $g(x) = x^2 - 2$. Compare the graph to the graph of $f(x) = x^2$.

SOLUTION

Step 1 Make a table of values.

x	−2	−1	0	1	2
$g(x)$	2	−1	−2	−1	2

Step 2 Plot the ordered pairs.

Step 3 Draw a smooth curve through the points.

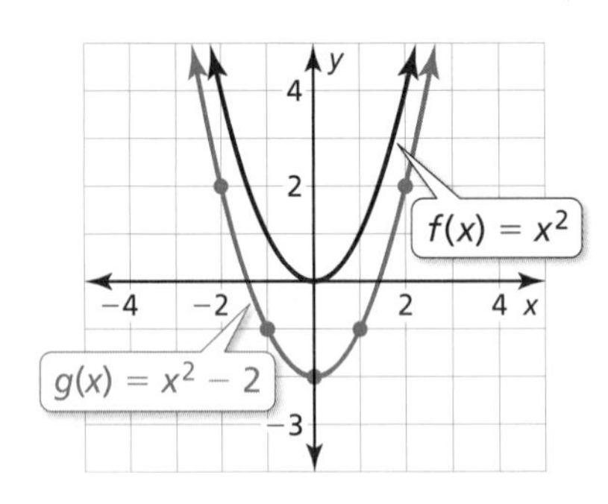

REMEMBER

The graph of $y = f(x) + k$ is a vertical translation, and the graph of $y = f(x - h)$ is a horizontal translation of the graph of f.

▶ Both graphs open up and have the same axis of symmetry, $x = 0$. The vertex of the graph of g, $(0, -2)$, is below the vertex of the graph of f, $(0, 0)$, because the graph of g is a vertical translation 2 units down of the graph of f.

Monitoring Progress

Help in English and Spanish at *BigIdeasMath.com*

Graph the function. Compare the graph to the graph of $f(x) = x^2$.

1. $g(x) = x^2 - 5$

2. $h(x) = x^2 + 3$

EXAMPLE 2 Graphing $y = ax^2 + c$

Graph $g(x) = 4x^2 + 1$. Compare the graph to the graph of $f(x) = x^2$.

SOLUTION

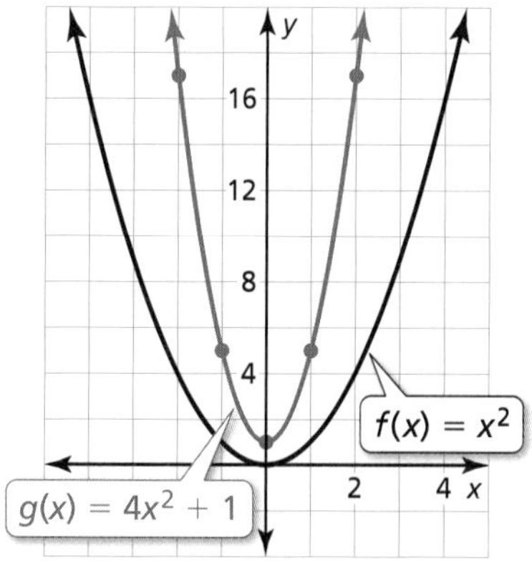

Step 1 Make a table of values.

x	-2	-1	0	1	2
$g(x)$	17	5	1	5	17

Step 2 Plot the ordered pairs.

Step 3 Draw a smooth curve through the points.

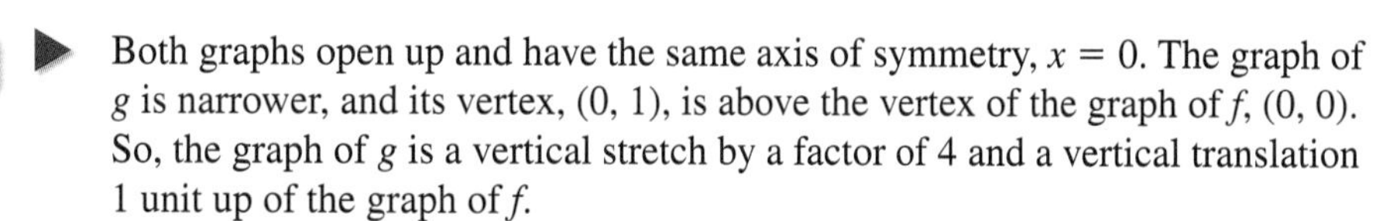

Both graphs open up and have the same axis of symmetry, $x = 0$. The graph of g is narrower, and its vertex, $(0, 1)$, is above the vertex of the graph of f, $(0, 0)$. So, the graph of g is a vertical stretch by a factor of 4 and a vertical translation 1 unit up of the graph of f.

EXAMPLE 3 Translating the Graph of $y = ax^2 + c$

Let $f(x) = -0.5x^2 + 2$ and $g(x) = f(x) - 7$.

a. Describe the transformation from the graph of f to the graph of g. Then graph f and g in the same coordinate plane.

b. Write an equation that represents g in terms of x.

SOLUTION

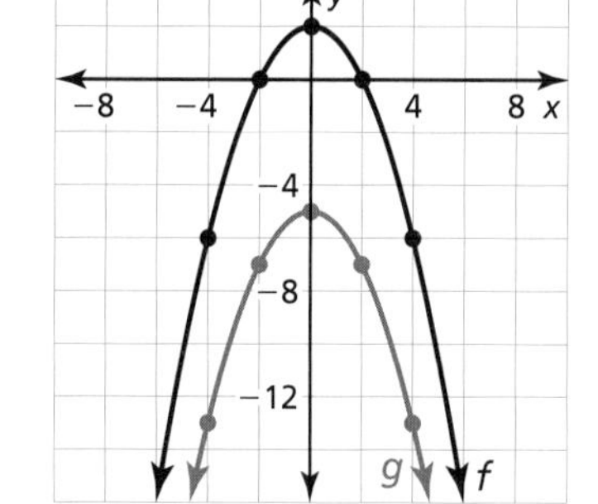

a. The function g is of the form $y = f(x) + k$, where $k = -7$. So, the graph of g is a vertical translation 7 units down of the graph of f.

x	-4	-2	0	2	4
$f(x)$	-6	0	2	0	-6
$g(x)$	-13	-7	-5	-7	-13

($f(x)$ row: $-0.5x^2 + 2$; $g(x)$ row: $f(x) - 7$)

b. $g(x) = f(x) - 7$ — Write the function g.

$= -0.5x^2 + 2 - 7$ — Substitute for $f(x)$.

$= -0.5x^2 - 5$ — Subtract.

So, the equation $g(x) = -0.5x^2 - 5$ represents g in terms of x.

Monitoring Progress

Help in English and Spanish at *BigIdeasMath.com*

Graph the function. Compare the graph to the graph of $f(x) = x^2$.

3. $g(x) = 2x^2 - 5$

4. $h(x) = -\frac{1}{4}x^2 + 4$

5. Let $f(x) = 3x^2 - 1$ and $g(x) = f(x) + 3$.

  a. Describe the transformation from the graph of f to the graph of g. Then graph f and g in the same coordinate plane.

  b. Write an equation that represents g in terms of x.

Solving Real-Life Problems

A **zero of a function** f is an x-value for which $f(x) = 0$. A zero of a function is an x-intercept of the graph of the function.

EXAMPLE 4 Solving a Real-Life Problem

The function $f(t) = -16t^2 + s_0$ represents the approximate height (in feet) of a falling object t seconds after it is dropped from an initial height s_0 (in feet). An egg is dropped from a height of 64 feet.

a. After how many seconds does the egg hit the ground?

b. Suppose the initial height is adjusted by k feet. How will this affect part (a)?

64 ft

SOLUTION

1. **Understand the Problem** You know the function that models the height of a falling object and the initial height of an egg. You are asked to find how many seconds it takes the egg to hit the ground when dropped from the initial height. Then you need to describe how a change in the initial height affects how long it takes the egg to hit the ground.

2. **Make a Plan** Use the initial height to write a function that models the height of the egg. Use a table to graph the function. Find the zero(s) of the function to answer the question. Then explain how vertical translations of the graph affect the zero(s) of the function.

3. **Solve the Problem**

COMMON ERROR

The graph in Step 1 shows the height of the object over time, not the path of the object.

a. The initial height is 64 feet. So, the function $f(t) = -16t^2 + 64$ represents the height of the egg t seconds after it is dropped. The egg hits the ground when $f(t) = 0$.

Step 1 Make a table of values and sketch the graph.

t	0	1	2
$f(t)$	64	48	0

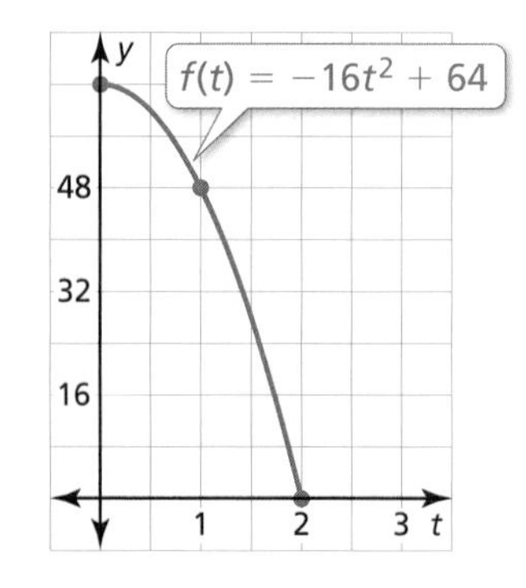

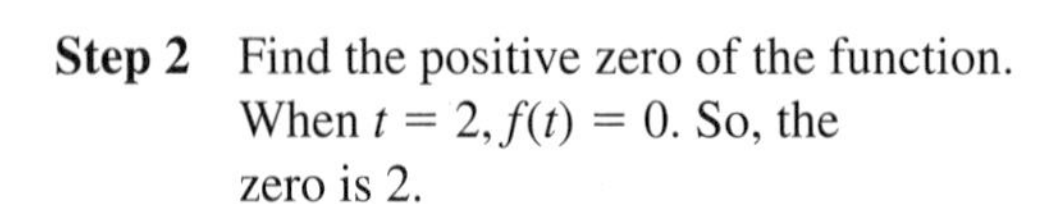

Step 2 Find the positive zero of the function. When $t = 2$, $f(t) = 0$. So, the zero is 2.

▶ The egg hits the ground 2 seconds after it is dropped.

b. When the initial height is adjusted by k feet, the graph of f is translated up k units when $k > 0$ or down $|k|$ units when $k < 0$. So, the x-intercept of the graph of f will move right when $k > 0$ or left when $k < 0$.

▶ When $k > 0$, the egg will take more than 2 seconds to hit the ground. When $k < 0$, the egg will take less than 2 seconds to hit the ground.

4. **Look Back** To check that the egg hits the ground 2 seconds after it is dropped, you can solve $0 = -16t^2 + 64$ by factoring.

Monitoring Progress

Help in English and Spanish at *BigIdeasMath.com*

6. Explain why only nonnegative values of t are used in Example 4.

7. **WHAT IF?** The egg is dropped from a height of 100 feet. After how many seconds does the egg hit the ground?

8.2 Exercises

Dynamic Solutions available at *BigIdeasMath.com*

Vocabulary and Core Concept Check

1. **VOCABULARY** State the vertex and axis of symmetry of the graph of $y = ax^2 + c$.

2. **WRITING** How does the graph of $y = ax^2 + c$ compare to the graph of $y = ax^2$?

Monitoring Progress and Modeling with Mathematics

In Exercises 3–6, graph the function. Compare the graph to the graph of $f(x) = x^2$. *(See Example 1.)*

3. $g(x) = x^2 + 6$
4. $h(x) = x^2 + 8$
5. $p(x) = x^2 - 3$
6. $q(x) = x^2 - 1$

In Exercises 7–12, graph the function. Compare the graph to the graph of $f(x) = x^2$. *(See Example 2.)*

7. $g(x) = -x^2 + 3$
8. $h(x) = -x^2 - 7$
9. $s(x) = 2x^2 - 4$
10. $t(x) = -3x^2 + 1$
11. $p(x) = -\frac{1}{3}x^2 - 2$
12. $q(x) = \frac{1}{2}x^2 + 6$

In Exercises 13–16, describe the transformation from the graph of f to the graph of g. Then graph f and g in the same coordinate plane. Write an equation that represents g in terms of x. *(See Example 3.)*

13. $f(x) = 3x^2 + 4$
    $g(x) = f(x) + 2$
14. $f(x) = \frac{1}{2}x^2 + 1$
    $g(x) = f(x) - 4$
15. $f(x) = -\frac{1}{4}x^2 - 6$
    $g(x) = f(x) - 3$
16. $f(x) = 4x^2 - 5$
    $g(x) = f(x) + 7$

17. **ERROR ANALYSIS** Describe and correct the error in comparing the graphs.

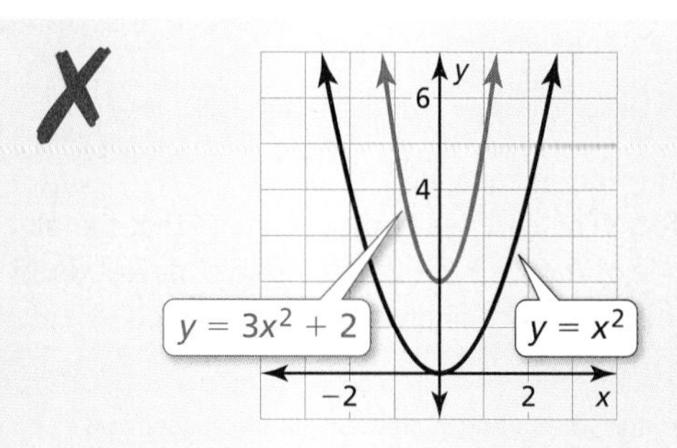

The graph of $y = 3x^2 + 2$ is a vertical shrink by a factor of 3 and a translation 2 units up of the graph of $y = x^2$.

18. **ERROR ANALYSIS** Describe and correct the error in graphing and comparing $f(x) = x^2$ and $g(x) = x^2 - 10$.

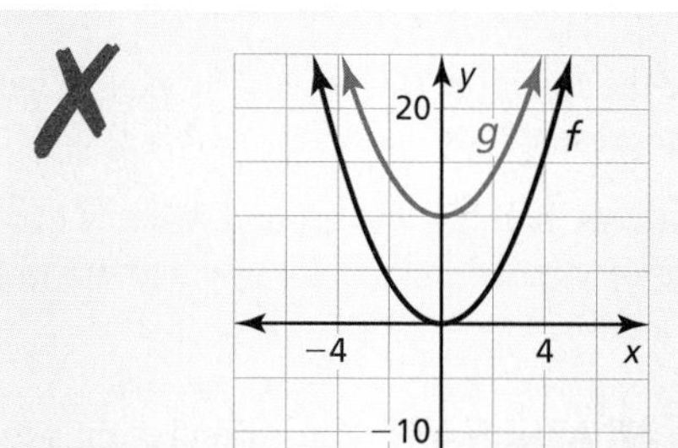

Both graphs open up and have the same axis of symmetry. However, the vertex of the graph of g, (0, 10), is 10 units above the vertex of the graph of f, (0, 0).

In Exercises 19–26, find the zeros of the function.

19. $y = x^2 - 1$
20. $y = x^2 - 36$
21. $f(x) = -x^2 + 25$
22. $f(x) = -x^2 + 49$
23. $f(x) = 4x^2 - 16$
24. $f(x) = 3x^2 - 27$
25. $f(x) = -12x^2 + 3$
26. $f(x) = -8x^2 + 98$

27. **MODELING WITH MATHEMATICS** A water balloon is dropped from a height of 144 feet. *(See Example 4.)*
    a. After how many seconds does the water balloon hit the ground?
    b. Suppose the initial height is adjusted by k feet. How does this affect part (a)?

28. **MODELING WITH MATHEMATICS** The function $y = -16x^2 + 36$ represents the height y (in feet) of an apple x seconds after falling from a tree. Find and interpret the x- and y-intercepts.

In Exercises 29–32, sketch a parabola with the given characteristics.

29. The parabola opens up, and the vertex is (0, 3).

30. The vertex is (0, 4), and one of the x-intercepts is 2.

31. The related function is increasing when $x < 0$, and the zeros are -1 and 1.

32. The highest point on the parabola is (0, -5).

33. DRAWING CONCLUSIONS You and your friend both drop a ball at the same time. The function $h(x) = -16x^2 + 256$ represents the height (in feet) of your ball after x seconds. The function $g(x) = -16x^2 + 300$ represents the height (in feet) of your friend's ball after x seconds.

a. Write the function $T(x) = h(x) - g(x)$. What does $T(x)$ represent?

b. When your ball hits the ground, what is the height of your friend's ball? Use a graph to justify your answer.

34. MAKING AN ARGUMENT Your friend claims that in the equation $y = ax^2 + c$, the vertex changes when the value of a changes. Is your friend correct? Explain your reasoning.

35. MATHEMATICAL CONNECTIONS The area A (in square feet) of a square patio is represented by $A = x^2$, where x is the length of one side of the patio. You add 48 square feet to the patio, resulting in a total area of 192 square feet. What are the dimensions of the original patio? Use a graph to justify your answer.

36. HOW DO YOU SEE IT? The graph of $f(x) = ax^2 + c$ is shown. Points A and B are the same distance from the vertex of the graph of f. Which point is closer to the vertex of the graph of f as c increases?

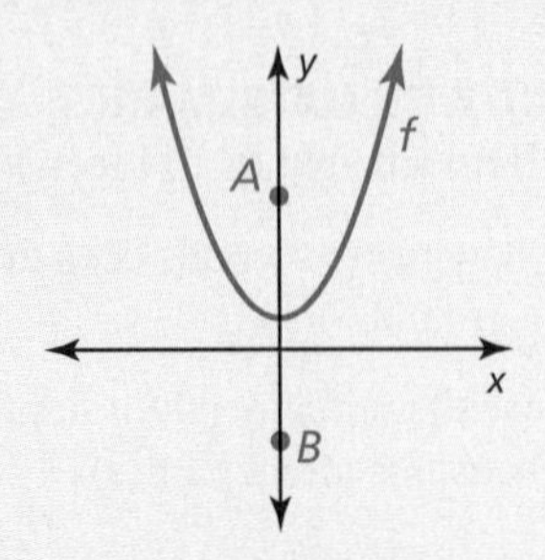

37. REASONING Describe two algebraic methods you can use to find the zeros of the function $f(t) = -16t^2 + 400$. Check your answer by graphing.

38. PROBLEM SOLVING The paths of water from three different garden waterfalls are given below. Each function gives the height h (in feet) and the horizontal distance d (in feet) of the water.

Waterfall 1 $h = -3.1d^2 + 4.8$

Waterfall 2 $h = -3.5d^2 + 1.9$

Waterfall 3 $h = -1.1d^2 + 1.6$

a. Which waterfall drops water from the highest point?

b. Which waterfall follows the narrowest path?

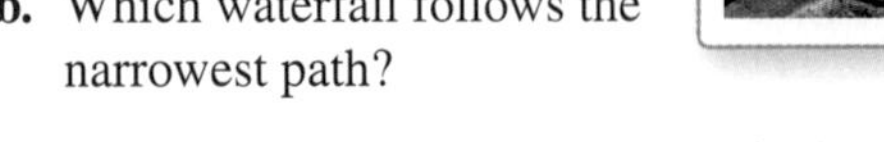

c. Which waterfall sends water the farthest?

39. WRITING EQUATIONS Two acorns fall to the ground from an oak tree. One falls 45 feet, while the other falls 32 feet.

a. For each acorn, write an equation that represents the height h (in feet) as a function of the time t (in seconds).

b. Describe how the graphs of the two equations are related.

40. THOUGHT PROVOKING One of two classic problems in calculus is to find the area under a curve. Approximate the area of the region bounded by the parabola and the x-axis. Show your work.

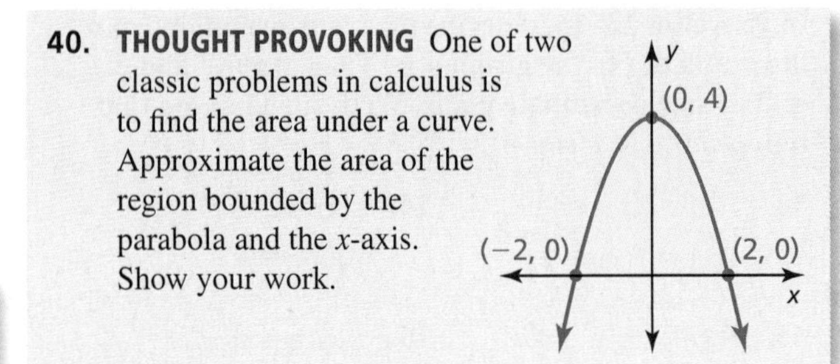

41. CRITICAL THINKING A cross section of the parabolic surface of the antenna shown can be modeled by $y = 0.012x^2$, where x and y are measured in feet. The antenna is moved up so that the outer edges of the dish are 25 feet above the x-axis. Where is the vertex of the cross section located? Explain.

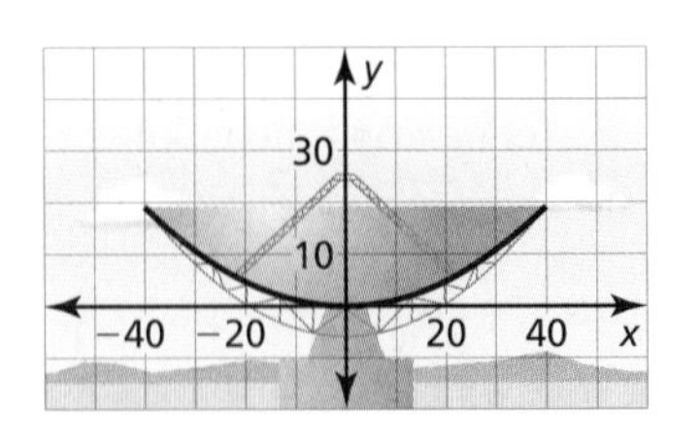

Maintaining Mathematical Proficiency

Reviewing what you learned in previous grades and lessons

Evaluate the expression when $a = 4$ and $b = -3$. *(Skills Review Handbook)*

42. $\frac{a}{4b}$ **43.** $-\frac{b}{2a}$ **44.** $\frac{a - b}{3a + b}$ **45.** $-\frac{b + 2a}{ab}$

8.3 Graphing $f(x) = ax^2 + bx + c$

Essential Question How can you find the vertex of the graph of $f(x) = ax^2 + bx + c$?

EXPLORATION 1 Comparing x-Intercepts with the Vertex

Work with a partner.

a. Sketch the graphs of $y = 2x^2 - 8x$ and $y = 2x^2 - 8x + 6$.

b. What do you notice about the x-coordinate of the vertex of each graph?

c. Use the graph of $y = 2x^2 - 8x$ to find its x-intercepts. Verify your answer by solving $0 = 2x^2 - 8x$.

d. Compare the value of the x-coordinate of the vertex with the values of the x-intercepts.

EXPLORATION 2 Finding x-Intercepts

Work with a partner.

a. Solve $0 = ax^2 + bx$ for x by factoring.

b. What are the x-intercepts of the graph of $y = ax^2 + bx$?

c. Copy and complete the table to verify your answer.

x	$y = ax^2 + bx$
0	
$-\frac{b}{a}$	

CONSTRUCTING VIABLE ARGUMENTS

To be proficient in math, you need to make conjectures and build a logical progression of statements.

EXPLORATION 3 Deductive Reasoning

Work with a partner. Complete the following logical argument.

The x-intercepts of the graph of $y = ax^2 + bx$ are 0 and $-\frac{b}{a}$.

The vertex of the graph of $y = ax^2 + bx$ occurs when $x =$ ______.

The vertices of the graphs of $y = ax^2 + bx$ and $y = ax^2 + bx + c$ have the same x-coordinate.

The vertex of the graph of $y = ax^2 + bx + c$ occurs when $x =$ ______.

Communicate Your Answer

4. How can you find the vertex of the graph of $f(x) = ax^2 + bx + c$?

5. Without graphing, find the vertex of the graph of $f(x) = x^2 - 4x + 3$. Check your result by graphing.

8.3 Lesson

Core Vocabulary

maximum value, *p. 433*
minimum value, *p. 433*

Previous
independent variable
dependent variable

What You Will Learn

- Graph quadratic functions of the form $f(x) = ax^2 + bx + c$.
- Find maximum and minimum values of quadratic functions.

Graphing $f(x) = ax^2 + bx + c$

Core Concept

Graphing $f(x) = ax^2 + bx + c$

- The graph opens up when $a > 0$, and the graph opens down when $a < 0$.
- The y-intercept is c.
- The x-coordinate of the vertex is $-\frac{b}{2a}$.
- The axis of symmetry is $x = -\frac{b}{2a}$.

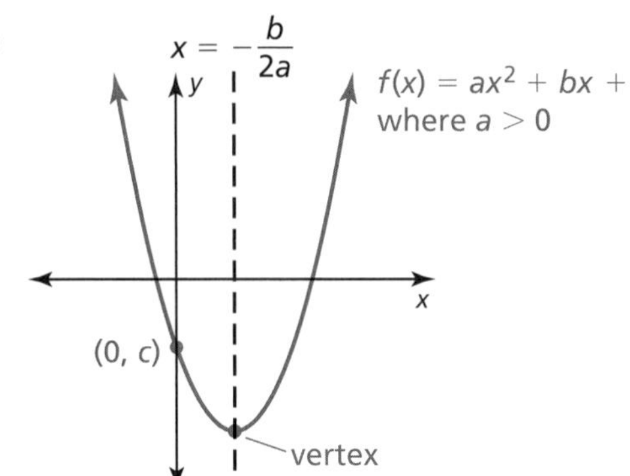

EXAMPLE 1 Finding the Axis of Symmetry and the Vertex

Find (a) the axis of symmetry and (b) the vertex of the graph of $f(x) = 2x^2 + 8x - 1$.

SOLUTION

a. Find the axis of symmetry when $a = 2$ and $b = 8$.

$x = -\frac{b}{2a}$ — Write the equation for the axis of symmetry.

$x = -\frac{8}{2(2)}$ — Substitute 2 for a and 8 for b.

$x = -2$ — Simplify.

▶ The axis of symmetry is $x = -2$.

b. The axis of symmetry is $x = -2$, so the x-coordinate of the vertex is -2. Use the function to find the y-coordinate of the vertex.

$f(x) = 2x^2 + 8x - 1$ — Write the function.

$f(-2) = 2(-2)^2 + 8(-2) - 1$ — Substitute -2 for x.

$= -9$ — Simplify.

▶ The vertex is $(-2, -9)$.

Check

$f(x) = 2x^2 + 8x - 1$

X=−2 Y=−9

Monitoring Progress  Help in English and Spanish at *BigIdeasMath.com*

Find (a) the axis of symmetry and (b) the vertex of the graph of the function.

1. $f(x) = 3x^2 - 2x$ **2.** $g(x) = x^2 + 6x + 5$ **3.** $h(x) = -\frac{1}{2}x^2 + 7x - 4$

EXAMPLE 2 Graphing $f(x) = ax^2 + bx + c$

COMMON ERROR

Be sure to include the negative sign before the fraction when finding the axis of symmetry.

Graph $f(x) = 3x^2 - 6x + 5$. Describe the domain and range.

SOLUTION

Step 1 Find and graph the axis of symmetry.

$$x = -\frac{b}{2a} = -\frac{(-6)}{2(3)} = 1$$ Substitute and simplify.

Step 2 Find and plot the vertex.

The axis of symmetry is $x = 1$, so the x-coordinate of the vertex is 1. Use the function to find the y-coordinate of the vertex.

$f(x) = 3x^2 - 6x + 5$ Write the function.

$f(1) = 3(1)^2 - 6(1) + 5$ Substitute 1 for x.

$= 2$ Simplify.

So, the vertex is (1, 2).

Step 3 Use the y-intercept to find two more points on the graph.

Because $c = 5$, the y-intercept is 5. So, (0, 5) lies on the graph. Because the axis of symmetry is $x = 1$, the point (2, 5) also lies on the graph.

Step 4 Draw a smooth curve through the points.

$f(x) = 3x^2 - 6x + 5$

▶ The domain is all real numbers. The range is $y \geq 2$.

REMEMBER

The domain is the set of all possible input values of the independent variable x. The range is the set of all possible output values of the dependent variable y.

Monitoring Progress

Help in English and Spanish at *BigIdeasMath.com*

Graph the function. Describe the domain and range.

4. $h(x) = 2x^2 + 4x + 1$ **5.** $k(x) = x^2 - 8x + 7$ **6.** $p(x) = -5x^2 - 10x - 2$

Finding Maximum and Minimum Values

Core Concept

Maximum and Minimum Values

The y-coordinate of the vertex of the graph of $f(x) = ax^2 + bx + c$ is the **maximum value** of the function when $a < 0$ or the **minimum value** of the function when $a > 0$.

$f(x) = ax^2 + bx + c, a < 0$

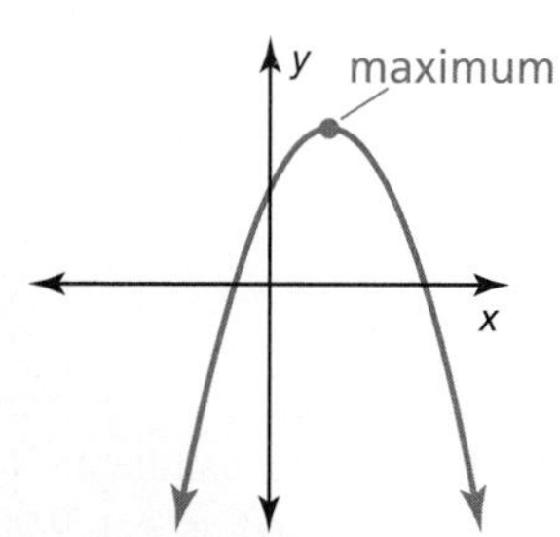

$f(x) = ax^2 + bx + c, a > 0$

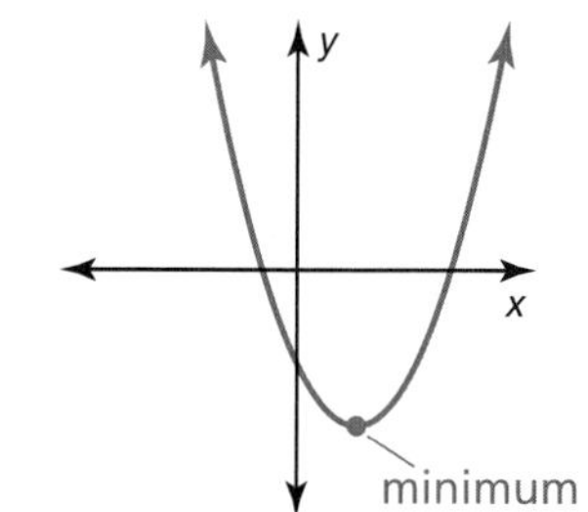

EXAMPLE 3 Finding a Maximum or Minimum Value

Tell whether the function $f(x) = -4x^2 - 24x - 19$ has a minimum value or a maximum value. Then find the value.

SOLUTION

For $f(x) = -4x^2 - 24x - 19$, $a = -4$ and $-4 < 0$. So, the parabola opens down and the function has a maximum value. To find the maximum value, find the y-coordinate of the vertex.

First, find the x-coordinate of the vertex. Use $a = -4$ and $b = -24$.

$$x = -\frac{b}{2a} = -\frac{(-24)}{2(-4)} = -3$$ Substitute and simplify.

Then evaluate the function when $x = -3$ to find the y-coordinate of the vertex.

$$f(-3) = -4(-3)^2 - 24(-3) - 19$$ Substitute -3 for x.

$$= 17$$ Simplify.

▶ The maximum value is 17.

EXAMPLE 4 Finding a Minimum Value

The suspension cables between the two towers of the Mackinac Bridge in Michigan form a parabola that can be modeled by $y = 0.000098x^2 - 0.37x + 552$, where x and y are measured in feet. What is the height of the cable above the water at its lowest point?

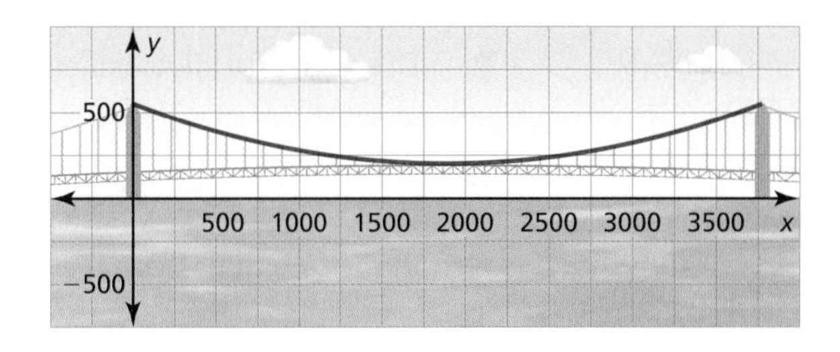

SOLUTION

The lowest point of the cable is at the vertex of the parabola. Find the x-coordinate of the vertex. Use $a = 0.000098$ and $b = -0.37$.

$$x = -\frac{b}{2a} = -\frac{(-0.37)}{2(0.000098)} \approx 1888$$ Substitute and use a calculator.

Substitute 1888 for x in the equation to find the y-coordinate of the vertex.

$$y = 0.000098(1888)^2 - 0.37(1888) + 552 \approx 203$$

▶ The cable is about 203 feet above the water at its lowest point.

Monitoring Progress

Help in English and Spanish at *BigIdeasMath.com*

Tell whether the function has a minimum value or a maximum value. Then find the value.

7. $g(x) = 8x^2 - 8x + 6$

8. $h(x) = -\frac{1}{4}x^2 + 3x + 1$

9. The cables between the two towers of the Tacoma Narrows Bridge in Washington form a parabola that can be modeled by $y = 0.00016x^2 - 0.46x + 507$, where x and y are measured in feet. What is the height of the cable above the water at its lowest point?

EXAMPLE 5 Modeling with Mathematics

A group of friends is launching water balloons. The function $f(t) = -16t^2 + 80t + 5$ represents the height (in feet) of the first water balloon t seconds after it is launched. The height of the second water balloon t seconds after it is launched is shown in the graph. Which water balloon went higher?

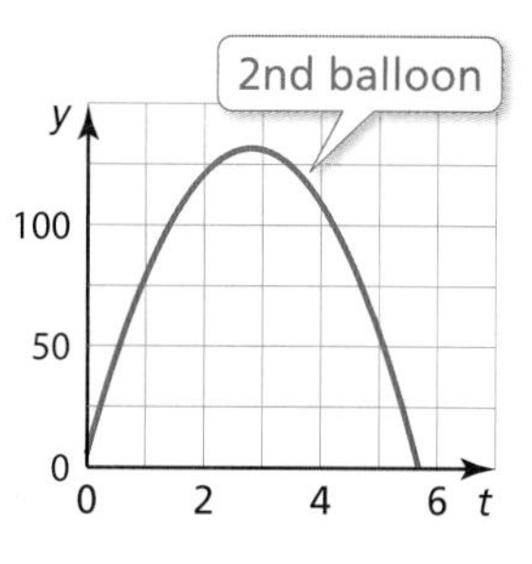

SOLUTION

1. **Understand the Problem** You are given a function that represents the height of the first water balloon. The height of the second water balloon is represented graphically. You need to find and compare the maximum heights of the water balloons.

2. **Make a Plan** To compare the maximum heights, represent both functions graphically. Use a graphing calculator to graph $f(t) = -16t^2 + 80t + 5$ in an appropriate viewing window. Then visually compare the heights of the water balloons.

3. **Solve the Problem** Enter the function $f(t) = -16t^2 + 80t + 5$ into your calculator and graph it. Compare the graphs to determine which function has a greater maximum value.

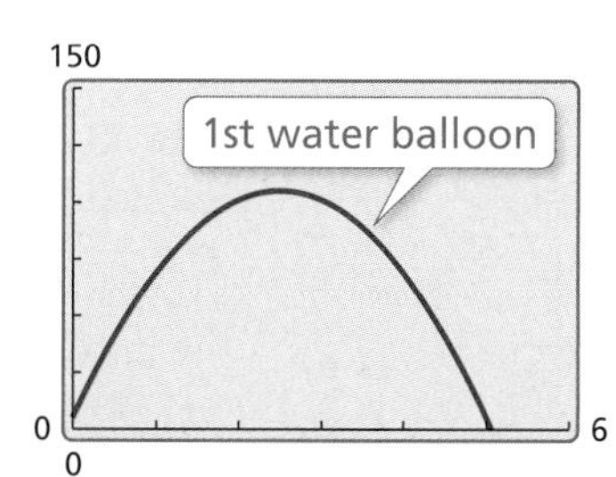

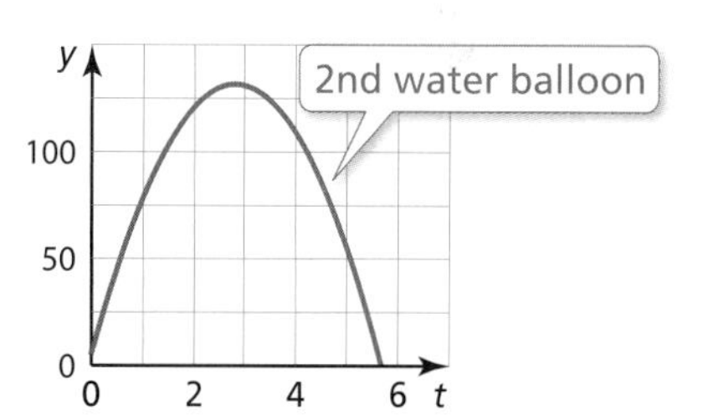

You can see that the second water balloon reaches a height of about 125 feet, while the first water balloon reaches a height of only about 100 feet.

▶ So, the second water balloon went higher.

4. **Look Back** Use the *maximum* feature to determine that the maximum value of $f(t) = -16t^2 + 80t + 5$ is 105. Use a straightedge to represent a height of 105 feet on the graph that represents the second water balloon to clearly see that the second water balloon went higher.

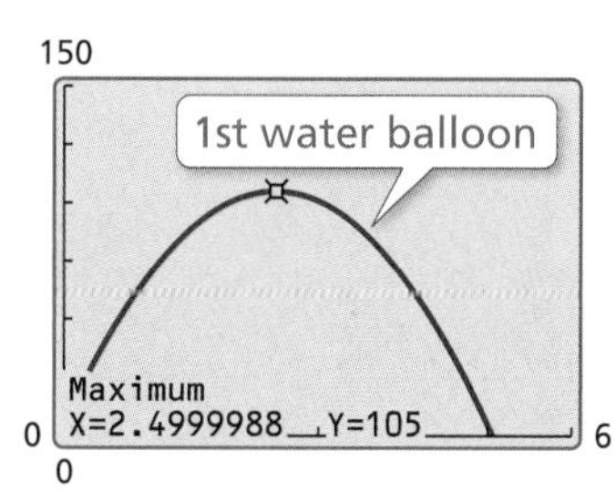

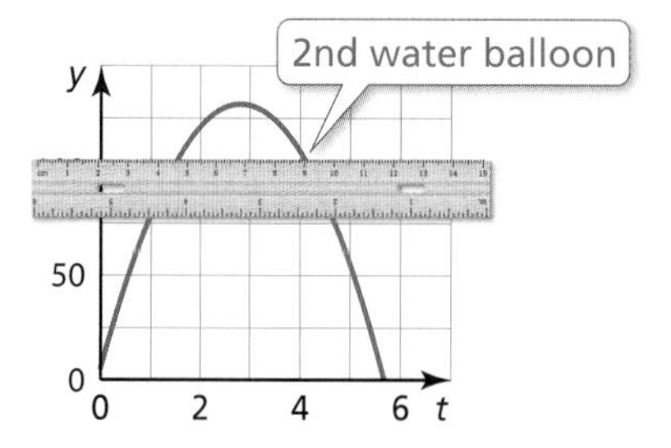

MODELING WITH MATHEMATICS

Because time cannot be negative, use only nonnegative values of t.

Monitoring Progress

Help in English and Spanish at *BigIdeasMath.com*

10. Which balloon is in the air longer? Explain your reasoning.

11. Which balloon reaches its maximum height faster? Explain your reasoning.

8.3 Exercises

Dynamic Solutions available at *BigIdeasMath.com*

Vocabulary and Core Concept Check

1. **VOCABULARY** Explain how you can tell whether a quadratic function has a maximum value or a minimum value without graphing the function.

2. **DIFFERENT WORDS, SAME QUESTION** Consider the quadratic function $f(x) = -2x^2 + 8x + 24$. Which is different? Find "both" answers.

What is the maximum value of the function?	What is the greatest number in the range of the function?
What is the y-coordinate of the vertex of the graph of the function?	What is the axis of symmetry of the graph of the function?

Monitoring Progress and Modeling with Mathematics

In Exercises 3–6, find the vertex, the axis of symmetry, and the y-intercept of the graph.

3. 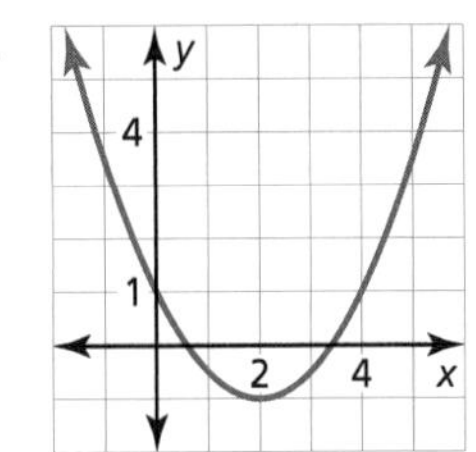

4. 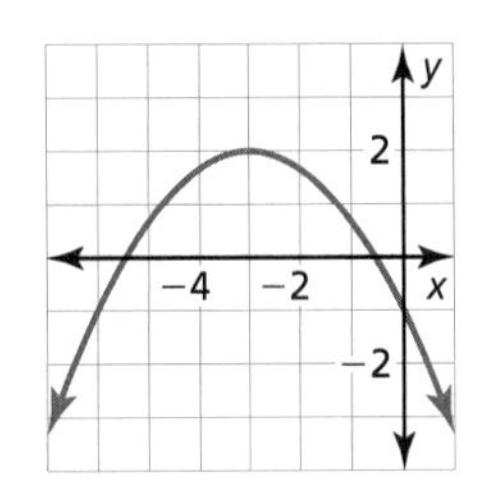

5. 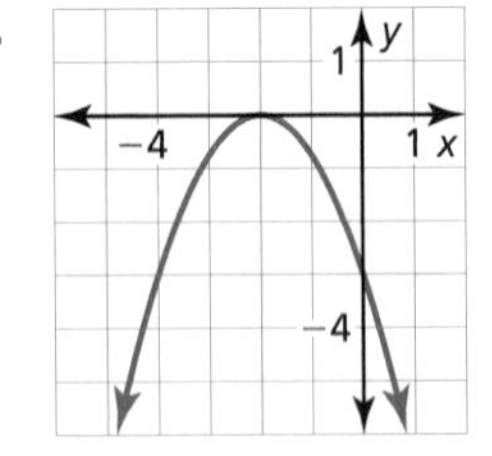

6. 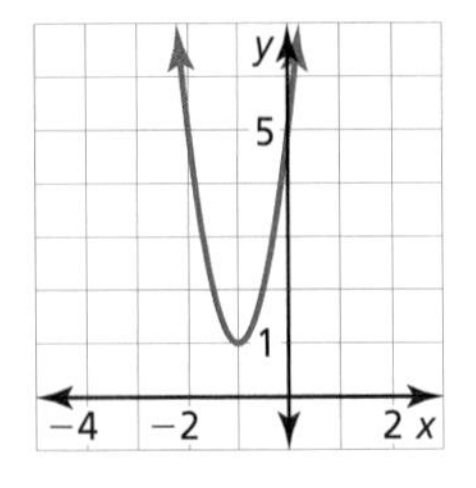

In Exercises 7–12, find (a) the axis of symmetry and (b) the vertex of the graph of the function. *(See Example 1.)*

7. $f(x) = 2x^2 - 4x$
8. $y = 3x^2 + 2x$
9. $y = -9x^2 - 18x - 1$
10. $f(x) = -6x^2 + 24x - 20$
11. $f(x) = \frac{2}{5}x^2 - 4x + 14$
12. $y = -\frac{3}{4}x^2 + 9x - 18$

In Exercises 13–18, graph the function. Describe the domain and range. *(See Example 2.)*

13. $f(x) = 2x^2 + 12x + 4$
14. $y = 4x^2 + 24x + 13$
15. $y = -8x^2 - 16x - 9$
16. $f(x) = -5x^2 + 20x - 7$
17. $y = \frac{2}{3}x^2 - 6x + 5$
18. $f(x) = -\frac{1}{2}x^2 - 3x - 4$

19. **ERROR ANALYSIS** Describe and correct the error in finding the axis of symmetry of the graph of $y = 3x^2 - 12x + 11$.

✗ $x = -\frac{b}{2a} = \frac{-12}{2(3)} = -2$

The axis of symmetry is $x = -2$.

20. **ERROR ANALYSIS** Describe and correct the error in graphing the function $f(x) = x^2 + 4x + 3$.

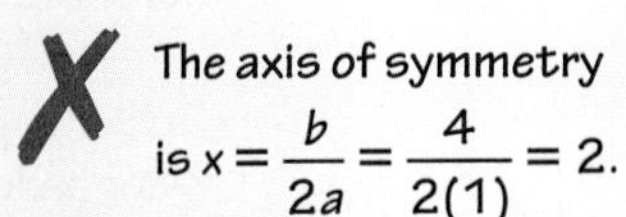
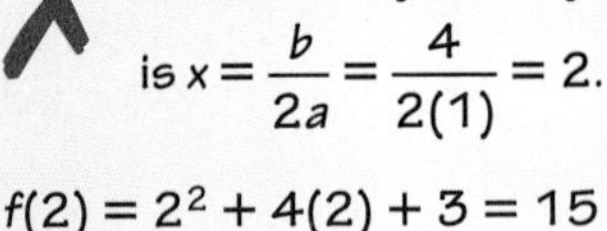

The axis of symmetry
is x = b/2a = 4/2(1) = 2.
f(2) = 2² + 4(2) + 3 = 15
So, the vertex is (2, 15).

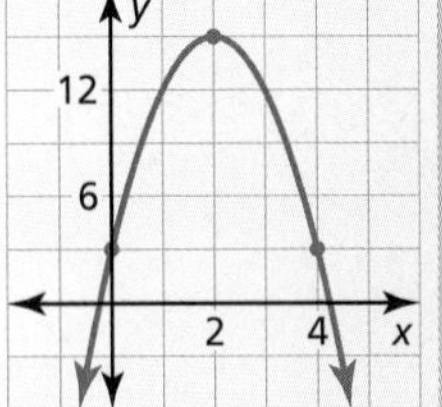

The y-intercept is 3. So, the points (0, 3) and (4, 3) lie on the graph.

In Exercises 21–26, tell whether the function has a minimum value or a maximum value. Then find the value. *(See Example 3.)*

21. $y = 3x^2 - 18x + 15$
22. $f(x) = -5x^2 + 10x + 7$
23. $f(x) = -4x^2 + 4x - 2$
24. $y = 2x^2 - 10x + 13$
25. $y = -\frac{1}{2}x^2 - 11x + 6$
26. $f(x) = \frac{1}{5}x^2 - 5x + 27$

27. **MODELING WITH MATHEMATICS** The function shown represents the height h (in feet) of a firework t seconds after it is launched. The firework explodes at its highest point. *(See Example 4.)*

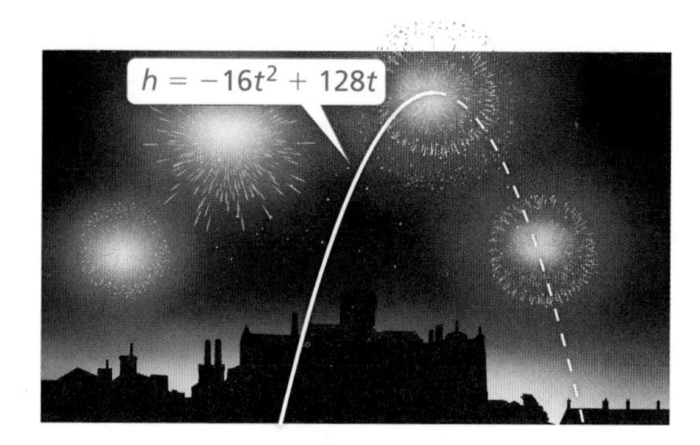

a. When does the firework explode?

b. At what height does the firework explode?

28. **MODELING WITH MATHEMATICS** The function $h(t) = -16t^2 + 16t$ represents the height (in feet) of a horse t seconds after it jumps during a steeplechase.

a. When does the horse reach its maximum height?

b. Can the horse clear a fence that is 3.5 feet tall? If so, by how much?

c. How long is the horse in the air?

29. **MODELING WITH MATHEMATICS** The cable between two towers of a suspension bridge can be modeled by the function shown, where x and y are measured in feet. The cable is at road level midway between the towers.

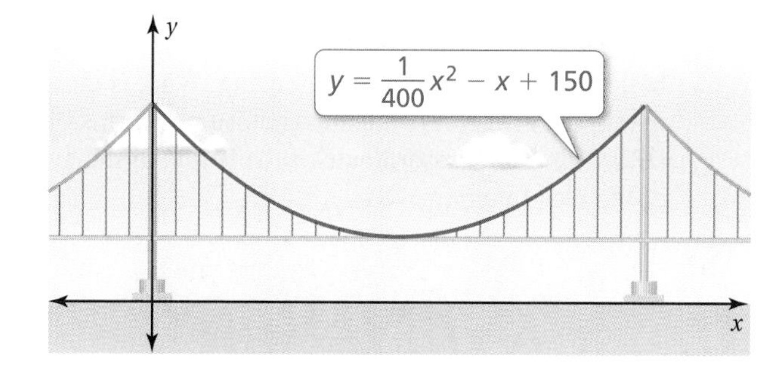

a. How far from each tower shown is the lowest point of the cable?

b. How high is the road above the water?

c. Describe the domain and range of the function shown.

30. **REASONING** Find the axis of symmetry of the graph of the equation $y = ax^2 + bx + c$ when $b = 0$. Can you find the axis of symmetry when $a = 0$? Explain.

31. **ATTENDING TO PRECISION** The vertex of a parabola is $(3, -1)$. One point on the parabola is $(6, 8)$. Find another point on the parabola. Justify your answer.

32. **MAKING AN ARGUMENT** Your friend claims that it is possible to draw a parabola through any two points with different x-coordinates. Is your friend correct? Explain.

USING TOOLS In Exercises 33–36, use the *minimum* or *maximum* feature of a graphing calculator to approximate the vertex of the graph of the function.

33. $y = 0.5x^2 + \sqrt{2}x - 3$

34. $y = -6.2x^2 + 4.8x - 1$

35. $y = -\pi x^2 + 3x$

36. $y = 0.25x^2 - 5^{2/3}x + 2$

37. **MODELING WITH MATHEMATICS** The opening of one aircraft hangar is a parabolic arch that can be modeled by the equation $y = -0.006x^2 + 1.5x$, where x and y are measured in feet. The opening of a second aircraft hangar is shown in the graph. *(See Example 5.)*

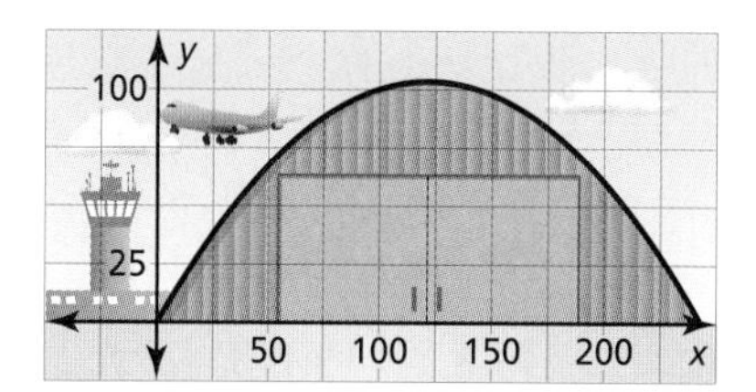

a. Which aircraft hangar is taller?

b. Which aircraft hangar is wider?

38. **MODELING WITH MATHEMATICS** An office supply store sells about 80 graphing calculators per month for \$120 each. For each \$6 decrease in price, the store expects to sell eight more calculators. The revenue from calculator sales is given by the function $R(n) =$ (unit price)(units sold), or $R(n) = (120 - 6n)(80 + 8n)$, where n is the number of \$6 price decreases.

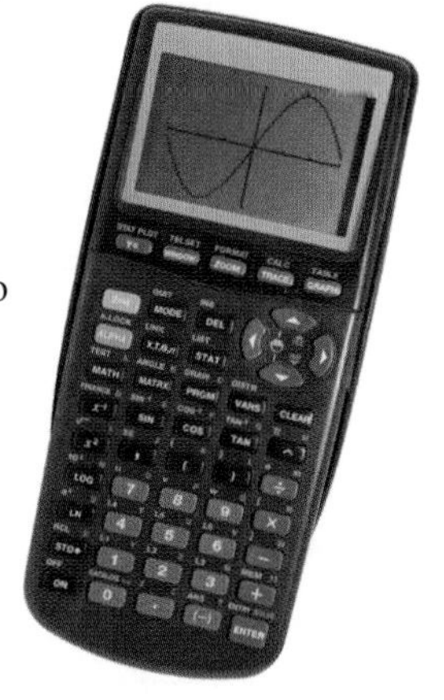

a. How much should the store charge to maximize monthly revenue?

b. Using a different revenue model, the store expects to sell five more calculators for each \$4 decrease in price. Which revenue model results in a greater maximum monthly revenue? Explain.

MATHEMATICAL CONNECTIONS In Exercises 39 and 40, (a) find the value of x that maximizes the area of the figure and (b) find the maximum area.

39.

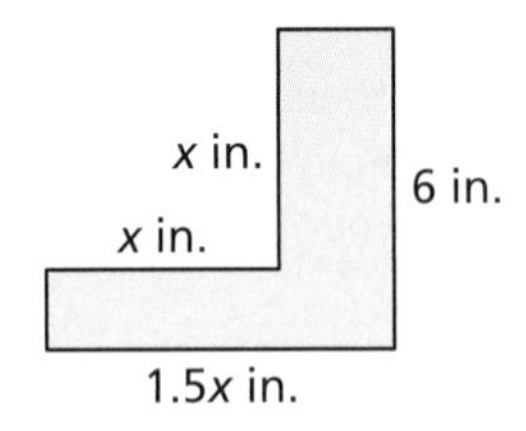

40.

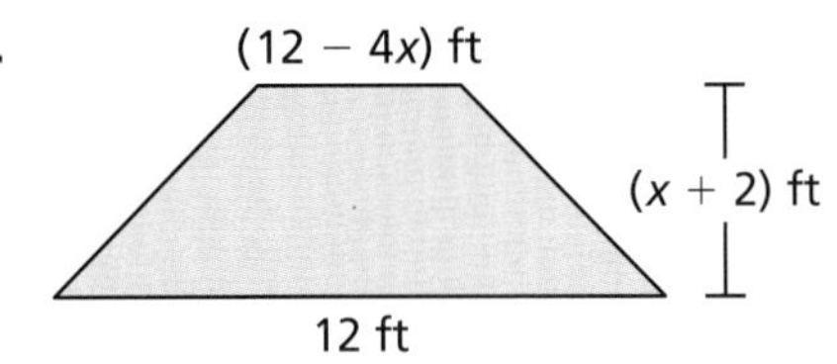

41. WRITING Compare the graph of $g(x) = x^2 + 4x + 1$ with the graph of $h(x) = x^2 - 4x + 1$.

42. HOW DO YOU SEE IT? During an archery competition, an archer shoots an arrow. The arrow follows the parabolic path shown, where x and y are measured in meters.

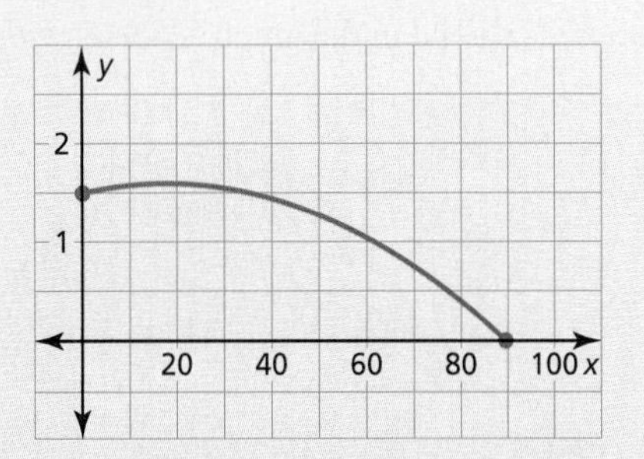

a. What is the initial height of the arrow?

b. Estimate the maximum height of the arrow.

c. How far does the arrow travel?

43. USING TOOLS The graph of a quadratic function passes through (3, 2), (4, 7), and (9, 2). Does the graph open up or down? Explain your reasoning.

44. REASONING For a quadratic function f, what does $f\left(-\frac{b}{2a}\right)$ represent? Explain your reasoning.

45. PROBLEM SOLVING Write a function of the form $y = ax^2 + bx$ whose graph contains the points (1, 6) and (3, 6).

46. CRITICAL THINKING Parabolas A and B contain the points shown. Identify characteristics of each parabola, if possible. Explain your reasoning.

Parabola A	
x	y
2	3
6	4

Parabola B	
x	y
1	4
3	−4
5	4

47. MODELING WITH MATHEMATICS At a basketball game, an air cannon launches T-shirts into the crowd. The function $y = -\frac{1}{8}x^2 + 4x$ represents the path of a T-shirt. The function $3y = 2x - 14$ represents the height of the bleachers. In both functions, y represents vertical height (in feet) and x represents horizontal distance (in feet). At what height does the T-shirt land in the bleachers?

48. THOUGHT PROVOKING One of two classic problems in calculus is finding the slope of a *tangent line* to a curve. An example of a tangent line, which just touches the parabola at one point, is shown.

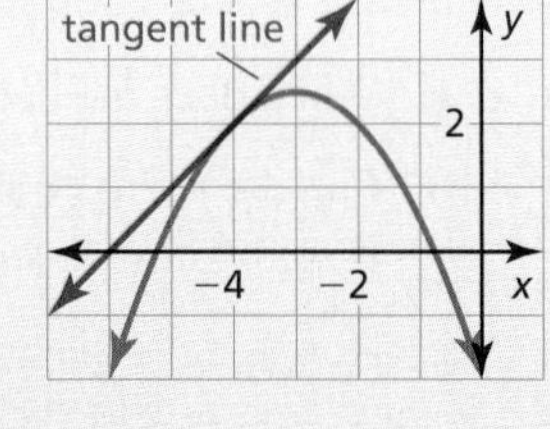

Approximate the slope of the tangent line to the graph of $y = x^2$ at the point (1, 1). Explain your reasoning.

49. PROBLEM SOLVING The owners of a dog shelter want to enclose a rectangular play area on the side of their building. They have k feet of fencing. What is the maximum area of the outside enclosure in terms of k? (*Hint:* Find the y-coordinate of the vertex of the graph of the area function.)

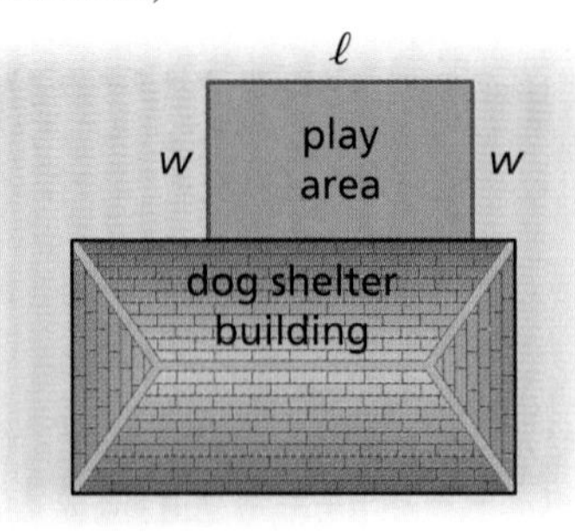

Maintaining Mathematical Proficiency

Reviewing what you learned in previous grades and lessons

Describe the transformation(s) from the graph of $f(x) = |x|$ to the graph of the given function. *(Section 3.7)*

50. $q(x) = |x + 6|$ **51.** $h(x) = -0.5|x|$ **52.** $g(x) = |x - 2| + 5$ **53.** $p(x) = 3|x + 1|$

8.1–8.3 What Did You Learn?

Core Vocabulary

quadratic function, *p. 420*
parabola, *p. 420*
vertex, *p. 420*
axis of symmetry, *p. 420*
zero of a function, *p. 428*
maximum value, *p. 433*
minimum value, *p. 433*

Core Concepts

Section 8.1

Characteristics of Quadratic Functions, *p. 420*
Graphing $f(x) = ax^2$ When $a > 0$, *p. 421*
Graphing $f(x) = ax^2$ When $a < 0$, *p. 421*

Section 8.2

Graphing $f(x) = ax^2 + c$, *p. 426*

Section 8.3

Graphing $f(x) = ax^2 + bx + c$, *p. 432*
Maximum and Minimum Values, *p. 433*

Mathematical Practices

1. Explain your plan for solving Exercise 18 on page 423.
2. How does graphing a function in Exercise 27 on page 429 help you answer the questions?
3. What definition and characteristics of the graph of a quadratic function did you use to answer Exercise 44 on page 438?

Study Skills
Learning Visually
• Draw a picture of a word problem before writing a verbal model. You do not have to be an artist.
• When making a review card for a word problem, include a picture. This will help you recall the information while taking a test.
• Make sure your notes are visually neat for easy recall.

8.1–8.3 Quiz

Identify characteristics of the quadratic function and its graph. *(Section 8.1)*

1. 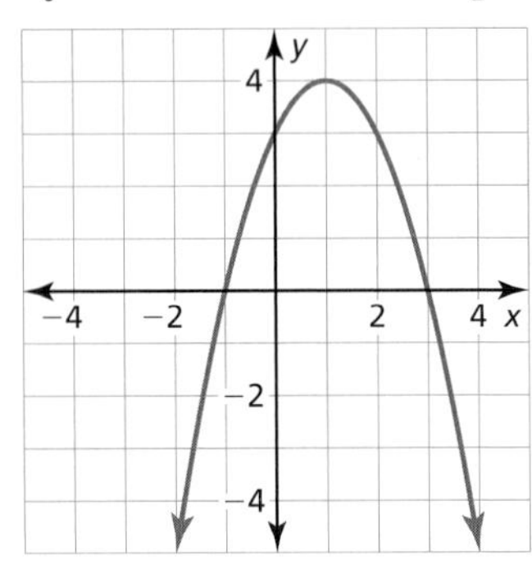

2. 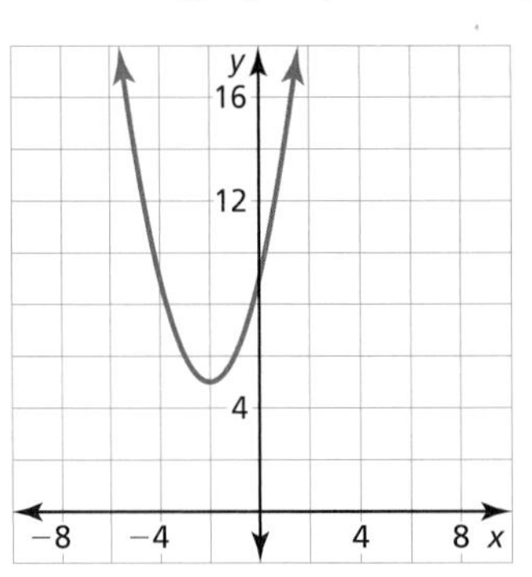

Graph the function. Compare the graph to the graph of $f(x) = x^2$. *(Section 8.1 and Section 8.2)*

3. $h(x) = -x^2$

4. $p(x) = 2x^2 + 2$

5. $r(x) = 4x^2 - 16$

6. $b(x) = 8x^2$

7. $g(x) = \frac{2}{5}x^2$

8. $m(x) = -\frac{1}{2}x^2 - 4$

Describe the transformation from the graph of f to the graph of g. Then graph f and g in the same coordinate plane. Write an equation that represents g in terms of x. *(Section 8.2)*

9. $f(x) = 2x^2 + 1$; $g(x) = f(x) + 2$

10. $f(x) = -3x^2 + 12$; $g(x) = f(x) - 9$

11. $f(x) = \frac{1}{2}x^2 - 2$; $g(x) = f(x) - 6$

12. $f(x) = 5x^2 - 3$; $g(x) = f(x) + 1$

Graph the function. Describe the domain and range. *(Section 8.3)*

13. $f(x) = -4x^2 - 4x + 7$

14. $f(x) = 2x^2 + 12x + 5$

15. $y = x^2 + 4x - 5$

16. $y = -3x^2 + 6x + 9$

Tell whether the function has a minimum value or a maximum value. Then find the value. *(Section 8.3)*

17. $f(x) = 5x^2 + 10x - 3$

18. $f(x) = -\frac{1}{2}x^2 + 2x + 16$

19. $y = -x^2 + 4x + 12$

20. $y = 2x^2 + 8x + 3$

21. The distance y (in feet) that a coconut falls after t seconds is given by the function $y = 16t^2$. Use a graph to determine how many seconds it takes for the coconut to fall 64 feet. *(Section 8.1)*

22. The function $y = -16t^2 + 25$ represents the height y (in feet) of a pinecone t seconds after falling from a tree. *(Section 8.2)*

a. After how many seconds does the pinecone hit the ground?

b. A second pinecone falls from a height of 36 feet. Which pinecone hits the ground in the least amount of time? Explain.

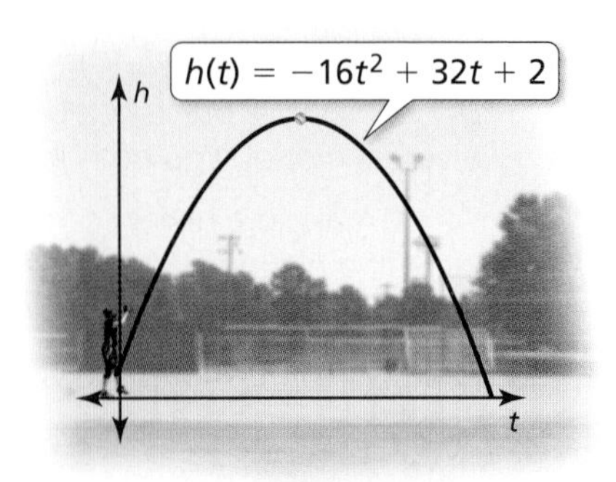

23. The function shown models the height (in feet) of a softball t seconds after it is pitched in an underhand motion. Describe the domain and range. Find the maximum height of the softball. *(Section 8.3)*

8.4 Graphing $f(x) = a(x - h)^2 + k$

Essential Question How can you describe the graph of $f(x) = a(x - h)^2$?

EXPLORATION 1 Graphing $y = a(x - h)^2$ When $h > 0$

Work with a partner. Sketch the graphs of the functions in the same coordinate plane. How does the value of h affect the graph of $y = a(x - h)^2$?

a. $f(x) = x^2$ and $g(x) = (x - 2)^2$

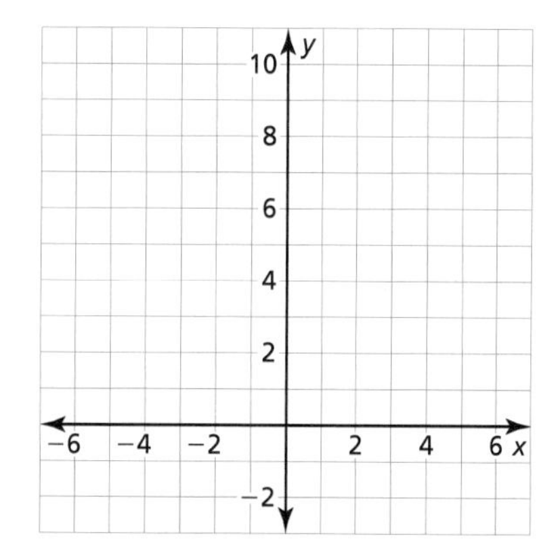

b. $f(x) = 2x^2$ and $g(x) = 2(x - 2)^2$

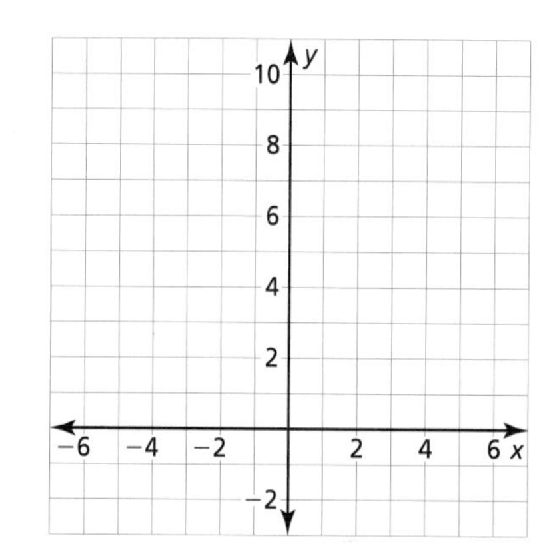

EXPLORATION 2 Graphing $y = a(x - h)^2$ When $h < 0$

Work with a partner. Sketch the graphs of the functions in the same coordinate plane. How does the value of h affect the graph of $y = a(x - h)^2$?

a. $f(x) = -x^2$ and $g(x) = -(x + 2)^2$

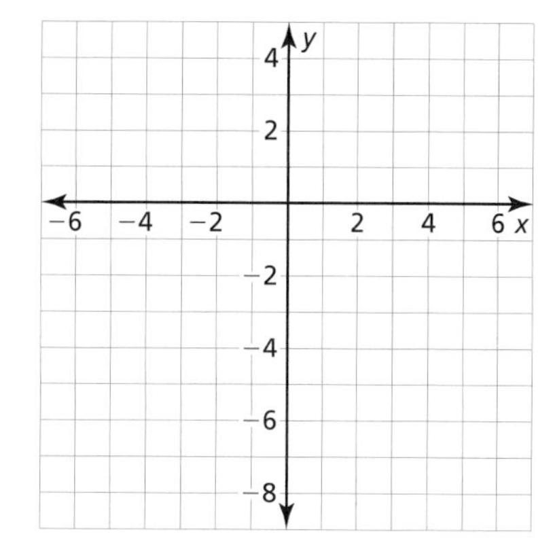

b. $f(x) = -2x^2$ and $g(x) = -2(x + 2)^2$

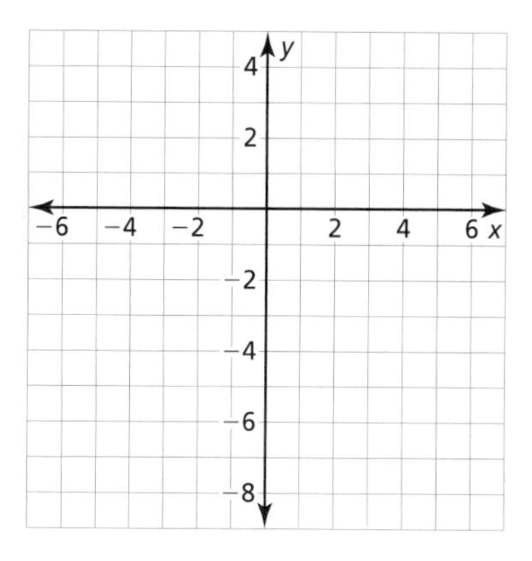

USING TOOLS STRATEGICALLY

To be proficient in math, you need to consider the available tools, such as a graphing calculator, when solving a mathematical problem.

Communicate Your Answer

3. How can you describe the graph of $f(x) = a(x - h)^2$?

4. Without graphing, describe the graph of each function. Use a graphing calculator to check your answer.

a. $y = (x - 3)^2$

b. $y = (x + 3)^2$

c. $y = -(x - 3)^2$

8.4 Lesson

What You Will Learn

- Identify even and odd functions.
- Graph quadratic functions of the form $f(x) = a(x - h)^2$.
- Graph quadratic functions of the form $f(x) = a(x - h)^2 + k$.
- Model real-life problems using $f(x) = a(x - h)^2 + k$.

Core Vocabulary

Previous
reflection

Identifying Even and Odd Functions

Core Concept

Even and Odd Functions

A function $y = f(x)$ is **even** when $f(-x) = f(x)$ for each x in the domain of f. The graph of an even function is symmetric about the y-axis.

A function $y = f(x)$ is **odd** when $f(-x) = -f(x)$ for each x in the domain of f. The graph of an odd function is symmetric about the origin. A graph is *symmetric about the origin* when it looks the same after reflections in the x-axis and then in the y-axis.

STUDY TIP

The graph of an odd function looks the same after a 180° rotation about the origin.

EXAMPLE 1 Identifying Even and Odd Functions

Determine whether each function is *even*, *odd*, or *neither*.

a. $f(x) = 2x$ **b.** $g(x) = x^2 - 2$ **c.** $h(x) = 2x^2 + x - 2$

SOLUTION

a. $f(x) = 2x$ Write the original function.

$f(-x) = 2(-x)$ Substitute $-x$ for x.

$= -2x$ Simplify.

$= -f(x)$ Substitute $f(x)$ for $2x$.

▶ Because $f(-x) = -f(x)$, the function is odd.

b. $g(x) = x^2 - 2$ Write the original function.

$g(-x) = (-x)^2 - 2$ Substitute $-x$ for x.

$= x^2 - 2$ Simplify.

$= g(x)$ Substitute $g(x)$ for $x^2 - 2$.

▶ Because $g(-x) = g(x)$, the function is even.

c. $h(x) = 2x^2 + x - 2$ Write the original function.

$h(-x) = 2(-x)^2 + (-x) - 2$ Substitute $-x$ for x.

$= 2x^2 - x - 2$ Simplify.

▶ Because $h(x) = 2x^2 + x - 2$ and $-h(x) = -2x^2 - x + 2$, you can conclude that $h(-x) \neq h(x)$ and $h(-x) \neq -h(x)$. So, the function is neither even nor odd.

STUDY TIP

Most functions are neither even nor odd.

Monitoring Progress  Help in English and Spanish at *BigIdeasMath.com*

Determine whether the function is *even*, *odd*, or *neither*.

1. $f(x) = 5x$ **2.** $g(x) = 2^x$ **3.** $h(x) = 2x^2 + 3$

Graphing $f(x) = a(x - h)^2$

Core Concept

Graphing $f(x) = a(x - h)^2$

- When $h > 0$, the graph of $f(x) = a(x - h)^2$ is a horizontal translation h units right of the graph of $f(x) = ax^2$.
- When $h < 0$, the graph of $f(x) = a(x - h)^2$ is a horizontal translation $|h|$ units left of the graph of $f(x) = ax^2$.

The vertex of the graph of $f(x) = a(x - h)^2$ is $(h, 0)$, and the axis of symmetry is $x = h$.

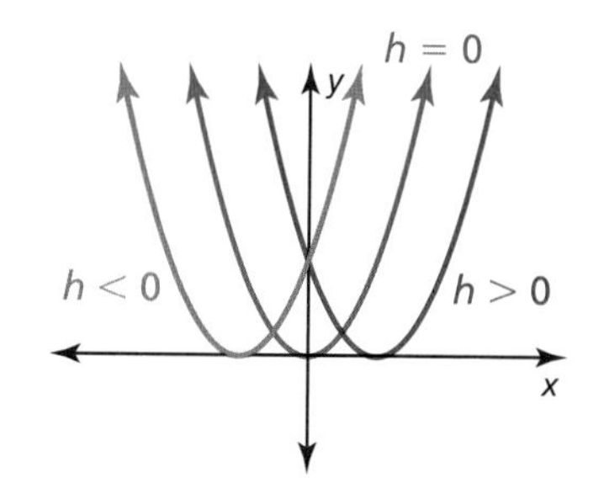

EXAMPLE 2 **Graphing $y = a(x - h)^2$**

Graph $g(x) = \frac{1}{2}(x - 4)^2$. Compare the graph to the graph of $f(x) = x^2$.

ANOTHER WAY

In Step 3, you could instead choose two x-values greater than the x-coordinate of the vertex.

SOLUTION

Step 1 Graph the axis of symmetry. Because $h = 4$, graph $x = 4$.

Step 2 Plot the vertex. Because $h = 4$, plot $(4, 0)$.

Step 3 Find and plot two more points on the graph. Choose two x-values less than the x-coordinate of the vertex. Then find $g(x)$ for each x-value.

When $x = 0$:

$$g(0) = \frac{1}{2}(0 - 4)^2 = 8$$

When $x = 2$:

$$g(2) = \frac{1}{2}(2 - 4)^2 = 2$$

So, plot $(0, 8)$ and $(2, 2)$.

Step 4 Reflect the points plotted in Step 3 in the axis of symmetry. So, plot $(8, 8)$ and $(6, 2)$.

Step 5 Draw a smooth curve through the points.

STUDY TIP

From the graph, you can see that $f(x) = x^2$ is an even function. However, $g(x) = \frac{1}{2}(x - 4)^2$ is neither even nor odd.

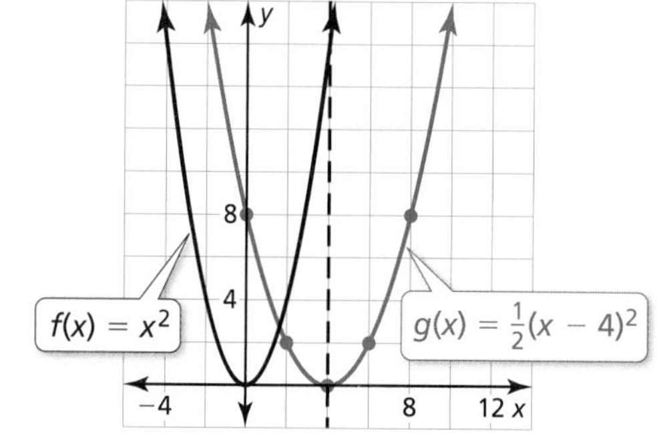

▶ Both graphs open up. The graph of g is wider than the graph of f. The axis of symmetry $x = 4$ and the vertex $(4, 0)$ of the graph of g are 4 units right of the axis of symmetry $x = 0$ and the vertex $(0, 0)$ of the graph of f. So, the graph of g is a translation 4 units right and a vertical shrink by a factor of $\frac{1}{2}$ of the graph of f.

Monitoring Progress  Help in English and Spanish at *BigIdeasMath.com*

Graph the function. Compare the graph to the graph of $f(x) = x^2$.

4. $g(x) = 2(x + 5)^2$

5. $h(x) = -(x - 2)^2$

Graphing $f(x) = a(x - h)^2 + k$

Core Concept

Graphing $f(x) = a(x - h)^2 + k$

The **vertex form** of a quadratic function is $f(x) = a(x - h)^2 + k$, where $a \neq 0$. The graph of $f(x) = a(x - h)^2 + k$ is a translation h units horizontally and k units vertically of the graph of $f(x) = ax^2$.

The vertex of the graph of $f(x) = a(x - h)^2 + k$ is (h, k), and the axis of symmetry is $x = h$.

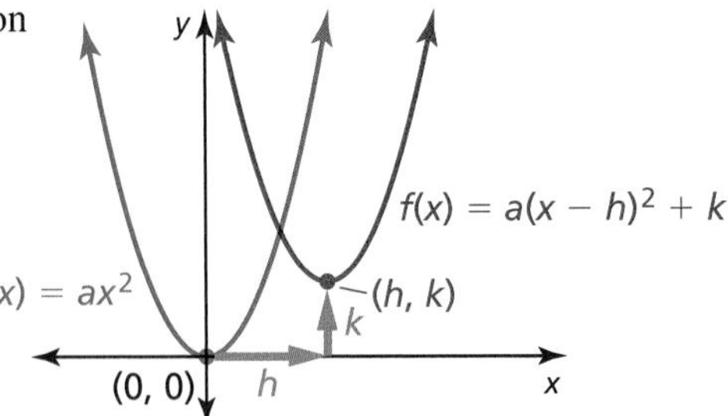

EXAMPLE 3 **Graphing $y = a(x - h)^2 + k$**

Graph $g(x) = -2(x + 2)^2 + 3$. Compare the graph to the graph of $f(x) = x^2$.

SOLUTION

Step 1 Graph the axis of symmetry. Because $h = -2$, graph $x = -2$.

Step 2 Plot the vertex. Because $h = -2$ and $k = 3$, plot $(-2, 3)$.

Step 3 Find and plot two more points on the graph. Choose two x-values less than the x-coordinate of the vertex. Then find $g(x)$ for each x-value. So, plot $(-4, -5)$ and $(-3, 1)$.

x	-4	-3
$g(x)$	-5	1

Step 4 Reflect the points plotted in Step 3 in the axis of symmetry. So, plot $(-1, 1)$ and $(0, -5)$.

Step 5 Draw a smooth curve through the points.

▶ The graph of g opens down and is narrower than the graph of f. The vertex of the graph of g, $(-2, 3)$, is 2 units left and 3 units up of the vertex of the graph of f, $(0, 0)$. So, the graph of g is a vertical stretch by a factor of 2, a reflection in the x-axis, and a translation 2 units left and 3 units up of the graph of f.

EXAMPLE 4 **Transforming the Graph of $y = a(x - h)^2 + k$**

Consider function g in Example 3. Graph $f(x) = g(x + 5)$.

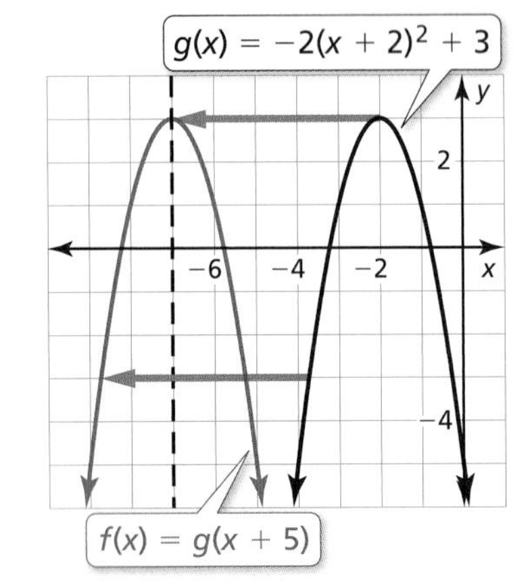

SOLUTION

The function f is of the form $y = g(x - h)$, where $h = -5$. So, the graph of f is a horizontal translation 5 units left of the graph of g. To graph f, subtract 5 from the x-coordinates of the points on the graph of g.

Monitoring Progress

Help in English and Spanish at *BigIdeasMath.com*

Graph the function. Compare the graph to the graph of $f(x) = x^2$.

6. $g(x) = 3(x - 1)^2 + 6$

7. $h(x) = \frac{1}{2}(x + 4)^2 - 2$

8. Consider function g in Example 3. Graph $f(x) = g(x) - 3$.

Modeling Real-Life Problems

EXAMPLE 5 Modeling with Mathematics

Water fountains are usually designed to give a specific visual effect. For example, the water fountain shown consists of streams of water that are shaped like parabolas. Notice how the streams are designed to land on the underwater spotlights. Write and graph a quadratic function that models the path of a stream of water with a maximum height of 5 feet, represented by a vertex of (3, 5), landing on a spotlight 6 feet from the water jet, represented by (6, 0).

SOLUTION

1. **Understand the Problem** You know the vertex and another point on the graph that represents the parabolic path. You are asked to write and graph a quadratic function that models the path.

2. **Make a Plan** Use the given points and the vertex form to write a quadratic function. Then graph the function.

3. **Solve the Problem**
   Use the vertex form, vertex (3, 5), and point (6, 0) to find the value of a.

$f(x) = a(x - h)^2 + k$	Write the vertex form of a quadratic function.
$f(x) = a(x - 3)^2 + 5$	Substitute 3 for h and 5 for k.
$0 = a(6 - 3)^2 + 5$	Substitute 6 for x and 0 for $f(x)$.
$0 = 9a + 5$	Simplify.
$-\frac{5}{9} = a$	Solve for a.

So, $f(x) = -\frac{5}{9}(x - 3)^2 + 5$ models the path of a stream of water. Now graph the function.

Step 1 Graph the axis of symmetry. Because $h = 3$, graph $x = 3$.

Step 2 Plot the vertex, (3, 5).

Step 3 Find and plot two more points on the graph. Because the x-axis represents the water surface, the graph should only contain points with nonnegative values of $f(x)$. You know that (6, 0) is on the graph. To find another point, choose an x-value between $x = 3$ and $x = 6$. Then find the corresponding value of $f(x)$.

$$f(4.5) = -\frac{5}{9}(4.5 - 3)^2 + 5 = 3.75$$

So, plot (6, 0) and (4.5, 3.75).

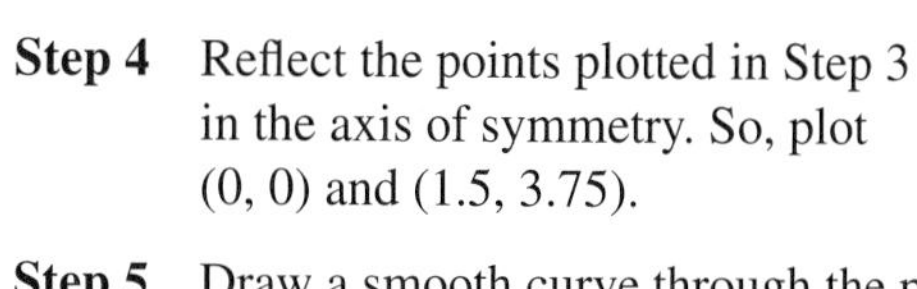

Step 4 Reflect the points plotted in Step 3 in the axis of symmetry. So, plot (0, 0) and (1.5, 3.75).

Step 5 Draw a smooth curve through the points.

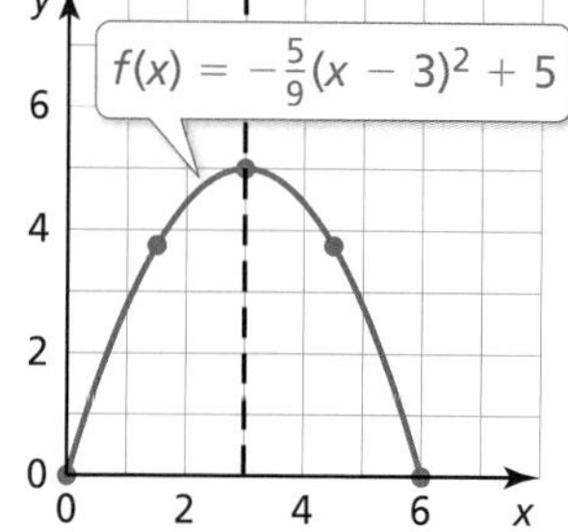

8
Maximum
X=3 Y=5
0 8
0

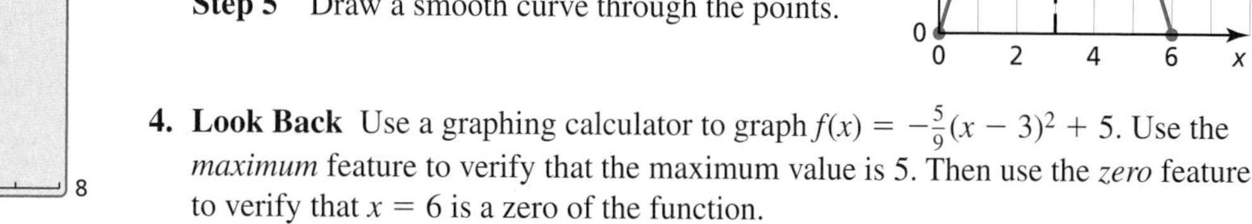

4. **Look Back** Use a graphing calculator to graph $f(x) = -\frac{5}{9}(x - 3)^2 + 5$. Use the *maximum* feature to verify that the maximum value is 5. Then use the *zero* feature to verify that $x = 6$ is a zero of the function.

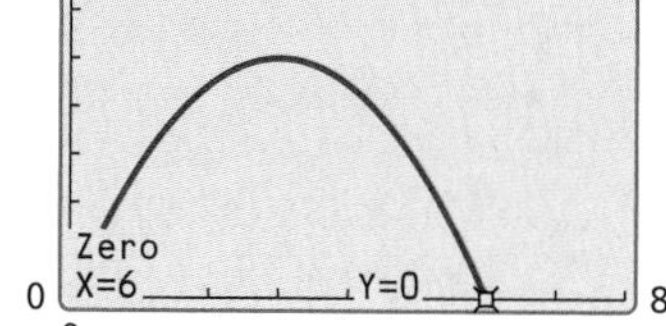

Monitoring Progress Help in English and Spanish at *BigIdeasMath.com*

9. **WHAT IF?** The vertex is (3, 6). Write and graph a quadratic function that models the path.

8.4 Exercises

Dynamic Solutions available at *BigIdeasMath.com*

Vocabulary and Core Concept Check

1. **VOCABULARY** Compare the graph of an even function with the graph of an odd function.

2. **OPEN-ENDED** Write a quadratic function whose graph has a vertex of (1, 2).

3. **WRITING** Describe the transformation from the graph of $f(x) = ax^2$ to the graph of $g(x) = a(x - h)^2 + k$.

4. **WHICH ONE DOESN'T BELONG?** Which function does *not* belong with the other three? Explain your reasoning.

$f(x) = 8(x + 4)^2$ $f(x) = (x - 2)^2 + 4$ $f(x) = 2(x + 0)^2$ $f(x) = 3(x + 1)^2 + 1$

Monitoring Progress and Modeling with Mathematics

In Exercises 5–12, determine whether the function is *even*, *odd*, or *neither*. *(See Example 1.)*

5. $f(x) = 4x + 3$
6. $g(x) = 3x^2$
7. $h(x) = 5^x + 2$
8. $m(x) = 2x^2 - 7x$
9. $p(x) = -x^2 + 8$
10. $f(x) = -\frac{1}{2}x$
11. $n(x) = 2x^2 - 7x + 3$
12. $r(x) = -6x^2 + 5$

In Exercises 13–18, determine whether the function represented by the graph is *even*, *odd*, or *neither*.

13.
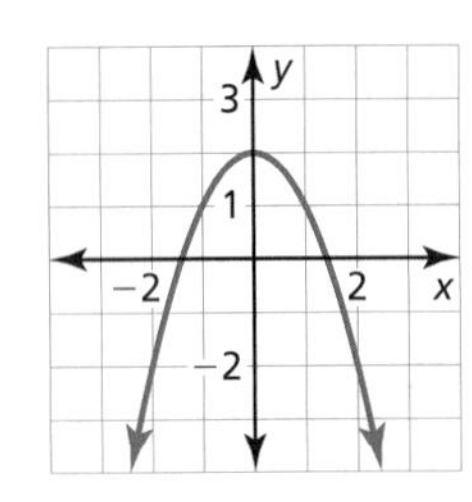

14.
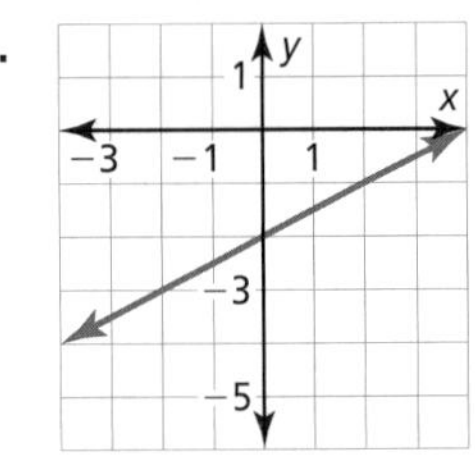

15.
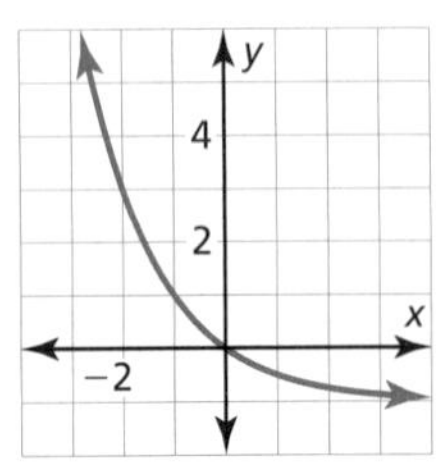

16.
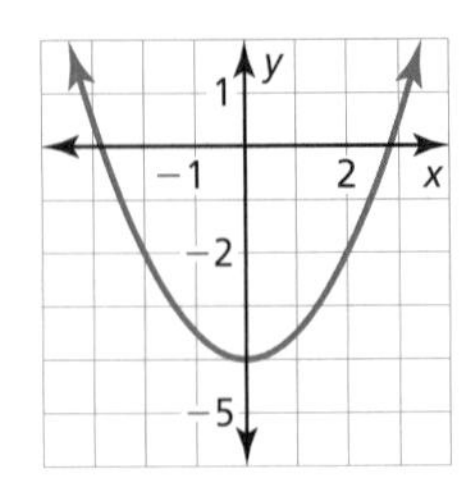

17.
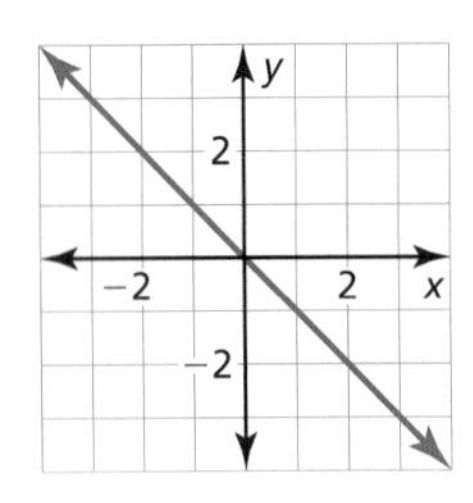

18.
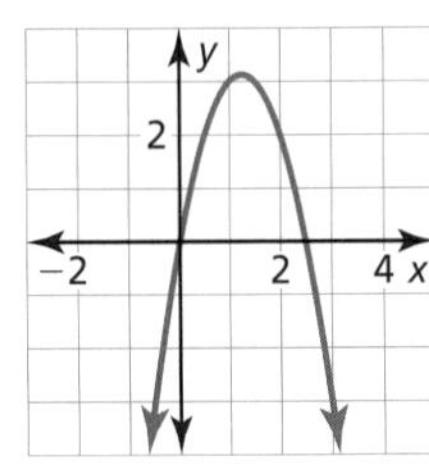

In Exercises 19–22, find the vertex and the axis of symmetry of the graph of the function.

19. $f(x) = 3(x + 1)^2$
20. $f(x) = \frac{1}{4}(x - 6)^2$
21. $y = -\frac{1}{8}(x - 4)^2$
22. $y = -5(x + 9)^2$

In Exercises 23–28, graph the function. Compare the graph to the graph of $f(x) = x^2$. *(See Example 2.)*

23. $g(x) = 2(x + 3)^2$
24. $p(x) = 3(x - 1)^2$
25. $r(x) = \frac{1}{4}(x + 10)^2$
26. $n(x) = \frac{1}{3}(x - 6)^2$
27. $d(x) = \frac{1}{5}(x - 5)^2$
28. $q(x) = 6(x + 2)^2$

29. **ERROR ANALYSIS** Describe and correct the error in determining whether the function $f(x) = x^2 + 3$ is even, odd, or neither.

$f(x) = x^2 + 3$
$f(-x) = (-x)^2 + 3$
$= x^2 + 3$
$= f(x)$
So, $f(x)$ is an odd function.

30. **ERROR ANALYSIS** Describe and correct the error in finding the vertex of the graph of the function.

$y = -(x + 8)^2$
Because $h = -8$, the vertex is $(0, -8)$.

In Exercises 31–34, find the vertex and the axis of symmetry of the graph of the function.

31. $y = -6(x + 4)^2 - 3$ **32.** $f(x) = 3(x - 3)^2 + 6$

33. $f(x) = -4(x + 3)^2 + 1$ **34.** $y = -(x - 6)^2 - 5$

In Exercises 35–38, match the function with its graph.

35. $y = -(x + 1)^2 - 3$ **36.** $y = -\frac{1}{2}(x - 1)^2 + 3$

37. $y = \frac{1}{3}(x - 1)^2 + 3$ **38.** $y = 2(x + 1)^2 - 3$

A.
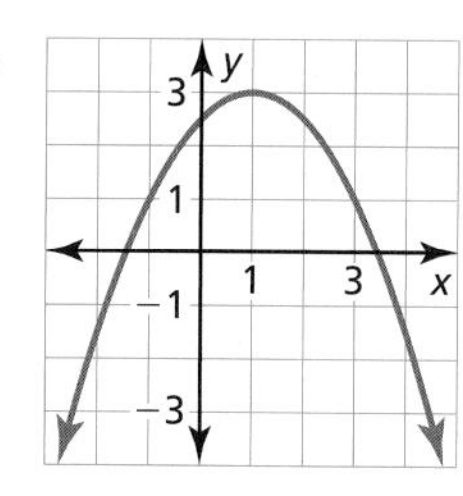

B.
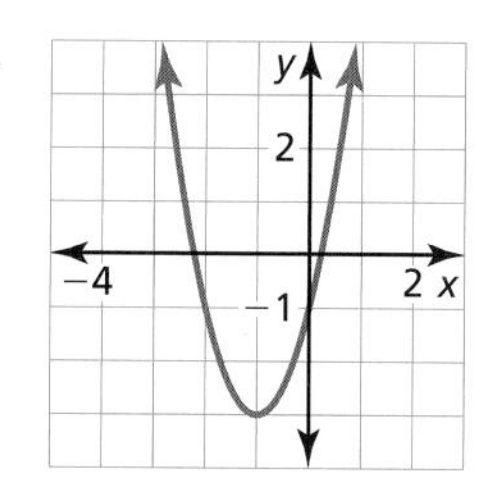

C.
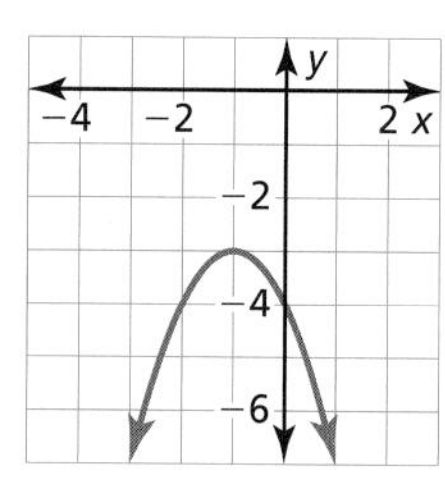

D.
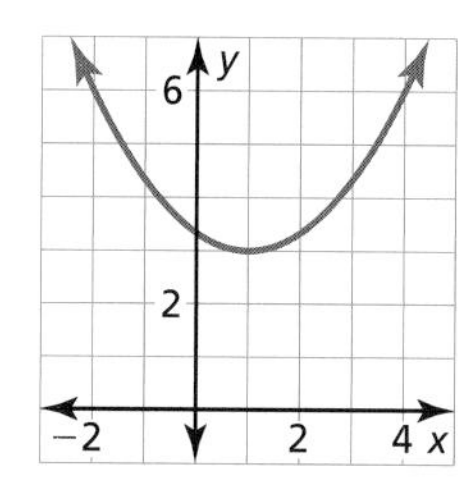

In Exercises 39–44, graph the function. Compare the graph to the graph of $f(x) = x^2$. *(See Example 3.)*

39. $h(x) = (x - 2)^2 + 4$ **40.** $g(x) = (x + 1)^2 - 7$

41. $r(x) = 4(x - 1)^2 - 5$ **42.** $n(x) = -(x + 4)^2 + 2$

43. $g(x) = -\frac{1}{3}(x + 3)^2 - 2$ **44.** $r(x) = \frac{1}{2}(x - 2)^2 - 4$

In Exercises 45–48, let $f(x) = (x - 2)^2 + 1$. Match the function with its graph.

45. $g(x) = f(x - 1)$ **46.** $r(x) = f(x + 2)$

47. $h(x) = f(x) + 2$ **48.** $p(x) = f(x) - 3$

A.
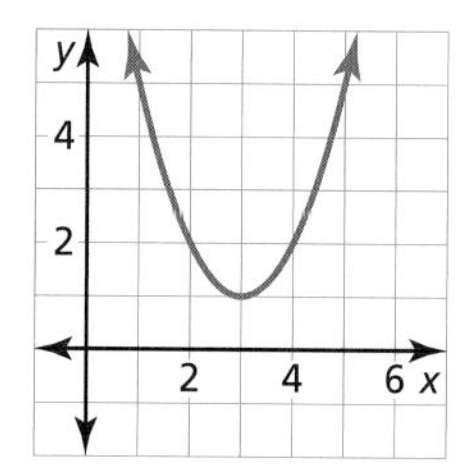

B.
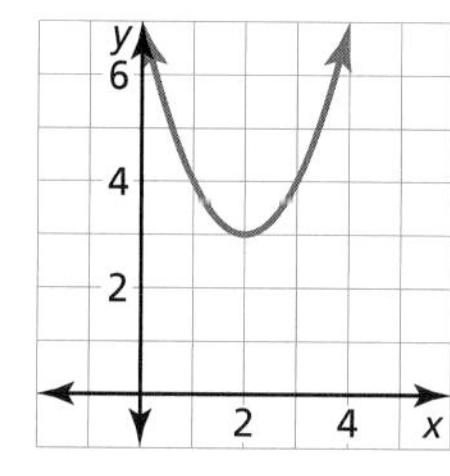

C.
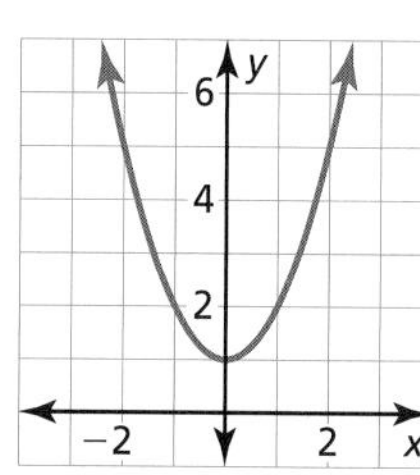

D.
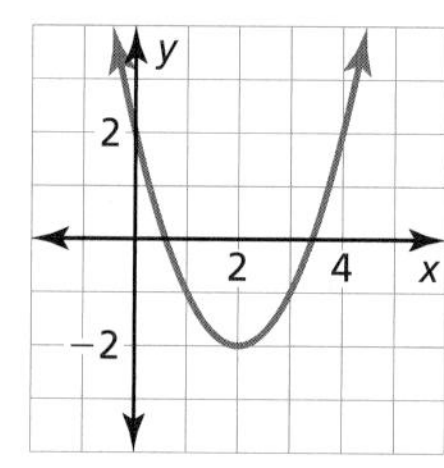

In Exercises 49–54, graph g. *(See Example 4.)*

49. $f(x) = 2(x - 1)^2 + 1$; $g(x) = f(x + 3)$

50. $f(x) = -(x + 1)^2 + 2$; $g(x) = \frac{1}{2}f(x)$

51. $f(x) = -3(x + 5)^2 - 6$; $g(x) = 2f(x)$

52. $f(x) = 5(x - 3)^2 - 1$; $g(x) = f(x) - 6$

53. $f(x) = (x + 3)^2 + 5$; $g(x) = f(x - 4)$

54. $f(x) = -2(x - 4)^2 - 8$; $g(x) = -f(x)$

55. MODELING WITH MATHEMATICS The height (in meters) of a bird diving to catch a fish is represented by $h(t) = 5(t - 2.5)^2$, where t is the number of seconds after beginning the dive.

a. Graph h.

b. Another bird's dive is represented by $r(t) = 2h(t)$. Graph r.

c. Compare the graphs. Which bird starts its dive from a greater height? Explain.

56. MODELING WITH MATHEMATICS A kicker punts a football. The height (in yards) of the football is represented by $f(x) = -\frac{1}{9}(x - 30)^2 + 25$, where x is the horizontal distance (in yards) from the kicker's goal line.

a. Graph f. Describe the domain and range.

b. On the next possession, the kicker punts the football. The height of the football is represented by $g(x) = f(x + 5)$. Graph g. Describe the domain and range.

c. Compare the graphs. On which possession does the kicker punt closer to his goal line? Explain.

In Exercises 57–62, write a quadratic function in vertex form whose graph has the given vertex and passes through the given point.

57. vertex: (1, 2); passes through (3, 10)

58. vertex: (−3, 5); passes through (0, −14)

59. vertex: (−2, −4); passes through (−1, −6)

60. vertex: (1, 8); passes through (3, 12)

61. vertex: (5, −2); passes through (7, 0)

62. vertex: (−5, −1); passes through (−2, 2)

63. **MODELING WITH MATHEMATICS** A portion of a roller coaster track is in the shape of a parabola. Write and graph a quadratic function that models this portion of the roller coaster with a maximum height of 90 feet, represented by a vertex of (25, 90), passing through the point (50, 0). *(See Example 5.)*

64. **MODELING WITH MATHEMATICS** A flare is launched from a boat and travels in a parabolic path until reaching the water. Write and graph a quadratic function that models the path of the flare with a maximum height of 300 meters, represented by a vertex of (59, 300), landing in the water at the point (119, 0).

In Exercises 65–68, rewrite the quadratic function in vertex form.

65. $y = 2x^2 - 8x + 4$

66. $y = 3x^2 + 6x - 1$

67. $f(x) = -5x^2 + 10x + 3$

68. $f(x) = -x^2 - 4x + 2$

69. **REASONING** Can a function be symmetric about the x-axis? Explain.

70. **HOW DO YOU SEE IT?** The graph of a quadratic function is shown. Determine which symbols to use to complete the vertex form of the quadratic function. Explain your reasoning.

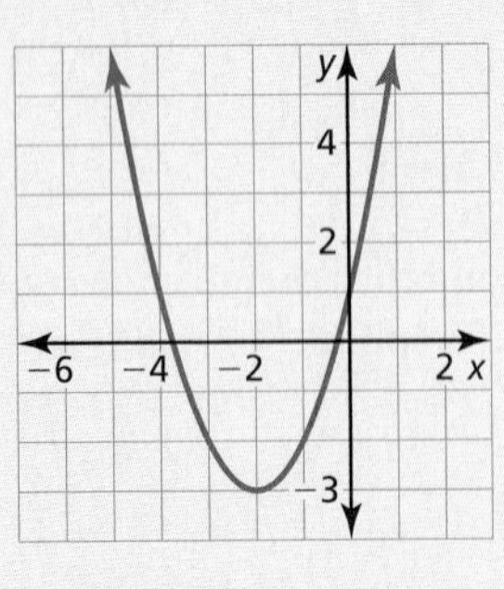

$y = a(x \quad 2)^2 \quad 3$

In Exercises 71–74, describe the transformation from the graph of f to the graph of h. Write an equation that represents h in terms of x.

71. $f(x) = -(x + 1)^2 - 2$
$h(x) = f(x) + 4$

72. $f(x) = 2(x - 1)^2 + 1$
$h(x) = f(x - 5)$

73. $f(x) = 4(x - 2)^2 + 3$
$h(x) = 2f(x)$

74. $f(x) = -(x + 5)^2 - 6$
$h(x) = \frac{1}{3}f(x)$

75. **REASONING** The graph of $y = x^2$ is translated 2 units right and 5 units down. Write an equation for the function in vertex form and in standard form. Describe advantages of writing the function in each form.

76. **THOUGHT PROVOKING** Which of the following are true? Justify your answers.

a. Any constant multiple of an even function is even.

b. Any constant multiple of an odd function is odd.

c. The sum or difference of two even functions is even.

d. The sum or difference of two odd functions is odd.

e. The sum or difference of an even function and an odd function is odd.

77. **COMPARING FUNCTIONS** A cross section of a birdbath can be modeled by $y = \frac{1}{81}(x - 18)^2 - 4$, where x and y are measured in inches. The graph shows the cross section of another birdbath.

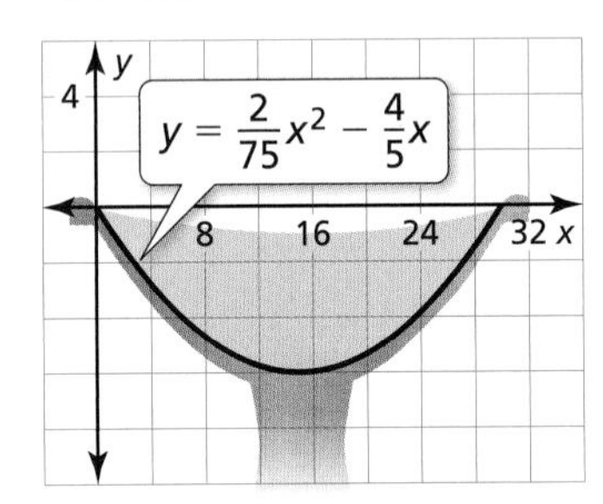

a. Which birdbath is deeper? Explain.

b. Which birdbath is wider? Explain.

78. **REASONING** Compare the graphs of $y = 2x^2 + 8x + 8$ and $y = x^2$ without graphing the functions. How can factoring help you compare the parabolas? Explain.

79. **MAKING AN ARGUMENT** Your friend says all absolute value functions are even because of their symmetry. Is your friend correct? Explain.

Maintaining Mathematical Proficiency

Reviewing what you learned in previous grades and lessons

Solve the equation. *(Section 7.4)*

80. $x(x - 1) = 0$

81. $(x + 3)(x - 8) = 0$

82. $(3x - 9)(4x + 12) = 0$

8.5 Using Intercept Form

Essential Question What are some of the characteristics of the graph of $f(x) = a(x - p)(x - q)$?

EXPLORATION 1 **Using Zeros to Write Functions**

Work with a partner. Each graph represents a function of the form $f(x) = (x - p)(x - q)$ or $f(x) = -(x - p)(x - q)$. Write the function represented by each graph. Explain your reasoning.

a.
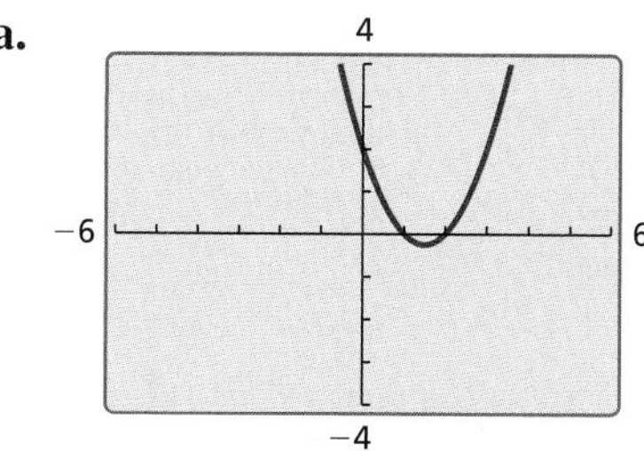

b.
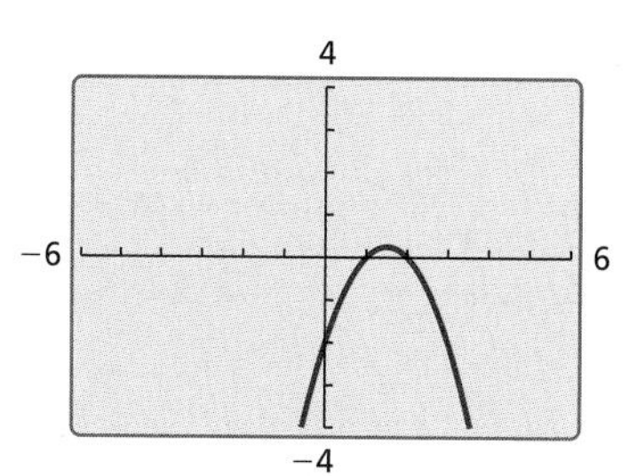

c.
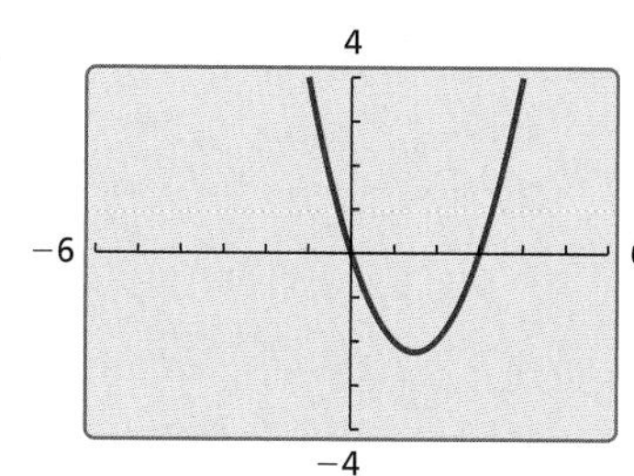

d.
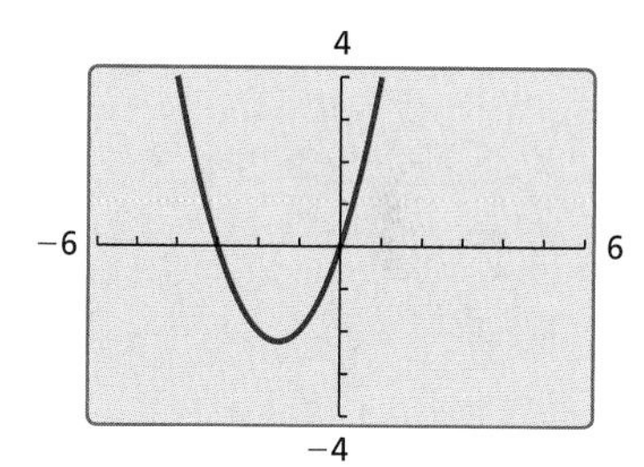

e.
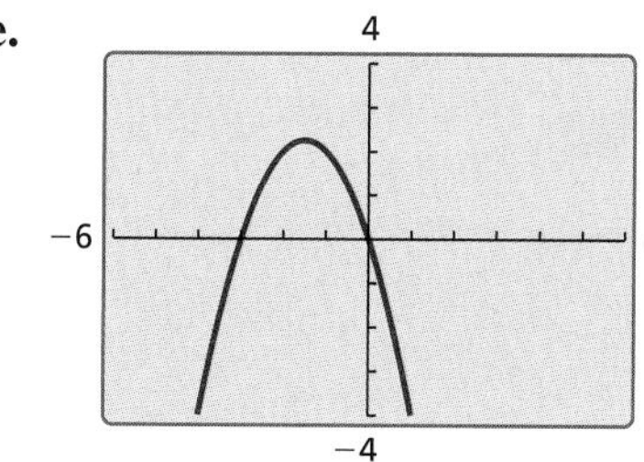

f.
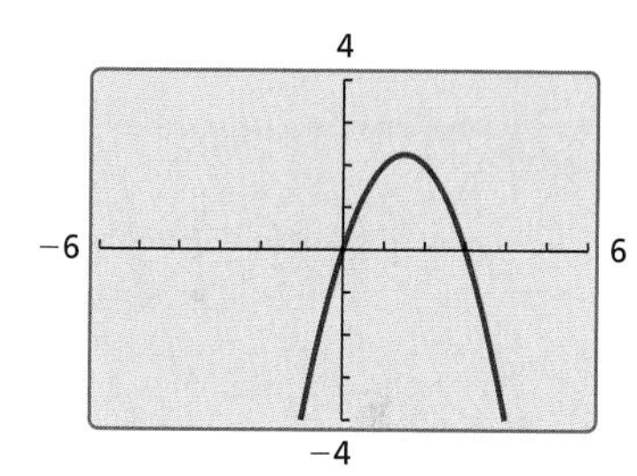

g.
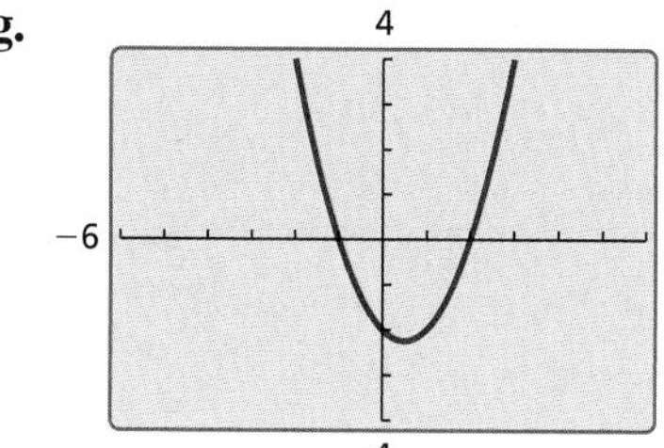

h.
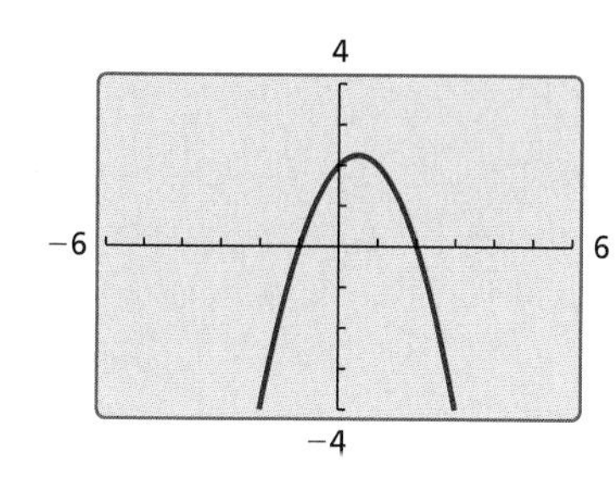

CONSTRUCTING VIABLE ARGUMENTS

To be proficient in math, you need to justify your conclusions and communicate them to others.

Communicate Your Answer

2. What are some of the characteristics of the graph of $f(x) = a(x - p)(x - q)$?

3. Consider the graph of $f(x) = a(x - p)(x - q)$.

a. Does changing the sign of a change the x-intercepts? Does changing the sign of a change the y-intercept? Explain your reasoning.

b. Does changing the value of p change the x-intercepts? Does changing the value of p change the y-intercept? Explain your reasoning.

8.5 Lesson

Core Vocabulary

intercept form, *p. 450*

What You Will Learn

- Graph quadratic functions of the form $f(x) = a(x - p)(x - q)$.
- Use intercept form to find zeros of functions.
- Use characteristics to graph and write quadratic functions.
- Use characteristics to graph and write cubic functions.

Graphing $f(x) = a(x - p)(x - q)$

You have already graphed quadratic functions written in several different forms, such as $f(x) = ax^2 + bx + c$ (standard form) and $g(x) = a(x - h)^2 + k$ (vertex form). Quadratic functions can also be written in **intercept form**, $f(x) = a(x - p)(x - q)$, where $a \neq 0$. In this form, the polynomial that defines a function is in factored form and the x-intercepts of the graph can be easily determined.

Core Concept

Graphing $f(x) = a(x - p)(x - q)$

- The x-intercepts are p and q.
- The axis of symmetry is halfway between $(p, 0)$ and $(q, 0)$. So, the axis of symmetry is $x = \frac{p + q}{2}$.
- The graph opens up when $a > 0$, and the graph opens down when $a < 0$.

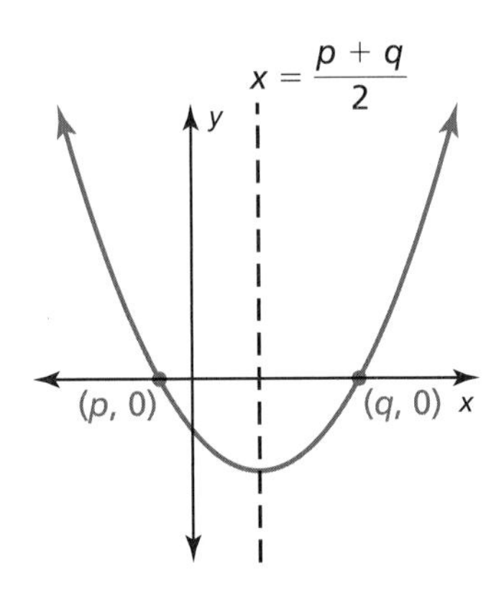

EXAMPLE 1 Graphing $f(x) = a(x - p)(x - q)$

Graph $f(x) = -(x + 1)(x - 5)$. Describe the domain and range.

SOLUTION

Step 1 Identify the x-intercepts. Because the x-intercepts are $p = -1$ and $q = 5$, plot $(-1, 0)$ and $(5, 0)$.

Step 2 Find and graph the axis of symmetry.

$$x = \frac{p + q}{2} = \frac{-1 + 5}{2} = 2$$

Step 3 Find and plot the vertex.

The x-coordinate of the vertex is 2. To find the y-coordinate of the vertex, substitute 2 for x and simplify.

$$f(2) = -(2 + 1)(2 - 5) = 9$$

So, the vertex is $(2, 9)$.

Step 4 Draw a parabola through the vertex and the points where the x-intercepts occur.

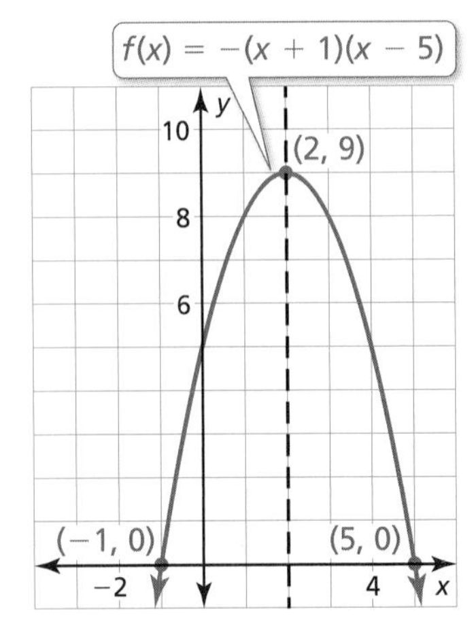

▶ The domain is all real numbers. The range is $y \leq 9$.

EXAMPLE 2 Graphing a Quadratic Function

Graph $f(x) = 2x^2 - 8$. Describe the domain and range.

SOLUTION

Step 1 Rewrite the quadratic function in intercept form.

$f(x) = 2x^2 - 8$ — Write the function.

$= 2(x^2 - 4)$ — Factor out common factor.

$= 2(x + 2)(x - 2)$ — Difference of two squares pattern

Step 2 Identify the x-intercepts. Because the x-intercepts are $p = -2$ and $q = 2$, plot $(-2, 0)$ and $(2, 0)$.

Step 3 Find and graph the axis of symmetry.

$$x = \frac{p + q}{2} = \frac{-2 + 2}{2} = 0$$

Step 4 Find and plot the vertex.

The x-coordinate of the vertex is 0.
The y-coordinate of the vertex is

$$f(0) = 2(0)^2 - 8 = -8.$$

So, the vertex is $(0, -8)$.

Step 5 Draw a parabola through the vertex and the points where the x-intercepts occur.

(−2, 0) y (2, 0)
−4 4 x
−2
−4
(0, −8) $f(x) = 2x^2 - 8$

▶ The domain is all real numbers. The range is $y \geq -8$.

Monitoring Progress

Help in English and Spanish at *BigIdeasMath.com*

Graph the quadratic function. Label the vertex, axis of symmetry, and x-intercepts. Describe the domain and range of the function.

1. $f(x) = (x + 2)(x - 3)$
2. $g(x) = -2(x - 4)(x + 1)$
3. $h(x) = 4x^2 - 36$

REMEMBER
Functions have zeros, and *graphs* have x-intercepts.

Using Intercept Form to Find Zeros of Functions

In Section 8.2, you learned that a zero of a function is an x-value for which $f(x) = 0$. You can use the intercept form of a function to find the zeros of the function.

EXAMPLE 3 Finding Zeros of a Function

Find the zeros of $f(x) = (x - 1)(x + 2)$.

SOLUTION

To find the zeros, determine the x-values for which $f(x)$ is 0.

$f(x) = (x - 1)(x + 2)$ — Write the function.

$0 = (x - 1)(x + 2)$ — Substitute 0 for $f(x)$.

$x - 1 = 0$ *or* $x + 2 = 0$ — Zero-Product Property

$x = 1$ *or* $x = -2$ — Solve for x.

▶ So, the zeros of the function are -2 and 1.

Check

6
−5 4
−4

Factors and Zeros

For any factor $x - n$ of a polynomial, n is a zero of the function defined by the polynomial.

EXAMPLE 4 Finding Zeros of Functions

Find the zeros of each function.

a. $f(x) = -2x^2 - 10x - 12$ **b.** $h(x) = (x - 1)(x^2 - 16)$

SOLUTION

Write each function in intercept form to identify the zeros.

a. $f(x) = -2x^2 - 10x - 12$ Write the function.

$= -2(x^2 + 5x + 6)$ Factor out common factor.

$= -2(x + 3)(x + 2)$ Factor the trinomial.

▶ So, the zeros of the function are -3 and -2.

b. $h(x) = (x - 1)(x^2 - 16)$ Write the function.

$= (x - 1)(x + 4)(x - 4)$ Difference of two squares pattern

▶ So, the zeros of the function are -4, 1, and 4.

LOOKING FOR STRUCTURE

The function in Example 4(b) is called a *cubic function*. You can extend the concept of intercept form to cubic functions. You will graph a cubic function in Example 7.

Monitoring Progress  Help in English and Spanish at *BigIdeasMath.com*

Find the zero(s) of the function.

4. $f(x) = (x - 6)(x - 1)$ **5.** $g(x) = 3x^2 - 12x + 12$ **6.** $h(x) = x(x^2 - 1)$

Using Characteristics to Graph and Write Quadratic Functions

EXAMPLE 5 Graphing a Quadratic Function Using Zeros

Use zeros to graph $h(x) = x^2 - 2x - 3$.

SOLUTION

The function is in standard form. You know that the parabola opens up ($a > 0$) and the y-intercept is -3. So, begin by plotting $(0, -3)$.

Notice that the polynomial that defines the function is factorable. So, write the function in intercept form and identify the zeros.

$h(x) = x^2 - 2x - 3$ Write the function.

$= (x + 1)(x - 3)$ Factor the trinomial.

The zeros of the function are -1 and 3. So, plot $(-1, 0)$ and $(3, 0)$. Draw a parabola through the points.

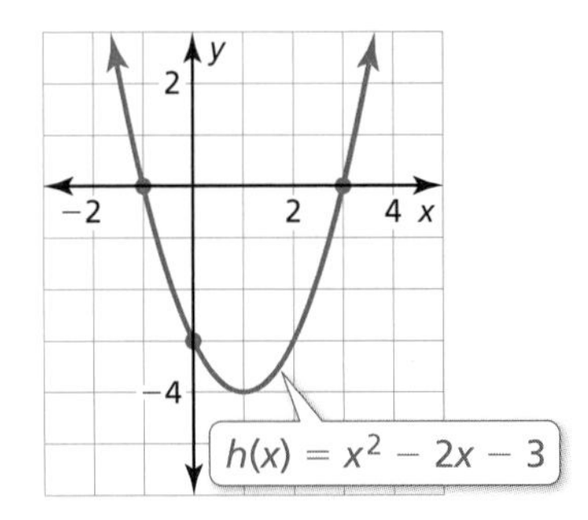

ATTENDING TO PRECISION

To sketch a more precise graph, make a table of values and plot other points on the graph.

EXAMPLE 6 Writing Quadratic Functions

STUDY TIP

In part (a), many possible functions satisfy the given condition. The value a can be *any* nonzero number. To allow easier calculations, let $a = 1$. By letting $a = 2$, the resulting function would be $f(x) = 2x^2 + 12x + 22$.

Write a quadratic function in standard form whose graph satisfies the given condition(s).

a. vertex: $(-3, 4)$

b. passes through $(-9, 0)$, $(-2, 0)$, and $(-4, 20)$

SOLUTION

a. Because you know the vertex, use vertex form to write a function.

$f(x) = a(x - h)^2 + k$	Vertex form
$= 1(x + 3)^2 + 4$	Substitute for a, h, and k.
$= x^2 + 6x + 9 + 4$	Find the product $(x + 3)^2$.
$= x^2 + 6x + 13$	Combine like terms.

b. The given points indicate that the x-intercepts are -9 and -2. So, use intercept form to write a function.

$f(x) = a(x - p)(x - q)$	Intercept form
$= a(x + 9)(x + 2)$	Substitute for p and q.

Use the other given point, $(-4, 20)$, to find the value of a.

$20 = a(-4 + 9)(-4 + 2)$	Substitute -4 for x and 20 for $f(x)$.
$20 = a(5)(-2)$	Simplify.
$-2 = a$	Solve for a.

Use the value of a to write the function.

$f(x) = -2(x + 9)(x + 2)$	Substitute -2 for a.
$= -2x^2 - 22x - 36$	Simplify.

Monitoring Progress  Help in English and Spanish at *BigIdeasMath.com*

Use zeros to graph the function.

7. $f(x) = (x - 1)(x - 4)$

8. $g(x) = x^2 + x - 12$

Write a quadratic function in standard form whose graph satisfies the given condition(s).

9. x-intercepts: -1 and 1

10. vertex: $(8, 8)$

11. passes through $(0, 0)$, $(10, 0)$, and $(4, 12)$

12. passes through $(-5, 0)$, $(4, 0)$, and $(3, -16)$

Using Characteristics to Graph and Write Cubic Functions

In Example 4, you extended the concept of intercept form to cubic functions.

$$f(x) = a(x - p)(x - q)(x - r),\ a \neq 0 \qquad \text{Intercept form of a cubic function}$$

The x-intercepts of the graph of f are p, q, and r.

EXAMPLE 7 Graphing a Cubic Function Using Zeros

Use zeros to graph $f(x) = x^3 - 4x$.

SOLUTION

Notice that the polynomial that defines the function is factorable. So, write the function in intercept form and identify the zeros.

$f(x) = x^3 - 4x$ Write the function.

$= x(x^2 - 4)$ Factor out x.

$= x(x + 2)(x - 2)$ Difference of two squares pattern

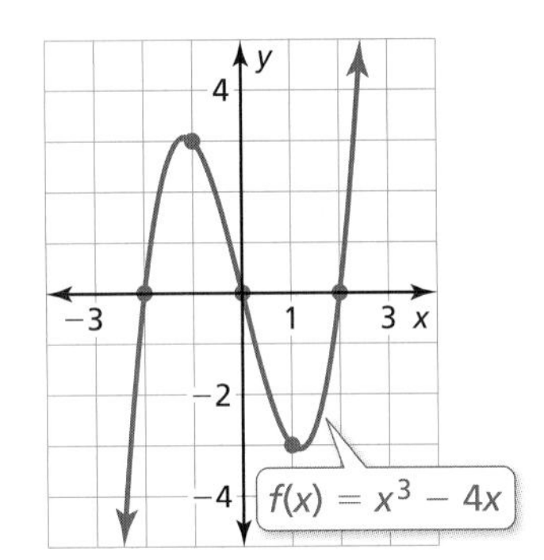

The zeros of the function are -2, 0, and 2. So, plot $(-2, 0)$, $(0, 0)$, and $(2, 0)$.

To help determine the shape of the graph, find points between the zeros.

x	-1	1
f(x)	3	-3

Plot $(-1, 3)$ and $(1, -3)$. Draw a smooth curve through the points.

EXAMPLE 8 Writing a Cubic Function

The graph represents a cubic function. Write the function.

SOLUTION

From the graph, you can see that the x-intercepts are 0, 2, and 5. Use intercept form to write a function.

$f(x) = a(x - p)(x - q)(x - r)$ Intercept form

$= a(x - 0)(x - 2)(x - 5)$ Substitute for p, q, and r.

$= a(x)(x - 2)(x - 5)$ Simplify.

Use the other given point, $(3, 12)$, to find the value of a.

$12 = a(3)(3 - 2)(3 - 5)$ Substitute 3 for x and 12 for $f(x)$.

$-2 = a$ Solve for a.

Use the value of a to write the function.

$f(x) = -2(x)(x - 2)(x - 5)$ Substitute -2 for a.

$= -2x^3 + 14x^2 - 20x$ Simplify.

▶ The function represented by the graph is $f(x) = -2x^3 + 14x^2 - 20x$.

Monitoring Progress

Help in English and Spanish at *BigIdeasMath.com*

Use zeros to graph the function.

13. $g(x) = (x - 1)(x - 3)(x + 3)$

14. $h(x) = x^3 - 6x^2 + 5x$

15. The zeros of a cubic function are -3, -1, and 1. The graph of the function passes through the point $(0, -3)$. Write the function.

8.5 Exercises

Dynamic Solutions available at *BigIdeasMath.com*

Vocabulary and Core Concept Check

1. **COMPLETE THE SENTENCE** The values p and q are __________ of the graph of the function $f(x) = a(x - p)(x - q)$.

2. **WRITING** Explain how to find the maximum value or minimum value of a quadratic function when the function is given in intercept form.

Monitoring Progress and Modeling with Mathematics

In Exercises 3–6, find the x-intercepts and axis of symmetry of the graph of the function.

3. 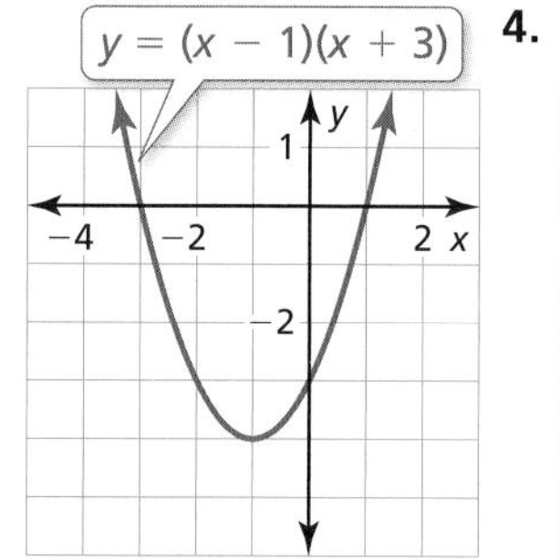

4. 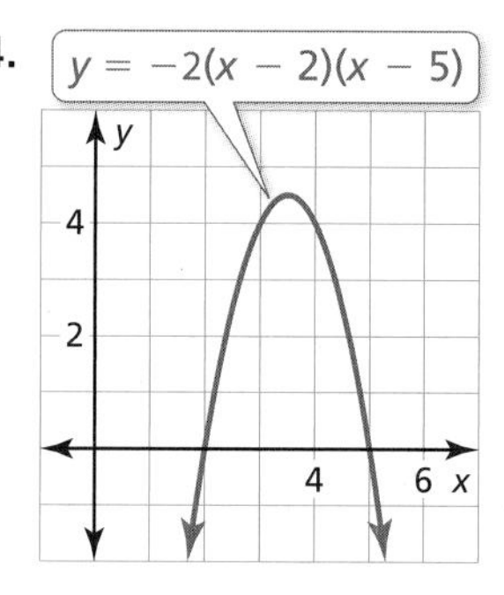

5. $f(x) = -5(x + 7)(x - 5)$

6. $g(x) = \frac{2}{3}x(x + 8)$

In Exercises 7–12, graph the quadratic function. Label the vertex, axis of symmetry, and x-intercepts. Describe the domain and range of the function. *(See Example 1.)*

7. $f(x) = (x + 4)(x + 1)$

8. $y = (x - 2)(x + 2)$

9. $y = -(x + 6)(x - 4)$

10. $h(x) = -4(x - 7)(x - 3)$

11. $g(x) = 5(x + 1)(x + 2)$

12. $y = -2(x - 3)(x + 4)$

In Exercises 13–20, graph the quadratic function. Label the vertex, axis of symmetry, and x-intercepts. Describe the domain and range of the function. *(See Example 2.)*

13. $y = x^2 - 9$

14. $f(x) = x^2 - 8x$

15. $h(x) = -5x^2 + 5x$

16. $y = 3x^2 - 48$

17. $q(x) = x^2 + 9x + 14$

18. $p(x) = x^2 + 6x - 27$

19. $y = 4x^2 - 36x + 32$

20. $y = -2x^2 - 4x + 30$

In Exercises 21–30, find the zero(s) of the function. *(See Examples 3 and 4.)*

21. $y = -2(x - 2)(x - 10)$

22. $f(x) = \frac{1}{3}(x + 5)(x - 1)$

23. $g(x) = x^2 + 5x - 24$

24. $y = x^2 - 17x + 52$

25. $y = 3x^2 - 15x - 42$

26. $g(x) = -4x^2 - 8x - 4$

27. $f(x) = (x + 5)(x^2 - 4)$

28. $h(x) = (x^2 - 36)(x - 11)$

29. $y = x^3 - 49x$

30. $y = x^3 - x^2 - 9x + 9$

In Exercises 31–36, match the function with its graph.

31. $y = (x + 5)(x + 3)$

32. $y = (x + 5)(x - 3)$

33. $y = (x - 5)(x + 3)$

34. $y = (x - 5)(x - 3)$

35. $y = (x + 5)(x - 5)$

36. $y = (x + 3)(x - 3)$

A. 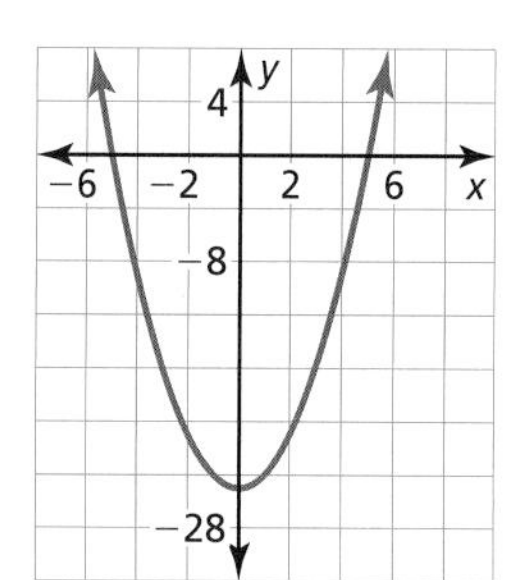

B. 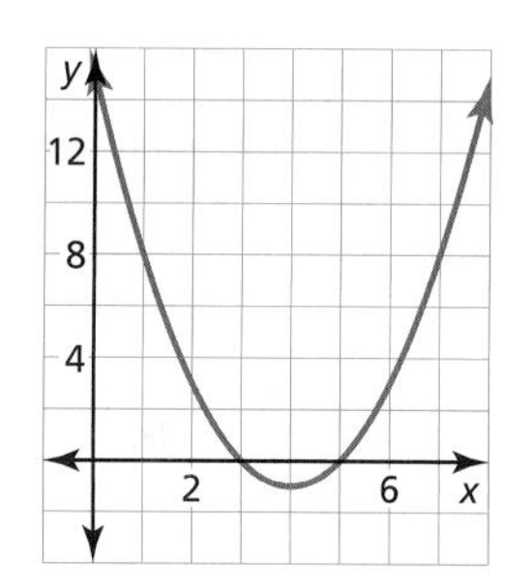

C. 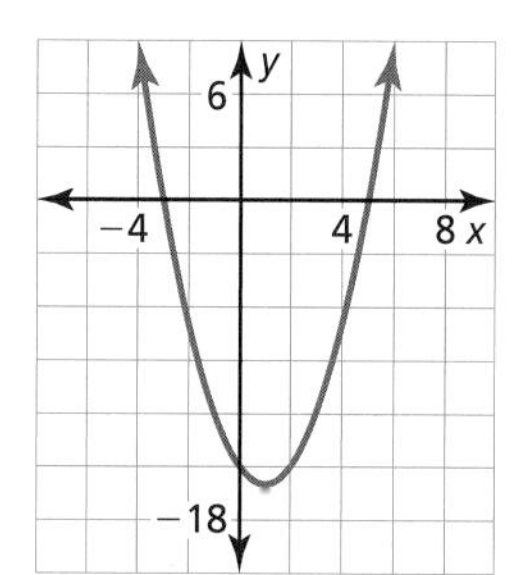

D. 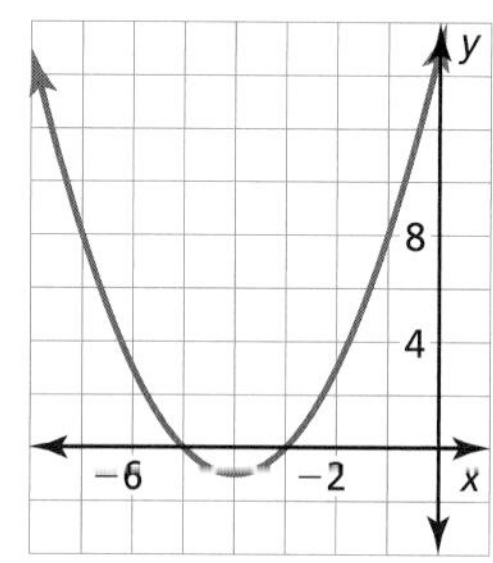

E. 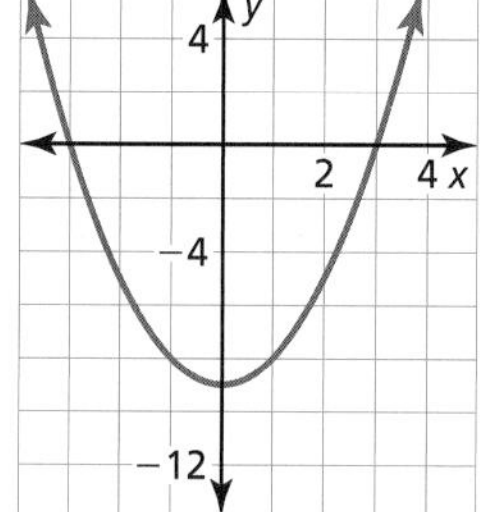

F. 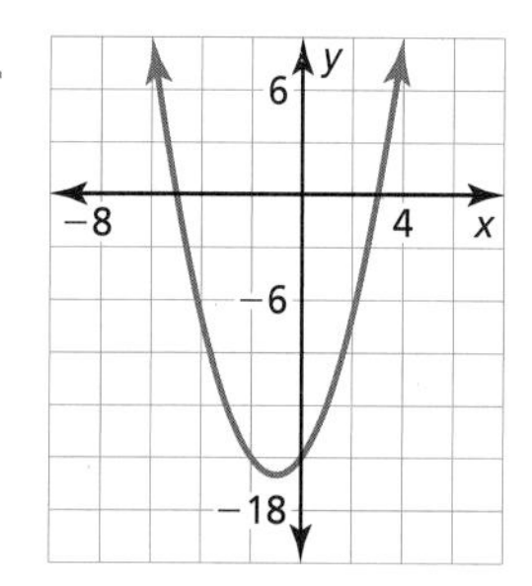

In Exercises 37–42, use zeros to graph the function. *(See Example 5.)*

37. $f(x) = (x + 2)(x - 6)$ **38.** $g(x) = -3(x + 1)(x + 7)$

39. $y = x^2 - 11x + 18$ **40.** $y = x^2 - x - 30$

41. $y = -5x^2 - 10x + 40$ **42.** $h(x) = 8x^2 - 8$

ERROR ANALYSIS In Exercises 43 and 44, describe and correct the error in finding the zeros of the function.

43.

$y = 5(x + 3)(x - 2)$

The zeros of the function are 3 and −2.

44.

$y = (x + 4)(x^2 - 9)$

The zeros of the function are −4 and 9.

In Exercises 45–56, write a quadratic function in standard form whose graph satisfies the given condition(s). *(See Example 6.)*

45. vertex: (7, −3) **46.** vertex: (4, 8)

47. x-intercepts: 1 and 9 **48.** x-intercepts: −2 and −5

49. passes through (−4, 0), (3, 0), and (2, −18)

50. passes through (−5, 0), (−1, 0), and (−4, 3)

51. passes through (7, 0)

52. passes through (0, 0) and (6, 0)

53. axis of symmetry: $x = -5$

54. y increases as x increases when $x < 4$; y decreases as x increases when $x > 4$.

55. range: $y \geq -3$ **56.** range: $y \leq 10$

In Exercises 57–60, write the quadratic function represented by the graph.

57.

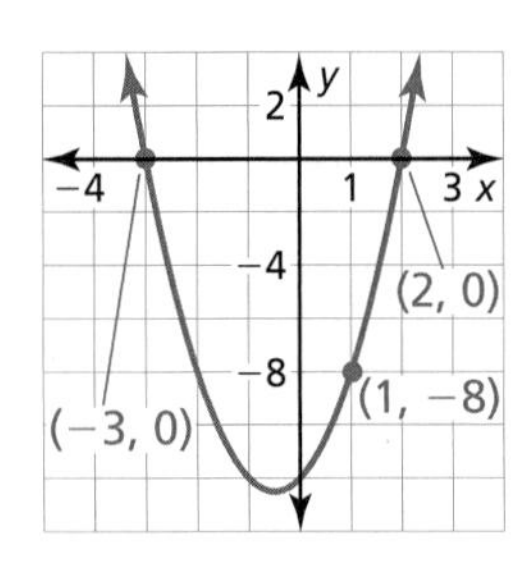

58.

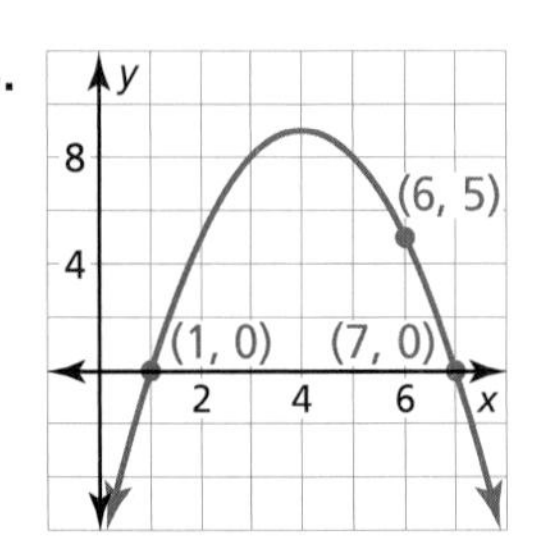

59.

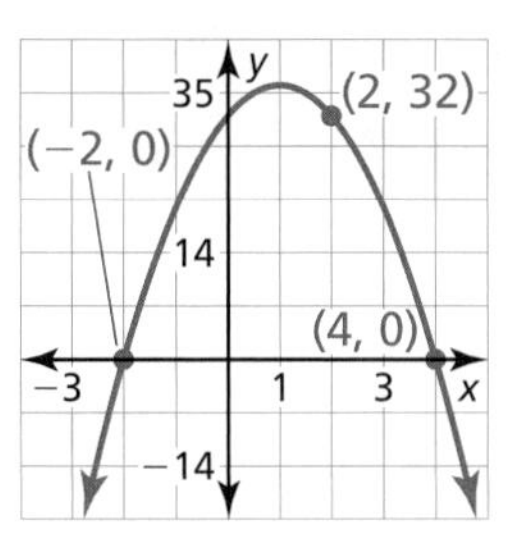

60.

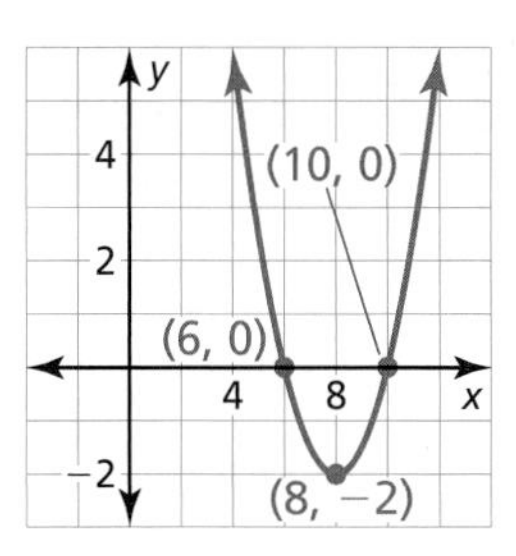

In Exercises 61–68, use zeros to graph the function. *(See Example 7.)*

61. $y = 5x(x + 2)(x - 6)$ **62.** $f(x) = -x(x + 9)(x + 3)$

63. $h(x) = (x - 2)(x + 2)(x + 7)$

64. $y = (x + 1)(x - 5)(x - 4)$

65. $f(x) = 3x^3 - 48x$ **66.** $y = -2x^3 + 20x^2 - 50x$

67. $y = -x^3 - 16x^2 - 28x$

68. $g(x) = 6x^3 + 30x^2 - 36x$

In Exercises 69–72, write the cubic function represented by the graph. *(See Example 8.)*

69.

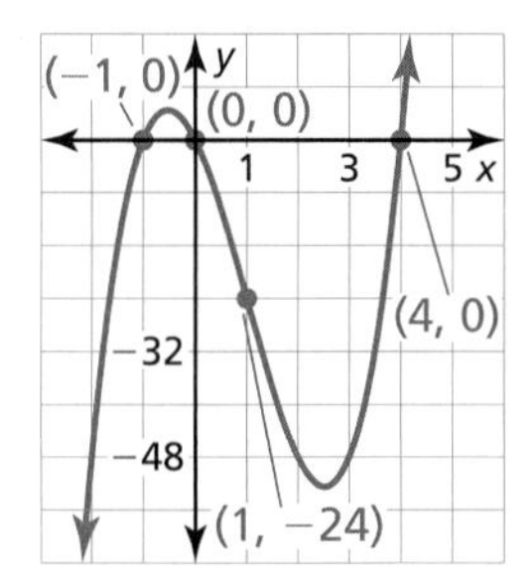

70.

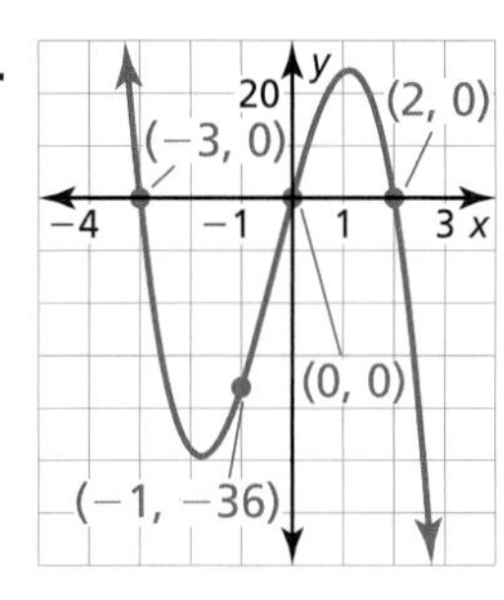

71.

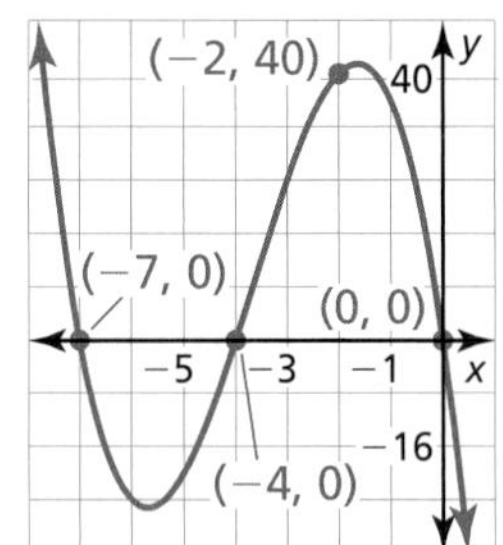

72.

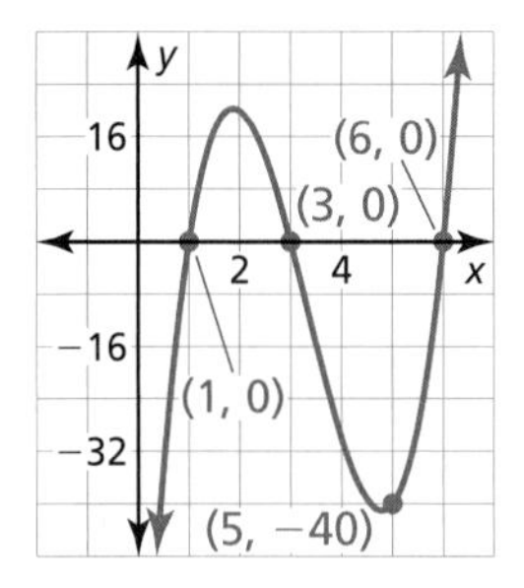

In Exercises 73–76, write a cubic function whose graph satisfies the given condition(s).

73. x-intercepts: −2, 3, and 8

74. x-intercepts: −7, −5, and 0

75. passes through (1, 0) and (7, 0)

76. passes through (0, 6)

In Exercises 77–80, all the zeros of a function are given. Use the zeros and the other point given to write a quadratic or cubic function represented by the table.

77.

x	y
0	0
2	30
7	0

78.

x	y
−3	0
1	−72
4	0

79.

x	y
−4	0
−3	0
0	−180
3	0

80.

x	y
−8	0
−6	−36
−3	0
0	0

In Exercises 81–84, sketch a parabola that satisfies the given conditions.

81. x-intercepts: −4 and 2; range: $y \geq -3$

82. axis of symmetry: $x = 6$; passes through (4, 15)

83. range: $y \leq 5$; passes through (0, 2)

84. x-intercept: 6; y-intercept: 1; range: $y \geq -4$

85. MODELING WITH MATHEMATICS Satellite dishes are shaped like parabolas to optimally receive signals. The cross section of a satellite dish can be modeled by the function shown, where x and y are measured in feet. The x-axis represents the top of the opening of the dish.

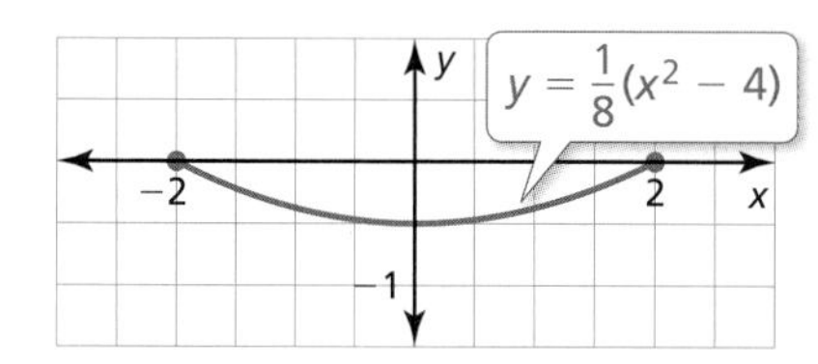

a. How wide is the satellite dish?

b. How deep is the satellite dish?

c. Write a quadratic function in standard form that models the cross section of a satellite dish that is 6 feet wide and 1.5 feet deep.

86. MODELING WITH MATHEMATICS A professional basketball player's shot is modeled by the function shown, where x and y are measured in feet.

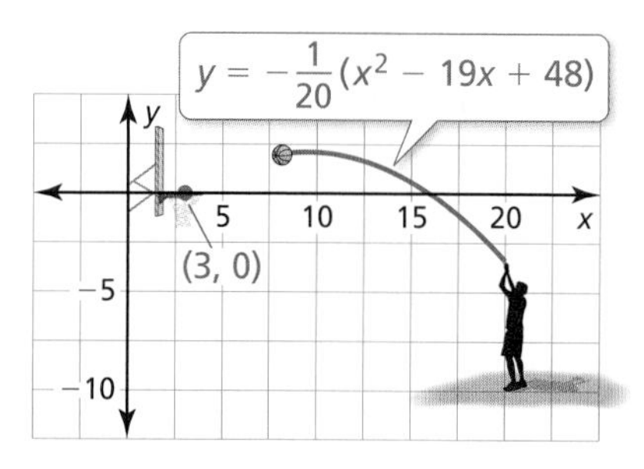

a. Does the player make the shot? Explain.

b. The basketball player releases another shot from the point (13, 0) and makes the shot. The shot also passes through the point (10, 1.4). Write a quadratic function in standard form that models the path of the shot.

USING STRUCTURE In Exercises 87–90, match the function with its graph.

87. $y = -x^2 + 5x$

88. $y = x^2 - x - 12$

89. $y = x^3 - 2x^2 - 8x$

90. $y = x^3 - 4x^2 - 11x + 30$

A.

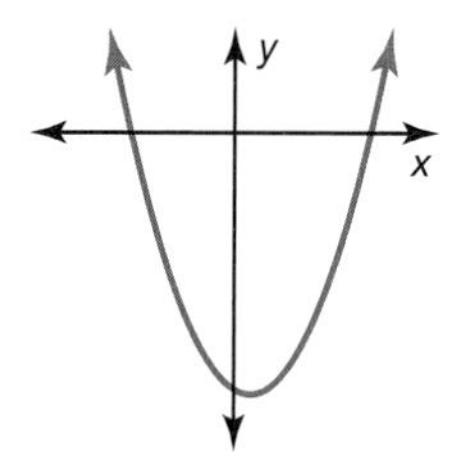

B.

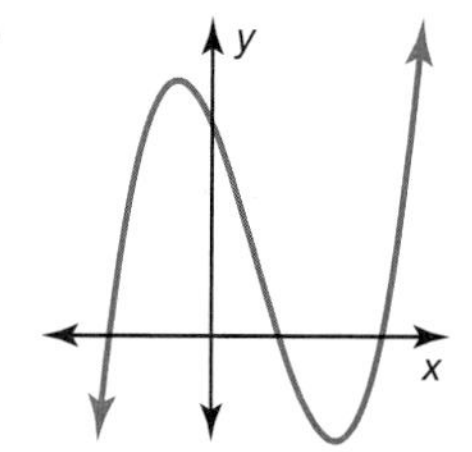

C.

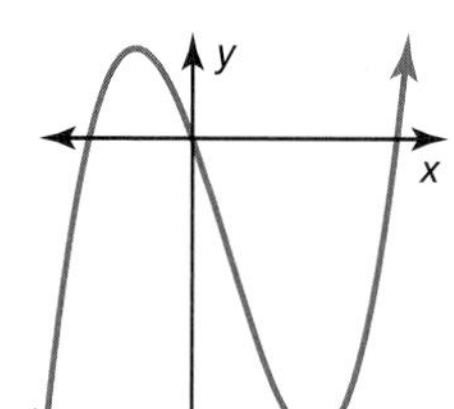

D.

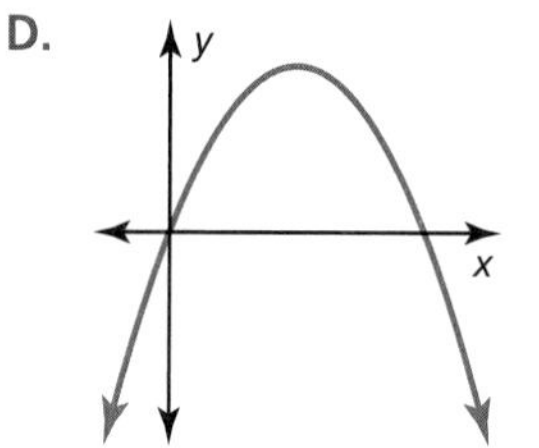

91. CRITICAL THINKING Write a quadratic function represented by the table, if possible. If not, explain why.

x	−5	−3	−1	1
y	0	12	4	0

92. HOW DO YOU SEE IT? The graph shows the parabolic arch that supports the roof of a convention center, where x and y are measured in feet.

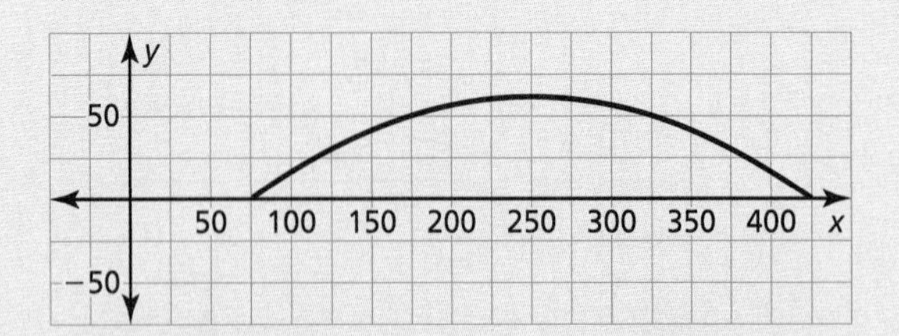

a. The arch can be represented by a function of the form $f(x) = a(x - p)(x - q)$. Estimate the values of p and q.

b. Estimate the width and height of the arch. Explain how you can use your height estimate to calculate a.

ANALYZING EQUATIONS In Exercises 93 and 94, (a) rewrite the quadratic function in intercept form and (b) graph the function using any method. Explain the method you used.

93. $f(x) = -3(x + 1)^2 + 27$

94. $g(x) = 2(x - 1)^2 - 2$

95. WRITING Can a quadratic function with exactly one real zero be written in intercept form? Explain.

96. MAKING AN ARGUMENT Your friend claims that any quadratic function can be written in standard form and in vertex form. Is your friend correct? Explain.

97. PROBLEM SOLVING Write the function represented by the graph in intercept form.

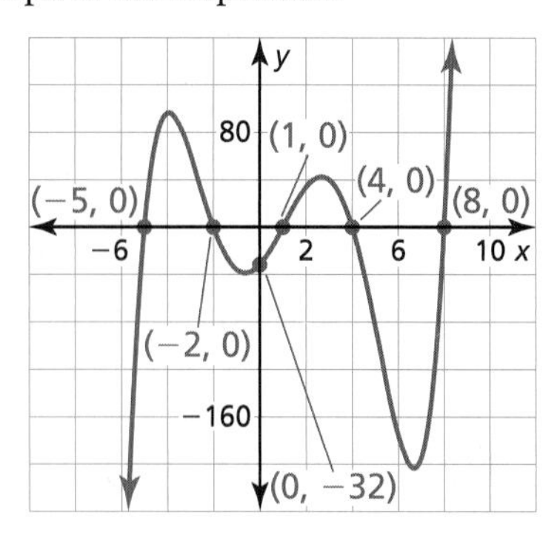

98. THOUGHT PROVOKING Sketch the graph of each function. Explain your procedure.

a. $f(x) = (x^2 - 1)(x^2 - 4)$

b. $g(x) = x(x^2 - 1)(x^2 - 4)$

99. REASONING Let k be a constant. Find the zeros of the function $f(x) = kx^2 - k^2x - 2k^3$ in terms of k.

PROBLEM SOLVING In Exercises 100 and 101, write a system of two quadratic equations whose graphs intersect at the given points. Explain your reasoning.

100. $(-4, 0)$ and $(2, 0)$

101. $(3, 6)$ and $(7, 6)$

Maintaining Mathematical Proficiency

Reviewing what you learned in previous grades and lessons

The scatter plot shows the amounts x (in grams) of fat and the numbers y of calories in 12 burgers at a fast-food restaurant. *(Section 4.4)*

102. How many calories are in the burger that contains 12 grams of fat?

103. How many grams of fat are in the burger that contains 600 calories?

104. What tends to happen to the number of calories as the number of grams of fat increases?

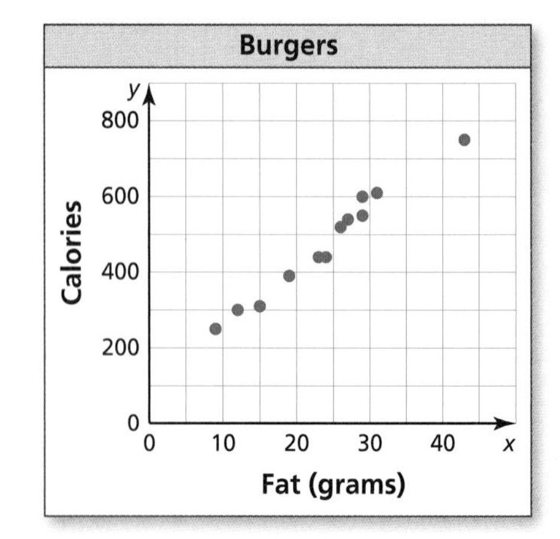

Determine whether the sequence is *arithmetic*, *geometric*, or *neither*. Explain your reasoning. *(Section 6.6)*

105. 3, 11, 21, 33, 47, . . .

106. −2, −6, −18, −54, . . .

107. 26, 18, 10, 2, −6, . . .

108. 4, 5, 9, 14, 23, . . .

8.6 Comparing Linear, Exponential, and Quadratic Functions

Essential Question How can you compare the growth rates of linear, exponential, and quadratic functions?

EXPLORATION 1 **Comparing Speeds**

Work with a partner. Three cars start traveling at the same time. The distance traveled in t minutes is y miles. Complete each table and sketch all three graphs in the same coordinate plane. Compare the speeds of the three cars. Which car has a constant speed? Which car is accelerating the most? Explain your reasoning.

t	$y = t$
0	
0.2	
0.4	
0.6	
0.8	
1.0	

t	$y = 2^t - 1$
0	
0.2	
0.4	
0.6	
0.8	
1.0	

t	$y = t^2$
0	
0.2	
0.4	
0.6	
0.8	
1.0	

COMPARING PREDICTIONS

To be proficient in math, you need to visualize the results of varying assumptions, explore consequences, and compare predictions with data.

EXPLORATION 2 **Comparing Speeds**

Work with a partner. Analyze the speeds of the three cars over the given time periods. The distance traveled in t minutes is y miles. Which car eventually overtakes the others?

t	$y = t$
1.0	
1.5	
2.0	
2.5	
3.0	
3.5	
4.0	
4.5	
5.0	

t	$y = 2^t - 1$
1.0	
1.5	
2.0	
2.5	
3.0	
3.5	
4.0	
4.5	
5.0	

t	$y = t^2$
1.0	
1.5	
2.0	
2.5	
3.0	
3.5	
4.0	
4.5	
5.0	

Communicate Your Answer

3. How can you compare the growth rates of linear, exponential, and quadratic functions?

4. Which function has a growth rate that is eventually much greater than the growth rates of the other two functions? Explain your reasoning.

8.6 Lesson

Core Vocabulary

average rate of change, *p. 462*

Previous

slope

What You Will Learn

- Choose functions to model data.
- Write functions to model data.
- Compare functions using average rates of change.
- Solve real-life problems involving different function types.

Choosing Functions to Model Data

So far, you have studied linear functions, exponential functions, and quadratic functions. You can use these functions to model data.

Core Concept

Linear, Exponential, and Quadratic Functions

Linear Function	Exponential Function	Quadratic Function
$y = mx + b$	$y = ab^x$	$y = ax^2 + bx + c$

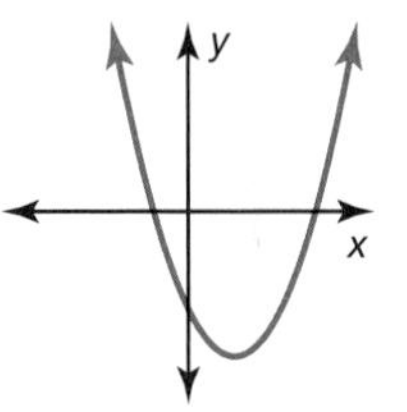

EXAMPLE 1 Using Graphs to Identify Functions

Plot the points. Tell whether the points appear to represent a *linear*, an *exponential*, or a *quadratic* function.

a. $(4, 4), (2, 0), (0, 0), \left(1, -\frac{1}{2}\right), (-2, 4)$

b. $(0, 1), (2, 4), (4, 7), (-2, -2), (-4, -5)$

c. $(0, 2), (2, 8), (1, 4), (-1, 1), \left(-2, \frac{1}{2}\right)$

SOLUTION

a.

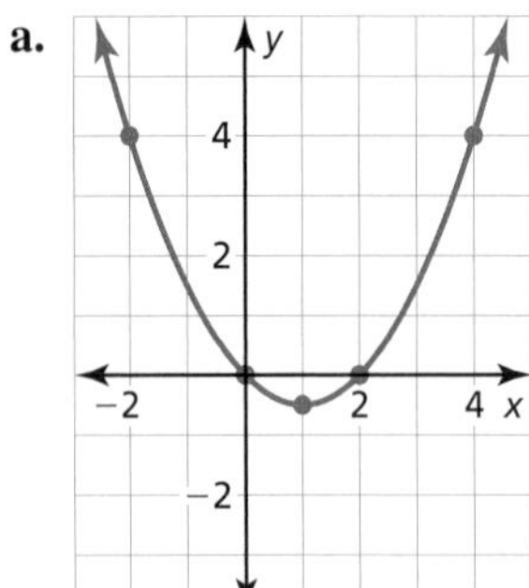

▶ quadratic

b.

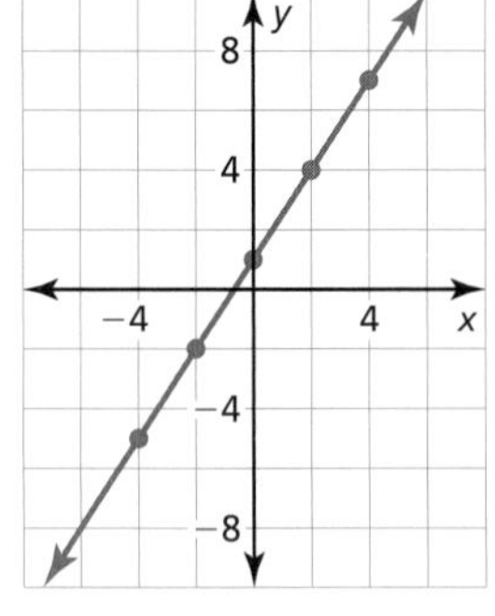

▶ linear

c.

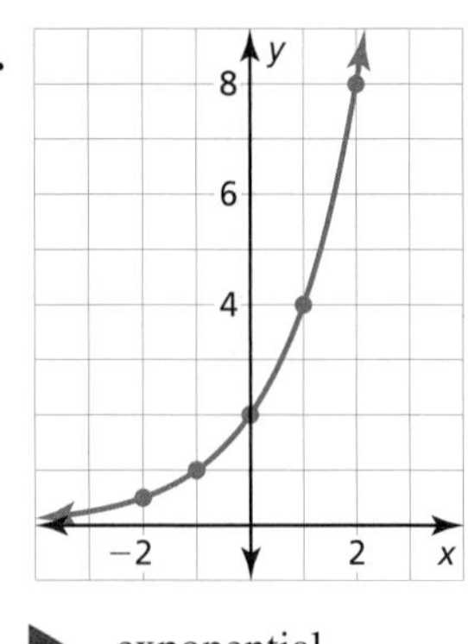

▶ exponential

Monitoring Progress

Help in English and Spanish at *BigIdeasMath.com*

Plot the points. Tell whether the points appear to represent a *linear*, an *exponential*, or a *quadratic* function.

1. $(-1, 5), (2, -1), (0, -1), (3, 5), (1, -3)$

2. $(-1, 2), (-2, 8), (-3, 32), \left(0, \frac{1}{2}\right), \left(1, \frac{1}{8}\right)$

3. $(-3, 5), (0, -1), (2, -5), (-4, 7), (1, -3)$

Core Concept

Differences and Ratios of Functions

You can use patterns between consecutive data pairs to determine which type of function models the data. The differences of consecutive y-values are called *first differences*. The differences of consecutive first differences are called *second differences*.

- **Linear Function** The first differences are constant.
- **Exponential Function** Consecutive y-values have a common *ratio*.
- **Quadratic Function** The second differences are constant.

In all cases, the differences of consecutive x-values need to be constant.

STUDY TIP

The first differences for exponential and quadratic functions are *not* constant.

EXAMPLE 2 Using Differences or Ratios to Identify Functions

Tell whether each table of values represents a *linear*, an *exponential*, or a *quadratic* function.

a.

x	−3	−2	−1	0	1
y	11	8	5	2	−1

b.

x	−2	−1	0	1	2
y	1	2	4	8	16

c.

x	−2	−1	0	1	2
y	−1	−2	−1	2	7

STUDY TIP

First determine that the differences of consecutive x-values are constant. Then check whether the first differences are constant or consecutive y-values have a common ratio. If neither of these is true, check whether the second differences are constant.

SOLUTION

a.

	+1	+1	+1	+1	
x	−3	−2	−1	0	1
y	11	8	5	2	−1
	−3	−3	−3	−3	

▶ The first differences are constant. So, the table represents a linear function.

b.

	+1	+1	+1	+1	
x	−2	−1	0	1	2
y	1	2	4	8	16
	×2	×2	×2	×2	

▶ Consecutive y-values have a common ratio. So, the table represents an exponential function.

c.

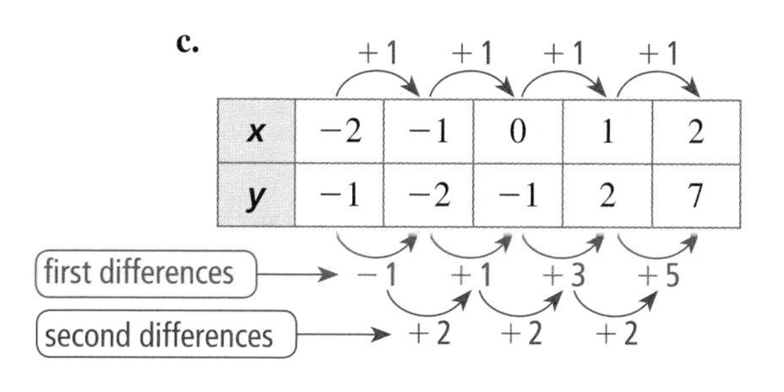

x	−2	−1	0	1	2
y	−1	−2	−1	2	7

▶ The second differences are constant. So, the table represents a quadratic function.

Monitoring Progress  Help in English and Spanish at *BigIdeasMath.com*

4. Tell whether the table of values represents a *linear*, an *exponential*, or a *quadratic* function.

x	−1	0	1	2	3
y	1	3	9	27	81

Writing Functions to Model Data

EXAMPLE 3 Writing a Function to Model Data

x	2	4	6	8	10
y	12	0	−4	0	12

Tell whether the table of values represents a *linear*, an *exponential*, or a *quadratic* function. Then write the function.

SOLUTION

Step 1 Determine which type of function the table of values represents.

The second differences are constant. So, the table represents a quadratic function.

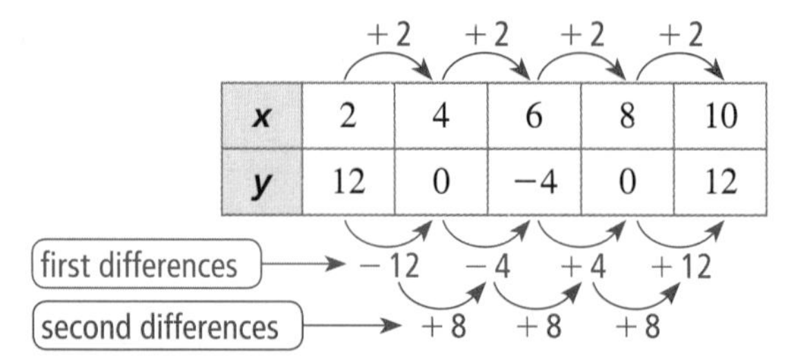

Step 2 Write an equation of the quadratic function. Using the table, notice that the x-intercepts are 4 and 8. So, use intercept form to write a function.

$y = a(x - 4)(x - 8)$ Substitute for p and q in intercept form.

Use another point from the table, such as (2, 12), to find a.

$12 = a(2 - 4)(2 - 8)$ Substitute 2 for x and 12 for y.

$1 = a$ Solve for a.

Use the value of a to write the function.

$y = (x - 4)(x - 8)$ Substitute 1 for a.

$= x^2 - 12x + 32$ Use the FOIL Method and combine like terms.

▶ So, the quadratic function is $y = x^2 - 12x + 32$.

STUDY TIP

To check your function in Example 3, substitute the other points from the table to verify that they satisfy the function.

Monitoring Progress  Help in English and Spanish at *BigIdeasMath.com*

5. Tell whether the table of values represents a *linear*, an *exponential*, or a *quadratic* function. Then write the function.

x	−1	0	1	2	3
y	16	8	4	2	1

Comparing Functions Using Average Rates of Change

For nonlinear functions, the rate of change is not constant. You can compare two nonlinear functions over the same interval using their *average rates of change*. The **average rate of change** of a function $y = f(x)$ between $x = a$ and $x = b$ is the slope of the line through $(a, f(a))$ and $(b, f(b))$.

$$\text{average rate of change} = \frac{\text{change in } y}{\text{change in } x} = \frac{f(b) - f(a)}{b - a}$$

Exponential Function

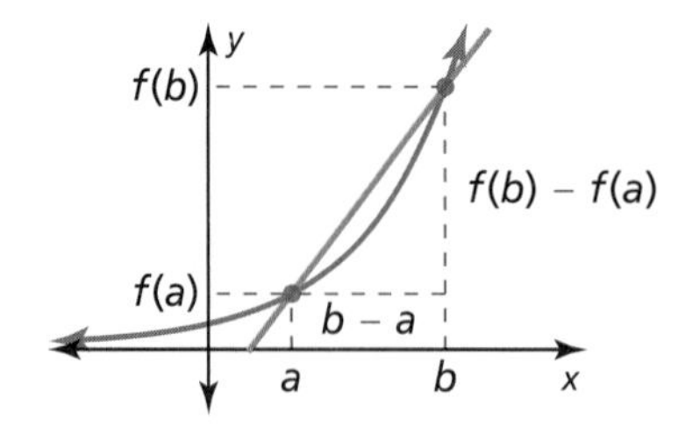

Quadratic Function

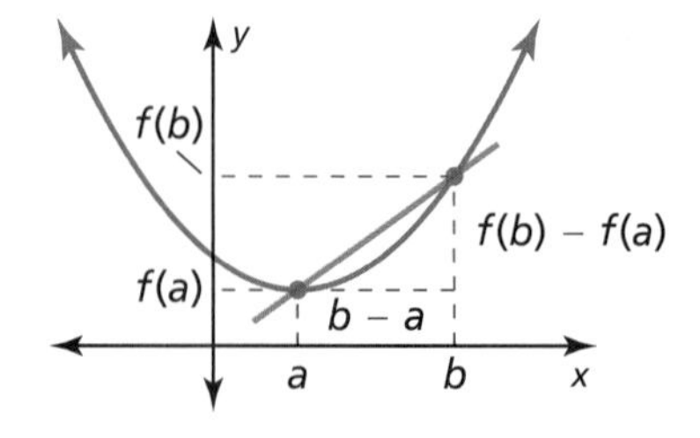

Core Concept

Comparing Functions Using Average Rates of Change

- As a and b increase, the average rate of change between $x = a$ and $x = b$ of an increasing exponential function $y = f(x)$ will eventually exceed the average rate of change between $x = a$ and $x = b$ of an increasing quadratic function $y = g(x)$ or an increasing linear function $y = h(x)$. So, as x increases, $f(x)$ will eventually exceed $g(x)$ or $h(x)$.
- As a and b increase, the average rate of change between $x = a$ and $x = b$ of an increasing quadratic function $y = g(x)$ will eventually exceed the average rate of change between $x = a$ and $x = b$ of an increasing linear function $y = h(x)$. So, as x increases, $g(x)$ will eventually exceed $h(x)$.

STUDY TIP

You can explore these concepts using a graphing calculator.

EXAMPLE 4 Using and Interpreting Average Rates of Change

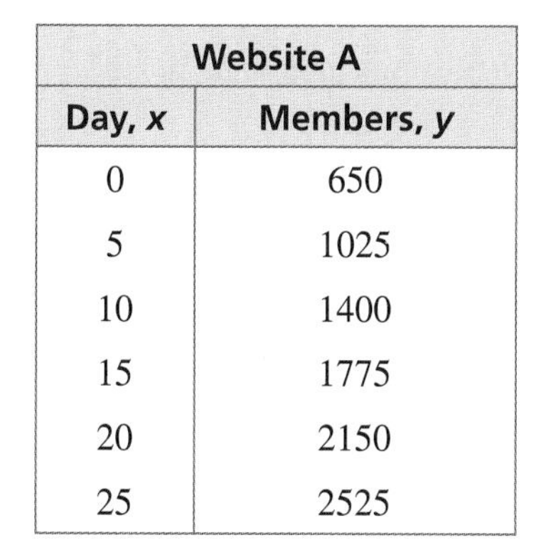

Website A	
Day, x	Members, y
0	650
5	1025
10	1400
15	1775
20	2150
25	2525

Two social media websites open their memberships to the public. (a) Compare the websites by calculating and interpreting the average rates of change from Day 10 to Day 20. (b) Predict which website will have more members after 50 days. Explain.

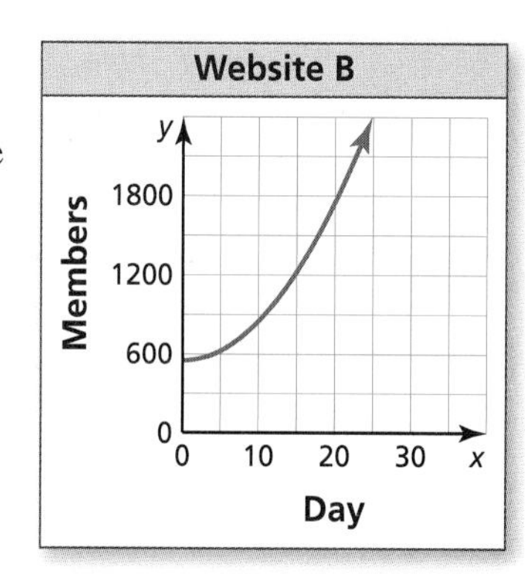

SOLUTION

a. Calculate the average rates of change by using the points whose x-coordinates are 10 and 20.

Website A: Use (10, 1400) and (20, 2150).

$$\text{average rate of change} = \frac{f(b) - f(a)}{b - a} = \frac{2150 - 1400}{20 - 10} = \frac{750}{10} = 75$$

Website B: Use the graph to estimate the points when $x = 10$ and $x = 20$. Use (10, 850) and (20, 1800).

$$\text{average rate of change} = \frac{f(b) - f(a)}{b - a} \approx \frac{1800 - 850}{20 - 10} = \frac{950}{10} = 95$$

▶ From Day 10 to Day 20, Website A membership increases at an average rate of 75 people per day, and Website B membership increases at an average rate of about 95 people per day. So, Website B membership is growing faster.

b. Using the table, membership increases and the average rates of change are constant. So, Website A membership can be represented by an increasing linear function. Using the graph, membership increases and the average rates of change are increasing. It appears that Website B membership can be represented by an increasing exponential or quadratic function.

After 25 days, the memberships of both websites are about equal and the average rate of change of Website B exceeds the average rate of change of Website A. So, Website B will have more members after 50 days.

Monitoring Progress  Help in English and Spanish at *BigIdeasMath.com*

6. Compare the websites in Example 4 by calculating and interpreting the average rates of change from Day 0 to Day 10.

Solving Real-Life Problems

EXAMPLE 5 Comparing Different Function Types

In 1900, Littleton had a population of 1000 people. Littleton's population increased by 50 people each year. In 1900, Tinyville had a population of 500 people. Tinyville's population increased by 5% each year.

a. In what year were the populations about equal?

b. Suppose Littleton's initial population doubled to 2000 and maintained a constant rate of increase of 50 people each year. Did Tinyville's population still catch up to Littleton's population? If so, in which year?

c. Suppose Littleton's rate of increase doubled to 100 people each year, in addition to doubling the initial population. Did Tinyville's population still catch up to Littleton's population? Explain.

SOLUTION

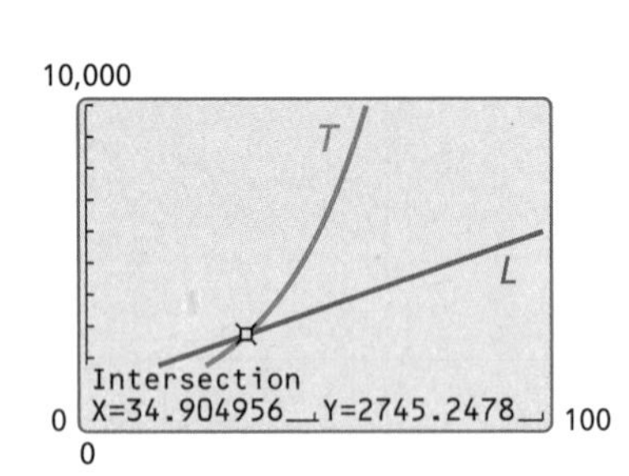

a. Let x represent the number of years since 1900. Write a function to model the population of each town.

Littleton: $L(x) = 50x + 1000$ Linear function

Tinyville: $T(x) = 500(1.05)^x$ Exponential function

Use a graphing calculator to graph each function in the same viewing window. Use the *intersect* feature to find the value of x for which $L(x) \approx T(x)$. The graphs intersect when $x \approx 34.9$.

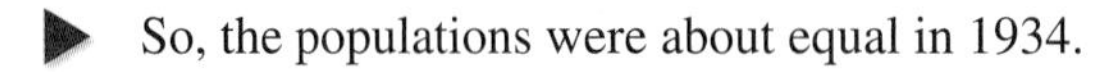

▶ So, the populations were about equal in 1934.

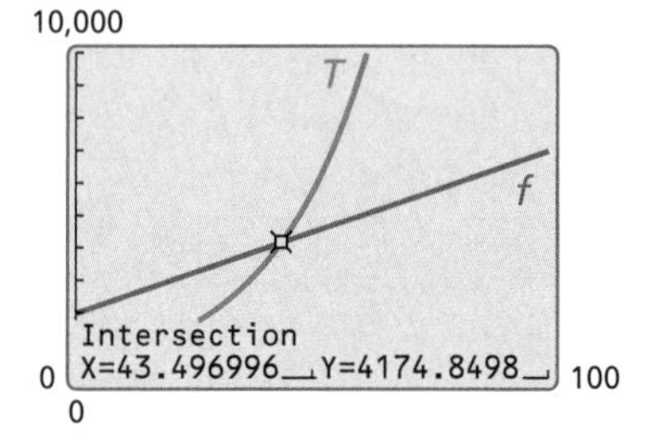

b. Littleton's new population function is $f(x) = 50x + 2000$. Use a graphing calculator to graph f and T in the same viewing window. Use the *intersect* feature to find the value of x for which $f(x) \approx T(x)$. The graphs intersect when $x \approx 43.5$.

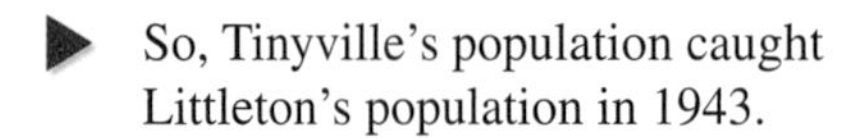

▶ So, Tinyville's population caught Littleton's population in 1943.

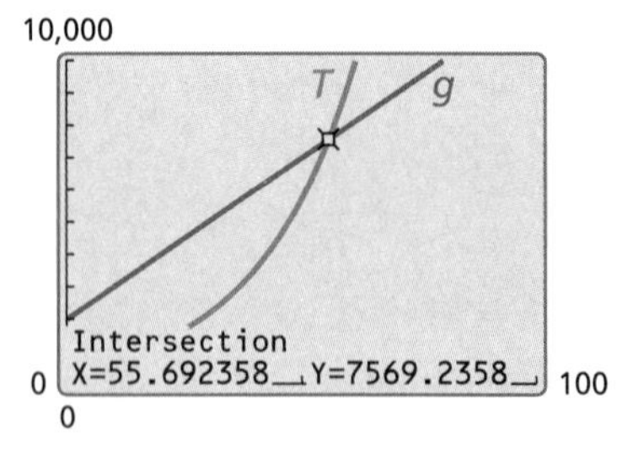

c. Littleton's new population function is $g(x) = 100x + 2000$. Use a graphing calculator to graph g and T in the same viewing window. Use the *intersect* feature to find the value of x for which $g(x) \approx T(x)$. The graphs intersect when $x \approx 55.7$.

▶ So, Tinyville's population caught Littleton's population in 1955. Because Littleton's population shows linear growth and Tinyville's population shows exponential growth, Tinyville's population eventually exceeded Littleton's regardless of Littleton's constant rate or initial value.

Monitoring Progress

Help in English and Spanish at *BigIdeasMath.com*

7. WHAT IF? Tinyville's population increased by 8% each year. In what year were the populations about equal?

8.6 Exercises

Dynamic Solutions available at *BigIdeasMath.com*

Vocabulary and Core Concept Check

1. **WRITING** Name three types of functions that you can use to model data. Describe the equation and graph of each type of function.

2. **WRITING** How can you decide whether to use a linear, an exponential, or a quadratic function to model a data set?

3. **VOCABULARY** Describe how to find the average rate of change of a function $y = f(x)$ between $x = a$ and $x = b$.

4. **WHICH ONE DOESN'T BELONG?** Which graph does *not* belong with the other three? Explain your reasoning.

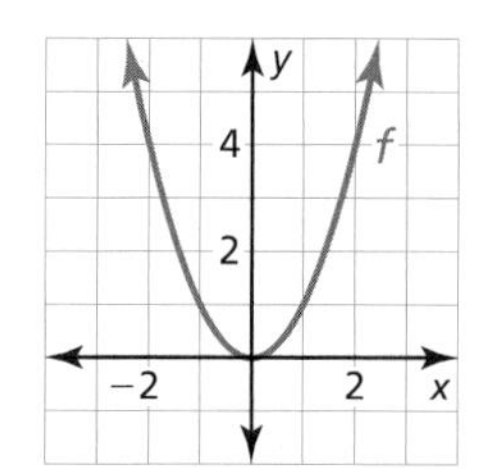

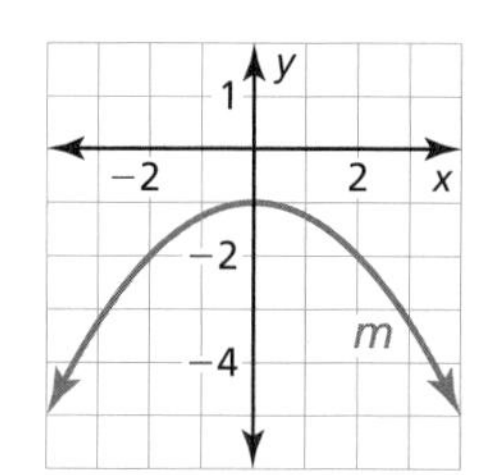

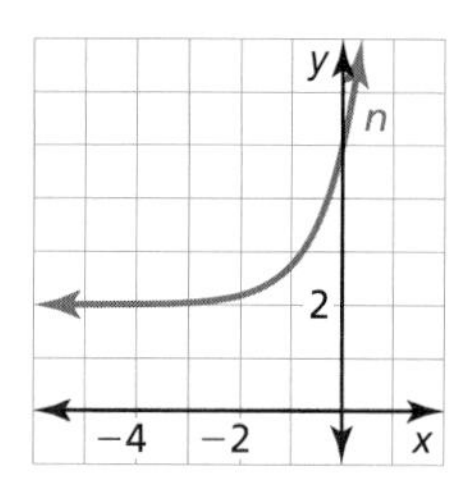

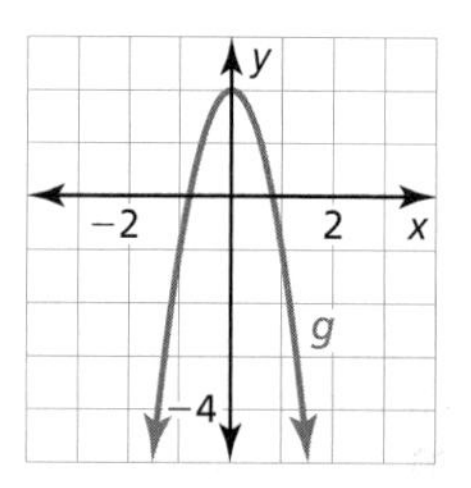

Monitoring Progress and Modeling with Mathematics

In Exercises 5–8, tell whether the points appear to represent a *linear*, an *exponential*, or a *quadratic* function.

5.

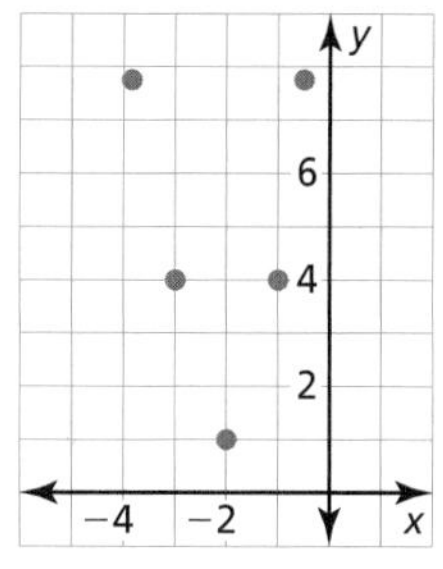

6.

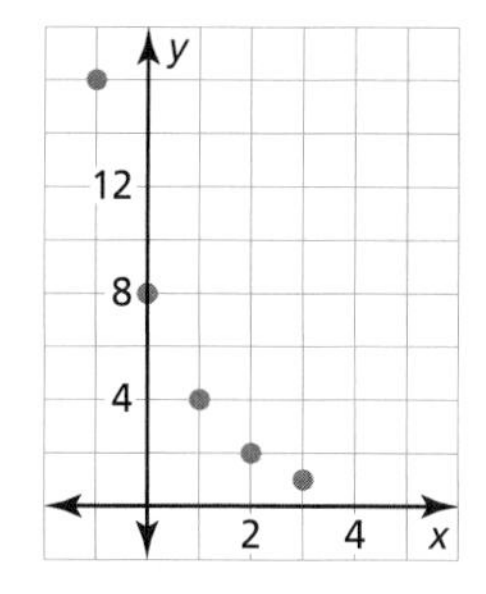

7.

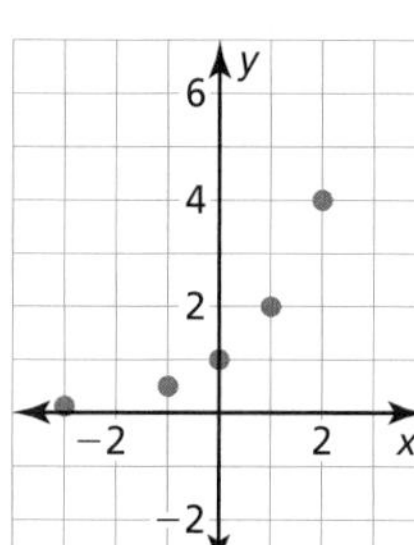

8.

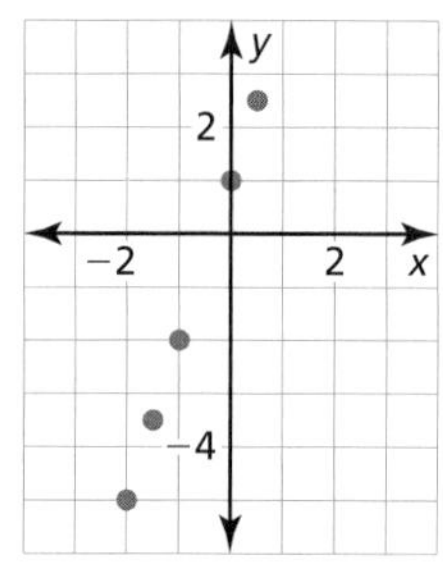

In Exercises 9–14, plot the points. Tell whether the points appear to represent a *linear*, an *exponential*, or a *quadratic* function. *(See Example 1.)*

9. $(-2, -1), (-1, 0), (1, 2), (2, 3), (0, 1)$

10. $\left(0, \frac{1}{4}\right), (1, 1), (2, 4), (3, 16), \left(-1, \frac{1}{16}\right)$

11. $(0, -3), (1, 0), (2, 9), (-2, 9), (-1, 0)$

12. $(-1, -3), (-3, 5), (0, -1), (1, 5), (2, 15)$

13. $(-4, -4), (-2, -3.4), (0, -3), (2, -2.6), (4, -2)$

14. $(0, 8), (-4, 0.25), (-3, 0.4), (-2, 1), (-1, 3)$

In Exercises 15–18, tell whether the table of values represents a *linear*, an *exponential*, or a *quadratic* function. *(See Example 2.)*

15.

x	−2	−1	0	1	2
y	0	0.5	1	1.5	2

16.

x	−1	0	1	2	3
y	0.2	1	5	25	125

17.

x	2	3	4	5	6
y	2	6	18	54	162

18.

x	−3	−2	−1	0	1
y	2	4.5	8	12.5	18

19. **MODELING WITH MATHEMATICS** A student takes a subway to a public library. The table shows the distances d (in miles) the student travels in t minutes. Let the time t represent the independent variable. Tell whether the data can be modeled by a *linear*, an *exponential*, or a *quadratic* function. Explain.

Time, t	0.5	1	3	5
Distance, d	0.335	0.67	2.01	3.35

20. **MODELING WITH MATHEMATICS** A store sells custom circular rugs. The table shows the costs c (in dollars) of rugs that have diameters of d feet. Let the diameter d represent the independent variable. Tell whether the data can be modeled by a *linear*, an *exponential*, or a *quadratic* function. Explain.

Diameter, d	3	4	5	6
Cost, c	63.90	113.60	177.50	255.60

In Exercises 21–26, tell whether the data represent a *linear*, an *exponential*, or a *quadratic* function. Then write the function. *(See Example 3.)*

21. $(-2, 8), (-1, 0), (0, -4), (1, -4), (2, 0), (3, 8)$

22. $(-3, 8), (-2, 4), (-1, 2), (0, 1), (1, 0.5)$

23.

x	−2	−1	0	1	2
y	4	1	−2	−5	−8

24.

x	−1	0	1	2	3
y	2.5	5	10	20	40

25.

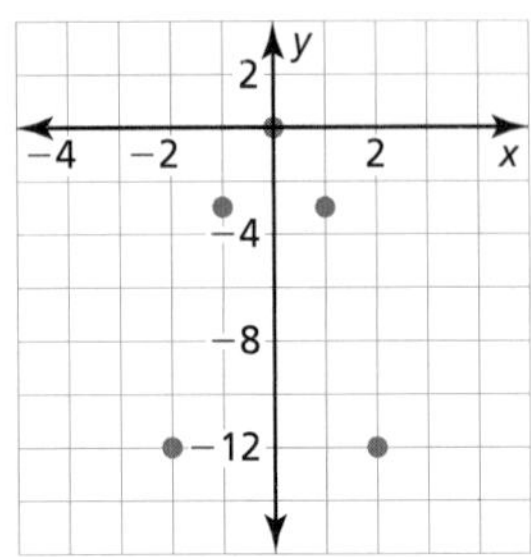

26.

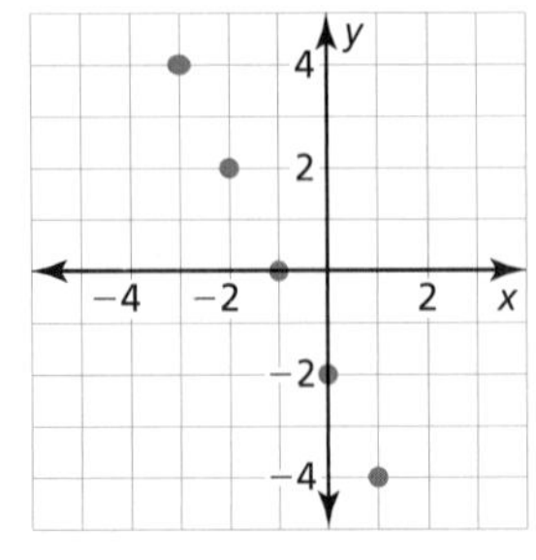

27. **ERROR ANALYSIS** Describe and correct the error in determining whether the table represents a linear, an exponential, or a quadratic function.

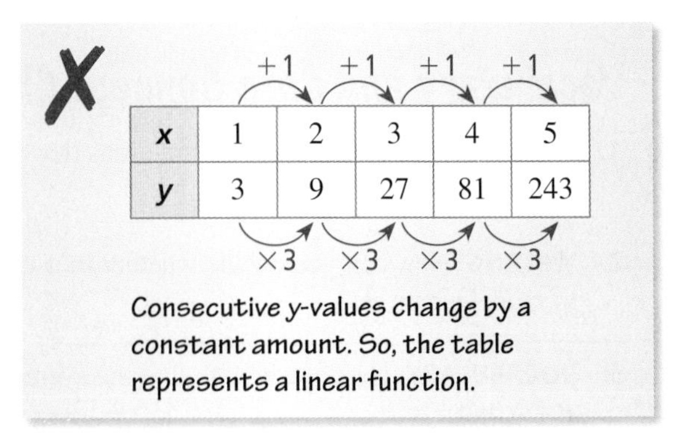

x	1	2	3	4	5
y	3	9	27	81	243

Consecutive y-values change by a constant amount. So, the table represents a linear function.

28. **ERROR ANALYSIS** Describe and correct the error in writing the function represented by the table.

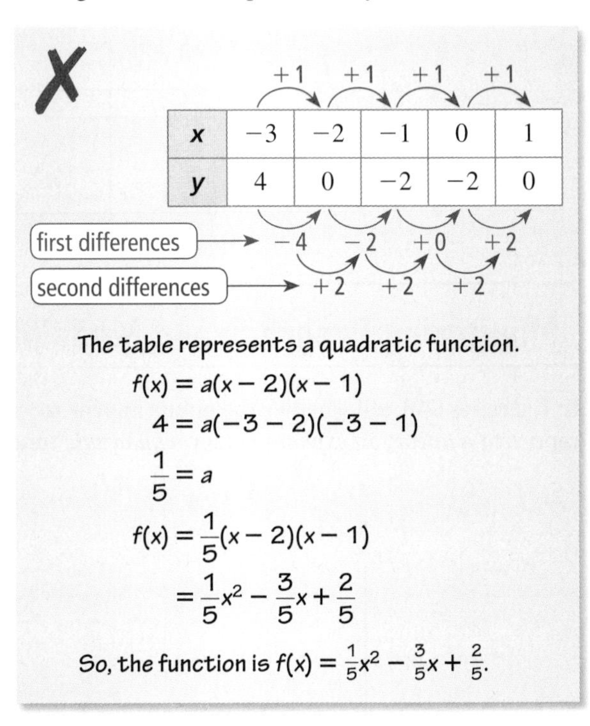

x	−3	−2	−1	0	1
y	4	0	−2	−2	0

The table represents a quadratic function.

$$f(x) = a(x - 2)(x - 1)$$
$$4 = a(-3 - 2)(-3 - 1)$$
$$\frac{1}{5} = a$$
$$f(x) = \frac{1}{5}(x - 2)(x - 1)$$
$$= \frac{1}{5}x^2 - \frac{3}{5}x + \frac{2}{5}$$

So, the function is $f(x) = \frac{1}{5}x^2 - \frac{3}{5}x + \frac{2}{5}$.

29. **REASONING** The table shows the numbers of people attending the first five football games at a high school.

Game, g	1	2	3	4	5
People, p	252	325	270	249	310

a. Plot the points. Let the game g represent the independent variable.

b. Can a linear, an exponential, or a quadratic function represent this situation? Explain.

30. **MODELING WITH MATHEMATICS** The table shows the breathing rates y (in liters of air per minute) of a cyclist traveling at different speeds x (in miles per hour).

Speed, x	20	21	22	23	24
Breathing rate, y	51.4	57.1	63.3	70.3	78.0

a. Plot the points. Let the speed x represent the independent variable. Then determine the type of function that best represents this situation.

b. Write a function that models the data.

c. Find the breathing rate of a cyclist traveling 18 miles per hour. Round your answer to the nearest tenth.

31. **ANALYZING RATES OF CHANGE** The function $f(t) = -16t^2 + 48t + 3$ represents the height (in feet) of a volleyball t seconds after it is hit into the air.

a. Copy and complete the table.

t	0	0.5	1	1.5	2	2.5	3
$f(t)$							

b. Plot the ordered pairs and draw a smooth curve through the points.

c. Describe where the function is increasing and decreasing.

d. Find the average rate of change for each 0.5-second interval in the table. What do you notice about the average rates of change when the function is increasing? decreasing?

32. **ANALYZING RELATIONSHIPS** The population of Town A in 1970 was 3000. The population of Town A increased by 20% every decade. Let x represent the number of decades since 1970. The graph shows the population of Town B. *(See Example 4.)*

a. Compare the populations of the towns by calculating and interpreting the average rates of change from 1990 to 2010.

b. Predict which town will have a greater population after 2030. Explain.

Town B
y
Population
6000
5000
4000
0
0 2 4 x
Decades since 1970

33. **ANALYZING RELATIONSHIPS** Three organizations are collecting donations for a cause. Organization A begins with one donation, and the number of donations quadruples each hour. The table shows the numbers of donations collected by Organization B. The graph shows the numbers of donations collected by Organization C.

Time (hours), t	Number of donations, y
0	0
1	4
2	8
3	12
4	16
5	20
6	24

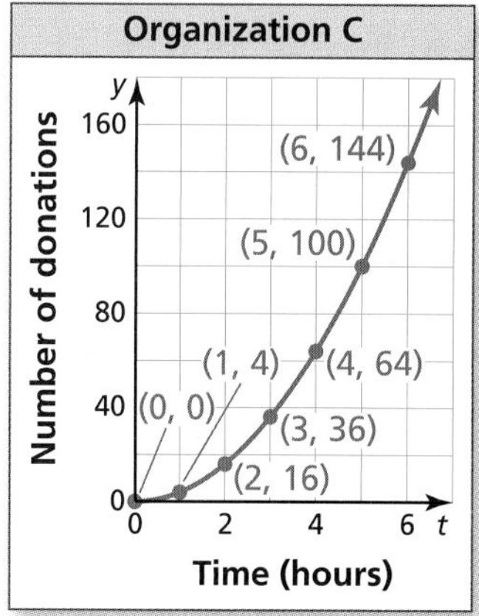

a. What type of function represents the numbers of donations collected by Organization A? B? C?

b. Find the average rates of change of each function for each 1-hour interval from $t = 0$ to $t = 6$.

c. For which function does the average rate of change increase most quickly? What does this tell you about the numbers of donations collected by the three organizations?

34. **COMPARING FUNCTIONS** The room expenses for two different resorts are shown. *(See Example 5.)*

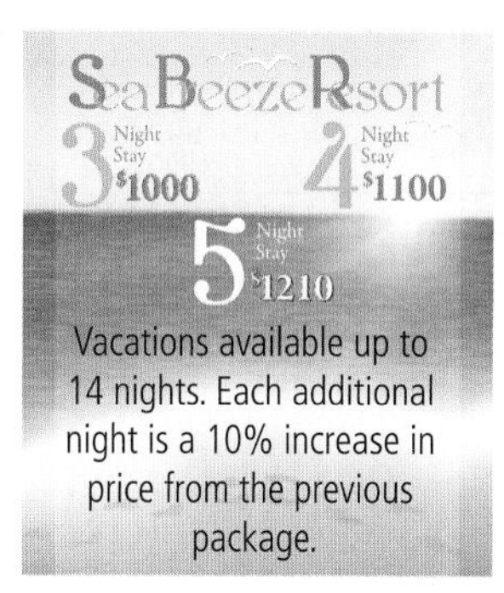

a. For what length of vacation does each resort cost about the same?

b. Suppose Blue Water Resort charges $1450 for the first three nights and $105 for each additional night. Would Sea Breeze Resort ever be more expensive than Blue Water Resort? Explain.

c. Suppose Sea Breeze Resort charges $1200 for the first three nights. The charge increases 10% for each additional night. Would Blue Water Resort ever be more expensive than Sea Breeze Resort? Explain.

35. REASONING Explain why the average rate of change of a linear function is constant and the average rate of change of a quadratic or exponential function is not constant.

36. HOW DO YOU SEE IT? Match each graph with its function. Explain your reasoning.

a.

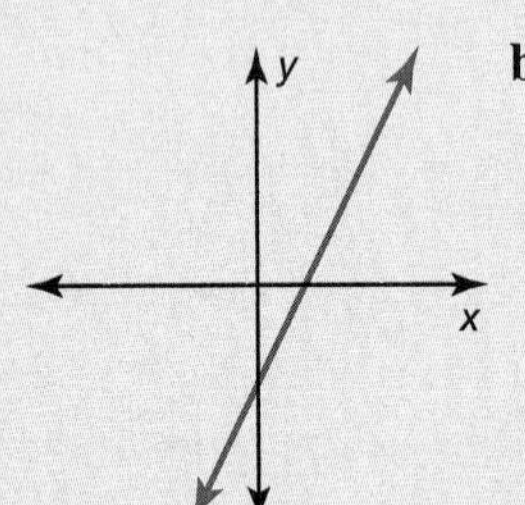

b.

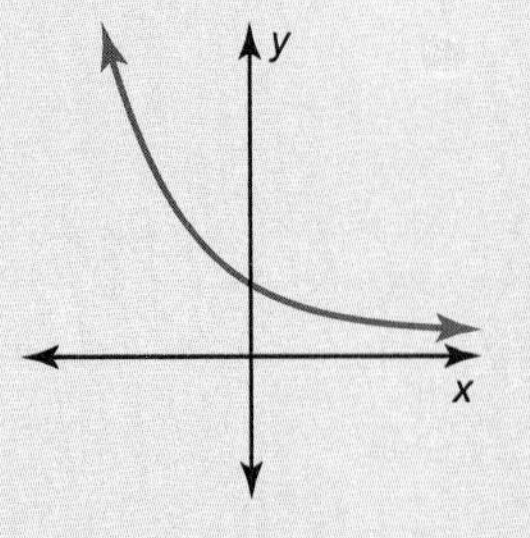

c.

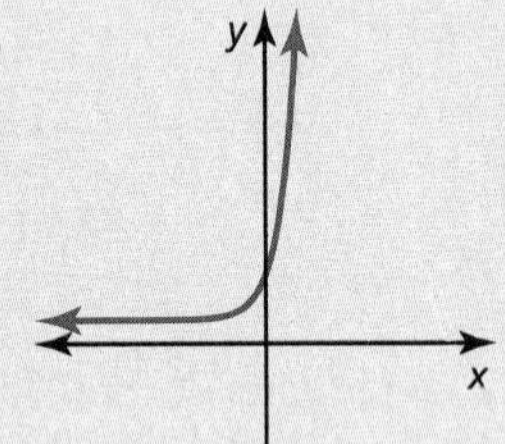

d.

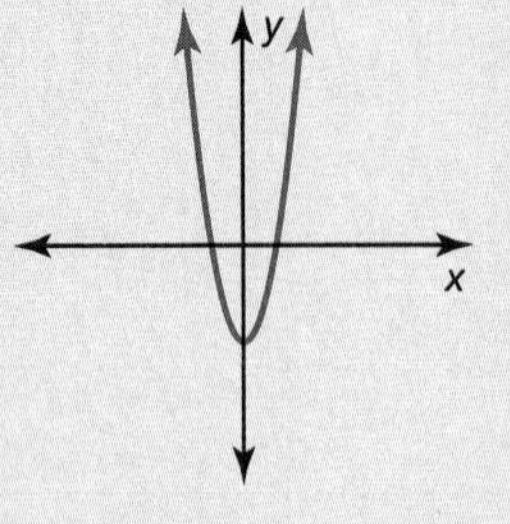

A. $y = 2x^2 - 4$

B. $y = 2(4)^x + 1$

C. $y = 2\left(\frac{3}{4}\right)^x + 1$

D. $y = 2x - 4$

37. CRITICAL THINKING In the ordered pairs below, the y-values are given in terms of n. Tell whether the ordered pairs represent a *linear*, an *exponential*, or a *quadratic* function. Explain.

$(1, 3n - 1)$, $(2, 10n + 2)$, $(3, 26n)$,
$(4, 51n - 7)$, $(5, 85n - 19)$

38. USING STRUCTURE Write a function that has constant second differences of 3.

39. CRITICAL THINKING Is the graph of a set of points enough to determine whether the points represent a linear, an exponential, or a quadratic function? Justify your answer.

40. THOUGHT PROVOKING Find four different patterns in the figure. Determine whether each pattern represents a *linear*, an *exponential*, or a *quadratic* function. Write a model for each pattern.

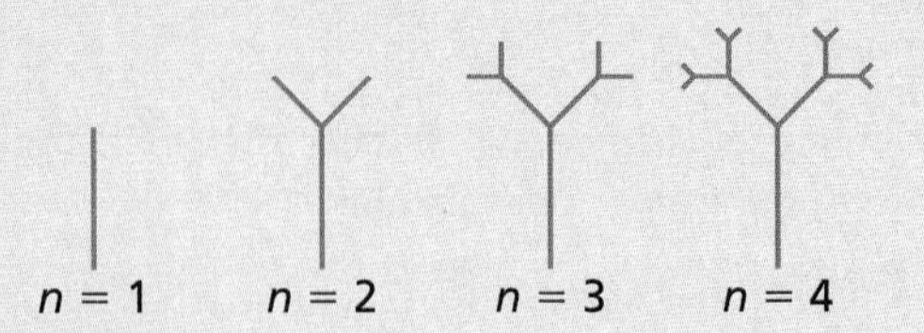

41. MAKING AN ARGUMENT Function p is an exponential function and function q is a quadratic function. Your friend says that after about $x = 3$, function q will always have a greater y-value than function p. Is your friend correct? Explain.

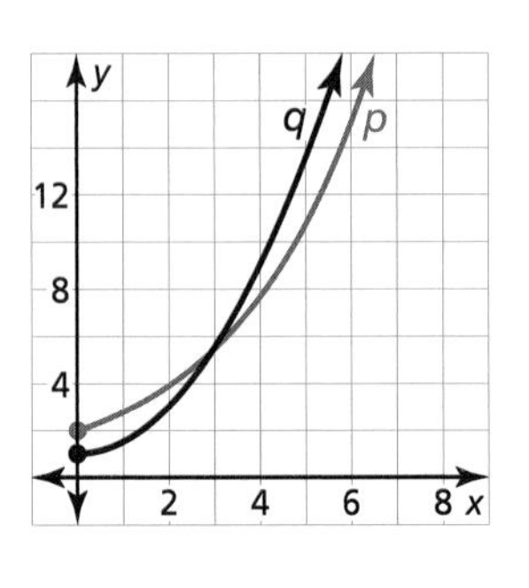

42. USING TOOLS The table shows the amount a (in billions of dollars) United States residents spent on pets or pet-related products and services each year for a 5-year period. Let the year x represent the independent variable. Using technology, find a function that models the data. How did you choose the model? Predict how much residents will spend on pets or pet-related products and services in Year 7.

Year, x	1	2	3	4	5
Amount, a	53.1	56.9	61.8	65.7	67.1

Maintaining Mathematical Proficiency

Reviewing what you learned in previous grades and lessons

Evaluate the expression. *(Section 6.2)*

43. $\sqrt{121}$

44. $\sqrt[3]{125}$

45. $\sqrt[3]{512}$

46. $\sqrt[5]{243}$

Find the product. *(Section 7.3)*

47. $(x + 8)(x - 8)$

48. $(4y + 2)(4y - 2)$

49. $(3a - 5b)(3a + 5b)$

50. $(-2r + 6s)(-2r - 6s)$

8.4–8.6 What Did You Learn?

Core Vocabulary

even function, *p. 442*
odd function, *p. 442*
vertex form (of a quadratic function), *p. 444*
intercept form, *p. 450*
average rate of change, *p. 462*

Core Concepts

Section 8.4

Even and Odd Functions, *p. 442*
Graphing $f(x) = a(x - h)^2$, *p. 443*
Graphing $f(x) = a(x - h)^2 + k$, *p. 444*
Writing Quadratic Functions of the Form $f(x) = a(x - h)^2 + k$, *p. 445*

Section 8.5

Graphing $f(x) = a(x - p)(x - q)$, *p. 450*
Factors and Zeros, *p. 452*
Using Characteristics to Graph and Write Quadratic Functions, *p. 452*
Using Characteristics to Graph and Write Cubic Functions, *p. 453*

Section 8.6

Linear, Exponential, and Quadratic Functions, *p. 460*
Differences and Ratios of Functions, *p. 461*
Writing Functions to Model Data, *p. 462*
Comparing Functions Using Average Rates of Change, *p. 463*

Mathematical Practices

1. How can you use technology to confirm your answer in Exercise 64 on page 448?
2. How did you use the structure of the equation in Exercise 85 on page 457 to solve the problem?
3. Describe why your answer makes sense considering the context of the data in Exercise 20 on page 466.

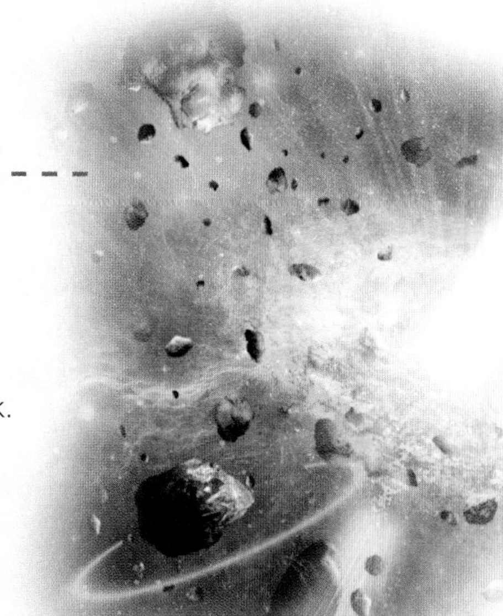

Performance Task

Asteroid Aim

Apps take a long time to design and program. One app in development is a game in which players shoot lasers at asteroids. They score points based on the number of hits per shot. The designer wants your feedback. Do you think students will like the game and want to play it? What changes would improve it?

To explore the answers to this question and more, go to ***BigIdeasMath.com***.

8 Chapter Review

Dynamic Solutions available at *BigIdeasMath.com*

8.1 Graphing $f(x) = ax^2$ (pp. 419–424)

Graph $g(x) = -4x^2$. Compare the graph to the graph of $f(x) = x^2$.

Step 1 Make a table of values.

x	-2	-1	0	1	2
$g(x)$	-16	-4	0	-4	-16

Step 2 Plot the ordered pairs.

Step 3 Draw a smooth curve through the points.

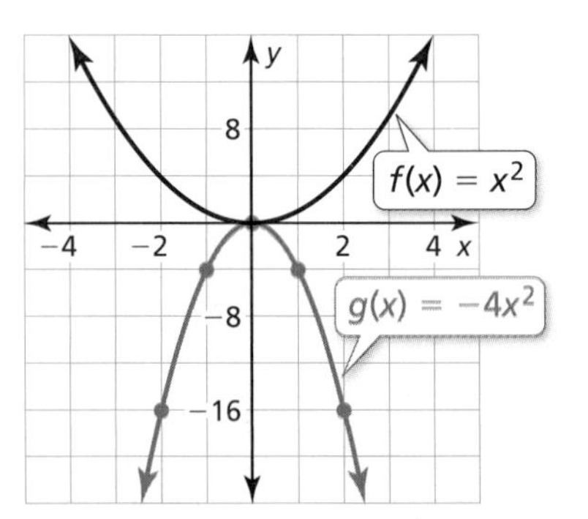

▶ The graphs have the same vertex, (0, 0), and the same axis of symmetry, $x = 0$, but the graph of g opens down and is narrower than the graph of f. So, the graph of g is a vertical stretch by a factor of 4 and a reflection in the x-axis of the graph of f.

Graph the function. Compare the graph to the graph of $f(x) = x^2$.

1. $p(x) = 7x^2$ **2.** $q(x) = \frac{1}{2}x^2$ **3.** $g(x) = -\frac{3}{4}x^2$ **4.** $h(x) = -6x^2$

5. Identify characteristics of the quadratic function and its graph.

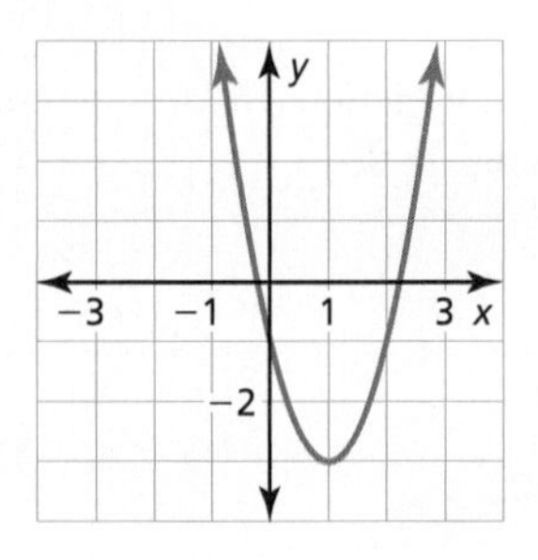

8.2 Graphing $f(x) = ax^2 + c$ (pp. 425–430)

Graph $g(x) = 2x^2 + 3$. Compare the graph to the graph of $f(x) = x^2$.

Step 1 Make a table of values.

x	-2	-1	0	1	2
$g(x)$	11	5	3	5	11

Step 2 Plot the ordered pairs.

Step 3 Draw a smooth curve through the points.

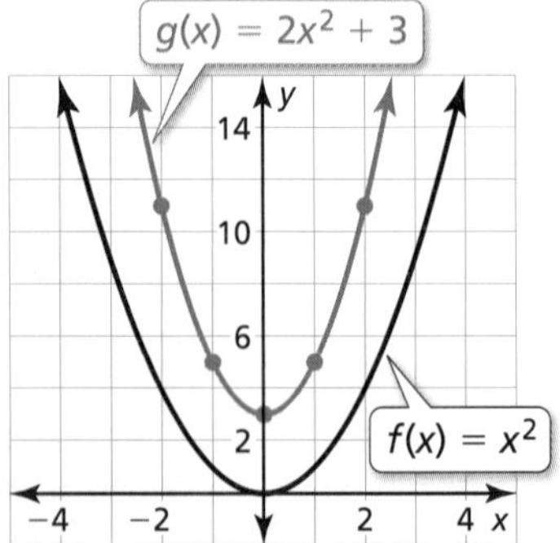

▶ Both graphs open up and have the same axis of symmetry, $x = 0$.
The graph of g is narrower, and its vertex, (0, 3), is above the vertex of the graph of f, (0, 0). So, the graph of g is a vertical stretch by a factor of 2 and a vertical translation 3 units up of the graph of f.

Graph the function. Compare the graph to the graph of $f(x) = x^2$.

6. $g(x) = x^2 + 5$ **7.** $h(x) = -x^2 - 4$ **8.** $m(x) = -2x^2 + 6$ **9.** $n(x) = \frac{1}{3}x^2 - 5$

8.3 Graphing $f(x) = ax^2 + bx + c$ (pp. 431–438)

Graph $f(x) = 4x^2 + 8x - 1$. Describe the domain and range.

Step 1 Find and graph the axis of symmetry: $x = -\frac{b}{2a} = -\frac{8}{2(4)} = -1$.

Step 2 Find and plot the vertex. The axis of symmetry is $x = -1$. So, the x-coordinate of the vertex is -1. The y-coordinate of the vertex is $f(-1) = 4(-1)^2 + 8(-1) - 1 = -5$. So, the vertex is $(-1, -5)$.

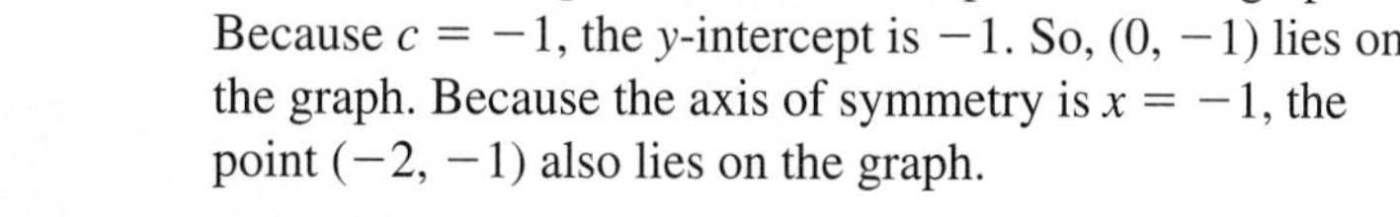

Step 3 Use the y-intercept to find two more points on the graph. Because $c = -1$, the y-intercept is -1. So, $(0, -1)$ lies on the graph. Because the axis of symmetry is $x = -1$, the point $(-2, -1)$ also lies on the graph.

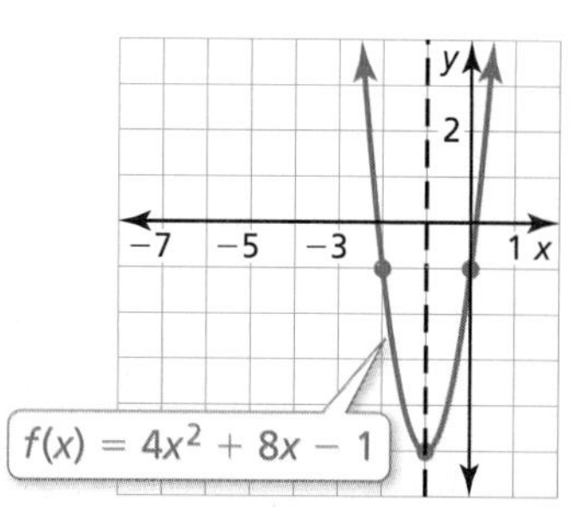

Step 4 Draw a smooth curve through the points.

▶ The domain is all real numbers. The range is $y \geq -5$.

Graph the function. Describe the domain and range.

10. $y = x^2 - 2x + 7$

11. $f(x) = -3x^2 + 3x - 4$

12. $y = \frac{1}{2}x^2 - 6x + 10$

13. The function $f(t) = -16t^2 + 88t + 12$ represents the height (in feet) of a pumpkin t seconds after it is launched from a catapult. When does the pumpkin reach its maximum height? What is the maximum height of the pumpkin?

8.4 Graphing $f(x) = a(x - h)^2 + k$ (pp. 441–448)

Determine whether $f(x) = 2x^2 + 4$ is *even*, *odd*, or *neither*.

$f(x) = 2x^2 + 4$	Write the original function.
$f(-x) = 2(-x)^2 + 4$	Substitute $-x$ for x.
$= 2x^2 + 4$	Simplify.
$= f(x)$	Substitute $f(x)$ for $2x^2 + 4$.

▶ Because $f(-x) = f(x)$, the function is even.

Determine whether the function is *even*, *odd*, or *neither*.

14. $w(x) = 5^x$

15. $r(x) = -8x$

16. $h(x) = 3x^2 - 2x$

Graph the function. Compare the graph to the graph of $f(x) = x^2$.

17. $h(x) = 2(x - 4)^2$

18. $g(x) = \frac{1}{2}(x - 1)^2 + 1$

19. $q(x) = -(x + 4)^2 + 7$

20. Consider the function $g(x) = -3(x + 2)^2 - 4$. Graph $h(x) = g(x - 1)$.

21. Write a quadratic function whose graph has a vertex of $(3, 2)$ and passes through the point $(4, 7)$.

8.5 Using Intercept Form (pp. 449–458)

Use zeros to graph $h(x) = x^2 - 7x + 6$.

The function is in standard form. The parabola opens up ($a > 0$), and the y-intercept is 6. So, plot (0, 6).

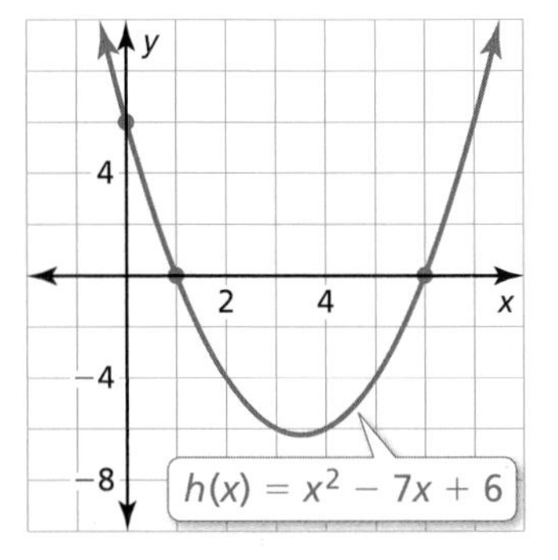

The polynomial that defines the function is factorable. So, write the function in intercept form and identify the zeros.

$h(x) = x^2 - 7x + 6$ Write the function.

$= (x - 6)(x - 1)$ Factor the trinomial.

The zeros of the function are 1 and 6. So, plot (1, 0) and (6, 0). Draw a parabola through the points.

Graph the quadratic function. Label the vertex, axis of symmetry, and x-intercepts. Describe the domain and range of the function.

22. $y = (x - 4)(x + 2)$ **23.** $f(x) = -3(x + 3)(x + 1)$ **24.** $y = x^2 - 8x + 15$

Use zeros to graph the function.

25. $y = -2x^2 + 6x + 8$ **26.** $f(x) = x^2 + x - 2$ **27.** $f(x) = 2x^3 - 18x$

28. Write a quadratic function in standard form whose graph passes through (4, 0) and (6, 0).

8.6 Comparing Linear, Exponential, and Quadratic Functions (pp. 459–468)

Tell whether the data represent a *linear*, an *exponential*, or a *quadratic* function.

a. $(-4, 1), (-3, -2), (-2, -3)$
$(-1, -2), (0, 1)$

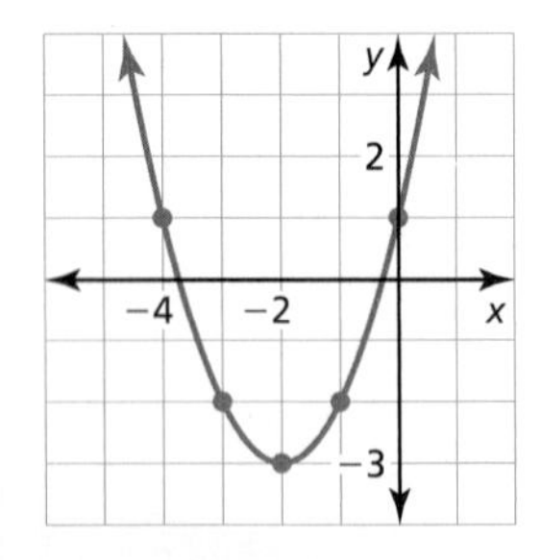

▶ The points appear to represent a quadratic function.

b.

x	−1	0	1	2	3
y	15	8	1	−6	−13

+1 +1 +1 +1

x	−1	0	1	2	3
y	15	8	1	−6	−13

−7 −7 −7 −7

▶ The first differences are constant. So, the table represents a linear function.

29. Tell whether the table of values represents a *linear*, an *exponential*, or *a quadratic* function. Then write the function.

x	−1	0	1	2	3
y	512	128	32	8	2

30. The balance y (in dollars) of your savings account after t years is represented by $y = 200(1.1)^t$. The beginning balance of your friend's account is \$250, and the balance increases by \$20 each year. (a) Compare the account balances by calculating and interpreting the average rates of change from $t = 2$ to $t = 7$. (b) Predict which account will have a greater balance after 10 years. Explain.

8 Chapter Test

Graph the function. Compare the graph to the graph of $f(x) = x^2$.

1. $h(x) = 2x^2 - 3$

2. $g(x) = -\frac{1}{2}x^2$

3. $p(x) = \frac{1}{2}(x + 1)^2 - 1$

4. Consider the graph of the function f.

   a. Find the domain, range, and zeros of the function.

   b. Write the function f in standard form.

   c. Compare the graph of f to the graph of $g(x) = x^2$.

   d. Graph $h(x) = f(x - 6)$.

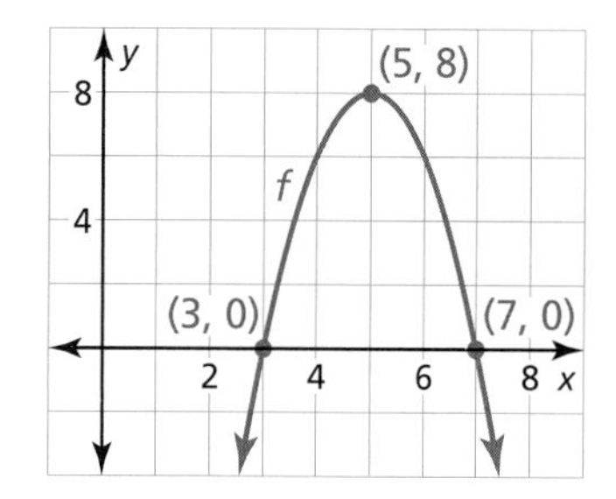

Use zeros to graph the function. Describe the domain and range of the function.

5. $f(x) = 2x^2 - 8x + 8$

6. $y = -(x + 5)(x - 1)$

7. $h(x) = 16x^2 - 4$

Tell whether the table of values represents a *linear*, an *exponential*, or a *quadratic* function. Explain your reasoning. Then write the function.

8.

x	-1	0	1	2	3
y	4	8	16	32	64

9.

x	-2	-1	0	1	2
y	-8	-2	0	-2	-8

Write a quadratic function in standard form whose graph satisfies the given conditions. Explain the process you used.

10. passes through $(-8, 0)$, $(-2, 0)$, and $(-6, 4)$

11. passes through $(0, 0)$, $(10, 0)$, and $(9, -27)$

12. is even and has a range of $y \geq 3$

13. passes through $(4, 0)$ and $(1, 9)$

14. The table shows the distances d (in miles) that Earth moves in its orbit around the Sun after t seconds. Let the time t be the independent variable. Tell whether the data can be modeled by a *linear*, an *exponential*, or a *quadratic* function. Explain. Then write a function that models the data.

Time, t	1	2	3	4	5
Distance, d	19	38	57	76	95

15. You are playing tennis with a friend. The path of the tennis ball after you return a serve can be modeled by the function $y = -0.005x^2 + 0.17x + 3$, where x is the horizontal distance (in feet) from where you hit the ball and y is the height (in feet) of the ball.

    a. What is the maximum height of the tennis ball?

    b. You are standing 30 feet from the net, which is 3 feet high. Will the ball clear the net? Explain your reasoning.

16. Find values of a, b, and c so that the function $f(x) = ax^2 + bx + c$ is (a) even, (b) odd, and (c) neither even nor odd.

17. Consider the function $f(x) = x^2 + 4$. Find the average rate of change from $x = 0$ to $x = 1$, from $x = 1$ to $x = 2$, and from $x = 2$ to $x = 3$. What do you notice about the average rates of change when the function is increasing?

8 Cumulative Assessment

1. Which function is represented by the graph?

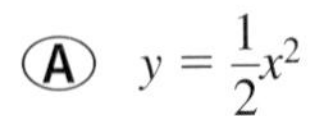
Ⓐ $y = \frac{1}{2}x^2$

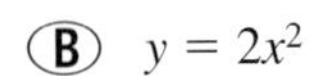
Ⓑ $y = 2x^2$

Ⓒ $y = -\frac{1}{2}x^2$

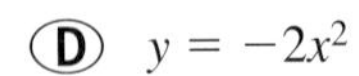
Ⓓ $y = -2x^2$

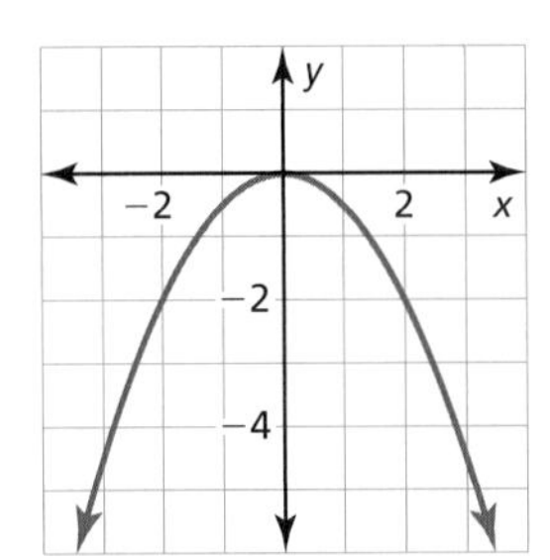

2. Find all numbers between 0 and 100 that are in the range of the function defined below.

$$f(1) = 1,\ f(2) = 1,\ f(n) = f(n-1) + f(n-2)$$

3. The function $f(t) = -16t^2 + v_0t + s_0$ represents the height (in feet) of a ball t seconds after it is thrown from an initial height s_0 (in feet) with an initial vertical velocity v_0 (in feet per second). The ball reaches its maximum height after $\frac{7}{8}$ second when it is thrown with an initial vertical velocity of ______ feet per second.

4. Classify each system of equations by the number of solutions.

$y = 6x + 9$
$y = -\frac{1}{6}x + 9$

$7x + 4y = 12$
$8y - 12 = -14x$

$2x + 4y = -2$
$10x + 4y = -2$

$3x + y = 5$
$-15 + 3y + 9x = 0$

$y - 2x = \frac{3}{2}$
$-3 + 2y = 4x$

$y = -3x + 5$
$y = -3x + 9$

5. Your friend claims that quadratic functions can have two, one, or no real zeros. Do you support your friend's claim? Use graphs to justify your answer.

6. Which polynomial represents the area (in square feet) of the shaded region of the figure?

Ⓐ $a^2 - x^2$

Ⓑ $x^2 - a^2$

Ⓒ $x^2 - 2ax + a^2$

Ⓓ $x^2 + 2ax + a^2$

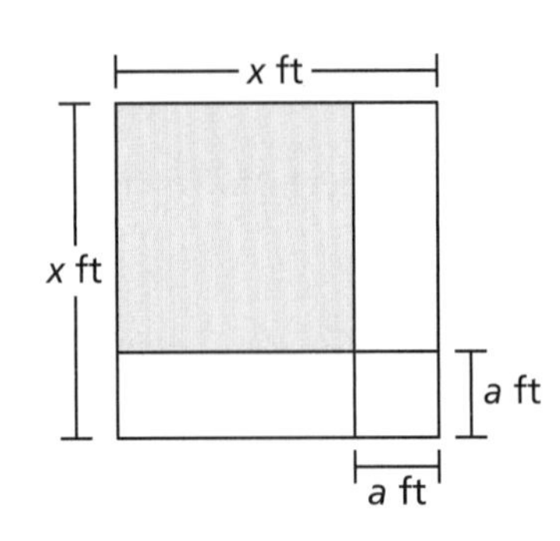

7. Consider the functions represented by the tables.

x	0	1	2	3
$p(x)$	-4	-16	-28	-40

x	1	2	3	4
$r(x)$	0	15	40	75

x	1	2	3	4
$s(x)$	72	36	18	9

x	1	3	5	7
$t(x)$	3	-5	-21	-45

a. Classify each function as *linear*, *exponential*, or *quadratic*.

b. Order the functions from least to greatest according to the average rates of change between $x = 1$ and $x = 3$.

8. Complete each function using the symbols + or −, so that the graph of the quadratic function satisfies the given conditions.

a. $f(x) = 5(x \;\square\; 3)^2 \;\square\; 4$; vertex: $(-3, 4)$

b. $g(x) = -(x \;\square\; 2)(x \;\square\; 8)$; x-intercepts: -8 and 2

c. $h(x) = \;\square\; 3x^2 \;\square\; 6$; range: $y \geq -6$

d. $j(x) = \;\square\; 4(x \;\square\; 1)(x \;\square\; 1)$; range: $y \leq 4$

9. The graph shows the amounts y (in dollars) that a referee earns for refereeing x high school volleyball games.

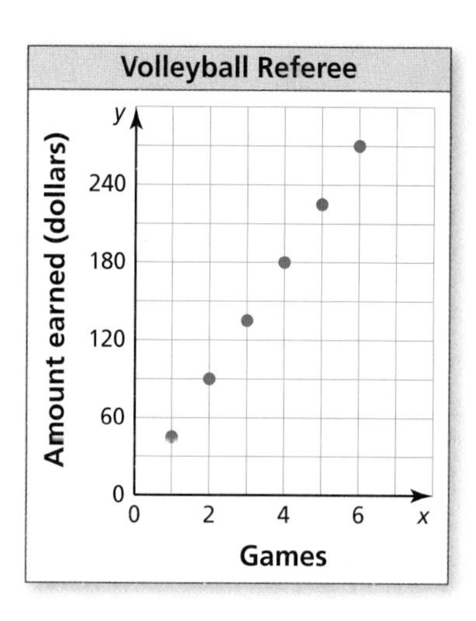

a. Does the graph represent a linear or nonlinear function? Explain.

b. Describe the domain of the function. Is the domain discrete or continuous?

c. Write a function that models the data.

d. Can the referee earn exactly \$500? Explain.

10. Which expressions are equivalent to $(b^{-5})^{-4}$?

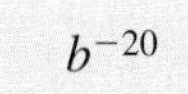

b^{-20}	$b^{-6}b^{-14}$	$(b^{-4})^{-5}$	b^{-9}	$(b^{-2})^{-7}$
$b^{-5}b^{-4}$	$(b^{10})^2$	b^{20}	$(b^{-12})^3$	$b^{12}b^8$

9 Solving Quadratic Equations

Dolphin *(p. 521)*

Half-pipe *(p. 513)*

Pond *(p. 501)*

Kicker *(p. 493)*

Parthenon *(p. 483)*

Maintaining Mathematical Proficiency

Factoring Perfect Square Trinomials

Example 1 **Factor $x^2 + 14x + 49$.**

$$x^2 + 14x + 49 = x^2 + 2(x)(7) + 7^2 \qquad \text{Write as } a^2 + 2ab + b^2.$$

$$= (x + 7)^2 \qquad \text{Perfect square trinomial pattern}$$

Factor the trinomial.

1. $x^2 + 10x + 25$ **2.** $x^2 - 20x + 100$ **3.** $x^2 + 12x + 36$

4. $x^2 - 18x + 81$ **5.** $x^2 + 16x + 64$ **6.** $x^2 - 30x + 225$

Solving Systems of Linear Equations by Graphing

Example 2 **Solve the system of linear equations by graphing.**

$y = 2x + 1$ Equation 1

$y = -\frac{1}{3}x + 8$ Equation 2

Step 1 Graph each equation.

Step 2 Estimate the point of intersection.
The graphs appear to intersect at (3, 7).

Step 3 Check your point from Step 2.

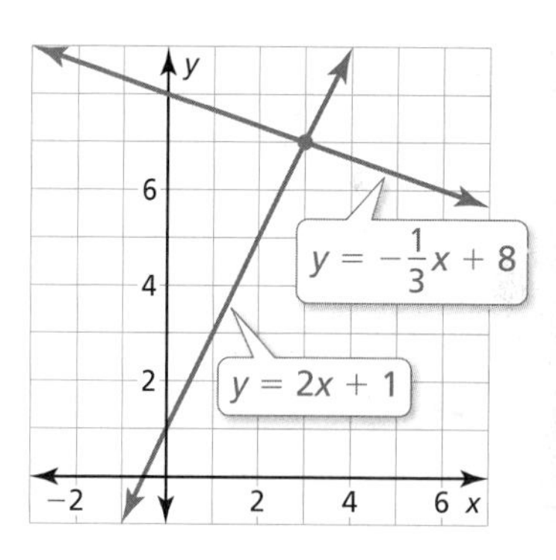

Equation 1	Equation 2
$y = 2x + 1$	$y = -\frac{1}{3}x + 8$
$7 \stackrel{?}{=} 2(3) + 1$	$7 \stackrel{?}{=} -\frac{1}{3}(3) + 8$
$7 = 7$ ✓	$7 = 7$ ✓

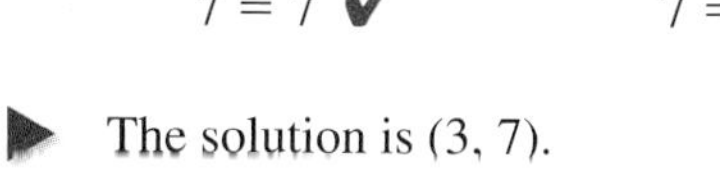

▶ The solution is (3, 7).

Solve the system of linear equations by graphing.

7. $y = -5x + 3$
$y = 2x - 4$

8. $y = \frac{3}{2}x - 2$
$y = -\frac{1}{4}x + 5$

9. $y = \frac{1}{2}x + 4$
$y = -3x - 3$

10. **ABSTRACT REASONING** What value of c makes $x^2 + bx + c$ a perfect square trinomial?

Mathematical Practices

Mathematically proficient students monitor their work and change course as needed.

Problem-Solving Strategies

Core Concept

Guess, Check, and Revise

When solving a problem in mathematics, it is often helpful to estimate a solution and then observe how close that solution is to being correct. For instance, you can use the guess, check, and revise strategy to find a decimal approximation of the square root of 2.

	Guess	Check	How to revise
1.	1.4	$1.4^2 = 1.96$	Increase guess.
2.	1.41	$1.41^2 = 1.9881$	Increase guess.
3.	1.415	$1.415^2 = 2.002225$	Decrease guess.

By continuing this process, you can determine that the square root of 2 is approximately 1.4142.

Approximating a Solution of an Equation

The graph of $y = x^2 + x - 1$ is shown. Approximate the positive solution of the equation $x^2 + x - 1 = 0$ to the nearest thousandth.

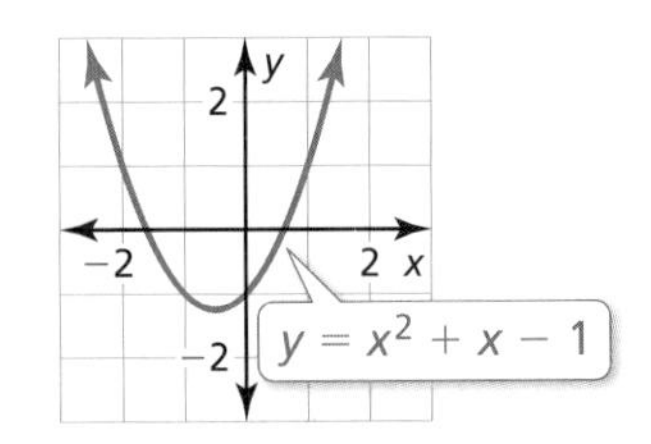

SOLUTION

Using the graph, you can make an initial estimate of the positive solution to be $x = 0.65$.

	Guess	Check	How to revise
1.	0.65	$0.65^2 + 0.65 - 1 = 0.0725$	Decrease guess.
2.	0.62	$0.62^2 + 0.62 - 1 = 0.0044$	Decrease guess.
3.	0.618	$0.618^2 + 0.618 - 1 = -0.000076$	Increase guess.
4.	0.6181	$0.6181^2 + 0.6181 - 1 \approx 0.00015$	The solution is between 0.618 and 0.6181.

▶ So, to the nearest thousandth, the positive solution of the equation is $x = 0.618$.

Monitoring Progress

1. Use the graph in Example 1 to approximate the negative solution of the equation $x^2 + x - 1 = 0$ to the nearest thousandth.

2. The graph of $y = x^2 + x - 3$ is shown. Approximate both solutions of the equation $x^2 + x - 3 = 0$ to the nearest thousandth.

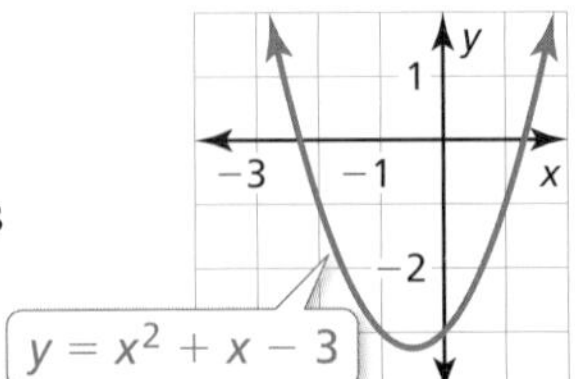

9.1 Properties of Radicals

Essential Question

How can you multiply and divide square roots?

EXPLORATION 1 Operations with Square Roots

Work with a partner. For each operation with square roots, compare the results obtained using the two indicated orders of operations. What can you conclude?

a. Square Roots and Addition

Is $\sqrt{36} + \sqrt{64}$ equal to $\sqrt{36 + 64}$?

In general, is $\sqrt{a} + \sqrt{b}$ equal to $\sqrt{a + b}$? Explain your reasoning.

b. Square Roots and Multiplication

Is $\sqrt{4} \cdot \sqrt{9}$ equal to $\sqrt{4 \cdot 9}$?

In general, is $\sqrt{a} \cdot \sqrt{b}$ equal to $\sqrt{a \cdot b}$? Explain your reasoning.

c. Square Roots and Subtraction

Is $\sqrt{64} - \sqrt{36}$ equal to $\sqrt{64 - 36}$?

In general, is $\sqrt{a} - \sqrt{b}$ equal to $\sqrt{a - b}$? Explain your reasoning.

d. Square Roots and Division

Is $\dfrac{\sqrt{100}}{\sqrt{4}}$ equal to $\sqrt{\dfrac{100}{4}}$?

In general, is $\dfrac{\sqrt{a}}{\sqrt{b}}$ equal to $\sqrt{\dfrac{a}{b}}$? Explain your reasoning.

REASONING ABSTRACTLY

To be proficient in math, you need to recognize and use counterexamples.

EXPLORATION 2 Writing Counterexamples

Work with a partner. A **counterexample** is an example that proves that a general statement is *not* true. For each general statement in Exploration 1 that is not true, write a counterexample different from the example given.

Communicate Your Answer

3. How can you multiply and divide square roots?

4. Give an example of multiplying square roots and an example of dividing square roots that are different from the examples in Exploration 1.

5. Write an algebraic rule for each operation.

a. the product of square roots

b. the quotient of square roots

9.1 Lesson

Core Vocabulary

Previous
radicand
perfect cube

What You Will Learn

- Use properties of radicals to simplify expressions.
- Simplify expressions by rationalizing the denominator.
- Perform operations with radicals.

Using Properties of Radicals

A **radical expression** is an expression that contains a radical. An expression involving a radical with index n is in **simplest form** when these three conditions are met.

- No radicands have perfect nth powers as factors other than 1.
- No radicands contain fractions.
- No radicals appear in the denominator of a fraction.

You can use the property below to simplify radical expressions involving square roots.

Core Concept

Product Property of Square Roots

Words The square root of a product equals the product of the square roots of the factors.

Numbers $\sqrt{9 \cdot 5} = \sqrt{9} \cdot \sqrt{5} = 3\sqrt{5}$

Algebra $\sqrt{ab} = \sqrt{a} \cdot \sqrt{b}$, where $a, b \geq 0$

STUDY TIP

There can be more than one way to factor a radicand. An efficient method is to find the greatest perfect square factor.

EXAMPLE 1 Using the Product Property of Square Roots

a. $\sqrt{108} = \sqrt{36 \cdot 3}$ — Factor using the greatest perfect square factor.

$= \sqrt{36} \cdot \sqrt{3}$ — Product Property of Square Roots

$= 6\sqrt{3}$ — Simplify.

b. $\sqrt{9x^3} = \sqrt{9 \cdot x^2 \cdot x}$ — Factor using the greatest perfect square factor.

$= \sqrt{9} \cdot \sqrt{x^2} \cdot \sqrt{x}$ — Product Property of Square Roots

$= 3x\sqrt{x}$ — Simplify.

STUDY TIP

In this course, whenever a variable appears in the radicand, assume that it has only *nonnegative* values.

Monitoring Progress  Help in English and Spanish at *BigIdeasMath.com*

Simplify the expression.

1. $\sqrt{24}$ **2.** $-\sqrt{80}$ **3.** $\sqrt{49x^3}$ **4.** $\sqrt{75n^5}$

Core Concept

Quotient Property of Square Roots

Words The square root of a quotient equals the quotient of the square roots of the numerator and denominator.

Numbers $\sqrt{\frac{3}{4}} = \frac{\sqrt{3}}{\sqrt{4}} = \frac{\sqrt{3}}{2}$

Algebra $\sqrt{\frac{a}{b}} = \frac{\sqrt{a}}{\sqrt{b}}$, where $a \geq 0$ and $b > 0$

EXAMPLE 2 Using the Quotient Property of Square Roots

a. $\sqrt{\frac{15}{64}} = \frac{\sqrt{15}}{\sqrt{64}}$ Quotient Property of Square Roots

$= \frac{\sqrt{15}}{8}$ Simplify.

b. $\sqrt{\frac{81}{x^2}} = \frac{\sqrt{81}}{\sqrt{x^2}}$ Quotient Property of Square Roots

$= \frac{9}{x}$ Simplify.

You can extend the Product and Quotient Properties of Square Roots to other radicals, such as cube roots. When using these *properties of cube roots*, the radicands may contain negative numbers.

EXAMPLE 3 Using Properties of Cube Roots

STUDY TIP

To write a cube root in simplest form, find factors of the radicand that are perfect cubes.

a. $\sqrt[3]{-128} = \sqrt[3]{-64 \cdot 2}$ Factor using the greatest perfect cube factor.

$= \sqrt[3]{-64} \cdot \sqrt[3]{2}$ Product Property of Cube Roots

$= -4\sqrt[3]{2}$ Simplify.

b. $\sqrt[3]{125x^7} = \sqrt[3]{125 \cdot x^6 \cdot x}$ Factor using the greatest perfect cube factors.

$= \sqrt[3]{125} \cdot \sqrt[3]{x^6} \cdot \sqrt[3]{x}$ Product Property of Cube Roots

$= 5x^2\sqrt[3]{x}$ Simplify.

c. $\sqrt[3]{\frac{y}{216}} = \frac{\sqrt[3]{y}}{\sqrt[3]{216}}$ Quotient Property of Cube Roots

$= \frac{\sqrt[3]{y}}{6}$ Simplify.

d. $\sqrt[3]{\frac{8x^4}{27y^3}} = \frac{\sqrt[3]{8x^4}}{\sqrt[3]{27y^3}}$ Quotient Property of Cube Roots

$= \frac{\sqrt[3]{8 \cdot x^3 \cdot x}}{\sqrt[3]{27 \cdot y^3}}$ Factor using the greatest perfect cube factors.

$= \frac{\sqrt[3]{8} \cdot \sqrt[3]{x^3} \cdot \sqrt[3]{x}}{\sqrt[3]{27} \cdot \sqrt[3]{y^3}}$ Product Property of Cube Roots

$= \frac{2x\sqrt[3]{x}}{3y}$ Simplify.

Monitoring Progress

Help in English and Spanish at *BigIdeasMath.com*

Simplify the expression.

5. $\sqrt{\frac{23}{9}}$ **6.** $-\sqrt{\frac{17}{100}}$ **7.** $\sqrt{\frac{36}{z^2}}$ **8.** $\sqrt{\frac{4x^2}{64}}$

9. $\sqrt[3]{54}$ **10.** $\sqrt[3]{16x^4}$ **11.** $\sqrt[3]{\frac{a}{-27}}$ **12.** $\sqrt[3]{\frac{25c^7d^3}{64}}$

Rationalizing the Denominator

When a radical is in the denominator of a fraction, you can multiply the fraction by an appropriate form of 1 to eliminate the radical from the denominator. This process is called **rationalizing the denominator**.

EXAMPLE 4 Rationalizing the Denominator

STUDY TIP

Rationalizing the denominator works because you multiply the numerator and denominator by the same nonzero number a, which is the same as multiplying by $\frac{a}{a}$, or 1.

a. $\frac{\sqrt{5}}{\sqrt{3n}} = \frac{\sqrt{5}}{\sqrt{3n}} \cdot \frac{\sqrt{3n}}{\sqrt{3n}}$ Multiply by $\frac{\sqrt{3n}}{\sqrt{3n}}$.

$= \frac{\sqrt{15n}}{\sqrt{9n^2}}$ Product Property of Square Roots

$= \frac{\sqrt{15n}}{\sqrt{9} \cdot \sqrt{n^2}}$ Product Property of Square Roots

$= \frac{\sqrt{15n}}{3n}$ Simplify.

b. $\frac{2}{\sqrt[3]{9}} = \frac{2}{\sqrt[3]{9}} \cdot \frac{\sqrt[3]{3}}{\sqrt[3]{3}}$ Multiply by $\frac{\sqrt[3]{3}}{\sqrt[3]{3}}$.

$= \frac{2\sqrt[3]{3}}{\sqrt[3]{27}}$ Product Property of Cube Roots

$= \frac{2\sqrt[3]{3}}{3}$ Simplify.

The binomials $a\sqrt{b} + c\sqrt{d}$ and $a\sqrt{b} - c\sqrt{d}$, where a, b, c, and d are rational numbers, are called **conjugates**. You can use conjugates to simplify radical expressions that contain a sum or difference involving square roots in the denominator.

EXAMPLE 5 Rationalizing the Denominator Using Conjugates

LOOKING FOR STRUCTURE

Notice that the product of two conjugates $a\sqrt{b} + c\sqrt{d}$ and $a\sqrt{b} - c\sqrt{d}$ does not contain a radical and is a *rational* number.

$(a\sqrt{b} + c\sqrt{d})(a\sqrt{b} - c\sqrt{d})$
$= (a\sqrt{b})^2 - (c\sqrt{d})^2$
$= a^2b - c^2d$

Simplify $\frac{7}{2 - \sqrt{3}}$.

SOLUTION

$\frac{7}{2 - \sqrt{3}} = \frac{7}{2 - \sqrt{3}} \cdot \frac{2 + \sqrt{3}}{2 + \sqrt{3}}$ The conjugate of $2 - \sqrt{3}$ is $2 + \sqrt{3}$.

$= \frac{7(2 + \sqrt{3})}{2^2 - (\sqrt{3})^2}$ Sum and difference pattern

$= \frac{14 + 7\sqrt{3}}{1}$ Simplify.

$= 14 + 7\sqrt{3}$ Simplify.

Monitoring Progress  Help in English and Spanish at *BigIdeasMath.com*

Simplify the expression.

13. $\frac{1}{\sqrt{5}}$ **14.** $\frac{\sqrt{10}}{\sqrt{3}}$ **15.** $\frac{7}{\sqrt{2x}}$ **16.** $\sqrt{\frac{2y^2}{3}}$

17. $\frac{5}{\sqrt[3]{32}}$ **18.** $\frac{8}{1 + \sqrt{3}}$ **19.** $\frac{\sqrt{13}}{\sqrt{5} - 2}$ **20.** $\frac{12}{\sqrt{2} + \sqrt{7}}$

EXAMPLE 6 Solving a Real-Life Problem

The distance d (in miles) that you can see to the horizon with your eye level h feet above the water is given by $d = \sqrt{\frac{3h}{2}}$. How far can you see when your eye level is 5 feet above the water?

SOLUTION

$$d = \sqrt{\frac{3(5)}{2}} \qquad \text{Substitute 5 for } h.$$

$$= \frac{\sqrt{15}}{\sqrt{2}} \qquad \text{Quotient Property of Square Roots}$$

$$= \frac{\sqrt{15}}{\sqrt{2}} \cdot \frac{\sqrt{2}}{\sqrt{2}} \qquad \text{Multiply by } \frac{\sqrt{2}}{\sqrt{2}}.$$

$$= \frac{\sqrt{30}}{2} \qquad \text{Simplify.}$$

▶ You can see $\frac{\sqrt{30}}{2}$, or about 2.74 miles.

EXAMPLE 7 Modeling with Mathematics

The ratio of the length to the width of a *golden rectangle* is $(1 + \sqrt{5}) : 2$. The dimensions of the face of the Parthenon in Greece form a golden rectangle. What is the height h of the Parthenon?

SOLUTION

1. **Understand the Problem** Think of the length and height of the Parthenon as the length and width of a golden rectangle. The length of the rectangular face is 31 meters. You know the ratio of the length to the height. Find the height h.
2. **Make a Plan** Use the ratio $(1 + \sqrt{5}) : 2$ to write a proposition and solve for h.
3. **Solve the Problem**

$$\frac{1 + \sqrt{5}}{2} = \frac{31}{h} \qquad \text{Write a proportion.}$$

$$h(1 + \sqrt{5}) = 62 \qquad \text{Cross Products Property}$$

$$h = \frac{62}{1 + \sqrt{5}} \qquad \text{Divide each side by } 1 + \sqrt{5}.$$

$$h = \frac{62}{1 + \sqrt{5}} \cdot \frac{1 - \sqrt{5}}{1 - \sqrt{5}} \qquad \text{Multiply the numerator and denominator by the conjugate.}$$

$$h = \frac{62 - 62\sqrt{5}}{-4} \qquad \text{Simplify.}$$

$$h \approx 19.16 \qquad \text{Use a calculator.}$$

▶ The height is about 19 meters.

4. **Look Back** $\frac{1 + \sqrt{5}}{2} \approx 1.62$ and $\frac{31}{19.16} \approx 1.62$. So, your answer is reasonable.

Monitoring Progress

Help in English and Spanish at *BigIdeasMath.com*

21. WHAT IF? In Example 6, how far can you see when your eye level is 35 feet above the water?

22. The dimensions of a dance floor form a golden rectangle. The shorter side of the dance floor is 50 feet. What is the length of the longer side of the dance floor?

Performing Operations with Radicals

Radicals with the same index and radicand are called **like radicals**. You can add and subtract like radicals the same way you combine like terms by using the Distributive Property.

STUDY TIP

Do not assume that radicals with different radicands cannot be added or subtracted. Always check to see whether you can simplify the radicals. In some cases, the radicals will become like radicals.

EXAMPLE 8 Adding and Subtracting Radicals

a. $5\sqrt{7} + \sqrt{11} - 8\sqrt{7} = 5\sqrt{7} - 8\sqrt{7} + \sqrt{11}$ — Commutative Property of Addition

$= (5 - 8)\sqrt{7} + \sqrt{11}$ — Distributive Property

$= -3\sqrt{7} + \sqrt{11}$ — Subtract.

b. $10\sqrt{5} + \sqrt{20} = 10\sqrt{5} + \sqrt{4 \cdot 5}$ — Factor using the greatest perfect square factor.

$= 10\sqrt{5} + \sqrt{4} \cdot \sqrt{5}$ — Product Property of Square Roots

$= 10\sqrt{5} + 2\sqrt{5}$ — Simplify.

$= (10 + 2)\sqrt{5}$ — Distributive Property

$= 12\sqrt{5}$ — Add.

c. $6\sqrt[3]{x} + 2\sqrt[3]{x} = (6 + 2)\sqrt[3]{x}$ — Distributive Property

$= 8\sqrt[3]{x}$ — Add.

EXAMPLE 9 Multiplying Radicals

Simplify $\sqrt{5}(\sqrt{3} - \sqrt{75})$.

SOLUTION

Method 1 $\sqrt{5}(\sqrt{3} - \sqrt{75}) = \sqrt{5} \cdot \sqrt{3} - \sqrt{5} \cdot \sqrt{75}$ — Distributive Property

$= \sqrt{15} - \sqrt{375}$ — Product Property of Square Roots

$= \sqrt{15} - 5\sqrt{15}$ — Simplify.

$= (1 - 5)\sqrt{15}$ — Distributive Property

$= -4\sqrt{15}$ — Subtract.

Method 2 $\sqrt{5}(\sqrt{3} - \sqrt{75}) = \sqrt{5}(\sqrt{3} - 5\sqrt{3})$ — Simplify $\sqrt{75}$.

$= \sqrt{5}[(1 - 5)\sqrt{3}]$ — Distributive Property

$= \sqrt{5}(-4\sqrt{3})$ — Subtract.

$= -4\sqrt{15}$ — Product Property of Square Roots

Monitoring Progress

Simplify the expression.

23. $3\sqrt{2} - \sqrt{6} + 10\sqrt{2}$

24. $4\sqrt{7} - 6\sqrt{63}$

25. $4\sqrt[3]{5x} - 11\sqrt[3]{5x}$

26. $\sqrt{3}(8\sqrt{2} + 7\sqrt{32})$

27. $(2\sqrt{5} - 4)^2$

28. $\sqrt[3]{-4}(\sqrt[3]{2} - \sqrt[3]{16})$

9.1 Exercises

Dynamic Solutions available at *BigIdeasMath.com*

Vocabulary and Core Concept Check

1. **COMPLETE THE SENTENCE** The process of eliminating a radical from the denominator of a radical expression is called ______________.

2. **VOCABULARY** What is the conjugate of the binomial $\sqrt{6} + 4$?

3. **WRITING** Are the expressions $\frac{1}{3}\sqrt{2x}$ and $\sqrt{\frac{2x}{9}}$ equivalent? Explain your reasoning.

4. **WHICH ONE DOESN'T BELONG?** Which expression does *not* belong with the other three? Explain your reasoning.

$-\frac{1}{3}\sqrt{6}$	$6\sqrt{3}$	$\frac{1}{6}\sqrt{3}$	$-3\sqrt{3}$

Monitoring Progress and Modeling with Mathematics

In Exercises 5–12, determine whether the expression is in simplest form. If the expression is not in simplest form, explain why.

5. $\sqrt{19}$

6. $\sqrt{\frac{1}{7}}$

7. $\sqrt{48}$

8. $\sqrt{34}$

9. $\frac{5}{\sqrt{2}}$

10. $\frac{3\sqrt{10}}{4}$

11. $\frac{1}{2 + \sqrt[3]{2}}$

12. $6 - \sqrt[3]{54}$

In Exercises 13–20, simplify the expression. *(See Example 1.)*

13. $\sqrt{20}$

14. $\sqrt{32}$

15. $\sqrt{128}$

16. $-\sqrt{72}$

17. $\sqrt{125b}$

18. $\sqrt{4x^2}$

19. $-\sqrt{81m^3}$

20. $\sqrt{48n^5}$

In Exercises 21–28, simplify the expression. *(See Example 2.)*

21. $\sqrt{\frac{4}{49}}$

22. $-\sqrt{\frac{7}{81}}$

23. $-\sqrt{\frac{23}{64}}$

24. $\sqrt{\frac{65}{121}}$

25. $\sqrt{\frac{a^3}{49}}$

26. $\sqrt{\frac{144}{k^2}}$

27. $\sqrt{\frac{100}{4x^2}}$

28. $\sqrt{\frac{25v^2}{36}}$

In Exercises 29–36, simplify the expression. *(See Example 3.)*

29. $\sqrt[3]{16}$

30. $\sqrt[3]{-108}$

31. $\sqrt[3]{-64x^5}$

32. $-\sqrt[3]{343n^2}$

33. $\sqrt[3]{\frac{6c}{-125}}$

34. $\sqrt[3]{\frac{8h^4}{27}}$

35. $-\sqrt[3]{\frac{81y^2}{1000x^3}}$

36. $\sqrt[3]{\frac{21}{-64a^3b^6}}$

ERROR ANALYSIS In Exercises 37 and 38, describe and correct the error in simplifying the expression.

37.

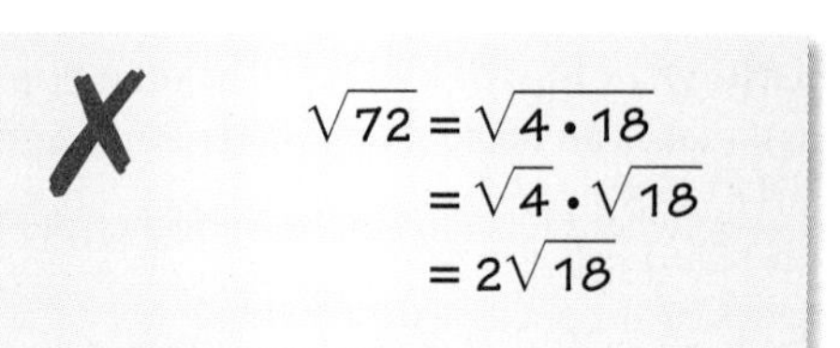

38.

✗

$$\sqrt[3]{\frac{128y^3}{125}} = \frac{\sqrt[3]{128y^3}}{125} = \frac{\sqrt[3]{64 \cdot 2 \cdot y^3}}{125} = \frac{\sqrt[3]{64} \cdot \sqrt[3]{2} \cdot \sqrt[3]{y^3}}{125} = \frac{4y\sqrt[3]{2}}{125}$$

In Exercises 39–44, write a factor that you can use to rationalize the denominator of the expression.

39. $\frac{4}{\sqrt{6}}$

40. $\frac{1}{\sqrt{13z}}$

41. $\frac{2}{\sqrt[3]{x^2}}$

42. $\frac{3m}{\sqrt[3]{4}}$

43. $\frac{\sqrt{2}}{\sqrt{5} - 8}$

44. $\frac{5}{\sqrt{3} + \sqrt{7}}$

In Exercises 45–54, simplify the expression.
(See Example 4.)

45. $\frac{2}{\sqrt{2}}$

46. $\frac{4}{\sqrt{3}}$

47. $\frac{\sqrt{5}}{\sqrt{48}}$

48. $\sqrt{\frac{4}{52}}$

49. $\frac{3}{\sqrt{a}}$

50. $\frac{1}{\sqrt{2x}}$

51. $\sqrt{\frac{3d^2}{5}}$

52. $\frac{\sqrt{8}}{\sqrt{3n^3}}$

53. $\frac{4}{\sqrt[3]{25}}$

54. $\sqrt[3]{\frac{1}{108y^2}}$

In Exercises 55–60, simplify the expression.
(See Example 5.)

55. $\frac{1}{\sqrt{7} + 1}$

56. $\frac{2}{5 - \sqrt{3}}$

57. $\frac{\sqrt{10}}{7 - \sqrt{2}}$

58. $\frac{\sqrt{5}}{6 + \sqrt{5}}$

59. $\frac{3}{\sqrt{5} - \sqrt{2}}$

60. $\frac{\sqrt{3}}{\sqrt{7} + \sqrt{3}}$

61. MODELING WITH MATHEMATICS The time t (in seconds) it takes an object to hit the ground is given by $t = \sqrt{\frac{h}{16}}$, where h is the height (in feet) from which the object was dropped. *(See Example 6.)*

a. How long does it take an earring to hit the ground when it falls from the roof of the building?

b. How much sooner does the earring hit the ground when it is dropped from two stories (22 feet) below the roof?

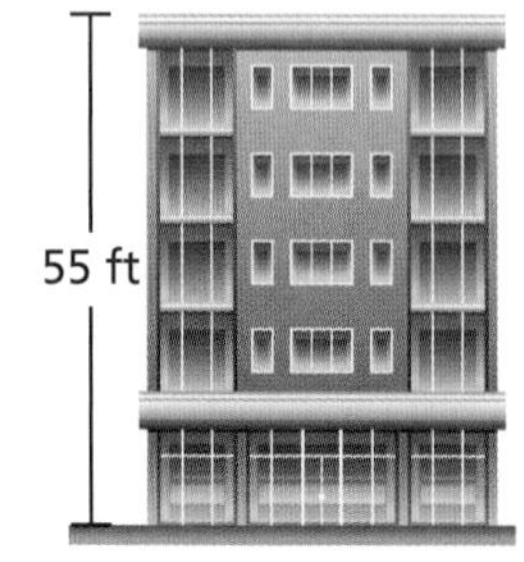

62. MODELING WITH MATHEMATICS The orbital period of a planet is the time it takes the planet to travel around the Sun. You can find the orbital period P (in Earth years) using the formula $P = \sqrt{d^3}$, where d is the average distance (in astronomical units, abbreviated AU) of the planet from the Sun.

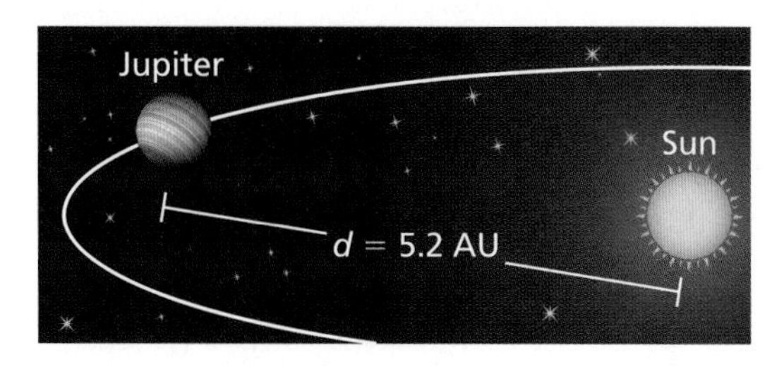

a. Simplify the formula.

b. What is Jupiter's orbital period?

63. MODELING WITH MATHEMATICS The electric current I (in amperes) an appliance uses is given by the formula $I = \sqrt{\frac{P}{R}}$, where P is the power (in watts) and R is the resistance (in ohms). Find the current an appliance uses when the power is 147 watts and the resistance is 5 ohms.

64. MODELING WITH MATHEMATICS You can find the average annual interest rate r (in decimal form) of a savings account using the formula $r = \sqrt{\frac{V_2}{V_0}} - 1$, where V_0 is the initial investment and V_2 is the balance of the account after 2 years. Use the formula to compare the savings accounts. In which account would you invest money? Explain.

Account	Initial investment	Balance after 2 years
1	\$275	\$293
2	\$361	\$382
3	\$199	\$214
4	\$254	\$272
5	\$386	\$406

In Exercises 65–68, evaluate the function for the given value of x. Write your answer in simplest form and in decimal form rounded to the nearest hundredth.

65. $h(x) = \sqrt{5x}; x = 10$

66. $g(x) = \sqrt{3x}; x = 60$

67. $r(x) = \sqrt{\dfrac{3x}{3x^2 + 6}}; x = 4$

68. $p(x) = \sqrt{\dfrac{x - 1}{5x}}; x = 8$

In Exercises 69–72, evaluate the expression when $a = -2$, $b = 8$, and $c = \frac{1}{2}$. Write your answer in simplest form and in decimal form rounded to the nearest hundredth.

69. $\sqrt{a^2 + bc}$

70. $-\sqrt{4c - 6ab}$

71. $-\sqrt{2a^2 + b^2}$

72. $\sqrt{b^2 - 4ac}$

73. MODELING WITH MATHEMATICS The text in the book shown forms a golden rectangle. What is the width w of the text? *(See Example 7.)*

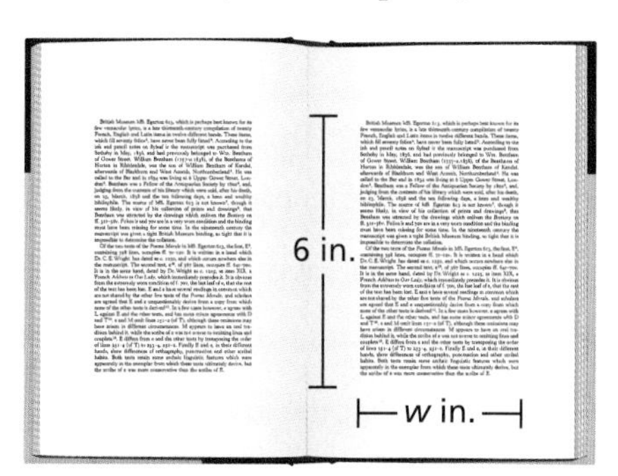

74. MODELING WITH MATHEMATICS The flag of Togo is approximately the shape of a golden rectangle. What is the width w of the flag?

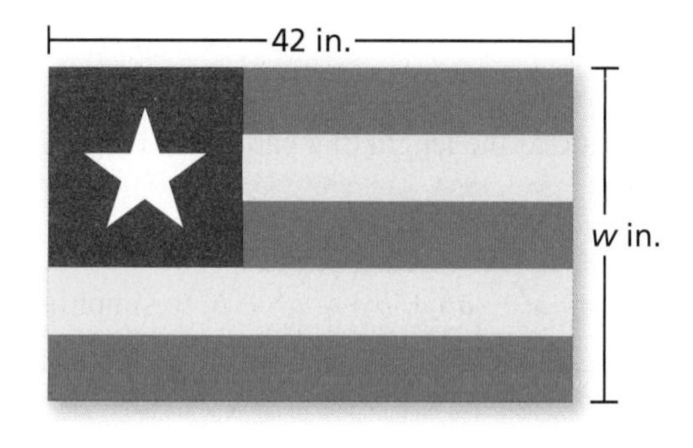

In Exercises 75–82, simplify the expression. *(See Example 8.)*

75. $\sqrt{3} - 2\sqrt{2} + 6\sqrt{2}$

76. $\sqrt{5} - 5\sqrt{13} - 8\sqrt{5}$

77. $2\sqrt{6} - 5\sqrt{54}$

78. $9\sqrt{32} + \sqrt{2}$

79. $\sqrt{12} + 6\sqrt{3} + 2\sqrt{6}$

80. $3\sqrt{7} - 5\sqrt{14} + 2\sqrt{28}$

81. $\sqrt[3]{-81} + 4\sqrt[3]{3}$

82. $6\sqrt[3]{128t} - 2\sqrt[3]{2t}$

In Exercises 83–90, simplify the expression. *(See Example 9.)*

83. $\sqrt{2}(\sqrt{45} + \sqrt{5})$

84. $\sqrt{3}(\sqrt{72} - 3\sqrt{2})$

85. $\sqrt{5}(2\sqrt{6x} - \sqrt{96x})$

86. $\sqrt{7y}(\sqrt{27y} + 5\sqrt{12y})$

87. $(4\sqrt{2} - \sqrt{98})^2$

88. $(\sqrt{3} + \sqrt{48})(\sqrt{20} - \sqrt{5})$

89. $\sqrt[3]{3}(\sqrt[3]{4} + \sqrt[3]{32})$

90. $\sqrt[3]{2}(\sqrt[3]{135} - 4\sqrt[3]{5})$

91. MODELING WITH MATHEMATICS The circumference C of the art room in a mansion is approximated by the formula $C \approx 2\pi\sqrt{\dfrac{a^2 + b^2}{2}}$. Approximate the circumference of the room.

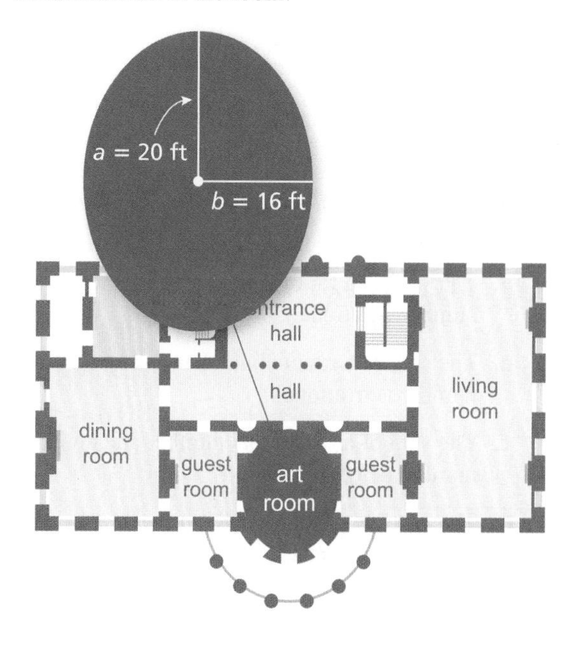

92. CRITICAL THINKING Determine whether each expression represents a *rational* or an *irrational* number. Justify your answer.

a. $4 + \sqrt{6}$

b. $\dfrac{\sqrt{48}}{\sqrt{3}}$

c. $\dfrac{8}{\sqrt{12}}$

d. $\sqrt{3} + \sqrt{7}$

e. $\dfrac{a}{\sqrt{10} - \sqrt{2}}$, where a is a positive integer

f. $\dfrac{2 + \sqrt{5}}{2b + \sqrt{5b^2}}$, where b is a positive integer

In Exercises 93–98, simplify the expression.

93. $\sqrt[5]{\dfrac{13}{5x^5}}$

94. $\sqrt[4]{\dfrac{10}{81}}$

95. $\sqrt[4]{256y}$

96. $\sqrt[5]{160x^6}$

97. $6\sqrt[4]{9} - \sqrt[5]{9} + 3\sqrt[4]{9}$

98. $\sqrt[5]{2}(\sqrt[4]{7} + \sqrt[5]{16})$

REASONING **In Exercises 99 and 100, use the table shown.**

	2	$\frac{1}{4}$	0	$\sqrt{3}$	$-\sqrt{3}$	π
2						
$\frac{1}{4}$						
0						
$\sqrt{3}$						
$-\sqrt{3}$						
π						

99. Copy and complete the table by (a) finding each sum $\left(2 + 2, 2 + \frac{1}{4}, \text{etc.}\right)$ and (b) finding each product $\left(2 \cdot 2, 2 \cdot \frac{1}{4}, \text{etc.}\right)$.

100. Use your answers in Exercise 99 to determine whether each statement is *always*, *sometimes*, or *never* true. Justify your answer.

a. The sum of a rational number and a rational number is rational.

b. The sum of a rational number and an irrational number is irrational.

c. The sum of an irrational number and an irrational number is irrational.

d. The product of a rational number and a rational number is rational.

e. The product of a nonzero rational number and an irrational number is irrational.

f. The product of an irrational number and an irrational number is irrational.

101. REASONING Let m be a positive integer. For what values of m will the simplified form of the expression $\sqrt{2^m}$ contain a radical? For what values will it *not* contain a radical? Explain.

102. HOW DO YOU SEE IT? The edge length s of a cube is an irrational number, the surface area is an irrational number, and the volume is a rational number. Give a possible value of s.

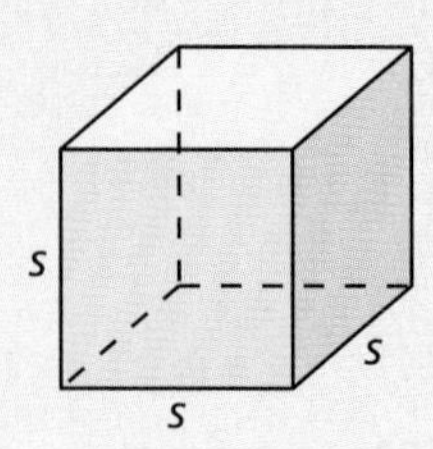

103. REASONING Let a and b be positive numbers. Explain why $\sqrt{ab}$ lies between a and b on a number line. (*Hint:* Let $a < b$ and multiply each side of $a < b$ by a. Then let $a < b$ and multiply each side by b.)

104. MAKING AN ARGUMENT Your friend says that you can rationalize the denominator of the expression $\frac{2}{4 + \sqrt[3]{5}}$ by multiplying the numerator and denominator by $4 - \sqrt[3]{5}$. Is your friend correct? Explain.

105. PROBLEM SOLVING The ratio of consecutive terms $\frac{a_n}{a_{n-1}}$ in the Fibonacci sequence gets closer and closer to the golden ratio $\frac{1 + \sqrt{5}}{2}$ as n increases. Find the term that precedes 610 in the sequence.

106. THOUGHT PROVOKING Use the golden ratio $\frac{1 + \sqrt{5}}{2}$ and the golden ratio conjugate $\frac{1 - \sqrt{5}}{2}$ for each of the following.

a. Show that the golden ratio and golden ratio conjugate are both solutions of $x^2 - x - 1 = 0$.

b. Construct a geometric diagram that has the golden ratio as the length of a part of the diagram.

107. CRITICAL THINKING Use the special product pattern $(a + b)(a^2 - ab + b^2) = a^3 + b^3$ to simplify the expression $\frac{2}{\sqrt[3]{x} + 1}$. Explain your reasoning.

Maintaining Mathematical Proficiency

Reviewing what you learned in previous grades and lessons

Graph the linear equation. Identify the *x*-intercept. *(Section 3.5)*

108. $y = x - 4$ **109.** $y = -2x + 6$ **110.** $y = -\frac{1}{3}x - 1$ **111.** $y = \frac{3}{2}x + 6$

Solve the equation. Check your solution. *(Section 6.5)*

112. $32 = 2^x$ **113.** $27^x = 3^{x-6}$ **114.** $\left(\frac{1}{6}\right)^{2x} = 216^{1-x}$ **115.** $625^x = \left(\frac{1}{25}\right)^{x+2}$

9.2 Solving Quadratic Equations by Graphing

Essential Question How can you use a graph to solve a quadratic equation in one variable?

Based on what you learned about the x-intercepts of a graph in Section 3.4, it follows that the x-intercept of the graph of the linear equation

$y = ax + b$ 2 variables

is the same value as the solution of

$ax + b = 0.$ 1 variable

You can use similar reasoning to solve *quadratic equations*.

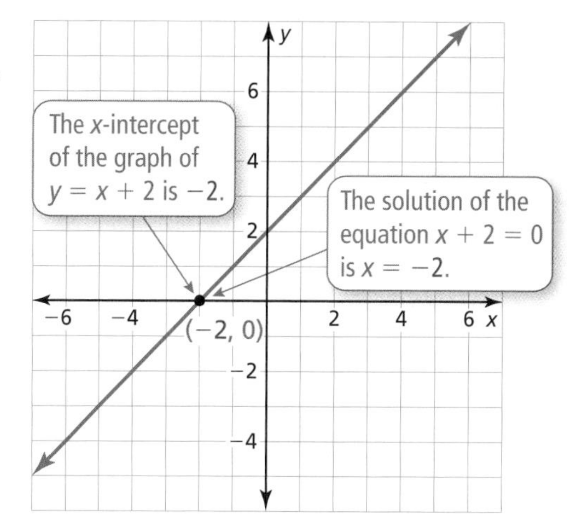

EXPLORATION 1 Solving a Quadratic Equation by Graphing

Work with a partner.

a. Sketch the graph of $y = x^2 - 2x$.

b. What is the definition of an x-intercept of a graph? How many x-intercepts does this graph have? What are they?

c. What is the definition of a solution of an equation in x? How many solutions does the equation $x^2 - 2x = 0$ have? What are they?

d. Explain how you can verify the solutions you found in part (c).

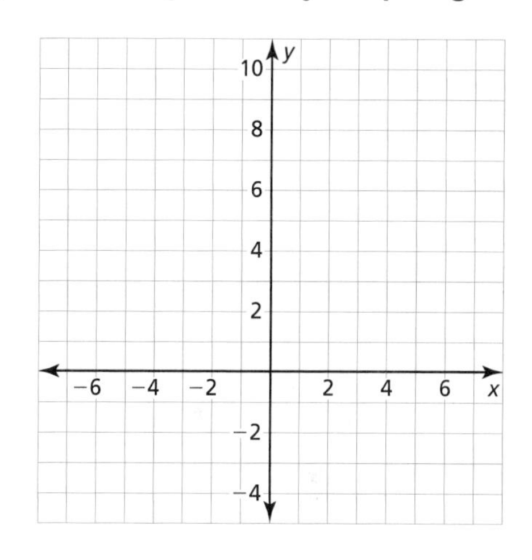

EXPLORATION 2 Solving Quadratic Equations by Graphing

Work with a partner. Solve each equation by graphing.

a. $x^2 - 4 = 0$

b. $x^2 + 3x = 0$

c. $-x^2 + 2x = 0$

d. $x^2 - 2x + 1 = 0$

e. $x^2 - 3x + 5 = 0$

f. $-x^2 + 3x - 6 = 0$

MAKING SENSE OF PROBLEMS

To be proficient in math, you need to check your answers to problems using a different method and continually ask yourself, "Does this make sense?"

Communicate Your Answer

3. How can you use a graph to solve a quadratic equation in one variable?

4. After you find a solution graphically, how can you check your result algebraically? Check your solutions for parts (a)–(d) in Exploration 2 algebraically.

5. How can you determine graphically that a quadratic equation has no solution?

9.2 Lesson

Core Vocabulary

quadratic equation, *p. 490*

Previous
x-intercept
root
zero of a function

What You Will Learn

- Solve quadratic equations by graphing.
- Use graphs to find and approximate the zeros of functions.
- Solve real-life problems using graphs of quadratic functions.

Solving Quadratic Equations by Graphing

A **quadratic equation** is a nonlinear equation that can be written in the standard form $ax^2 + bx + c = 0$, where $a \neq 0$.

In Chapter 7, you solved quadratic equations by factoring. You can also solve quadratic equations by graphing.

Core Concept

Solving Quadratic Equations by Graphing

Step 1 Write the equation in standard form, $ax^2 + bx + c = 0$.

Step 2 Graph the related function $y = ax^2 + bx + c$.

Step 3 Find the *x*-intercepts, if any.

The solutions, or *roots*, of $ax^2 + bx + c = 0$ are the *x*-intercepts of the graph.

EXAMPLE 1 Solving a Quadratic Equation: Two Real Solutions

Solve $x^2 + 2x = 3$ by graphing.

SOLUTION

Step 1 Write the equation in standard form.

$x^2 + 2x = 3$ Write original equation.

$x^2 + 2x - 3 = 0$ Subtract 3 from each side.

Step 2 Graph the related function $y = x^2 + 2x - 3$.

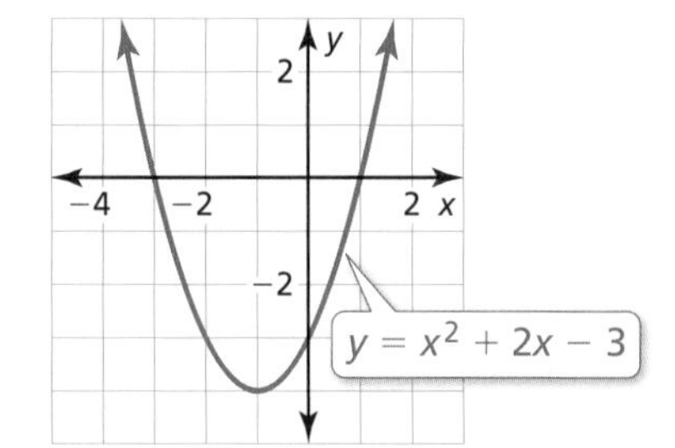

Step 3 Find the *x*-intercepts. The *x*-intercepts are -3 and 1.

▶ So, the solutions are $x = -3$ and $x = 1$.

Check

$x^2 + 2x = 3$	Original equation	$x^2 + 2x = 3$
$(-3)^2 + 2(-3) \stackrel{?}{=} 3$	Substitute.	$1^2 + 2(1) \stackrel{?}{=} 3$
$3 = 3$ ✓	Simplify.	$3 = 3$ ✓

Monitoring Progress  Help in English and Spanish at *BigIdeasMath.com*

Solve the equation by graphing. Check your solutions.

1. $x^2 - x - 2 = 0$ **2.** $x^2 + 7x = -10$ **3.** $x^2 + x = 12$

EXAMPLE 2 Solving a Quadratic Equation: One Real Solution

Solve $x^2 - 8x = -16$ by graphing.

SOLUTION

Step 1 Write the equation in standard form.

$$x^2 - 8x = -16 \qquad \text{Write original equation.}$$

$$x^2 - 8x + 16 = 0 \qquad \text{Add 16 to each side.}$$

Step 2 Graph the related function $y = x^2 - 8x + 16$.

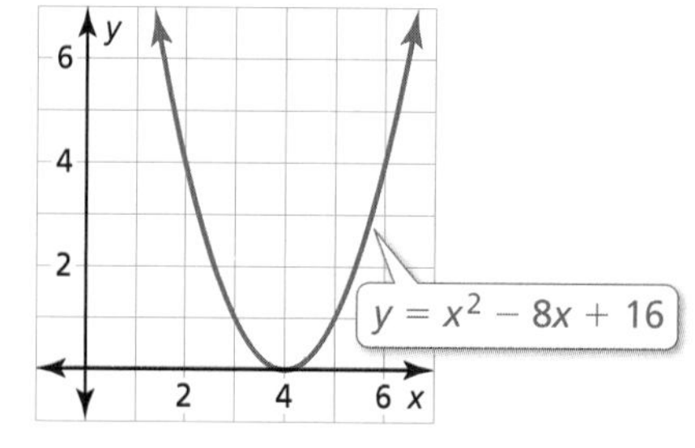

Step 3 Find the x-intercept. The only x-intercept is at the vertex, (4, 0).

▶ So, the solution is $x = 4$.

ANOTHER WAY

You can also solve the equation in Example 2 by factoring.

$$x^2 - 8x + 16 = 0$$

$$(x - 4)(x - 4) = 0$$

So, $x = 4$.

EXAMPLE 3 Solving a Quadratic Equation: No Real Solutions

Solve $-x^2 = 2x + 4$ by graphing.

SOLUTION

Method 1 Write the equation in standard form, $x^2 + 2x + 4 = 0$. Then graph the related function $y = x^2 + 2x + 4$, as shown at the left.

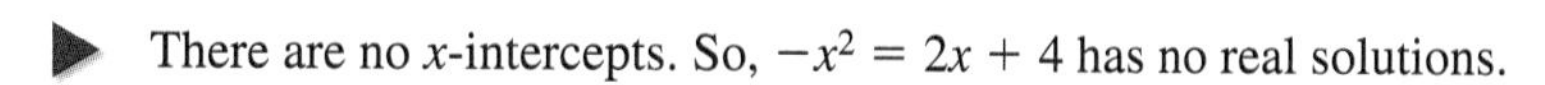

▶ There are no x-intercepts. So, $-x^2 = 2x + 4$ has no real solutions.

Method 2 Graph each side of the equation.

$$y = -x^2 \qquad \text{Left side}$$

$$y = 2x + 4 \qquad \text{Right side}$$

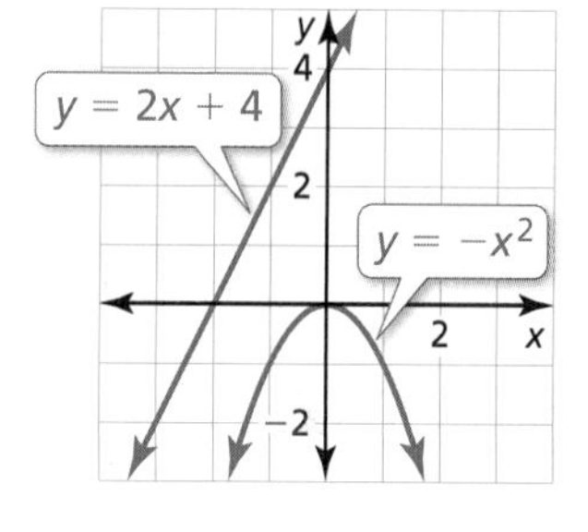

▶ The graphs do not intersect. So, $-x^2 = 2x + 4$ has no real solutions.

Monitoring Progress

Help in English and Spanish at *BigIdeasMath.com*

Solve the equation by graphing.

4. $x^2 + 36 = 12x$ **5.** $x^2 + 4x = 0$ **6.** $x^2 + 10x = -25$

7. $x^2 = 3x - 3$ **8.** $x^2 + 7x = -6$ **9.** $2x + 5 = -x^2$

Concept Summary

Number of Solutions of a Quadratic Equation

A quadratic equation has:

- two real solutions when the graph of its related function has two x-intercepts.
- one real solution when the graph of its related function has one x-intercept.
- no real solutions when the graph of its related function has no x-intercepts.

Finding Zeros of Functions

Recall that a zero of a function is an x-intercept of the graph of the function.

EXAMPLE 4 Finding the Zeros of a Function

The graph of $f(x) = (x - 3)(x^2 - x - 2)$ is shown. Find the zeros of f.

SOLUTION

The x-intercepts are -1, 2, and 3.

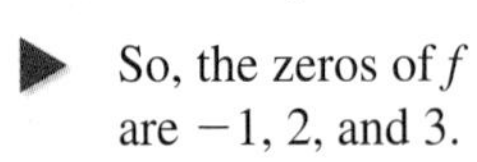

▶ So, the zeros of f are -1, 2, and 3.

Check

$f(-1) = (-1 - 3)[(-1)^2 - (-1) - 2] = 0$ ✓

$f(2) = (2 - 3)(2^2 - 2 - 2) = 0$ ✓

$f(3) = (3 - 3)(3^2 - 3 - 2) = 0$ ✓

$f(x) = (x - 3)(x^2 - x - 2)$

The zeros of a function are not necessarily integers. To approximate zeros, analyze the signs of function values. When two function values have different signs, a zero lies between the x-values that correspond to the function values.

EXAMPLE 5 Approximating the Zeros of a Function

The graph of $f(x) = x^2 + 4x + 1$ is shown. Approximate the zeros of f to the nearest tenth.

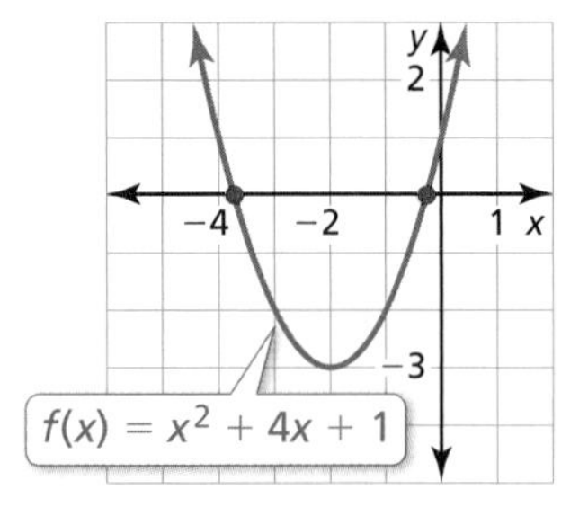

SOLUTION

There are two x-intercepts: one between -4 and -3, and another between -1 and 0.

Make tables using x-values between -4 and -3, and between -1 and 0. Use an increment of 0.1. Look for a change in the signs of the function values.

x	-3.9	-3.8	-3.7	-3.6	-3.5	-3.4	-3.3	-3.2	-3.1
$f(x)$	0.61	0.24	-0.11	-0.44	-0.75	-1.04	-1.31	-1.56	-1.79

change in signs

x	-0.9	-0.8	-0.7	-0.6	-0.5	-0.4	-0.3	-0.2	-0.1
$f(x)$	-1.79	-1.56	-1.31	-1.04	-0.75	-0.44	-0.11	0.24	0.61

The function values that are closest to 0 correspond to x-values that best approximate the zeros of the function.

▶ In each table, the function value closest to 0 is -0.11. So, the zeros of f are about -3.7 and -0.3.

ANOTHER WAY

You could approximate one zero using a table and then use the axis of symmetry to find the other zero.

Monitoring Progress  Help in English and Spanish at *BigIdeasMath.com*

10. Graph $f(x) = x^2 + x - 6$. Find the zeros of f.

11. Graph $f(x) = -x^2 + 2x + 2$. Approximate the zeros of f to the nearest tenth.

Seconds, t	Height, h
0	2
1	61
2	88
3	83
4	46
5	−23

Solving Real-Life Problems

EXAMPLE 6 Real-Life Application

A football player kicks a football 2 feet above the ground with an initial vertical velocity of 75 feet per second. The function $h = -16t^2 + 75t + 2$ represents the height h (in feet) of the football after t seconds. (a) Find the height of the football each second after it is kicked. (b) Use the results of part (a) to estimate when the height of the football is 50 feet. (c) Using a graph, after how many seconds is the football 50 feet above the ground?

SOLUTION

a. Make a table of values starting with $t = 0$ seconds using an increment of 1. Continue the table until a function value is negative.

▶ The height of the football is 61 feet after 1 second, 88 feet after 2 seconds, 83 feet after 3 seconds, and 46 feet after 4 seconds.

b. From part (a), you can estimate that the height of the football is 50 feet between 0 and 1 second and between 3 and 4 seconds.

▶ Based on the function values, it is reasonable to estimate that the height of the football is 50 feet slightly less than 1 second and slightly less than 4 seconds after it is kicked.

c. To determine when the football is 50 feet above the ground, find the t-values for which $h = 50$. So, solve the equation $-16t^2 + 75t + 2 = 50$ by graphing.

Step 1 Write the equation in standard form.

$$-16t^2 + 75t + 2 = 50 \quad \text{Write the equation.}$$

$$-16t^2 + 75t - 48 = 0 \quad \text{Subtract 50 from each side.}$$

Step 2 Use a graphing calculator to graph the related function $h = -16t^2 + 75t - 48$.

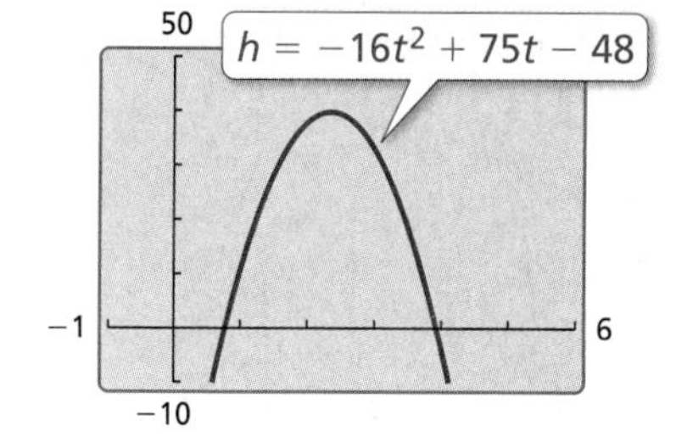

REMEMBER

Equations have *solutions*, or *roots*. Graphs have *x-intercepts*. Functions have *zeros*.

Step 3 Use the *zero* feature to find the zeros of the function.

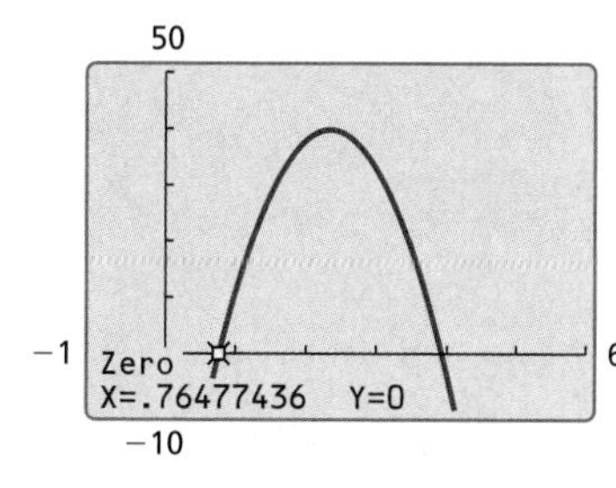

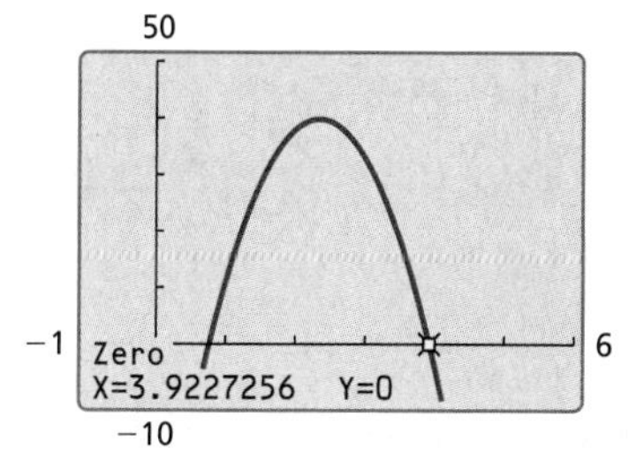

▶ The football is 50 feet above the ground after about 0.8 second and about 3.9 seconds, which supports the estimates in part (b).

Monitoring Progress  Help in English and Spanish at *BigIdeasMath.com*

12. WHAT IF? After how many seconds is the football 65 feet above the ground?

9.2 Exercises

Dynamic Solutions available at *BigIdeasMath.com*

Vocabulary and Core Concept Check

1. **VOCABULARY** What is a quadratic equation?

2. **WHICH ONE DOESN'T BELONG?** Which equation does *not* belong with the other three? Explain your reasoning.

$x^2 + 5x = 20$	$x^2 + x - 4 = 0$	$x^2 - 6 = 4x$	$7x + 12 = x^2$

3. **WRITING** How can you use a graph to find the number of solutions of a quadratic equation?

4. **WRITING** How are solutions, roots, x-intercepts, and zeros related?

Monitoring Progress and Modeling with Mathematics

In Exercises 5–8, use the graph to solve the equation.

5. $-x^2 + 2x + 3 = 0$

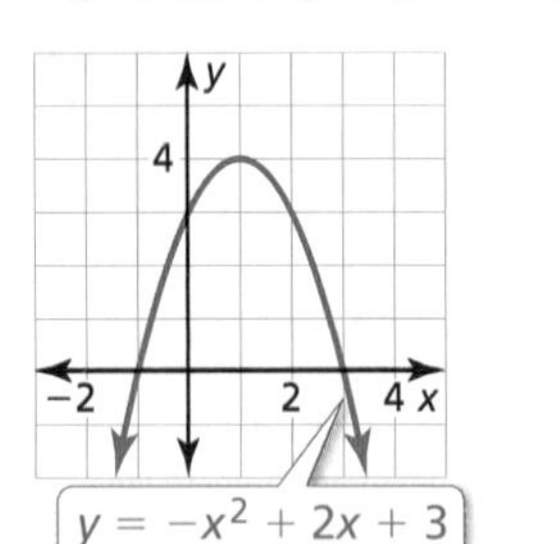

6. $x^2 - 6x + 8 = 0$

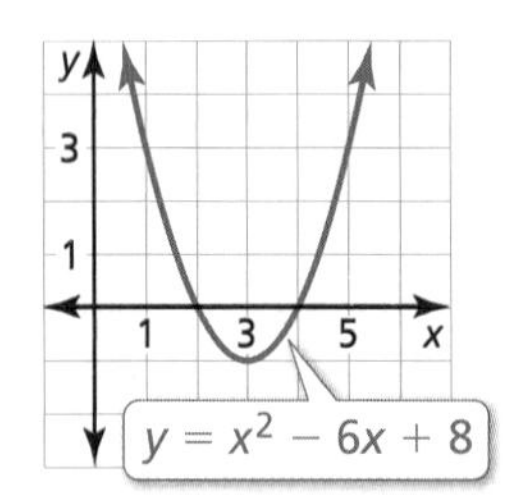

7. $x^2 + 8x + 16 = 0$

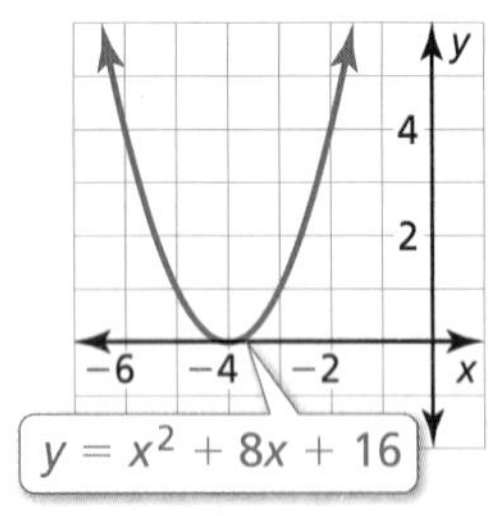

8. $-x^2 - 4x - 6 = 0$

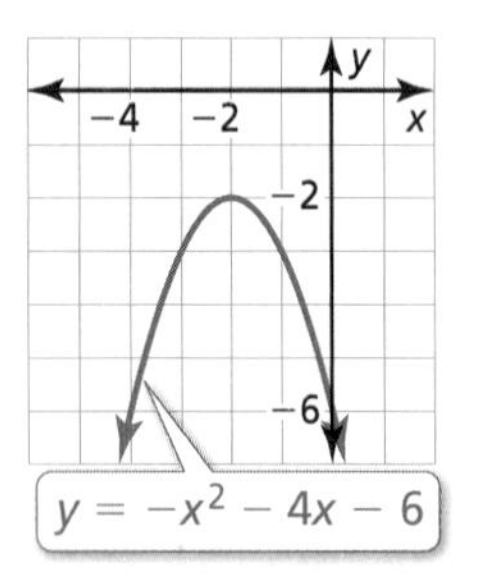

In Exercises 9–12, write the equation in standard form.

9. $4x^2 = 12$
10. $-x^2 = 15$
11. $2x - x^2 = 1$
12. $5 + x = 3x^2$

In Exercises 13–24, solve the equation by graphing. *(See Examples 1, 2, and 3.)*

13. $x^2 - 5x = 0$
14. $x^2 - 4x + 4 = 0$
15. $x^2 - 2x + 5 = 0$
16. $x^2 - 6x - 7 = 0$
17. $x^2 = 6x - 9$
18. $-x^2 = 8x + 20$
19. $x^2 = -1 - 2x$
20. $x^2 = -x - 3$
21. $4x - 12 = -x^2$
22. $5x - 6 = x^2$
23. $x^2 - 2 = -x$
24. $16 + x^2 = -8x$

25. **ERROR ANALYSIS** Describe and correct the error in solving $x^2 + 3x = 18$ by graphing.

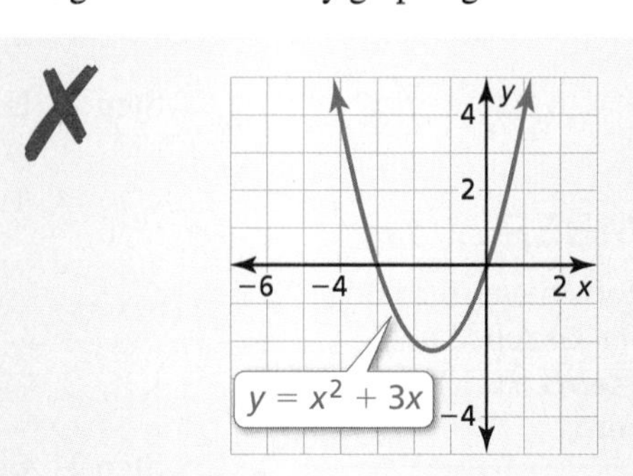

26. **ERROR ANALYSIS** Describe and correct the error in solving $x^2 + 6x + 9 = 0$ by graphing.

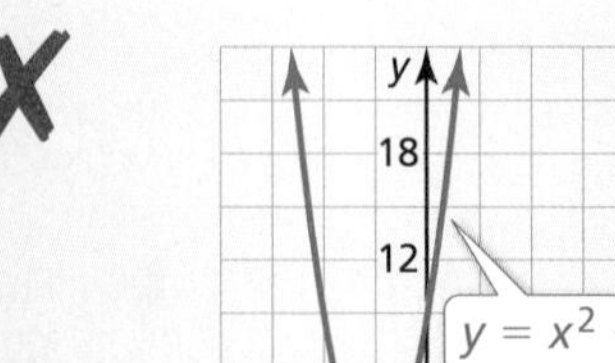
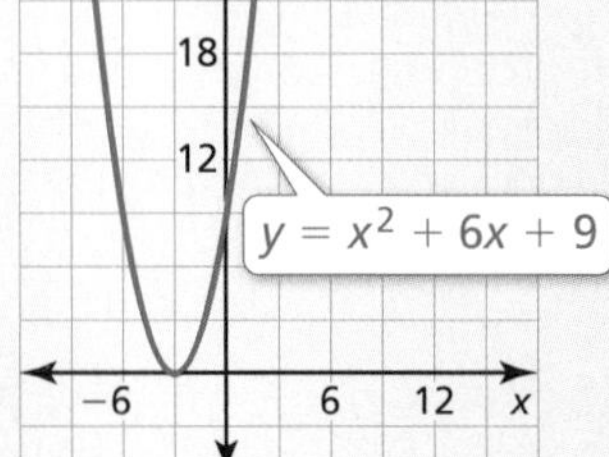

27. **MODELING WITH MATHEMATICS** The height y (in yards) of a flop shot in golf can be modeled by $y = -x^2 + 5x$, where x is the horizontal distance (in yards).

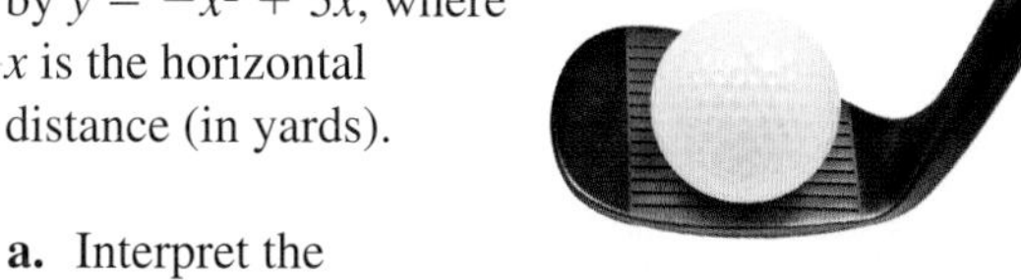

  a. Interpret the x-intercepts of the graph of the equation.

  b. How far away does the golf ball land?

28. **MODELING WITH MATHEMATICS** The height h (in feet) of an underhand volleyball serve can be modeled by $h = -16t^2 + 30t + 4$, where t is the time (in seconds).

  a. Do both t-intercepts of the graph of the function have meaning in this situation? Explain.

  b. No one receives the serve. After how many seconds does the volleyball hit the ground?

In Exercises 29–36, solve the equation by using Method 2 from Example 3.

29. $x^2 = 10 - 3x$

30. $2x - 3 = x^2$

31. $5x - 7 = x^2$

32. $x^2 = 6x - 5$

33. $x^2 + 12x = -20$

34. $x^2 + 8x = 9$

35. $-x^2 - 5 = -2x$

36. $-x^2 - 4 = -4x$

In Exercises 37–42, find the zero(s) of f. *(See Example 4.)*

37.

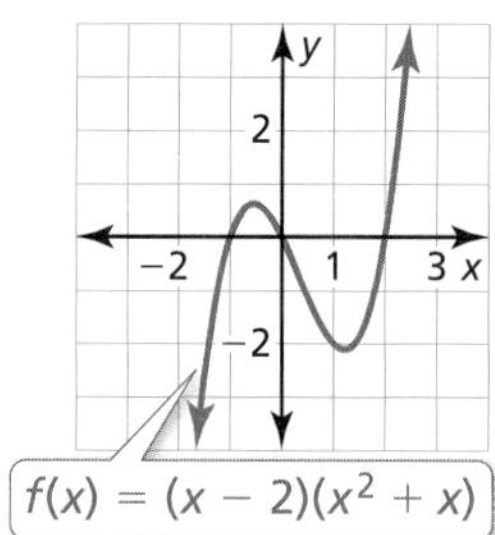

38.

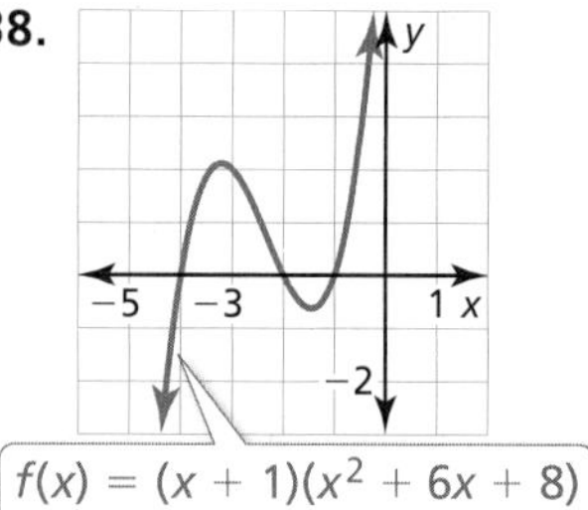

39.

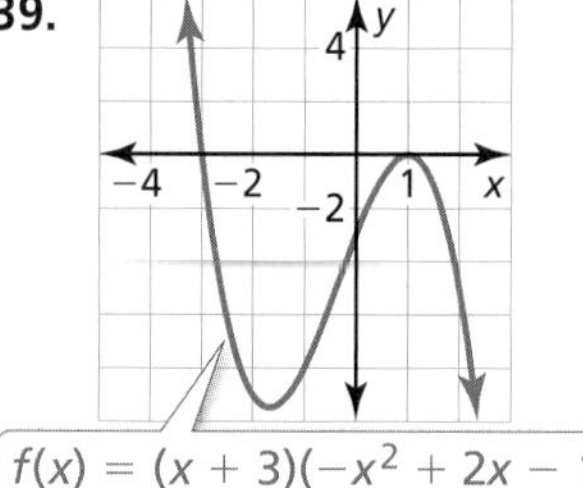

40.

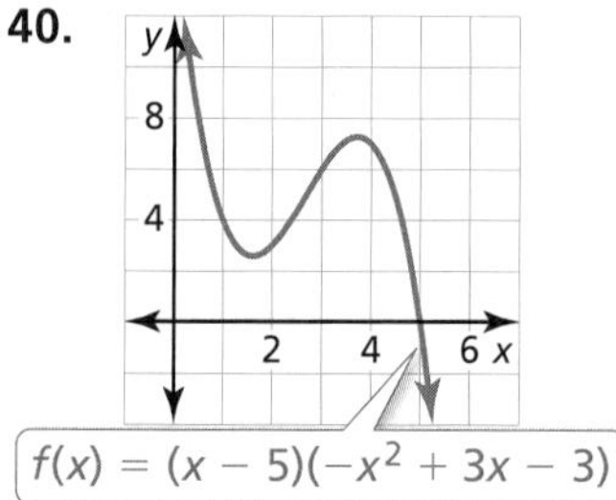

41.

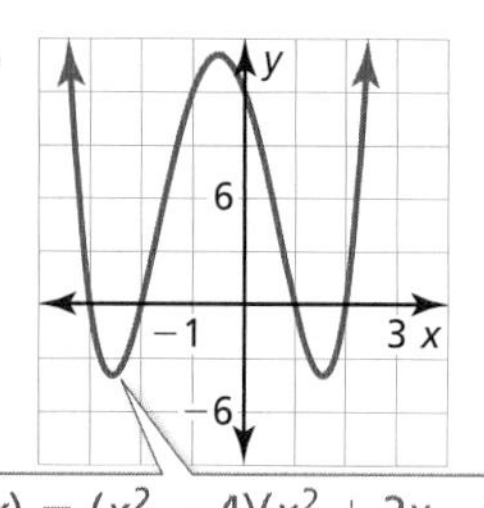

42.

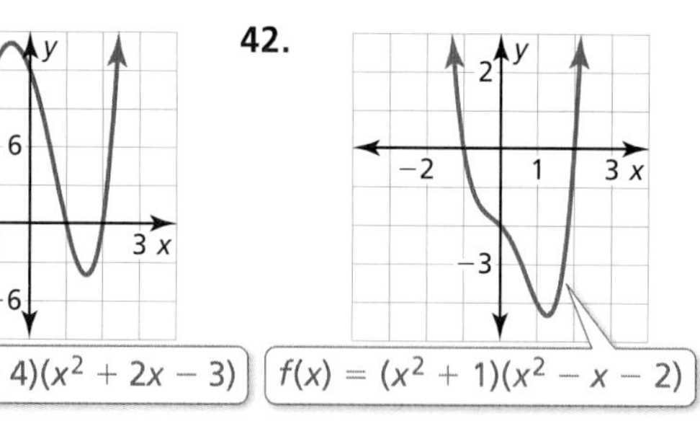

In Exercises 43–46, approximate the zeros of f to the nearest tenth. *(See Example 5.)*

43.

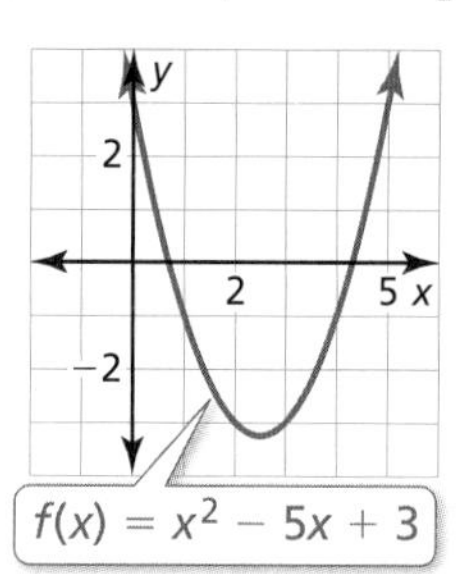

44.

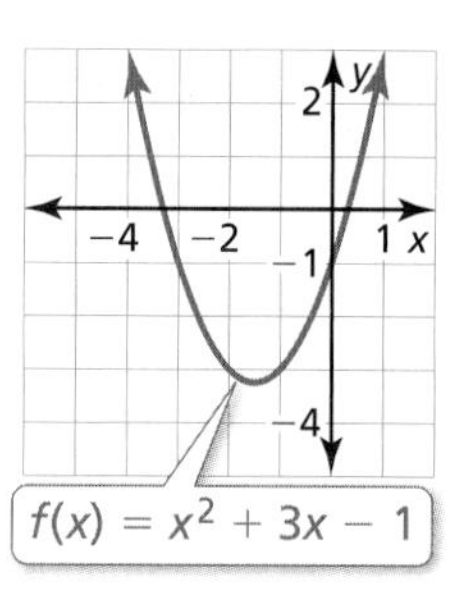

45.

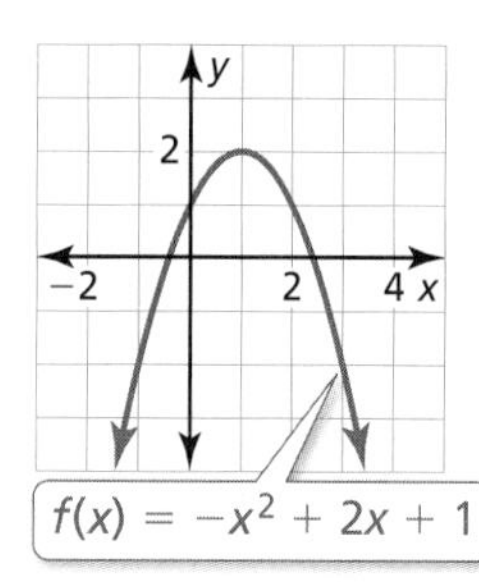

46.

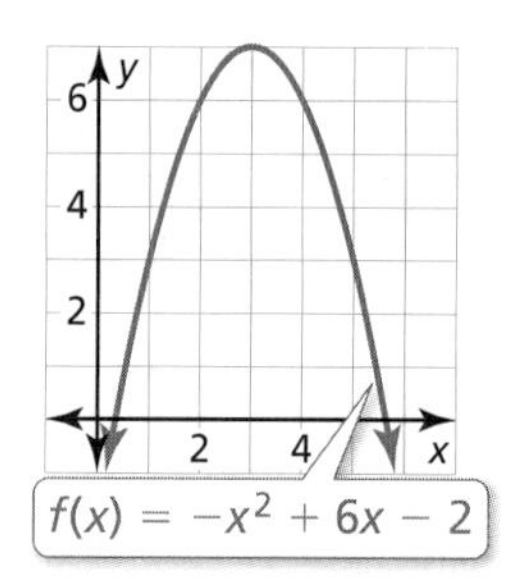

In Exercises 47–52, graph the function. Approximate the zeros of the function to the nearest tenth, if necessary.

47. $f(x) = x^2 + 6x + 1$

48. $f(x) = x^2 - 3x + 2$

49. $y = -x^2 + 4x - 2$

50. $y = -x^2 + 9x - 6$

51. $f(x) = \frac{1}{2}x^2 + 2x - 5$

52. $f(x) = -3x^2 + 4x + 3$

53. **MODELING WITH MATHEMATICS** At a Civil War reenactment, a cannonball is fired into the air with an initial vertical velocity of 128 feet per second. The release point is 6 feet above the ground. The function $h = -16t^2 + 128t + 6$ represents the height h (in feet) of the cannonball after t seconds. *(See Example 6.)*

  a. Find the height of the cannonball each second after it is fired.

  b. Use the results of part (a) to estimate when the height of the cannonball is 150 feet.

  c. Using a graph, after how many seconds is the cannonball 150 feet above the ground?

54. MODELING WITH MATHEMATICS You throw a softball straight up into the air with an initial vertical velocity of 40 feet per second. The release point is 5 feet above the ground. The function $h = -16t^2 + 40t + 5$ represents the height h (in feet) of the softball after t seconds.

a. Find the height of the softball each second after it is released.

b. Use the results of part (a) to estimate when the height of the softball is 15 feet.

c. Using a graph, after how many seconds is the softball 15 feet above the ground?

MATHEMATICAL CONNECTIONS In Exercises 55 and 56, use the given surface area S of the cylinder to find the radius r to the nearest tenth.

55. $S = 225$ ft^2

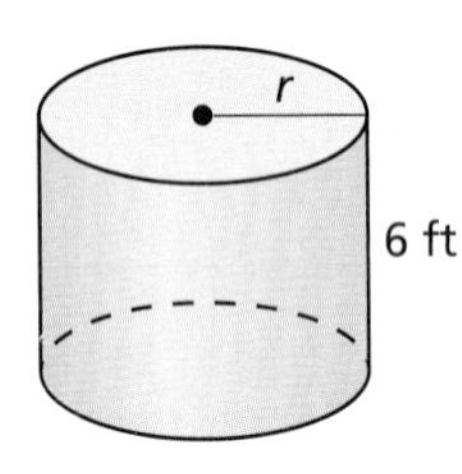

56. $S = 750$ m^2

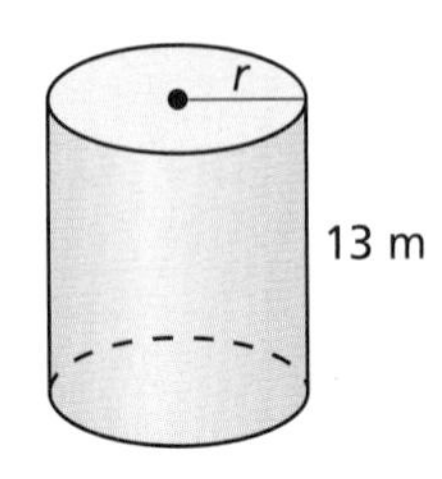

57. WRITING Explain how to approximate zeros of a function when the zeros are not integers.

58. HOW DO YOU SEE IT? Consider the graph shown.

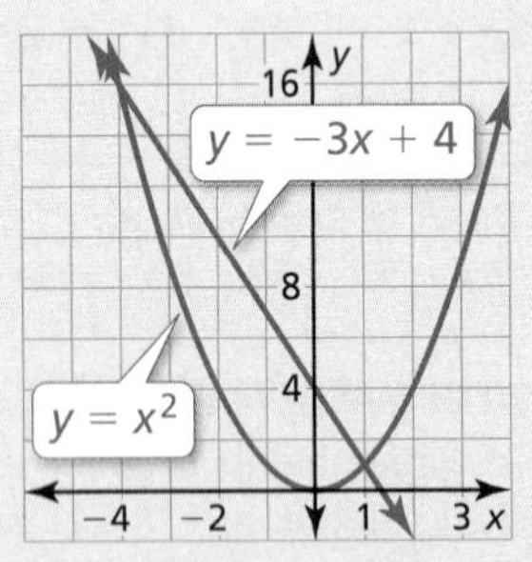

a. How many solutions does the quadratic equation $x^2 = -3x + 4$ have? Explain.

b. Without graphing, describe what you know about the graph of $y = x^2 + 3x - 4$.

59. COMPARING METHODS Example 3 shows two methods for solving a quadratic equation. Which method do you prefer? Explain your reasoning.

60. THOUGHT PROVOKING How many different parabolas have -2 and 2 as x-intercepts? Sketch examples of parabolas that have these two x-intercepts.

61. MODELING WITH MATHEMATICS To keep water off a road, the surface of the road is shaped like a parabola. A cross section of the road is shown in the diagram. The surface of the road can be modeled by $y = -0.0017x^2 + 0.041x$, where x and y are measured in feet. Find the width of the road to the nearest tenth of a foot.

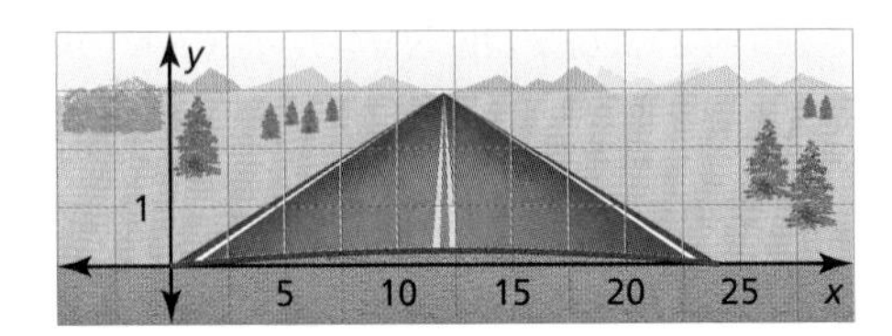

62. MAKING AN ARGUMENT A stream of water from a fire hose can be modeled by $y = -0.003x^2 + 0.58x + 3$, where x and y are measured in feet. A firefighter is standing 57 feet from a building and is holding the hose 3 feet above the ground. The bottom of a window of the building is 26 feet above the ground. Your friend claims the stream of water will pass through the window. Is your friend correct? Explain.

REASONING In Exercises 63–65, determine whether the statement is *always*, *sometimes*, or *never* true. Justify your answer.

63. The graph of $y = ax^2 + c$ has two x-intercepts when a is negative.

64. The graph of $y = ax^2 + c$ has no x-intercepts when a and c have the same sign.

65. The graph of $y = ax^2 + bx + c$ has more than two x-intercepts when $a \neq 0$.

Maintaining Mathematical Proficiency

Reviewing what you learned in previous grades and lessons

Determine whether the table represents an *exponential growth function*, an *exponential decay function*, or *neither*. Explain. *(Section 6.4)*

66.

x	-1	0	1	2
y	18	3	$\frac{1}{2}$	$\frac{1}{12}$

67.

x	0	1	2	3
y	2	8	32	128

9.3 Solving Quadratic Equations Using Square Roots

Essential Question

How can you determine the number of solutions of a quadratic equation of the form $ax^2 + c = 0$?

EXPLORATION 1 The Number of Solutions of $ax^2 + c = 0$

Work with a partner. Solve each equation by graphing. Explain how the number of solutions of $ax^2 + c = 0$ relates to the graph of $y = ax^2 + c$.

a. $x^2 - 4 = 0$

b. $2x^2 + 5 = 0$

c. $x^2 = 0$

d. $x^2 - 5 = 0$

EXPLORATION 2 Estimating Solutions

Work with a partner. Complete each table. Use the completed tables to estimate the solutions of $x^2 - 5 = 0$. Explain your reasoning.

a.

x	$x^2 - 5$
2.21	
2.22	
2.23	
2.24	
2.25	
2.26	

b.

x	$x^2 - 5$
−2.21	
−2.22	
−2.23	
−2.24	
−2.25	
−2.26	

ATTENDING TO PRECISION

To be proficient in math, you need to calculate accurately and express numerical answers with a level of precision appropriate for the problem's context.

EXPLORATION 3 Using Technology to Estimate Solutions

Work with a partner. Two equations are equivalent when they have the same solutions.

a. Are the equations $x^2 - 5 = 0$ and $x^2 = 5$ equivalent? Explain your reasoning.

b. Use the square root key on a calculator to estimate the solutions of $x^2 - 5 = 0$. Describe the accuracy of your estimates in Exploration 2.

c. Write the exact solutions of $x^2 - 5 = 0$.

Communicate Your Answer

4. How can you determine the number of solutions of a quadratic equation of the form $ax^2 + c = 0$?

5. Write the exact solutions of each equation. Then use a calculator to estimate the solutions.

a. $x^2 - 2 = 0$

b. $3x^2 - 18 = 0$

c. $x^2 = 8$

9.3 Lesson

Core Vocabulary

Previous
square root
zero of a function

What You Will Learn

- Solve quadratic equations using square roots.
- Approximate the solutions of quadratic equations.

Solving Quadratic Equations Using Square Roots

Earlier in this chapter, you studied properties of square roots. Now you will use square roots to solve quadratic equations of the form $ax^2 + c = 0$. First isolate x^2 on one side of the equation to obtain $x^2 = d$. Then solve by taking the square root of each side.

Core Concept

Solutions of $x^2 = d$

- When $d > 0$, $x^2 = d$ has two real solutions, $x = \pm\sqrt{d}$.
- When $d = 0$, $x^2 = d$ has one real solution, $x = 0$.
- When $d < 0$, $x^2 = d$ has no real solutions.

ANOTHER WAY

You can also solve $3x^2 - 27 = 0$ by factoring.

$$3(x^2 - 9) = 0$$
$$3(x - 3)(x + 3) = 0$$
$$x = 3 \text{ or } x = -3$$

EXAMPLE 1 Solving Quadratic Equations Using Square Roots

a. Solve $3x^2 - 27 = 0$ using square roots.

$3x^2 - 27 = 0$	Write the equation.
$3x^2 = 27$	Add 27 to each side.
$x^2 = 9$	Divide each side by 3.
$x = \pm\sqrt{9}$	Take the square root of each side.
$x = \pm 3$	Simplify.

▶ The solutions are $x = 3$ and $x = -3$.

b. Solve $x^2 - 10 = -10$ using square roots.

$x^2 - 10 = -10$	Write the equation.
$x^2 = 0$	Add 10 to each side.
$x = 0$	Take the square root of each side.

▶ The only solution is $x = 0$.

c. Solve $-5x^2 + 11 = 16$ using square roots.

$-5x^2 + 11 = 16$	Write the equation.
$-5x^2 = 5$	Subtract 11 from each side.
$x^2 = -1$	Divide each side by -5.

▶ The square of a real number cannot be negative. So, the equation has no real solutions.

STUDY TIP

Each side of the equation $(x-1)^2 = 25$ is a square. So, you can still solve by taking the square root of each side.

EXAMPLE 2 Solving a Quadratic Equation Using Square Roots

Solve $(x-1)^2 = 25$ using square roots.

SOLUTION

$(x-1)^2 = 25$ — Write the equation.

$x - 1 = \pm 5$ — Take the square root of each side.

$x = 1 \pm 5$ — Add 1 to each side.

▶ So, the solutions are $x = 1 + 5 = 6$ and $x = 1 - 5 = -4$.

Check

Use a graphing calculator to check your answer. Rewrite the equation as $(x-1)^2 - 25 = 0$. Graph the related function $f(x) = (x-1)^2 - 25$ and find the zeros of the function. The zeros are -4 and 6.

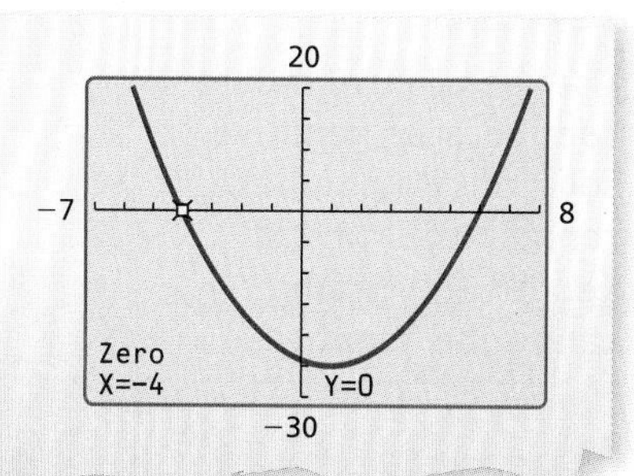

Monitoring Progress  Help in English and Spanish at *BigIdeasMath.com*

Solve the equation using square roots.

1. $-3x^2 = -75$ **2.** $x^2 + 12 = 10$ **3.** $4x^2 - 15 = -15$

4. $(x+7)^2 = 0$ **5.** $4(x-3)^2 = 9$ **6.** $(2x+1)^2 = 36$

Approximating Solutions of Quadratic Equations

EXAMPLE 3 Approximating Solutions of a Quadratic Equation

Solve $4x^2 - 13 = 15$ using square roots. Round the solutions to the nearest hundredth.

SOLUTION

$4x^2 - 13 = 15$ — Write the equation.

$4x^2 = 28$ — Add 13 to each side.

$x^2 = 7$ — Divide each side by 4.

$x = \pm\sqrt{7}$ — Take the square root of each side.

$x \approx \pm 2.65$ — Use a calculator.

▶ The solutions are $x \approx -2.65$ and $x \approx 2.65$.

Check

Graph each side of the equation and find the points of intersection. The x-values of the points of intersection are about -2.65 and 2.65.

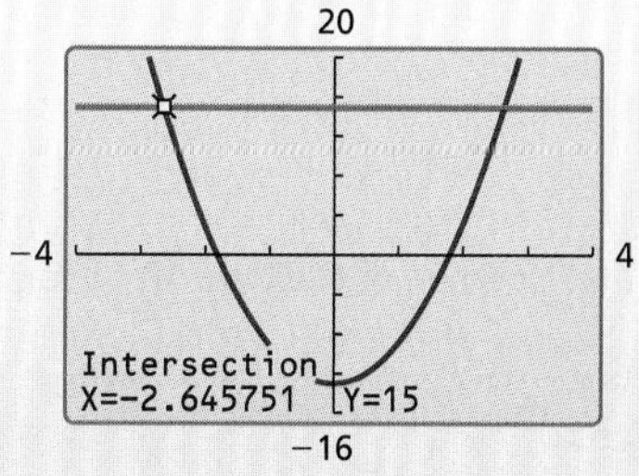

20
−4
4
Intersection
X=-2.645751
Y=15
−16

Monitoring Progress  Help in English and Spanish at *BigIdeasMath.com*

Solve the equation using square roots. Round your solutions to the nearest hundredth.

7. $x^2 + 8 = 19$ **8.** $5x^2 - 2 = 0$ **9.** $3x^2 - 30 = 4$

EXAMPLE 4 Solving a Real-Life Problem

A touch tank has a height of 3 feet. Its length is three times its width. The volume of the tank is 270 cubic feet. Find the length and width of the tank.

SOLUTION

The length ℓ is three times the width w, so $\ell = 3w$. Write an equation using the formula for the volume of a rectangular prism.

INTERPRETING MATHEMATICAL RESULTS

Use the positive square root because negative solutions do not make sense in this context. Length and width cannot be negative.

$V = \ell wh$	Write the formula.
$270 = 3w(w)(3)$	Substitute 270 for V, $3w$ for ℓ, and 3 for h.
$270 = 9w^2$	Multiply.
$30 = w^2$	Divide each side by 9.
$\pm\sqrt{30} = w$	Take the square root of each side.

The solutions are $\sqrt{30}$ and $-\sqrt{30}$. Use the positive solution.

▶ So, the width is $\sqrt{30} \approx 5.5$ feet and the length is $3\sqrt{30} \approx 16.4$ feet.

EXAMPLE 5 Rearranging and Evaluating a Formula

The area A of an equilateral triangle with side length s is given by the formula $A = \frac{\sqrt{3}}{4}s^2$. Solve the formula for s. Then approximate the side length of the traffic sign that has an area of 390 square inches.

SOLUTION

ANOTHER WAY

Notice that you can rewrite the formula as $s = \frac{2}{3^{1/4}}\sqrt{A}$, or $s \approx 1.52\sqrt{A}$. This can help you efficiently find the value of s for various values of A.

Step 1 Solve the formula for s.

$A = \frac{\sqrt{3}}{4}s^2$	Write the formula.
$\frac{4A}{\sqrt{3}} = s^2$	Multiply each side by $\frac{4}{\sqrt{3}}$.
$\sqrt{\frac{4A}{\sqrt{3}}} = s$	Take the positive square root of each side.

Step 2 Substitute 390 for A in the new formula and evaluate.

$$s = \sqrt{\frac{4A}{\sqrt{3}}} = \sqrt{\frac{4(390)}{\sqrt{3}}} = \sqrt{\frac{1560}{\sqrt{3}}} \approx 30$$ Use a calculator.

▶ The side length of the traffic sign is about 30 inches.

Monitoring Progress  Help in English and Spanish at *BigIdeasMath.com*

10. WHAT IF? In Example 4, the volume of the tank is 315 cubic feet. Find the length and width of the tank.

11. The surface area S of a sphere with radius r is given by the formula $S = 4\pi r^2$. Solve the formula for r. Then find the radius of a globe with a surface area of 804 square inches.

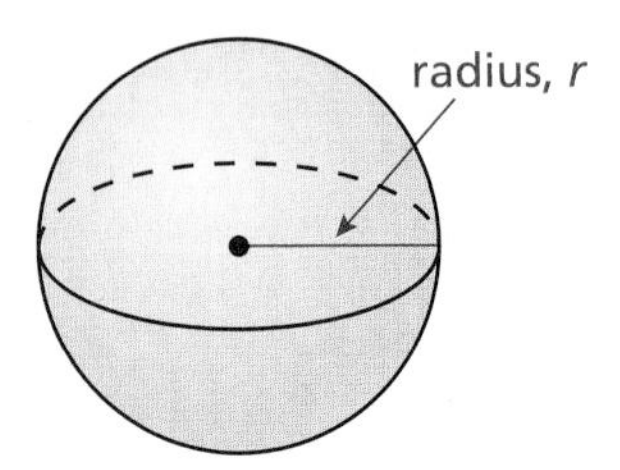

9.3 Exercises

Dynamic Solutions available at *BigIdeasMath.com*

Vocabulary and Core Concept Check

1. **COMPLETE THE SENTENCE** The equation $x^2 = d$ has ____ real solutions when $d > 0$.

2. **DIFFERENT WORDS, SAME QUESTION** Which is different? Find "both" answers.

Solve $x^2 = 144$ using square roots.	Solve $x^2 - 144 = 0$ using square roots.
Solve $x^2 + 146 = 2$ using square roots.	Solve $x^2 + 2 = 146$ using square roots.

Monitoring Progress and Modeling with Mathematics

In Exercises 3–8, determine the number of real solutions of the equation. Then solve the equation using square roots.

3. $x^2 = 25$
4. $x^2 = -36$
5. $x^2 = -21$
6. $x^2 = 400$
7. $x^2 = 0$
8. $x^2 = 169$

In Exercises 9–18, solve the equation using square roots. *(See Example 1.)*

9. $x^2 - 16 = 0$
10. $x^2 + 6 = 0$
11. $3x^2 + 12 = 0$
12. $x^2 - 55 = 26$
13. $2x^2 - 98 = 0$
14. $-x^2 + 9 = 9$
15. $-3x^2 - 5 = -5$
16. $4x^2 - 371 = 29$
17. $4x^2 + 10 = 11$
18. $9x^2 - 35 = 14$

In Exercises 19–24, solve the equation using square roots. *(See Example 2.)*

19. $(x + 3)^2 = 0$
20. $(x - 1)^2 = 4$
21. $(2x - 1)^2 = 81$
22. $(4x + 5)^2 = 9$
23. $9(x + 1)^2 = 16$
24. $4(x - 2)^2 = 25$

In Exercises 25–30, solve the equation using square roots. Round your solutions to the nearest hundredth. *(See Example 3.)*

25. $x^2 + 6 = 13$
26. $x^2 + 11 = 24$
27. $2x^2 - 9 = 11$
28. $5x^2 + 2 = 6$
29. $-21 = 15 - 2x^2$
30. $2 = 4x^2 - 5$

31. **ERROR ANALYSIS** Describe and correct the error in solving the equation $2x^2 - 33 = 39$ using square roots.

✗

$$2x^2 - 33 = 39$$
$$2x^2 = 72$$
$$x^2 = 36$$
$$x = 6$$

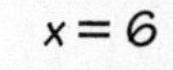

The solution is $x = 6$.

32. **MODELING WITH MATHEMATICS** An in-ground pond has the shape of a rectangular prism. The pond has a depth of 24 inches and a volume of 72,000 cubic inches. The length of the pond is two times its width. Find the length and width of the pond. *(See Example 4.)*

33. **MODELING WITH MATHEMATICS** A person sitting in the top row of the bleachers at a sporting event drops a pair of sunglasses from a height of 24 feet. The function $h = -16x^2 + 24$ represents the height h (in feet) of the sunglasses after x seconds. How long does it take the sunglasses to hit the ground?

34. **MAKING AN ARGUMENT** Your friend says that the solution of the equation $x^2 + 4 = 0$ is $x = 0$. Your cousin says that the equation has no real solutions. Who is correct? Explain your reasoning.

35. **MODELING WITH MATHEMATICS** The design of a square rug for your living room is shown. You want the area of the inner square to be 25% of the total area of the rug. Find the side length x of the inner square.

36. **MATHEMATICAL CONNECTIONS** The area A of a circle with radius r is given by the formula $A = \pi r^2$. *(See Example 5.)*

   a. Solve the formula for r.

   b. Use the formula from part (a) to find the radius of each circle.

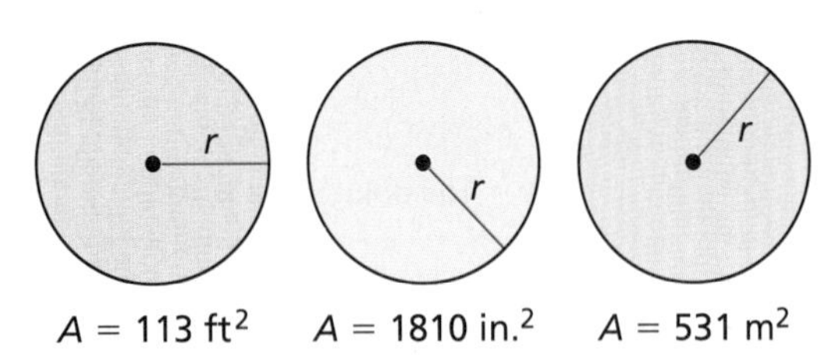

   c. Explain why it is beneficial to solve the formula for r before finding the radius.

37. **WRITING** How can you approximate the roots of a quadratic equation when the roots are not integers?

38. **WRITING** Given the equation $ax^2 + c = 0$, describe the values of a and c so the equation has the following number of solutions.

   a. two real solutions

   b. one real solution

   c. no real solutions

39. **REASONING** Without graphing, where do the graphs of $y = x^2$ and $y = 9$ intersect? Explain.

40. **HOW DO YOU SEE IT?** The graph represents the function $f(x) = (x - 1)^2$. How many solutions does the equation $(x - 1)^2 = 0$ have? Explain.

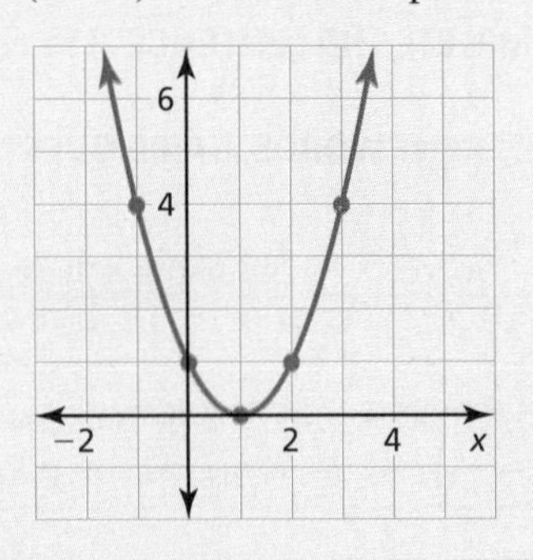

41. **REASONING** Solve $x^2 = 1.44$ without using a calculator. Explain your reasoning.

42. **THOUGHT PROVOKING** The quadratic equation

$$ax^2 + bx + c = 0$$

can be rewritten in the following form.

$$\left(x + \frac{b}{2a}\right)^2 = \frac{b^2 - 4ac}{4a^2}$$

Use this form to write the solutions of the equation.

43. **REASONING** An equation of the graph shown is $y = \frac{1}{2}(x - 2)^2 + 1$. Two points on the parabola have y-coordinates of 9. Find the x-coordinates of these points.

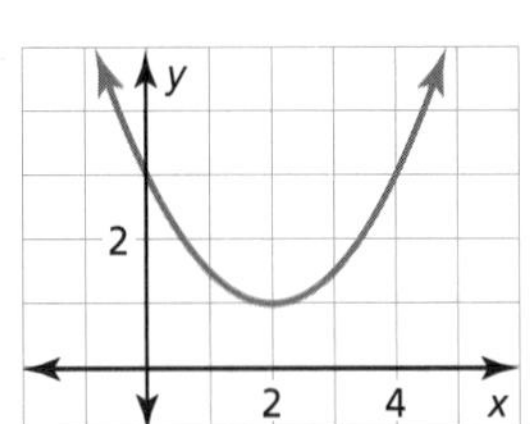

44. **CRITICAL THINKING** Solve each equation without graphing.

   a. $x^2 - 12x + 36 = 64$

   b. $x^2 + 14x + 49 = 16$

Maintaining Mathematical Proficiency

Reviewing what you learned in previous grades and lessons

Factor the polynomial. *(Section 7.7)*

45. $x^2 + 8x + 16$

46. $x^2 - 4x + 4$

47. $x^2 - 14x + 49$

48. $x^2 + 18x + 81$

49. $x^2 + 12x + 36$

50. $x^2 - 22x + 121$

9.1–9.3 What Did You Learn?

Core Vocabulary

counterexample, *p. 479*
radical expression, *p. 480*
simplest form of a radical, *p. 480*
rationalizing the denominator, *p. 482*
conjugates, *p. 482*
like radicals, *p. 484*
quadratic equation, *p. 490*

Core Concepts

Section 9.1

Product Property of Square Roots, *p. 480*
Quotient Property of Square Roots, *p. 480*
Rationalizing the Denominator, *p. 482*
Performing Operations with Radicals, *p. 484*

Section 9.2

Solving Quadratic Equations by Graphing, *p. 490*
Number of Solutions of a Quadratic Equation, *p. 491*
Finding Zeros of Functions, *p. 492*

Section 9.3

Solutions of $x^2 = d$, *p. 498*
Approximating Solutions of Quadratic Equations, *p. 499*

Mathematical Practices

1. For each part of Exercise 100 on page 488 that is *sometimes* true, list all examples and counterexamples from the table that represent the sum or product being described.
2. Which Examples can you use to help you solve Exercise 54 on page 496?
3. Describe how solving a simpler equation can help you solve the equation in Exercise 41 on page 502.

Study Skills

Keeping a Positive Attitude

Do you ever feel frustrated or overwhelmed by math? You're not alone. Just take a deep breath and assess the situation. Try to find a productive study environment, review your notes and the examples in the textbook, and ask your teacher or friends for help.

9.1–9.3 Quiz

Simplify the expression. *(Section 9.1)*

1. $\sqrt{112x^3}$

2. $\sqrt{\frac{18}{81}}$

3. $\sqrt[3]{-625}$

4. $\frac{12}{\sqrt{32}}$

5. $\frac{4}{\sqrt{11}}$

6. $\sqrt{\frac{144}{13}}$

7. $\sqrt[3]{\frac{54x^4}{343y^6}}$

8. $\sqrt{\frac{4x^2}{28y^4z^5}}$

9. $\frac{6}{5 + \sqrt{3}}$

10. $2\sqrt{5} + 7\sqrt{10} - 3\sqrt{20}$

11. $\frac{10}{\sqrt{8} - \sqrt{10}}$

12. $\sqrt{6}(7\sqrt{12} - 4\sqrt{3})$

Use the graph to solve the equation. *(Section 9.2)*

13. $x^2 - 2x - 3 = 0$

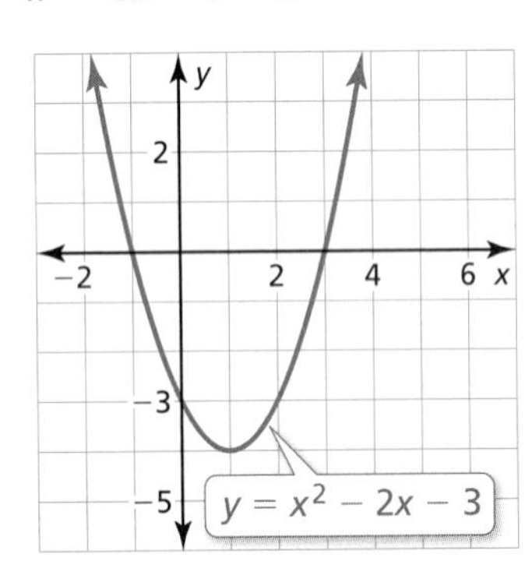

14. $x^2 - 2x + 3 = 0$

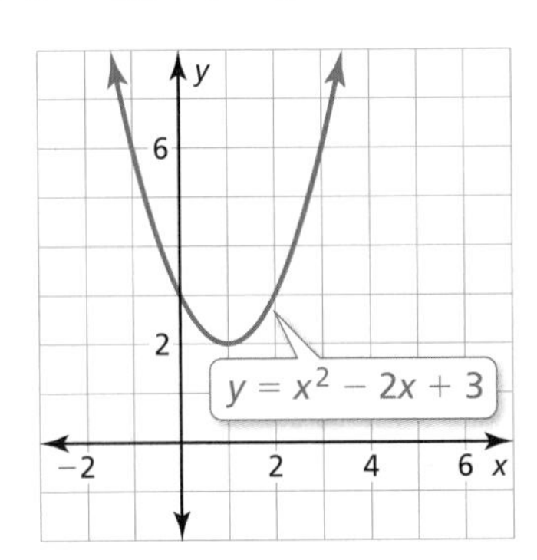

15. $x^2 + 10x + 25 = 0$

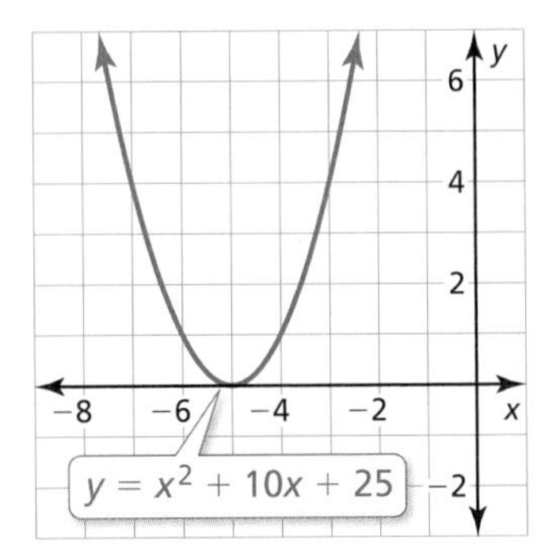

Solve the equation by graphing. *(Section 9.2)*

16. $x^2 + 9x + 14 = 0$

17. $x^2 - 7x = 8$

18. $x + 4 = -x^2$

Solve the equation using square roots. *(Section 9.3)*

19. $4x^2 = 64$

20. $-3x^2 + 6 = 10$

21. $(x - 8)^2 = 1$

22. Explain how to determine the number of real solutions of $x^2 = 100$ without solving. *(Section 9.3)*

23. The length of a rectangular prism is four times its width. The volume of the prism is 380 cubic meters. Find the length and width of the prism. *(Section 9.3)*

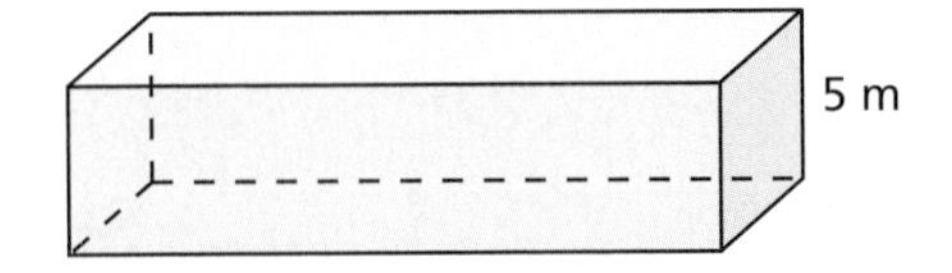

24. You cast a fishing lure into the water from a height of 4 feet above the water. The height h (in feet) of the fishing lure after t seconds can be modeled by the equation $h = -16t^2 + 24t + 4$. *(Section 9.2)*

a. After how many seconds does the fishing lure reach a height of 12 feet?

b. After how many seconds does the fishing lure hit the water?

9.4 Solving Quadratic Equations by Completing the Square

Essential Question How can you use "completing the square" to solve a quadratic equation?

EXPLORATION 1 Solving by Completing the Square

Work with a partner.

a. Write the equation modeled by the algebra tiles. This is the equation to be solved.

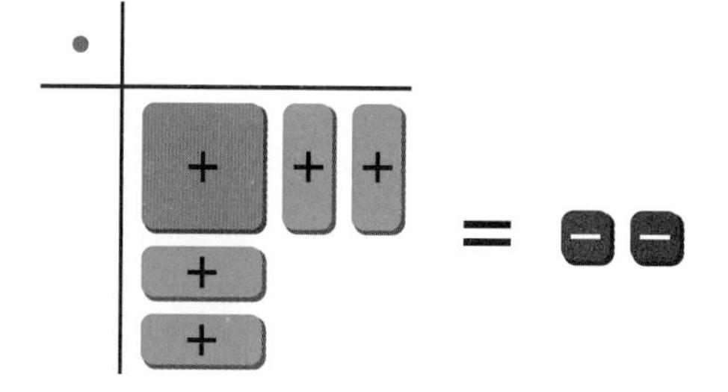

b. Four algebra tiles are added to the left side to "complete the square." Why are four algebra tiles also added to the right side?

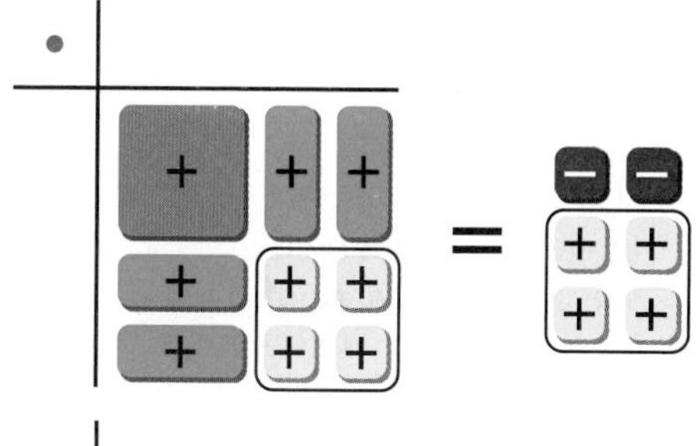

c. Use algebra tiles to label the dimensions of the square on the left side and simplify on the right side.

d. Write the equation modeled by the algebra tiles so that the left side is the square of a binomial. Solve the equation using square roots.

MAKING SENSE OF PROBLEMS

To be proficient in math, you need to explain to yourself the meaning of a problem. After that, you need to look for entry points to its solution.

EXPLORATION 2 Solving by Completing the Square

Work with a partner.

a. Write the equation modeled by the algebra tiles.

b. Use algebra tiles to "complete the square."

c. Write the solutions of the equation.

d. Check each solution in the original equation.

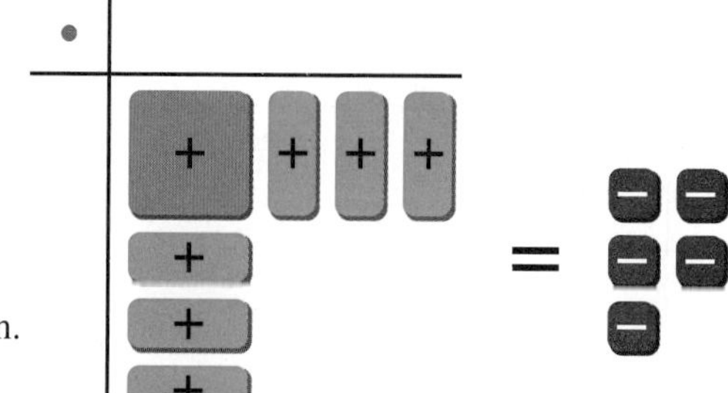

Communicate Your Answer

3. How can you use "completing the square" to solve a quadratic equation?

4. Solve each quadratic equation by completing the square.

a. $x^2 - 2x = 1$ **b.** $x^2 - 4x = -1$ **c.** $x^2 + 4x = -3$

9.4 Lesson

Core Vocabulary

completing the square, *p. 506*

Previous
perfect square trinomial
coefficient
maximum value
minimum value
vertex form of a quadratic function

What You Will Learn

- Complete the square for expressions of the form $x^2 + bx$.
- Solve quadratic equations by completing the square.
- Find and use maximum and minimum values.
- Solve real-life problems by completing the square.

Completing the Square

For an expression of the form $x^2 + bx$, you can add a constant c to the expression so that $x^2 + bx + c$ is a perfect square trinomial. This process is called **completing the square**.

Core Concept

Completing the Square

Words To complete the square for an expression of the form $x^2 + bx$, follow these steps.

Step 1 Find one-half of b, the coefficient of x.

Step 2 Square the result from Step 1.

Step 3 Add the result from Step 2 to $x^2 + bx$.

Factor the resulting expression as the square of a binomial.

Algebra $x^2 + bx + \left(\frac{b}{2}\right)^2 = \left(x + \frac{b}{2}\right)^2$

JUSTIFYING STEPS

In each diagram below, the combined area of the shaded regions is $x^2 + bx$. Adding $\left(\frac{b}{2}\right)^2$ completes the square in the second diagram.

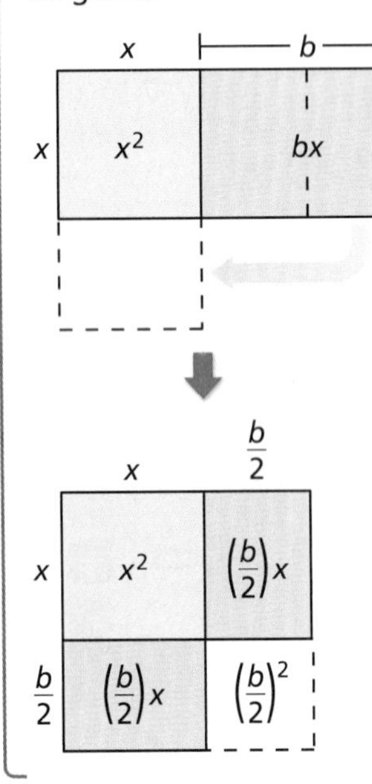

EXAMPLE 1 Completing the Square

Complete the square for each expression. Then factor the trinomial.

a. $x^2 + 6x$ **b.** $x^2 - 9x$

SOLUTION

a. Step 1 Find one-half of b. $\frac{b}{2} = \frac{6}{2} = 3$

Step 2 Square the result from Step 1. $3^2 = 9$

Step 3 Add the result from Step 2 to $x^2 + bx$. $x^2 + 6x + 9$

▶ $x^2 + 6x + 9 = (x + 3)^2$

b. Step 1 Find one-half of b. $\frac{b}{2} = \frac{-9}{2}$

Step 2 Square the result from Step 1. $\left(\frac{-9}{2}\right)^2 = \frac{81}{4}$

Step 3 Add the result from Step 2 to $x^2 + bx$. $x^2 - 9x + \frac{81}{4}$

▶ $x^2 - 9x + \frac{81}{4} = \left(x - \frac{9}{2}\right)^2$

Monitoring Progress  Help in English and Spanish at *BigIdeasMath.com*

Complete the square for the expression. Then factor the trinomial.

1. $x^2 + 10x$ **2.** $x^2 - 4x$ **3.** $x^2 + 7x$

Solving Quadratic Equations by Completing the Square

The method of completing the square can be used to solve any quadratic equation. To solve a quadratic equation by completing the square, you must write the equation in the form $x^2 + bx = d$.

COMMON ERROR

When completing the square to solve an equation, be sure to add $\left(\frac{b}{2}\right)^2$ to each side of the equation.

EXAMPLE 2 Solving a Quadratic Equation: $x^2 + bx = d$

Solve $x^2 - 16x = -15$ by completing the square.

SOLUTION

$x^2 - 16x = -15$	Write the equation.
$x^2 - 16x + (-8)^2 = -15 + (-8)^2$	Complete the square by adding $\left(\frac{-16}{2}\right)^2$, or $(-8)^2$, to each side.
$(x - 8)^2 = 49$	Write the left side as the square of a binomial.
$x - 8 = \pm 7$	Take the square root of each side.
$x = 8 \pm 7$	Add 8 to each side.

▶ The solutions are $x = 8 + 7 = 15$ and $x = 8 - 7 = 1$.

Check

$x^2 - 16x = -15$	Original equation	$x^2 - 16x = -15$
$15^2 - 16(15) \stackrel{?}{=} -15$	Substitute.	$1^2 - 16(1) \stackrel{?}{=} -15$
$-15 = -15$ ✓	Simplify.	$-15 = -15$ ✓

COMMON ERROR

Before you complete the square, be sure that the coefficient of the x^2-term is 1.

EXAMPLE 3 Solving a Quadratic Equation: $ax^2 + bx + c = 0$

Solve $2x^2 + 20x - 8 = 0$ by completing the square.

SOLUTION

$2x^2 + 20x - 8 = 0$	Write the equation.
$2x^2 + 20x = 8$	Add 8 to each side.
$x^2 + 10x = 4$	Divide each side by 2.
$x^2 + 10x + 5^2 = 4 + 5^2$	Complete the square by adding $\left(\frac{10}{2}\right)^2$, or 5^2, to each side.
$(x + 5)^2 = 29$	Write the left side as the square of a binomial.
$x + 5 = \pm\sqrt{29}$	Take the square root of each side.
$x = -5 \pm \sqrt{29}$	Subtract 5 from each side.

▶ The solutions are $x = -5 + \sqrt{29} \approx 0.39$ and $x = -5 - \sqrt{29} \approx -10.39$.

Monitoring Progress  Help in English and Spanish at *BigIdeasMath.com*

Solve the equation by completing the square. Round your solutions to the nearest hundredth, if necessary.

4. $x^2 - 2x = 3$ **5.** $m^2 + 12m = -8$ **6.** $3g^2 - 24g + 27 = 0$

Finding and Using Maximum and Minimum Values

One way to find the maximum or minimum value of a quadratic function is to write the function in vertex form by completing the square. Recall that the vertex form of a quadratic function is $y = a(x - h)^2 + k$, where $a \neq 0$. The vertex of the graph is (h, k).

EXAMPLE 4 Finding a Minimum Value

Find the minimum value of $y = x^2 + 4x - 1$.

SOLUTION

Write the function in vertex form.

$y = x^2 + 4x - 1$	Write the function.
$y + 1 = x^2 + 4x$	Add 1 to each side.
$y + 1 + 4 = x^2 + 4x + 4$	Complete the square for $x^2 + 4x$.
$y + 5 = x^2 + 4x + 4$	Simplify the left side.
$y + 5 = (x + 2)^2$	Write the right side as the square of a binomial.
$y = (x + 2)^2 - 5$	Write in vertex form.

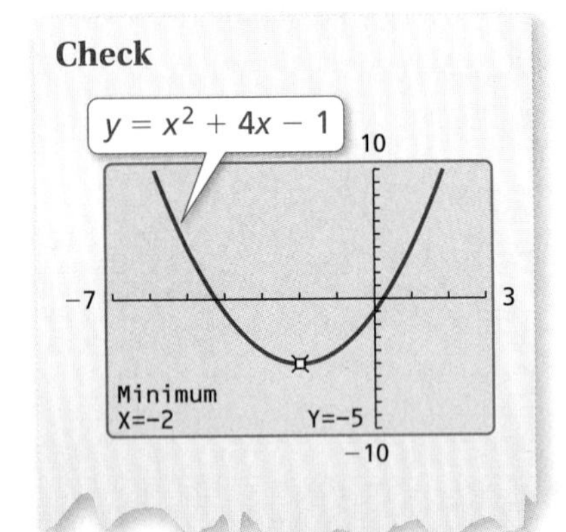

The vertex is $(-2, -5)$. Because a is positive ($a = 1$), the parabola opens up and the y-coordinate of the vertex is the minimum value.

▶ So, the function has a minimum value of -5.

EXAMPLE 5 Finding a Maximum Value

Find the maximum value of $y = -x^2 + 2x + 7$.

SOLUTION

Write the function in vertex form.

STUDY TIP

Adding 1 inside the parentheses results in subtracting 1 from the right side of the equation.

$y = -x^2 + 2x + 7$	Write the function.
$y - 7 = -x^2 + 2x$	Subtract 7 from each side.
$y - 7 = -(x^2 - 2x)$	Factor out −1.
$y - 7 - 1 = -(x^2 - 2x + 1)$	Complete the square for $x^2 - 2x$.
$y - 8 = -(x^2 - 2x + 1)$	Simplify the left side.
$y - 8 = -(x - 1)^2$	Write $x^2 - 2x + 1$ as the square of a binomial.
$y = -(x - 1)^2 + 8$	Write in vertex form.

The vertex is $(1, 8)$. Because a is negative ($a = -1$), the parabola opens down and the y-coordinate of the vertex is the maximum value.

▶ So, the function has a maximum value of 8.

Monitoring Progress  Help in English and Spanish at *BigIdeasMath.com*

Determine whether the quadratic function has a maximum or minimum value. Then find the value.

7. $y = -x^2 - 4x + 4$ **8.** $y = x^2 + 12x + 40$ **9.** $y = x^2 - 2x - 2$

EXAMPLE 6 Interpreting Forms of Quadratic Functions

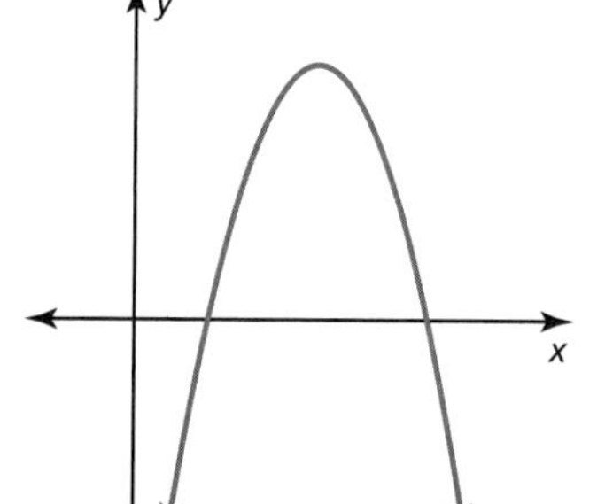

$f(x) = -\frac{1}{2}(x + 4)^2 + 8$

$g(x) = -(x - 5)^2 + 9$

$m(x) = (x - 3)(x - 12)$

$p(x) = -(x - 2)(x - 8)$

Which of the functions could be represented by the graph? Explain.

SOLUTION

You do not know the scale of either axis. To eliminate functions, consider the characteristics of the graph and information provided by the form of each function. The graph appears to be a parabola that opens down, which means the function has a maximum value. The vertex of the graph is in the first quadrant. Both x-intercepts are positive.

- The graph of f opens down because $a < 0$, which means f has a maximum value. However, the vertex $(-4, 8)$ of the graph of f is in the second quadrant. So, the graph does not represent f.
- The graph of g opens down because $a < 0$, which means g has a maximum value. The vertex $(5, 9)$ of the graph of g is in the first quadrant. By solving $0 = -(x - 5)^2 + 9$, you see that the x-intercepts of the graph of g are 2 and 8. So, the graph could represent g.
- The graph of m has two positive x-intercepts. However, its graph opens up because $a > 0$, which means m has a minimum value. So, the graph does not represent m.
- The graph of p has two positive x-intercepts, and its graph opens down because $a < 0$. This means that p has a maximum value and the vertex must be in the first quadrant. So, the graph could represent p.

▶ The graph could represent function g or function p.

EXAMPLE 7 Real-Life Application

The function $y = -16x^2 + 96x$ represents the height y (in feet) of a model rocket x seconds after it is launched. (a) Find the maximum height of the rocket. (b) Find and interpret the axis of symmetry.

SOLUTION

a. To find the maximum height, identify the maximum value of the function.

$$y = -16x^2 + 96x \qquad \text{Write the function.}$$

$$y = -16(x^2 - 6x) \qquad \text{Factor out } -16.$$

$$y - 144 = -16(x^2 - 6x + 9) \qquad \text{Complete the square for } x^2 - 6x.$$

$$y = -16(x - 3)^2 + 144 \qquad \text{Write in vertex form.}$$

STUDY TIP

Adding 9 inside the parentheses results in subtracting 144 from the right side of the equation.

▶ Because the maximum value is 144, the model rocket reaches a maximum height of 144 feet.

b. The vertex is $(3, 144)$. So, the axis of symmetry is $x = 3$. On the left side of $x = 3$, the height increases as time increases. On the right side of $x = 3$, the height decreases as time increases.

Monitoring Progress  Help in English and Spanish at *BigIdeasMath.com*

Determine whether the function could be represented by the graph in Example 6. Explain.

10. $h(x) = (x - 8)^2 + 10$

11. $n(x) = -2(x - 5)(x - 20)$

12. WHAT IF? Repeat Example 7 when the function is $y = -16x^2 + 128x$.

Solving Real-Life Problems

EXAMPLE 8 Modeling with Mathematics

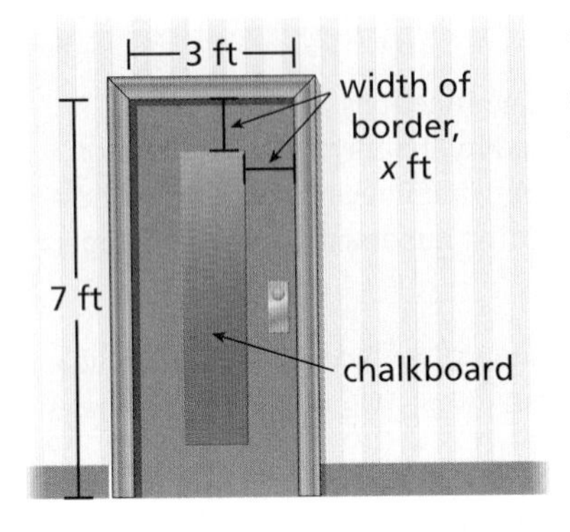

You decide to use chalkboard paint to create a chalkboard on a door. You want the chalkboard to cover 6 square feet and to have a uniform border, as shown. Find the width of the border to the nearest inch.

SOLUTION

1. **Understand the Problem** You know the dimensions (in feet) of the door from the diagram. You also know the area (in square feet) of the chalkboard and that it will have a uniform border. You are asked to find the width of the border to the nearest inch.

2. **Make a Plan** Use a verbal model to write an equation that represents the area of the chalkboard. Then solve the equation.

3. **Solve the Problem**

   Let x be the width (in feet) of the border, as shown in the diagram.

Area of chalkboard (square feet)	=	Length of chalkboard (feet)	•	Width of chalkboard (feet)
6	=	$(7 - 2x)$	•	$(3 - 2x)$

$6 = (7 - 2x)(3 - 2x)$ Write the equation.

$6 = 21 - 20x + 4x^2$ Multiply the binomials.

$-15 = 4x^2 - 20x$ Subtract 21 from each side.

$-\frac{15}{4} = x^2 - 5x$ Divide each side by 4.

$-\frac{15}{4} + \frac{25}{4} = x^2 - 5x + \frac{25}{4}$ Complete the square for $x^2 - 5x$.

$\frac{5}{2} = x^2 - 5x + \frac{25}{4}$ Simplify the left side.

$\frac{5}{2} = \left(x - \frac{5}{2}\right)^2$ Write the right side as the square of a binomial.

$\pm\sqrt{\frac{5}{2}} = x - \frac{5}{2}$ Take the square root of each side.

$\frac{5}{2} \pm \sqrt{\frac{5}{2}} = x$ Add $\frac{5}{2}$ to each side.

The solutions of the equation are $x = \frac{5}{2} + \sqrt{\frac{5}{2}} \approx 4.08$ and $x = \frac{5}{2} - \sqrt{\frac{5}{2}} \approx 0.92$.

It is not possible for the width of the border to be 4.08 feet because the width of the door is 3 feet. So, the width of the border is about 0.92 foot.

$0.92 \text{ ft} \cdot \frac{12 \text{ in.}}{1 \text{ ft}} = 11.04 \text{ in.}$ Convert 0.92 foot to inches.

▶ The width of the border should be about 11 inches.

4. **Look Back** When the width of the border is slightly less than 1 foot, the length of the chalkboard is slightly more than 5 feet and the width of the chalkboard is slightly more than 1 foot. Multiplying these dimensions gives an area close to 6 square feet. So, an 11-inch border is reasonable.

Monitoring Progress  Help in English and Spanish at *BigIdeasMath.com*

13. **WHAT IF?** You want the chalkboard to cover 4 square feet. Find the width of the border to the nearest inch.

9.4 Exercises

Dynamic Solutions available at *BigIdeasMath.com*

Vocabulary and Core Concept Check

1. **COMPLETE THE SENTENCE** The process of adding a constant c to the expression $x^2 + bx$ so that $x^2 + bx + c$ is a perfect square trinomial is called ______________.

2. **VOCABULARY** Explain how to complete the square for an expression of the form $x^2 + bx$.

3. **WRITING** Is it more convenient to complete the square for $x^2 + bx$ when b is odd or when b is even? Explain.

4. **WRITING** Describe how you can use the process of completing the square to find the maximum or minimum value of a quadratic function.

Monitoring Progress and Modeling with Mathematics

In Exercises 5–10, find the value of c that completes the square.

5. $x^2 - 8x + c$
6. $x^2 - 2x + c$
7. $x^2 + 4x + c$
8. $x^2 + 12x + c$
9. $x^2 - 15x + c$
10. $x^2 + 9x + c$

In Exercises 11–16, complete the square for the expression. Then factor the trinomial. *(See Example 1.)*

11. $x^2 - 10x$
12. $x^2 - 40x$
13. $x^2 + 16x$
14. $x^2 + 22x$
15. $x^2 + 5x$
16. $x^2 - 3x$

In Exercises 17–22, solve the equation by completing the square. Round your solutions to the nearest hundredth, if necessary. *(See Example 2.)*

17. $x^2 + 14x = 15$
18. $x^2 - 6x = 16$
19. $x^2 - 4x = -2$
20. $x^2 + 2x = 5$
21. $x^2 - 5x = 8$
22. $x^2 + 11x = -10$

23. **MODELING WITH MATHEMATICS** The area of the patio is 216 square feet.

    a. Write an equation that represents the area of the patio.

    b. Find the dimensions of the patio by completing the square.

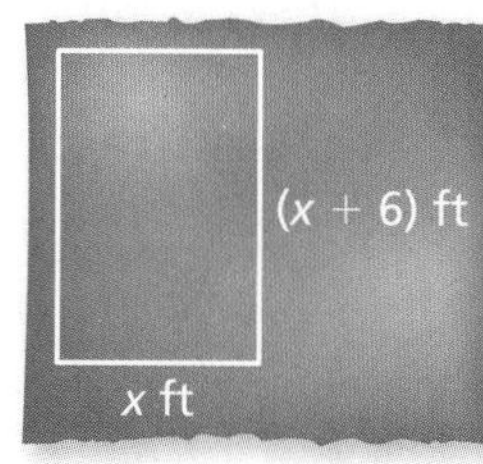

24. **MODELING WITH MATHEMATICS** Some sand art contains sand and water sealed in a glass case, similar to the one shown. When the art is turned upside down, the sand and water fall to create a new picture. The glass case has a depth of 1 centimeter and a volume of 768 cubic centimeters.

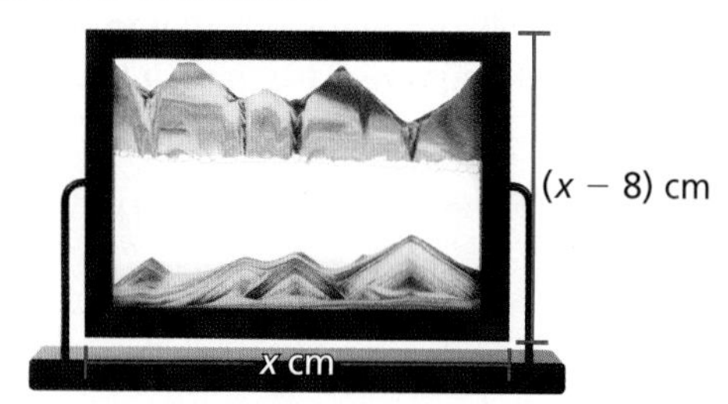

    a. Write an equation that represents the volume of the glass case.

    b. Find the dimensions of the glass case by completing the square.

In Exercises 25–32, solve the equation by completing the square. Round your solutions to the nearest hundredth, if necessary. *(See Example 3.)*

25. $x^2 - 8x + 15 = 0$
26. $x^2 + 4x - 21 = 0$
27. $2x^2 + 20x + 44 = 0$
28. $3x^2 - 18x + 12 = 0$
29. $-3x^2 - 24x + 17 = -40$
30. $-5x^2 - 20x + 35 = 30$
31. $2x^2 - 14x + 10 = 26$
32. $4x^2 + 12x - 15 = 5$

33. ERROR ANALYSIS Describe and correct the error in solving $x^2 + 8x = 10$ by completing the square.

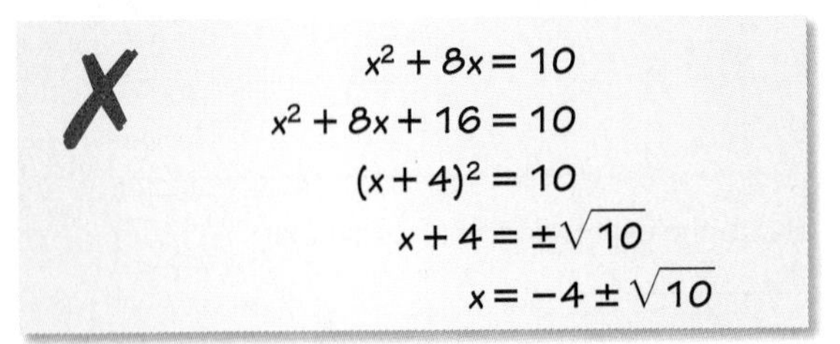

$x^2 + 8x = 10$
$x^2 + 8x + 16 = 10$
$(x + 4)^2 = 10$
$x + 4 = \pm\sqrt{10}$
$x = -4 \pm \sqrt{10}$

34. ERROR ANALYSIS Describe and correct the error in the first two steps of solving $2x^2 - 2x - 4 = 0$ by completing the square.

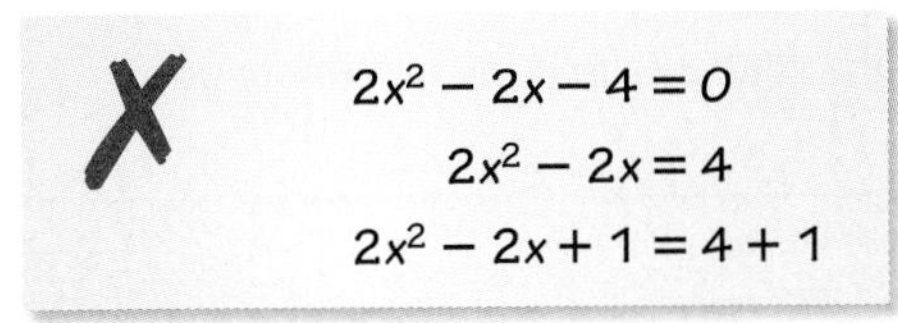

$2x^2 - 2x - 4 = 0$
$2x^2 - 2x = 4$
$2x^2 - 2x + 1 = 4 + 1$

35. NUMBER SENSE Find all values of b for which $x^2 + bx + 25$ is a perfect square trinomial. Explain how you found your answer.

36. REASONING You are completing the square to solve $3x^2 + 6x = 12$. What is the first step?

In Exercises 37–40, write the function in vertex form by completing the square. Then match the function with its graph.

37. $y = x^2 + 6x + 3$ **38.** $y = -x^2 + 8x - 12$

39. $y = -x^2 - 4x - 2$ **40.** $y = x^2 - 2x + 4$

A.

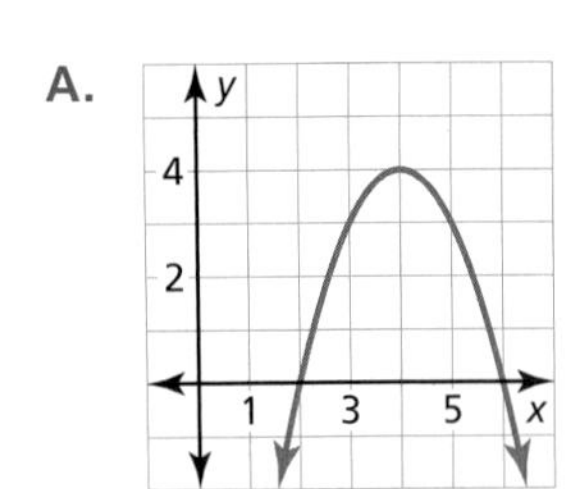

B.

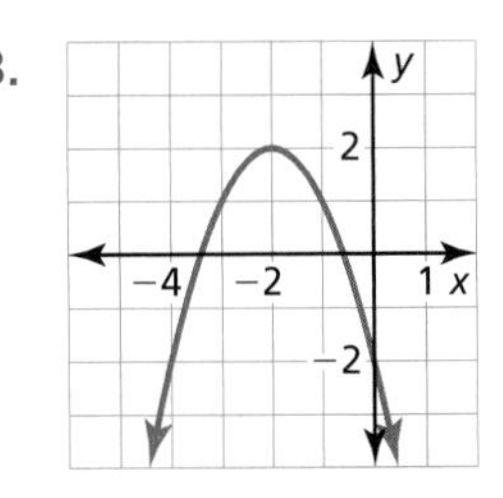

C.

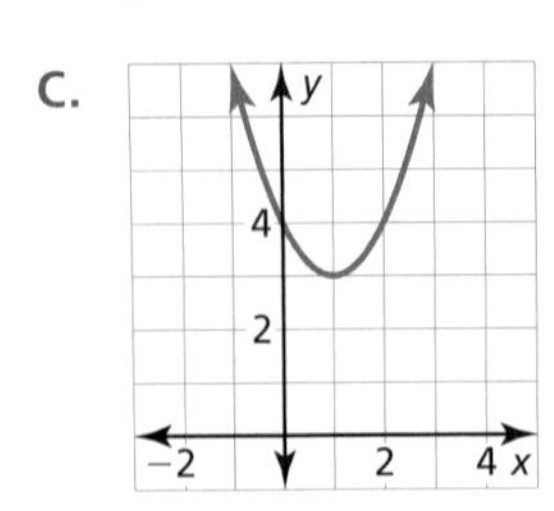

D.

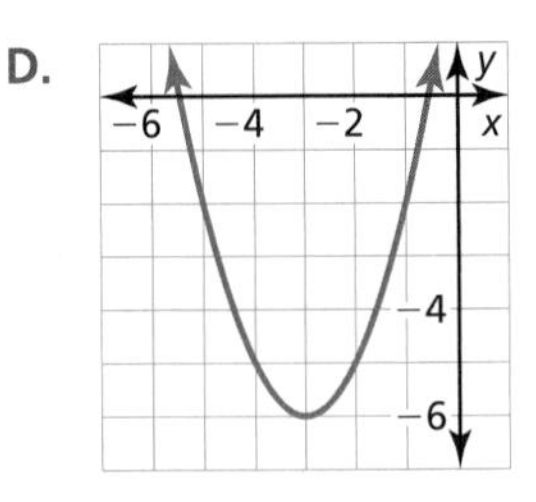

In Exercises 41–46, determine whether the quadratic function has a maximum or minimum value. Then find the value. *(See Examples 4 and 5.)*

41. $y = x^2 - 4x - 2$ **42.** $y = x^2 + 6x + 10$

43. $y = -x^2 - 10x - 30$ **44.** $y = -x^2 + 14x - 34$

45. $f(x) = -3x^2 - 6x - 9$ **46.** $f(x) = 4x^2 - 28x + 32$

In Exercises 47–50, determine whether the graph could represent the function. Explain.

47. $y = -(x + 8)(x + 3)$ **48.** $y = (x - 5)^2$

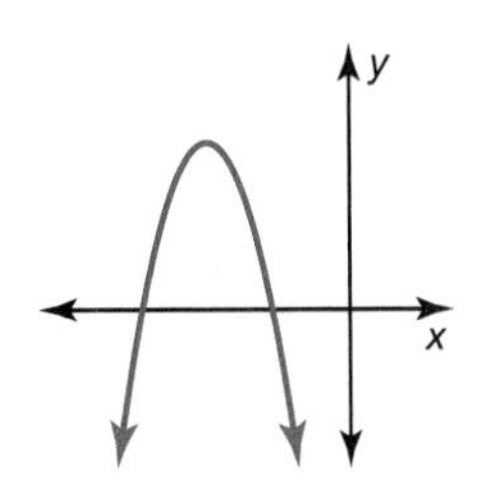

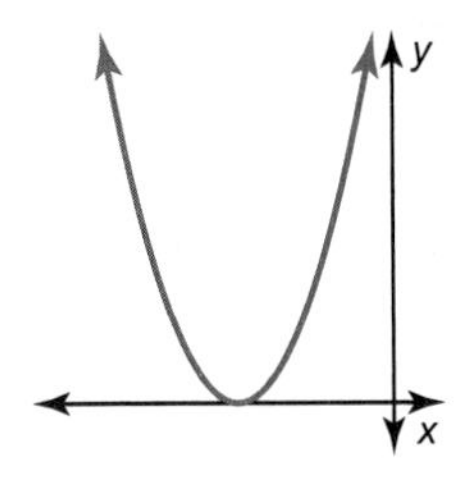

49. $y = \frac{1}{4}(x + 2)^2 - 4$ **50.** $y = -2(x - 1)(x + 2)$

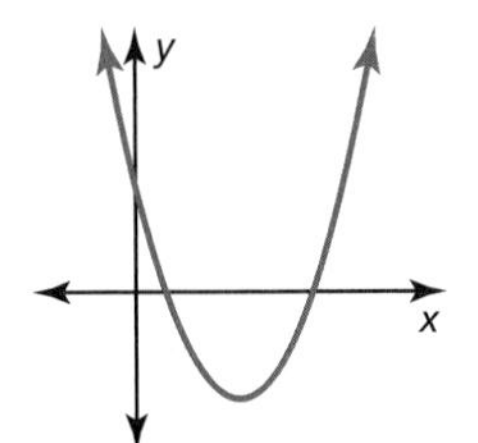

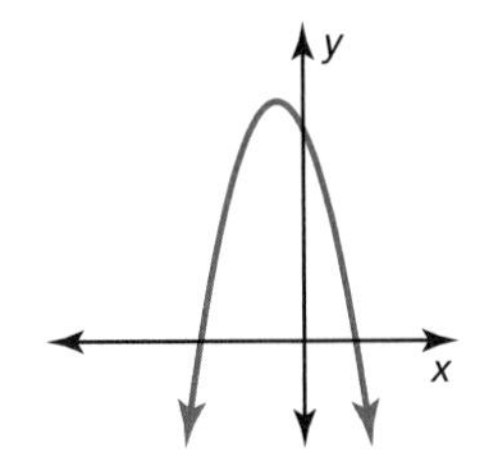

In Exercises 51 and 52, determine which of the functions could be represented by the graph. Explain. *(See Example 6.)*

51.

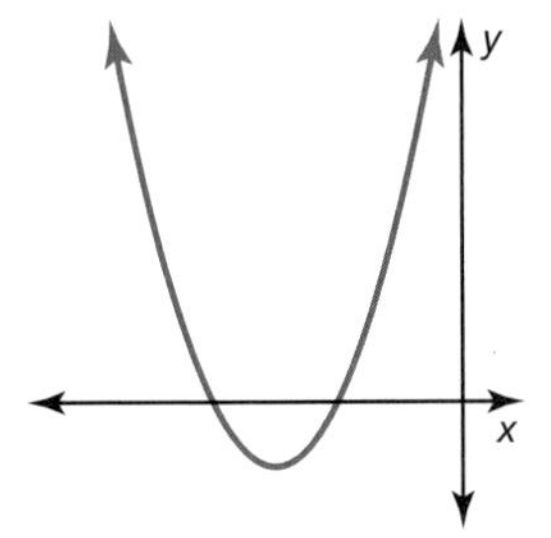

$h(x) = (x + 2)^2 + 3$

$f(x) = 2(x + 3)^2 - 2$

$g(x) = -\frac{1}{2}(x - 8)(x - 4)$

$m(x) = (x + 2)(x + 4)$

52.

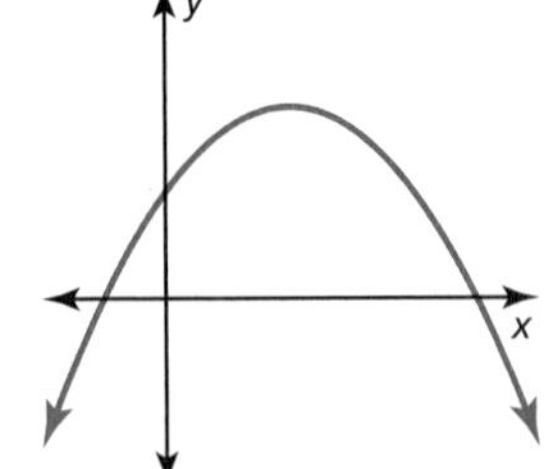

$r(x) = -\frac{1}{3}(x - 5)(x + 1)$

$p(x) = -2(x - 2)(x - 6)$

$q(x) = (x + 1)^2 + 4$

$n(x) = -(x - 2)^2 + 9$

53. MODELING WITH MATHEMATICS The function $h = -16t^2 + 48t$ represents the height h (in feet) of a kickball t seconds after it is kicked from the ground. *(See Example 7.)*

a. Find the maximum height of the kickball.

b. Find and interpret the axis of symmetry.

54. MODELING WITH MATHEMATICS You throw a stone from a height of 16 feet with an initial vertical velocity of 32 feet per second. The function $h = -16t^2 + 32t + 16$ represents the height h (in feet) of the stone after t seconds.

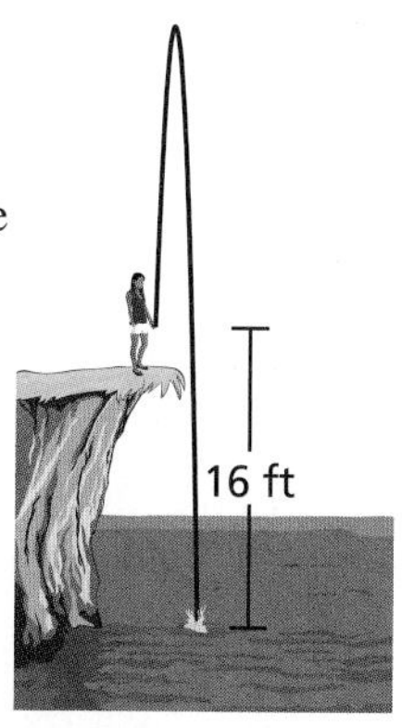

a. Find the maximum height of the stone.

b. Find and interpret the axis of symmetry.

55. MODELING WITH MATHEMATICS You are building a rectangular brick patio surrounded by a crushed stone border with a uniform width, as shown. You purchase patio bricks to cover 140 square feet. Find the width of the border. *(See Example 8.)*

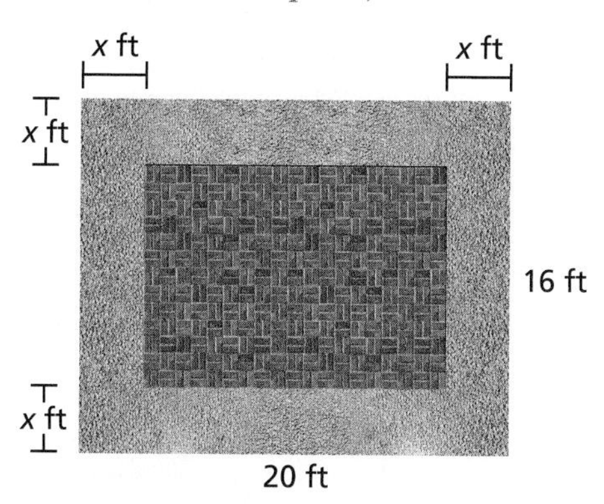

56. MODELING WITH MATHEMATICS You are making a poster that will have a uniform border, as shown. The total area of the poster is 722 square inches. Find the width of the border to the nearest inch.

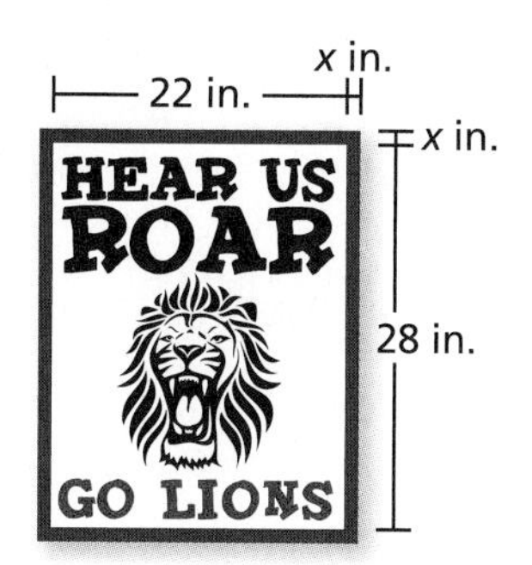

MATHEMATICAL CONNECTIONS **In Exercises 57 and 58, find the value of x. Round your answer to the nearest hundredth, if necessary.**

57. $A = 108 \text{ m}^2$

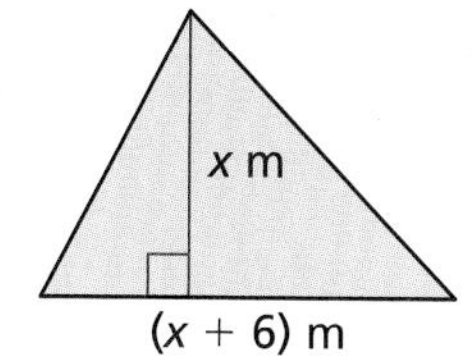

58. $A = 288 \text{ in.}^2$

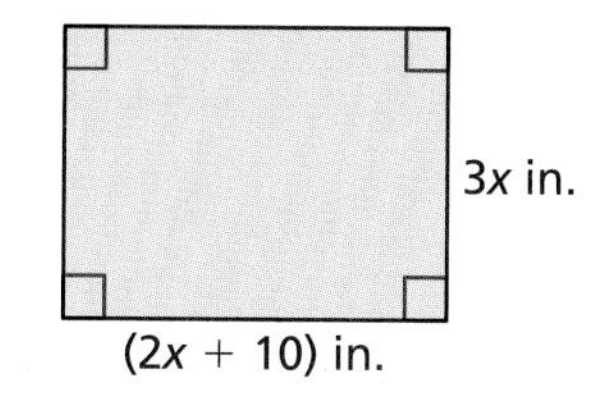

In Exercises 59–62, solve the equation by completing the square. Round your solutions to the nearest hundredth, if necessary.

59. $0.5x^2 + x - 2 = 0$

60. $0.75x^2 + 1.5x = 4$

61. $\frac{8}{3}x - \frac{2}{3}x^2 = -\frac{5}{6}$

62. $\frac{1}{4}x^2 + \frac{1}{2}x - \frac{5}{4} = 0$

63. PROBLEM SOLVING The distance d (in feet) that it takes a car to come to a complete stop can be modeled by $d = 0.05s^2 + 2.2s$, where s is the speed of the car (in miles per hour). A car has 168 feet to come to a complete stop. Find the maximum speed at which the car can travel.

64. PROBLEM SOLVING During a "big air" competition, snowboarders launch themselves from a half-pipe, perform tricks in the air, and land back in the half-pipe. The height h (in feet) of a snowboarder above the bottom of the half-pipe can be modeled by $h = -16t^2 + 24t + 16.4$, where t is the time (in seconds) after the snowboarder launches into the air. The snowboarder lands 3.2 feet lower than the height of the launch. How long is the snowboarder in the air? Round your answer to the nearest tenth of a second.

Cross section of a half-pipe

65. PROBLEM SOLVING You have 80 feet of fencing to make a rectangular horse pasture that covers 750 square feet. A barn will be used as one side of the pasture, as shown.

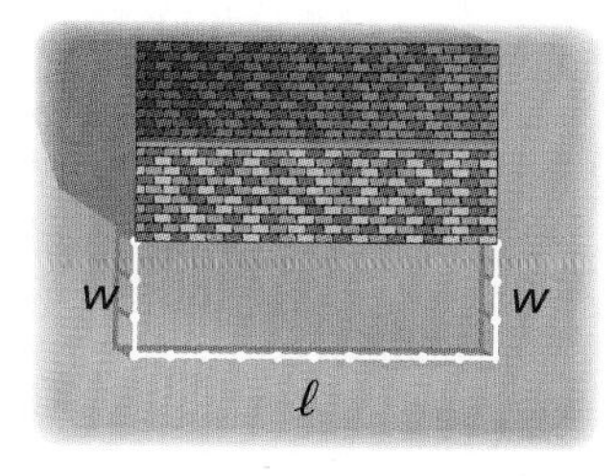

a. Write equations for the amount of fencing to be used and the area enclosed by the fencing.

b. Use substitution to solve the system of equations from part (a). What are the possible dimensions of the pasture?

66. HOW DO YOU SEE IT? The graph represents the quadratic function $y = x^2 - 4x + 6$.

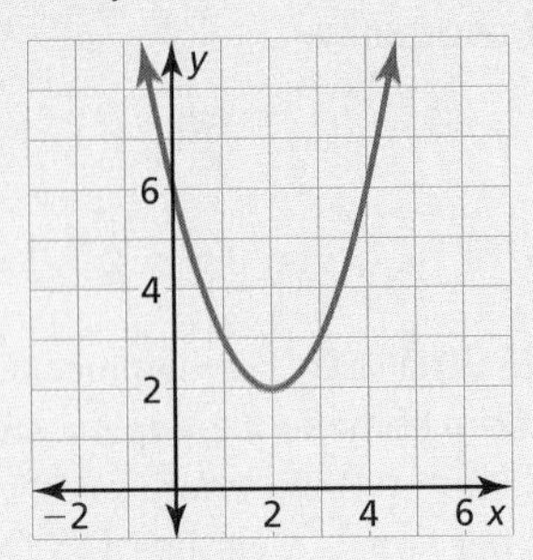

a. Use the graph to estimate the x-values for which $y = 3$.

b. Explain how you can use the method of completing the square to check your estimates in part (a).

67. COMPARING METHODS Consider the quadratic equation $x^2 + 12x + 2 = 12$.

a. Solve the equation by graphing.

b. Solve the equation by completing the square.

c. Compare the two methods. Which do you prefer? Explain.

68. THOUGHT PROVOKING Sketch the graph of the equation $x^2 - 2xy + y^2 - x - y = 0$. Identify the graph.

69. REASONING The product of two consecutive even integers that are positive is 48. Write and solve an equation to find the integers.

70. REASONING The product of two consecutive odd integers that are negative is 195. Write and solve an equation to find the integers.

71. MAKING AN ARGUMENT You purchase stock for \$16 per share. You sell the stock 30 days later for \$23.50 per share. The price y (in dollars) of a share during the 30-day period can be modeled by $y = -0.025x^2 + x + 16$, where x is the number of days after the stock is purchased. Your friend says you could have sold the stock earlier for \$23.50 per share. Is your friend correct? Explain.

72. REASONING You are solving the equation $x^2 + 9x = 18$. What are the advantages of solving the equation by completing the square instead of using other methods you have learned?

73. PROBLEM SOLVING You are knitting a rectangular scarf. The pattern results in a scarf that is 60 inches long and 4 inches wide. However, you have enough yarn to knit 396 square inches. You decide to increase the dimensions of the scarf so that you will use all your yarn. The increase in the length is three times the increase in the width. What are the dimensions of your scarf?

74. WRITING How many solutions does $x^2 + bx = c$ have when $c < -\left(\frac{b}{2}\right)^2$? Explain.

Maintaining Mathematical Proficiency

Reviewing what you learned in previous grades and lessons

Write a recursive rule for the sequence. *(Section 6.7)*

75.

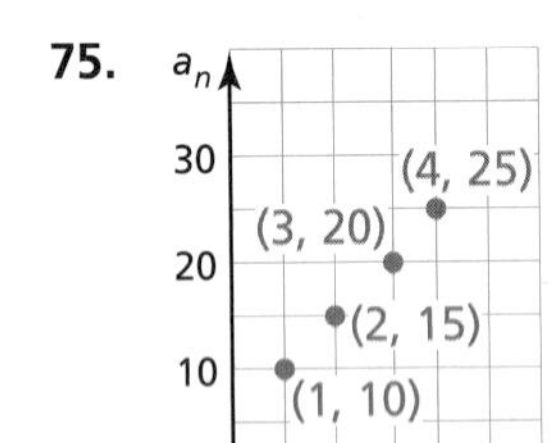

76.

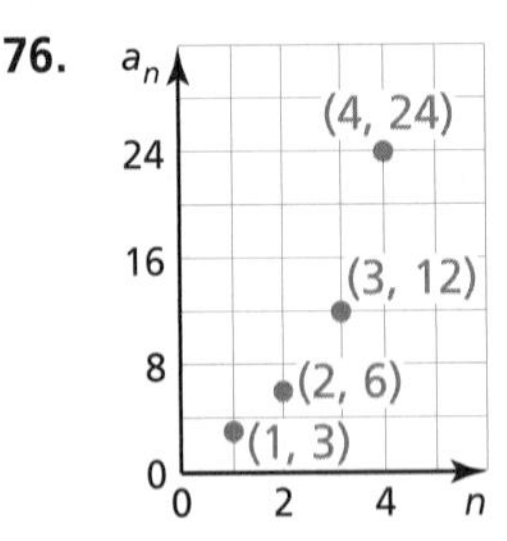

77.

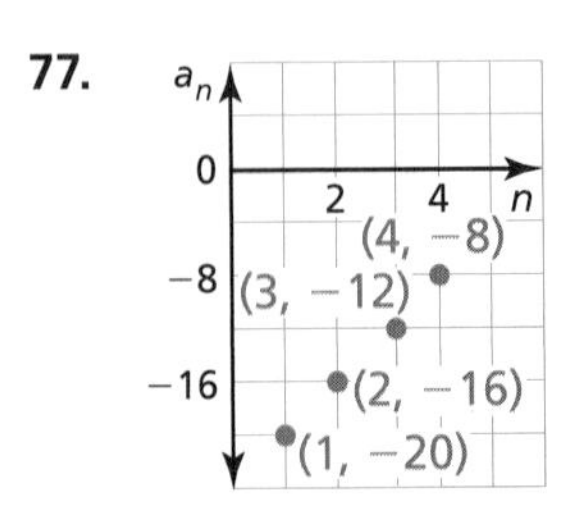

Simplify the expression $\sqrt{b^2 - 4ac}$ for the given values. *(Section 9.1)*

78. $a = 3, b = -6, c = 2$

79. $a = -2, b = 4, c = 7$

80. $a = 1, b = 6, c = 4$

9.5 Solving Quadratic Equations Using the Quadratic Formula

Essential Question How can you derive a formula that can be used to write the solutions of any quadratic equation in standard form?

EXPLORATION 1 Deriving the Quadratic Formula

Work with a partner. The following steps show a method of solving $ax^2 + bx + c = 0$. Explain what was done in each step.

$$ax^2 + bx + c = 0$$ ← 1. Write the equation.

$$4a^2x^2 + 4abx + 4ac = 0$$ ← 2. What was done?

$$4a^2x^2 + 4abx + 4ac + b^2 = b^2$$ ← 3. What was done?

$$4a^2x^2 + 4abx + b^2 = b^2 - 4ac$$ ← 4. What was done?

$$(2ax + b)^2 = b^2 - 4ac$$ ← 5. What was done?

$$2ax + b = \pm\sqrt{b^2 - 4ac}$$ ← 6. What was done?

$$2ax = -b \pm \sqrt{b^2 - 4ac}$$ ← 7. What was done?

Quadratic Formula: $$x = \frac{-b \pm \sqrt{b^2 - 4ac}}{2a}$$ ← 8. What was done?

EXPLORATION 2 Deriving the Quadratic Formula by Completing the Square

Work with a partner.

a. Solve $ax^2 + bx + c = 0$ by completing the square. (*Hint:* Subtract c from each side, divide each side by a, and then proceed by completing the square.)

b. Compare this method with the method in Exploration 1. Explain why you think $4a$ and b^2 were chosen in Steps 2 and 3 of Exploration 1.

Communicate Your Answer

3. How can you derive a formula that can be used to write the solutions of any quadratic equation in standard form?

4. Use the Quadratic Formula to solve each quadratic equation.

a. $x^2 + 2x - 3 = 0$ **b.** $x^2 - 4x + 4 = 0$ **c.** $x^2 + 4x + 5 = 0$

5. Use the Internet to research *imaginary numbers*. How are they related to quadratic equations?

USING TOOLS STRATEGICALLY

To be proficient in math, you need to identify relevant external mathematical resources.

9.5 Lesson

Core Vocabulary

Quadratic Formula, *p. 516*
discriminant, *p. 518*

What You Will Learn

- Solve quadratic equations using the Quadratic Formula.
- Interpret the discriminant.
- Choose efficient methods for solving quadratic equations.

Using the Quadratic Formula

By completing the square for the quadratic equation $ax^2 + bx + c = 0$, you can develop a formula that gives the solutions of any quadratic equation in standard form. This formula is called the **Quadratic Formula**.

Core Concept

Quadratic Formula

The real solutions of the quadratic equation $ax^2 + bx + c = 0$ are

$$x = \frac{-b \pm \sqrt{b^2 - 4ac}}{2a} \qquad \text{Quadratic Formula}$$

where $a \neq 0$ and $b^2 - 4ac \geq 0$.

EXAMPLE 1 Using the Quadratic Formula

Solve $2x^2 - 5x + 3 = 0$ using the Quadratic Formula.

SOLUTION

$$x = \frac{-b \pm \sqrt{b^2 - 4ac}}{2a} \qquad \text{Quadratic Formula}$$

$$= \frac{-(-5) \pm \sqrt{(-5)^2 - 4(2)(3)}}{2(2)} \qquad \text{Substitute 2 for } a, -5 \text{ for } b, \text{ and 3 for } c.$$

$$= \frac{5 \pm \sqrt{1}}{4} \qquad \text{Simplify.}$$

$$= \frac{5 \pm 1}{4} \qquad \text{Evaluate the square root.}$$

▶ So, the solutions are $x = \frac{5 + 1}{4} = \frac{3}{2}$ and $x = \frac{5 - 1}{4} = 1$.

STUDY TIP

You can use the roots of a quadratic equation to factor the related expression. In Example 1, you can use 1 and $\frac{3}{2}$ to factor $2x^2 - 5x + 3$ as $(x - 1)(2x - 3)$.

Check

$2x^2 - 5x + 3 = 0$	Original equation	$2x^2 - 5x + 3 = 0$
$2\left(\frac{3}{2}\right)^2 - 5\left(\frac{3}{2}\right) + 3 \stackrel{?}{=} 0$	Substitute.	$2(1)^2 - 5(1) + 3 \stackrel{?}{=} 0$
$\frac{9}{2} - \frac{15}{2} + 3 \stackrel{?}{=} 0$	Simplify.	$2 - 5 + 3 \stackrel{?}{=} 0$
$0 = 0$ ✓	Simplify.	$0 = 0$ ✓

Monitoring Progress

Help in English and Spanish at *BigIdeasMath.com*

Solve the equation using the Quadratic Formula. Round your solutions to the nearest tenth, if necessary.

1. $x^2 - 6x + 5 = 0$

2. $\frac{1}{2}x^2 + x - 10 = 0$

3. $-3x^2 + 2x + 7 = 0$

4. $4x^2 - 4x = -1$

EXAMPLE 2 Modeling With Mathematics

The number y of Northern Rocky Mountain wolf breeding pairs x years since 1990 can be modeled by the function $y = 0.20x^2 + 1.8x - 3$. When were there about 35 breeding pairs?

Wolf Breeding Pairs

$y = 0.20x^2 + 1.8x - 3$

Number of breeding pairs: 0, 15, 30, 45, 60, 75, 90, 105, 120

Years since 1990: 0, 3, 6, 9, 12, 15, 18

SOLUTION

1. **Understand the Problem** You are given a quadratic function that represents the number of wolf breeding pairs for years after 1990. You need to use the model to determine when there were 35 wolf breeding pairs.

2. **Make a Plan** To determine when there were 35 wolf breeding pairs, find the x-values for which $y = 35$. So, solve the equation $35 = 0.20x^2 + 1.8x - 3$.

3. **Solve the Problem**

$35 = 0.20x^2 + 1.8x - 3$ Write the equation.

$0 = 0.20x^2 + 1.8x - 38$ Write in standard form.

$x = \dfrac{-b \pm \sqrt{b^2 - 4ac}}{2a}$ Quadratic Formula

$= \dfrac{-1.8 \pm \sqrt{1.8^2 - 4(0.2)(-38)}}{2(0.2)}$ Substitute 0.2 for a, 1.8 for b, and -38 for c.

$= \dfrac{-1.8 \pm \sqrt{33.64}}{0.4}$ Simplify.

$= \dfrac{-1.8 \pm 5.8}{0.4}$ Simplify.

The solutions are $x = \dfrac{-1.8 + 5.8}{0.4} = 10$ and $x = \dfrac{-1.8 - 5.8}{0.4} = -19$.

INTERPRETING MATHEMATICAL RESULTS

You can ignore the solution $x = -19$ because -19 represents the year 1971, which is not in the given time period.

▶ Because x represents the number of years since 1990, x is greater than or equal to zero. So, there were about 35 breeding pairs 10 years after 1990, in 2000.

4. **Look Back** Use a graphing calculator to graph the equations $y = 0.20x^2 + 1.8x - 3$ and $y = 35$. Then use the *intersect* feature to find the point of intersection. The graphs intersect at (10, 35).

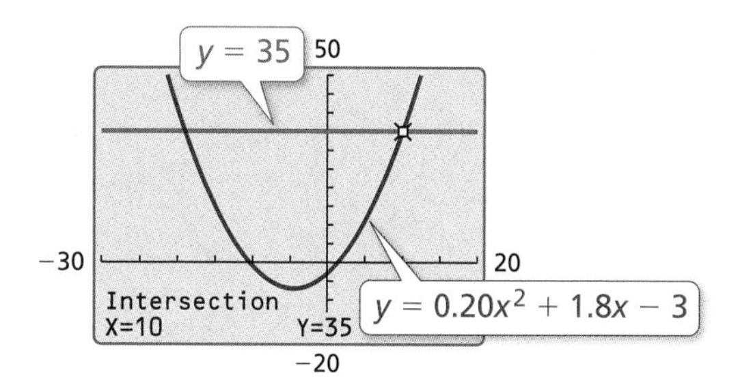

Monitoring Progress  Help in English and Spanish at *BigIdeasMath.com*

5. **WHAT IF?** When were there about 60 wolf breeding pairs?

6. The number y of bald eagle nesting pairs in a state x years since 2000 can be modeled by the function $y = 0.34x^2 + 13.1x + 51$.

   a. When were there about 160 bald eagle nesting pairs?

   b. How many bald eagle nesting pairs were there in 2000?

Interpreting the Discriminant

The expression $b^2 - 4ac$ in the Quadratic Formula is called the **discriminant**.

$$x = \frac{-b \pm \sqrt{b^2 - 4ac}}{2a} \longleftarrow \text{discriminant}$$

Because the discriminant is under the radical symbol, you can use the value of the discriminant to determine the number of real solutions of a quadratic equation and the number of x-intercepts of the graph of the related function.

Core Concept

Interpreting the Discriminant

$b^2 - 4ac > 0$

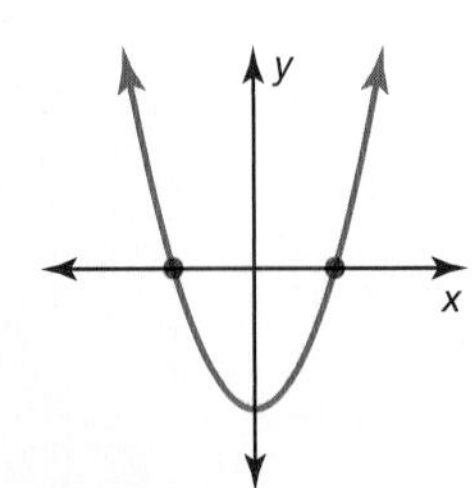

- two real solutions
- two x-intercepts

$b^2 - 4ac = 0$

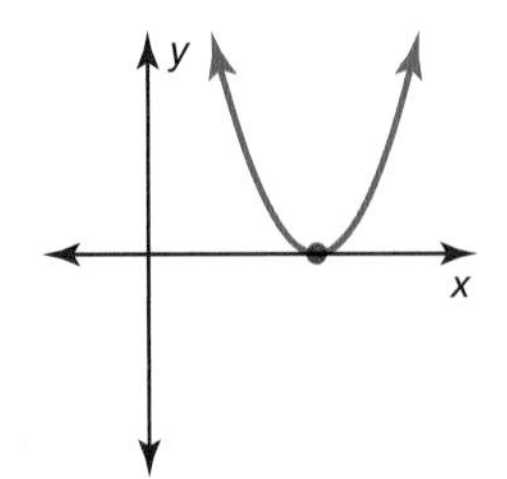

- one real solution
- one x-intercept

$b^2 - 4ac < 0$

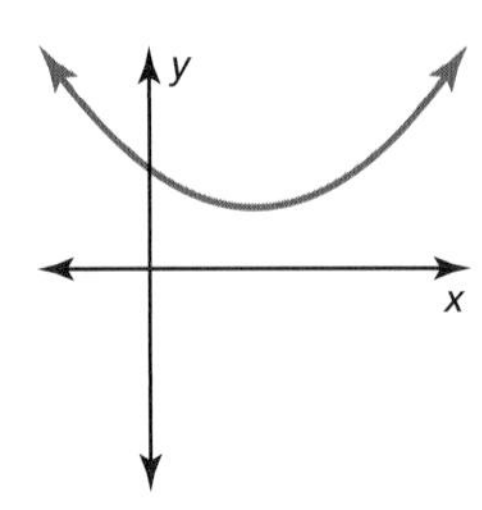

- no real solutions
- no x-intercepts

STUDY TIP

The solutions of a quadratic equation may be real numbers or *imaginary numbers*. You will study imaginary numbers in a future course.

EXAMPLE 3 Determining the Number of Real Solutions

a. Determine the number of real solutions of $x^2 + 8x - 3 = 0$.

$b^2 - 4ac = 8^2 - 4(1)(-3)$	Substitute 1 for a, 8 for b, and -3 for c.
$= 64 + 12$	Simplify.
$= 76$	Add.

▶ The discriminant is greater than 0. So, the equation has two real solutions.

b. Determine the number of real solutions of $9x^2 + 1 = 6x$.

Write the equation in standard form: $9x^2 - 6x + 1 = 0$.

$b^2 - 4ac = (-6)^2 - 4(9)(1)$	Substitute 9 for a, -6 for b, and 1 for c.
$= 36 - 36$	Simplify.
$= 0$	Subtract.

▶ The discriminant is 0. So, the equation has one real solution.

Monitoring Progress

Help in English and Spanish at *BigIdeasMath.com*

Determine the number of real solutions of the equation.

7. $-x^2 + 4x - 4 = 0$

8. $6x^2 + 2x = -1$

9. $\frac{1}{2}x^2 = 7x - 1$

EXAMPLE 4 Finding the Number of x-Intercepts of a Parabola

Find the number of x-intercepts of the graph of $y = 2x^2 + 3x + 9$.

SOLUTION

Determine the number of real solutions of $0 = 2x^2 + 3x + 9$.

$b^2 - 4ac = 3^2 - 4(2)(9)$ Substitute 2 for *a*, 3 for *b*, and 9 for *c*.

$= 9 - 72$ Simplify.

$= -63$ Subtract.

Because the discriminant is less than 0, the equation has no real solutions.

▶ So, the graph of $y = 2x^2 + 3x + 9$ has no x-intercepts.

Check

Use a graphing calculator to check your answer. Notice that the graph of $y = 2x^2 + 3x + 9$ has no x-intercepts.

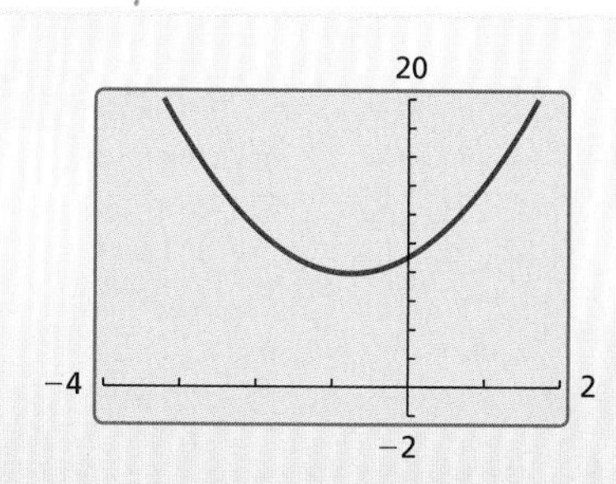

Monitoring Progress

Help in English and Spanish at *BigIdeasMath.com*

Find the number of x-intercepts of the graph of the function.

10. $y = -x^2 + x - 6$ **11.** $y = x^2 - x$ **12.** $f(x) = x^2 + 12x + 36$

Choosing an Efficient Method

The table shows five methods for solving quadratic equations. For a given equation, it may be more efficient to use one method instead of another. Some advantages and disadvantages of each method are shown.

Core Concept

Methods for Solving Quadratic Equations

Method	Advantages	Disadvantages
Factoring *(Lessons 7.5–7.8)*	• Straightforward when the equation can be factored easily	• Some equations are not factorable.
Graphing *(Lesson 9.2)*	• Can easily see the number of solutions • Use when approximate solutions are sufficient. • Can use a graphing calculator	• May not give exact solutions
Using Square Roots *(Lesson 9.3)*	• Use to solve equations of the form $x^2 = d$.	• Can only be used for certain equations
Completing the Square *(Lesson 9.4)*	• Best used when $a = 1$ and b is even	• May involve difficult calculations
Quadratic Formula *(Lesson 9.5)*	• Can be used for any quadratic equation • Gives exact solutions	• Takes time to do calculations

EXAMPLE 5 Choosing a Method

Solve the equation using any method. Explain your choice of method.

a. $x^2 - 10x = 1$ **b.** $2x^2 - 13x - 24 = 0$ **c.** $x^2 + 8x + 12 = 0$

SOLUTION

a. The coefficient of the x^2-term is 1, and the coefficient of the x-term is an even number. So, solve by completing the square.

$x^2 - 10x = 1$ Write the equation.

$x^2 - 10x + 25 = 1 + 25$ Complete the square for $x^2 - 10x$.

$(x - 5)^2 = 26$ Write the left side as the square of a binomial.

$x - 5 = \pm\sqrt{26}$ Take the square root of each side.

$x = 5 \pm \sqrt{26}$ Add 5 to each side.

▶ So, the solutions are $x = 5 + \sqrt{26} \approx 10.1$ and $x = 5 - \sqrt{26} \approx -0.1$.

b. The equation is not easily factorable, and the numbers are somewhat large. So, solve using the Quadratic Formula.

$x = \dfrac{-b \pm \sqrt{b^2 - 4ac}}{2a}$ Quadratic Formula

$= \dfrac{-(-13) \pm \sqrt{(-13)^2 - 4(2)(-24)}}{2(2)}$ Substitute 2 for a, -13 for b, and -24 for c.

$= \dfrac{13 \pm \sqrt{361}}{4}$ Simplify.

$= \dfrac{13 \pm 19}{4}$ Evaluate the square root.

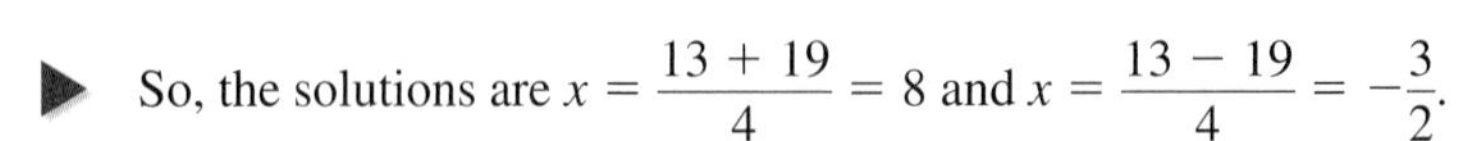

▶ So, the solutions are $x = \dfrac{13 + 19}{4} = 8$ and $x = \dfrac{13 - 19}{4} = -\dfrac{3}{2}$.

Check

Graph the related function $f(x) = x^2 + 8x + 12$ and find the zeros. The zeros are -6 and -2.

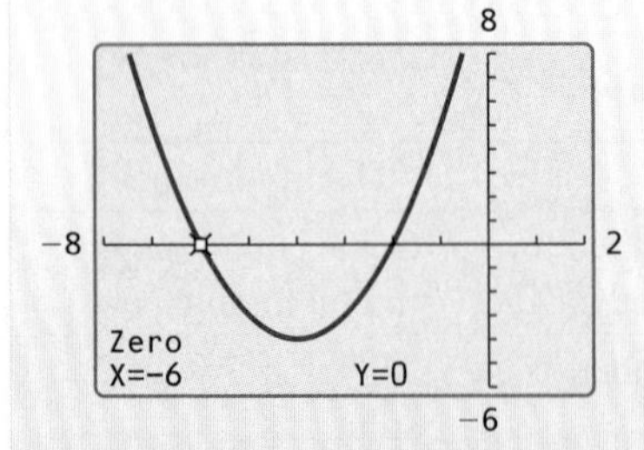

c. The equation is easily factorable. So, solve by factoring.

$x^2 + 8x + 12 = 0$ Write the equation.

$(x + 2)(x + 6) = 0$ Factor the polynomial.

$x + 2 = 0$ or $x + 6 = 0$ Zero-Product Property

$x = -2$ or $x = -6$ Solve for x.

▶ The solutions are $x = -2$ and $x = -6$.

Monitoring Progress

Help in English and Spanish at *BigIdeasMath.com*

Solve the equation using any method. Explain your choice of method.

13. $x^2 + 11x - 12 = 0$

14. $9x^2 - 5 = 4$

15. $5x^2 - x - 1 = 0$

16. $x^2 = 2x - 5$

9.5 Exercises

Dynamic Solutions available at *BigIdeasMath.com*

Vocabulary and Core Concept Check

1. **VOCABULARY** What formula can you use to solve any quadratic equation? Write the formula.

2. **VOCABULARY** In the Quadratic Formula, what is the discriminant? What does the value of the discriminant determine?

Monitoring Progress and Modeling with Mathematics

In Exercises 3–8, write the equation in standard form. Then identify the values of a, b, and c that you would use to solve the equation using the Quadratic Formula.

3. $x^2 = 7x$
4. $x^2 - 4x = -12$
5. $-2x^2 + 1 = 5x$
6. $3x + 2 = 4x^2$
7. $4 - 3x = -x^2 + 3x$
8. $-8x - 1 = 3x^2 + 2$

In Exercises 9–22, solve the equation using the Quadratic Formula. Round your solutions to the nearest tenth, if necessary. *(See Example 1.)*

9. $x^2 - 12x + 36 = 0$
10. $x^2 + 7x + 16 = 0$
11. $x^2 - 10x - 11 = 0$
12. $2x^2 - x - 1 = 0$
13. $2x^2 - 6x + 5 = 0$
14. $9x^2 - 6x + 1 = 0$
15. $6x^2 - 13x = -6$
16. $-3x^2 + 6x = 4$
17. $1 - 8x = -16x^2$
18. $x^2 - 5x + 3 = 0$
19. $x^2 + 2x = 9$
20. $5x^2 - 2 = 4x$
21. $2x^2 + 9x + 7 = 3$
22. $8x^2 + 8 = 6 - 9x$

23. **MODELING WITH MATHEMATICS** A dolphin jumps out of the water, as shown in the diagram. The function $h = -16t^2 + 26t$ models the height h (in feet) of the dolphin after t seconds. After how many seconds is the dolphin at a height of 5 feet? *(See Example 2.)*

24. **MODELING WITH MATHEMATICS** The amount of trout y (in tons) caught in a lake from 1995 to 2014 can be modeled by the equation $y = -0.08x^2 + 1.6x + 10$, where x is the number of years since 1995.

    a. When were about 15 tons of trout caught in the lake?

    b. Do you think this model can be used to determine the amounts of trout caught in future years? Explain your reasoning.

In Exercises 25–30, determine the number of real solutions of the equation. *(See Example 3.)*

25. $x^2 - 6x + 10 = 0$
26. $x^2 - 5x - 3 = 0$
27. $2x^2 - 12x = -18$
28. $4x^2 = 4x - 1$
29. $-\frac{1}{4}x^2 + 4x = -2$
30. $-5x^2 + 8x = 9$

In Exercises 31–36, find the number of x-intercepts of the graph of the function. *(See Example 4.)*

31. $y = x^2 + 5x - 1$
32. $y = 4x^2 + 4x + 1$
33. $y = -6x^2 + 3x - 4$
34. $y = -x^2 + 5x + 13$
35. $f(x) = 4x^2 + 3x - 6$
36. $f(x) = 2x^2 + 8x + 8$

In Exercises 37–44, solve the equation using any method. Explain your choice of method. *(See Example 5.)*

37. $-10x^2 + 13x = 4$
38. $x^2 - 3x - 40 = 0$
39. $x^2 + 6x = 5$
40. $-5x^2 = -25$
41. $x^2 + x - 12 = 0$
42. $x^2 - 4x + 1 = 0$
43. $4x^2 - x = 17$
44. $x^2 + 6x + 9 = 16$

45. ERROR ANALYSIS Describe and correct the error in solving the equation $3x^2 - 7x - 6 = 0$ using the Quadratic Formula.

✗

$$x = \frac{-7 \pm \sqrt{(-7)^2 - 4(3)(-6)}}{2(3)}$$

$$= \frac{-7 \pm \sqrt{121}}{6}$$

$$x = \frac{2}{3} \text{ and } x = -3$$

46. ERROR ANALYSIS Describe and correct the error in solving the equation $-2x^2 + 9x = 4$ using the Quadratic Formula.

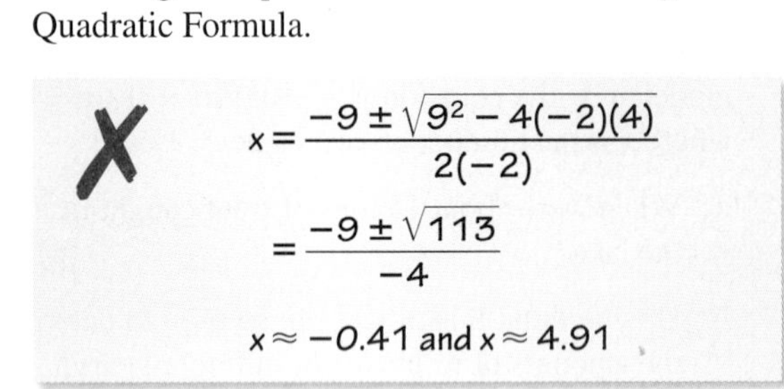

47. MODELING WITH MATHEMATICS A fountain shoots a water arc that can be modeled by the graph of the equation $y = -0.006x^2 + 1.2x + 10$, where x is the horizontal distance (in feet) from the river's north shore and y is the height (in feet) above the river. Does the water arc reach a height of 50 feet? If so, about how far from the north shore is the water arc 50 feet above the water?

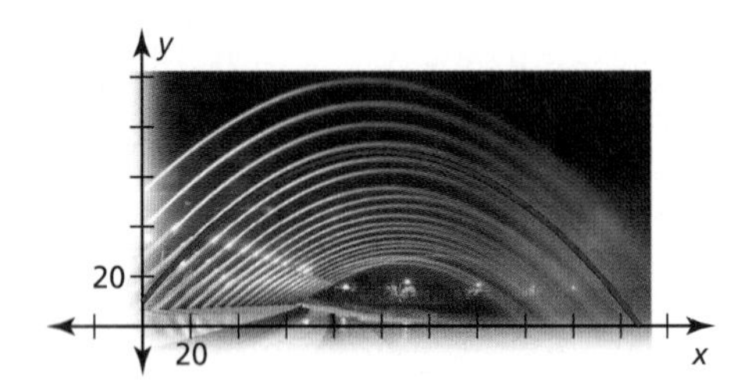

48. MODELING WITH MATHEMATICS Between the months of April and September, the number y of hours of daylight per day in Seattle, Washington, can be modeled by $y = -0.00046x^2 + 0.076x + 13$, where x is the number of days since April 1.

a. Do any of the days between April and September in Seattle have 17 hours of daylight? If so, how many?

b. Do any of the days between April and September in Seattle have 14 hours of daylight? If so, how many?

49. MAKING AN ARGUMENT Your friend uses the discriminant of the equation $2x^2 - 5x - 2 = -11$ and determines that the equation has two real solutions. Is your friend correct? Explain your reasoning.

50. MODELING WITH MATHEMATICS The frame of the tent shown is defined by a rectangular base and two parabolic arches that connect the opposite corners of the base. The graph of $y = -0.18x^2 + 1.6x$ models the height y (in feet) of one of the arches x feet along the diagonal of the base. Can a child who is 4 feet tall walk under one of the arches without having to bend over? Explain.

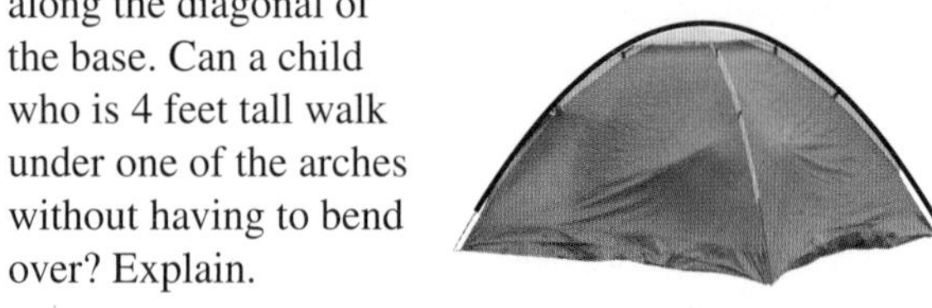

MATHEMATICAL CONNECTIONS In Exercises 51 and 52, use the given area A of the rectangle to find the value of x. Then give the dimensions of the rectangle.

51. $A = 91 \text{ m}^2$

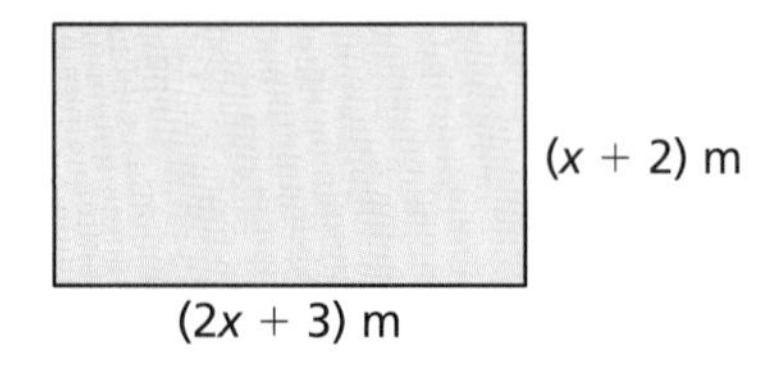

52. $A = 209 \text{ ft}^2$

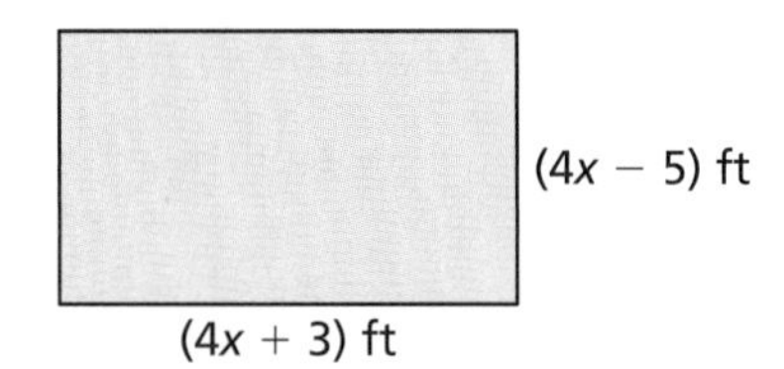

COMPARING METHODS In Exercises 53 and 54, solve the equation by (a) graphing, (b) factoring, and (c) using the Quadratic Formula. Which method do you prefer? Explain your reasoning.

53. $x^2 + 4x + 4 = 0$

54. $3x^2 + 11x + 6 = 0$

55. REASONING How many solutions does the equation $ax^2 + bx + c = 0$ have when a and c have different signs? Explain your reasoning.

56. REASONING When the discriminant is a perfect square, are the solutions of $ax^2 + bx + c = 0$ rational or irrational? (Assume a, b, and c are integers.) Explain your reasoning.

REASONING In Exercises 57–59, give a value of c for which the equation has (a) two solutions, (b) one solution, and (c) no solutions.

57. $x^2 - 2x + c = 0$

58. $x^2 - 8x + c = 0$

59. $4x^2 + 12x + c = 0$

60. REPEATED REASONING You use the Quadratic Formula to solve an equation.

a. You obtain solutions that are integers. Could you have used factoring to solve the equation? Explain your reasoning.

b. You obtain solutions that are fractions. Could you have used factoring to solve the equation? Explain your reasoning.

c. Make a generalization about quadratic equations with rational solutions.

61. MODELING WITH MATHEMATICS The fuel economy y (in miles per gallon) of a car can be modeled by the equation $y = -0.013x^2 + 1.25x + 5.6$, where $5 \le x \le 75$ and x is the speed (in miles per hour) of the car. Find the speed(s) at which you can travel and have a fuel economy of 32 miles per gallon.

62. MODELING WITH MATHEMATICS The depth d (in feet) of a river can be modeled by the equation $d = -0.25t^2 + 1.7t + 3.5$, where $0 \le t \le 7$ and t is the time (in hours) after a heavy rain begins. When is the river 6 feet deep?

ANALYZING EQUATIONS In Exercises 63–68, tell whether the vertex of the graph of the function lies above, below, or on the x-axis. Explain your reasoning without using a graph.

63. $y = x^2 - 3x + 2$

64. $y = 3x^2 - 6x + 3$

65. $y = 6x^2 - 2x + 4$

66. $y = -15x^2 + 10x - 25$

67. $f(x) = -3x^2 - 4x + 8$

68. $f(x) = 9x^2 - 24x + 16$

69. REASONING NASA creates a weightless environment by flying a plane in a series of parabolic paths. The height h (in feet) of a plane after t seconds in a parabolic flight path can be modeled by $h = -11t^2 + 700t + 21{,}000$. The passengers experience a weightless environment when the height of the plane is greater than or equal to 30,800 feet. For approximately how many seconds do passengers experience weightlessness on such a flight? Explain.

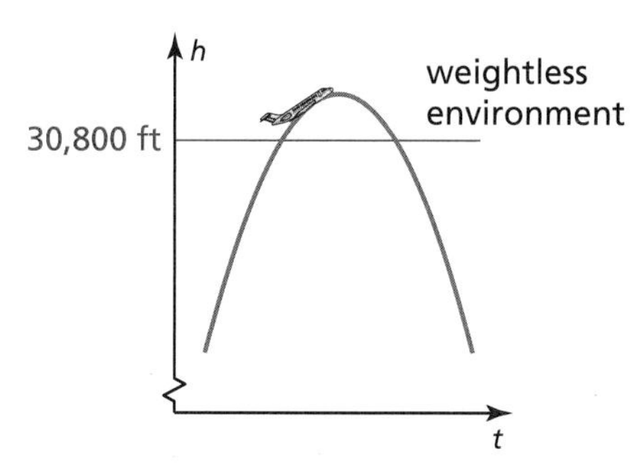

70. WRITING EQUATIONS Use the numbers to create a quadratic equation with the solutions $x = -1$ and $x = -\frac{1}{4}$.

$$\_\_\_x^2 + \_\_\_x + \_\_\_ = 0$$

−5	−4	−3	−2	−1
1	2	3	4	5

71. PROBLEM SOLVING A rancher constructs two rectangular horse pastures that share a side, as shown. The pastures are enclosed by 1050 feet of fencing. Each pasture has an area of 15,000 square feet.

a. Show that $y = 350 - \frac{4}{3}x$.

b. Find the possible lengths and widths of each pasture.

72. PROBLEM SOLVING A kicker punts a football from a height of 2.5 feet above the ground with an initial vertical velocity of 45 feet per second.

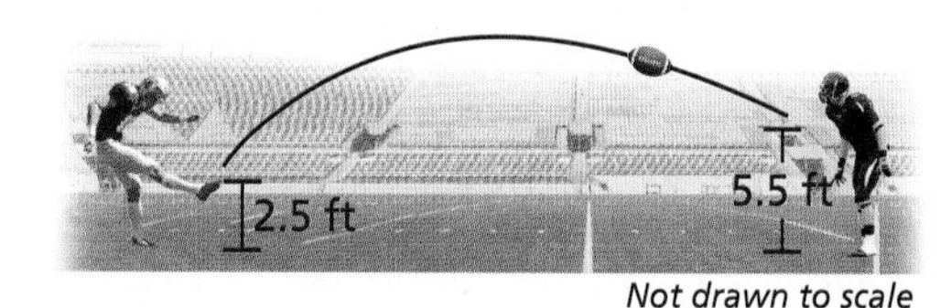

Not drawn to scale

a. Write an equation that models this situation using the function $h = -16t^2 + v_0t + s_0$, where h is the height (in feet) of the football, t is the time (in seconds) after the football is punted, v_0 is the initial vertical velocity (in feet per second), and s_0 is the initial height (in feet).

b. The football is caught 5.5 feet above the ground, as shown in the diagram. Find the amount of time that the football is in the air.

73. CRITICAL THINKING The solutions of the quadratic equation $ax^2 + bx + c = 0$ are $x = \frac{-b + \sqrt{b^2 - 4ac}}{2a}$ and $x = \frac{-b - \sqrt{b^2 - 4ac}}{2a}$. Find the mean of the solutions. How is the mean of the solutions related to the graph of $y = ax^2 + bx + c$? Explain.

74. HOW DO YOU SEE IT? Match each graph with its discriminant. Explain your reasoning.

A.

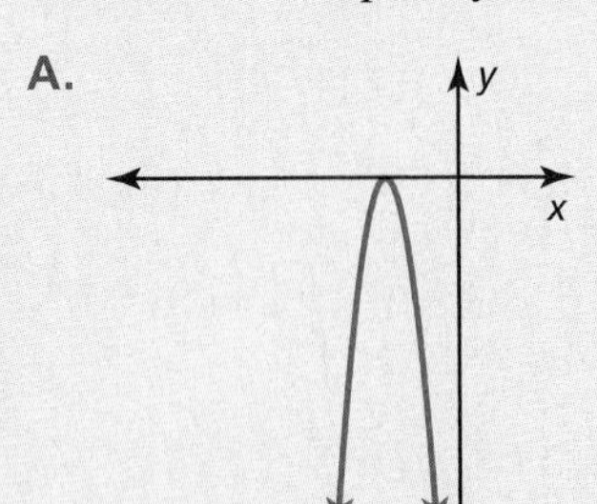

B.

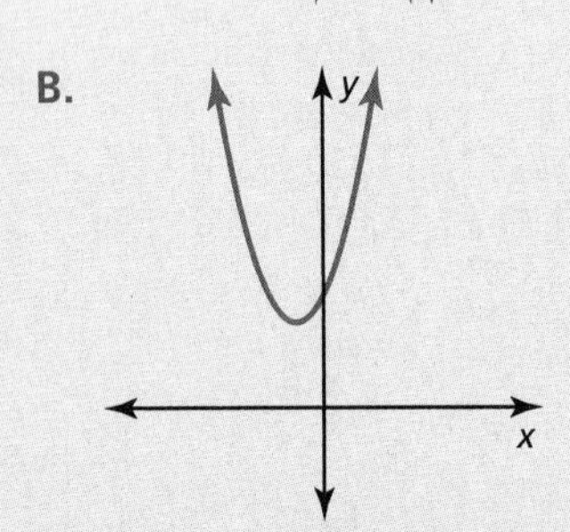

C.

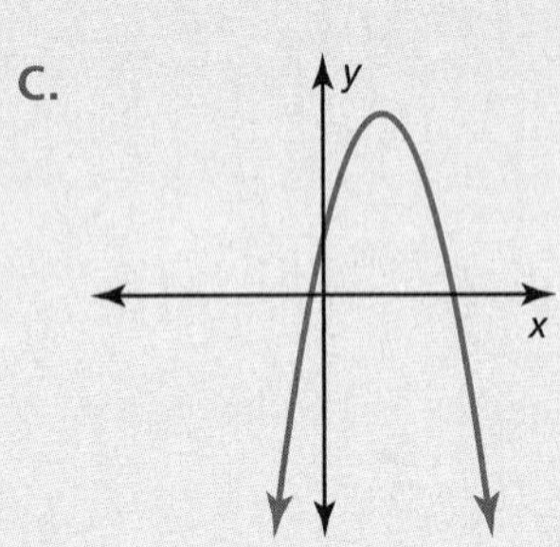

a. $b^2 - 4ac > 0$

b. $b^2 - 4ac = 0$

c. $b^2 - 4ac < 0$

75. CRITICAL THINKING You are trying to hang a tire swing. To get the rope over a tree branch that is 15 feet high, you tie the rope to a weight and throw it over the branch. You release the weight at a height s_0 of 5.5 feet. What is the minimum initial vertical velocity v_0 needed to reach the branch? (*Hint*: Use the equation $h = -16t^2 + v_0t + s_0$.)

76. THOUGHT PROVOKING Consider the graph of the standard form of a quadratic function $y = ax^2 + bx + c$. Then consider the Quadratic Formula as given by

$$x = -\frac{b}{2a} \pm \frac{\sqrt{b^2 - 4ac}}{2a}.$$

Write a graphical interpretation of the two parts of this formula.

77. ANALYZING RELATIONSHIPS Find the sum and product of $\frac{-b + \sqrt{b^2 - 4ac}}{2a}$ and $\frac{-b - \sqrt{b^2 - 4ac}}{2a}$. Then write a quadratic equation whose solutions have a sum of 2 and a product of $\frac{1}{2}$.

78. WRITING A FORMULA Derive a formula that can be used to find solutions of equations that have the form $ax^2 + x + c = 0$. Use your formula to solve $-2x^2 + x + 8 = 0$.

79. MULTIPLE REPRESENTATIONS If p is a solution of a quadratic equation $ax^2 + bx + c = 0$, then $(x - p)$ is a factor of $ax^2 + bx + c$.

a. Copy and complete the table for each pair of solutions.

Solutions	Factors	Quadratic equation
3, 4	$(x - 3), (x - 4)$	$x^2 - 7x + 12 = 0$
$-1, 6$		
0, 2		
$-\frac{1}{2}, 5$		

b. Graph the related function for each equation. Identify the zeros of the function.

CRITICAL THINKING In Exercises 80–82, find all values of k for which the equation has (a) two solutions, (b) one solution, and (c) no solutions.

80. $2x^2 + x + 3k = 0$

81. $x^2 - 4kx + 36 = 0$

82. $kx^2 + 5x - 16 = 0$

Maintaining Mathematical Proficiency

Reviewing what you learned in previous grades and lessons

Solve the system of linear equations using any method. Explain why you chose the method. *(Section 5.1, Section 5.2, and Section 5.3)*

83. $y = -x + 4$
$y = 2x - 8$

84. $x = 16 - 4y$
$3x + 4y = 8$

85. $2x - y = 7$
$2x + 7y = 31$

86. $3x - 2y = -20$
$x + 1.2y = 6.4$

9.6 Solving Nonlinear Systems of Equations

Essential Question How can you solve a system of two equations when one is linear and the other is quadratic?

EXPLORATION 1 Solving a System of Equations

Work with a partner. Solve the system of equations by graphing each equation and finding the points of intersection.

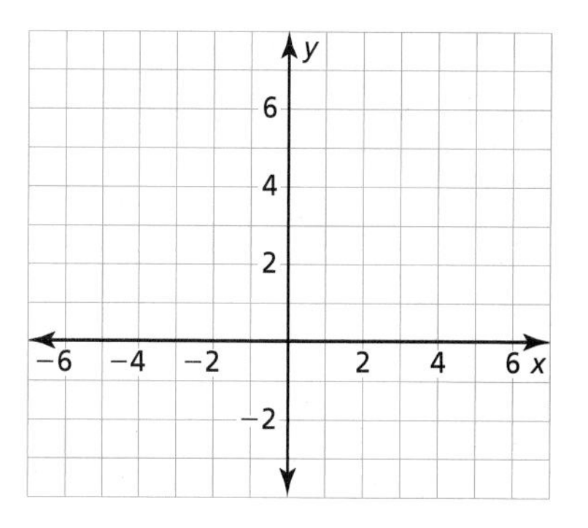

System of Equations

$y = x + 2$ Linear

$y = x^2 + 2x$ Quadratic

EXPLORATION 2 Analyzing Systems of Equations

Work with a partner. Match each system of equations with its graph. Then solve the system of equations.

a. $y = x^2 - 4$
$y = -x - 2$

b. $y = x^2 - 2x + 2$
$y = 2x - 2$

c. $y = x^2 + 1$
$y = x - 1$

d. $y = x^2 - x - 6$
$y = 2x - 2$

A.

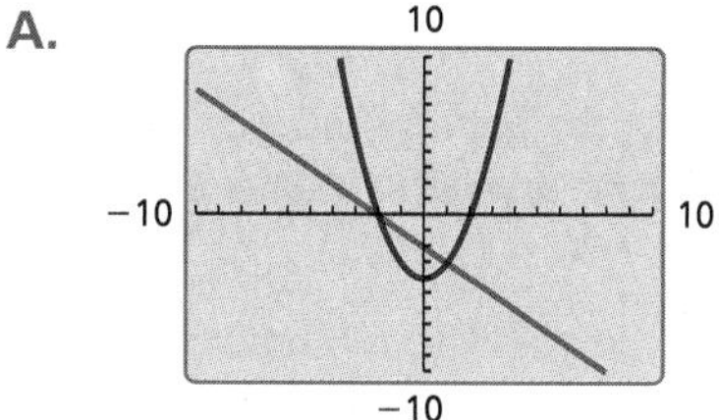

B.

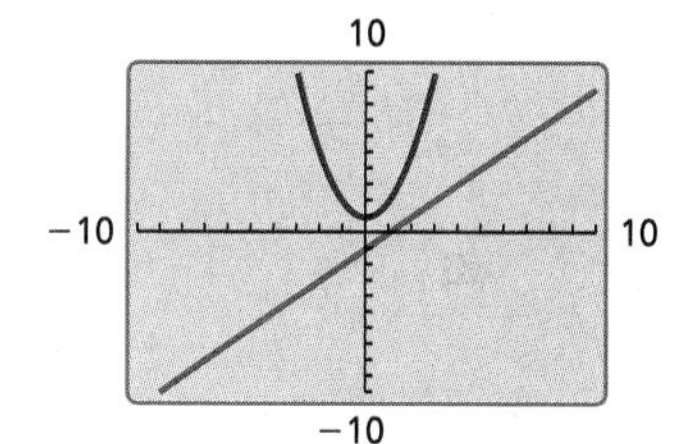

C.

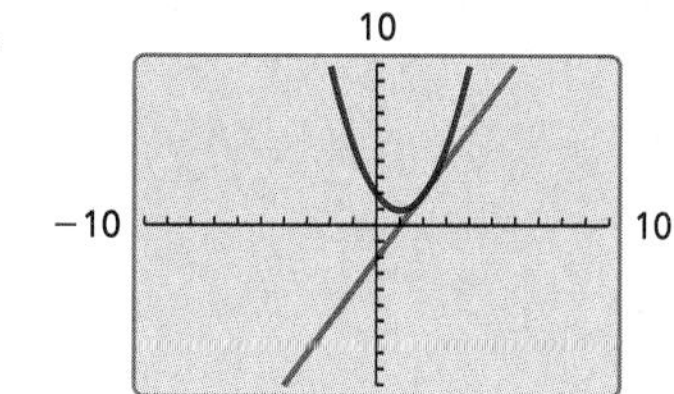

D.

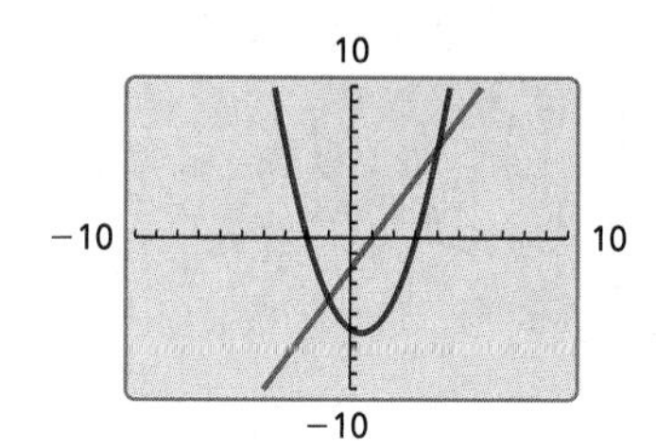

MAKING SENSE OF PROBLEMS

To be proficient in math, you need to analyze givens, relationships, and goals.

Communicate Your Answer

3. How can you solve a system of two equations when one is linear and the other is quadratic?

4. Write a system of equations (one linear and one quadratic) that has (a) no solutions, (b) one solution, and (c) two solutions. Your systems should be different from those in Explorations 1 and 2.

9.6 Lesson

Core Vocabulary

system of nonlinear equations, p. 526

Previous
system of linear equations

What You Will Learn

- Solve systems of nonlinear equations by graphing.
- Solve systems of nonlinear equations algebraically.
- Approximate solutions of nonlinear systems and equations.

Solving Nonlinear Systems by Graphing

The methods for solving systems of linear equations can also be used to solve *systems of nonlinear equations*. A **system of nonlinear equations** is a system in which at least one of the equations is nonlinear.

When a nonlinear system consists of a linear equation and a quadratic equation, the graphs can intersect in zero, one, or two points. So, the system can have zero, one, or two solutions, as shown.

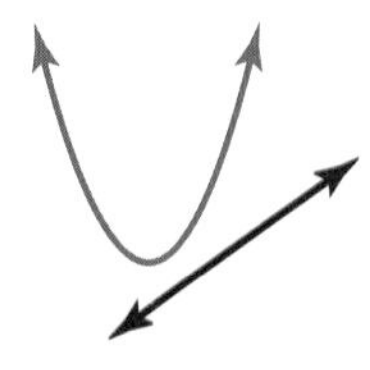

No solutions

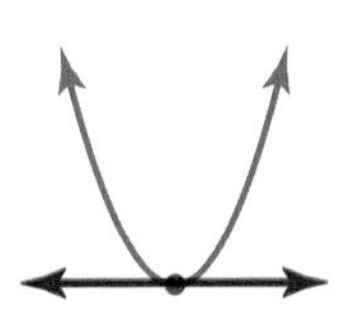

One solution

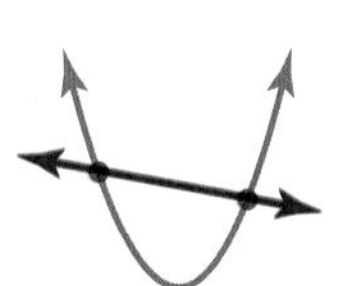

Two solutions

EXAMPLE 1 Solving a Nonlinear System by Graphing

Solve the system by graphing.

$y = 2x^2 + 5x - 1$ Equation 1

$y = x - 3$ Equation 2

SOLUTION

Step 1 Graph each equation.

Step 2 Estimate the point of intersection. The graphs appear to intersect at $(-1, -4)$.

Step 3 Check the point from Step 2 by substituting the coordinates into each of the original equations.

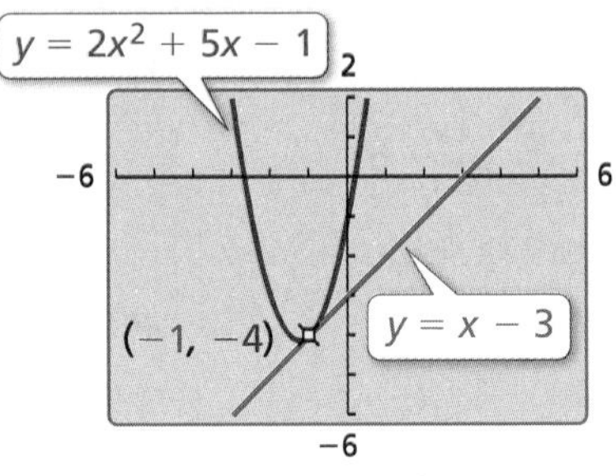

Equation 1

$y = 2x^2 + 5x - 1$

$-4 \stackrel{?}{=} 2(-1)^2 + 5(-1) - 1$

$-4 = -4$ ✓

Equation 2

$y = x - 3$

$-4 \stackrel{?}{=} -1 - 3$

$-4 = -4$ ✓

▶ The solution is $(-1, -4)$.

Monitoring Progress

Help in English and Spanish at *BigIdeasMath.com*

Solve the system by graphing.

1. $y = x^2 + 4x - 4$
$y = 2x - 5$

2. $y = -x + 6$
$y = -2x^2 - x + 3$

3. $y = 3x - 15$
$y = \frac{1}{2}x^2 - 2x - 7$

REMEMBER

The algebraic procedures that you use to solve nonlinear systems are similar to the procedures that you used to solve linear systems in Sections 5.2 and 5.3.

Solving Nonlinear Systems Algebraically

EXAMPLE 2 Solving a Nonlinear System by Substitution

Solve the system by substitution.

$y = x^2 + x - 1$ Equation 1

$y = -2x + 3$ Equation 2

SOLUTION

Step 1 The equations are already solved for y.

Step 2 Substitute $-2x + 3$ for y in Equation 1 and solve for x.

$-2x + 3 = x^2 + x - 1$ Substitute $-2x + 3$ for y in Equation 1.

$3 = x^2 + 3x - 1$ Add $2x$ to each side.

$0 = x^2 + 3x - 4$ Subtract 3 from each side.

$0 = (x + 4)(x - 1)$ Factor the polynomial.

$x + 4 = 0$ or $x - 1 = 0$ Zero-Product Property

$x = -4$ or $x = 1$ Solve for x.

Step 3 Substitute -4 and 1 for x in Equation 2 and solve for y.

$y = -2(-4) + 3$ Substitute for x in Equation 2. $y = -2(1) + 3$

$= 11$ Simplify. $= 1$

▶ So, the solutions are $(-4, 11)$ and $(1, 1)$.

Check

Use a graphing calculator to check your answer. Notice that the graphs have two points of intersection at $(-4, 11)$ and $(1, 1)$.

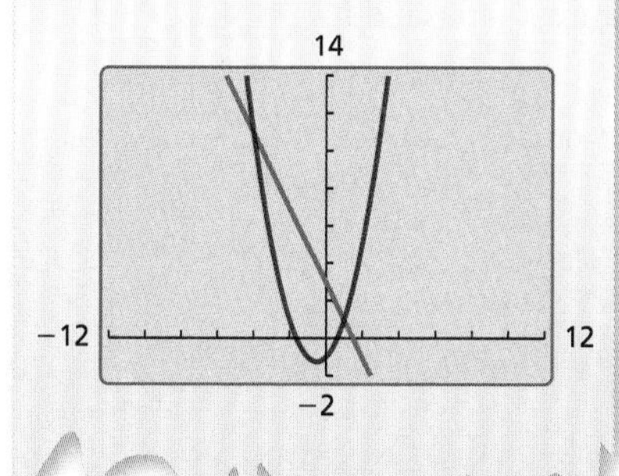

EXAMPLE 3 Solving a Nonlinear System by Elimination

Solve the system by elimination.

$y = x^2 - 3x - 2$ Equation 1

$y = -3x - 8$ Equation 2

SOLUTION

Step 1 Because the coefficients of the y-terms are the same, you do not need to multiply either equation by a constant.

Step 2 Subtract Equation 2 from Equation 1.

$y = x^2 - 3x - 2$ Equation 1

$y = \quad -3x - 8$ Equation 2

$0 = x^2 \quad + 6$ Subtract the equations.

Step 3 Solve for x.

$0 = x^2 + 6$ Resulting equation from Step 2

$-6 = x^2$ Subtract 6 from each side.

▶ The square of a real number cannot be negative. So, the system has no real solutions.

Check

Use a graphing calculator to check your answer. The graphs do not intersect.

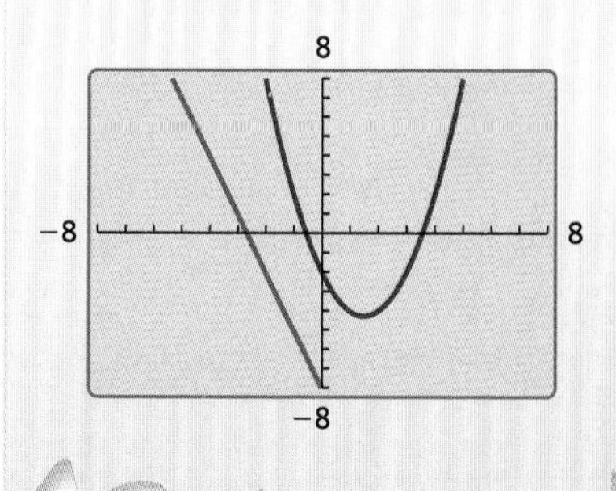

Monitoring Progress  Help in English and Spanish at *BigIdeasMath.com*

Solve the system by substitution.

4. $y = x^2 + 9$
$y = 9$

5. $y = -5x$
$y = x^2 - 3x - 3$

6. $y = -3x^2 + 2x + 1$
$y = 5 - 3x$

Solve the system by elimination.

7. $y = x^2 + x$
$y = x + 5$

8. $y = 9x^2 + 8x - 6$
$y = 5x - 4$

9. $y = 2x + 5$
$y = -3x^2 + x - 4$

Approximating Solutions

When you cannot find the exact solution(s) of a system of equations, you can analyze output values to approximate the solution(s).

EXAMPLE 4 Approximating Solutions of a Nonlinear System

Approximate the solution(s) of the system to the nearest thousandth.

$y = \frac{1}{2}x^2 + 3$ Equation 1

$y = 3^x$ Equation 2

SOLUTION

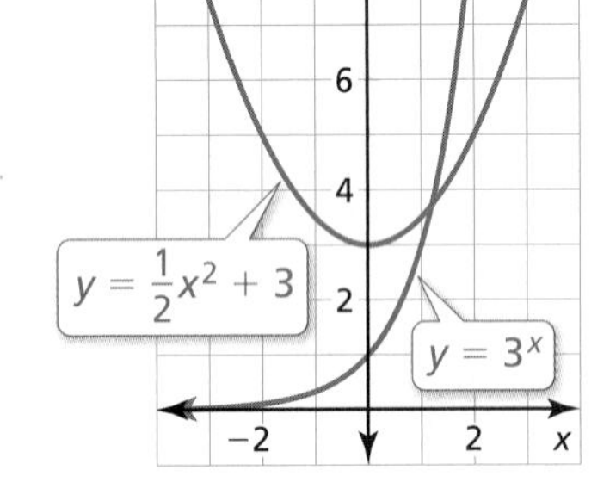

Sketch a graph of the system. You can see that the system has one solution between $x = 1$ and $x = 2$.

Substitute 3^x for y in Equation 1 and rewrite the equation.

$3^x = \frac{1}{2}x^2 + 3$ Substitute 3^x for y in Equation 1.

$3^x - \frac{1}{2}x^2 - 3 = 0$ Rewrite the equation.

Because you do not know how to solve this equation algebraically, let $f(x) = 3^x - \frac{1}{2}x^2 - 3$. Then evaluate the function for x-values between 1 and 2.

$f(1.1) \approx -0.26$
$f(1.2) \approx 0.02$
Because $f(1.1) < 0$ and $f(1.2) > 0$, the zero is between 1.1 and 1.2.

REMEMBER
The function values that are closest to 0 correspond to x-values that best approximate the zeros of the function.

$f(1.2)$ is closer to 0 than $f(1.1)$, so decrease your guess and evaluate $f(1.19)$.

$f(1.19) \approx -0.012$ Because $f(1.19) < 0$ and $f(1.2) > 0$, the zero is between 1.19 and 1.2. So, increase guess.

$f(1.191) \approx -0.009$ Result is negative. Increase guess.

$f(1.192) \approx -0.006$ Result is negative. Increase guess.

$f(1.193) \approx -0.003$ Result is negative. Increase guess.

$f(1.194) \approx -0.0002$ Result is negative. Increase guess.

$f(1.195) \approx 0.003$ Result is positive.

Because $f(1.194)$ is closest to 0, $x \approx 1.194$.

Substitute $x = 1.194$ into one of the original equations and solve for y.

$y = \frac{1}{2}x^2 + 3 = \frac{1}{2}(1.194)^2 + 3 \approx 3.713$

▶ So, the solution of the system is about (1.194, 3.713).

REMEMBER

When entering the equations, be sure to use an appropriate viewing window that shows all the points of intersection. For this system, an appropriate viewing window is $-4 \le x \le 4$ and $-4 \le y \le 4$.

Recall from Section 5.5 that you can use systems of equations to solve equations with variables on both sides.

EXAMPLE 5 Approximating Solutions of an Equation

Solve $-2(4)^x + 3 = 0.5x^2 - 2x$.

SOLUTION

You do not know how to solve this equation algebraically. So, use each side of the equation to write the system $y = -2(4)^x + 3$ and $y = 0.5x^2 - 2x$.

Method 1 Use a graphing calculator to graph the system. Then use the *intersect* feature to find the coordinates of each point of intersection.

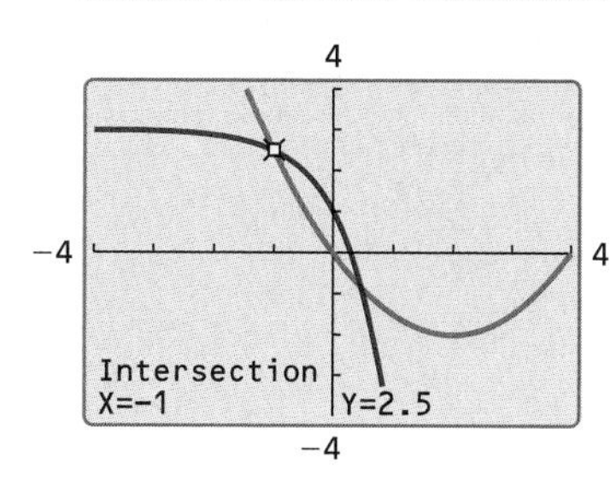

One point of intersection is $(-1, 2.5)$.

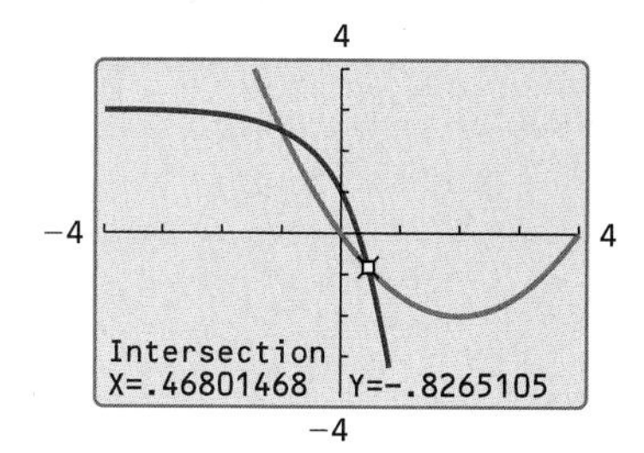

The other point of intersection is about $(0.47, -0.83)$.

▶ So, the solutions of the equation are $x = -1$ and $x \approx 0.47$.

Method 2 Use the *table* feature to create a table of values for the equations. Find the x-values for which the corresponding y-values are approximately equal.

X	Y1	Y2
-1.03	2.5204	2.5905
-1.02	2.5137	2.5602
-1.01	2.5069	2.5301
-1	2.5	2.5
-.99	2.493	2.4701
-.98	2.4859	2.4402
-.97	2.4788	2.4105
X=-1		

When $x = -1$, the corresponding y-values are 2.5.

X	Y1	Y2
.44	-.6808	-.7832
.45	-.7321	-.7988
.46	-.7842	-.8142
.47	-.8371	-.8296
.48	-.8906	-.8448
.49	-.9449	-.86
.50	-1	-.875
X=.47		

When $x = 0.47$, the corresponding y-values are approximately -0.83.

STUDY TIP

You can use the differences between the corresponding y-values to determine the best approximation of a solution.

▶ So, the solutions of the equation are $x = -1$ and $x \approx 0.47$.

Monitoring Progress  Help in English and Spanish at *BigIdeasMath.com*

Use the method in Example 4 to approximate the solution(s) of the system to the nearest thousandth.

10. $y = 4^x$
$y = x^2 + x + 3$

11. $y = 4x^2 - 1$
$y = -2(3)^x + 4$

12. $y = x^2 + 3x$
$y = -x^2 + x + 10$

Solve the equation. Round your solution(s) to the nearest hundredth.

13. $3^x - 1 = x^2 - 2x + 5$

14. $4x^2 + x = -2\left(\frac{1}{2}\right)^x + 5$

9.6 Exercises

Dynamic Solutions available at *BigIdeasMath.com*

Vocabulary and Core Concept Check

1. **VOCABULARY** Describe how to use substitution to solve a system of nonlinear equations.

2. **WRITING** How is solving a system of nonlinear equations similar to solving a system of linear equations? How is it different?

Monitoring Progress and Modeling with Mathematics

In Exercises 3–6, match the system of equations with its graph. Then solve the system.

3. $y = x^2 - 2x + 1$
   $y = x + 1$

4. $y = x^2 + 3x + 2$
   $y = -x - 3$

5. $y = x - 1$
   $y = -x^2 + x - 1$

6. $y = -x + 3$
   $y = -x^2 - 2x + 5$

A. 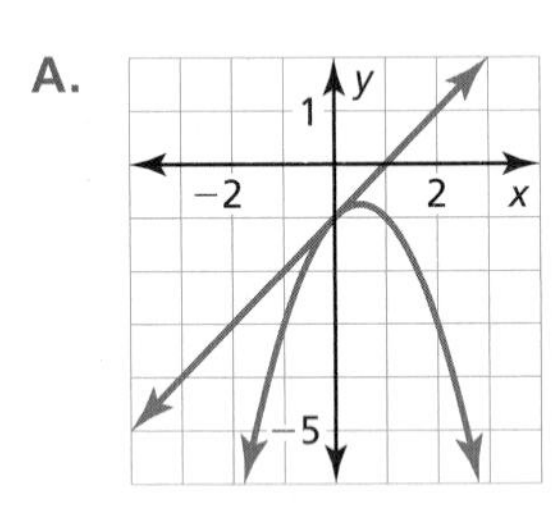

B. 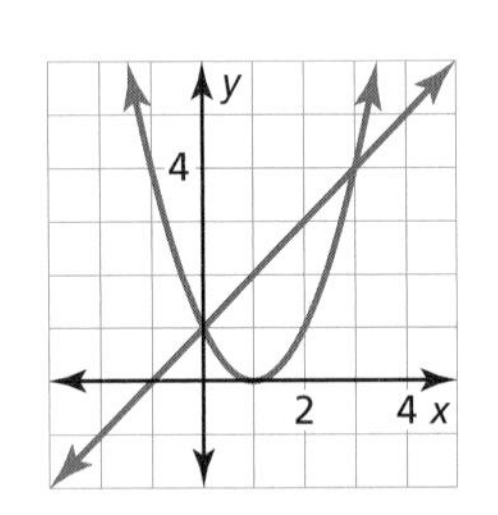

C. 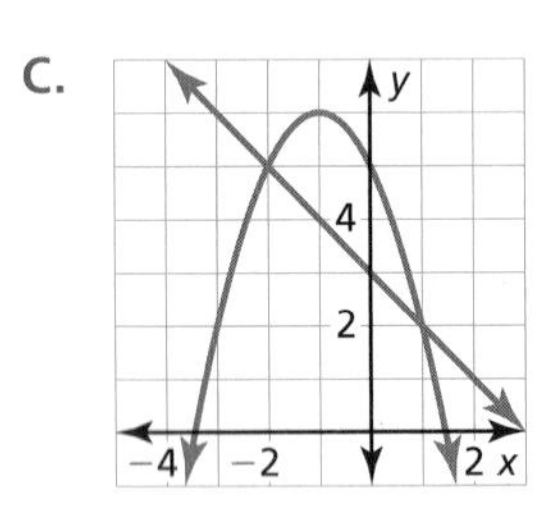

D. 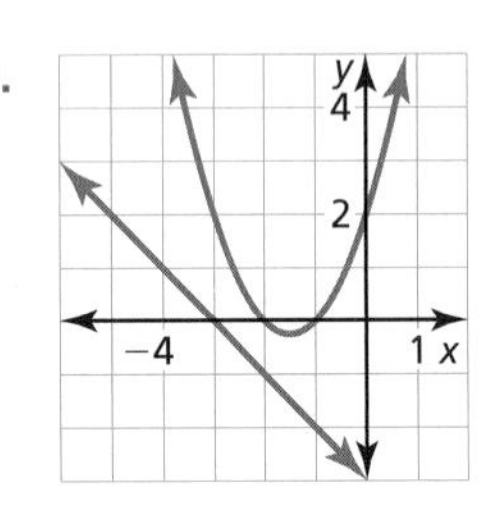

In Exercises 7–12, solve the system by graphing. *(See Example 1.)*

7. $y = 3x^2 - 2x + 1$
   $y = x + 7$

8. $y = x^2 + 2x + 5$
   $y = -2x - 5$

9. $y = -2x^2 - 4x$
   $y = 2$

10. $y = \frac{1}{2}x^2 - 3x + 4$
    $y = x - 2$

11. $y = \frac{1}{3}x^2 + 2x - 3$
    $y = 2x$

12. $y = 4x^2 + 5x - 7$
    $y = -3x + 5$

In Exercises 13–18, solve the system by substitution. *(See Example 2.)*

13. $y = x - 5$
    $y = x^2 + 4x - 5$

14. $y = -3x^2$
    $y = 6x + 3$

15. $y = -x + 7$
    $y = -x^2 - 2x - 1$

16. $y = -x^2 + 7$
    $y = 2x + 4$

17. $y - 5 = -x^2$
    $y = 5$

18. $y = 2x^2 + 3x - 4$
    $y - 4x = 2$

In Exercises 19–26, solve the system by elimination. *(See Example 3.)*

19. $y = x^2 - 5x - 7$
    $y = -5x + 9$

20. $y = -3x^2 + x + 2$
    $y = x + 4$

21. $y = -x^2 - 2x + 2$
    $y = 4x + 2$

22. $y = -2x^2 + x - 3$
    $y = 2x - 2$

23. $y = 2x - 1$
    $y = x^2$

24. $y = x^2 + x + 1$
    $y = -x - 2$

25. $y + 2x = 0$
    $y = x^2 + 4x - 6$

26. $y = 2x - 7$
    $y + 5x = x^2 - 2$

27. **ERROR ANALYSIS** Describe and correct the error in solving the system of equations by graphing.

$y = x^2 - 3x + 4$
$y = 2x + 4$

The only solution of the system is (0, 4).

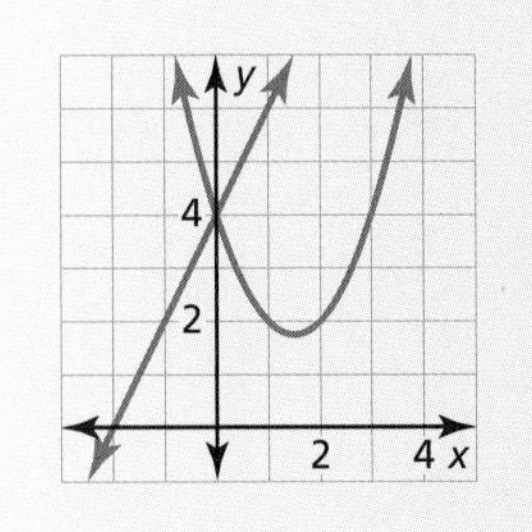

28. ERROR ANALYSIS Describe and correct the error in solving for one of the variables in the system.

> ✗
> $y = 3x^2 - 6x + 4$
> $y = 4$
>
> $y = 3(4)^2 - 6(4) + 4$ Substitute.
> $y = 28$ Simplify.

In Exercises 29–32, use the table to describe the locations of the zeros of the quadratic function *f*.

29.

x	-4	-3	-2	-1	0	1
$f(x)$	-2	2	4	4	2	-2

30.

x	-1	0	1	2	3	4
$f(x)$	11	5	1	-1	-1	1

31.

x	-4	-3	-2	-1	0	1
$f(x)$	3	-1	-1	3	11	23

32.

x	1	2	3	4	5	6
$f(x)$	-25	-9	1	5	3	-5

In Exercises 33–38, use the method in Example 4 to approximate the solution(s) of the system to the nearest thousandth. *(See Example 4.)*

33. $y = x^2 + 2x + 3$
$y = 3^x$

34. $y = 2^x + 5$
$y = x^2 - 3x + 1$

35. $y = 2(4)^x - 1$
$y = 3x^2 + 8x$

36. $y = -x^2 - 4x - 4$
$y = -5^x - 2$

37. $y = -x^2 - x + 5$
$y = 2x^2 + 6x - 3$

38. $y = 2x^2 + x - 8$
$y = x^2 - 5$

In Exercises 39–46, solve the equation. Round your solution(s) to the nearest hundredth. *(See Example 5.)*

39. $3x + 1 = x^2 + 7x - 1$

40. $-x^2 + 2x = -2x + 5$

41. $x^2 - 6x + 4 = -x^2 - 2x$

42. $2x^2 + 8x + 10 = -x^2 - 2x + 5$

43. $-4\left(\frac{1}{2}\right)^x = -x^2 - 5$

44. $1.5(2)^x - 3 = -x^2 + 4x$

45. $8^{x-2} + 3 = 2\left(\frac{3}{2}\right)^x$

46. $-0.5(4)^x = 5^x - 6$

47. COMPARING METHODS Solve the system in Exercise 37 using substitution. Compare the exact solutions to the approximated solutions.

48. COMPARING METHODS Solve the system in Exercise 38 using elimination. Compare the exact solutions to the approximated solutions.

49. MODELING WITH MATHEMATICS The attendances y for two movies can be modeled by the following equations, where x is the number of days since the movies opened.

$y = -x^2 + 35x + 100$ Movie A

$y = -5x + 275$ Movie B

When is the attendance for each movie the same?

50. MODELING WITH MATHEMATICS You and a friend are driving boats on the same lake. Your path can be modeled by the equation $y = -x^2 - 4x - 1$, and your friend's path can be modeled by the equation $y = 2x + 8$. Do your paths cross each other? If so, what are the coordinates of the point(s) where the paths meet?

51. MODELING WITH MATHEMATICS The arch of a bridge can be modeled by $y = -0.002x^2 + 1.06x$, where x is the distance (in meters) from the left pylons and y is the height (in meters) of the arch above the water. The road can be modeled by the equation $y = 52$. To the nearest meter, how far from the left pylons are the two points where the road intersects the arch of the bridge?

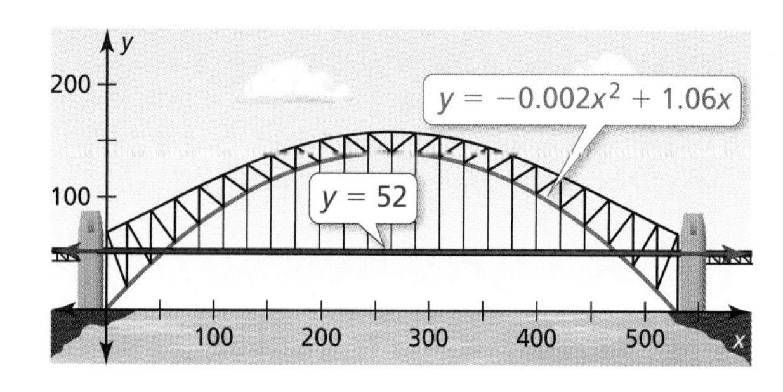

52. MAKING AN ARGUMENT Your friend says that a system of equations consisting of a linear equation and a quadratic equation can have zero, one, two, or infinitely many solutions. Is your friend correct? Explain.

COMPARING METHODS **In Exercises 53 and 54, solve the system of equations by (a) graphing, (b) substitution, and (c) elimination. Which method do you prefer? Explain your reasoning.**

53. $y = 4x + 3$
$y = x^2 + 4x - 1$

54. $y = x^2 - 5$
$y = -x + 7$

55. MODELING WITH MATHEMATICS The function $y = -x^2 + 65x + 256$ models the number y of subscribers to a website, where x is the number of days since the website launched. The number of subscribers to a competitor's website can be modeled by a linear function. The websites have the same number of subscribers on Days 1 and 34.

a. Write a linear function that models the number of subscribers to the competitor's website.

b. Solve the system to verify the function from part (a).

56. HOW DO YOU SEE IT? The diagram shows the graphs of two equations in a system that has one solution.

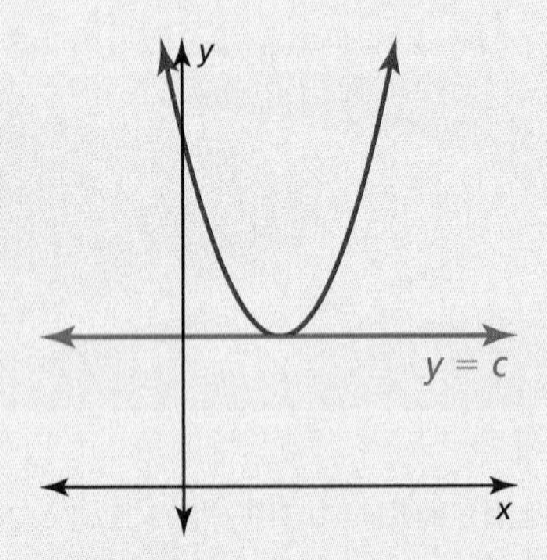

a. How many solutions will the system have when you change the linear equation to $y = c + 2$?

b. How many solutions will the system have when you change the linear equation to $y = c - 2$?

57. WRITING A system of equations consists of a quadratic equation whose graph opens up and a quadratic equation whose graph opens down. Describe the possible numbers of solutions of the system. Sketch examples to justify your answer.

58. PROBLEM SOLVING The population of a country is 2 million people and increases by 3% each year. The country's food supply is sufficient to feed 3 million people and increases at a constant rate that feeds 0.25 million additional people each year.

a. When will the country first experience a food shortage?

b. The country doubles the rate at which its food supply increases. Will food shortages still occur? If so, in what year?

59. ANALYZING GRAPHS Use the graphs of the linear and quadratic functions.

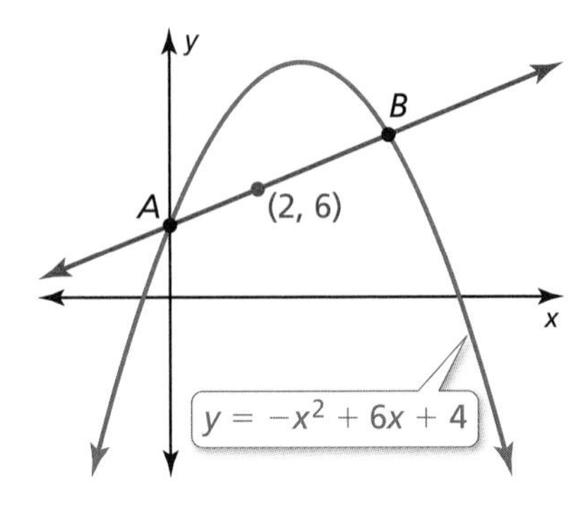

a. Find the coordinates of point A.

b. Find the coordinates of point B.

60. THOUGHT PROVOKING Is it possible for a system of two quadratic equations to have exactly three solutions? exactly four solutions? Explain your reasoning. (*Hint*: Rotations of the graphs of quadratic equations still represent quadratic equations.)

61. PROBLEM SOLVING Solve the system of three equations shown.

$y = 2x - 8$
$y = x^2 - 4x - 3$
$y = -3(2)^x$

62. PROBLEM SOLVING Find the point(s) of intersection, if any, of the line $y = -x - 1$ and the circle $x^2 + y^2 = 41$.

Maintaining Mathematical Proficiency

Reviewing what you learned in previous grades and lessons

Graph the system of linear inequalities. *(Section 5.7)*

63. $y > 2x$
$y > -x + 4$

64. $y \geq 4x + 1$
$y \leq 7$

65. $y - 3 \leq -2x$
$y + 5 < 3x$

66. $x + y > -6$
$2y \leq 3x + 4$

Graph the function. Describe the domain and range. *(Section 8.3)*

67. $y = 3x^2 + 2$

68. $y = -x^2 - 6x$

69. $y = -2x^2 + 12x - 7$

70. $y = 5x^2 + 10x - 3$

9.4–9.6 What Did You Learn?

Core Vocabulary

completing the square, *p. 506*
Quadratic Formula, *p. 516*
discriminant, *p. 518*
system of nonlinear equations, *p. 526*

Core Concepts

Section 9.4

Completing the Square, *p. 506*

Section 9.5

Quadratic Formula, *p. 516*
Interpreting the Discriminant, *p. 518*

Section 9.6

Solving Systems of Nonlinear Equations, *p. 526*

Mathematical Practices

1. How does your answer to Exercise 74 on page 514 help create a shortcut when solving some quadratic equations by completing the square?
2. What logical progression led you to your answer in Exercise 55 on page 522?
3. Compare the methods used to solve Exercise 53 on page 532. Discuss the similarities and differences among the methods.

Performance Task

Form Matters

Each form of a quadratic function has its pros and cons. Using one form, you can easily find the vertex, but the zeros are more difficult to find. Using another form, you can easily find the y-intercept, but the vertex is more difficult to find. Which form would you use in different situations? How can you convert one form into another?

To explore the answers to these questions and more, go to ***BigIdeasMath.com***.

9 Chapter Review

Dynamic Solutions available at *BigIdeasMath.com*

9.1 Properties of Radicals (pp. 479–488)

a. Simplify $\sqrt[3]{27x^{10}}$.

$\sqrt[3]{27x^{10}} = \sqrt[3]{27 \cdot x^9 \cdot x}$ — Factor using the greatest perfect cube factors.

$= \sqrt[3]{27} \cdot \sqrt[3]{x^9} \cdot \sqrt[3]{x}$ — Product Property of Cube Roots

$= 3x^3\sqrt[3]{x}$ — Simplify.

b. Simplify $\dfrac{12}{3 + \sqrt{5}}$.

$\dfrac{12}{3+\sqrt{5}} = \dfrac{12}{3+\sqrt{5}} \cdot \dfrac{3-\sqrt{5}}{3-\sqrt{5}}$ — The conjugate of $3 + \sqrt{5}$ is $3 - \sqrt{5}$.

$= \dfrac{12(3-\sqrt{5})}{3^2 - (\sqrt{5})^2}$ — Sum and difference pattern

$= \dfrac{36 - 12\sqrt{5}}{4}$ — Simplify.

$= 9 - 3\sqrt{5}$ — Simplify.

Simplify the expression.

1. $\sqrt{72p^7}$ **2.** $\sqrt{\dfrac{45}{7y}}$ **3.** $\sqrt[3]{\dfrac{125x^{11}}{4}}$ **4.** $\dfrac{8}{\sqrt{6}+2}$

5. $4\sqrt{3} + 5\sqrt{12}$ **6.** $15\sqrt[3]{2} - 2\sqrt[3]{54}$ **7.** $(3\sqrt{7} + 5)^2$ **8.** $\sqrt{6}(\sqrt{18} + \sqrt{8})$

9.2 Solving Quadratic Equations by Graphing (pp. 489–496)

Solve $x^2 + 3x = 4$ by graphing.

Step 1 Write the equation in standard form.

$x^2 + 3x = 4$ — Write original equation.

$x^2 + 3x - 4 = 0$ — Subtract 4 from each side.

Step 2 Graph the related function $y = x^2 + 3x - 4$.

Step 3 Find the *x*-intercepts. The *x*-intercepts are -4 and 1.

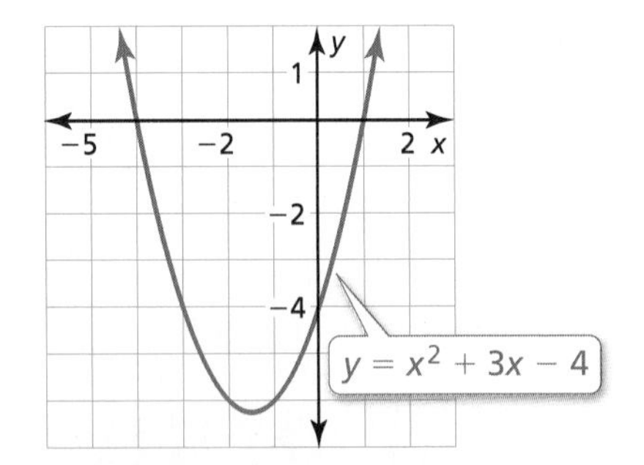

▶ So, the solutions are $x = -4$ and $x = 1$.

Solve the equation by graphing.

9. $x^2 - 9x + 18 = 0$ **10.** $x^2 - 2x = -4$

11. $-8x - 16 = x^2$

12. The graph of $f(x) = (x + 1)(x^2 + 2x - 3)$ is shown. Find the zeros of f.

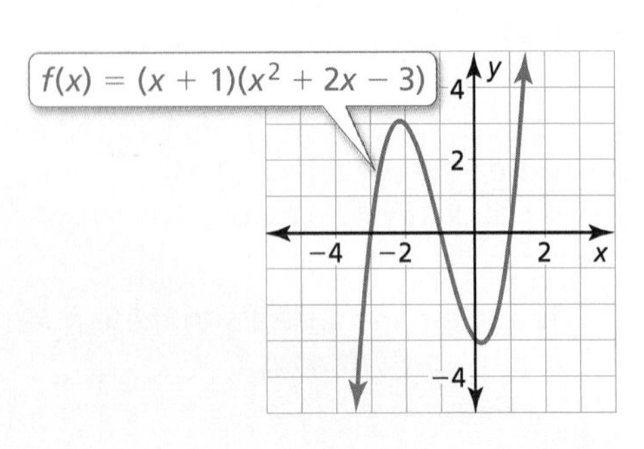

13. Graph $f(x) = x^2 + 2x - 5$. Approximate the zeros of f to the nearest tenth.

9.3 Solving Quadratic Equations Using Square Roots (pp. 497–502)

A sprinkler sprays water that covers a circular region of 90π square feet. Find the diameter of the circle.

Write an equation using the formula for the area of a circle.

$A = \pi r^2$	Write the formula.
$90\pi = \pi r^2$	Substitute 90π for A.
$90 = r^2$	Divide each side by π.
$\pm\sqrt{90} = r$	Take the square root of each side.
$\pm 3\sqrt{10} = r$	Simplify.

A diameter cannot be negative, so use the positive square root. The diameter is twice the radius. So, the diameter is $6\sqrt{10}$.

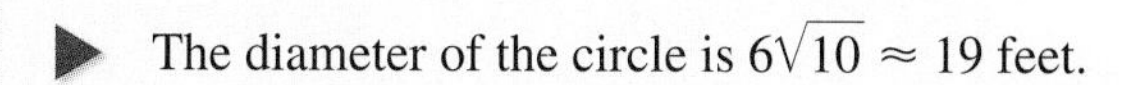

▶ The diameter of the circle is $6\sqrt{10} \approx 19$ feet.

Solve the equation using square roots. Round your solutions to the nearest hundredth, if necessary.

14. $x^2 + 5 = 17$

15. $x^2 - 14 = -14$

16. $(x + 2)^2 = 64$

17. $4x^2 + 25 = -75$

18. $(x - 1)^2 = 0$

19. $19 = 30 - 5x^2$

9.4 Solving Quadratic Equations by Completing the Square (pp. 505–514)

Solve $x^2 - 6x + 4 = 11$ by completing the square.

$x^2 - 6x + 4 = 11$	Write the equation.
$x^2 - 6x = 7$	Subtract 4 from each side.
$x^2 - 6x + (-3)^2 = 7 + (-3)^2$	Complete the square by adding $\left(\frac{-6}{2}\right)^2$, or $(-3)^2$, to each side.
$(x - 3)^2 = 16$	Write the left side as the square of a binomial.
$x - 3 = \pm 4$	Take the square root of each side.
$x = 3 \pm 4$	Add 3 to each side.

▶ The solutions are $x = 3 + 4 = 7$ and $x = 3 - 4 = -1$.

Solve the equation by completing the square. Round your solutions to the nearest hundredth, if necessary.

20. $x^2 + 6x - 40 = 0$

21. $x^2 + 2x + 5 = 4$

22. $2x^2 - 4x = 10$

Determine whether the quadratic function has a maximum or minimum value. Then find the value.

23. $y = -x^2 + 6x - 1$

24. $f(x) = x^2 + 4x + 11$

25. $y = 3x^2 - 24x + 15$

26. The width w of a credit card is 3 centimeters shorter than the length ℓ. The area is 46.75 square centimeters. Find the perimeter.

9.5 Solving Quadratic Equations Using the Quadratic Formula (pp. 515–524)

Solve $-3x^2 + x = -8$ using the Quadratic Formula.

$-3x^2 + x = -8$	Write the equation.
$-3x^2 + x + 8 = 0$	Write in standard form.
$x = \dfrac{-b \pm \sqrt{b^2 - 4ac}}{2a}$	Quadratic Formula
$x = \dfrac{-1 \pm \sqrt{1^2 - 4(-3)(8)}}{2(-3)}$	Substitute -3 for a, 1 for b, and 8 for c.
$x = \dfrac{-1 \pm \sqrt{97}}{-6}$	Simplify.

▶ So, the solutions are $x = \dfrac{-1 + \sqrt{97}}{-6} \approx -1.5$ and $x = \dfrac{-1 - \sqrt{97}}{-6} \approx 1.8$.

Solve the equation using the Quadratic Formula. Round your solutions to the nearest tenth, if necessary.

27. $x^2 + 2x - 15 = 0$ **28.** $2x^2 - x + 8 = 16$ **29.** $-5x^2 + 10x = 5$

Find the number of x-intercepts of the graph of the function.

30. $y = -x^2 + 6x - 9$ **31.** $y = 2x^2 + 4x + 8$ **32.** $y = -\frac{1}{2}x^2 + 2x$

9.6 Solving Nonlinear Systems of Equations (pp. 525–532)

Solve the system by substitution. $y = x^2 - 5$ Equation 1
$y = -x + 1$ Equation 2

Step 1 The equations are already solved for y.

Step 2 Substitute $-x + 1$ for y in Equation 1 and solve for x.

$-x + 1 = x^2 - 5$	Substitute $-x + 1$ for y in Equation 1.
$1 = x^2 + x - 5$	Add x to each side.
$0 = x^2 + x - 6$	Subtract 1 from each side.
$0 = (x + 3)(x - 2)$	Factor the polynomial.
$x + 3 = 0$ or $x - 2 = 0$	Zero-Product Property
$x = -3$ or $x = 2$	Solve for x.

Step 3 Substitute -3 and 2 for x in Equation 2 and solve for y.

$y = -(-3) + 1$	Substitute for x in Equation 2.	$y = -2 + 1$
$= 4$	Simplify.	$= -1$

▶ So, the solutions are $(-3, 4)$ and $(2, -1)$.

Solve the system using any method.

33. $y = x^2 - 2x - 4$
$y = -5$

34. $y = x^2 - 9$
$y = 2x + 5$

35. $y = 2\left(\frac{1}{2}\right)^x - 5$
$y = -x^2 - x + 4$

9 Chapter Test

Solve the equation using any method. Explain your choice of method.

1. $x^2 - 121 = 0$

2. $x^2 - 6x = 10$

3. $-2x^2 + 3x + 7 = 0$

4. $x^2 - 7x + 12 = 0$

5. $5x^2 + x - 4 = 0$

6. $(4x + 3)^2 = 16$

7. Describe how you can use the method of completing the square to determine whether the function $f(x) = 2x^2 + 4x - 6$ can be represented by the graph shown.

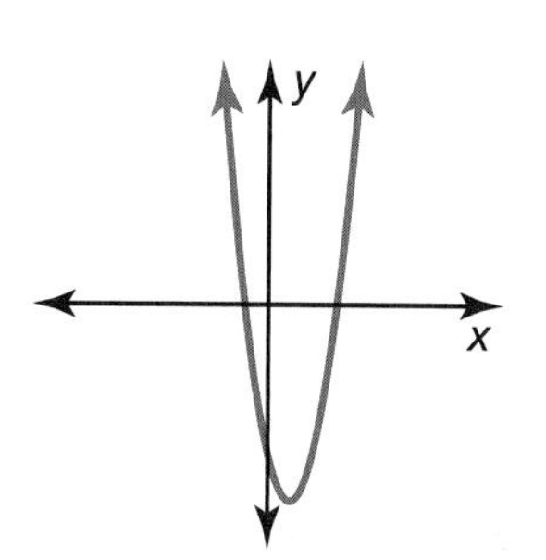

8. Write an expression involving radicals in which a conjugate can be used to simplify the expression.

Solve the system using any method.

9. $y = x^2 - 4x - 2$
$y = -4x + 2$

10. $y = -5x^2 + x - 1$
$y = -7$

11. $y = \frac{1}{2}(4)^x + 1$
$y = x^2 - 2x + 4$

12. A skier leaves an 8-foot-tall ramp with an initial vertical velocity of 28 feet per second. The function $h = -16t^2 + 28t + 8$ represents the height h (in feet) of the skier after t seconds. The skier has a perfect landing. How many points does the skier earn?

Criteria	Scoring
Maximum height	1 point per foot
Time in air	5 points per second
Perfect landing	25 points

13. An amusement park ride lifts seated riders 265 feet above the ground. The riders are then dropped and experience free fall until the brakes are activated 105 feet above the ground. The function $h = -16t^2 + 265$ represents the height h (in feet) of the riders t seconds after they are dropped. How long do the riders experience free fall? Round your solution to the nearest hundredth.

14. Write an expression in simplest form that represents the area of the painting shown.

$\frac{36}{\sqrt{3}}$ in.

$\sqrt{30x^7}$ in.

15. Explain how you can determine the number of times the graph of $y = 5x^2 - 10x + 5$ intersects the x-axis without graphing or solving an equation.

16. Consider the quadratic equation $ax^2 + bx + c = 0$. Find values of a, b, and c so that the graph of its related function has (a) two x-intercepts, (b) one x-intercept, and (c) no x-intercepts.

17. The numbers y of two types of bacteria after x hours are represented by the models below.

$y = 3x^2 + 8x + 20$ Type A
$y = 27x + 60$ Type B

a. When are there 400 Type A bacteria?

b. When are the number of Type A and Type B bacteria the same?

c. When are there more Type A bacteria than Type B? When are there more Type B bacteria than Type A? Use a graph to support your answer.

9 Cumulative Assessment

1. The graphs of four quadratic functions are shown. Determine whether the discriminants of the equations $f(x) = 0$, $g(x) = 0$, $h(x) = 0$, and $j(x) = 0$ are positive, negative, or zero.

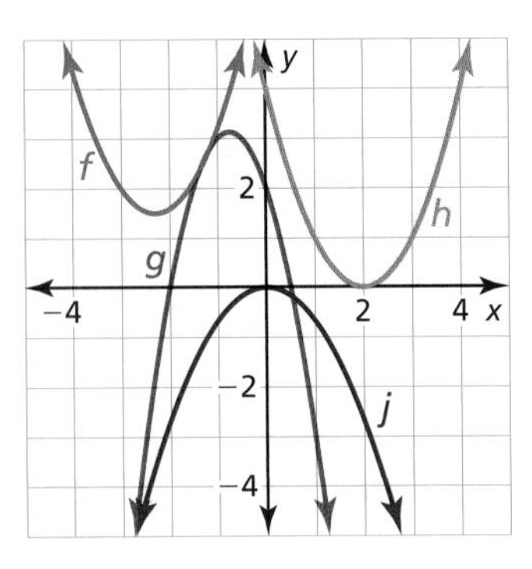

2. The function $f(x) = a(1.08)^x$ represents the total amount of money (in dollars) in Account A after x years. The function $g(x) = 600(b)^x$ represents the total amount of money (in dollars) in Account B after x years. Fill in values for a and b so that each statement is true.

   a. When $a =$ _____ and $b =$ _____, Account B has a greater initial amount and increases at a faster rate than Account A.

   b. When $a =$ _____ and $b =$ _____, Account B has a lesser initial amount than Account A but increases at a faster rate than Account A.

   c. When $a =$ _____ and $b =$ _____, Account B and Account A have the same initial amount, and Account B increases at a slower rate than Account A.

3. Your friend claims to be able to find the radius r of each figure, given the surface area S. Do you support your friend's claim? Justify your answer.

   a.

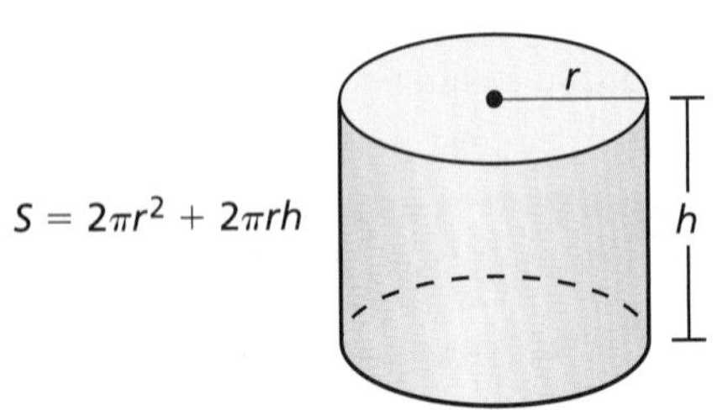

   b.

$S = 4\pi r^2$

r

4. The tables represent the numbers of items sold at a concession stand on days with different average temperatures. Determine whether the data represented by each table show a *positive*, a *negative*, or *no* correlation.

Temperature (°F), *x*	14	27	32	41	48	62	73
Cups of hot chocolate, *y*	35	28	22	9	4	2	1

Temperature (°F), *x*	14	27	32	41	48	62	73
Bottles of sports drink, *y*	8	12	13	16	19	27	29

5. Which graph shows exponential growth?

Ⓐ

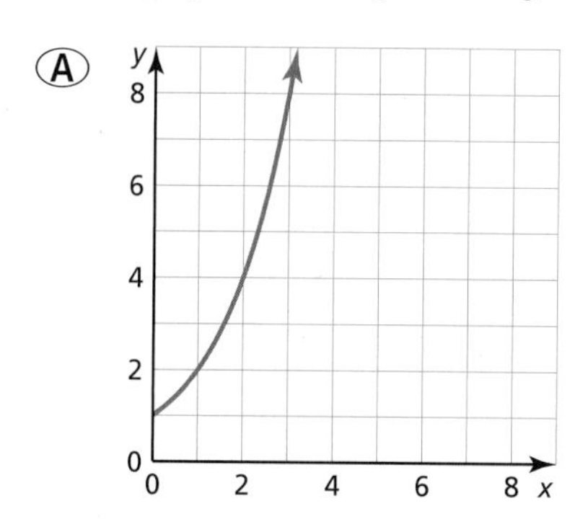

Ⓑ

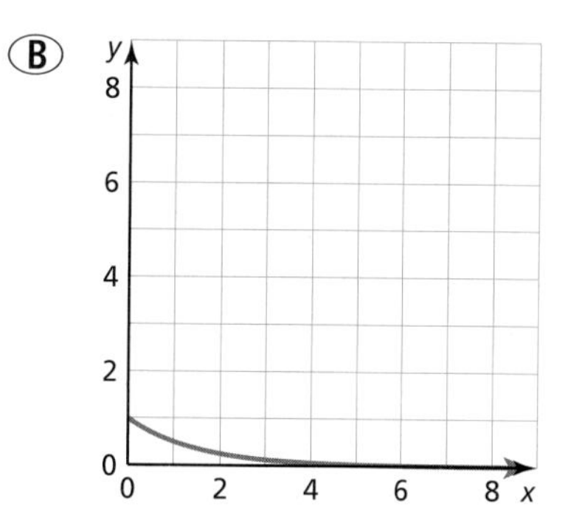

Ⓒ

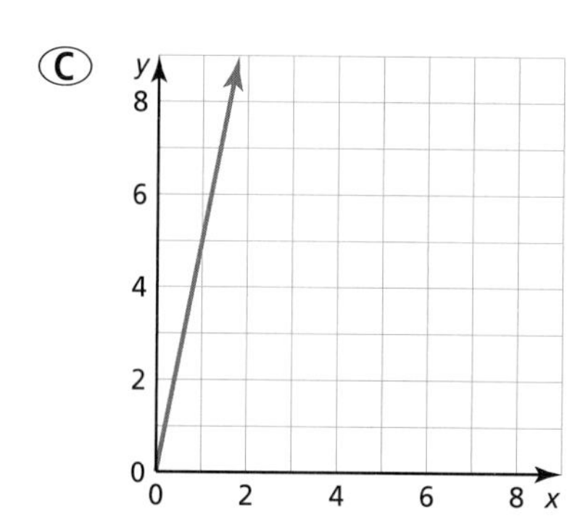

Ⓓ

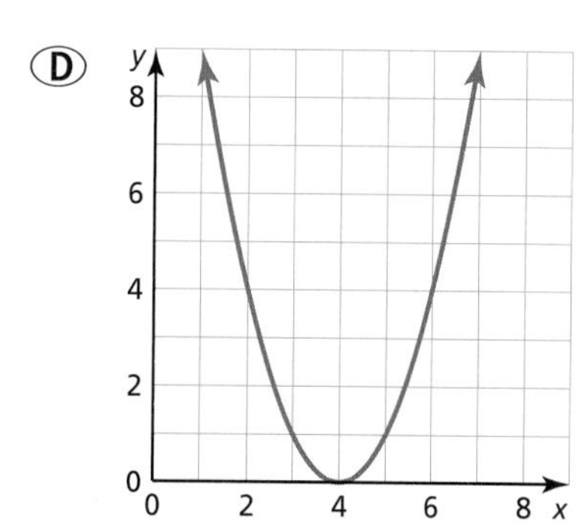

6. Which statement best describes the solution(s) of the system of equations?

$$y = x^2 + 2x - 8$$
$$y = 5x + 2$$

Ⓐ The graphs intersect at one point, $(-2, -8)$. So, there is one solution.

Ⓑ The graphs intersect at two points, $(-2, -8)$ and $(5, 27)$. So, there are two solutions.

Ⓒ The graphs do not intersect. So, there is no solution.

Ⓓ The graph of $y = x^2 + 2x - 8$ has two x-intercepts. So, there are two solutions.

7. Which expressions are in simplest form?

$x\sqrt{45x}$	$\dfrac{16}{\sqrt{5}}$	$\sqrt[3]{\dfrac{4}{9}}$	$16\sqrt{5}$	$3x\sqrt{5x}$
$\dfrac{\sqrt[3]{x^4}}{2}$	$\dfrac{4\sqrt{7}}{3}$	$\dfrac{\sqrt{16}}{5}$	$2\sqrt[3]{x^2}$	$3\dfrac{\sqrt{7}}{\sqrt{x}}$

8. The domain of the function shown is all integers in the interval $-3 < x \leq 3$. Find all the ordered pairs that are solutions of the equation $y = f(x)$.

$$f(x) = 4x - 5$$

10 Radical Functions and Equations

Crow Feeding Habits *(p. 573)*

Trapeze Artist *(p. 565)*

Asian Elephant *(p. 554)*

Firefighting *(p. 549)*

Tsunami *(p. 547)*

Maintaining Mathematical Proficiency

Evaluating Expressions Involving Square Roots

Example 1 **Evaluate $-4(\sqrt{121} - 16)$.**

$$-4(\sqrt{121} - 16) = -4(11 - 16) \quad \text{Evaluate the square root.}$$
$$= -4(-5) \quad \text{Subtract.}$$
$$= 20 \quad \text{Multiply.}$$

Evaluate the expression.

1. $7\sqrt{25} + 10$

2. $-8 - \sqrt{\frac{64}{16}}$

3. $5\left(\frac{\sqrt{81}}{3} - 7\right)$

4. $-2(3\sqrt{9} + 13)$

Transforming Linear Functions

Example 2 **Graph $f(x) = x$ and $g(x) = -3x - 4$. Describe the transformations from the graph of f to the graph of g.**

Note that you can rewrite g as $g(x) = -3f(x) - 4$.

Step 1 There is no horizontal translation from the graph of f to the graph of g.

Step 2 Stretch the graph of f vertically by a factor of 3 to get the graph of $h(x) = 3x$.

Step 3 Reflect the graph of h in the x-axis to get the graph of $r(x) = -3x$.

Step 4 Translate the graph of r vertically 4 units down to get the graph of $g(x) = -3x - 4$.

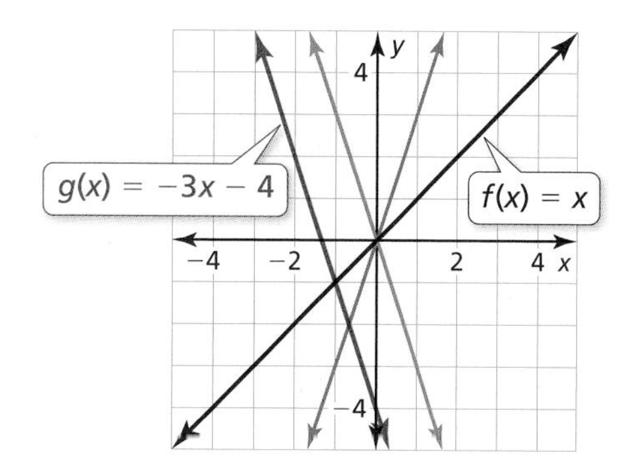

Graph f and g. Describe the transformations from the graph of f to the graph of g.

5. $f(x) = x$; $g(x) = 2x - 2$

6. $f(x) = x$; $g(x) = \frac{1}{3}x + 5$

7. $f(x) = x$; $g(x) = -x + 3$

8. **ABSTRACT REASONING** Let a and b represent constants, where $b \geq 0$. Describe the transformations from the graph of $m(x) = ax + b$ to the graph of $n(x) = -2ax - 4b$.

Mathematically proficient students distinguish correct reasoning from flawed reasoning.

Logical Reasoning

Core Concept

Logical Reasoning and Proof by Contradiction

Mathematics is a logical system that is built from only a few assumptions and undefined terms. The assumptions are called *axioms* or *postulates*. After starting with a collection of axioms and undefined terms, the remainder of mathematics is logically built using careful definitions and theorems (or rules), which are based on the axioms or on previously proven theorems.

To write an indirect proof, or *proof by contradiction*, identify the statement you want to prove. Assume temporarily that this statement is false by assuming that its opposite is true. Then reason logically until you reach a contradiction. Point out that the original statement must be true because the contradiction proves the temporary assumption false.

EXAMPLE 1 Understanding a Proof

A number is *rational* when it can be written as the ratio a/b of two integers, where $b \neq 0$. Use proof by contradiction to prove that $\sqrt{2}$ is not a rational number.

SOLUTION

Assume that $\sqrt{2}$ can be written as the ratio of two integers (in simplest form) and show that this assumption leads to a contradiction.

$\sqrt{2} = \frac{a}{b}$ Assume $\sqrt{2}$ is rational.

$2 = \frac{a^2}{b^2}$ Square each side.

$a^2 = 2b^2$ Multiply each side by b^2 and interchange left and right sides.

This implies that a^2 is even, which is only true when a is even (divisible by 2). This implies that a^2 is divisible by 2^2, or 4. So, $2b^2$ is also divisible by 4, meaning that b^2 is divisible by 2, b^2 is even, and b is even. Because both a and b are even, they have a common factor (of at least 2).

This contradicts the assumption that the ratio a/b is written in simplest form. So, the initial assumption that $\sqrt{2}$ is rational must be false. Therefore, $\sqrt{2}$ is not a rational number.

Monitoring Progress

1. Which of the following square roots are rational numbers? Explain your reasoning.

   $\sqrt{0}, \sqrt{1}, \sqrt{3}, \sqrt{4}, \sqrt{5}, \sqrt{6}, \sqrt{7}, \sqrt{8}, \sqrt{9}$

2. The sequence of steps shown appears to prove that $1 = 0$. What is wrong with this argument?

   $x = 1$ Let $x = 1$.

   $x - 1 = 0$ Subtract 1 from each side.

   $x(x - 1) = 0$ Multiply each side by x.

   $x = 0$ Divide each side by $(x - 1)$.

10.1 Graphing Square Root Functions

Essential Question What are some of the characteristics of the graph of a square root function?

EXPLORATION 1 Graphing Square Root Functions

Work with a partner.

- Make a table of values for each function.
- Use the table to sketch the graph of each function.
- Describe the domain of each function.
- Describe the range of each function.

a. $y = \sqrt{x}$

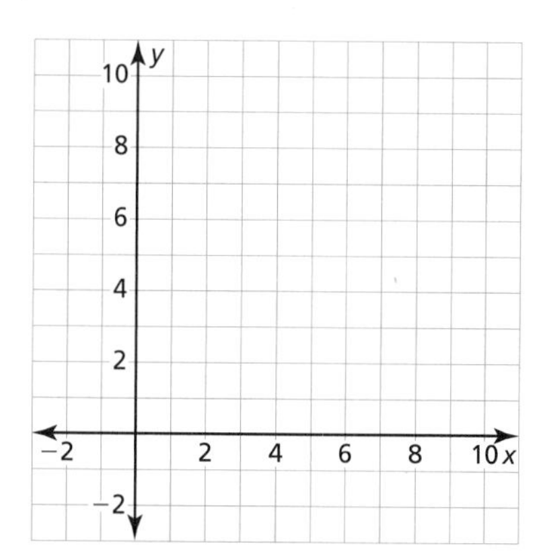

b. $y = \sqrt{x + 2}$

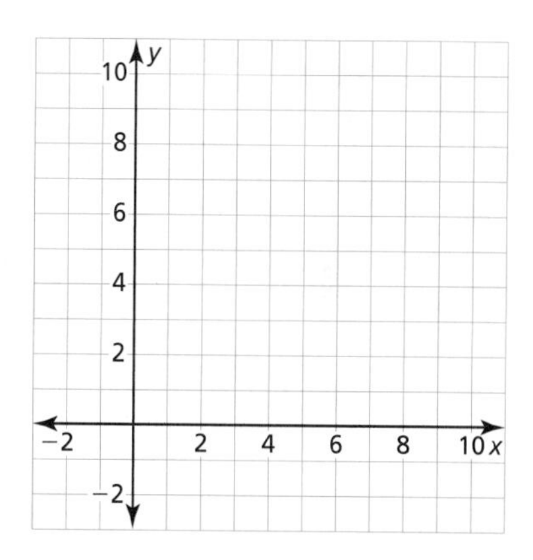

LOOKING FOR A PATTERN

To be proficient in math, you need to look closely to discern a pattern or structure.

EXPLORATION 2 Writing Square Root Functions

Work with a partner. Write a square root function, $y = f(x)$, that has the given values. Then use the function to complete the table.

a.

x	$f(x)$
−4	0
−3	
−2	
−1	$\sqrt{3}$
0	2
1	

b.

x	$f(x)$
−4	1
−3	
−2	
−1	$1 + \sqrt{3}$
0	3
1	

Communicate Your Answer

3. What are some of the characteristics of the graph of a square root function?

4. Graph each function. Then compare the graph to the graph of $f(x) = \sqrt{x}$.

a. $g(x) = \sqrt{x - 1}$ **b.** $g(x) = \sqrt{x} - 1$ **c.** $g(x) = 2\sqrt{x}$ **d.** $g(x) = -2\sqrt{x}$

10.1 Lesson

Core Vocabulary

square root function, *p. 544*
radical function, *p. 545*

Previous
radicand
transformation
average rate of change

What You Will Learn

- Graph square root functions.
- Compare square root functions using average rates of change.
- Solve real-life problems involving square root functions.

Graphing Square Root Functions

Core Concept

Square Root Functions

A **square root function** is a function that contains a square root with the independent variable in the radicand. The parent function for the family of square root functions is $f(x) = \sqrt{x}$. The domain of f is $x \geq 0$, and the range of f is $y \geq 0$.

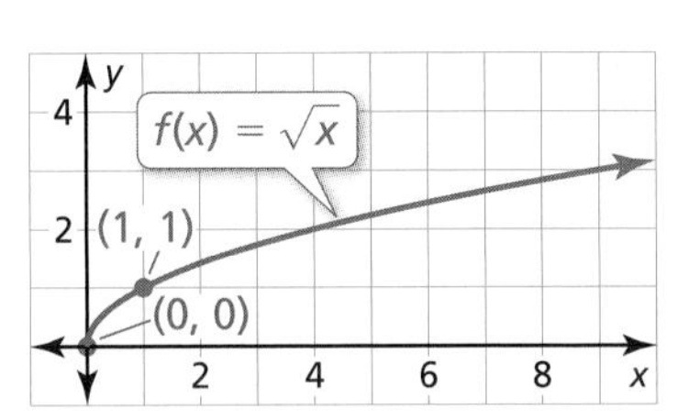

STUDY TIP

The graph of $f(x) = \sqrt{x}$ starts at (0, 0) and increases on the entire domain. So, the minimum value of f is 0.

The value of the radicand in a square root function cannot be negative. So, the domain of a square root function includes x-values for which the radicand is greater than or equal to 0.

EXAMPLE 1 Describing the Domain of a Square Root Function

Describe the domain of $f(x) = 3\sqrt{x - 5}$.

SOLUTION

The radicand cannot be negative. So, $x - 5$ is greater than or equal to 0.

$x - 5 \geq 0$ Write an inequality for the domain.

$x \geq 5$ Add 5 to each side.

▶ The domain is the set of real numbers greater than or equal to 5.

EXAMPLE 2 Graphing a Square Root Function

Graph $f(x) = \sqrt{x} + 3$. Describe the range of the function.

SOLUTION

Step 1 Use the domain of f, $x \geq 0$, to make a table of values.

x	0	1	4	9	16
$f(x)$	3	4	5	6	7

Step 2 Plot the ordered pairs.

Step 3 Draw a smooth curve through the points, starting at (0, 3).

▶ From the graph, you can see that the range of f is $y \geq 3$.

Monitoring Progress

Help in English and Spanish at *BigIdeasMath.com*

Describe the domain of the function.

1. $f(x) = 10\sqrt{x}$
2. $y = \sqrt{2x} + 7$
3. $h(x) = \sqrt{-x + 1}$

Graph the function. Describe the range.

4. $g(x) = \sqrt{x} - 4$
5. $y = \sqrt{x + 5}$
6. $n(x) = 5\sqrt{x}$

STUDY TIP

You will study another type of radical function in the next section.

A **radical function** is a function that contains a radical expression with the independent variable in the radicand. A square root function is a type of radical function.

You can transform graphs of radical functions in the same way you transformed graphs of functions previously. In Example 2, notice that the graph of f is a vertical translation of the graph of the parent square root function.

Core Concept

Transformation	$f(x)$ Notation	Examples	
Horizontal Translation Graph shifts left or right.	$f(x - h)$	$g(x) = \sqrt{x - 2}$ $g(x) = \sqrt{x + 3}$	2 units right 3 units left
Vertical Translation Graph shifts up or down.	$f(x) + k$	$g(x) = \sqrt{x} + 7$ $g(x) = \sqrt{x} - 1$	7 units up 1 unit down
Reflection Graph flips over x- or y-axis.	$f(-x)$ $-f(x)$	$g(x) = \sqrt{-x}$ $g(x) = -\sqrt{x}$	in the y-axis in the x-axis
Horizontal Stretch or Shrink Graph stretches away from or shrinks toward y-axis.	$f(ax)$	$g(x) = \sqrt{3x}$ $g(x) = \sqrt{\frac{1}{2}x}$	shrink by a factor of $\frac{1}{3}$ stretch by a factor of 2
Vertical Stretch or Shrink Graph stretches away from or shrinks toward x-axis.	$a \cdot f(x)$	$g(x) = 4\sqrt{x}$ $g(x) = \frac{1}{5}\sqrt{x}$	stretch by a factor of 4 shrink by a factor of $\frac{1}{5}$

EXAMPLE 3 Comparing Graphs of Square Root Functions

Graph $g(x) = -\sqrt{x - 2}$. Compare the graph to the graph of $f(x) = \sqrt{x}$.

SOLUTION

Step 1 Use the domain of g, $x \geq 2$, to make a table of values.

x	2	3	4	5	6
$g(x)$	0	-1	-1.4	-1.7	-2

Step 2 Plot the ordered pairs.

Step 3 Draw a smooth curve through the points, starting at (2, 0).

$f(x) = \sqrt{x}$

$g(x) = -\sqrt{x - 2}$

▶ The graph of g is a translation 2 units right and a reflection in the x-axis of the graph of f.

Monitoring Progress  Help in English and Spanish at *BigIdeasMath.com*

Graph the function. Compare the graph to the graph of $f(x) = \sqrt{x}$.

7. $h(x) = \sqrt{\frac{1}{4}x}$ **8.** $g(x) = \sqrt{x} - 6$ **9.** $m(x) = -3\sqrt{x}$

To graph a square root function of the form $y = a\sqrt{x-h} + k$, where $a \neq 0$, start at (h, k).

EXAMPLE 4 Graphing $y = a\sqrt{x-h} + k$

Let $g(x) = -2\sqrt{x-3} - 2$. (a) Describe the transformations from the graph of $f(x) = \sqrt{x}$ to the graph of g. (b) Graph g.

REMEMBER

The graph of $y = a \cdot f(x-h) + k$ can be obtained from the graph of $y = f(x)$ using the steps you learned in Section 3.6.

SOLUTION

a. Step 1 Translate the graph of f horizontally 3 units right to get the graph of $t(x) = \sqrt{x-3}$.

Step 2 Stretch the graph of t vertically by a factor of 2 to get the graph of $h(x) = 2\sqrt{x-3}$.

Step 3 Reflect the graph of h in the x-axis to get the graph of $r(x) = -2\sqrt{x-3}$.

Step 4 Translate the graph of r vertically 2 units down to get the graph of $g(x) = -2\sqrt{x-3} - 2$.

b. Step 1 Use the domain, $x \geq 3$, to make a table of values.

x	3	4	7	12
$g(x)$	−2	−4	−6	−8

Step 2 Plot the ordered pairs.

Step 3 Start at $(h, k) = (3, -2)$ and draw a smooth curve through the points.

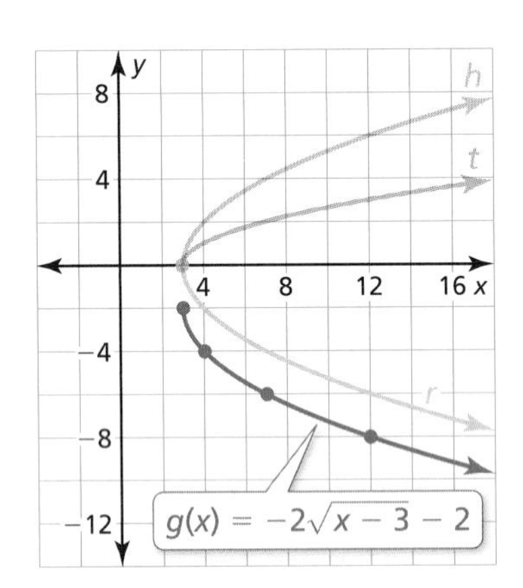

Monitoring Progress

Help in English and Spanish at *BigIdeasMath.com*

10. Let $g(x) = \frac{1}{2}\sqrt{x+4} + 1$. Describe the transformations from the graph of $f(x) = \sqrt{x}$ to the graph of g. Then graph g.

Comparing Average Rates of Change

EXAMPLE 5 Comparing Square Root Functions

The model $v(d) = \sqrt{2gd}$ represents the velocity v (in meters per second) of a free-falling object on the moon, where g is the constant 1.6 meters per second squared and d is the distance (in meters) the object has fallen. The velocity of a free-falling object on Earth is shown in the graph. Compare the velocities by finding and interpreting their average rates of change over the interval $d = 0$ to $d = 10$.

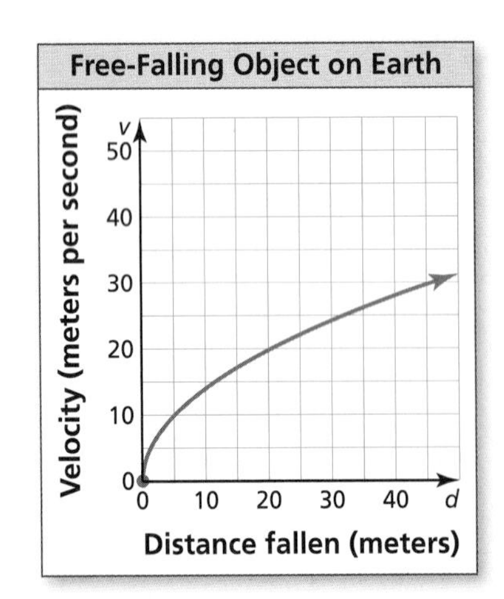

SOLUTION

To calculate the average rates of change, use points whose d-coordinates are 0 and 10.

Earth: Use the graph to estimate. Use (0, 0) and (10, 14).

$$\frac{v(10) - v(0)}{10 - 0} \approx \frac{14 - 0}{10} = 1.4 \qquad \text{Average rate of change on Earth}$$

Moon: Evaluate v when $d = 0$ and $d = 10$.

$$v(0) = \sqrt{2(1.6)(0)} = 0 \quad \text{and} \quad v(10) = \sqrt{2(1.6)(10)} = \sqrt{32} \approx 5.7$$

Use (0, 0) and $(10, \sqrt{32})$.

$$\frac{v(10) - v(0)}{10 - 0} = \frac{\sqrt{32} - 0}{10} \approx 0.57 \qquad \text{Average rate of change on the moon}$$

▶ From 0 to 10 meters, the velocity of a free-falling object increases at an average rate of about 1.4 meters per second per meter on Earth and about 0.57 meter per second per meter on the moon.

Monitoring Progress  Help in English and Spanish at *BigIdeasMath.com*

11. In Example 5, compare the velocities by finding and interpreting their average rates of change over the interval $d = 30$ to $d = 40$.

Solving Real-Life Problems

EXAMPLE 6 Real-Life Application

The velocity v (in meters per second) of a tsunami can be modeled by the function $v(x) = \sqrt{9.8x}$, where x is the water depth (in meters). (a) Use a graphing calculator to graph the function. At what depth does the velocity of the tsunami exceed 200 meters per second? (b) What happens to the average rate of change of the velocity as the water depth increases?

SOLUTION

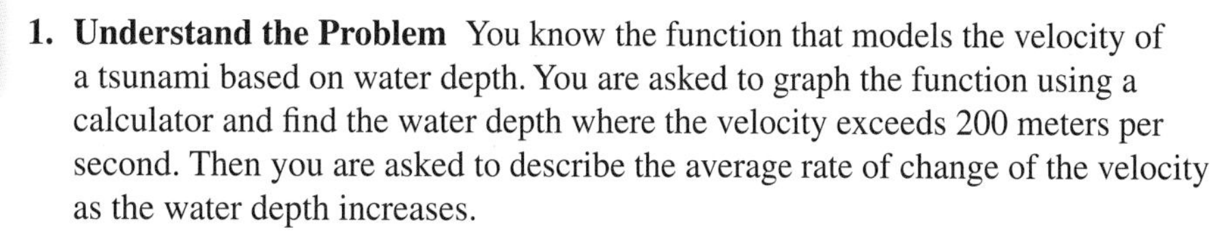

1. **Understand the Problem** You know the function that models the velocity of a tsunami based on water depth. You are asked to graph the function using a calculator and find the water depth where the velocity exceeds 200 meters per second. Then you are asked to describe the average rate of change of the velocity as the water depth increases.

2. **Make a Plan** Graph the function using a calculator. Use the *trace* feature to find the value of x when $v(x)$ is about 200. Then calculate and compare average rates of change of the velocity over different intervals.

3. **Solve the Problem**

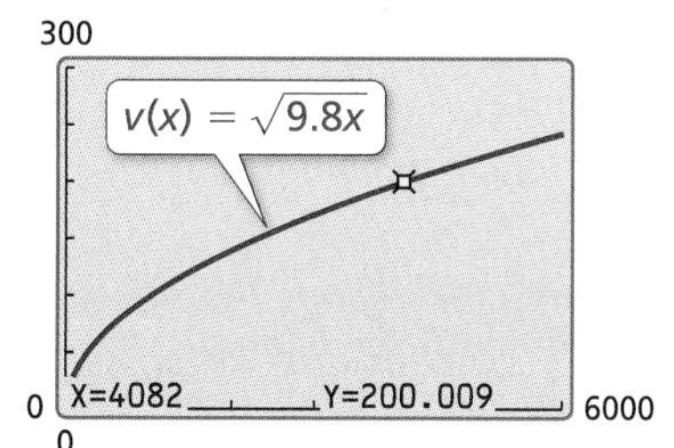

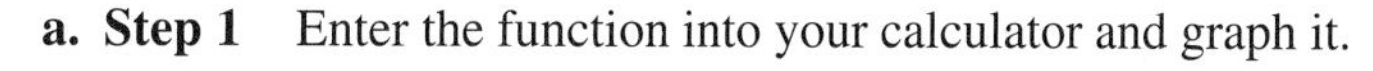

a. Step 1 Enter the function into your calculator and graph it.

Step 2 Use the *trace* feature to find the value of x when $v(x) \approx 200$.

▶ The velocity exceeds 200 meters per second at a depth of about 4082 meters.

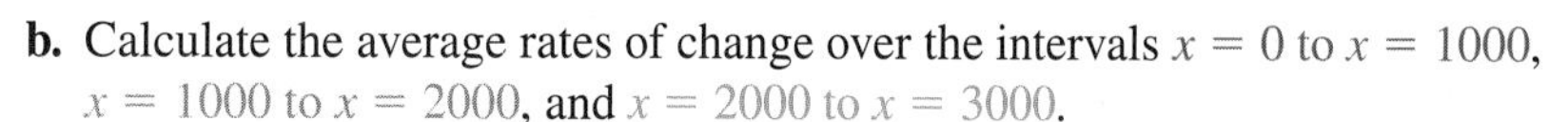

b. Calculate the average rates of change over the intervals $x = 0$ to $x = 1000$, $x = 1000$ to $x = 2000$, and $x = 2000$ to $x = 3000$.

$$\frac{v(1000) - v(0)}{1000 - 0} = \frac{\sqrt{9800} - 0}{1000} \approx 0.099 \qquad \text{0 to 1000 meters}$$

$$\frac{v(2000) - v(1000)}{2000 - 1000} = \frac{\sqrt{19{,}600} - \sqrt{9800}}{1000} \approx 0.041 \qquad \text{1000 to 2000 meters}$$

$$\frac{v(3000) - v(2000)}{3000 - 2000} = \frac{\sqrt{29{,}400} - \sqrt{19{,}600}}{1000} \approx 0.031 \qquad \text{2000 to 3000 meters}$$

▶ The average rate of change of the velocity decreases as the water depth increases.

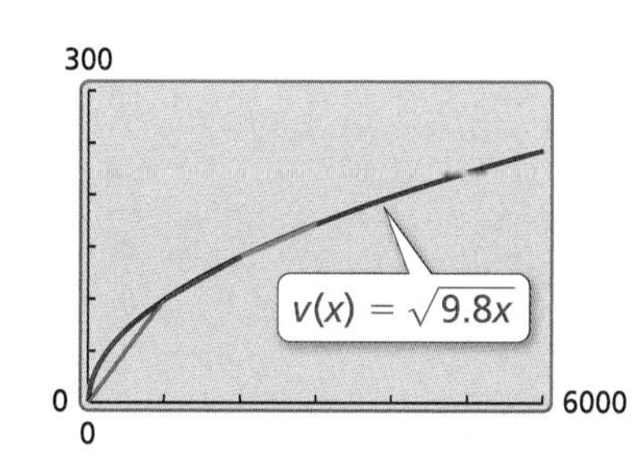

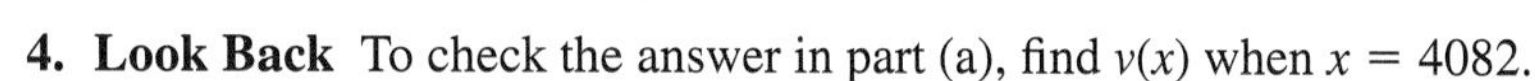

4. **Look Back** To check the answer in part (a), find $v(x)$ when $x = 4082$.

$v(4082) = \sqrt{9.8(4082)} \approx 200$ ✓

In part (b), the slopes of the line segments (shown at the left) that represent the average rates of change over the intervals are decreasing. So, the answer to part (b) is reasonable.

Monitoring Progress  Help in English and Spanish at *BigIdeasMath.com*

12. WHAT IF? At what depth does the velocity of the tsunami exceed 100 meters per second?

10.1 Exercises

Dynamic Solutions available at *BigIdeasMath.com*

Vocabulary and Core Concept Check

1. **COMPLETE THE SENTENCE** A ________ is a function that contains a radical expression with the independent variable in the radicand.

2. **VOCABULARY** Is $y = 2x\sqrt{5}$ a square root function? Explain.

3. **WRITING** How do you describe the domain of a square root function?

4. **REASONING** Is the graph of $g(x) = 1.25\sqrt{x}$ a vertical stretch or a vertical shrink of the graph of $f(x) = \sqrt{x}$? Explain.

Monitoring Progress and Modeling with Mathematics

In Exercises 5–14, describe the domain of the function. *(See Example 1.)*

5. $y = 8\sqrt{x}$
6. $y = \sqrt{4x}$
7. $y = 4 + \sqrt{-x}$
8. $y = \sqrt{-\frac{1}{2}x} + 1$
9. $h(x) = \sqrt{x - 4}$
10. $p(x) = \sqrt{x + 7}$
11. $f(x) = \sqrt{-x + 8}$
12. $g(x) = \sqrt{-x - 1}$
13. $m(x) = 2\sqrt{x + 4}$
14. $n(x) = \frac{1}{2}\sqrt{-x} - 2$

In Exercises 15–18, match the function with its graph. Describe the range.

15. $y = \sqrt{x - 3}$
16. $y = 3\sqrt{x}$
17. $y = \sqrt{x} - 3$
18. $y = \sqrt{-x + 3}$

A.

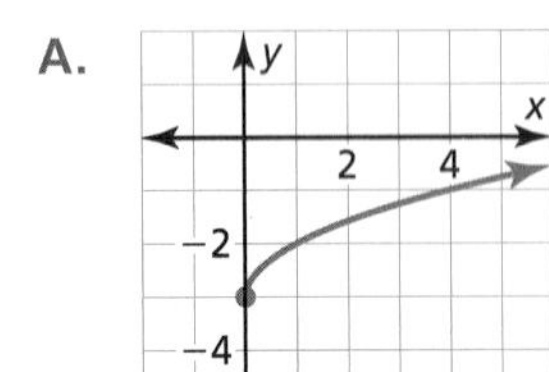

B.

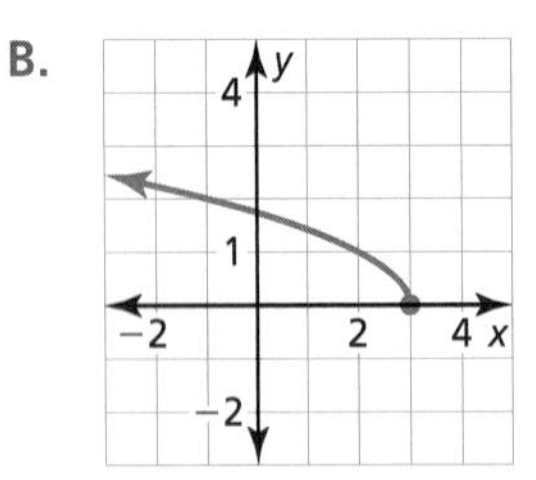

C.

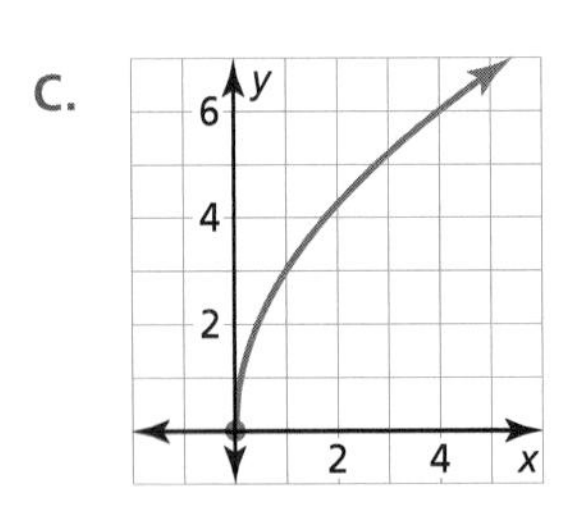

D.

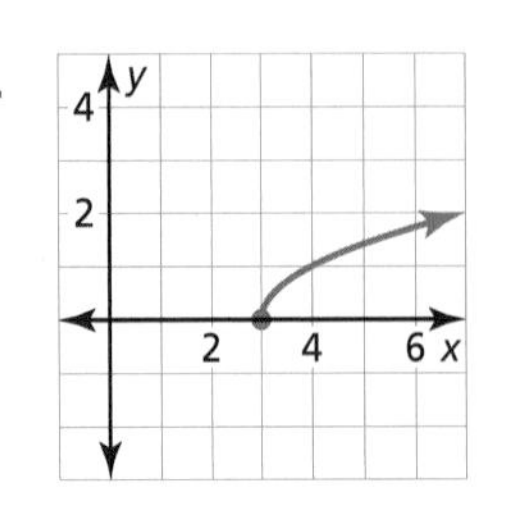

In Exercises 19–26, graph the function. Describe the range. *(See Example 2.)*

19. $y = \sqrt{3x}$
20. $y = 4\sqrt{-x}$
21. $y = \sqrt{x} + 5$
22. $y = -2 + \sqrt{x}$
23. $f(x) = -\sqrt{x - 3}$
24. $g(x) = \sqrt{x + 4}$
25. $h(x) = \sqrt{x + 2} - 2$
26. $f(x) = -\sqrt{x - 1} + 3$

In Exercises 27–34, graph the function. Compare the graph to the graph of $f(x) = \sqrt{x}$. *(See Example 3.)*

27. $g(x) = \frac{1}{4}\sqrt{x}$
28. $r(x) = \sqrt{2x}$
29. $h(x) = \sqrt{x + 3}$
30. $q(x) = \sqrt{x} + 8$
31. $p(x) = \sqrt{-\frac{1}{3}x}$
32. $g(x) = -5\sqrt{x}$
33. $m(x) = -\sqrt{x} - 6$
34. $n(x) = -\sqrt{x - 4}$

35. **ERROR ANALYSIS** Describe and correct the error in graphing the function $y = \sqrt{x + 1}$.

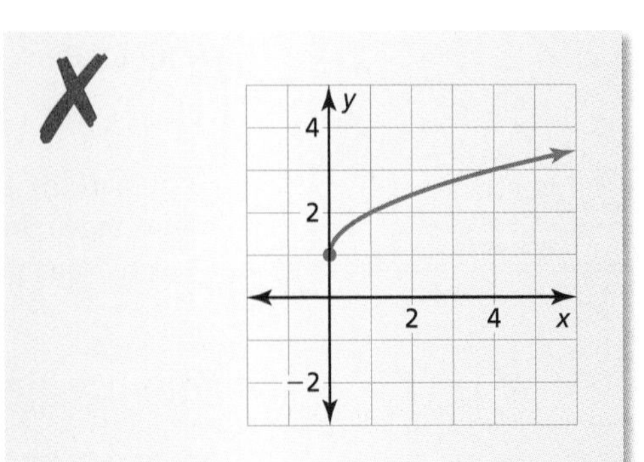

36. **ERROR ANALYSIS** Describe and correct the error in comparing the graph of $g(x) = -\frac{1}{4}\sqrt{x}$ to the graph of $f(x) = \sqrt{x}$.

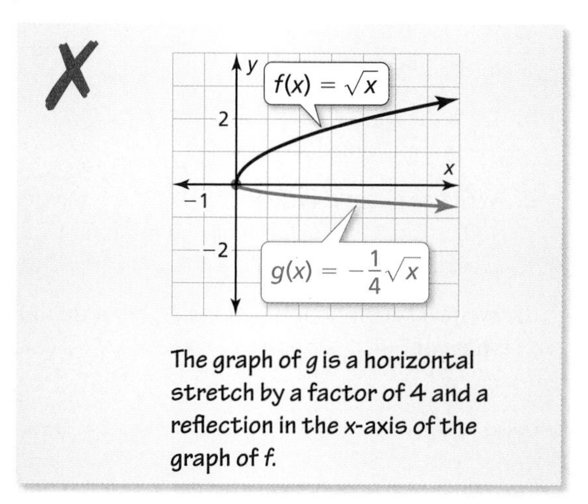

f(x) = √x
g(x) = −1/4 √x
The graph of g is a horizontal stretch by a factor of 4 and a reflection in the x-axis of the graph of f.

In Exercises 37–44, describe the transformations from the graph of $f(x) = \sqrt{x}$ to the graph of h. Then graph h. *(See Example 4.)*

37. $h(x) = 4\sqrt{x+2} - 1$

38. $h(x) = \frac{1}{2}\sqrt{x-6} + 3$

39. $h(x) = 2\sqrt{-x} - 6$

40. $h(x) = -\sqrt{x-3} - 2$

41. $h(x) = \frac{1}{3}\sqrt{x+3} - 3$

42. $h(x) = 2\sqrt{x-1} + 4$

43. $h(x) = -2\sqrt{x-1} + 5$

44. $h(x) = -5\sqrt{x+2} - 1$

45. **COMPARING FUNCTIONS** The model $S(d) = \sqrt{30df}$ represents the speed S (in miles per hour) of a van before it skids to a stop, where f is the drag factor of the road surface and d is the length (in feet) of the skid marks. The drag factor of Road Surface A is 0.75. The graph shows the speed of the van on Road Surface B. Compare the speeds by finding and interpreting their average rates of change over the interval $d = 0$ to $d = 15$. *(See Example 5.)*

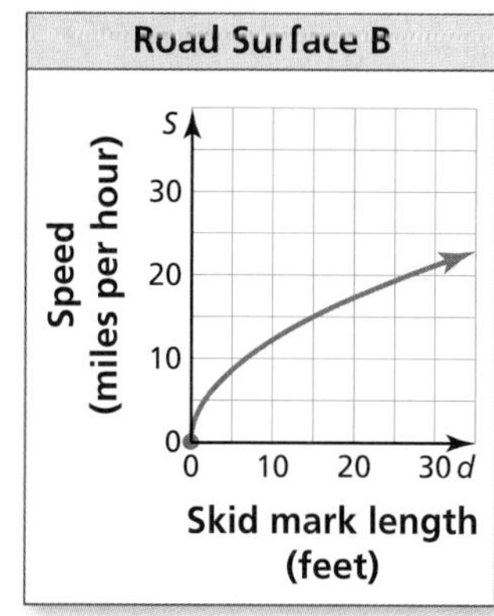

46. **COMPARING FUNCTIONS** The velocity v (in meters per second) of an object in motion is given by $v(E) = \sqrt{\frac{2E}{m}}$, where E is the kinetic energy of the object (in joules) and m is the mass of the object (in kilograms). The mass of Object A is 4 kilograms. The graph shows the velocity of Object B. Compare the velocities of the objects by finding and interpreting the average rates of change over the interval $E = 0$ to $E = 6$.

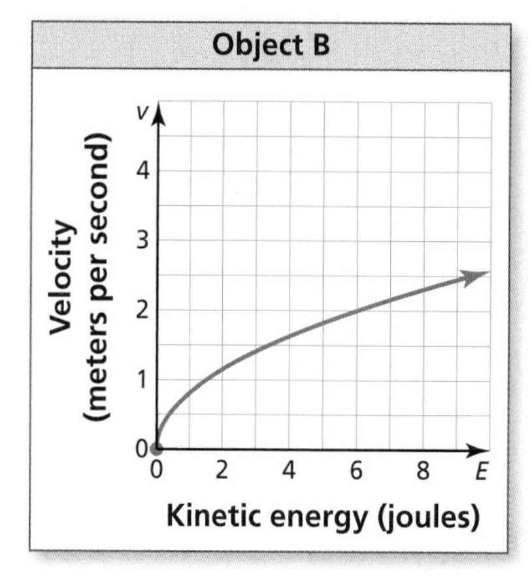

47. **OPEN-ENDED** Consider the graph of $y = \sqrt{x}$.

a. Write a function that is a vertical translation of the graph of $y = \sqrt{x}$.

b. Write a function that is a reflection of the graph of $y = \sqrt{x}$.

48. **REASONING** Can the domain of a square root function include negative numbers? Can the range include negative numbers? Explain your reasoning.

49. **PROBLEM SOLVING** The nozzle pressure of a fire hose allows firefighters to control the amount of water they spray on a fire. The flow rate f (in gallons per minute) can be modeled by the function $f = 120\sqrt{p}$, where p is the nozzle pressure (in pounds per square inch). *(See Example 6.)*

a. Use a graphing calculator to graph the function. At what pressure does the flow rate exceed 300 gallons per minute?

b. What happens to the average rate of change of the flow rate as the pressure increases?

50. PROBLEM SOLVING The speed s (in meters per second) of a long jumper before jumping can be modeled by the function $s = 10.9\sqrt{h}$, where h is the maximum height (in meters from the ground) of the jumper.

a. Use a graphing calculator to graph the function. A jumper is running 9.2 meters per second. Estimate the maximum height of the jumper.

b. Suppose the runway and pit are raised on a platform slightly higher than the ground. How would the graph of the function be transformed?

51. MATHEMATICAL CONNECTIONS The radius r of a circle is given by $r = \sqrt{\frac{A}{\pi}}$, where A is the area of the circle.

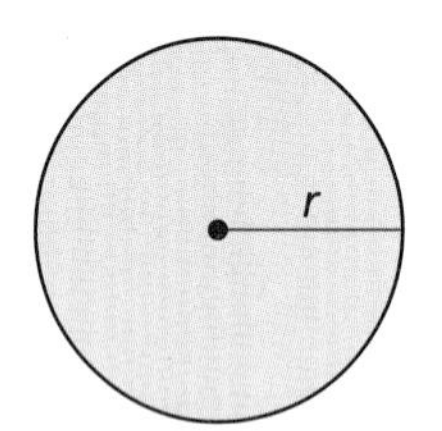

a. Describe the domain of the function. Use a graphing calculator to graph the function.

b. Use the *trace* feature to approximate the area of a circle with a radius of 5.4 inches.

52. REASONING Consider the function $f(x) = 8a\sqrt{x}$.

a. For what value of a will the graph of f be identical to the graph of the parent square root function?

b. For what values of a will the graph of f be a vertical stretch of the graph of the parent square root function?

c. For what values of a will the graph of f be a vertical shrink and a reflection of the graph of the parent square root function?

53. REASONING The graph represents the function $f(x) = \sqrt{x}$.

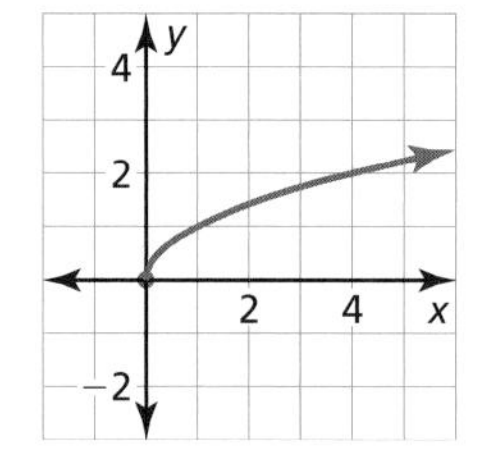

a. What is the minimum value of the function?

b. Does the function have a maximum value? Explain.

c. Write a square root function that has a maximum value. Does the function have a minimum value? Explain.

d. Write a square root function that has a minimum value of −4.

54. HOW DO YOU SEE IT? Match each function with its graph. Explain your reasoning.

A. $f(x) = \sqrt{x} + 2$ **B.** $m(x) = f(x) - 4$

C. $n(x) = f(-x)$ **D.** $p(x) = f(3x)$

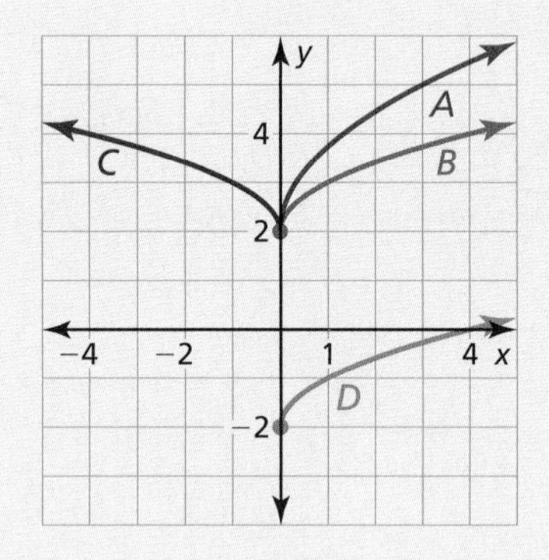

55. REASONING Without graphing, determine which function's graph rises more steeply, $f(x) = 5\sqrt{x}$ or $g(x) = \sqrt{5x}$. Explain your reasoning.

56. THOUGHT PROVOKING Use a graphical approach to find the solutions of $x - 1 = \sqrt{5x - 9}$. Show your work. Verify your solutions algebraically.

57. OPEN-ENDED Write a radical function that has a domain of all real numbers greater than or equal to −5 and a range of all real numbers less than or equal to 3.

Maintaining Mathematical Proficiency

Reviewing what you learned in previous grades and lessons

Evaluate the expression. *(Section 6.2)*

58. $\sqrt[3]{343}$ **59.** $\sqrt[3]{-64}$ **60.** $-\sqrt[3]{-\frac{1}{27}}$

Factor the polynomial. *(Section 7.5)*

61. $x^2 + 7x + 6$ **62.** $d^2 - 11d + 28$ **63.** $y^2 - 3y - 40$

10.2 Graphing Cube Root Functions

Essential Question What are some of the characteristics of the graph of a cube root function?

EXPLORATION 1 Graphing Cube Root Functions

Work with a partner.

- Make a table of values for each function. Use positive and negative values of x.
- Use the table to sketch the graph of each function.
- Describe the domain of each function.
- Describe the range of each function.

a. $y = \sqrt[3]{x}$

b. 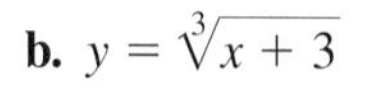 $y = \sqrt[3]{x + 3}$

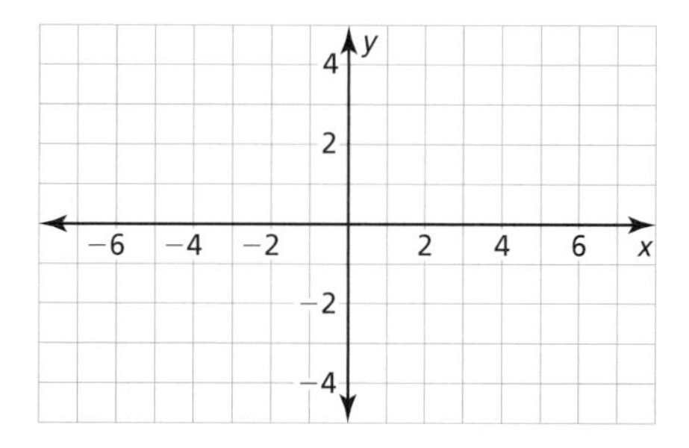

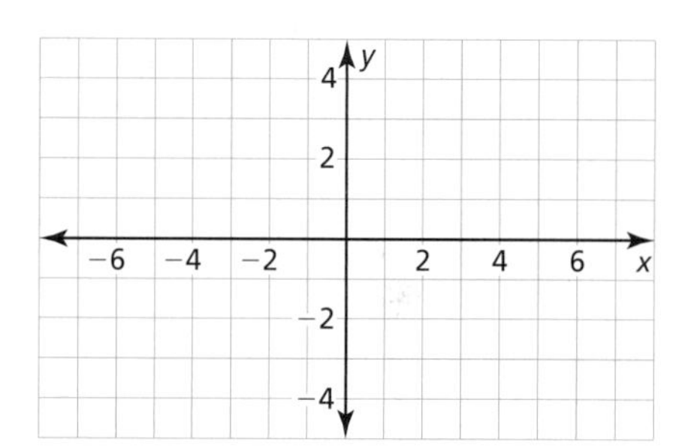

LOOKING FOR REGULARITY IN REPEATED REASONING

To be proficient in math, you need to notice whether calculations are repeated and look for both general methods and shortcuts.

EXPLORATION 2 Writing Cube Root Functions

Work with a partner. Write a cube root function, $y = f(x)$, that has the given values. Then use the function to complete the table.

a.

x	$f(x)$
-4	0
-3	
-2	
-1	$\sqrt[3]{3}$
0	

x	$f(x)$
1	
2	
3	
4	2
5	

b.

x	$f(x)$
-4	1
-3	
-2	
-1	$1 + \sqrt[3]{3}$
0	

x	$f(x)$
1	
2	
3	
4	3
5	

Communicate Your Answer

3. What are some of the characteristics of the graph of a cube root function?

4. Graph each function. Then compare the graph to the graph of $f(x) = \sqrt[3]{x}$.

a. $g(x) = \sqrt[3]{x - 1}$

b. $g(x) = \sqrt[3]{x} - 1$

c. $g(x) = 2\sqrt[3]{x}$

d. $g(x) = -2\sqrt[3]{x}$

10.2 Lesson

Core Vocabulary

cube root function, *p. 552*

Previous
radical function
index

What You Will Learn

- Graph cube root functions.
- Compare cube root functions using average rates of change.
- Solve real-life problems involving cube root functions.

Graphing Cube Root Functions

Core Concept

Cube Root Functions

A **cube root function** is a radical function with an index of 3. The parent function for the family of cube root functions is $f(x) = \sqrt[3]{x}$. The domain and range of f are all real numbers.

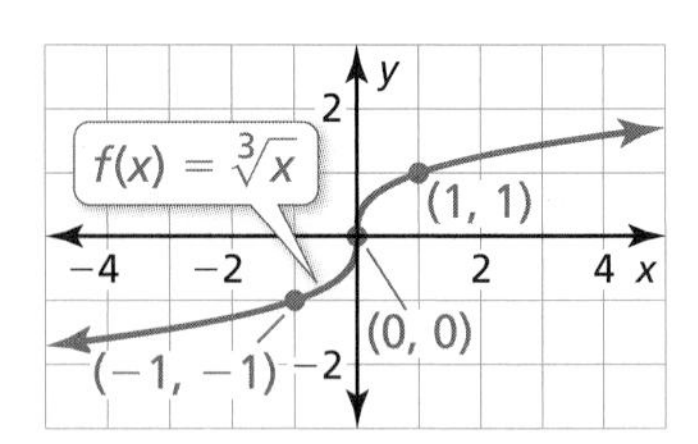

The graph of $f(x) = \sqrt[3]{x}$ increases on the entire domain.

You can transform graphs of cube root functions in the same way you transformed graphs of square root functions.

LOOKING FOR STRUCTURE

Use x-values so that the cube root of the radicand is an integer. This makes it easier to perform the calculations and plot the points.

EXAMPLE 1 Comparing Graphs of Cube Root Functions

Graph $h(x) = \sqrt[3]{x} - 4$. Compare the graph to the graph of $f(x) = \sqrt[3]{x}$.

SOLUTION

Step 1 Make a table of values.

x	−8	−1	0	1	8
$h(x)$	−6	−5	−4	−3	−2

Step 2 Plot the ordered pairs.

Step 3 Draw a smooth curve through the points.

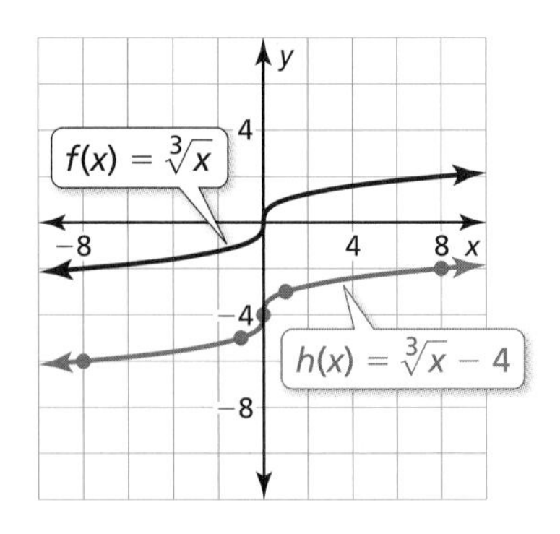

- The graph of h is a translation 4 units down of the graph of f.

Monitoring Progress

Help in English and Spanish at *BigIdeasMath.com*

Graph the function. Compare the graph to the graph of $f(x) = \sqrt[3]{x}$.

1. $h(x) = \sqrt[3]{x} + 3$

2. $m(x) = \sqrt[3]{x - 5}$

3. $g(x) = 4\sqrt[3]{x}$

EXAMPLE 2 Comparing Graphs of Cube Root Functions

Graph $g(x) = -\sqrt[3]{x+2}$. Compare the graph to the graph of $f(x) = \sqrt[3]{x}$.

SOLUTION

Step 1 Make a table of values.

x	-10	-3	-2	-1	6
$g(x)$	2	1	0	-1	-2

Step 2 Plot the ordered pairs.

Step 3 Draw a smooth curve through the points.

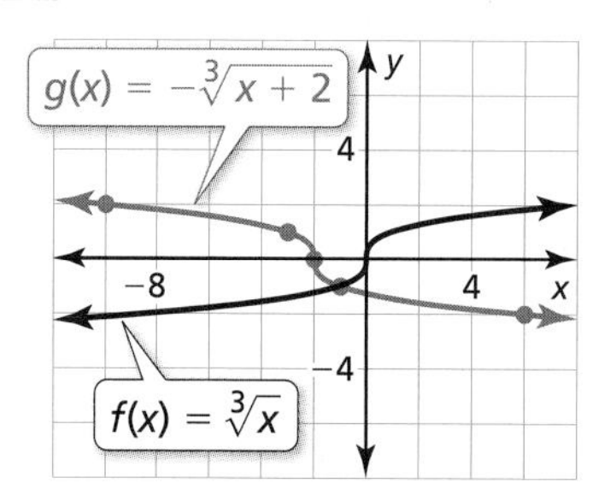

▶ The graph of g is a translation 2 units left and a reflection in the x-axis of the graph of f.

REMEMBER

The graph of $y = a \cdot f(x - h) + k$ can be obtained from the graph of $y = f(x)$ using the steps you learned in Section 3.6.

EXAMPLE 3 Graphing $y = a\sqrt[3]{x - h} + k$

Let $g(x) = 2\sqrt[3]{x-3} + 4$. (a) Describe the transformations from the graph of $f(x) = \sqrt[3]{x}$ to the graph of g. (b) Graph g.

SOLUTION

a. Step 1 Translate the graph of f horizontally 3 units right to get the graph of $t(x) = \sqrt[3]{x-3}$.

Step 2 Stretch the graph of t vertically by a factor of 2 to get the graph of $h(x) = 2\sqrt[3]{x-3}$.

Step 3 Because $a > 0$, there is no reflection.

Step 4 Translate the graph of h vertically 4 units up to get the graph of $g(x) = 2\sqrt[3]{x-3} + 4$.

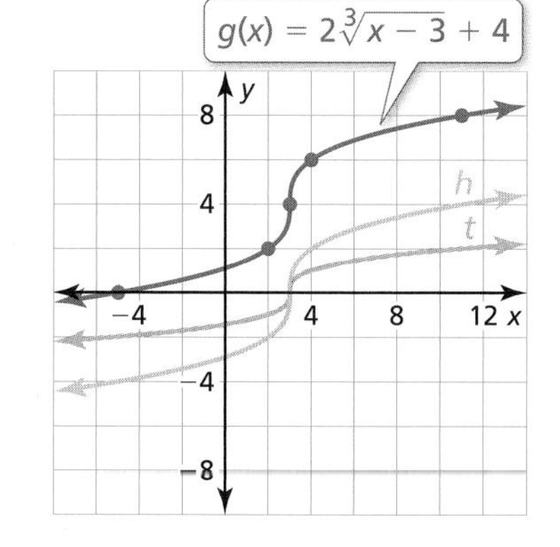

b. Step 1 Make a table of values.

x	-5	2	3	4	11
$g(x)$	0	2	4	6	8

Step 2 Plot the ordered pairs.

Step 3 Draw a smooth curve through the points.

Monitoring Progress

Help in English and Spanish at *BigIdeasMath.com*

Graph the function. Compare the graph to the graph of $f(x) = \sqrt[3]{x}$.

4. $g(x) = \sqrt[3]{0.5x} + 5$

5. $h(x) = 4\sqrt[3]{x-1}$

6. $n(x) = \sqrt[3]{4-x}$

7. Let $g(x) = -\frac{1}{2}\sqrt[3]{x+2} - 4$. Describe the transformations from the graph of $f(x) = \sqrt[3]{x}$ to the graph of g. Then graph g.

Comparing Average Rates of Change

EXAMPLE 4 Comparing Cube Root Functions

The graph of cube root function m is shown. Compare the average rate of change of m to the average rate of change of $h(x) = \sqrt[3]{\frac{1}{4}x}$ over the interval $x = 0$ to $x = 8$.

SOLUTION

To calculate the average rates of change, use points whose x-coordinates are 0 and 8.

Function m: Use the graph to estimate. Use (0, 0) and (8, 8).

$$\frac{m(8) - m(0)}{8 - 0} \approx \frac{8 - 0}{8} = 1 \qquad \text{Average rate of change of } m$$

Function h: Evaluate h when $x = 0$ and $x = 8$.

$$h(0) = \sqrt[3]{\frac{1}{4}(0)} = 0 \quad \text{and} \quad h(8) = \sqrt[3]{\frac{1}{4}(8)} = \sqrt[3]{2} \approx 1.3$$

Use (0, 0) and $\left(8, \sqrt[3]{2}\right)$.

$$\frac{h(8) - h(0)}{8 - 0} = \frac{\sqrt[3]{2} - 0}{8} \approx 0.16 \qquad \text{Average rate of change of } h$$

▶ The average rate of change of m is $1 \div \frac{\sqrt[3]{2}}{8} \approx 6.3$ times greater than the average rate of change of h over the interval $x = 0$ to $x = 8$.

Monitoring Progress  Help in English and Spanish at *BigIdeasMath.com*

8. In Example 4, compare the average rates of change over the interval $x = 2$ to $x = 10$.

Solving Real-Life Problems

EXAMPLE 5 Real-Life Application

The shoulder height h (in centimeters) of a male Asian elephant can be modeled by the function $h = 62.5\sqrt[3]{t} + 75.8$, where t is the age (in years) of the elephant. Use a graphing calculator to graph the function. Estimate the age of an elephant whose shoulder height is 200 centimeters.

SOLUTION

Step 1 Enter $y_1 = 62.5\sqrt[3]{t} + 75.8$ and $y_2 = 200$ into your calculator and graph the equations. Choose a viewing window that shows the point where the graphs intersect.

Step 2 Use the *intersect* feature to find the x-coordinate of the intersection point.

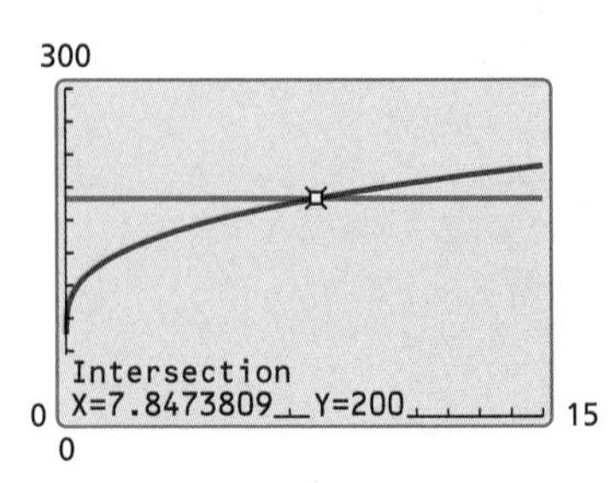

▶ The two graphs intersect at about (8, 200). So, the elephant is about 8 years old.

Monitoring Progress  Help in English and Spanish at *BigIdeasMath.com*

9. WHAT IF? Estimate the age of an elephant whose shoulder height is 175 centimeters.

10.2 Exercises

Dynamic Solutions available at *BigIdeasMath.com*

Vocabulary and Core Concept Check

1. **COMPLETE THE SENTENCE** The __________ of the radical in a cube root function is 3.

2. **WRITING** Describe the domain and range of the function $f(x) = \sqrt[3]{x-4} + 1$.

Monitoring Progress and Modeling with Mathematics

In Exercises 3–6, match the function with its graph.

3. $y = \sqrt[3]{x+2}$

4. $y = \sqrt[3]{x-2}$

5. $y = \sqrt[3]{x} + 2$

6. $y = \sqrt[3]{x} - 2$

A. 

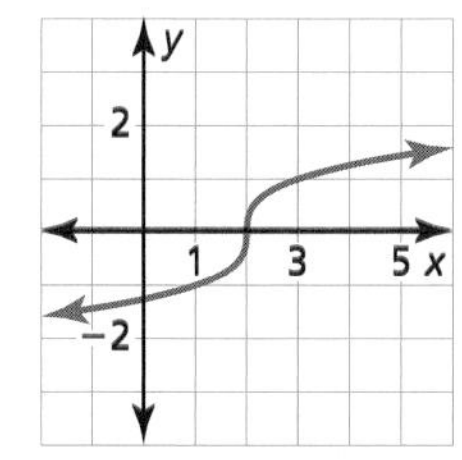

B.

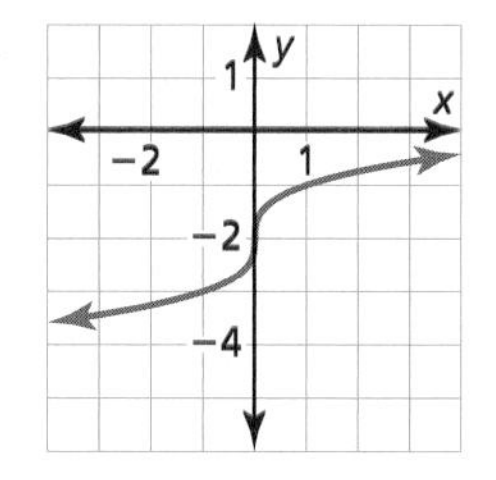

C.

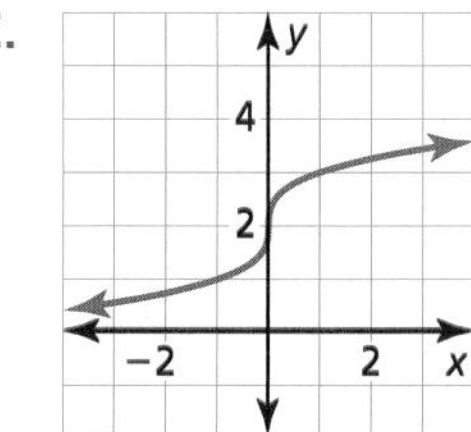

D.

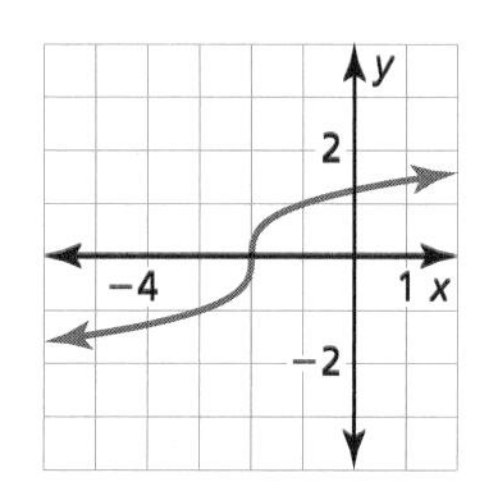

In Exercises 7–12, graph the function. Compare the graph to the graph of $f(x) = \sqrt[3]{x}$. *(See Example 1.)*

7. $h(x) = \sqrt[3]{x-4}$

8. $g(x) = \sqrt[3]{x+1}$

9. $m(x) = \sqrt[3]{x} + 5$

10. $q(x) = \sqrt[3]{x} - 3$

11. $p(x) = 6\sqrt[3]{x}$

12. $j(x) = \sqrt[3]{\frac{1}{2}x}$

In Exercises 13–16, compare the graphs. Find the value of h, k, or a.

13.

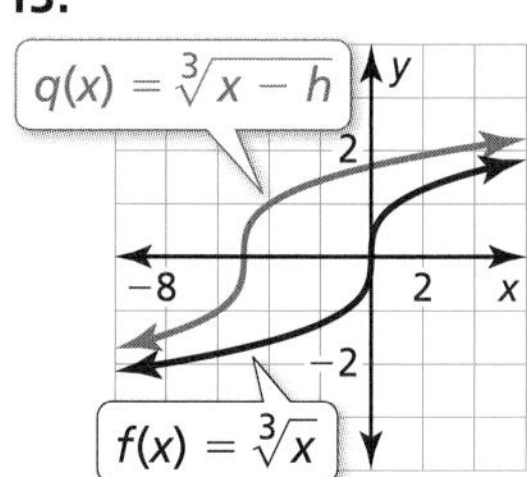

14.

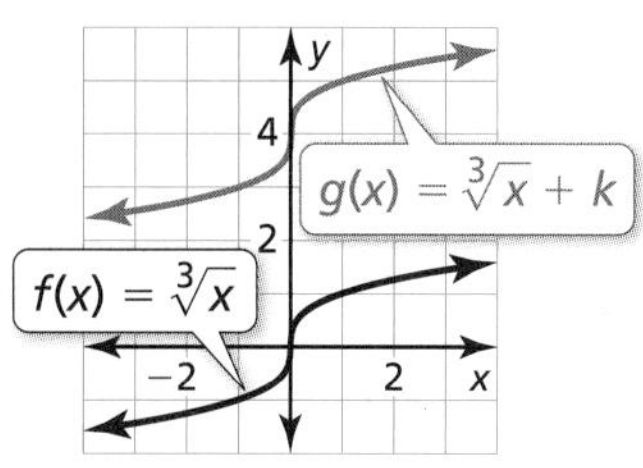

15.

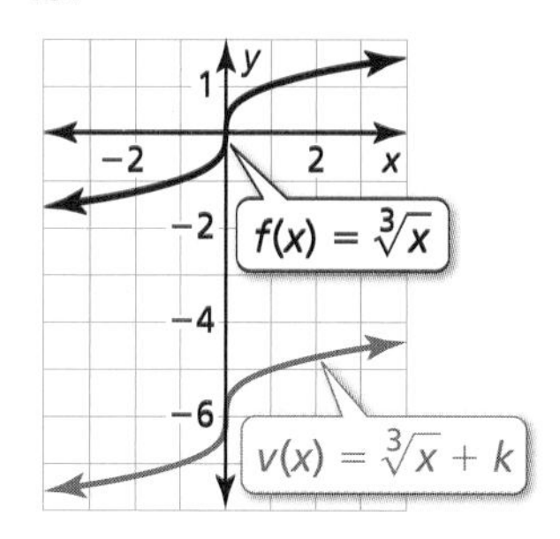

16.

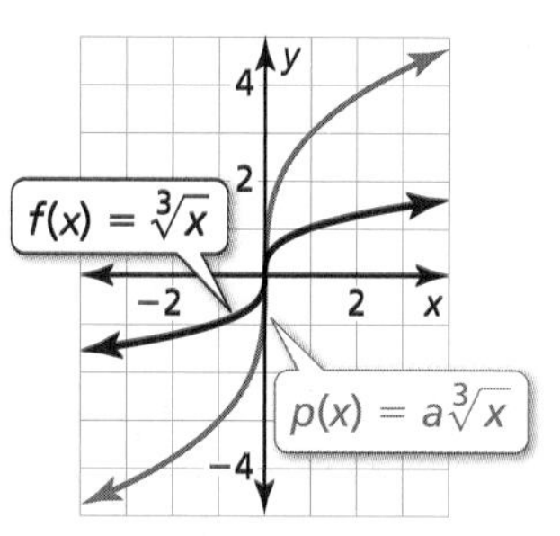

In Exercises 17–26, graph the function. Compare the graph to the graph of $f(x) = \sqrt[3]{x}$. *(See Example 2.)*

17. $r(x) = -\sqrt[3]{x-2}$

18. $h(x) = -\sqrt[3]{x} + 3$

19. $k(x) = 5\sqrt[3]{x+1}$

20. $j(x) = 0.5\sqrt[3]{x-4}$

21. $g(x) = 4\sqrt[3]{x} - 3$

22. $m(x) = 3\sqrt[3]{x} + 7$

23. $n(x) = \sqrt[3]{-8x} - 1$

24. $v(x) = \sqrt[3]{5x} + 2$

25. $q(x) = \sqrt[3]{2(x+3)}$

26. $p(x) = \sqrt[3]{3(1-x)}$

In Exercises 27–32, describe the transformations from the graph of $f(x) = \sqrt[3]{x}$ to the graph of the given function. Then graph the given function. *(See Example 3.)*

27. $g(x) = \sqrt[3]{x-4} + 2$

28. $n(x) = \sqrt[3]{x+1} - 3$

29. $j(x) = -5\sqrt[3]{x+3} + 2$

30. $k(x) = 6\sqrt[3]{x-9} - 5$

31. $v(x) = \frac{1}{3}\sqrt[3]{x-1} + 7$

32. $h(x) = -\frac{3}{2}\sqrt[3]{x+4} - 3$

33. **ERROR ANALYSIS** Describe and correct the error in graphing the function $f(x) = \sqrt[3]{x-3}$.

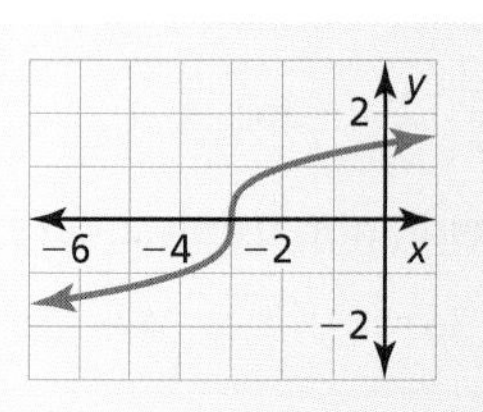

34. ERROR ANALYSIS Describe and correct the error in graphing the function $h(x) = \sqrt[3]{x} + 1$.

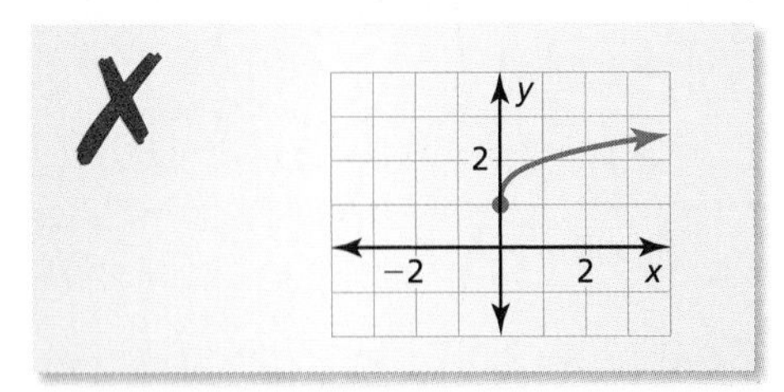

35. COMPARING FUNCTIONS The graph of cube root function q is shown. Compare the average rate of change of q to the average rate of change of $f(x) = 3\sqrt[3]{x}$ over the interval $x = 0$ to $x = 6$. *(See Example 4.)*

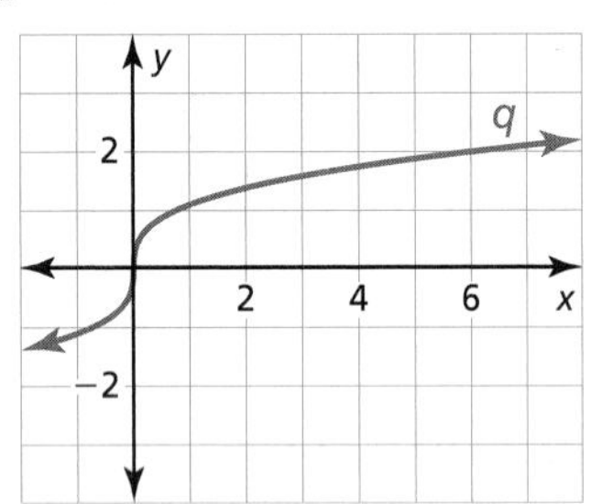

36. COMPARING FUNCTIONS The graphs of two cube root functions are shown. Compare the average rates of change of the two functions over the interval $x = -2$ to $x = 2$.

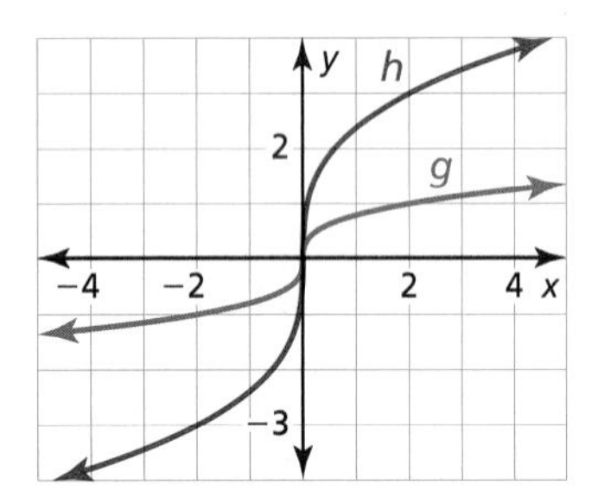

37. MODELING WITH MATHEMATICS For a drag race car that weighs 1600 kilograms, the velocity v (in kilometers per hour) reached by the end of a drag race can be modeled by the function $v = 23.8\sqrt[3]{p}$, where p is the car's power (in horsepower). Use a graphing calculator to graph the function. Estimate the power of a 1600-kilogram car that reaches a velocity of 220 kilometers per hour. *(See Example 5.)*

38. MODELING WITH MATHEMATICS The radius r of a sphere is given by the function $r = \sqrt[3]{\frac{3}{4\pi}V}$, where V is the volume of the sphere. Use a graphing calculator to graph the function. Estimate the volume of a spherical beach ball with a radius of 13 inches.

39. MAKING AN ARGUMENT Your friend says that all cube root functions are odd functions. Is your friend correct? Explain.

40. HOW DO YOU SEE IT? The graph represents the cube root function $f(x) = \sqrt[3]{x}$.

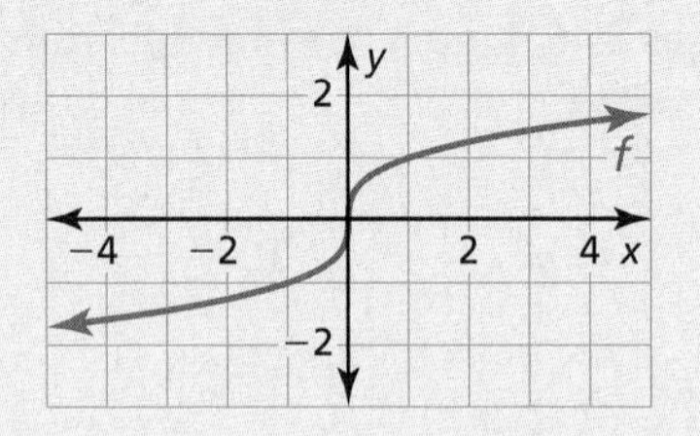

a. On what interval is f negative? positive?

b. On what interval, if any, is f decreasing? increasing?

c. Does f have a maximum or minimum value? Explain.

d. Find the average rate of change of f over the interval $x = -1$ to $x = 1$.

41. PROBLEM SOLVING Write a cube root function that passes through the point (3, 4) and has an average rate of change of -1 over the interval $x = -5$ to $x = 2$.

42. THOUGHT PROVOKING Write the cube root function represented by the graph. Use a graphing calculator to check your answer.

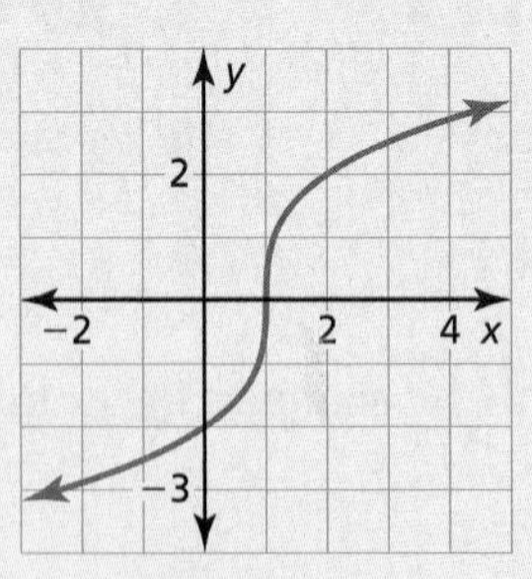

Maintaining Mathematical Proficiency

Reviewing what you learned in previous grades and lessons

Factor the polynomial. *(Section 7.6)*

43. $3x^2 + 12x - 36$ **44.** $2x^2 - 11x + 9$ **45.** $4x^2 + 7x - 15$

Solve the equation using square roots. *(Section 9.3)*

46. $x^2 - 36 = 0$ **47.** $5x^2 + 20 = 0$ **48.** $(x + 4)^2 = 81$ **49.** $25(x - 2)^2 = 9$

10.1–10.2 What Did You Learn?

Core Vocabulary

square root function, *p. 544*
radical function, *p. 545*
cube root function, *p. 552*

Core Concepts

Section 10.1

Square Root Functions, *p. 544*
Transformations of Square Root Functions, *p. 545*
Comparing Square Root Functions Using Average Rates of Change, *p. 546*

Section 10.2

Cube Root Functions, *p. 552*
Comparing Cube Root Functions Using Average Rates of Change, *p. 554*

Mathematical Practices

1. In Exercise 45 on page 549, what information are you given? What relationships are present? What is your goal?
2. What units of measure did you use in your answer to Exercise 38 on page 556? Explain your reasoning.

Study Skills

Making Note Cards

Invest in three different colors of note cards. Use one color for each of the following: vocabulary words, rules, and calculator keystrokes.

- Using the first color of note cards, write a vocabulary word on one side of a card. On the other side, write the definition and an example. If possible, put the definition in your own words.
- Using the second color of note cards, write a rule on one side of a card. On the other side, write an explanation and an example.
- Using the third color of note cards, write a calculation on one side of a card. On the other side, write the keystrokes required to perform the calculation.

Use the note cards as references while completing your homework. Quiz yourself once a day.

10.1–10.2 Quiz

Describe the domain of the function. *(Section 10.1)*

1. $y = \sqrt{x - 3}$

2. $f(x) = 15\sqrt{x}$

3. $y = \sqrt{3 - x}$

Graph the function. Describe the range. Compare the graph to the graph of $f(x) = \sqrt{x}$. *(Section 10.1)*

4. $g(x) = \sqrt{x} + 5$

5. $n(x) = \sqrt{x - 4}$

6. $r(x) = -\sqrt{x - 2} + 1$

Graph the function. Compare the graph to the graph of $f(x) = \sqrt[3]{x}$. *(Section 10.2)*

7. $b(x) = \sqrt[3]{x + 2}$

8. $h(x) = -3\sqrt[3]{x} - 6$

9. $q(x) = \sqrt[3]{-4 - x}$

Compare the graphs. Find the value of *h*, *k*, or *a*. *(Section 10.1 and Section 10.2)*

10.

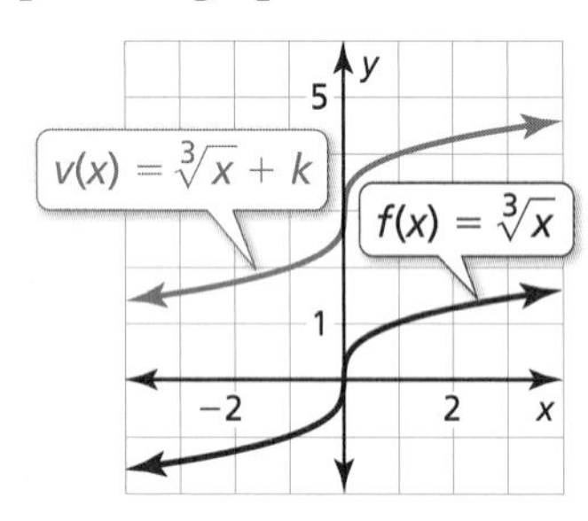

11.

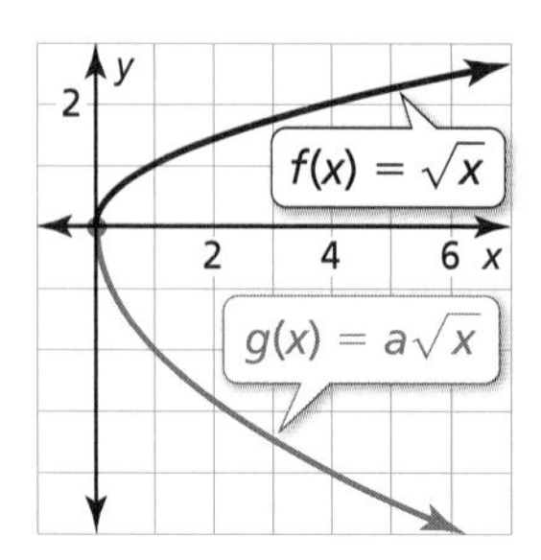

12.

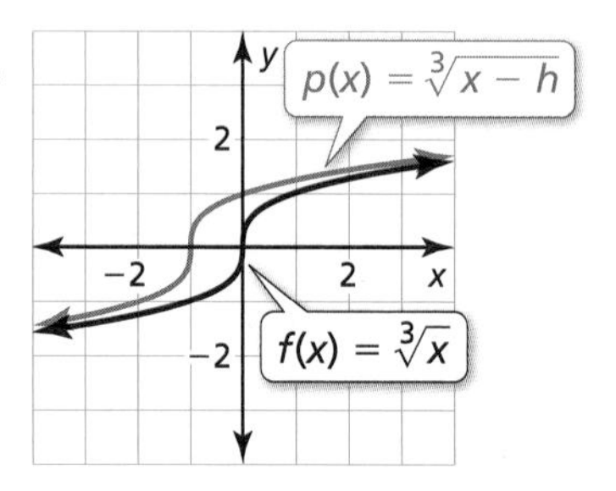

Describe the transformations from the graph of *f* to the graph of *h*. Then graph *h*. *(Section 10.1 and Section 10.2)*

13. $f(x) = \sqrt{x}$; $h(x) = -3\sqrt{x + 2} + 6$

14. $f(x) = \sqrt[3]{x}$; $h(x) = \frac{1}{2}\sqrt[3]{x} - 3$

15. The time t (in seconds) it takes a dropped object to fall h feet is given by $t = \frac{1}{4}\sqrt{h}$. *(Section 10.1)*

a. Use a graphing calculator to graph the function. Describe the domain and range.

b. It takes about 7.4 seconds for a stone dropped from the New River Gorge Bridge in West Virginia to reach the water below. About how high is the bridge above the New River?

16. The radius r of a sphere is given by the function $r = \sqrt[3]{\frac{3}{4\pi}V}$, where V is the volume of the sphere. Spaceship Earth is a spherical structure at Walt Disney World that has an inner radius of about 25 meters. Use a graphing calculator to graph the function. Estimate the volume of Spaceship Earth. *(Section 10.2)*

17. The graph of square root function g is shown. Compare the average rate of change of g to the average rate of change of $h(x) = \sqrt[3]{\frac{3}{2}x}$ over the interval $x = 0$ to $x = 3$. *(Section 10.1 and Section 10.2)*

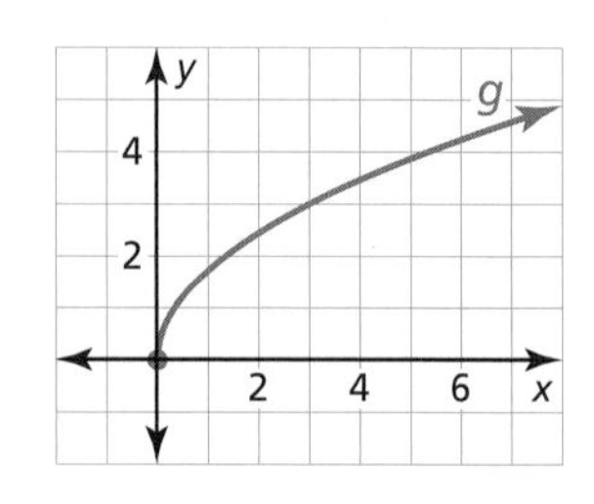

10.3 Solving Radical Equations

Essential Question How can you solve an equation that contains square roots?

EXPLORATION 1 Analyzing a Free-Falling Object

Work with a partner. The table shows the time t (in seconds) that it takes a free-falling object (with no air resistance) to fall d feet.

d (feet)	t (seconds)
0	0.00
32	1.41
64	2.00
96	2.45
128	2.83
160	3.16
192	3.46
224	3.74
256	4.00
288	4.24
320	4.47

MODELING WITH MATHEMATICS

To be proficient in math, you need to routinely interpret your mathematical results in the context of the situation and reflect on whether the results make sense.

a. Use the data in the table to sketch the graph of t as a function of d. Use the coordinate plane below.

b. Use your graph to estimate the time it takes the object to fall 240 feet.

c. The relationship between d and t is given by the function

$$t = \sqrt{\frac{d}{16}}.$$

Use this function to check your estimate in part (b).

d. It takes 5 seconds for the object to hit the ground. How far did it fall? Explain your reasoning.

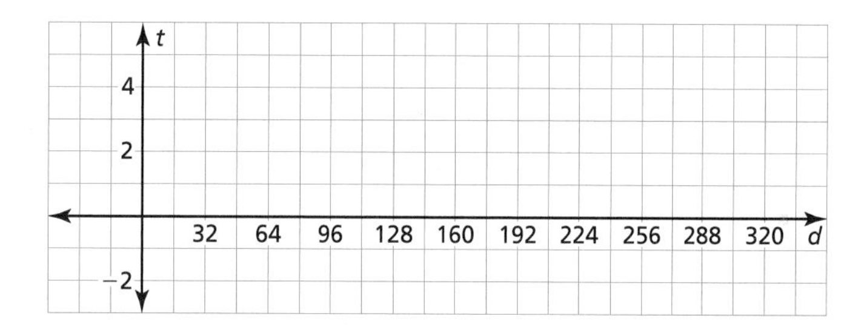

EXPLORATION 2 Solving a Square Root Equation

Work with a partner. The speed s (in feet per second) of the free-falling object in Exploration 1 is given by the function

$$s = \sqrt{64d}.$$

Find the distance the object has fallen when it reaches each speed.

a. $s = 8$ ft/sec **b.** $s = 16$ ft/sec **c.** $s = 24$ ft/sec

Communicate Your Answer

3. How can you solve an equation that contains square roots?

4. Use your answer to Question 3 to solve each equation.

a. $5 = \sqrt{x + 20}$ **b.** $4 = \sqrt{x - 18}$

c. $\sqrt{x} + 2 = 3$ **d.** $-3 = -2\sqrt{x}$

10.3 Lesson

Core Vocabulary

radical equation, *p. 560*

Previous
radical
radical expression
extraneous solution

What You Will Learn

- Solve radical equations.
- Identify extraneous solutions.
- Solve real-life problems involving radical equations.

Solving Radical Equations

A **radical equation** is an equation that contains a radical expression with a variable in the radicand. To solve a radical equation involving a square root, first use properties of equality to isolate the radical on one side of the equation. Then use the following property to eliminate the radical and solve for the variable.

Core Concept

Squaring Each Side of an Equation

Words If two expressions are equal, then their squares are also equal.

Algebra If $a = b$, then $a^2 = b^2$.

EXAMPLE 1 Solving Radical Equations

Solve each equation.

a. $\sqrt{x} + 5 = 13$

b. $3 - \sqrt{x} = 0$

SOLUTION

a. $\sqrt{x} + 5 = 13$ Write the equation.

$\sqrt{x} = 8$ Subtract 5 from each side.

$(\sqrt{x})^2 = 8^2$ Square each side of the equation.

$x = 64$ Simplify.

The solution is $x = 64$.

Check

$\sqrt{x} + 5 = 13$

$\sqrt{64} + 5 \stackrel{?}{=} 13$

$8 + 5 \stackrel{?}{=} 13$

$13 = 13$ ✓

b. $3 - \sqrt{x} = 0$ Write the equation.

$3 = \sqrt{x}$ Add $\sqrt{x}$ to each side.

$3^2 = (\sqrt{x})^2$ Square each side of the equation.

$9 = x$ Simplify.

The solution is $x = 9$.

Check

$3 - \sqrt{x} = 0$

$3 - \sqrt{9} \stackrel{?}{=} 0$

$3 - 3 \stackrel{?}{=} 0$

$0 = 0$ ✓

Monitoring Progress  Help in English and Spanish at *BigIdeasMath.com*

Solve the equation. Check your solution.

1. $\sqrt{x} = 6$
2. $\sqrt{x} - 7 = 3$
3. $\sqrt{y} + 15 = 22$
4. $1 - \sqrt{c} = -2$

EXAMPLE 2 Solving a Radical Equation

$4\sqrt{x+2} + 3 = 19$	Original equation
$4\sqrt{x+2} = 16$	Subtract 3 from each side.
$\sqrt{x+2} = 4$	Divide each side by 4.
$(\sqrt{x+2})^2 = 4^2$	Square each side of the equation.
$x + 2 = 16$	Simplify.
$x = 14$	Subtract 2 from each side.

▶ The solution is $x = 14$.

Check

$4\sqrt{x+2} + 3 = 19$

$4\sqrt{14+2} + 3 \stackrel{?}{=} 19$

$4\sqrt{16} + 3 \stackrel{?}{=} 19$

$4(4) + 3 \stackrel{?}{=} 19$

$19 = 19$ ✓

EXAMPLE 3 Solving an Equation with Radicals on Both Sides

Solve $\sqrt{2x-1} = \sqrt{x+4}$.

SOLUTION

Method 1

$\sqrt{2x-1} = \sqrt{x+4}$	Write the equation.
$(\sqrt{2x-1})^2 = (\sqrt{x+4})^2$	Square each side of the equation.
$2x - 1 = x + 4$	Simplify.
$x = 5$	Solve for x.

▶ The solution is $x = 5$.

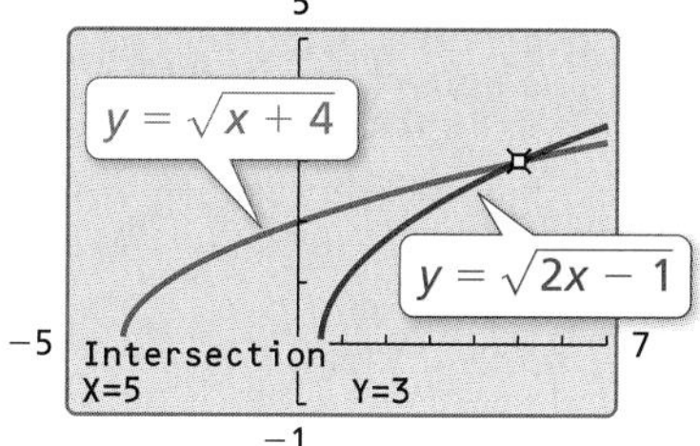

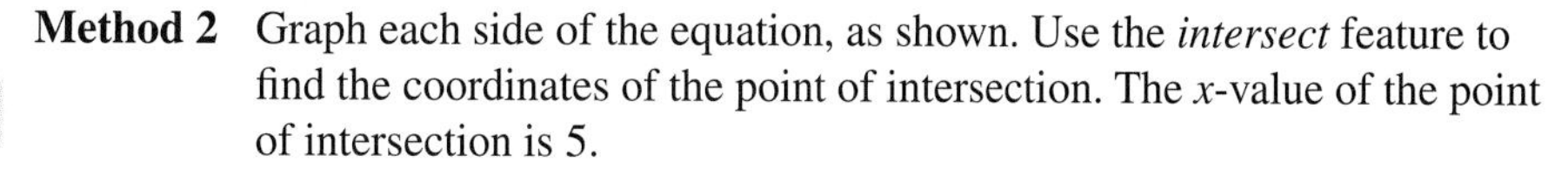

Method 2 Graph each side of the equation, as shown. Use the *intersect* feature to find the coordinates of the point of intersection. The x-value of the point of intersection is 5.

▶ So, the solution is $x = 5$.

EXAMPLE 4 Solving a Radical Equation Involving a Cube Root

Solve $\sqrt[3]{5x-2} = 12$.

SOLUTION

$\sqrt[3]{5x-2} = 12$	Write the equation.
$(\sqrt[3]{5x-2})^3 = 12^3$	Cube each side of the equation.
$5x - 2 = 1728$	Simplify.
$x = 346$	Solve for x.

▶ The solution is $x = 346$.

LOOKING FOR STRUCTURE

You can extend the concept taught in Examples 1–3 to solve a radical equation involving a cube root. Instead of squaring each side of the equation, you *cube* each side to eliminate the radical.

Monitoring Progress

Help in English and Spanish at *BigIdeasMath.com*

Solve the equation. Check your solution.

5. $\sqrt{x+4} + 7 = 11$ **6.** $15 = 6 + \sqrt{3w-9}$ **7.** $\sqrt{3x+1} = \sqrt{4x-7}$

8. $\sqrt{n} = \sqrt{5n-1}$ **9.** $\sqrt[3]{y} - 4 = 1$ **10.** $\sqrt[3]{3c+7} = 10$

Identifying Extraneous Solutions

Squaring each side of an equation can sometimes introduce an extraneous solution.

ATTEND TO PRECISION

To understand how extraneous solutions can be introduced, consider the equation $\sqrt{x} = -2$. This equation has no real solution, however, you obtain $x = 4$ after squaring each side.

EXAMPLE 5 Identifying an Extraneous Solution

Solve $x = \sqrt{x + 6}$.

SOLUTION

$x = \sqrt{x+6}$	Write the equation.
$x^2 = \left(\sqrt{x+6}\right)^2$	Square each side of the equation.
$x^2 = x + 6$	Simplify.
$x^2 - x - 6 = 0$	Subtract x and 6 from each side.
$(x - 3)(x + 2) = 0$	Factor.
$x - 3 = 0$ or $x + 2 = 0$	Zero-Product Property
$x = 3$ or $x = -2$	Solve for x.

Check Check each solution in the original equation.

$3 \stackrel{?}{=} \sqrt{3+6}$	Substitute for x.	$-2 \stackrel{?}{=} \sqrt{-2+6}$
$3 \stackrel{?}{=} \sqrt{9}$	Simplify.	$-2 \stackrel{?}{=} \sqrt{4}$
$3 = 3$ ✓	Simplify.	$-2 \neq 2$ ✗

STUDY TIP

Be sure to always substitute your solutions into the original equation to check for extraneous solutions.

▶ Because $x = -2$ does not satisfy the original equation, it is an extraneous solution. The only solution is $x = 3$.

EXAMPLE 6 Identifying an Extraneous Solution

Solve $13 + \sqrt{5n} = 3$.

SOLUTION

$13 + \sqrt{5n} = 3$	Write the equation.
$\sqrt{5n} = -10$	Subtract 13 from each side.
$\left(\sqrt{5n}\right)^2 = (-10)^2$	Square each side of the equation.
$5n = 100$	Simplify.
$n = 20$	Divide each side by 5.

Check

$13 + \sqrt{5n} = 3$

$13 + \sqrt{5(20)} \stackrel{?}{=} 3$

$13 + \sqrt{100} \stackrel{?}{=} 3$

$23 \neq 3$ ✗

▶ Because $n = 20$ does not satisfy the original equation, it is an extraneous solution. So, the equation has no solution.

Monitoring Progress  Help in English and Spanish at *BigIdeasMath.com*

Solve the equation. Check your solution(s).

11. $\sqrt{4 - 3x} = x$ **12.** $\sqrt{3m} + 10 = 1$ **13.** $p + 1 = \sqrt{7p + 15}$

Solving Real-Life Problems

EXAMPLE 7 Modeling with Mathematics

STUDY TIP

The period of a pendulum is the amount of time it takes for the pendulum to swing back and forth.

The period P (in seconds) of a pendulum is given by the function $P = 2\pi\sqrt{\frac{L}{32}}$, where L is the pendulum length (in feet). A pendulum has a period of 4 seconds. Is this pendulum twice as long as a pendulum with a period of 2 seconds? Explain your reasoning.

SOLUTION

1. **Understand the Problem** You are given a function that represents the period P of a pendulum based on its length L. You need to find and compare the values of L for two values of P.

2. **Make a Plan** Substitute $P = 2$ and $P = 4$ into the function and solve for L. Then compare the values.

3. **Solve the Problem**

$P = 2\pi\sqrt{\frac{L}{32}}$	Write the function.	$P = 2\pi\sqrt{\frac{L}{32}}$
$2 = 2\pi\sqrt{\frac{L}{32}}$	Substitute for P.	$4 = 2\pi\sqrt{\frac{L}{32}}$
$\frac{2}{2\pi} = \sqrt{\frac{L}{32}}$	Divide each side by 2π.	$\frac{4}{2\pi} = \sqrt{\frac{L}{32}}$
$\frac{1}{\pi} = \sqrt{\frac{L}{32}}$	Simplify.	$\frac{2}{\pi} = \sqrt{\frac{L}{32}}$
$\frac{1}{\pi^2} = \frac{L}{32}$	Square each side and simplify.	$\frac{4}{\pi^2} = \frac{L}{32}$
$\frac{32}{\pi^2} = L$	Multiply each side by 32.	$\frac{128}{\pi^2} = L$
$3.24 \approx L$	Use a calculator.	$12.97 \approx L$

▶ No, the length of the pendulum with a period of 4 seconds is $\frac{128}{\pi^2} \div \frac{32}{\pi^2} = 4$ times longer than the length of a pendulum with a period of 2 seconds.

4. **Look Back** Use the *trace* feature of a graphing calculator to check your solutions.

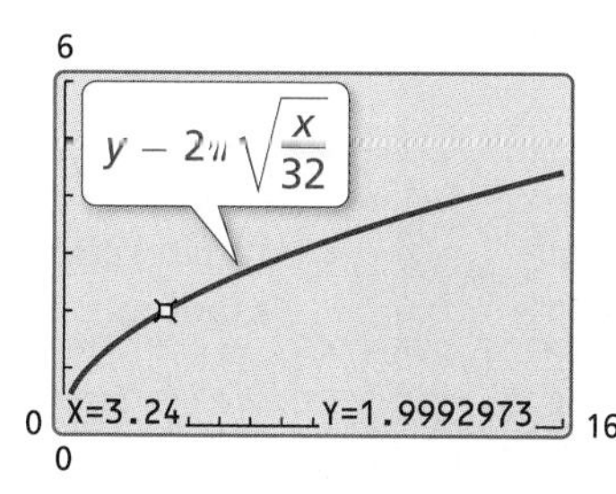

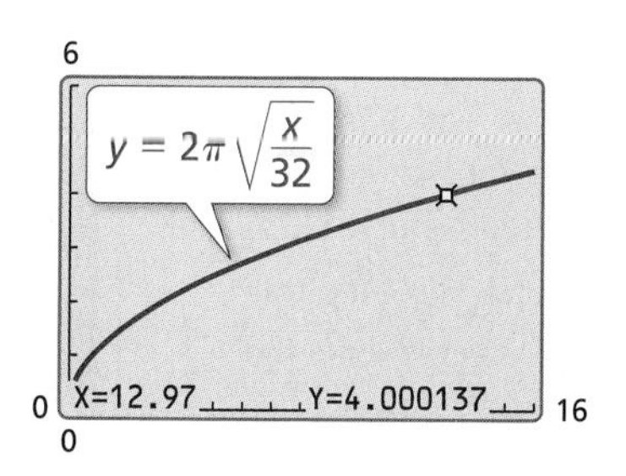

Monitoring Progress  Help in English and Spanish at *BigIdeasMath.com*

14. What is the length of a pendulum that has a period of 2.5 seconds?

10.3 Exercises

Dynamic Solutions available at *BigIdeasMath.com*

Vocabulary and Core Concept Check

1. **VOCABULARY** Why should you check every solution of a radical equation?

2. **WHICH ONE DOESN'T BELONG?** Which equation does *not* belong with the other three? Explain your reasoning.

$\sqrt{x} + 6 = 10$ | $2\sqrt{x+3} = 32$ | $x\sqrt{3} - 5 = 4$ | $\sqrt{x-1} = 16$

Monitoring Progress and Modeling with Mathematics

In Exercises 3–12, solve the equation. Check your solution. *(See Example 1.)*

3. $\sqrt{x} = 9$
4. $\sqrt{y} = 4$
5. $7 = \sqrt{m} - 5$
6. $\sqrt{p} - 7 = -1$
7. $\sqrt{c} + 12 = 23$
8. $\sqrt{x} + 6 = 8$
9. $4 - \sqrt{a} = 2$
10. $-8 = 7 - \sqrt{r}$
11. $3\sqrt{y} - 18 = -3$
12. $2\sqrt{q} + 5 = 11$

In Exercises 13–20, solve the equation. Check your solution. *(See Example 2.)*

13. $\sqrt{a-3} + 5 = 9$
14. $\sqrt{b+7} - 5 = -2$
15. $2\sqrt{x+4} = 16$
16. $5\sqrt{y-2} = 10$
17. $-1 = \sqrt{5r+1} - 7$
18. $2 = \sqrt{4s-4} - 4$
19. $7 + 3\sqrt{3p-9} = 25$
20. $19 - 4\sqrt{3c-11} = 11$

21. **MODELING WITH MATHEMATICS** The Cave of Swallows is a natural open-air pit cave in the state of San Luis Potosí, Mexico. The 1220-foot-deep cave was a popular destination for BASE jumpers. The function $t = \frac{1}{4}\sqrt{d}$ represents the time t (in seconds) that it takes a BASE jumper to fall d feet. How far does a BASE jumper fall in 3 seconds?

22. **MODELING WITH MATHEMATICS** The edge length s of a cube with a surface area of A is given by $s = \sqrt{\frac{A}{6}}$. What is the surface area of a cube with an edge length of 4 inches?

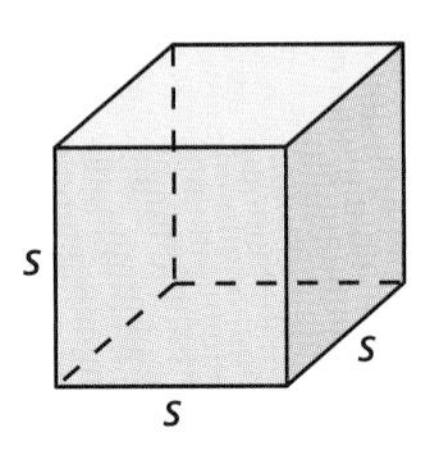

In Exercises 23–26, use the graph to solve the equation.

23. $\sqrt{2x+2} = \sqrt{x+3}$

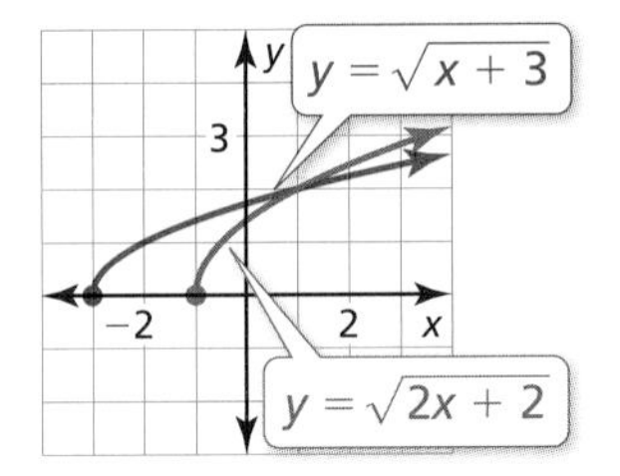

24. $\sqrt{3x+1} = \sqrt{4x-4}$

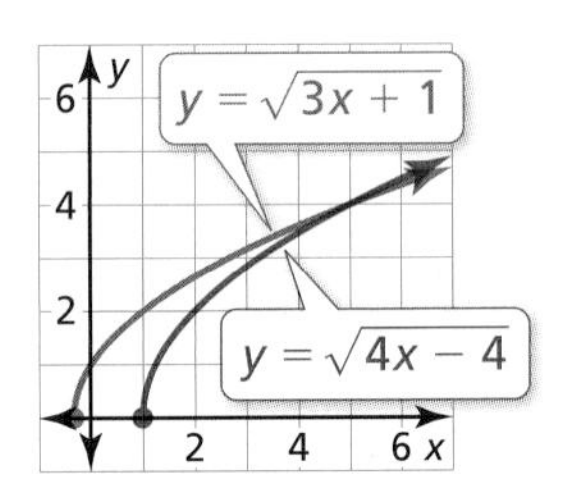

25. $\sqrt{x+2} - \sqrt{2x} = 0$

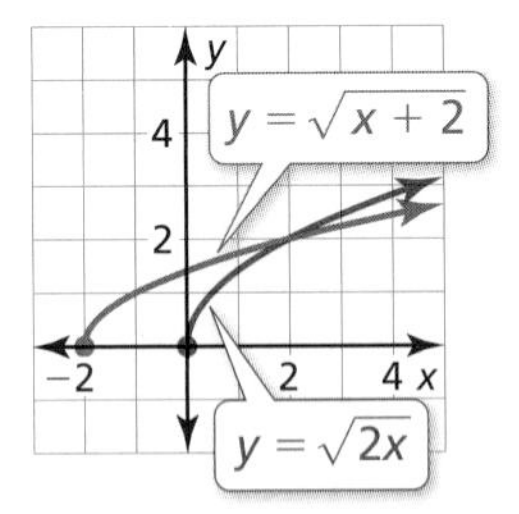

26. $\sqrt{x+5} - \sqrt{3x+7} = 0$

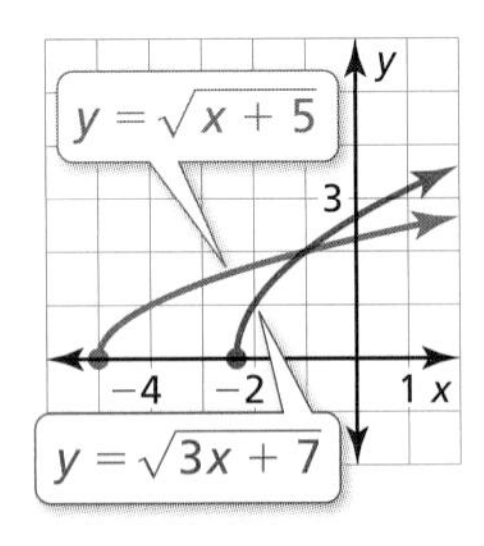

In Exercises 27–34, solve the equation. Check your solution. *(See Example 3.)*

27. $\sqrt{2x-9} = \sqrt{x}$
28. $\sqrt{y+1} = \sqrt{4y-8}$
29. $\sqrt{3g+1} = \sqrt{7g-19}$
30. $\sqrt{8h-7} = \sqrt{6h+7}$
31. $\sqrt{\frac{p}{2} - 2} = \sqrt{p-8}$
32. $\sqrt{2v-5} = \sqrt{\frac{v}{3} + 5}$
33. $\sqrt{2c+1} - \sqrt{4c} = 0$
34. $\sqrt{5r} - \sqrt{8r-2} = 0$

MATHEMATICAL CONNECTIONS In Exercises 35 and 36, find the value of x.

35. Perimeter $= 22$ cm

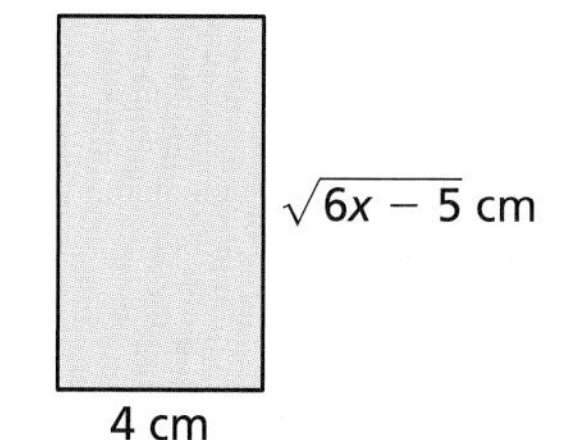

36. Area $= \sqrt{5x - 4}$ ft^2

$\sqrt{3x + 12}$ ft

2 ft

In Exercises 37–44, solve the equation. Check your solution. *(See Example 4.)*

37. $\sqrt[3]{x} = 4$

38. $\sqrt[3]{y} = 2$

39. $6 = \sqrt[3]{8g}$

40. $\sqrt[3]{r + 19} = 3$

41. $\sqrt[3]{2s + 9} = -3$

42. $-5 = \sqrt[3]{10x + 15}$

43. $\sqrt[3]{y + 6} = \sqrt[3]{5y - 2}$

44. $\sqrt[3]{7j - 2} = \sqrt[3]{j + 4}$

In Exercises 45–48, determine which solution, if any, is an extraneous solution.

45. $\sqrt{6x - 5} = x; x = 5, x = 1$

46. $\sqrt{2y + 3} = y; y = -1, y = 3$

47. $\sqrt{12p + 16} = -2p; p = -1, p = 4$

48. $-3g = \sqrt{-18 - 27g}; g = -2, g = -1$

In Exercises 49–58, solve the equation. Check your solution(s). *(See Examples 5 and 6.)*

49. $y = \sqrt{5y - 4}$

50. $\sqrt{-14 - 9x} = x$

51. $\sqrt{1 - 3a} = 2a$

52. $2q = \sqrt{10q - 6}$

53. $9 + \sqrt{5p} = 4$

54. $\sqrt{3n} - 11 = -5$

55. $\sqrt{2m + 2} - 3 = 1$

56. $15 + \sqrt{4b} - 8 = 13$

57. $r + 4 = \sqrt{-4r - 19}$

58. $\sqrt{3 - s} = s - 1$

ERROR ANALYSIS In Exercises 59 and 60, describe and correct the error in solving the equation.

59.

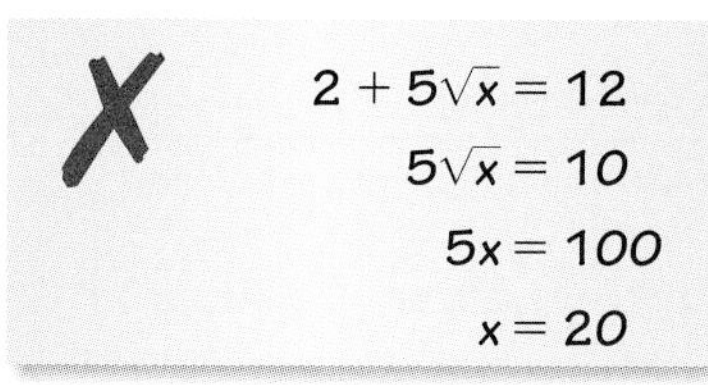

$2 + 5\sqrt{x} = 12$

$5\sqrt{x} = 10$

$5x = 100$

$x = 20$

60.

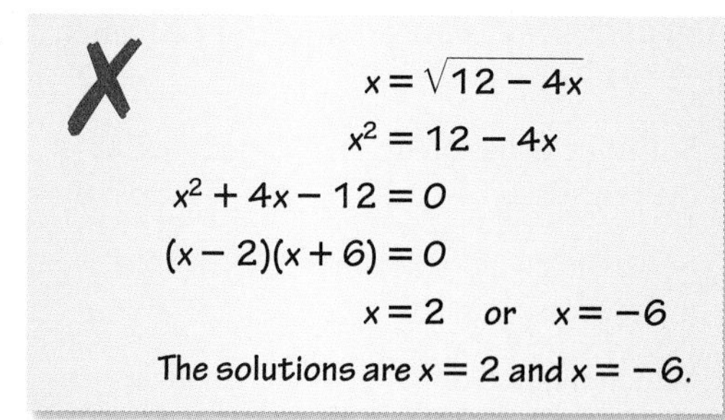

$x = \sqrt{12 - 4x}$

$x^2 = 12 - 4x$

$x^2 + 4x - 12 = 0$

$(x - 2)(x + 6) = 0$

$x = 2$ or $x = -6$

The solutions are $x = 2$ and $x = -6$.

61. REASONING Explain how to use mental math to solve $\sqrt{2x} + 5 = 1$.

62. WRITING Explain how you would solve $\sqrt[4]{m + 4} - \sqrt[4]{3m} = 0$.

63. MODELING WITH MATHEMATICS The formula $V = \sqrt{PR}$ relates the voltage V (in volts), power P (in watts), and resistance R (in ohms) of an electrical circuit. The hair dryer shown is on a 120-volt circuit. Is the resistance of the hair dryer half as much as the resistance of the same hair dryer on a 240-volt circuit? Explain your reasoning. *(See Example 7.)*

64. MODELING WITH MATHEMATICS The time t (in seconds) it takes a trapeze artist to swing back and forth is represented by the function $t = 2\pi\sqrt{\frac{r}{32}}$, where r is the rope length (in feet). It takes the trapeze artist 6 seconds to swing back and forth. Is this rope $\frac{3}{2}$ as long as the rope used when it takes the trapeze artist 4 seconds to swing back and forth? Explain your reasoning.

REASONING In Exercises 65–68, determine whether the statement is *true* or *false*. If it is false, explain why.

65. If $\sqrt{a} = b$, then $(\sqrt{a})^2 = b^2$.

66. If $\sqrt{a} = \sqrt{b}$, then $a = b$.

67. If $a^2 = b^2$, then $a = b$.

68. If $a^2 = \sqrt{b}$, then $a^4 = (\sqrt{b})^2$.

69. **COMPARING METHODS** Consider the equation $x + 2 = \sqrt{2x - 3}$.

a. Solve the equation by graphing. Describe the process.

b. Solve the equation algebraically. Describe the process.

c. Which method do you prefer? Explain your reasoning.

70. **HOW DO YOU SEE IT?** The graph shows two radical functions.

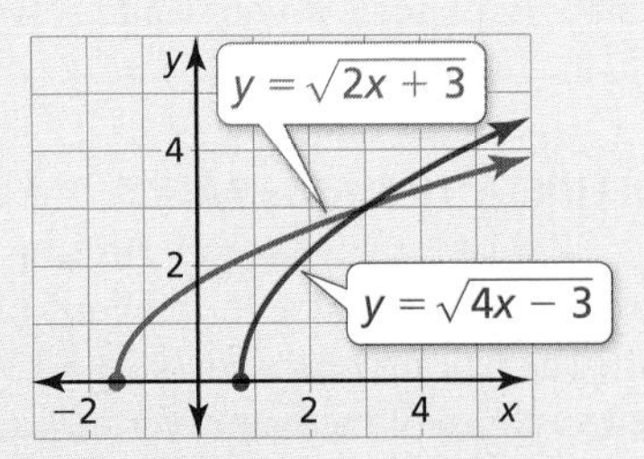

a. Write an equation whose solution is the x-coordinate of the point of intersection of the graphs.

b. Use the graph to solve the equation.

71. **MATHEMATICAL CONNECTIONS** The slant height s of a cone with a radius of r and a height of h is given by $s = \sqrt{r^2 + h^2}$. The slant heights of the two cones are equal. Find the radius of each cone.

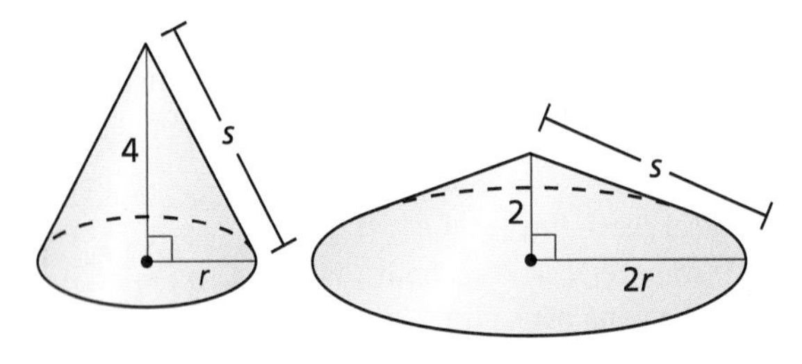

72. **CRITICAL THINKING** How is squaring $\sqrt{x + 2}$ different from squaring $\sqrt{x} + 2$?

USING STRUCTURE In Exercises 73–78, solve the equation. Check your solution.

73. $\sqrt{m + 15} = \sqrt{m} + \sqrt{5}$

74. $2 - \sqrt{x + 1} = \sqrt{x + 2}$

75. $\sqrt{5y + 9} + \sqrt{5y} = 9$

76. $\sqrt{2c - 8} - \sqrt{2c} - 4 = 0$

77. $2\sqrt{1 + 4h} - 4\sqrt{h} - 2 = 0$

78. $\sqrt{20 - 4z} + 2\sqrt{-z} = 10$

79. **OPEN-ENDED** Write a radical equation that has a solution of $x = 5$.

80. **OPEN-ENDED** Write a radical equation that has $x = 3$ and $x = 4$ as solutions.

81. **MAKING AN ARGUMENT** Your friend says the equation $\sqrt{(2x + 5)^2} = 2x + 5$ is always true, because after simplifying the left side of the equation, the result is an equation with infinitely many solutions. Is your friend correct? Explain.

82. **THOUGHT PROVOKING** Solve the equation $\sqrt[3]{x + 1} = \sqrt{x - 3}$. Show your work and explain your steps.

83. **MODELING WITH MATHEMATICS** The frequency f (in cycles per second) of a string of an electric guitar is given by the equation $f = \frac{1}{2\ell}\sqrt{\frac{T}{m}}$, where ℓ is the length of the string (in meters), T is the string's tension (in newtons), and m is the string's mass per unit length (in kilograms per meter). The high E string of an electric guitar is 0.64 meter long with a mass per unit length of 0.000401 kilogram per meter.

a. How much tension is required to produce a frequency of about 330 cycles per second?

b. Would you need more or less tension to create the same frequency on a string with greater mass per unit length? Explain.

Maintaining Mathematical Proficiency

Reviewing what you learned in previous grades and lessons

Find the product. *(Section 7.2)*

84. $(x + 8)(x - 2)$

85. $(3p - 1)(4p + 5)$

86. $(s + 2)(s^2 + 3s - 4)$

Graph the function. Compare the graph to the graph of $f(x) = x^2$. *(Section 8.1)*

87. $r(x) = 3x^2$

88. $g(x) = \frac{3}{4}x^2$

89. $h(x) = -5x^2$

10.4 Inverse of a Function

Essential Question How are a function and its inverse related?

EXPLORATION 1 Exploring Inverse Functions

Work with a partner. The functions f and g are *inverses* of each other. Compare the tables of values of the two functions. How are the functions related?

x	0	0.5	1	1.5	2	2.5	3	3.5
$f(x)$	0	0.25	1	2.25	4	6.25	9	12.25

x	0	0.25	1	2.25	4	6.25	9	12.25
$g(x)$	0	0.5	1	1.5	2	2.5	3	3.5

EXPLORATION 2 Exploring Inverse Functions

Work with a partner.

a. Plot the two sets of points represented by the tables in Exploration 1. Use the coordinate plane below.

b. Connect each set of points with a smooth curve.

c. Describe the relationship between the two graphs.

d. Write an equation for each function.

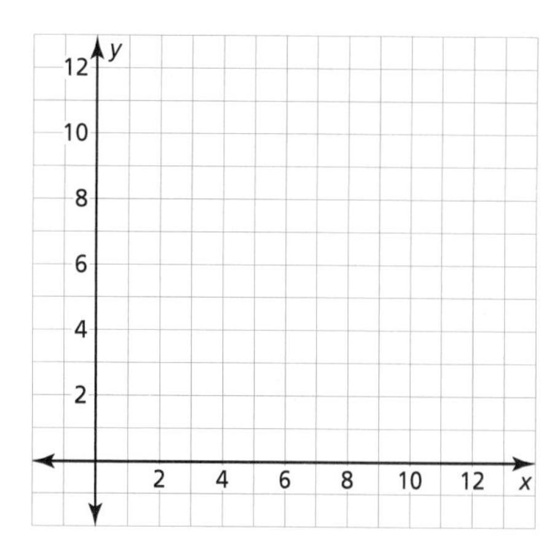

Communicate Your Answer

3. How are a function and its inverse related?

4. A table of values for a function f is given. Create a table of values for a function g, the inverse of f.

x	0	1	2	3	4	5	6	7
$f(x)$	1	2	3	4	5	6	7	8

5. Sketch the graphs of $f(x) = x + 4$ and its inverse in the same coordinate plane. Then write an equation of the inverse of f. Explain your reasoning.

ATTENDING TO PRECISION

To be proficient in math, you need to communicate precisely with others.

10.4 Lesson

Core Vocabulary

inverse relation, p. 568
inverse function, p. 569

Previous
input
output
inverse operations
reflection
line of reflection

What You Will Learn

- Find inverses of relations.
- Explore inverses of functions.
- Find inverses of functions algebraically.
- Find inverses of nonlinear functions.

Finding Inverses of Relations

Recall that a relation pairs inputs with outputs. An **inverse relation** switches the input and output values of the original relation.

Core Concept

Inverse Relation

When a relation contains (a, b), the inverse relation contains (b, a).

EXAMPLE 1 Finding Inverses of Relations

Find the inverse of each relation.

a. $(-4, 7), (-2, 4), (0, 1), (2, -2), (4, -5)$ — Switch the coordinates of each ordered pair.

$(7, -4), (4, -2), (1, 0), (-2, 2), (-5, 4)$ — Inverse relation

b.

Input	−1	0	1	2	3	4
Output	5	10	15	20	25	30

Switch the inputs and outputs.

Inverse relation:

Input	5	10	15	20	25	30
Output	−1	0	1	2	3	4

Monitoring Progress

Help in English and Spanish at *BigIdeasMath.com*

Find the inverse of the relation.

1. $(-3, -4), (-2, 0), (-1, 4), (0, 8), (1, 12), (2, 16), (3, 20)$

2.

Input	−2	−1	0	1	2
Output	4	1	0	1	4

Exploring Inverses of Functions

Throughout this book, you have used given inputs to find corresponding outputs of $y = f(x)$ for various types of functions. You have also used given outputs to find corresponding inputs. Now you will solve equations of the form $y = f(x)$ for x to obtain a formula for finding the input given a specific output of the function f.

EXAMPLE 2 Writing a Formula for the Input of a Function

Let $f(x) = 2x + 1$. Solve $y = f(x)$ for x. Then find the input when the output is -3.

SOLUTION

$y = 2x + 1$ Set y equal to $f(x)$.

$y - 1 = 2x$ Subtract 1 from each side.

$\frac{y-1}{2} = x$ Divide each side by 2.

Find the input when $y = -3$.

$x = \frac{-3-1}{2}$ Substitute -3 for y.

$= \frac{-4}{2}$ Subtract.

$= -2$ Divide.

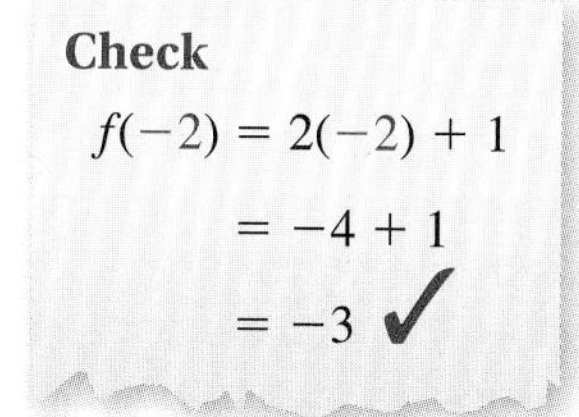

▶ So, the input is -2 when the output is -3.

Monitoring Progress  Help in English and Spanish at *BigIdeasMath.com*

Solve $y = f(x)$ for x. Then find the input when the output is 4.

3. $f(x) = x - 6$ **4.** $f(x) = \frac{1}{2}x + 3$ **5.** $f(x) = 4x^2$

UNDERSTANDING MATHEMATICAL TERMS

The term *inverse functions* does not refer to a new type of function. Rather, it describes any pair of functions that are inverses.

In Example 2, notice the steps involved after substituting for x in $y = 2x + 1$ and after substituting for y in $x = \frac{y-1}{2}$.

$y = 2x + 1$		$x = \frac{y-1}{2}$
Step 1 Multiply by 2.		**Step 1** Subtract 1.
Step 2 Add 1.	inverse operations in the reverse order	**Step 2** Divide by 2.

Notice that these steps *undo* each other. **Inverse functions** are functions that undo each other. In Example 2, you can use the equation solved for x to write the inverse of f by switching the roles of x and y.

$f(x) = 2x + 1$ original function $g(x) = \frac{x-1}{2}$ inverse function

Because an inverse function interchanges the input and output values of the original function, the domain and range are also interchanged.

LOOKING FOR A PATTERN

Notice that the graph of the inverse function g is a reflection of the graph of the original function f. The line of reflection is $y = x$.

Original function: $f(x) = 2x + 1$

x	-2	-1	0	1	2
y	-3	-1	1	3	5

Inverse function: $g(x) = \frac{x-1}{2}$

x	-3	-1	1	3	5
y	-2	-1	0	1	2

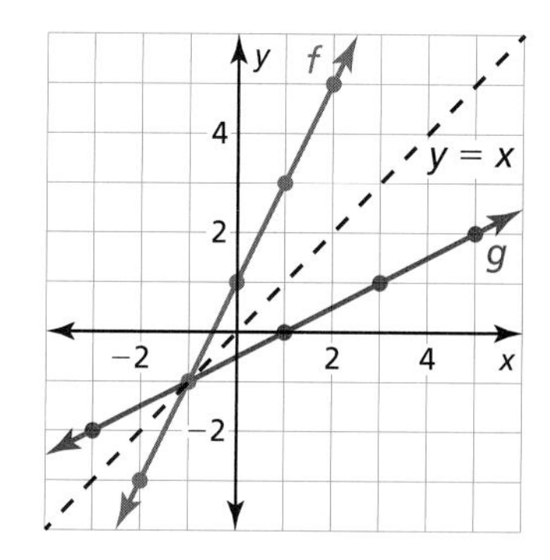

Finding Inverses of Functions Algebraically

Core Concept

Finding Inverses of Functions Algebraically

Step 1 Set y equal to $f(x)$.

Step 2 Switch x and y in the equation.

Step 3 Solve the equation for y.

STUDY TIP

On the previous page, you solved a function for x and switched the roles of x and y to find the inverse function. You can also find the inverse function by switching x and y first, and then solving for y.

EXAMPLE 3 **Finding the Inverse of a Linear Function**

Find the inverse of $f(x) = 4x - 9$.

SOLUTION

Method 1 Use the method above.

Step 1	$f(x) = 4x - 9$	Write the function.
	$y = 4x - 9$	Set y equal to $f(x)$.
Step 2	$x = 4y - 9$	Switch x and y in the equation.
Step 3	$x + 9 = 4y$	Add 9 to each side.
	$\dfrac{x+9}{4} = y$	Divide each side by 4.

▶ The inverse of f is $g(x) = \dfrac{x+9}{4}$, or $g(x) = \dfrac{1}{4}x + \dfrac{9}{4}$.

Check

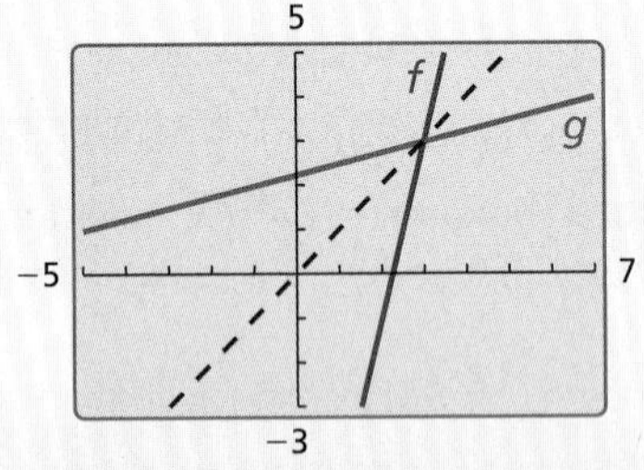

The graph of g appears to be a reflection of the graph of f in the line $y = x$. ✓

Method 2 Use inverse operations in the reverse order.

$f(x) = 4x - 9$ Multiply the input x by 4 and then subtract 9.

To find the inverse, apply inverse operations in the reverse order.

$g(x) = \dfrac{x+9}{4}$ Add 9 to the input x and then divide by 4.

▶ The inverse of f is $g(x) = \dfrac{x+9}{4}$, or $g(x) = \dfrac{1}{4}x + \dfrac{9}{4}$.

Monitoring Progress

Help in English and Spanish at *BigIdeasMath.com*

Find the inverse of the function. Then graph the function and its inverse.

6. $f(x) = 6x$ **7.** $f(x) = -x + 5$ **8.** $f(x) = \frac{1}{4}x - 1$

Finding Inverses of Nonlinear Functions

The inverse of the linear function in Example 3 is also a function. The inverse of a function, however, is *not* always a function. The graph of $f(x) = x^2$ is shown along with its reflection in the line $y = x$. Notice that the graph of the inverse of $f(x) = x^2$ does not pass the Vertical Line Test. So, the inverse is *not* a function.

When the domain of $f(x) = x^2$ is *restricted* to only nonnegative real numbers, the inverse of f *is* a function, as shown in the next example.

EXAMPLE 4 Finding the Inverse of a Quadratic Function

Find the inverse of $f(x) = x^2$, $x \geq 0$. Then graph the function and its inverse.

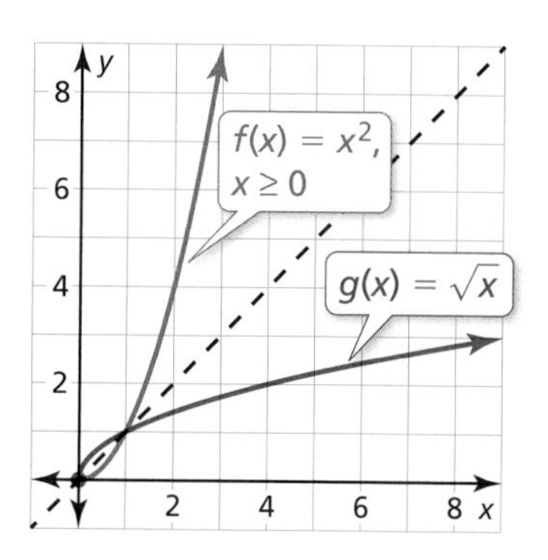

SOLUTION

$f(x) = x^2$	Write the function.
$y = x^2$	Set y equal to $f(x)$.
$x = y^2$	Switch x and y in the equation.
$\pm\sqrt{x} = y$	Take square root of each side.

▶ Because the domain of f is restricted to nonnegative values of x, the range of the inverse must also be restricted to nonnegative values. So, the inverse of f is $g(x) = \sqrt{x}$.

You can use the graph of a function f to determine whether the inverse of f is a function by applying the *Horizontal Line Test*.

Core Concept

Horizontal Line Test

The inverse of a function f is also a function if and only if no horizontal line intersects the graph of f more than once.

EXAMPLE 5 Finding the Inverse of a Radical Function

Consider the function $f(x) = \sqrt{x + 2}$. Determine whether the inverse of f is a function. Then find the inverse.

SOLUTION

Graph the function f. Because no horizontal line intersects the graph more than once, the inverse of f is a function. Find the inverse.

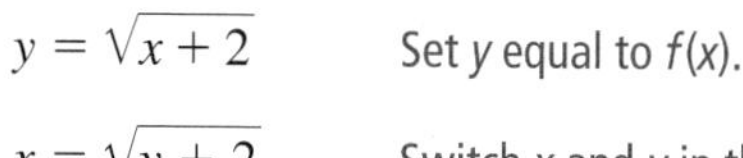

$y = \sqrt{x + 2}$	Set y equal to $f(x)$.
$x = \sqrt{y + 2}$	Switch x and y in the equation.
$x^2 = \left(\sqrt{y + 2}\right)^2$	Square each side.
$x^2 = y + 2$	Simplify.
$x^2 - 2 = y$	Subtract 2 from each side.

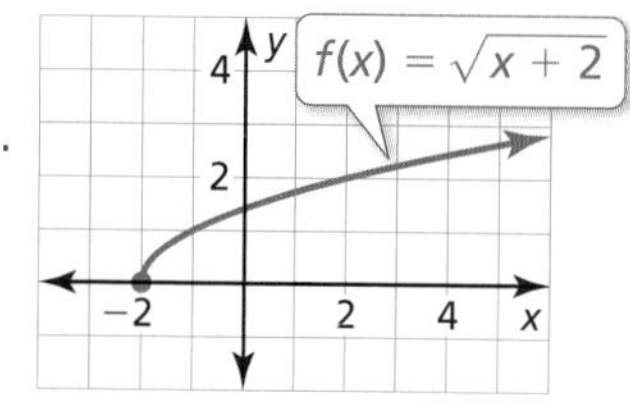

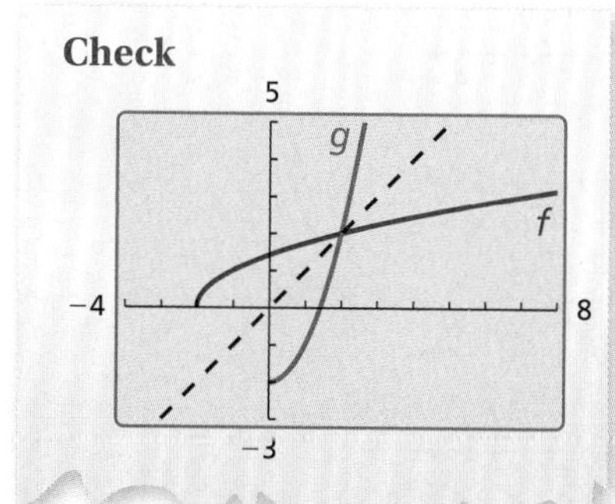

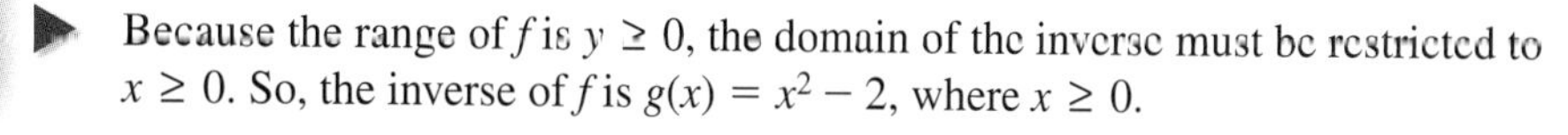

▶ Because the range of f is $y \geq 0$, the domain of the inverse must be restricted to $x \geq 0$. So, the inverse of f is $g(x) = x^2 - 2$, where $x \geq 0$.

Monitoring Progress  Help in English and Spanish at *BigIdeasMath.com*

Find the inverse of the function. Then graph the function and its inverse.

9. $f(x) = -x^2, x \leq 0$

10. $f(x) = 4x^2 + 3, x \geq 0$

11. Is the inverse of $f(x) = \sqrt{2x - 1}$ a function? Find the inverse.

10.4 Exercises

Dynamic Solutions available at *BigIdeasMath.com*

Vocabulary and Core Concept Check

1. **COMPLETE THE SENTENCE** A relation contains the point $(-3, 10)$. The __________ contains the point $(10, -3)$.

2. **DIFFERENT WORDS, SAME QUESTION** Consider the function f represented by the graph. Which is different? Find "both" answers.

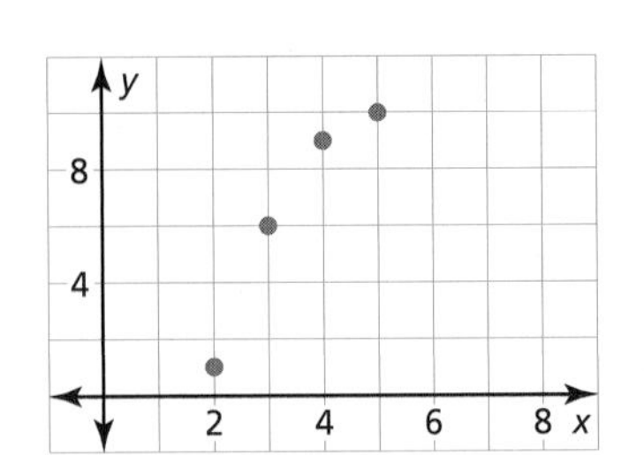

Graph the inverse of the function.

Reflect the graph of the function in the x-axis.

Reflect the graph of the function in the line $y = x$.

Switch the inputs and outputs of the function and graph the resulting function.

Monitoring Progress and Modeling with Mathematics

In Exercises 3–8, find the inverse of the relation. *(See Example 1.)*

3. $(1, 0), (3, -8), (4, -3), (7, -5), (9, -1)$

4. $(2, 1), (4, -3), (6, 7), (8, 1), (10, -4)$

5.

Input	−5	−5	0	5	10
Output	8	6	0	6	8

6.

Input	−12	−8	−5	−3	−2
Output	2	5	−1	10	−2

7. Input → Output

- −3 → 9
- −1 → 1
- 1 → 1
- 3 → 4

8. Input → Output

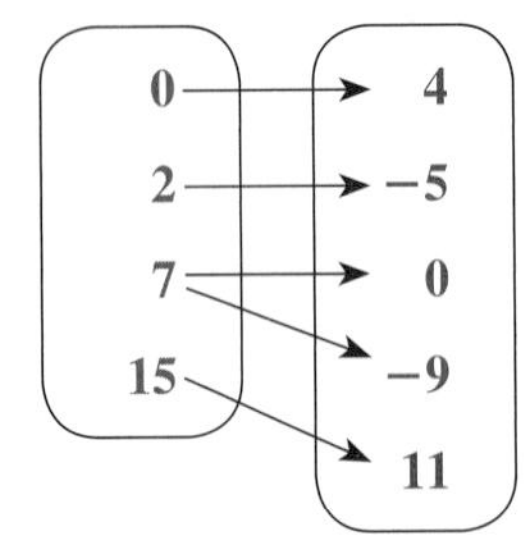

In Exercises 9–14, solve $y = f(x)$ for x. Then find the input when the output is 2. *(See Example 2.)*

9. $f(x) = x + 5$

10. $f(x) = 2x - 3$

11. $f(x) = \frac{1}{4}x - 1$

12. $f(x) = \frac{2}{3}x + 4$

13. $f(x) = 9x^2$

14. $f(x) = \frac{1}{2}x^2 - 7$

In Exercises 15 and 16, graph the inverse of the function by reflecting the graph in the line $y = x$. Describe the domain and range of the inverse.

15.

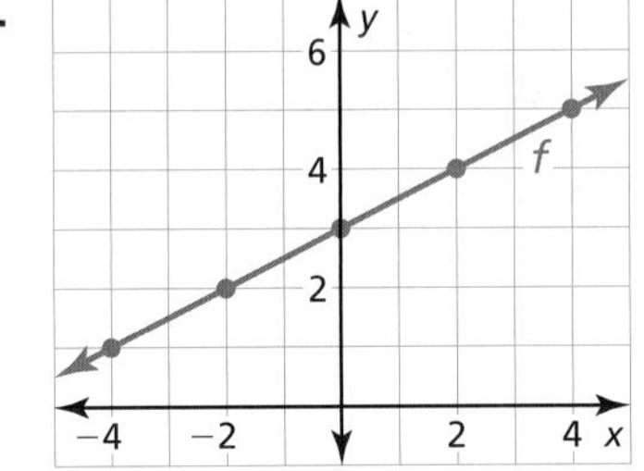

16.

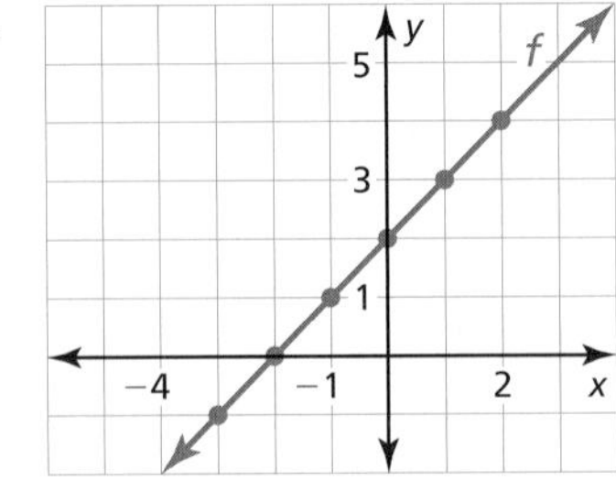

In Exercises 17–22, find the inverse of the function. Then graph the function and its inverse. *(See Example 3.)*

17. $f(x) = 4x - 1$

18. $f(x) = -2x + 5$

19. $f(x) = -3x - 2$

20. $f(x) = 2x + 3$

21. $f(x) = \frac{1}{3}x + 8$

22. $f(x) = -\frac{3}{2}x + \frac{7}{2}$

In Exercises 23–28, find the inverse of the function. Then graph the function and its inverse. *(See Example 4.)*

23. $f(x) = 4x^2, x \geq 0$

24. $f(x) = -\frac{1}{25}x^2, x \leq 0$

25. $f(x) = -x^2 + 10, x \leq 0$

26. $f(x) = 2x^2 + 6, x \geq 0$

27. $f(x) = \frac{1}{9}x^2 + 2, x \geq 0$

28. $f(x) = -4x^2 - 8, x \leq 0$

In Exercises 29–32, use the Horizontal Line Test to determine whether the inverse of f is a function.

29.

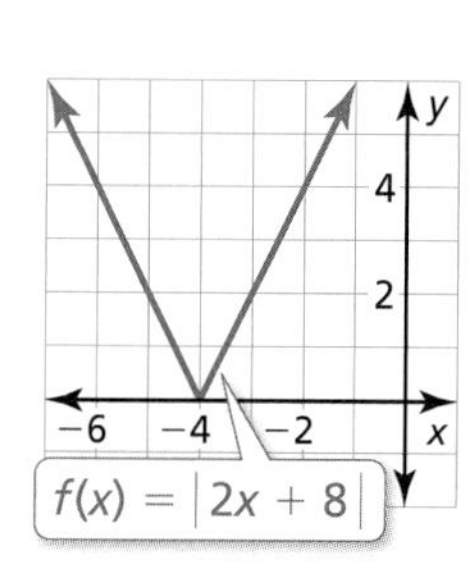

30.

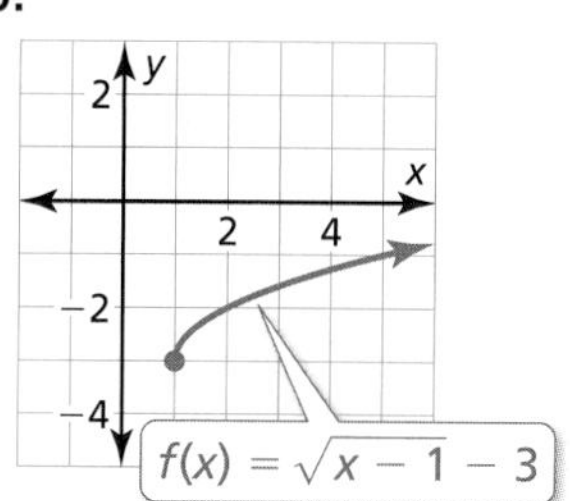

31.

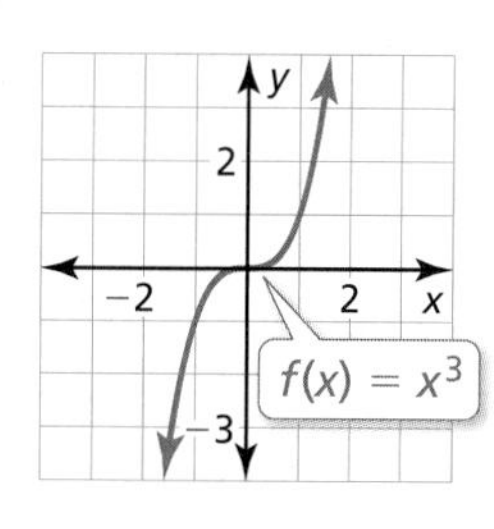

32.

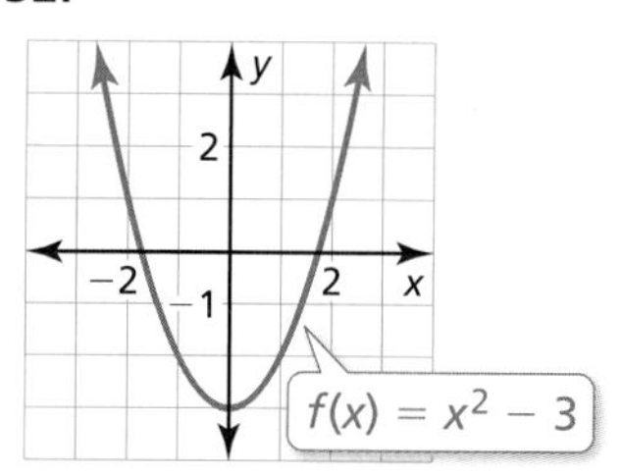

In Exercises 33–42, determine whether the inverse of f is a function. Then find the inverse. *(See Example 5.)*

33. $f(x) = \sqrt{x + 3}$

34. $f(x) = \sqrt{x - 5}$

35. $f(x) = \sqrt{2x - 6}$

36. $f(x) = \sqrt{4x + 1}$

37. $f(x) = 3\sqrt{x - 8}$

38. $f(x) = -\frac{1}{4}\sqrt{5x + 2}$

39. $f(x) = -\sqrt{3x + 5} - 2$

40. $f(x) = 2\sqrt{x - 7} + 6$

41. $f(x) = 2x^2$

42. $f(x) = |x|$

43. ERROR ANALYSIS Describe and correct the error in finding the inverse of the function $f(x) = 3x + 5$.

$$y = 3x + 5$$
$$y - 5 = 3x$$
$$\frac{y - 5}{3} = x$$

The inverse of f is $g(x) = \frac{y - 5}{3}$, or $g(x) = \frac{y}{3} - \frac{5}{3}$.

44. ERROR ANALYSIS Describe and correct the error in finding and graphing the inverse of the function $f(x) = \sqrt{x - 3}$.

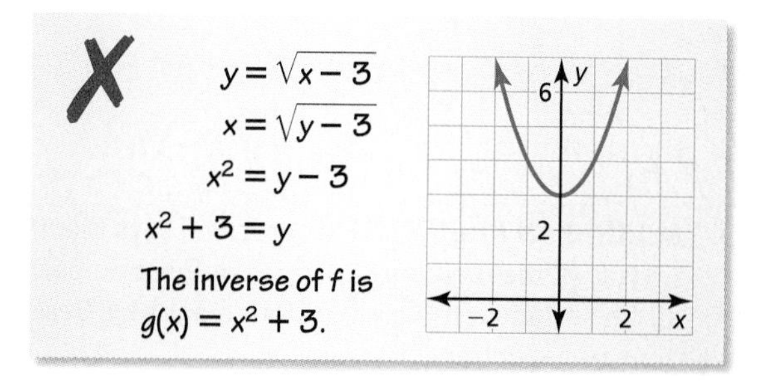

45. MODELING WITH MATHEMATICS The euro is the unit of currency for the European Union. On a certain day, the number E of euros that could be obtained for D U.S. dollars was represented by the formula shown.

$$E = 0.74683D$$

Solve the formula for D. Then find the number of U.S. dollars that could be obtained for 250 euros on that day.

46. MODELING WITH MATHEMATICS A crow is flying at a height of 50 feet when it drops a walnut to break it open. The height h (in feet) of the walnut above ground can be modeled by $h = -16t^2 + 50$, where t is the time (in seconds) since the crow dropped the walnut. Solve the equation for t. After how many seconds will the walnut be 15 feet above the ground?

MATHEMATICAL CONNECTIONS In Exercises 47 and 48, s is the side length of an equilateral triangle. Solve the formula for s. Then evaluate the new formula for the given value.

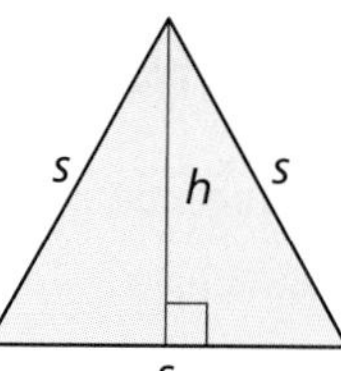

47. Height: $h = \frac{\sqrt{3}s}{2}$; $h = 16$ in.

48. Area: $A = \frac{\sqrt{3}s^2}{4}$; $A = 11$ ft²

In Exercises 49–54, find the inverse of the function. Then graph the function and its inverse.

49. $f(x) = 2x^3$

50. $f(x) = x^3 - 4$

51. $f(x) = (x - 5)^3$

52. $f(x) = 8(x + 2)^3$

53. $f(x) = 4\sqrt[3]{x}$

54. $f(x) = -\sqrt[3]{x - 1}$

55. MAKING AN ARGUMENT Your friend says that the inverse of the function $f(x) = 3$ is a function because all linear functions pass the Horizontal Line Test. Is your friend correct? Explain.

56. HOW DO YOU SEE IT? Pair the graph of each function with the graph of its inverse.

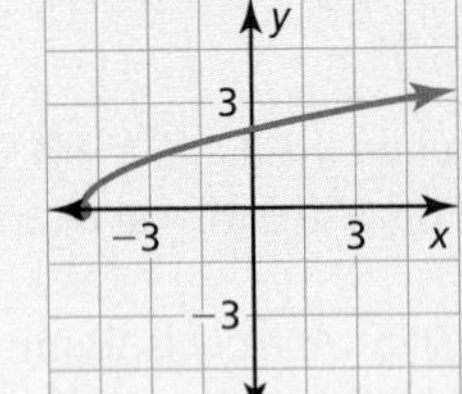

B.

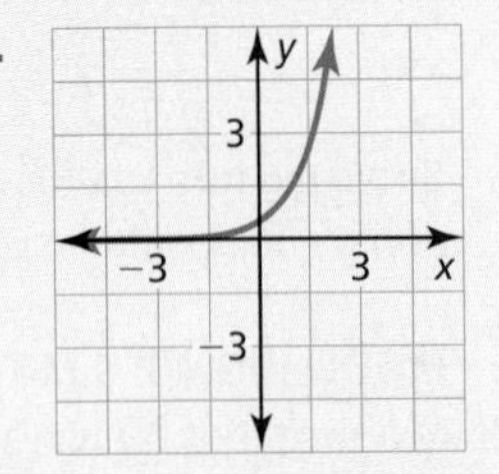

C.

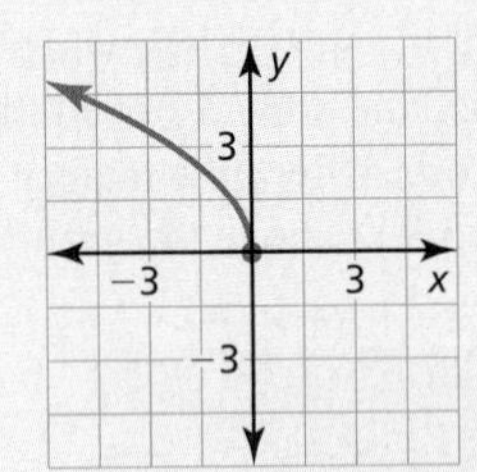

D.

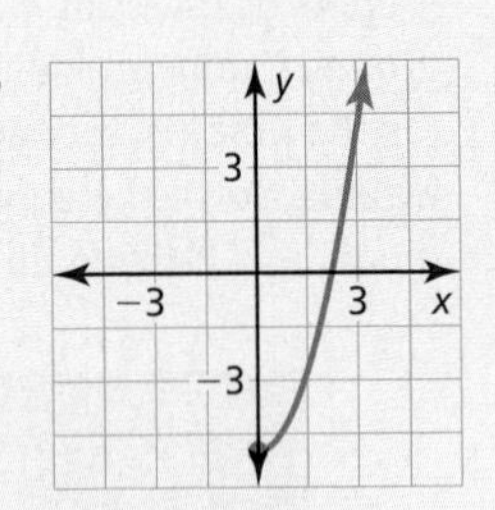

E.

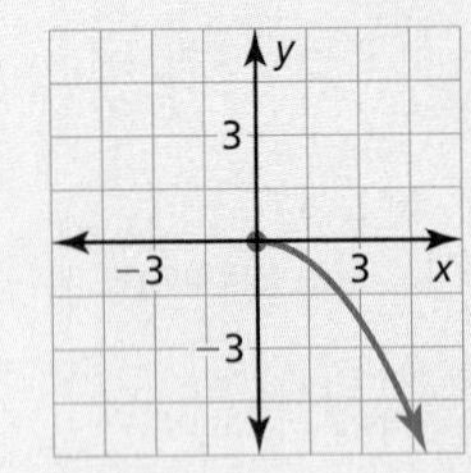

F.

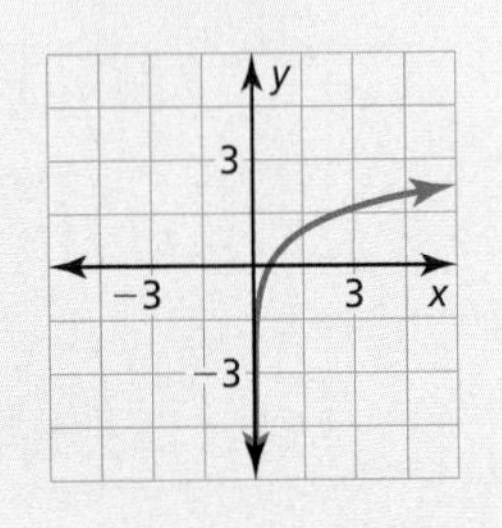

57. WRITING Describe changes you could make to the function $f(x) = x^2 - 5$ so that its inverse is a function. Describe the domain and range of the new function and its inverse.

58. CRITICAL THINKING Can an even function with at least two values in its domain have an inverse that is a function? Explain.

59. OPEN-ENDED Write a function such that the graph of its inverse is a line with a slope of 4.

60. CRITICAL THINKING Consider the function $g(x) = -x$.

a. Graph $g(x) = -x$ and explain why it is its own inverse.

b. Graph other linear functions that are their own inverses. Write equations of the lines you graph.

c. Use your results from part (b) to write a general equation that describes the family of linear functions that are their own inverses.

61. REASONING Show that the inverse of any linear function $f(x) = mx + b$, where $m \neq 0$, is also a linear function. Write the slope and y-intercept of the graph of the inverse in terms of m and b.

62. THOUGHT PROVOKING The graphs of $f(x) = x^3 - 3x$ and its inverse are shown. Find the greatest interval $-a \leq x \leq a$ for which the inverse of f is a function. Write an equation of the inverse function.

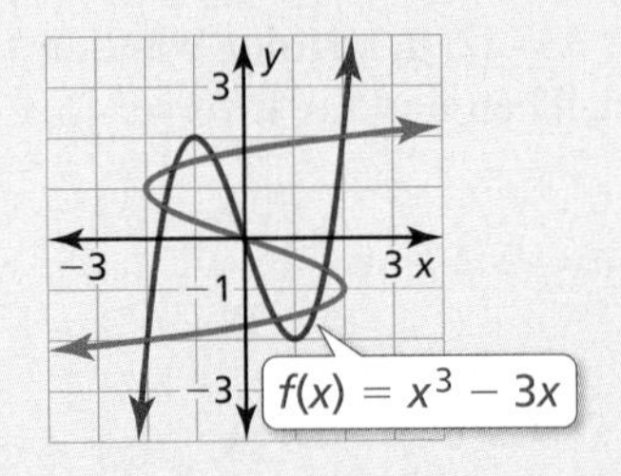

63. REASONING Is the inverse of $f(x) = 2|x + 1|$ a function? Are there any values of a, h, and k for which the inverse of $f(x) = a|x - h| + k$ is a function? Explain your reasoning.

Maintaining Mathematical Proficiency

Reviewing what you learned in previous grades and lessons

Find the sum or difference. *(Section 7.1)*

64. $(2x - 9) - (6x + 5)$

65. $(8y + 1) + (-y - 12)$

66. $(t^2 - 4t - 4) + (7t^2 + 12t + 3)$

67. $(-3d^2 + 10d - 8) - (7d^2 - d - 6)$

Graph the function. Compare the graph to the graph of $f(x) = x^2$. *(Section 8.2)*

68. $g(x) = x^2 + 6$

69. $h(x) = -x^2 - 2$

70. $p(x) = -4x^2 + 5$

71. $q(x) = \frac{1}{3}x^2 - 1$

10.3–10.4 What Did You Learn?

Core Vocabulary

radical equation, *p. 560*
inverse relation, *p. 568*
inverse function, *p. 569*

Core Concepts

Section 10.3

Squaring Each Side of an Equation, *p. 560*
Identifying Extraneous Solutions, *p. 562*

Section 10.4

Inverse Relation, *p. 568*
Finding Inverses of Functions Algebraically, *p. 570*
Finding Inverses of Nonlinear Functions, *p. 570*
Horizontal Line Test, *p. 571*

Mathematical Practices

1. Could you also solve Exercises 37–44 on page 565 by graphing? Explain.
2. What external resources could you use to check the reasonableness of your answer in Exercise 45 on page 573?

Performance Task

Medication and the Mosteller Formula

When taking medication, it is critical to take the correct dosage. For children in particular, body surface area (BSA) is a key component in calculating that dosage. The Mosteller Formula is commonly used to approximate body surface area. How will you use this formula to calculate BSA for the optimum dosage?

To explore the answers to this question and more, go to ***BigIdeasMath.com***.

10 Chapter Review

Dynamic Solutions available at *BigIdeasMath.com*

10.1 Graphing Square Root Functions (pp. 543–550)

a. Describe the domain of $f(x) = 4\sqrt{x + 2}$.

The radicand cannot be negative. So, $x + 2$ is greater than or equal to 0.

$x + 2 \geq 0$ — Write an inequality for the domain.

$x \geq -2$ — Subtract 2 from each side.

▶ The domain is the set of real numbers greater than or equal to -2.

b. Graph $g(x) = \sqrt{x} - 1$. Describe the range. Compare the graph to the graph of $f(x) = \sqrt{x}$.

Step 1 Use the domain of g, $x \geq 0$, to make a table of values.

x	0	1	4	9	16
$g(x)$	−1	0	1	2	3

Step 2 Plot the ordered pairs.

Step 3 Draw a smooth curve through the points, starting at $(0, -1)$.

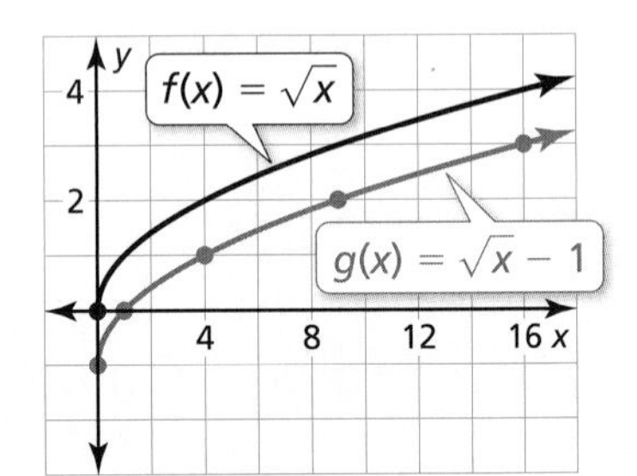

▶ The range of g is $y \geq -1$. The graph of g is a translation 1 unit down of the graph of f.

Graph the function. Describe the domain and range. Compare the graph to the graph of $f(x) = \sqrt{x}$.

1. $g(x) = \sqrt{x} + 7$
2. $h(x) = \sqrt{x - 6}$
3. $r(x) = -\sqrt{x + 3} - 1$
4. Let $g(x) = \frac{1}{4}\sqrt{x - 6} + 2$. Describe the transformations from the graph of $f(x) = \sqrt{x}$ to the graph of g. Then graph g.

10.2 Graphing Cube Root Functions (pp. 551–556)

Graph $g(x) = -\sqrt[3]{x - 2}$. Compare the graph to the graph of $f(x) = \sqrt[3]{x}$.

Step 1 Make a table of values.

x	−6	1	2	3	10
$g(x)$	2	1	0	−1	−2

Step 2 Plot the ordered pairs.

Step 3 Draw a smooth curve through the points.

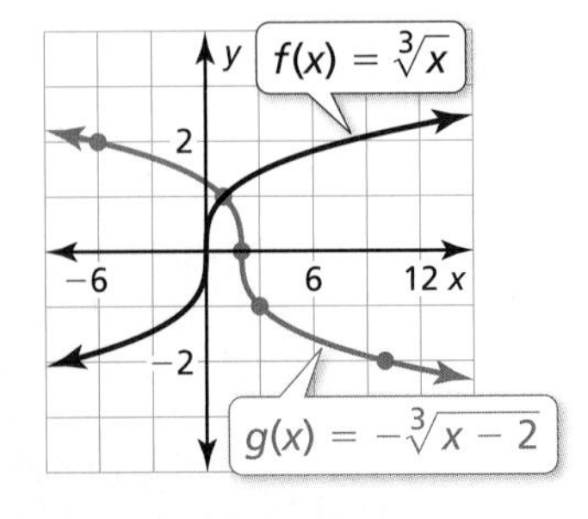

▶ The graph of g is a translation 2 units right and a reflection in the x-axis of the graph of f.

Graph the function. Compare the graph to the graph of $f(x) = \sqrt[3]{x}$.

5. $g(x) = \sqrt[3]{x} + 4$

6. $h(x) = -8\sqrt[3]{x}$

7. $s(x) = \sqrt[3]{-2(x - 3)}$

8. Let $g(x) = -3\sqrt[3]{x + 2} - 1$. Describe the transformations from the graph of $f(x) = \sqrt[3]{x}$ to the graph of g. Then graph g.

9. The graph of cube root function r is shown. Compare the average rate of change of r to the average rate of change of $p(x) = \sqrt[3]{\frac{1}{2}x}$ over the interval $x = 0$ to $x = 8$.

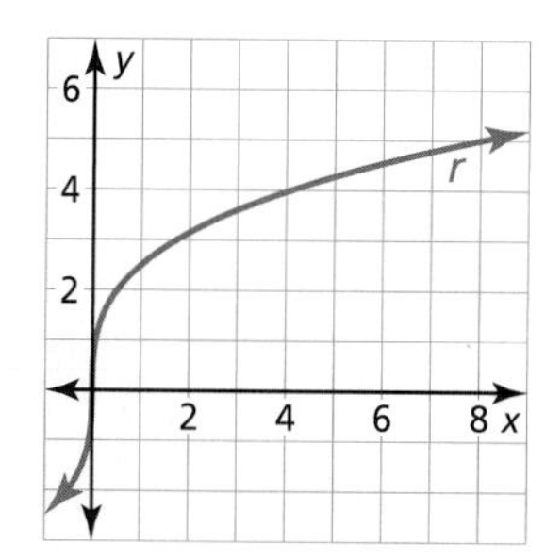

10.3 Solving Radical Equations (pp. 559–566)

Solve $\sqrt{10 - 3x} = x$.

$\sqrt{10 - 3x} = x$	Write the equation.
$\left(\sqrt{10 - 3x}\right)^2 = x^2$	Square each side of the equation.
$10 - 3x = x^2$	Simplify.
$0 = x^2 + 3x - 10$	Subtract 10 and add $3x$ to each side.
$0 = (x - 2)(x + 5)$	Factor.
$x - 2 = 0$ or $x + 5 = 0$	Zero-Product Property
$x = 2$ or $x = -5$	Solve for x.

Check Check each solution in the original equation.

$\sqrt{10 - 3(2)} \stackrel{?}{=} 2$	Substitute for x.	$\sqrt{10 - 3(-5)} \stackrel{?}{=} -5$
$\sqrt{4} \stackrel{?}{=} 2$	Simplify.	$\sqrt{25} \stackrel{?}{=} -5$
$2 = 2$ ✓	Simplify.	$5 \neq -5$ ✗

▶ Because $x = -5$ does not satisfy the original equation, it is an extraneous solution. The only solution is $x = 2$.

Solve the equation. Check your solution(s).

10. $8 + \sqrt{x} = 18$

11. $\sqrt[3]{x - 1} = 3$

12. $\sqrt{5x - 9} = \sqrt{4x}$

13. $x = \sqrt{3x + 4}$

14. $8\sqrt{x - 5} + 34 = 58$

15. $\sqrt{5x} + 6 = 5$

16. The radius r of a cylinder is represented by the function $r = \sqrt{\frac{V}{\pi h}}$, where V is the volume and h is the height of the cylinder. What is the volume of the cylindrical can?

10.4 Inverse of a Function (pp. 567–574)

a. Find the inverse of the relation.

Input	-4	-2	0	2	4	6
Output	-3	0	3	6	9	12

Switch the inputs and outputs.

Inverse relation:

Input	-3	0	3	6	9	12
Output	-4	-2	0	2	4	6

b. Find the inverse of $f(x) = \sqrt{x - 4}$. Then graph the function and its inverse.

$y = \sqrt{x - 4}$ Set y equal to $f(x)$.

$x = \sqrt{y - 4}$ Switch x and y in the equation.

$x^2 = \left(\sqrt{y - 4}\right)^2$ Square each side.

$x^2 = y - 4$ Simplify.

$x^2 + 4 = y$ Add 4 to each side.

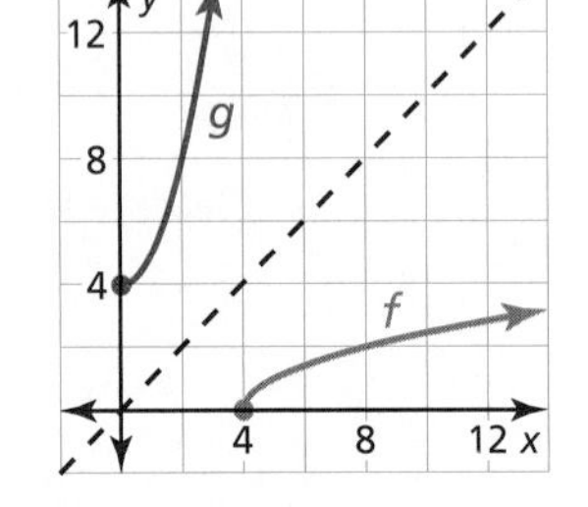

Because the range of f is $y \geq 0$, the domain of the inverse must be restricted to $x \geq 0$.

▶ So, the inverse of f is $g(x) = x^2 + 4$, where $x \geq 0$.

Find the inverse of the relation.

17. $(1, -10)$, $(3, -4)$, $(5, 4)$, $(7, 14)$, $(9, 26)$

18.

Input	-4	-2	0	2	4
Output	6	3	0	-3	-6

Find the inverse of the function. Then graph the function and its inverse.

19. $f(x) = -5x + 10$

20. $f(x) = 3x^2 - 1, x \geq 0$

21. $f(x) = \frac{1}{2}\sqrt{2x + 6}$

22. Consider the function $f(x) = x^2 + 4$. Use the Horizontal Line Test to determine whether the inverse of f is a function.

23. In bowling, a handicap is an adjustment to a bowler's score to even out differences in ability levels. In a particular league, you can find a bowler's handicap h by using the formula $h = 0.8(210 - a)$, where a is the bowler's average. Solve the formula for a. Then find a bowler's average when the bowler's handicap is 28.

10 Chapter Test

Find the inverse of the function.

1. $f(x) = 5x - 8$
2. $f(x) = 2\sqrt{x + 3} - 1$
3. $f(x) = -\frac{1}{3}x^2 + 4, x \geq 0$

Graph the function f. Describe the domain and range. Compare the graph of f to the graph of g.

4. $f(x) = -\sqrt{x + 6};\ g(x) = \sqrt{x}$
5. $f(x) = \sqrt{x - 3} + 2;\ g(x) = \sqrt{x}$
6. $f(x) = \sqrt[3]{x} - 5;\ g(x) = \sqrt[3]{x}$
7. $f(x) = -2\sqrt[3]{x + 1};\ g(x) = \sqrt[3]{x}$

Solve the equation. Check your solution(s).

8. $9 - \sqrt{x} = 3$
9. $\sqrt{2x - 7} - 3 = 6$
10. $\sqrt{8x - 21} = \sqrt{18 - 5x}$
11. $x + 5 = \sqrt{7x + 53}$

12. When solving the equation $x - 5 = \sqrt{ax + b}$, you obtain $x = 2$ and $x = 8$. Explain why at least one of these solutions must be extraneous.

Describe the transformations from the graph of $f(x) = \sqrt[3]{x}$ to the graph of the given function. Then graph the given function.

13. $h(x) = 4\sqrt[3]{x - 1} + 5$
14. $w(x) = -\sqrt[3]{x + 7} - 2$

15. The velocity v (in meters per second) of a roller coaster at the bottom of a hill is given by $v = \sqrt{19.6h}$, where h is the height (in meters) of the hill. (a) Use a graphing calculator to graph the function. Describe the domain and range. (b) How tall must the hill be for the velocity of the roller coaster at the bottom of the hill to be at least 28 meters per second? (c) What happens to the average rate of change of the velocity as the height of the hill increases?

16. The speed s (in meters per second) of sound through air is given by $s = 20\sqrt{T + 273}$, where T is the temperature (in degrees Celsius).
    a. What is the temperature when the speed of sound through air is 340 meters per second?
    b. How long does it take you to hear the wolf howl when the temperature is $-17°C$?

17. How can you restrict the domain of the function $f(x) = (x - 3)^2$ so that the inverse of f is a function?

18. Write a radical function that has a domain of all real numbers less than or equal to 0 and a range of all real numbers greater than or equal to 9.

10 Cumulative Assessment

1. Fill in the function so that it is represented by the graph.

$f(x) = \square\sqrt{x - \square} + \square$

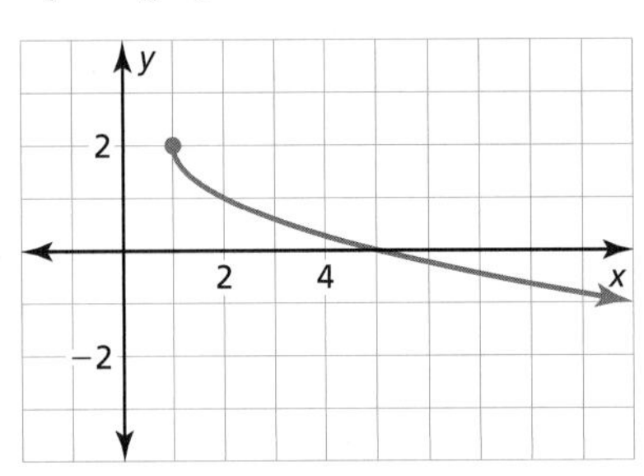

2. Consider the equation $y = mx + b$. Fill in values for m and b so that each statement is true.

   a. When $m =$ ______ and $b =$ ______, the graph of the equation passes through the point $(-1, 4)$.

   b. When $m =$ ______ and $b =$ ______, the graph of the equation has a positive slope and passes through the point $(-2, -5)$.

   c. When $m =$ ______ and $b =$ ______, the graph of the equation is perpendicular to the graph of $y = 4x - 3$ and passes through the point $(1, 6)$.

3. Which graph represents the inverse of the function $f(x) = 2x + 4$?

Ⓐ 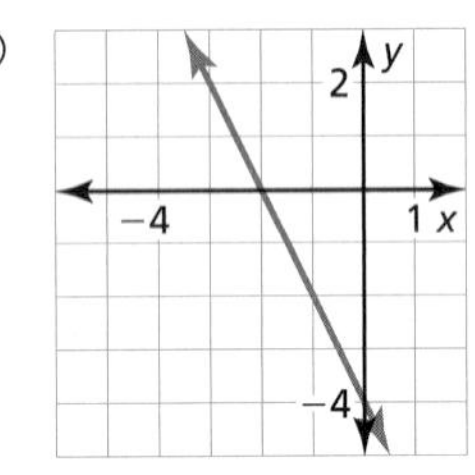

Ⓑ 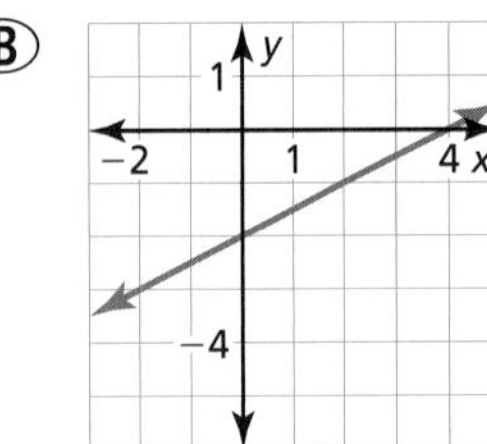

Ⓒ 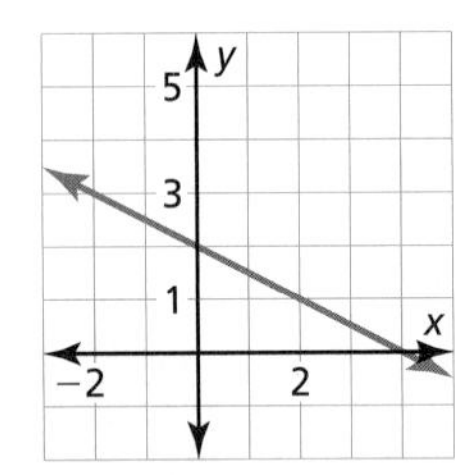

Ⓓ 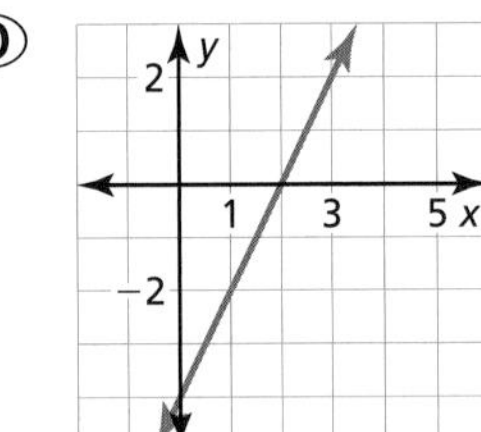

4. Consider the equation $x = \sqrt{ax + b}$. Student A claims this equation has one real solution. Student B claims this equation has two real solutions. Use the numbers to answer parts (a)–(c).

−4	−3	−2	−1	0	1	2	3	4

   a. Choose values for a and b to create an equation that supports Student A's claim.

   b. Choose values for a and b to create an equation that supports Student B's claim.

   c. Choose values for a and b to create an equation that does not support either student's claim.

5. Which equation represents the nth term of the sequence 3, 12, 48, 192, . . .?

Ⓐ $a_n = 3(4)^{n-1}$ Ⓑ $a_n = 3(9)^{n-1}$

Ⓒ $a_n = 9n - 6$ Ⓓ $a_n = 9n + 3$

6. Consider the function $f(x) = \frac{1}{2}\sqrt[3]{x + 3}$. The graph represents function g. Select all the statements that are true.

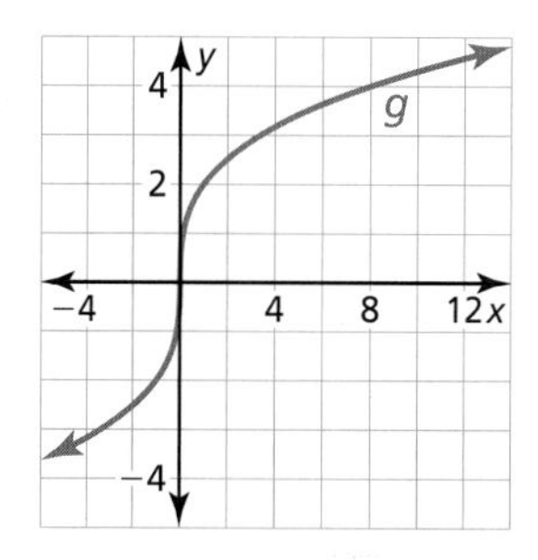

- The x-intercept of the graph of f is greater than the x-intercept of the graph of g.
- The graph of g is always increasing.
- The average rate of change of g decreases as x increases.
- The average rate of change of f increases as x increases.
- The average rate of change of g is greater than the average rate of change of f over the interval $x = 0$ to $x = 8$.

7. Place each function into one of the three categories.

No zeros	**One zero**	**Two zeros**
$f(x) = 3x^2 + 4x + 2$ $f(x) = -x^2 + 2x$	$f(x) = 4x^2 - 8x + 4$ $f(x) = x^2 - 3x - 21$	$f(x) = 7x^2$ $f(x) = -6x^2 - 5$

8. You are making a tabletop with a tiled center and a uniform mosaic border.

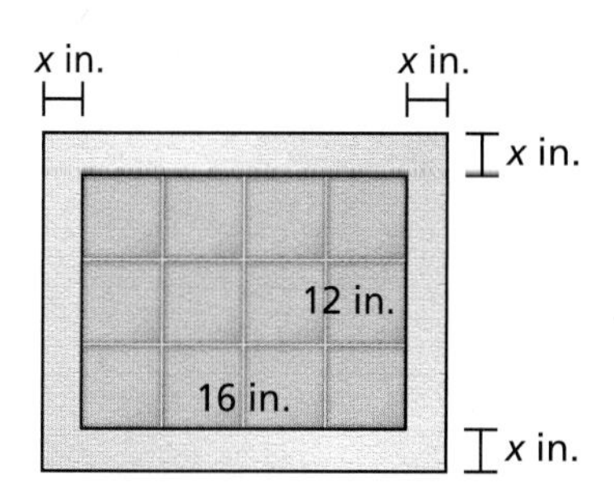

a. Write the polynomial in standard form that represents the perimeter of the tabletop.

b. Write the polynomial in standard form that represents the area of the tabletop.

c. The perimeter of the tabletop is less than 80 inches, and the area of tabletop is at least 252 square inches. Select all the possible values of x.

0.5 | 1 | 1.5 | 2 | 2.5 | 3 | 3.5 | 4

11 Data Analysis and Displays

Watching Sports on TV *(p. 616)*

Shoes *(p. 603)*

Backpacking *(p. 594)*

Bowling Scores *(p. 591)*

Altitudes of Airplanes *(p. 589)*

Maintaining Mathematical Proficiency

Displaying Data

Example 1 **The frequency table shows the numbers of books that 12 people read last month. Display the data in a histogram.**

Number of books	Frequency
0–1	6
2–3	4
4–5	0
6–7	2

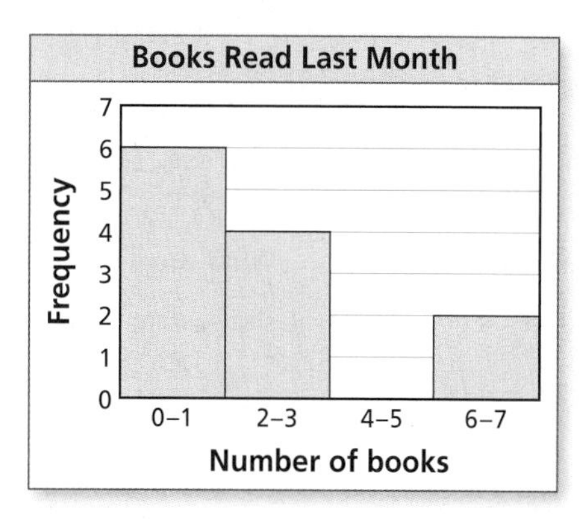

Example 2 **The table shows the results of a survey. Display the data in a circle graph.**

Class trip location	Water park	Museum	Zoo	Other
Students	25	11	5	4

A total of 45 students took the survey.

Water park:

$\frac{25}{45} \cdot 360° = 200°$

Museum:

$\frac{11}{45} \cdot 360° = 88°$

Zoo:

$\frac{5}{45} \cdot 360° = 40°$

Other:

$\frac{4}{45} \cdot 360° = 32°$

Class Trip Locations

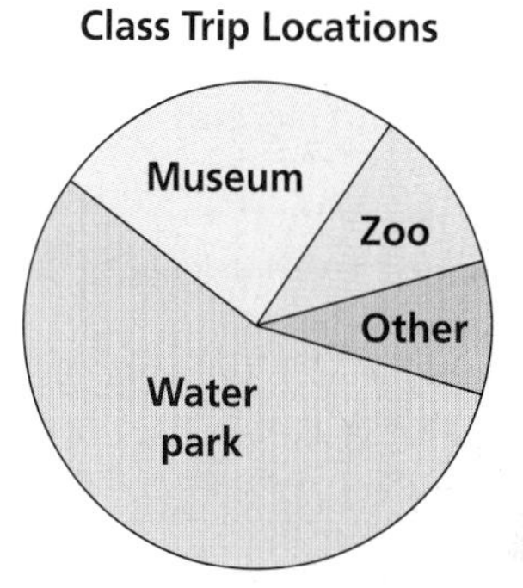

The table shows the results of a survey. Display the data in a histogram.

1.

After-school activities	Frequency
0–1	11
2–3	8
4–5	6
6–7	1

2.

Pets	Frequency
0–1	10
2–3	18
4–5	2

The table shows the results of a survey. Display the data in a circle graph.

3.

Favorite subject	Math	Science	English	History
Students	8	5	7	4

4. **ABSTRACT REASONING** Twenty people respond "yes" or "no" to a survey question. Let a and b represent the frequencies of the responses. What must be true about the sum of a and b? What must be true about the sum when "maybe" is an option for the response?

Mathematical Practices

Mathematically proficient students use diagrams and graphs to show relationships between data. They also analyze data to draw conclusions.

Using Data Displays

Core Concept

Displaying Data Graphically

When solving a problem involving data, it is helpful to display the data graphically. This can be done in a variety of ways.

Data Display	What does it do?
Pictograph	shows data using pictures
Bar Graph	shows data in specific categories
Circle Graph	shows data as parts of a whole
Line Graph	shows how data change over time
Histogram	shows frequencies of data values in intervals of the same size
Stem-and-Leaf Plot	orders numerical data and shows how they are distributed
Box-and-Whisker Plot	shows the variability of a data set using quartiles
Dot Plot	shows the number of times each value occurs in a data set
Scatter Plot	shows the relationship between two data sets using ordered pairs in a coordinate plane

A |
B |
C |
D |

1 | 0 2 3 6
2 | 1 1 5
3 | 9
4 | 0 6

Monitoring Progress

1. The table shows the estimated populations of males and females by age in the United States in 2012. Use a spreadsheet, graphing calculator, or some other form of technology to make two different displays for the data.
2. Explain why you chose each type of data display in Monitoring Progress Question 1. What conclusions can you draw from your data displays?

U.S. Population by Age and Gender

Ages (years)	Males	Females
0–14	31,242,542	29,901,556
15–29	33,357,203	31,985,028
30–44	30,667,513	30,759,902
45–59	31,875,279	33,165,976
60–74	19,737,347	22,061,730
75–89	6,999,292	10,028,195

11.1 Measures of Center and Variation

Essential Question How can you describe the variation of a data set?

EXPLORATION 1 Describing the Variation of Data

Work with a partner. The graphs show the weights of the players on a professional football team and a professional baseball team.

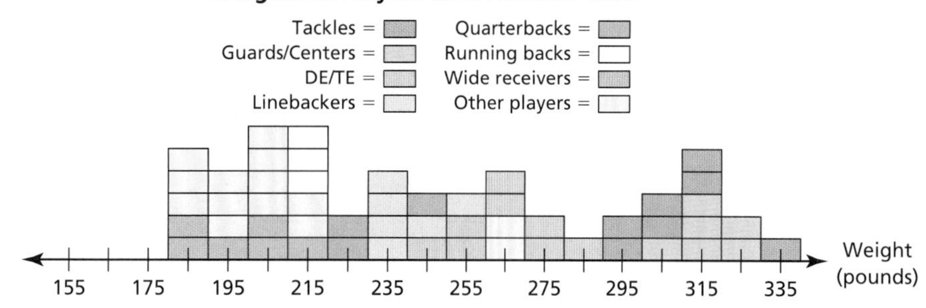

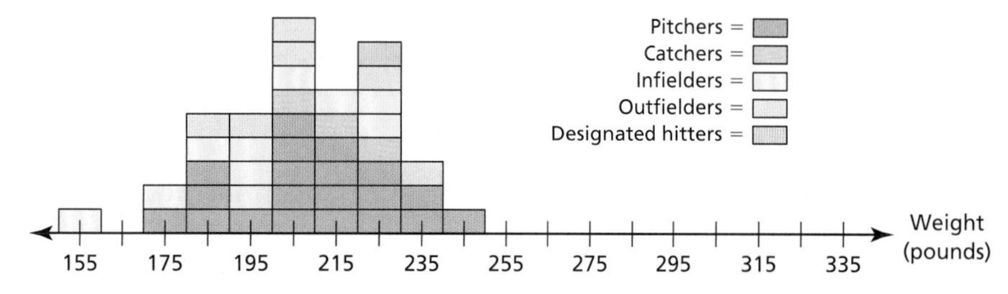

CONSTRUCTING VIABLE ARGUMENTS

To be proficient in math, you need to reason inductively about data, making plausible arguments that take into account the context from which the data arose.

a. Describe the data in each graph in terms of how much the weights vary from the mean. Explain your reasoning.

b. Compare how much the weights of the players on the football team vary from the mean to how much the weights of the players on the baseball team vary from the mean.

c. Does there appear to be a correlation between the body weights and the positions of players in professional football? in professional baseball? Explain.

EXPLORATION 2 Describing the Variation of Data

Work with a partner. The weights (in pounds) of the players on a professional basketball team by position are as follows.

Power forwards: 235, 255, 295, 245; small forwards: 235, 235;
centers: 255, 245, 325; point guards: 205, 185, 205; shooting guards: 205, 215, 185

Make a graph that represents the weights and positions of the players. Does there appear to be a correlation between the body weights and the positions of players in professional basketball? Explain your reasoning.

Communicate Your Answer

3. How can you describe the variation of a data set?

11.1 Lesson

Core Vocabulary

measure of center, *p. 586*
mean, *p. 586*
median, *p. 586*
mode, *p. 586*
outlier, *p. 587*
measure of variation, *p. 587*
range, *p. 587*
standard deviation, *p. 588*
data transformation, *p. 589*

What You Will Learn

- Compare the mean, median, and mode of a data set.
- Find the range and standard deviation of a data set.
- Identify the effects of transformations on data.

Comparing the Mean, Median, and Mode

A **measure of center** is a measure that represents the center, or typical value, of a data set. The *mean*, *median*, and *mode* are measures of center.

Core Concept

Mean

The **mean** of a numerical data set is the sum of the data divided by the number of data values. The symbol $\overline{x}$ represents the mean. It is read as "*x*-bar."

Median

The **median** of a numerical data set is the middle number when the values are written in numerical order. When a data set has an even number of values, the median is the mean of the two middle values.

Mode

The **mode** of a data set is the value or values that occur most often. There may be one mode, no mode, or more than one mode.

EXAMPLE 1 Comparing Measures of Center

Students' Hourly Wages	
\$16.50	\$8.25
\$8.75	\$8.45
\$8.65	\$8.25
\$9.10	\$9.25

An amusement park hires students for the summer. The students' hourly wages are shown in the table.

a. Find the mean, median, and mode of the hourly wages.

b. Which measure of center best represents the data? Explain.

SOLUTION

a. Mean $\overline{x} = \dfrac{16.5 + 8.75 + 8.65 + 9.1 + 8.25 + 8.45 + 8.25 + 9.25}{8} = 9.65$

Median 8.25, 8.25, 8.45, $\underbrace{8.65, 8.75}$, 9.10, 9.25, 16.50 — Order the data.

$\dfrac{17.4}{2} = 8.7$ — Mean of two middle values

Mode 8.25, 8.25, 8.45, 8.65, 8.75, 9.10, 9.25, 16.50 — 8.25 occurs most often.

▶ The mean is \$9.65, the median is \$8.70, and the mode is \$8.25.

b. The median best represents the data. The mode is less than most of the data, and the mean is greater than most of the data.

STUDY TIP

Mode is the only measure of center that can represent a nonnumerical data set.

Monitoring Progress  Help in English and Spanish at *BigIdeasMath.com*

1. WHAT IF? The park hires another student at an hourly wage of \$8.45.
(a) How does this additional value affect the mean, median, and mode? Explain.
(b) Which measure of center best represents the data? Explain.

An **outlier** is a data value that is much greater than or much less than the other values in a data set.

EXAMPLE 2 Removing an Outlier

Consider the data in Example 1. (a) Identify the outlier. How does the outlier affect the mean, median, and mode? (b) Describe one possible explanation for the outlier.

SOLUTION

a. The value \$16.50 is much greater than the other wages. It is the outlier. Find the mean, median, and mode without the outlier.

STUDY TIP

Outliers usually have the greatest effect on the mean.

Mean $\bar{x} = \dfrac{8.75 + 8.65 + 9.1 + 8.25 + 8.45 + 8.25 + 9.25}{7} \approx 8.67$

Median 8.25, 8.25, 8.45, 8.65, 8.75, 9.10, 9.25 — The middle value is 8.65.

Mode 8.25, 8.25, 8.45, 8.65, 8.75, 9.10, 9.25 — The mode is 8.25.

▶ When you remove the outlier, the mean decreases \$9.65 − \$8.67 = \$0.98, the median decreases \$8.70 − \$8.65 = \$0.05, and the mode is the same.

b. The outlier could be a student who is hired to maintain the park's website, while the other students could be game attendants.

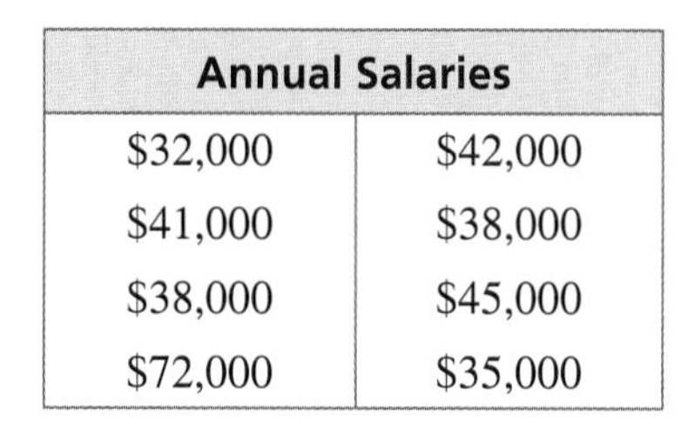

Annual Salaries	
\$32,000	\$42,000
\$41,000	\$38,000
\$38,000	\$45,000
\$72,000	\$35,000

Monitoring Progress  Help in English and Spanish at *BigIdeasMath.com*

2. The table shows the annual salaries of the employees of an auto repair service. (a) Identify the outlier. How does the outlier affect the mean, median, and mode? (b) Describe one possible explanation for the outlier.

Finding the Range and Standard Deviation

A **measure of variation** is a measure that describes the spread, or distribution, of a data set. One measure of variation is the *range*. The **range** of a data set is the difference of the greatest value and the least value.

EXAMPLE 3 Finding a Range

Two reality cooking shows select 12 contestants each. The ages of the contestants are shown in the tables. Find the range of the ages for each show. Compare your results.

Show A	
Ages	
20	29
19	22
25	27
27	29
30	20
21	31

Show B	
Ages	
25	19
20	27
22	25
27	22
48	21
32	24

SOLUTION

Show A 19, 20, 20, 21, 22, 25, 27, 27, 29, 29, 30, 31 — Order the data.

So, the range is 31 − 19, or 12 years.

Show B 19, 20, 21, 22, 22, 24, 25, 25, 27, 27, 32, 48 — Order the data.

So, the range is 48 − 19, or 29 years.

▶ The range of the ages for Show A is 12 years, and the range of the ages for Show B is 29 years. So, the ages for Show B are more spread out.

Monitoring Progress  Help in English and Spanish at *BigIdeasMath.com*

3. After the first week, the 25-year-old is voted off Show A and the 48-year-old is voted off Show B. How does this affect the range of the ages of the remaining contestants on each show in Example 3? Explain.

A disadvantage of using the range to describe the spread of a data set is that it uses only two data values. A measure of variation that uses all the values of a data set is the *standard deviation*.

Core Concept

Standard Deviation

The **standard deviation** of a numerical data set is a measure of how much a typical value in the data set differs from the mean. The symbol σ represents the standard deviation. It is read as "sigma." It is given by

$$\sigma = \sqrt{\frac{(x_1 - \overline{x})^2 + (x_2 - \overline{x})^2 + \cdots + (x_n - \overline{x})^2}{n}}$$

where n is the number of values in the data set. The deviation of a data value x is the difference of the data value and the mean of the data set, $x - \overline{x}$.

Step 1 Find the mean, $\overline{x}$.

Step 2 Find the deviation of each data value, $x - \overline{x}$.

Step 3 Square each deviation, $(x - \overline{x})^2$.

Step 4 Find the mean of the squared deviations. This is called the *variance*.

Step 5 Take the square root of the variance.

REMEMBER

An ellipsis " . . . " indicates that a pattern continues.

A small standard deviation means that the data are clustered around the mean. A large standard deviation means that the data are more spread out.

EXAMPLE 4 Finding a Standard Deviation

Find the standard deviation of the ages for Show A in Example 3. Use a table to organize your work. Interpret your result.

SOLUTION

x	$\overline{x}$	$x - \overline{x}$	$(x - \overline{x})^2$
20	25	−5	25
29	25	4	16
19	25	−6	36
22	25	−3	9
25	25	0	0
27	25	2	4
27	25	2	4
29	25	4	16
30	25	5	25
20	25	−5	25
21	25	−4	16
31	25	6	36

Step 1 Find the mean, $\overline{x}$.

$$\overline{x} = \frac{300}{12} = 25$$

Step 2 Find the deviation of each data value, $x - \overline{x}$, as shown.

Step 3 Square each deviation, $(x - \overline{x})^2$, as shown.

Step 4 Find the mean of the squared deviations, or variance.

$$\frac{(x_1 - \overline{x})^2 + (x_2 - \overline{x})^2 + \cdots + (x_n - \overline{x})^2}{n} = \frac{25 + 16 + \cdots + 36}{12} = \frac{212}{12} \approx 17.7$$

Step 5 Use a calculator to take the square root of the variance.

$$\sqrt{\frac{(x_1 - \overline{x})^2 + (x_2 - \overline{x})^2 + \cdots + (x_n - \overline{x})^2}{n}} = \sqrt{\frac{212}{12}} \approx 4.2$$

▶ The standard deviation is about 4.2. This means that the typical age of a contestant on Show A differs from the mean by about 4.2 years.

Monitoring Progress  Help in English and Spanish at *BigIdeasMath.com*

4. Find the standard deviation of the ages for Show B in Example 3. Interpret your result.

5. Compare the standard deviations for Show A and Show B. What can you conclude?

Effects of Data Transformations

A **data transformation** is a procedure that uses a mathematical operation to change a data set into a different data set.

STUDY TIP

The standard deviation stays the same because the amount by which each data value deviates from the mean stays the same.

Core Concept

Data Transformations Using Addition

When a real number k is added to each value in a numerical data set

- the measures of center of the new data set can be found by adding k to the original measures of center.
- the measures of variation of the new data set are the *same* as the original measures of variation.

Data Transformations Using Multiplication

When each value in a numerical data set is multiplied by a real number k, where $k > 0$, the measures of center and variation can be found by multiplying the original measures by k.

EXAMPLE 5 Real-Life Application

Consider the data in Example 1. (a) Find the mean, median, mode, range, and standard deviation when each hourly wage increases by \$0.50. (b) Find the mean, median, mode, range, and standard deviation when each hourly wage increases by 10%.

Students' Hourly Wages	
\$17.00	\$8.75
\$9.25	\$8.95
\$9.15	\$8.75
\$9.60	\$9.75

SOLUTION

a. Method 1 Make a new table by adding \$0.50 to each hourly wage. Find the mean, median, mode, range, and standard deviation of the new data set.

▶ Mean: \$10.15 Median: \$9.20 Mode: \$8.75
Range: \$8.25 Standard deviation: \$2.61

Method 2 Find the mean, median, mode, range, and standard deviation of the original data set.

Mean: \$9.65 Median: \$8.70 Mode: \$8.25 From Example 1
Range: \$8.25 Standard deviation: \$2.61

Add \$0.50 to the mean, median, and mode. The range and standard deviation are the same as the original range and standard deviation.

▶ Mean: \$10.15 Median: \$9.20 Mode: \$8.75
Range: \$8.25 Standard deviation: \$2.61

b. Increasing by 10% means to multiply by 1.1. So, multiply the original mean, median, mode, range, and standard deviation from Method 2 of part (a) by 1.1.

▶ Mean: \$10.62 Median: \$9.57 Mode: \$9.08
Range: \$9.08 Standard deviation: \$2.87

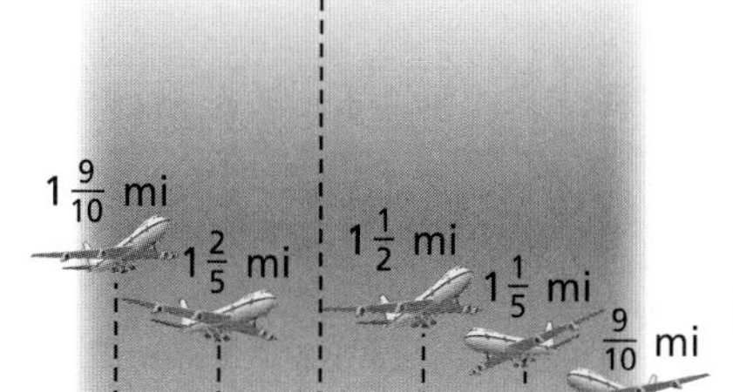

Monitoring Progress  Help in English and Spanish at *BigIdeasMath.com*

6. Find the mean, median, mode, range, and standard deviation of the altitudes of the airplanes when each altitude increases by $1\frac{1}{2}$ miles.

11.1 Exercises

Dynamic Solutions available at *BigIdeasMath.com*

Vocabulary and Core Concept Check

1. **VOCABULARY** In a data set, what does a measure of center represent? What does a measure of variation describe?

2. **WRITING** Describe how removing an outlier from a data set affects the mean of the data set.

3. **OPEN-ENDED** Create a data set that has more than one mode.

4. **REASONING** What is an advantage of using the range to describe a data set? Why do you think the standard deviation is considered a more reliable measure of variation than the range?

Monitoring Progress and Modeling with Mathematics

In Exercises 5–8, (a) find the mean, median, and mode of the data set and (b) determine which measure of center best represents the data. Explain. *(See Example 1.)*

5. 3, 5, 1, 5, 1, 1, 2, 3, 15

6. 12, 9, 17, 15, 10

7. 13, 30, 16, 19, 20, 22, 25, 31

8. 14, 15, 3, 15, 14, 14, 18, 15, 8, 16

9. **ANALYZING DATA** The table shows the lengths of nine movies.

Movie Lengths (hours)		
$1\frac{1}{3}$	$1\frac{2}{3}$	2
3	$2\frac{1}{3}$	$1\frac{2}{3}$
2	2	$1\frac{2}{3}$

   a. Find the mean, median, and mode of the lengths.

   b. Which measure of center best represents the data? Explain.

10. **ANALYZING DATA** The table shows the daily changes in the value of a stock over 12 days.

Changes in Stock Value (dollars)			
1.05	2.03	−13.78	−2.41
2.64	0.67	4.02	1.39
0.66	−0.28	−3.01	2.20

   a. Find the mean, median, and mode of the changes in stock value.

   b. Which measure of center best represents the data? Explain.

   c. On the 13th day, the value of the stock increases by \$4.28. How does this additional value affect the mean, median, and mode? Explain.

In Exercises 11–14, find the value of x.

11. 2, 8, 9, 7, 6, x; The mean is 6.

12. 12.5, −10, −7.5, x; The mean is 11.5.

13. 9, 10, 12, x, 20, 25; The median is 14.

14. 30, 45, x, 100; The median is 51.

15. **ANALYZING DATA** The table shows the masses of eight polar bears. *(See Example 2.)*

Masses (kilograms)			
455	262	471	358
364	553	62	351

   a. Identify the outlier. How does the outlier affect the mean, median, and mode?

   b. Describe one possible explanation for the outlier.

16. **ANALYZING DATA** The sizes of emails (in kilobytes) in your inbox are 2, 3, 5, 2, 1, 46, 3, 7, 2, and 1.

   a. Identify the outlier. How does the outlier affect the mean, median, and mode?

   b. Describe one possible explanation for the outlier.

17. **ANALYZING DATA** The scores of two golfers are shown. Find the range of the scores for each golfer. Compare your results. *(See Example 3.)*

Golfer A		Golfer B	
83	88	89	87
84	95	93	95
91	89	92	94
90	87	88	91
98	95	89	92

18. ANALYZING DATA The graph shows a player's monthly home run totals in two seasons. Find the range of the number of home runs for each season. Compare your results.

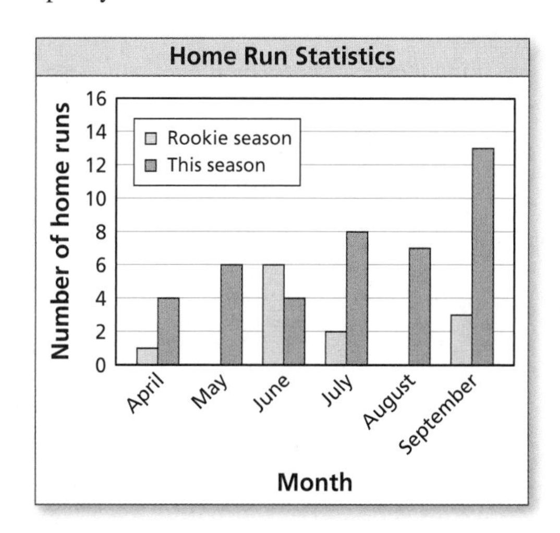

In Exercises 19–22, find (a) the range and (b) the standard deviation of the data set.

19. 40, 35, 45, 55, 60

20. 141, 116, 117, 135, 126, 121

21. 0.5, 2.0, 2.5, 1.5, 1.0, 1.5

22. 8.2, 10.1, 2.6, 4.8, 2.4, 5.6, 7.0, 3.3

23. ANALYZING DATA Consider the data in Exercise 17. *(See Example 4.)*

a. Find the standard deviation of the scores of Golfer A. Interpret your result.

b. Find the standard deviation of the scores of Golfer B. Interpret your result.

c. Compare the standard deviations for Golfer A and Golfer B. What can you conclude?

24. ANALYZING DATA Consider the data in Exercise 18.

a. Find the standard deviation of the monthly home run totals in the player's rookie season. Interpret your result.

b. Find the standard deviation of the monthly home run totals in this season. Interpret your result.

c. Compare the standard deviations for the rookie season and this season. What can you conclude?

In Exercises 25 and 26, find the mean, median, and mode of the data set after the given transformation.

25. In Exercise 5, each data value increases by 4.

26. In Exercise 6, each data value increases by 20%.

27. TRANSFORMING DATA Find the values of the measures shown when each value in the data set increases by 14. *(See Example 5.)*

Mean: 62 Median: 55 Mode: 49

Range: 46 Standard deviation: 15.5

28. TRANSFORMING DATA Find the values of the measures shown when each value in the data set is multiplied by 0.5.

Mean: 320 Median: 300 Mode: none

Range: 210 Standard deviation: 70.6

29. ERROR ANALYSIS Describe and correct the error in finding the median of the data set.

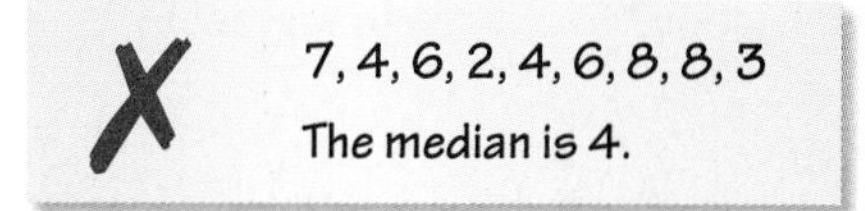

30. ERROR ANALYSIS Describe and correct the error in finding the range of the data set after the given transformation.

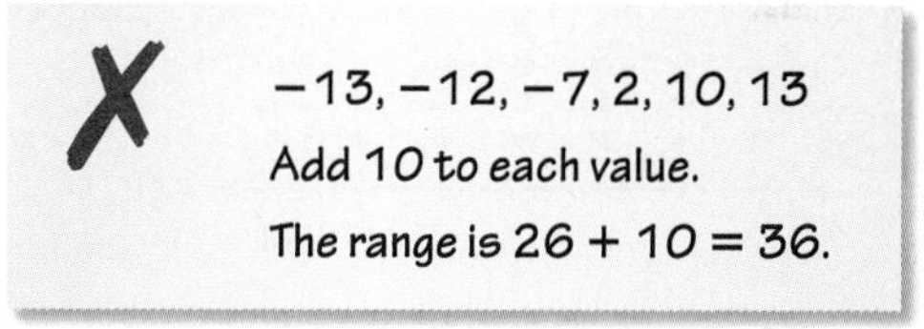

31. PROBLEM SOLVING In a bowling match, the team with the greater mean score wins. The scores of the members of two bowling teams are shown.

Team A: 172, 130, 173, 212
Team B: 136, 184, 168, 192

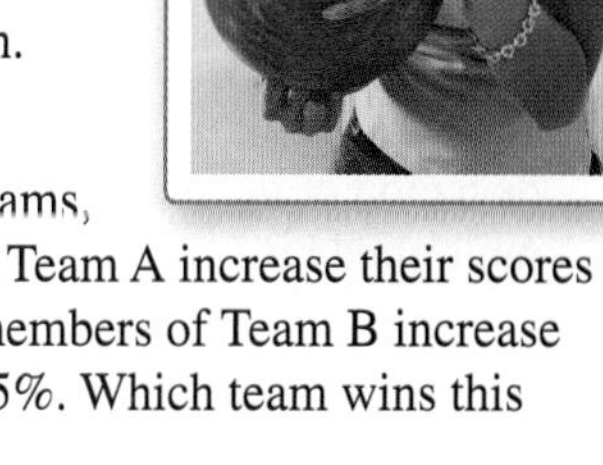

a. Which team wins the match? If the team with the greater median score wins, is the result the same? Explain.

b. Which team is more consistent? Explain.

c. In another match between the two teams, all the members of Team A increase their scores by 15 and all the members of Team B increase their scores by 12.5%. Which team wins this match? Explain.

32. MAKING AN ARGUMENT Your friend says that when two data sets have the same range, you can assume the data sets have the same standard deviation, because both range and standard deviation are measures of variation. Is your friend correct? Explain.

33. ANALYZING DATA The table shows the results of a survey that asked 12 students about their favorite meal. Which measure of center (mean, median, or mode) can be used to describe the data? Explain.

Favorite Meal			
spaghetti	pizza	steak	hamburger
steak	taco	pizza	chili
pizza	chicken	fish	spaghetti

34. HOW DO YOU SEE IT? The dot plots show the ages of the members of three different adventure clubs. Without performing calculations, which data set has the greatest standard deviation? Which has the least standard deviation? Explain your reasoning.

Ⓐ

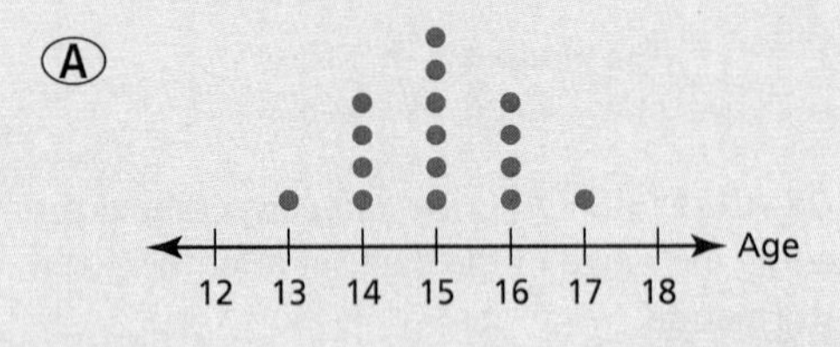

Ⓑ

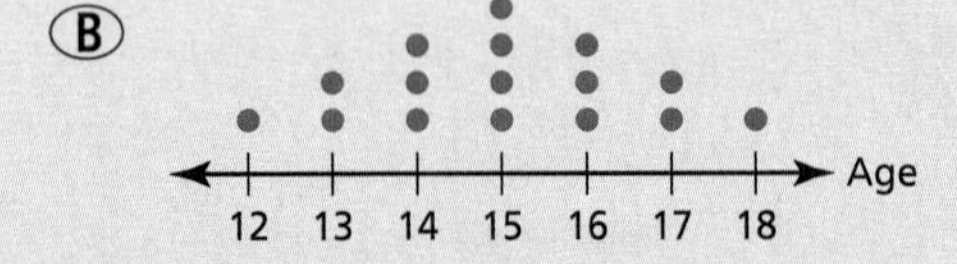

Ⓒ

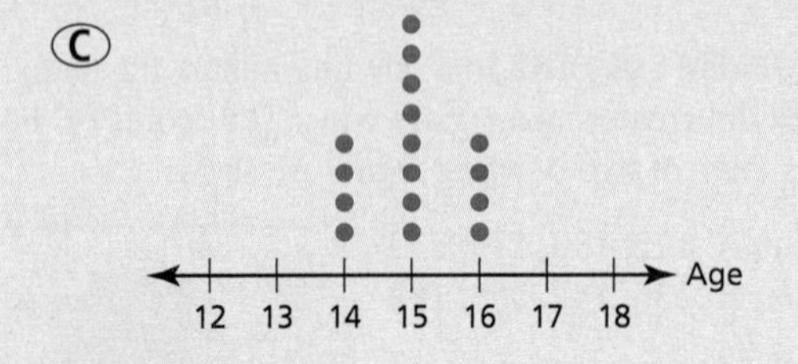

35. REASONING A data set is described by the measures shown.

Mean: 27 Median: 32 Mode: 18

Range: 41 Standard deviation: 9

Find the mean, median, mode, range, and standard deviation of the data set when each data value is multiplied by 3 and then increased by 8.

36. CRITICAL THINKING Can the standard deviation of a data set be 0? Can it be negative? Explain.

37. USING TOOLS Measure the heights (in inches) of the students in your class.

a. Find the mean, median, mode, range, and standard deviation of the heights.

b. A new student who is 7 feet tall joins your class. How would you expect this student's height to affect the measures in part (a)? Verify your answer.

38. THOUGHT PROVOKING To find the arithmetic mean of n numbers, divide the sum of the numbers by n. To find the geometric mean of n numbers $a_1, a_2, a_3, \ldots,$ a_n, take the nth root of the product of the numbers.

$$\text{geometric mean} = \sqrt[n]{a_1 \cdot a_2 \cdot a_3 \cdot \ldots \cdot a_n}$$

Compare the arithmetic mean to the geometric mean of n numbers.

39. PROBLEM SOLVING The circle graph shows the distribution of the ages of 200 students in a college Psychology I class.

College Student Ages

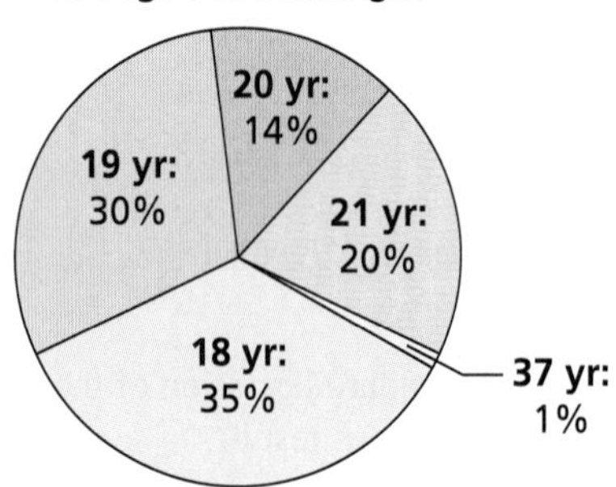

a. Find the mean, median, and mode of the students' ages.

b. Identify the outliers. How do the outliers affect the mean, median, and mode?

c. Suppose all 200 students take the same Psychology II class exactly 1 year later. Draw a new circle graph that shows the distribution of the ages of this class and find the mean, median, and mode of the students' ages.

Maintaining Mathematical Proficiency

Reviewing what you learned in previous grades and lessons

Solve the inequality. *(Section 2.4)*

40. $6x + 1 \leq 4x - 9$ **41.** $-3(3y - 2) < 1 - 9y$ **42.** $2(5c - 4) \geq 5(2c + 8)$ **43.** $4(3 - w) > 3(4w - 4)$

Evaluate the function for the given value of *x*. *(Section 6.3)*

44. $f(x) = 4^x; x = 3$ **45.** $f(x) = 7^x; x = -2$ **46.** $f(x) = 5(2)^x; x = 6$ **47.** $f(x) = -2(3)^x; x = 4$

11.2 Box-and-Whisker Plots

Essential Question How can you use a box-and-whisker plot to describe a data set?

EXPLORATION 1 Drawing a Box-and-Whisker Plot

Work with a partner. The numbers of first cousins of the students in a ninth-grade class are shown. A *box-and-whisker plot* is one way to represent the data visually.

Numbers of First Cousins			
3	10	18	8
9	3	0	32
23	19	13	8
6	3	3	10
12	45	1	5
13	24	16	14

a. Order the data on a strip of grid paper with 24 equally spaced boxes.

Fold the paper in half to find the median.

b. Fold the paper in half again to divide the data into four groups. Because there are 24 numbers in the data set, each group should have 6 numbers. Find the least value, the greatest value, the first quartile, and the third quartile.

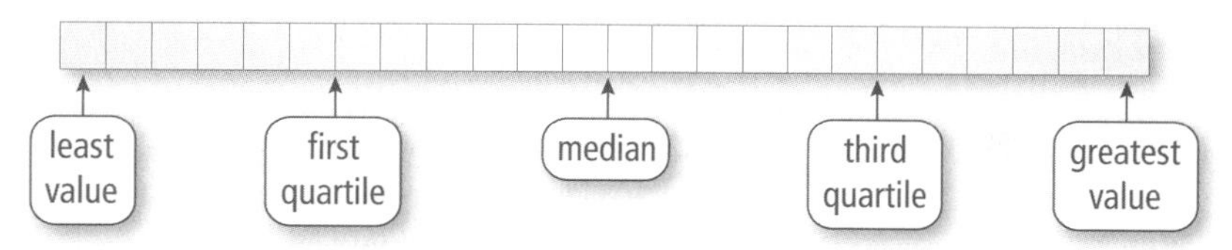

c. Explain how the box-and-whisker plot shown represents the data set.

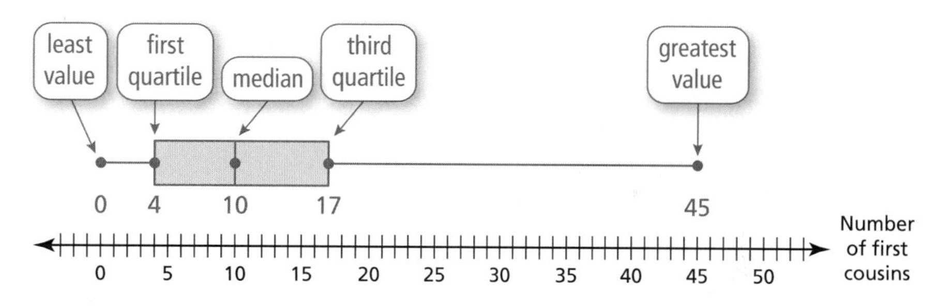

MODELING WITH MATHEMATICS

To be proficient in math, you need to identify important quantities in a practical situation.

Communicate Your Answer

2. How can you use a box-and-whisker plot to describe a data set?

3. Interpret each box-and-whisker plot.

a. body mass indices (BMI) of students in a ninth-grade class

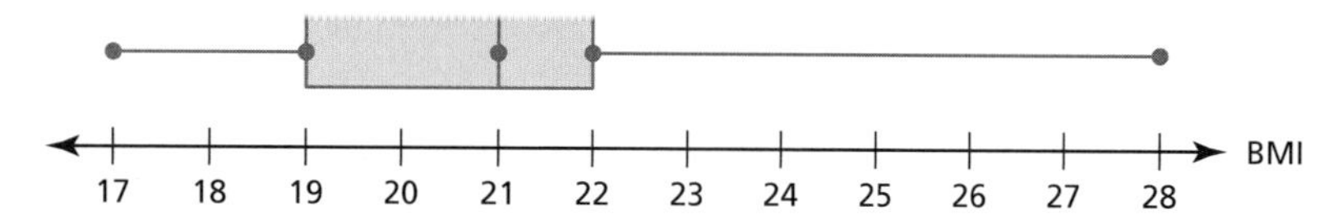

b. heights of roller coasters at an amusement park

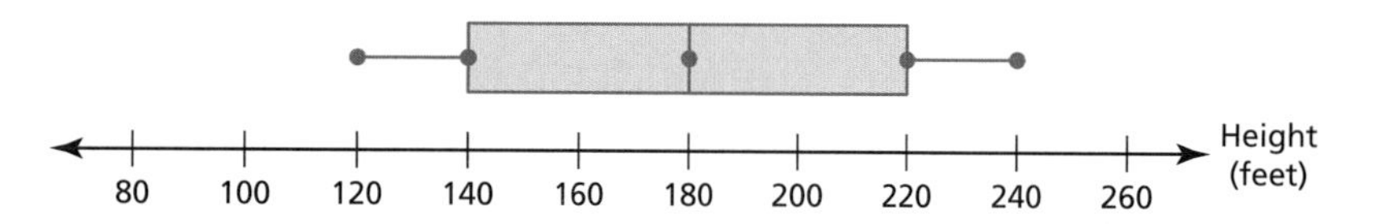

11.2 Lesson

Core Vocabulary

box-and-whisker plot, *p. 594*
quartile, *p. 594*
five-number summary, *p. 594*
interquartile range, *p. 595*

What You Will Learn

- Use box-and-whisker plots to represent data sets.
- Interpret box-and-whisker plots.
- Use box-and-whisker plots to compare data sets.

Using Box-and-Whisker Plots to Represent Data Sets

Core Concept

Box-and-Whisker Plot

A **box-and-whisker plot** shows the variability of a data set along a number line using the least value, the greatest value, and the *quartiles* of the data. **Quartiles** divide the data set into four equal parts. The median (second quartile, Q2) divides the data set into two halves. The median of the lower half is the first quartile, Q1. The median of the upper half is the third quartile, Q3.

STUDY TIP

Sometimes, the first quartile is called the *lower quartile* and the third quartile is called the *upper quartile*.

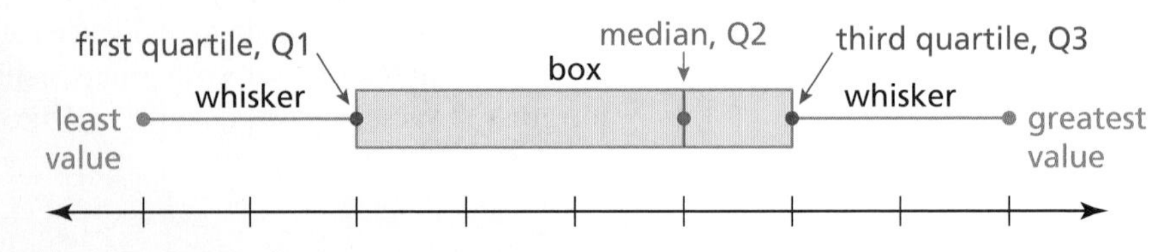

The five numbers that make up a box-and-whisker plot are called the **five-number summary** of the data set.

EXAMPLE 1 Making a Box-and-Whisker Plot

Make a box-and-whisker plot that represents the ages of the members of a backpacking expedition in the mountains.

24, 30, 30, 22, 25, 22, 18, 25, 28, 30, 25, 27

SOLUTION

Step 1 Order the data. Find the median and the quartiles.

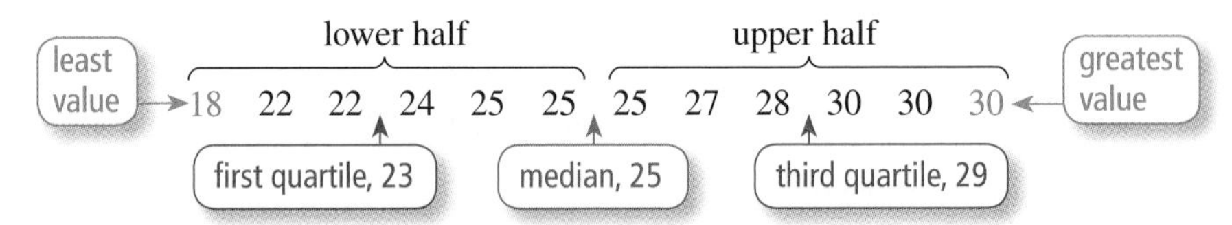

Step 2 Draw a number line that includes the least and greatest values. Graph points above the number line for the five-number summary.

Step 3 Draw a box using Q1 and Q3. Draw a line through the median. Draw whiskers from the box to the least and greatest values.

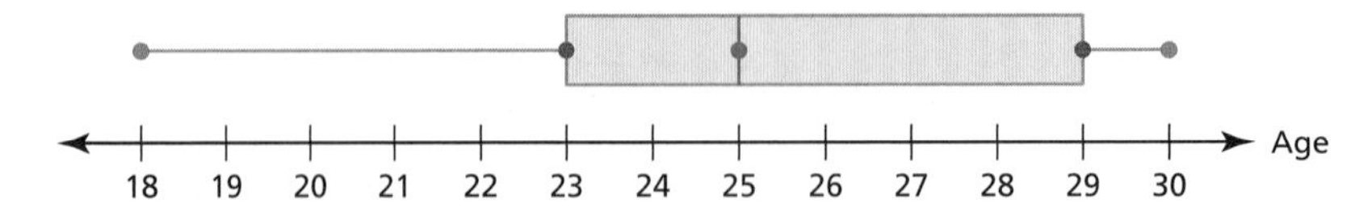

Monitoring Progress  Help in English and Spanish at *BigIdeasMath.com*

1. A basketball player scores 14, 16, 20, 5, 22, 30, 16, and 28 points during a tournament. Make a box-and-whisker plot that represents the data.

Interpreting Box-and-Whisker Plots

The figure shows how data are distributed in a box-and-whisker plot.

STUDY TIP

A long whisker or box indicates that the data are more spread out.

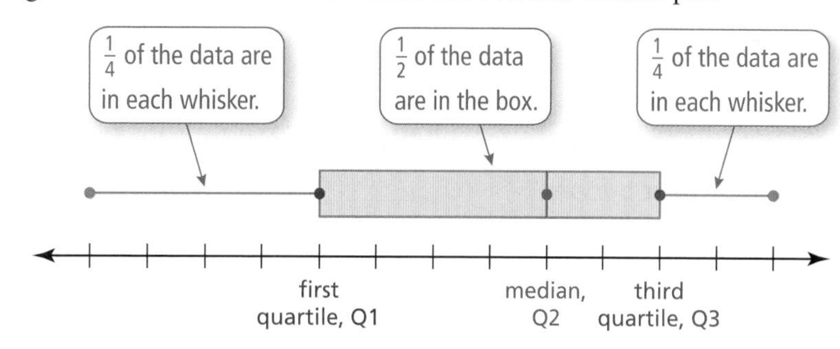

Another measure of variation for a data set is the **interquartile range** (IQR), which is the difference of the third quartile, Q3, and the first quartile, Q1. It represents the range of the middle half of the data.

EXAMPLE 2 Interpreting a Box-and-Whisker Plot

The box-and-whisker plot represents the lengths (in seconds) of the songs played by a rock band at a concert.

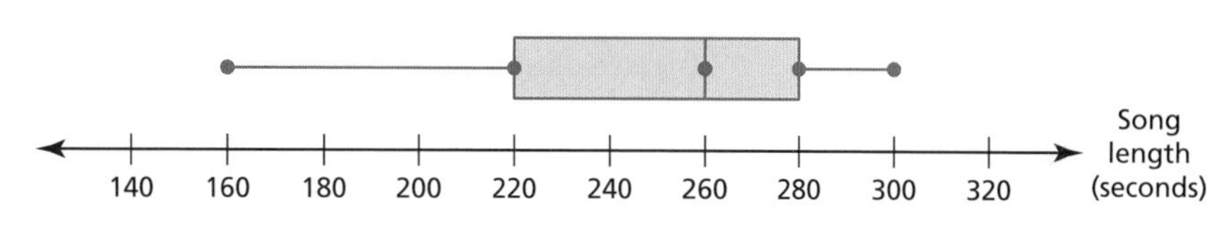

a. Find and interpret the range of the data.

b. Describe the distribution of the data.

c. Find and interpret the interquartile range of the data.

d. Are the data more spread out below Q1 or above Q3? Explain.

SOLUTION

a. The least value is 160. The greatest value is 300.

▶ So, the range is $300 - 160 = 140$ seconds. This means that the song lengths vary by no more than 140 seconds.

b. Each whisker represents 25% of the data. The box represents 50% of the data. So,

- 25% of the song lengths are between 160 and 220 seconds.
- 50% of the song lengths are between 220 and 280 seconds.
- 25% of the song lengths are between 280 and 300 seconds.

c. $\text{IQR} = \text{Q3} - \text{Q1} = 280 - 220 = 60$

▶ So, the interquartile range is 60 seconds. This means that the middle half of the song lengths vary by no more than 60 seconds.

d. The left whisker is longer than the right whisker.

▶ So, the data below Q1 are more spread out than data above Q3.

Monitoring Progress  Help in English and Spanish at *BigIdeasMath.com*

Use the box-and-whisker plot in Example 1.

2. Find and interpret the range and interquartile range of the data.

3. Describe the distribution of the data.

Using Box-and-Whisker Plots to Compare Data Sets

A box-and-whisker plot shows the shape of a distribution.

STUDY TIP

If you can draw a line through the median of a box-and-whisker plot, and each side is approximately a mirror image of the other, then the distribution is symmetric.

Shapes of Box-and-Whisker Plots

Skewed left	Symmetric	Skewed right
• The left whisker is longer than the right whisker. • Most of the data are on the right side of the plot.	• The whiskers are about the same length. • The median is in the middle of the plot.	• The right whisker is longer than the left whisker. • Most of the data are on the left side of the plot.

EXAMPLE 3 Comparing Box-and-Whisker Plots

The double box-and-whisker plot represents the test scores for your class and your friend's class.

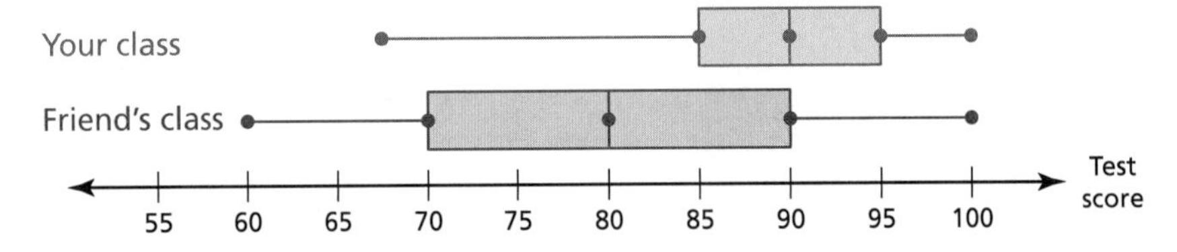

a. Identify the shape of each distribution.

b. Which test scores are more spread out? Explain.

SOLUTION

a. For your class, the left whisker is longer than the right whisker, and most of the data are on the right side of the plot. For your friend's class, the whisker lengths are equal, and the median is in the middle of the plot.

▶ So, the distribution for your class is skewed left, and the distribution for your friend's class is symmetric.

b. The range and interquartile range of the test scores in your friend's class are greater than the range and interquartile range in your class.

▶ So, the test scores in your friend's class are more spread out.

Monitoring Progress  Help in English and Spanish at *BigIdeasMath.com*

4. The double box-and-whisker plot represents the surfboard prices at Shop A and Shop B. Identify the shape of each distribution. Which shop's prices are more spread out? Explain.

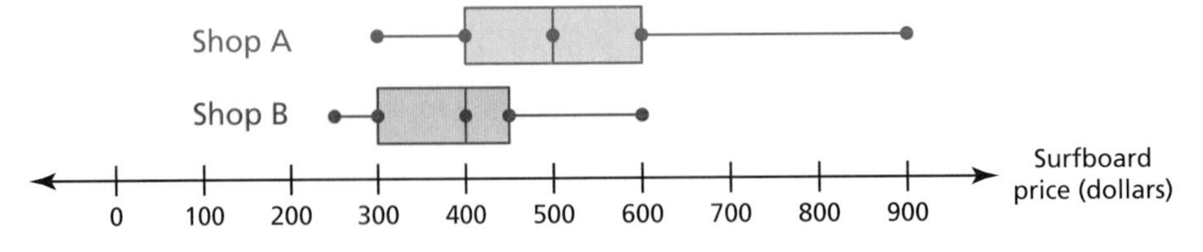

11.2 Exercises

Dynamic Solutions available at *BigIdeasMath.com*

Vocabulary and Core Concept Check

1. **WRITING** Describe how to find the first quartile of a data set.

2. **DIFFERENT WORDS, SAME QUESTION** Consider the box-and-whisker plot shown. Which is different? Find "both" answers.

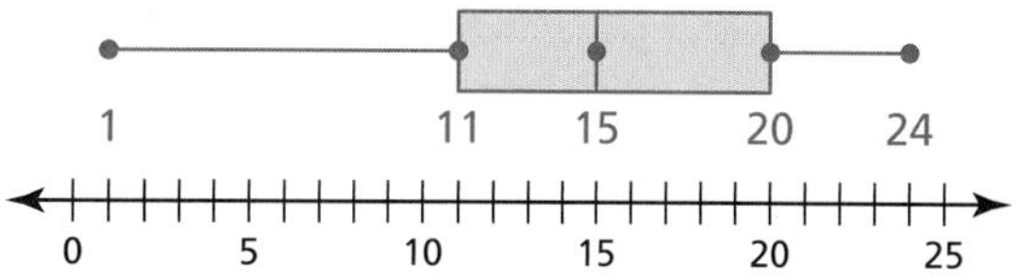

Find the interquartile range of the data.

Find the range of the middle half of the data.

Find the difference of the greatest value and the least value of the data set.

Find the difference of the third quartile and the first quartile.

Monitoring Progress and Modeling with Mathematics

In Exercises 3–8, use the box-and-whisker plot to find the given measure.

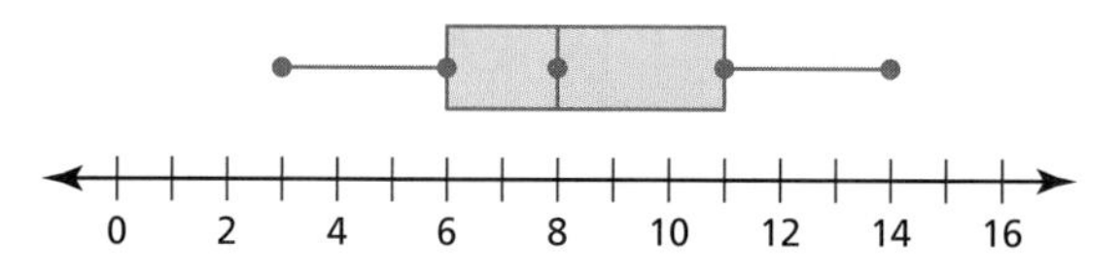

3. least value
4. greatest value
5. third quartile
6. first quartile
7. median
8. range

In Exercises 9–12, make a box-and-whisker plot that represents the data. *(See Example 1.)*

9. Hours of television watched: 0, 3, 4, 5, 2, 4, 6, 5

10. Cat lengths (in inches): 16, 18, 20, 25, 17, 22, 23, 21

11. Elevations (in feet): −2, 0, 5, −4, 1, −3, 2, 0, 2, −3, 6

12. MP3 player prices (in dollars): 124, 95, 105, 110, 95, 124, 300, 190, 114

13. **ANALYZING DATA** The dot plot represents the numbers of hours students spent studying for an exam. Make a box-and-whisker plot that represents the data.

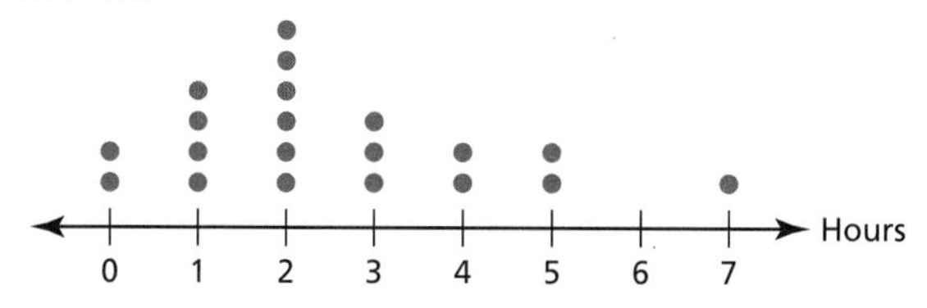

14. **ANALYZING DATA** The stem-and-leaf plot represents the lengths (in inches) of the fish caught on a fishing trip. Make a box-and-whisker plot that represents the data.

Stem	Leaf
0	6 7 8 8 9
1	0 0 2 2 3 4 4 7
2	1 2

Key: 1 | 0 = 10 inches

15. **ANALYZING DATA** The box-and-whisker plot represents the prices (in dollars) of the entrées at a restaurant. *(See Example 2.)*

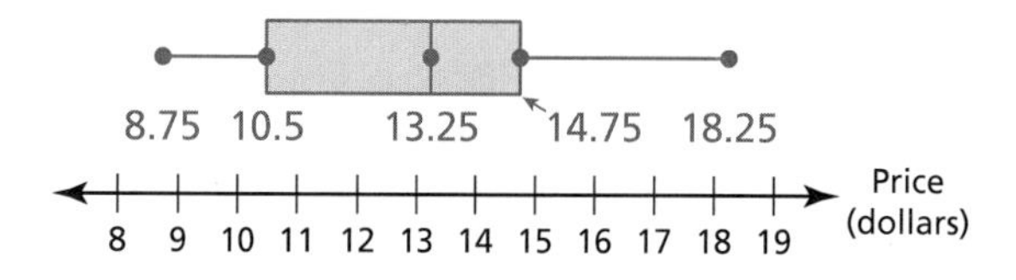

a. Find and interpret the range of the data.

b. Describe the distribution of the data.

c. Find and interpret the interquartile range of the data.

d. Are the data more spread out below Q1 or above Q3? Explain.

16. ANALYZING DATA A baseball player scores 101 runs in a season. The box-and-whisker plot represents the numbers of runs the player scores against different opposing teams.

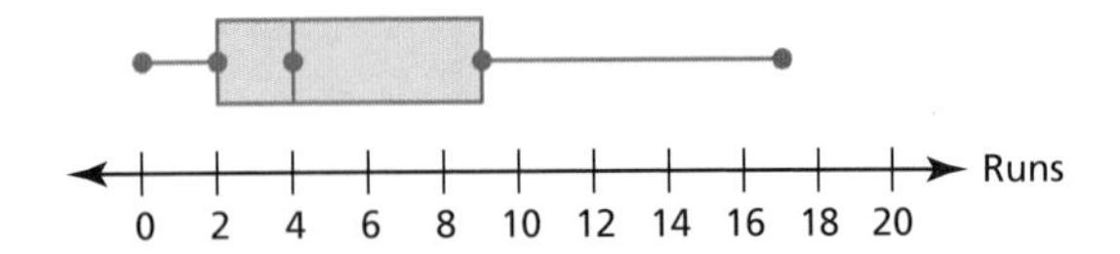

a. Find and interpret the range and interquartile range of the data.

b. Describe the distribution of the data.

c. Are the data more spread out between Q1 and Q2 or between Q2 and Q3? Explain.

17. ANALYZING DATA The double box-and-whisker plot represents the monthly car sales for a year for two sales representatives. *(See Example 3.)*

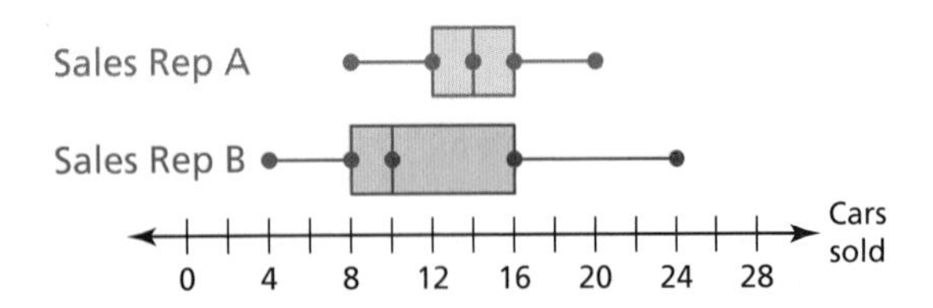

a. Identify the shape of each distribution.

b. Which representative's sales are more spread out? Explain.

c. Which representative had the single worst sales month during the year? Explain.

18. ERROR ANALYSIS Describe and correct the error in describing the box-and-whisker plot.

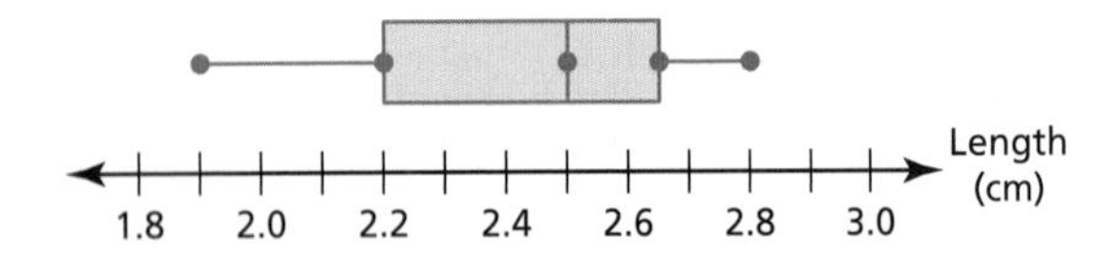

The distribution is skewed left. So, most of the data are on the left side of the plot.

19. WRITING Given the numbers 36 and 12, identify which number is the range and which number is the interquartile range of a data set. Explain.

20. HOW DO YOU SEE IT? The box-and-whisker plot represents a data set. Determine whether each statement is always true. Explain your reasoning.

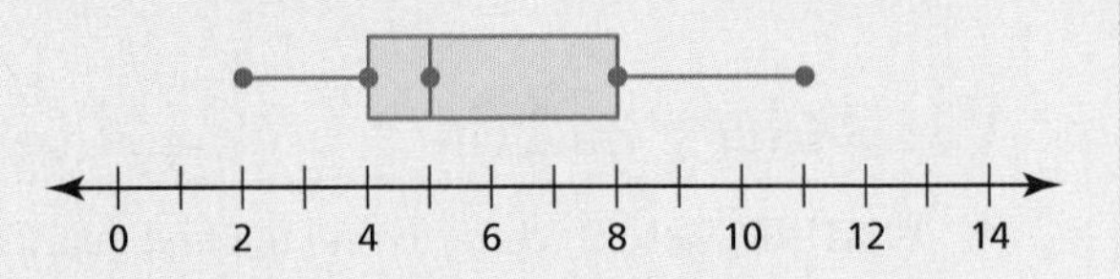

a. The data set contains the value 11.

b. The data set contains the value 6.

c. The distribution is skewed right.

d. The mean of the data is 5.

21. ANALYZING DATA The double box-and-whisker plot represents the battery lives (in hours) of two brands of cell phones.

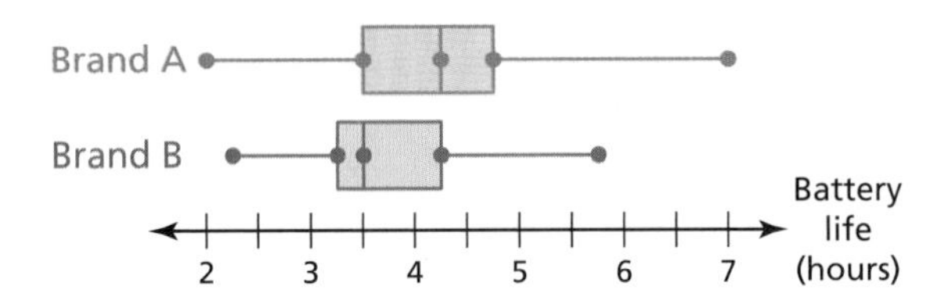

a. Identify the shape of each distribution.

b. What is the range of the upper 75% of each brand?

c. Compare the interquartile ranges of the two data sets.

d. Which brand do you think has a greater standard deviation? Explain.

e. You need a cell phone that has a battery life of more than 3.5 hours most of the time. Which brand should you buy? Explain.

22. THOUGHT PROVOKING Create a data set that can be represented by the box-and-whisker plot shown. Justify your answer.

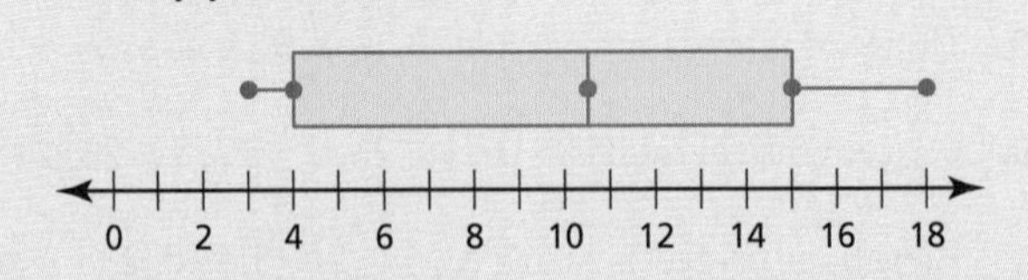

23. CRITICAL THINKING Two data sets have the same median, the same interquartile range, and the same range. Is it possible for the box-and-whisker plots of the data sets to be different? Justify your answer.

Maintaining Mathematical Proficiency

Reviewing what you learned in previous grades and lessons

Use zeros to graph the function. *(Section 8.5)*

24. $f(x) = -2(x + 9)(x - 3)$

25. $y = 3(x - 5)(x + 5)$

26. $y = 4x^2 - 16x - 48$

27. $h(x) = -x^2 + 5x + 14$

11.3 Shapes of Distributions

Essential Question How can you use a histogram to characterize the basic shape of a distribution?

EXPLORATION 1 Analyzing a Famous Symmetric Distribution

Work with a partner. A famous data set was collected in Scotland in the mid-1800s. It contains the chest sizes, measured in inches, of 5738 men in the Scottish Militia. Estimate the percent of the chest sizes that lie within (a) 1 standard deviation of the mean, (b) 2 standard deviations of the mean, and (c) 3 standard deviations of the mean. Explain your reasoning.

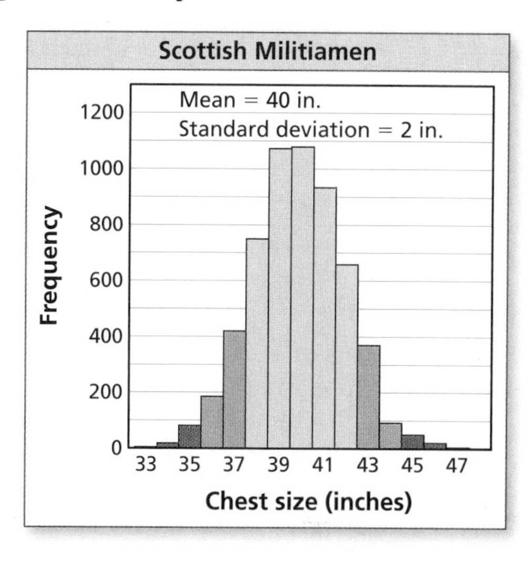

EXPLORATION 2 Comparing Two Symmetric Distributions

Work with a partner. The graphs show the distributions of the heights of 250 adult American males and 250 adult American females.

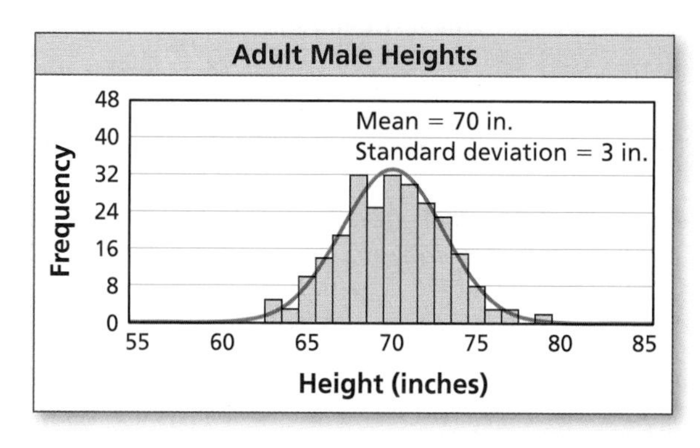

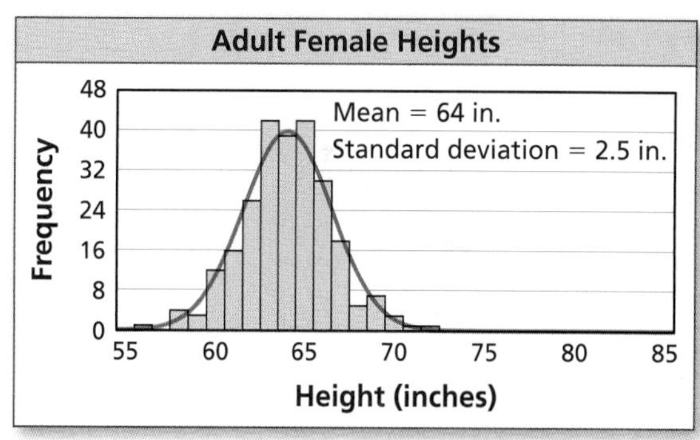

a. Which data set has a smaller standard deviation? Explain what this means in the context of the problem.

b. Estimate the percent of male heights between 67 inches and 73 inches.

ATTENDING TO PRECISION

To be proficient in math, you need to express numerical answers with a level of precision appropriate for the problem's context.

Communicate Your Answer

3. How can you use a histogram to characterize the basic shape of a distribution?

4. All three distributions in Explorations 1 and 2 are roughly symmetric. The histograms are called "bell-shaped."

a. What are the characteristics of a symmetric distribution?

b. Why is a symmetric distribution called "bell-shaped?"

c. Give two other real-life examples of symmetric distributions.

11.3 Lesson

Core Vocabulary

Previous
histogram
frequency table

What You Will Learn

- Describe the shapes of data distributions.
- Use the shapes of data distributions to choose appropriate measures.
- Compare data distributions.

Describing the Shapes of Data Distributions

Recall that a histogram is a bar graph that shows the frequency of data values in intervals of the same size. A histogram is another useful data display that shows the shape of a distribution.

STUDY TIP

If all the bars of a histogram are about the same height, then the distribution is a *flat*, or *uniform*, distribution. A uniform distribution is also symmetric.

Core Concept

Symmetric and Skewed Distributions

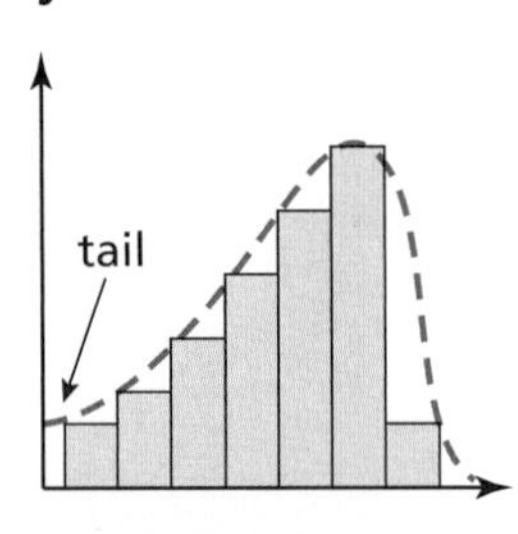

Skewed left

- The "tail" of the graph extends to the left.
- Most of the data are on the right.

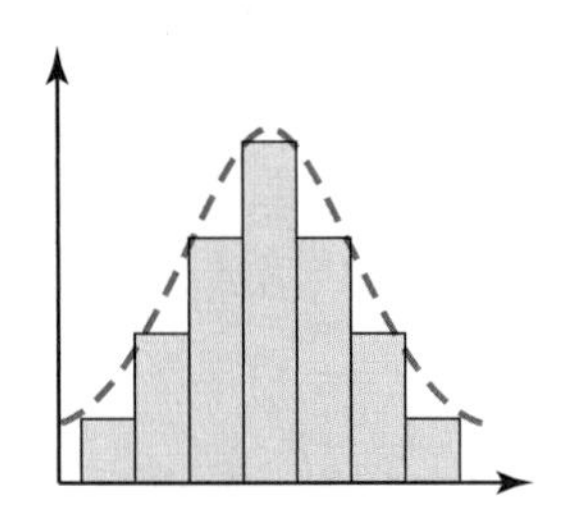

Symmetric

- The data on the right of the distribution are approximately a mirror image of the data on the left of the distribution.

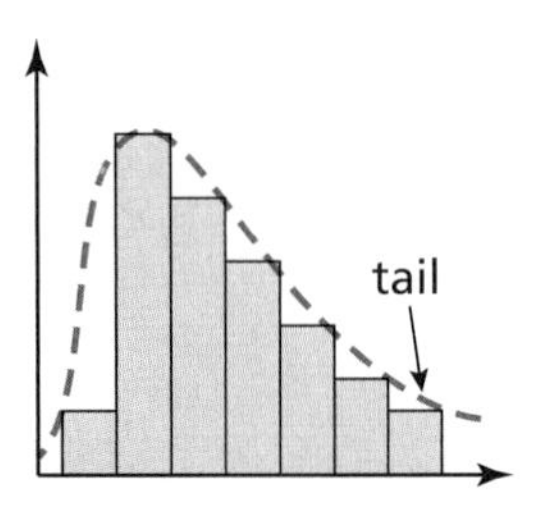

Skewed right

- The "tail" of the graph extends to the right.
- Most of the data are on the left.

EXAMPLE 1 Describing the Shape of a Distribution

Number of tickets sold	Frequency
1–8	5
9–16	9
17–24	16
25–32	25
33–40	20
41–48	8
49–56	7

The frequency table shows the numbers of raffle tickets sold by students in your grade. Display the data in a histogram. Describe the shape of the distribution.

SOLUTION

Step 1 Draw and label the axes.

Step 2 Draw a bar to represent the frequency of each interval.

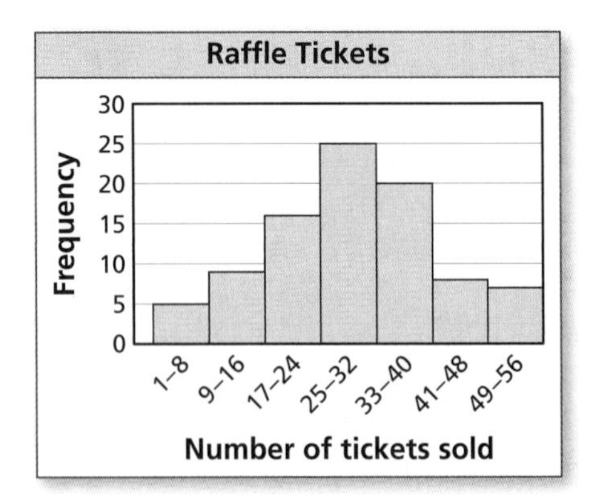

The data on the right of the distribution are approximately a mirror image of the data on the left of the distribution.

▶ So, the distribution is symmetric.

Monitoring Progress  Help in English and Spanish at *BigIdeasMath.com*

Number of pounds	Frequency
1–10	7
11–20	8
21–30	10
31–40	16
41–50	34
51–60	15

1. The frequency table shows the numbers of pounds of aluminum cans collected by classes for a fundraiser. Display the data in a histogram. Describe the shape of the distribution.

Choosing Appropriate Measures

Use the shape of a distribution to choose the most appropriate measure of center and measure of variation to describe the data set.

STUDY TIP

When a distribution is symmetric, the mean and median are about the same. When a distribution is skewed, the mean will be in the direction in which the distribution is skewed while the median will be less affected.

Core Concept

Choosing Appropriate Measures

When a data distribution is symmetric,

- use the mean to describe the center and
- use the standard deviation to describe the variation.

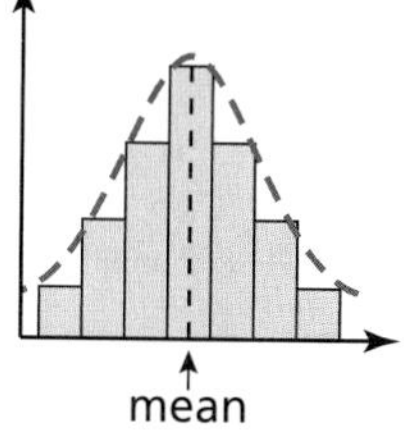

When a data distribution is skewed,

- use the median to describe the center and
- use the five-number summary to describe the variation.

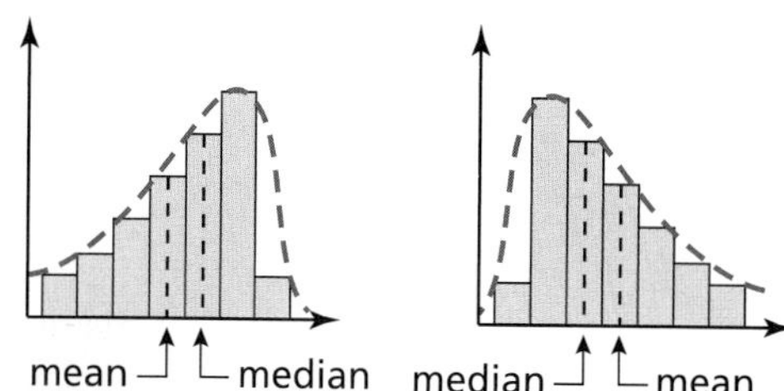

EXAMPLE 2 Choosing Appropriate Measures

Speeds (mi/h)		
32	44	39
53	38	48
56	41	42
50	50	55
55	45	49
51	53	52
54	60	55
52	50	52
55	40	60
45	58	47

A police officer measures the speeds (in miles per hour) of 30 motorists. The results are shown in the table at the left. (a) Display the data in a histogram using six intervals beginning with 31–35. (b) Which measures of center and variation best represent the data? (c) The speed limit is 45 miles per hour. How would you interpret these results?

SOLUTION

a. Make a frequency table using the described intervals. Then use the frequency table to make a histogram.

Speed (mi/h)	Frequency
31–35	1
36–40	3
41–45	5
46–50	6
51–55	11
56–60	4

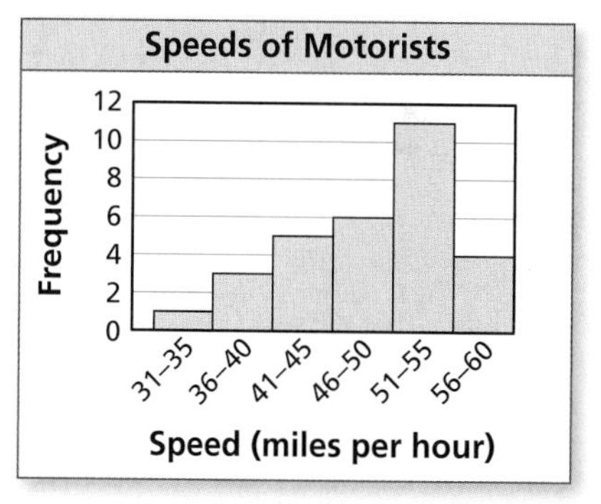

b. Because most of the data are on the right and the tail of the graph extends to the left, the distribution is skewed left. So, use the median to describe the center and the five-number summary to describe the variation.

c. Using the frequency table and the histogram, you can see that most of the speeds are more than 45 miles per hour. So, most of the motorists were speeding.

Monitoring Progress

Help in English and Spanish at *BigIdeasMath.com*

Email Attachments Sent				
74	105	98	68	64
85	75	60	48	51
65	55	58	45	38
64	52	65	30	70
72	5	45	77	83
42	25	95	16	120

2. You record the numbers of email attachments sent by 30 employees of a company in 1 week. Your results are shown in the table. (a) Display the data in a histogram using six intervals beginning with 1–20. (b) Which measures of center and variation best represent the data? Explain.

Comparing Data Distributions

EXAMPLE 3 Comparing Data Distributions

Emoticons are graphic symbols that represent facial expressions. They are used to convey a person's mood in a text message. The double histogram shows the distributions of emoticon messages sent by a group of female students and a group of male students during 1 week. Compare the distributions using their shapes and appropriate measures of center and variation.

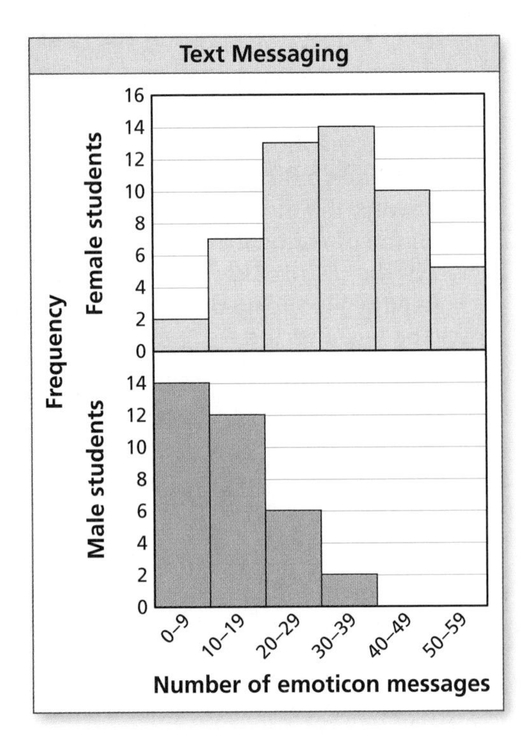

SOLUTION

Because the data on the right of the distribution for the female students are approximately a mirror image of the data on the left of the distribution, the distribution is symmetric. So, the mean and standard deviation best represent the distribution for female students.

Because most of the data are on the left of the distribution for the male students and the tail of the graph extends to the right, the distribution is skewed right. So, the median and five-number summary best represent the distribution for male students.

The mean of the female data set is probably in the 30–39 interval, while the median of the male data set is in the 10–19 interval. So, a typical female student is much more likely to use emoticons than a typical male student.

The data for the female students is more variable than the data for the male students. This means that the use of emoticons tends to differ more from one female student to the next.

Monitoring Progress  Help in English and Spanish at *BigIdeasMath.com*

3. Compare the distributions using their shapes and appropriate measures of center and variation.

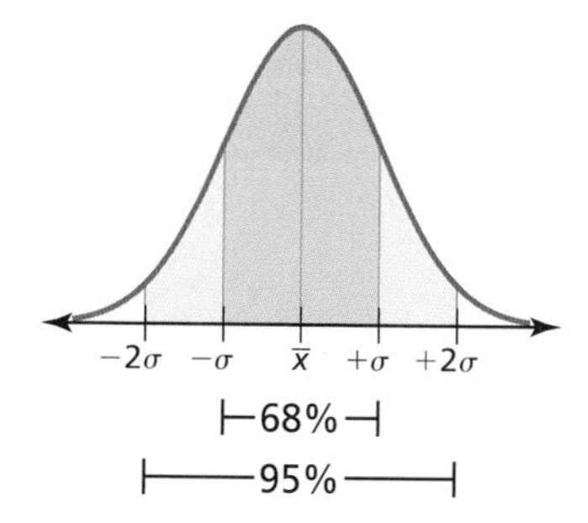

Many real-life data sets have distributions that are bell-shaped and approximately symmetric about the mean. In a future course, you will study this type of distribution in detail. For now, the following rules can help you see how valuable the standard deviation can be as a measure of variation.

- About 68% of the data lie within 1 standard deviation of the mean.
- About 95% of the data lie within 2 standard deviations of the mean.
- Data values that are more than 2 standard deviations from the mean are considered unusual.

Because the data are symmetric, you can deduce that 34% of the data lie within 1 standard deviation to the left of the mean, and 34% of the data lie within 1 standard deviation to the right of the mean.

EXAMPLE 4 Comparing Data Distributions

The table shows the results of a survey that asked men and women how many pairs of shoes they own.

a. Make a double box-and-whisker plot that represents the data. Describe the shape of each distribution.

b. Compare the number of pairs of shoes owned by men to the number of pairs of shoes owned by women.

c. About how many of the women surveyed would you expect to own between 10 and 18 pairs of shoes?

	Men	Women
Survey size	35	40
Minimum	2	5
Maximum	17	24
1st Quartile	5	12
Median	7	14
3rd Quartile	10	17
Mean	8	14
Standard deviation	3	4

SOLUTION

a.

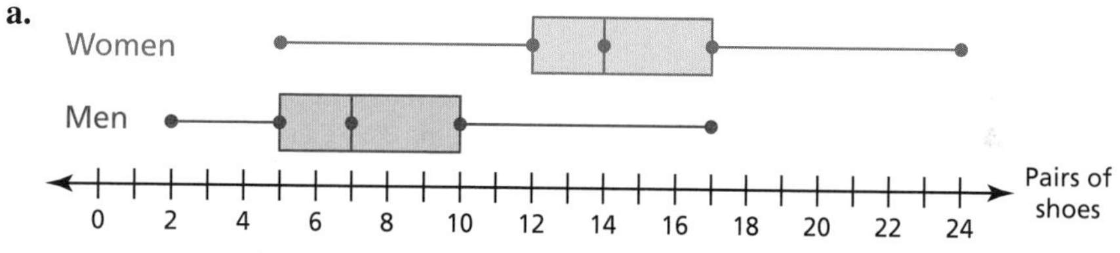

▶ The distribution for men is skewed right, and the distribution for women is symmetric.

b. The centers and spreads of the two data sets are quite different from each other. The mean for women is twice the median for men, and there is more variability in the number of pairs of shoes owned by women.

c. Assuming the symmetric distribution is bell-shaped, you know about 68% of the data lie within 1 standard deviation of the mean. Because the mean is 14 and the standard deviation is 4, the interval from 10 to 18 represents about 68% of the data. So, you would expect about $0.68 \cdot 40 \approx 27$ of the women surveyed to own between 10 and 18 pairs of shoes.

Monitoring Progress  Help in English and Spanish at *BigIdeasMath.com*

4. Why is the mean greater than the median for the men?

5. If 50 more women are surveyed, about how many more would you expect to own between 10 and 18 pairs of shoes?

11.3 Exercises

Dynamic Solutions available at *BigIdeasMath.com*

Vocabulary and Core Concept Check

1. **VOCABULARY** Describe how data are distributed in a symmetric distribution, a distribution that is skewed left, and a distribution that is skewed right.

2. **WRITING** How does the shape of a distribution help you decide which measures of center and variation best describe the data?

Monitoring Progress and Modeling with Mathematics

3. **DESCRIBING DISTRIBUTIONS** The frequency table shows the numbers of hours that students volunteer per month. Display the data in a histogram. Describe the shape of the distribution. *(See Example 1.)*

Number of volunteer hours	1–2	3–4	5–6	7–8	9–10	11–12	13–14
Frequency	1	5	12	20	15	7	2

4. **DESCRIBING DISTRIBUTIONS** The frequency table shows the results of a survey that asked people how many hours they spend online per week. Display the data in a histogram. Describe the shape of the distribution.

Hours online	Frequency
0–3	5
4–7	7
8–11	12
12–15	14
16–19	26
20–23	45
24–27	33

In Exercises 5 and 6, describe the shape of the distribution of the data. Explain your reasoning.

5.

Stem	Leaf
1	1 1 3 4 8
2	2 3 4 7 8
3	1 2 4 9
4	0 3 2
5	7 9
6	6

Key: 3 | 1 = 31

6.

Stem	Leaf
5	0 0 1
6	3 6 7 9
7	1 4 5 8 9
8	2 4 5 5 7
9	4 6 8 9
10	1 3 4

Key: 6 | 3 = 63

In Exercises 7 and 8, determine which measures of center and variation best represent the data. Explain your reasoning.

7.

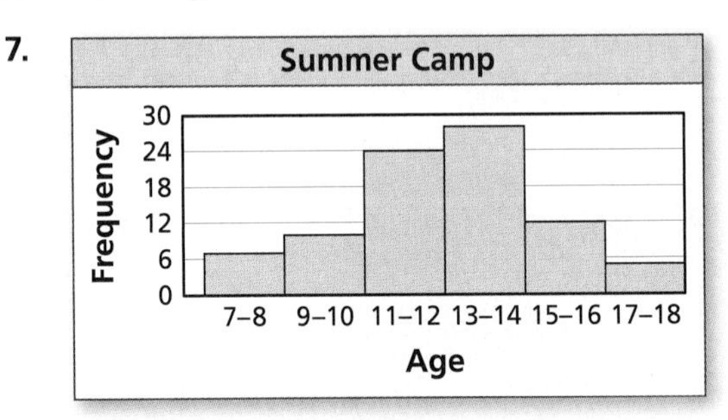

8.

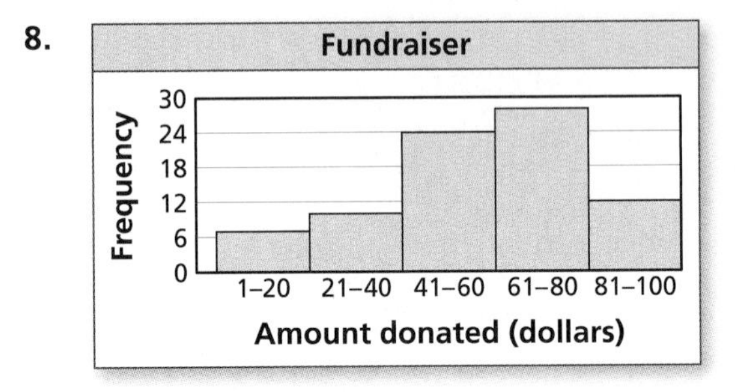

9. **ANALYZING DATA** The table shows the last 24 ATM withdrawals at a bank. *(See Example 2.)*

   a. Display the data in a histogram using seven intervals beginning with 26–50.

   b. Which measures of center and variation best represent the data? Explain.

   c. The bank charges a fee for any ATM withdrawal less than \$150. How would you interpret the data?

ATM Withdrawals (dollars)		
120	100	70
60	40	80
150	80	50
120	60	175
30	50	50
60	200	30
100	150	110
70	40	100

10. ANALYZING DATA Measuring an IQ is an inexact science. However, IQ scores have been around for years in an attempt to measure human intelligence. The table shows some of the greatest known IQ scores.

a. Display the data in a histogram using five intervals beginning with 151–166.

b. Which measures of center and variation best represent the data? Explain.

c. The distribution of IQ scores for the human population is symmetric. What happens to the shape of the distribution in part (a) as you include more and more IQ scores from the human population in the data set?

IQ Scores		
170	190	180
160	180	210
154	170	180
195	230	160
170	186	180
225	190	170

ERROR ANALYSIS In Exercises 11 and 12, describe and correct the error in the statements about the data displayed in the histogram.

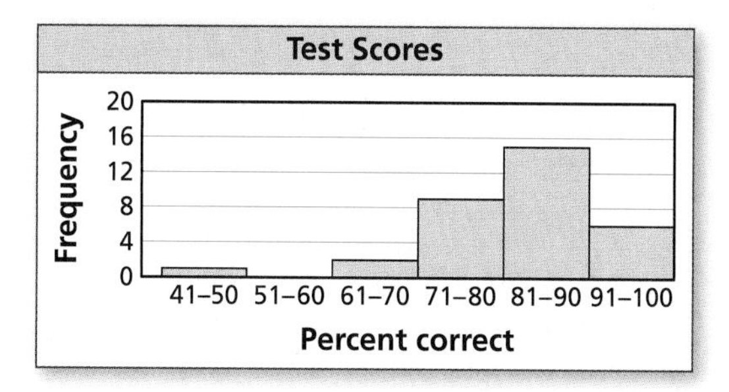

11. ✗ *Most of the data are on the right. So, the distribution is skewed right.*

12. ✗ *Because the distribution is skewed, use the standard deviation to describe the variation of the data.*

13. USING TOOLS For a large data set, would you use a stem-and-leaf plot or a histogram to show the distribution of the data? Explain.

14. REASONING For a symmetric distribution, why is the mean used to describe the center and the standard deviation used to describe the variation? For a skewed distribution, why is the median used to describe the center and the five-number summary used to describe the variation?

15. COMPARING DATA SETS The double histogram shows the distributions of daily high temperatures for two towns over a 50-day period. Compare the distributions using their shapes and appropriate measures of center and variation. *(See Example 3.)*

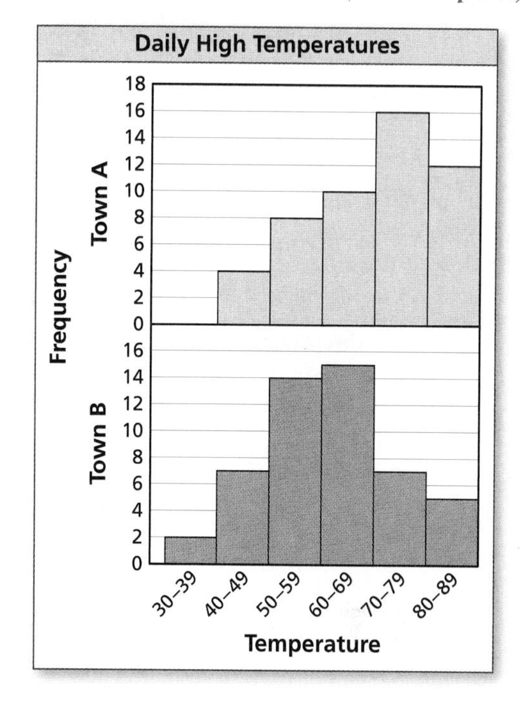

16. COMPARING DATA SETS The frequency tables show the numbers of entrées in certain price ranges (in dollars) at two different restaurants. Display the data in a double histogram. Compare the distributions using their shapes and appropriate measures of center and variation.

Restaurant A	
Price range	Frequency
8–10	5
11–13	9
14–16	12
17–19	4
20–22	3
23–25	0

Restaurant B	
Price range	Frequency
8–10	0
11–13	2
14–16	5
17–19	7
20–22	8
23–25	6

17. OPEN-ENDED Describe a real-life data set that has a distribution that is skewed right.

18. OPEN-ENDED Describe a real-life data set that has a distribution that is skewed left.

19. **COMPARING DATA SETS** The table shows the results of a survey that asked freshmen and sophomores how many songs they have downloaded on their MP3 players. *(See Example 4.)*

	Freshmen	Sophomores
Survey size	45	54
Minimum	250	360
Maximum	2150	2400
1st Quartile	800	780
Median	1200	2000
3rd Quartile	1600	2200
Mean	1150	1650
Standard deviation	420	480

a. Make a double box-and-whisker plot that represents the data. Describe the shape of each distribution.

b. Compare the number of songs downloaded by freshmen to the number of songs downloaded by sophomores.

c. About how many of the freshmen surveyed would you expect to have between 730 and 1570 songs downloaded on their MP3 players?

d. If you survey 100 more freshmen, about how many would you expect to have downloaded between 310 and 1990 songs on their MP3 players?

20. **COMPARING DATA SETS** You conduct the same survey as in Exercise 19 but use a different group of freshmen. The results are as follows.
Survey size: 60; minimum: 200; maximum: 2400; 1st quartile: 640; median: 1670; 3rd quartile: 2150; mean: 1480; standard deviation: 500

a. Compare the number of songs downloaded by this group of freshmen to the number of songs downloaded by sophomores.

b. Why is the median greater than the mean for this group of freshmen?

21. **REASONING** A data set has a symmetric distribution. Every value in the data set is doubled. Describe the shape of the new distribution. Are the measures of center and variation affected? Explain.

22. **HOW DO YOU SEE IT?** Match the distribution with the corresponding box-and-whisker plot.

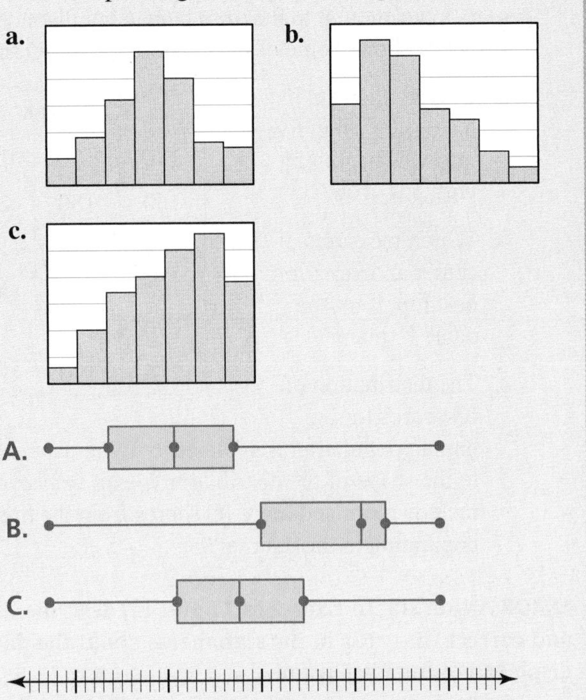

23. **REASONING** You record the following waiting times at a restaurant.

Waiting Times (minutes)									
26	38	15	8	22	42	25	20	17	18
40	35	24	31	42	29	25	0	30	13

a. Display the data in a histogram using five intervals beginning with 0–9. Describe the shape of the distribution.

b. Display the data in a histogram using 10 intervals beginning with 0–4. What happens when the number of intervals is increased?

c. Which histogram best represents the data? Explain your reasoning.

24. **THOUGHT PROVOKING** The shape of a *bimodal* distribution is shown. Describe a real-life example of a bimodal distribution.

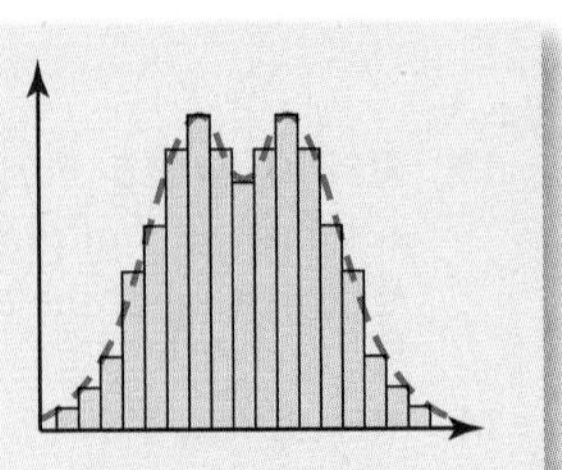

Maintaining Mathematical Proficiency

Reviewing what you learned in previous grades and lessons

Find the domain of the function. *(Section 10.1)*

25. $f(x) = \sqrt{x + 6}$

26. $f(x) = \sqrt{2x}$

27. $f(x) = \frac{1}{4}\sqrt{x} - 7$

Core Vocabulary

measure of center, *p. 586*
mean, *p. 586*
median, *p. 586*
mode, *p. 586*
outlier, *p. 587*
measure of variation, *p. 587*
range, *p. 587*
standard deviation, *p. 588*
data transformation, *p. 589*
box-and-whisker plot, *p. 594*
quartile, *p. 594*
five-number summary, *p. 594*
interquartile range, *p. 595*

Core Concepts

Section 11.1

Measures of Center, *p. 586*
Measures of Variation, *p. 587*
Data Transformations Using Addition, *p. 589*
Data Transformations Using Multiplication, *p. 589*

Section 11.2

Box-and-Whisker Plot, *p. 594*
Shapes of Box-and-Whisker Plots, *p. 596*

Section 11.3

Symmetric and Skewed Distributions, *p. 600*
Choosing Appropriate Measures, *p. 601*

Mathematical Practices

1. Exercises 15 and 16 on page 590 are similar. For each data set, is the outlier much greater than or much less than the rest of the data values? Compare how the outliers affect the means. Explain why this makes sense.
2. In Exercise 18 on page 605, provide a possible reason for why the distribution is skewed left.

Study Skills

Studying for Finals

- Form a study group of three or four students several weeks before the final exam.
- Find out what material you must know for the final exam, even if your teacher has not yet covered it.
- Ask for a practice final exam or create one yourself and have your teacher look at it.
- Have each group member take the practice final exam.
- Decide when the group is going to meet and what you will cover during each session.
- During the sessions, make sure you stay on track.

11.1–11.3 Quiz

Find the mean, median, and mode of the data set. Which measure of center best represents the data? Explain. *(Section 11.1)*

1.

Hours Spent on Project		
$3\frac{1}{2}$	5	$2\frac{1}{2}$
3	$3\frac{1}{2}$	$\frac{1}{2}$

2.

Waterfall Height (feet)		
1000	1267	1328
1200	1180	1000
2568	1191	1100

Find the range and standard deviation of each data set. Then compare your results. *(Section 11.1)*

3. Absent students during a week of school
Female: 6, 2, 4, 3, 4
Male: 5, 3, 6, 6, 9

4. Numbers of points scored
Juniors: 19, 15, 20, 10, 14, 21, 18, 15
Seniors: 22, 19, 29, 32, 15, 26, 30, 19

Make a box-and-whisker plot that represents the data. *(Section 11.2)*

5. Ages of family members:
60, 15, 25, 20, 55, 70, 40, 30

6. Minutes of violin practice:
20, 50, 60, 40, 40, 30, 60, 40, 50, 20, 20, 35

7. Display the data in a histogram. Describe the shape of the distribution. *(Section 11.3)*

Quiz score	0–2	3–5	6–8	9–11	12–14
Frequency	1	3	6	16	4

8. The table shows the prices of eight mountain bikes in a sporting goods store. *(Section 11.1 and Section 11.2)*

Price (dollars)	98	119	95	211	130	98	100	125

a. Find the mean, median, mode, range, and standard deviation of the prices.

b. Identify the outlier. How does the outlier affect the mean, median, and mode?

c. Make a box-and-whisker plot that represents the data. Find and interpret the interquartile range of the data. Identify the shape of the distribution.

d. Find the mean, median, mode, range, and standard deviation of the prices when the store offers a 5% discount on all mountain bikes.

9. The table shows the times of 20 presentations. *(Section 11.3)*

a. Display the data in a histogram using five intervals beginning with 3–5.

b. Which measures of center and variation best represent the data? Explain.

c. The presentations are supposed to be 10 minutes long. How would you interpret these results?

Time (minutes)			
9	7	10	12
10	11	8	10
10	17	11	5
9	10	4	12
6	14	8	10

11.4 Two-Way Tables

Essential Question How can you read and make a two-way table?

EXPLORATION 1 Reading a Two-Way Table

Work with a partner. You are the manager of a sports shop. The two-way tables show the numbers of soccer T-shirts in stock at your shop at the beginning and end of the selling season. (a) Complete the totals for the rows and columns in each table. (b) How would you alter the number of T-shirts you order for next season? Explain your reasoning.

Beginning of season		T-Shirt Size					
		S	M	L	XL	XXL	Total
Color	blue/white	5	6	7	6	5	
	blue/gold	5	6	7	6	5	
	red/white	5	6	7	6	5	
	black/white	5	6	7	6	5	
	black/gold	5	6	7	6	5	
	Total						145

End of season		T-Shirt Size					
		S	M	L	XL	XXL	Total
Color	blue/white	5	4	1	0	2	
	blue/gold	3	6	5	2	0	
	red/white	4	2	4	1	3	
	black/white	3	4	1	2	1	
	black/gold	5	2	3	0	2	
	Total						

MODELING WITH MATHEMATICS

To be proficient in math, you need to identify important quantities and map their relationships using tools such as graphs and two-way tables.

EXPLORATION 2 Making a Two-Way Table

Work with a partner. The three-dimensional bar graph shows the numbers of hours students work at part-time jobs.

a. Make a two-way table showing the data. Use estimation to find the entries in your table.

b. Write two observations that summarize the data in your table.

Part-Time Jobs of Students at a High School

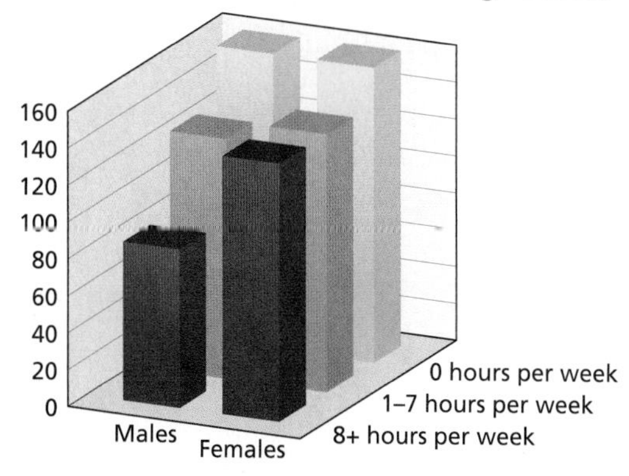

Communicate Your Answer

3. How can you read and make a two-way table?

11.4 Lesson

Core Vocabulary

two-way table, p. 610
joint frequency, p. 610
marginal frequency, p. 610
joint relative frequency, p. 611
marginal relative frequency, p. 611
conditional relative frequency, p. 612

What You Will Learn

- Find and interpret marginal frequencies.
- Make two-way tables.
- Find relative and conditional relative frequencies.
- Use two-way tables to recognize associations in data.

Finding and Interpreting Marginal Frequencies

A **two-way table** is a frequency table that displays data collected from one source that belong to two different categories. One category of data is represented by rows, and the other is represented by columns. For instance, the two-way table below shows the results of a survey that asked freshmen and sophomores whether they access the Internet using a mobile device, such as a smartphone.

The two categories of data are *class* and *mobile access*. Class is further divided into *freshman* and *sophomore*, and mobile access is further divided into *yes* and *no*.

Each entry in the table is called a **joint frequency**. The sums of the rows and columns in a two-way table are called **marginal frequencies**.

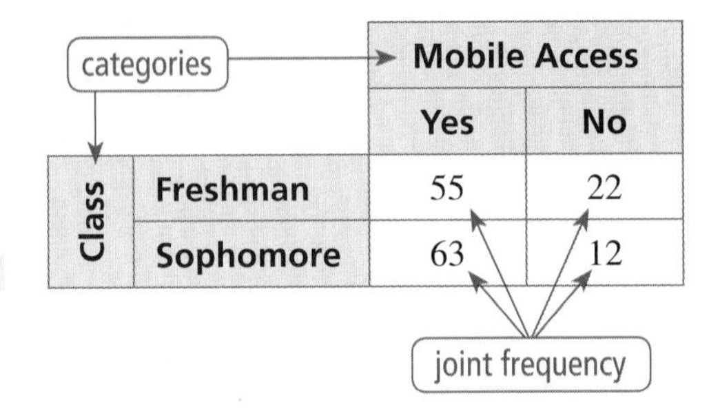

categories
Mobile Access
Yes
No
Class
Freshman
55
22
Sophomore
63
12
joint frequency

REMEMBER

The frequency of an event is the number of times the event occurs.

EXAMPLE 1 **Finding and Interpreting Marginal Frequencies**

Find and interpret the marginal frequencies for the two-way table above.

SOLUTION

Create a new column and a new row for the marginal frequencies. Then add the entries in each row and column.

		Mobile Access		
		Yes	No	Total
Class	Freshman	55	22	77
	Sophomore	63	12	75
	Total	118	34	152

77 freshmen responded.
75 sophomores responded.
152 students were surveyed.
118 students access the Internet using a mobile device.
34 students do not access the Internet using a mobile device.

STUDY TIP

The sum of the "total" row should be equal to the sum of the "total" column. Place this sum of the marginal frequencies at the bottom right of your two-way table.

Monitoring Progress

Help in English and Spanish at *BigIdeasMath.com*

1. You conduct a technology survey to publish on your school's website. You survey students in the school cafeteria about the technological devices they own. The results are shown in the two-way table. Find and interpret the marginal frequencies.

		Tablet Computer	
		Yes	No
Cell Phone	Yes	34	124
	No	18	67

Making Two-Way Tables

EXAMPLE 2 **Making a Two-Way Table**

You conduct a survey that asks 286 students in your freshman class whether they play a sport or a musical instrument. One hundred eighteen of the students play a sport, and 64 of those students play an instrument. Ninety-three of the students do not play a sport or an instrument. Organize the results in a two-way table. Include the marginal frequencies.

SOLUTION

Step 1 Determine the two categories for the table: *sport* and *instrument*.

Step 2 Use the given joint and marginal frequencies to fill in parts of the table.

		Instrument		
		Yes	No	Total
Sport	Yes	64		118
	No		93	
	Total			286

Step 3 Use reasoning to find the missing joint and marginal frequencies. For instance, you can conclude that there are $286 - 118 = 168$ students who do not play a sport, and $118 - 64 = 54$ students who play a sport but do not play an instrument.

		Instrument		
		Yes	No	Total
Sport	Yes	64	54	118
	No	75	93	168
	Total	139	147	286

Monitoring Progress  Help in English and Spanish at *BigIdeasMath.com*

2. You survey students about whether they are getting a summer job. Seventy-five males respond, with 18 of them responding "no." Fifty-seven females respond, with 45 of them responding "yes." Organize the results in a two-way table. Include the marginal frequencies.

Finding Relative and Conditional Relative Frequencies

You can display entries of a two-way table as frequency counts (as in Examples 1 and 2) or as *relative frequencies*.

Relative Frequencies

A **joint relative frequency** is the ratio of a frequency that is not in the "total" row or the "total" column to the total number of values or observations.

A **marginal relative frequency** is the sum of the joint relative frequencies in a row or a column.

When finding relative frequencies in a two-way table, you can use the corresponding decimals or percents.

EXAMPLE 3 Finding Relative Frequencies

The two-way table shows the results of a survey that asked college-bound high school students whether they plan to major in a medical field. Make a two-way table that shows the joint and marginal relative frequencies.

		Major in Medical Field	
		Yes	No
Class	Junior	124	219
	Senior	101	236

SOLUTION

There are 124 + 219 + 101 + 236 = 680 students in the survey. To find the joint relative frequencies, divide each frequency by 680. Then find the sum of each row and each column to find the marginal relative frequencies.

STUDY TIP

The sum of the marginal relative frequencies in the "total" row and the "total" column should each be equal to 1.

		Major in Medical Field		
		Yes	No	Total
Class	Junior	$\frac{124}{680} \approx 0.18$	$\frac{219}{680} \approx 0.32$	0.50
	Senior	$\frac{101}{680} \approx 0.15$	$\frac{236}{680} \approx 0.35$	0.50
	Total	0.33	0.67	1

About 50% of the students are juniors.

About 35% of the students are seniors and are not planning to major in a medical field.

Core Concept

Conditional Relative Frequencies

A **conditional relative frequency** is the ratio of a joint relative frequency to the marginal relative frequency. You can find a conditional relative frequency using a row total or a column total of a two-way table.

EXAMPLE 4 Finding Conditional Relative Frequencies

Use the survey results in Example 3 to make a two-way table that shows the conditional relative frequencies based on the column totals.

SOLUTION

Use the marginal relative frequency of each *column* to calculate the conditional relative frequencies.

STUDY TIP

When you use column totals, the sum of the conditional relative frequencies for each column should be equal to 1.

		Major in Medical Field	
		Yes	No
Class	Junior	$\frac{0.18}{0.33} \approx 0.55$	$\frac{0.32}{0.67} \approx 0.48$
	Senior	$\frac{0.15}{0.33} \approx 0.45$	$\frac{0.35}{0.67} \approx 0.52$

Given that a student is not planning to major in a medical field, the conditional relative frequency that he or she is a junior is about 48%.

Monitoring Progress

Help in English and Spanish at *BigIdeasMath.com*

3. Use the survey results in Monitoring Progress Question 2 to make a two-way table that shows the joint and marginal relative frequencies. What percent of students are not getting a summer job?

4. Use the survey results in Example 3 to make a two-way table that shows the conditional relative frequencies based on the row totals. Given that a student is a senior, what is the conditional relative frequency that he or she is planning to major in a medical field?

Recognizing Associations in Data

EXAMPLE 5 Recognizing Associations in Data

You survey students and find that 40% exercise regularly, 35% eat fruits and vegetables each day, and 52% do not exercise and do not eat fruits and vegetables each day. Is there an association between exercising regularly and eating fruits and vegetables each day?

SOLUTION

Use the given information to make a two-way table. Use reasoning to find the missing joint and marginal relative frequencies.

		Exercise Regularly		
		Yes	No	Total
Eats Fruit/ Vegetables	Yes	27%	8%	35%
	No	13%	52%	65%
	Total	40%	60%	100%

Use conditional relative frequencies based on the column totals to determine whether there is an association. Of the students who exercise regularly, 67.5% eat fruits and vegetables each day. Of the students who do not exercise regularly, only about 13% eat fruits and vegetables each day. It appears that students who exercise regularly are more likely to eat more fruits and vegetables than students who do not exercise regularly.

		Exercise Regularly	
		Yes	No
Eats Fruit/ Vegetables	Yes	$\frac{0.27}{0.4} = 0.675$	$\frac{0.08}{0.6} \approx 0.133$
	No	$\frac{0.13}{0.4} = 0.325$	$\frac{0.52}{0.6} \approx 0.867$

► So, there is an association between exercising regularly and eating fruits and vegetables each day.

You can also find the conditional relative frequencies by dividing each joint frequency by its corresponding column total or row total.

EXAMPLE 6 Recognizing Associations in Data

The two-way table shows the results of a survey that asked students whether they share a computer at home with other family members. Is there an association between age and sharing a computer?

		Age			
		12–13	14–15	16–17	18–19
Share a Computer	Yes	40	47	42	22
	No	10	25	36	34

SOLUTION

Use conditional relative frequencies based on column totals to determine whether there is an association. Based on this sample, 80% of students ages 12–13 share a computer and only about 39% of students ages 18–19 share a computer.

		Age			
		12–13	14–15	16–17	18–19
Share a Computer	Yes	$\frac{40}{50} = 0.8$	$\frac{47}{72} \approx 0.65$	$\frac{42}{78} \approx 0.54$	$\frac{22}{56} \approx 0.39$
	No	$\frac{10}{50} \approx 0.2$	$\frac{25}{72} \approx 0.35$	$\frac{36}{78} \approx 0.46$	$\frac{34}{56} \approx 0.61$

► The table shows that as age increases, students are less likely to share a computer with other family members. So, there is an association.

Monitoring Progress  Help in English and Spanish at *BigIdeasMath.com*

5. Using the results of the survey in Monitoring Progress Question 1, is there an association between owning a tablet computer and owning a cell phone? Explain your reasoning.

11.4 Exercises

Dynamic Solutions available at *BigIdeasMath.com*

Vocabulary and Core Concept Check

1. **COMPLETE THE SENTENCE** Each entry in a two-way table is called a(n) __________.

2. **WRITING** When is it appropriate to use a two-way table to organize data?

3. **VOCABULARY** Explain the relationship between joint relative frequencies, marginal relative frequencies, and conditional relative frequencies.

4. **WRITING** Describe two ways you can find conditional relative frequencies.

Monitoring Progress and Modeling with Mathematics

You conduct a survey that asks 346 students whether they buy lunch at school. In Exercises 5–8, use the results of the survey shown in the two-way table.

		Buy Lunch at School	
		Yes	No
Class	Freshman	92	86
	Sophomore	116	52

5. How many freshmen were surveyed?

6. How many sophomores were surveyed?

7. How many students buy lunch at school?

8. How many students do not buy lunch at school?

In Exercises 9 and 10, find and interpret the marginal frequencies. *(See Example 1.)*

9.

		Set Academic Goals	
		Yes	No
Gender	Male	64	168
	Female	54	142

10.

		Cat	
		Yes	No
Dog	Yes	104	208
	No	186	98

11. **USING TWO-WAY TABLES** You conduct a survey that asks students whether they plan to participate in school spirit week. The results are shown in the two-way table. Find and interpret the marginal frequencies.

		Participate in Spirit Week		
		Yes	No	Undecided
Class	Freshman	112	56	54
	Sophomore	92	68	32

12. **USING TWO-WAY TABLES** You conduct a survey that asks college-bound high school seniors about the type of degree they plan to receive. The results are shown in the two-way table. Find and interpret the marginal frequencies.

		Type of Degree		
		Associate's	Bachelor's	Master's
Gender	Male	58	126	42
	Female	62	118	48

USING STRUCTURE In Exercises 13 and 14, complete the two-way table.

13.

		Traveled on an Airplane		
		Yes	No	Total
Class	Freshman		62	
	Sophomore	184		
	Total	274		352

14.

		Plan to Attend School Dance		
		Yes	No	Total
Gender	Male	38		
	Female		24	112
	Total			196

15. **MAKING TWO-WAY TABLES** You conduct a survey that asks 245 students in your school whether they have taken a Spanish or a French class. One hundred nine of the students have taken a Spanish class, and 45 of those students have taken a French class. Eighty-two of the students have not taken a Spanish or a French class. Organize the results in a two-way table. Include the marginal frequencies. *(See Example 2.)*

16. **MAKING TWO-WAY TABLES** A car dealership has 98 cars on its lot. Fifty-five of the cars are new. Of the new cars, 36 are domestic cars. There are 15 used foreign cars on the lot. Organize this information in a two-way table. Include the marginal frequencies.

In Exercises 17 and 18, make a two-way table that shows the joint and marginal relative frequencies. *(See Example 3.)*

17.

		Exercise Preference	
		Aerobic	Anaerobic
Gender	Male	88	104
	Female	96	62

18.

		Meat	
		Turkey	Ham
Bread	White	452	146
	Wheat	328	422

19. **USING TWO-WAY TABLES** Refer to Exercise 17. What percent of students prefer aerobic exercise? What percent of students are males who prefer anaerobic exercise?

20. **USING TWO-WAY TABLES** Refer to Exercise 18. What percent of the sandwiches are on wheat bread? What percent of the sandwiches are turkey on white bread?

ERROR ANALYSIS In Exercises 21 and 22, describe and correct the error in using the two-way table.

		Participate in Fundraiser	
		Yes	No
Class	Freshman	187	85
	Sophomore	123	93

21.

One hundred eighty-seven freshmen responded to the survey.

22.

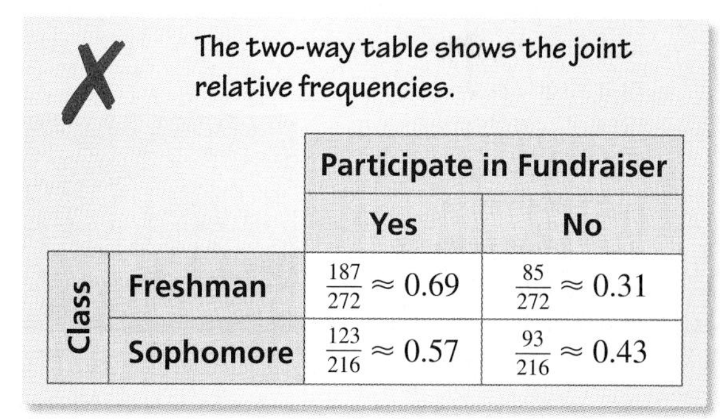

The two-way table shows the joint relative frequencies.

		Participate in Fundraiser	
		Yes	No
Class	Freshman	$\frac{187}{272} \approx 0.69$	$\frac{85}{272} \approx 0.31$
	Sophomore	$\frac{123}{216} \approx 0.57$	$\frac{93}{216} \approx 0.43$

23. **USING TWO-WAY TABLES** A company is hosting an event for its employees to celebrate the end of the year. It asks the employees whether they prefer a lunch event or a dinner event. It also asks whether they prefer a catered event or a potluck. The results are shown in the two-way table. Make a two-way table that shows the conditional relative frequencies based on the row totals. Given that an employee prefers a lunch event, what is the conditional relative frequency that he or she prefers a catered event? *(See Example 4.)*

		Menu	
		Potluck	Catered
Meal	Lunch	36	48
	Dinner	44	72

24. USING TWO-WAY TABLES The two-way table shows the results of a survey that asked students about their preference for a new school mascot. Make a two-way table that shows the conditional relative frequencies based on the column totals. Given that a student prefers a hawk as a mascot, what is the conditional relative frequency that he or she prefers a cartoon mascot?

		Type		
		Tiger	Hawk	Dragon
Style	Realistic	67	74	51
	Cartoon	58	18	24

25. ANALYZING TWO-WAY TABLES You survey college-bound seniors and find that 85% plan to live on campus, 35% plan to have a car while at college, and 5% plan to live off campus and not have a car. Is there an association between living on campus and having a car at college? Explain. *(See Example 5.)*

26. ANALYZING TWO-WAY TABLES You survey students and find that 70% watch sports on TV, 48% participate in a sport, and 16% do not watch sports on TV or participate in a sport. Is there an association between participating in a sport and watching sports on TV? Explain.

27. ANALYZING TWO-WAY TABLES The two-way table shows the results of a survey that asked adults whether they participate in recreational skiing. Is there an association between age and recreational skiing? *(See Example 6.)*

		Age				
		21–30	31–40	41–50	51–60	61–70
Ski	Yes	87	93	68	37	20
	No	165	195	148	117	125

28. ANALYZING TWO-WAY TABLES Refer to Exercise 12. Is there an association between gender and type of degree? Explain.

29. WRITING Compare Venn diagrams and two-way tables.

30. HOW DO YOU SEE IT? The graph shows the results of a survey that asked students about their favorite movie genre.

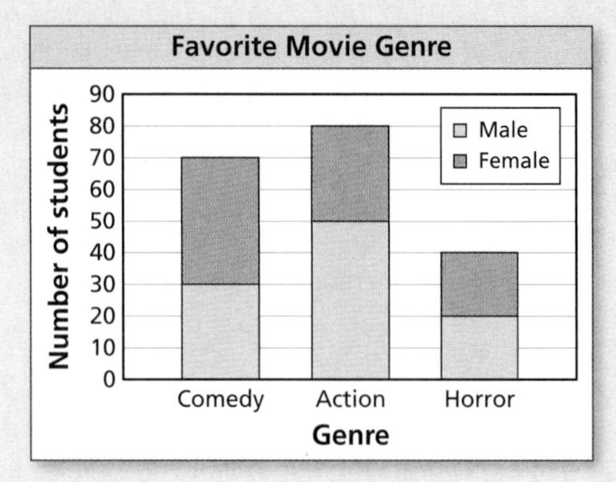

a. Display the given information in a two-way table.

b. Which of the data displays do you prefer? Explain.

31. PROBLEM SOLVING A box office sells 1809 tickets to a play, 800 of which are for the main floor. The tickets consist of $2x + y$ adult tickets on the main floor, $x - 40$ child tickets on the main floor, $x + 2y$ adult tickets in the balcony, and $3x - y - 80$ child tickets in the balcony.

a. Organize this information in a two-way table.

b. Find the values of x and y.

c. What percent of tickets are adult tickets?

d. What percent of child tickets are balcony tickets?

32. THOUGHT PROVOKING Compare "one-way tables" and "two-way tables." Is it possible to have a "three-way table?" If so, give an example of a three-way table.

Maintaining Mathematical Proficiency

Reviewing what you learned in previous grades and lessons

Tell whether the table of values represents a *linear*, an *exponential*, or a *quadratic* function. *(Section 8.6)*

33.

x	0	1	2	3	4
y	144	24	4	$\frac{2}{3}$	$\frac{1}{9}$

34.

x	−1	0	1	2	3
y	3	0	−1	0	3

11.5 Choosing a Data Display

Essential Question How can you display data in a way that helps you make decisions?

EXPLORATION 1 Displaying Data

Work with a partner. Analyze the data and then create a display that best represents the data. Explain your choice of data display.

a. A group of schools in New England participated in a 2-month study and reported 3962 animals found dead along roads.

birds: 307	mammals: 2746	amphibians: 145
reptiles: 75	unknown: 689	

b. The data below show the numbers of black bears killed on a state's roads from 1993 to 2012.

1993: 30	2000: 47	2007: 99
1994: 37	2001: 49	2008: 129
1995: 46	2002: 61	2009: 111
1996: 33	2003: 74	2010: 127
1997: 43	2004: 88	2011: 141
1998: 35	2005: 82	2012: 135
1999: 43	2006: 109	

c. A 1-week study along a 4-mile section of road found the following weights (in pounds) of raccoons that had been killed by vehicles.

13.4	14.8	17.0	12.9	21.3	21.5	16.8	14.8
15.2	18.7	18.6	17.2	18.5	9.4	19.4	15.7
14.5	9.5	25.4	21.5	17.3	19.1	11.0	12.4
20.4	13.6	17.5	18.5	21.5	14.0	13.9	19.0

d. A yearlong study by volunteers in California reported the following numbers of animals killed by motor vehicles.

raccoons: 1693	gray squirrels: 715
skunks: 1372	cottontail rabbits: 629
ground squirrels: 845	barn owls: 486
opossum: 763	jackrabbits: 466
deer: 761	gopher snakes: 363

USING TOOLS STRATEGICALLY

To be proficient in math, you need to identify relevant external mathematical resources.

Communicate Your Answer

2. How can you display data in a way that helps you make decisions?

3. Use the Internet or some other reference to find examples of the following types of data displays.

bar graph	circle graph	scatter plot
stem-and-leaf plot	pictograph	line graph
box-and-whisker plot	histogram	dot plot

11.5 Lesson

Core Vocabulary

qualitative (categorical) data, *p. 618*
quantitative data, *p. 618*
misleading graph, *p. 620*

What You Will Learn

- Classify data as quantitative or qualitative.
- Choose and create appropriate data displays.
- Analyze misleading graphs.

Classifying Data

Data sets can consist of two types of data: *qualitative* or *quantitative.*

Core Concept

Types of Data

Qualitative data, or **categorical data**, consist of labels or nonnumerical entries that can be separated into different categories. When using qualitative data, operations such as adding or finding a mean do not make sense.

Quantitative data consist of numbers that represent counts or measurements.

STUDY TIP

Just because a frequency count can be shown for a data set does not make it quantitative. A frequency count can be shown for both qualitative and quantitative data.

EXAMPLE 1 Classifying Data

Tell whether the data are *qualitative* or *quantitative*.

a. prices of used cars at a dealership

b. jersey numbers on a basketball team

c. lengths of songs played at a concert

d. zodiac signs of students in your class

SOLUTION

a. Prices are numerical entries. So, the data are quantitative.

b. Jersey numbers are numerical, but they are labels. It does not make sense to compare them, and you cannot measure them. So, the data are qualitative.

c. Song lengths are numerical measurements. So, the data are quantitative.

d. Zodiac signs are nonnumerical entries that can be separated into different categories. So, the data are qualitative.

Monitoring Progress

Help in English and Spanish at *BigIdeasMath.com*

Tell whether the data are *qualitative* or *quantitative*. Explain your reasoning.

1. telephone numbers in a directory

2. ages of patients at a hospital

3. lengths of videos on a website

4. types of flowers at a florist

Qualitative and quantitative data can be collected from the same data source, as shown below. You can use these types of data together to obtain a more accurate description of a population.

Data Source	Quantitative Data	Qualitative Data
a student	How much do you earn per hour at your job? \$10.50	What is your occupation? painter
a house	How many square feet of living space is in the house? 2500 ft^2	In what city is the house located? Chicago

Choosing and Creating Appropriate Data Displays

As shown on page 584, you have learned a variety of ways to display data sets graphically. Choosing an appropriate data display can depend on whether the data are qualitative or quantitative.

EXAMPLE 2 Choosing and Creating a Data Display

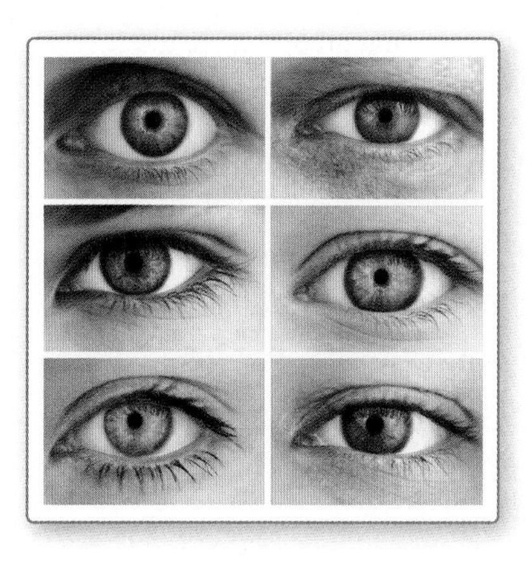

Analyze the data and then create a display that best represents the data. Explain your reasoning.

a.

Eye Color Survey	
Color	**Number of students**
brown	63
blue	37
hazel	25
green	10
gray	3
amber	2

b.

Speeds of Vehicles (mi/h)			
Interstate A		**Interstate B**	
65	67	67	72
68	71	70	78
72	70	65	71
68	65	71	80
65	68	84	81
75	82	77	79
68	59	69	70
62	68	66	69
75	80	73	75
77	75	84	79

SOLUTION

a. A circle graph is one appropriate way to display this qualitative data. It shows data as parts of a whole.

Step 1 Find the angle measure for each section of the circle graph by multiplying the fraction of students who have each eye color by 360°. Notice that there are $63 + 37 + 25 + 10 + 3 + 2 = 140$ students in the survey.

Brown: $\frac{63}{140} \cdot 360^\circ \approx 162^\circ$ **Blue:** $\frac{37}{140} \cdot 360^\circ \approx 95^\circ$ **Hazel:** $\frac{25}{140} \cdot 360^\circ \approx 64^\circ$

Green: $\frac{10}{140} \cdot 360^\circ \approx 26^\circ$ **Gray:** $\frac{3}{140} \cdot 360^\circ \approx 8^\circ$ **Amber:** $\frac{2}{140} \cdot 360^\circ \approx 5^\circ$

Step 2 Use a protractor to draw the angle measures found in Step 1 on a circle. Then label each section and title the circle graph, as shown.

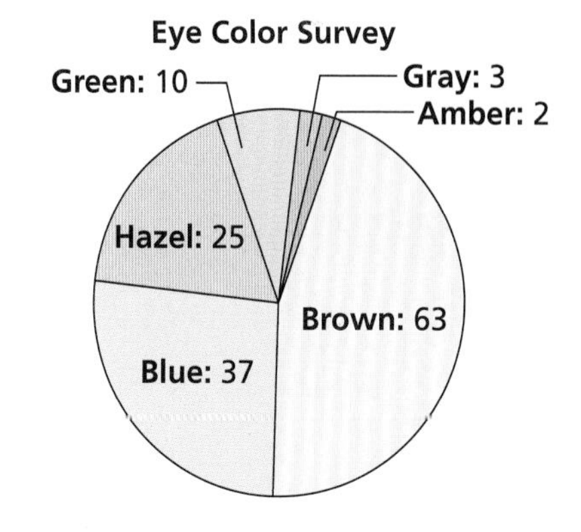

b. A double box-and-whisker plot is one appropriate way to display this quantitative data. Use the five-number summary of each data set to create a double box-and-whisker plot.

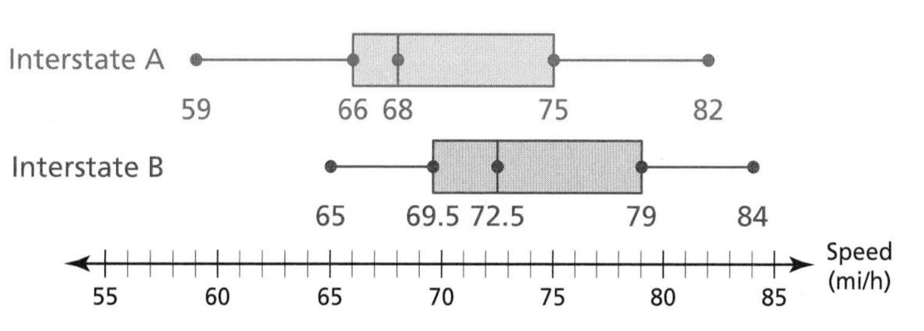

Monitoring Progress  Help in English and Spanish at *BigIdeasMath.com*

5. Display the data in Example 2(a) in another way.

6. Display the data in Example 2(b) in another way.

Analyzing Misleading Graphs

Just as there are several ways to display data accurately using graphs, there are several ways to display data that are misleading. A **misleading graph** is a statistical graph that is not drawn appropriately. This may occur when the creator of a graph wants to give viewers the impression that results are better than they actually are. Below are some questions you can ask yourself when analyzing a statistical graph that will help you recognize when a graph is trying to deceive or mislead.

- Does the graph have a title?
- Are the numbers of the scale evenly spaced?
- Does the scale begin at zero? If not, is there a break?
- Does the graph need a key?
- Are all the axes or sections of the graph labeled?
- Are all the components of the graph, such as the bars, the same size?

EXAMPLE 3 Analyzing Misleading Graphs

Describe how each graph is misleading. Then explain how someone might misinterpret the graph.

a.

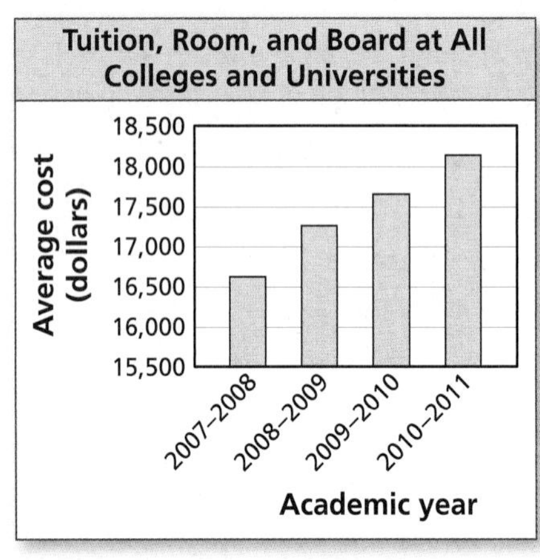

b.

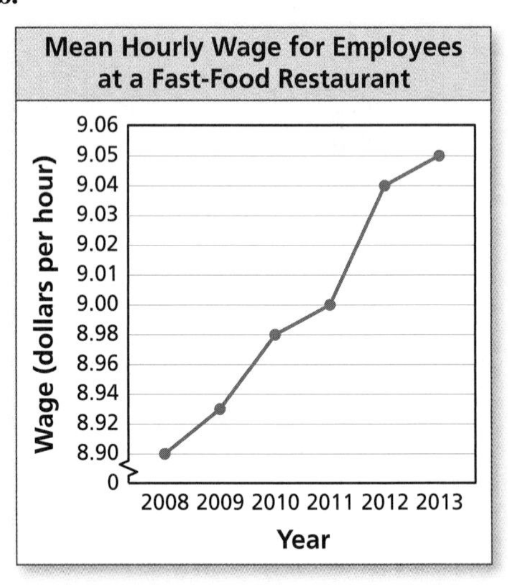

SOLUTION

a. The scale on the vertical axis of the graph starts at \$15,500 and does not have a break. This makes it appear that the average cost increased rapidly for the years given.

Someone might believe that the average cost more than doubled from 2007 to 2011, when actually, it increased by only about \$1500.

b. The scale on the vertical axis has very small increments that are not equal.

Someone might believe that the greatest increase in the mean hourly wage occurred from 2011 to 2012, when the greatest increase actually occurred from 2009 to 2010.

Monitoring Progress Help in English and Spanish at *BigIdeasMath.com*

7. Redraw the graphs in Example 3 so they are not misleading.

11.5 Exercises

Dynamic Solutions available at *BigIdeasMath.com*

Vocabulary and Core Concept Check

1. **OPEN-ENDED** Describe two ways that a line graph can be misleading.

2. **WHICH ONE DOESN'T BELONG?** Which data set does *not* belong with the other three? Explain your reasoning.

ages of people attending a concert	heights of skyscrapers in a city
populations of counties in a state	breeds of dogs at a pet store

Monitoring Progress and Modeling with Mathematics

In Exercises 3–8, tell whether the data are *qualitative* or *quantitative*. Explain your reasoning. *(See Example 1.)*

3. brands of cars in a parking lot

4. weights of bears at a zoo

5. budgets of feature films

6. file formats of documents on a computer

7. shoe sizes of students in your class

8. street addresses in a phone book

In Exercises 9–12, choose an appropriate data display for the situation. Explain your reasoning.

9. the number of students in a marching band each year

10. a comparison of students' grades (out of 100) in two different classes

11. the favorite sports of students in your class

12. the distribution of teachers by age

In Exercises 13–16, analyze the data and then create a display that best represents the data. Explain your reasoning. *(See Example 2.)*

13.

Ages of World Cup Winners	
2010 Men's World Cup Winner (Spain)	2011 Women's World Cup Winner (Japan)
29 24 23 30 32 26 28 30 26 23 32 28 22 28 24 21 27 22 25 21 24 24 27	36 27 24 20 27 23 29 26 25 32 27 27 22 25 24 23 24 28 20 18 24

14.

Average Precipitation (inches)			
January	1.1	July	4.0
February	1.5	August	4.4
March	2.2	September	4.2
April	3.7	October	3.5
May	5.1	November	2.1
June	5.5	December	1.8

15.

Grades (out of 100) on a Test				
96	74	97	80	62
84	88	53	77	75
89	81	52	85	63
87	95	59	83	100

16.

Colors of Cars that Drive by Your House			
white	25	green	3
red	12	silver/gray	27
yellow	1	blue	6
black	21	brown/biege	5

17. **DISPLAYING DATA** Display the data in Exercise 13 in another way.

18. **DISPLAYING DATA** Display the data in Exercise 14 in another way.

19. **DISPLAYING DATA** Display the data in Exercise 15 in another way.

20. **DISPLAYING DATA** Display the data in Exercise 16 in another way.

In Exercises 21–24, describe how the graph is misleading. Then explain how someone might misinterpret the graph. *(See Example 3.)*

21.

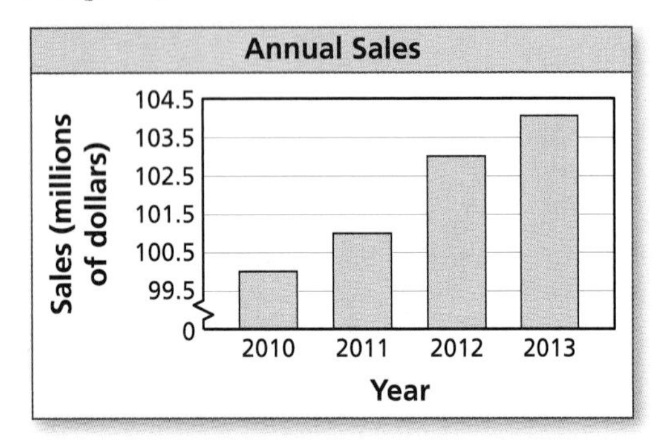

22.

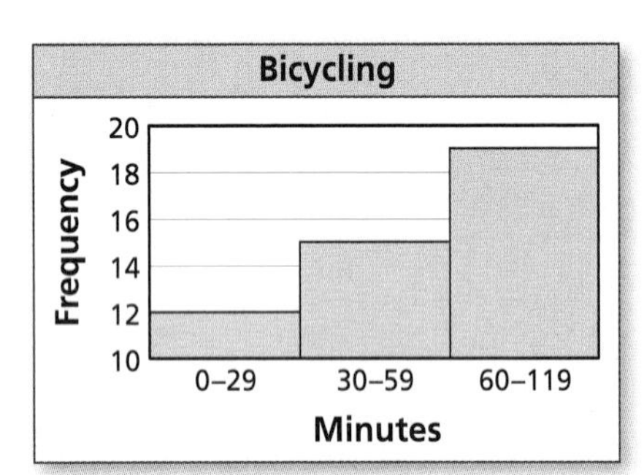

23.

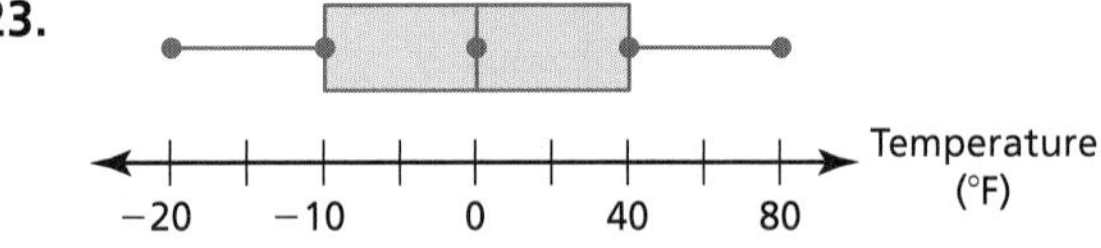

24.

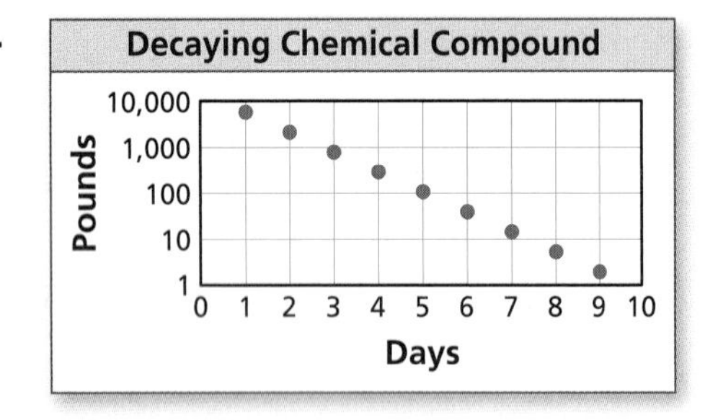

25. **DISPLAYING DATA** Redraw the graph in Exercise 21 so it is not misleading.

26. **DISPLAYING DATA** Redraw the graph in Exercise 22 so it is not misleading.

27. **MAKING AN ARGUMENT** A data set gives the ages of voters for a city election. Classmate A says the data should be displayed in a bar graph, while Classmate B says the data would be better displayed in a histogram. Who is correct? Explain.

28. **HOW DO YOU SEE IT?** The manager of a company sees the graph shown and concludes that the company is experiencing a decline. What is missing from the graph? Explain why the manager may be mistaken.

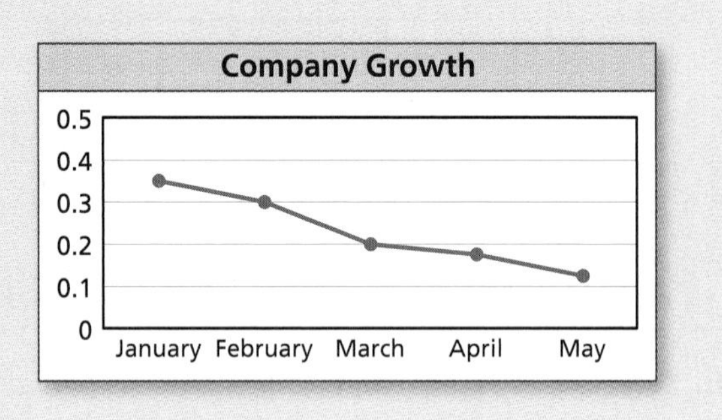

29. **REASONING** A survey asked 100 students about the sports they play. The results are shown in the circle graph.

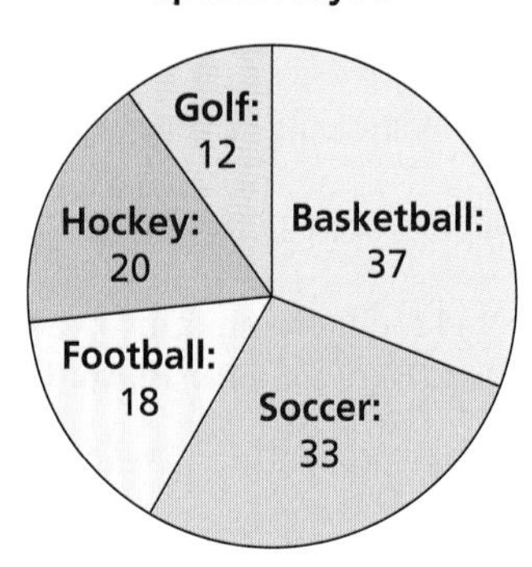

a. Explain why the graph is misleading.

b. What type of data display would be more appropriate for the data? Explain.

30. **THOUGHT PROVOKING** Use a spreadsheet program to create a type of data display that is not used in this section.

31. **REASONING** What type of data display shows the mode of a data set?

Maintaining Mathematical Proficiency

Reviewing what you learned in previous grades and lessons

Determine whether the relation is a function. Explain. *(Section 3.1)*

32. $(-5, -1), (-6, 0), (-5, 1), (-2, 2), (3, 3)$

33. $(0, 1), (4, 0), (8, 1), (12, 2), (16, 3)$

11.4–11.5 What Did You Learn?

Core Vocabulary

two-way table, *p. 610*
joint frequency, *p. 610*
marginal frequency, *p. 610*
joint relative frequency, *p. 611*
marginal relative frequency, *p. 611*
conditional relative frequency, *p. 612*
qualitative (categorical) data, *p. 618*
quantitative data, *p. 618*
misleading graph, *p. 620*

Core Concepts

Section 11.4

Joint and Marginal Frequencies, *p. 610*
Making Two-Way Tables, *p. 611*
Relative Frequencies, *p. 611*
Conditional Relative Frequencies, *p. 612*
Recognizing Associations in Data, *p. 613*

Section 11.5

Types of Data, *p. 618*
Choosing and Creating Appropriate Data Displays, *p. 619*
Analyzing Misleading Graphs, *p. 620*

Mathematical Practices

1. Consider the data given in the two-way table for Exercises 5–8 on page 614. Your sophomore friend responded to the survey. Is your friend more likely to have responded "yes" or "no" to buying a lunch? Explain.
2. Use your answer to Exercise 28 on page 622 to explain why it is important for a company manager to see accurate graphs.

Performance Task

College Student Study Time

Data from a small survey at a state university could provide insight into the amount of study time necessary to be successful in college. Based on the information you find when you organize the data, what advice should you give your peers? How will you support your conclusions?

To explore the answers to these questions and more, go to ***BigIdeasMath.com***.

11 Chapter Review

Dynamic Solutions available at *BigIdeasMath.com*

11.1 Measures of Center and Variation *(pp. 585–592)*

The table shows the number of miles you ran each day for 10 days. Find the mean, median, and mode of the distances.

Miles Run	
3.5	4.1
4.0	4.3
4.4	4.5
3.9	2.0
4.3	5.0

Mean $\bar{x} = \dfrac{3.5 + 4.0 + 4.4 + 3.9 + 4.3 + 4.1 + 4.3 + 4.5 + 2.0 + 5.0}{10} = 4$

Median 2.0, 3.5, 3.9, 4.0, 4.1, 4.3, 4.3, 4.4, 4.5, 5.0 Order the data.

$\dfrac{8.4}{2} = 4.2$ Mean of two middle values

Mode 2.0, 3.5, 3.9, 4.0, 4.1, 4.3, 4.3, 4.4, 4.5, 5.0 4.3 occurs most often.

▶ The mean is 4 miles, the median is 4.2 miles, and the mode is 4.3 miles.

1. Use the data in the example above. You run 4.0 miles on Day 11. How does this additional value affect the mean, median, and mode? Explain.
2. Use the data in the example above. You run 10.0 miles on Day 11. How does this additional value affect the mean, median, and mode? Explain.

Find the mean, median, and mode of the data.

3.

Ski Resort Temperatures (°F)		
11	3	3
0	−9	−2
10	10	10

4.

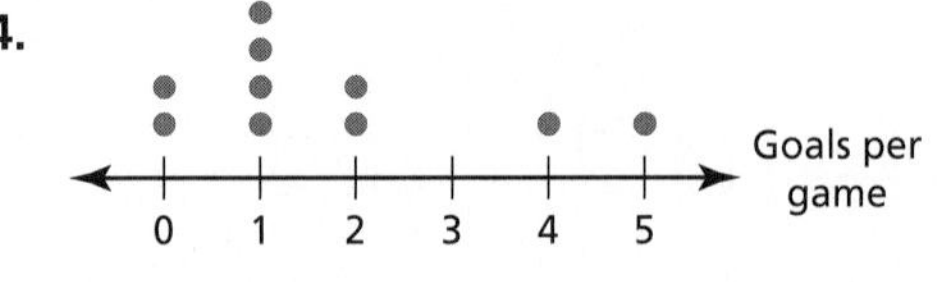

Find the range and standard deviation of each data set. Then compare your results.

5.

Bowling Scores			
Player A		Player B	
205	190	228	205
185	200	172	181
210	219	154	240
174	203	235	235
194	230	168	192

6.

Tablet Prices			
Store A		Store B	
\$140	\$180	\$225	\$310
\$200	\$250	\$260	\$190
\$150	\$190	\$190	\$285
\$250	\$160	\$160	\$240

Find the values of the measures shown after the given transformation.

Mean: 109 Median: 104 Mode: 96 Range: 45 Standard deviation: 3.6

7. Each value in the data set increases by 25.
8. Each value in the data set is multiplied by 0.6.

11.2 Box-and-Whisker Plots (pp. 593–598)

Make a box-and-whisker plot that represents the weights (in pounds) of pumpkins sold at a market.

16, 20, 11, 15, 10, 8, 8, 19, 11, 9, 16, 9

Step 1 Order the data. Find the median and the quartiles.

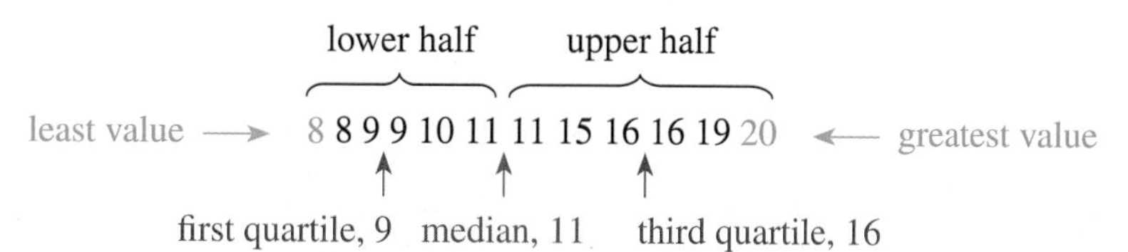

Step 2 Draw a number line that includes the least and greatest values. Graph points above the number line for the five-number summary.

Step 3 Draw a box using Q_1 and Q_3. Draw a line through the median. Draw whiskers from the box to the least and greatest values.

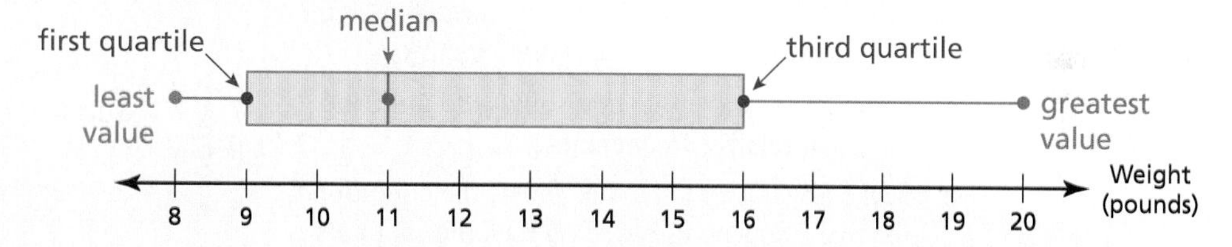

Make a box-and-whisker plot that represents the data. Identify the shape of the distribution.

9. Ages of volunteers at a hospital:
14, 17, 20, 16, 17, 14, 21, 18, 22

10. Masses (in kilograms) of lions:
120, 230, 180, 210, 200, 200, 230, 160

11.3 Shapes of Distributions (pp. 599–606)

The histogram shows the amounts of money a group of adults have in their pockets. Describe the shape of the distribution. Which measures of center and variation best represent the data?

▶ The distribution is skewed left. So, use the median to describe the center and the five-number summary to describe the variation.

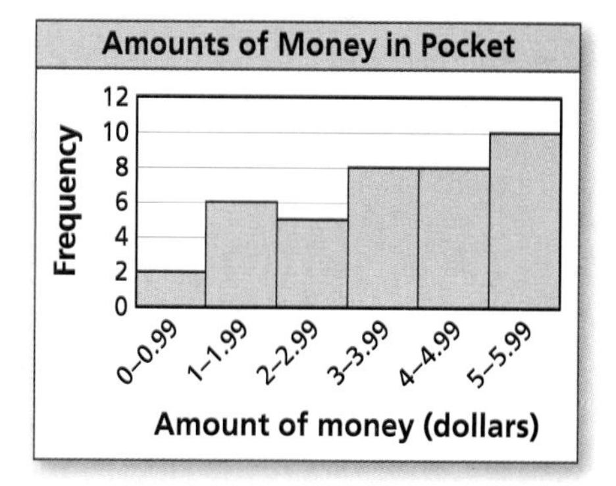

Amount	Frequency
0–0.99	9
1–1.99	10
2–2.99	9
3–3.99	7
4–4.99	4
5–5.99	1

11. The frequency table shows the amounts (in dollars) of money the students in a class have in their pockets.

a. Display the data in a histogram. Describe the shape of the distribution.

b. Which measures of center and variation best represent the data?

c. Compare this distribution with the distribution shown above using their shapes and appropriate measures of center and variation.

11.4 Two-Way Tables (pp. 609–616)

You conduct a survey that asks 130 students about whether they have an after-school job. Sixty males respond, 38 of which have a job. Twenty-six females do not have a job. Organize the results in a two-way table. Find and interpret the marginal frequencies.

		After-School Job		
		Yes	No	Total
Gender	Males	38	22	60
	Females	44	26	70
	Total	82	48	130

60 males responded.

70 females responded.

130 students were surveyed.

82 students have a job.

48 students do not have a job.

12. The two-way table shows the results of a survey that asked shoppers at a mall about whether they like the new food court.

a. Make a two-way table that shows the joint and marginal relative frequencies.

b. Make a two-way table that shows the conditional relative frequencies based on the column totals.

		Food Court	
		Like	Dislike
Shoppers	Adults	21	79
	Teenagers	96	4

11.5 Choosing a Data Display (pp. 617–622)

Analyze the data and then create a display that best represents the data.

Ages of U.S. Presidents at Inauguration										
57	61	57	57	58	57	61	54	68	51	49
64	50	48	65	52	56	46	54	49	51	47
55	55	54	42	51	56	55	51	54	51	60
62	43	55	56	61	52	69	64	46	54	47

A stem-and-leaf plot is one appropriate way to display this quantitative data. It orders numerical data and shows how they are distributed.

Ages of U.S. Presidents at Inauguration

Stem	Leaf
4	2 3 6 6 7 7 8 9 9
5	0 1 1 1 1 1 2 2 4 4 4 4 4 5 5 5 5 6 6 6 7 7 7 7 8
6	0 1 1 1 2 4 4 5 8 9

Key: $5|0 = 50$

13. Analyze the data in the table at the right and then create a display that best represents the data. Explain your reasoning.

Perfect Attendance	
Class	Number of students
freshman	84
sophomore	42
junior	67
senior	31

Tell whether the data are *qualitative* or *quantitative*. Explain.

14. heights of the members of a basketball team

15. grade level of students in an elementary school

11 Chapter Test

Describe the shape of the data distribution. Then determine which measures of center and variation best represent the data.

1.

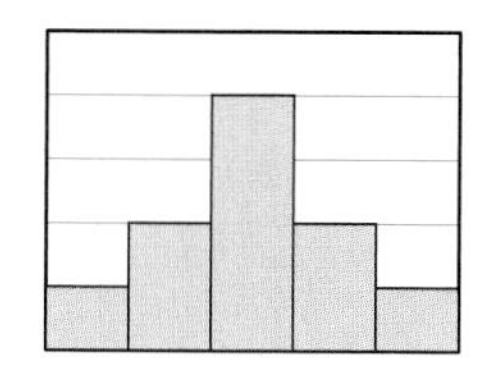

2.

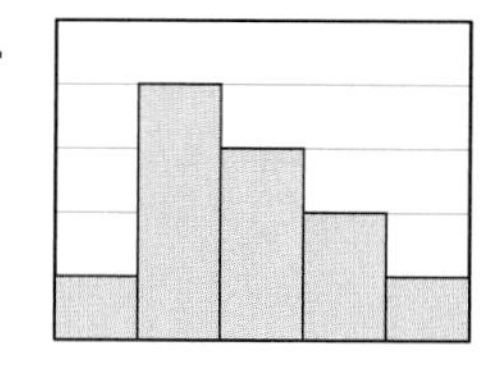

3.

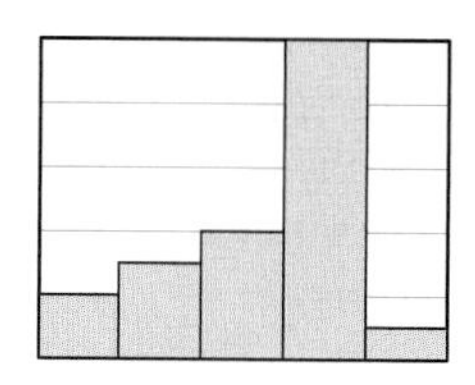

4. Determine whether each statement is *always*, *sometimes*, or *never* true. Explain your reasoning.

a. The sum of the marginal relative frequencies in the "total" row and the "total" column of a two-way table should each be equal to 1.

b. In a box-and-whisker plot, the length of the box to the left of the median and the length of the box to the right of the median are equal.

c. Qualitative data are numerical.

5. Find the mean, median, mode, range, and standard deviation of the prices.

Prices of Shirts at a Clothing Store			
\$15.50	\$18.90	\$10.60	\$12.25
\$7.80	\$23.50	\$9.75	\$21.70

6. Repeat Exercise 5 when all the shirts in the clothing store are 20% off.

7. Which data display best represents the data, a histogram or a stem-and-leaf plot? Explain.

15, 21, 18, 10, 12, 11, 17, 18, 16, 12, 20, 12, 17, 16

8. The tables show the battery lives (in hours) of two brands of laptops.

a. Make a double box-and-whisker plot that represents the data.

b. Identify the shape of each distribution.

c. Which brand's battery lives are more spread out? Explain.

d. Compare the distributions using their shapes and appropriate measures of center and variation.

Brand A	
20.75	18.5
13.5	16.25
8.5	13.5
14.5	15.5
11.5	16.75

Brand B	
10.5	12.5
9.5	10.25
9.0	9.75
8.5	8.5
9.0	7.0

Preferred method of exercise	Number of students
walking	20
jogging	28
biking	17
swimming	11
lifting weights	10
dancing	14

9. The table shows the results of a survey that asked students their preferred method of exercise. Analyze the data and then create a display that best represents the data. Explain your reasoning.

10. You conduct a survey that asks 271 students in your class whether they are attending the class field trip. One hundred twenty-one males respond, 92 of which are attending the field trip. Thirty-one females are not attending the field trip.

a. Organize the results in a two-way table. Find and interpret the marginal frequencies.

b. What percent of females are attending the class field trip?

11 Cumulative Assessment

1. You ask all the students in your grade whether they have a cell phone. The results are shown in the two-way table. Your friend claims that a greater percent of males in your grade have cell phones than females. Do you support your friend's claim? Justify your answer.

		Cell Phones	
		Yes	No
Gender	Male	27	12
	Female	31	17

2. Use the graphs of the functions to answer each question.

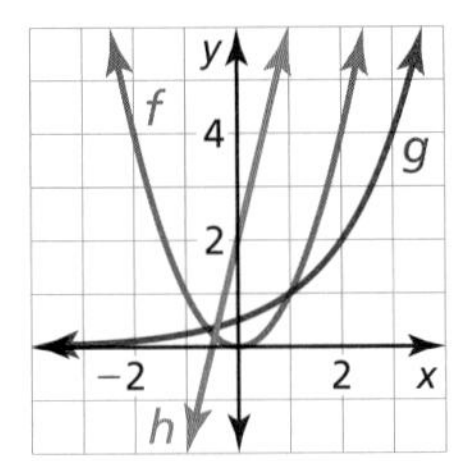

   a. Are there any values of x greater than 0 where $f(x) > h(x)$? Explain.

   b. Are there any values of x greater than 1 where $g(x) > f(x)$? Explain.

   c. Are there any values of x greater than 0 where $g(x) > h(x)$? Explain.

3. Classify the shape of each distribution as *symmetric*, *skewed left*, or *skewed right*.

   a. 

   b. 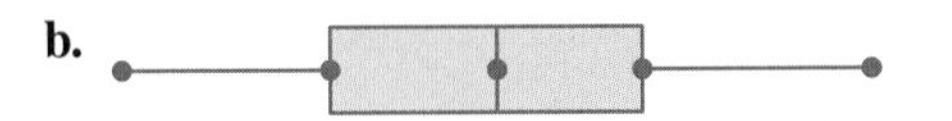

   c. 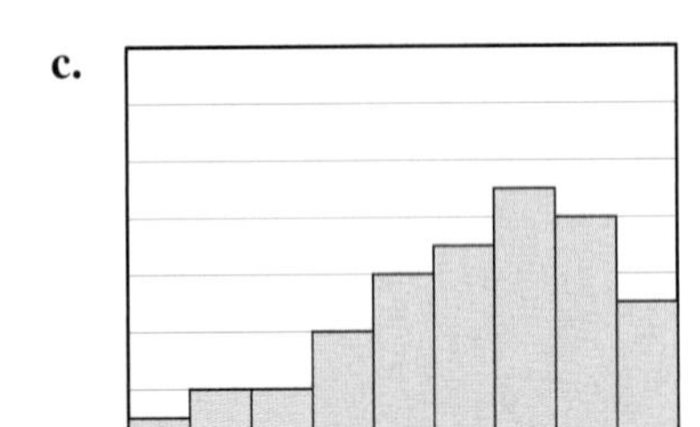

   d. 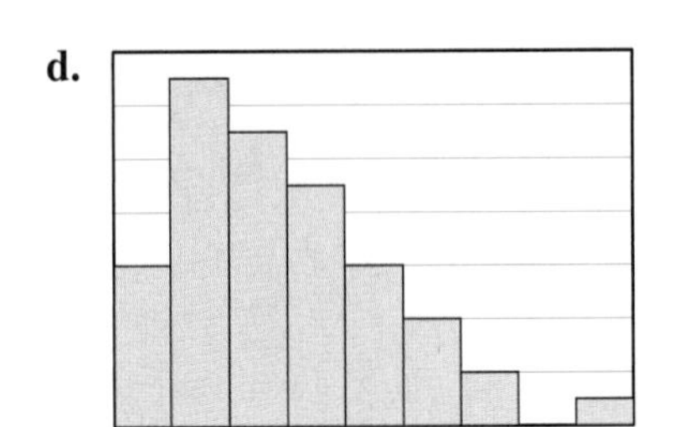

4. Complete the equation so that the solutions of the system of equations are $(-2, 4)$ and $(1, -5)$.

$$y = \square x + \square$$
$$y = 2x^2 - x - 6$$

5. Pair each function with its inverse.

$y = -3x^2, x \geq 0$	$y = -x + 7$	$y = 2x - 4$	$y = \sqrt{-\frac{1}{3}x}$
$y = \frac{1}{2}x + 2$	$y = x^2 - 5, x \geq 0$	$y = \sqrt{x + 5}$	$y = -x + 7$

6. The box-and-whisker plot represents the lengths (in minutes) of project presentations at a science fair. Find the interquartile range of the data. What does this represent in the context of the situation?

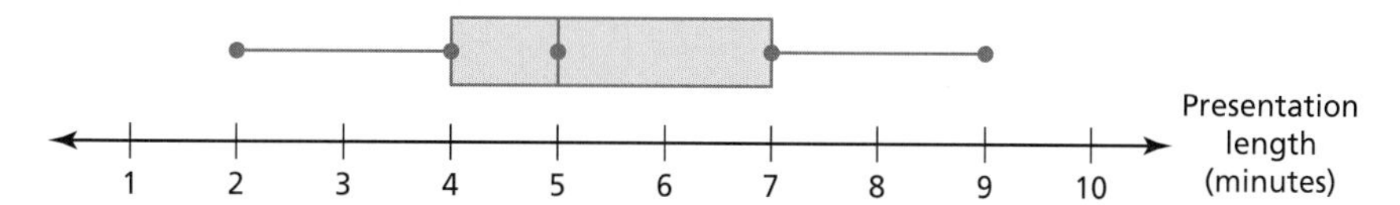

Ⓐ 7; The middle half of the presentation lengths vary by no more than 7 minutes.

Ⓑ 3; The presentation lengths vary by no more than 3 minutes.

Ⓒ 3; The middle half of the presentation lengths vary by no more than 3 minutes.

Ⓓ 7; The presentation lengths vary by no more than 7 minutes.

7. Scores in a video game can be between 0 and 100. Use the data set shown to fill in a value for x so that each statement is true.

a. When $x =$ _____, the mean of the scores is 45.5.

b. When $x =$ _____, the median of the scores is 47.

c. When $x =$ _____, the mode of the scores is 63.

d. When $x =$ _____, the range of the scores is 71.

Video Game Scores	
36	28
48	x
42	57
63	52

8. Select all the numbers that are in the range of the function shown.

$$y = \begin{cases} x^2 + 4x + 7, & \text{if } x \leq -1 \\ \frac{1}{2}x + 2, & \text{if } x > -1 \end{cases}$$

0	$\frac{1}{2}$	1	$1\frac{1}{2}$	2	$2\frac{1}{2}$	3	$3\frac{1}{2}$	4

9. A traveler walks and takes a shuttle bus to get to a terminal of an airport. The function $y = D(x)$ represents the traveler's distance (in feet) after x minutes.

a. Estimate and interpret $D(2)$.

b. Use the graph to find the solution of the equation $D(x) = 3500$. Explain the meaning of the solution.

c. How long does the traveler wait for the shuttle bus?

d. How far does the traveler ride on the shuttle bus?

e. What is the total distance that the traveler walks before and after riding the shuttle bus?

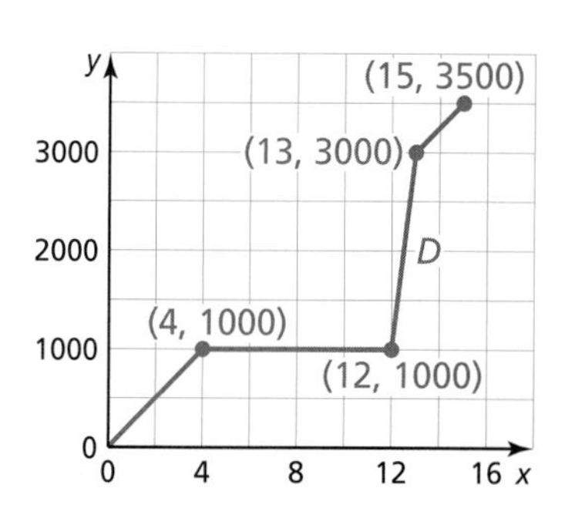

Selected Answers

Chapter 1

Chapter 1 Maintaining Mathematical Proficiency *(p. 1)*

1. −7 **2.** −13 **3.** 8 **4.** 32 **5.** −7 **6.** 2
7. 22 **8.** 5 **9.** −4 **10.** −24 **11.** 63
12. −28 **13.** 4 **14.** −8 **15.** −4 **16.** 48
17. 6 **18.** 12

19. a. If the signs are the same, add the absolute values and attach the sign. If the signs are different, subtract the absolute values and attach the sign of the number with the greatest absolute value; *Sample answer:* $-6 + 2 = -4$

b. Add the opposite; *Sample answer:* $5 - (-3) = 5 + 3 = 8$

c. Multiply the absolute values. If the signs are the same, then the product is positive. If the signs are different, then the product is negative; *Sample answer:* $(-6)(-4) = 24$

d. Divide the absolute values. If the signs are the same, then the quotient is positive. If the signs are different, then the quotient is negative; *Sample answer:* $-15 \div 3 = -5$

1.1 Vocabulary and Core Concept Check *(p. 8)*

1. + and −; × and ÷

3. Division Property of Equality; Divide each side by 14.

1.1 Monitoring Progress and Modeling with Mathematics *(pp. 8–10)*

5. $x = 3$; Subtract 5 from each side.

7. $y = 7$; Add 4 to each side.

9. $w = -7$; Subtract 3 from each side.

11. $p = -3$; Add 11 to each side.

13. $r = 18$; Add 8 to each side. **15.** $p - 12.95 = 44$; \$56.95

17. $x + 100 + 120 + 100 = 360$; $x = 40$

19. $x + 76 + 92 + 122 = 360$; $x = 70$

21. $g = 4$; Divide each side by 5.

23. $p = 15$; Multiply each side by 5.

25. $r = -8$; Divide each side by −8.

27. $x = 48$; Multiply each side by 6.

29. $s = -6$; Divide each side by 9. **31.** $t = -1$

33. $m = 14$ **35.** $a = 5.6$ **37.** $j = -18$

39. Subtract −0.8 from each side, not add; $r = 12.6 - (-0.8)$; $r = 13.4$

41. C; Multiplying the number of eggs in each carton by the number of cartons will give the total number of eggs; 9 cartons

43. $9.5 = 1.9w$; 5 ft **45.** Multiplication Property of Equality

47. a. $4p = 30.40$; \$7.60

b. no; Each CD costs \$9.50 at the regular price, so 3 CDs would cost \$28.50 which is greater than \$25.

49. a. 5; 10 **b.** −2; 9

51. 71; Because $\frac{1}{6}$ of the girls is 6, there are 36 girls. Because $\frac{2}{7}$ of the boys is 10, there are 35 boys; $36 + 35 = 71$

53. $B = 12\pi \text{ in.}^2$ **55.** $B = 9\pi \text{ m}^2$

57. a. 132 hits

b. no; Dividing by a larger number of at-bats decreases the value of the average.

1.1 Maintaining Mathematical Proficiency *(p. 10)*

59. $\frac{5}{6}x + \frac{15}{4}$ **61.** $8p + 16q + 24$

63. 0.02 **65.** 298.26

1.2 Vocabulary and Core Concept Check *(p. 16)*

1. like terms

1.2 Monitoring Progress and Modeling with Mathematics *(pp. 16–18)*

3. $w = 4$ **5.** $q = 1$ **7.** $z = -32$ **9.** $h = 4$

11. $y = 4$ **13.** $v = 3$ **15.** 6 min **17.** $z = 3$

19. $m = 3$ **21.** $c = 5$ **23.** $x = 29$

25. $k = 45$; 45°, 90°, 45°

27. $b = 90$; 90°, 135°, 90°, 90°, 135°

29. $2n + 13 = 75$; $n = 31$ **31.** $8 + \frac{n}{3} = -2$; $n = -30$

33. $6(n + 15) = -42$; $n = -22$

35. $30(8.75) + 11t = 400$; 12.5 h

37. $1.08(2t + 2.50) + 3 = 13.80$; \$3.75

39. Distributive Property; Simplify; Combine like terms; Subtract 6 from each side; Divide each side by 3.

41. In the third step, the right side should be 8 × 4, not 8 ÷ 4; $x - 2 = 32$; $x = 34$

43. $2y + 2 \cdot \frac{11}{8}y = 190$, $y = 40$; 55 in. by 40 in.

45. $x = \frac{15}{16}$; *Sample answer:* method 1; There are no fractions until the last step.

47. no; Solving the equation $0.25(d + 8) + 0.10d = 2.80$ results in the number of dimes not being a whole number.

49. 16, 18, 20; The next consecutive even integers after $2n$ are $2n + 2$ and $2n + 4$. Solve the equation $2n + (2n + 2) + (2n + 4) = 54$. Then substitute the solution into the expressions for the integers.

51. $x = -\frac{7}{b}$ **53.** $x = \frac{12.5 + b}{a}$ **55.** $x = -\frac{8}{b}$

1.2 Maintaining Mathematical Proficiency *(p. 18)*

57. $m + 5$ **59.** −2 **61.** b **63.** a **65.** a

1.3 Vocabulary and Core Concept Check *(p. 23)*

1. no; Solving the equation gives a statement that is never true, not one that is always true.

1.3 Monitoring Progress and Modeling with Mathematics *(pp. 23–24)*

3. $x = 3$ **5.** $p = 7$ **7.** $t = -1$ **9.** $x = \frac{1}{2}$

11. $g = -4$ **13.** $x = -3$ **15.** $y = -12$ **17.** 2 h

19. no solution **21.** $h = 3$; one solution

23. infinitely many solutions

25. In the second step, you should add $3c$ to each side; $8c - 6 = 4$, $8c = 10$, $c = \frac{5}{4}$

27. $60 + 42.95x = 25 + 49.95x$; 5th month

29. $r = -2$

31. $x = 10$; $S = 62.5\pi$ cm^2 or about 196.35 cm^2; $V = 62.5\pi$ cm^3 or about 196.35 cm^3

33. 4 sec **35.** $a = 5$; Both sides simplify to $10x + 15$.

37. 11, 12

39. **a.** *Sample answer:* $3x + 12 = 2x + x$; simplifies to a statement that is never true

b. *Sample answer:* $5x + 3 = 2x + 3 + 3x$; simplifies to a statement that is always true

1.3 Maintaining Mathematical Proficiency *(p. 24)*

41. $-4, |2|, |-4|, 5, 9$ **43.** $-19, -18, |-18|, |22|, |-24|$

1.4 Vocabulary and Core Concept Check *(p. 32)*

1. an apparent solution that must be rejected because it does not satisfy the original equation

1.4 Monitoring Progress and Modeling with Mathematics *(pp. 32–34)*

3. 9 **5.** 0 **7.** -35 **9.** 9

11. $w = -6, w = 6$;

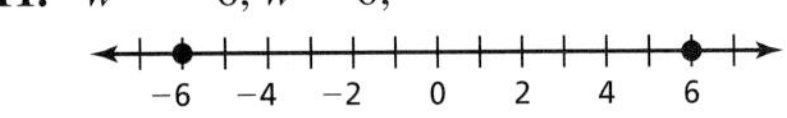

13. no solution

15. $m = -10, m = 4$;

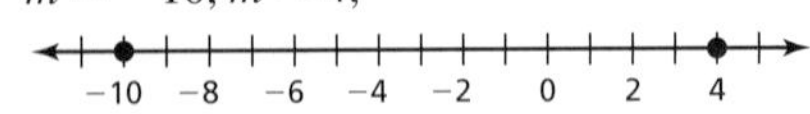

17. $d = -5, d = 5$;

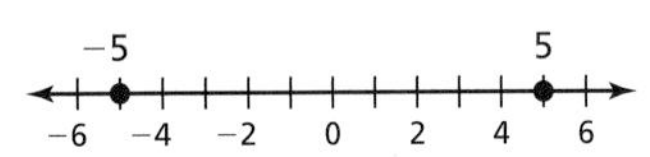

19. $b = -3.5, b = 6$;

21. no solution **23.** $s = 20$;

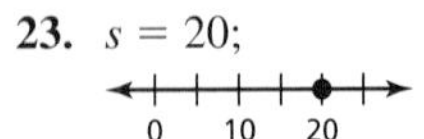

25. **a.** 91.4 94.5

91 92 93 94 95

b. $|d - 92{,}950{,}000| = 1{,}550{,}000$

27. B **29.** C **31.** $|x - 13| = 5$

33. $|x - 5.5| = 3.5$ **35.** $n = 3, n = 5$ **37.** $b = 3, b = 5$

39. $p = \frac{2}{3}, p = 10$ **41.** $h = 0.25$ **43.** $f = -1$

45. 5 sec, 7.5 sec

47. **a.** $|x - 32| = 5$; 27%, 37%

b. no; $\frac{1}{3}\left(33\frac{1}{3}\%\right)$ falls within the range of possible values.

49. The absolute value cannot be negative. So, there is no solution.

51. No solution: $|x - 2| + 6 = 0$, $|x - 6| - 5 = -9$
One solution: $|x - 1| + 4 = 4$, $|x + 5| - 8 = -8$
Two solutions: $|x + 8| + 2 = 7$, $|x + 3| - 1 = 0$

53. always; Square roots of the same number have the same absolute value.

55. sometimes; The equation will only have two solutions if p is positive.

57. Absolute value equations will have no solution when the absolute value is equal to a negative number, one solution when the absolute value is equal to zero, and two solutions when the absolute value is equal to a positive number; *Sample answer:* $|x + 12| = -2$ has no solution, $|x + 12| = 0$ has one solution, and $|x + 12| = 2$ has two solutions.

59. $x = 1, x = -5$; Interpret $|x + 2|$ as a single quantity and solve for it, then solve the resulting absolute value equation.

61. 1; If $c = d$, then the absolute value expression will be equal to 0; 2; If $c > d$, then $d - c$ will be negative. When this value is divided by a negative value of a, the result will be positive.

1.4 Maintaining Mathematical Proficiency *(p. 34)*

63. Division Property of Equality **65.** 6 in. **67.** 9 cm

1.5 Vocabulary and Core Concept Check *(p. 40)*

1. no; It only has one variable.

1.5 Monitoring Progress and Modeling with Mathematics *(pp. 40–42)*

3. $y = 13 + 3x$ **5.** $y = -13 + 9x$ **7.** $y = 9x - 45$

9. $y = x - 3$ **11.** $y = 18x + 12$ **13.** $x = \frac{1}{12}y$

15. $x = \dfrac{a}{2 + 6z}$ **17.** $x = \dfrac{y - 6}{4 + r}$ **19.** $x = \dfrac{r}{s + t}$

21. $x = \dfrac{y - 12}{-5 - 4k}$

23. **a.** $x = \dfrac{C - 60}{85}$ **b.** 3 trips; 5 trips

25. The equation is not solved for x because there is still a term with x on both sides; $x = y - x + 6$; $2x = y + 6$; $x = \dfrac{y + 6}{2}$

27. $C = R - P$ **29.** $b_2 = \dfrac{2A}{h} - b_1$

31. $C = A\left(\dfrac{R}{5} + 0.3\right)$ **33.** **a.** $r = \dfrac{L - S}{L}$ **b.** 0.4

35. 6.25 yr

37. **a.** $P = 2x + 2\pi r$ **b.** $x = \frac{1}{2}P - \pi r$ **c.** 173 ft

39. **a.** $r = \dfrac{C}{2\pi}$ **b.** 1.1 ft; 1.3 ft; 1.4 ft

c. First find the radius using the formula from part (a), then substitute this into the formula for the area of a circle.

41. no; 70°F is about 21.1°C, which is greater than 20°C.

43. $A = \dfrac{5}{2}bh$; $h = \dfrac{2A}{5b}$ **45.** $a = \dfrac{b + c}{bx - 1}$

1.5 Maintaining Mathematical Proficiency *(p. 42)*

47. 35 **49.** 47

51. $x = -2, x = 8$; **53.** no solution

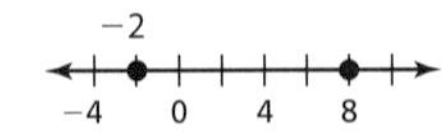

Chapter 1 Review *(pp. 44–46)*

1. $z = -9$; Subtract 3 from each side.

2. $t = -13$; Divide each side by -0.2.

3. $n = 10$; Multiply each side by -5. **4.** $y = -9$

5. $b = -5$ **6.** $n = 6$ **7.** $z = -5$ **8.** $x = 18$

9. $w = \frac{25}{4}$ **10.** $x = 10$; 110°, 50°, 20°

11. $x = 126$; 126°, 96°, 126°, 96°, 96° **12.** $n = -4$

13. infinitely many solutions **14.** no solution

15. $y = 14, y = -20$ **16.** $w = 3, w = -\frac{1}{5}$ **17.** $x = -1$

18. $|v - 84.5| = 10.5$ **19.** $y = \frac{1}{2}x - 5$ **20.** $y = 2x - 2$

21. $y = \dfrac{a}{9 + 3x}$ **22. a.** $h = \dfrac{3V}{B}$ **b.** 18 cm

23. a. $K = \frac{5}{9}(F - 32) + 273.15$ **b.** about 355.37 K

Chapter 2

Chapter 2 Maintaining Mathematical Proficiency *(p. 51)*

1. **2.**

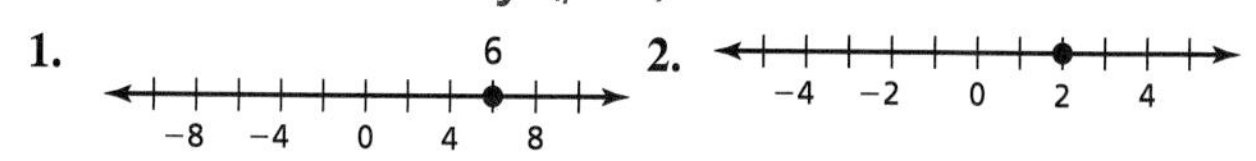

3. **4.**

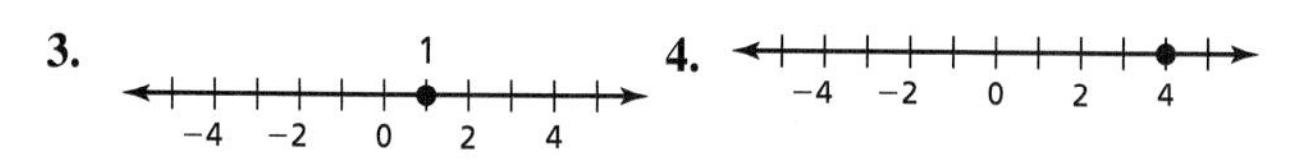

5. **6.**

7. < **8.** < **9.** < **10.** > **11.** =

12. < **13.** $-b < -a$

2.1 Vocabulary and Core Concept Check *(p. 58)*

1. inequality

3. Draw an open circle when a number is not part of the solution. Draw a closed circle when a number is part of the solution. Draw an arrow to the left or right to show that the graph continues in that direction.

2.1 Monitoring Progress and Modeling with Mathematics *(pp. 58–60)*

5. $x > 3$ **7.** $15 \leq \frac{t}{5}$ **9.** $\frac{1}{2}y > 22$ **11.** $13 \geq v - 1$

13. $w \geq 1.7$ **15.** no **17.** yes **19.** yes

21. yes **23.** yes

25. a. $h < 107$ in.

b. no; A height of 9 feet is equal to 108 inches, which is not less than 107 inches.

27. Because -1 is not less than -4, the final result is not true; $-1 \not< -4$; 8 is not in the solution set.

29. **31.**

33. **35.**

37. $x < 7$ **39.** $1.3 \leq z$

41. $x \leq 4$ **43.** $x > 3$

45. C; The temperature must be at least 2°F warmer, so the increase is represented by $x \geq 2$.

47. $\ell < 4800$

49. *Sample answer:* You spend \$23 on admission and x dollars on snacks, and you can spend no more than \$31 total.

51. $0.90x \leq 24$; yes; Because $0.9(25) = \$22.50$, which is less than \$24, the inequality is true.

53. *Sample answer:* A temperature above the freezing point of water can be represented by $T > 0$ if the temperature is in degrees Celsius, or by $T > 32$ if the temperature is in degrees Fahrenheit.

55. $x < 14$ m **57.** $x < 3$ cm

59. a. $r > \dfrac{40}{7}$ (about 5.71)

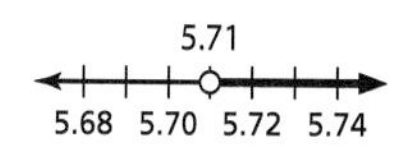

b. no; The graph includes speeds beyond the maximum speed a human can run.

2.1 Maintaining Mathematical Proficency *(p. 60)*

61. $y = 14$ **63.** $y = -1$ **65.** $x = \dfrac{s - 2r}{3}$

67. $x = n - \frac{1}{2}$

2.2 Vocabulary and Core Concept Check *(p. 65)*

1. Subtraction Property of Inequality

2.2 Monitoring Progress and Modeling with Mathematics *(pp. 65–66)*

3. subtract 11 **5.** add 9

7. $x < -1$ **9.** $m \leq 7$

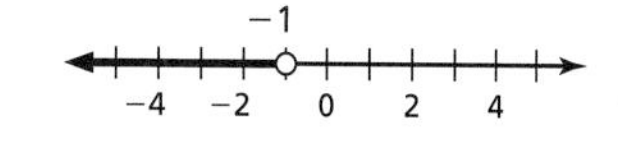

11. $r < 1$ **13.** $w > -2$

15. $h \geq 8$ **17.** $j < 2$

19. $p \leq 17$

21. $n + 8 > 11; n > 3$ **23.** $n - 9 < 4; n < 13$

25. a. $38 + w \leq 50; w \leq 12$

b. no; The total being added is 14 pounds, which is not a solution of the inequality found in part (a).

27. The graph is going in the wrong direction.

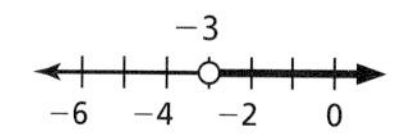

29. 33 or more goals

31. A; Subtract 3 from each side; D; Add b, $-x$, and -3 to each side.

33. $6.4 + 4.9 + 4.1 + x \leq 18.7; x \leq 3.3$

35. no; *Sample answer:* 3, 7, 8, 9, and 12; There are infinitely many solutions. Check 8 and a few numbers greater than and less than 8.

37. a. $x \geq 6$

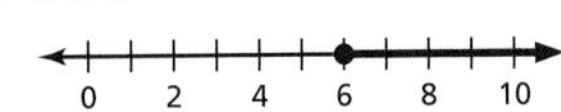

b. $x < 17.7$

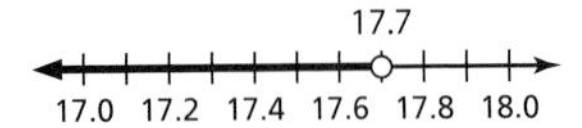

2.2 Maintaining Mathematical Proficiency *(p. 66)*

39. -63 **41.** 9 **43.** $x = 4$ **45.** $s = -104$

2.3 Vocabulary and Core Concept Check *(p. 71)*

1. When solving $2x < -8$, the inequality symbol is not reversed when dividing each side by 2. When solving $-2x < 8$, the inequality is reversed when dividing each side by -2.

2.3 Monitoring Progress and Modeling with Mathematics *(p. 71–72)*

3. $x < 2$

5. $n \geq -2$

7. $x > -4$

9. $w \leq 25$

11. $t > -2$

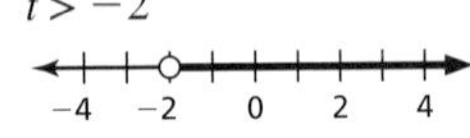

13. $z \geq 5$

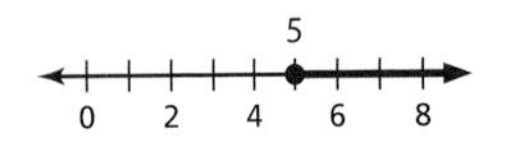

15. $n \leq -3$

17. $m < 32$

19. $5p \leq 12, p \leq 2.4$ **21.** $y > 12$

23. $x \leq -9$ **25.** $x > \frac{3}{8}$

27. The inequality should not be reversed when multiplying each side by $\frac{3}{2}$; $\frac{3}{2} \cdot (-6) > \frac{3}{2} \cdot \frac{2}{3}x$; $-\frac{18}{2} > x$; $-9 > x$; $x < -9$; The solution is $x < -9$.

29. $(14 \cdot 14)c \leq 700$; ft, ft, dollars

$196c \leq 700$; ft^2, dollars

$c \leq 3.57$; dollars/ft^2

31. a. $d \leq 6.3(2), d \leq 12.6$

b. yes; The distance traveled in 4 hours would be no more than 25.2 miles, which is less than the distance required for a marathon.

33. more than 300 million pennies

35. a. $A > B$ or $B < A$ **b.** $-A < -B$ or $-B > -A$

c. As numbers move farther away from zero, their absolute value becomes larger. $A > B$ and $|A| > |B|$. $-A < -B$ and $|A| > |B|$.

37. $\frac{C}{2\pi} > 5, C > 10\pi$

39. $36p \geq 12, p \geq \frac{1}{3}$

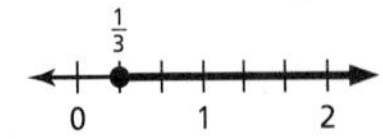

2.3 Maintaining Mathematical Proficiency *(p. 72)*

41. $y = -4$ **43.** $z = 6$ **45.** $\frac{16}{30}$ **47.** $\frac{2}{3}$

2.4 Vocabulary and Core Concept Check *(p. 77)*

1. *Sample answer:* The same steps can be applied when solving multi-step inequalities and multi-step equations, except that when each side of an inequality is divided by a negative number, the inequality must be reversed.

2.4 Monitoring Progress and Modeling with Mathematics *(pp. 77–78)*

3. B **5.** C

7. $x > 5$

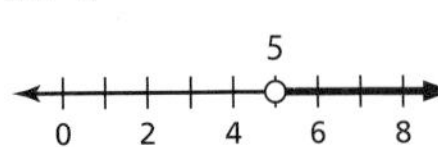

9. $v \leq 2$

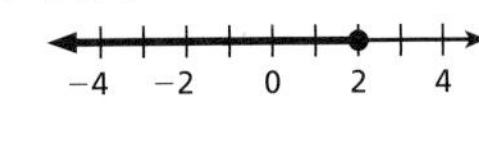

11. $w > 2$

13. $p < -32$

15. $a \geq -3$

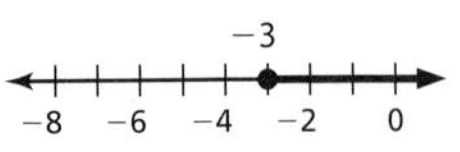

17. $m > 3$ **19.** $d > -2$ **21.** all real numbers

23. no solution **25.** no solution **27.** all real numbers

29. In the first step, you need to use the Distributive Property on the left side; $x + 24 \geq 12$; $x \geq -12$

31. $20n + 100 \leq 320, n \leq 11$ **33.** $12(2x - 3) > 60; x > 4$

35. 7 stories; Using the Pythagorean Theorem, the 74-foot ladder can reach at most 70 feet. Solving the inequality $10n - 8 \leq 70$ gives $n \leq 7.8$, so the ladder cannot quite reach the 8th story.

37. $r \geq 3$ **39.** $a = 4$

2.4 Maintaining Mathematical Proficiency *(p. 78)*

41. $6y \leq 10$ **43.** $\frac{r}{7} \leq 18$

2.5 Vocabulary and Core Concept Check *(p. 85)*

1. The graph of $-6 \leq x \leq -4$ shows a single segment between -6 and -4. The graph of $x \leq -6$ or $x \geq -4$ shows two opposite rays with endpoints at -6 and -4.

2.5 Monitoring Progress and Modeling with Mathematics *(pp. 85–86)*

3. $-3 < x \leq 2$ **5.** $x \leq -7$ or $x \geq -4$

7. $2 < p < 6$

9. $m > -7\frac{2}{3}$ or $m \leq -10$

11. $-2500 \leq e \leq -100$

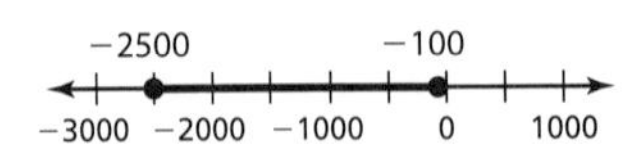

13. $1 < x \leq 6$

15. $v < -5$ or $v > 5$

17. $r < 2$ or $r \geq 7$

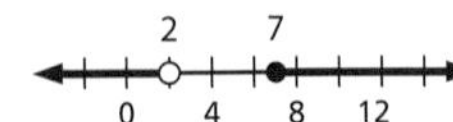

19. $-10 < x < 5$

21. In the second step, 3 should have been subtracted from 4 on the left side;
$1 < -2x < 6$; $-\frac{1}{2} > x > -3$

23. $-20 \le \frac{5}{9}(F - 32) \le -15$, $-4 \le F \le 5$ **25.** no solution

27. $y > 7$ **29.** all real numbers

31. $\le$

33. $7 + 5 > x \Rightarrow x < 12$, $7 + x > 5 \Rightarrow x > -2$, $5 + x > 7 \Rightarrow x > 2$; no; A value of 1 does not make the inequality $x > 2$ true.

2.5 Maintaining Mathematical Proficiency *(p. 86)*

35. $d = 54$, $d = -54$ **37.** $r = 3$, $r = \frac{1}{2}$

39. 4; The data values are clustered close together.

2.6 Vocabulary and Core Concept Check *(p. 91)*

1. yes; Because the absolute value must be positive, all values will be greater than -6.

2.6 Monitoring Progress and Modeling with Mathematics *(pp. 91–92)*

3. $-3 < x < 3$ **5.** $d < -12$ *or* $d > -6$

7. all real numbers **9.** no solution

11. $t \le -\frac{2}{3}$ *or* $t \ge 3$ **13.** $m > 20$ *or* $m < 8$

15. $-4 \le w \le -\frac{4}{3}$ **17.** $f < 3$ *or* $f > 3$

19. $|w - 500| \le 30$; 470 to 530 words

21. did not rewrite the absolute value inequality as a compound inequality; $-20 < x - 5 < 20$; $-15 < x < 25$

23. $|n| < 6$; $-6 < n < 6$ **25.** $|\frac{1}{2}n - 14| \le 5$; $18 \le n \le 38$

27. the gasket with a weight of 0.53 lb

29. $|\frac{1}{2} \cdot 4(x + 6) - 2 \cdot 6| < 2$; $-1 < x < 1$ **31.** true

33. false; It has to be a solution of $x + 3 \le -8$ *or* $x + 3 \ge 8$.

35. no; If n is 0, the statement is false.

37. no solution; An absolute value cannot be negative; all real numbers; An absolute value must be positive or zero, so it will always be greater than any negative number.

39. The solution of $|x| < 5$ is the set of values that make both of 2 inequalities true, the solution of $|x| > 5$ is the set of values that make either of 2 inequalities true.

2.6 Maintaining Mathematical Proficiency *(p. 92)*

41 and 43.

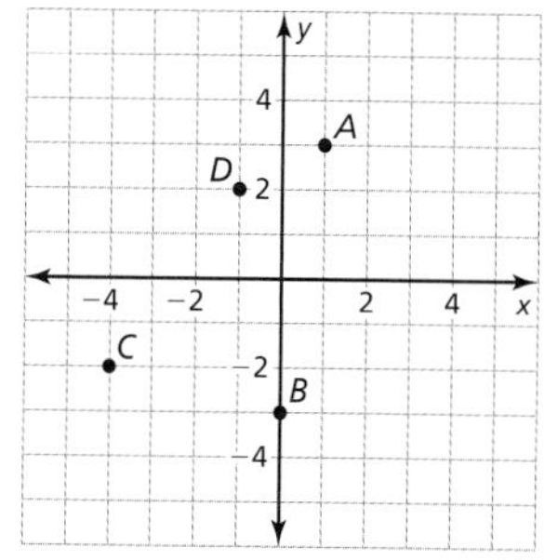

41. Quadrant I **43.** Quadrant III **45.** 1, 6, 11, 16, 21

Chapter 2 Review *(pp. 94–96)*

1. $d - 2 < -1$ **2.** $5h \le 10$

3. **4.**

5. **6.** $p < 6$

7. $r < -2$ **8.** $m \le 8.8$

9. $x > -7$ **10.** $g \ge -20$

11. $n \ge -4$ **12.** $s \le -88$

13. $q > 18$ **14.** $k < -5$

15. $x > 5$ **16.** $b > -26$

17. $n \ge 1$ **18.** $s \le \frac{14}{3}$

19. no solution **20.** all real numbers

21. $-6 < x \le 8$ **22.** $-2 \le z \le 6$

23. $r < -20$ *or* $r \ge -5$ **24.** $m \ge 10$ *or* $m \le -10$

25. no solution **26.** $3 \le f \le 9$

Selected Answers

27. $b < -12$ *or* $b > -4$

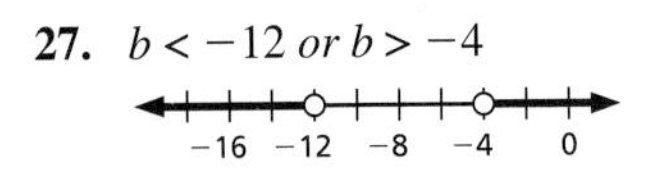

28. $-\frac{7}{3} < g < 1$

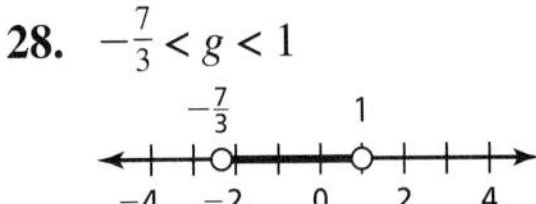

29. all real numbers

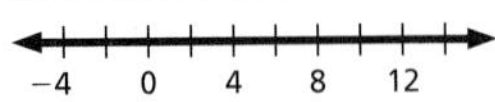

30. $|h - 106| \leq 7$; 99 cm to 113 cm

Chapter 3

Chapter 3 Maintaining Mathematical Proficiency *(p. 101)*

1–6.

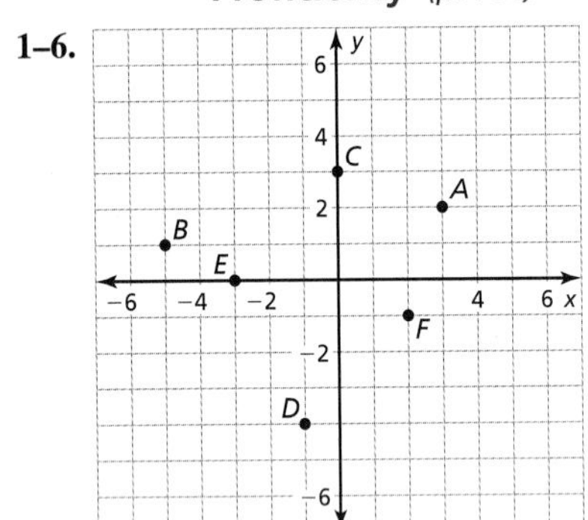

1. in Quadrant I
2. in Quadrant II
3. on positive y-axis
4. in Quadrant III
5. on negative x-axis
6. in Quadrant IV
7. 17
8. -7
9. 68
10. 34
11. 40
12. -6
13. Start at the origin. Move a units right and b units up. Plot the point; Start at the origin. Move a units left and b units up. Plot the point; Start at the origin. Move a units right and b units down. Plot the point; Start at the origin. Move a units left and b units down. Plot the point.

3.1 Vocabulary and Core Concept Check *(p. 108)*

1. The independent variable can be any value in the domain, but the dependent variable depends on the values of the independent variable.

3.1 Monitoring Progress and Modeling with Mathematics *(pp. 108–110)*

3. function; Every input has exactly one output.
5. not a function; The input 2 has two outputs, 3 and 2.
7. not a function; The input 16 has two outputs, -2 and 2, and the input 1 has two outputs, -1 and 1.
9. function; No vertical line can be drawn through more than one point on the graph.
11. not a function; A vertical line can be drawn through more than one point on the graph in many places, such as (4, 0) and (4, 6).
13. domain: $-2, -1, 0, 1, 2$; range: $-2, 0, 2$
15. domain: $-4 \leq x \leq 2$; range: $2 \leq y \leq 6$
17. **a.** y is the dependent variable and x is the independent variable.
    b. 500, 525, 550, 575, 600, 625
19. A function can have one output paired with two inputs, but cannot have one input paired with more than one output; The relation is a function. No input is paired with more than one output.
21. The amount of time is the dependent variable and the number of quarters is the independent variable.
23. **a.** *Sample answer:* The balance of the savings account is \$100 in month 0 and increases by \$25 per month through month 4.
    b. (0, 100), (1, 125), (2, 150), (3, 175), (4, 200)
    c.

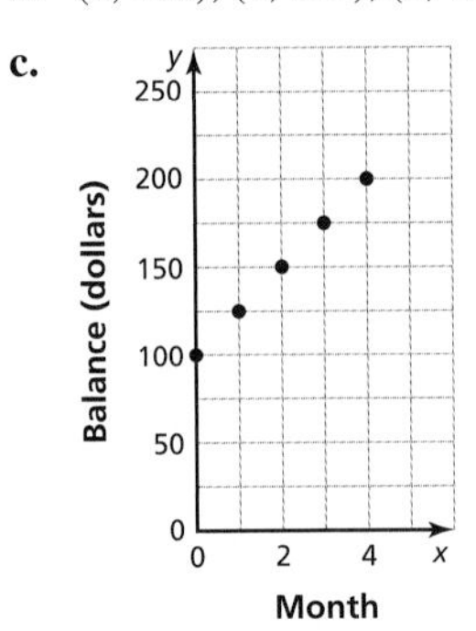

25. -2
27. **a.** Each letter-number combination is paired with exactly one food or drink item.
    b. The food item is the dependent variable and the letter-number combination is the independent variable.
    c. domain: A1, A2, A3, B1, B2, B3, B4, C1, C2, C3, C4; range: popcorn, nuts, pretzels, protein bar, granola bar, cereal, energy bar, orange juice, water, milk
29. no; A vertical line does not represent a function.
31. no; Items that cost the same to make could be sold for different prices.
33. yes; Each student has exactly one homeroom teacher.
35. true
37. false; More than one input can have the same output in a function, so reversing the values may not produce a function.
39. **a.** $P = h + 23$
    b. P is the dependent variable and h is the independent variable.
    c. domain: $3 < h < 23$; range: $26 < P < 46$
41. domain: all real numbers; range: $y \leq 0$
43. domain: all real numbers; range: $y \leq 4$

3.1 Maintaining Mathematical Proficiency *(p. 110)*

45. $3 \geq x$ **47.** $w + 4 > -12$ **49.** 81 **51.** 32

3.2 Vocabulary and Core Concept Check *(p. 117)*

1. $y = mx + b$
3. Discrete domains consist of only certain numbers in an interval. Continuous domains consist of all numbers in an interval.

3.2 Monitoring Progress and Modeling with Mathematics *(pp. 117–120)*

5. nonlinear; The graph is not a line.

7. linear; The graph is a line.

9. nonlinear; The graph is not a line.

11. linear; As x increases by 1, y increases by 5. The rate of change is constant.

13. nonlinear; As x increases by 4, y decreases by different amounts. The rate of change is not constant.

15. The increase in y needs to be done by adding or subtracting the same amount to be linear, not multiplying; As x increases by 2, y increases by different amounts. The rate of change is not constant. So, the function is nonlinear.

17. nonlinear; It cannot be rewritten in the form $y = mx + b$.

19. linear; It can be rewritten as $y = -1x + 2$.

21. linear; It can be rewritten as $y = 18x + 12$.

23. linear; It can be rewritten as $y = 9x - 13$.

25. A, C, F; None of these can be rewritten in the form $y = mx + b$.

27. 2, 4, 6; discrete; The graph consists of individual points.

29. discrete; The number of bags must be a whole number.

31. continuous; The time can be any value greater than or equal to 0.

33. There is no point with an x-value of 2.5; 2.5 is not in the domain.

35. **a.** 0, 1, 2, 3, 4, 5, 6; discrete; The number of books must be a whole number.

b.

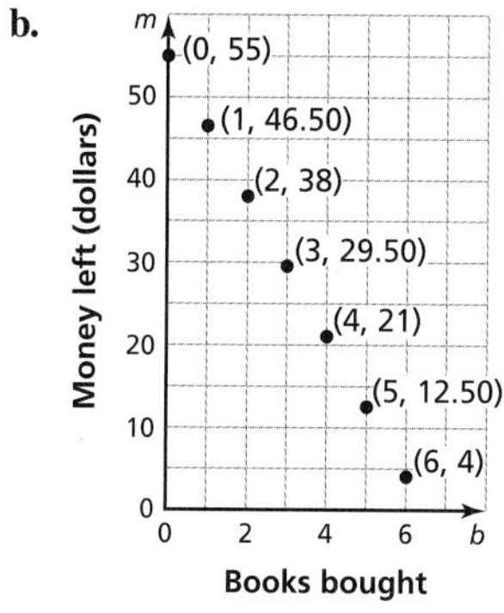

37. **a.** yes; As t increases by 2, d increases by 0.434. The rate of change is constant.

b. $t \geq 0$; continuous; The time can be any value greater than or equal to 0.

c.

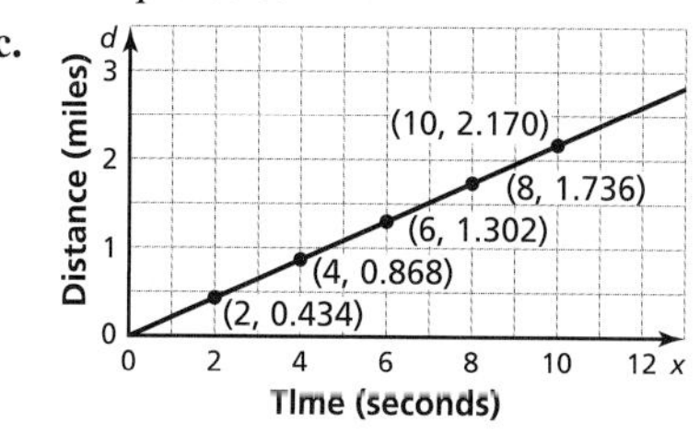

39. *Sample answer:* The number of hours on a parking meter is a function of the number of tokens used, 4 tokens for 1 hour and a maximum time of 2 hours; discrete; The number of tokens used must be 0, 4, or 8.

41. *Sample answer:* The depth (in feet) of a scuba diver returning to the surface of an ocean as a function of the time; continuous; The time can be any value from 0 to 30.

43. **a.** 51.00 **b.** \$10.20

45. nonlinear; $V = 9s^2$ cannot be written in linear form.

47. linear; The formula can be written in the form $V = (4\pi)h + 0$.

49. **a.** the total gallons of water in one jug of each type; continuous

b. the total number of jugs of both types; discrete

c. the total gallons of water in all the jugs of the first type; continuous

d. the total gallons of water in all the jugs of both types; continuous

51. linear; As x increases by 1, y increases by 4. The rate of change is constant.

53. *Sample answer:* how long it takes an ice cube to melt as a function of its temperature in degrees Celsius

3.2 Maintaining Mathematical Proficiency *(p. 120)*

55. no; The graph does not pass through the origin.

57. no; The graph does not show a constant rate of change.

59. 14 **61.** 4

3.3 Vocabulary and Core Concept Check *(p. 125)*

1. function notation

3.3 Monitoring Progress and Modeling with Mathematics *(pp. 125–126)*

3. 4; 6; 11 **5.** 13; 9; -1

7. -11; -3; 17 **9.** 11; 7; -3

11. **a.** There are no customers in the restaurant at 8 A.M.

b. There are the same number of customers in the restaurant at 11 A.M. as there are at 4 P.M.

c. There are 29 customers in the restaurant n hours after 8 A.M.

d. There are fewer customers in the restaurant at 9 P.M. than there are at 8 P.M.

13. $x = -9$ **15.** $x = -2$ **17.** $x = -2$ **19.** $x = 5$

21. **a.** \$77.50 **b.** 8 tickets

23.

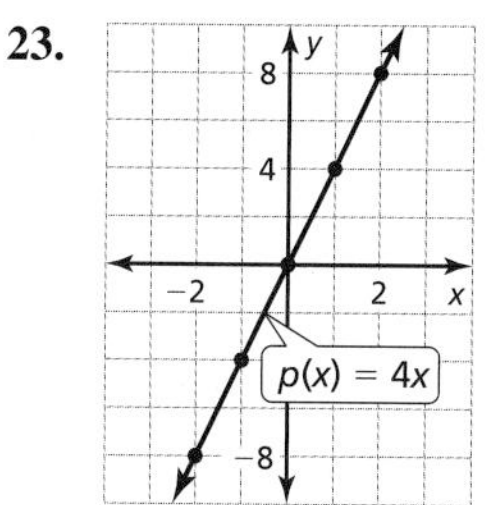

25.

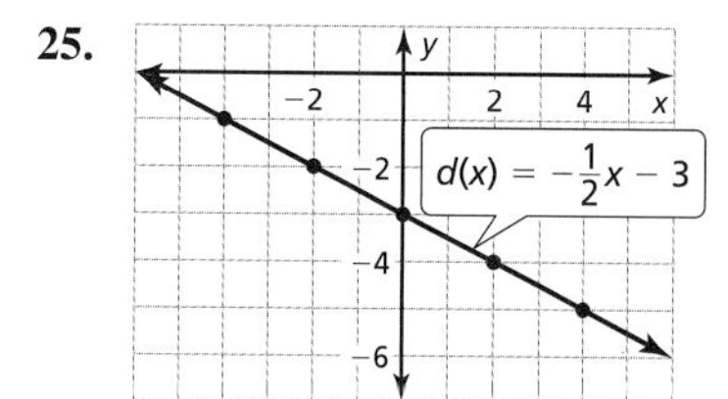

27.

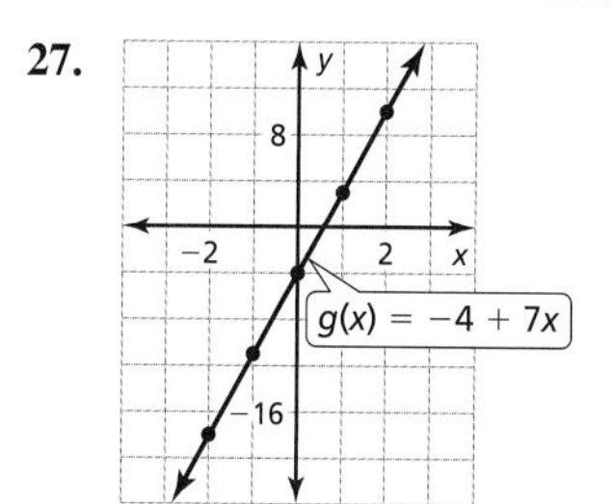

Selected Answers

29. the tablet computer; The graph of the tablet computer shows that when $p(t) = 0$, t is 6, which is greater than 5.

31. no; Because the function rule is unknown, it is unknown whether an increase in x would result in an increase or decrease in the value of the function.

33. **a.** $d(r) = 2r$; 10; The diameter of the circle is 10 feet.

b. $A(r) = \pi r^2$; 25π or about 78.5; The area of the circle is 25π square feet.

c. $C(r) = 2\pi r$; 10π or about 31.4; The circumference of the circle is 10π feet.

35. **a.** (5, 9)

b. (5, −3)

3.3 Maintaining Mathematical Proficiency *(p. 126)*

37. $9 \le x \le 17$;

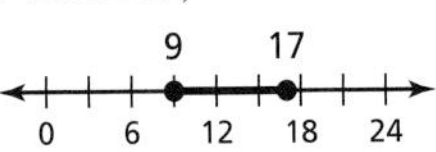

39. $-3 < k < -\frac{1}{3}$;

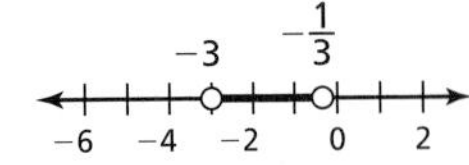

41. $-1 \le y < 3$;

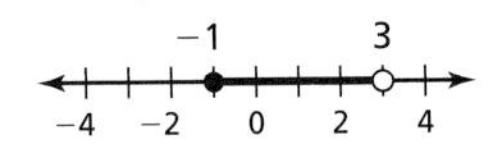

3.4 Vocabulary and Core Concept Check *(p. 133)*

1. They are both places where a graph crosses an axis; The x-intercept is where a graph crosses the x-axis. The y-intercept is where a graph crosses the y-axis.

3.4 Monitoring Progress and Modeling with Mathematics *(p. 133–134)*

3.

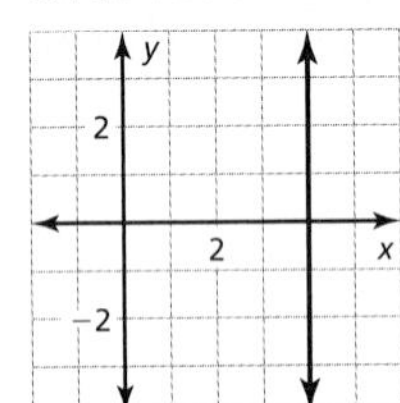

5.

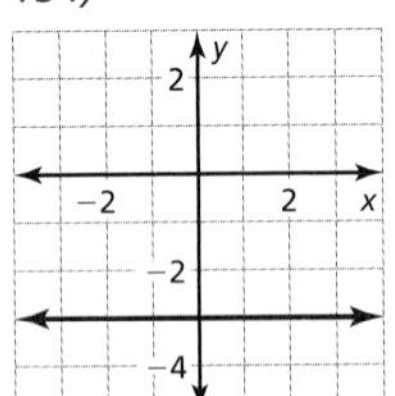

7. 6; 4 **9.** 4; −2 **11.** $\frac{2}{3}$; $-\frac{1}{3}$

13.

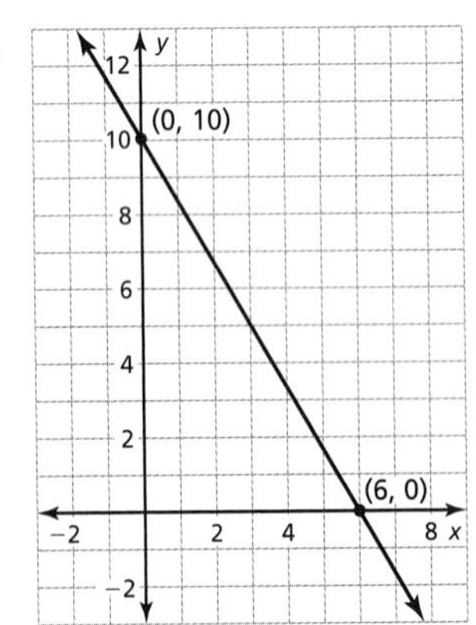

15.

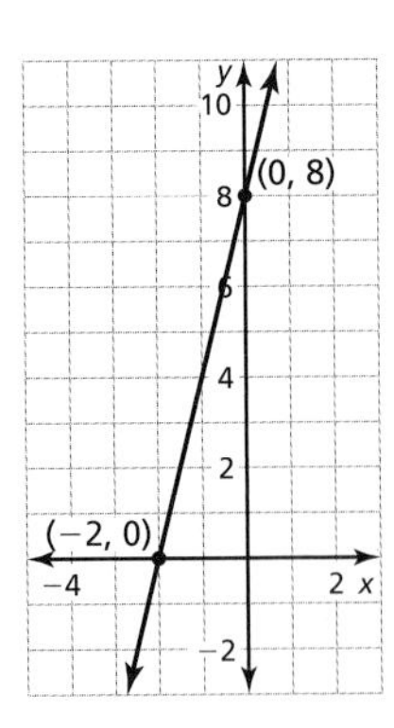

17.

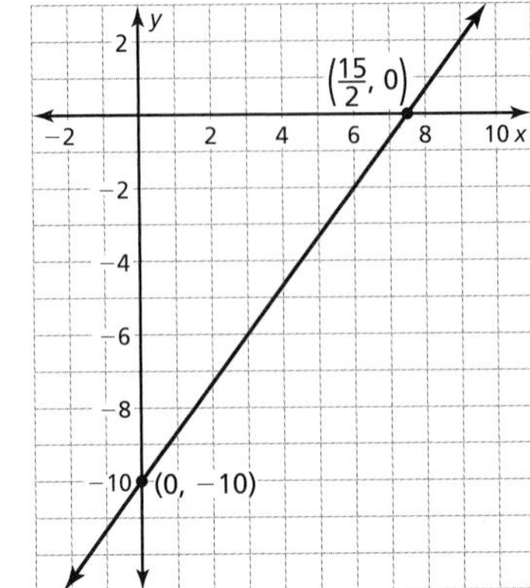

19.

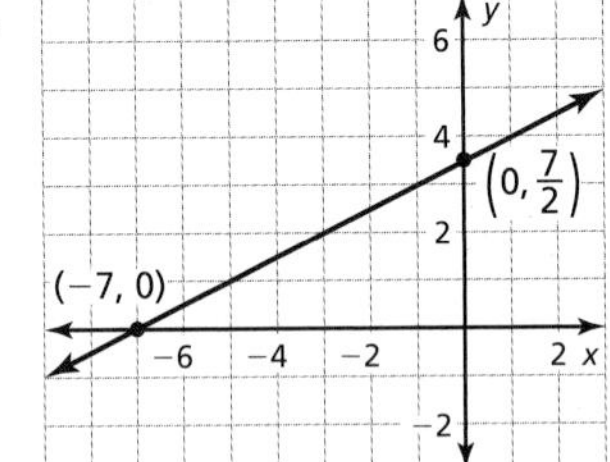

21.

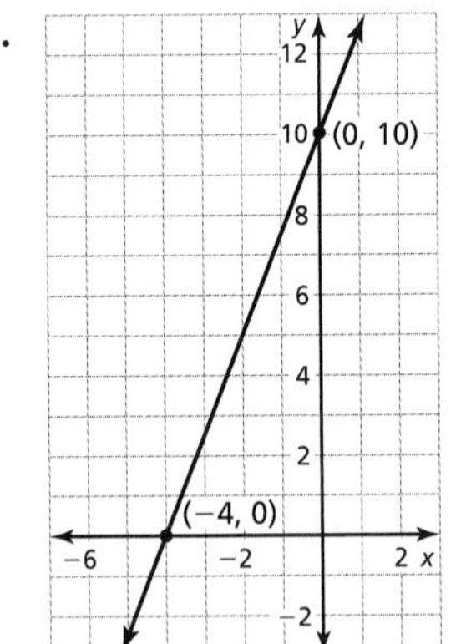

23. **a.**

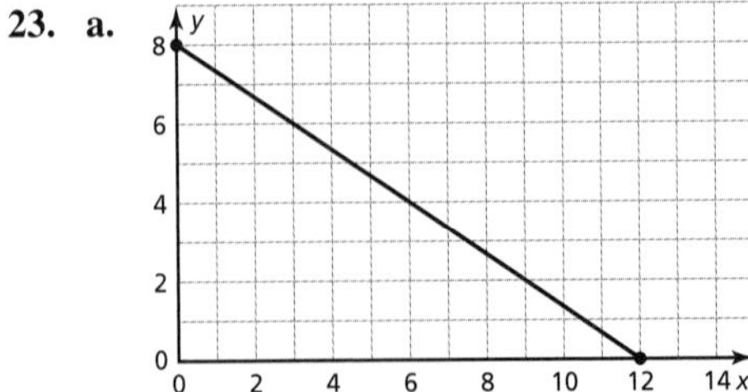

The x-intercept shows they can take 12 cars and 0 vans. The y-intercept shows they can take 8 vans and 0 cars.

b. *Sample answers:* 12 cars, 0 vans; 9 cars, 2 vans; 6 cars, 4 vans; 0 cars, 8 vans

25. The equations locate 2 separate intercepts, not 1. The intercepts are 8 and 2.

27. no; You have to substitute 0 for y to find the x-intercept, not for x.

29. A **31.** D

33.

square; The graphs are 2 horizontal lines and 2 vertical lines, which will intersect at right angles. The length of each side is 3 units.

35. -3; 6

37. yes; $x = a$ can be written as $1x + 0y = a$, and $y = b$ can be written as $0x + 1y = b$.

3.4 Maintaining Mathematical Proficiency *(p. 134)*

39. $\frac{1}{2}$ **41.** $-\frac{4}{5}$

3.5 Vocabulary and Core Concept Check *(p. 141)*

1. slope

3. $y = mx + b$; The equation gives the slope m and the y-intercept b.

3.5 Monitoring Progress and Modeling with Mathematics *(pp. 141–144)*

5. negative; $-\frac{3}{5}$ **7.** zero; 0 **9.** $\frac{1}{2}$ **11.** undefined

13. $m = 60$; The bus is traveling at a speed of 60 miles per hour.

15. slope: -3; y-intercept: 2 **17.** slope: 6; y-intercept: 0

19. slope: 2; y-intercept: 4 **21.** slope: 5; y-intercept: 8

23. To be in slope-intercept form the equation needs to be solved for y, not x; $y = -\frac{1}{4}x$; The slope is $-\frac{1}{4}$ and the y-intercept is 0.

25.

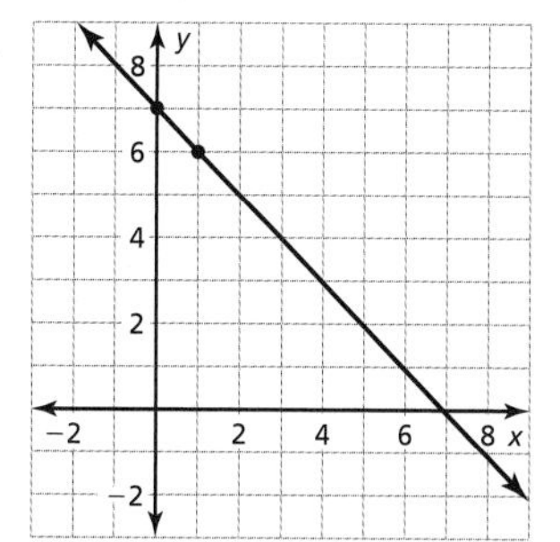

x-intercept: 7

27.

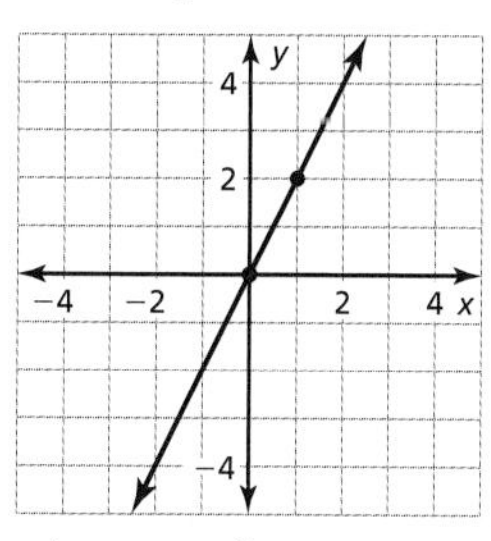

x-intercept: 0

29.

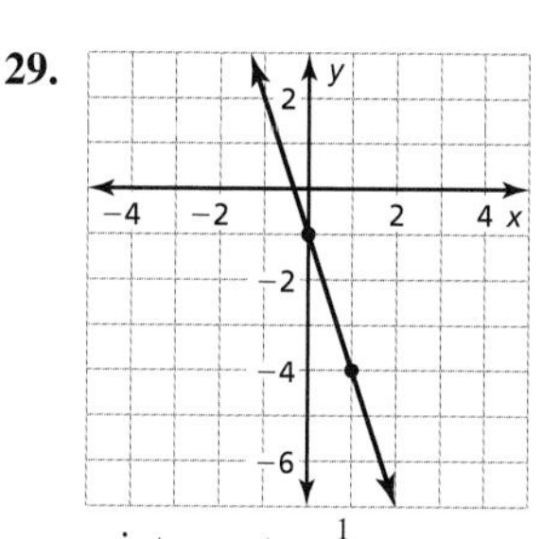

x-intercept: $-\frac{1}{3}$

31.

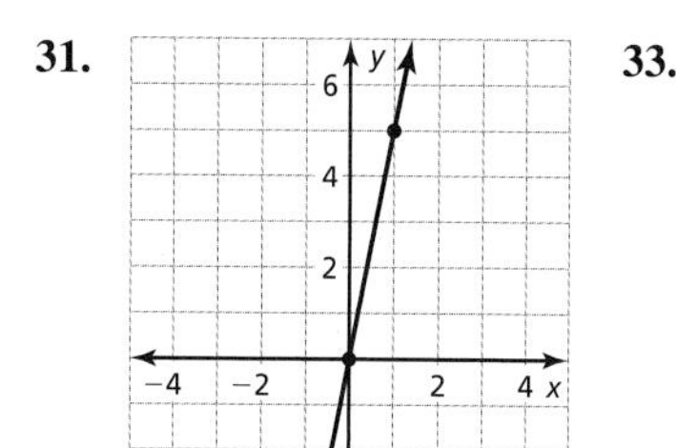

x-intercept: 0

33.

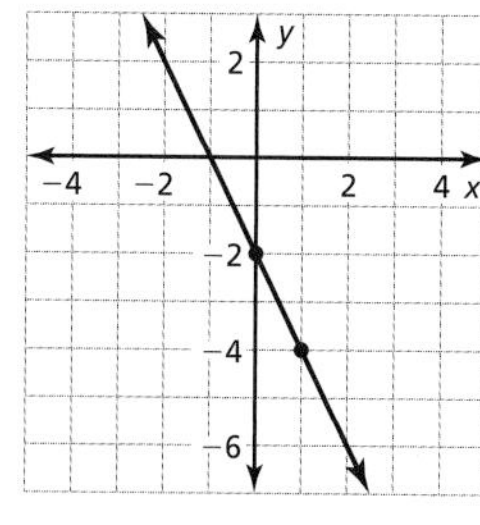

slope: -2; y-intercept: -2; x-intercept: -1

35.

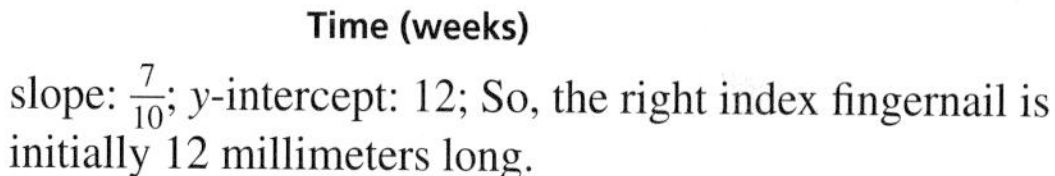

slope: $\frac{7}{10}$; y-intercept: 12; So, the right index fingernail is initially 12 millimeters long.

37. a.

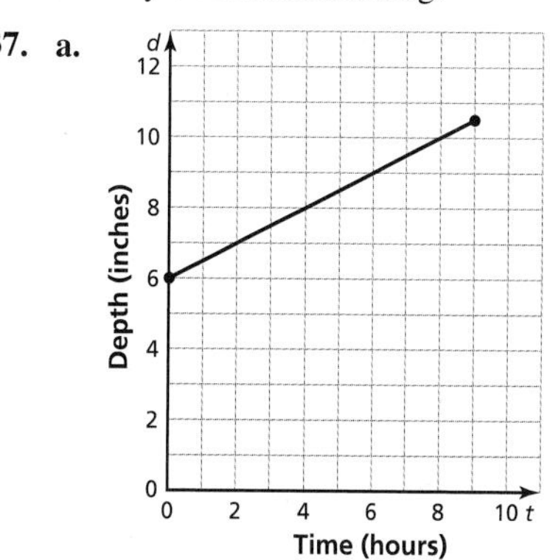

domain: $0 \le t \le 9$; range: $6 \le d \le 10\frac{1}{2}$

b. slope: $\frac{1}{2}$; So, $\frac{1}{2}$ inch of snow falls every hour during the storm; d-intercept: 6; So, there were 6 inches of snow already on the ground at the start of the storm.

39.

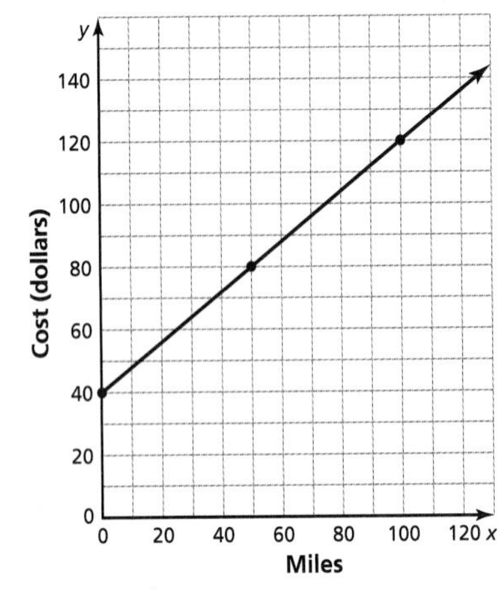

The slope of $\frac{4}{5}$ is greater than the slope in Exercise 38. The y-intercept of 40 is less than the c-intercept in Exercise 38.

41. The slope was interpreted as the y-intercept, and the y-intercept was interpreted as the slope.

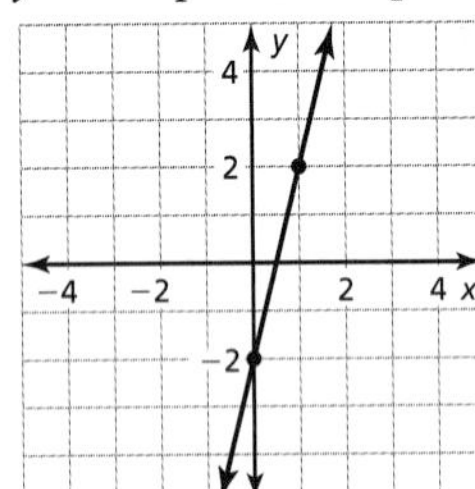

43. **a.**

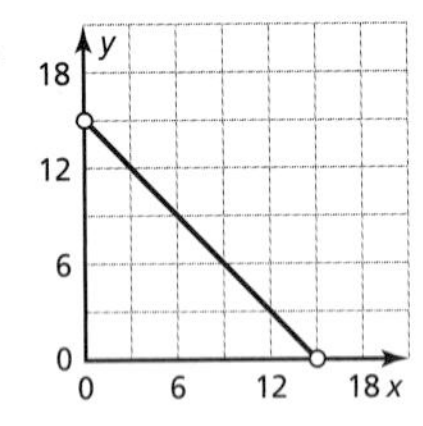

b. The slope of each graph is the same. The open circles in the graph in part (a) are closer to the origin.

45. **a.** $y = \frac{1}{3}x + 5$

b. $y = \frac{7}{4}x - \frac{1}{4}$; $y = 2x - 4$

c. $y = -3x + 8$

d. $y = -4x - 9$; $y = -x - \frac{4}{3}$

47. *Sample answer:* the ratio of the rise to the run; the ratio of the change in y to the change in x

49. **a.**

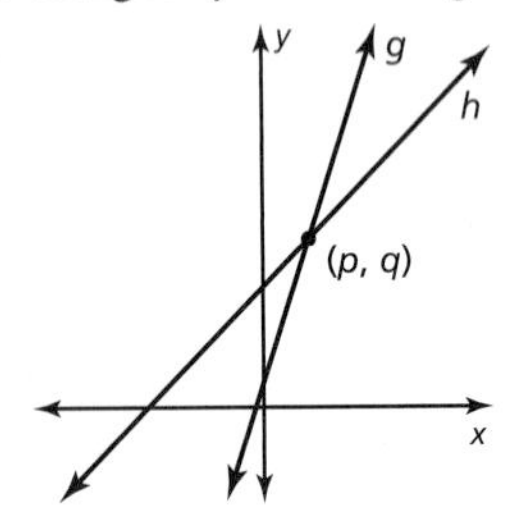

b. a is the y-intercept of the graph of g; b is the y-intercept of the graph of h

c. 8 units greater

51. $k = \frac{1}{8}$

53.
$$\frac{y_2 - y_1}{x_2 - x_1} = \frac{(mx_2 + b) - (mx_1 + b)}{x_2 - x_1} = \frac{mx_2 + b - mx_1 - b}{x_2 - x_1} = \frac{mx_2 - mx_1}{x_2 - x_1} = \frac{m(x_2 - x_1)}{x_2 - x_1} = m$$

3.5 Maintaining Mathematical Proficiency *(p. 144)*

55. $X'(0, 0)$, $Y'(4, 2)$, $Z'(4, 0)$

57. nonlinear; It cannot be rewritten in the form $y = mx + b$.

59. linear; It can be rewritten as $y = -3x + 12$.

3.6 Vocabulary and Core Concept Check *(p. 151)*

1. The graphs of all other nonconstant linear functions are transformations of the graph of $f(x) = x$.

3. causes a horizontal stretch or shrink; causes a vertical stretch or shrink

3.6 Monitoring Progress and Modeling with Mathematics *(pp. 151–154)*

5. The graph of g is a vertical translation 2 units up of the graph of f.

7. The graph of g is a vertical translation 3 units down of the graph of f.

9. The graph of g is a horizontal translation 5 units left of the graph of f.

11. The graph of f is a horizontal translation 5 units right of the graph of d.

13. The graph of h is a reflection in the x-axis of the graph of f.

15. The graph of h is a reflection in the y-axis of the graph of f.

17. The graph of r is a vertical stretch of the graph of f by a factor of 2.

19. The graph of r is a horizontal stretch of the graph of f by a factor of 2.

21. The graph of r is a vertical stretch of the graph of f by a factor of 3.

23. The graph of h is a horizontal shrink of the graph of f by a factor of $\frac{1}{3}$.

25. The graph of h is a vertical shrink of the graph of f by a factor of $\frac{1}{6}$.

27. The graph of h is a horizontal shrink of the graph of f by a factor of $\frac{1}{5}$.

29. The graph of g is a vertical shrink of the graph of f by a factor of $\frac{1}{4}$.

31. The graph of g is a horizontal translation 2 units right of the graph of f.

33. The graph of g is a vertical stretch of the graph of f by a factor of 6.

35. $g(x) = f(x - 2)$ **37.** $g(x) = 4f(x)$

39. $f(x - 2)$ is a translation to the right, not to the left.

41.

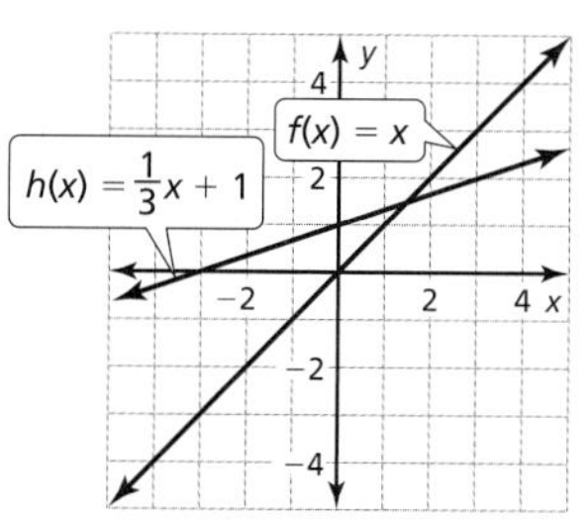

The transformations are a vertical shrink by a factor of $\frac{1}{3}$ then a vertical translation 1 unit up.

43.

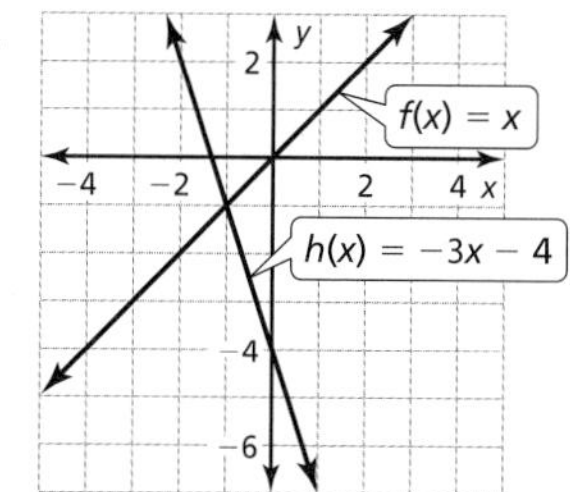

The transformations are a vertical stretch by a factor of 3, then a reflection in the x-axis, then a vertical translation 4 units down.

45.

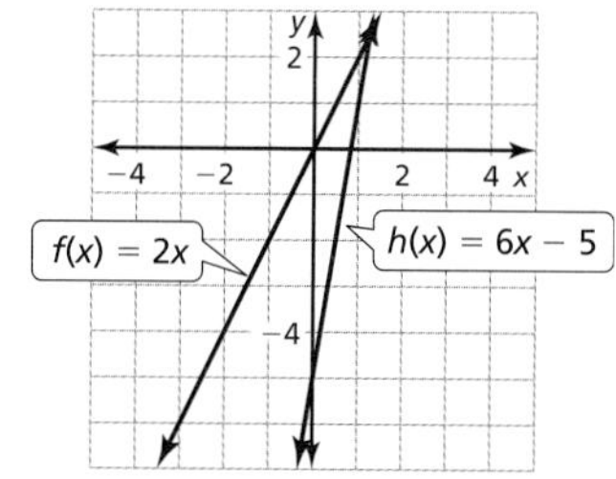

The transformations are a vertical stretch by a factor of 3 then a vertical translation 5 units down.

47. The graph of d is a reflection in the y-axis of the graph of t.

49. **a.** stretch **b.** right **c.** 1

51. B and C, A and F; Adding 4 to the y-coordinate of each point on B gives the corresponding y-coordinate on C. Adding 2 to the y-coordinate of each point on F gives you the corresponding y-coordinate on A.

53. **a.** The new graph is a vertical translation 10 units up of the old graph.

b. The new graph is a horizontal stretch of the old graph by a factor of 2.

55.

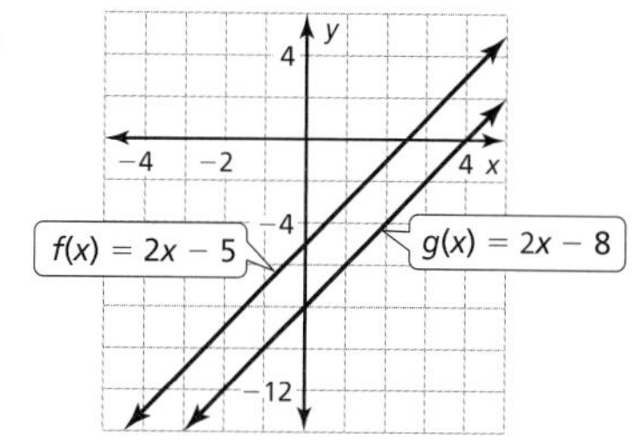

$g(x) = f(x) - 3$; The graph of g is a vertical translation 3 units down of the graph of f.

57.

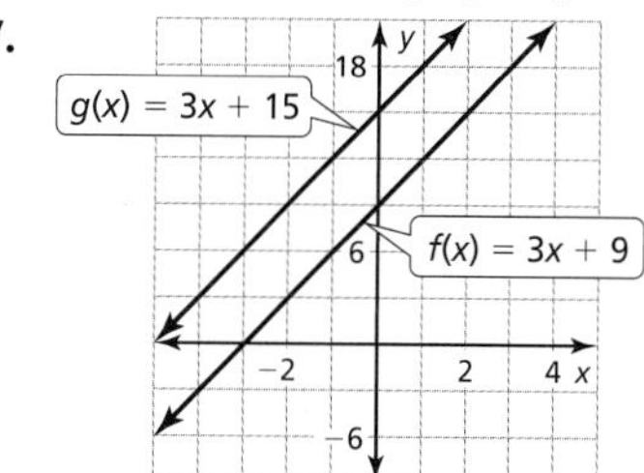

$g(x) = f(x) + 6$; The graph of g is a vertical translation 6 units up of the graph of f.

59.

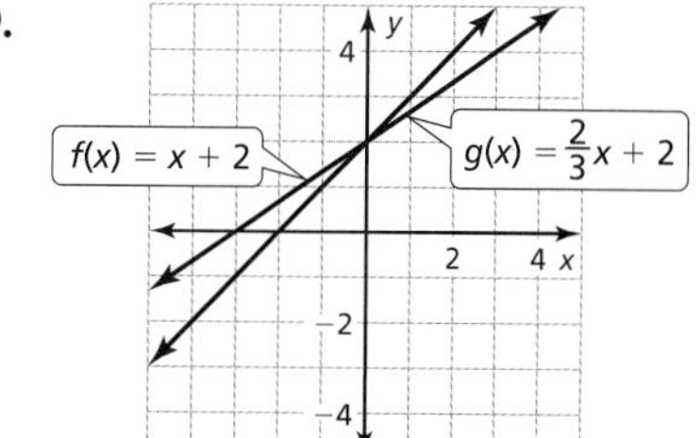

$g(x) = f\left(\frac{2}{3}x\right)$; The graph of g is a horizontal stretch of the graph of f by a factor of $\frac{3}{2}$.

61. Translate the graph of $f(x) = x$ horizontally 5 units left.

63. $r = 2$ **65.** $r = 2$

67. when the slope is 1; A slope of 1 occurs when the ratio of the vertical change to the horizontal change is 1, meaning the vertical change and horizontal change are the same.

3.6 Maintaining Mathematical Proficiency *(p. 154)*

69. $w = \frac{P - 2\ell}{2}$ or $w = \frac{1}{2}P - \ell$

71. $x < -10$ or $x > 6$

73. no solution

3.7 Vocabulary and Core Concept Check *(p. 160)*

1. vertex

3. *Sample answer:* $g(x) = a|x|$ is a vertical stretch or shrink, $g(x) = |x - h|$ is a horizontal translation, and $g(x) = |x| + k$ is a vertical translation.

3.7 Monitoring Progress and Modeling with Mathematics *(pp. 160–162)*

5.

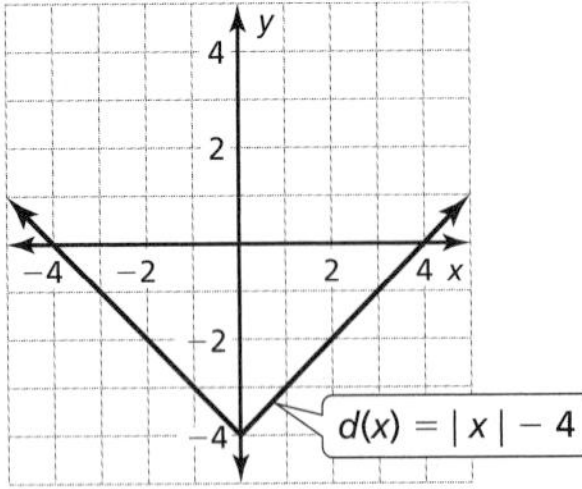

The graph of d is a vertical translation 4 units down of the graph of f; domain: all real numbers; range: $y \geq -4$

7.

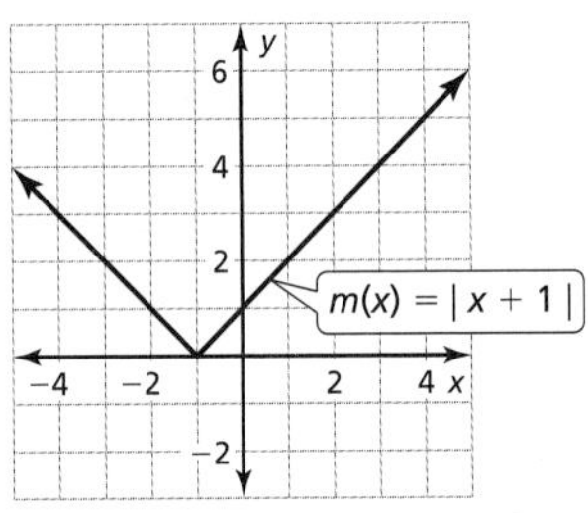

The graph of m is a horizontal translation 1 unit left of the graph of f; domain: all real numbers; range: $y \geq 0$

9.

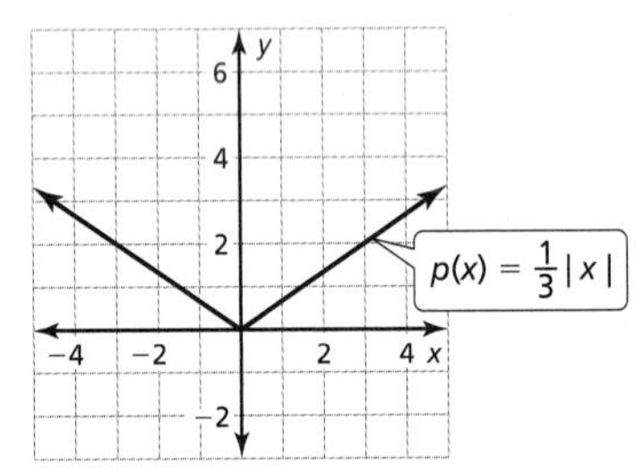

The graph of p is a vertical shrink of the graph of f by a factor of $\frac{1}{3}$; domain: all real numbers; range: $y \geq 0$

11.

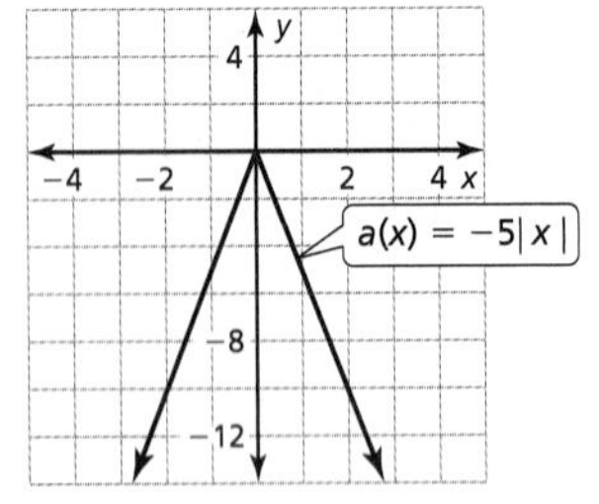

The graph of a is a vertical stretch of the graph of f by a factor of 5 and a reflection in the x-axis; domain: all real numbers; range: $y \leq 0$

13.

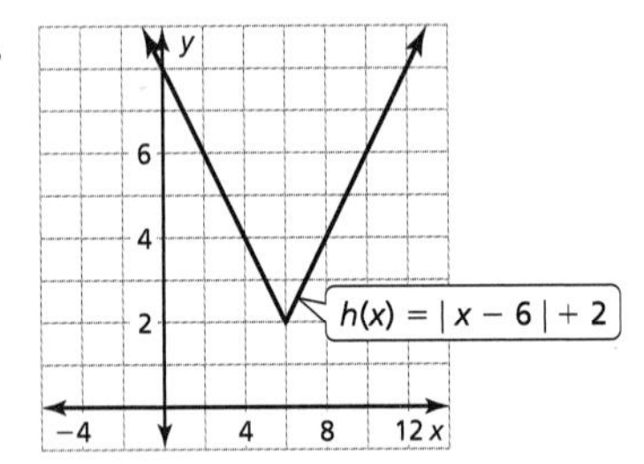

The graph of h is a vertical translation 2 units up of the graph of f.

15.

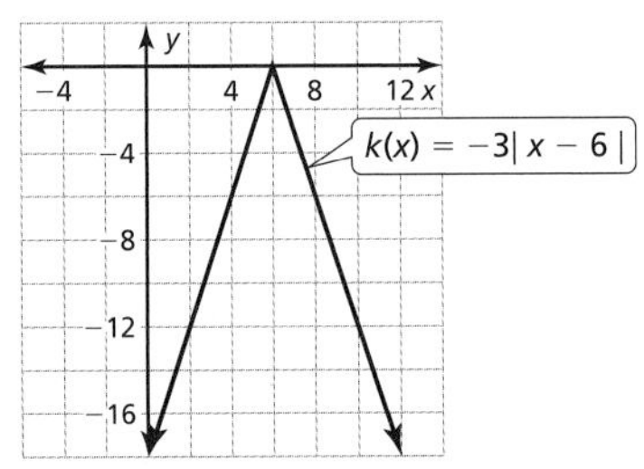

The graph of k is a vertical stretch of the graph of f by a factor of 3 and a reflection in the x-axis.

17.

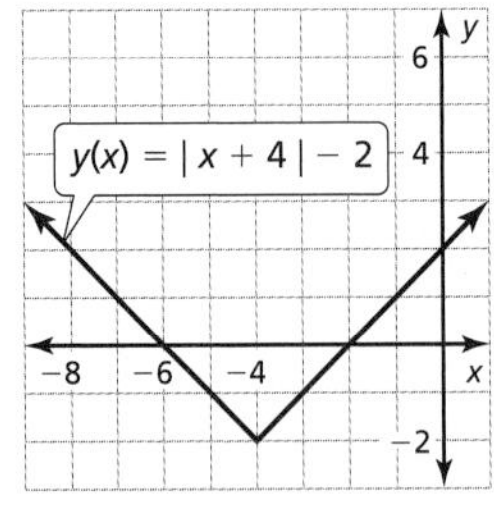

The graph of y is a horizontal translation 1 unit left of the graph of f.

19. The graph of g is a vertical translation 3 units down of the graph of f; $k = -3$

21. The graph of p is a vertical stretch of the graph of f by a factor of 3 and a reflection in the x-axis; $a = -3$

23. $h(x) = |x| - 7$ **25.** $h(x) = \frac{1}{4}|x|$

27.

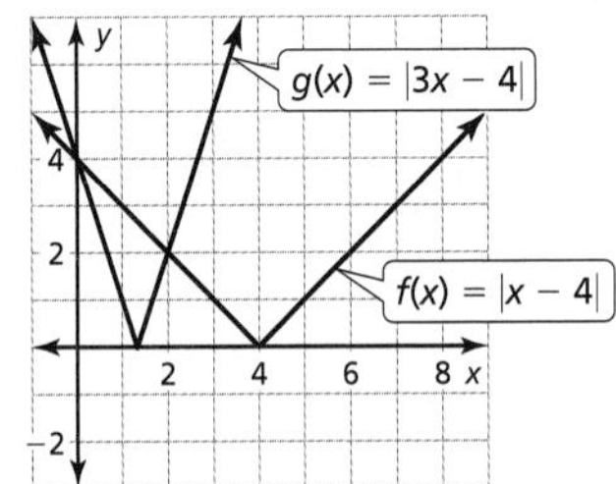

The graph of g is a horizontal shrink of the graph of f by a factor of $\frac{1}{3}$.

29.

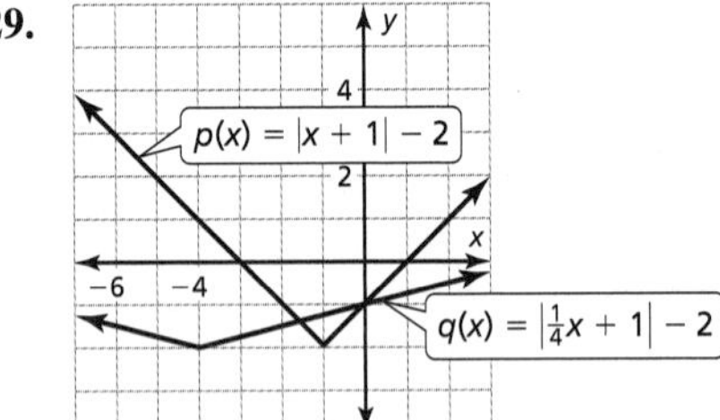

The graph of q is a horizontal stretch of the graph of p by a factor of 4.

31.

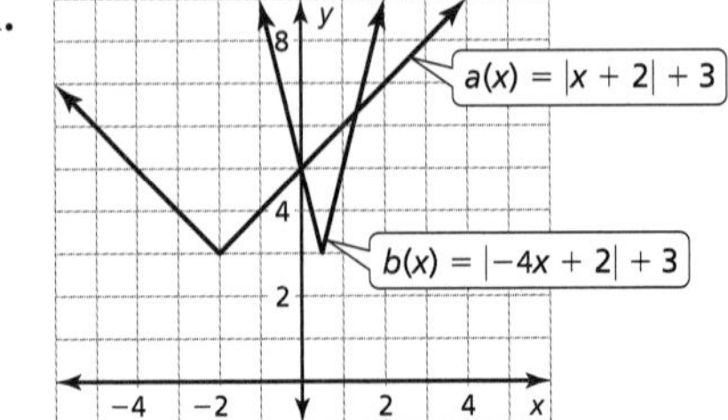

The graph of b is a horizontal shrink of the graph of a by a factor of $\frac{1}{4}$ and a reflection in the y-axis.

33. The transformations are a horizontal translation 2 units left then a vertical translation 6 units down.

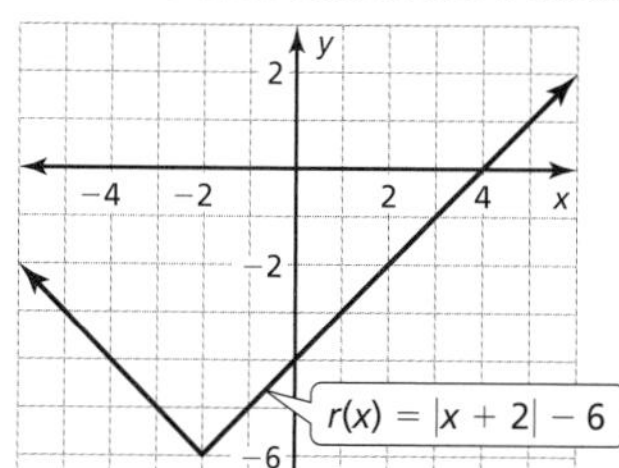

35. The transformations are a horizontal translation 3 units right, then a reflection in the x-axis, then a vertical translation 5 units up.

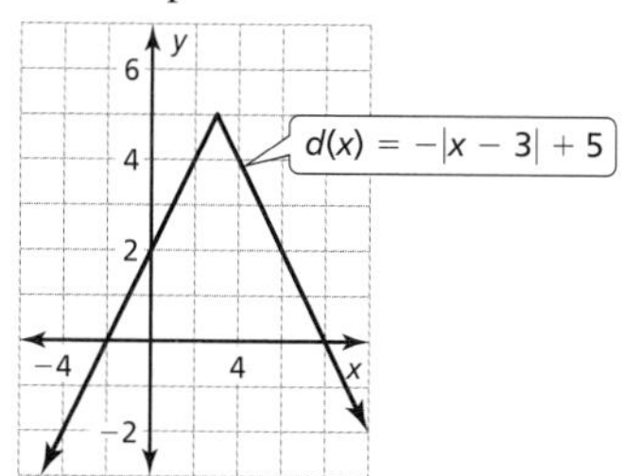

37. The transformations are a horizontal translation 4 units left, then a vertical shrink by a factor of $\frac{1}{2}$, then a vertical translation 1 unit down.

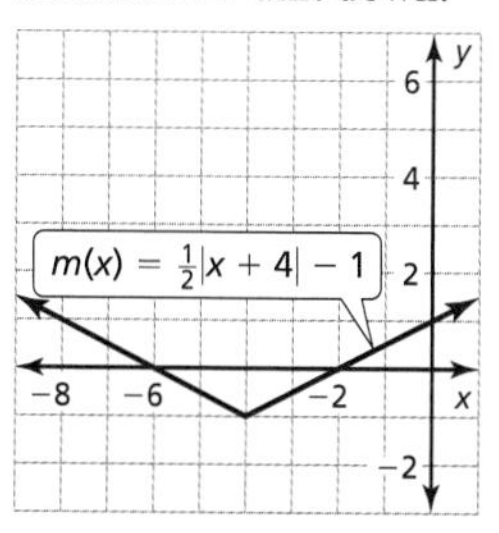

39. The transformations are a horizontal translation 1 unit left, then a reflection in the y-axis, then a vertical translation 5 units down.

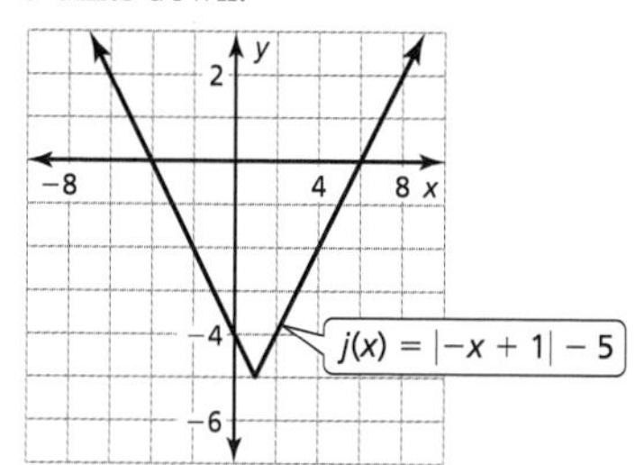

41. **a.**

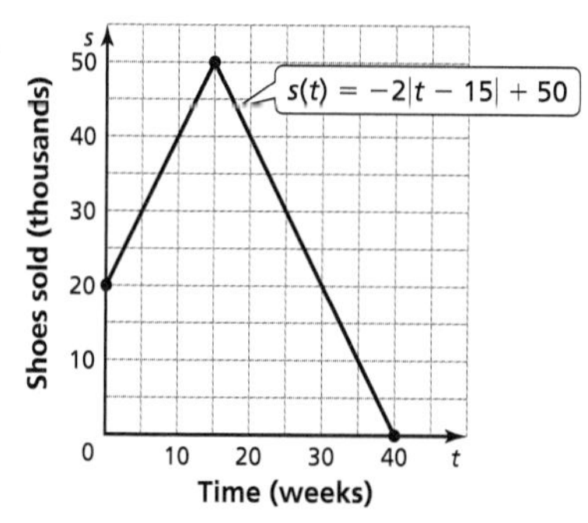

b. 50,000 pairs of shoes

43. **a.** $A'\left(-\frac{1}{2}, -2\right)$; $B'(1, -5)$; $C'(-4, -7)$

b. $A'\left(\frac{5}{2}, 3\right)$; $B'(4, 0)$; $C'(-1, -2)$

c. $A'\left(-\frac{1}{2}, -3\right)$; $B'(1, 0)$; $C'(-4, 2)$

d. $A'\left(-\frac{1}{2}, 12\right)$; $B'(1, 0)$; $C'(-4, -8)$

45. The graph should have a horizontal shift right, not left.

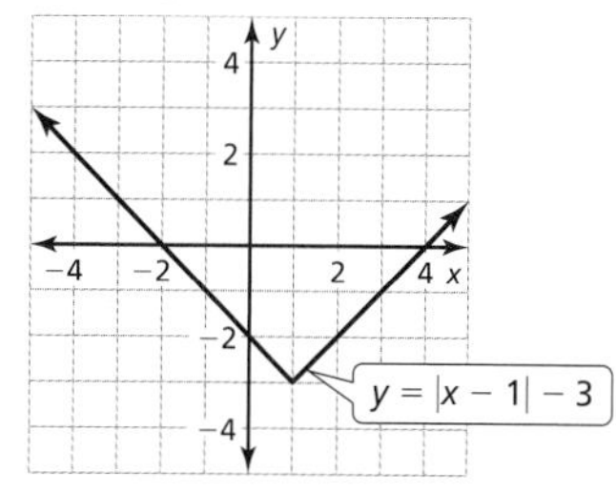

47. *Sample answer:* $f(x) = -|x| + 2$

49. The graph of p is a horizontal translation 6 units right of the graph of $f(x) = |x|$. The graph of q is a vertical translation 6 units down of the graph of $f(x) = |x|$.

51.

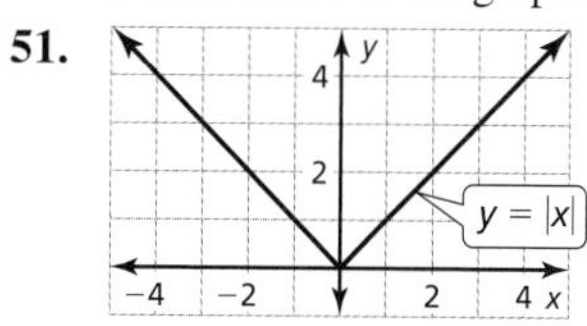

$y = -x$; $y = x$

53.

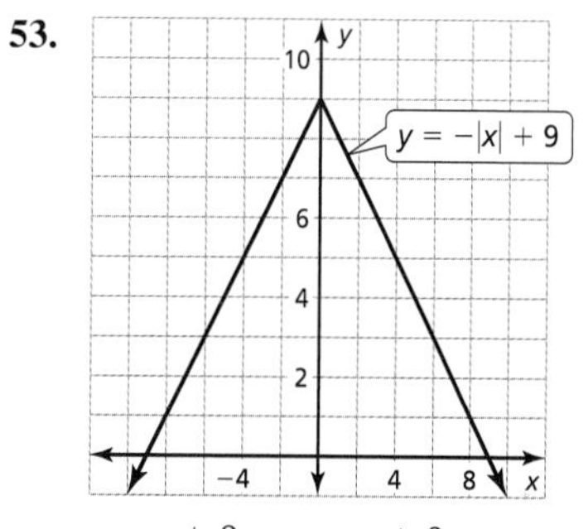

$y = x + 9$; $y = -x + 9$

55.

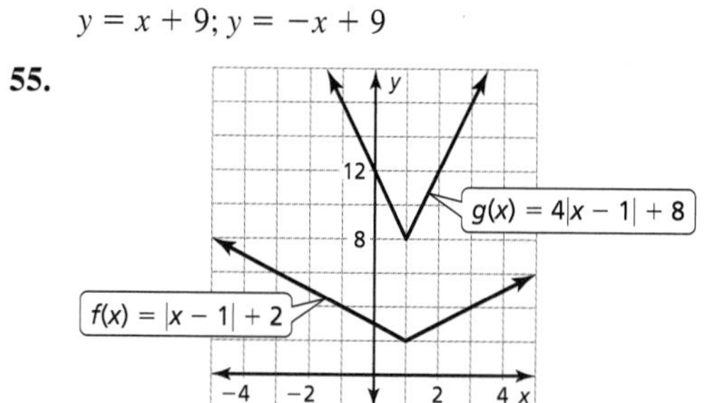

The graph of g is a vertical stretch by a factor of 4 of the graph of f.

57.

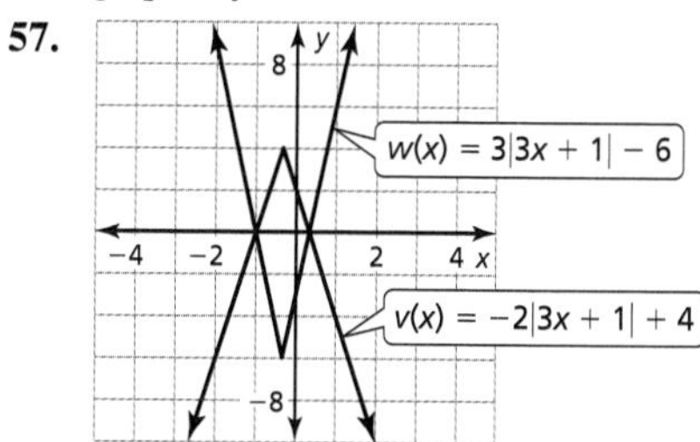

The graph of w is a vertical stretch of the graph of v by a factor of $\frac{3}{2}$ and a reflection in the x-axis.

Selected Answers

59. The transformations are a vertical translation 4 units down, then a reflection in the x-axis, then a vertical shrink by a factor of $\frac{1}{2}$, then a horizontal translation 1 unit right; *Sample answer:* Determine the transformations from h to g, then perform the opposite transformations in reverse order.

61.

$x = 0, x = -4$

63. $\left(\frac{h}{a}, k\right)$

3.7 Maintaining Mathematical Proficiency *(p. 162)*

65. no solution **67.** $x > -\frac{3}{4}$ **69.** $\frac{1}{3}$

Chapter 3 Review *(pp. 164–168)*

1. function; Every input has exactly one output.

2. not a function; A vertical line can be drawn through more than one point on the graph in many places.

3. function; Every input has exactly one output.

4. **a.** The amount of money in the bank account is the dependent variable and the hours you babysat is the independent variable.

b. domain: $0 \leq x \leq 4$; range: $100 \leq y \leq 140$

5. linear; As x increases by 5, y decreases by 3. The rate of change is constant.

6. nonlinear; The graph is not a line.

7. **a.** 0, 1, 2, 3, 4, 5, 6, 7; discrete; The number of tickets bought must be a whole number.

b.

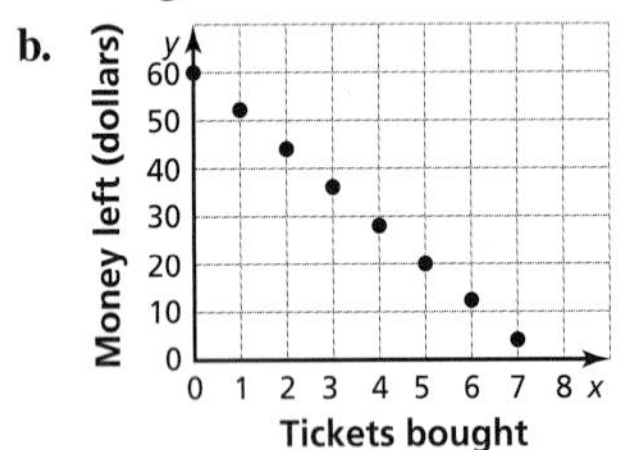

8. 5, 8, 13 **9.** 13, 4, −11 **10.** $x = 7$ **11.** $x = -4$

12.

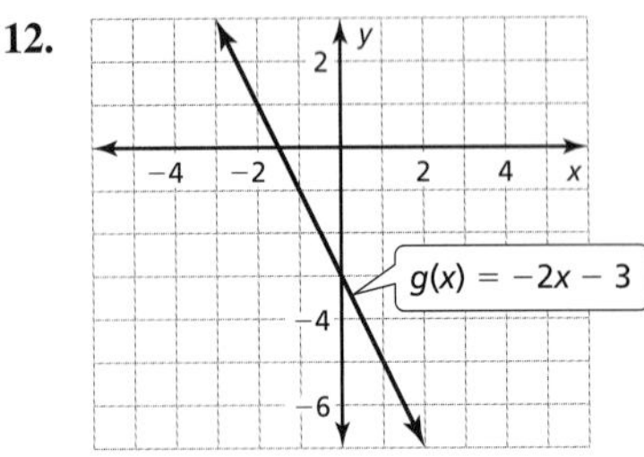

13.

14.

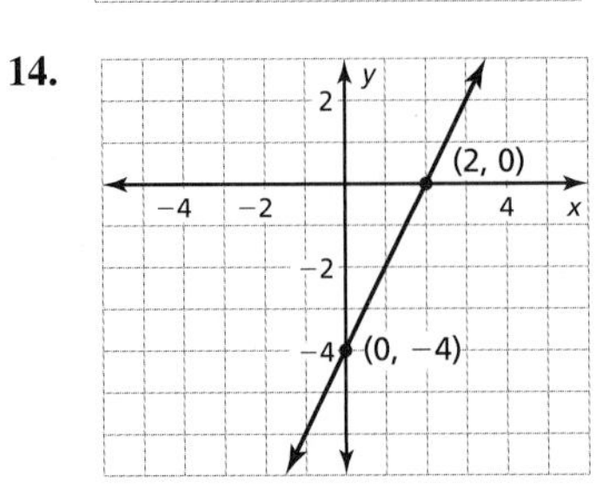

15.

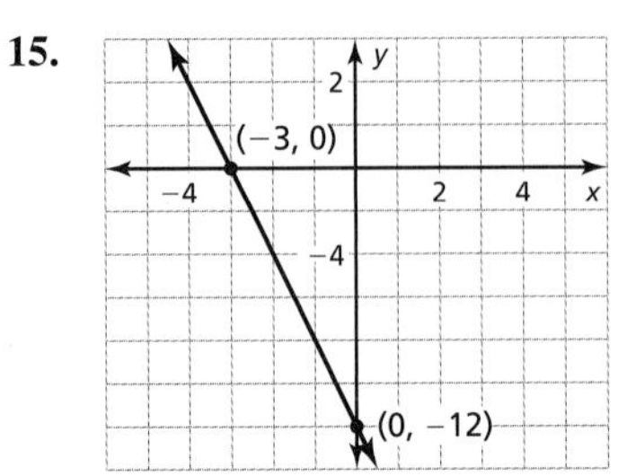

16.

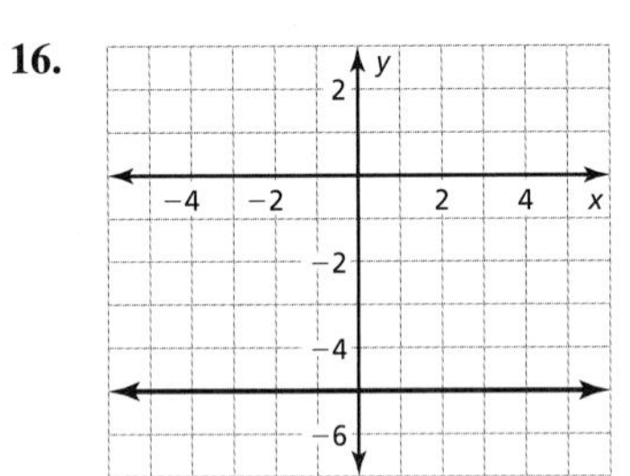

17.

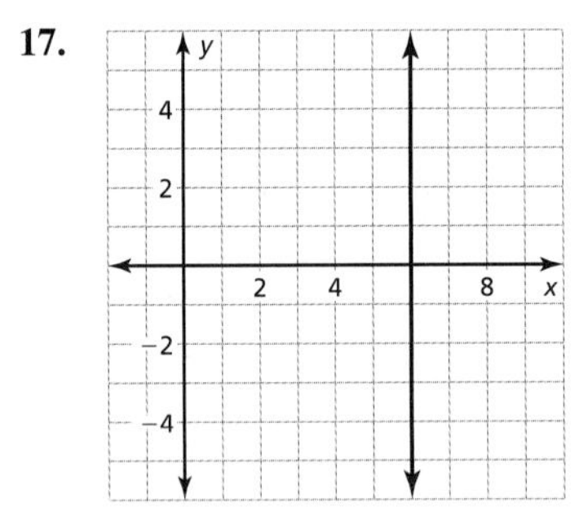

18. $\frac{6}{5}$ **19.** undefined **20.** 0

21.

x-intercept: −2

22.

x-intercept: 2

23.

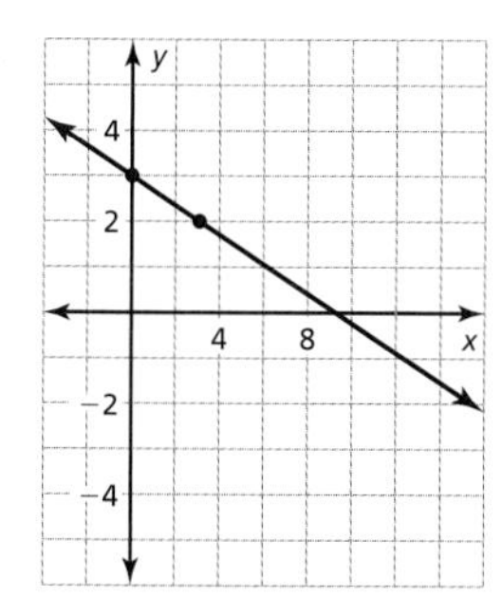

x-intercept: 9

24.

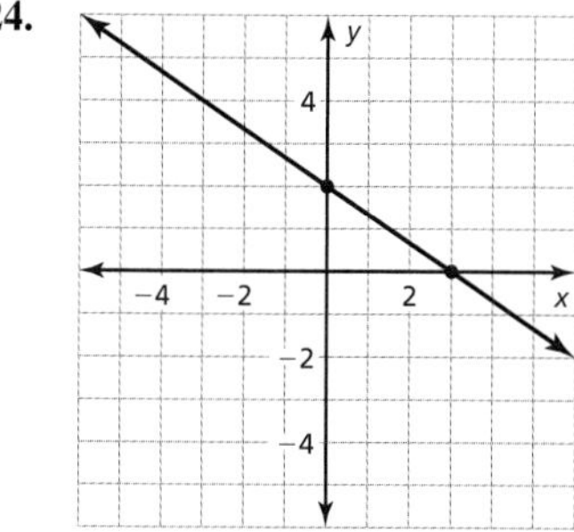

slope: $-\frac{2}{3}$; y-intercept: 2; x-intercept: 3

25.

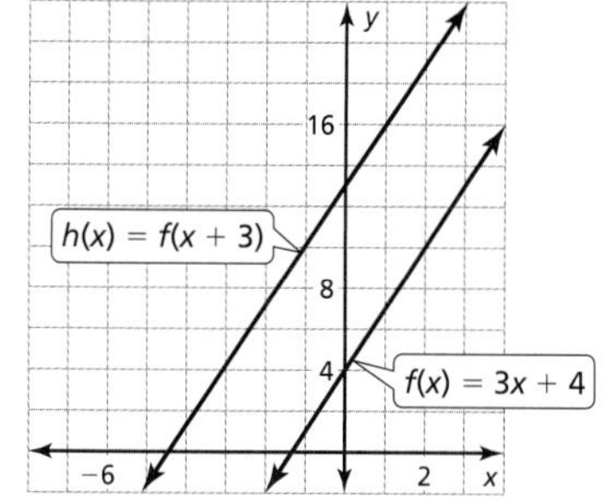

The graph of h is a horizontal translation 3 units left of the graph of f.

26.

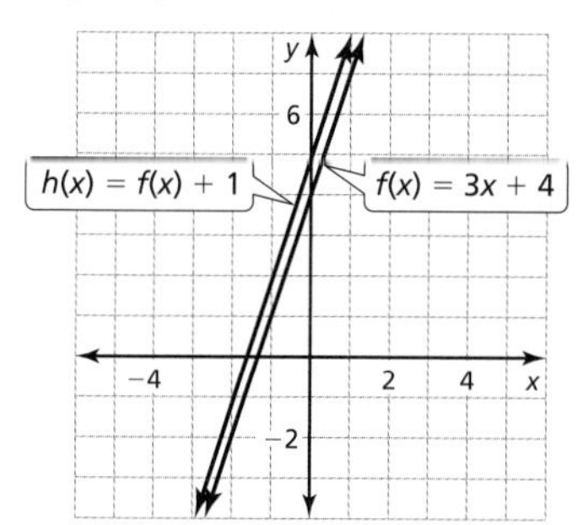

The graph of h is a vertical translation 1 unit up of the graph of f.

27.

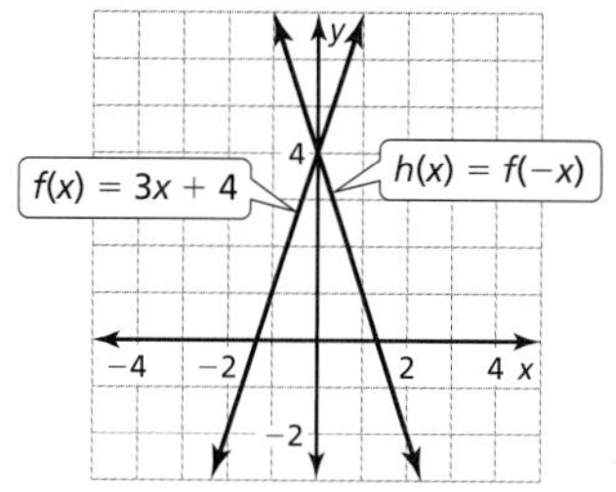

The graph of h is a reflection in the y-axis of the graph of f.

28.

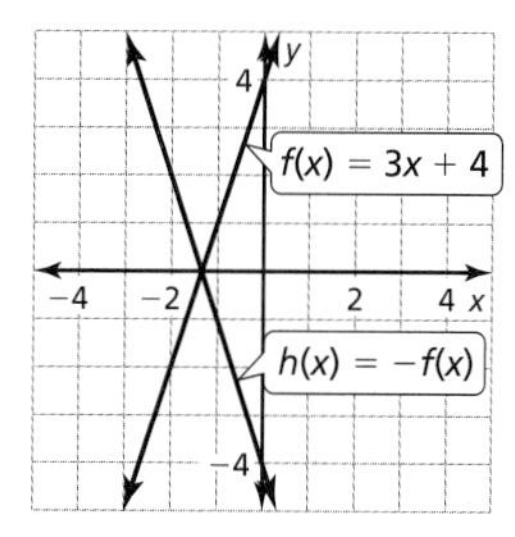

The graph of h is a reflection in the x-axis of the graph of f.

29.

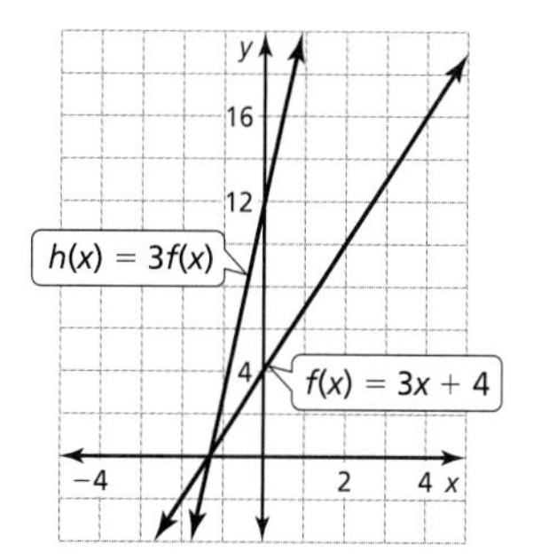

The graph of h is a vertical stretch of the graph of f by a factor of 3.

30.

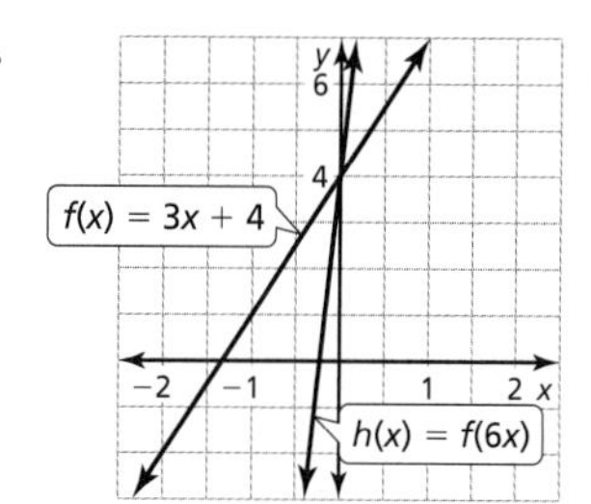

The graph of h is a horizontal shrink of the graph of f by a factor of $\frac{1}{6}$.

31.

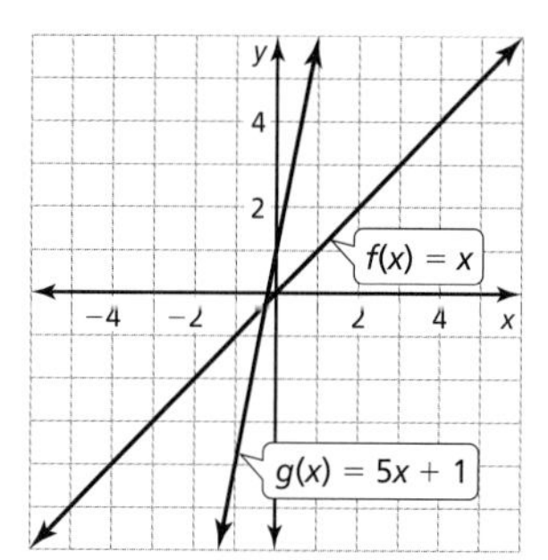

The transformations are a vertical stretch by a factor of 5, then a vertical translation 1 unit up.

32.

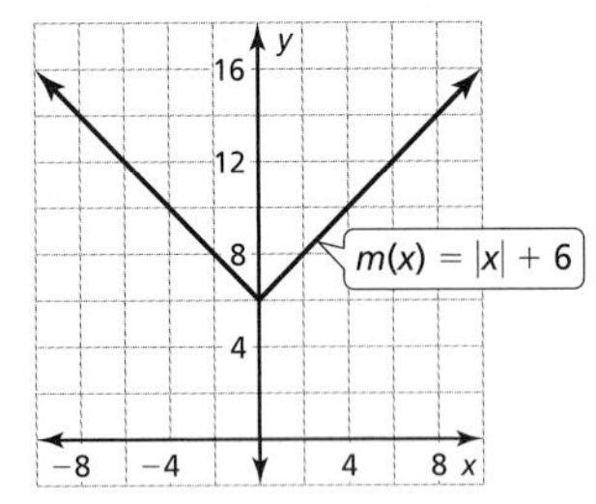

The graph of m is a vertical translation 6 units up of the graph of f; domain: all real numbers; range: $y \geq 6$

33.

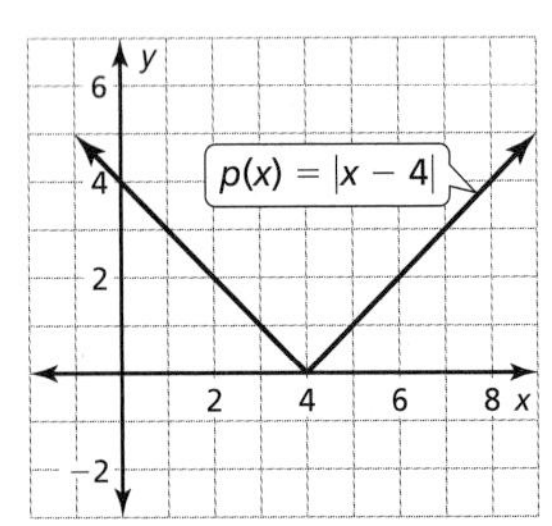

The graph of p is a horizontal translation 4 units right of the graph of f; domain: all real numbers; range: $y \geq 0$

34.

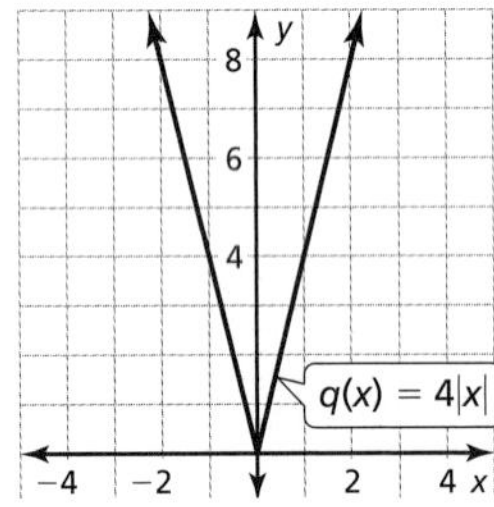

The graph of q is a vertical stretch of the graph of f by a factor of 4; domain: all real numbers; range: $y \geq 0$

35.

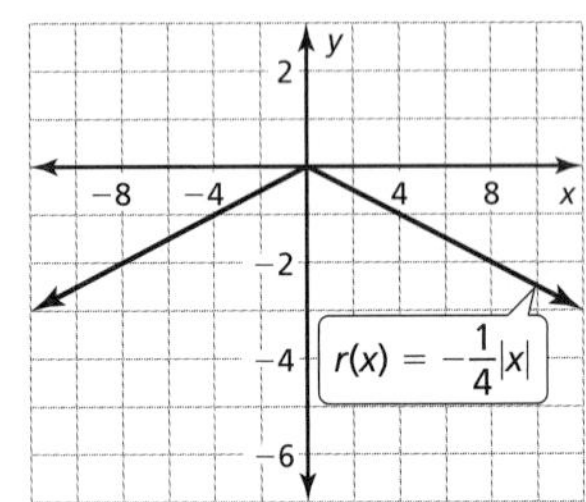

The graph of r is a vertical shrink of the graph of f by a factor of $\frac{1}{4}$ and a reflection in the x-axis; domain: all real numbers; range: $y \leq 0$

36.

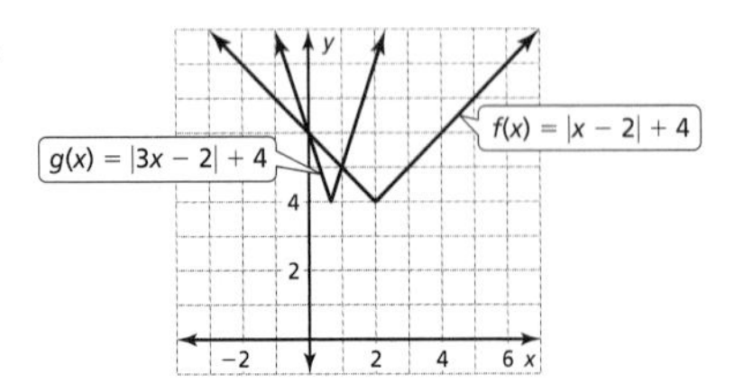

The graph of g is a horizontal shrink of the graph of f by a factor of $\frac{1}{3}$.

37. a. The transformations are a horizontal translation 1 unit right, then a vertical shrink by a factor of $\frac{1}{3}$, then a vertical translation 2 units down.

b.

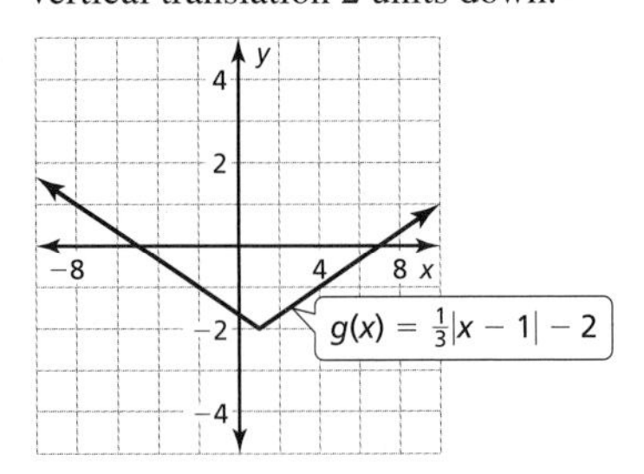

Chapter 4

Chapter 4 Maintaining Mathematical Proficiency *(p. 173)*

1. $(-5, -2)$ **2.** $(2, 0)$ **3.** C **4.** E

5. $y = -5 + x$ **6.** $y = -\frac{1}{3} - 2x$ **7.** $y = 4x - 5$

8. $y = \frac{1}{4}x + 7$ **9.** $y = 4x - \frac{1}{2}$ **10.** $y = -\frac{3}{4}x - 1$

11. Points in Quadrant I move to Quadrant III, points in Quadrant II move to Quadrant IV, points in Quadrant III move to Quadrant I, and points in Quadrant IV move to Quadrant II.

4.1 Vocabulary and Core Concept Check *(p. 179)*

1. linear model

4.1 Monitoring Progress and Modeling with Mathematics *(pp. 179–180)*

3. $y = 2x + 9$ **5.** $y = -3x$ **7.** $y = \frac{2}{3}x - 8$

9. $y = \frac{1}{3}x + 2$ **11.** $y = -\frac{4}{3}x$ **13.** $y = -3x + 10$

15. $y = -4$ **17.** $y = \frac{8}{3}x + 5$ **19.** $f(x) = x + 2$

21. $f(x) = -\frac{1}{4}x - 2$ **23.** $f(x) = -5x - 4$

25. $f(x) = -2x + 1$

27. The slope and the y-intercept were substituted incorrectly; $y = 2x + 7$

29. a. $y = -0.005t + 3.91$

b. 3.71 min; 3.61 min

31. no; Because the x-coordinates are the same, the line is vertical and has an undefined slope.

33. $y = -\frac{A}{B}x + \frac{C}{B}$; slope is $\frac{6}{5}$; y-intercept is $\frac{9}{5}$

35. $y = \frac{5}{3}x - 1$

37. slope $= \frac{(b + m) - b}{1 - 0} = m$; y-intercept $= b$; $y = mx + b$;

Substitute -1 for x in $y = mx + b$ and verify that $y = b - m$.

4.1 Maintaining Mathematical Proficiency *(p. 180)*

39. $y = \frac{1}{4}$ **41.** $n = 0$

43.

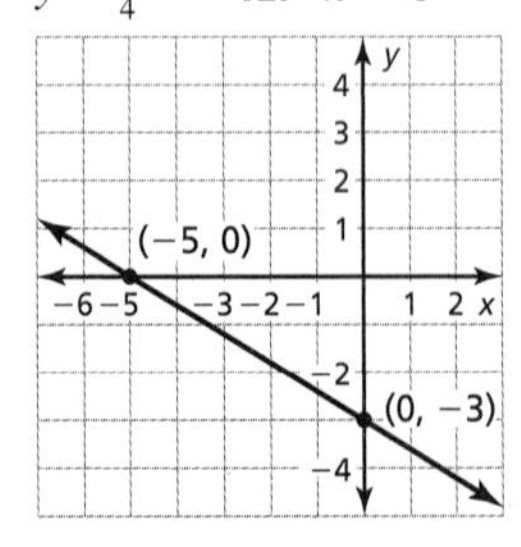

45.

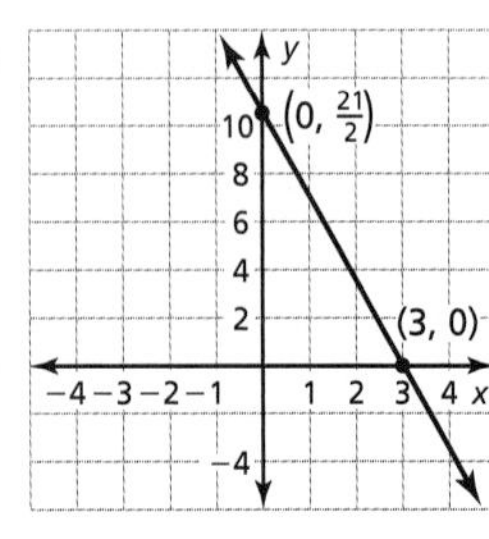

4.2 Vocabulary and Core Concept Check *(p. 185)*

1. $-2; (-5, 5)$

4.2 Monitoring Progress and Modeling with Mathematics *(pp. 185–186)*

3. $y - 1 = 2(x - 2)$ **5.** $y + 4 = -6(x - 7)$

7. $y = -3(x - 9)$ **9.** $y - 6 = \frac{3}{2}(x + 6)$

11. $y = 2x - 5$ **13.** $y = -\frac{1}{2}x + 1$ **15.** $y = -2x + 16$

17. $y = 2x - 13$ **19.** $y = -9$ **21.** $f(x) = -3x + 4$

23. $f(x) = -\frac{1}{2}x$ **25.** $f(x) = \frac{1}{4}x + \frac{7}{4}$

27. no; y does not increase at a constant rate.

29. yes; y increases at a constant rate; $y = 0.2x + 1.2$

31. In point-slope form, the slope is multiplied by the quantity, $x - x_1$; $y - y_1 = m(x - x_1)$;
$y - 4 = -3(x - 5)$; $y - 4 = -3x + 15$; $y = -3x + 19$;
A function is $g(x) = -3x + 19$.

33. a. $C = 80n + 145$

b. \$865

35. *Sample answer:* Plot the point (4, 1), then use the slope of $\frac{3}{2}$ to find a second point on the graph and draw a line through the points; Rewrite the equation in slope-intercept form, then use the y-intercept, -5, and the slope of $\frac{3}{2}$ to graph the equation.

37. *Sample answer:* point-slope form; The value of the y-intercept is unknown.

39. *Sample answer:* The graph of $y - k = m(x - h)$ is a translation h units to the right and k units upward of the graph of $y = mx$.

4.2 Maintaining Mathematical Proficiency *(p. 186)*

41. $\frac{1}{5}$ **43.** $-\frac{7}{2}$

4.3 Vocabulary and Core Concept Check *(p. 191)*

1. parallel

4.3 Monitoring Progress and Modeling with Mathematics *(pp. 191–192)*

3. lines a and b; They have the same slope.

5. lines a and c; They have the same slope.

7. none; None of the lines have the same slope.

9. $y = 2x + 5$ **11.** $y = \frac{1}{3}x - 4$

13. None are parallel or perpendicular; None of the lines have the same slope or slopes that are negative reciprocals of each other.

15. None are parallel; Lines b and c are perpendicular; None of the lines have the same slope and the slope of line b is the negative reciprocal of the slope of line c.

17. Lines a and b are parallel; Line c is perpendicular to lines a and b; Lines a and b have the same slope and the slope of line c is the negative reciprocal of the slopes of lines a and b.

19. $y = -2x + 24$ **21.** $y = -\frac{1}{4}x + \frac{9}{4}$

23. a. $y = -4x + 19$ **b.** $y = \frac{1}{4}x + 2$

25. Parallel lines have the same slope, not negative reciprocal slopes; $y - 3 = \frac{1}{4}(x - 1)$; $y - 3 = \frac{1}{4}x - \frac{1}{4}$; $y = \frac{1}{4}x + \frac{11}{4}$

27. $y = -\frac{3}{4}x + \frac{3}{2}$

29. a. yes; Opposite sides are parallel.

b. no; Adjacent sides are not perpendicular.

31. no; The lines that form the angle are not perpendicular.

33. never; Perpendicular lines have opposite reciprocal slopes, so one must be positive and the other must be negative.

35. sometimes; They are perpendicular when the slopes are negative reciprocals, otherwise they will not be perpendicular.

4.3 Maintaining Mathematical Proficiency *(p. 192)*

37. function; Each input value is paired with exactly one output value.

4.4 Vocabulary and Core Concept Check *(p. 199)*

1. increase

4.4 Monitoring Progress and Modeling with Mathematics *(pp. 199–200)*

3. 6 **5.** 7 **7. a.** \$1100 **b.** 12 GB **c.** increases

9. positive **11.** no

13.

no

15. a. *Sample answer:* $y = -0.3x + 35$

b. *Sample answer:* The slope of -0.3 means the birthrate is decreasing by about 3 births per 1000 people every 10 years. The y-intercept of 35 means in 1960 the birth rate was about 35 births per 1000 people.

17. *Sample answer:*

Weight of car (pounds), x	2400	2500	2900	3000
Gas mileage (mpg), y	39	38	25	32

Weight of car (pounds), x	3400	3500	3700	5100
Gas mileage (mpg), y	30	24	21	16

19. a. *Sample answer:*

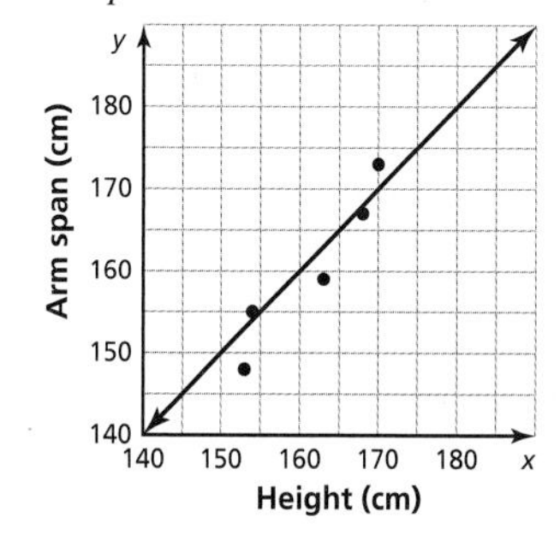

b. *Sample answer:* The slope of 1 means a person's arm span increases by about 1 centimeter for every 1 centimeter increase in height. The y-intercept of 0 has no meaning in this context because the height cannot be 0.

21. *Sample answer:* When the data are from two sets such as age and time.

23. no; The data points do not have a linear trend.

4.4 Maintaining Mathematical Proficiency *(p. 200)*

25. $-18, 0, 24$ **27.** $-23, -8, 12$

Selected Answers

4.5 Vocabulary and Core Concept Check (p. 206)

1. when the actual y-value is greater than the y-value from the model; when the actual y-value is less than the y-value from the model.

3. Interpolation is using a graph or its equation to approximate a value between two known values, and extrapolation is using a graph or its equation to predict a value outside the range of known values.

4.5 Monitoring Progress and Modeling with Mathematics (pp. 206–208)

5. no; The residual points are not evenly dispersed about the horizontal axis.

7. yes; The residual points are evenly dispersed about the horizontal axis.

9. yes; The residual points are evenly dispersed about the horizontal axis.

11. $y = 2.1x - 8$; $r = 0.980$; strong positive correlation

13. $y = 1.4x + 16$; $r = 0.999$; strong positive correlation

15. The slope and y-intercept were reversed; $y = -4.47x + 23.16$

17. **a.** $y = 381x - 566$

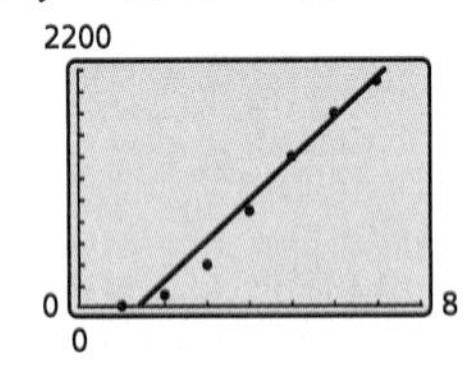

b. $r = 0.989$; strong positive correlation

c. The slope of 381 means the number of people who reported an earthquake increased by about 381 each minute after the earthquake ended. The y-intercept of -566 has no meaning in this context because the number of people cannot be negative.

19. **a.** $y = -0.2x + 20$

b. $r = -0.968$; strong negative correlation

c. The slope of -0.2 means the selling price decreases by about \$200 for every increase in mileage of 1000 miles. The y-intercept of 19.7 has no meaning in this context because a used car cannot have 0 mileage.

d. 22,500 mi **e.** \$18,800

21. There is a negative correlation and a causal relationship because the more time you spend talking on the phone, the less charge there is left in the battery.

23. A correlation is unlikely. The number of hats you own is not related to the size of your head.

25. *Sample answer:* ACT math score and SAT math score

27. **a.** $y = -0.08x + 3.8$; $r = -0.965$; strong negative correlation

b. The slope of -0.08 means the GPA decreases by about 0.08 for every hour spent watching television in a week. The y-intercept of 3.8 means that a student who watches no television has a GPA of about 3.8.

c. 2.7

d. *Sample answer:* no; The time spent watching television does not determine GPA.

29. **a.** 2863 people; 5149 people

b. relatively accurate at 9 min, but not accurate at 15 min.

31. **a.** $y = 513.5x - 298$; $r = 0.993$; strong positive correlation

b. no; The year does not determine the number of text messages sent.

c. 25.5; -128; 117.5; 50; -63.5

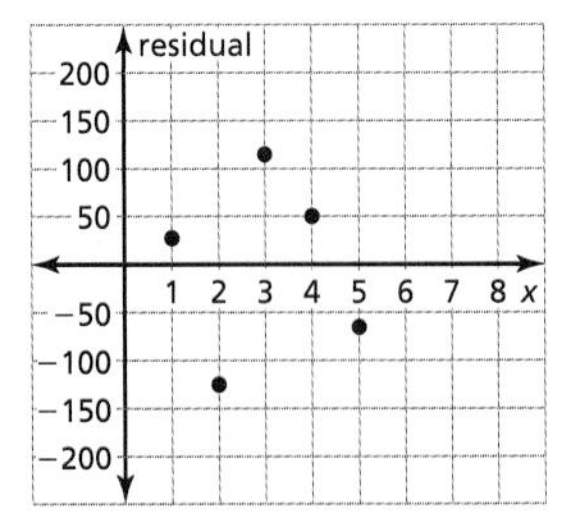

The equation $y = 513.5x - 298$ is a good fit.

d. *Sample answer:* part (a); The correlation coefficient is a single value, which is easily interpreted where as interpreting the scatter plot of the residuals is more subjective.

4.5 Maintaining Mathematical Proficiency (p. 208)

33. linear; The rate of change is constant.

4.6 Vocabulary and Core Concept Check (p. 214)

1. The graph of an arithmetic sequence is a graph of a linear function whose domain is the set of positive integers.

4.6 Monitoring Progress and Modeling with Mathematics (pp. 214–216)

3. 15, 28, 41 **5.** 5 **7.** 4 **9.** -1.5

11. 31, 34, 37 **13.** 36, 41, 46 **15.** 0.1, -0.2, -0.5

17.

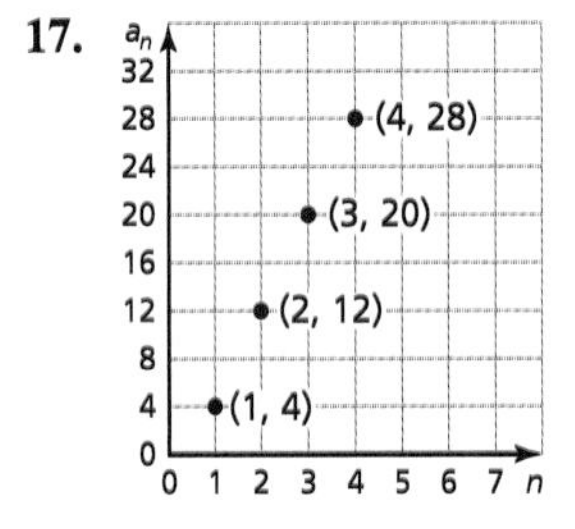

19.

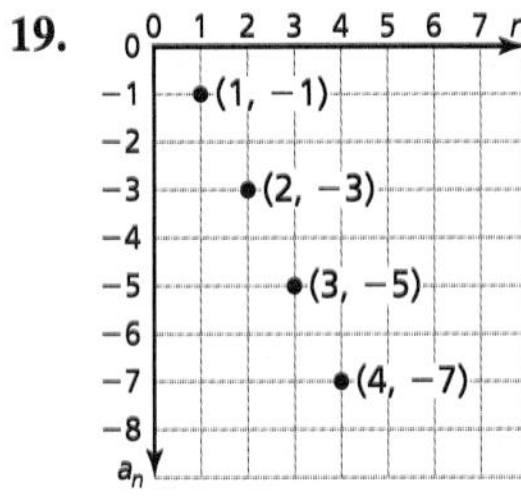

21.

a_n 16 14 12 10 8 6 4 2 0; 0 1 2 3 4 5 6 7 n; $(4, 13\frac{1}{2})$; (3, 9); $(2, 4\frac{1}{2})$; (1, 0)

23. not an arithmetic sequence; Consecutive terms do not have a common difference.

25. arithmetic sequence; Consecutive terms have a common difference of -15.

27. arithmetic sequence; 13 **29.** not an arithmetic sequence

31. 4, 6, 8, 10; yes; Consecutive terms have a common difference of 2.

33. $a_n = n - 6$; 4 **35.** $a_n = \frac{1}{2}n$; 5

37. $a_n = -10n + 20$; -80

39. -1 is added each time, not 1; The common difference is -1.

41. 7.5, 12, 16.5

43. a.

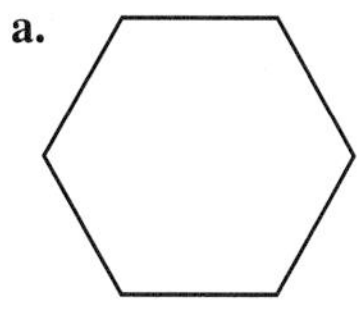
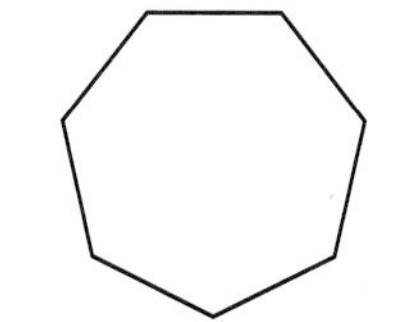
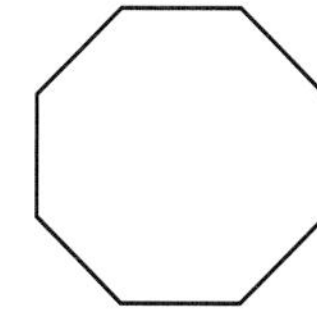

b. a regular 22-sided polygon

45. a. $f(n) = 5n$

b.

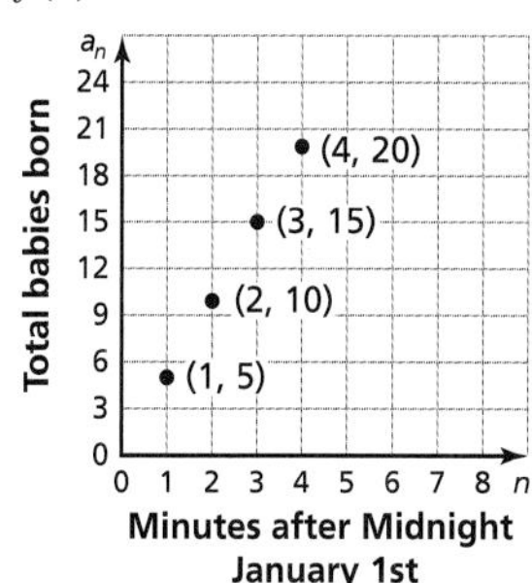

c. 20 min

47. arithmetic sequence; $f(n) = 4n - 3$; 117

49. discrete; discrete

51. *Sample answer:* $-12, -15, -18, -21$; $a_n = -3n - 9$; 12, 9, 6, 3; $a_n = -3n + 15$

53. 119 **55.** $f(n) = 6n + 17$

57. a. arithmetic sequence; Consecutive terms have a common difference of $2x$.

b. not an arithmetic sequence; Consecutive terms do not have a common difference.

4.6 Maintaining Mathematical Proficiency *(p. 216)*

59. $b > 19$

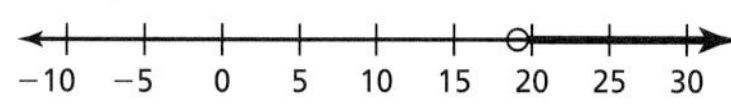

61. $y \leq -4$

63.

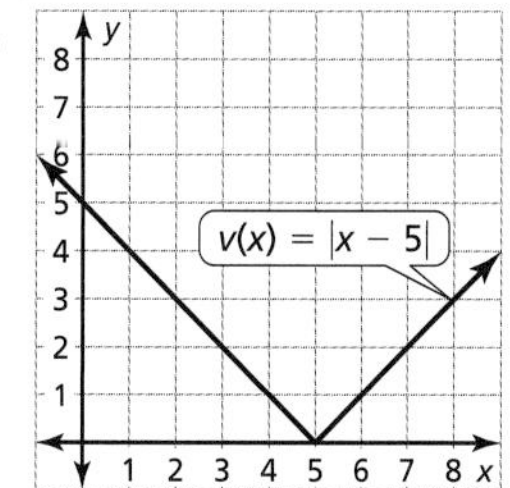

The graph of v is a horizontal translation 5 units right of the graph of f; domain: all real numbers; range: $y \geq 0$

65.

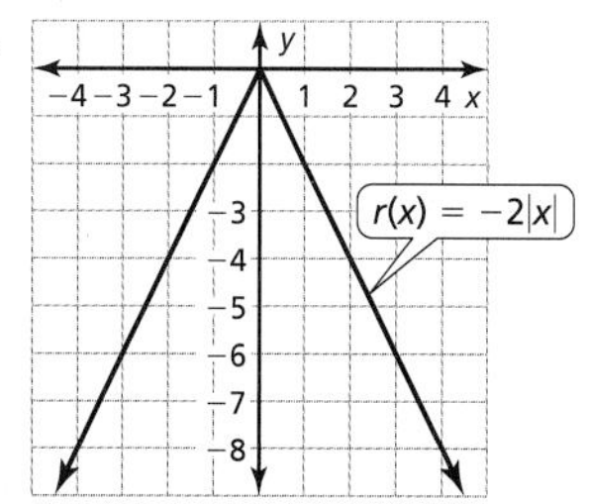

The graph of r is a vertical stretch of the graph of f by a factor of 2 and a reflection in the x-axis; domain: all real numbers; range: $y \leq 0$

4.7 Vocabulary and Core Concept Check *(p. 222)*

1. A piecewise function is a function defined by two or more equations. A step function is a special piecewise function defined by a constant value over each part of its domain.

4.7 Monitoring Progress and Modeling with Mathematics *(pp. 222–224)*

3. -16 **5.** 3 **7.** 8 **9.** 3 **11.** -1

13. 240 mi

15.

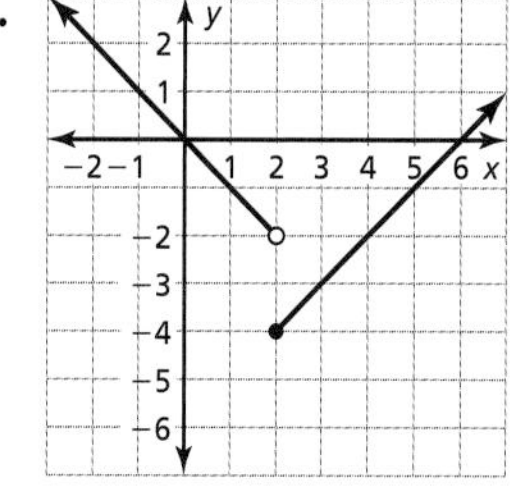

domain: all real numbers; range: $y \geq -4$

17.

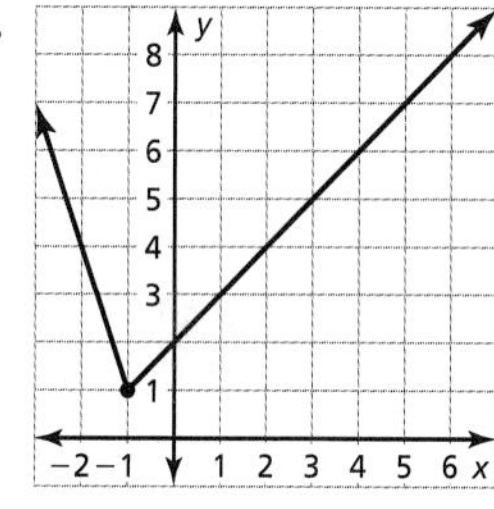

domain: all real numbers; range: $y \geq 1$

19.

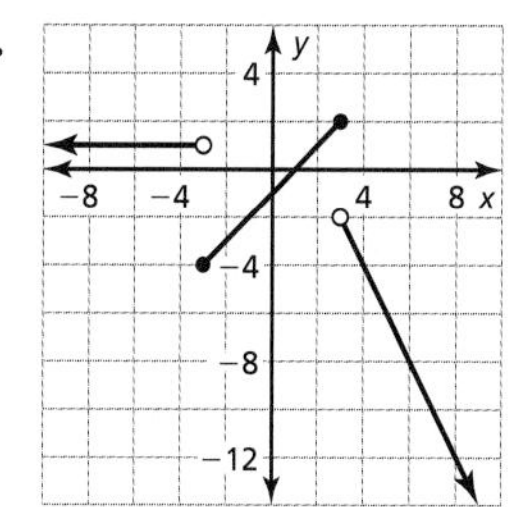

domain: all real numbers; range: $y \leq 2$

21. The wrong expression was used to evaluate the function; $f(5) = 5 + 8 = 13$

23. $f(x) = \begin{cases} x + 2, & \text{if } x < 0 \\ 2, & \text{if } x \geq 0 \end{cases}$

25. $f(x) = \begin{cases} -x, & \text{if } x < 4 \\ -x + 1, & \text{if } x \geq 4 \end{cases}$

27. $f(x) = \begin{cases} 1, & \text{if } x \leq -2 \\ 2x, & \text{if } -2 < x \leq 0 \\ -\frac{1}{2}x + 2, & \text{if } x > 0 \end{cases}$

29. $f(x) = \begin{cases} -5, & \text{if } -5 \leq x < -3 \\ -3, & \text{if } -3 \leq x < -1 \\ -1, & \text{if } -1 \leq x < 1 \end{cases}$

Selected Answers

31. 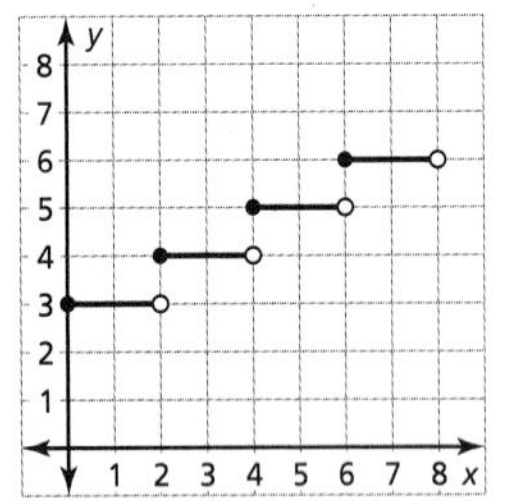

domain: $0 \le x < 8$;
range: 3, 4, 5, 6

33. 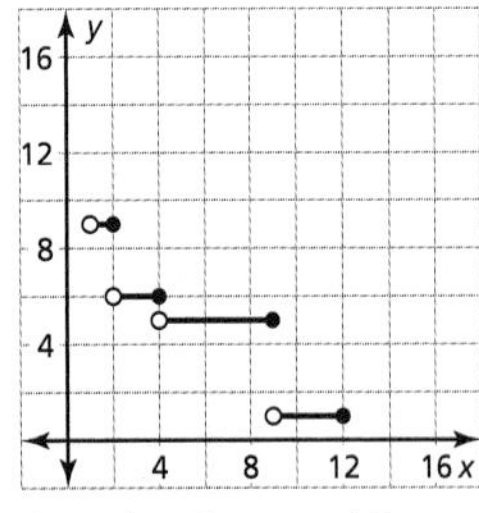

domain: $1 < x \le 12$;
range: 1, 5, 6, 9

35. $C(p) = \begin{cases} 180, & \text{if } 0 < p \le 5 \\ 210, & \text{if } 5 < p \le 6 \\ 240, & \text{if } 6 < p \le 7 \\ 270, & \text{if } 7 < p \le 8 \\ 300, & \text{if } 8 < p \le 9 \end{cases}$

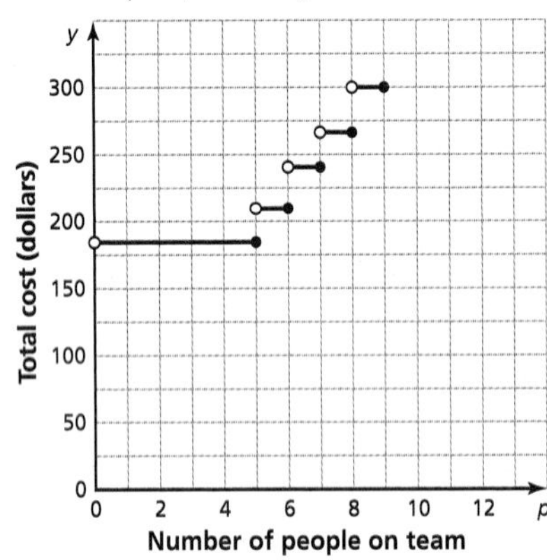

37. $y = \begin{cases} -x + 1, & \text{if } x < 0 \\ x + 1, & \text{if } x \ge 0 \end{cases}$ **39.** $y = \begin{cases} -x + 2, & \text{if } x < 2 \\ x - 2, & \text{if } x \ge 2 \end{cases}$

41. $y = \begin{cases} -2x - 6, & \text{if } x < -3 \\ 2x + 6, & \text{if } x \ge -3 \end{cases}$

43. $y = \begin{cases} 5x - 40, & \text{if } x < 8 \\ -5x + 40, & \text{if } x \ge 8 \end{cases}$

45. $y = \begin{cases} x - 1, & \text{if } x < 3 \\ -x + 5, & \text{if } x \ge 3 \end{cases}$

47. **a.** $f(x) = 2|x - 3|$ **b.** $f(x) = \begin{cases} -2x + 6, & \text{if } x < 3 \\ 2x - 6, & \text{if } x \ge 3 \end{cases}$

49. **a.** -4 **b.** 6

51. 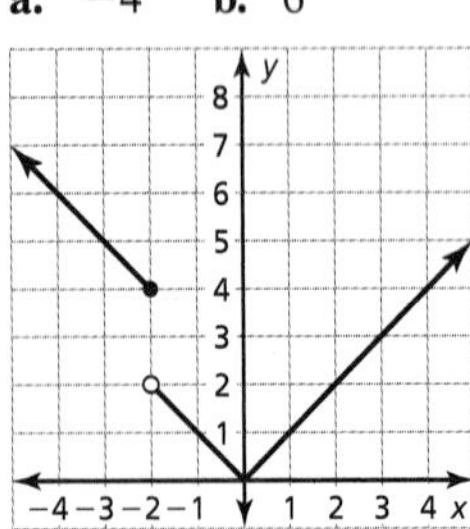

domain: all real numbers; range: $y \ge 0$

53. 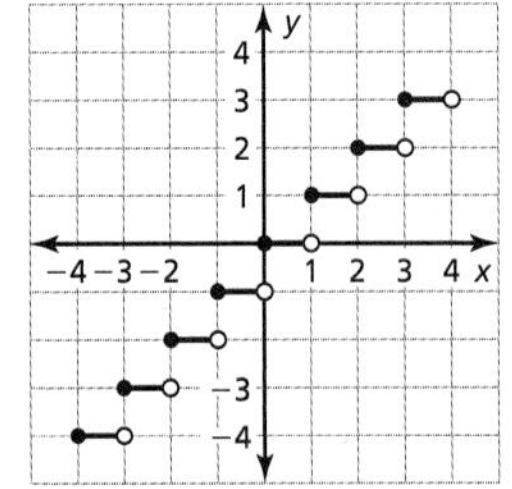

yes; yes; It is a function defined by two or more equations and it is defined by a constant value over each part of its domain.

55. **a.** $f(x) = \begin{cases} x, & \text{if } 0 \le x \le 2 \\ 2x - 2, & \text{if } 2 < x \le 8 \\ x + 6, & \text{if } 8 < x \le 9 \end{cases}$

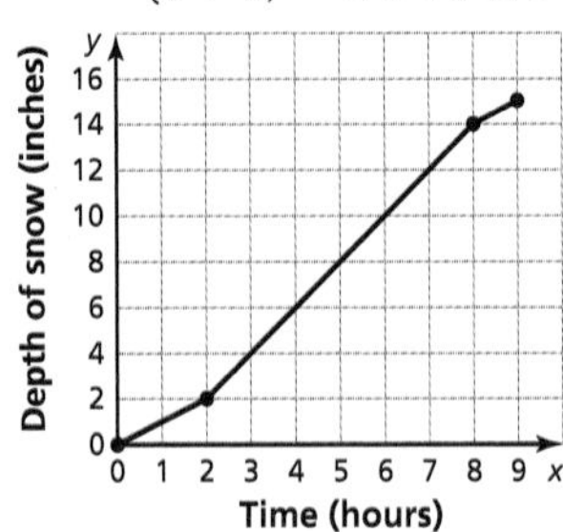

b. no; The total accumulation is 15 inches.

4.7 Maintaining Mathematical Proficiency *(p. 224)*

57. $t \le 4$ or $t \ge 18$

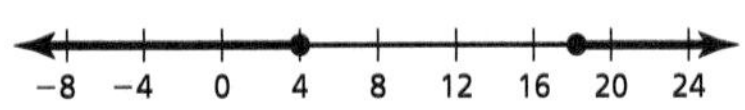

59. 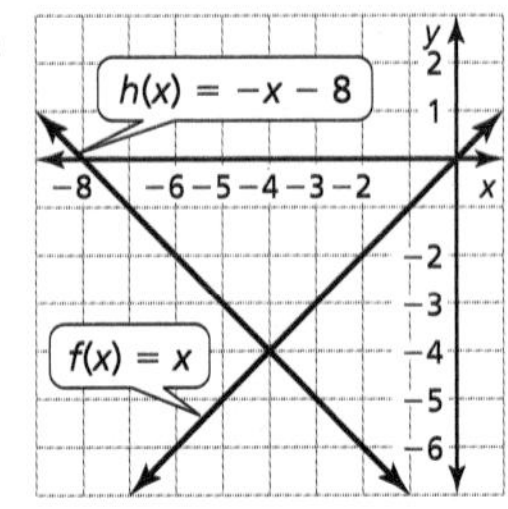

The graph of h is a reflection in the x-axis and a vertical translation 8 units down of the graph of f.

Chapter 4 Review *(pp. 226–228)*

1. $y = -\frac{1}{2}x + 1$ **2.** $y - 7 = -(x - 4)$

3. $f(x) = x - 5$ **4.** $f(x) = -4$ **5.** $f(x) = -\frac{5}{3}x + 18$

6. Lines a and b are parallel; None of the lines are perpendicular; Lines a and b have the same slope and none of them have negative reciprocal slopes.

7. None of the lines are parallel; Lines b and c are perpendicular; None of the lines have the same slope and the slope of line b is the negative reciprocal of the slope of line c.

8. $y = -4x + 9$ **9.** $y = \frac{1}{2}x - 4$ **10.** $4h$

11. *Sample answer:* $y = \frac{1}{5}x + 2$; The slope of $\frac{1}{5}$ means the roasting time increases by about $\frac{1}{5}$ hour for each pound the weight of the turkey increases. The y-intercept of 2 has no meaning in this context because the weight of the turkey cannot be 0 pounds.

12.

13. a. 65 in. **b.** 6.5

14. *Sample answer:* no; Height does not determine shoe size.

15. $a_n = -n + 12$; -18 **16.** $a_n = 6n$; 180

17. $a_n = 3n - 12$; 78 **18. a.** 3 **b.** -10

19.

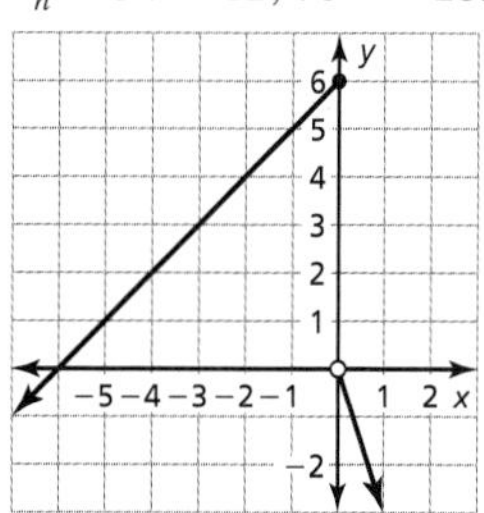

domain: all real numbers;
range: $y \leq 6$

20.

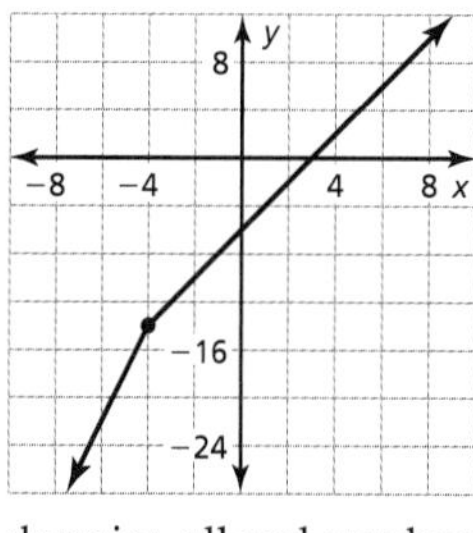

domain: all real numbers;
range: all real numbers

21. $y = \begin{cases} -x + 15, & \text{if } x < 0 \\ x + 15, & \text{if } x \geq 0 \end{cases}$

22. $y = \begin{cases} -4x - 20, & \text{if } x < -5 \\ 4x + 20, & \text{if } x \geq -5 \end{cases}$

23. $y = \begin{cases} -2x - 7, & \text{if } x < -2 \\ 2x + 1, & \text{if } x \geq -2 \end{cases}$

24. $f(x) = \begin{cases} 65, & \text{if } 0 < x \leq 1 \\ 100, & \text{if } 1 < x \leq 2 \\ 135, & \text{if } 2 < x \leq 3 \end{cases}$

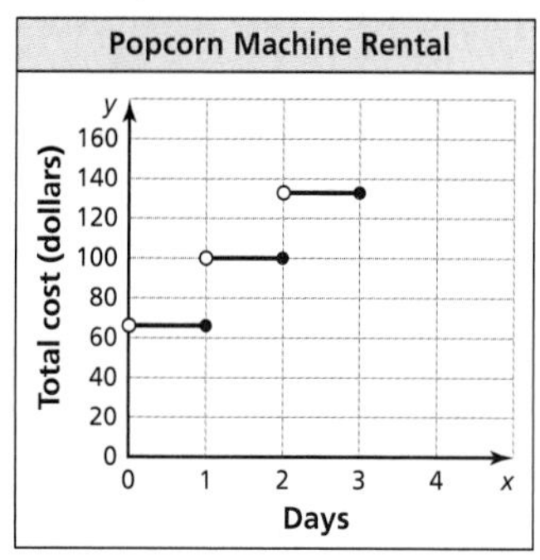

Chapter 5

Chapter 5 Maintaining Mathematical Proficiency *(p. 233)*

1.

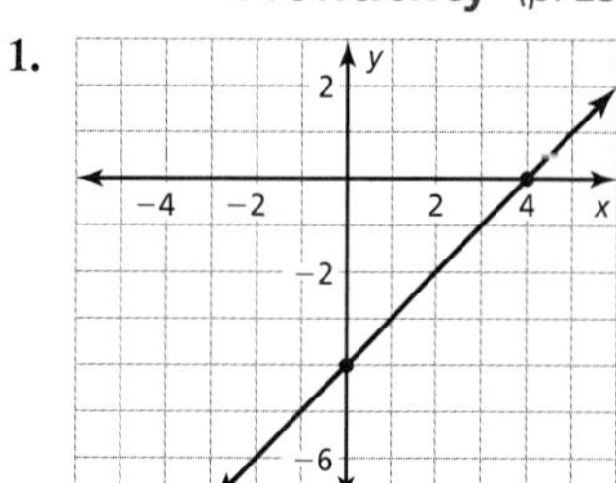

2.

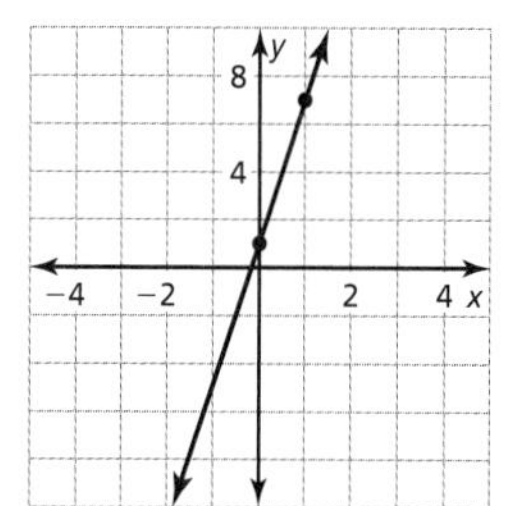

3.

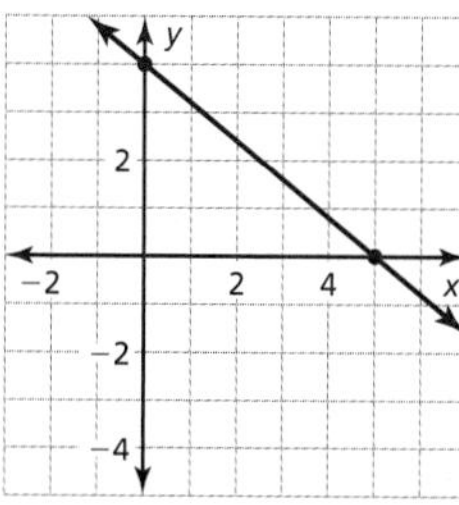

4.

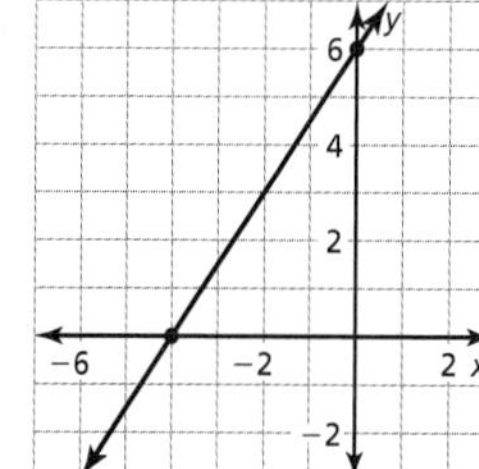

5. $m > 5$

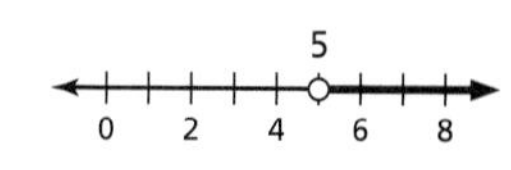

6. $t \leq -4$

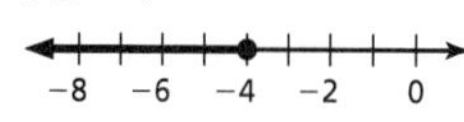

7. $a \leq 9$

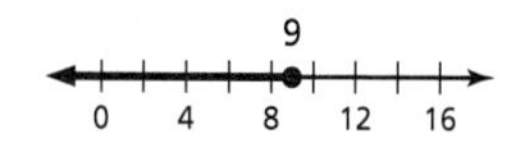

8. $z > 3$

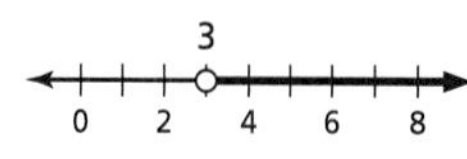

9. $k < 6$

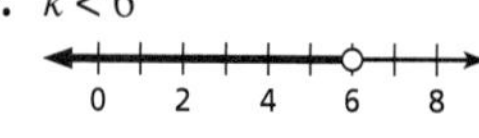

10. $w \geq -3$

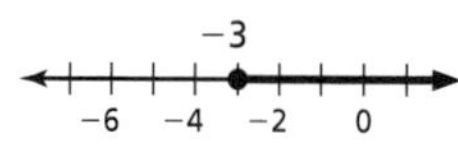

11. The two lines intersect.

5.1 Vocabulary and Core Concept Check *(p. 239)*

1. yes; They are two equations in the same variables.

5.1 Monitoring Progress and Modeling with Mathematics *(pp. 239–240)*

3. yes **5.** no **7.** yes **9.** $(1, -3)$ **11.** $(-4, 5)$

13. $(3, 4)$ **15.** $(-9, -1)$ **17.** $(-1, 2)$ **19.** $(-4, 0)$

21. The graph of the second equation should be $y = \frac{2}{3}x - 1$.

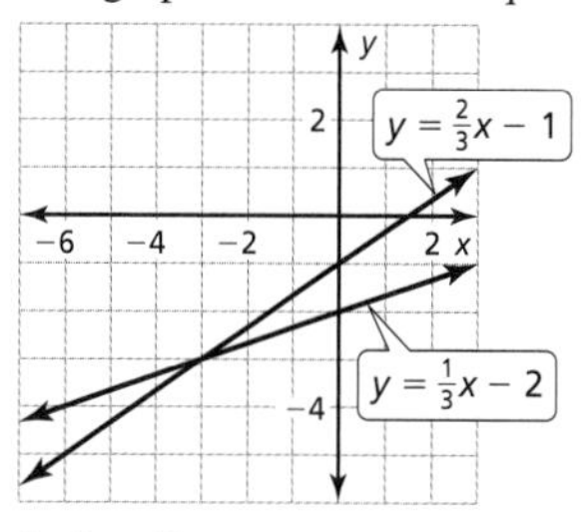

$(-3, -3)$

23. $(10, 5)$ **25.** $(1.2, 1.4)$

27. 30 min on the elliptical trainer, 10 min on the stationary bike

29. $A = 18x - 18$; $P = 6x + 6$; $(2, 18)$; The value of the perimeter and the area of the rectangle will both be 18 when $x = 2$.

31. a. $x = 3$ **b.** $(3, 5)$

c. The equation in part (a) shows that the 2 values of y in the system in part (b) are equal. The x-coordinate of the solution of the system in part (b) is the solution of the equation in part (a).

Selected Answers

33. a. $y = 5x$ and $y = 3x + 3$

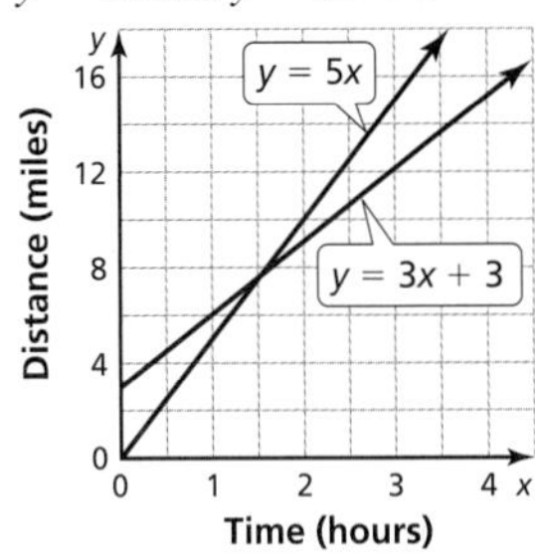

b. no; The point of intersection does not occur at $x = 1$.

5.1 Maintaining Mathematical Proficiency *(p. 240)*

35. $y = 2x + 3$

5.2 Vocabulary and Core Concept Check *(p. 245)*

1. Solve one of the equations for one of the variables. Substitute the expression for that variable into the other equation to find the value of the other variable. Substitute this value into one of the original equations to find the value of the remaining variable.

5.2 Monitoring Progress and Modeling with Mathematics *(pp. 245–246)*

3. *Sample answer:* $x - 2y = 0$; This equation can be solved for x easily.
5. *Sample answer:* $5x - y = 5$; This equation can be solved for y easily.
7. *Sample answer:* $x - y = -3$; This equation can be solved for x easily.
9. $(5, 3)$ 11. $(-4, 5)$ 13. $(6, 7)$ 15. $(5, -8)$
17. In Step 2, the expression for y needs to be substituted in the other equation; $8x + 2(5x - 4) = -12$, $8x + 10x - 8 = -12$, $18x - 8 = -12$, $18x = -4$, $x = -\frac{2}{9}$
19. $x + y = 180$ and $x = 3y$; 135 acres of corn, 45 acres of wheat
21. *Sample answer:* $y = x + 2$ and $x + 2y = 13$
23. *Sample answer:* $x = \frac{1}{3}y$ and $x + y = -16$
25. 8 five-point problems, 30 two-point problems
27. a. $x + y + 90 = 180$ b. $x = 67$, $y = 23$
29. $a = 4$, $b = 5$
31. *Sample answer:* $y = -3x + 4$ and $x + y = 6$
33. 144 pop songs, 48 rock songs, 80 hip-hop songs 35. 47

5.2 Maintaining Mathematical Proficiency *(p. 246)*

37. -13 39. $3d + 5$ 41. $64v$

5.3 Vocabulary and Core Concept Check *(p. 251)*

1. *Sample answer:* $x + 3y = 5$ and $-x + 4y = 10$

5.3 Monitoring Progress and Modeling with Mathematics *(pp. 251–252)*

3. $(1, 6)$ 5. $(4, 5)$ 7. $(-1, 2)$ 9. $(0, -10)$
11. $(1, 1)$ 13. $(8, 3)$ 15. $(-7, -12)$ 17. $(5, -3)$
19. $5x + x \neq 4x$; $6x = 24$, $x = 4$
21. $x + 5y = 22.45$ and $x + 7y = 25.45$; \$14.95 fee, \$1.50 per quart of oil
23. $(2, -1)$; *Sample answer:* elimination because y has the same coefficient in both equations
25. $(4, 6)$; *Sample answer:* substitution because the second equation is already solved for y
27. 4 and –4; These values yield x terms with either the same or opposite coefficients.
29. no; *Sample answer:* The system $8x - 5y = 11$ and $4x - 3y = 5$ can be solved by elimination in fewer steps than it can be solved by substitution.
31. a. $2\ell + 2w = 18$ and $6\ell + 4w = 46$; length: 5 in.; width: 4 in.
    b. length: 15 in.; width: 8 in.
33. 4.5 qt of 100% fruit juice, 1.5 qt of 20% fruit juice
35. $(-5, 4, 2)$; *Sample answer:* Subtract Equation 2 from Equation 1. The resulting equation only has 1 variable, y, so use it to solve for y. Substitute this result in Equation 3 and solve for x. Substitute the values of x and y in Equation 2 and solve for z.

5.3 Maintaining Mathematical Proficiency *(p. 252)*

37. $t = \frac{3}{8}$; one solution
39. all real numbers; infinitely many solutions
41. $y = 5x + 6$

5.4 Vocabulary and Core Concept Check *(p. 257)*

1. no; Two lines cannot intersect at exactly two points.

5.4 Monitoring Progress and Modeling with Mathematics *(pp. 257–258)*

3. F; no solution 5. B; infinitely many solutions
7. D; no solution 9. $(0, -4)$
11. infinitely many solutions 13. infinitely many solutions
15. no solution
17. infinitely many solutions; The lines have the same slope and the same y-intercept, so they are the same line.
19. one solution; The lines have different slopes, so they will intersect.
21. no solution; The lines have the same slope but different y-intercepts, so they are parallel.
23. The lines are not parallel, so they must intersect.

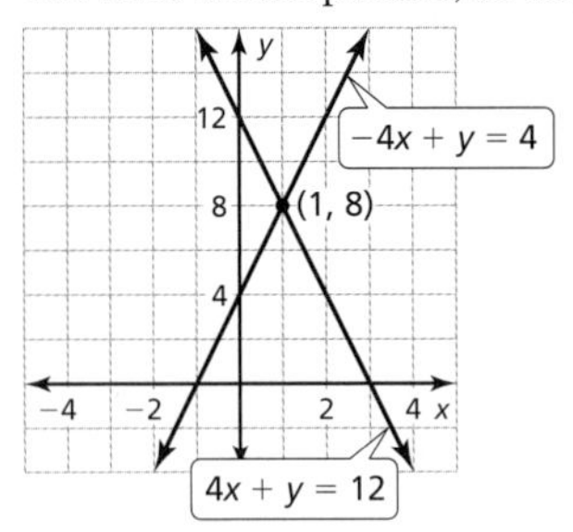

$(1, 8)$

25. $3f + 4a = 6$ and $\frac{9}{2}f + 6a = 9$; infinitely many solutions
27. yes; The system of equations $150c + 80b = 22{,}860$ and $170c + 100b = 27{,}280$ has a solution.
29. one; The lines have different slopes, so they will intersect at exactly one point.
31. a. never; The y-intercepts are different, so they can never be equations for the same line.
    b. sometimes; When $a = b$, the lines are parallel and there is no solution.
    c. always; When $a < b$, the slopes are different and the lines intersect at one point.

5.4 Maintaining Mathematical Proficiency *(p. 258)*

33. $x = -6$, $x = -2$ 35. $x = 1$, $x = 5$

5.5 Vocabulary and Core Concept Check *(p. 265)*

1. $x = 6$

5.5 Monitoring Progress and Modeling with Mathematics *(pp. 265–266)*

3. $x = 1$ **5.** $x = -3$ **7.** $x = -2$ **9.** $x = -3$

11. $x = 2$ **13.** $x = 3$ **15.** $x = 2$; one solution

17. all real numbers; infinitely many solutions

19. no solution **21.** $x = -2, x = 1$ **23.** $x = 3, x = -1$

25. $x = 2, x = -2$ **27.** $x = 2$ **29.** $x = -3, x = 1$

31. $x = -2$ **33.** 75 guests **35.** 30 seconds

37. *Sample answer:* $m = 1, b = 8$ **39.** $x = 10$

41. a. negative **b.** positive

5.5 Maintaining Mathematical Proficiency *(p. 266)*

43. −6 −4 −2 0 2

45.

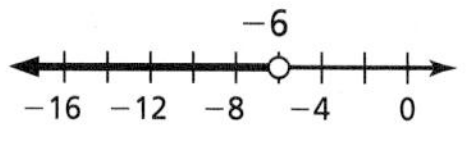

47. The graph of g is a reflection in the x-axis of the graph of f.

49. The graph of g is a horizontal translation 1 unit right of the graph of f.

5.6 Vocabulary and Core Concept Check *(p. 271)*

1. Substitute the values into the inequality and verify that the statement is true.

5.6 Monitoring Progress and Modeling with Mathematics *(pp. 271–272)*

3. yes **5.** no **7.** yes **9.** no **11.** no

13. yes **15.** no

17. no; (12, 14) is not a solution of the inequality.

19.

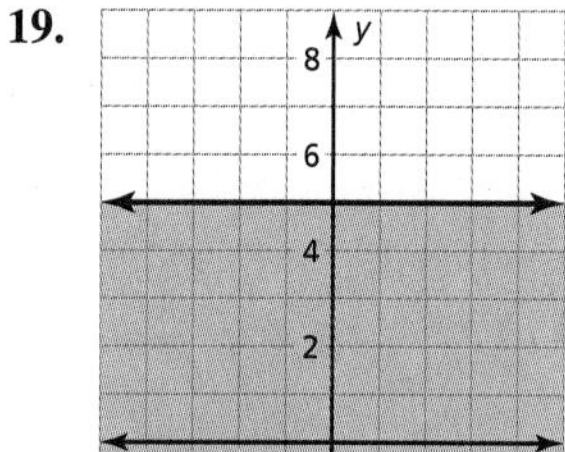

21.

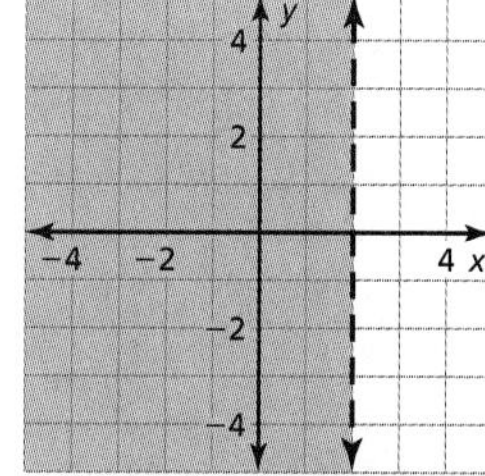

23.

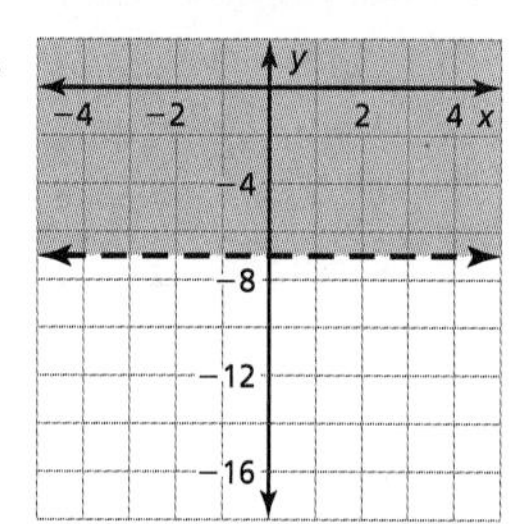

25.

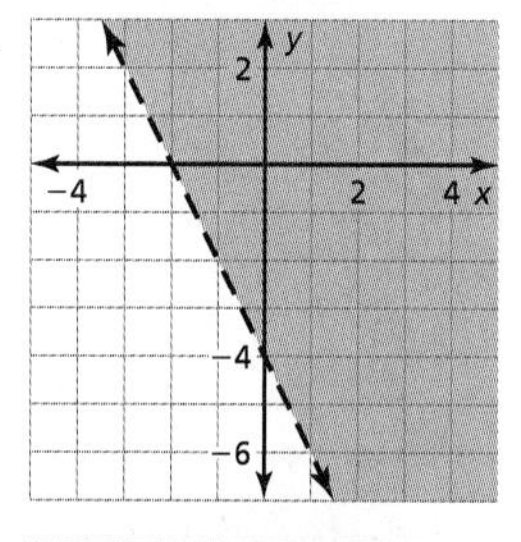

27.

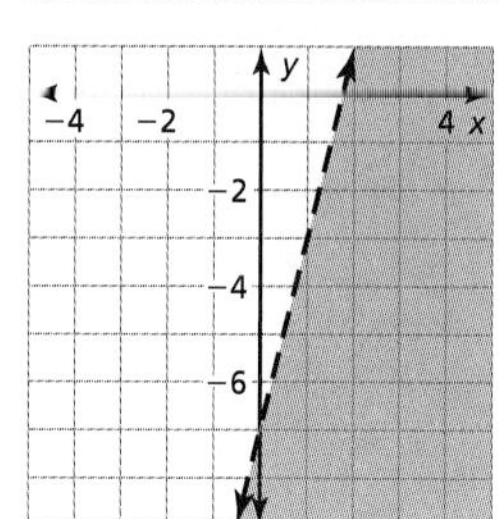

29.

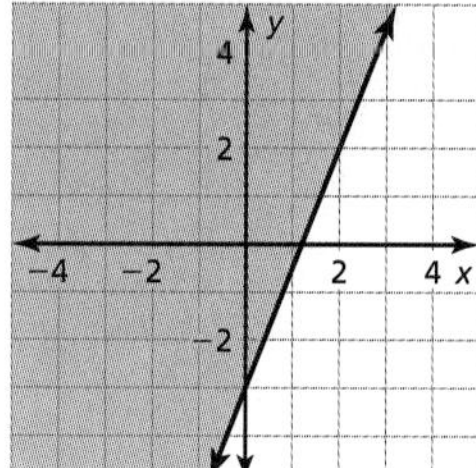

31. The line should be dashed.

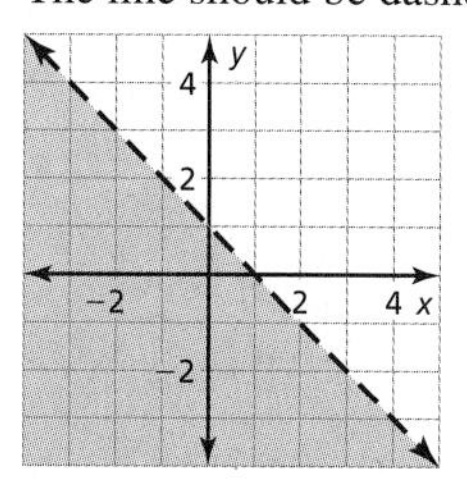

33. $0.75x + 2.25y \leq 20$

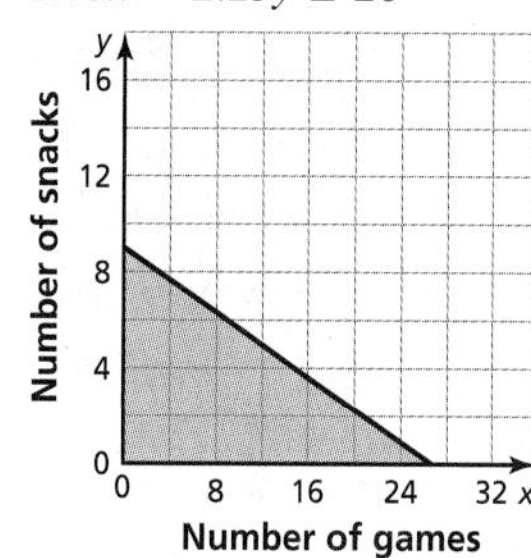

Sample answer: (12, 4), You can play 12 games and buy 4 snacks; (5, 2), You can play 5 games and buy 2 snacks.

35. $y > 2x + 1$ **37.** $y \leq -\frac{1}{2}x - 2$

39. a. $75x + 40y \leq 1800$

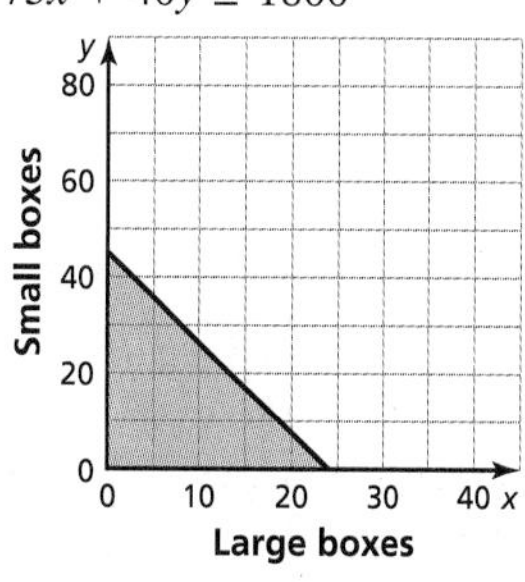

b. *Sample answer:* The number of boxes must be a positive whole number. It is also unlikely for 1 person to carry 45 boxes onto an elevator.

41. A test point on the boundary line is not in either half-plane, so it will not indicate which half-plane to shade.

43. no; You cannot use (0, 0) as a test point when it is on the boundary line.

45. $y < 3x + 5$

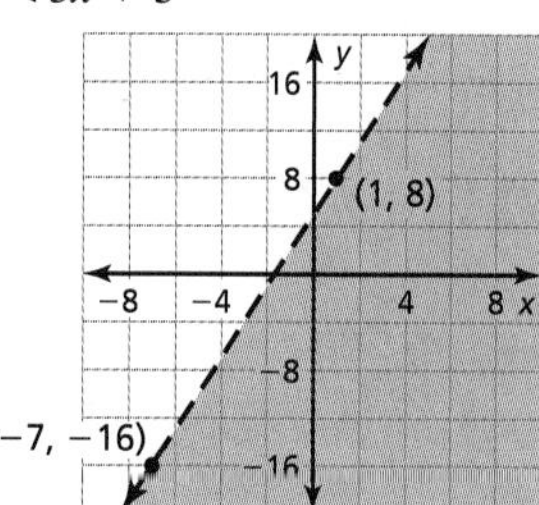

5.6 Maintaining Mathematical Proficiency *(p. 272)*

47. −20, −23, −26

5.7 Vocabulary and Core Concept Check *(p. 278)*

1. Substitute the values into both inequalities and verify that both inequalities are true.

5.7 Monitoring Progress and Modeling with Mathematics *(pp. 278–280)*

3. no **5.** no **7.** yes **9.** no

11.

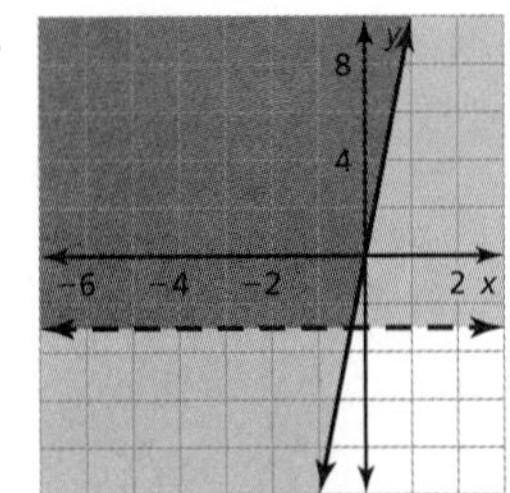

13.

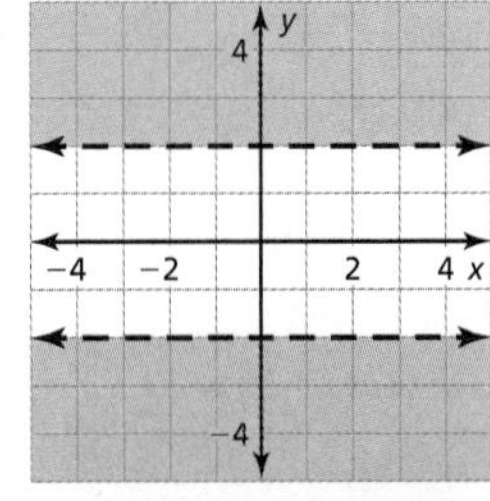

15.

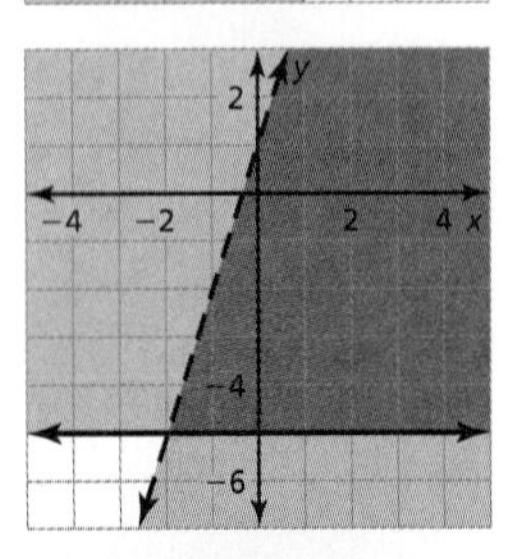

17.

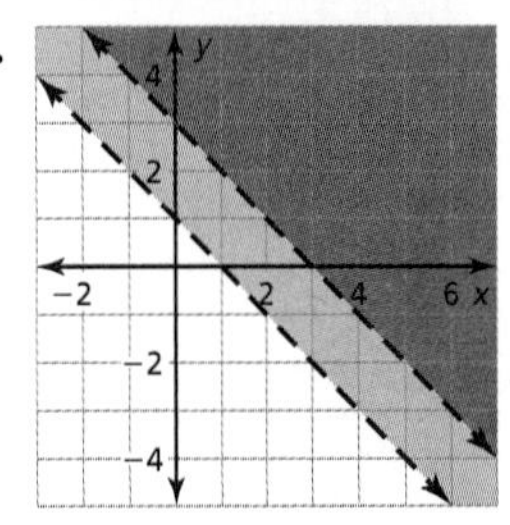

19.

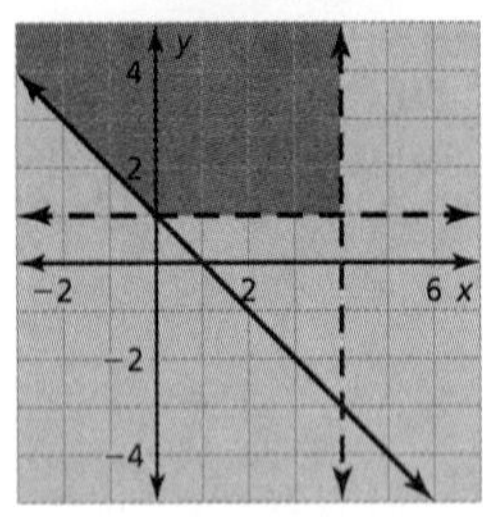

21. $x \geq -1$ and $y < 3$ **23.** $y \geq \frac{2}{3}x - 2$ and $y \geq -3x + 2$

25. $y > -2x - 1$ and $y < -2x - 3$

27. The wrong regions are shaded for both inequalities.

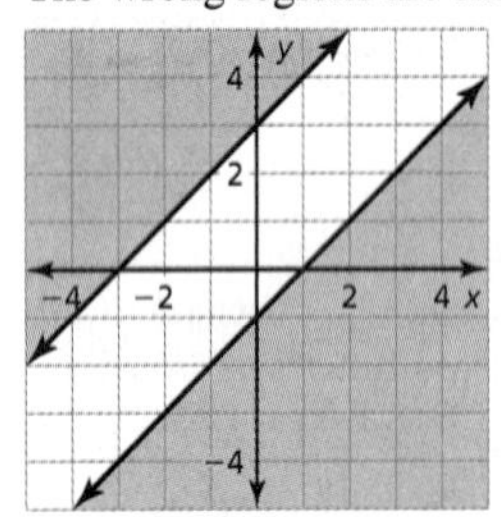

29. **a.** $4x + 3y \leq 21$ and $x + y \geq 3$

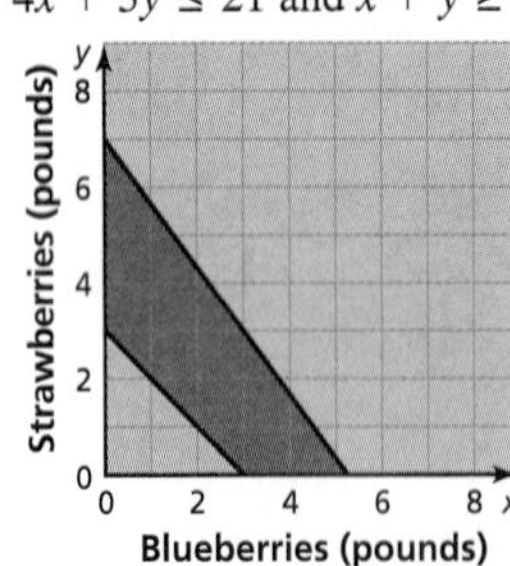

b. *Sample answer:* (2, 4), You can buy 2 pounds of blueberries and 4 pounds of strawberries.

c. yes

31. **a.** $x \leq 15$, $y \leq 10$, and $x + y \leq 20$

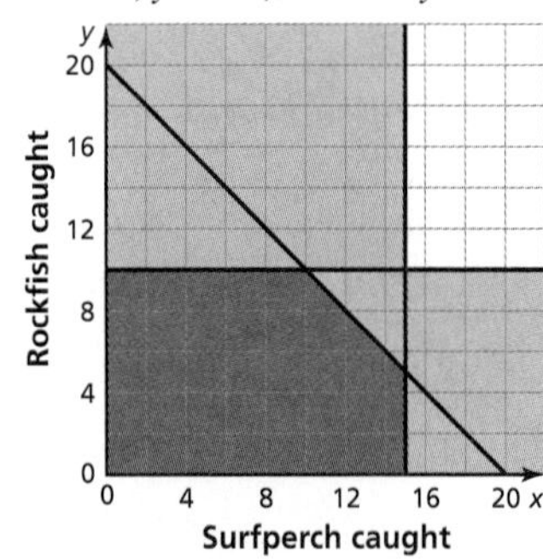

b. yes

33. **a.** $x \geq -1, x \leq 6, y \geq -3$, and $y \leq 1$ **b.** 28 square units

35. *Sample answer:* \$300 on savings, \$500 on housing

37. Both can be done by graphing; Solving a system of linear inequalities requires finding overlapping half-planes, solving a system of linear equations does not.

39. $y > -x$; $y < x$; and $x > 0$

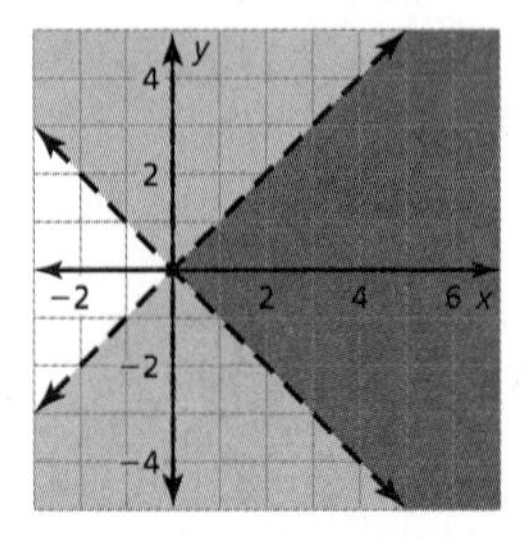

41. no; The solution of each inequality is a half-plane, and so the intersection can be at most a half-plane.

43. *Sample answer:* $x > 2$ and $y < -4$

45. **a.** *Sample answer:* $-4x + 2y < 6$

b. *Sample answer:* $-2x + y > 3$

47. *Sample answer:* $x \geq 4, x \leq 4, y \geq 5$, and $y \leq 5$

5.7 Maintaining Mathematical Proficiency *(p. 280)*

49. 4^5 **51.** x^6 **53.** $y = -3x + 5$ **55.** $y = \frac{4}{3}x$

Chapter 5 Review *(pp. 282–284)*

1. (2, −5) **2.** (1, −1) **3.** (4, −1) **4.** (−2, −3)

5. (2, 1) **6.** (2, 0) **7.** 8 brushes, 4 tubes of paint

8. (2, −8) **9.** (−2, 5) **10.** (4, 5) **11.** no solution

12. no solution **13.** infinitely many solutions

14. $x = -3$ **15.** $x = -5$ **16.** $x = 13, x = 1$

17.

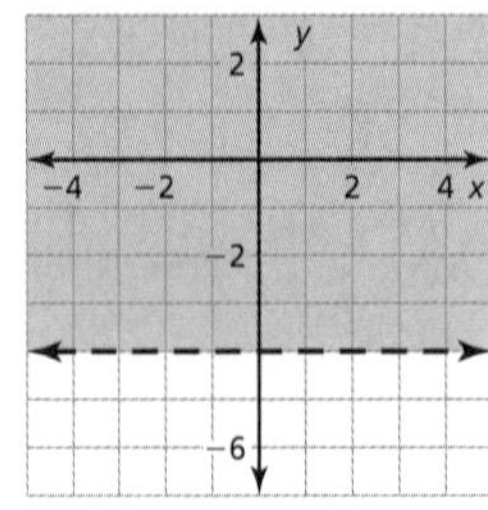

18.

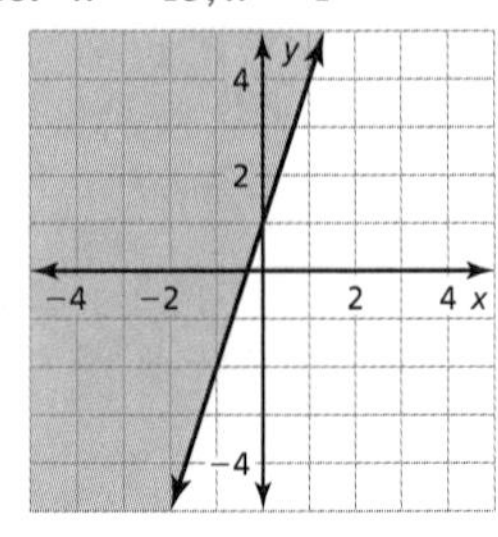

19.

20. 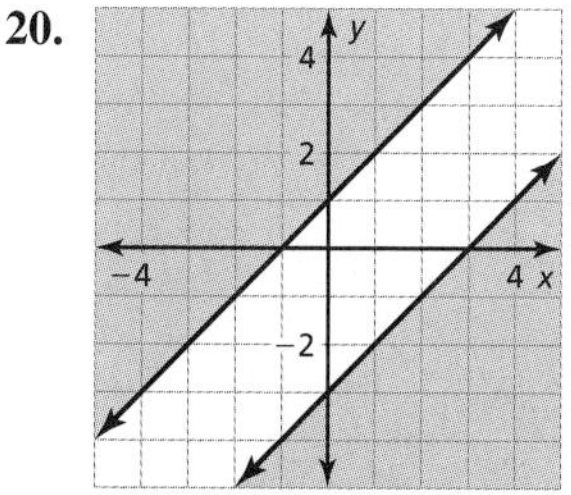

21. 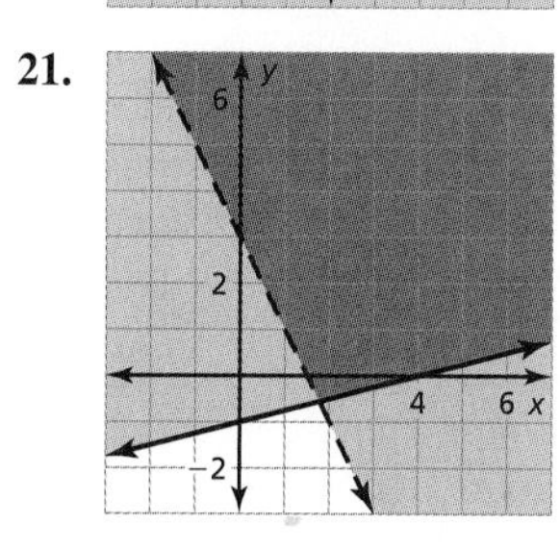

22. 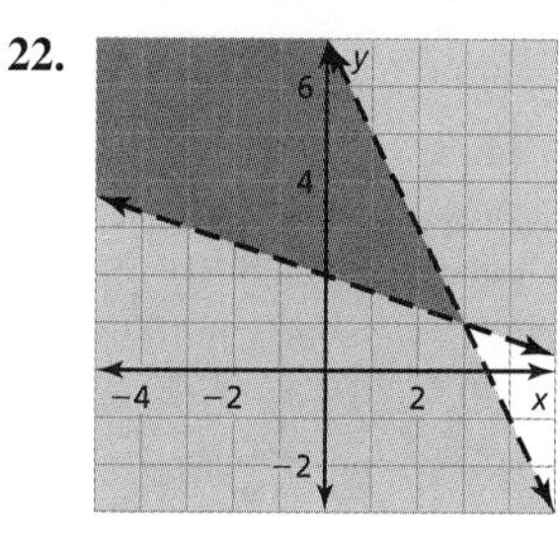

Chapter 6

Chapter 6 Maintaining Mathematical Proficiency (p. 289)

1. −9 **2.** 106 **3.** −12 **4.** 8 **5.** −2
6. −5 **7.** ±11 **8.** $a_n = 2n + 10$
9. $a_n = -3n + 9$ **10.** $a_n = -7n + 29$
11. yes; no; The product of two perfect squares can be represented by $m^2n^2 = (mm)(nn) = (mn)(mn) = (mn)^2$. If m and n are integers, their product is also an integer, so $(mn)^2$ is an integer. There are many counterexamples illustrating that the quotient of two perfect squares does not have to be a perfect square, such as $9 \div 4$.

6.1 Vocabulary and Core Concept Check (p. 296)

1. Product of Powers Property, Power of a Power Property, definition of negative exponents; Use the Product of Powers Property to simplify the expression inside the parentheses to 4^4. Then, use the Power of a Power Property to simplify the entire expression to 4^{-8}. Then, use the definition of negative exponents to produce the final answer, $\frac{1}{4^8} = \frac{1}{65{,}536}$.
3. The Quotient of Powers Property is used when dividing powers that have the same base. The answer is the common base raised to the difference of the exponents of the numerator and denominator.

6.1 Monitoring Progress and Modeling with Mathematics (pp. 296–298)

5. 1 **7.** $\frac{1}{625}$ **9.** $\frac{1}{16}$ **11.** $-\frac{4}{3}$ **13.** $\frac{1}{x^7}$
15. $\frac{9}{y^3}$ **17.** $\frac{1}{4m^3}$ **19.** $\frac{b^7}{64}$ **21.** $\frac{32x^7}{y^6}$ **23.** 625
25. 6561 **27.** p^{24} **29.** $\frac{1}{216}$ **31.** x^2 **33.** 10^{-2} m
35. The product has a base of 2, not $2 \cdot 2$; $2^4 \cdot 2^5 = 2^9$
37. $-125z^3$ **39.** $\frac{n^2}{36}$ **41.** $\frac{1}{243s^{40}}$ **43.** $\frac{36}{w^6}$
45. B, C, D; These expressions simplify to be the volume of the sphere, which is $\frac{32\pi s^6}{3}$.
47. $\frac{16y^{28}}{81x^{12}}$ **49.** $\frac{n^{10}}{144m^4}$ **51.** 4.5×10^{-3}; 0.0045
53. 4×10^2; 400
55. about 4.113×10^4 lb/acre; about 41,130 lb/acre
57. **a.** Power of a Product Property
b. Express $\frac{(6x)^3}{(2x)^3}$ as $\left(\frac{6x}{2x}\right)^3$. Simplify the expression inside the parentheses to produce $(3)^3$, so the volume is 27 times greater.
59. $(2ab)^3$ **61.** $(2w^3z^2)^6$
63. **a.** $\left(\frac{1}{6}\right)^n$ **b.** $\frac{1}{1296}$
c. $\frac{1}{32}$; The probability of flipping heads once is $\frac{1}{2}$, and $\left(\frac{1}{2}\right)^5$ is $\frac{1}{32}$.
65. $x = 8, y = -1$; Using the Quotient of Powers Property, you can conclude from the first equation that $x - y = 9$. Using the Product of Powers Property and the Quotient of Powers Property, you can conclude from the second equation that $x + 2 - 3y = 13$. Use these equations to solve a system of linear equations.
67. no; The mass of the seed from the double coconut palm is 10 kilograms.
69. **a.** When $a > 1$ and $n < 0$, $a^n < a^{-n}$ because a^n will be less than 1 and a^{-n} will be greater than 1. When $a > 1$ and $n = 0$, $a^n = a^{-n} = 1$, because any number to the zero power is 1. When $a > 1$ and $n > 0$, $a^n > a^{-n}$ because a^n will be greater than 1 and a^{-n} will be less than 1.
b. When $0 < a < 1$ and $n < 0$, $a^n > a^{-n}$, because a^n will be greater than 1 and a^{-n} will be less than 1. When $0 < a < 1$ and $n = 0$, $a^n = a^{-n} = 1$, because any number to the zero power is 1. When $0 < a < 1$ and $n > 0$, $a^n < a^{-n}$ because a^n will be less than 1 and a^{-n} will be greater than 1.

6.1 Maintaining Mathematical Proficiency (p. 298)

71. −10
73. natural number, whole number, integer, rational number, real number
75. irrational number, real number

6.2 Vocabulary and Core Concept Check (p. 303)

1. Find the fourth root of 81, or what real number multiplied by itself four times produces 81.

6.2 Monitoring Progress and Modeling with Mathematics (pp. 303–304)

3. $10^{1/2}$ **5.** $\sqrt[3]{15}$ **7.** ±6 **9.** 10 **11.** $s = 4$ in.
13. 4 **15.** −7 **17.** 2 **19.** $8^{4/5}$ **21.** $(\sqrt[7]{-4})^2$
23. 8 **25.** not a real number **27.** −32
29. The numerator and denominator are reversed; $(\sqrt[3]{2})^4 = 2^{4/3}$
31. $\frac{1}{10}$ **33.** $\frac{1}{9}$ **35.** 6 ft² **37.** about 1 in.
39. Write the radicand, a, as the base and write the exponent as a fraction with the power, m, as the numerator and the index, n, as the denominator.
41. about 5.5% **43.** −1, 0, and 1 **45.** $(xy)^{1/2}$
47. $2xy^2$ **49.** about 1.38 ft
51. always; Power of a Power Property
53. always; definition of rational exponent
55. sometimes; false if $x = 0$ because division by 0 is undefined

6.2 Maintaining Mathematical Proficiency *(p. 304)*

57. $f(-3) = -16; f(0) = -10; f(8) = 6$

59. $h(-3) = 16; h(0) = 13; h(8) = 5$

6.3 Vocabulary and Core Concept Check *(p. 310)*

1. *Sample answer:*

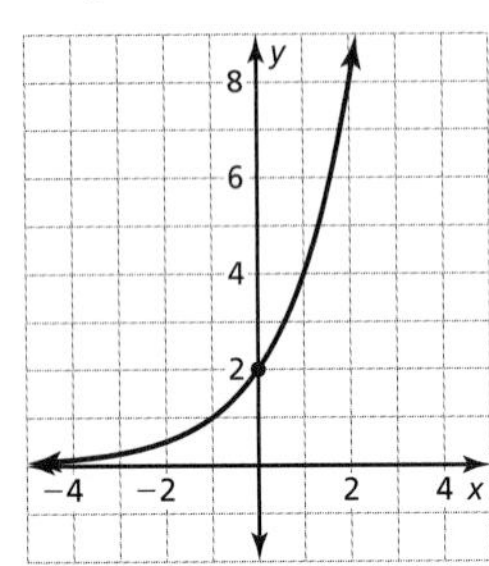

3. The graph of $y = 2(5)^x$ is a vertical stretch by a factor of 2 of the graph of $y = 5^x$. The y-intercept of $y = 2(5)^x$, 2, is above the y-intercept of $y = 5^x$, 1.

6.3 Monitoring Progress and Modeling with Mathematics *(pp. 310–312)*

5. yes; It fits the pattern $y = ab^x$.

7. no; The exponent is a constant.

9. no; Although it fits the pattern $y = ab^x$, b cannot be negative.

11. linear; As x increases by 1, y increases by 2. The rate of change is constant.

13. exponential; As x increases by 1, y is multiplied by 4.

15. 9 **17.** -100 **19.** 72 **21.** C **23.** A

25.

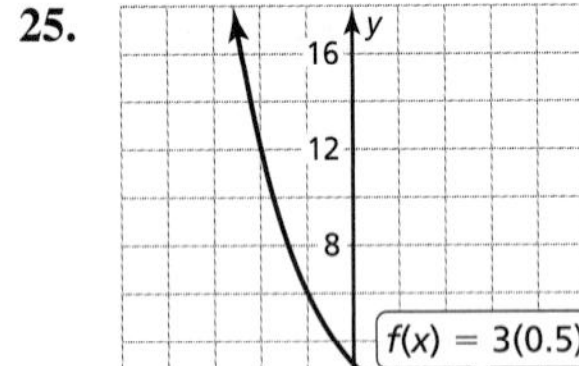

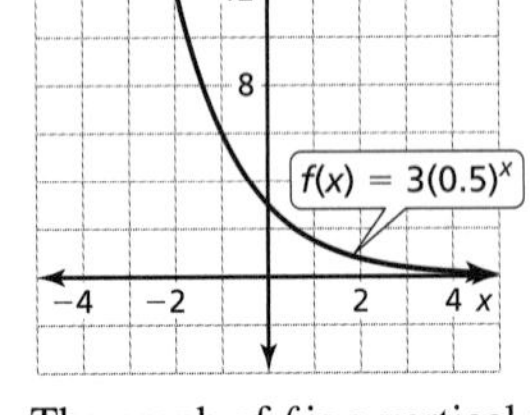

The graph of f is a vertical stretch by a factor of 3 of the graph of the parent function. The y-intercept of the graph of f, 3, is above the y-intercept of the graph of the parent function, 1; domain: all real numbers, range: $y > 0$

27.

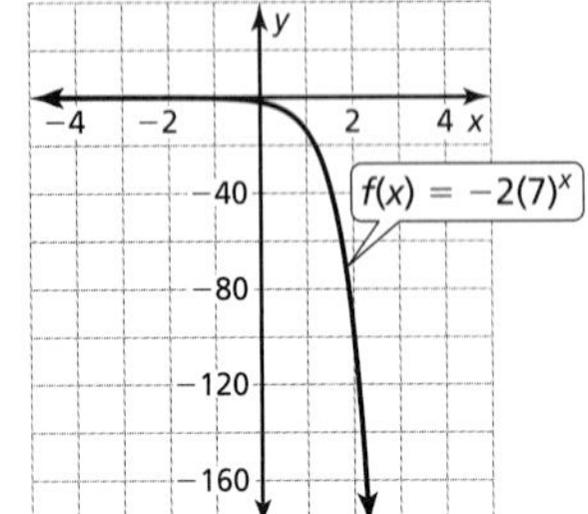

The graph of f is a vertical stretch by a factor of 2 and a reflection in the x-axis of the graph of the parent function. The y-intercept of the graph of f, -2, is below the y-intercept of the graph of the parent function, 1; domain: all real numbers, range: $y < 0$

29.

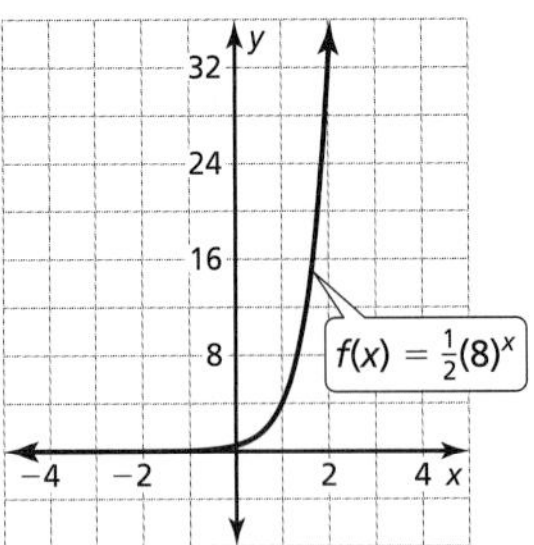

The graph of f is a vertical shrink by a factor of $\frac{1}{2}$ of the graph of the parent function. The y-intercept of the graph of f, $\frac{1}{2}$, is below the y-intercept of the graph of the parent function, 1; domain: all real numbers, range: $y > 0$

31.

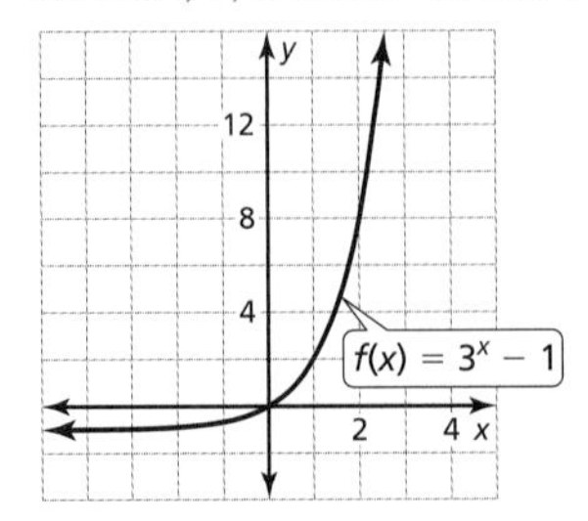

domain: all real numbers, range: $y > -1$

33.

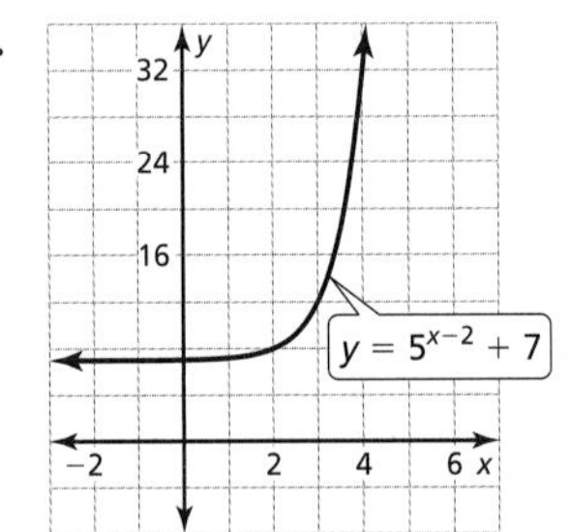

domain: all real numbers, range: $y > 7$

35.

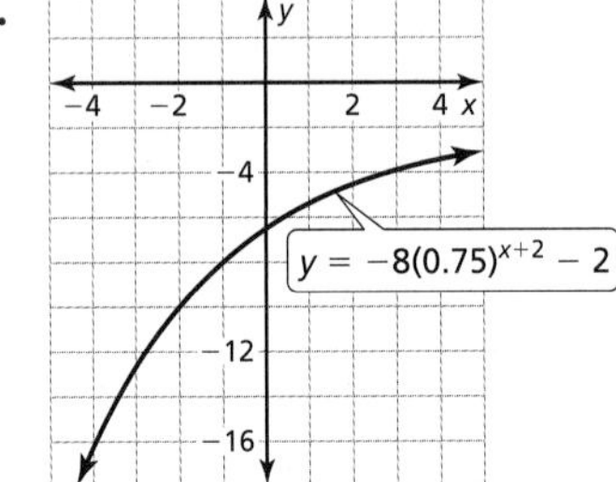

domain: all real numbers, range: $y < -2$

37. The graph of g is a vertical shrink by a factor of $\frac{1}{2}$ of the graph of f; $a = \frac{1}{2}$

39. The graph of g is a horizontal translation 4 units right of the graph of f; $h = 4$

41. need to simplify the power before multiplying; $6(0.5)^{-2} = 6(4) = 24$

43. 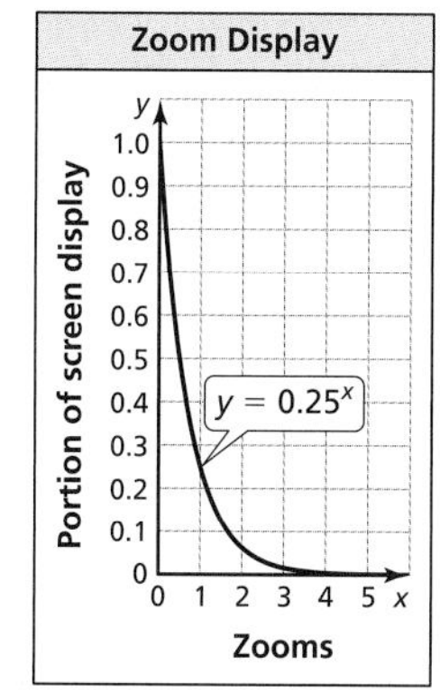

The value of f is less than the value of g over the entire interval.

45. a. 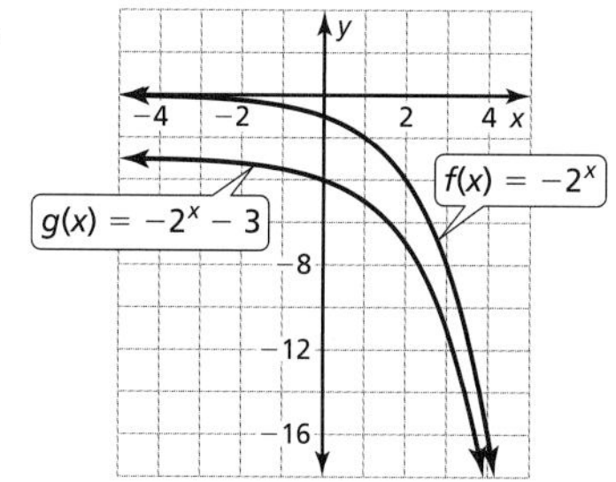

domain: $x \geq 0$, range: $0 < y \leq 1$

b. The y-intercept is 1. This means that when you do not zoom in, 100%, or all, of the original screen display is seen.

c. $6\frac{1}{4}\%$

47. $y = 2(7)^x$ **49.** $y = -\frac{1}{2}(2)^x$

51. a. $y = 40\left(\frac{3}{2}\right)^x$ **b.** about 304 visitors

53. 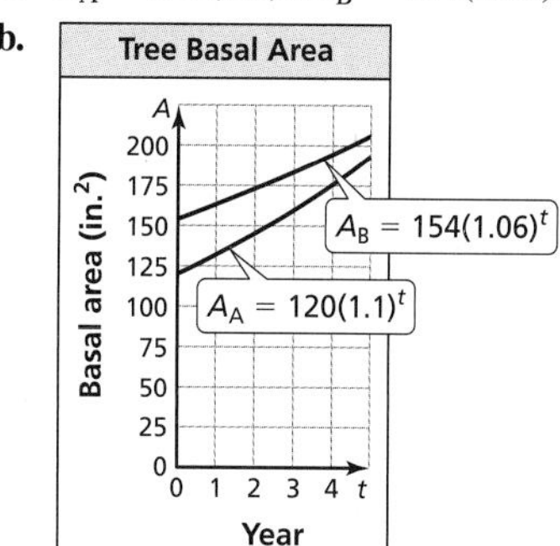

The y-intercept of g is 3 units below the y-intercept of f. The domain of both functions is all real numbers. The range of g is $y < -3$ and the range of f is $y < 0$.

55. When a is positive, it causes a vertical stretch or shrink on the graph. When a is negative, it causes a vertical stretch or shrink and a reflection in the x-axis on the graph.

57. $g(x) = 5^{x-3} + 4$ **59.** -192

61. $\frac{f(x+k)}{f(x)} = \frac{ab^{x+k}}{ab^x} = \frac{b^{x+k}}{b^x} = b^{(x+k)-x} = b^k$

63. *Sample answer:* $y = 8(2)^x$

6.3 Maintaining Mathematical Proficiency *(p. 312)*

65. 0.35 **67.** 2.5

6.4 Vocabulary and Core Concept Check *(p. 319)*

1. rate of growth

3. Exponential growth occurs when a quantity increases by the same factor over equal intervals of time. Exponential decay occurs when a quantity decreases by the same factor over equal intervals of time.

6.4 Monitoring Progress and Modeling with Mathematics *(pp. 319–322)*

5. $a = 350$, $r = 75\%$; about 5744.6

7. $a = 25$, $r = 20\%$; about 62.2

9. $a = 1500$, $r = 7.4\%$; about 2143.4

11. $a = 6.72$, $r = 100\%$; about 215.0

13. $y = 10{,}000(1.65)^t$ **15.** $y = 210{,}000(1.125)^t$

17. a. $y = 315{,}000(1.02)^t$ **b.** about 468,000

19. $a = 575$, $r = 60\%$; about 36.8

21. $a = 240$, $r = 25\%$; about 101.3

23. $a = 700$, $r = 0.5\%$; about 689.6

25. $a = 1$, $r = 12.5\%$; about 0.7 **27.** $y = 100{,}000(0.98)^t$

29. $y = 100(0.905)^t$

31. The growth rate is $1 + 1.5$, not just 1.5; $b(t) = 10(2.5)^t$; $b(8) = 10(2.5)^8 \approx 15{,}258.8$; After 8 hours, there are about 15,259 bacteria in the culture.

33. exponential decay; As x increases by 1, y is multiplied by $\frac{1}{5}$.

35. neither; As x increases by 1, y decreases by 6.

37. exponential growth; As x increases by 5, y is multiplied by 4.

39. a. exponential decay **b.** about $15,155

41. exponential decay; 20% **43.** exponential decay; 5%

45. exponential growth; 6% **47.** exponential growth; 25%

49. $y \approx 1.52(0.9)^t$; exponential decay

51. $y \approx 2(1.69)^t$; exponential growth

53. $x(t) \approx (1.20)^t$; exponential growth

55. $b(t) \approx 0.67(0.55)^t$; exponential decay

57. $y = 2000(1.0125)^{4t}$ **59.** $y = 6200(1.007)^{12t}$

61. a. $A_A = 120(1.1)^t$; $A_B = 154(1.06)^t$

b. Tree Basal Area

Basal area (in.²) vs. Year: $A_B = 154(1.06)^t$, $A_A = 120(1.1)^t$

The basal area of tree B is larger than the basal area of tree A, but the difference between the basal areas is decreasing.

63. a. $y = 25{,}000(1.055)^t$ **b.** about 0.45%

c. 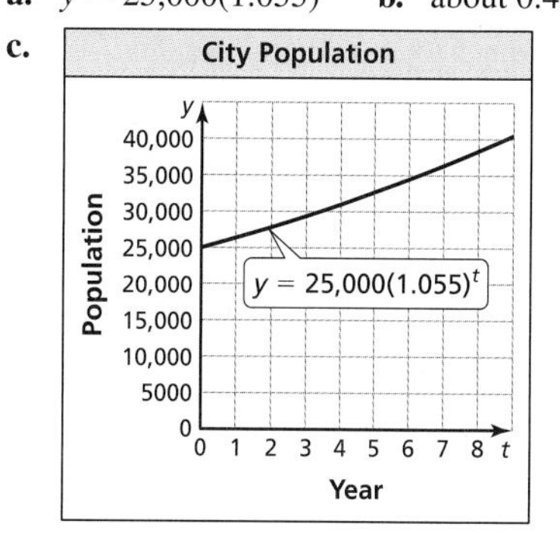

about 30,971

65. a. $y \approx 800(0.9943)^{60t} \approx 800(0.9943^{60})^t \approx 800(0.7097)^t$;
$y \approx 800(0.843)^{2t} \approx 800(0.843^2)^t \approx 800(0.7106)^t$

b. All three functions indicate the initial amount of ibuprofen in a person's bloodstream is 800 mg. The first function indicates the amount of ibuprofen in a person's bloodstream decreases by about 29% each hour. The second function indicates the amount of ibuprofen in a person's bloodstream decreases by about 0.57% each minute. The third function indicates the amount of ibuprofen in a person's bloodstream decreases by about 15.7% each half-hour.

67. 200%; The growth factor is 3, which is also $r + 1$, so r, the growth rate, is 2, or 200%.

69. *Sample answer:* $y = 5(1)^x$; One to any power is 1, so this is a constant function equivalent to $y = 5$.

71. no; The discount is 20% of the preceding day's price, not always the original price, so the amount of the discount is less each day.

6.4 Maintaining Mathematical Proficiency *(p. 322)*

73. $x = -3$ 75. $r = 5$ 77. slope: $\frac{1}{4}$; y-intercept: 7

79. slope: $-\frac{1}{2}$; y-intercept: 4

6.5 Vocabulary and Core Concept Check *(p. 329)*

1. Rewrite each side of the equation using the same base, then set the exponents equal to each other and solve. If it is impossible to rewrite each side of an exponential equation using the same base, you can solve the equation by graphing each side and finding the point(s) of intersection.

6.5 Monitoring Progress and Modeling with Mathematics *(pp. 329–330)*

3. $x = 2$ 5. $x = 4$ 7. $x = 6$ 9. $x = -5$

11. $x = 3$ 13. $x = -3$ 15. $x = -2$ 17. $x = 3$

19. The exponents are not equal because there is not a common base; $5^{3x+2} = (5^2)^{x-8}$; $5^{3x+2} = 5^{2x-16}$; $3x + 2 = 2x - 16$; $x = -18$

21. C; $x \approx 2.58$ 23. B; $x \approx -0.89$ 25. $x \approx -0.61$

27. no solution 29. $x \approx -7.30$ and $x \approx -2.75$

31. $x \approx 2.06$ 33. no solution 35. $x \approx -1.77$

37. $x = -2$ 39. $x = -3$ 41. 3 times

43. all real numbers 45. no solution

47. Any number to the zero power is 1, so $x - 4 = 0$, and $x = 4$.

49. $800 = 500(1.06)^x$; about 8.07 years

51. One to any power is 1, even if the powers are not equal; *Sample answer:* $1^3 = 1^5$, but $3 \neq 5$.

53. $x = \frac{5}{2}$ 55. $x = -\frac{5}{3}$ 57. $x = 18$

59. yes; If x was 0, b would equal 1. By the definition of negative exponents, $\left(\frac{1}{a}\right)^x = (a^{-1})^x$. By the Power of a Power Property, $(a^{-1})^x = a^{-x}$. a must be raised to a positive exponent to stay positive. So, x must be negative.

6.5 Maintaining Mathematical Proficiency *(p. 330)*

61. no 63. yes, 10

6.6 Vocabulary and Core Concept Check *(p. 336)*

1. The first sequence is an arithmetic sequence with a common difference of 2. The second sequence is a geometric sequence with a common ratio of 2.

6.6 Monitoring Progress and Modeling with Mathematics *(pp. 336–338)*

3. 3 5. -8 7. $\frac{3}{4}$

9. arithmetic; There is a common difference of 8.

11. neither; There is no common difference or common ratio.

13. geometric; There is a common ratio of $\frac{1}{8}$.

15. geometric; There is a common ratio of 5.

17. neither; There is no common difference or common ratio.

19. 1280; 5120; 20,480

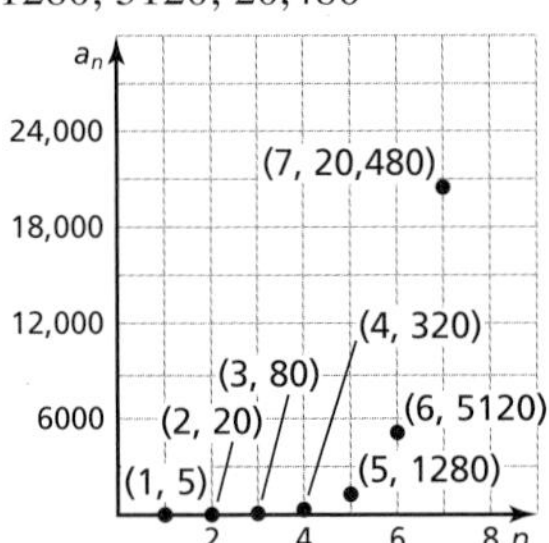

21. $1, -\frac{1}{3}, \frac{1}{9}$

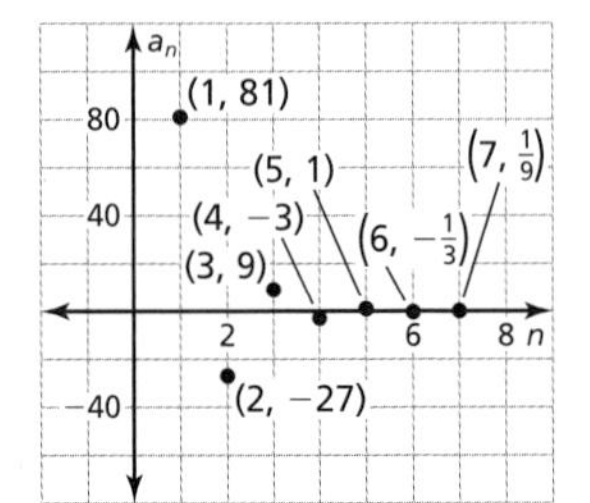

23. $\frac{1}{8}, \frac{1}{32}, \frac{1}{128}$

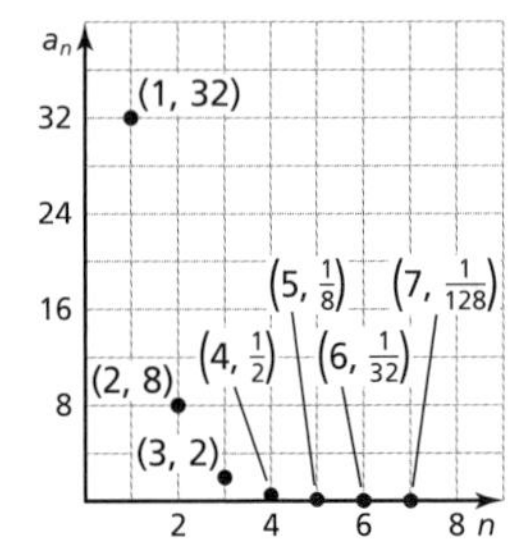

25. $a_n = 2(4)^{n-1}$; 2048 27. $a_n = -\frac{1}{8}(2)^{n-1}$; -4

29. $a_n = 7640(0.1)^{n-1}$; 0.0764

31. $a_n = 0.5(-6)^{n-1}$; -3888

33. 16 teams; 8 teams; 4 teams

35. The common factor is $-\frac{1}{2}$, not -2;

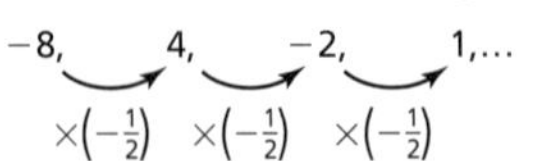

The next three terms are $-\frac{1}{2}$, $\frac{1}{4}$, and $-\frac{1}{8}$.

37. a. $a_n = 625\left(\frac{4}{5}\right)^{n-1}$ b. 5 swings

39. a. $a_n = 9^{n-1}$

b. a large square containing 387,420,489 small squares

41. $a_n = 64\left(\frac{1}{2}\right)^{n-1}$; geometric; There is a common ratio of $\frac{1}{2}$.

43. Graphs of arithmetic sequences form a linear pattern. Graphs of geometric sequences form an exponential pattern when the common ratio is positive, and a pattern of points alternating between quadrants I and IV when the common ratio is negative.

45. yes; yes; It is an arithmetic sequence with a common difference of 0, and a geometric sequence with a common ratio of 1.

47. 59,049

49. dependent; Each term is calculated from the preceding term.

51. a. 2048, 1536, 1152, 864, 648 **b.** $a_n = 2048\left(\frac{3}{4}\right)^{n-1}$

c. Theoretically, all of the soup is never gone, because an exponential function approaches 0, but never actually equals 0. However, in real life, eventually there will not be enough soup to serve.

53. *Sample answer:* $1, -2, 4, -8, \ldots$

6.6 Maintaining Mathematical Proficiency *(p. 338)*

55. yes; The residual points are evenly dispersed about the horizontal axis.

6.7 Vocabulary and Core Concept Check *(p. 344)*

1. recursive equation

6.7 Monitoring Progress and Modeling with Mathematics *(pp. 344–346)*

3. geometric **5.** arithmetic

7. 0, 2, 4, 6, 8, 10

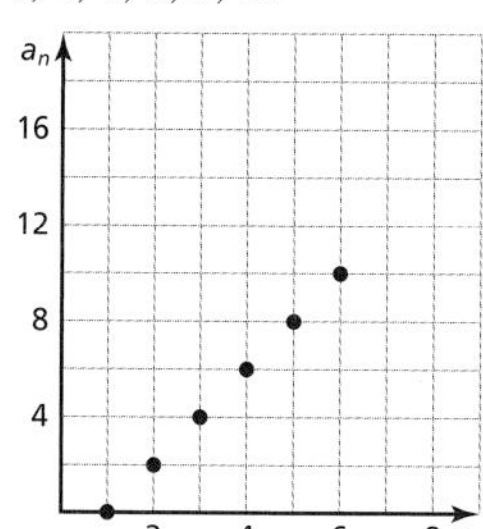

9. 2, 6, 18, 54, 162, 486

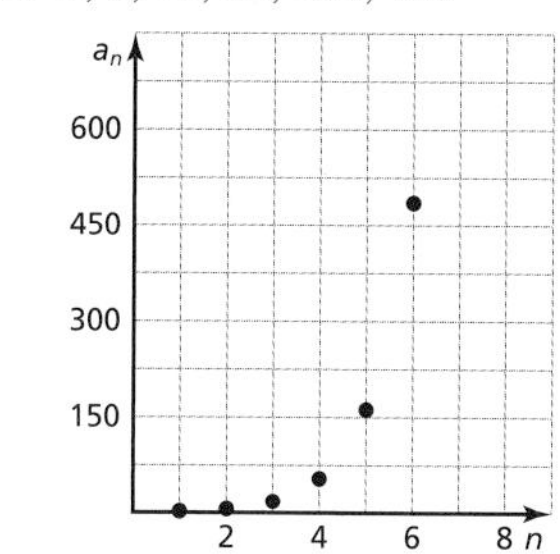

11. $80, -40, 20, -10, 5, -\frac{5}{2}$

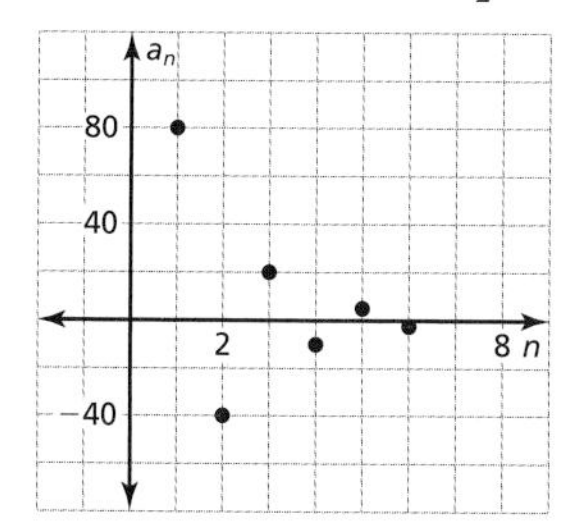

13. $a_1 = 7, a_n = a_{n-1} + 9$ **15.** $a_1 = 243, a_n = \frac{1}{3}a_{n-1}$

17. $a_1 = 0, a_n = a_{n-1} - 3$ **19.** $a_1 = -1, a_n = 4a_{n-1}$

21. $a_1 = 1$ cell, $a_n = (2a_{n-1})$ cells **23.** $a_n = 3n - 6$

25. $a_n = 16(0.5)^{n-1}$ **27.** $a_n = 17n - 13$

29. $a_1 = 7, a_n = 3a_{n-1}$ **31.** $a_1 = 4.5, a_n = a_{n-1} + 1.5$

33. $a_1 = 1, a_n = -5a_{n-1}$

35.

$a_1 = 5, a_n = a_{n-1} + 15$;
$a_n = 15n - 10$

37.

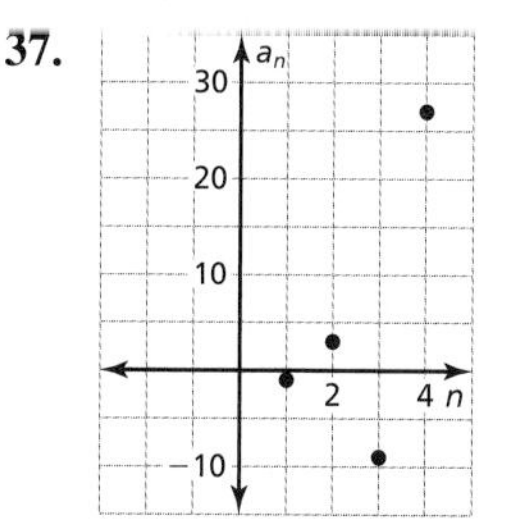

$a_1 = -1, a_n = -3a_{n-1}$;
$a_n = -(-3)^{n-1}$

39. $a_1 = 1, a_2 = 3, a_n = a_{n-2} + a_{n-1}$; 18, 29

41. $a_1 = 2, a_2 = 4, a_n = a_{n-1} - a_{n-2}$; 2, 4

43. $a_1 = 1, a_2 = 3, a_n = (a_n - 1)(a_{n-2})$; 243, 6561

45. The common difference is −12, not 12;
$a_n = 6 + (n - 1)(-12)$; $a_n = 6 - 12n + 12$; $a_n = 18 - 12n$

47. 10; 31; 66 **49.** −8; 8; −8 **51.** 15; −15; −10

53. a. 3, 5, 7, 9, 11, 13, 15, 17, 19, 21

b. 3; 12; 48; 192; 768; 3072; 12,288; 49,152; 196,608; 786,432

c. 4, 7, 3, −4, −7, −3, 4, 7, 3, −4

55. 5, 19, 61, 187, 565; neither; There is no common difference or common ratio.

57. a. Substituting $n - 1$ for n in the explicit rule that defines an arithmetic sequence results in this expression.

b. Write the equation; Identity Property of Addition; Additive Inverse Property; Associative Property of Addition; Distributive Property; Substitution Property of Equality

59. $a_1 = 3, a_n = a_{n-1} + 2^n$

6.7 Maintaining Mathematical Proficiency *(p. 346)*

61. $-6y - 5$ **63.** $8m + 3$ **65.** $f(x) = -\frac{1}{2}x - 1$

67. $f(x) = 2x - 7$

Chapter 6 Review *(pp. 348–350)*

1. $\frac{1}{y^2}$ **2.** $\frac{1}{x^3}$ **3.** y^6 **4.** $\frac{25y^8}{4x^4}$ **5.** 2 **6.** −3

7. 125 **8.** not a real number

9.

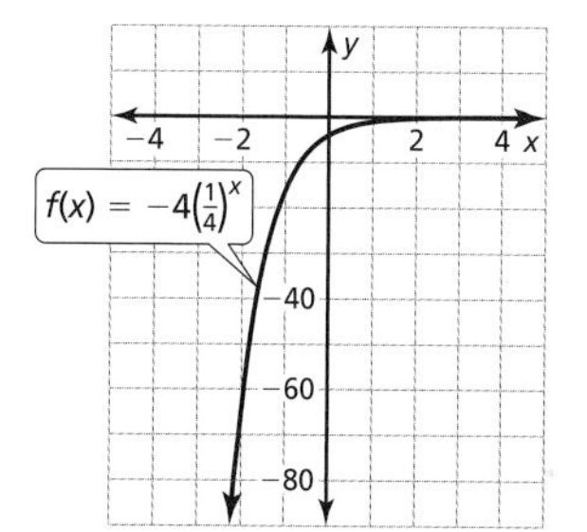

domain: all real numbers, range: $y < 0$

10.

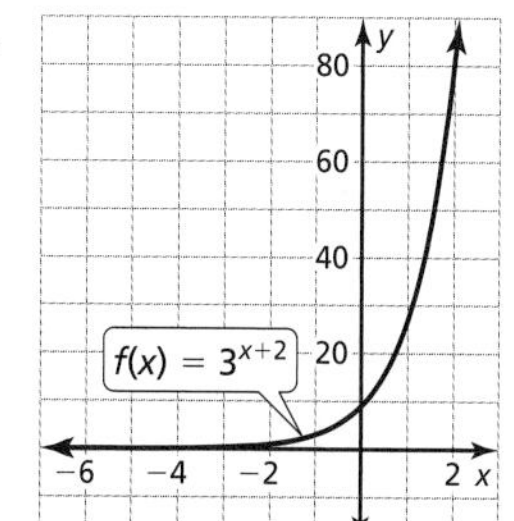

domain: all real numbers, range: $y > 0$

11.

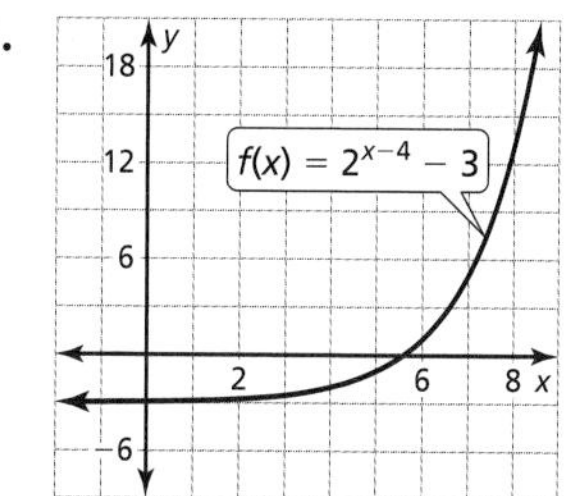

domain: all real numbers, range: $y > -3$

12. $f(x) = 2\left(\frac{1}{2}\right)^x$

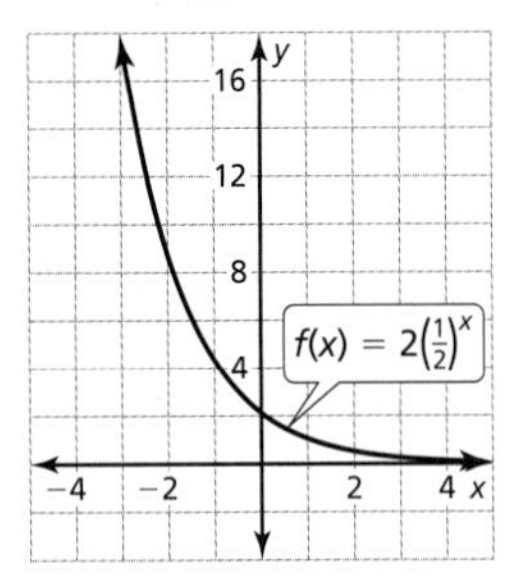

The graph of f is a vertical stretch by a factor of 2 of the graph of g.

13. exponential growth; As x increases by 1, y is multiplied by 2.

14. exponential decay; As x increases by 1, y is multiplied by $\frac{2}{3}$.

15. $f(t) \approx 7.81(1.25)^t$, exponential growth; 25%

16. $y \approx 1.59^t$, exponential growth; about 59%

17. $f(t) \approx 12.05(0.84)^t$, exponential decay; 16%

18. **a.** $y = 750(1.0125)^{4t}$ **b.** \$914.92

19. **a.** $y = 1500(0.86)^t$ **b.** about 1.2%

c.

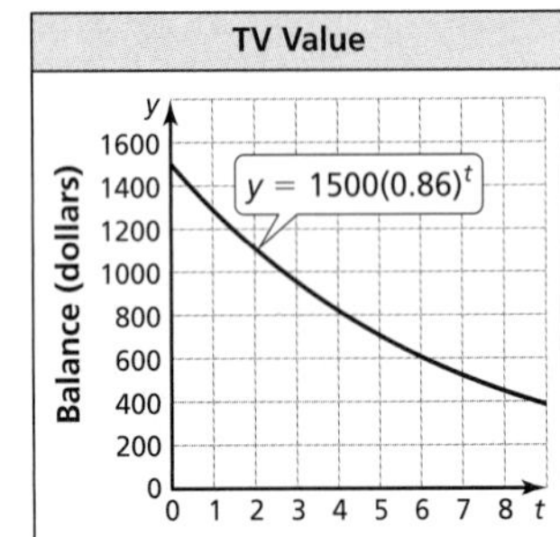

about \$950

20. $x = 1$ **21.** $x = 2$ **22.** no solution

23. $x \approx -2.23$ **24.** $x = -4$ **25.** $x = 5$

26. geometric; There is a common ratio of 4; 768, 3072, 12,288

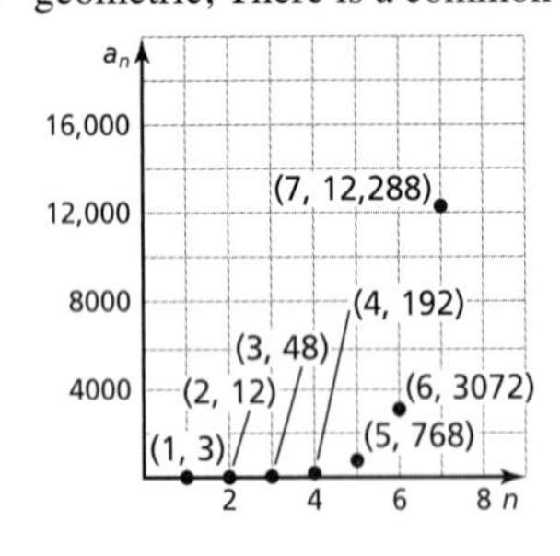

27. neither; There is no common ratio or common difference.

28. geometric; There is a common ratio of $-\frac{1}{5}$; $\frac{3}{5}$, $-\frac{3}{25}$, $\frac{3}{125}$

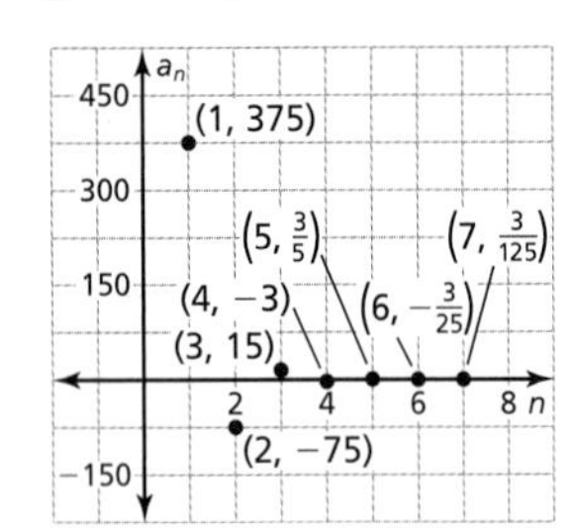

29. $a_n = 4^{n-1}$; 65,536 **30.** $a_n = 5(-2)^{n-1}$; 1280

31. $a_n = 486\left(\frac{1}{3}\right)^{n-1}$; $\frac{2}{27}$

32. 4, 9, 14, 19, 24, 29

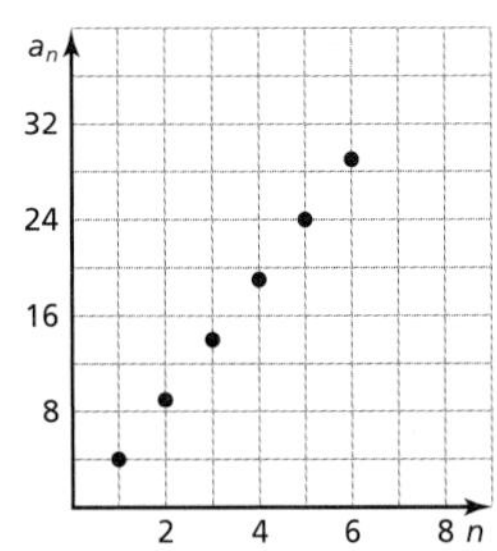

33. −4, 12, −36, 108, −324, 972

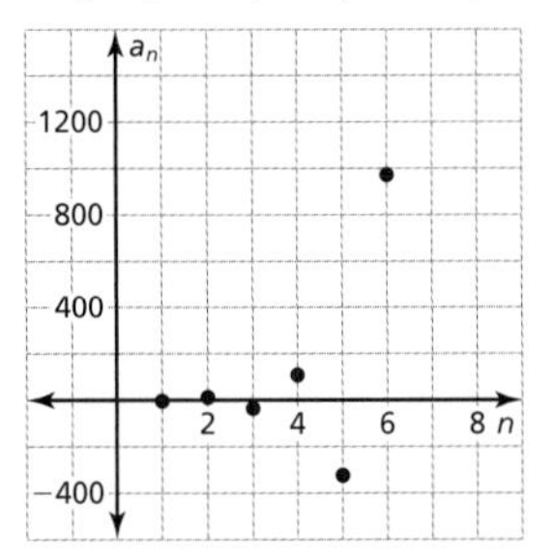

34. 32, 8, 2, $\frac{1}{2}$, $\frac{1}{8}$, $\frac{1}{32}$

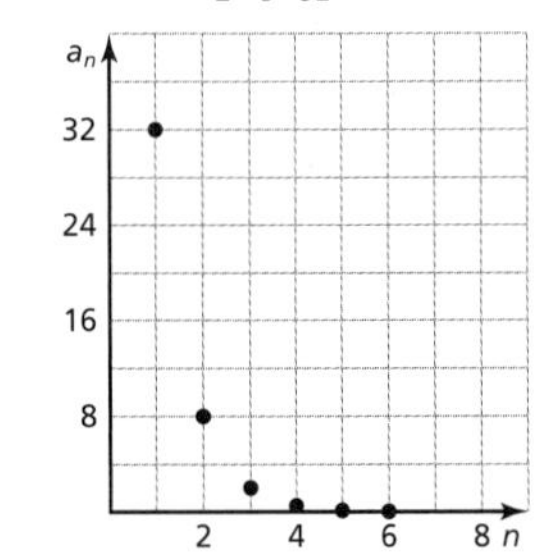

35. $a_1 = 3, a_n = a_{n-1} + 5$ **36.** $a_1 = 3, a_n = 2a_{n-1}$

37. $a_1 = 7, a_2 = 6, a_n = a_{n-2} + a_{n-1}$

38.

$a_1 = 8, a_n = 5a_{n-1}$; $a_n = 8(5)^{n-1}$

Chapter 7

Chapter 7 Maintaining Mathematical Proficiency *(p. 355)*

1. $5x - 7$ **2.** $-5r + 5$ **3.** $2t - 1$ **4.** $3s + 2$

5. $9m - 21$ **6.** $3h + 26$ **7.** 4 **8.** 21 **9.** 27

10. 12 **11.** 4 **12.** 1

13. no; The number 1 is a factor of every integer.

7.1 Vocabulary and Core Concept Check *(p. 362)*

1. when the exponents of the terms decrease from left to right

3. Determine if performing the operation on any two numbers in the set always results in a number that is also in the set.

7.1 Monitoring Progress and Modeling with Mathematics (pp. 362–364)

5. 1 **7.** 2 **9.** 9 **11.** 11

13. $2c^4 + 6c^2 - c$; 4; 2; trinomial

15. $3p^2 + 7$; 2; 3; binomial **17.** $3t^8$; 8; 3; monomial

19. $-\frac{5}{7}r^8 + 2r^5 + \pi r^2$; 8; $-\frac{5}{7}$; trinomial

21. It is the product of a number, $\frac{4}{3}\pi$, and a variable with a whole number exponent, r^3; 3

23. $3y + 10$ **25.** $n^2 - 8n + 5$ **27.** $6g^2 - 9g + 4$

29. $a^3 - 5a^2 + 4a + 5$ **31.** $-2d - 8$

33. $-2y^2 + 2y + 18$ **35.** $k^3 - k^2 - 7k + 14$

37. $t^4 + 8t^2 + 8t - 12$

39. When writing the subtraction as addition, the last term of the polynomial was not multiplied by -1;
$= (x^2 + x) + (-2x^2 + 3x) = (x^2 - 2x^2) + (x + 3x)$
$= -x^2 + 4x$

41. $3b + 2$ **43.** $s^2 - 12st$ **45.** $2c^2 - 2cd - 4d^2$

47. always; A polynomial is a monomial or a sum of monomials, and each monomial is a term of the polynomial.

49. sometimes; The two terms in the binomial can be of any degree.

51. $-16t^2 - 45t + 200$; 139 ft

53. **a.** $-46t + 92$

b. The constant term 92 indicates that the distance between the two balls is 92 feet when they begin. The coefficient of the linear term -46 indicates that the two balls become 46 feet closer to each other each second.

55. $12x - 3$

57. yes; Addition is commutative and associative, so you can add in any order.

59. **a.** no; The product of two negative integers is always a positive integer.

b. yes; The sum of two whole numbers is always a whole number.

61. **a.** $19x^2 - 12x$ **b.** \$14,310

7.1 Maintaining Mathematical Proficiency (p. 364)

63. $34y - 34$

7.2 Vocabulary and Core Concept Check (p. 369)

1. *Sample answer:* Distribute one of the binomials over each term in the other binomial and simplify; Write each binomial as a sum of terms and make a table of products.

7.2 Monitoring Progress and Modeling with Mathematics (pp. 369–370)

3. $x^2 + 4x + 3$ **5.** $z^2 - 2z - 15$ **7.** $g^2 - 9g + 14$

9. $3m^2 + 28m + 9$ **11.** $x^2 + 5x + 6$

13. $h^2 - 17h + 72$ **15.** $12k^2 + 23k - 9$

17. $8j^2 - 26j + 21$

19. t also should be multiplied by $t + 5$;
$= t(t + 5) - 2(t + 5) = t^2 + 5t - 2t - 10 = t^2 + 3t - 10$

21. $b^2 + 10b + 21$ **23.** $k^2 + 4k - 5$ **25.** $q^2 - \frac{1}{2}q - \frac{3}{16}$

27. $3r^2 - 29r + 18$ **29.** $w^3 + 8w^2 + 15w$

31. $2x^2 + x - 45$ **33.** $\frac{1}{2}x^2 + \frac{11}{2}x + 15$

35. $x^3 + 7x^2 + 14x + 8$ **37.** $y^3 + 11y^2 + 22y - 6$

39. $-5b^3 + 15b^2 + 24b - 16$ **41.** $18e^3 - 27e^2 + 37e + 7$

43. **a.** $(40x^2 + 240x + 200)$ ft^2 **b.** 57,600 ft^2

45. The degree of the product is the sum of the degrees of each binomial.

47. no; FOIL would leave out the products that include the middle terms of the two trinomials.

49. yes; You are both multiplying the same binomials, and neither the order in which you multiply nor the method used will make a difference.

51. **a.** They have the same signs.

b. They have opposite signs.

7.2 Maintaining Mathematical Proficiency (p. 370)

53. $y = \begin{cases} -6x + 18, & \text{if } x < 3 \\ 6x - 18, & \text{if } x \geq 3 \end{cases}$ **55.** 10^{11} **57.** $\frac{1}{27z^{18}}$

7.3 Vocabulary and Core Concept Check (p. 375)

1. Substitute the first term of the binomial for a and the second term of the binomial for b in the square of a binomial pattern, then simplify.

7.3 Monitoring Progress and Modeling with Mathematics (pp. 375–376)

3. $x^2 + 16x + 64$ **5.** $4f^2 - 4f + 1$ **7.** $49t^2 - 56t + 16$

9. $4a^2 + 4ab + b^2$ **11.** $x^2 + 8x + 16$

13. $49n^2 - 70n + 25$ **15.** $t^2 - 49$ **17.** $16x^2 - 1$

19. $64 - 9a^2$ **21.** $p^2 - 100q^2$ **23.** $y^2 - 16$

25. 384 **27.** 1764 **29.** 930.25

31. The middle term in the square of a binomial pattern was not included; $= k^2 + 2(k)(4) + 4^2 = k^2 + 8k + 16$

33. **a.** $(x^2 + 100x + 2500)$ ft^2 **b.** 4225 ft^2; 1725 ft^2

35. **a.** 25% **b.** $(0.5N + 0.5a)^2 = 0.25N^2 + 0.5Na + 0.25a^2$

37. $(x + 11)(x - 11)$; $x^2 - 121$ fits the product side of the sum and difference pattern, so working backwards, a and b are the square roots of a^2 and b^2.

39. $x^4 - 1$ **41.** $4m^4 - 20m^2n^2 + 25n^4$

43. no; $\left(4\frac{1}{3}\right)^2$ can be written as $\left(4 + \frac{1}{3}\right)^2$, however using the square of a binomial pattern results in $16 + \frac{8}{3} + \frac{1}{9}$, which is $18\frac{7}{9}$, not $16\frac{1}{9}$.

45. $k = 64$ **47.** *Sample answer:* $a = 3, b = 4$

7.3 Maintaining Mathematical Proficiency (p. 376)

49. $9(r + 3)$ **51.** $5(3x - 2y)$

7.4 Vocabulary and Core Concept Check (p. 381)

1. Set $3x = 0$ and $x - 6 = 0$, then solve both of the equations to get the solutions $x = 0$ and $x = 6$.

7.4 Monitoring Progress and Modeling with Mathematics (pp. 381–382)

3. $x = 0, x = -7$ **5.** $t = 0, t = 5$ **7.** $s = 9, s = 1$

9. $a = 3, a = -5$ **11.** $m = -\frac{4}{5}$ **13.** $g = \frac{3}{2}, g = 7$

15. $z = 0, z = -2, z = 1$ **17.** $r = 4, r = -8$

19. $c = 3, c = -1, c = 6$ **21.** $x = 8, x = -8$

23. $x = 14, x = 5$ **25.** $5z(z + 9)$ **27.** $3y^2(y - 3)$

29. $n^5(5n + 2)$ **31.** $p = 0, p = \frac{1}{4}$ **33.** $c = 0, c = -\frac{5}{2}$

35. $n = 0, n = 3$

37. also need to set $6x = 0$ and solve; $6x = 0$ or $x + 5 = 0$; $x = 0$ or $x = -5$; The roots are $x = 0$ and $x = -5$.

39. 20 ft

41. $x = 0, x = 0.3$ sec; The roots represent the times when the penguin is at water level. $x = 0$ is when it leaves the water, and $x = 0.3$ second is when it returns to the water after the leap.

43. 2; x-intercepts occur when $y = 0$ and the equation has 2 roots when $y = 0$.

45. no; Roots will occur if $x^2 + 3 = 0$ or $x^4 + 1 = 0$. However, solving these equations results in $x^2 = -3$ or $x^4 = -1$, and even powers of any number cannot be negative.

47. **a.** $x = -y, x = \frac{1}{2}y$ **b.** $x = \pm y, x = -4y$

7.4 Maintaining Mathematical Proficiency *(p. 382)*

49. 1, 10; 2, 5 **51.** 1, 30; 2, 15; 3, 10; 5, 6

7.5 Vocabulary and Core Concept Check *(p. 389)*

1. They have opposite signs; When factoring $x^2 + bx + c = (x + p)(x + q)$, if c is negative, p and q must have opposite signs.

7.5 Monitoring Progress and Modeling with Mathematics *(pp. 389–390)*

3. $(x + 1)(x + 7)$ **5.** $(n + 4)(n + 5)$ **7.** $(h + 2)(h + 9)$

9. $(v - 1)(v - 4)$ **11.** $(d - 2)(d - 3)$

13. $(w - 8)(w - 9)$ **15.** $(x - 1)(x + 4)$

17. $(n - 2)(n + 6)$ **19.** $(y - 6)(y + 8)$

21. $(x + 4)(x - 5)$ **23.** $(t + 2)(t - 8)$

25. **a.** $(x - 5)$ ft **b.** 16 ft

27. $4 + 12$ is not 14; $= (x + 6)(x + 8)$

29. $m = -1, m = -2$ **31.** $x = 2, x = -7$

33. $t = -3, t = -12$ **35.** $a = 5, a = -10$

37. $m = 4, m = 11$ **39.** 100 in.^{2}

41. yes; p and q must be factors of -12 that have a sum of b, and -12 has 6 sets of integer factors, -1 and 12, -2 and 6, -3 and 4, -4 and 3, -6 and 2, and -12 and 1.

43. length: 11 ft, width: 4 ft

45. $x^2 - 2x - 24 = 0$; Multiply $(x + 4)(x - 6)$.

47. **a.** $-x^2 + 38x$ **b.** $-x^2 + 38x = 280$; 10 m

49. $(r + 3s)(r + 4s)$ **51.** $(x + 5y)(x - 7y)$

7.5 Maintaining Mathematical Proficiency *(p. 390)*

53. $z = -17$ **55.** $k = 0$

7.6 Vocabulary and Core Concept Check *(p. 395)*

1. 3

7.6 Monitoring Progress and Modeling with Mathematics *(pp. 395–396)*

3. $3(x - 1)(x + 2)$ **5.** $4(k + 3)(k + 4)$

7. $7(b - 4)(b - 5)$ **9.** $(3h + 2)(h + 3)$

11. $(2x - 1)(3x - 1)$ **13.** $(n + 2)(3n - 1)$

15. $2(g - 2)(4g + 3)$ **17.** $-(t - 3)(3t - 2)$

19. $-(c - 5)(4c + 1)$ **21.** $-(3w - 4)(5w + 7)$

23. need to factor 2 out of every term; $= 2(x^2 - x - 12) = 2(x + 3)(x - 4)$

25. $x = -2, x = 3$ **27.** $n = -\frac{5}{3}, n = \frac{3}{4}$

29. $x = -\frac{7}{2}, x = 5$ **31.** $x = -1, x = \frac{5}{7}$

33. **a.** $(5x - 2)$ ft

b. Substitute 3 for x into the expression for the area $15x^2 - x - 2$, then simplify; Substitute 3 for x into the expressions for the length $(5x - 2)$ and width $(3x + 1)$, simplify each, then multiply these two numbers.

35. length: 70 m, width: 31 m

37. *Sample answer:* $6x^2 + 3x$

39. when no combination of factors of a and c produce the correct middle term; *Sample answer:* $2x^2 + x + 1$

41. $\pm 9, \pm 12, \pm 21$ **43.** 3.5 in. **45.** $(k + 2j)(4k - j)$

47. $-(a - 2b)(6a - 7b)$

7.6 Maintaining Mathematical Proficiency *(p. 396)*

49. ± 8 **51.** -15 **53.** $(-1, -4)$ **55.** $(0, -7)$

7.7 Vocabulary and Core Concept Check *(p. 401)*

1. yes; The square roots of the first and last terms are y and 8, and the middle term is $2 \cdot y \cdot 8$, so it fits the pattern.

7.7 Monitoring Progress and Modeling with Mathematics *(pp. 401–402)*

3. $(m + 7)(m - 7)$ **5.** $(8 + 9d)(8 - 9d)$

7. $9(5a + 2b)(5a - 2b)$ **9.** 63 **11.** 900 **13.** 600

15. $(h + 6)^2$ **17.** $(y - 11)^2$ **19.** $(a - 14)^2$

21. $(5n + 2)^2$

23. should follow the difference of two squares pattern; $= (n + 8)(n - 8)$

25. **a.** $(d + 4)$ cm **b.** $(4d + 16)$ cm

27. $z = -2, z = 2$ **29.** $k = 8$ **31.** $n = 3$

33. $y = -\frac{1}{4}$ **35.** $3(z + 3)(z - 3)$ **37.** $4(y - 2)^2$

39. $2(5y + 6)^2$ **41.** 1.25 sec

43. **a.** no; $w^2 + 18w + 81$ **b.** no; $y^2 - 10y + 25$

45. Square each binomial, then combine like terms; Use the difference of two squares pattern with each binomial as one of the terms, then simplify; *Sample answer:* The difference of two squares pattern; You do not need to square any binomials.

47. **a.** $(9x^2 - 144)$ in.^{2}

b. 5 in.; Setting the polynomial in part (a) equal to 81 and solving results in $x = -5$ or $x = 5$. Length cannot be negative, so 5 is the solution.

7.7 Maintaining Mathematical Proficiency *(p. 402)*

49. $2 \cdot 5^2$ **51.** $5 \cdot 17$

53.

55.

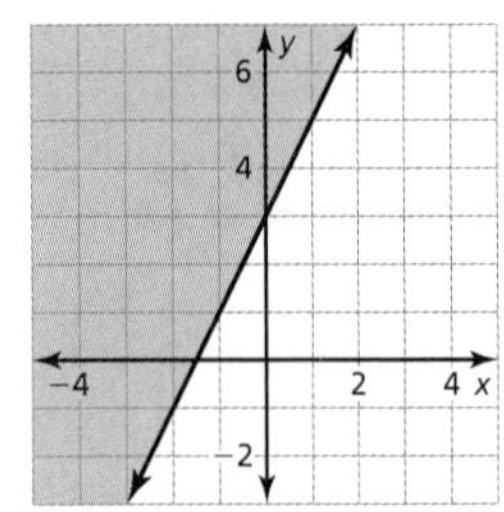

7.8 Vocabulary and Core Concept Check *(p. 407)*

1. It is written as a product of unfactorable polynomials with integer coefficients.

7.8 Monitoring Progress and Modeling with Mathematics *(pp. 407–408)*

3. $(x + 1)(x^2 + 2)$ **5.** $(z - 4)(3z^2 + 2)$

7. $(x + y)(x + 8)$ **9.** $(m - 3)(m + n)$

11. $2x(x + 1)(x - 1)$ **13.** unfactorable **15.** $6g(g - 2)^2$
17. $3r^3(r + 6)(r - 5)$ **19.** $-4c^2(c^2 - 2c + 7)$
21. $(b - 5)(b + 2)(b - 2)$ **23.** $n = 0, n = 2, n = 4$
25. $x = -1, x = -2, x = 2$ **27.** $s = 0, s = 2, s = -2$
29. $x = 0, x = -9, x = 9$ **31.** $x = 0, x = 4$
33. In the second group, factor out -6 instead of 6; $= a^2(a + 8) - 6(a + 8) = (a + 8)(a^2 - 6)$
35. **a.** $(4w^2 + 16w)$ in.^{3}
b. length: 4 in., width: 4 in., height: 8 in.
37. $(x + 2y)(x + 1)(x - 1)$ **39.** $(4s - 1)(s + 3t)$
41. no; The factors of the polynomial are $x^2 + 3$ and $x + 2$. Using the Zero-Product Property, $x + 2 = 0$ will give 1 real solution, but $x^2 + 3 = 0$ has no real solutions.
43. **a.** *Sample answer:* $x^3 + x^2 + x + 2$
b. *Sample answer:* $x^3 + x^2 + x + 1$
45. $3z, 2z + 3, 2z - 3$ **47.** radius: 5, height: 8
49. **a.** *Sample answer:* $w = 40$; When $w = 40$, factoring out $5x$ will leave a perfect square trinomial, so there will be 2 factors.
b. *Sample answer:* $w = 50$; When $w = 50$, factoring out $5x$ will leave a factorable trinomial that is not a perfect square, so there will be 3 factors.

7.8 Maintaining Mathematical Proficiency *(p. 408)*

51. (2, 3) **53.** (−6, −2)

55.

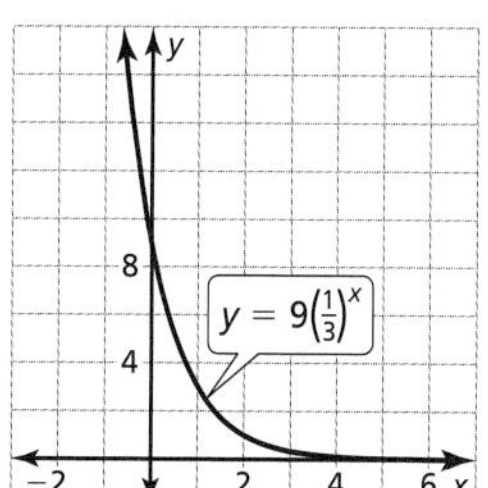

domain: all real numbers, range: $y > 0$

57.

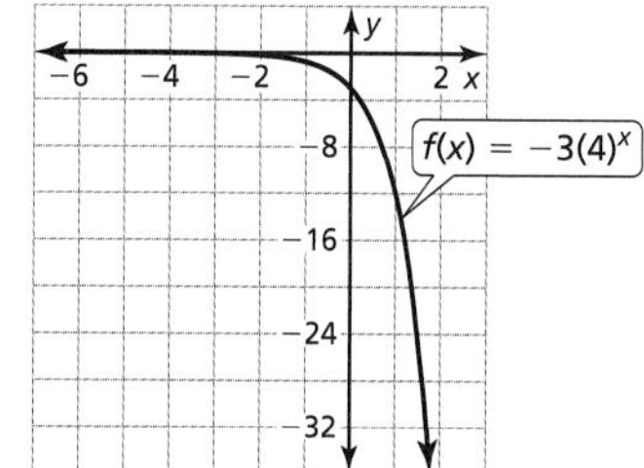

domain: all real numbers, range: $y < 0$

Chapter 7 Review *(pp. 410–412)*

1. $2x^2 + 6$; 2; 2; binomial
2. $5p^6 - 3p^3 - 4$; 6; 5; trinomial
3. $9x^7 + 13x^5 - 6x^2$; 7; 9; trinomial
4. $8y^3 - 12y$; 3; 8; binomial **5.** $4a + 6$
6. $3x^2 + 6x + 10$ **7.** $-2y^2 + 6y + 4$
8. $-6p^2 - 12p + 7$ **9.** $x^2 + 2x - 24$
10. $3y^2 - 7y - 40$ **11.** $x^3 + 11x^2 + 28x$
12. $-12y^3 + 7y^2 + 20y - 7$ **13.** $x^2 - 81$ **14.** $4y^2 - 16$
15. $p^2 + 8p + 16$ **16.** $4d^2 - 4d + 1$
17. $x = 0, x = -5$ **18.** $z = -3, z = 7$ **19.** $b = -13$
20. $y = 0, y = 9, y = -4$ **21.** $(p + 7)(p - 5)$
22. $(b + 8)(b + 10)$ **23.** $(z + 3)(z - 7)$
24. $(x - 7)(x - 4)$ **25.** $(t + 6)(3t - 2)$
26. $-(y + 4)(5y + 2)$ **27.** $(2x + 1)(3x + 7)$
28. $-(y - 2)(2y - 3)$ **29.** $(z + 9)(3z - 1)$
30. $(2a - 3)(5a + 1)$ **31.** $(x + 3)(x - 3)$
32. $(y + 10)(y - 10)$ **33.** $(z - 3)^2$ **34.** $(m + 8)^2$
35. $n(n + 3)(n - 3)$ **36.** $(x - 3)(x + 4a)$
37. $2x^2(x^2 + x - 10)$ **38.** $x = 0, x = 6, x = -3$
39. $x = -\frac{3}{2}, x = \frac{3}{2}$ **40.** $z = -3, z = 5, z = -5$
41. length: 12 ft, width: 4 ft, height: 2 ft

Chapter 8

Chapter 8 Maintaining Mathematical Proficiency *(p. 417)*

1.

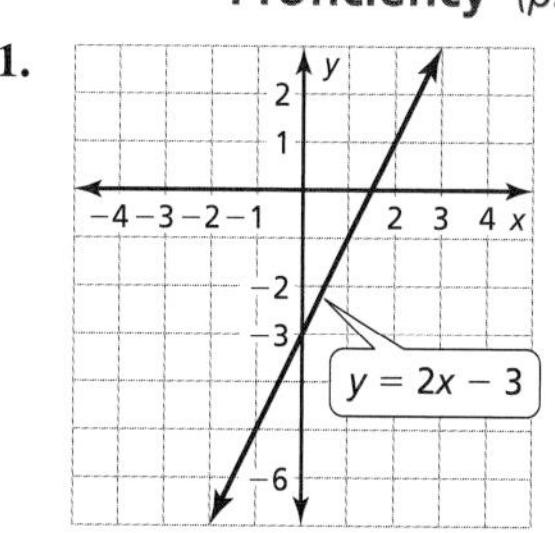

2.

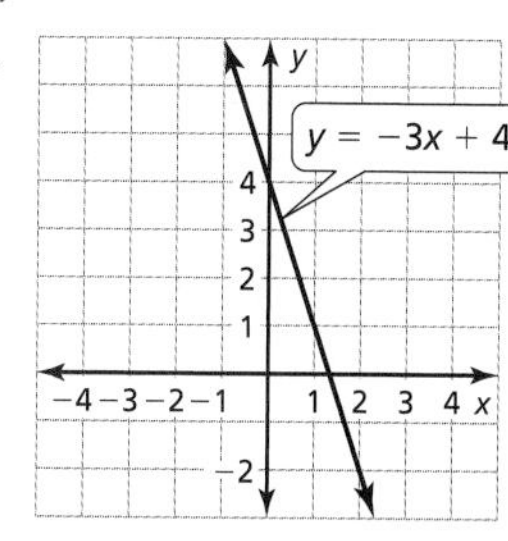

3.

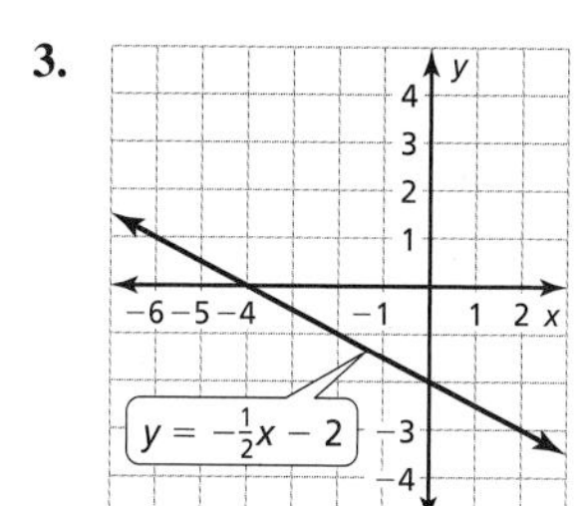

4.

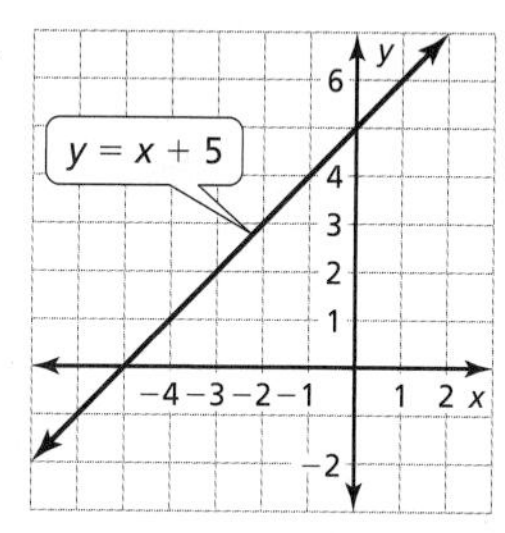

5. 11 **6.** 8 **7.** −11 **8.** −7
9. 3 **10.** −26
11. $a, 4a, 9a, 16a, 25a$; *Sample answer:* The coefficient of each difference is the next consecutive odd integer; $36a$

8.1 Vocabulary and Core Concept Check *(p. 423)*

1. parabola

8.1 Monitoring Progress and Modeling with Mathematics *(pp. 423–424)*

3. The vertex is $(1, -1)$. The axis of symmetry is $x = 1$. The domain is all real numbers. The range is $y \leq -1$. When $x < 1$, y increases as x increases. When $x > 1$, y increases as x decreases.

5.

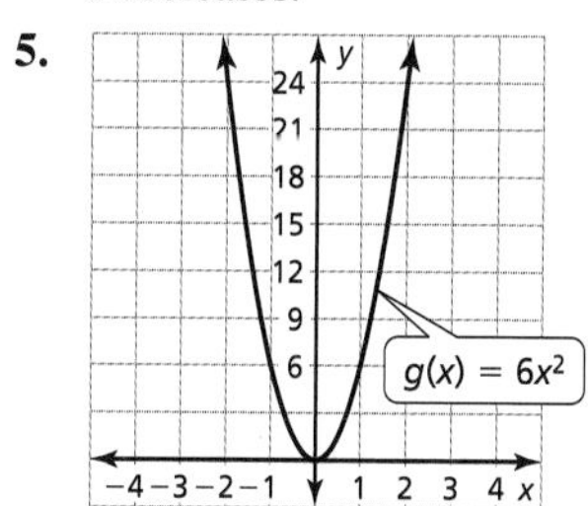

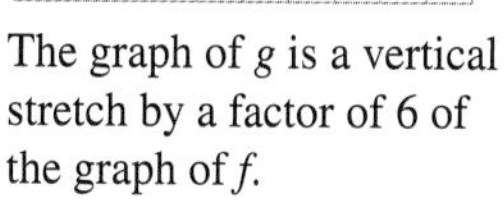

The graph of g is a vertical stretch by a factor of 6 of the graph of f.

7.

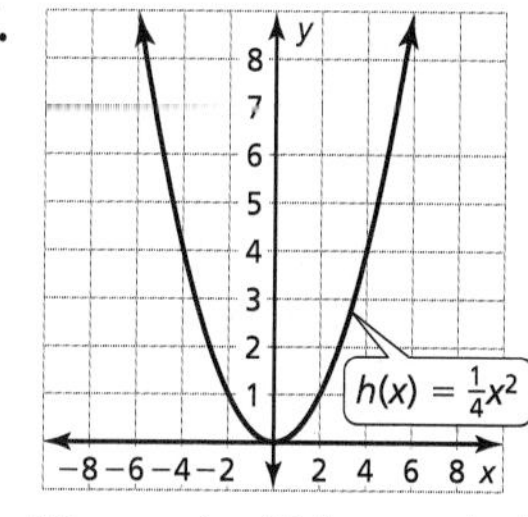

The graph of h is a vertical shrink by a factor of $\frac{1}{4}$ of the graph of f.

9. 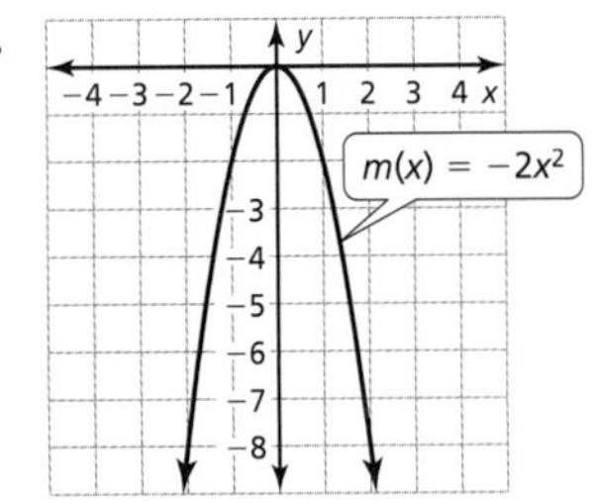

The graph of m is a vertical stretch by a factor of 2 and a reflection in the x-axis of the graph of f.

11. 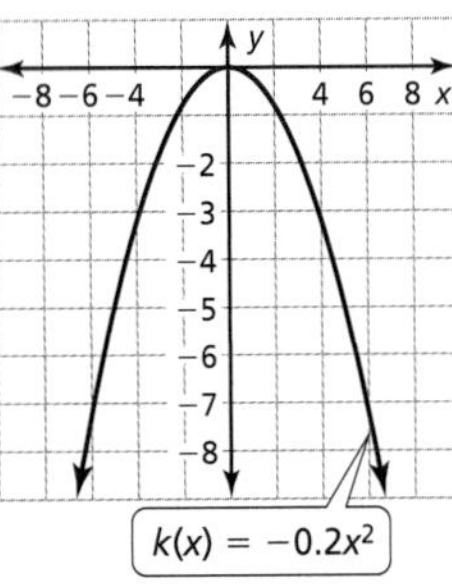

The graph of k is a vertical shrink by a factor of 0.2 and a reflection in the x-axis of the graph of f.

13. 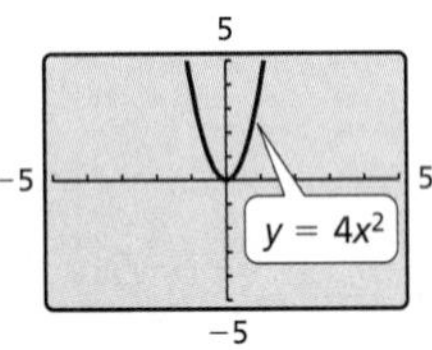

The graph of $y = 4x^2$ is a reflection in the x-axis of the graph of $y = -4x^2$.

15. 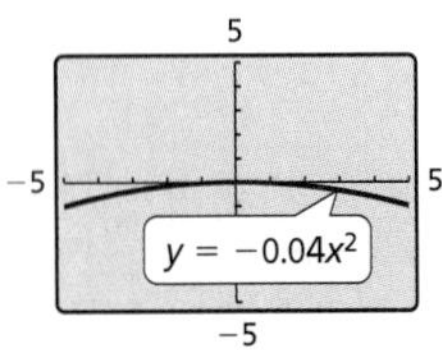

The graph of $y = -0.04x^2$ is a vertical shrink by a factor of $\frac{1}{100}$ of the graph of $y = -4x^2$.

17. The graph of $y = 0.5x^2$ should be wider than the graph of $y = x^2$.

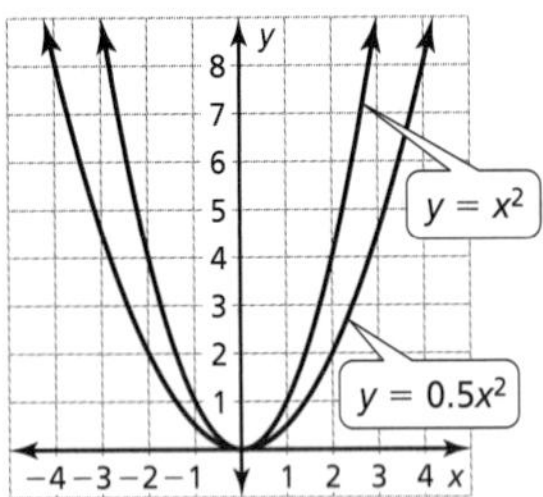

The graphs have the same vertex and the same axis of symmetry. The graph of $y = 0.5x^2$ is wider than the graph of $y = x^2$.

19. **a.** domain: $d \geq 0$, range: $z \geq 0$

b. 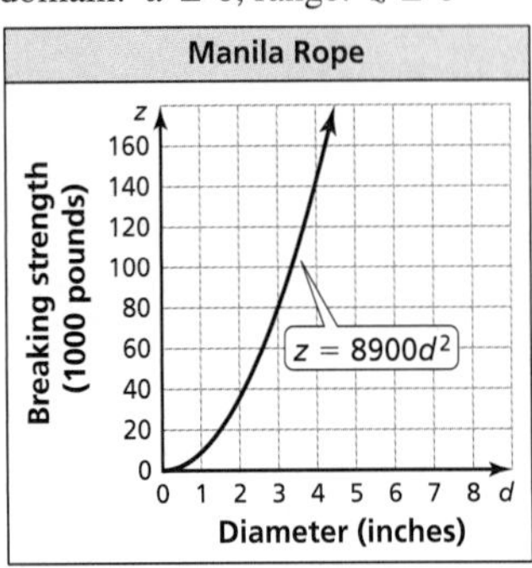

c. no; *Sample answer:* The relationship is quadratic, so a rope with 4 times the diameter will have $4^2 = 16$ times the breaking strength.

21. f is increasing when $x > 0$. g is increasing when $x < 0$.

23. f; $a = \frac{3}{4}$

25. *Sample answer:* The vertex of a parabola that opens up is the minimum point, so its y-coordinate is the minimum value of y. The graph passes through $(6, -3)$, so 2 is not the minimum value of y.

27. always; *Sample answer:* When $|a| > 1$, the graph of f will be a vertical stretch of the graph of g, so it will be narrower.

29. never; *Sample answer:* When $|a| > |d|$, the graph of f will be a vertical stretch of the graph of g, so it will be narrower, not wider.

31. **a.** 8 cm

b. no; *Sample answer:* A faster rotational speed would increase the depth. The diagram shown has a depth of 3.2 centimeters. A model of $y = 0.1x^2$ would only have a depth of 1.6 centimeters, so it would have a slower rotational speed.

8.1 Maintaining Mathematical Proficiency *(p. 424)*

33. 3 **35.** 11

8.2 Vocabulary and Core Concept Check *(p. 429)*

1. $(0, c)$, $x = 0$

8.2 Monitoring Progress and Modeling with Mathematics *(pp. 429–430)*

3. 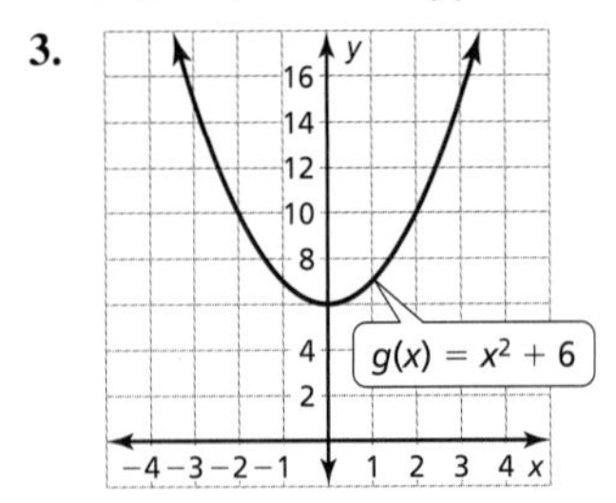

The graph of g is a vertical translation 6 units up of the graph of f.

5. 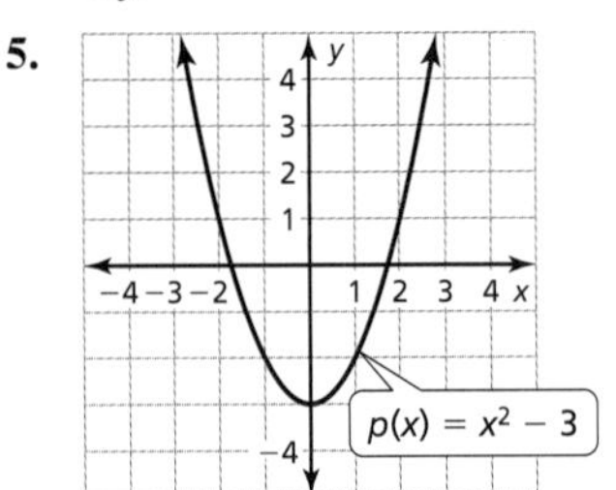

The graph of p is a vertical translation 3 units down of the graph of f.

7. 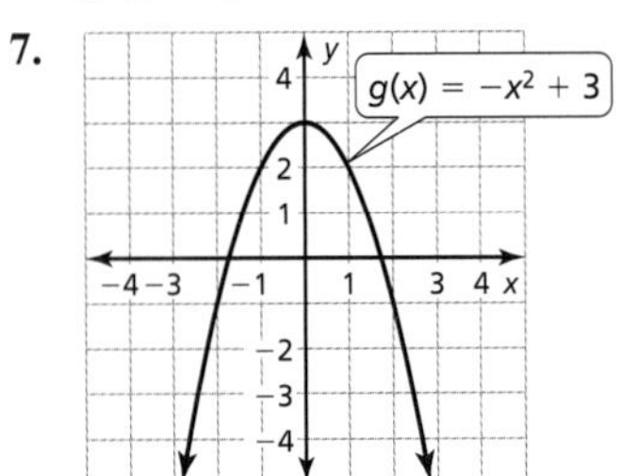

The graph of g is a reflection in the x-axis, and a vertical translation 3 units up of the graph of f.

9.

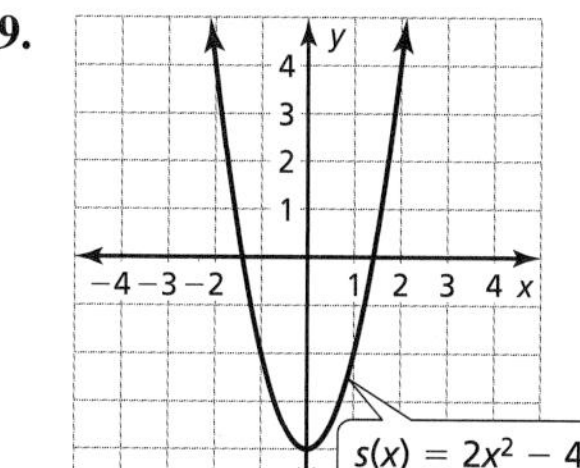

The graph of s is a vertical stretch by a factor of 2 and a vertical translation 4 units down of the graph of f.

11.

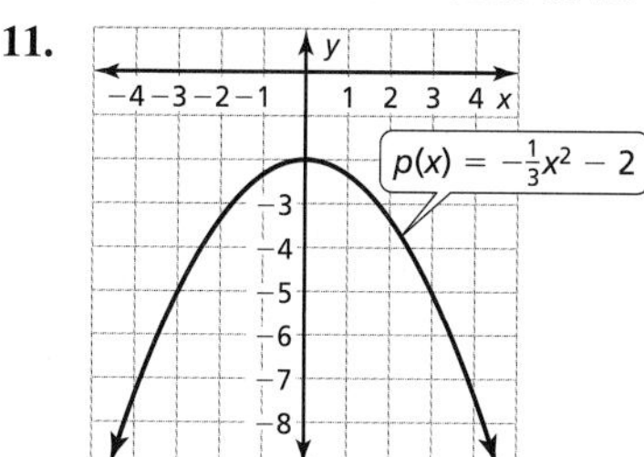

The graph of p is a vertical shrink by a factor of $\frac{1}{3}$, a reflection in the x-axis, and a vertical translation 2 units down of the graph of f.

13. The graph of g is a vertical translation 2 units up of the graph of f.

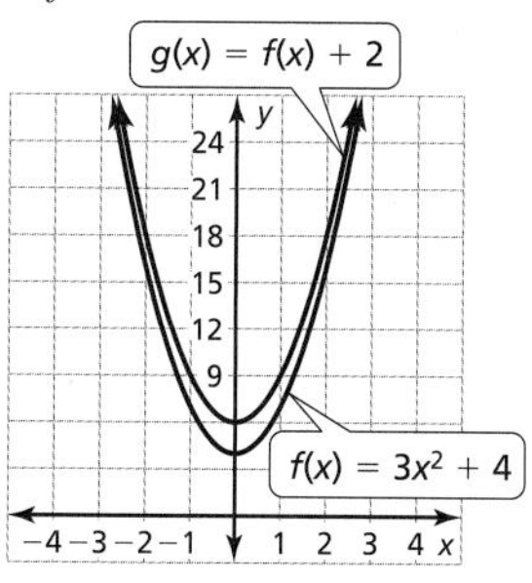

$g(x) = 3x^2 + 6$

15. The graph of g is a vertical translation 3 units down of the graph of f.

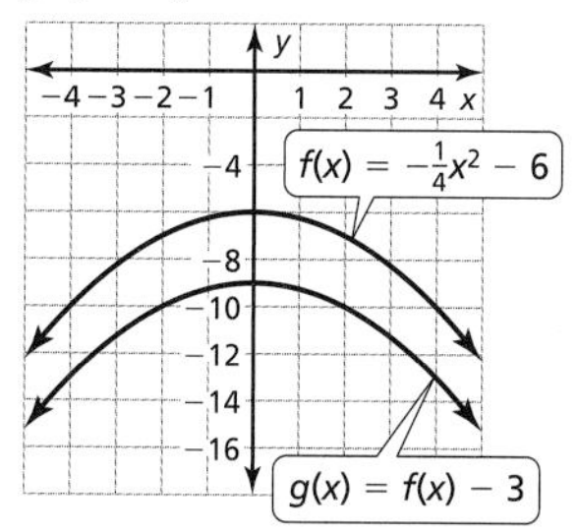

$g(x) = -\frac{1}{4}x^2 - 9$

17. The graph of $y = 3x^2 + 2$ is narrower, so it should be a stretch not a shrink; The graph of $y = 3x^2 + 2$ is a vertical stretch by a factor of 3 and a translation 2 units up of the graph of $y = x^2$.

19. $x = 1, x = -1$ **21.** $x = 5, x = -5$

23. $x = 2, x = -2$ **25.** $x = \frac{1}{2}, x = -\frac{1}{2}$

27. a. 3 sec

b. When $k > 0$, the water balloon will take more than 3 seconds to hit the ground. When $k < 0$, the water balloon will take less than 3 seconds to hit the ground.

29. *Sample answer:*

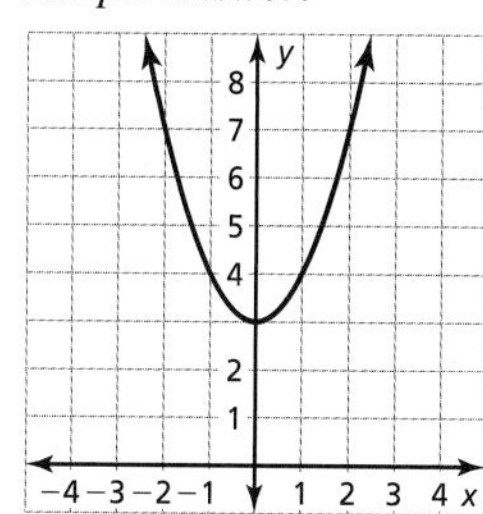

31. *Sample answer:*

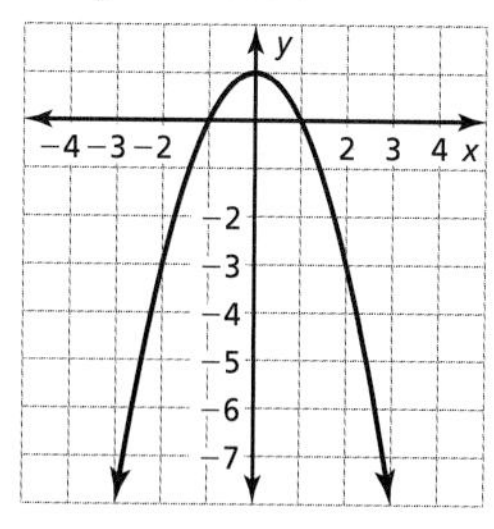

33. a. $T(x) = -44$; the distance between the balls

b. 44 ft

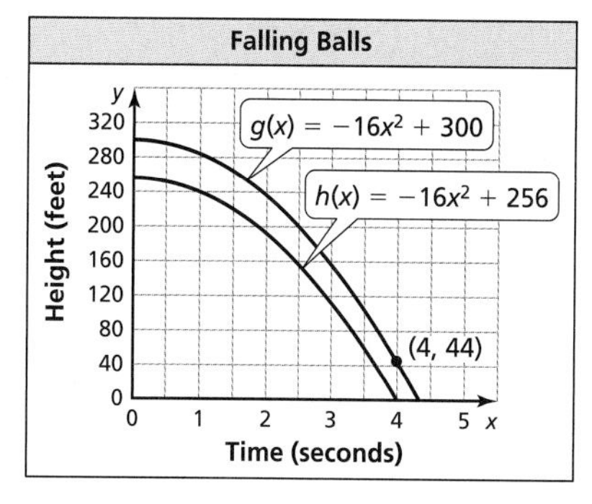

35. 12 ft by 12 ft

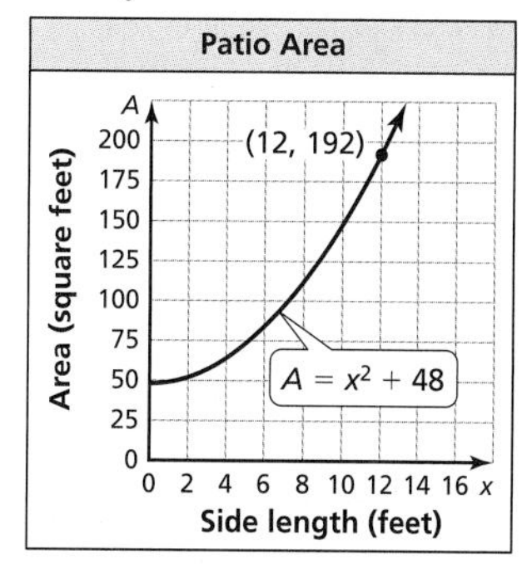

37. Graph the function and determine the x-intercepts; Set the function equal to 0, factor $-16t^2 + 400$, and apply the Zero-Product Property.

39. a. $h = -16t^2 + 45$; $h = -16t^2 + 32$

b. The graph of $h = -16t^2 + 32$ is a vertical translation 13 units down of the graph of $h = -16t^2 + 45$.

41. (0, 5.8); *Sample answer:* The outer edges are located 40 feet from the center. Substituting this into $y = 0.012x^2$ indicates they are 19.2 feet above the x-axis. To be 25 feet above the x-axis, they must be vertically translated up 5.8 feet.

8.2 Maintaining Mathematical Proficiency *(p. 430)*

43. $\frac{3}{8}$ **45.** $\frac{5}{12}$

8.3 Vocabulary and Core Concept Check *(p. 436)*

1. *Sample answer:* If the leading coefficient is positive, the graph has a minimum value. If the leading coefficient is negative, the graph has a maximum value.

8.3 Monitoring Progress and Modeling with Mathematics *(pp. 436–438)*

3. $(2, -1)$; $x = 2$; 1 **5.** $(-2, 0)$; $x = -2$; -3

7. a. $x = 1$ **b.** $(1, -2)$ **9. a.** $x = -1$ **b.** $(-1, 8)$

11. a. $x = 5$ **b.** $(5, 4)$

13. 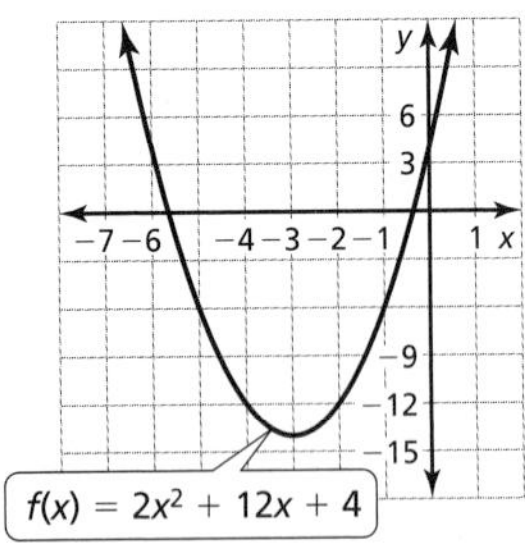

domain: all real numbers, range: $y \geq -14$

15. 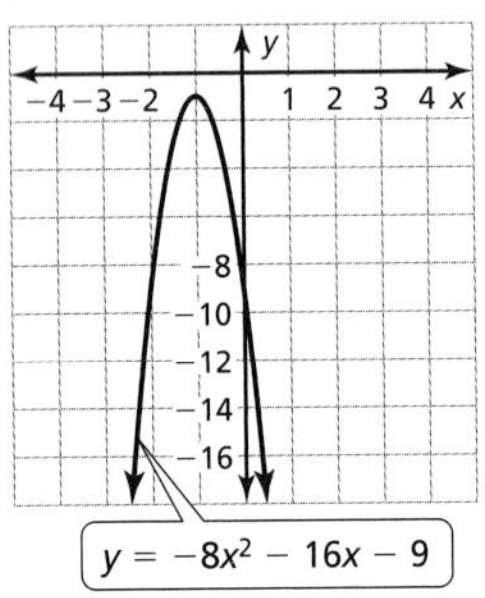

domain: all real numbers, range: $y \leq -1$

17. 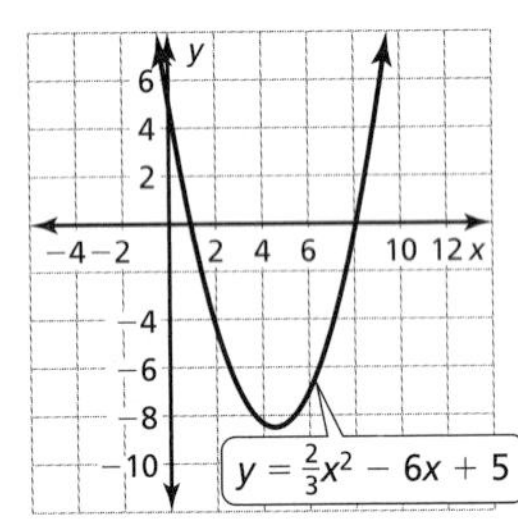

domain: all real numbers, range: $y \geq -\frac{17}{2}$

19. There should be two negatives in the substitution, one from the formula and one because b is -12;

$x = -\frac{b}{2a} = -\frac{-12}{2(3)} = 2$; The axis of symmetry is $x = 2$.

21. minimum value; -12 **23.** maximum value; -1

25. maximum value; $66\frac{1}{2}$ **27. a.** 4 sec **b.** 256 ft

29. **a.** 200 ft **b.** 50 ft

c. domain: $0 \leq x \leq 400$, range: $50 \leq y \leq 150$

31. (0, 8); *Sample answer:* Because the axis of symmetry is $x = 3$, the point (0, 8) would also lie on the graph.

33. $(-1.41, -4)$ **35.** (0.48, 0.72)

37. **a.** second aircraft hangar **b.** first aircraft hangar

39. **a.** $x = 4.5$ **b.** 20.25 in.^{2}

41. The graph of g is a reflection in the y-axis of the graph of h.

43. down; *Sample answer:* Because (3, 2) and (9, 2) have the same y-coordinate, any point with an x-coordinate between 3 and 9 lies on the part of the parabola between these two points that passes through the vertex. Because 7 is greater than 2, the vertex must be above these 2 points, so the parabola opens down.

45. $y = -2x^2 + 8x$ **47.** 14 ft **49.** $\frac{k^2}{8}$ ft^2

8.3 Maintaining Mathematical Proficiency *(p. 438)*

51. The graph of h is a vertical shrink by a factor of 0.5 and a reflection in the x-axis of the graph of f.

53. The graph of p is a vertical stretch by a factor of 3 and a horizontal translation 1 unit left of the graph of f.

8.4 Vocabulary and Core Concept Check *(p. 446)*

1. *Sample answer:* The graph of an even function is symmetric about the y-axis. The graph of an odd function is symmetric about the origin.

3. The graph of g is a horizontal translation h units right if h is positive or $|h|$ units left if h is negative, and a vertical translation k units up if k is positive or $|k|$ units down if k is negative of the graph of f.

8.4 Monitoring Progress and Modeling with Mathematics *(pp. 446–448)*

5. neither **7.** neither **9.** even **11.** neither

13. even **15.** neither **17.** odd

19. $(-1, 0)$; $x = -1$ **21.** (4, 0); $x = 4$

23. 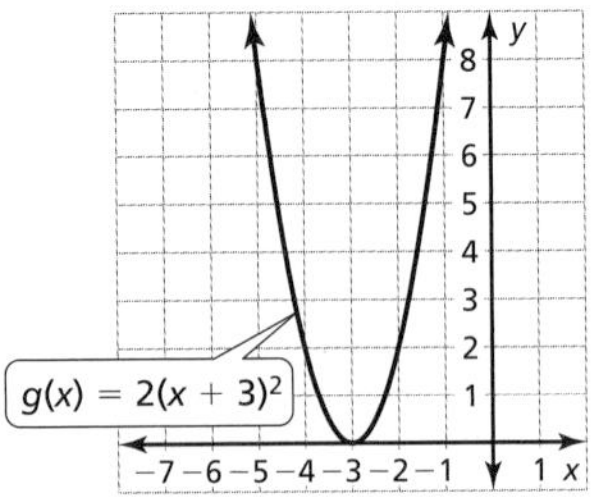

The graph of g is a horizontal translation 3 units left and a vertical stretch by a factor of 2 of the graph of f.

25. 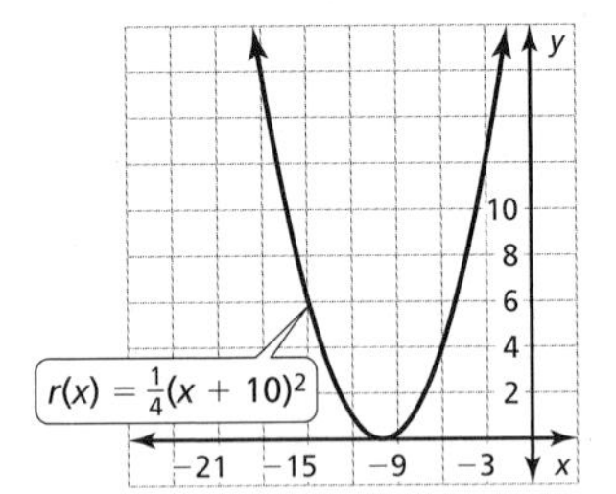

The graph of r is a horizontal translation 10 units left and a vertical shrink by a factor of $\frac{1}{4}$ of the graph of f.

27. 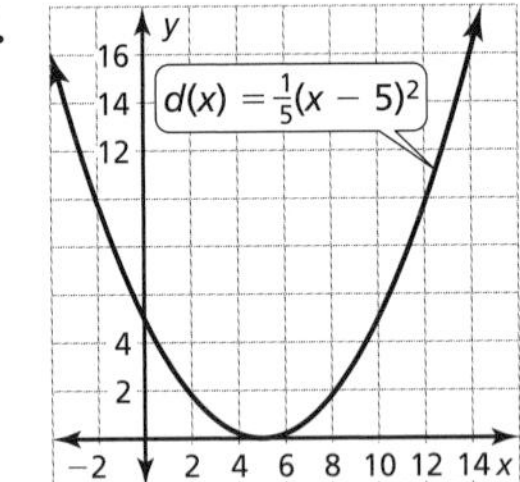

The graph of d is a horizontal translation 5 units right and a vertical shrink by a factor of $\frac{1}{5}$ of the graph of f.

29. If $f(-x) = f(x)$ the function is even; So, $f(x)$ is an even function.

31. $(-4, -3)$; $x = -4$ **33.** $(-3, 1)$; $x = -3$

35. C **37.** D

39. 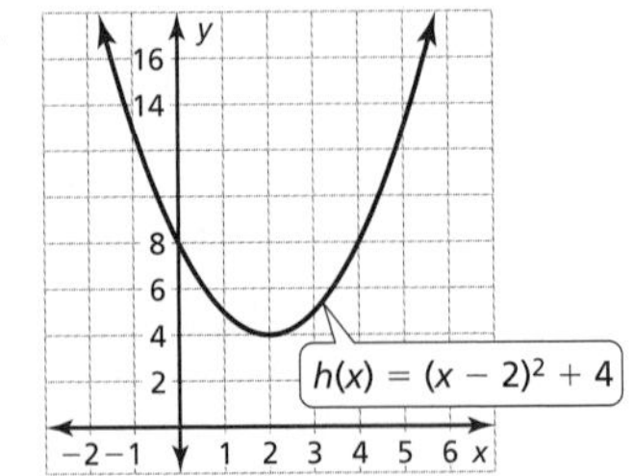

The graph of h is a translation 2 units right and 4 units up of the graph of f.

41. $r(x) = 4(x - 1)^2 - 5$

The graph of r is a vertical stretch by a factor of 4, and a translation 1 unit right and 5 units down of the graph of f.

43.

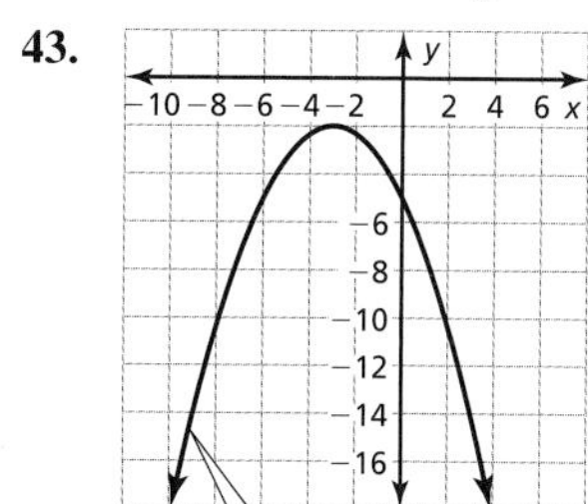

The graph of g is a vertical shrink by a factor of $\frac{1}{3}$, a reflection in the x-axis, and a translation 3 units left and 2 units down of the graph of f.

45. A **47.** B

49.

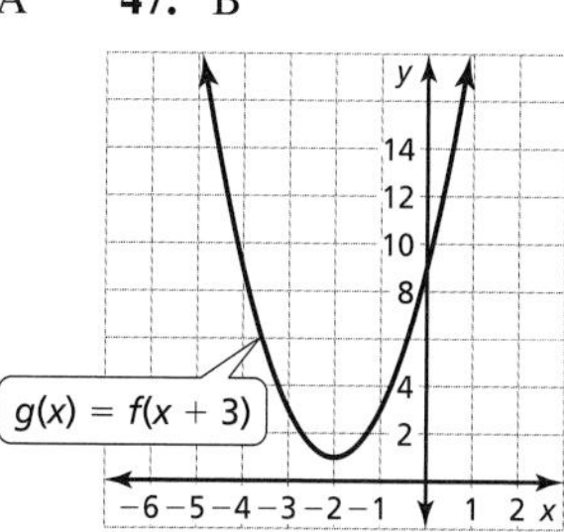

51.

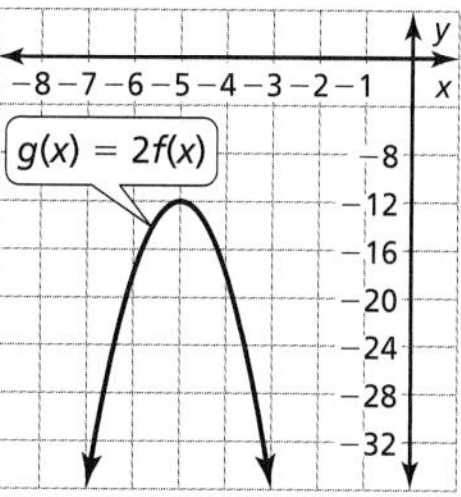

53.

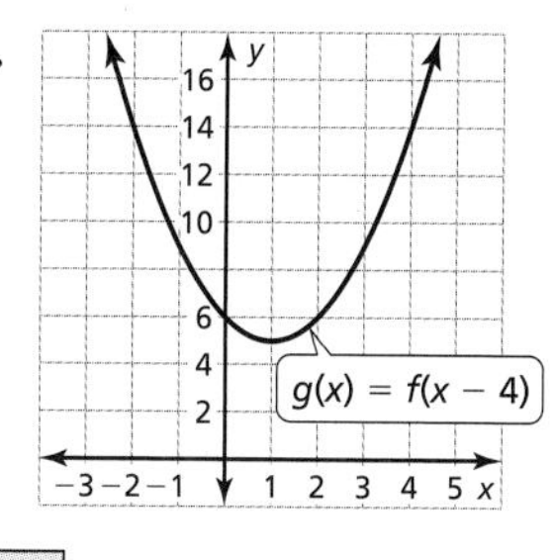

55. a.

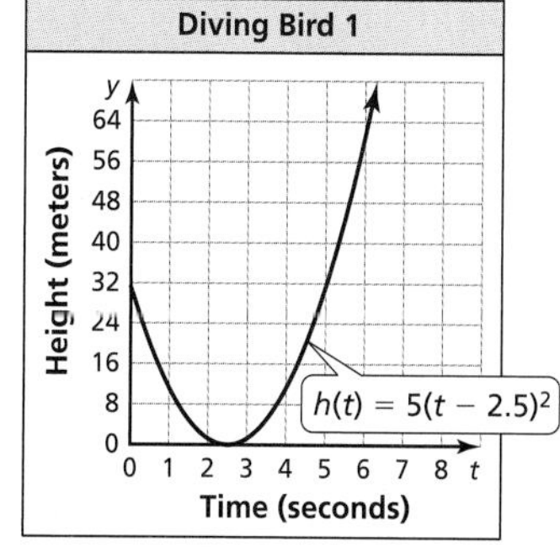

b. Diving Bird 2

$r(t) = 2h(t)$

c. The graph of r is a vertical stretch by a factor of 2 of the graph of h; the second bird; *Sample answer:* Because $r(t)$ is twice $h(t)$, the second bird starts at a height twice as high as the first bird.

57. $f(x) = 2(x - 1)^2 + 2$ **59.** $f(x) = -2(x + 2)^2 - 4$

61. $f(x) = \frac{1}{2}(x - 5)^2 - 2$

63. $f(x) = -\frac{18}{125}(x - 25)^2 + 90$

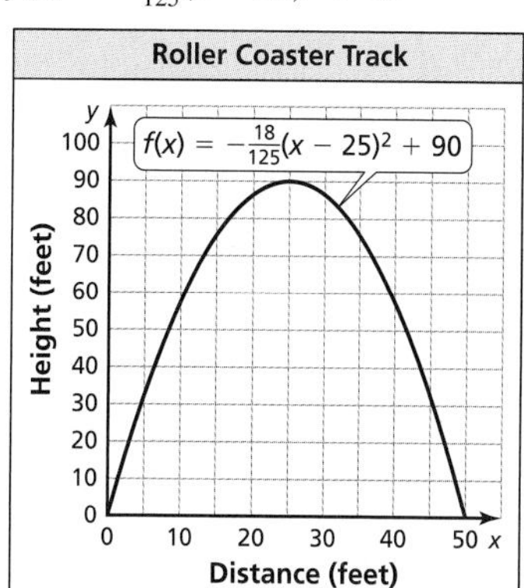

65. $y = 2(x - 2)^2 - 4$ **67.** $f(x) = -5(x - 1)^2 + 8$

69. no; *Sample answer:* The graph would not pass the vertical line test.

71. The graph of h is a vertical translation 4 units up of the graph of f; $h(x) = -(x + 1)^2 + 2$

73. The graph of h is a vertical stretch by a factor of 2 of the graph of f; $h(x) = 8(x - 2)^2 + 6$

75. $y = (x - 2)^2 - 5$; $y = x^2 - 4x - 1$; *Sample answer:* The vertex, $(2, -5)$, can be quickly determined from the vertex form; The y-intercept, -1, can be quickly determined from the standard form.

77. a. the second birdbath; *Sample answer:* The first birdbath has a depth of 4 inches and the second birdbath has a depth of 6 inches.

b. the first birdbath; *Sample answer:* The first birdbath has a width of 36 inches and the second birdbath has a width of 30 inches.

79. no; *Sample answer:* If the absolute value function includes a horizontal translation, it will not be symmetric about the y-axis.

8.4 Maintaining Mathematical Proficiency *(p. 448)*

81. $x = -3, x = 8$

8.5 Vocabulary and Core Concept Check *(p. 455)*

1. x-intercepts

8.5 Monitoring Progress and Modeling with Mathematics *(pp. 455–458)*

3. $-3, 1; x = -1$ **5.** $-7, 5; x = -1$

7.

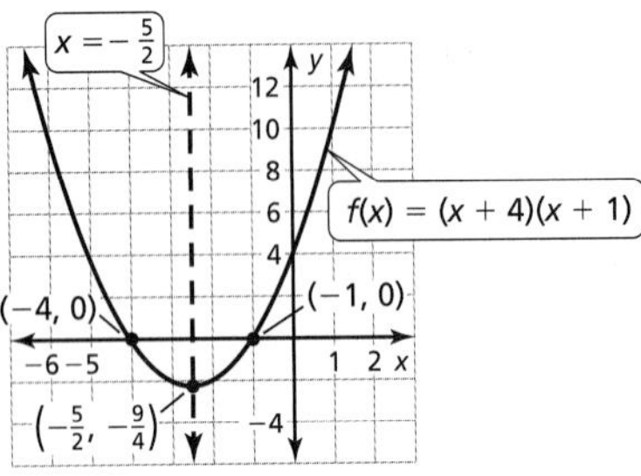

domain: all real numbers, range: $y \geq -\frac{9}{4}$

9.

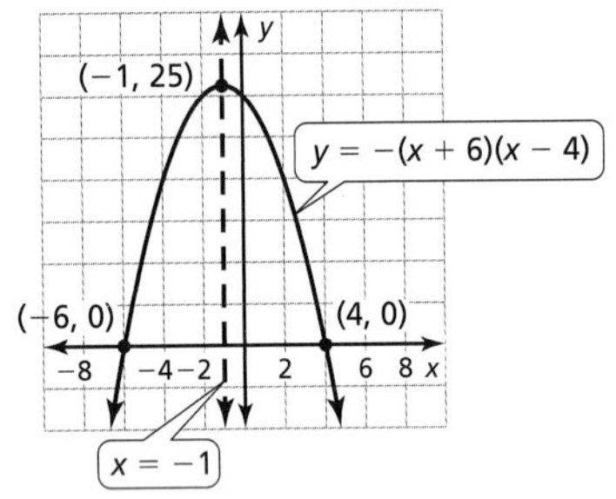

domain: all real numbers, range: $y \leq 25$

11.

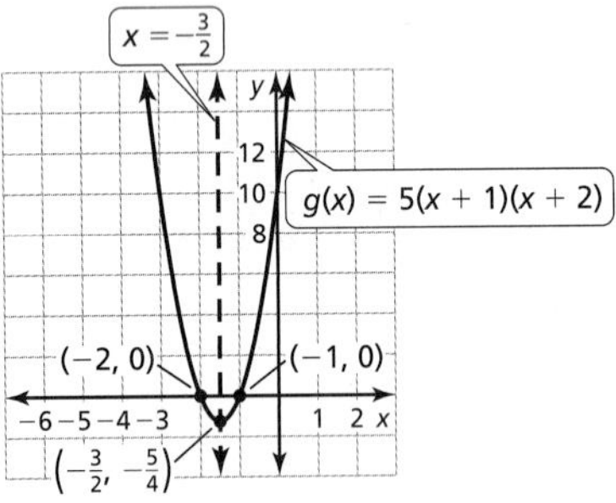

domain: all real numbers, range: $y \geq -\frac{5}{4}$

13.

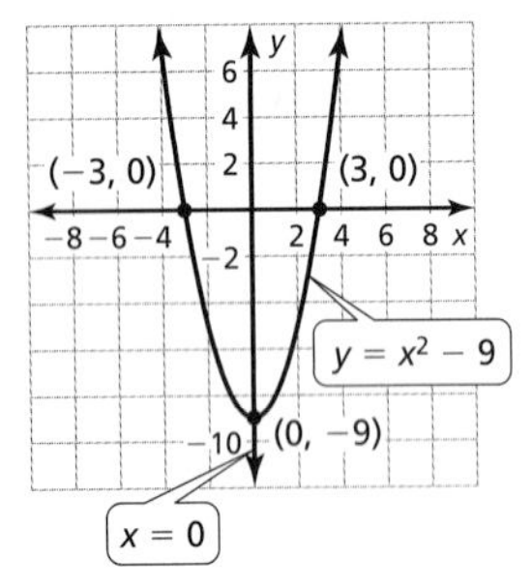

domain: all real numbers, range: $y \geq -9$

15.

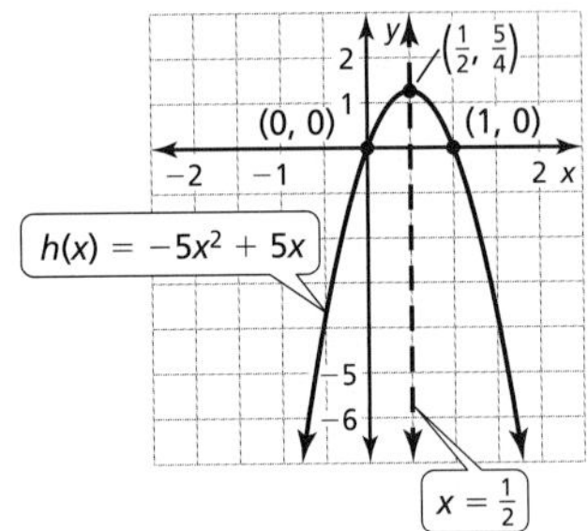

domain: all real numbers, range: $y \leq \frac{5}{4}$

17.

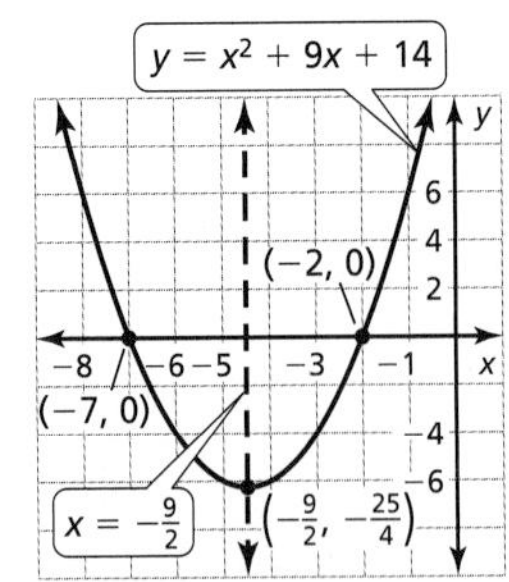

domain: all real numbers, range: $y \geq -\frac{25}{4}$

19.

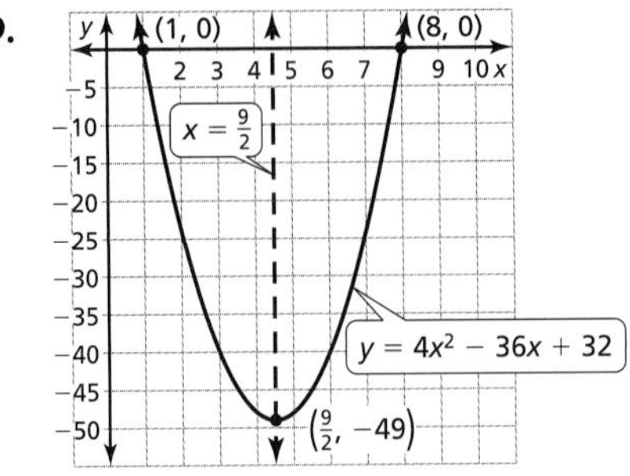

domain: all real numbers, range: $y \geq -49$

21. 2, 10 **23.** $-8, 3$ **25.** $-2, 7$ **27.** $-5, -2, 2$

29. $-7, 0, 7$ **31.** D **33.** C **35.** A

37.

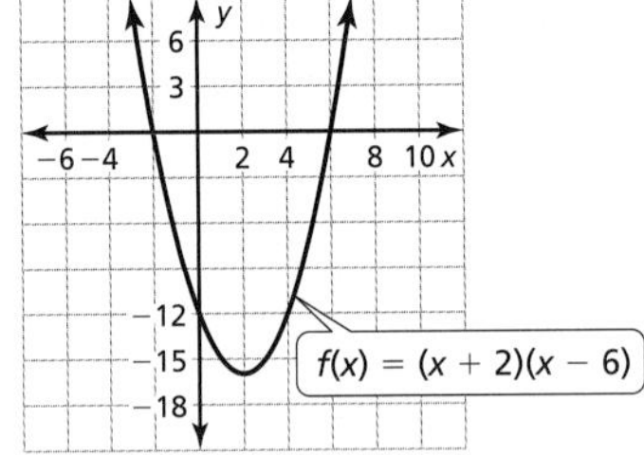

39.

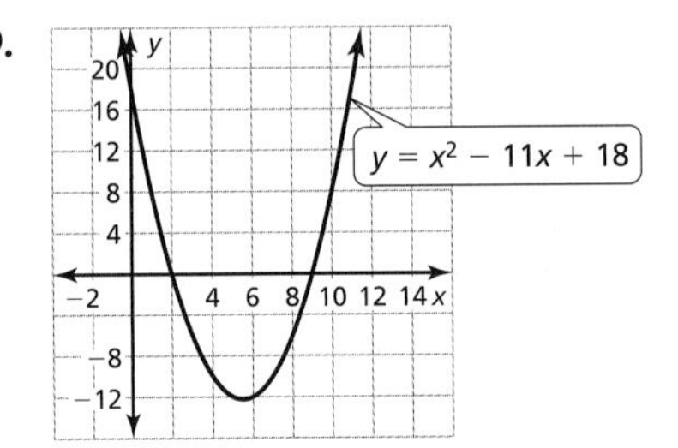

41.

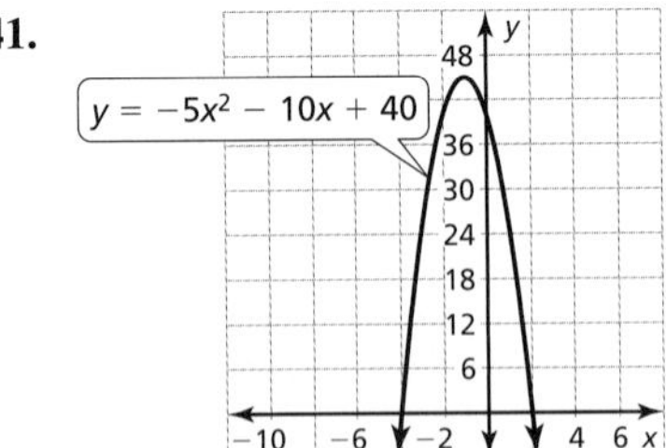

43. The factors need to be set equal to 0 and solved to find the zeros; $x + 3 = 0$ or $x - 2 = 0$; $x = -3$ or $x = 2$; The zeros of the function are -3 and 2.

45. *Sample answer:* $f(x) = x^2 - 14x + 46$

47. *Sample answer:* $f(x) = x^2 - 10x + 9$

49. $f(x) = 3x^2 + 3x - 36$

51. *Sample answer:* $f(x) = x^2 - 7x$

53. *Sample answer:* $f(x) = x^2 + 10x + 25$

55. *Sample answer:* $f(x) = x^2 - 3$

57. $f(x) = 2x^2 + 2x - 12$ **59.** $f(x) = -4x^2 + 8x + 32$

61.

63.

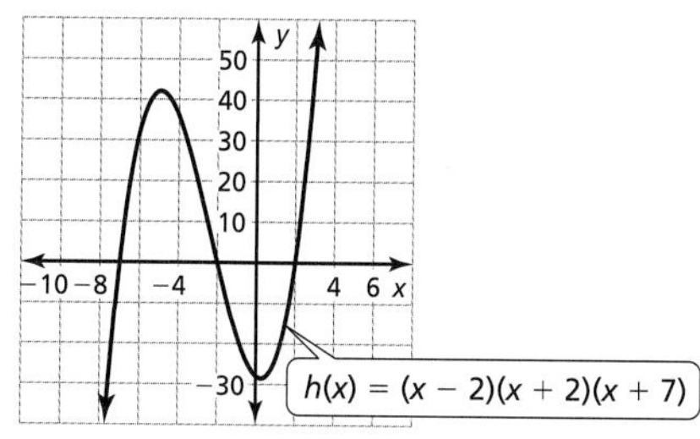

65.

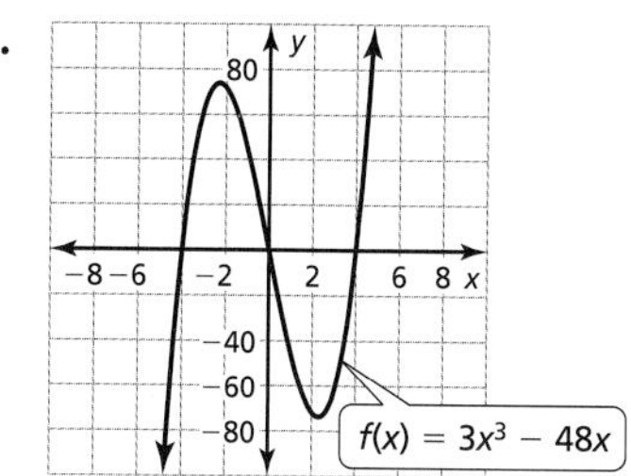

67.

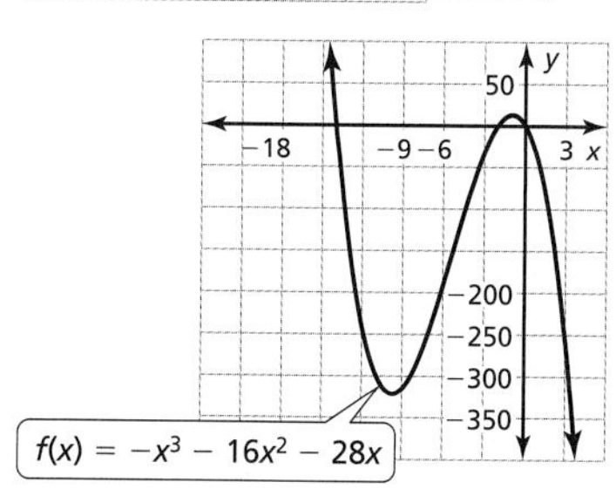

69. $f(x) = 4x^3 - 12x^2 - 16x$ **71.** $f(x) = -2x^3 - 22x^2 - 56x$

73. *Sample answer:* $f(x) = x^3 - 9x^2 + 2x + 48$

75. *Sample answer:* $f(x) = x^3 - 8x^2 + 7x$

77. $f(x) = -3x^2 + 21x$ **79.** $f(x) = 5x^3 + 20x^2 - 45x - 180$

81.

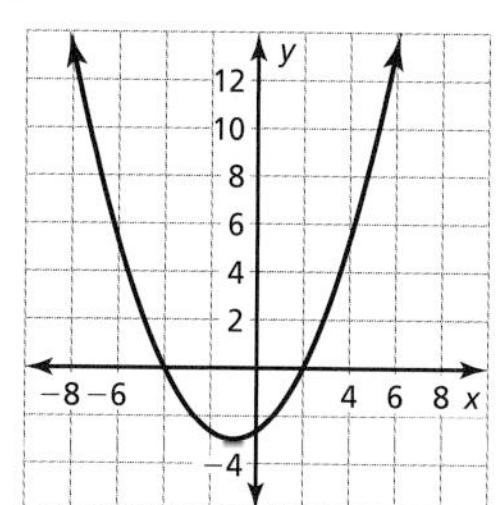

83. *Sample answer:*

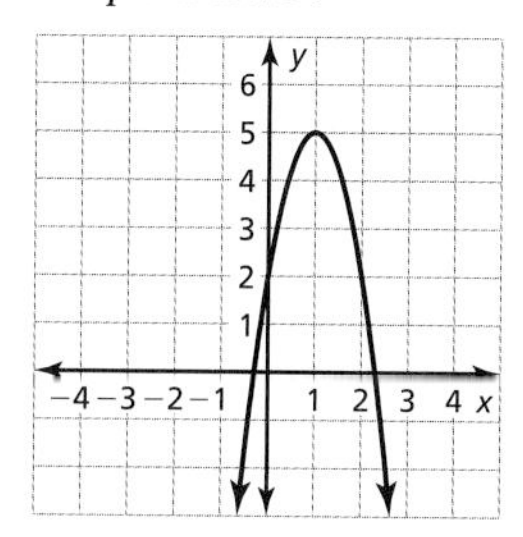

85. a. 4 ft **b.** $\frac{1}{2}$ ft **c.** $y = \frac{1}{6}(x^2 - 9)$

87. D **89.** C

91. not possible; *Sample answer:* Because -5 and 1 are the x-intercepts, the axis of symmetry is $x = -2$. The points $(-3, 12)$ and $(-1, 4)$ are the same horizontal distance from the axis of symmetry, so for both of them to lie on the parabola they would have to have the same y-coordinate.

93. a. $f(x) = -3(x + 4)(x - 2)$

b.

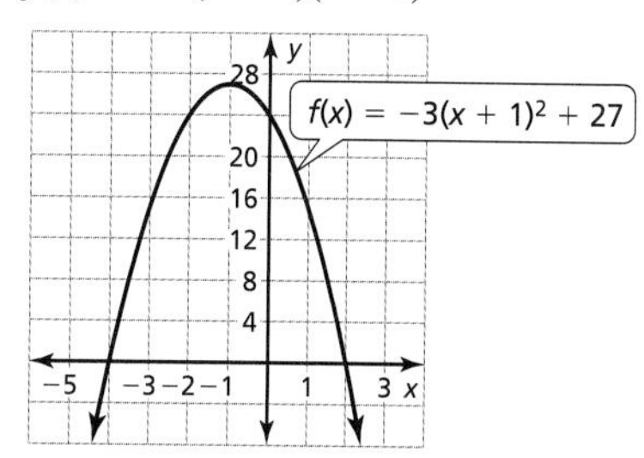

Sample answer: Plot the vertex $(-1, 27)$, which can be determined from the vertex form. Then plot the x-intercepts $(-4, 0)$ and $(2, 0)$, which can be determined from the intercept form. Draw a smooth curve through these points.

95. yes; *Sample answer:* Let p and q be real zeros of a quadratic function. When $p = q$, the function has exactly one real zero and can be written as $y = a(x - p)(x - p)$.

97. $f(x) = \frac{1}{10}(x + 5)(x + 2)(x - 1)(x - 4)(x - 8)$

99. $-k$, $2k$

101. *Sample answer:* $y = (x - 5)^2 + 2$ and $y = -(x - 5)^2 + 10$; The two given points have the same y-coordinate, so the axis of symmetry of any parabola passing through them would be halfway between, which is $x = 5$. The vertex form can be used with $h = 5$ and any selected value of k, where $k \neq 6$, to find quadratic equations that would pass through these points. Any two of these equations would form a system with these two points as solutions.

8.5 Maintaining Mathematical Proficiency *(p. 458)*

103. 29 g

105. neither; There is no common difference or common ratio.

107. arithmetic; There is a common difference of -8.

8.6 Vocabulary and Core Concept Check *(p. 465)*

1. linear, exponential, quadratic; $y = mx + b$, $y = ab^x$, $y = ax^2 + bx + c$; a straight line, a continuously increasing or decreasing curve, a parabola

3. Find the slope of the line through $(a, f(a))$ and $(b, f(b))$.

8.6 Monitoring Progress and Modeling with Mathematics *(pp. 465–468)*

5. quadratic **7.** exponential

9.

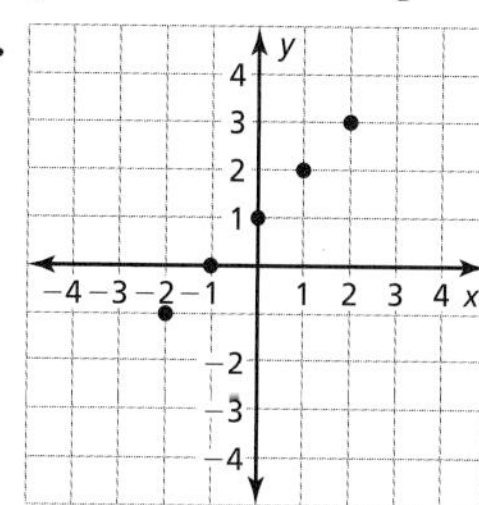

linear

11.

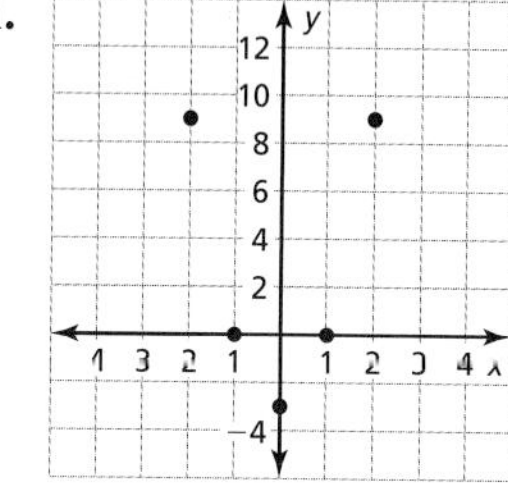

quadratic

13. 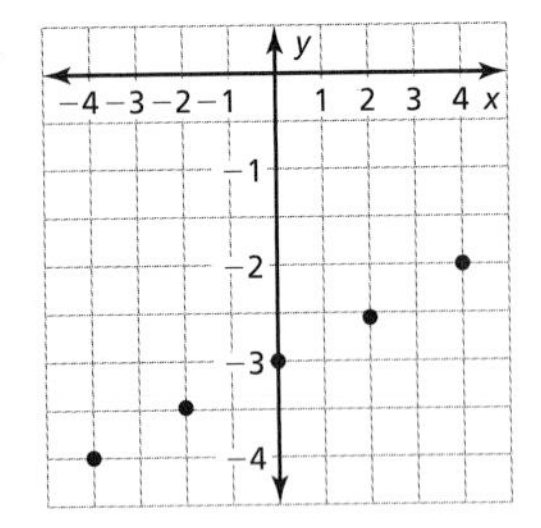

linear

15. linear **17.** exponential

19. linear; The first differences are constant.

21. quadratic; $y = 2x^2 - 2x - 4$ **23.** linear; $y = -3x - 2$

25. quadratic; $y = -3x^2$

27. Consecutive y-values have a constant ratio. They do not change by a constant amount; Consecutive y-values change by a constant ratio. So, the table represents an exponential function.

29. a. 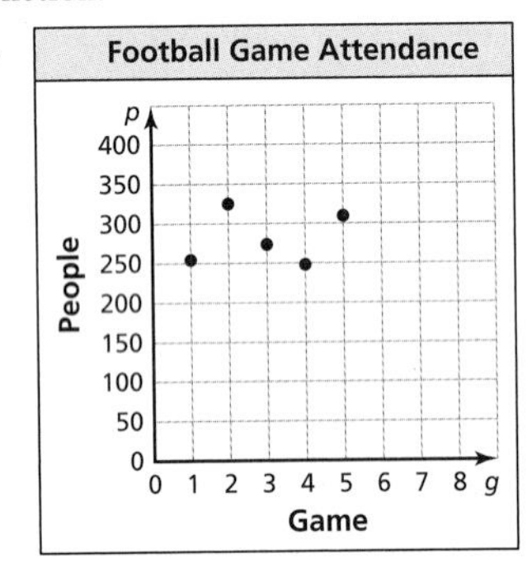

b. no; The points do not appear to follow the shape of any of these types of functions.

31. a. 3, 23, 35, 39, 35, 23, 3

b. 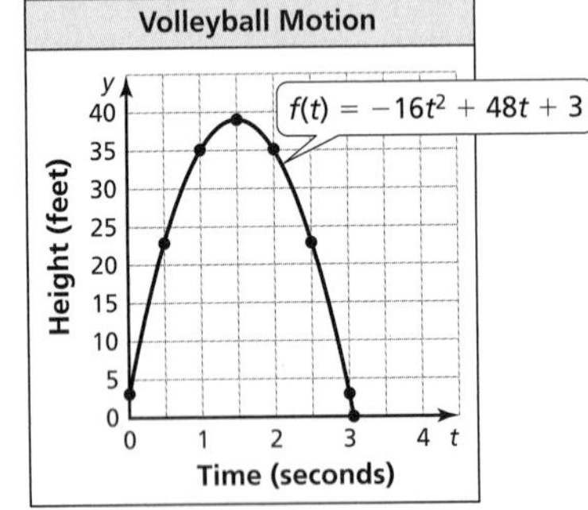

c. The function is increasing between 0 and 1.5 seconds, and the function is decreasing between 1.5 seconds and about 3 seconds.

d. 40, 24, 8, −8, −24, −40; The average rate of change decreases when the function is increasing, and the average rate of change increases in the negative direction when the function is decreasing.

33. a. exponential; linear; quadratic

b. Organization A: 3, 12, 48, 192, 768, 3072; Organization B: 4, 4, 4, 4, 4, 4; Organization C: 4, 12, 20, 28, 36, 44

c. Organization A; Organization A will have the most donations, followed by Organization C, then Organization B.

35. The average rate of change of a linear function is constant because the dependent variable of a linear function increases by the same amount for each constant change in the independent variable. The average rate of change of a quadratic or exponential function is not constant because the dependent variable of a quadratic or exponential function changes by a different amount for each constant change in the independent variable.

37. quadratic; The second differences have a constant value of $9n - 5$.

39. no; *Sample answer:* There may not be enough points to clearly determine the shape of the graph.

41. no; The graph of an exponential growth function will always eventually have greater y-values than the graph of a quadratic function.

8.6 Maintaining Mathematical Proficiency *(p. 468)*

43. 11 **45.** 8 **47.** $x^2 - 64$ **49.** $9a^2 - 25b^2$

Chapter 8 Review *(pp. 470–472)*

1. 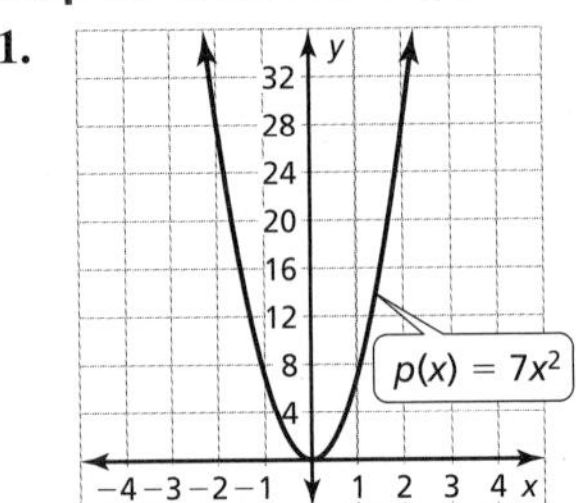

The graph of p is a vertical stretch by a factor of 7 of the graph of f.

2. 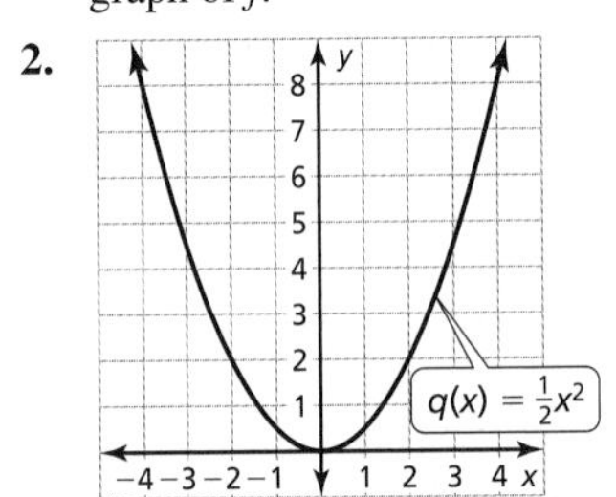

The graph of q is a vertical shrink by a factor of $\frac{1}{2}$ of the graph of f.

3. 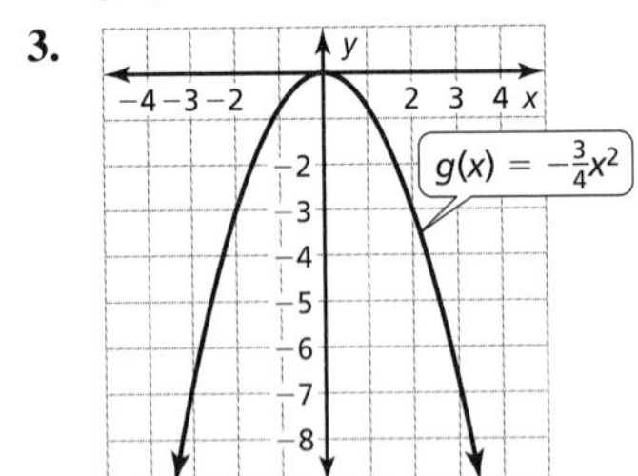

The graph of g is a vertical shrink by a factor of $\frac{3}{4}$ and a reflection in the x-axis of the graph of f.

4. 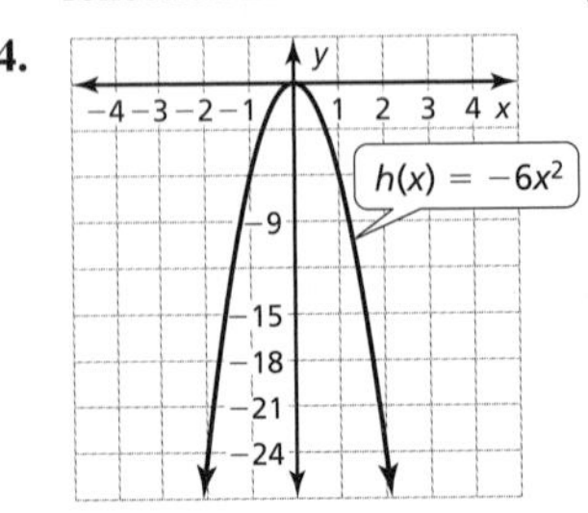

The graph of h is a vertical stretch by a factor of 6 and a reflection in the x-axis of the graph of f.

5. The vertex is $(1, -3)$. The axis of symmetry is $x = 1$. The domain is all real numbers. The range is $y \geq -3$. When $x < 1$, y increases as x decreases. When $x > 1$, y increases as x increases.

6. $g(x) = x^2 + 5$

The graph of g is a vertical translation 5 units up of the graph of f.

7.

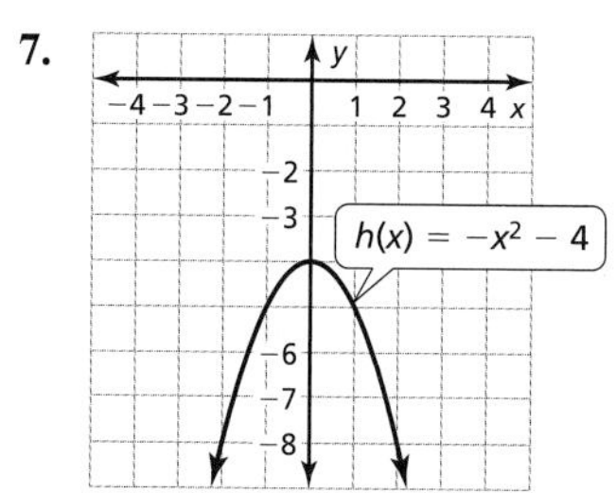

The graph of h is a reflection in the x-axis, and a vertical translation 4 units down of the graph of f.

8.

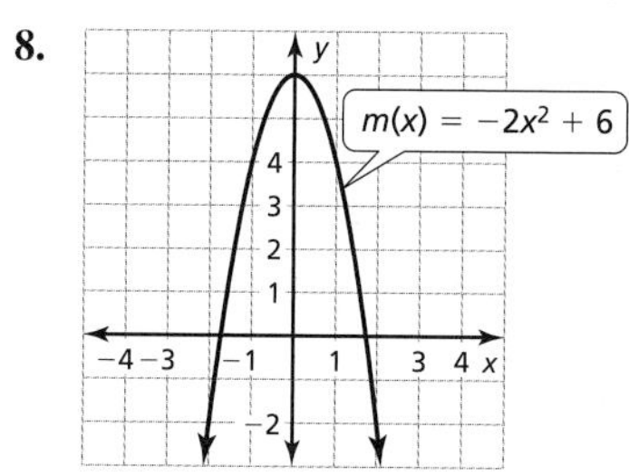

The graph of m is a vertical stretch by a factor of 2, a reflection in the x-axis, and a vertical translation 6 units up of the graph of f.

9.

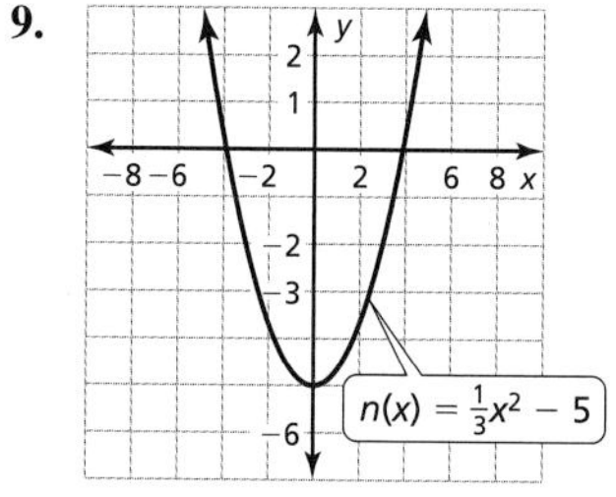

The graph of n is a vertical shrink by a factor of $\frac{1}{3}$ and a vertical translation 5 units down of the graph of f.

10.

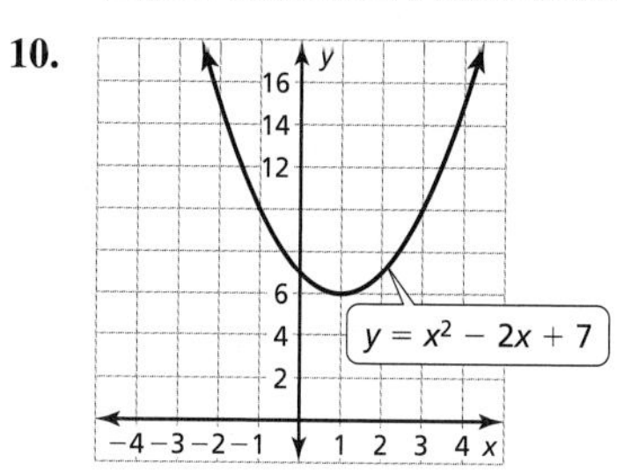

domain: all real numbers, range: $y \geq 6$

11.

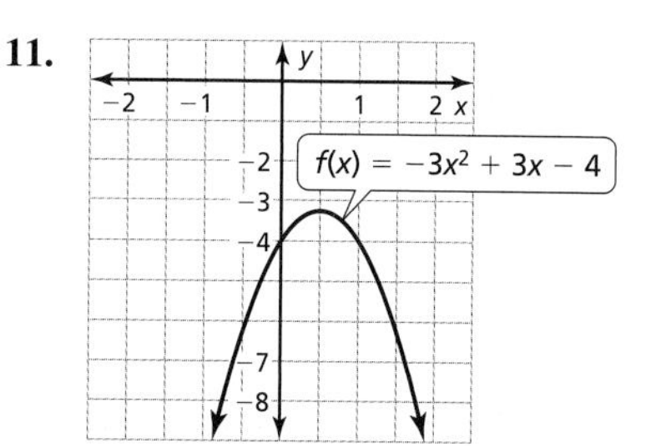

domain: all real numbers, range: $y \leq -\frac{13}{4}$

12.

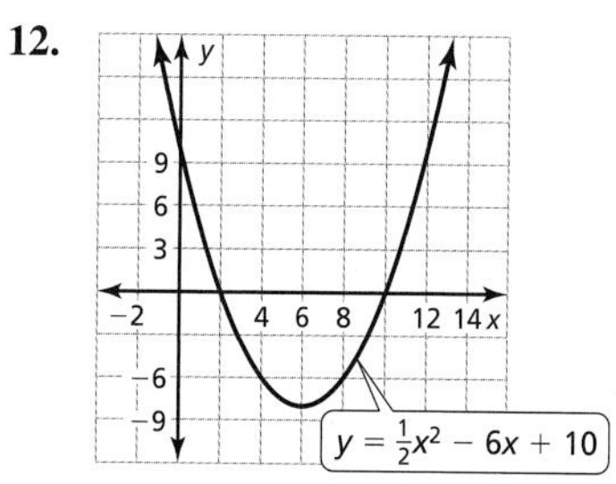

domain: all real numbers, range: $y \geq -8$

13. 2.75 sec; 133 ft **14.** neither **15.** odd

16. neither

17.

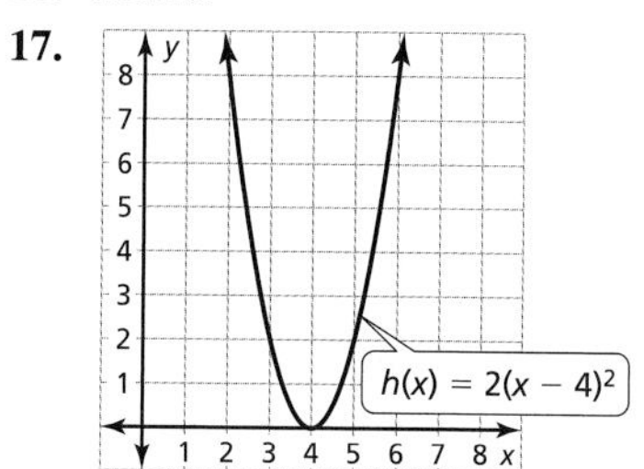

The graph of h is a vertical stretch by a factor of 2 and a horizontal translation 4 units right of the graph of f.

18.

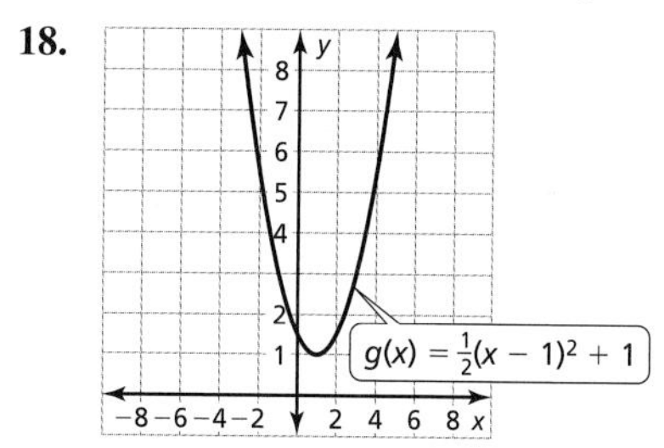

The graph of g is a vertical shrink by a factor of $\frac{1}{2}$, and a translation 1 unit right and 1 unit up of the graph of f.

19.

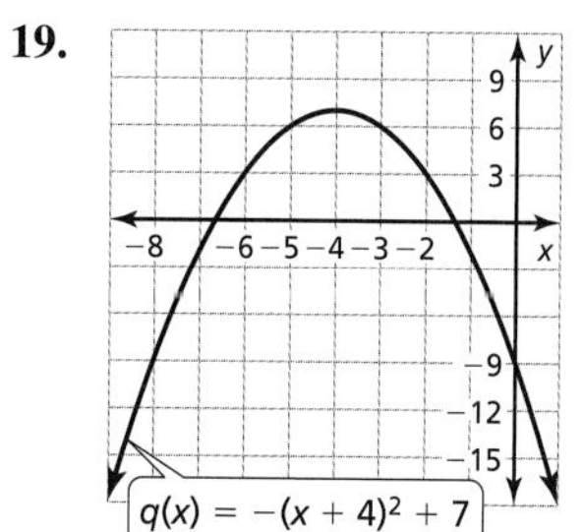

The graph of q is a reflection in the x-axis, and a translation 4 units left and 7 units up of the graph of f.

20. 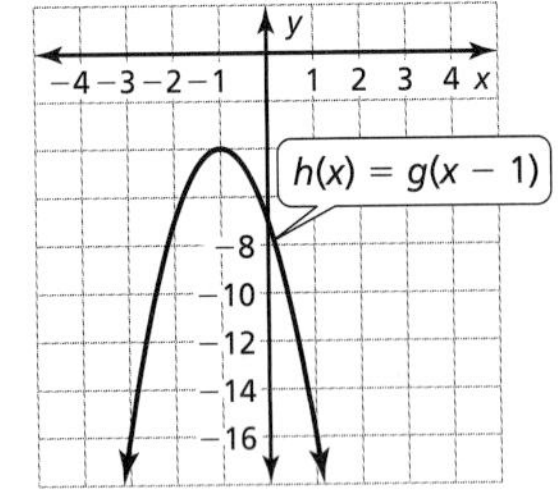

21. $f(x) = 5(x - 3)^2 + 2$

22. 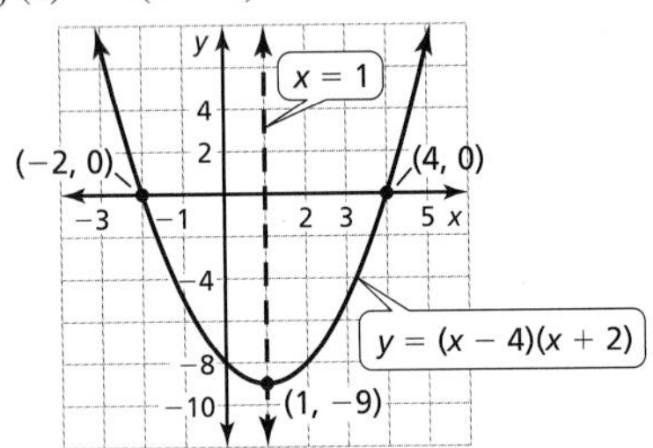

domain: all real numbers, range: $y \geq -9$

23. 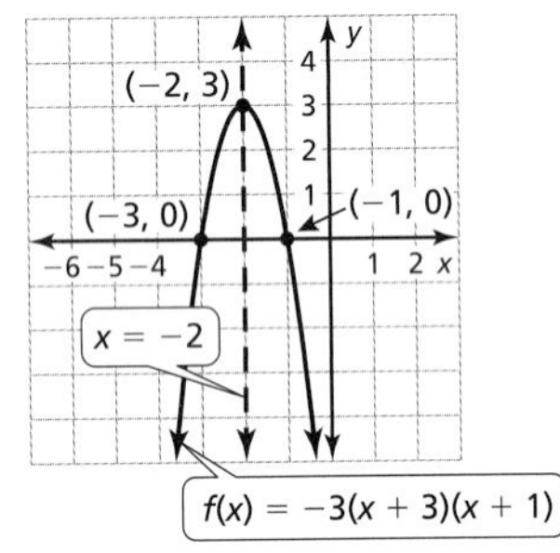

domain: all real numbers, range: $y \leq 3$

24. 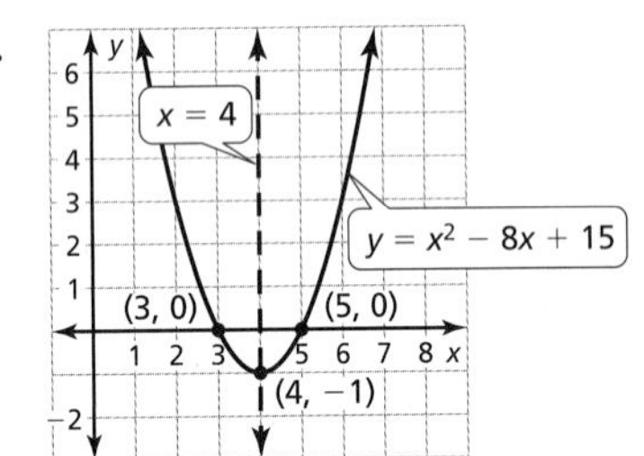

domain: all real numbers, range: $y \geq -1$

25. 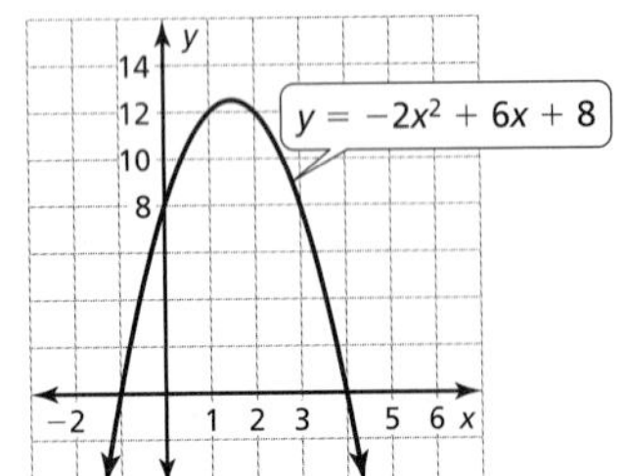

26. 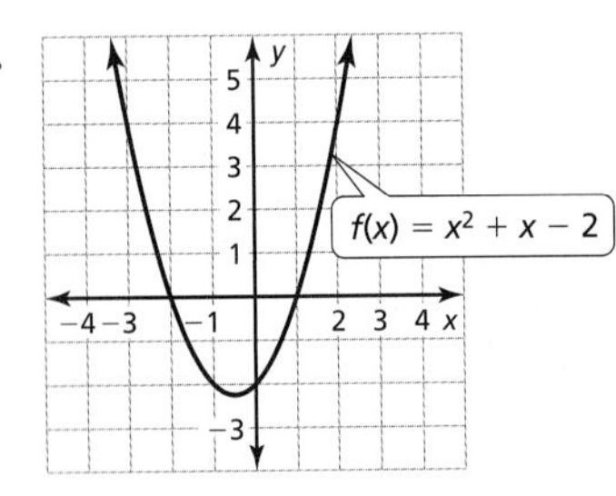

27. 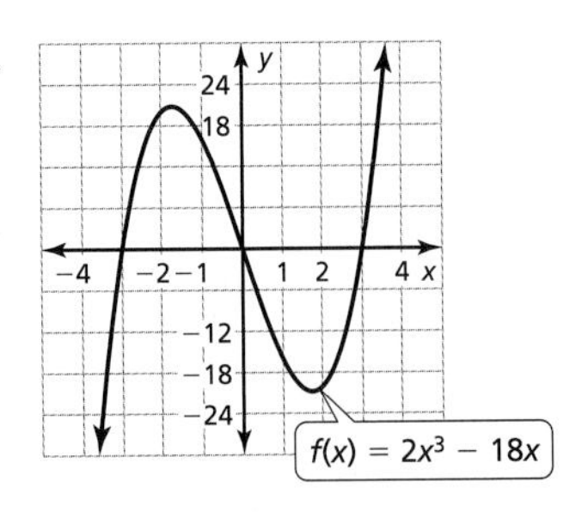

28. *Sample answer:* $x^2 - 10x + 24$

29. exponential; $y = 128\left(\frac{1}{4}\right)^x$

30. **a.** From year 2 to year 7, your account increased at an average rate of about \$29.55 per year and your friend's account increased at an average rate of \$20 per year. So, your account is growing faster.

b. your account; After 7 years, the balances are about equal and the average rate of change of your balance, which shows exponential growth, exceeds the average rate of change of your friend's balance, which shows linear growth. So, your account will have a greater balance after 10 years.

Chapter 9

Chapter 9 Maintaining Mathematical Proficiency *(p. 477)*

1. $(x + 5)^2$ **2.** $(x - 10)^2$ **3.** $(x + 6)^2$ **4.** $(x - 9)^2$
5. $(x + 8)^2$ **6.** $(x - 15)^2$ **7.** $(1, -2)$ **8.** $(4, 4)$
9. $(-2, 3)$ **10.** $\left(\frac{b}{2}\right)^2$

9.1 Vocabulary and Core Concept Check *(p. 485)*

1. rationalizing the denominator

3. yes; $\sqrt{\frac{2x}{9}} = \sqrt{\frac{1}{9} \cdot 2x}, = \sqrt{\frac{1}{9}} \cdot \sqrt{2x} = \frac{1}{3}\sqrt{2x}$

9.1 Monitoring Progress and Modeling with Mathematics *(pp. 485–488)*

5. yes
7. no; The radicand has a perfect square factor of 16.
9. no; A radical appears in the denominator of a fraction.
11. no; A radical appears in the denominator of a fraction.
13. $2\sqrt{5}$ **15.** $8\sqrt{2}$ **17.** $5\sqrt{5b}$ **19.** $-9m\sqrt{m}$
21. $\frac{2}{7}$ **23.** $-\frac{\sqrt{23}}{8}$ **25.** $\frac{a\sqrt{a}}{7}$ **27.** $\frac{5}{x}$ **29.** $2\sqrt[3]{2}$
31. $-4x\sqrt[3]{x^2}$ **33.** $-\frac{\sqrt[3]{6c}}{5}$ **35.** $-\frac{3\sqrt[3]{3y^2}}{10x}$
37. The radicand 18 has a perfect square factor of 9; $\sqrt{72} = \sqrt{36 \cdot 2} = 6\sqrt{2}$
39. $\frac{\sqrt{6}}{\sqrt{6}}$ **41.** $\frac{\sqrt[3]{x}}{\sqrt[3]{x}}$ **43.** $\frac{\sqrt{5} + 8}{\sqrt{5} + 8}$ **45.** $\sqrt{2}$
47. $\frac{\sqrt{15}}{12}$ **49.** $\frac{3\sqrt{a}}{a}$ **51.** $\frac{d\sqrt{15}}{5}$ **53.** $\frac{4\sqrt[3]{5}}{5}$
55. $\frac{\sqrt{7} - 1}{6}$ **57.** $\frac{7\sqrt{10} + 2\sqrt{5}}{47}$ **59.** $\sqrt{5} + \sqrt{2}$
61. **a.** about 1.85 sec **b.** about 0.41 sec
63. about 5.42 amperes **65.** $5\sqrt{2}$, about 7.07

67. $\frac{\sqrt{2}}{3}$, about 0.47 69. $2\sqrt{2}$, about 2.83

71. $-6\sqrt{2}$, about -8.49 73. about 3.71 in.

75. $\sqrt{3} + 4\sqrt{2}$ 77. $-13\sqrt{6}$ 79. $8\sqrt{3} + 2\sqrt{6}$

81. $\sqrt[3]{3}$ 83. $4\sqrt{10}$ 85. $-2\sqrt{30x}$ 87. 18

89. $3\sqrt[3]{12}$ 91. about 114 ft 93. $\frac{\sqrt[5]{8125}}{5x}$ 95. $4\sqrt[4]{y}$

97. $9\sqrt[4]{9} - \sqrt[5]{9}$

99. a. $4, 2\frac{1}{4}, 2, 2 + \sqrt{3}, 2 - \sqrt{3}, 2 + \pi$;

$2\frac{1}{4}, \frac{1}{2}, \frac{1}{4}, \frac{1}{4} + \sqrt{3}, \frac{1}{4} - \sqrt{3}, \frac{1}{4} + \pi$;

$2, \frac{1}{4}, 0, \sqrt{3}, -\sqrt{3}, \pi$;

$2 + \sqrt{3}, \frac{1}{4} + \sqrt{3}, \sqrt{3}, 2\sqrt{3}, 0, \pi + \sqrt{3}$;

$2 - \sqrt{3}, \frac{1}{4} - \sqrt{3}, -\sqrt{3}, 0, -2\sqrt{3}, \pi - \sqrt{3}$;

$2 + \pi, \frac{1}{4} + \pi, \pi, \pi + \sqrt{3}, \pi - \sqrt{3}, 2\pi$

b. $4, \frac{1}{2}, 0, 2\sqrt{3}, -2\sqrt{3}, 2\pi$;

$\frac{1}{2}, \frac{1}{16}, 0, \frac{\sqrt{3}}{4}, -\frac{\sqrt{3}}{4}, \frac{\pi}{4}$;

$0, 0, 0, 0, 0, 0$;

$2\sqrt{3}, \frac{\sqrt{3}}{4}, 0, 3, -3, \pi\sqrt{3}$;

$-2\sqrt{3}, -\frac{\sqrt{3}}{4}, 0, -3, 3, -\pi\sqrt{3}$;

$2\pi, \frac{\pi}{4}, 0, \pi\sqrt{3}, -\pi\sqrt{3}, \pi^2$

101. odd; even; When m is even, 2^m is a perfect square.

103. $a^2 < ab < b^2$ when $a < b$. 105. 377

107. $\frac{2\sqrt[3]{x^2} - 2\sqrt[3]{x} + 2}{x + 1}$; Multiplying the numerator and denominator by $\sqrt[3]{x^2} - \sqrt[3]{x} + 1$ rationalizes the denominator.

9.1 Maintaining Mathematical Proficiency *(p. 488)*

109.

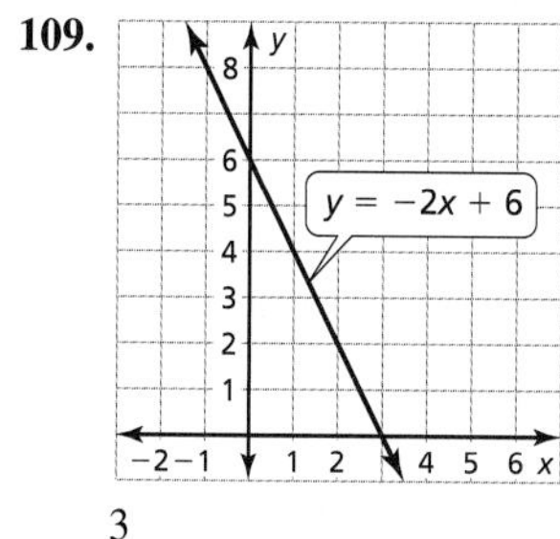

3

111.

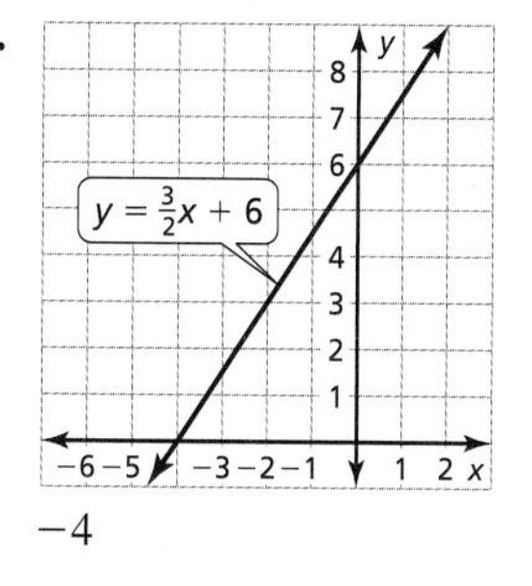

-4

113. $x = -3$ 115. $x = -\frac{2}{3}$

9.2 Vocabulary and Core Concept Check *(p. 494)*

1. an equation that can be written in the standard form $ax^2 + bx + c = 0$, where $a \neq 0$

3. The number of x-intercepts is the number of solutions.

9.2 Monitoring Progress and Modeling with Mathematics *(pp. 494–496)*

5. $x = 3, x = -1$ 7. $x = -4$

9. $4x^2 - 12 = 0$ or $-4x^2 + 12 = 0$

11. $x^2 - 2x + 1 = 0$ or $-x^2 + 2x - 1 = 0$ 13. $x = 0, x = 5$

15. no solution 17. $x = 3$ 19. $x = -1$

21. $x = -6, x = 2$ 23. $x = -2, x = 1$

25. The equation needs to be in standard form; $x^2 + 3x - 18 = 0$

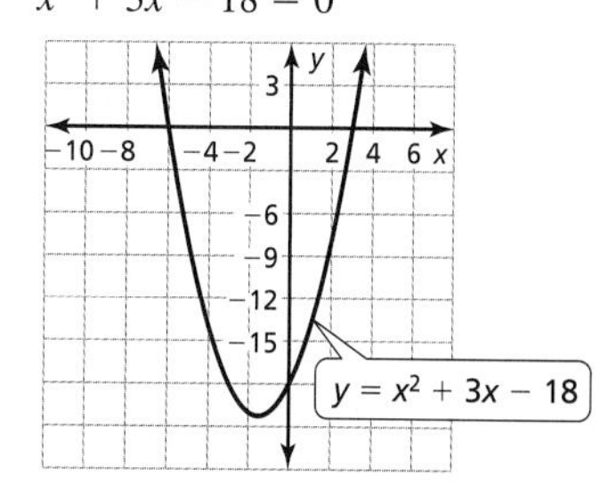

The solutions are $x = -6$ and $x = 3$.

27. a. The x-intercepts are the distances at which the height is 0 yards.

b. 5 yd

29. $x = -5, x = 2$ 31. no solution

33. $x = -10, x = -2$ 35. no solution 37. $2, 0, -1$

39. $-3, 1$ 41. $-2, 2, -3, 1$ 43. about 4.3, about 0.7

45. about 2.4, about -0.4

47.

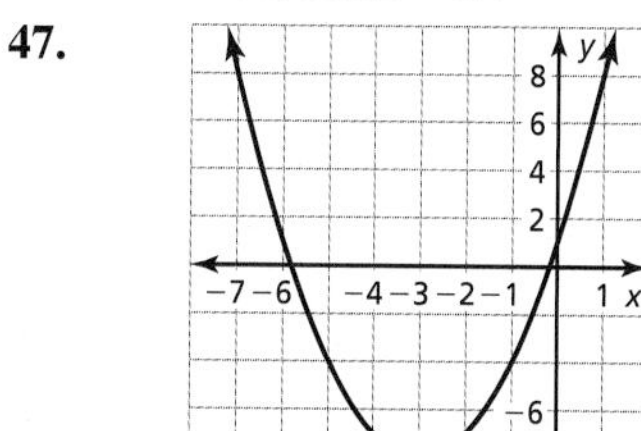
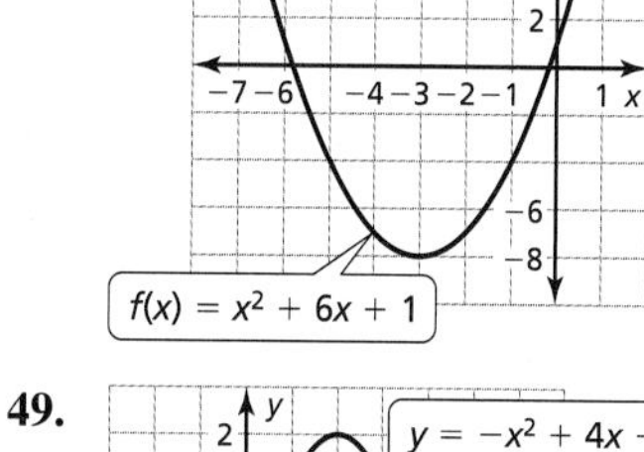

about -0.2, about -5.8

49.

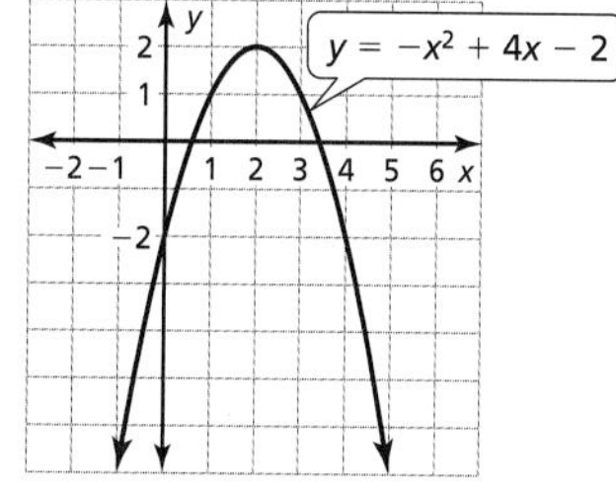

about 3.4, about 0.6

51.

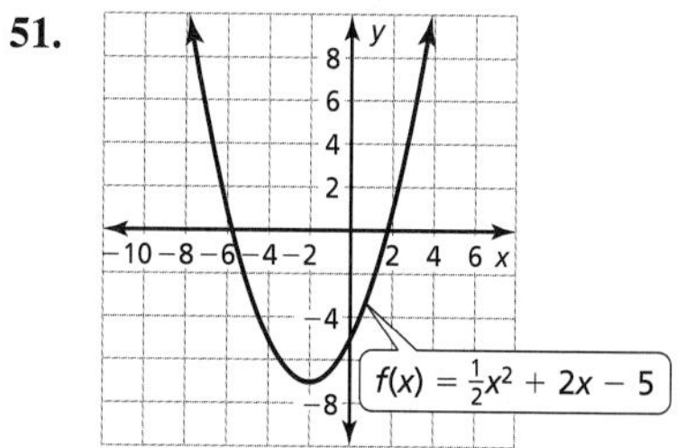

about 1.7, about -5.7

53. a. 118 ft, 198 ft, 246 ft, 262 ft, 246 ft, 198 ft, 118 ft, 6 ft

b. about 1.5 sec, about 6.5 sec

c. about 1.4 sec, about 6.6 sec

55. about 3.7 ft

57. Graph the function to determine which integers the solutions are between. Then make tables using x-values between the integers with an interval of 0.1. Look for a change of sign in the function values, then select the value closest to zero.

59. *Sample answer:* Method 1; Only one graph needs to be drawn.

61. about 24.1 ft

63. sometimes; $y = -2x^2 + 1$ has two x-intercepts, but $y = -2x^2 + (-1)$ has no x-intercepts.

65. never; The graph of $y = ax^2 + bx + c$ has at most two x-intercepts.

9.2 Maintaining Mathematical Proficiency *(p. 496)*

67. exponential growth function; As x increases by 1, y is multiplied by 4.

9.3 Vocabulary and Core Concept Check *(p. 501)*

1. two

9.3 Monitoring Progress and Modeling with Mathematics *(pp. 501–502)*

3. 2; $x = 5, x = -5$ **5.** 0; no real solutions

7. 1; $x = 0$ **9.** $x = 4, x = -4$ **11.** no real solutions

13. $x = 7, x = -7$ **15.** $x = 0$ **17.** $x = \frac{1}{2}, x = -\frac{1}{2}$

19. $x = -3$ **21.** $x = 5, x = -4$ **23.** $x = \frac{1}{3}, x = -\frac{7}{3}$

25. $x \approx 2.65, x \approx -2.65$ **27.** $x \approx 3.16, x \approx -3.16$

29. $x \approx 4.24, x \approx -4.24$

31. The number 36 has both a positive and negative square root; $x = \pm 6$

33. about 1.2 sec **35.** 3 ft

37. *Sample answer:* Use a calculator.

39. $(3, 9), (-3, 9)$; When $y = 9, x = \pm 3$.

41. $x = 1.2, x = -1.2$; $1.2^2 = 1.44$ **43.** $x = 6, x = -2$

9.3 Maintaining Mathematical Proficiency *(p. 502)*

45. $(x + 4)^2$ **47.** $(x - 7)^2$ **49.** $(x + 6)^2$

9.4 Vocabulary and Core Concept Check *(p. 511)*

1. completing the square

3. even; When b is even, $\frac{b}{2}$ is an integer.

9.4 Monitoring Progress and Modeling with Mathematics *(pp. 511–514)*

5. 16 **7.** 4 **9.** $\frac{225}{4}$ **11.** $x^2 - 10x + 25; (x - 5)^2$

13. $x^2 + 16x + 64; (x + 8)^2$ **15.** $x^2 + 5x + \frac{25}{4}; \left(x + \frac{5}{2}\right)^2$

17. $x = 1, x = -15$ **19.** $x \approx 3.41, x \approx 0.59$

21. $x \approx 6.27, x \approx -1.27$

23. a. $x^2 + 6x = 216$ **b.** width: 12 ft, length: 18 ft

25. $x = 5, x = 3$ **27.** $x \approx -3.27, x \approx -6.73$

29. $x \approx 1.92, x \approx -9.92$ **31.** $x = 8, x = -1$

33. The number 16 should be added to each side of the equation; $x^2 + 8x + 16 = 10 + 16$; $x = -4 \pm \sqrt{26}$

35. $b = 10, b = -10$; In a perfect square trinomial $c = \left(\frac{b}{2}\right)^2$, so $b = \pm 2\sqrt{c}$.

37. $y = (x + 3)^2 - 6$; D **39.** $y = -(x + 2)^2 + 2$; B

41. minimum value; -6 **43.** maximum value; -5

45. maximum value; -6

47. yes; The graph has two negative x-intercepts and it opens down.

49. no; The x-intercepts are both positive.

51. f, m; The graph has two negative x-intercepts and it opens up.

53. a. 36 ft

b. $x = \frac{3}{2}$; On the left side of $x = \frac{3}{2}$, the height increases as time increases. On the right side of $x = \frac{3}{2}$, the height decreases as time increases.

55. 3 ft **57.** 12 **59.** $x \approx 1.24, x \approx -3.24$

61. $x \approx 4.29, x \approx -0.29$ **63.** 40 mi/h

65. a. $\ell + 2w = 80, \ell w = 750$

b. length: 30 ft, width: 25 ft; length: 50 ft, width: 15 ft

67. a. $x \approx 0.8, x \approx -12.8$ **b.** $x \approx 0.78, x \approx -12.78$

c. *Sample answer:* completing the square; The result is more accurate.

69. $x(x + 2) = 48$, 6 and 8

71. yes; Substituting 23.50 for y in the model, the stock is worth \$23.50 ten days and thirty days after the stock is purchased.

73. length: 66 in., width: 6 in.

9.4 Maintaining Mathematical Proficiency *(p. 514)*

75. $a_1 = 10, a_n = a_{n-1} + 5$ **77.** $a_1 = -20, a_n = a_{n-1} + 4$

79. $6\sqrt{2}$

9.5 Vocabulary and Core Concept Check *(p. 521)*

1. the quadratic formula; $x = \dfrac{-b \pm \sqrt{b^2 - 4ac}}{2a}$

9.5 Monitoring Progress and Modeling with Mathematics *(pp. 521–524)*

3. $x^2 - 7x = 0; a = 1, b = -7, c = 0$ or $-x^2 + 7x = 0; a = -1, b = 7, c = 0$

5. $-2x^2 - 5x + 1 = 0; a = -2, b = -5, c = 1$ or $2x^2 + 5x - 1 = 0; a = 2, b = 5, c = -1$

7. $x^2 - 6x + 4 = 0; a = 1, b = -6, c = 4$ or $-x^2 + 6x - 4 = 0; a = -1, b = 6, c = -4$

9. $x = 6$ **11.** $x = 11, x = -1$ **13.** no real solutions

15. $x = \frac{3}{2}, x = \frac{2}{3}$ **17.** $x = \frac{1}{4}$ **19.** $x \approx 2.2, x \approx -4.2$

21. $x = -\frac{1}{2}, x = -4$ **23.** about 0.2 sec, about 1.4 sec

25. no real solutions **27.** one real solution

29. two real solutions **31.** two x-intercepts

33. no x-intercepts **35.** two x-intercepts

37. $x = \frac{1}{2}, x = \frac{4}{5}$; *Sample answer:* The equation is not easily factorable and $a \neq 1$, so solve using the quadratic formula.

39. $x \approx 0.74, x \approx -6.74$; *Sample answer:* $a = 1$ and b is even, so solve by completing the square.

41. $x = -4, x = 3$; *Sample answer:* The equation is easily factorable, so solve by factoring.

43. $x \approx 2.19, x \approx -1.94$; *Sample answer:* The equation cannot be factored and $a \neq 1$, so solve using the quadratic formula.

45. $-b$ should be $-(-7)$, not -7; $x = \dfrac{-(-7) \pm \sqrt{(-7)^2 - 4(3)(-6)}}{2(3)}$; $x = 3$ and $x = -\dfrac{2}{3}$

47. yes; about 42 ft, about 158 ft

49. no; The discriminant is -47, so the equation has no real solutions.

51. 5; length: 13 m, width: 7 m

53. a–c. $x = -2$

Sample answer: factoring; The equation is easily factorable.

55. 2; When a and c have different signs, ac is negative, so the discriminant is positive.

57. a. *Sample answer:* $\frac{1}{2}$ **b.** 1 **c.** *Sample answer:* 2

59. a. *Sample answer:* 1 **b.** 9 **c.** *Sample answer:* 10

61. about 31 mi/h, about 65 mi/h

63. below the x-axis; The discriminant is positive and $a > 0$.

65. above the x-axis; The discriminant is negative and $a > 0$.

67. above the x-axis; The discriminant is positive and $a < 0$.

69. about 22 sec; The height is 30,800 feet after about 20.8 seconds and after about 42.8 seconds.

71. a. $4x + 3y = 1050$; $3y = 1050 - 4x$; $y = 350 - \frac{4}{3}x$

b. length: about 54 ft, width: about 278 ft; length: about 209 ft, width: about 72 ft

73. $-\dfrac{b}{2a}$; The mean of the solutions is the x-coordinate of the vertex; The mean of the solutions is equal to the graph's axis of symmetry, which is where the vertex lies.

75. about 24.7 ft/sec

77. $\dfrac{-b}{a}$; $\dfrac{c}{a}$; *Sample answer:* $2x^2 - 4x + 1 = 0$

79. a. $(x + 1)$, $(x - 6)$; $x^2 - 5x - 6 = 0$;
x, $(x - 2)$; $x^2 - 2x = 0$;
$\left(x + \frac{1}{2}\right)$, $(x - 5)$; $x^2 - \frac{9}{2}x - \frac{5}{2} = 0$

b.

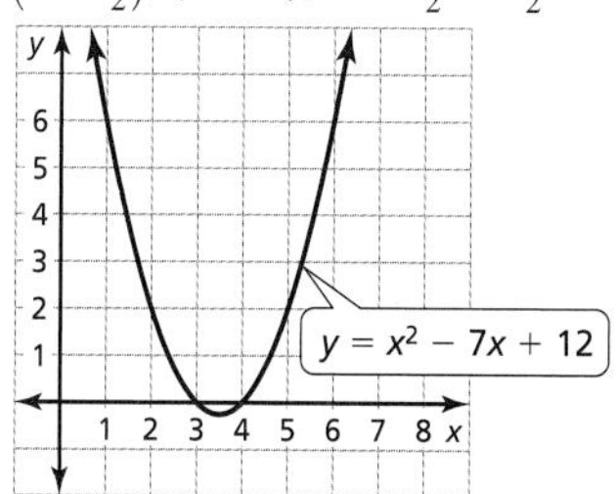

3, 4

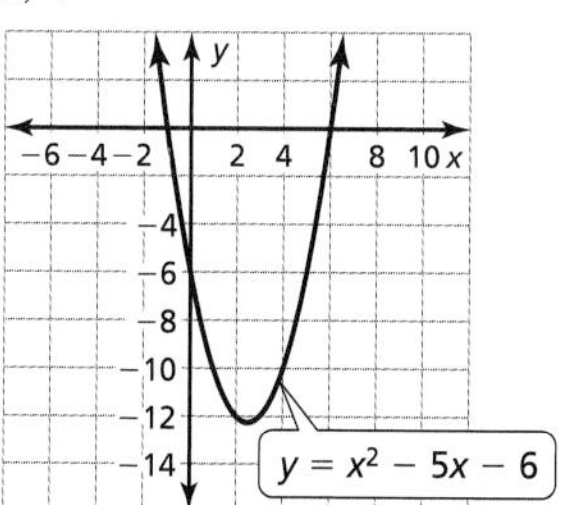

−1, 6

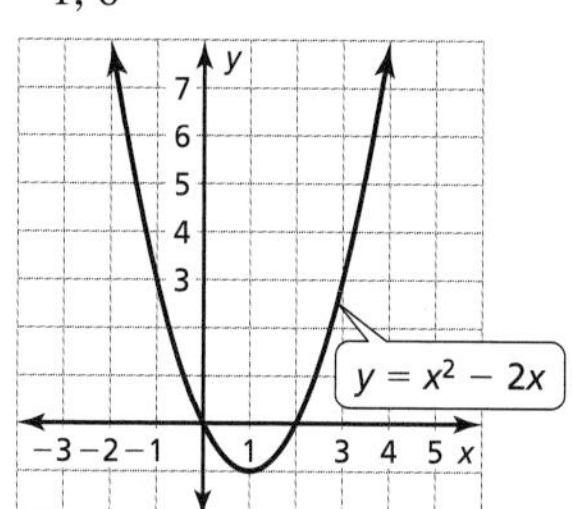

0, 2

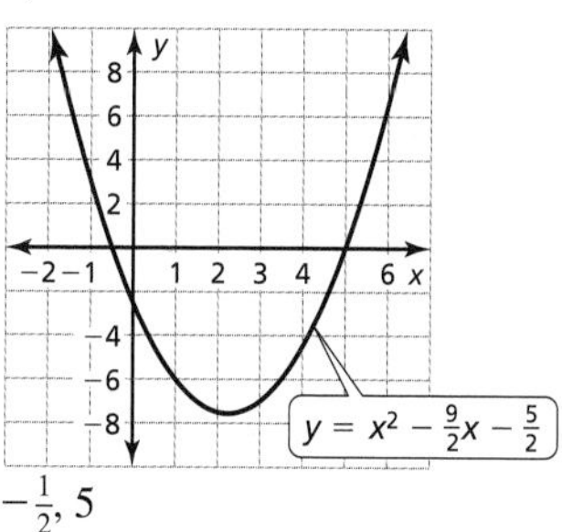

$-\frac{1}{2}$, 5

81. a. $k < -3$ or $k > 3$ **b.** $k = -3$ or $k = 3$

c. $-3 < k < 3$

9.5 Maintaining Mathematical Proficiency *(p. 524)*

83. (4, 0); *Sample answer:* substitution because both equations are solved for y

85. (5, 3); *Sample answer:* elimination because one pair of like terms has the same coefficient

9.6 Vocabulary and Core Concept Check *(p. 530)*

1. Solve one of the equations for one of the variables and substitute into the other equation and solve.

9.6 Monitoring Progress and Modeling with Mathematics *(pp. 530–532)*

3. B; (0, 1), (3, 4) **5.** A; (0, −1) **7.** (2, 9), (−1, 6)

9. (−1, 2) **11.** (3, 6), (−3, −6)

13. (0, −5), (−3, −8) **15.** no solutions **17.** (0, 5)

19. (4, −11), (−4, 29) **21.** (0, 2), (−6, −22) **23.** (1, 1)

25. about (0.87, −1.74), about (−6.87, 13.74)

27. The graph does not show both solutions.

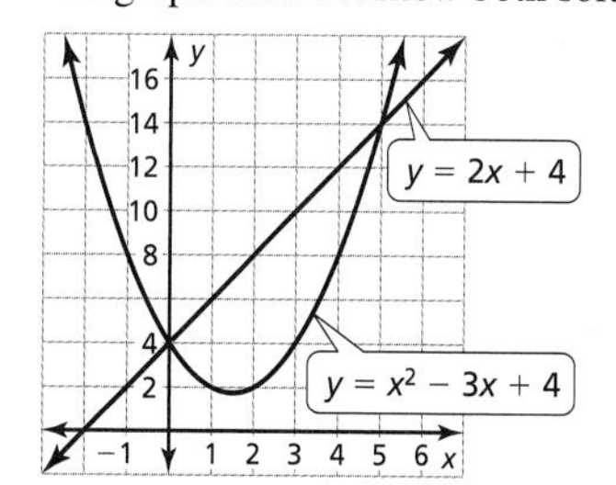

The solutions are (0, 4) and (5, 14).

29. between −4 and −3, between 0 and 1

31. between −4 and −3, between −2 and −1

33. about (2.350, 13.225)

35. about (−2.543, −0.941), about (0.185, 1.586), about (1.854, 25.152)

37. about (0.840, 3.454), about (−3.174, −1.898)

39. $x \approx 0.45$, $x \approx -4.45$ **41.** no solutions

43. $x \approx -0.36$ **45.** $x \approx 1.13$, $x \approx 2.40$

47. $\left(\dfrac{-7 + \sqrt{145}}{6}, \dfrac{7 + 2\sqrt{145}}{9}\right)$, $\left(\dfrac{-7 - \sqrt{145}}{6}, \dfrac{7 - 5\sqrt{145}}{9}\right)$; They are about the same.

49. 5 days and 35 days after the movie opened

51. about 55 m, about 475 m

53. a–c. (2, 11), (−2, −5)

Sample answer: elimination; The resulting equation can be written in the form $x^2 = d$.

55. a. $y = 30x + 290$ **b.** (1, 320), (34, 1310)

57. The possible number of solutions is 0, 1, or 2.

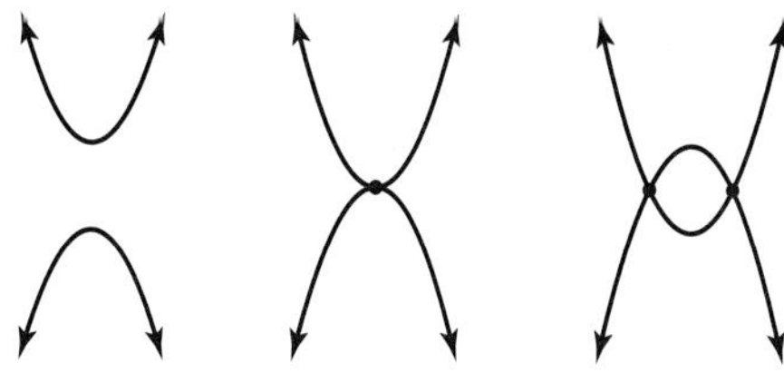

59. a. (0, 4) **b.** (5, 9) **61.** (1, −6)

9.6 Maintaining Mathematical Proficiency *(p. 532)*

63.

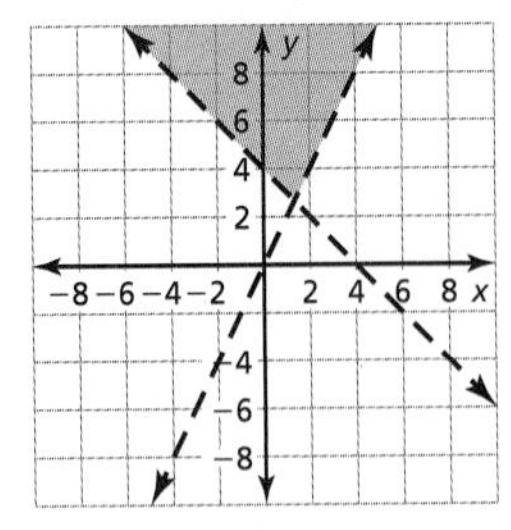

65.

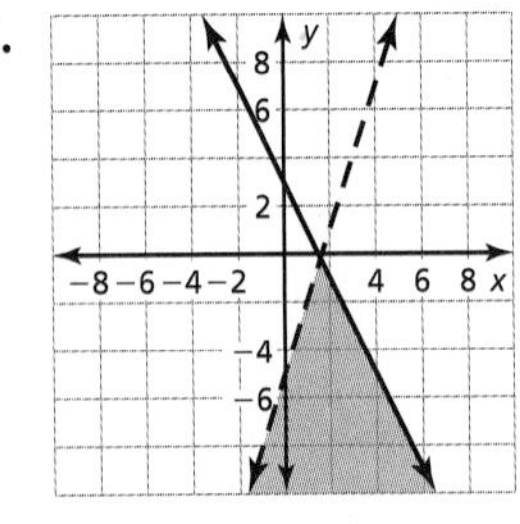

67.

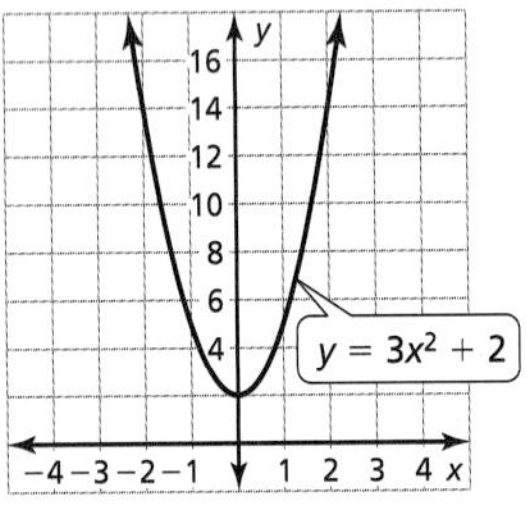

domain: all real numbers; range: $y \geq 2$

69.

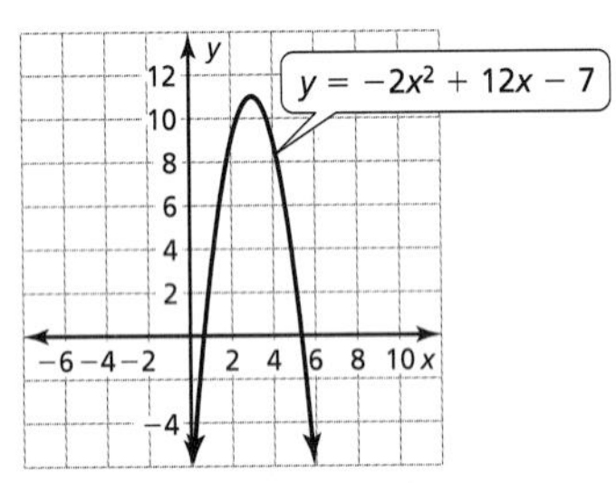

domain: all real numbers; range: $y \leq 11$

Chapter 9 Review *(pp. 534–536)*

1. $6p^3\sqrt{2p}$ **2.** $\frac{3\sqrt{35y}}{7y}$ **3.** $\frac{5x^3\sqrt[3]{2x^2}}{2}$ **4.** $4\sqrt{6} - 8$

5. $14\sqrt{3}$ **6.** $9\sqrt[3]{2}$ **7.** $88 + 30\sqrt{7}$ **8.** $10\sqrt{3}$

9. $x = 6, x = 3$ **10.** no solutions **11.** $x = -4$

12. $-3, -1, 1$

13.

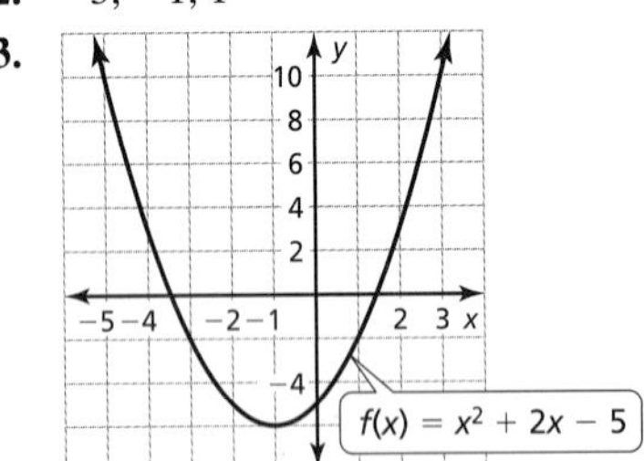

$-3.4, 1.4$

14. $x \approx 3.46, x \approx -3.46$ **15.** $x = 0$ **16.** $x = 6, x = -10$

17. no solutions **18.** $x = 1$ **19.** $x \approx 1.48, x \approx -1.48$

20. $x = 4, x = -10$ **21.** $x = -1$

22. $x \approx 3.45, x \approx -1.45$ **23.** maximum value; 8

24. minimum value; 7 **25.** minimum value; -33

26. 28 cm **27.** $x = 3, x = -5$ **28.** $x \approx 2.3, x \approx -1.8$

29. $x = 1$ **30.** one x-intercept **31.** no x-intercepts

32. two x-intercepts **33.** $(1, -5)$

34. about $(4.87, 14.75)$, about $(-2.87, -0.75)$

35. about $(-1.88, 2.35)$, about $(2.48, -4.64)$

Chapter 10

Chapter 10 Maintaining Mathematical Proficiency *(p. 541)*

1. 45 **2.** -10 **3.** -20 **4.** -44

5.

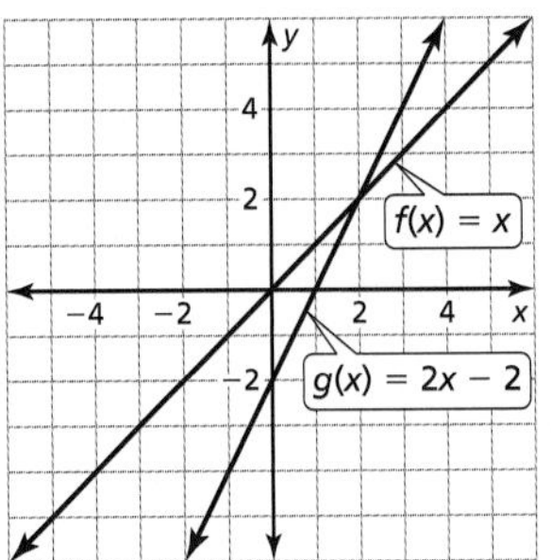

The graph of g is a vertical stretch by a factor of 2 then a vertical translation 2 units down of the graph of f.

6.

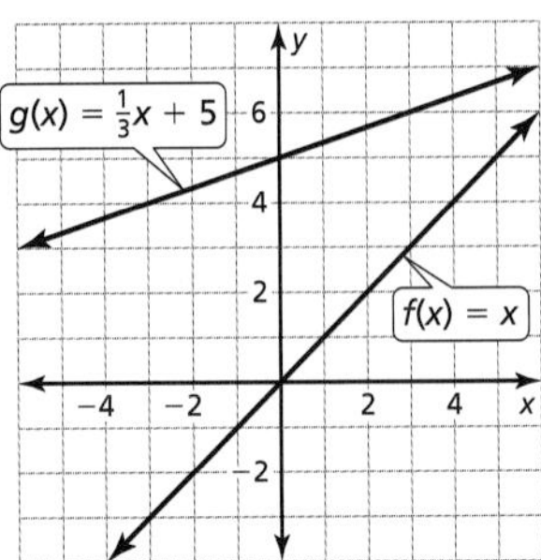

The graph of g is a vertical shrink by a factor of $\frac{1}{3}$ then a vertical translation 5 units up of the graph of f.

7.

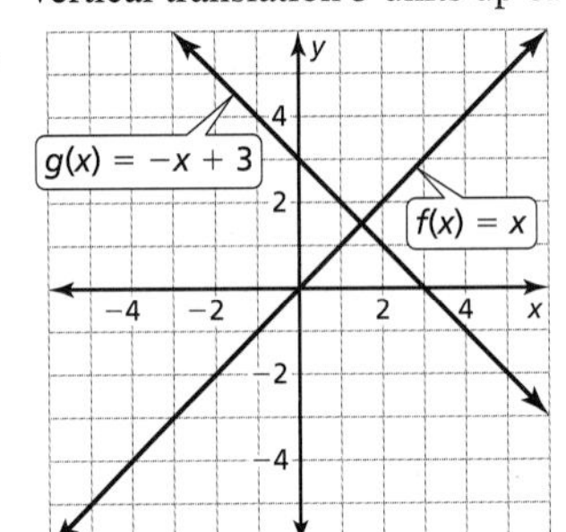

The graph of g is a reflection in the x-axis then a vertical translation 3 units up of the graph of f.

8. The graph of n is a vertical stretch by a factor of 2, a reflection in the x-axis, then a vertical translation $2b$ units down of the graph of m.

10.1 Vocabulary and Core Concept Check *(p. 548)*

1. radical function

3. Describe the x-values for which the radicand is greater than or equal to 0.

10.1 Monitoring Progress and Modeling with Mathematics *(pp. 548–550)*

5. $x \geq 0$ **7.** $x \leq 0$ **9.** $x \geq 4$ **11.** $x \leq 8$

13. $x \geq -4$ **15.** D; $y \geq 0$ **17.** A; $y \geq -3$

19.

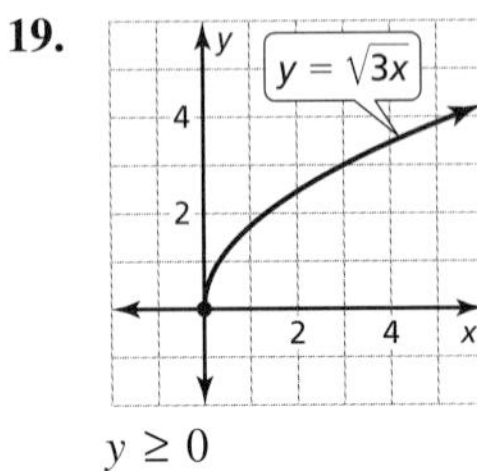

$y \geq 0$

21.

$y \geq 5$

23. 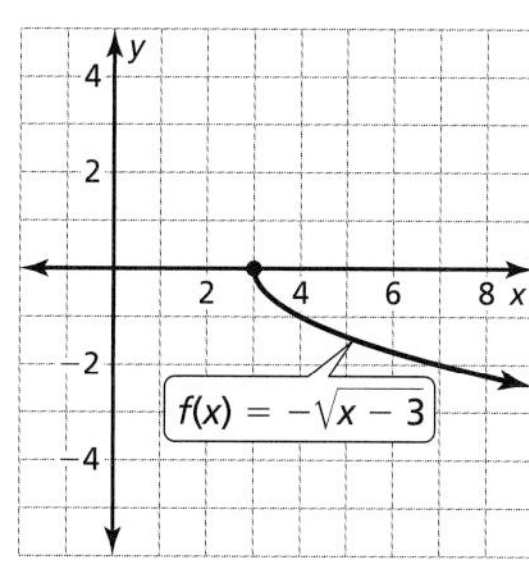

$y \leq 0$

25. 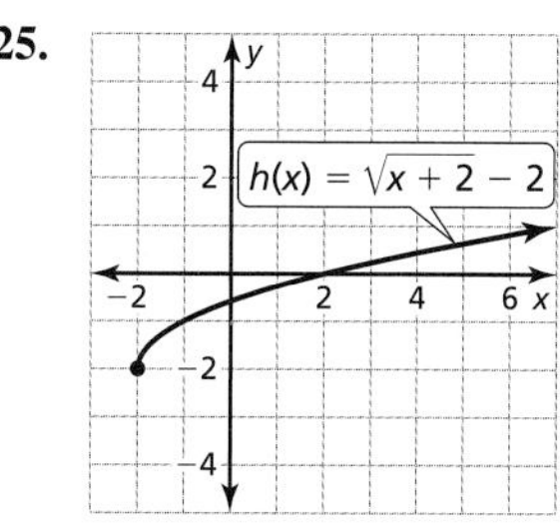

$y \geq -2$

27. 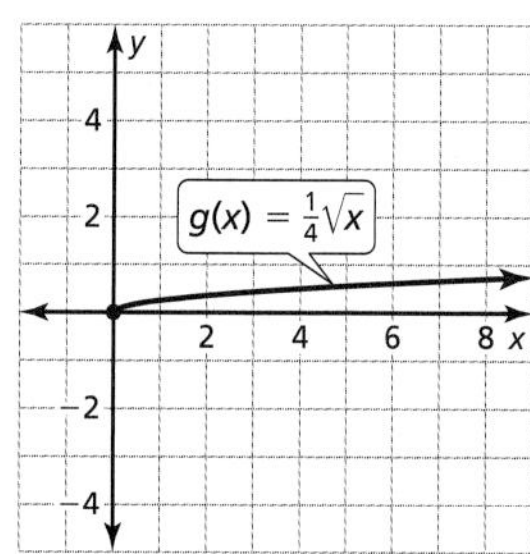

The graph of g is a vertical shrink by a factor of $\frac{1}{4}$ of the graph of f.

29. 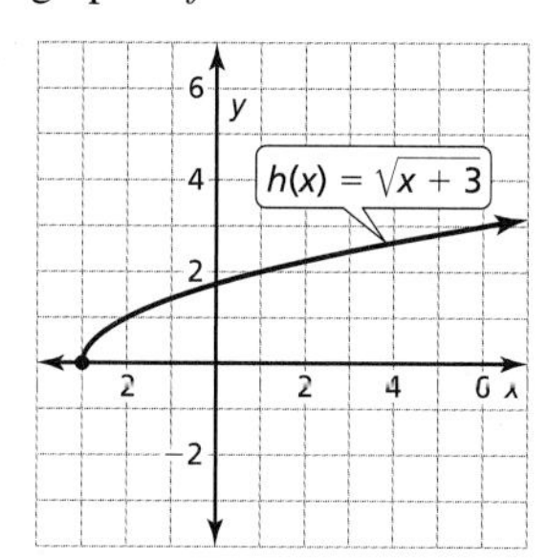

The graph of h is a translation 3 units left of the graph of f.

31. 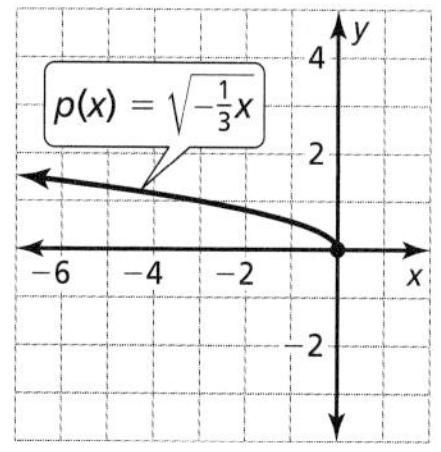

The graph of p is a horizontal stretch by a factor of 3 and a reflection in the y-axis of the graph of f.

33. 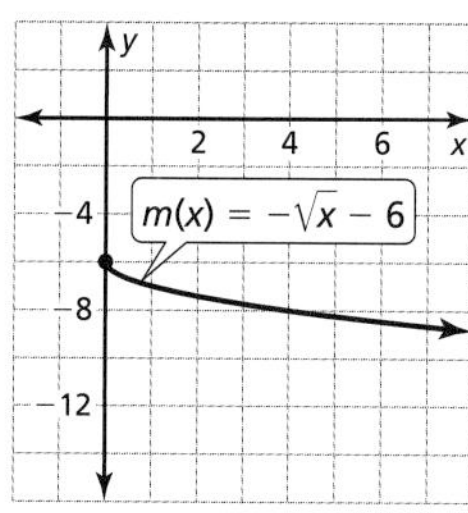

The graph of m is a reflection in the x-axis and a translation 6 units down of the graph of f.

35. The graph of $y = \sqrt{x}$ is translated 1 unit up instead of 1 unit left.

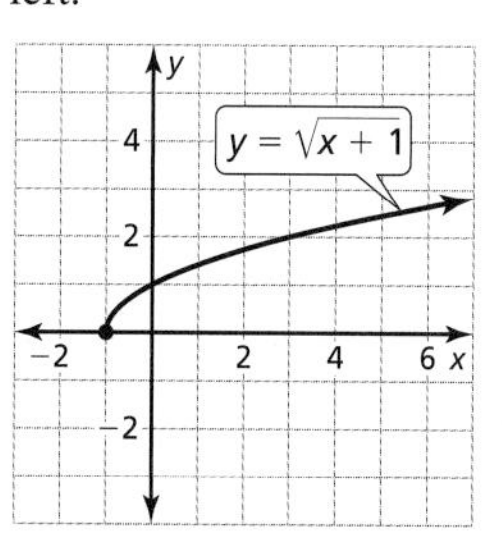

37. Translate 2 units left, stretch vertically by a factor of 4, and translate 1 unit down to obtain the graph of h.

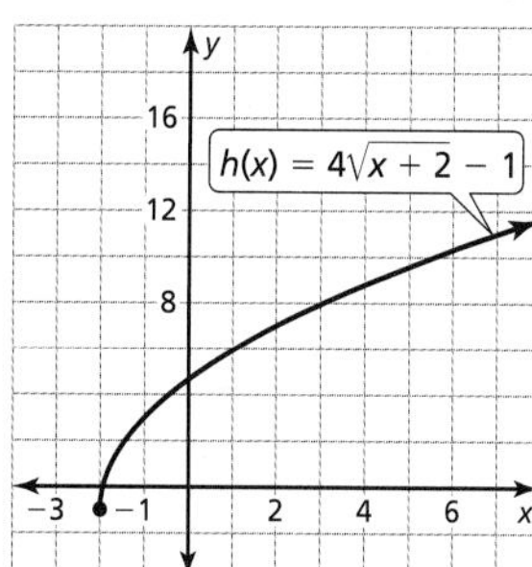

39. Reflect in the y-axis, stretch vertically by a factor of 2, and translate 6 units down to obtain the graph of h.

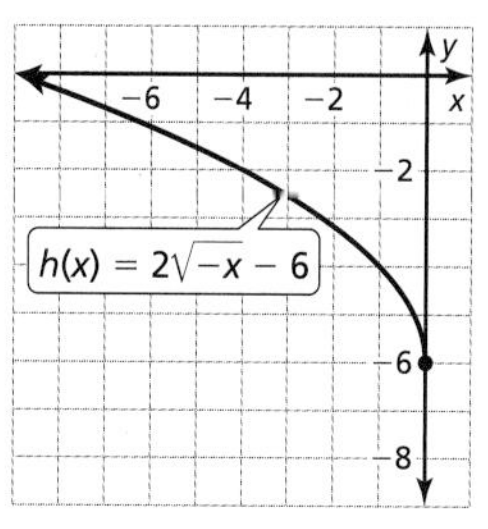

Selected Answers

41. Translate 3 units left, shrink vertically by a factor of $\frac{1}{3}$, and translate 3 units down to obtain the graph of h.

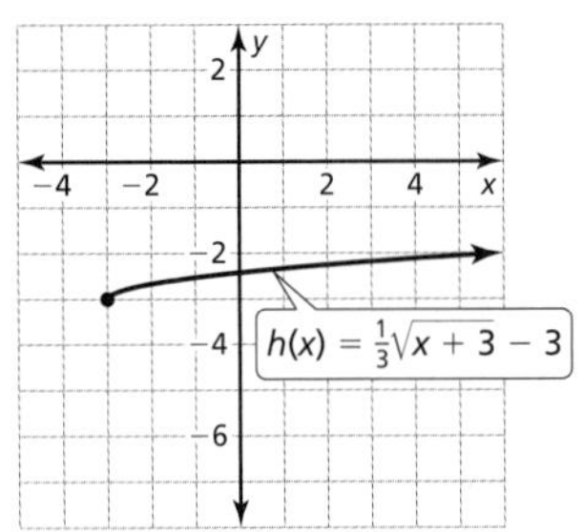

43. Translate 1 unit right, stretch vertically by a factor of 2, reflect in the x-axis, and translate 5 units up to obtain the graph of h.

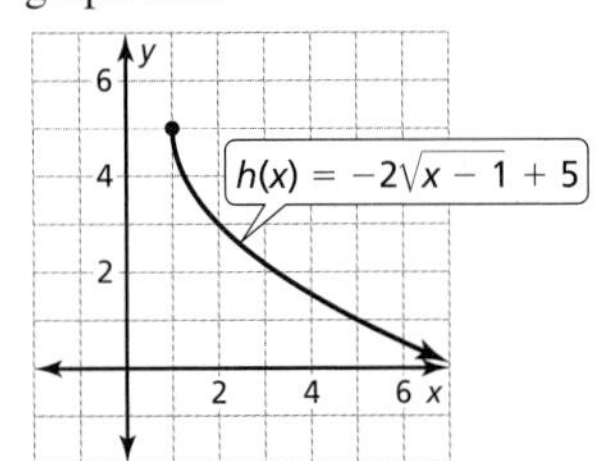

45. As skid mark lengths increase from 0 to 15 feet, the initial van speeds increase at an average rate of about 1.22 miles per hour per foot on Road Surface A and about 1.00 mile per hour per foot on Road Surface B.

47. **a.** *Sample answer:* $y = \sqrt{x} + 1$

b. *Sample answer:* $y = -\sqrt{x}$

49. **a.**

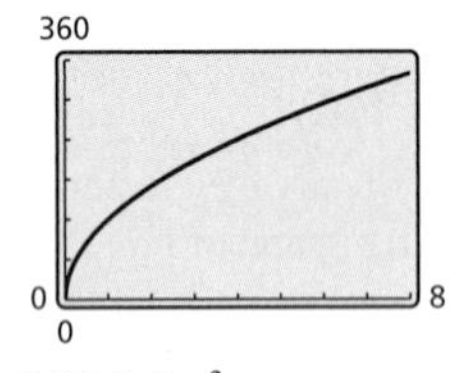

6.25 lb/in.^{2}

b. decreases

51. **a.** $A > 0$

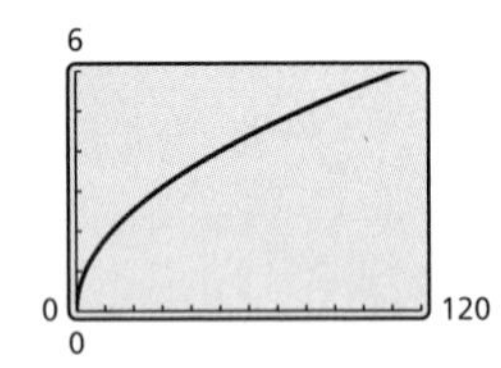

b. about 91.61 in.^{2}

53. **a.** 0 **b.** no; $f(x)$ increases as x increases.

c. *Sample answer:* $f(x) = -\sqrt{x}$; no; $f(x)$ decreases as x increases.

d. *Sample answer:* $f(x) = -4 + \sqrt{x}$

55. $f(x) = 5\sqrt{x}$; The graph of $f(x) = 5\sqrt{x}$ is a vertical stretch of the graph of $y = \sqrt{x}$ by a factor of 5, and the graph of $g(x) = \sqrt{5x} = \sqrt{5} \cdot \sqrt{x}$ is a vertical stretch of the graph of $y = \sqrt{x}$ by a factor of $\sqrt{5}$.

57. *Sample answer:* $f(x) = 3 - \sqrt{x + 5}$

10.1 Maintaining Mathematical Proficiency *(p. 550)*

59. -4 61. $(x + 1)(x + 6)$ 63. $(y - 8)(y + 5)$

10.2 Vocabulary and Core Concept Check *(p. 555)*

1. index

10.2 Monitoring Progress and Modeling with Mathematics *(pp. 555–556)*

3. D 5. C

7.

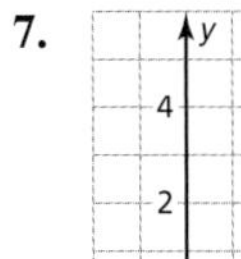

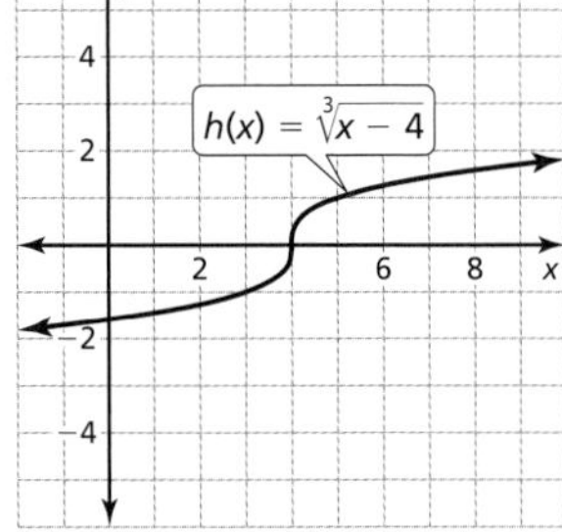

The graph of h is a translation 4 units right of the graph of f.

9.

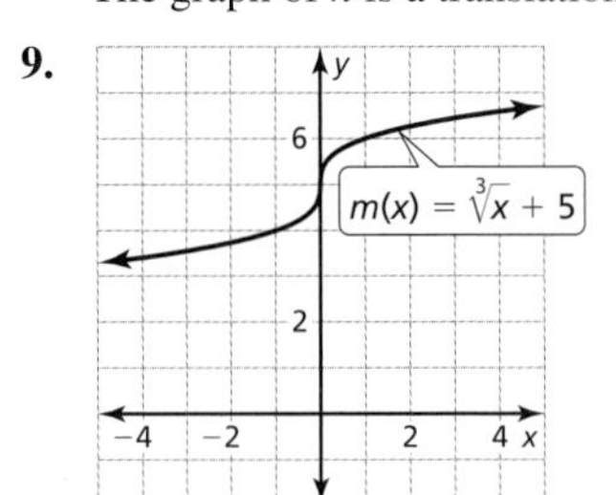

The graph of m is a translation 5 units up of the graph of f.

11.

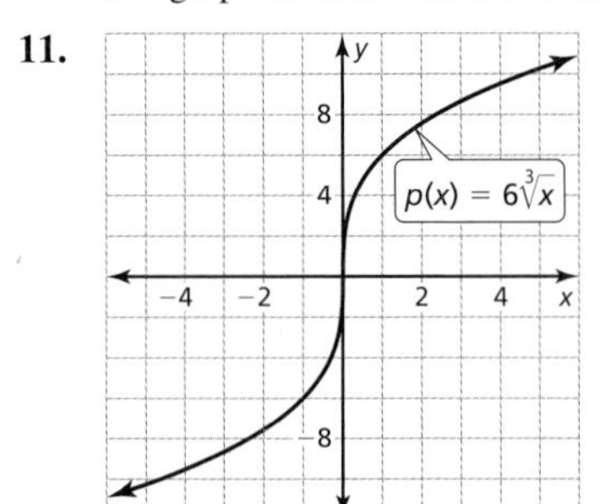

The graph of p is a vertical stretch by a factor of 6 of the graph of f.

13. The graph of q is a translation 5 units left of the graph of f; $h = -5$

15. The graph of v is a translation 6 units down of the graph of f; $k = -6$

17.

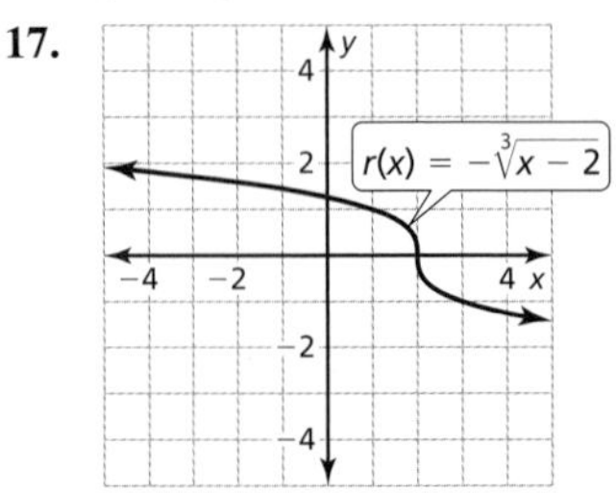

The graph of r is a reflection in the x-axis and a translation 2 units right of the graph of f.

19.

$k(x) = 5\sqrt[3]{x+1}$

The graph of k is a vertical stretch by a factor of 5 and a translation 1 unit left of the graph of f.

21.

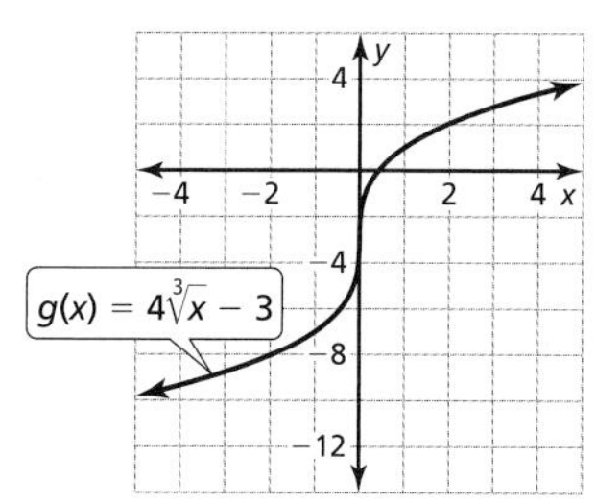

The graph of g is a vertical stretch by a factor of 4 and a translation 3 units down of the graph of f.

23.

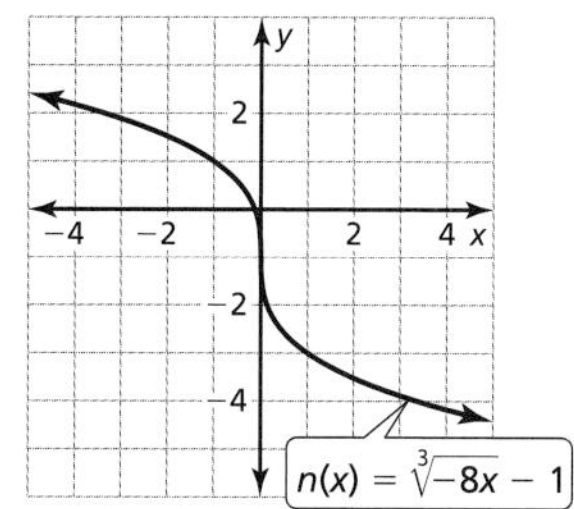

The graph of n is a horizontal shrink by a factor of $\frac{1}{8}$, a reflection in the y-axis, and a translation 1 unit down of the graph of f.

25.

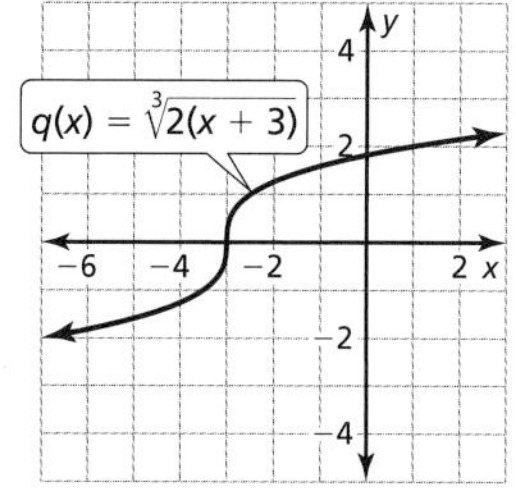

The graph of q is a horizontal shrink by a factor of $\frac{1}{2}$ and a translation 3 units left of the graph of f.

27. Translate 4 units right and 2 units up to obtain the graph of g.

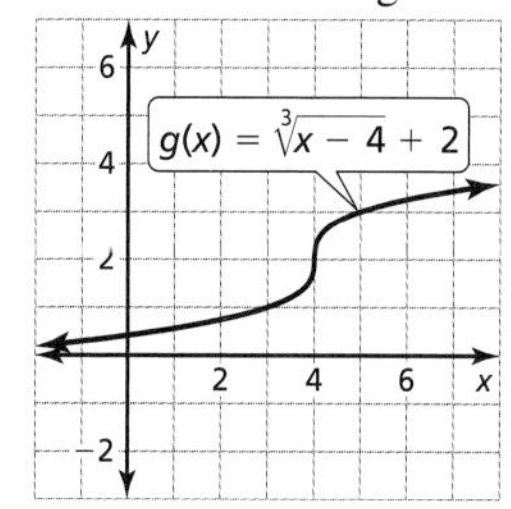

29. Translate 3 units left, stretch vertically by a factor of 5, reflect in the x-axis, and translate 2 units up to obtain the graph of j.

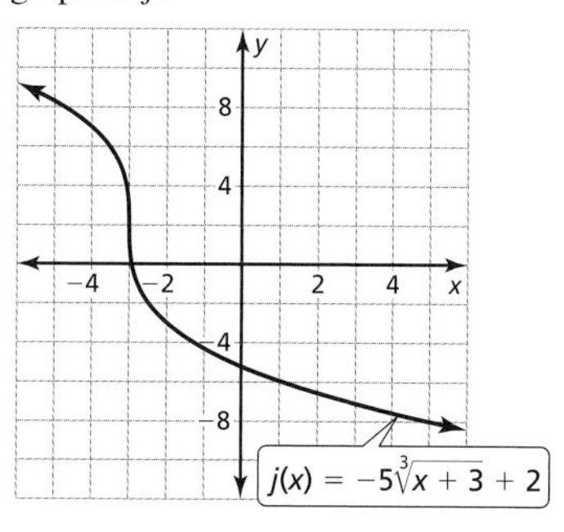

31. Translate 1 unit right, shrink vertically by a factor of $\frac{1}{3}$, and translate 7 units up to obtain the graph of v.

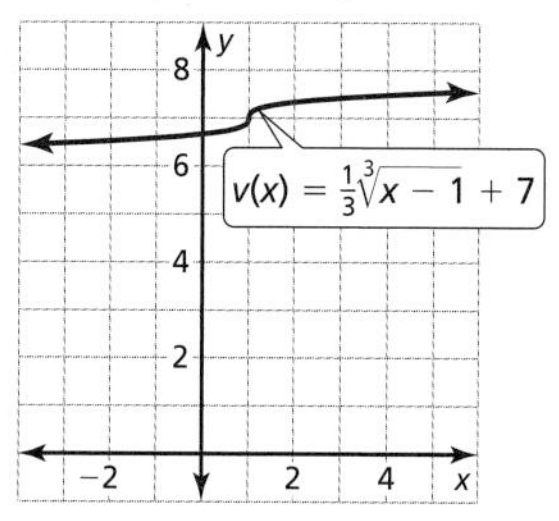

33. The graph should be a translation 3 units right, not a translation 3 units left, of the graph of $y = \sqrt[3]{x}$.

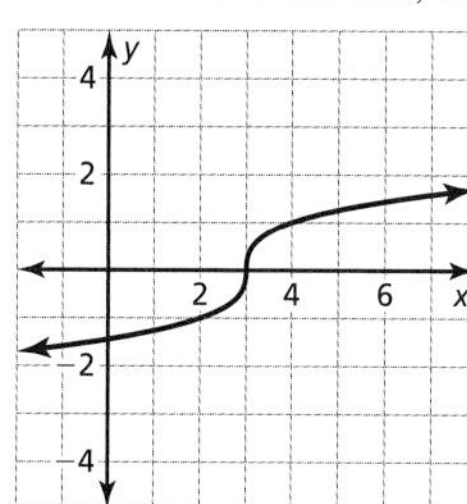

35. $\dfrac{2}{3\sqrt[3]{6}} \approx 0.37$ times greater

37.

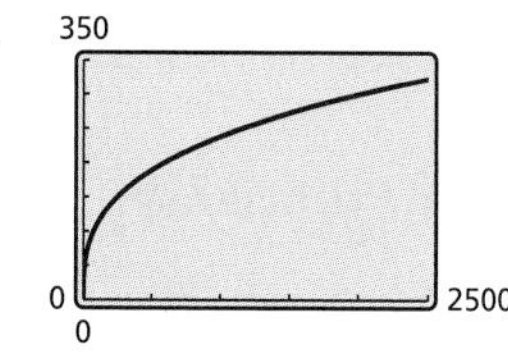

about 790 hp

39. no; A function $f(x)$ is an odd function when $f(-x) = -f(x)$ for each x in the domain of f. *Sample answer:* A cube root function, such as $f(x) = \sqrt[3]{x} + 1$, is not an odd function because $f(-x) = \sqrt[3]{-x} + 1 = -\sqrt[3]{x} + 1$ and $-f(x) = -(\sqrt[3]{x} + 1) = -\sqrt[3]{x} - 1$. $f(-x) \neq -f(x)$.

41. *Sample answer:* $f(x) = -7\sqrt[3]{x-3} + 4$

10.2 Maintaining Mathematical Proficiency *(p. 556)*

43. $3(x + 6)(x - 2)$ **45.** $(x + 3)(4x - 5)$

47. no real solutions **49.** $x = \frac{7}{5}, x = \frac{13}{5}$

10.3 Vocabulary and Core Concept Check *(p. 564)*

1. Raising each side of a radical equation to an exponent can introduce extraneous solutions.

10.3 Monitoring Progress and Modeling with Mathematics (pp. 564–566)

3. $x = 81$ **5.** $m = 144$ **7.** $c = 121$ **9.** $a = 4$
11. $y = 25$ **13.** $a = 19$ **15.** $x = 60$ **17.** $r = 7$
19. $p = 15$ **21.** 144 ft **23.** $x = 1$ **25.** $x = 2$
27. $x = 9$ **29.** $g = 5$ **31.** $p = 12$ **33.** $c = \frac{1}{2}$
35. $x = 9$ **37.** $x = 64$ **39.** $g = 27$ **41.** $s = -18$
43. $y = 2$ **45.** neither **47.** $p = 4$ **49.** $y = 1, y = 4$
51. $a = \frac{1}{4}$ **53.** no solution **55.** $m = 7$

57. no solution

59. An error was made in squaring both sides because the radical was not isolated. The third equation should be $\sqrt{x} = 2$, and the fourth equation should be $x = 4$.

61. This equation is equivalent to $\sqrt{2x} = -4$, which has no solution because the positive square root of a number cannot be negative.

63. no; The resistance is $\frac{7.68}{30.72} = \frac{1}{4}$ as much for a 120-volt circuit than for a 240-volt circuit.

65. true

67. false; when $b = -a$, then $a^2 = (-a)^2$, but $a \neq -a$

69. **a.** no solution; Graph $y = x + 2$ and $y = \sqrt{2x - 3}$ on the same set of coordinate axes. The graphs do not intersect, so there is no solution.

b. no solution; Square each side, simplify, and subtract $2x - 3$ from each side of the equation. Then, use the quadratic formula, which gives no real solutions.

c. *Sample answer:* the graphical approach, because it gives a visual picture of the situation

71. first cone: $r = 2$ units; second cone: $2r = 4$ units

73. $m = 5$ **75.** $y = \frac{16}{5}$ **77.** $h = 0$

79. *Sample answer:* $\sqrt{x - 4} = 1$

81. no; Because $\sqrt{(2x + 5)^2}$ cannot be negative, the equation is true only for $x \geq -\frac{5}{2}$, where $2x + 5$ is nonnegative.

83. **a.** about 71.5 N

b. more; To keep the same ratio $\frac{T}{m}$, if m is increased, then T must be increased also.

10.3 Maintaining Mathematical Proficiency (p. 566)

85. $12p^2 + 11p - 5$

87.

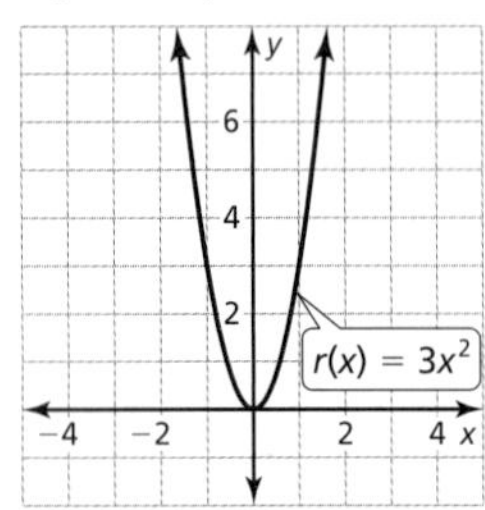

Both graphs open up and have the same vertex, (0, 0), and the same axis of symmetry, $x = 0$. The graph of r is narrower than the graph of f because the graph of r is a vertical stretch by a factor of 3 of the graph of f.

89.

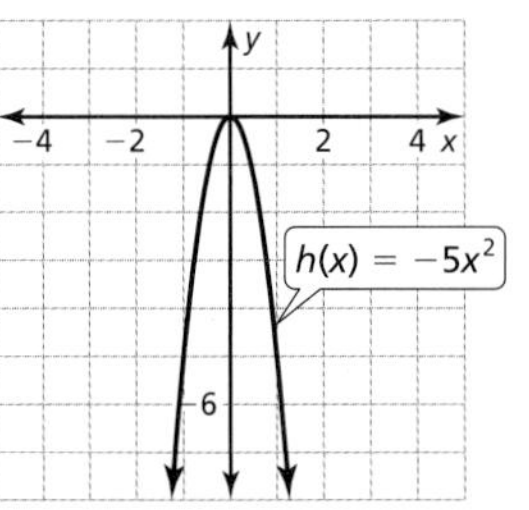

Both the graphs have the same vertex, (0, 0), and the same axis of symmetry, $x = 0$, but the graph of h opens down and is narrower than the graph of f, because the graph of h is a vertical stretch by a factor of 5 and a reflection in the x-axis of the graph of f.

10.4 Vocabulary and Core Concept Check (p. 572)

1. inverse relation

10.4 Monitoring Progress and Modeling with Mathematics (pp. 572–574)

3. (0, 1), (−8, 3), (−3, 4), (−5, 7), (−1, 9)

5.

Input	8	6	0	6	8
Output	−5	−5	0	5	10

7.

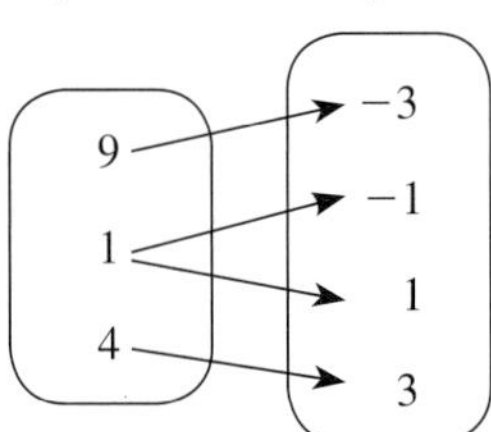

9. $x = y - 5$; -3 **11.** $x = 4y + 4$; 12

13. $x = \pm\frac{\sqrt{y}}{3}$; $\pm\frac{\sqrt{2}}{3}$

15.

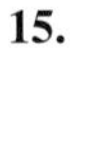

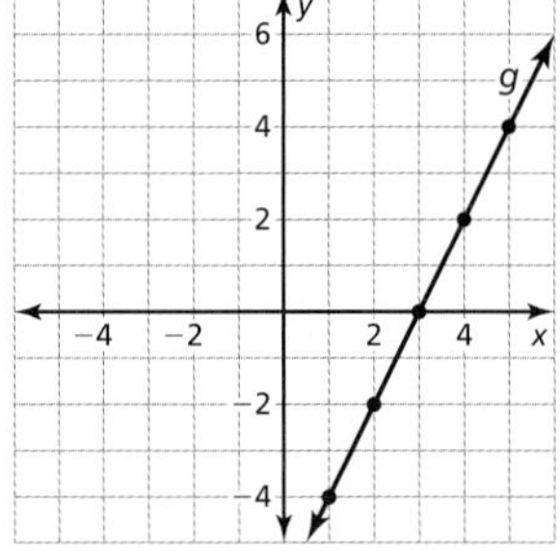

domain of inverse: all real numbers; range of inverse: all real numbers

17. $g(x) = \frac{x + 1}{4}$

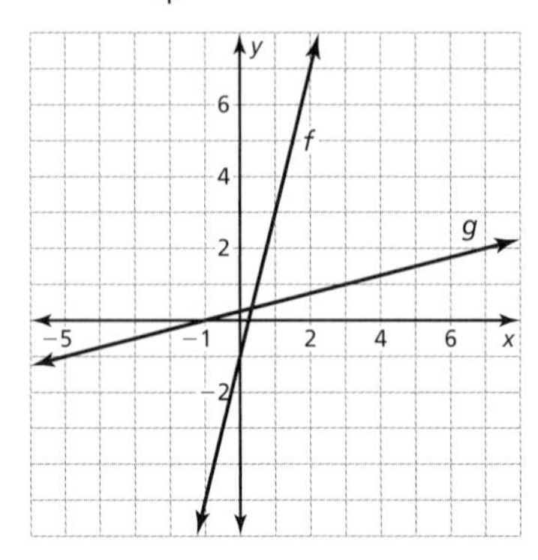

19. $g(x) = -\dfrac{x + 2}{3}$

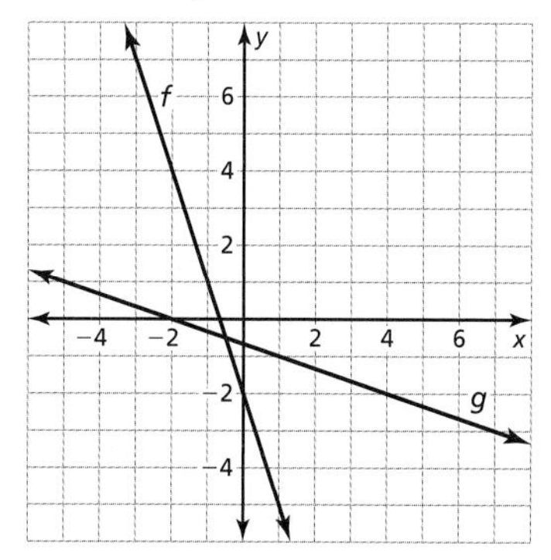

21. $g(x) = 3x - 24$

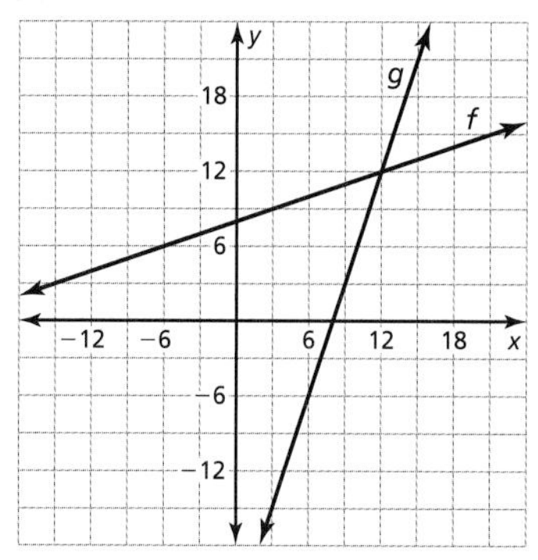

23. $g(x) = \dfrac{\sqrt{x}}{2}$

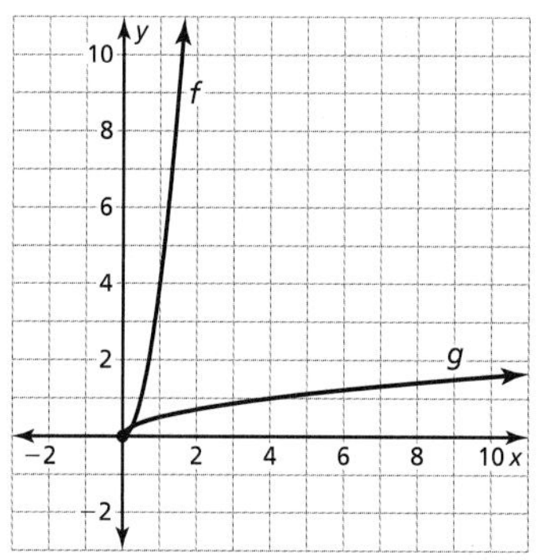

25. $g(x) = -\sqrt{10 - x}$

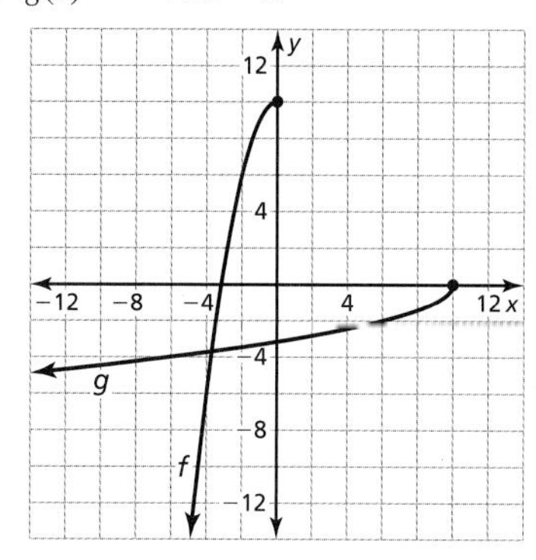

27. $g(x) = 3\sqrt{x - 2}$

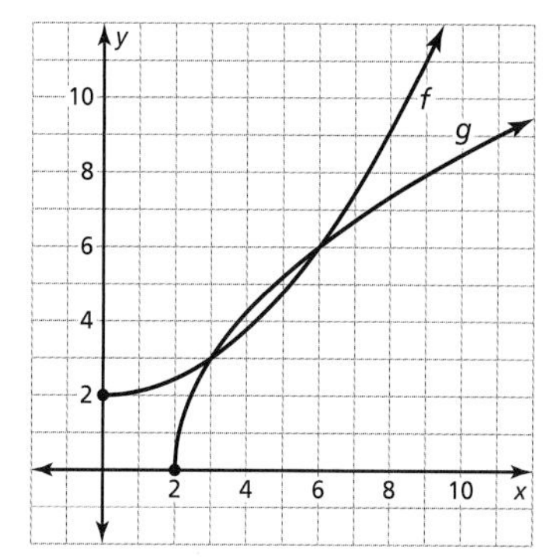

29. not a function **31.** function

33. yes; $g(x) = x^2 - 3, x \geq 0$ **35.** yes; $g(x) = \dfrac{x^2 + 6}{2}, x \geq 0$

37. yes; $g(x) = \dfrac{x^2}{9} + 8, x \geq 0$

39. yes; $g(x) = \dfrac{(-x - 2)^2 - 5}{3}; x \leq -2$ **41.** no; $y = \pm\sqrt{\dfrac{x}{2}}$

43. x and y were not switched in the equation; The inverse of f is $g(x) = \dfrac{x - 5}{3}$ or $g(x) = \dfrac{x}{3} - \dfrac{5}{3}$.

45. $D = 1.33899E$; about 334.75 U.S. dollars

47. $s = \dfrac{2\sqrt{3}}{3}h$; $s \approx 18.48$ in.

49. $g(x) = \sqrt[3]{\dfrac{x}{2}}$

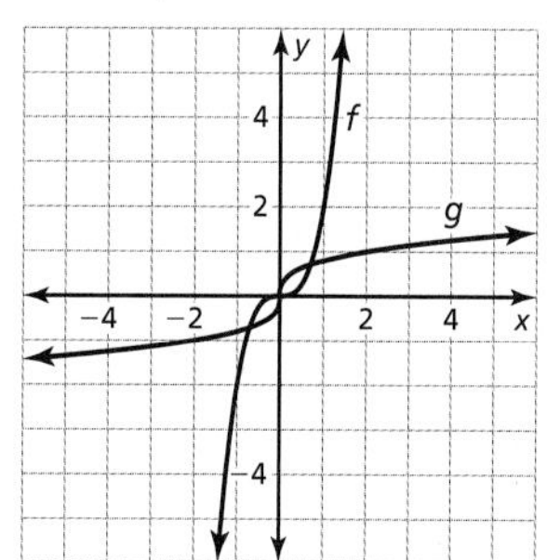

51. $g(x) = \sqrt[3]{x} + 5$

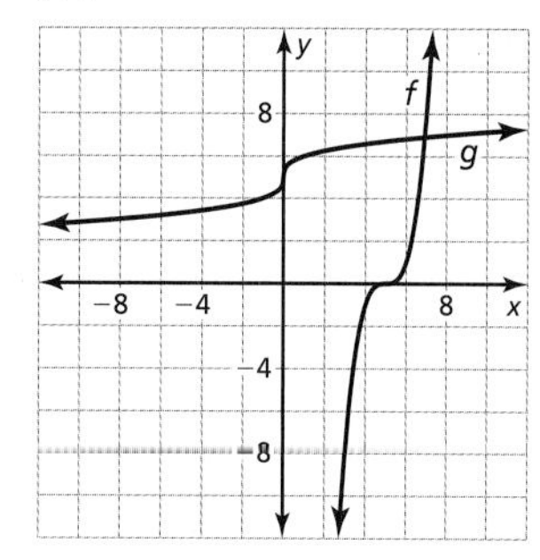

53. $g(x) = \left(\frac{x}{4}\right)^3$

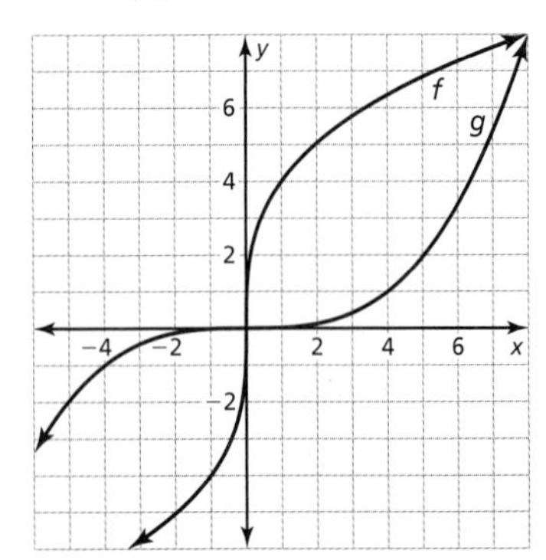

55. no; The inverse of $f(x) = 3$ is the vertical line $x = 3$, which does not pass the Horizontal Line Test and is not a function.

57. *Sample answer:* Restrict the domain to nonnegative values of x; new function: $x \geq 0$; $y \geq -5$; inverse function: $x \geq -5$; $y \geq 0$

59. *Sample answer:* $f(x) = \frac{1}{4}x$

61. The inverse of $f(x) = mx + b$, where $b \neq 0$, is

$g(x) = \frac{x - b}{m} = \frac{1}{m}x - \frac{b}{m}$, which is a linear function. The graph of the inverse function g has a slope of $\frac{1}{m}$ and a y-intercept of $-\frac{b}{m}$.

63. no; no; For any values of h, k, and a, a horizontal line intersects the graph of f more than once.

10.4 Maintaining Mathematical Proficiency *(p. 574)*

65. $7y - 11$

67. $-10d^2 + 11d - 2$

69.

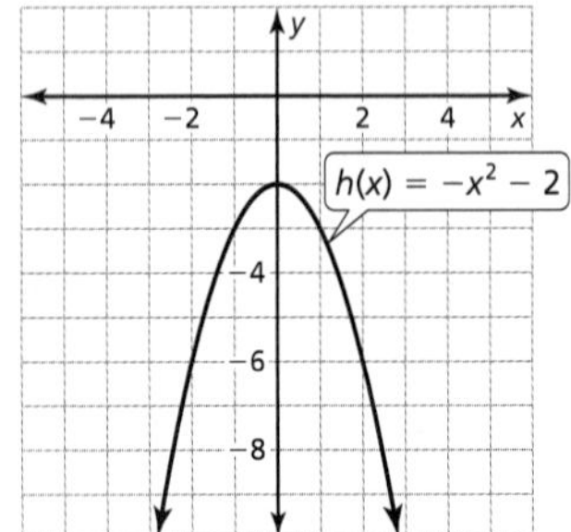

Both the graph of f and the graph of h have the same axis of symmetry, $x = 0$, but the graph of f opens up and the graph of h opens down. The vertex of the graph of f is (0, 0) and the vertex of the graph of h is (0, −2). So, the graph of h is a reflection in the x-axis and a vertical translation 2 units down of the graph of f.

71.

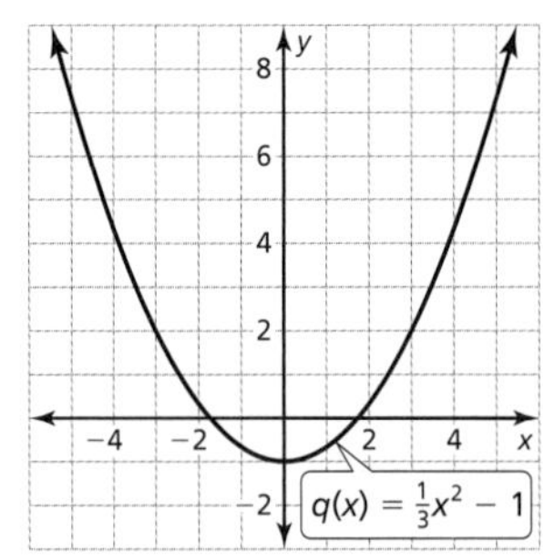

Both the graph of f and the graph of q open up and have the same axis of symmetry, $x = 0$, but the graph of q is wider than the graph of f. Also, the vertex (0, −1) of the graph of q is below the vertex (0, 0) of the graph of f. The graph of q is a vertical shrink by a factor of $\frac{1}{3}$ and a vertical translation 1 unit down of the graph of f.

Chapter 10 Review *(pp. 576–578)*

1.

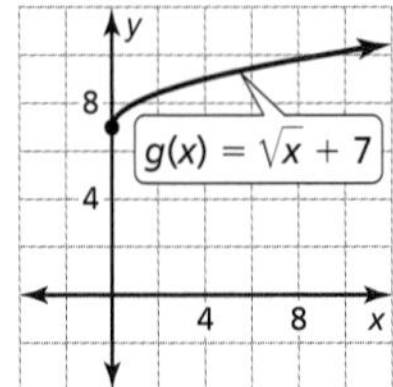

$x \geq 0$; $y \geq 7$; The graph of g is a translation 7 units up of the graph of f.

2.

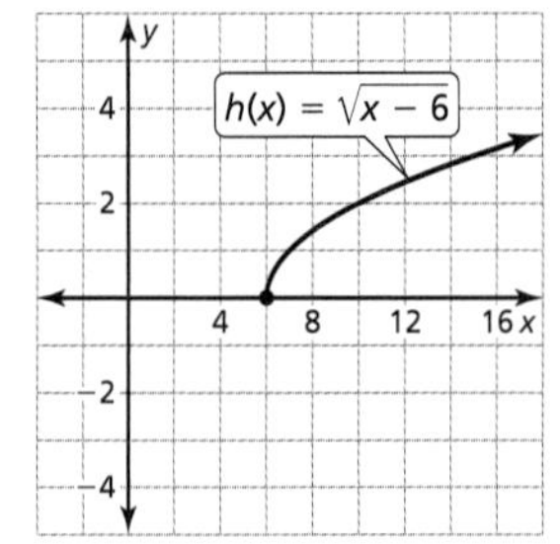

$x \geq 6$; $y \geq 0$; The graph of h is a translation 6 units right of the graph of f.

3.

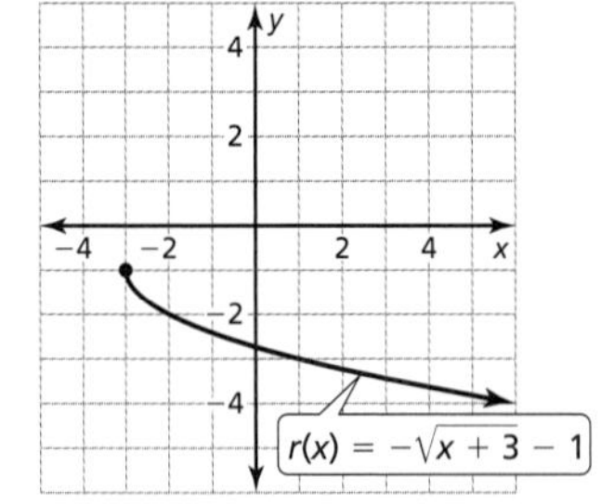

$x \geq -3$; $y \leq -1$; The graph of r is a translation 3 units left, a reflection in the x-axis, and a translation 1 unit down of the graph of f.

4. Translate 6 units right, shrink vertically by a factor of $\frac{1}{4}$, and translate 2 units up to obtain the graph of g.

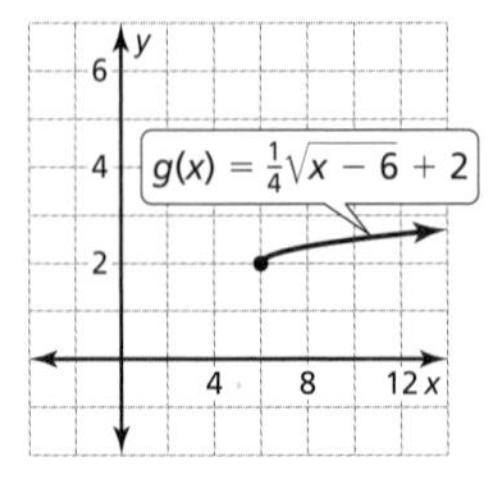

5. $g(x) = \sqrt[3]{x} + 4$

The graph of g is a translation 4 units up of the graph of f.

6. 

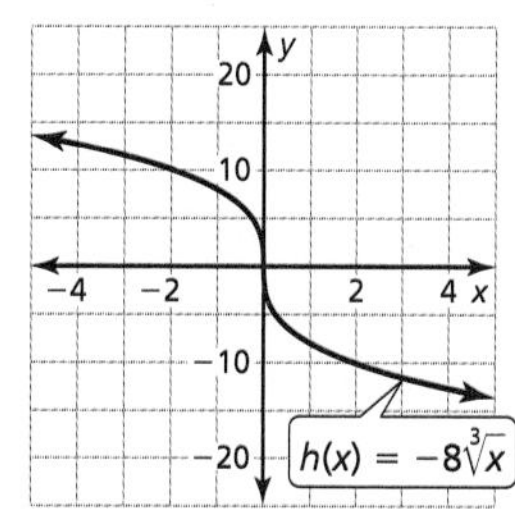

The graph of h is a vertical stretch by a factor of 8 and a reflection in the x-axis of the graph of f.

7.

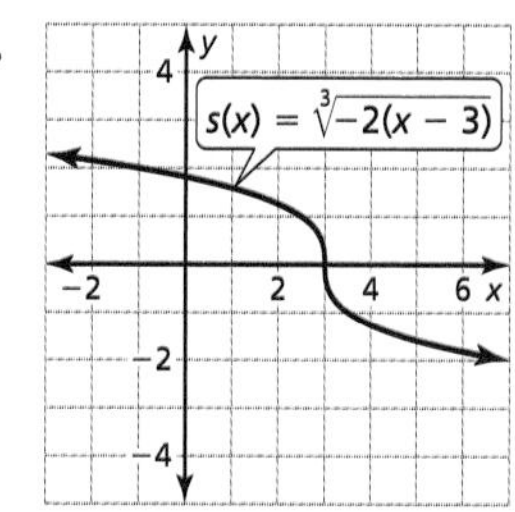

The graph of s is a reflection in the y-axis, a translation 3 units right, and a horizontal shrink by a factor of $\frac{1}{2}$ of the graph of f.

8. Stretch vertically by a factor of 3, reflect in the x-axis, translate 2 units left, and translate 1 unit down to obtain the graph of g.

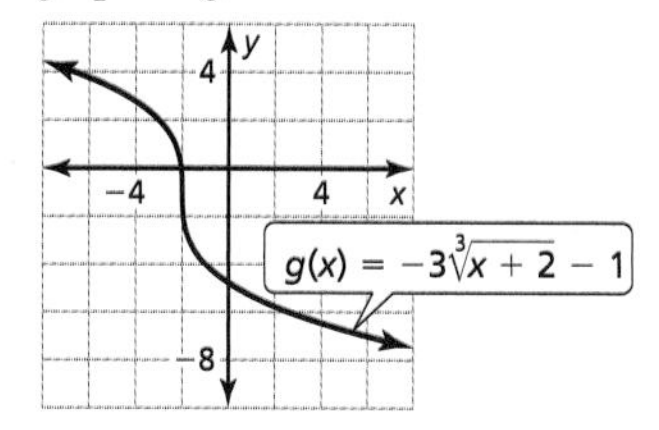

9. The average rate of change of r is about $0.625 \div \left(\frac{\sqrt[3]{4}}{8}\right) \approx 3.15$ times greater than the average rate of change of p over the interval $x = 0$ to $x = 8$.

10. $x = 100$ **11.** $x = 28$ **12.** $x = 9$ **13.** $x = 4$

14. $x = 14$ **15.** no solution **16.** $16\pi \approx 50.3$ in.^{3}

17. $(-10, 1), (-4, 3), (4, 5), (14, 7), (26, 9)$

18.

Input	6	3	0	−3	−6
Output	−4	−2	0	2	4

19. $g(x) = \frac{10 - x}{5}$

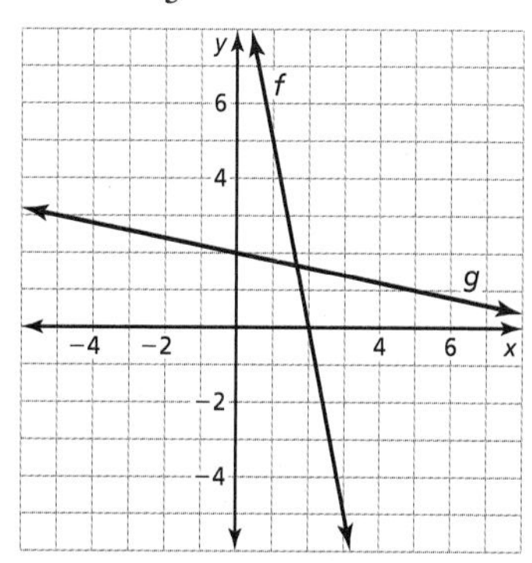

20. $g(x) = \frac{1}{3}\sqrt{3(x+1)}$

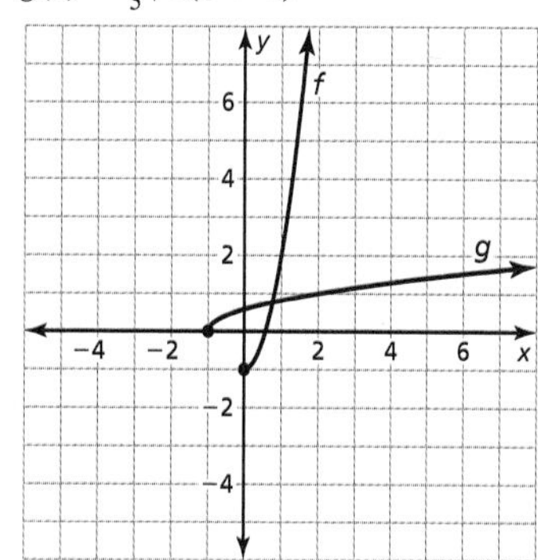

21. $g(x) = 2x^2 - 3, x \geq 0$

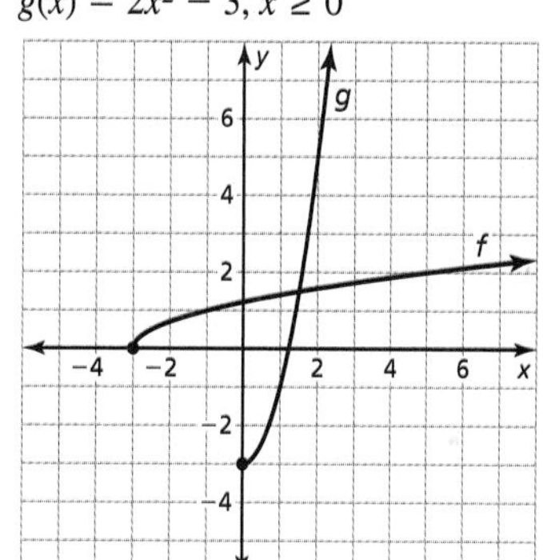

22. not a function

23. $a = 210 - 1.25h$; 175

Chapter 11

Chapter 11 Maintaining Mathematical Proficiency (p. 583)

1.

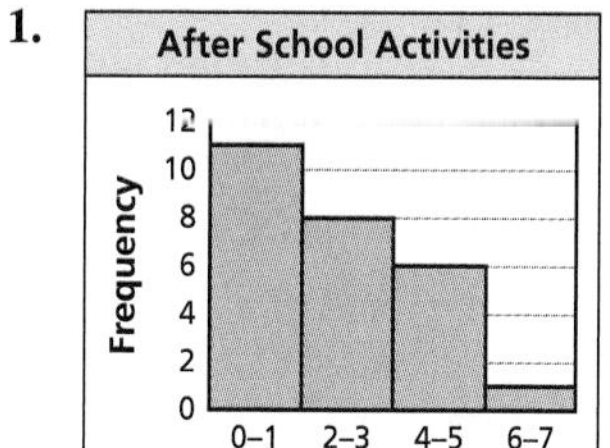

2.

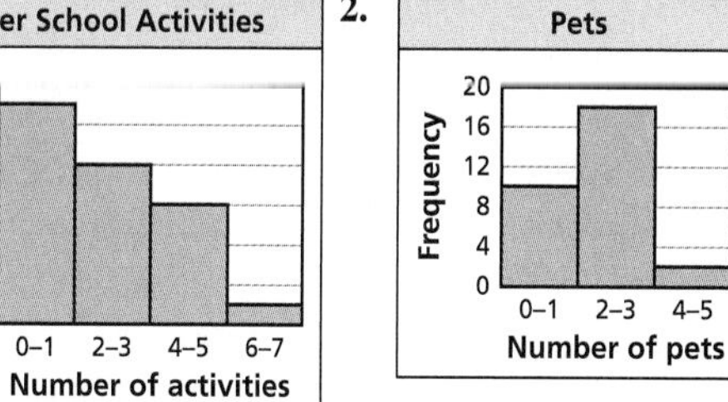

3. Students' Favorite Subjects

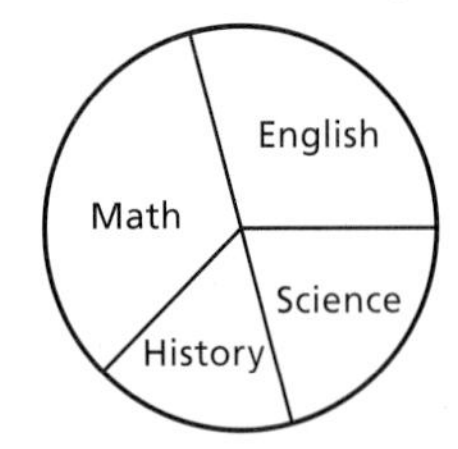

4. $a + b = 20$; $a + b \leq 20$

11.1 Vocabulary and Core Concept Check *(p. 590)*

1. the center, or typical value; the distribution of the data

3. *Sample answer:* 3, 4, 4, 7, 9, 9

11.1 Monitoring Progress and Modeling with Mathematics *(pp. 590–592)*

5. a. mean: 4, median: 3, mode: 1

b. median; The mean is greater than most of the data and the mode is less than most of the data.

7. a. mean: 22, median: 21, mode: none

b. mean; There are no outliers.

9. a. mean: about 1.96, median: 2, modes: $1\frac{2}{3}$ and 2

b. median; The data are evenly distributed.

11. 4 **13.** 16

15. a. 62; The outlier decreases the mean and median and does not affect the mode.

b. *Sample answer:* The outlier could be the mass of a baby polar bear.

17. Golfer A: 15; Golfer B: 8; The range for Golfer A is greater.

19. a. 25 **b.** about 9.27 **21. a.** 2 **b.** about 0.65

23. a. about 4.6; The typical score differs from the mean by about 4.6 points.

b. about 2.5; The typical score differs from the mean by about 2.5 points.

c. The standard deviation for Golfer A is greater, so the scores are more spread out.

25. mean: 8, median: 7, mode: 5

27. mean: 76, median: 69, mode: 63, range: 46, standard deviation: 15.5

29. The numbers are not in numerical order; The median is 6.

31. a. Team A; no; Team B has a greater median.

b. Team B; The range and standard deviation are less than Team A's.

c. Team B; The means are 186.75 and 191.25.

33. mode; The data are not numeric.

35. mean: 89, median: 104, mode: 62, range: 123, standard deviation: 27

37. a. Answers will vary.

b. *Sample answer:* The mean, median, range, and standard deviation increase.

39. a. mean: 19.37, median: 19, mode: 18

b. 37, 37; They increase the mean and median but do not affect the mode.

c. College Student Ages

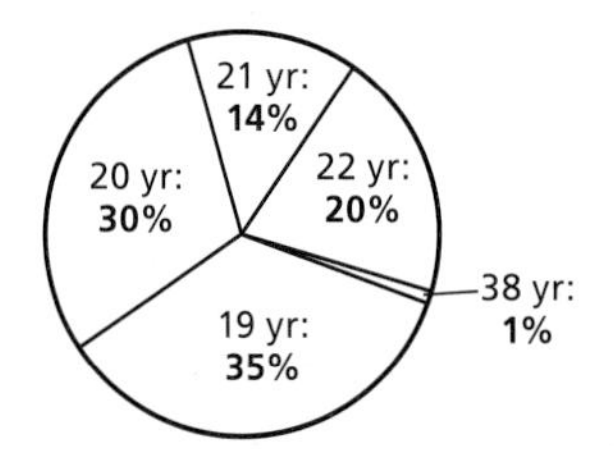

mean: 20.37, median: 20, mode: 19

11.1 Maintaining Mathematical Proficiency *(p. 592)*

41. no solution **43.** $w < \frac{3}{2}$ **45.** $\frac{1}{49}$ **47.** -162

11.2 Vocabulary and Core Concept Check *(p. 597)*

1. Order the data and find the median of the lower half.

11.2 Monitoring Progress and Modeling with Mathematics *(pp. 597–598)*

3. 3 **5.** 11 **7.** 8

9.

Hours
0 1 2 3 4 5 6 7

11.

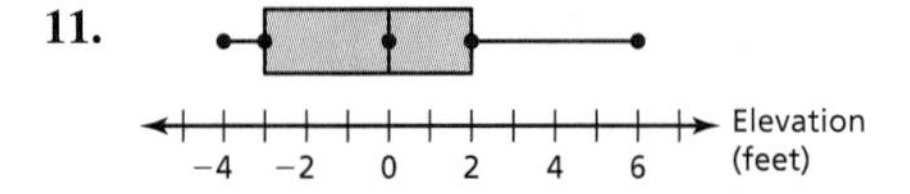

13.

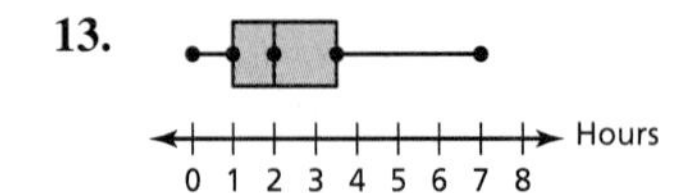

15. a. 9.5; The prices vary by no more than $9.50.

b. 25% of the prices are between $8.75 and $10.50, 50% of the prices are between $10.50 and $14.75, and 25% of the prices are between $14.75 and $18.25.

c. 4.25; The middle half of the prices vary by no more than $4.25.

d. above Q3; the whisker is longer

17. a. Sales Rep A: symmetric; Sales Rep B: skewed right

b. Sales Rep B; The range and interquartile range are greater.

c. Sales Rep B; The least value is 4.

19. range: 36, interquartile range: 12; The range is greater.

21. a. Both distributions are skewed right.

b. Brand A: 3.5, Brand B: 2.5

c. The interquartile range of Brand A is greater.

d. Brand A; The range is greater.

e. Brand A; 75% of the battery lives are greater than 3.5 hours.

23. yes; *Sample answer:*

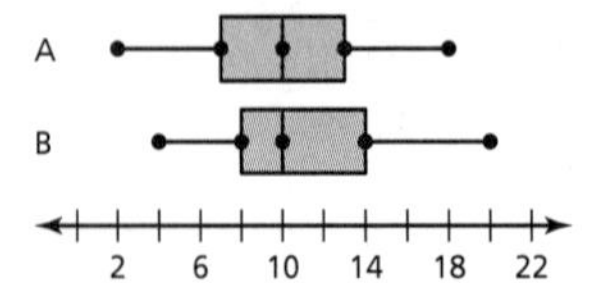

11.2 Maintaining Mathematical Proficiency *(p. 598)*

25.

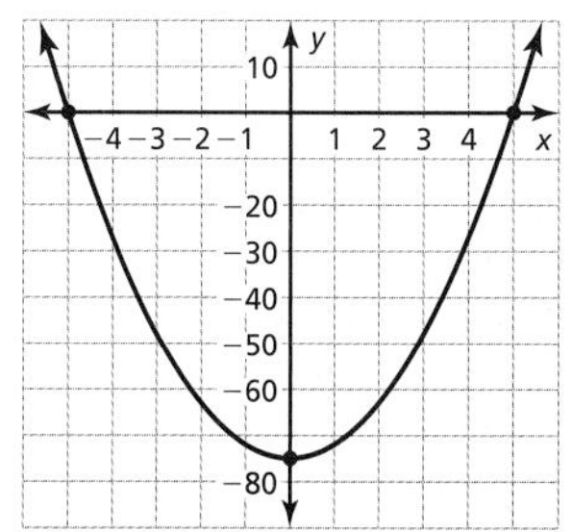

27.

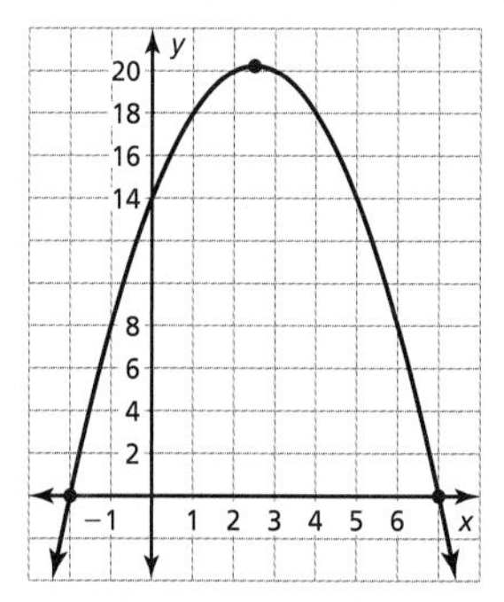

11.3 Vocabulary and Core Concept Check *(p. 604)*

1. The data in a symmetric distribution are evenly distributed to the left and right of the highest bar, most of the data in a distribution that is skewed left are on the right, and most of the data in a distribution that is skewed right are on the left.

11.3 Monitoring Progress and Modeling with Mathematics *(pp. 604–606)*

3.

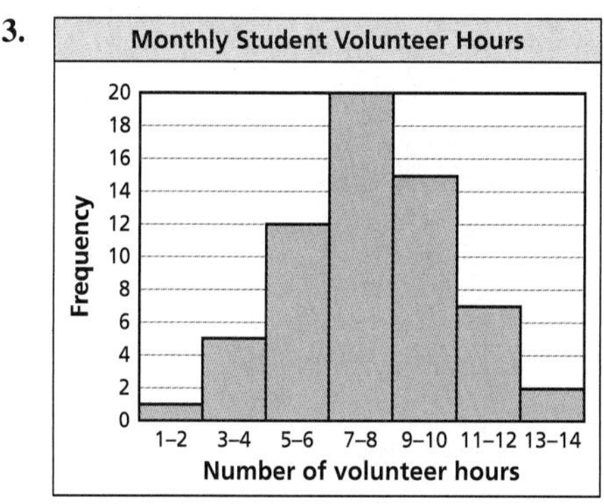

symmetric

5. skewed right; Most of the data are on the first two stems.

7. mean, standard deviation; The distribution is symmetric.

9. a.

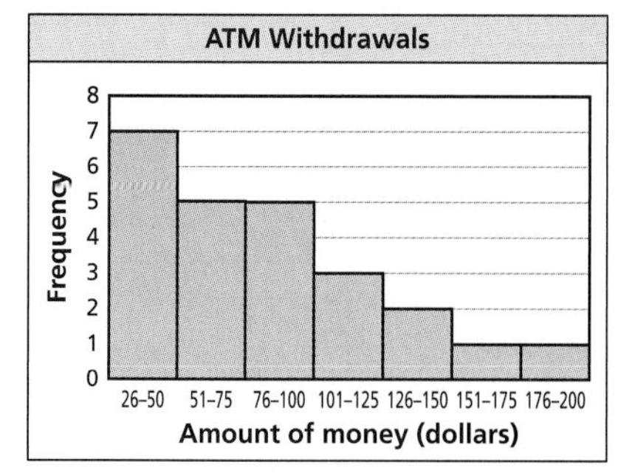

b. median, five-number summary; The distribution is skewed right.

c. Most people were charged a fee.

11. When most of the data are on the right, the distribution is skewed left, not right.

13. histogram; You do not need to list every value.

15. Town A usually has a greater daily high temperature, and the daily high temperature tends to differ more for town B.

17. *Sample answer:* salaries of employees at a large company

19. a.

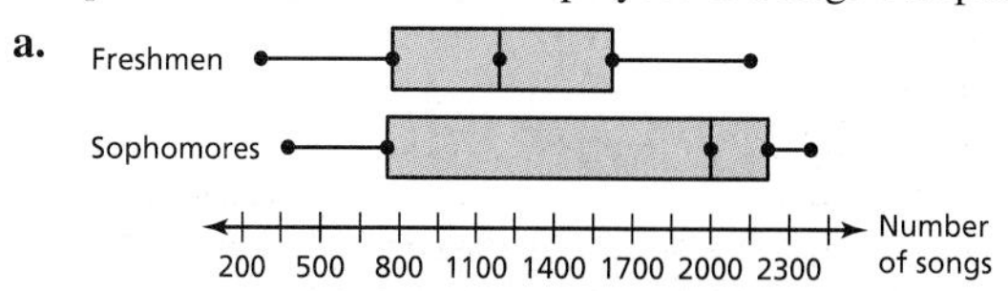

The distribution of data for freshmen is symmetric. The distribution of data for sophomores is skewed left.

b. The number of songs downloaded by freshmen tends to be less than the number of songs downloaded by sophomores.

c. about 31 freshmen **d.** about 95 freshmen

21. symmetric; yes; The measures of center and variation double.

23. a.

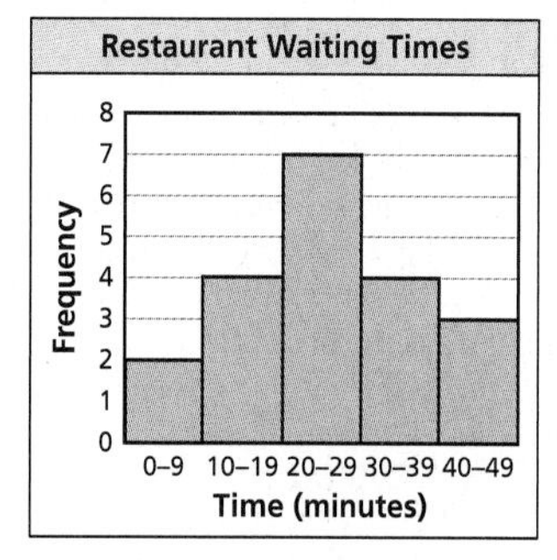

approximately symmetric

b.

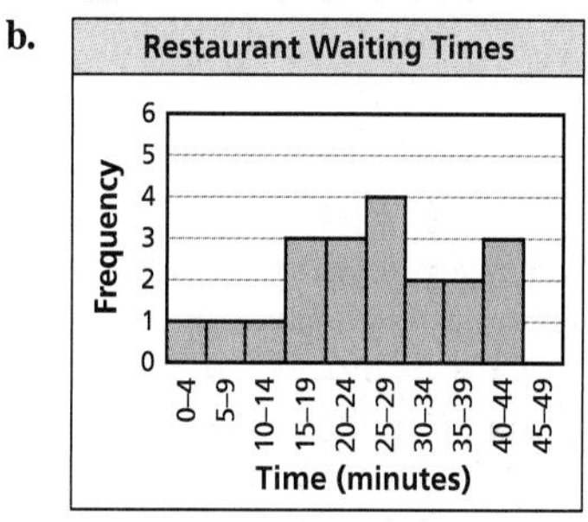

The distribution becomes skewed left.

c. part (b); More intervals show the spread of the data better.

11.3 Maintaining Mathematical Proficiency *(p. 606)*

25. $x \geq -6$ **27.** $x \geq 7$

11.4 Vocabulary and Core Concept Check *(p. 614)*

1. joint frequency

3. A marginal relative frequency is the sum of the joint relative frequencies in a row or column. A conditional relative frequency is the ratio of a joint relative frequency to the marginal relative frequency.

11.4 Monitoring Progress and Modeling with Mathematics *(pp. 614–616)*

5. 178 **7.** 208

9. 118 students have set academic goals, 310 students have not set academic goals, 232 males responded, 196 females responded, 428 students were surveyed.

11. 204 students plan to participate, 124 students do not plan to participate, 86 students are undecided, 222 freshmen responded, 192 sophomores responded, 414 students were surveyed.

13. 90; 152; 16; 200; 78

15.

		Spanish Class		
		Yes	No	Total
French Class	Yes	45	54	99
	No	64	82	146
	Total	109	136	245

17.

		Exercise Preference		
		Aerobic	Anaerobic	Total
Gender	Male	0.25	0.30	0.55
	Female	0.27	0.18	0.45
	Total	0.52	0.48	1

19. 52%; 30%

21. This number is only the number of freshmen who participated in the fundraiser; 272 freshmen responded to the survey.

23.

		Menu	
		Potluck	Catered
Meal	Lunch	0.43	0.57
	Dinner	0.38	0.62

about 57%

25. yes; About 71% of students who have a car and about 92% of students who do not have a car live on campus.

27. yes; About 35% of adults ages 21–30 and about 14% of adults ages 61–70 participate in recreational skiing.

29. Venn diagrams can be used to display the information in a two-way table that has two rows and two columns.

31. **a.**

		Seat Location		
		Main Floor	Balcony	Total
Ticket Type	Adult	$2x + y$	$x + 2y$	$3x + 3y$
	Child	$x - 40$	$3x - y - 80$	$4x - y - 120$
	Total	800	1009	1809

b. $x = 249; y = 93$ **c.** about 57% **d.** about 73%

11.4 Maintaining Mathematical Proficiency *(p. 616)*

33. exponential

11.5 Vocabulary and Core Concept Check *(p. 621)*

1. *Sample answer:* The numbers on the scale(s) could be unevenly spaced; The scale(s) may not begin at zero.

11.5 Monitoring Progress and Modeling with Mathematics *(pp. 621–622)*

3. qualitative; Brands of cars are nonnumerical.

5. quantitative; Budgets are numerical values.

7. quantitative; Shoe sizes are numerical values.

9. *Sample answer:* line graph; It shows data values over time.

11. *Sample answer:* bar graph; It shows data in each specific category.

13. *Sample answer:*

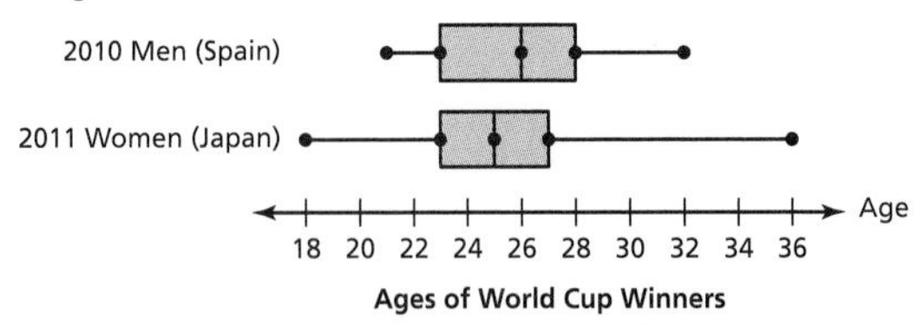

A double box-and-whisker plot shows the distributions of the data.

15. *Sample answer:*

Grades (out of 100) on a Test

Stem	Leaf
5	2 3 9
6	2 3
7	4 5 7
8	0 1 3 4 5 7 8 9
9	5 6 7
10	0

Key: 5 | 3 = 53 out of 100

A stem-and-leaf plot shows how the grades are distributed.

17. *Sample answer:*

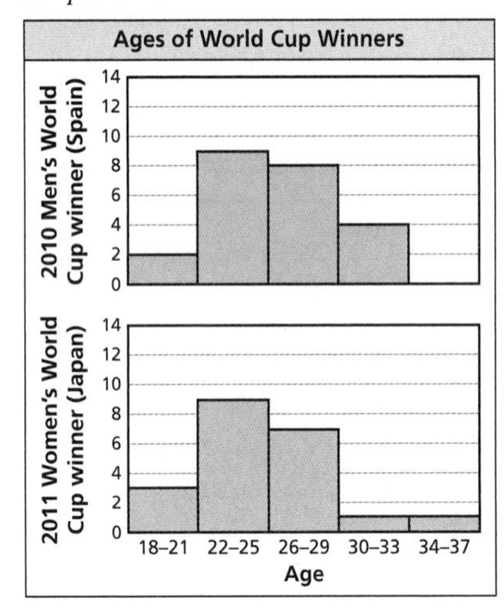

19. *Sample answer:*

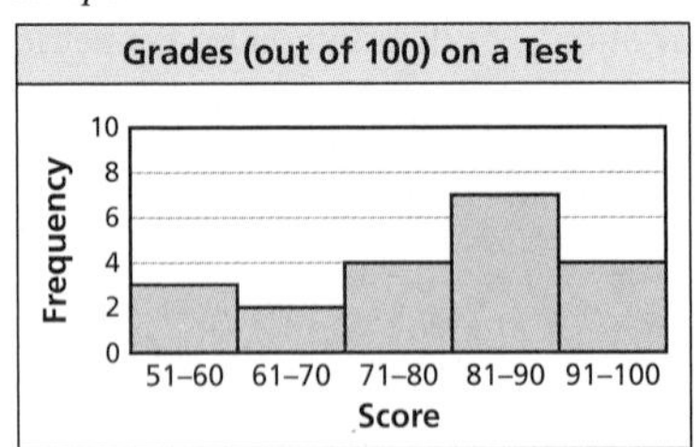

21. The scale on the vertical axis has very small increments; Someone might believe that the annual sales more than tripled from 2010 to 2013.

23. The increments on the scale are not equal; Someone might believe that the temperatures are evenly distributed.

25. *Sample answer:*

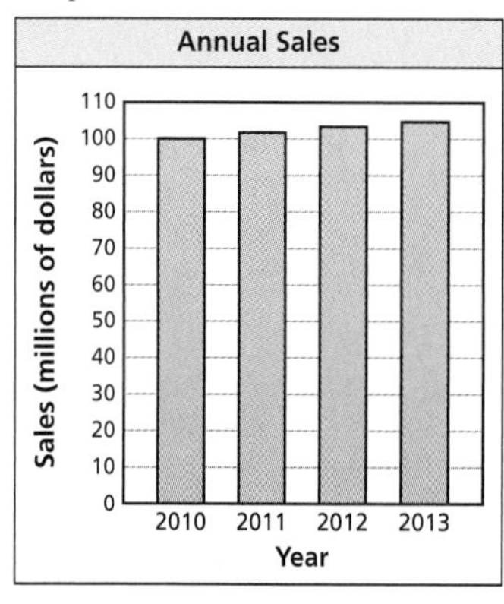

27. Classmate B; The data are quantitative, not qualitative.

29. **a.** There are 120 responses, so the regions do not represent numbers of students.

b. *Sample answer:* bar graph; A bar graph shows data in categories.

31. *Sample answer:* dot plot

11.5 Maintaining Mathematical Proficiency *(p. 622)*

33. yes; Every input has exactly one output.

Chapter 11 Review *(pp. 624–626)*

1. mean: 4, same; median: 4.1, decreases; modes: 4.0 and 4.3; 4 is the mean and is less than the median.

2. mean: 4.5, increases; median: 4.3, increases; mode: 4.3; 10.0 is greater than the mean and median.

3. mean: 4, median: 3, mode: 10

4. mean: 1.7, median: 1, mode: 1

5. Player A: range = 56, standard deviation ≈ 15.56; Player B: range = 86, standard deviation ≈ 30.33; The scores for Player B are more spread out.

6. Store A: range = $110, standard deviation ≈ $39.37; Store B: range = $150, standard deviation ≈ $48.09; The prices at Store B are more spread out.

7. mean: 134, median: 129, mode: 121, range: 45, standard deviation: 3.6

8. mean: 65.4, median: 62.4, mode: 57.6, range: 27, standard deviation: 2.16

9.

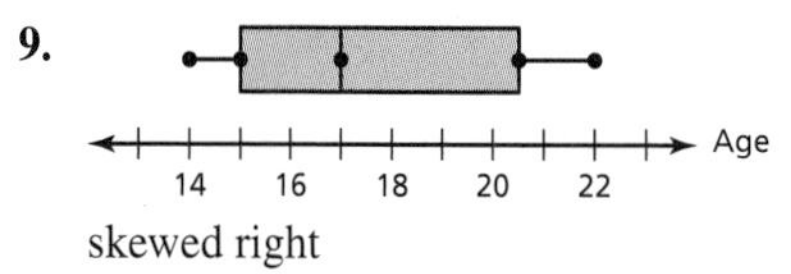

skewed right

10.

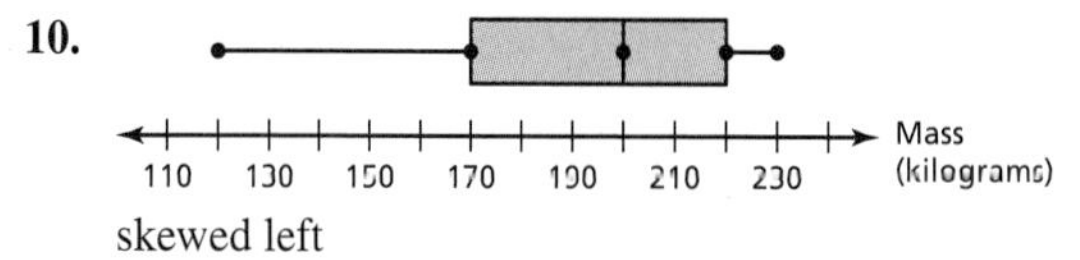

skewed left

11. **a.**

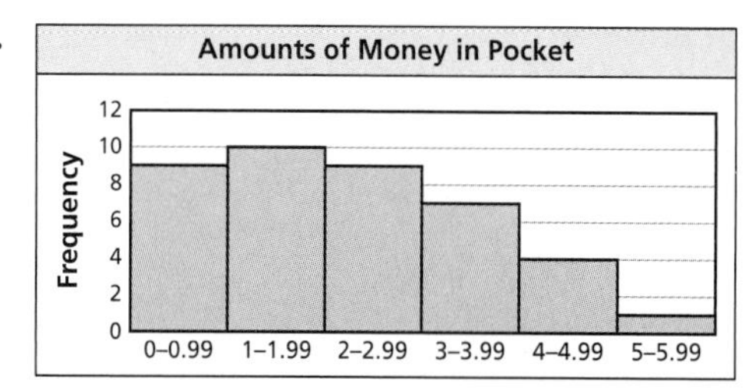

skewed right

b. median, five-number summary

c. The adults typically have more money in their pockets and the amounts vary more.

12. **a.**

		Food Court		
		Like	Dislike	Total
Shoppers	Adults	0.10	0.40	0.50
	Teenagers	0.48	0.02	0.50
	Total	0.58	0.42	1

b.

		Food Court	
		Like	Dislike
Shoppers	Adults	0.18	0.95
	Teenagers	0.82	0.05

13. *Sample answer:*

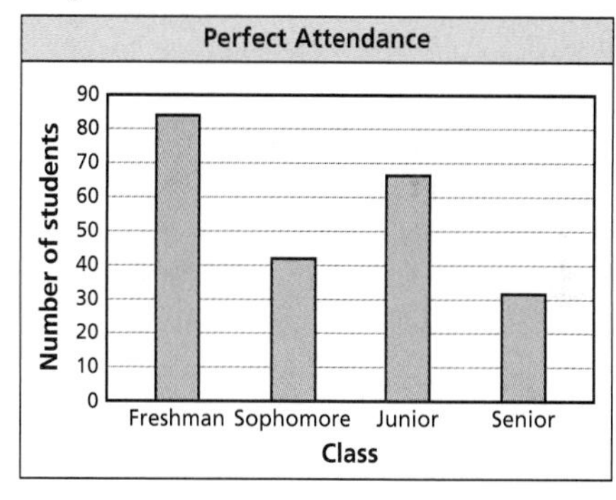

A bar graph shows data in specific categories.

14. quantitative; Heights are numerical values.

15. qualitative; It does not make sense to measure grade levels.

Selected Answers

English-Spanish Glossary

English / Spanish

A

absolute deviation *(p. 90)* The absolute value of the difference of a number x and a given value

desviación absoluta *(p. 90)* El valor absoluto de la diferencia de un número x y un valor dado

absolute value equation *(p. 28)* An equation that contains an absolute value expression

ecuación de valor absoluto *(p. 28)* Una ecuación que contiene una expresión de valor absoluto

absolute value function *(p. 156)* A function that contains an absolute value expression

función de valor absoluto *(p. 156)* Una función que contiene una expresión de valor absoluto

absolute value inequality *(p. 88)* An inequality that contains an absolute value expression

desigualdad de valor absoluto *(p. 88)* Una desigualdad que contiene una expresión de valor absoluto

arithmetic sequence *(p. 210)* An ordered list of numbers in which the difference between each pair of consecutive terms is the same

secuencia aritmética *(p. 210)* Una lista ordenada en donde la diferencia entre cada par de términos consecutivos es la misma

average rate of change *(p. 462)* The slope of the line through $(a, f(a))$ and $(b, f(b))$ of a function $y = f(x)$ between $x = a$ and $x = b$

tasa de variación media *(p. 462)* La pendiente de la línea a través de $(a, f(a))$ y $(b, f(b))$ de una función $y = f(x)$ entre $x = a$ y $x = b$

axis of symmetry *(p. 420)* The vertical line that divides a parabola into two symmetric parts

eje de simetría *(p. 420)* La línea vertical que divide una parábola en dos partes simétricas

B

binomial *(p. 359)* A polynomial with two terms

binomio *(p. 359)* Un polinomio con dos términos

box-and-whisker plot *(p. 594)* A graph that shows the variability of a data set along a number line using the least value, the greatest value, and the quartiles of the data

diagrama de cajas y bigotes *(p. 594)* Un gráfico que muestra la variabilidad de un conjunto de datos a lo largo de una línea de números usando el valor menor, el valor mayor y los cuartiles de los datos

C

categorical data *(p. 618)* Data that consist of labels or nonnumerical entries that can be separated into different categories

datos categóricos *(p. 618)* Datos que consisten en etiquetas o valores no numéricos que pueden separarse en categorías distintas

causation *(p. 205)* When a change in one variable causes a change in another variable

causalidad *(p. 205)* Cuando un cambio en una variable causa un cambio en otra variable

closed *(p. 360)* When an operation performed on any two numbers in the set results in a number that is also in the set

operación interna *(p. 360)* Cuando una operación efectuada en dos números cualesquiera del conjunto da como resultado un número que también está en el conjunto

common difference *(p. 210)* The difference between each pair of consecutive terms in an arithmetic sequence

diferencia común *(p. 210)* La diferencia entre cada par de términos consecutivos en una secuencia aritmética

common ratio *(p. 332)* The ratio between each pair of consecutive terms in a geometric sequence

razón común *(p. 332)* La razón entre cada par de términos consecutivos en una secuencia geométrica

completing the square *(p. 506)* To add a constant c to an expression of the form $x^2 + bx$ so that $x^2 + bx + c$ is a perfect square trinomial

completando el cuadrado *(p. 506)* Agregar una constante c a una expresión de la forma $x^2 + bx$ para que $x^2 + bx + c$ sea un trinomio de cuadrado perfecto

compound inequality *(p. 82)* An inequality formed by joining two inequalities with the word "and" or the word "or"

desigualdad compuesta *(p. 82)* Una desigualdad formada por dos desigualdades con la palabra "y" o la palabra "o"

compound interest *(p. 317)* The interest earned on the principal and on previously earned interest

interés compuesto *(p. 317)* El interés obtenido en el principal y en intereses previamente obtenidos

conditional relative frequency *(p. 612)* The ratio of a joint relative frequency to the marginal relative frequency

frecuencia relativa condicional *(p. 612)* La razón de una frecuencia relativa conjunta a una frecuencia relativa marginal

conjecture *(p. 3)* An unproven statement about a general mathematical concept

conjetura *(p. 3)* Una afirmación no comprobada sobre un concepto matemático general

conjugates *(p. 482)* Binomials of the form $a\sqrt{b} + c\sqrt{d}$ and $a\sqrt{b} - c\sqrt{d}$, where a, b, c, and d are rational numbers

conjugados *(p. 482)* Binomios de la forma $a\sqrt{b} + c\sqrt{d}$ y $a\sqrt{b} - c\sqrt{d}$, donde a, b, c y d son números racionales

constant function *(p. 138)* A linear equation written in the form $y = 0x + b$ or $y = b$

función constante *(p. 138)* Una ecuación lineal escrita bajo la forma $y = 0x + b$ o $y = b$

continuous domain *(p. 114)* A set of input values that consists of all numbers in an interval

dominio continuo *(p. 114)* Un conjunto de valores que consiste en todos los números en un intervalo

correlation *(p. 197)* A relationship between data sets

correlación *(p. 197)* Una relación entre conjuntos de datos

correlation coefficient *(p. 203)* A value r that tells how closely the equation of the line of best fit models the data

coeficiente de correlación *(p. 203)* Un valor r que indica cuán fielmente la ecuación de la línea de mejor ajuste modela los datos

counterexample *(p. 479)* An example that proves that a general statement is not true

contraejemplo *(p. 479)* Un ejemplo que prueba que una afirmación general no es verdadera

cube root function *(p. 552)* A radical function with an index of 3

función de raíz cúbica *(p. 552)* Una función radical con un índice de 3

D

data transformation *(p. 589)* A procedure that uses a mathematical operation to change a data set into a different data set

transformación de datos *(p. 589)* Un procedimiento que usa una operación matemática para cambiar un conjunto de datos en un conjunto de datos diferentes

degree of a monomial *(p. 358)* The sum of the exponents of the variables in the monomial

grado de un monomio *(p. 358)* La suma de los exponentes de las variables en el monomio

degree of a polynomial *(p. 359)* The greatest degree of the terms in a polynomial

grado de un polinomio *(p. 359)* El grado mayor de los términos en un polinomio

dependent variable *(p. 107)* The variable that represents the output values of a function

variable dependiente *(p. 107)* La variable que representa los resultados de una función

discrete domain *(p. 114)* A set of input values that consists of only certain numbers in an interval

dominio discreto *(p. 114)* Un conjunto de valores que consiste en solamente ciertos números en un intervalo

discriminant *(p. 518)* The expression $b^2 - 4ac$ in the Quadratic Formula

discriminante *(p. 518)* La expresión $b^2 - 4ac$ en la Fórmula Cuadrática

domain *(p. 106)* The set of all possible input values of a function

dominio *(p. 106)* El conjunto de todos los valores posibles de una función

E

equation *(p. 4)* A statement that two expressions are equal

ecuación *(p. 4)* Una afirmación de que dos expresiones son iguales

equivalent equations *(p. 4)* Equations that have the same solution(s)

ecuaciones equivalentes *(p. 4)* Ecuaciones que tiene(n) la(s) misma(s) solución(es)

equivalent inequalities *(p. 62)* Inequalities that have the same solutions

desigualdades equivalentes *(p. 62)* Desigualdades que tiene(n) la(s) misma(s) solución(es)

even function *(p. 442)* A function $y = f(x)$ is even when $f(-x) = f(x)$ for each x in the domain of f.

par función *(p. 442)* Una función $y = f(x)$ es par cuando $f(-x) = f(x)$ para cada x en el dominio de f.

explicit rule *(p. 340)* A rule to define arithmetic and geometric sequences that gives a_n as a function of the term's position number n in the sequence

regla explícita *(p. 340)* Una regla para definir secuencias aritméticas y geométricas que da a_n como función del número de posición del término n en la secuencia

exponential decay *(p. 315)* When a quantity decreases by the same factor over equal intervals of time

decaimiento exponencial *(p. 315)* Cuando una cantidad disminuye por el mismo factor sobre intervalos iguales de tiempo

exponential decay function *(p. 315)* A function of the form $y = a(1 - r)^t$, where $a > 0$ and $0 < r < 1$

función de decaimiento exponencial *(p. 315)* Una función de la forma $y = a(1 - r)^t$, donde $a > 0$ y $0 < r < 1$

exponential equation *(p. 326)* An equation in which variable expressions occur as exponents

ecuación exponencial *(p. 326)* Una ecuación en donde las expresiones de una variable ocurren como exponentes

exponential function *(p. 306)* A nonlinear function of the form $y = ab^x$, where $a \neq 0$, $b \neq 1$, and $b > 0$

función exponencial *(p. 306)* Una función no lineal de la forma $y = ab^x$, donde $a \neq 0$, $b \neq 1$, y $b > 0$

exponential growth *(p. 314)* When a quantity increases by the same factor over equal intervals of time

crecimiento exponencial *(p. 314)* Cuando una cantidad se incrementa por el mismo factor sobre intervalos iguales de tiempo

exponential growth function *(p. 314)* A function of the form $y = a(1 + r)^t$, where $a > 0$ and $r > 0$

función de crecimiento exponencial *(p. 314)* Una función de la forma $y = a(1 + r)^t$, donde $a > 0$ y $r > 0$

extraneous solution *(p. 31)* An apparent solution that must be rejected because it does not satisfy the original equation

solución extraña *(p. 31)* Una solución aparente que debe ser rechazada porque no satisface la ecuación original

extrapolation *(p. 205)* To predict a value outside the range of known values using a graph or its equation

extrapolación *(p. 205)* Predecir un valor fuera del rango de valores conocidos usando un gráfico o su ecuación

factored completely *(p. 404)* A polynomial that is written as a product of unfactorable polynomials with integer coefficients

factorizado completamente *(p. 404)* Un polinomio que se escribe como un producto de polinomios no factorizables con coeficientes de números enteros

factored form *(p. 378)* A polynomial that is written as a product of factors

forma factorizada *(p. 378)* Un polinomio que se escribe como un producto de factores

factoring by grouping *(p. 404)* To use the Distributive Property to factor a polynomial with four terms

factorización por agrupamiento *(p. 404)* Uso de la propiedad distributiva para factorizar un polinomio con cuatro términos

family of functions *(p. 146)* A group of functions with similar characteristics

familia de funciones *(p. 146)* Un grupo de funciones con características similares

five-number summary *(p. 594)* The five numbers that make up a box-and-whisker plot

resumen de cinco números *(p. 594)* Los cinco números que componen un diagrama de cajas y bigotes

FOIL Method *(p. 367)* A shortcut for multiplying two binomials by finding the sum of the products of the first terms, outer terms, inner terms, and last terms

método FOIL o de multiplicación de binomios *(p. 367)* Un método rápido para multiplicar dos binomios encontrando la suma de los productos de los primeros términos, los segundos términos, los terceros términos y los últimos términos

formula *(p. 37)* A literal equation that shows how one variable is related to one or more other variables

fórmula *(p. 37)* Una ecuación literal que muestra cómo una variable está relacionada con una o más variables

function *(p. 104)* A relation that pairs each input with exactly one output

función *(p. 104)* Una relación que asocia cada valor exactamente con un resultado

function notation *(p. 122)* Another name for y denoted as $f(x)$ and read as "the value of f at x" or "f of x"

notación de función *(p. 122)* Otro nombre para y que se denota como $f(x)$ y que se lee como "el valor de f en x" o "f de x"

geometric sequence *(p. 332)* An ordered list of numbers in which the ratio between each pair of consecutive terms is the same

secuencia geométrica *(p. 332)* Una lista ordenada en donde la razón entre cada par de términos consecutivos es la misma

graph of an inequality *(p. 56)* A graph that shows the solution set of an inequality on a number line

gráfico de una desigualdad *(p. 56)* Un gráfico que muestra el conjunto de soluciones de una desigualdad en una línea numérica

graph of a linear inequality *(p. 268)* The graph in two variables that shows all the solutions of the inequality in a coordinate plane

gráfico de una desigualdad lineal *(p. 268)* El gráfico en dos variables que muestra todas las soluciones de la desigualdad en un plano coordenado

graph of a system of linear inequalities *(p. 275)* The graph of all the solutions of the system of linear inequalities

gráfico de un sistema de desigualdades lineales *(p. 275)* El gráfico de todas las soluciones del sistema de las desigualdades lineales

half-planes *(p. 268)* Two regions of the coordinate plane divided by a boundary line

semiplanos *(p. 268)* Dos regiones del plano coordenado divididas por una línea limítrofe

horizontal shrink *(p. 148)* A transformation that causes the graph of a function to shrink toward the y-axis when all the x-coordinates are multiplied by a factor a, where $a > 1$

reducción horizontal *(p. 148)* Una transformación que hace que el gráfico de una función se reduzca hacia el eje y cuando todas las coordenadas x se multiplican por un factor a, donde $a > 1$

horizontal stretch *(p. 148)* A transformation that causes the graph of a function to stretch away from the y-axis when all the x-coordinates are multiplied by a factor a, where $0 < a < 1$

ampliación horizontal *(p. 148)* Una transformación que hace que el gráfico de una función se amplíe desde el eje y cuando todas las coordenadas x se multiplican por un factor a, donde $0 < a < 1$

I

identity *(p. 21)* An equation that is true for all values of the variable

identidad *(p. 21)* Una ecuación que es verdadera para todos los valores de la variable

independent variable *(p. 107)* The variable that represents the input values of a function

variable independiente *(p. 107)* La variable que representa los valores de una función

index of a radical *(p. 300)* The value of n in the radical $\sqrt[n]{a}$

índice de un radical *(p. 300)* El valor de n en el radical $\sqrt[n]{a}$

inequality *(p. 54)* A mathematical sentence that compares expressions

desigualdad *(p. 54)* Una oración matemática que compara expresiones

intercept form *(p. 450)* A quadratic function written in the form $f(x) = a(x - p)(x - q)$, where $a \neq 0$

forma intersección *(p. 450)* Una ecuación cuadrática escrita en la forma $f(x) = a(x - p)(x - q)$, donde $a \neq 0$

interpolation *(p. 205)* To approximate a value between two known values using a graph or its equation

interpolación *(p. 205)* Aproximar un valor entre dos valores conocidos usando un gráfico o su ecuación

interquartile range *(p. 595)* A measure of variation for a data set, which is the difference of the third quartile and the first quartile

rango de intercuartiles *(p. 595)* Una medida de variación para un conjunto de datos, el cual es la diferencia del tercer cuartil y el primer cuartil

inverse function *(p. 569)* Functions that undo each other

función inversa *(p. 569)* Funciones que se anulan entre sí

inverse operations *(p. 4)* Two operations that undo each other, such as addition and subtraction

operaciones inversas *(p. 4)* Dos operaciones que se anulan entre sí, como la suma y la resta

inverse relation *(p. 568)* When the input and output values of the original relation are switched

relación inversa *(p. 568)* Cuando los valores y los resultados de la relación original se intercambian

joint frequency *(p. 610)* Each entry in a two-way table

frecuencia conjunta *(p. 610)* Cada valor en una tabla de doble entrada

joint relative frequency *(p. 611)* The ratio of a frequency that is not in the "total" row or the "total" column to the total number of values or observations

frecuencia relativa conjunta *(p. 611)* La razón de una frecuencia que no está en la hilera "total" o columna "total" del número total de valores u observaciones

leading coefficient *(p. 359)* The coefficient of the first term of a polynomial written in standard form

coeficiente principal *(p. 359)* El coeficiente del primer término de un polinomio escrito en forma estándar

like radicals *(p. 484)* Radicals with the same index and radicand

radicales semejantes *(p. 484)* Radicales con el mismo índice y radicando

line of best fit *(p. 203)* A line that best models a set of data

línea de mejor ajuste *(p. 203)* Una línea que mejor modela un conjunto de datos

line of fit *(p. 198)* A line drawn on a scatter plot that is close to most of the data points

línea de ajuste *(p. 198)* Una línea dibujada en un diagrama de dispersión que está cerca a la mayoría de los puntos de datos

linear equation in one variable *(p. 4)* An equation that can be written in the form $ax + b = 0$, where a and b are constants and $a \neq 0$

ecuación lineal en una variable *(p. 4)* Una ecuación que puede escribirse en la forma $ax + b = 0$, donde a y b son constantes y $a \neq 0$

linear equation in two variables *(p. 112)* An equation that can be written in the form $y = mx + b$, where m and b are constants

ecuación lineal en dos variables *(p. 112)* Una ecuación que puede escribirse en la forma $y = mx + b$, donde m y b son constantes

linear function *(p. 112)* A function whose graph is a nonvertical line

función lineal *(p. 112)* Una función cuyo gráfico es una línea no vertical

linear inequality in two variables *(p. 268)* An inequality written in the form $ax + by < c$, $ax + by \leq c$, $ax + by > c$, or $ax + by \geq c$, where a, b, and c are real numbers

desigualdad lineal en dos variables *(p. 268)* Una desigualdad escrita en la forma $ax + by < c$, $ax + by \leq c$, $ax + by > c$, o $ax + by \geq c$, donde a, b y c son números reales

linear model *(p. 178)* A linear function that models a real-life situation

modelo lineal *(p. 178)* Una función lineal que modela una situación de la vida real

linear regression *(p. 203)* A method that graphing calculators use to find a precise line of fit that models a set of data

regresión lineal *(p. 203)* Un método que usan las calculadoras gráficas para encontrar una línea precisa de ajuste que modela un conjunto de datos

literal equation *(p. 36)* An equation that has two or more variables

ecuación literal *(p. 36)* Una ecuación que tiene dos o más variables

M

marginal frequency *(p. 610)* The sums of the rows and columns in a two-way table

frecuencia marginales *(p. 610)* Las sumas de las hileras y columnas en una tabla de doble entrada

marginal relative frequency *(p. 611)* The sum of the joint relative frequencies in a row or a column

frecuencia relativa marginal *(p. 611)* La suma de las frecuencias relativas conjuntas en una hilera o columna

maximum value *(p. 433)* The y-coordinate of the vertex of the graph of $f(x) = ax^2 + bx + c$, where $a < 0$

valor máximo *(p. 433)* La coordenada y del vértice del gráfico de $f(x) = ax^2 + bx + c$, donde $a < 0$

mean *(p. 586)* The sum of a numerical data set divided by the number of data values

media *(p. 586)* La suma de un conjunto de datos numéricos divididos entre el número de valores de datos

measure of center *(p. 586)* A measure that represents the center, or typical value, of a data set

medida del centro *(p. 586)* Una medida que representa el centro, o valor típico, de un conjunto de datos

measure of variation *(p. 587)* A measure that describes the spread, or distribution, of a data set

medida de variación *(p. 587)* Una medida que describe la extensión, o distribución, de un conjunto de datos

median *(p. 586)* The middle number of a numerical data set when the values are written in numerical order

mediana *(p. 586)* El número medio de un conjunto de datos numéricos cuando los valores se escriben en orden numérico

minimum value *(p. 433)* The y-coordinate of the vertex of the graph of $f(x) = ax^2 + bx + c$, where $a > 0$

valor mínimo *(p. 433)* La coordenada y del vértice del gráfico de $f(x) = ax^2 + bx + c$, donde $a > 0$

misleading graph *(p. 620)* A statistical graph that is not drawn appropriately

gráfico engañoso *(p. 620)* Un gráfico estadístico que no está dibujado apropiadamente

mode *(p. 586)* The value or values that occur most often in a data set

modo *(p. 586)* El valor o valores que ocurren con mayor frecuencia en un conjunto de datos

monomial *(p. 358)* A number, a variable, or the product of a number and one or more variables with whole number exponents

monomio *(p. 358)* Un número, una variable o el producto de un número y una o más variables con exponentes en números enteros

N

nonlinear function *(p. 112)* A function that does not have a constant rate of change and whose graph is not a line

función no lineal *(p. 112)* Una función que no tiene una tasa constante de cambio y cuyo gráfico no es una línea

nth root of a *(p. 300)* For an integer n greater than 1, if $b^n = a$, then b is an nth root of a.

raíz de orden n de a *(p. 300)* Para un número entero n mayor que 1, si $b^n = a$, entonces b es una raíz de orden n de a.

O

odd function *(p. 442)* A function $y = f(x)$ is odd when $f(-x) = -f(x)$ for each x in the domain of f.

impar funcíon *(p. 442)* Una función $y = f(x)$ es impar cuando $f(-x) = -f(x)$ para cada x en el dominio de f.

outlier *(p. 587)* A data value that is much greater than or much less than the other values in a data set

valor atípico *(p. 587)* Una valor de datos que es mucho mayor o mucho menor que los otros valores en un conjunto de datos

P

parabola *(p. 420)* The U-shaped graph of a quadratic function

parábola *(p. 420)* El gráfico en forma de U de una función cuadrática

parallel lines *(p. 188)* Two lines in the same plane that never intersect

líneas paralelas *(p. 188)* Dos líneas en el mismo plano que nunca se intersectan

parent function *(p. 146)* The most basic function in a family of functions

función principal *(p. 146)* La función más básica en una familia de funciones

perpendicular lines *(p. 189)* Two lines in the same plane that intersect to form right angles

líneas perpendiculares *(p. 189)* Dos líneas en el mismo plano que se intersectan para formar ángulos rectos

piecewise function *(p. 218)* A function defined by two or more equations

función por tramos *(p. 218)* Una función definida por dos o más ecuaciones

point-slope form *(p. 182)* A linear equation written in the form $y - y_1 = m(x - x_1)$

forma punto-pendiente *(p. 182)* Una ecuación lineal escrita en la forma $y - y_1 = m(x - x_1)$

polynomial *(p. 359)* A monomial or a sum of monomials

polinomio *(p. 359)* Un monomio o una suma de monomios

Q

quadratic equation *(p. 490)* A nonlinear equation that can be written in the standard form $ax^2 + bx + c = 0$, where $a \neq 0$

ecuación cuadrática *(p. 490)* Una ecuación no lineal que puede escribirse en la forma estándar $ax^2 + bx + c = 0$, donde $a \neq 0$

Quadratic Formula *(p. 516)* The real solutions of the quadratic equation $ax^2 + bx + c = 0$ are

$x = \dfrac{-b \pm \sqrt{b^2 - 4ac}}{2a}$, where $a \neq 0$ and $b^2 - 4ac \geq 0$.

fórmula cuadrática *(p. 516)* Las soluciones reales de la ecuación cuadrática $ax^2 + bx + c = 0$ son

$x = \dfrac{-b \pm \sqrt{b^2 - 4ac}}{2a}$, donde $a \neq 0$ y $b^2 - 4ac \geq 0$.

quadratic function *(p. 420)* A nonlinear function that can be written in the standard form $y = ax^2 + bx + c$, where $a \neq 0$

función cuadrática *(p. 420)* Una ecuación no lineal que puede escribirse en la forma estándar $y = ax^2 + bx + c$, donde $a \neq 0$

qualitative data *(p. 618)* Data that consist of labels or nonnumerical entries that can be separated into different categories

datos cualitativos *(p. 618)* Datos que consisten en etiquetas o valores no numéricos que pueden separarse en categorías distintas

quantitative data *(p. 618)* Data that consist of numbers that represent counts or measurements

datos cuantitativos *(p. 618)* Datos que consisten en números que representan conteos o medidas

quartiles *(p. 594)* Values of a box-and-whisker plot that divide a data set into four equal parts

cuartiles *(p. 594)* Valores de un diagrama de cajas y bigotes que dividen un conjunto de datos en cuatro partes iguales

R

radical *(p. 300)* An expression of the form $\sqrt[n]{a}$

radical *(p. 300)* Una expresión de la forma $\sqrt[n]{a}$

radical equation *(p. 560)* An equation that contains a radical expression with a variable in the radicand

ecuación radical *(p. 560)* Una ecuación que contiene una expresión radical con una variable en el radicando

radical expression *(p. 480)* An expression that contains a radical

expresión radical *(p. 480)* Una expresión que contiene un radical

radical function *(p. 545)* A function that contains a radical expression with the independent variable in the radicand

función radical *(p. 545)* Una ecuación que contiene una expresión radical con la variable independiente en el radicando

range of a data set *(p. 587)* The difference of the greatest value and the least value of a data set

rango de un conjunto de datos *(p. 587)* La diferencia del valor mayor y el valor menor de un conjunto de datos

range of a function *(p. 106)* The set of all possible output values of a function

rango de una función *(p. 106)* El conjunto de todos los resultados posibles de una función

rationalizing the denominator *(p. 482)* To eliminate a radical from the denominator of a fraction by multiplying by an appropriate form of 1

racionalización del denominador *(p. 482)* Eliminar un radical del denominador de una fracción multiplicando por una forma apropiada de 1

recursive rule *(p. 340)* A rule to define arithmetic and geometric sequences that gives the beginning term(s) of a sequence and a recursive equation that tells how a_n is related to one or more preceding terms

regla recursiva *(p. 340)* Una regla para definir las secuencias aritméticas y geométricas que da el(los) primer(os) término(s) de una secuencia y una ecuación recursiva que indica cómo se relaciona a_n a uno o más términos precedentes

reflection *(p. 147)* A transformation that flips a graph over a line called the *line of reflection*

reflexión *(p. 147)* Una transformación que voltea un gráfico sobre una línea llamada la *línea de reflexión*

relation *(p. 104)* A pairing of inputs with outputs

relación *(p. 104)* Una pareja de valores con resultados

repeated roots *(p. 379)* Two or more roots of an equation that are the same number

raíces repetidas *(p. 379)* Dos o más raíces de una ecuación que son el mismo número

residual *(p. 202)* The difference of the y-value of a data point and the corresponding y-value found using the line of fit

residual *(p. 202)* La diferencia del valor y de un punto de datos y el valor y correspondiente usando la línea de ajuste

rise *(p. 136)* The change in y between any two points on a line

desplazamiento vertical *(p. 136)* El cambio en y entre cualquier dos puntos en una recta

roots *(p. 378)* The solutions of a polynomial equation

raíces *(p. 378)* Las soluciones deuna ecuación de polinomios

rule *(p. 3)* A proven statement about a general mathematical concept

regla *(p. 3)* Una afirmación comprobada sobre un concepto matemático general

run *(p. 136)* The change in x between any two points on a line

desplazamiento horizontal *(p. 136)* El cambio en x entre cualquier dos puntos en una recta

S

scatter plot *(p. 196)* A graph that shows the relationship between two data sets

diagrama de dispersión *(p. 196)* Un gráfico que muestra la relación entre dos conjuntos de datos

sequence *(p. 210)* An ordered list of numbers

secuencia *(p. 210)* Una lista ordenada de números

simplest form of a radical *(p. 480)* An expression involving a radical with index n that has no radicands with perfect nth powers as factors other than 1, no radicands that contain fractions, and no radicals that appear in the denominator of a fraction

mínima expresión de un radical *(p. 480)* Una expresión que conlleva un radical con índice n que no tiene radicandos con potencias perfectas de orden n como factores distintos a 1, que no tiene radicandos que contengan fracciones y que no tiene radicales que aparezcan en el denominador de una fracción

slope *(p. 136)* The rate of change between any two points on a line

pendiente *(p. 136)* La tasa de cambio entre dos puntos cualesquiera en una línea

slope-intercept form *(p. 138)* A linear equation written in the form $y = mx + b$

forma intersección-pendiente *(p. 138)* Una ecuación lineal escrita en la forma $y = mx + b$

solution of an equation *(p. 4)* A value that makes an equation true

solución de una ecuación *(p. 4)* Un valor que hace que una ecuación sea verdadera

solution of an inequality *(p. 55)* A value that makes an inequality true

solución de una desigualdad *(p. 55)* Un valor que hace que una desigualdad sea verdadera

solution of a linear equation in two variables *(p. 114)* An ordered pair (x, y) that makes an equation true

solución de una ecuación lineal en dos variables *(p. 114)* Un par ordenado (x, y) que hace que una ecuación sea verdadera

solution of a linear inequality in two variables *(p. 268)* An ordered pair (x, y) that makes the inequality true

solución de una desigualdad lineal en dos variables *(p. 268)* Un par ordenado (x, y) que hace que la desigualdad sea verdadera

solution set *(p. 55)* The set of all solutions of an inequality

conjunto solución *(p. 55)* El conjunto de todas las soluciones de una desigualdad

solution of a system of linear equations *(p. 236)* An ordered pair that is a solution of each equation in the system

solución de un sistema de ecuaciones lineales *(p. 236)* Un par ordenado que es una solución de cada ecuación en el sistema

solution of a system of linear inequalities *(p. 274)* An ordered pair that is a solution of each inequality in the system

solución de un sistema de desigualdades lineales *(p. 274)* Un par ordenado que es una solución de cada desigualdad en el sistema

square root function *(p. 544)* A function that contains a square root with the independent variable in the radicand

función de raíz cuadrada *(p. 544)* Una función que contiene una raíz cuadrada con la variable independiente en el radicando

standard deviation *(p. 588)* A measure of how much a typical value in a numerical data set differs from the mean

desviación estándar *(p. 588)* Una medida de cuánto difiere un valor típico en un conjunto de datos numéricos de la media

standard form of a linear equation *(p. 130)* A linear equation written in the form $Ax + By = C$, where A, B, and C are real numbers and A and B are not both zero

forma estándar de una ecuación lineal *(p. 130)* Una ecuación lineal escrita en la forma $Ax + By = C$, donde A, B y C son números reales y ni A ni B son cero

standard form of a polynomial *(p. 359)* A polynomial in one variable written with the exponents of the terms decreasing from left to right

forma estándar de un polinomio *(p. 359)* Un polinomio en una variable escrita con los exponentes de los términos decreciendo de izquierda a derecha

step function *(p. 220)* A piecewise function defined by a constant value over each part of its domain

función escalón *(p. 220)* Una función por tramos definida por un valor constante sobre cada parte de su dominio

system of linear equations *(p. 236)* A set of two or more linear equations in the same variables

sistema de ecuaciones lineales *(p. 236)* Un conjunto de dos o más ecuaciones lineales en las mismas variables

system of linear inequalities *(p. 274)* A set of two or more linear inequalities in the same variables

sistema de desigualdades lineales *(p. 274)* Un conjunto de dos o más desigualdades lineales en las mismas variables

system of nonlinear equations *(p. 526)* A system in which at least one of the equations is nonlinear

sistema de ecuaciones no lineales *(p. 526)* Un sistema en donde al menos una de las ecuaciones no es lineal

term of a sequence *(p. 210)* Each number in a sequence

término de una sucesión *(p. 210)* Cada número en una secuencia

theorem *(p. 3)* A proven statement about a general mathematical concept

teorema *(p. 3)* Una declaración comprobada acerca de un concepto matemático general

transformation *(p. 146)* A change in the size, shape, position, or orientation of a graph

transformación *(p. 146)* Un cambio en el tamaño, forma, posición u orientación de un gráfico

translation *(p. 146)* A transformation that shifts a graph horizontally and/or vertically but does not change the size, shape, or orientation of the graph

traslación *(p. 146)* Una transformación que desplaza un gráfico horizontal y/o verticalmente pero no cambia el tamaño, forma u orientación del gráfico

trinomial *(p. 359)* A polynomial with three terms

trinomio *(p. 359)* Un polinomio con tres términos

two-way table *(p. 610)* A frequency table that displays data collected from one source that belong to two different categories

tabla de doble entrada *(p. 610)* Una tabla de frecuencia que muestra los datos recogidos de una fuente que pertenece a dos categorías distintas

vertex *(p. 156)* The point where a graph changes direction

vértice *(p. 156)* El punto en donde un gráfico cambia de dirección

vertex form of an absolute value function *(p. 158)* An absolute value function written in the form $f(x) = a|x - h| + k$, where $a \neq 0$

forma de vértice de una función de valor absoluto *(p. 158)* Una función de valor absoluto escrita en la forma $f(x) = a|x - h| + k$, donde $a \neq 0$

vertex form of a quadratic function *(p. 444)* A quadratic function written in the form $f(x) = a(x - h)^2 + k$, where $a \neq 0$

forma de vértice de una función cuadrática *(p. 444)* Una función cuadrática escrita en la forma $f(x) = a(x - h)^2 + k$, donde $a \neq 0$

vertex of a parabola *(p. 420)* The lowest point on a parabola that opens up or the highest point on a parabola that opens down

vértice de una parábola *(p. 420)* El punto más bajo de una parábola que se abre hacia arriba o el punto más alto de una parábola que se abre hacia abajo

vertical shrink *(p. 148)* A transformation that causes the graph of a function to shrink toward the x-axis when all the y-coordinates are multiplied by a factor a, where $0 < a < 1$

reducción vertical *(p. 148)* Una transformación que hace que el gráfico de una función se reduzca hacia el eje x cuando todas las coordenadas y se multiplican por un factor a, donde $0 < a < 1$

vertical stretch *(p. 148)* A transformation that causes the graph of a function to stretch away from the x-axis when all the y-coordinates are multiplied by a factor a, where $a > 1$

ampliación vertical *(p. 148)* Una transformación que hace que el gráfico de una función se amplíe desde el eje x cuando todas las coordenadas y se multiplican por un factor a, donde $a > 1$

x-intercept *(p. 131)* The x-coordinate of a point where the graph crosses the x-axis

intersección *x* *(p. 131)* La coordenada x de un punto donde el gráfico cruza el eje x

y-intercept *(p. 131)* The y-coordinate of a point where the graph crosses the y-axis

intersección *y* *(p. 131)* La coordenada y de un punto donde el gráfico cruza el eje y

zero of a function *(p. 428)* An x-value of a function f for which $f(x) = 0$

cero de una función *(p. 428)* Un valor x de una función f para el cual $f(x) = 0$

Zero-Product Property *(p. 378)* If the product of two real numbers is 0, then at least one of the numbers is 0.

propiedad de producto cero *(p. 378)* Si el producto de dos números reales es 0, entonces al menos uno de los números es 0.

Index

A

B

Index

Index

S

Reference

Properties

Properties of Equality

Addition Property of Equality
If $a = b$, then $a + c = b + c$.

Subtraction Property of Equality
If $a = b$, then $a - c = b - c$.

Multiplication Property of Equality
If $a = b$, then $a \cdot c = b \cdot c, c \neq 0$.

Division Property of Equality
If $a = b$, then $a \div c = b \div c, c \neq 0$.

Properties of Inequality

Addition Property of Inequality
If $a > b$, then $a + c > b + c$.
If $a < b$, then $a + c < b + c$.

Subtraction Property of Inequality
If $a > b$, then $a - c > b - c$.
If $a < b$, then $a - c < b - c$.

Multiplication Property of Inequality ($c > 0$)
If $a > b$ and $c > 0$, then $ac > bc$.
If $a < b$ and $c > 0$, then $ac < bc$.

Division Property of Inequality ($c > 0$)
If $a > b$ and $c > 0$, then $\frac{a}{c} > \frac{b}{c}$.
If $a < b$ and $c > 0$, then $\frac{a}{c} < \frac{b}{c}$.

Multiplication Property of Inequality ($c < 0$)
If $a > b$ and $c < 0$, then $ac < bc$.
If $a < b$ and $c < 0$, then $ac > bc$.

Division Property of Inequality ($c < 0$)
If $a > b$ and $c < 0$, then $\frac{a}{c} < \frac{b}{c}$.
If $a < b$ and $c < 0$, then $\frac{a}{c} > \frac{b}{c}$.

* The Properties of Inequality are also true for $\geq$ and $\leq$.

Properties of Exponents

Zero Exponent
$a^0 = 1$, where $a \neq 0$

Negative Exponent
$a^{-n} = \frac{1}{a^n}$, where $a \neq 0$

Product of Powers Property
$a^m \cdot a^n = a^{m+n}$

Quotient of Powers Property
$\frac{a^m}{a^n} = a^{m-n}$, where $a \neq 0$

Power of a Power Property
$(a^m)^n = a^{mn}$

Power of a Product Property
$(ab)^m = a^m b^m$

Power of a Quotient Property
$\left(\frac{a}{b}\right)^m = \frac{a^m}{b^m}$, where $b \neq 0$

Rational Exponents
$a^{m/n} = (a^{1/n})^m = (\sqrt[n]{a})^m$

Properties of Absolute Value

$|a| \geq 0$

$|-a| = |a|$

$|ab| = |a||b|$

$\left|\frac{a}{b}\right| = \frac{|a|}{|b|}, b \neq 0$

Properties of Radicals

Product Property of Square Roots
$\sqrt{ab} = \sqrt{a} \cdot \sqrt{b}$, where $a, b \geq 0$

Quotient Property of Square Roots
$\sqrt{\frac{a}{b}} = \frac{\sqrt{a}}{\sqrt{b}}$, where $a \geq 0$ and $b > 0$

Other Properties

Property of Equality for Exponential Equations
If $b > 0$ and $b \neq 1$, then $b^x = b^y$ if and only if $x = y$.

Zero-Product Property
If a and b are real numbers and $ab = 0$, then $a = 0$ or $b = 0$.

Patterns

Square of a Binomial Pattern
$(a+b)^2 = a^2 + 2ab + b^2$
$(a-b)^2 = a^2 - 2ab + b^2$

Sum and Difference Pattern
$(a+b)(a-b) = a^2 - b^2$

Difference of Two Squares Pattern
$a^2 - b^2 = (a+b)(a-b)$

Perfect Square Trinomial Pattern
$a^2 + 2ab + b^2 = (a+b)^2$
$a^2 - 2ab + b^2 = (a-b)^2$

Formulas

Slope
$m = \dfrac{y_2 - y_1}{x_2 - x_1}$

Slope-intercept form
$y = mx + b$

Point-slope form
$y - y_1 = m(x - x_1)$

Standard form of a linear equation
$Ax + By = C$, where A and B are not both 0

Vertex form of an absolute value function
$f(x) = a|x - h| + k$, where $a \neq 0$

Standard form of a quadratic function
$f(x) = ax^2 + bx + c$, where $a \neq 0$

Vertex form of a quadratic function
$f(x) = a(x - h)^2 + k$, where $a \neq 0$

Intercept form of a quadratic function
$f(x) = a(x - p)(x - q)$, where $a \neq 0$

Quadratic Formula
$x = \dfrac{-b \pm \sqrt{b^2 - 4ac}}{2a}$, where $a \neq 0$ and $b^2 - 4ac \geq 0$

Exponential growth
$y = a(1 + r)^t$, where $a > 0$ and $r > 0$

Exponential decay
$y = a(1 - r)^t$, where $a > 0$ and $0 < r < 1$

Explicit rule for an arithmetic sequence
$a_n = a_1 + (n - 1)d$

Explicit rule for a geometric sequence
$a_n = a_1 r^{n-1}$

Recursive equation for an arithmetic sequence
$a_n = a_{n-1} + d$

Recursive equation for a geometric sequence
$a_n = r \cdot a_{n-1}$

Standard deviation

$$\sigma = \sqrt{\frac{(x_1 - \overline{x})^2 + (x_2 - \overline{x})^2 + \cdots + (x_n - \overline{x})^2}{n}}$$

Perimeter, Area, and Volume Formulas

Square

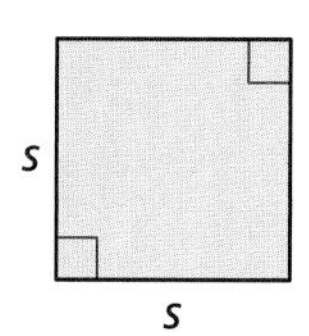

$P = 4s$
$A = s^2$

Rectangle

w
ℓ

$P = 2\ell + 2w$
$A = \ell w$

Triangle

a
h
c
b

$P = a + b + c$
$A = \frac{1}{2}bh$

Circle

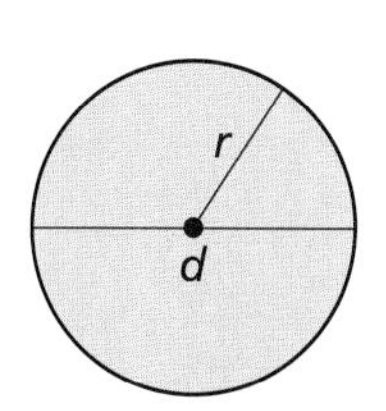

$C = \pi d$ or $C = 2\pi r$
$A = \pi r^2$

Parallelogram

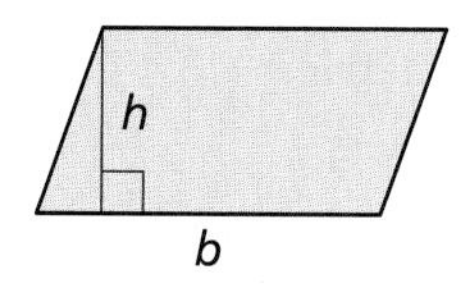

$A = bh$

Trapezoid

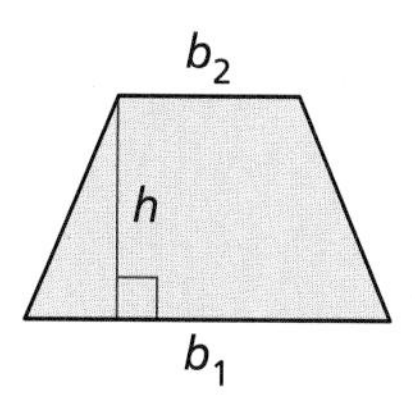

$A = \frac{1}{2}h(b_1 + b_2)$

Rhombus/Kite

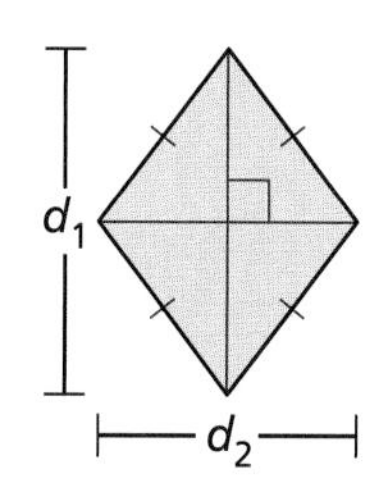

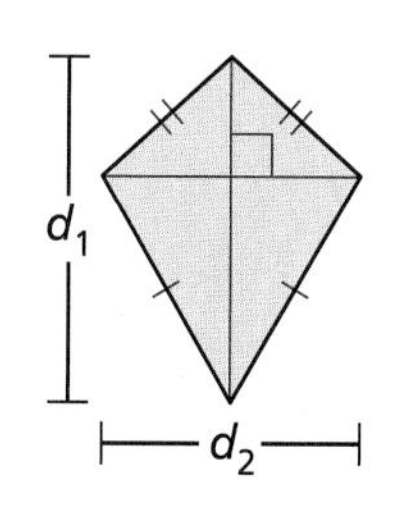

$A = \frac{1}{2}d_1d_2$

Regular n-gon

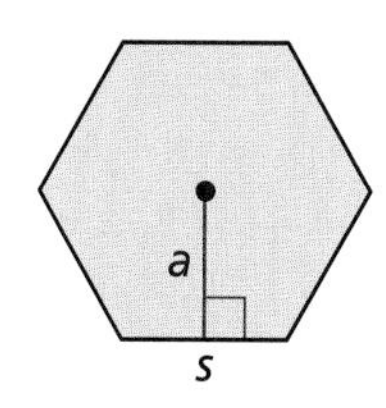

$A = \frac{1}{2}aP$ or $A = \frac{1}{2}a \cdot ns$

Prism

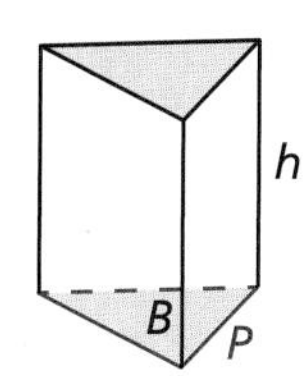

$L = Ph$
$S = 2B + Ph$
$V = Bh$

Cylinder

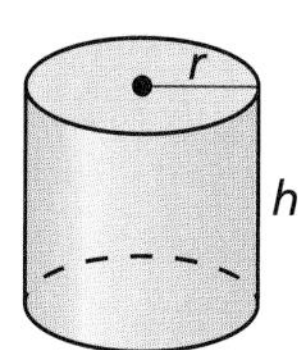

$L = 2\pi rh$
$S = 2\pi r^2 + 2\pi rh$
$V = \pi r^2h$

Pyramid

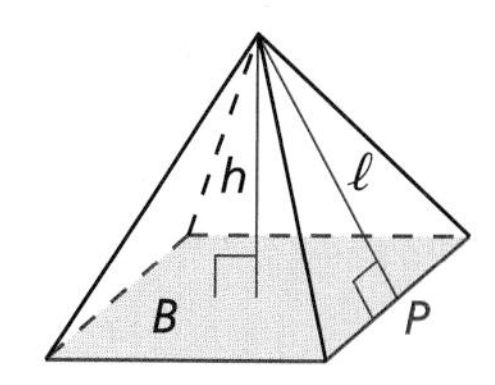

$L = \frac{1}{2}P\ell$
$S = B + \frac{1}{2}P\ell$
$V = \frac{1}{3}Bh$

Cone

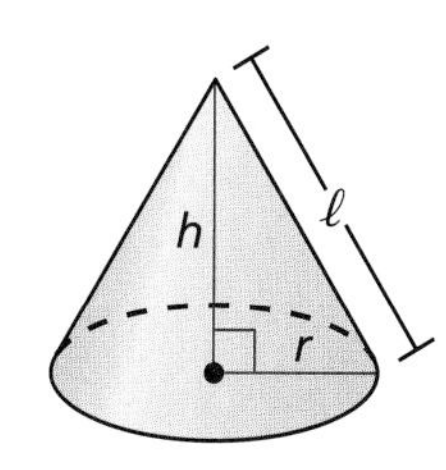

$L = \pi r\ell$

$S = \pi r^2 + \pi r\ell$

$V = \frac{1}{3}\pi r^2h$

Sphere

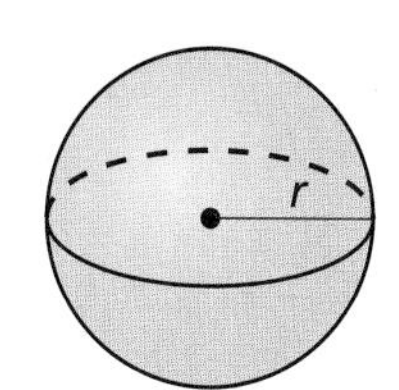

$S = 4\pi r^2$

$V = \frac{4}{3}\pi r^3$

Reference

Other Formulas

Pythagorean Theorem

$a^2 + b^2 = c^2$

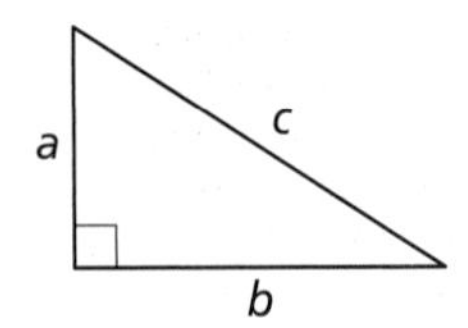

Simple Interest

$I = Prt$

Compound Interest

$y = P\left(1 + \frac{r}{n}\right)^{nt}$

Distance

$d = rt$

Conversions

U.S. Customary

1 foot = 12 inches
1 yard = 3 feet
1 mile = 5280 feet
1 mile = 1760 yards
1 acre = 43,560 square feet
1 cup = 8 fluid ounces
1 pint = 2 cups
1 quart = 2 pints
1 gallon = 4 quarts
1 gallon = 231 cubic inches
1 pound = 16 ounces
1 ton = 2000 pounds

U.S. Customary to Metric

1 inch = 2.54 centimeters
1 foot ≈ 0.3 meter
1 mile ≈ 1.61 kilometers
1 quart ≈ 0.95 liter
1 gallon ≈ 3.79 liters
1 cup ≈ 237 milliliters
1 pound ≈ 0.45 kilogram
1 ounce ≈ 28.3 grams
1 gallon ≈ 3785 cubic centimeters

Time

1 minute = 60 seconds
1 hour = 60 minutes
1 hour = 3600 seconds
1 year = 52 weeks

Temperature

$C = \frac{5}{9}(F - 32)$

$F = \frac{9}{5}C + 32$

Metric

1 centimeter = 10 millimeters
1 meter = 100 centimeters
1 kilometer = 1000 meters
1 liter = 1000 milliliters
1 kiloliter = 1000 liters
1 milliliter = 1 cubic centimeter
1 liter = 1000 cubic centimeters
1 cubic millimeter = 0.001 milliliter
1 gram = 1000 milligrams
1 kilogram = 1000 grams

Metric to U.S. Customary

1 centimeter ≈ 0.39 inch
1 meter ≈ 3.28 feet
1 meter ≈ 39.37 inches
1 kilometer ≈ 0.62 mile
1 liter ≈ 1.06 quarts
1 liter ≈ 0.26 gallon
1 kilogram ≈ 2.2 pounds
1 gram ≈ 0.035 ounce
1 cubic meter ≈ 264 gallons

Credits

Front Matter

viii Stephanie Lupoli/Shutterstock.com; **ix** © Anthony Berenyi | Dreamstime.com; **x** zhangyang13576997233/Shutterstock.com; **xi** WDG Photo/Shutterstock.com; **xii** eye-blink/Shutterstock.com; **xiii** Amnartk/Shutterstock.com; **xiv** anyaivanova/Shutterstock.com; **xv** Johnny Adolphson/Shutterstock.com; **xvi** ©iStockphoto.com/MinistryOfJoy; **xvii** Marafona/Shutterstock.com; **xviii** ©iStockphoto.com/Cesar Okada; **xix** Monkey Business Images/Shutterstock.com

Chapter 1

0 *top left* Doug Meek/Shutterstock.com; *top right* aceshot1/Shutterstock.com; *center left* Stephanie Lupoli/Shutterstock.com; *bottom right* Stefan Schurr/Shutterstock.com; *bottom left* Ververidis Vasilis/Shutterstock.com; **6** Ververidis Vasilis/Shutterstock.com; **9** cowardlion/Shutterstock.com; **14** Stefan Schurr/Shutterstock.com; **15** Eric Isselee/Shutterstock.com; **22** Stephanie Lupoli/Shutterstock.com; **25** Aleksander Erin/Shutterstock.com; **26** Invisible Studio/Shutterstock.com, avNY/Shutterstock.com, barragan/Shutterstock.com; **29** aceshot1/Shutterstock.com; **32** Jagodka/Shutterstock.com; **38** Petr84/Shutterstock.com; **40** Maxim Tupikov/Shutterstock.com; **41** *bottom left* Vladimir Dokovski/Shutterstock.com; *top right* Potapov Alexander/Shutterstock.com; **42** John Blanton/Shutterstock.com; **43** ollyy/Shutterstock.com

Chapter 2

50 *top left* DDCoral/Shutterstock.com; *top right* Christopher Boswell/Shutterstock.com; *center left* Catalin Petolea/Shutterstock.com; *bottom right* © Anthony Berenyi | Dreamstime.com; *bottom left* © Archhunter.de; **58** Gaieva Tetiana/Shutterstock.com; **59** © Archhunter.de; **64** © Anthony Berenyi | Dreamstime.com; **65** ©iStockphoto.com/NoDerog; **70** Oleksiy Mark/Shutterstock.com; **77** ©iStockphoto.com/fotoVoyager; **78** terekhov igor/Shutterstock.com; **79** Aleksander Erin/Shutterstock.com; **84** *top left* Umberto Shtanzman/Shutterstock.com; *bottom left* marekuliasz/Shutterstock.com; **85** Christopher Boswell/Shutterstock.com; **86** Niyazz/Shutterstock.com; **91** DDCoral/Shutterstock.com; **93** Lucky Business/Shutterstock.com; **97** Dmitry Naumov/Shutterstock.com

Chapter 3

100 *top left* © Sandra Sims | Dreamstime.com; *top right* Yiorgos GR/Shutterstock.com; *center left* zhangyang13576997233/Shutterstock.com; *bottom right* Myotis/Shutterstock.com; *bottom left* guroldinneden/Shutterstock.com; **107** Monika Gniot/Shutterstock.com; **109** *top left* guroldinneden/Shutterstock.com; *bottom right* ©iStockphoto.com/sorbetto; **116** Myotis/Shutterstock.com; **118** © Beaucroft | Dreamstime.com; **119** Vixit/Shutterstock.com; **120** Coprid/Shutterstock.com; **125** zhangyang13576997233/Shutterstock.com; **127** Aleksander Erin/Shutterstock.com; **134** Pertusinas/Shutterstock.com; **140** © Sandra Sims | Dreamstime.com; **142** Beneda Miroslav/Shutterstock.com; **151** Jacek Chabraszewski/Shutterstock.com; **153** ecco/Shutterstock.com, bazzier/Shutterstock.com; **161** Pakhnyushcha/Shutterstock.com; **163** ©iStockphoto.com/NejroN

Chapter 4

172 *top left* © Shannon Fagan | Dreamstime.com; *top right* © Joseph Helfenberger | Dreamstime.com; *center left* Asmus/Shutterstock.com; *bottom right* © Rmarmion | Dreamstime.com; *bottom left* WDG Photo/Shutterstock.com; **178** WDG Photo/Shutterstock.com; **180** bikeriderlondon/Shutterstock.com; **184** © Rmarmion | Dreamstime.com; **190** Asmus/Shutterstock.com; **192** ©iStockphoto.com/mecaleha; **193** Aleksander Erin/Shutterstock.com; **204** © Joseph Helfenberger | Dreamstime.com; **207** Adisa/Shutterstock.com; **215** NatUlrich/Shutterstock.com; **220** © Shannon Fagan | Dreamstime.com; **222** manish mansinh/Shutterstock.com; **225** and **231** Africa Studio/Shutterstock.com

Chapter 5

232 *top left* eye-blink/Shutterstock.com; *top right* Mike Neale/Shutterstock.com; *center left* majeczka/Shutterstock.com; *bottom right* Igor Bulgarin/Shutterstock.com; *bottom left* urbanlight/Shutterstock.com; **238** urbanlight/Shutterstock.com; **240** *top left* Tatuasha/Shutterstock.com; *top right* Rudy Umans/Shutterstock.com; *center left* 7yonov/Shutterstock.com; **244** Igor Bulgarin/Shutterstock.com; **245** BluIz60/Shutterstock.com; **250** VoodooDot/Shutterstock.com, Catherinecml/Shutterstock.com; **251** Pixachi/Shutterstock.com; **258** Viktoria/Shutterstock.com, t_lidiya/Shutterstock.com, ecco/Shutterstock.com, Yeko Photo Studio/Shutterstock.com; **259** Aleksander Erin/Shutterstock.com; **260** *left* Irina Rogova/Shutterstock.com; *right* Kostyantyn Ivanyshen/Shutterstock.com; **264** Ijzendoorn/Shutterstock.com; **266** Mike Neale/Shutterstock.com; **270** Tatyana Vyc/Shutterstock.com; **277** fotohunter/Shutterstock.com; **279** *center left* Marek R. Swadzba/Shutterstock.com; *Exercise 31 left* Chad King / NOAA MBNMS, *right* eye-blink/Shutterstock.com; **281** Melpomene/Shutterstock.com; **285** *left* Sashkin/Shutterstock.com; *right* Boris Sosnovyy/Shutterstock.com

Chapter 6

288 *top left* Amnartk/Shutterstock.com; *top right* ©iStockphoto.com/Sam Burt Photography; *center left* © Ivan Cholakov | Dreamstime.com; *bottom right* U.S. Department of Energy; *bottom left* karl umbriaco/Shutterstock.com; **290** Kasia/Shutterstock.com; **296** *top right* Vitaly Korovin/Shutterstock.com; *center right* Jiri Hera/Shutterstock.com; **297** *center left* Mark III Photonics/Shutterstock.com; *top right* NASA; **302** Glenda M. Powers/Shutterstock.com, siamionau pavel/Shutterstock.com; **304** *top left* ©iStockphoto.com/katkov, ©iStockphoto.com/Andrew Johnson; *center right* DTR; **311** karl umbriaco/Shutterstock.com; **314** locote/Shutterstock.com; **318** risteski goce/Shutterstock.com; **319** Sergey Goruppa/Shutterstock.com; **320** rudall30/Shutterstock.com; **321** *center left* bornholm/Shutterstock.com, Masalski Maksim/Shutterstock.com; *top right* Tupungato/Shutterstock.com; *center right* U.S. Department of Energy; **322** mariematata/Shutterstock.com; **323** Aleksander Erin/Shutterstock.com; **324** © Leerobin | Dreamstime.com; **330** © Ivan Cholakov | Dreamstime.com; **337** Alexander Lukin/Shutterstock.com; **338** ©iStockphoto.com/Sam Burt Photography; **339** Eric Isselee/Shutterstock.com, Vasyl Helevachuk/Shutterstock.com; **341** Steshkin Yevgeniy/Shutterstock.com, mayer kleinostheim/Shutterstock.com, ©iStockphoto.com/Matej Michelizza, Refat/Shutterstock.com; **343** Photo by Alvesgaspar/Joaquim Alves Gaspar, Lisboa, Portugal, modified by RDBury.; **344** *top right* jovan vitanovski/Shutterstock.com; *bottom right* Verdateo/Shutterstock.com; **345** aquatic creature/Shutterstock.com; **347** mangostock/Shutterstock.com

Chapter 7

354 *top left* anyaivanova/Shutterstock.com; *top right* jaroslava V/ Shutterstock.com; *center left* luchschen/Shutterstock.com; *bottom right* Rudy Balasko/Shutterstock.com; *bottom left* Andresr/Shutterstock.com; **361** ©iStockphoto.com/edge69; **362** ecco/Shutterstock.com; **363** *center left* Johanna Goodyear/Shutterstock.com; *Exercise 51* ©iStockphoto.com/ edge69; *Exercise 52* Matt Antonino/Shutterstock.com; *Exercise 53* ©iStockphoto.com/lumpynoodles; **368** Fejas/Shutterstock.com; **370** Andresr/Shutterstock.com; **374** Eric Isselee/Shutterstock.com, Erik Lam/Shutterstock.com; **376** Tamara Kulikova/Shutterstock.com; **382** Rudy Balasko/Shutterstock.com; **383** Aleksander Erin/Shutterstock.com; **384** *center right* Tuja/Shutterstock.com; *bottom right* ©iStockphoto.com/ tirc83; **389** deer boy/Shutterstock.com; **390** *top left* luchschen/ Shutterstock.com; *center left* cobalt88/Shutterstock.com; **394** jaroslava V/ Shutterstock.com; **400** G Tipene/Shutterstock.com; **401** Mark Stout Photography/Shutterstock.com; **402** *top left* MarijaPiliponyte/ Shutterstock.com; *center left* Sergiy Telesh/Shutterstock.com; **406** Dirk Ercken/Shutterstock.com; **407** Anna Breitenberger/Shutterstock.com; **409** Denise Lett/Shutterstock.com

Chapter 8

416 *top left* Johnny Adolphson/Shutterstock.com; *top right* Serg64/ Shutterstock.com; *center left* Susana Ortega/Shutterstock.com; *bottom right* MishAl/Shutterstock.com; *bottom left* Ron Zmiri/Shutterstock.com; **423** Charles Brutlag/Shuttterstock.com; **424** Wolna/Shutterstock.com, ©iStockphoto.com/PLAINVIEW; **428** ©iStockphoto.com/edge69; **430** Ron Zmiri/Shutterstock.com; **437** *top left* James Thew/ Shutterstock.com; *bottom right* James Hoenstine/Shutterstock.com; **439**Aleksander Erin/Shutterstock.com; **440** Rob Byron/Shutterstock.com; **445** dompr/Shutterstock.com; **447** Paul St. Clair/Shutterstock.com; **448** Susana Ortega/Shutterstock.com; **457** Serg64/Shutterstock.com; **464** Johnny Adolphson/Shutterstock.com; **466** Zarja/Shutterstock.com; **467** *top left* homydesign/Shutterstock.com; *bottom right* BlueOrange Studio/ Shutterstock.com, Nikolamirejovska/Shutterstock.com; **469** Alin Brotea/ Shutterstock.com

Chapter 9

476 *top left* © Kjersti Joergensen | Dreamstime.com; *top right* NatalieJean/ Shutterstock.com; *center left* Zhukov Oleg/Shutterstock.com; *bottom right* Richard Paul Kane/Shutterstock.com; *bottom left* ©iStockphoto.com/ MinistryOfJoy; **483** *top left* Radovan Spurny/Shutterstock.com, Worldpics/ Shutterstock.com, wong yu liang/Shutterstock.com; *center left* ©iStockphoto.com/MinistryOfJoy; **486** Jon Le-Bon/Shutterstock.com; **487** Dan Gerber/Shutterstock.com; **493** Richard Paul Kane/ Shutterstock.com; **495** *top left* Ben Haslam/Haslam Photography/ Shutterstock.com; *bottom right* Don Williamson/Shutterstock.com; **496** Steve Cukrov/Shutterstock.com; **501** Zhukov Oleg/Shutterstock.com; **502** Baker Alhashki/Shutterstock.com; **503** Aleksander Erin/ Shutterstock.com; **504** ©iStockphoto.com/mikdam; **513** *center left* graphixmania/Shutterstock.com; *bottom left* Petrovic Igor/Shutterstock.com; *center right* © Kaleb Timberlake | Dreamstime.com; **514** *top right* Ermek/ Shutterstock.com; *center right* © South12th | Dreamstime.com; **521** © Kjersti Joergensen | Dreamstime.com; **522** *center left* ©iStockphoto.com/edhor; *top right* Vereshchagin Dmitry/Shutterstock.com; **523** bikeriderlondon/Shutterstock.com, Aspen Photo/Shutterstock.com, Ruth Peterkin/Shutterstock.com; **531** luigi nifosi/Shutterstock.com; **533** wavebreakmedia/Shutterstock.com; **537** Artazum and Iriana Shiyan/ Shutterstock.com

Chapter 10

540 *top left* ©iStockphoto.com/Andrew_Howe; *top right* © Lostafichuk | Dreamstime.com; *center left* BsChan/Shutterstock.com; *bottom right* Johnny Habell/Shutterstock.com; *bottom left* Marafona/Shutterstock.com; **547** Marafona/Shutterstock.com; **549** *bottom left* Eduard Härkönen/ Shutterstock.com; *bottom right* Johnny Habell/Shutterstock.com; **550** ©iStockphoto.com/technotr; **554** BsChan/Shutterstock.com; **557** Aleksander Erin/Shutterstock.com; **558** Henryk Sadura/ Shutterstock.com; **563** Julia Baturina/Shutterstock.com; **564** Stubbs; **565** *center right* Neamov/Shutterstock.com, Subbotina Anna/ Shutterstock.com, Coprid/Shutterstock.com; *bottom right* © Lostafichuk | Dreamstime.com; **566** mekCar/Shutterstock.com; **573** Eric Isselee/ Shutterstock.com, mycola/Shutterstock.com; **575** Nikola Bilic/ Shutterstock.com; **578** © Andres Rodriguez | Dreamstime.com

Chapter 11

582 *top left* ©iStockphoto.com/David Sucsy; *top right* ©iStockphoto.com/ Cesar Okada; *center left* My Good Images/Shutterstock.com; *bottom right* bikeriderlondon/Shutterstock.com; *bottom left* Iryna Rasko/ Shutterstock.com; **591** bikeriderlondon/Shutterstock.com; **594** My Good Images/Shutterstock.com; **595** Stokkete/Shutterstock.com; **597** Krasowit/ Shutterstock.com; **602** Kostenyukova Nataliya/Shutterstock.com; **603** ©iStockphoto.com/Cesar Okada; **604** Monkey Business Images/ Shutterstock.com; **607** Aleksander Erin/Shutterstock.com; **608** Alexandr79/ Shutterstock.com; **615** Kenneth William Caleno/Shutterstock.com; **616** *center left* ©iStockphoto.com/David Sucsy; *center right* Johann Helgason/Shutterstock.com; **619** Taiga/Shutterstock.com; **623** Lucky Business/Shutterstock.com